Western World

HOLT PEOPLE, PLACES, AND CHANGE

An Introduction to World Studies

HOLT, RINEHART AND WINSTON

A Harcourt Education Company

Austin • Orlando • Chicago • New York • Toronto • London • San Diego

THE AUTHORS

Prof. Robert J. Sager is Chair of Earth Sciences at Pierce College in Lakewood, Washington. Prof. Sager received his B.S. in geology and geography and M.S. in geography from the University of Wisconsin and holds a J.D. in international law from Western State University College of Law. He is the coauthor of several geography and earth science textbooks and has written many articles and educational media programs on the geography of the Pacific. Prof. Sager has received several National Science Foundation study grants and has twice been a recipient of the University of Texas NISOD National Teaching Excellence Award. He is a founding member of the Southern California Geographic Alliance and former president of the Association of Washington Geographers.

Prof. David M. Helgren is Director of the Center for Geographic Education at San Jose State University in California, where he is also Chair of the Department of Geography. Prof. Helgren received his Ph.D. in geography from the University of Chicago. He is the coauthor of several geography textbooks and has written many articles on the geography of Africa. Awards from the National Geographic Society, the National Science Foundation, and the L. S. B. Leakey Foundation have supported his many field research projects. Prof. Helgren is a former president of the California Geographical Society and a founder of the Northern California Geographic Alliance.

Prof. Alison S. Brooks is Professor of Anthropology at George Washington University and a Research Associate in Anthropology at the Smithsonian Institution. She received her A.B., M.A., and Ph.D. in Anthropology from Harvard University. Since 1964, she has carried out ethnological and archaeological research in Africa, Europe, and Asia and is the author of more than 300 scholarly and popular publications. She has served as a consultant to Smithsonian exhibits and to National Geographic, Public Broadcasting, the Discovery Channel, and other public media. In addition, she is a founder and editor of *Anthro Notes: The National Museum of Natural History's Bulletin for Teachers* and has received numerous grants and awards to develop and lead in-service training institutes for teachers in grades 5–12. She served as the American Anthropological Association's representative to the NCSS task force on developing Scope and Sequence guidelines for Social Studies Education in grades K–12.

While the details of the young people's stories in the chapter openers are real, their identities have been changed to protect their privacy.

Cover and Title Page Photo Credits: (child image) AlaskaStock Images; (bkgd) Image Copyright © 2003 PhotoDisc, Inc./HRW

Printed in the United States of America

ISBN 0-03-053612-X

1 2 3 4 5 6 7 8 9   032   05 04 03 02

CONTENT REVIEWERS

EDUCATIONAL REVIEWERS

It's All About

INTERVIEWS with young people from around the world begin each chapter, making real world peer connections for your students. Plus students can read the interviews in their entirety on our Web site.

The first step to success in the social studies classroom is capturing and sustaining the interest of your students. *HOLT PEOPLE, PLACES, AND CHANGE* is designed to be open and friendly to all students, so that they develop an enthusiasm for learning and an appreciation for their world.

HOLT PEOPLE, PLACES, AND CHANGE offers
- Built-in Reading Support
- Technology with Instructional Value
- Standardized Testing Strategies and Skill Building
- The Best Teacher's Management System in the Industry

RELEVANCE

CNNfyi.com™ is designed to give students in grades 6–12 access to the news about people, places, and environments around the globe while offering "real-world" articles, career and college resources, and online activities.

In-Text Features that Put Geography into Perspective

- Case Study
- Connecting to Art
- Connecting to History
- Connecting to Literature
- Connecting to Math
- Connecting to Science
- Connecting to Technology
- Daily Life
- Focus on Culture

- Focus on Economy
- Focus on Environment
- Focus on Government
- Focus on Regions
- Geo Skills
- Hands On Geography
- Our Amazing Planet
- Why It Matters

Reading for

At Holt, we don't assume that students know how or have any desire to make sense of what they're reading, and we develop our programs based on that assumption. We don't just ask students questions about content, we give them strategies to get to that content. Through design, research, and the help of experts like Dr. Judith Irvin, we make sure students' reading needs are covered with our programs.

Helping Students Make Sense of What They're Reading

An Essay by Dr. Judith Irvin, Ph.D.

Who in middle and high schools helps students become more successful at reading and writing informational text? When I ask this question of a school faculty, the Language Arts/English teachers point to the social studies and science teachers because they are the ones with this type of textbook. The social studies and science teachers point to the Language Arts/English teachers because they are the ones that "do" words.

I advocate teachers taking an active role in helping students learn how to use text structure and context to understand what they read. Through consistent and systematic instruction that includes modeling of effective reading behavior, teachers can assist students in becoming better readers while at the same time helping them learn more content material.

The strategies in this book are designed to assist students with getting started, maintaining focus with reading, and organizing information for later retrieval. They engage students in learning material, provide the vehicle for them to organize and reorganize concepts, and extend their understanding through writing.

When teachers combine the teaching of reading and the teaching of content together into meaningful, systematic, and corrected instruction, students can apply what they have learned to understanding increasingly more difficult and complex texts as they progress through the school years.

READING STRATEGIES FOR THE SOCIAL STUDIES CLASSROOM

by Dr. Judith Irvin, Ph.D., Reading Education

HOLT, RINEHART AND WINSTON
Reading Strategies for the Social Studies Classroom
Dr. Judith Irvin

Additional Reading Support

- **Graphic Organizer Activities**
- **Guided Reading Strategies**
- **Main Idea Activities for English Language Learners and Special-Needs Students**
- **Audio CD Program**

MEANING

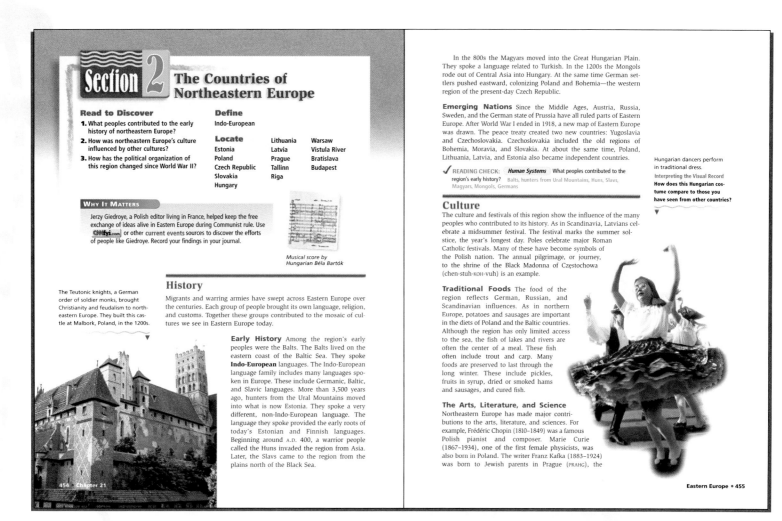

Section 2 — The Countries of Northeastern Europe

Read to Discover
1. What peoples contributed to the early history of northeastern Europe?
2. How was northeastern Europe's culture influenced by other cultures?
3. How has the political organization of this region changed since World War II?

Define
Indo-European

Locate
Estonia
Poland
Czech Republic
Slovakia
Hungary
Lithuania
Latvia
Prague
Tallinn
Riga
Warsaw
Vistula River
Bratislava
Budapest

WHY IT MATTERS
Jerzy Giedroyc, a Polish editor living in France, helped keep the free exchange of ideas alive in Eastern Europe during Communist rule. Use CNNfyi.com or other current events sources to discover the efforts of people like Giedroyc. Record your findings in your journal.

Musical score by Hungarian Béla Bartók

The Teutonic knights, a German order of soldier monks, brought Christianity and feudalism to northeastern Europe. They built this castle at Malbork, Poland, in the 1200s.

History

Migrants and warring armies have swept across Eastern Europe over the centuries. Each group of people brought its own language, religion, and customs. Together these groups contributed to the mosaic of cultures we see in Eastern Europe today.

Early History Among the region's early peoples were the Balts. The Balts lived on the eastern coast of the Baltic Sea. They spoke **Indo-European** languages. The Indo-European language family includes many languages spoken in Europe. These include Germanic, Baltic, and Slavic languages. More than 3,500 years ago, hunters from the Ural Mountains moved into what is now Estonia. They spoke a very different, non-Indo-European language. The language they spoke provided the early roots of today's Estonian and Finnish languages. Beginning around A.D. 400, a warrior people called the Huns invaded the region from Asia. Later, the Slavs came to the region from the plains north of the Black Sea.

454 • Chapter 21

In the 800s the Magyars moved into the Great Hungarian Plain. They spoke a language related to Turkish. In the 1200s the Mongols rode out of Central Asia into Hungary. At the same time German settlers pushed eastward, colonizing Poland and Bohemia—the western region of the present-day Czech Republic.

Emerging Nations Since the Middle Ages, Austria, Russia, Sweden, and the German state of Prussia have all ruled parts of Eastern Europe. After World War I ended in 1918, a new map of Eastern Europe was drawn. The peace treaty created two new countries: Yugoslavia and Czechoslovakia. Czechoslovakia included the old regions of Bohemia, Moravia, and Slovakia. At about the same time, Poland, Lithuania, Latvia, and Estonia also became independent countries.

✓ READING CHECK: **Human Systems** What peoples contributed to the region's early history? Balts, hunters from Ural Mountains, Huns, Slavs, Magyars, Mongols, Germans

Culture

The culture and festivals of this region show the influence of the many peoples who contributed to its history. As in Scandinavia, Latvians celebrate a midsummer festival. The festival marks the summer solstice, the year's longest day. Poles celebrate major Roman Catholic festivals. Many of these have become symbols of the Polish nation. The annual pilgrimage, or journey, to the shrine of the Black Madonna of Częstochowa (chen-stuh-KOH-vuh) is an example.

Traditional Foods The food of the region reflects German, Russian, and Scandinavian influences. As in northern Europe, potatoes and sausages are important in the diets of Poland and the Baltic countries. Although the region has only limited access to the sea, the fish of lakes and rivers are often the center of a meal. These fish often include trout and carp. Many foods are preserved to last through the long winter. These include pickles, fruits in syrup, dried or smoked hams and sausages, and cured fish.

The Arts, Literature, and Science Northeastern Europe has made major contributions to the arts, literature, and sciences. For example, Frédéric Chopin (1810–1849) was a famous Polish pianist and composer. Marie Curie (1867–1934), one of the first female physicists, was also born in Poland. The writer Franz Kafka (1883–1924) was born to Jewish parents in Prague (PRAHG), the

Hungarian dancers perform in traditional dress.
Interpreting the Visual Record
How does this Hungarian costume compare to those you have seen from other countries?

Eastern Europe • 455

Successful Readers must have:

1 AN ENGAGING NARRATIVE

Great care is taken in selecting and presenting content in a way that students will find motivating and engaging. Features such as **Youth Interviews** help students connect their own lives to the lives and cultures of other students around the world.

2 A FORECAST OF WHAT THEY WILL LEARN

Read to Discover questions give students insight into the content they will cover in the chapter to come. In features such as **Why It Matters,** students gain insight into regional issues.

3 VOCABULARY DEFINED IN CONTEXT

Important new terms are identified at the beginning of every section and are defined in context so students will develop an understanding of the contextual meaning of all terms.

4 STRATEGIES FOR UNDERSTANDING WHAT THEY READ

Through the design of the text, students are led through the content using built-in reading strategies. For example, **Reading Checks** in the text are used as a comprehension tool. The checks remind students to stop and engage with what they have read, functioning as a "Tutor in the Text."

Get Your Students

Your students love activities that get them involved with the content. That's why Holt offers active-learning resources that link directly to program content and provide a multitude of different lessons for large-group, small-group, and individual projects.

CREATIVE TEACHING STRATEGIES

These innovative teaching strategies can be utilized at various points in your lesson. The wide range of cooperative-learning activities, including learning stations and simulations, motivate your students and help them develop critical-thinking skills.

HANDS-ON GEOGRAPHY ACTIVITIES

From hands-on study of world cultures to hands-on practice with skill building, the following booklets cover it all. *Cultures of the World Activities* is a stand-alone booklet containing recipes, games, and craft activities. *Geography for Life Activities with Answer Key* contains a a problem-solving activity for each chapter, reflecting the skills and knowledge called for in the **National Geography Standards**.

GEOGRAPHY APPLICATIONS

For use in geography as well as earth science courses, here are two stand-alone booklets that organize applications with special relevance. *Environmental and Global Issues Activities* contains activities related to current environmental and global issues. *Lab Activities for Geography and Earth Science* contains laboratory activities related to physical geography and earth science.

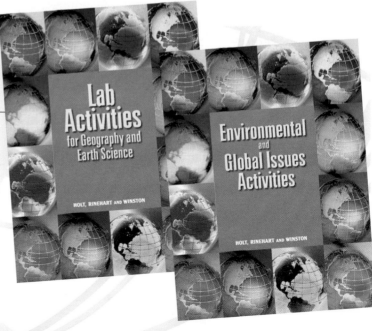

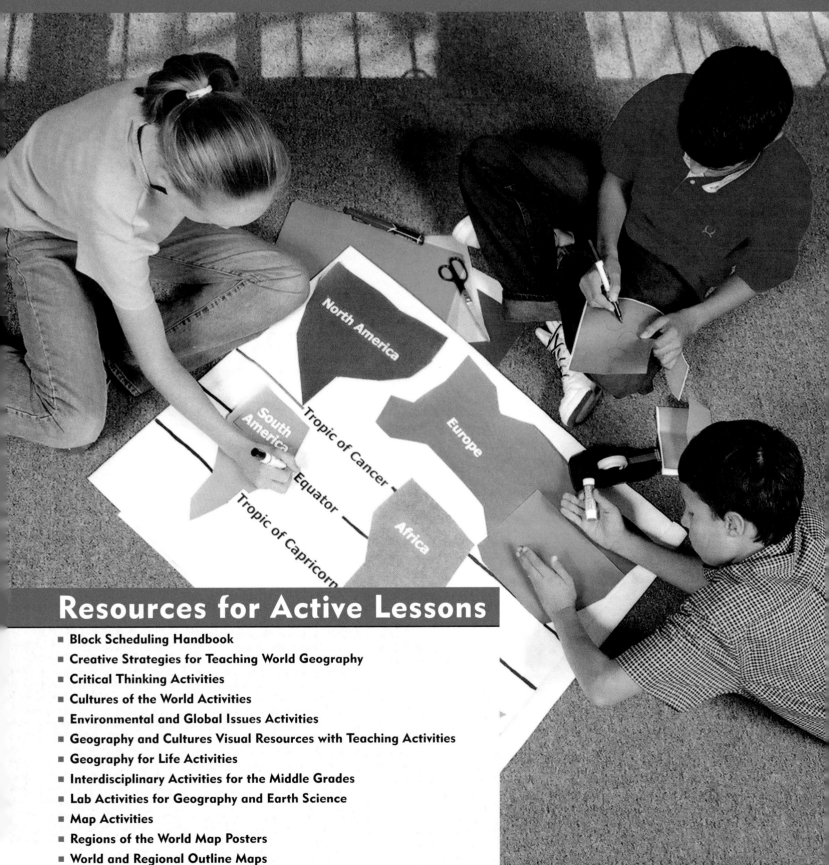

Involved in Learning

Resources for Active Lessons

- Block Scheduling Handbook
- Creative Strategies for Teaching World Geography
- Critical Thinking Activities
- Cultures of the World Activities
- Environmental and Global Issues Activities
- Geography and Cultures Visual Resources with Teaching Activities
- Geography for Life Activities
- Interdisciplinary Activities for the Middle Grades
- Lab Activities for Geography and Earth Science
- Map Activities
- Regions of the World Map Posters
- World and Regional Outline Maps
- World History and Geography Document-Based Questions Activities

Joining Forces

CNN® Presents

Geography: Yesterday and Today

- Mapping Change
- Birth of an Island
- Global Warming
- Farming with Saltwater
- World Population—Hitting Six Billion
- What's Cooking for the Holidays?
- Do Aspen Trees Hold the Key?
- Tradition and Change—The James Bay Project
- Baseball, Mexican Style
- Preserving Heritage
- The Power of Poetry
- Destination Brazil
- Harnessing Nature
- The Fishing Life
- Europe's Immigration Challenge
- The Mountain People of Yo
- Castles for Sale
- Lake Baykal
- Cossacks Return to Power
- Dateline Kazakhstan
- Jordan's Water Crisis
- Kuwait—Restoring the Environment
- Cairo—Selling the Suburbs
- The Women of Nigeria
- The Nairobi National Park
- The Medicine Tree
- Mining in South Africa the Traditional Way
- The Yangtze (Chang) River Dam
- The Tokyo Grave Crisis
- The Highway to Prosperity
- The Borobudur Temple of Indonesia
- Indian Pop Music
- The Tengboche Monastery
- The Great Barrier Reef
- Dateline Antarctica

World Cultures: Yesterday and Today

- The Kennewick Man
- Restoring the Sphinx
- The Sacred Ganges
- Treasures from China
- Crete: Past and Present
- The Ancient Theaters of Epidaurus
- In the Shadow of Mt. Vesuvius
- The Silk Road
- Pilgrimage to Mecca
- Mother Teresa: A Devoted Life
- The Bells of Agnone
- Buddhism in Mongolia
- The Inca's Frozen Past
- Divinci for Sale
- Worlds Meet in the Americas
- Restoring St. Petersburg
- The New Globe Theater
- A Key to the Bastille
- Hong Kong: Past and Present
- Riverboats Return
- Endangered Edison
- Honoring Women's Suffrage
- Fabergé: The Czar's Jeweler
- The Farms of Zimbabwe
- Recovering Russia's Royalty
- The Killer Flu of 1918
- The Great Depression
- Japan's Royal Family
- Rosie the Riveter
- NATO at 50
- North Korea Turns 50
- A Kenyan Reggae Festival
- Dateline Iraq
- Mexico's Jewish Community
- The History of Haiti
- The Walls That Divide
- Mission to Mars

to Enrich Your Classroom

CNNfyi.com

At **CNNfyi.com**, students will love exploring news stories written by experienced journalists as well as student bureau reporters. Stories link to homework help and lesson plans.

CNN PRESENTS VIDEO LIBRARY

The **CNN PRESENTS** video collection tackles the issue of making content relevant to students head on. Real-world news stories enable students to see the connections between classroom curriculum and today's issues and events around the nation and the world.

CNN PRESENTS...

- **America: Yesterday and Today, Beginnings to 1914**
- **America: Yesterday and Today, 1850 to Present**
- **America: Yesterday and Today, Modern Times**
- **Geography: Yesterday and Today**

- **World Cultures: Yesterday and Today**
- **American Government**
- **Economics**
- **September 11, 2001, Part One**
- **September 11, 2001, Part Two**

Holt is proud to team up with CNN/TURNER LEARNING® to provide you and your students with exceptional current and historical news videos and online resources that add depth and relevance to your daily instruction. This information collection takes your classroom to the far corners of the globe without students ever leaving their desks!

Your Multitalented Classroom

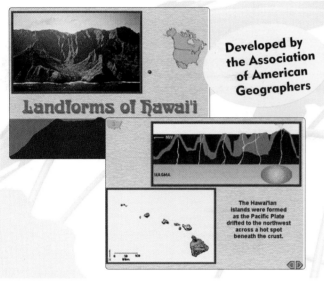

Landforms of Hawai'i

The Hawai'ian islands were formed as the Pacific Plate drifted to the northwest across a hot spot beneath the crust.

Developed by the Association of American Geographers

ACTIVITIES AND READINGS IN THE GEOGRAPHY OF THE WORLD

Integrate real geography into the topic you're studying with *Activities and Readings in the Geography of the World (ARGWorld).* This CD–ROM features world geography case studies with a multitude of activities that focus around geographical themes, population geography, economic geography, political geography, and environmental issues. Case studies will help teachers address the **National Geography Standards**.

HOLT RESEARCHER ONLINE: WORLD HISTORY AND CULTURES

New and online—students can access this outstanding research tool at **www.hrw.com**. A fully searchable database provides biographies, nation profiles and statistics, a glossary, and powerful graphic capabilities.

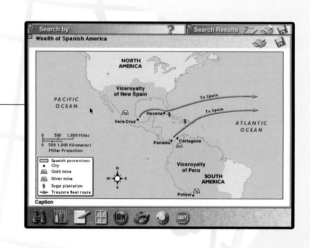

GLOBAL SKILL BUILDER CD–ROM

This CD-ROM is a comprehensive program containing interactive lessons that motivate your students to strengthen their map, graph, and computer skills. A handy *User's Guide and Teacher's Manual* provides student project sheets for each lesson along with optional suggestions for using the Internet to help complete the activity.

Guided Tour
1 Navigating the Internet
2 Mapping the Earth
3 Understanding Map Projections
4 Determining Absolute and Relative Location
5 Understanding Time Zones
6 Using Map Legends and Symbols
7 Identifying Different Types of Maps
8 Comparing Maps

3 Understanding Map Projections

Northern Hemisphere: Flat-Plane Projection

This map of the Northern Hemisphere is a flat-plane projection. Because the North Pole is the point of contact, directions and distances from the North Pole are accurate.

This projection shows true area, but not true shapes. A flat-plane projection is a good tool for pilots and ship navigators because it indicates true direction.

Global Skill Builder
CD-ROM for Macintosh and Windows

needs Multimedia Tools

THE WORLD TODAY VIDEODISC PROGRAM

This unique resource offers a stimulating outlook on world geography by showing your students the different ways geographers organize the world, and challenging them to contemplate and discuss significant world issues. Compelling video segments with in-depth content cover contemporary culture in every major world region.

PEOPLE, PLACES, AND CHANGE AUDIO CD PROGRAM

The **Audio CD Program** provides in-depth audio section summaries and self-check activity sheets to help those students who respond to auditory learning. Available in English and Spanish.

Audio CD Program

Other Multimedia Products

- **CNN Presents Geography: Modern Times**
- **CNN Presents Geography: Yesterday and Today**
- **CNN Presents World Cultures: Yesterday and Today**
- **Holt Researcher Online: World History and Cultures**

Technology with

go.hrw.com FOR TEACHERS

Throughout the *Annotated Teacher's Edition*, you'll find **Internet Connect** boxes that take you to specific chapter activities, links, current events, and more that correlate directly to the section you are teaching. Through **go.hrw.com** you'll find a wealth of teaching resources at your fingertips for fun, interactive lessons.

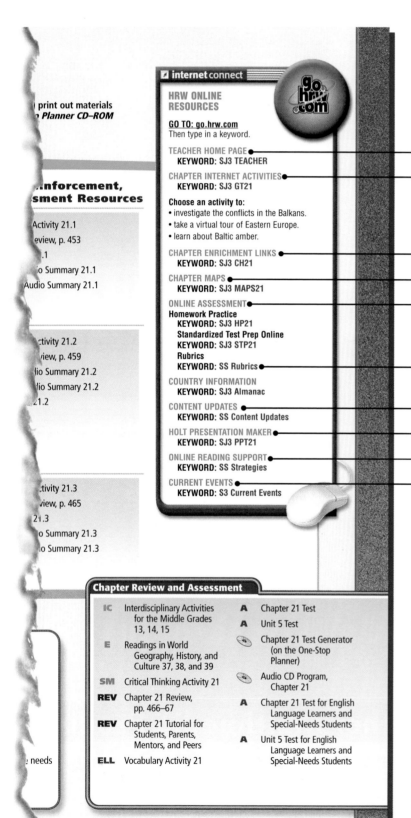

internet connect

HRW ONLINE RESOURCES

GO TO: go.hrw.com
Then type in a keyword.

TEACHER HOME PAGE
KEYWORD: SJ3 TEACHER

CHAPTER INTERNET ACTIVITIES
KEYWORD: SJ3 GT21

Choose an activity to:
• investigate the conflicts in the Balkans.
• take a virtual tour of Eastern Europe.
• learn about Baltic amber.

CHAPTER ENRICHMENT LINKS
KEYWORD: SJ3 CH21

CHAPTER MAPS
KEYWORD: SJ3 MAPS21

ONLINE ASSESSMENT
Homework Practice
KEYWORD: SJ3 HP21
Standardized Test Prep Online
KEYWORD: SJ3 STP21
Rubrics
KEYWORD: SS Rubrics

COUNTRY INFORMATION
KEYWORD: SJ3 Almanac

CONTENT UPDATES
KEYWORD: SS Content Updates

HOLT PRESENTATION MAKER
KEYWORD: SJ3 PPT21

ONLINE READING SUPPORT
KEYWORD: SS Strategies

CURRENT EVENTS
KEYWORD: S3 Current Events

DIRECT LAUNCH TO CHAPTER ACTIVITIES

GUIDED ONLINE ACTIVITIES

LINKS FOR EVERY SECTION

MAPS AND CHARTS

INTERACTIVE PRACTICE AND REVIEW

RUBRICS FOR SUBJECTIVE GRADING

UP-TO-DATE INFORMATION

CLASSROOM PRESENTATION SUPPORT

PRACTICE FOR READING SUCCESS

WEB RESOURCES FOR CURRENT ISSUES

print out materials
Planner CD–ROM

...nforcement,
...ssment Resources

Activity 21.1
...eview, p. 453
...1
io Summary 21.1
Audio Summary 21.1

...ctivity 21.2
...view, p. 459
...io Summary 21.2
...lio Summary 21.2
...21.2

...tivity 21.3
...view, p. 465
...21.3
...o Summary 21.3
...io Summary 21.3

needs

Chapter Review and Assessment

IC	Interdisciplinary Activities for the Middle Grades 13, 14, 15	**A**	Chapter 21 Test
		A	Unit 5 Test
E	Readings in World Geography, History, and Culture 37, 38, and 39		Chapter 21 Test Generator (on the One-Stop Planner)
SM	Critical Thinking Activity 21		Audio CD Program, Chapter 21
REV	Chapter 21 Review, pp. 466–67	**A**	Chapter 21 Test for English Language Learners and Special-Needs Students
REV	Chapter 21 Tutorial for Students, Parents, Mentors, and Peers	**A**	Unit 5 Test for English Language Learners and Special-Needs Students
ELL	Vocabulary Activity 21		

449B

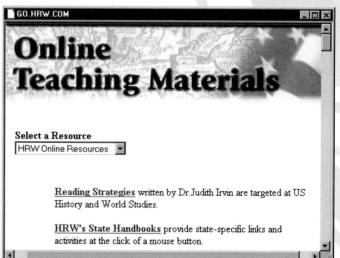

GO.HRW.COM

Online Teaching Materials

Select a Resource
HRW Online Resources ▾

Reading Strategies written by Dr. Judith Irvin are targeted at US History and World Studies.

HRW's State Handbooks provide state-specific links and activities at the click of a mouse button.

Instructional Value

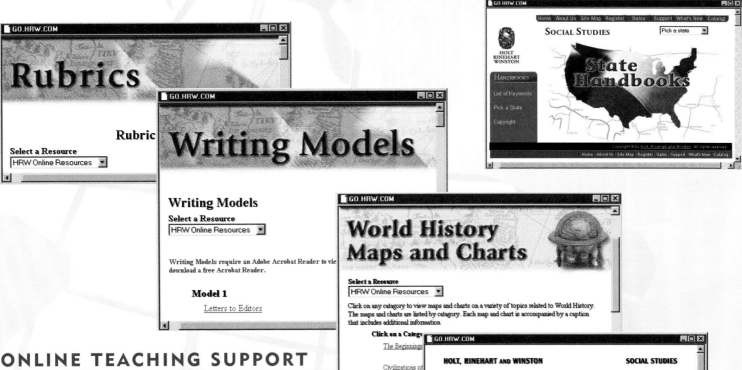

ONLINE TEACHING SUPPORT

Teacher materials on **go.hrw.com** offer you multiple resources for keeping content current. From **World History Maps and Charts** to **State Handbooks,** we've got it all.

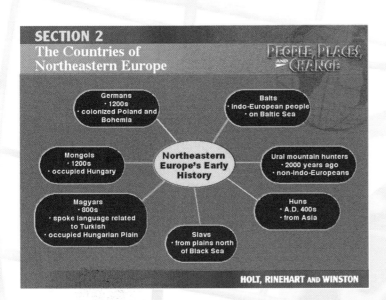

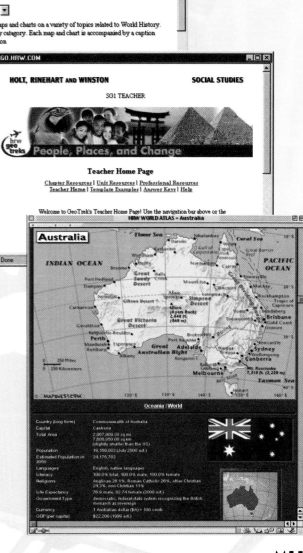

CLASSROOM PRESENTATION SUPPORT

Lecture notes and animated graphic organizers help add visual support to your classroom presentations.

Technology that

go.hrw.com FOR STUDENTS

Your students can access interactive activities, homework help, up-to-date maps, and more when they visit **go.hrw.com** and type in the keywords they find in their text.

ONLINE WORLD TRAVEL

When you log on to **go.hrw.com**, you and your students gain passage to **GeoTreks**—a site with guided Internet activities that integrate program content, spark imaginations, and promote online research skills. You'll find:

Interactive templates for creating newspapers, postcards, travel brochures, guided research reports, and more

GeoMaps—Interactive satellite maps of the world's regions for content review

Drag-and-drop exercises to review chapter content in short, fun activities

Chapter Web Links for prescreened, age-appropriate Web sites and current events

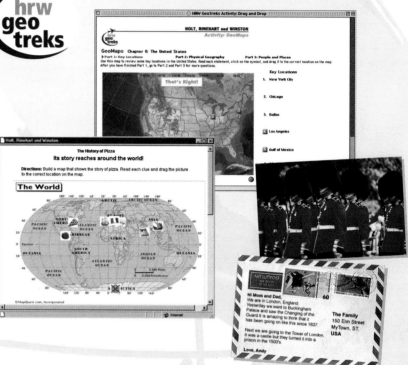

HOMEWORK PRACTICE

This helpful tool allows students to practice and review content by chapter anywhere there is a computer.

HRW ONLINE ATLAS AND HISTORICAL MAPS

The helpful online atlas contains over 300 well-rendered and clearly labeled country and state maps. Available in English and Spanish, these maps are continually updated so you can rest assured that you and your students have the latest and most accurate geographical content.

Online historical maps provide fascinating visual "snapshots" of the past. Students will relish the chance to explore medieval European trade routes, explorers' routes, ancient African kingdoms, and more.

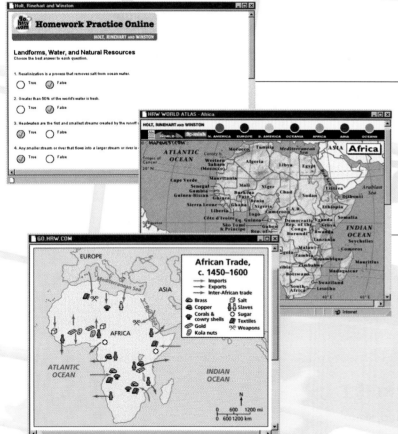

M12

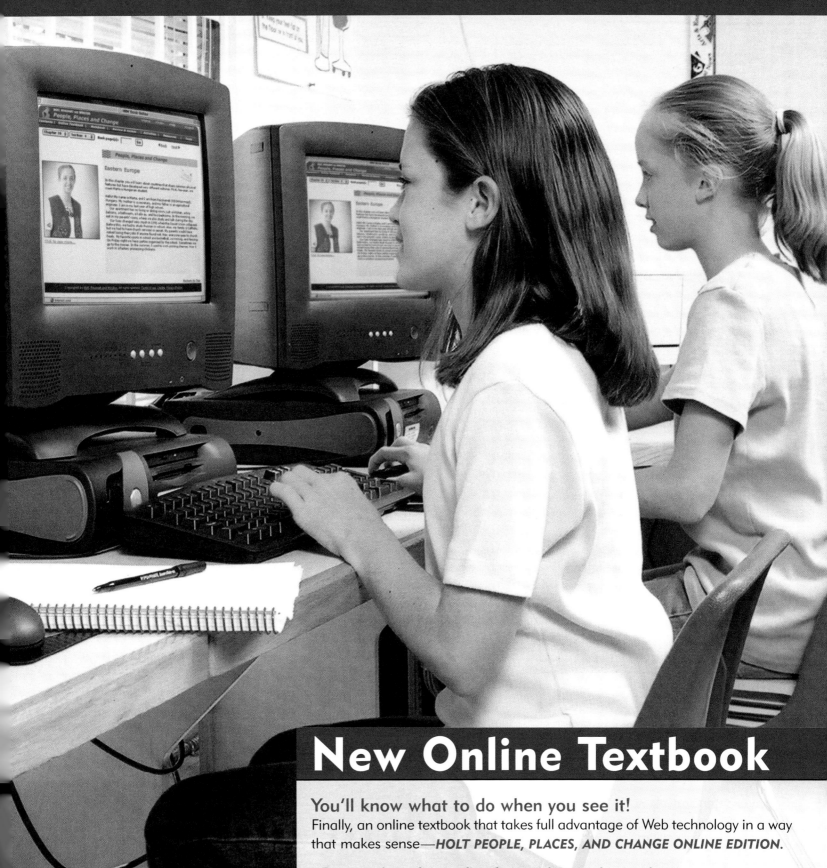

Delivers Content

New Online Textbook

You'll know what to do when you see it!
Finally, an online textbook that takes full advantage of Web technology in a way that makes sense—*HOLT PEOPLE, PLACES, AND CHANGE ONLINE EDITION.*

- **Entire student edition online, formatted to match printed text**
- **User-friendly navigation**
- **Hot links to interactive activities, practice, and assessment**
- **Student Notebook for online responses**

Unique Teacher's

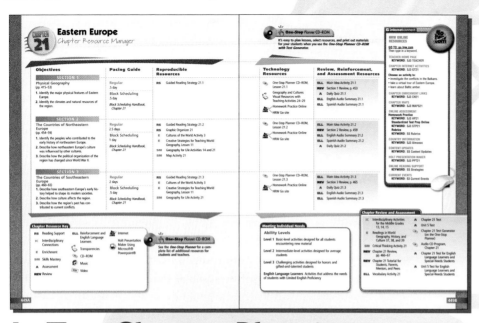

In-Text Chapter Planning

TEACHER TO TEACHER

These strategies are offered in the columns of your *Annotated Teacher's Edition* and provide you with valuable, classroom-tested ideas and activities that have been developed and successfully applied by your peers.

FOOD FESTIVAL

Kolaches are popular Czech or Polish pastries. For a shortcut version, use frozen bread dough. Or, make a sweetened yeast dough from scratch. After the dough has risen, roll it out, cut it into circles 2–3 inches across, and indent the center of each. Add a filling of fruit preserves, cottage cheese with egg and sugar, or a sweetened poppyseed paste. Let rise again. Sprinkle with a streusel topping of sugar, cinnamon, butter, and a little flour. Bake at 375° for 15–20 minutes. There are many *kolache* recipes on the Internet. *Kolacky* and *kolachke* are alternate spellings.

OBJECTIVE-BASED LESSON CYCLE

With lively activities and presentation strategies such as **Let's Get Started, Building Vocabulary,** and **Graphic Organizers,** your step-by-step lesson cycle makes planning your lessons easy and productive.

Side-Column Annotations that Spark Curiosity

- **Across the Curriculum: Art**
- **Across the Curriculum: History**
- **Across the Curriculum: Literature**
- **Across the Curriculum: Math**
- **Across the Curriculum: Science**
- **Across the Curriculum: Technology**
- **Cooperative Learning**
- **Cultural Kaleidoscope**
- **Daily Life**
- **Eye on Earth**
- **Geography sidelight**
- **Global Perspectives**
- **Historical Geography**
- **Linking Past to Present**
- **National Geography Standards**
- **People in the Profile**
- **Using Illustrations**

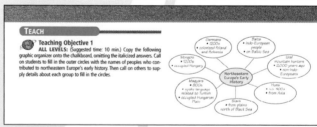

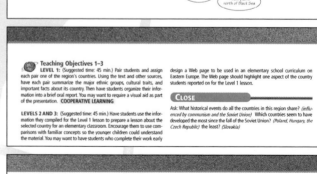

Management System

Everything you need is on one disc!

One-Stop Planner
with Test Generator

ONE-STOP PLANNER® CD–ROM WITH TEST GENERATOR

Holt brings you the most user-friendly management system in the industry with the *One-Stop Planner CD–ROM with Test Generator.* Plan and manage your lessons from this single disc containing all the teaching resources for *Holt People, Places, and Change,* valuable planning and assessment tools, and more.

- **Editable lesson plans**
- **Classroom Lecture Notes and Animated Graphic Organizers**
- **Easy-to-use test generator**
- **Previews of all teaching and video resources**
- **Easy printing feature**
- **Direct launch to go.hrw.com**

BLOCK SCHEDULING HANDBOOK

This is more than a pacing guide—it provides daily lesson plans that suggest practical ways to cover more than one textbook section in an extended class period and ways to make interdisciplinary connections.

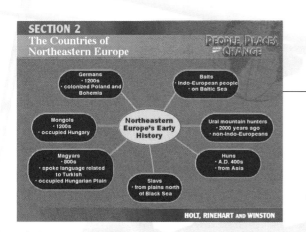

PRESENTATIONS THAT BENEFIT LEARNING

Classroom presentations and lecture notes can be accessed with ease when you use Holt's **Presentation** tool found on the *One-Stop Planner CD–ROM.* This resource helps you spice up your presentations and gives you ideas to build on. You'll find Microsoft® PowerPoint® presentations that include lecture notes and animated graphic organizers for each chapter and section of your text.

Assessment for

INTERACTIVE
PRACTICE
ACTIVITIES
FOR EACH SECTION

Section Review 1

Homework Practice Online
Keyword: SJ3 HP21

Define and explain: oil shale, lignite, amber

Working with Sketch Maps On a map of Eastern Europe that you draw or that your teacher provides, label the following: Baltic Sea, Adriatic Sea, Black Sea, Danube River, Dinaric Alps, Balkan Mountains, and Carpathian Mountains.

Reading for the Main Idea

1. *Places and Regions* On which three major seas do the countries of Eastern Europe have coasts?

2. *Environment and Society* What types of mineral and energy resources are available in this region? How does this influence individual economies?

Critical Thinking

3. **Making Generalizations and Predictions** Would this region be suitable for agriculture? Why?

4. **Identifying Cause and Effect** How did Communist rule contribute to the pollution problems of this region?

Organizing What You Know

5. **Summarizing** Copy the following graphic organizer. Use it to summarize the physical features, climate, and resources of the heartland, the Baltics, and the Balkans. Then write and answer one question about the region's geography based on the chart.

Region	Physical features	Climate	Resources

THE SUPERIOR TEST GENERATOR THAT REALLY WORKS!

M16

CHAPTER 21 — Reviewing What You Know

Building Vocabulary

On a separate sheet of paper, write sentences to define each of the following words.

1. oil shale
2. lignite
3. amber
4. Indo-European
5. Roma

Reviewing the Main Ideas

1. *Places and Regions* What are the major landforms of Eastern Europe?

2. *Human Systems* What groups influenced the culture of Eastern Europe? How can these influences be seen in modern society?

3. *Environment and Society* What were some environmental effects of Communist economic policies in Eastern Europe?

4. *Human Systems* What political and economic systems were most common in Eastern Europe before the 1990s? How and why have these systems changed?

5. *Human Systems* List the countries that broke away from Yugoslavia in the early 1990s. What are the greatest sources of tension in these countries?

Understanding Environment and Society

Economic Geography

During the Communist era, the Council for Economic Assistance, or COMECON, played an important role in planning the economies of countries in the region. Create a presentation comparing the practices of this organization to the practices now in place in the region. Consider the following:

• How COMECON organized the distribution of goods and services.
• How countries in the region now organize distribution.

Thinking Critically

1. **Drawing Inferences and Conclusions** How has Eastern Europe's location influenced the diets of the region's people?

2. **Analyzing Information** What is Eastern Europe's most important river for transportation and trade? How can you tell that the river is important to economic development?

3. **Identifying Cause and Effect** What geographic factors help make Warsaw the transportation and communication center of Poland? Imagine that Warsaw is located along the Baltic coast of Poland or near the German border. How might Warsaw have developed differently?

4. **Comparing/Contrasting** Compare and contrast the breakups of Yugoslavia and Czechoslovakia.

5. **Summarizing** How has political change affected the economies of Eastern European countries?

466 • Chapter 21

CRITICAL-THINKING
REINFORCEMENT

Every Student

Building Social Studies Skills

Map ACTIVITY

On a separate sheet of paper, match the letters on the map with their correct labels.

Baltic Sea
Adriatic Sea
Black Sea
Danube River
Dinaric Alps

Balkan Mountains
Carpathian Mountains

Mental Mapping Skills ACTIVITY

Using the chapter map or a globe as a guide, draw a freehand map of Eastern Europe and label the following:

Bosnia and Herzegovina
Croatia
Czech Republic
Estonia

Hungary
Macedonia
Poland
Yugoslavia

WRITING ACTIVITY

Imagine that you are a teenager living in Romania and want to write a family memoir of life in Romania. Include accounts of life for your grandparents under strict Soviet rule and life for your parents during the Soviet Union's breakup. Also describe your life in free Romania. Be sure to use standard grammar, spelling, sentence structure, and punctuation.

Alternative Assessment

Portfolio ACTIVITY

Learning About Your Local Geography

Cooperative Project Ask international agencies or search the Internet for help in contacting a boy or girl in Eastern Europe. As a group, write a letter to the teen, telling about your daily lives.

internet connect

Internet Activity: go.hrw.com
KEYWORD: SJ3 GT21

Choose a topic to explore Eastern Europe:
- Investigate the conflicts in the Balkans.
- Take a virtual tour of Eastern Europe.
- Learn about Baltic amber.

Eastern Europe • 467

ACCESS ONLINE RUBRICS
FOR GRADING PROJECTS AND
PORTFOLIO ASSIGNMENTS

WORLD HISTORY AND GEOGRAPHY DOCUMENT-BASED QUESTIONS ACTIVITIES

This resource provides a wide variety of primary sources and thought-provoking questions to help students develop intelligent, well-formed opinions. Important historical and geographical themes are grouped together, allowing for scaffolded instruction.

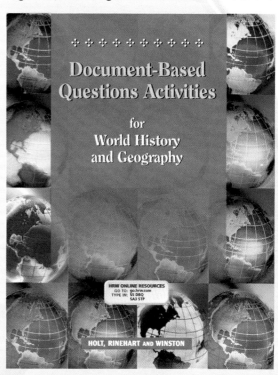

✦ ✦ ✦ ✦ ✦ ✦ ✦ ✦ ✦ ✦

Document-Based Questions Activities

for
World History and Geography

HRW ONLINE RESOURCES
GO TO: go.hrw.com
TYPE IN: SS DBQ
SA3 STP

HOLT, RINEHART AND WINSTON

Additional Print and Technology Assessment Resources

- **Daily Quizzes**
- **Chapter Tutorials for Students, Parents, Mentors, and Peers**
- **Chapter and Unit Tests**
- **Chapter and Unit Tests for English Language Learners and Special-Needs Students**
- **Alternative Assessment Handbook**
- **Test Generator (located on the One-Stop Planner)**

PEOPLE, PLACES, AND CHANGE

CONTENTS

- Mapping the Earth
- Mapmaking
- Map Essentials
- Working with Maps
- Using Graphs, Diagrams, Charts, and Tables
- Reading a Time-Zone Map
- Writing About Geography
- Doing Research
- Analyzing Primary and Secondary Sources
- Critical Thinking
- Decision-Making and Problem-Solving Skills
- Becoming a Strategic Reader
- Standardized Test-Taking Strategies

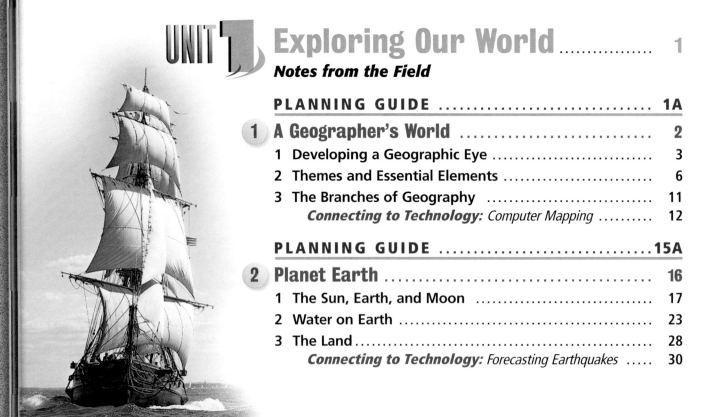

UNIT 1 Exploring Our World 1
Notes from the Field

UNIT 2 Gaining a Historical Perspective 96

Notes from the Field

UNIT 3 United States and Canada 206

Notes from the Field

UNIT 4 Middle and South America 280

Notes from the Field

UNIT 5 Europe 380

Notes from the Field

UNIT 6 **Russia and Northern Eurasia** 472

Notes from the Field

REFERENCE SECTION

FEATURES

FEATURES

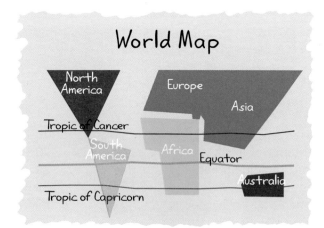

World Map

North America
Europe
Asia
Tropic of Cancer
South America
Africa
Equator
Australia
Tropic of Capricorn

MAPS

FEATURES

DIAGRAMS, CHARTS, and TABLES

FEATURES

DIAGRAMS, CHARTS, and TABLES *continued*

How To Use Your Textbook

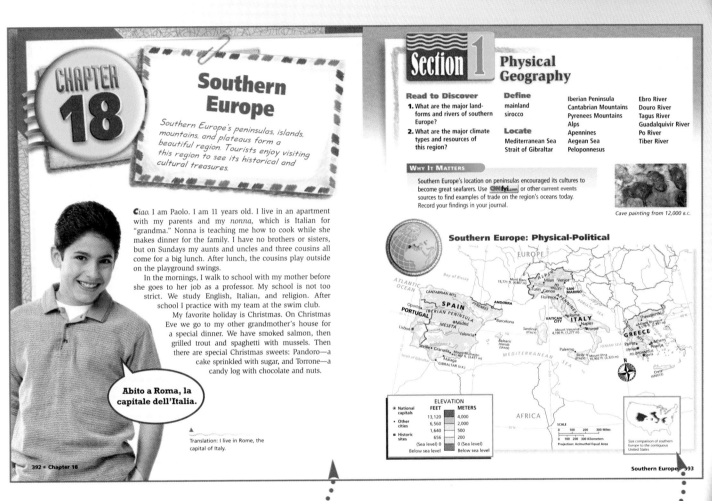

An interview with a student begins each regional chapter. These interviews give you a glimpse of what life is like for some people in the region you are about to study.

Chapter Map The map at the beginning of Section 1 in regional chapters shows you the countries you will read about. You can use this map to identify country names and capitals and to locate physical features. These chapter maps will also help you create sketch maps in section reviews.

Use these built-in tools to read for understanding.

Read to Discover questions begin each section of *Holt People, Places, and Change.* These questions serve as your guide as you read through the section. Keep them in mind as you explore the section content.

Why It Matters is an exciting way for you to make connections between what you are reading in your geography textbook and the world around you. Explore a topic that is relevant to our lives today by using **CNNfyi.com** .

Define and Locate terms are introduced at the beginning of each section. The Define terms include terms important to the study of geography and to the region you are studying. The Locate terms are important physical features or places from the region you are studying.

Interpreting the Visual Record features accompany many of the textbook's rich photographs. These features invite you to analyze the images so that you can learn more about their content and their links to what you are studying in the section. Other captions ask you to interpret maps, graphs, and charts.

Our Amazing Planet features provide interesting facts about the region you are studying. Here you will learn about the origins of place-names and fascinating tidbits like the size of South America's rain forests.

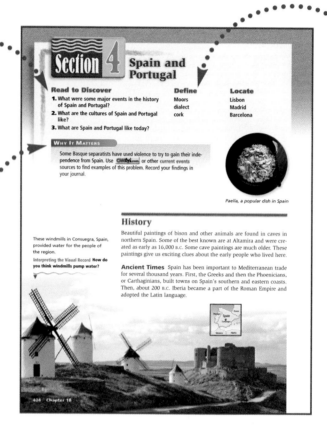

Reading Check questions appear often throughout the textbook to allow you to check your comprehension. As you read, pause for a moment to consider each Reading Check. If you have trouble answering the question, review the material that you just read.

Use these tools to pull together all of the information you have learned.

Critical Thinking activities in section and chapter reviews allow you to explore a topic in greater depth and to build your skills.

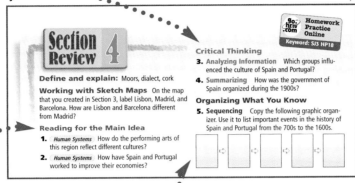

Homework Practice Online lets you log on to the HRW Go site to complete an interactive self-check of the material covered.

Reading for the Main Idea questions help review the main points you have studied in the section.

Graphic Organizers will help you pull together important information from the section.

Building Social Studies Skills activities help you develop the mapping and writing skills you need to study geography.

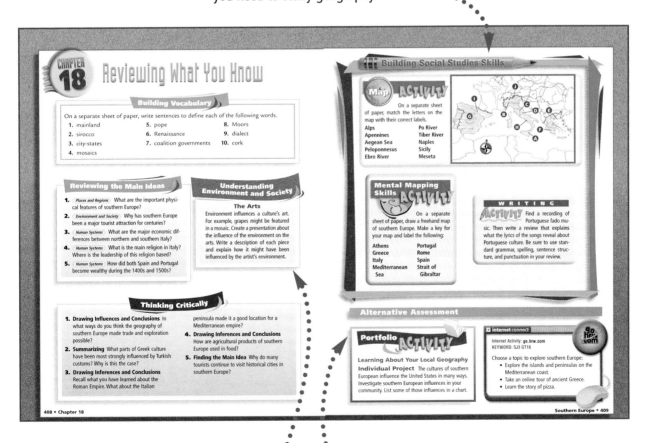

Understanding Environment and Society activities ask you to research and create a presentation expanding on an issue you have read about in the chapter.

Portfolio Activities are exciting and creative ways to explore your local geography and to make connections to the region you are studying.

Use these online tools to review and complete online activities.

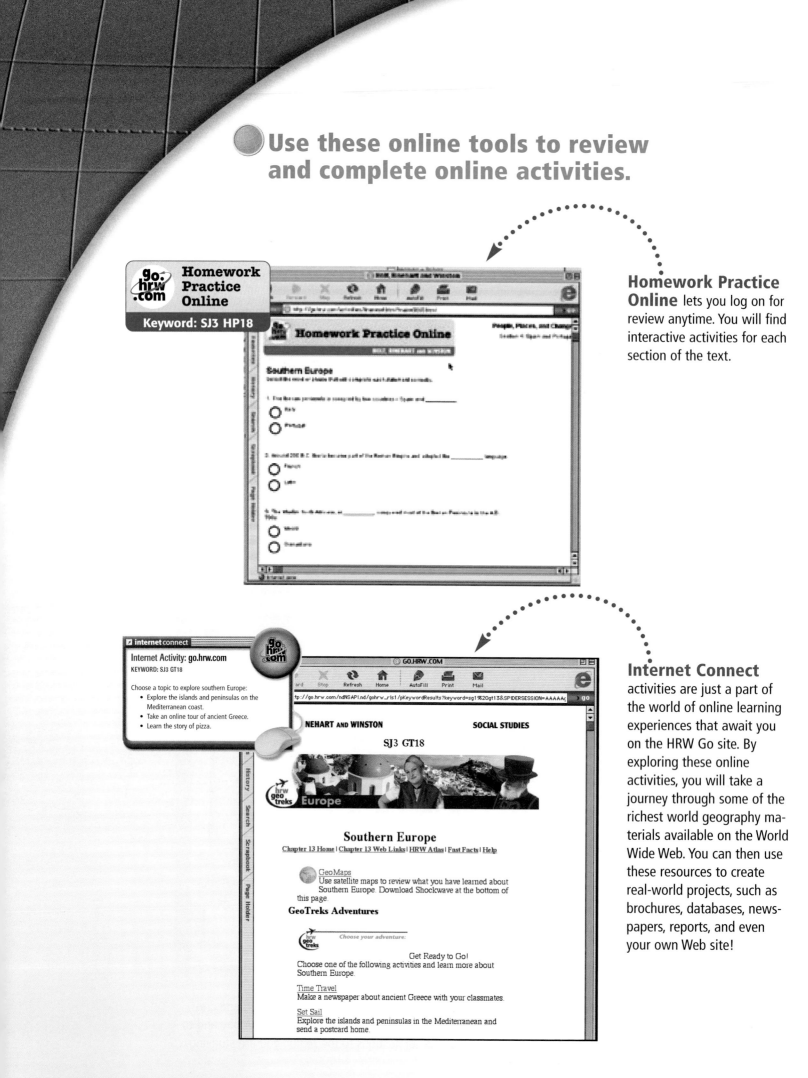

Homework Practice Online lets you log on for review anytime. You will find interactive activities for each section of the text.

Internet Connect activities are just a part of the world of online learning experiences that await you on the HRW Go site. By exploring these online activities, you will take a journey through some of the richest world geography materials available on the World Wide Web. You can then use these resources to create real-world projects, such as brochures, databases, newspapers, reports, and even your own Web site!

Why Geography Matters

Have you ever wondered. . .

why some places are deserts while other places get so much rain? What makes certain times of the year cooler than others? Why do some rivers run dry?

Maybe you live near mountains and wonder what processes created them. Do you know why the loss of huge forest areas in one part of the world can affect areas far away? Why does the United States have many different kinds of churches and other places of worship? Perhaps you are curious why Americans and people from other countries have such different points of view on many issues. The key to understanding questions and issues like these lies in the study of geography.

Geography and Your World

All you need to do is watch or read the news to see the importance of geography. You have probably seen news stories about the effects of floods, volcanic eruptions, and other natural events on people and places. You likely have also seen how conflict and cooperation shape the relations between peoples and countries around the world. The Why It Matters feature beginning every section of *Holt People, Places, and Change* uses the vast resources of **CNNfyi.com** or other current events sources to examine the importance of geography. Through this feature you will be able to draw connections between what you are studying in your geography textbook and events and conditions found around the world today.

The **CNNfyi.com** *Web site*

My fall semester project, growing a garden

Geography and Making Connections

When you think of the word *geography,* what comes to mind? Perhaps you simply picture people memorizing names of countries and capitals. Maybe you think of people studying maps to identify features like deserts, mountains, oceans, and rivers. These things are important, but the study of geography includes much more. Geography involves asking questions and solving problems. It focuses on looking at people and their ways of life as well as studying physical features like mountains, oceans, and rivers. Studying geography also means looking at why things are where they

are and at the relationships between human and physical features of Earth.

The study of geography helps us make connections between what was, what is, and what may be. It helps us understand the processes that have shaped the features we observe around us today, as well as the ways those features may be different tomorrow. In short, geography helps us understand the processes that have created a world that is home to more than 6 billion people and countless billions of other creatures.

Geography and You

Anyone can influence the geography of our world. For example, the actions of individuals affect local environments. Some individual actions might pollute the environment. Other actions might contribute to efforts to keep the environment clean and healthy. Various other things also influence geography. For example, governments create political divisions, such as countries and states. The borders between these divisions influence the human geography of regions by separating peoples, legal systems, and human activities.

Governments and businesses also plan and build structures like dams, railroads, and airports, which change the physical characteristics of places. As you might expect, some actions influence Earth's geography in negative ways, others in positive ways. Understanding geography helps us evaluate the consequences of our actions.

ATLAS
CONTENTS

WORLD: PHYSICAL

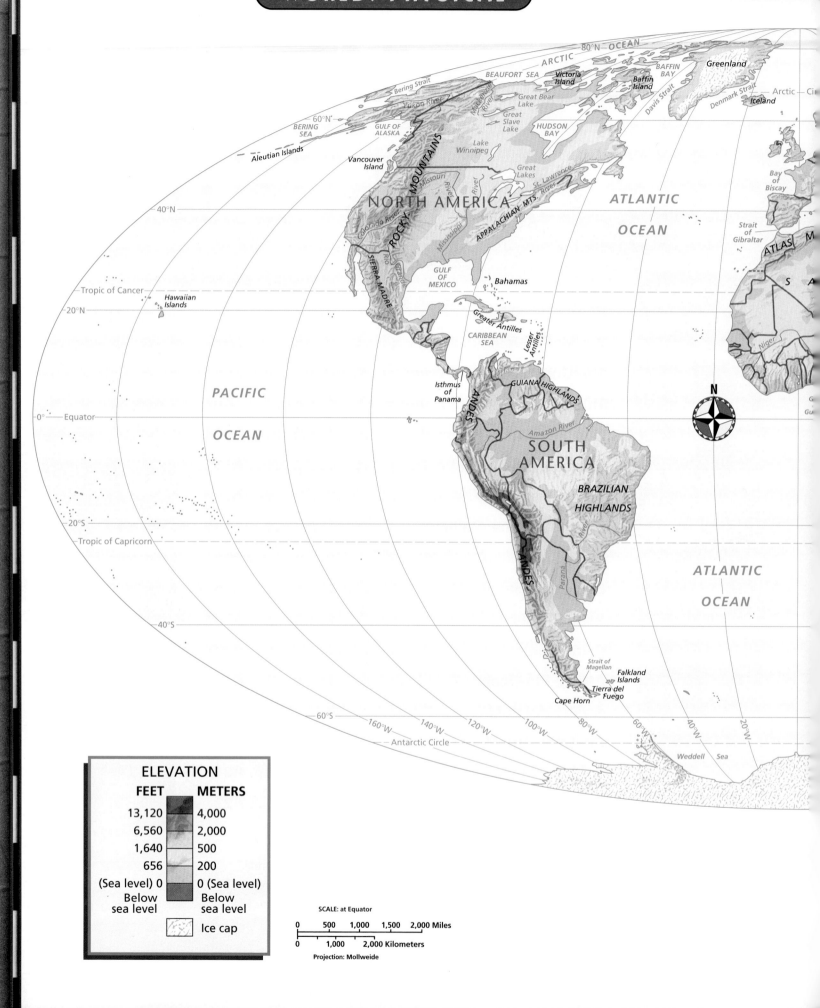

ARCTIC 80°N OCEAN

BEAUFORT SEA
Bering Strait
BERING SEA
60°N
GULF OF ALASKA
Aleutian Islands
Yukon River
Mackenzie River
Great Bear Lake
Great Slave Lake
HUDSON BAY
Victoria Island
Baffin Island
Baffin Bay
Davis Strait
Denmark Strait
Greenland
Iceland
Arctic Cir

Vancouver Island
ROCKY MOUNTAINS
Lake Winnipeg
Great Lakes
St. Lawrence River
APPALACHIAN MTS.
ATLANTIC OCEAN
Bay of Biscay
ATLAS
40°N
NORTH AMERICA
Colorado River
Missouri River
Mississippi
Strait of Gibraltar

Tropic of Cancer
20°N
Hawaiian Islands
SIERRA MADRE
Rio Grande
GULF OF MEXICO
Bahamas
Greater Antilles
CARIBBEAN SEA
Lesser Antilles
S A
Niger

PACIFIC
OCEAN
Isthmus of Panama
ANDES
GUIANA HIGHLANDS
N
Gu

0° Equator
Amazon River
SOUTH AMERICA

BRAZILIAN HIGHLANDS
River

20°S
Tropic of Capricorn
ANDES
Parana
River

ATLANTIC
OCEAN

40°S
Parana

Strait of Magellan
Falkland Islands
Tierra del Fuego
Cape Horn
60°S
160°W 140°W 120°W 100°W 80°W 60°W 40°W 20°W
Antarctic Circle
Weddell Sea

ELEVATION

FEET		METERS
13,120		4,000
6,560		2,000
1,640		500
656		200
(Sea level) 0		0 (Sea level)
Below sea level		Below sea level
	Ice cap	

SCALE: at Equator

0   500   1,000   1,500   2,000 Miles

0   1,000   2,000 Kilometers

Projection: Mollweide

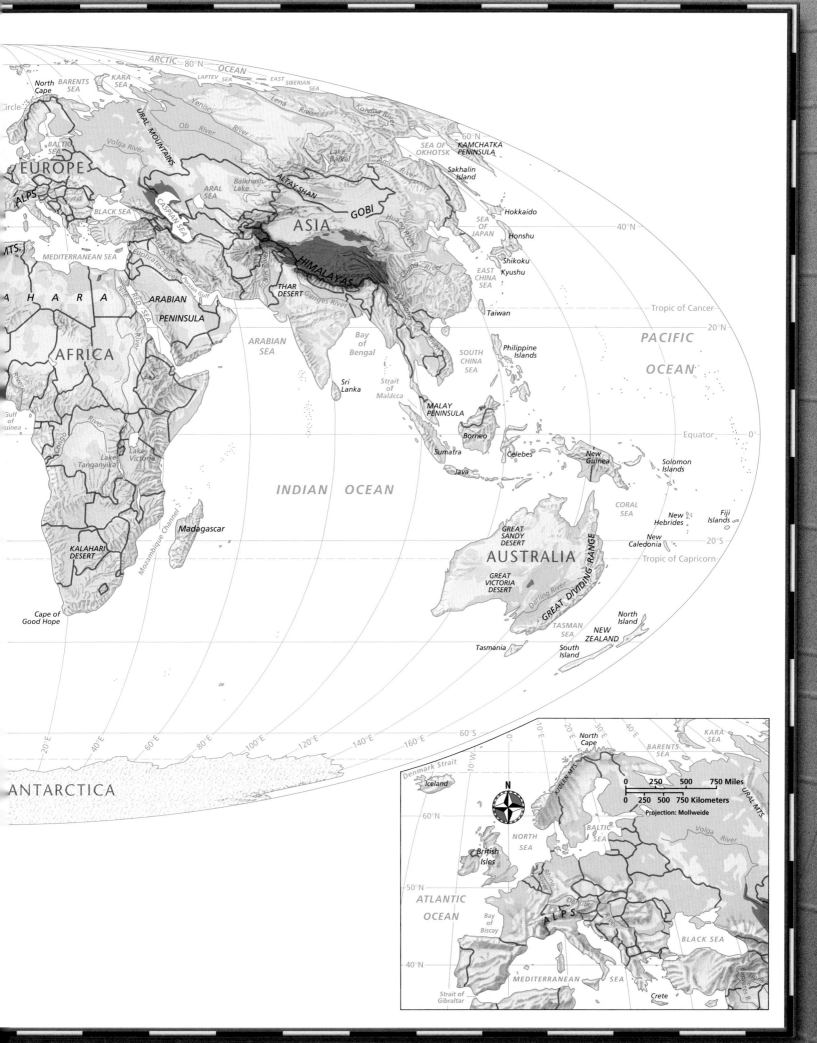

ARCTIC 80°N OCEAN

North
Cape
BARENTS
SEA
KARA
SEA
LAPTEV
SEA
EAST
SIBERIAN
SEA

Circle
Yenisey
River
Lena River
Kolyma River
60°N

URAL MOUNTAINS
Ob River
SEA OF
OKHOTSK
KAMCHATKA
PENINSULA

BALTIC
SEA
Volga River
Lake
Baikal
Amur River
Sakhalin
Island

EUROPE
ARAL
SEA
Balkhash
Lake
ALTAY SHAN
Hokkaido

ALPS
CASPIAN SEA
ASIA
GOBI
Honshu
40°N

BLACK SEA
Huang River
SEA
OF
JAPAN
Shikoku

MTS.
MEDITERRANEAN SEA
Tigris River
Euphrates River
Persian Gulf
HIMALAYAS
Chang River
Kyushu

SAHARA
Nile River
RED SEA
ARABIAN
PENINSULA
THAR
DESERT
Ganges River
Mekong River
EAST
CHINA
SEA
Tropic of Cancer

AFRICA
ARABIAN
SEA
Bay
of
Bengal
Taiwan
20°N

Gulf
of
Guinea
Congo River
Sri
Lanka
Strait
of
Malacca
SOUTH
CHINA
SEA
Philippine
Islands
PACIFIC
OCEAN

Lake
Tanganyika
Lake
Victoria
MALAY
PENINSULA
Borneo
New
Guinea
Equator 0°

Sumatra
Celebes
Solomon
Islands

INDIAN OCEAN
Java
CORAL
SEA
Fiji
Islands

Madagascar
GREAT
SANDY
DESERT
New
Hebrides
New
Caledonia
20°S

KALAHARI
DESERT
Mozambique Channel
AUSTRALIA
GREAT
VICTORIA
DESERT
Darling River
GREAT DIVIDING RANGE
Tropic of Capricorn

Cape of
Good Hope
TASMAN
SEA
NEW
ZEALAND
North
Island

20°E 40°E 60°E 80°E 100°E 120°E 140°E 160°E 60°S
Tasmania
South
Island

ANTARCTICA

Denmark Strait
Iceland
20°E 30°E 40°E
North
Cape
BARENTS
SEA
KARA
SEA

KJÖLEN MTS.
URAL MTS.
0 250 500 750 Miles
0 250 500 750 Kilometers

60°N
N
British
Isles
NORTH
SEA
BALTIC
SEA
Volga River
Projection: Mollweide

ATLANTIC
OCEAN
50°N
10°W
Rhine River
Rhône River
Danube River
BLACK SEA

40°N
Bay
of
Biscay
ALPS
Euphrates R.

MEDITERRANEAN SEA
Strait of
Gibraltar
Crete

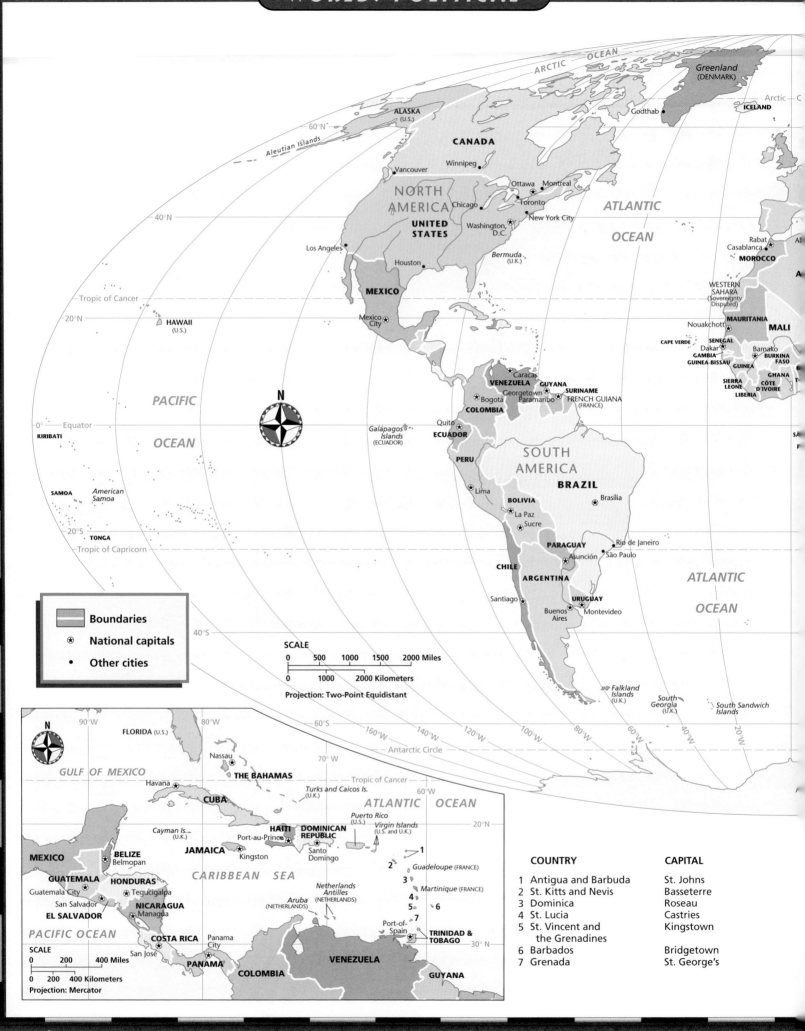

WORLD: POLITICAL

ARCTIC OCEAN

Greenland (DENMARK)

Arctic C

ICELAND

Godthab

ALASKA (U.S.)

60°N

Aleutian Islands

CANADA

Vancouver Winnipeg

40°N NORTH AMERICA Ottawa Montreal

Chicago Toronto

ATLANTIC

OCEAN

UNITED STATES Washington, D.C. New York City

Los Angeles Rabat
Casablanca

Houston Bermuda (U.K.) MOROCCO

MEXICO WESTERN SAHARA (Sovereignty Disputed)

Tropic of Cancer Nouakchott MAURITANIA MALI

20°N HAWAII (U.S.) Mexico City CAPE VERDE SENEGAL Bamako BURKINA FASO

Dakar GAMBIA GUINEA
GUINEA-BISSAU

SIERRA LEONE CÔTE GHANA
D'IVOIRE LIBERIA

Caracas SA

VENEZUELA GUYANA SURINAME

Georgetown FRENCH GUIANA (FRANCE)

Bogotá Paramaribo

PACIFIC COLOMBIA

N Quito

Galápagos ECUADOR SOUTH
AMERICA

0° Equator Islands (ECUADOR)

OCEAN PERU BRAZIL

KIRIBATI Lima Brasília

BOLIVIA

SAMOA American La Paz
Samoa Sucre

20°S TONGA Rio de Janeiro

Tropic of Capricorn PARAGUAY São Paulo

Asunción

CHILE ATLANTIC

ARGENTINA URUGUAY

Santiago Buenos Montevideo OCEAN
Aires

Boundaries

⊛ **National capitals**

• **Other cities**

40°S SCALE

0 500 1000 1500 2000 Miles

0 1000 2000 Kilometers Falkland South
Islands Georgia
(U.K.) (U.K.) South Sandwich
Islands

Projection: Two-Point Equidistant

60°S

Antarctic Circle

160°W 140°W 120°W 100°W 80°W 60°W 40°W 20°W

N

90°W 80°W 70°W FLORIDA (U.S.)

Nassau Tropic of Cancer

GULF OF MEXICO THE BAHAMAS Turks and Caicos Is. (U.K.) ATLANTIC OCEAN

Havana 60°W

CUBA Puerto Rico (U.S.) 20°N

Cayman Is. HAITI DOMINICAN Virgin Islands (U.S. and U.K.)

(U.K.) Port-au-Prince REPUBLIC 1

MEXICO JAMAICA Santo 2

BELIZE Kingston Domingo

Belmopan Guadeloupe (FRANCE)

CARIBBEAN SEA 3

GUATEMALA Netherlands Martinique (FRANCE)

HONDURAS Antilles 4

Guatemala City Tegucigalpa Aruba (NETHERLANDS) 6

San Salvador NICARAGUA (NETHERLANDS) 7

EL SALVADOR Managua Port-of-

PACIFIC OCEAN Spain TRINIDAD &
TOBAGO

COSTA RICA Panama 30°N

SCALE City

0 200 400 Miles San José PANAMA

0 200 400 Kilometers COLOMBIA VENEZUELA

Projection: Mercator GUYANA

COUNTRY	CAPITAL
1 Antigua and Barbuda	St. Johns
2 St. Kitts and Nevis	Basseterre
3 Dominica	Roseau
4 St. Lucia	Castries
5 St. Vincent and the Grenadines	Kingstown
6 Barbados	Bridgetown
7 Grenada	St. George's

ARCTIC OCEAN

RUSSIA

EUROPE

Moscow

KAZAKHSTAN
Astana

ASIA

Ulaanbaatar
MONGOLIA

Harbin

GEORGIA
Istanbul
Ankara
ARMENIA
TURKEY
Nicosia
CYPRUS
LEBANON
SYRIA
Beirut
Damascus
Jerusalem
ISRAEL
Amman
JORDAN
Cairo

TUNISIA
Tripoli
LIBYA
EGYPT

AFRICA

AZERBAIJAN
Baku
Tehran
IRAN
Baghdad
IRAQ
KUWAIT
BAHRAIN
QATAR
SAUDI
ARABIA
Riyadh

UZBEKISTAN
Tashkent
TURKMENISTAN
Ashgabat
TAJIKISTAN
Kabul
AFGHANISTAN
Islamabad
PAKISTAN
Karachi

Almaty
KYRGYZSTAN

CHINA

Beijing
Tianjin

NORTH
KOREA
P'yŏngyang
Seoul
Pusan
SOUTH
KOREA

JAPAN
Nagoya
Tokyo
Yokohama
Osaka

NIGER
CHAD
N'Djamena

NIGERIA
Abuja

CAMEROON
EQUATORIAL
GUINEA
GABON
REP.
OF THE
CONGO
RWANDA
DEMOCRATIC REP.
OF THE CONGO
Kinshasa

CENTRAL
AFRICAN
REPUBLIC

Khartoum
SUDAN
ERITREA
Asmara
Sanaa
YEMEN
DJIBOUTI
Addis Ababa
ETHIOPIA

UGANDA
KENYA
Nairobi
BURUNDI
TANZANIA
Dar es Salaam

SOMALIA

Delhi
New
Delhi
NEPAL Kathmandu
BHUTAN
INDIA
Mumbai
(Bombay)
Dhaka
BANGLADESH
Kolkata
(Calcutta)
MYANMAR
(BURMA)
LAOS
THAILAND
Bangkok
CAMBODIA
Phnom Penh
Ho Chi
Minh City
Chennai
(Madras)
SRI
LANKA
Colombo
Yangon
(Rangoon)

Wuhan
Chongqing

Guangzhou

Hong
Kong
Hanoi
VIETNAM

Shanghai

Taipei
TAIWAN

Tropic of Cancer

PACIFIC
OCEAN

Northern
Marianas
(U.S.)

Guam (U.S.)

Manila
PHILIPPINES

PALAU

BRUNEI
Kuala
Lumpur
MALAYSIA
Singapore
SINGAPORE

MALDIVES

SEYCHELLES

INDIAN OCEAN

Masqat
(Muscat)
OMAN
UNITED ARAB
EMIRATES

Equator

MARSHALL
ISLANDS

FEDERATED STATES
OF MICRONESIA

NAURU
KIRIBATI

INDONESIA

Jakarta
Surabaya

PAPUA
NEW
GUINEA
Port Moresby

SOLOMON
ISLANDS

TUVALU

Luanda
ANGOLA
ZAMBIA
Lusaka
NAMIBIA
Windhoek
BOTSWANA
Gaborone
Pretoria
Johannesburg
SWAZILAND
SOUTH
AFRICA
Cape Town
LESOTHO

MALAWI
MOZAMBIQUE
COMOROS
MADAGASCAR
Antananarivo
ZIMBABWE
Harare
Maputo

Réunion
(FRANCE)
MAURITIUS

AUSTRALIA

Tropic of Capricorn

New Caledonia
(FRANCE)
VANUATU
FIJI

20°S

Sydney
Canberra
Melbourne

NEW
ZEALAND
Wellington

Tasmania

ANTARCTICA

SCALE
0 250 500 750 Miles
0 250 500 750 Kilometers
Projection: Mollweide

ICELAND
Reykjavik
Arctic
Circle
N
NORWAY
SWEDEN
FINLAND
Helsinki
Oslo
Stockholm
10
St. Petersburg
RUSSIA
NORTH SEA
UNITED
KINGDOM
DENMARK
Copenhagen
9
8
Minsk
Moscow
Dublin
IRELAND
London
NETHERLANDS
Amsterdam
The Hague
Brussels
BELGIUM
GERMANY
Berlin
Warsaw
POLAND
BELARUS
Kiev
UKRAINE
ATLANTIC
OCEAN
Paris
LUXEMBOURG
FRANCE
Bern
SWITZERLAND
LIECHTENSTEIN
Vienna
AUSTRIA
2
HUNGARY
Budapest
1
3
4
5
7
MOLDOVA
Chişinău
ROMANIA
Bucharest
BULGARIA
Sofia
BLACK SEA
PORTUGAL
Lisbon
SPAIN
Madrid
ANDORRA
Corsica
(FRANCE)
MONACO
ITALY
SAN MARINO
VATICAN CITY
Rome
Sardinia
Balearic
Is. (SPAIN)
Tiranë
ALBANIA
6
GREECE
Athens
MEDITERRANEAN
SEA
Sicily
MALTA
Crete
Gibraltar (U.K.)

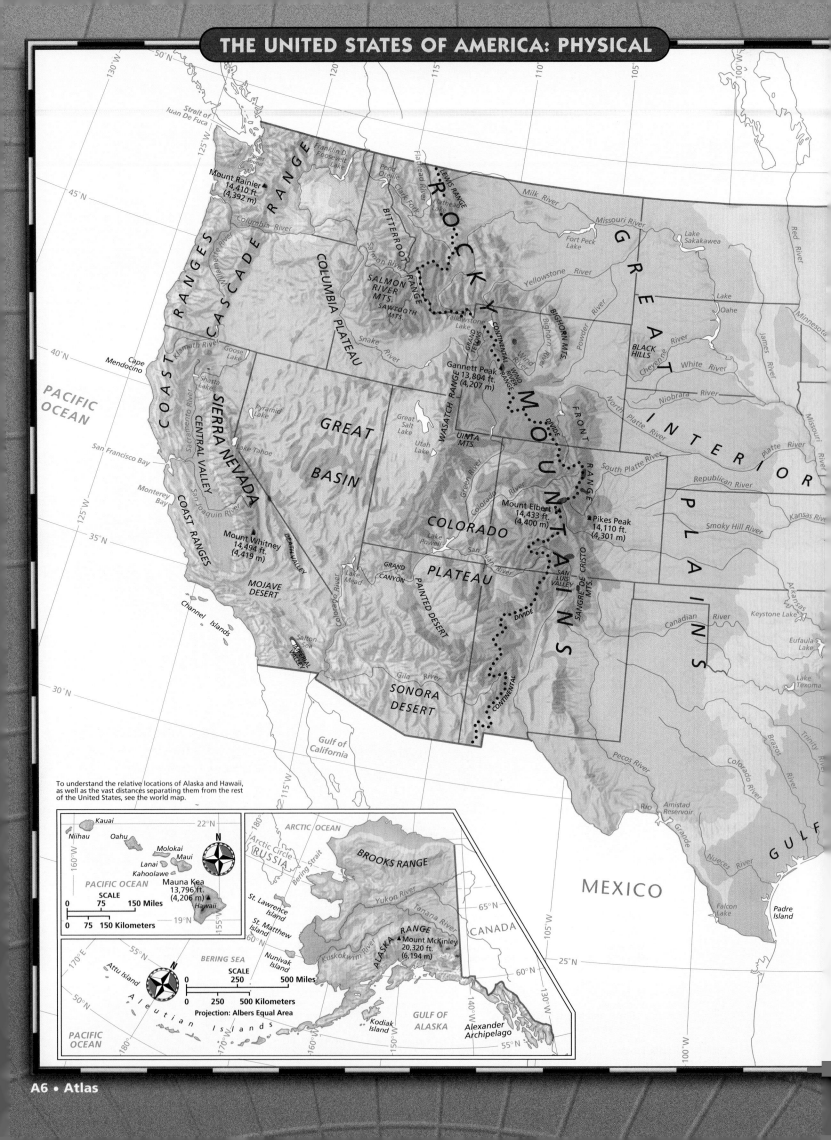

PACIFIC OCEAN

MEXICO

CANADA

COAST RANGES

CASCADE RANGE

COLUMBIA PLATEAU

SIERRA NEVADA

CENTRAL VALLEY

COAST RANGES

GREAT BASIN

SALMON RIVER MTS.

BITTERROOT RANGE

SAWTOOTH MTS.

ROCKY MOUNTAINS

GREAT INTERIOR PLAINS

WASATCH RANGE

UINTA MTS.

COLORADO PLATEAU

GRAND CANYON

PAINTED DESERT

MOJAVE DESERT

DEATH VALLEY

SONORA DESERT

IMPERIAL VALLEY

FRONT RANGE

SANGRE DE CRISTO MTS.

SAN LUIS VALLEY

BLACK HILLS

GRAND TETONS

CONTINENTAL DIVIDE

WIND RIVER RANGE

LEWIS RANGE

Mount Rainier 14,410 ft. (4,392 m)

Mount Whitney 14,494 ft. (4,419 m)

Gannett Peak 13,804 ft. (4,207 m)

Mount Elbert 14,433 ft. (4,400 m)

Pikes Peak 14,110 ft. (4,301 m)

Cape Mendocino

San Francisco Bay

Monterey Bay

Channel Islands

Strait of Juan De Fuca

Puget Sound

Franklin D. Roosevelt Lake

Pend Oreille

Flathead Lake

Flathead River

Clark Fork

Columbia River

Willamette River

Klamath River

Goose Lake

Shasta Lake

Pyramid Lake

Lake Tahoe

Sacramento River

San Joaquin River

Snake River

Salmon River

Great Salt Lake

Utah Lake

Lake Mead

Lake Powell

Colorado River

Green River

San Juan River

Gila River

Salton Sea

Gulf of California

Yellowstone Lake

Milk River

Missouri River

Fort Peck Lake

Yellowstone River

Bighorn River

Bighorn Mts.

Powder River

Lake Sakakawea

Lake Oahe

Red River

Cheyenne River

White River

Niobrara River

North Platte River

South Platte River

Republican River

Smoky Hill River

Kansas River

Arkansas River

Keystone Lake

Eufaula Lake

Lake Texoma

Canadian River

Pecos River

Brazos River

Colorado River

Trinity River

Nueces River

Rio Grande

Amistad Reservoir

Falcon Lake

Padre Island

Minnesota

James River

Platte River

Missouri River

GULF

MEXICO

To understand the relative locations of Alaska and Hawaii, as well as the vast distances separating them from the rest of the United States, see the world map.

Kauai

Niihau

Oahu

Molokai

Maui

Lanai

Kahoolawe

Mauna Kea 13,796 ft. (4,206 m)

Hawaii

PACIFIC OCEAN

SCALE
0 — 75 — 150 Miles
0 — 75 — 150 Kilometers

ARCTIC OCEAN

Arctic Circle

RUSSIA

Bering Strait

BROOKS RANGE

St. Lawrence Island

St. Matthew Island

Nunivak Island

Yukon River

Tanana River

Kuskokwim River

ALASKA RANGE

Mount McKinley 20,320 ft. (6,194 m)

CANADA

GULF OF ALASKA

Kodiak Island

Alexander Archipelago

BERING SEA

Attu Island

Aleutian Islands

PACIFIC OCEAN

SCALE
0 — 250 — 500 Miles
0 — 250 — 500 Kilometers

Projection: Albers Equal Area

CANADA

MESABI RANGE
Isle Royale
Lake Superior
Lake Huron
Lake Michigan
Lake Ontario
Lake Erie
Mississippi River
Wisconsin River
Des Moines River

P L A I N S

Lake of the Ozarks
ZARK PLATEAU
OZARK PLATEAU
River
White River
ACHITA MTS.
Red
Illinois River
Wabash River
Ohio River
Scioto River
Kentucky Lake
Lake Barkley
Cumberland River
Tennessee River
Mississippi River
Tombigbee River

ST. LAWRENCE SEAWAY
St. Lawrence River
LONGFELLOW MTS.
Penobscot River
St. John River
Lake Champlain
GREEN MTS.
WHITE MTS.
Connecticut River
ADIRONDACK MTS.
Finger Lakes
Hudson R.
Cape Cod
Long Island Sound
Long Island
CATSKILL MTS.
Allegheny River
Susquehanna River
Delaware R.
PLATEAU
MOUNTAINS
Delaware Bay
Potomac River
Monongahela R.
Kanawha River
Chesapeake Bay
James River
ALLEGHENY
APPALACHIAN
Roanoke River
Pamlico Sound
Cape Hatteras
CUMBERLAND PLATEAU
GREAT SMOKY MTS.
BLUE RIDGE Mountains
PIEDMONT
ATLANTIC COASTAL PLAIN

ATLANTIC OCEAN

Alabama R.
Coosa River
Chattahoochee River
Savannah River
Oconee River
Pearl River
Altamaha River
Sea Islands

C O A S T A L   P L A I N
oledo Bend Reservoir
abine River

Chandeleur Islands
Mississippi Delta

GULF OF MEXICO

Okefenokee Swamp
FLORIDA PENINSULA
Cape Canaveral
Lake Okeechobee
The Everglades
Cape Sable
Florida Key
Straits of Florida

THE BAHAMAS

CUBA

PACIFIC OCEAN

WASHINGTON
Seattle
Tacoma
Olympia ★
Portland
Salem ★
Eugene
OREGON
Puget Sound
Strait of Juan de Fuca
Spokane
Columbia River

Franklin D. Roosevelt Lake
Pend Oreille
IDAHO
Boise
Snake River

Flathead Lake
MONTANA
Helena ★
Billings
Fort Peck Lake
Yellowstone River
Missouri River

NORTH DAKOTA
Bismarck ★
Fargo
Lake Sakakawea
Red River
Lake Oahe

Yellowstone Lake
WYOMING
Casper
Cheyenne ★

SOUTH DAKOTA
Pierre ★
Sioux Falls
Minnesota River

Pocatello

Cape Mendocino
Shasta Lake
Goose Lake
Sacramento River
Pyramid Lake

NEVADA
Reno ★
Carson City ★
Lake Tahoe
Sacramento
Stockton
Modesto
San Joaquin R.
Oakland
San Francisco
San Francisco Bay
San Jose
Monterey Bay
Fresno
CALIFORNIA
Bakersfield
Los Angeles
Long Beach
Anaheim
Santa Ana
San Diego
Channel Islands
Salton Sea

Great Salt Lake
Salt Lake City ★
Utah Lake
Provo
UTAH
Green River
Lake Powell
Colorado River
Lake Mead
Las Vegas

COLORADO
Denver ★
Colorado Springs

NEBRASKA
Omaha
Lincoln ★
Platte River
Missouri River

KANSAS
Topeka ★
Wichita
Arkansas River

OKLAHOMA
Oklahoma City ★
Tulsa
Keystone Lake
Eufaula Lake
Canadian River
Lake Texoma

ARIZONA
Phoenix ★
Gila River
Tucson

NEW MEXICO
Santa Fe ★
Albuquerque
Amarillo
Lubbock

El Paso
Pecos River
Rio Grande

TEXAS
Abilene
Odessa
Fort Worth
Dallas
Waco
Colorado River
Brazos River
Austin
San Antonio
Houston
Corpus Christi
Laredo
Padre Island
Amistad Reservoir

Gulf of California

MEXICO

To understand the relative locations of Alaska and Hawaii as well as the vast distances separating them from the rest of the United States, see the world map.

Kauai
Niihau
Oahu
Honolulu
HAWAII
Molokai
Lanai
Kahoolawe
Maui
Hawaii
PACIFIC OCEAN
N
SCALE
0   75   150 Miles
0   75   150 Kilometers

ARCTIC OCEAN
Arctic Circle
RUSSIA
Bering Strait
Nome
Fairbanks
Yukon River
ALASKA
Anchorage
Juneau ★
CANADA
GULF OF ALASKA
Kodiak Island
Alexander Archipelago
St. Lawrence Island
St. Matthew Island
Nunivak Island
BERING SEA
Attu Island
Aleutian Islands
PACIFIC OCEAN
N
SCALE
0   250   500 Miles
0   250   500 Kilometers
Projection: Albers Equal Area

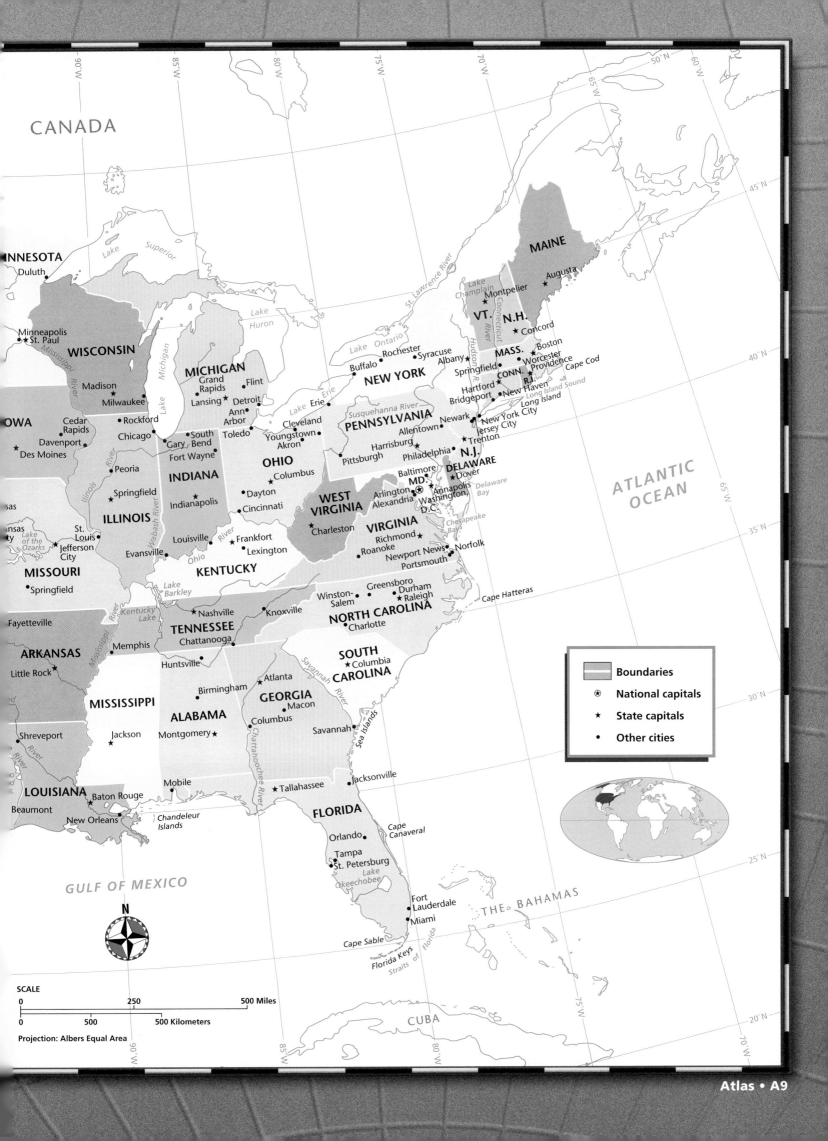

CANADA

MINNESOTA
• Duluth

• Minneapolis
★ St. Paul

WISCONSIN

Lake Superior

Lake Michigan

Lake Huron

Lake Ontario

St. Lawrence River

MAINE
• Augusta

Lake Champlain

• Montpelier
VT.   N.H.
★ Concord

MASS. • Boston
• Worcester
• Springfield Providence
CONN. R.I.
★ Hartford • New Haven   Cape Cod
• Bridgeport
Long Island Sound
Long Island

IOWA
• Cedar Rapids
• Davenport
• Des Moines

• Madison
• Milwaukee
• Rockford
• Chicago

MICHIGAN
Grand Rapids
• Flint
• Lansing • Detroit
• Ann Arbor
• South Bend
• Gary
Fort Wayne
• Toledo

OHIO
• Cleveland
• Youngstown
• Akron

• Columbus

NEW YORK
• Buffalo • Rochester
• Syracuse
• Albany

Lake Erie

PENNSYLVANIA
• Harrisburg
• Pittsburgh
Susquehanna River

Hudson River
Connecticut River

• Allentown
★ Harrisburg
• Philadelphia
• Newark
• New York City
• Jersey City
• Trenton
N.J.

ATLANTIC OCEAN

ILLINOIS
• Peoria
★ Springfield

INDIANA
★ Indianapolis
• Dayton
• Cincinnati

KANSAS

St. Louis
★ Jefferson City
MISSOURI
• Springfield

Lake of the Ozarks

KENTUCKY
• Louisville
★ Frankfort
• Lexington
Ohio River

Wabash River

WEST VIRGINIA
★ Charleston

Baltimore
MD.
DELAWARE
• Dover
Arlington ⊛ Annapolis
Alexandria   Washington, D.C.
Delaware Bay

VIRGINIA
• Richmond ★
• Roanoke
• Newport News • Norfolk
• Portsmouth
Chesapeake Bay

Lake Barkley

Kentucky Lake

TENNESSEE
★ Nashville
• Knoxville
• Chattanooga

NORTH CAROLINA
• Winston-Salem
• Greensboro • Durham
★ Raleigh
• Charlotte

Cape Hatteras

ARKANSAS
• Fayetteville
★ Little Rock

• Memphis
• Huntsville

SOUTH CAROLINA
★ Columbia

Savannah River

MISSISSIPPI

ALABAMA
• Birmingham
★ Montgomery

• Atlanta

GEORGIA
• Macon
• Columbus

• Savannah

Sea Islands

Mississippi River

• Jackson

Chattahoochee River

LOUISIANA
• Shreveport
• Beaumont
★ Baton Rouge
• New Orleans

• Mobile

Chandeleur Islands

★ Tallahassee
• Jacksonville

FLORIDA
• Orlando
Cape Canaveral

• Tampa
• St. Petersburg

Lake Okeechobee

GULF OF MEXICO

THE BAHAMAS

N

Cape Sable
Florida Keys
Straits of Florida

• Fort Lauderdale
• Miami

SCALE
0   250   500 Miles
0   500   500 Kilometers

Projection: Albers Equal Area

CUBA

Boundaries
⊛ National capitals
★ State capitals
• Other cities

NORTH AMERICA: PHYSICAL

ARCTIC OCEAN

EUROPE

ASIA

North Pole

POLAR ICE PACK

Queen Elizabeth Islands

Greenland

Ellesmere Island

Denmark Strait

Cape Farewell

BERING SEA

St. Lawrence Island

Nunivak Island

Bering Strait

BROOKS RANGE

Mt. McKinley 20,320 ft. (6,194 m)

ALASKA RANGE

Yukon River

BEAUFORT SEA

Banks Island

Victoria Island

Great Bear Lake

Mackenzie River

Baffin Island

Baffin Bay

Davis Strait

GULF OF ALASKA

YUKON PLATEAU

Kodiak Island

Alexander Archipelago

Queen Charlotte Islands

Great Slave Lake

Southampton Island

Coats Island

Mansel Island

Hudson Strait

LABRADOR SEA

Arctic Circle

Vancouver Island

PACIFIC OCEAN

Mount Rainier 14,410 ft. (4,392 m)

CASCADE RANGE

Fraser River

Columbia River

COAST RANGE

ROCKY MOUNTAINS

Peace River

Athabasca River

Lake Athabasca

Saskatchewan River

Nelson River

Lake Winnipeg

Hudson Bay

CANADIAN SHIELD

St. Lawrence River

GULF OF ST. LAWRENCE

Anticosti Island

Newfoundland

Prince Edward Island

Cape Breton Island

Cape Mendocino

Snake River

SIERRA NEVADA

CENTRAL VALLEY

GREAT BASIN

DEATH VALLEY

Mount Whitney 14,494 ft. (4,419 m)

COLORADO PLATEAU

Colorado River

Great Salt Lake

BLACK HILLS

Missouri River

GREAT PLAINS

Platte River

Arkansas River

Red River

Mississippi River

OZARK PLATEAU

INTERIOR PLAINS

Lake Superior

Lake Michigan

Lake Huron

Lake Ontario

L. Erie

Ohio River

Cumberland R.

Tennessee River

APPALACHIAN MOUNTAINS

PIEDMONT

Cape Cod

Long Island

ATLANTIC OCEAN

Bermuda

Cape Hatteras

ATLANTIC COASTAL PLAIN

Tropic of Cancer

Guadalupe Island

BAJA CALIFORNIA

GULF OF CALIFORNIA

SIERRA MADRE OCCIDENTAL

SIERRA MADRE ORIENTAL

Rio Grande

Brazos River

GULF COASTAL PLAIN

GULF OF MEXICO

FLORIDA PENINSULA

Cape Canaveral

Florida Keys

Straits of Florida

Bahamas

Cuba

Greater Antilles

Jamaica

Hispaniola

Puerto Rico

Lesser Antilles

Trinidad

CARIBBEAN SEA

Popocatépetl 17,887 ft. (5,452 m)

YUCATÁN PENINSULA

SIERRA MADRE DEL SUR

CENTRAL AMERICA

Lake Nicaragua

ISTHMUS OF PANAMA

SOUTH AMERICA

Equator 0°

N

ELEVATION

FEET	METERS
13,120	4,000
6,560	2,000
1,640	500
656	200
(Sea level) 0	0 (Sea level)
Below sea level	Below sea level
	Ice cap

SCALE

0   250   500   750   1,000 Miles

0   250   500   750   1,000 Kilometers

Projection: Azimuthal Equal Area

NORTH AMERICA: POLITICAL

ASIA

EUROPE

North Pole

ARCTIC OCEAN

Queen
Elizabeth
Islands

Ellesmere Island

Greenland
(DENMARK)

Arctic Circle

ICELAND

Denmark Strait

Beaufort
Sea

Point
Barrow

Banks
Island

Victoria
Island

Baffin
Bay

Baffin Island

Davis Strait

Cape
Farewell

ALASKA
(U.S.)

Yukon River

Great Bear
Lake

Mackenzie River

Great
Slave
Lake

Southampton
Island

Hudson Strait

LABRADOR
SEA

Anchorage

GULF OF
ALASKA

Kodiak
Island

Coats
Island

Mansel
Island

St. Lawrence
Island

BERING SEA

Nunivak
Island

Bering Strait

Alexander
Archipelago

Juneau

Queen
Charlotte
Islands

Peace River

Hudson
Bay

Anticosti
Island

Newfoundland

PACIFIC
OCEAN

Vancouver
Island

Edmonton

CANADA

Prince
Edward
Island

GULF OF
ST. LAWRENCE

St. Pierre and
Miquelon (FRANCE)

Vancouver

Calgary

Lake
Winnipeg

St. Lawrence R.

Quebec

Cape
Breton
Island

Seattle

Winnipeg

Lake
Superior

Montreal

Portland

Columbia River

Lake
Huron

Ottawa

Toronto

Lake
Ontario

Boston

Cape Cod

Snake
River

Minneapolis

Lake
Michigan

Detroit

Lake Erie

New York City

ATLANTIC
OCEAN

Cape Mendocino

Great
Salt
Lake

Salt Lake
City

Missouri River

Milwaukee

Cleveland

Philadelphia

Chicago

Columbus

Baltimore

San Francisco

Denver

Platte
River

Indianapolis

Washington, D.C.

San
Jose

Kansas City

St. Louis

Norfolk

UNITED STATES

Ohio R.

Bermuda
(U.K.)

Los Angeles

Colorado River

Memphis

Cape
Hatteras

San Diego

Phoenix

Atlanta

Tijuana

Red River

Birmingham

Dallas

Mississippi River

Jacksonville

Austin

Cape Canaveral

San
Antonio

Houston

New Orleans

Tropic of Cancer

Rio Grande

Miami

THE
BAHAMAS

Turks and Caicos
Islands (U.K.)

Florida
Keys

Nassau

Monterrey

GULF OF CALIFORNIA

GULF OF
MEXICO

DOMINICAN
REPUBLIC

Puerto Rico (U.S.)

ST. KITTS & NEVIS

ANTIGUA &
BARBUDA

Havana

San Juan

Guadeloupe
(FRANCE)

MEXICO

CUBA

DOMINICA

Guadalajara

Mexico
City

Mérida

Cayman Is.
(U.K.)

HAITI

Santo
Domingo

Virgin Is.
(U.S., U.K.)

Martinique (FRANCE)

BARBADOS

Puebla

Kingston

Port-au-
Prince

ST. LUCIA

Balsas R.

JAMAICA

ST. VINCENT AND
THE GRENADINES

GRENADA

Belmopan

Netherlands
Antilles
(NETHERLANDS)

BELIZE

CARIBBEAN SEA

GUATEMALA

HONDURAS

Aruba (NETHERLANDS)

TRINIDAD AND TOBAGO

Guatemala City

Tegucigalpa

San Salvador

NICARAGUA

EL SALVADOR

Managua

Panama
Canal

N

San José

Panama City

COSTA
RICA

PANAMA

SOUTH AMERICA

	Boundaries
⊛	National capitals
•	Other cities

SCALE

0   500   1000 Miles

0   500   1000 Kilometers

Projection: Azimuthal Equal Area

Equator 0°

Atlas • A11

SOUTH AMERICA: PHYSICAL

CENTRAL AMERICA

CARIBBEAN SEA

Panama Canal

GULF OF PANAMA

Margarita Island

Tobago
Trinidad

Lake Maracaibo

Orinoco River Delta

N

ATLANTIC OCEAN

LLANOS

Cauca River

Magdalena River

Meta River

Orinoco River

Angel Falls

GUIANA HIGHLANDS

Devil's Island
Cape Orange

Malpelo Island

▲ Mount Tolima
18,425 ft. (5,616 m)

Orinoco River

Amazon River Delta

Galápagos Islands

Equator 0°

▲ Mount Chimborazo
20,561 ft. (6,267 m)

GULF OF GUAYAQUIL

Marañón River

Caquetá River

Japurá River

Rio Negro

AMAZON BASIN

Amazon River

Equator 0°

ANDES

Amazon River

Juruá River

Purus River

River

Madeira River

Tapajós River

Xingu River

Tocantins River

Parnaíba River

▲ Mount Huascarán
22,205 ft. (6,768 m)

Ucayali River

BRAZILIAN HIGHLANDS

PACIFIC OCEAN

Lake Titicaca

▲ Ancohuma Peak
20,958 ft. (6,388 m)

Beni River

Mamoré River

MATO GROSSO PLATEAU

Araguaia River

São Francisco River

10°S

Lake Poopó

Pilcomayo River

ATACAMA DESERT

ANDES

CHACO

Paraguay River

BRAZILIAN PLATEAU

San Ambrosio Island

San Félix Island

Salado River

Paraná River

20°S

Tropic of Capricorn

Tropic of Capricorn

Juan Fernández Islands

Salado River

PAMPAS

Uruguay River

Río de la Plata

ATLANTIC OCEAN

▲ Mount Aconcagua
22,834 ft. (6,960 m)

Colorado River

30°S

ELEVATION

FEET		METERS
13,120		4,000
6,560		2,000
1,640		500
656		200
(Sea level) 0		0 (Sea level)
Below sea level		Below sea level

PATAGONIA

GULF OF SAN MATÍAS

Chiloé Island

CHONOS ARCHIPELAGO

GULF OF SAN JORGE

Cape Tres Puntas

40°S

Bahía Grande

SCALE

0   250   500   750   1,000 Miles

0   250   500   750   1,000 Kilometers

Projection: Azimuthal Equal Area

Strait of Magellan

Falkland Islands

South Georgia Islands

TIERRA DEL FUEGO

CAPE HORN

50°S

CENTRAL AMERICA

CARIBBEAN SEA

N

ATLANTIC OCEAN

Barranquilla
Cartagena
Caracas
Lake Maracaibo
VENEZUELA
Orinoco River
Medellín
Bogotá
COLOMBIA
Cali
Georgetown
Paramaribo
GUYANA
SURINAME
Cayenne
FRENCH GUIANA (FRANCE)

Malpelo Island (COLOMBIA)

Rio Negro
Amazon River
Equator 0°

Quito
ECUADOR
Guayaquil
Belém

Galápagos Islands (ECUADOR)

0° Equator

Amazon River

BRAZIL

Marañón River
Trujillo
PERU
Ucayali River
Recife

Callao
Lima
10°S

PACIFIC OCEAN

Lake Titicaca
BOLIVIA
La Paz
Arequipa
Lake Poopó
Sucre
Brasília
São Francisco River
Salvador

Belo Horizonte

Campinas
São Paulo
Rio de Janeiro

Paraná River
PARAGUAY
Asunción
Curitiba

Tropic of Capricorn

San Ambrosio Island (CHILE)
San Félix Island (CHILE)
CHILE
Pôrto Alegre

Córdoba
Uruguay River
URUGUAY

Juan Fernández Islands (CHILE)
Valparaíso
Santiago
Rosario
Buenos Aires
Montevideo
Rio de la Plata
ATLANTIC OCEAN

ARGENTINA

Boundaries

⊛ **National capitals**

• **Other cities**

SCALE
0   250   500   750   1000 Miles
0   250   500   250   1000 Kilometers
Projection: Azimuthal Equal Area

Strait of Magellan
Falkland Islands (U.K.)

Tierra del Fuego

South Georgia Island (U.K.)

ASIA

URAL MOUNTAINS

River

River

Pechora

Kama River

Volga River

CASPIAN SEA

Mt. Elbrus (5,642 m) 18,510 ft.

CAUCASUS MTS.

NORTHERN EUROPEAN PLAIN

Dvina

North Dvina

River

Don River

Rybinsk Reservoir

BARENTS SEA

White Sea

KOLA PENINSULA

Lake Onega

Lake Ladoga

Dnieper River

SEA OF AZOV

CRIMEAN PENINSULA

BLACK SEA

SOUTHWEST ASIA

North Cape

ARCTIC OCEAN

GULF OF BOTHNIA

GULF OF FINLAND

Daugava R.

Ozuina River

BALTIC SEA

PLAINS

Vistula River

Oder River

Nistru River

Dnestr River

Danube River

CARPATHIAN MTS.

TRANSYLVANIAN ALPS

BALKAN PENINSULA

AEGEAN SEA

SEA OF MARMARA

Rhodes

Crete

KJØLEN MOUNTAINS

Lake Vänern

Lake Vättern

Kattegat

Skagerrak

DINARIC ALPS

ADRIATIC SEA

APENNINES

Tiber River

TYRRHENIAN SEA

Sicily

Malta

MEDITERRANEAN SEA

NORWEGIAN SEA

N

Arctic Circle

Faeroe Islands

Shetland Islands

Orkney Islands

NORTH SEA

Hebrides

British Isles

IRISH SEA

PENNINES

Thames River

English Channel

Seine River

Rhine River

Elbe River

Danube River

Lake Geneva

ALPS

15,781 ft. Mont Blanc (4,810 m)

Rhône River

Po River

Corsica

Sardinia

Balearic Islands

AFRICA

Iceland

Bay of Biscay

Loire River

Garonne River

PYRENEES

Ebro River

IBERIAN PENINSULA

Douro River

Tagus River

Guadiana River

Guadalquivir

Cape Finisterre

Strait of Gibraltar

ATLANTIC OCEAN

ELEVATION

FEET	METERS
13,120	4,000
6,560	2,000
1,640	500
656	200
0 (Sea level)	0 (Sea level)
Below sea level	Below sea level

Ice cap

SCALE

0   250   500 Miles

0   500   500 Kilometers

Projection: Azimuthal Equal Area

70°E

60°E

50°E

40°E

30°E

20°E

10°E

0°

10°W

20°W

60°N

50°N

40°N

30°N

ASIA

URAL MOUNTAINS

ASIA

RUSSIA

Nizhny Novgorod

Ural River

Volga River

Don River

CASPIAN SEA

50°E

40°E

50°E

Moscow

BARENTS SEA

WHITE SEA

St. Petersburg

Dnieper River

30°E

40°E

SOUTHWEST ASIA

North Cape

FINLAND

Helsinki

GULF OF FINLAND

Tallinn ESTONIA

Riga LATVIA

BELARUS

Minsk

Kiev

UKRAINE

Chişinău

MOLDOVA

BLACK SEA

30°E

70°N

60°N

ARCTIC OCEAN

Arctic Circle

LITHUANIA

Vilnius

RUSSIA

Warsaw

POLAND

Krakow

SLOVAKIA

Bratislava

HUNGARY

Budapest

ROMANIA

Bucharest

Danube River

SERBIA

Belgrade

Sofia

BULGARIA

Skopje

MACEDONIA

GREECE

Athens

AEGEAN SEA

Rhodes

Crete

20°E

SWEDEN

BALTIC SEA

Stockholm

Göteborg

Berlin

Dresden

CZECH REPUBLIC

Prague

Vienna

AUSTRIA

LIECHTENSTEIN

SLOVENIA

Ljubljana

Zagreb

CROATIA

BOSNIA & HERZEGOVINA

Sarajevo

YUGOSLAVIA

MONTENEGRO

Tiranë

ALBANIA

SEA

MALTA

Valletta

20°E

NORWAY

Oslo

Bergen

NORTH SEA

DENMARK

Copenhagen

Hamburg

Elbe River

GERMANY

Cologne

Bonn

Munich

Luxembourg

Danube River

SWITZERLAND

Bern

Vaduz

Geneva

Lake Geneva

ALPS

Milan

San Marino

SAN MARINO

Po River

ITALY

Rome

VATICAN CITY

Naples

MONACO

Monaco

Corsica (FRANCE)

Sardinia (ITALY)

Sicily

MEDITERRANEAN

AFRICA

10°E

Faeroe Islands (DENMARK)

Shetland Islands

ICELAND

Reykjavik

SCOTLAND

Edinburgh

UNITED KINGDOM

Belfast

NORTHERN IRELAND

Dublin

IRELAND

WALES

ENGLAND

London

Liverpool

Thames R.

English Channel

Channel Islands (U.K.)

British Isles

NETHERLANDS

The Hague

Amsterdam

Brussels

BELGIUM

LUXEMBOURG

Paris

Seine River

FRANCE

Loire River

Rhine River

Rhône River

Lyons

Marseille

PYRENEES

ANDORRA

Andorra la Vella

Barcelona

Balearic Islands (SPAIN)

Valencia

SPAIN

Madrid

Seville

Gibraltar (U.K.)

Strait of Gibraltar

Tagus River

PORTUGAL

Lisbon

Bay of Biscay

ATLANTIC OCEAN

50°N

40°N

10°W

0°

20°W

N

Boundaries

⊛ National capitals

• Other cities

SCALE

0

250

500 Miles

0

250

500 Kilometers

Projection: Azimuthal Equal Area

EUROPE: POLITICAL

ASIA: PHYSICAL

ELEVATION

FEET	METERS
13,120	4,000
6,560	2,000
1,640	500
656	200
0 (Sea level)	0 (Sea level)
Below sea level	Below sea level

Ice cap

SCALE

0   500   1,000 Miles
0   500   1,000 Kilometers

Projection: Modified Oblique Conic

NORTH AMERICA
EUROPE
AFRICA
AUSTRALIA

PACIFIC OCEAN
INDIAN OCEAN
BERING SEA
SEA OF OKHOTSK
SEA OF JAPAN
YELLOW SEA
EAST CHINA SEA
SOUTH CHINA SEA
CELEBES SEA
BANDA SEA
ARAFURA SEA
JAVA SEA
ANDAMAN SEA
Bay of Bengal
ARABIAN SEA
GULF OF OMAN
PERSIAN GULF
GULF OF ADEN
RED SEA
MEDITERRANEAN SEA
BLACK SEA
SEA OF AZOV
CASPIAN SEA
BARENTS SEA
KARA SEA
LAPTEV SEA
GULF OF TONKIN
GULF OF THAILAND

Aleutian Islands
KAMCHATKA PENINSULA
CENTRAL RANGE
Sakhalin Island
Kuril Islands
Hokkaido
Honshu
Shikoku
Kyushu
Ryukyu Islands
Okinawa
Taiwan
Hainan
Luzon
Mindanao
Philippines
Moluccas
Celebes
Borneo
Bangka
Sumatra
Java
Mentawai Islands
Nicobar Islands
Andaman Islands
Lakshadweep Islands
Maldives
Sri Lanka
Socotra Island
Cyprus
New Siberian Islands
Wrangel Island
North Land
Franz Josef Land
Novaya Zemlya
Taymyr Peninsula
New Guinea
MAOKE MOUNTAIN

KOLYMA MTS.
CHERSKIY RANGE
VERKHOYANSKIY RANGE
STANOVOY MOUNTAINS
YABLONOVY RANGE
SAYAN MOUNTAINS
ALTAY SHAN
GREATER KHINGAN RANGE
MONGOLIAN PLATEAU
CENTRAL SIBERIAN PLATEAU
WEST SIBERIAN PLAIN
KAZAKH UPLANDS
URAL MOUNTAINS
SIBERIA
GOBI
TIAN SHAN
TARIM BASIN
TAKLIMAKAN DESERT
KUNLUN MOUNTAINS
PLATEAU OF TIBET
HIMALAYAS
HINDU KUSH
QIN LING
NORTH CHINA PLAIN
CHINA
GREAT WALL
BOHAI HILLS
INDOCHINA PENINSULA
MALAY PENINSULA
INDO-GANGETIC PLAIN
THAR DESERT
DECCAN PLATEAU
WESTERN GHATS
EASTERN GHATS
GREAT SALT DESERT
ZAGROS MTS.
CAUCASUS MTS.
ANATOLIAN PLATEAU
SYRIAN DESERT
AN-NAFUD
RUB' AL-KHALI
SINAI PENINSULA
URAL LOWLAND
USTYURT PLATEAU
KARA-KUM
KYZYL KUM

Mount Everest 29,035 ft. (8,850 m)
Mount Ararat 16,945 ft. (5,165 m)

Amur River
Aldan River
Lena River
Lower Tunguska River
Angara River
Yenisey River
Ob River
Irtysh River
Ishim River
Ural River
Syr Darya
Amu Darya
Lake Balkhash
Lake Baikal
Indus River
Ganges River
Brahmaputra River
Sutlej River
Godavari River
Huang River
Chang River
Xi River
Hong River
Mekong River
Chao Phraya River
Irrawaddy River
Salween River
Tigris River
Euphrates River
Bosporus
Strait of Hormuz
Korea Strait
Luzon Strait
Strait of Malacca

Tropic of Cancer
Arctic Circle
Equator

N

Atlas

Boundaries
National capitals ⊛
Other cities •

PACIFIC
OCEAN

AUSTRALIA

IRIAN JAYA

EAST TIMOR
(U.N.-Administered)

ARAFURA SEA

Aleutian Islands

BERING
SEA

SEA OF
OKHOTSK

Sakhalin
Island

Kuril Islands
(RUSSIA)

Sapporo

JAPAN

SEA OF JAPAN

Vladivostok

Tokyo
Yokohama
Osaka Nagoya
Kyoto
Hiroshima

NORTH KOREA
P'yongyang
SOUTH KOREA
Seoul
Pusan

Ryukyu
Islands
(JAPAN)

TAIWAN
Taipei

PHILIPPINES

Manila
SOUTH
CHINA
SEA

Luzon Strait

CELEBES
SEA

INDONESIA

JAVA Ujung Pandang
SEA
Jakarta Semarang
Bandung Surabaya

MALAYSIA
SINGAPORE
Singapore

BRUNEI
Bandar Seri
Begawan

Kuala Lumpur

Medan

SINGAPORE

Harbin
Changchun
Fushun
Dalian
Shenyang
Qingdao
Shanghai
Nanjing
Beijing
Tianjin

YELLOW
SEA

EAST
CHINA
SEA

Hong Kong
Macao
Guangzhou
Hainan
(CHINA)

VIETNAM
Hanoi

Ho Chi Minh City

CAMBODIA
Phnom
Penh

GULF OF
THAILAND

LAOS
Vientiane

THAILAND
Bangkok

MYANMAR
(BURMA)
Yangon
(Rangoon)

ANDAMAN
SEA

Andaman
Islands
(INDIA)

Nicobar Islands
(INDIA)

Yakutsk

Lena River

Irkutsk

Ulaanbaatar

MONGOLIA

Lake
Baykal

CHINA

Great Wall of China

Xi'an

Chengdu

Chongqing

Wuhan

Huang River

Yangtze River

Nu River

Mekong River

Mandalay

BHUTAN
Thimphu

BANGLADESH
Dhaka

Chittagong

NEPAL
Kathmandu

Irrawaddy River

Brahmaputra River

Ganges River

Kolkata
(Calcutta)

Nagpur

Bhopal

Ahmadabad

INDIA
Hyderabad

Chennai
(Madras)

SRI LANKA
Colombo

Bay of
Bengal

Lakshadweep
Islands
(INDIA)

Bangalore

Mumbai
(Bombay)

MALDIVES
Male

RUSSIA

Novosibirsk

Omsk

Ob River

Irtysh River

Yenisey River

Angara River

Amur River

KARA
SEA

LAPTEV
SEA

BARENTS
SEA

Ekaterinburg
Chelyabinsk

URAL MOUNTAINS

Astana

KAZAKHSTAN

Lake
Balkhash

Almaty

Bishkek
KYRGYZSTAN

UZBEKISTAN
Tashkent

TAJIKISTAN
Dushanbe

ARAL
SEA

TURKMENISTAN
Ashgabat

Kabul
AFGHANISTAN

Islamabad

Lahore
Faisalabad

PAKISTAN

Karachi

Delhi
New
Delhi
Jaipur

Indus River

Moscow

Arctic Circle

CASPIAN SEA

Ural River

GEORGIA
T'bilisi

ARMENIA
Yerevan

AZERBAIJAN
Baku

IRAN
Tehran
Mashhad
Isfahan
Shiraz
Tabriz

Mosul

IRAQ
Baghdad
Basra

Kuwait City
KUWAIT

BAHRAIN
Manama

QATAR
Doha

UNITED
ARAB
EMIRATES
Abu Dhabi

OMAN
Masqat
(Muscat)

PERSIAN GULF

Tigris River

Euphrates River

SAUDI
ARABIA
Riyadh

Mecca
Jidda

YEMEN
Sanaa

RED
SEA

GULF OF ADEN

Socotra
(YEMEN)

ARABIAN
SEA

BLACK SEA

TURKEY
Ankara

Istanbul
Izmir

CYPRUS
Nicosia

LEBANON
Beirut

SYRIA
Damascus
Aleppo

ISRAEL
Tel Aviv
Jerusalem

JORDAN
Amman

MEDITERRANEAN SEA

EUROPE

RUSSIA

AFRICA

INDIAN OCEAN

Equator

Tropic of Cancer

N

SCALE
0   500   1000 Miles
0   500   1000 Kilometers

Projection: Two-Point Equidistant

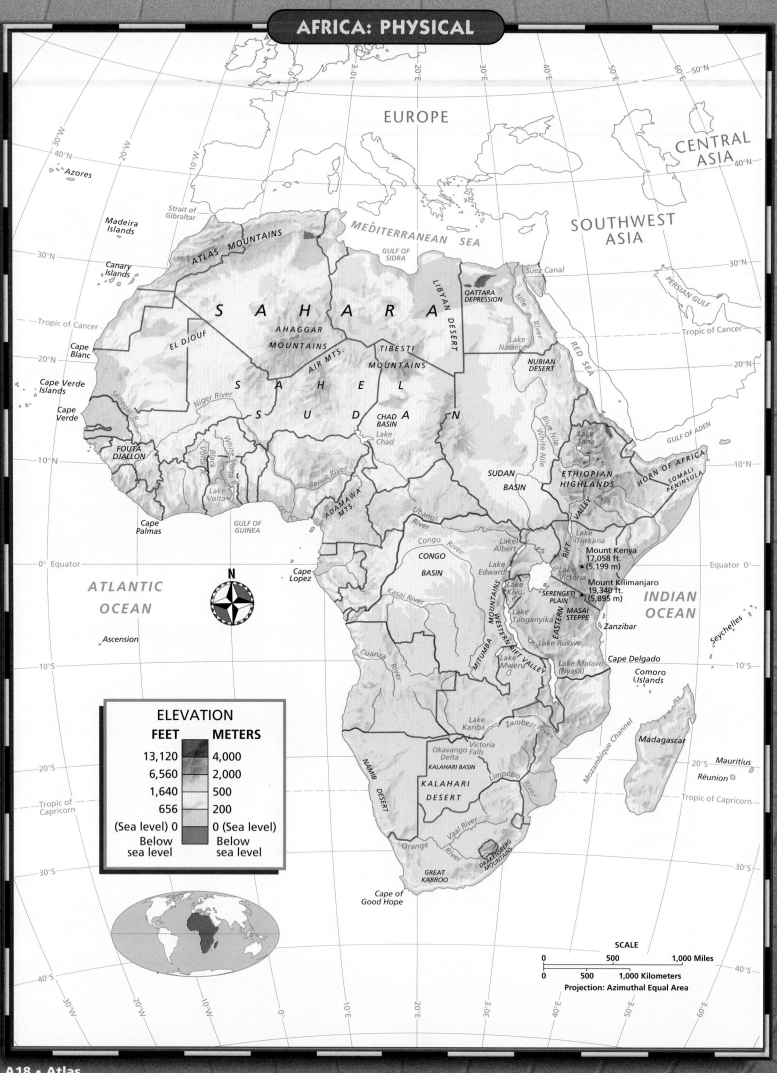

AFRICA: PHYSICAL

EUROPE

CENTRAL ASIA

SOUTHWEST ASIA

MEDITERRANEAN SEA

ATLANTIC OCEAN

INDIAN OCEAN

Azores

Madeira Islands

Canary Islands

Cape Blanc

Cape Verde Islands

Cape Verde

Strait of Gibraltar

ATLAS MOUNTAINS

GULF OF SIDRA

LIBYAN DESERT

QATTARA DEPRESSION

Suez Canal

Nile River

Lake Nasser

RED SEA

PERSIAN GULF

Tropic of Cancer

SAHARA

EL DJOUF

AHAGGAR MOUNTAINS

AIR MTS.

TIBESTI MOUNTAINS

NUBIAN DESERT

SAHEL

SUDAN

CHAD BASIN

Lake Chad

GULF OF ADEN

Lake Tana

Blue Nile

White Nile

ETHIOPIAN HIGHLANDS

HORN OF AFRICA

SOMALI PENINSULA

FOUTA DJALLON

Sénégal R.

Niger River

Black Volta R.

White Volta R.

Lake Volta

Benue River

Cape Palmas

GULF OF GUINEA

ADAMAWA MTS.

SUDAN BASIN

Cape Lopez

Ubangi River

Congo River

CONGO BASIN

Lake Albert

Lake Edward

Lake Kivu

Lake Victoria

Lake Turkana

RIFT VALLEY

Mount Kenya 17,058 ft. (5,199 m)

Mount Kilimanjaro 19,340 ft. (5,895 m)

SERENGETI PLAIN

MASAI STEPPE

Zanzibar

Equator

Ascension

Kasai River

WESTERN RIFT VALLEY

MITUMBA MOUNTAINS

EASTERN

Lake Tanganyika

Lake Rukwa

Lake Mweru

Cuanza River

Seychelles

Comoro Islands

Cape Delgado

NAMIB DESERT

Lake Kariba

Zambezi River

Okavango Delta

Victoria Falls

KALAHARI BASIN

KALAHARI DESERT

Limpopo River

Mozambique Channel

Madagascar

Mauritius

Réunion

Vaal River

Orange River

GREAT KARROO

DRAKENSBERG MOUNTAINS

Cape of Good Hope

ELEVATION

FEET	METERS
13,120	4,000
6,560	2,000
1,640	500
656	200
(Sea level) 0	0 (Sea level)
Below sea level	Below sea level

N

SCALE

0   500   1,000 Miles

0   500   1,000 Kilometers

Projection: Azimuthal Equal Area

AFRICA: POLITICAL

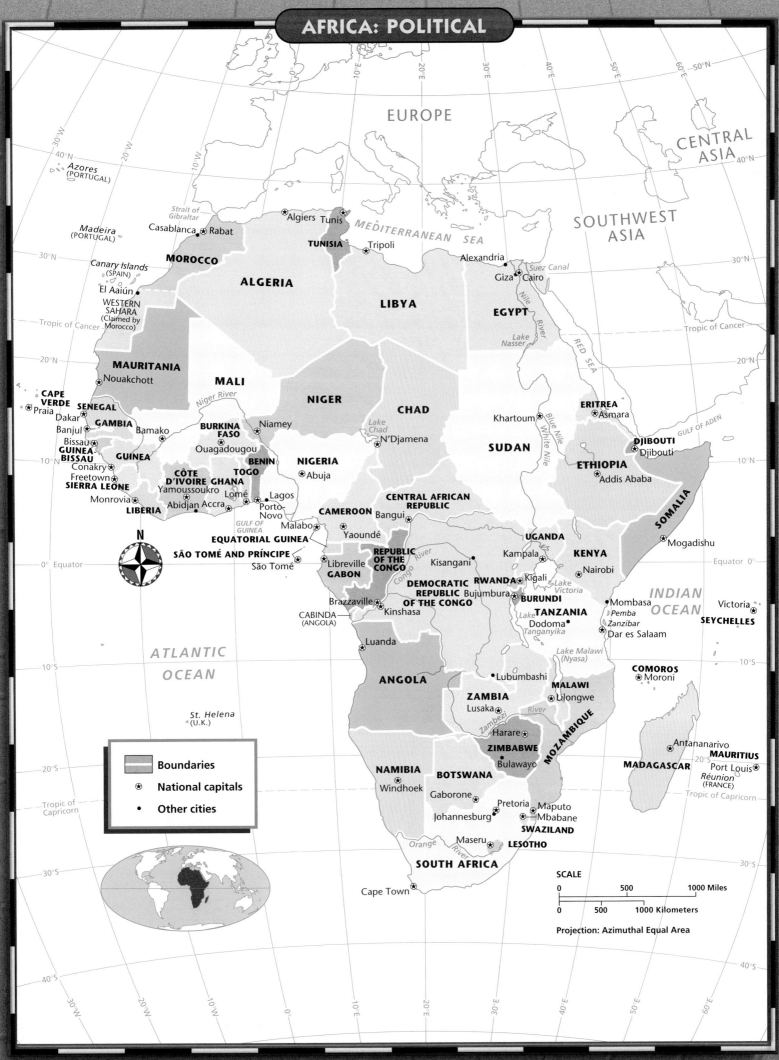

EUROPE

CENTRAL ASIA

SOUTHWEST ASIA

MEDITERRANEAN SEA

Azores (PORTUGAL)

Madeira (PORTUGAL)

Strait of Gibraltar
Algiers Tunis
Casablanca Rabat
TUNISIA
Tripoli

Alexandria
Giza Cairo
Suez Canal

Canary Islands (SPAIN)

MOROCCO

ALGERIA

LIBYA

EGYPT

El Aaiún
WESTERN SAHARA (Claimed by Morocco)

Tropic of Cancer

Nile River

Lake Nasser

RED SEA

Tropic of Cancer

MAURITANIA
Nouakchott

MALI

NIGER

CHAD

Khartoum

ERITREA
Asmara

GULF OF ADEN

CAPE VERDE
Praia

SENEGAL
Dakar

GAMBIA
Banjul

Bamako

BURKINA FASO
Ouagadougou

Niamey

Lake Chad

N'Djamena

SUDAN

White Nile
Blue Nile

DJIBOUTI
Djibouti

Niger River

ETHIOPIA
Addis Ababa

Bissau
GUINEA-BISSAU

GUINEA

Conakry
Freetown
SIERRA LEONE

Monrovia

LIBERIA

CÔTE D'IVOIRE
Yamoussoukro
Abidjan

GHANA
Accra

BENIN
TOGO
Lomé

NIGERIA
Abuja

Lagos
Porto-Novo

CENTRAL AFRICAN REPUBLIC

Bangui

CAMEROON

GULF OF GUINEA

Malabo

EQUATORIAL GUINEA

Yaoundé

SOMALIA

Mogadishu

UGANDA

Kampala

KENYA

Nairobi

SÃO TOMÉ AND PRÍNCIPE

São Tomé

Libreville

GABON

REPUBLIC OF THE CONGO

Congo River

Kisangani

RWANDA
Kigali

Lake Victoria

Equator

N

Equator

INDIAN OCEAN

Victoria

SEYCHELLES

Brazzaville
Kinshasa

DEMOCRATIC REPUBLIC OF THE CONGO

BURUNDI
Bujumbura

TANZANIA
Dodoma

Mombasa
Pemba
Zanzibar
Dar es Salaam

CABINDA (ANGOLA)

Lake Tanganyika

ATLANTIC OCEAN

Luanda

Lubumbashi

Lake Malawi (Nyasa)

COMOROS
Moroni

ANGOLA

ZAMBIA
Lusaka

MALAWI
Lilongwe

MOZAMBIQUE

St. Helena (U.K.)

Zambezi River

Antananarivo

MAURITIUS

NAMIBIA
Windhoek

BOTSWANA

Harare

ZIMBABWE
Bulawayo

MADAGASCAR

Port Louis
Réunion (FRANCE)

Tropic of Capricorn

Tropic of Capricorn

Gaborone

Pretoria
Johannesburg

Maputo
Mbabane
SWAZILAND

Orange River

Maseru
LESOTHO

SOUTH AFRICA

Legend

	Boundaries
⊛	National capitals
•	Other cities

Cape Town

SCALE

0   500   1000 Miles

0   500   1000 Kilometers

Projection: Azimuthal Equal Area

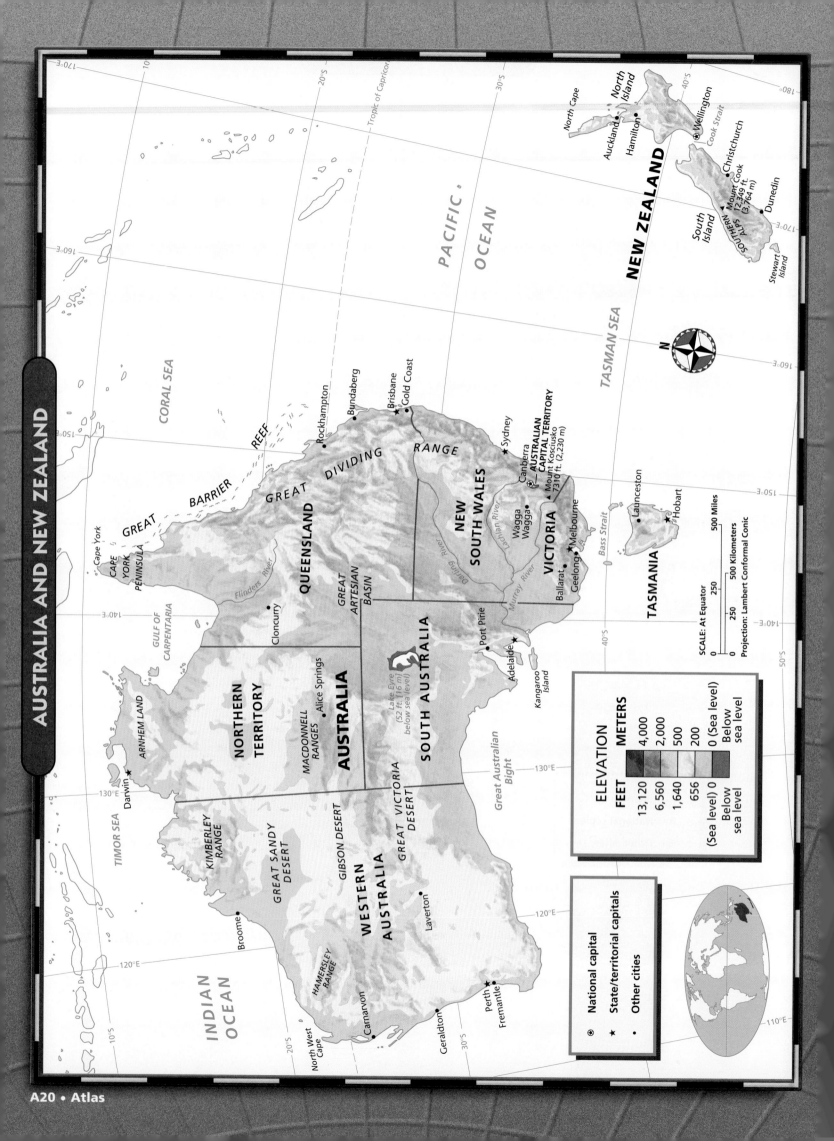

INDIAN OCEAN

TIMOR SEA

GULF OF CARPENTARIA

ARNHEM LAND

Darwin

KIMBERLEY RANGE

GREAT SANDY DESERT

GIBSON DESERT

NORTHERN TERRITORY

WESTERN AUSTRALIA

HAMERSLEY RANGE

GREAT VICTORIA DESERT

North West Cape

Carnarvon

Broome

Laverton

Geraldton

Perth
Fremantle

CAPE YORK PENINSULA

Cape York

CORAL SEA

GREAT BARRIER REEF

GREAT DIVIDING RANGE

Flinders River

Cloncurry

MACDONNELL RANGES • Alice Springs

AUSTRALIA

GREAT ARTESIAN BASIN

SOUTH AUSTRALIA

Lake Eyre
(52 ft. [16 m]
below sea level)

Port Pirie

Adelaide

Kangaroo Island

Great Australian Bight

QUEENSLAND

Rockhampton

Bundaberg

Brisbane
Gold Coast

NEW SOUTH WALES

Sydney

Darling River

Lachlan River

Wagga Wagga

Canberra
AUSTRALIAN CAPITAL TERRITORY
▲ Mount Kosciusko
7310 ft. (2,230 m)

VICTORIA

Murray River

Ballarat

Geelong • Melbourne

Bass Strait

TASMANIA

Launceston

Hobart

PACIFIC OCEAN

Tropic of Capricorn

N

TASMAN SEA

NEW ZEALAND

North Cape

North Island

Auckland
Hamilton

Wellington

Cook Strait

Christchurch

Mount Cook 12,349 ft.
(3,764 m)

SOUTHERN ALPS

South Island

Dunedin

Stewart Island

SCALE: At Equator

0   250   500 Miles

0   250   500 Kilometers

Projection: Lambert Conformal Conic

ELEVATION

FEET	METERS
13,120	4,000
6,560	2,000
1,640	500
656	200
(Sea level) 0	0 (Sea level)
Below sea level	Below sea level

⊛ National capital
★ State/territorial capitals
• Other cities

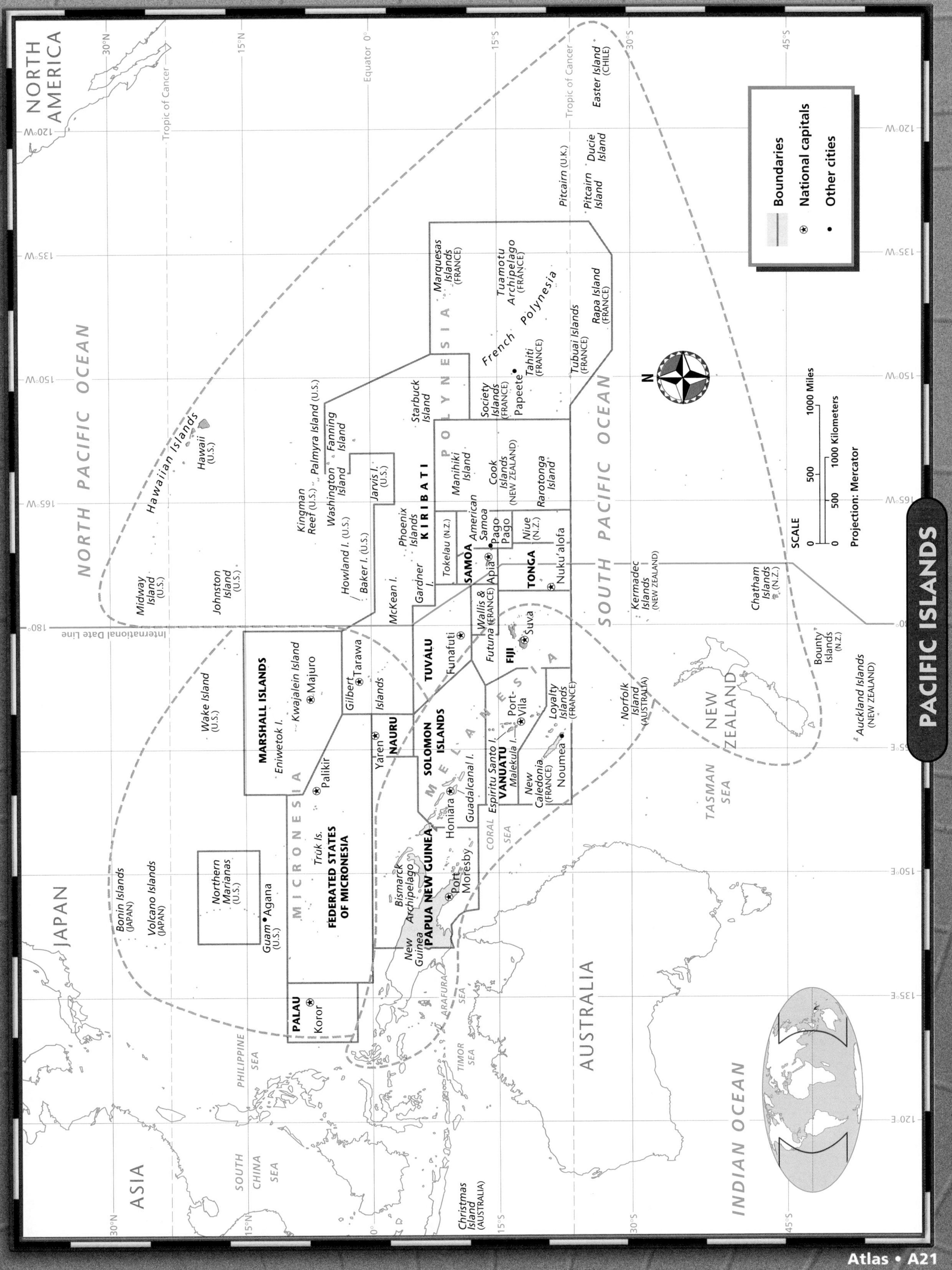

PACIFIC ISLANDS

NORTH AMERICA

ASIA

JAPAN

AUSTRALIA

NEW ZEALAND

NORTH PACIFIC OCEAN

SOUTH PACIFIC OCEAN

PHILIPPINE SEA

SOUTH CHINA SEA

TIMOR SEA

ARAFURA SEA

CORAL SEA

TASMAN SEA

INDIAN OCEAN

MICRONESIA

MELANESIA

POLYNESIA

Boundaries
⊛ National capitals
• Other cities

30°N
15°N
Equator 0°
15°S
30°S
45°S

120°W
135°W
150°W
165°W
180°
165°E
150°E
135°E
120°E

30°N
15°N
15°S
30°S
45°S

Tropic of Cancer
Tropic of Cancer
International Date Line

SCALE
0   500   1000 Miles
0   500   1000 Kilometers
Projection: Mercator

N

Hawaiian Islands
Hawaii (U.S.)

Midway Island (U.S.)
Johnston Island (U.S.)
Wake Island (U.S.)

Kingman Reef (U.S.)
Palmyra Island (U.S.)
Washington Island
Fanning Island
Howland I. (U.S.)
Baker I. (U.S.)
Jarvis I. (U.S.)
McKean I.
Gardner I.
Phoenix Islands

Starbuck Island

Marquesas Islands (FRANCE)
Tuamotu Archipelago (FRANCE)
French Polynesia
Society Islands (FRANCE)
Papeete
Tahiti (FRANCE)
Tubuai Islands (FRANCE)
Rapa Island (FRANCE)

Pitcairn (U.K.)
Pitcairn Island
Ducie Island

Easter Island (CHILE)

Manihiki Island
Cook Islands (NEW ZEALAND)
Rarotonga Island

KIRIBATI
Tarawa
Gilbert Islands

Tokelau (N.Z.)
SAMOA
Apia
American Samoa
Pago Pago
Niue (N.Z.)
TONGA
Nuku'alofa

Kermadec Islands (NEW ZEALAND)

Chatham Islands (N.Z.)

Bounty Islands (N.Z.)

Auckland Islands (NEW ZEALAND)

Norfolk Island (AUSTRALIA)

MARSHALL ISLANDS
Eniwetok I.
Kwajalein Island
⊛ Majuro

FEDERATED STATES OF MICRONESIA
Truk Is.
⊛ Palikir

Northern Marianas (U.S.)
Guam (U.S.)
• Agana

Bonin Islands (JAPAN)
Volcano Islands (JAPAN)

Christmas Island (AUSTRALIA)

PALAU
⊛ Koror

TUVALU
Funafuti ⊛

NAURU
⊛ Yaren

SOLOMON ISLANDS
Honiara ⊛
Guadalcanal I.

PAPUA NEW GUINEA
Port Moresby ⊛
New Guinea
Bismarck Archipelago

VANUATU
Espiritu Santo I.
Malekula I.
Port-Vila ⊛

New Caledonia (FRANCE)
Loyalty Islands (FRANCE)
Noumea

FIJI
Suva ⊛
Wallis & Futuna (FRANCE)

Atlas • A21

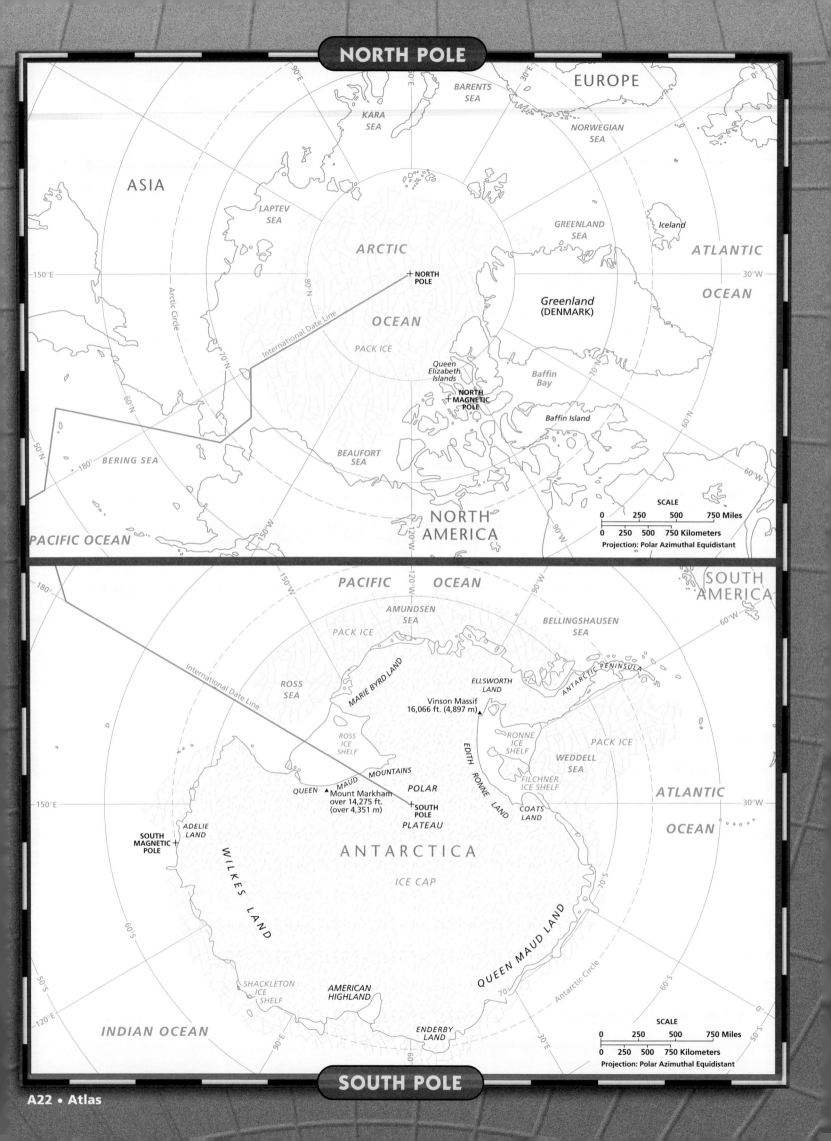

EUROPE

BARENTS
SEA

KARA
SEA

NORWEGIAN
SEA

ASIA

LAPTEV
SEA

GREENLAND
SEA

Iceland

ATLANTIC

ARCTIC

90°E

80°N

150°E

+ NORTH
POLE

30°W

OCEAN

International Date Line

70°N

Greenland
(DENMARK)

OCEAN

PACK ICE

Arctic Circle

60°N

Queen
Elizabeth
Islands

NORTH
+ MAGNETIC
POLE

Baffin
Bay

60°E

50°N

180°

BERING SEA

BEAUFORT
SEA

Baffin Island

PACIFIC OCEAN

150°W

120°W

NORTH
AMERICA

90°W

60°W

SCALE

| 0 | 250 | 500 | 750 Miles |

| 0 | 250 | 500 | 750 Kilometers |

Projection: Polar Azimuthal Equidistant

SOUTH
AMERICA

PACIFIC   OCEAN

120°W

180°

150°W

AMUNDSEN
SEA

90°W

BELLINGSHAUSEN
SEA

60°W

PACK ICE

International Date Line

ROSS
SEA

MARIE BYRD LAND

ELLSWORTH
LAND

ANTARCTIC PENINSULA

Vinson Massif
16,066 ft. (4,897 m)▲

ROSS
ICE
SHELF

RONNE
ICE
SHELF

PACK ICE

30°W

MOUNTAINS

QUEEN   MAUD   ▲ Mount Markham
over 14,275 ft.
(over 4,351 m)

POLAR

EDITH RONNE LAND

WEDDELL
SEA

ATLANTIC

+ SOUTH
POLE

FILCHNER
ICE SHELF

COATS
LAND

OCEAN

150°E

PLATEAU

30°W

SOUTH
MAGNETIC
POLE +

ADELIE
LAND

ANTARCTICA

ICE CAP

WILKES   LAND

QUEEN MAUD LAND

70°S

60°S

INDIAN OCEAN

SHACKLETON
ICE
SHELF

AMERICAN
HIGHLAND

70°

Antarctic Circle

60°S

120°E

ENDERBY
LAND

30°E

50°S

90°E

SCALE

| 0 | 250 | 500 | 750 Miles |

| 0 | 250 | 500 | 750 Kilometers |

Projection: Polar Azimuthal Equidistant

SKILLS HANDBOOK

CONTENTS

S tudying geography requires the ability to understand and use various tools. This Skills Handbook explains how to use maps, charts, and other graphics to help you learn about geography and the various regions of the world. Throughout this textbook, you will have the opportunity to improve these skills and build upon them.

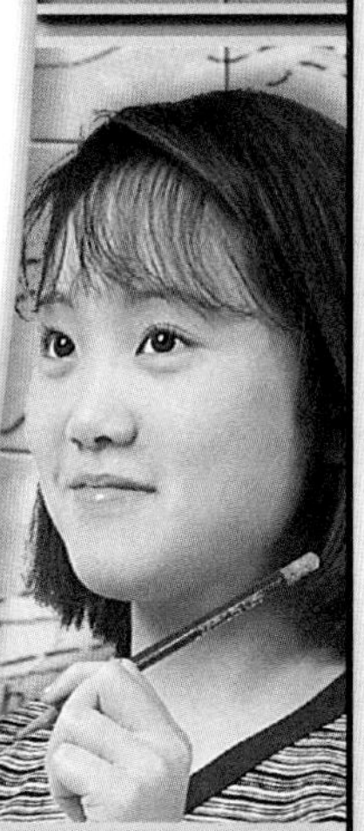

GEOGRAPHIC
Dictionary

- globe
- grid
- latitude
- equator
- parallels
- degrees
- minutes

- longitude
- prime meridian
- meridians
- hemispheres
- continents
- islands
- ocean

- map
- map projections
- compass rose
- scale
- legend

MAPPING
THE EARTH

The Globe

A **globe** is a scale model of Earth. It is useful for looking at the entire Earth or at large areas of Earth's surface.

The pattern of lines that circle the globe in east-west and north-south directions is called a **grid**. The intersection of these imaginary lines helps us find places on Earth.

The east-west lines in the grid are lines of **latitude**. These imaginary lines measure distance north and south of the **equator**. The equator is an imaginary line that circles the globe halfway between the North and South Poles. Lines of latitude are called **parallels** because they are always parallel to the equator. Parallels measure distance from the equator in **degrees**. The symbol for degrees is °. Degrees are further divided into **minutes**. The symbol for minutes is ´. There are 60 minutes in a degree. Parallels north of the equator are labeled with an *N*. Those south of the equator are labeled with an *S*.

The north-south lines are lines of **longitude**. These imaginary lines pass through the Poles. They measure distance east and west of the **prime meridian**. The prime meridian is an imaginary line that runs through Greenwich, England. It represents 0° longitude. Lines of longitude are called **meridians**.

Lines of latitude range from 0°, for locations on the equator, to 90°N or 90°S, for locations at the Poles. See **Figure 1**. Lines of longitude range from 0° on the prime meridian to 180° on a meridian in the mid-Pacific Ocean. Meridians west of the prime meridian to 180° are labeled with a *W*. Those east of the prime meridian to 180° are labeled with an *E*. See **Figure 2**.

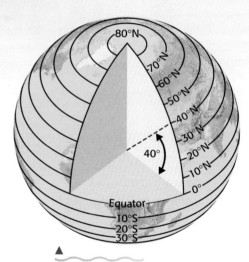

Figure 1: The east-west lines in the grid are lines of latitude.

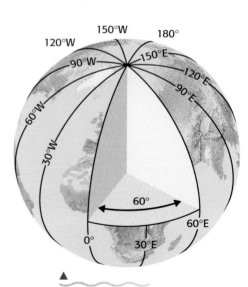

Figure 2: The north-south lines are lines of longitude.

NORTHERN HEMISPHERE

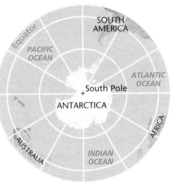

◀
Figure 3: The hemispheres

SOUTHERN HEMISPHERE

EASTERN HEMISPHERE

WESTERN HEMISPHERE

The equator divides the globe into two halves, called **hemispheres**. See **Figure 3**. The half north of the equator is the Northern Hemisphere. The southern half is the Southern Hemisphere. The prime meridian and the 180° meridian divide the world into the Eastern Hemisphere and the Western Hemisphere. The prime meridian separates parts of Europe and Africa into two different hemispheres. To prevent this, some mapmakers divide the Eastern and Western hemispheres at 20° W. This places all of Europe and Africa in the Eastern Hemisphere.

Our planet's land surface is organized into seven large landmasses, called **continents**. They are identified in **Figure 3**. Landmasses smaller than continents and completely surrounded by water are called **islands**. Geographers also organize Earth's water surface into parts. The largest is the world **ocean**. Geographers divide the world ocean into the Pacific Ocean, the Atlantic Ocean, the Indian Ocean, and the Arctic Ocean. Lakes and seas are smaller bodies of water.

YOUR TURN

1. Look at the Student Atlas map on page A4. What islands are located near the intersection of latitude 20° N and longitude 160° W?
2. Name the four hemispheres. In which hemispheres is the United States located?
3. Name the continents of the world.
4. Name the oceans of the world.

Answers
1. The Hawaiian Islands
2. The United States is located in the Northern and Western Hemispheres. The other two hemispheres are the Southern and Eastern.
3. North America, South America, Africa, Europe, Asia, Australia, Antarctica
4. Pacific, Atlantic, Indian, Arctic

MAPMAKING

A **map** is a flat diagram of all or part of Earth's surface. Mapmakers have different ways of showing our round Earth on flat maps. These different ways are called **map projections**. Because our planet is round, all flat maps lose some accuracy. Mapmakers must choose the type of map projection that is best for their purposes. Many map projections are one of three kinds: cylindrical, conic, or flat-plane.

Figure 4: If you remove the peel from the orange and flatten the peel, it will stretch and tear. The larger the piece of peel, the more its shape is distorted as it is flattened. Also distorted are the distances between points on the peel.

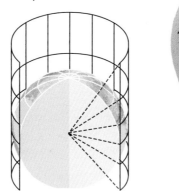

Figure 5A: Paper cylinder

Cylindrical projections are designed from a cylinder wrapped around the globe. See **Figure 5A**. The cylinder touches the globe only at the equator. The meridians are pulled apart and are parallel to each other instead of meeting at the Poles. This causes landmasses near the Poles to appear larger than they really are. **Figure 5B** is a Mercator projection, one type of cylindrical projection. The Mercator projection is useful for navigators because it shows true direction and shape. The Mercator projection for world maps, however, emphasizes the Northern Hemisphere. Africa and South America appear smaller than they really are.

Figure 5B: A Mercator projection, although accurate near the equator, distorts distances between regions of land. This projection also distorts the sizes of areas near the poles.

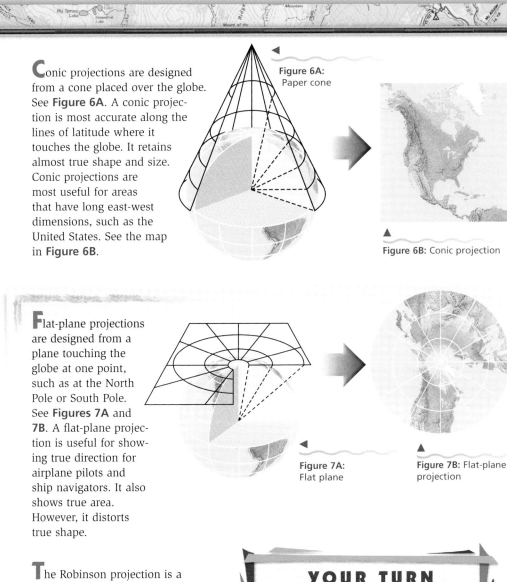

Conic projections are designed from a cone placed over the globe. See **Figure 6A**. A conic projection is most accurate along the lines of latitude where it touches the globe. It retains almost true shape and size. Conic projections are most useful for areas that have long east-west dimensions, such as the United States. See the map in **Figure 6B**.

Figure 6A: Paper cone

Figure 6B: Conic projection

Flat-plane projections are designed from a plane touching the globe at one point, such as at the North Pole or South Pole. See **Figures 7A** and **7B**. A flat-plane projection is useful for showing true direction for airplane pilots and ship navigators. It also shows true area. However, it distorts true shape.

Figure 7A: Flat plane

Figure 7B: Flat-plane projection

The Robinson projection is a compromise between size and shape distortions. It often is used for world maps, such as the map on page 76. The minor distortions in size at high latitudes on Robinson projections are balanced by realistic shapes at the middle and low latitudes.

YOUR TURN

1. What are three major kinds of map projections?
2. Why is a Robinson projection often used for world maps?
3. What kind of projection is a Mercator map?
4. When would a mapmaker choose to use a conic projection?

Your Turn

Answers
1. cylindrical, conic, flat-plane
2. The minor distortions in size at high latitudes are balanced by realistic shapes at the middle and low latitudes.
3. cylindrical
4. when mapping areas that have long east-west dimensions, such as the United States

MAP ESSENTIALS

In some ways, maps are like messages sent out in code. Mapmakers provide certain elements that help us translate these codes. These elements help us understand the message they are presenting about a particular part of the world. Of these elements, almost all maps have directional indicators, scales, and legends, or keys. **Figure 8**, a map of East Asia, has all three elements.

Figure 8: East and Southeast Asia—Physical

A directional indicator shows which directions are north, south, east, and west. Some mapmakers use a "north arrow," which points toward the North Pole. Remember, "north" is not always at the top of a map. The way a map is drawn and the location of directions on that map depend on the perspective of the mapmaker. Maps in this textbook indicate direction by using a **compass rose** (1). A compass rose has arrows that point to all four principal directions, as shown in **Figure 8**.

Mapmakers use scales to represent distances between points on a map. Scales may appear on maps in several different forms. The maps in this textbook provide a line **scale** (2). Scales give distances in miles and kilometers (km).

To find the distance between two points on the map in **Figure 8**, place a piece of paper so that the edge connects the two points. Mark the location of each point on the paper with a line or dot. Then, compare the distance between the two dots with the map's line scale. The number on the top of the scale gives the distance in miles. The number on the bottom gives the distance in kilometers. Because the distances are given in intervals, you will have to approximate the actual distance on the scale.

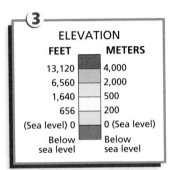

ELEVATION

FEET		METERS
13,120		4,000
6,560		2,000
1,640		500
656		200
(Sea level) 0		0 (Sea level)
Below sea level		Below sea level

The **legend** ③, or key, explains what the symbols on the map represent. Point symbols are used to specify the location of things, such as cities, that do not take up much space on a large-scale map. Some legends, such as the one in **Figure 8**, show which colors represent certain elevations. Other maps might have legends with symbols or colors that represent things such as roads. Legends can also show economic resources, land use, population density, and climate.

Size comparison of Canada to the contiguous United States

Physical maps at the beginning of each unit have size comparison maps ④. An outline of the mainland United States (not including Alaska and Hawaii) is compared to the area under study in that chapter. These size comparison maps help you understand the size of the areas you are studying in relation to the size of the United States.

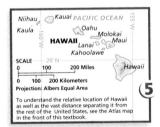

To understand the relative location of Hawaii as well as the vast distance separating it from the rest of the United States, see the Atlas map in the front of this textbook.

Inset maps are sometimes used to show a small part of a larger map. Mapmakers also use inset maps to show areas that are far away from the areas shown on the main map. Maps of the United States, for example, often include inset maps of Alaska and Hawaii ⑤. Those two states are too far from the other 48 states to accurately represent the true distance on the main map. Subject areas in inset maps can be drawn to a scale different from the scale used on the main map.

YOUR TURN

Look at the Student Atlas map on pages A4 and A5.
1. Locate the compass rose. What country is directly west of Madagascar in Africa?
2. What island country is located southeast of India?
3. Locate the distance scale. Using the inset map, find the approximate distance in miles and kilometers from Oslo, Norway, to Stockholm, Sweden.
4. What is the capital of Brazil? What other cities are shown in Brazil?

Your Turn

Answers
1. Mozambique
2. Sri Lanka
3. less than 500 miles (800 km)
4. Brasília is the capital. Rio de Janeiro and São Paulo are also shown.

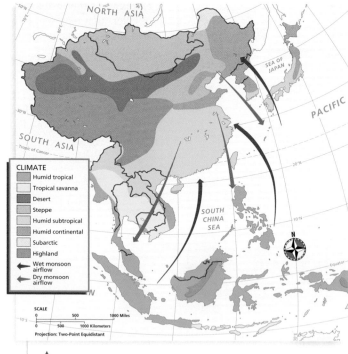
WORKING
WITH MAPS

The Atlas at the front of this textbook includes two kinds of maps: physical and political. At the beginning of most units in this textbook, you will find five kinds of maps. These physical, political, climate, population, and land use and resources maps provide different kinds of information about the region you will study in that unit. These maps are accompanied by questions. Some questions ask you to show how the information on each of the maps might be related.

Mapmakers often combine physical and political features into one map. Physical maps, such as the one in **Figure 8** on page S6, show important physical features in a region, including major mountains and mountain ranges, rivers, oceans and other bodies of water, deserts, and plains. Physical-political maps also show important political features, such as national borders, state and provincial boundaries, and capitals and other important cities. You will find a physical-political map at the beginning of most chapters.

Figure 9: East and Southeast Asia—Climate

Mapmakers use climate maps to show the most important weather patterns in certain areas. Climate maps throughout this textbook use color to show the various climate regions of the world. See **Figure 9**. Colors that identify climate types are found in a legend with each map. Boundaries between climate regions do not indicate an immediate change in the main weather conditions between two climate regions. Instead, boundaries show the general areas of gradual change between climate regions.

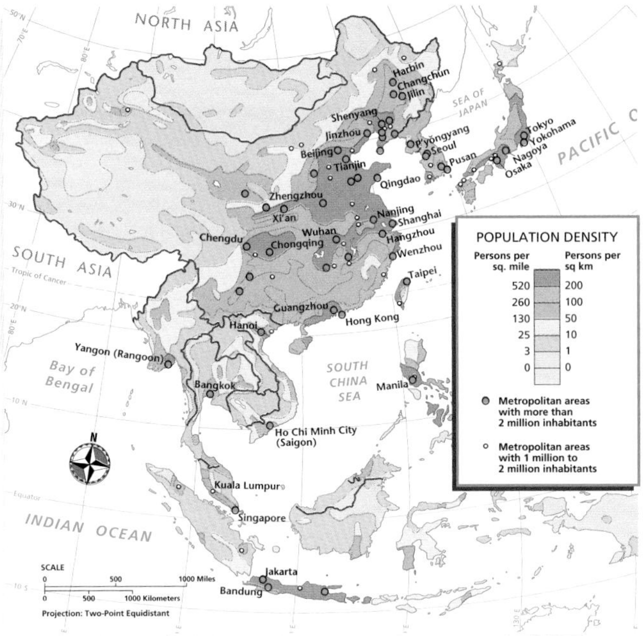

Figure 10: East and Southeast Asia—Population

Population maps show where people live in a particular region. They also show how crowded, or densely populated, regions are. Population maps throughout this textbook use color to show population density. See **Figure 10**. Each color represents a certain number of people living within a square mile or square kilometer. Population maps also use symbols to show metropolitan areas with populations of a particular size. These symbols and colors are shown in a legend.

Land Use and Resources maps show the important resources of a region. See **Figure 11**. Symbols and colors are used to show information about economic development, such as where industry is located or where farming is most common. The meanings of each symbol and color are shown in a legend.

Your Turn

Answers

1. to show the most important weather patterns in certain areas

2. more than 520 persons per square mile (200 per sq km)

3. oil

YOUR TURN

1. What is the purpose of a climate map?

2. Look at the population map. What is the population density of the area around Qingdao in northern China?

3. What energy resource is found near Ho Chi Minh City?

Figure 11: East and Southeast Asia—Land Use and Resources

USING

GRAPHS, DIAGRAMS, CHARTS, AND TABLES

Bar graphs are a visual way to present information. The bar graph in **Figure 12** shows the imports and exports of the countries of southern Europe. The amount of imports and exports in billions of dollars is listed on the left side of the graph. Along the bottom of the graph are the names of the countries of southern Europe. Above each country or group of countries is a vertical bar. The top of the bar corresponds to a number along the left side of the graph. For example, Italy imports $200 billion worth of goods.

Figure 12: Reading a bar graph

Often, line graphs are used to show such things as trends, comparisons, and size. The line graph in **Figure 13** shows the population growth of the world over time. The information on the left shows the number of people in billions. The years being studied are listed along the bottom. Lines connect points that show the population in billions at each year under study. This line graph projects population growth into the future.

Figure 13: Reading a line graph

A pie graph shows how a whole is divided into parts. In this kind of graph, a circle represents the whole. The wedges represent the parts. Bigger wedges represent larger parts of the whole. The pie graph in **Figure 14** shows the percentages of the world's coffee beans produced by various groups of countries. Brazil is the largest grower. It grows 25 percent of the world's coffee beans.

Major Producers of Coffee

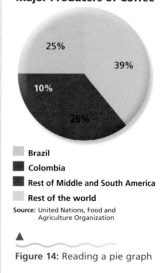

- Brazil
- Colombia
- Rest of Middle and South America
- Rest of the world

Source: United Nations, Food and Agriculture Organization

Figure 14: Reading a pie graph

Age structure diagrams show the number of males and females by age group. These diagrams are split into two sides, one for male and one for female. Along the bottom are numbers that show the number of males or females in the age groups. The age groups are listed on the side of the diagram. The wider the base of a country's diagram, the younger the population of that country. The wider the top of a country's diagram, the older the population.

Some countries have so many younger people that their age structure diagrams are shaped like pyramids. For this reason, these diagrams are sometimes called population pyramids. However, in some countries the population is more evenly distributed by age group. For example, see the age structure diagram for Germany in **Figure 15**. Germany's population is older. It is not growing as fast as countries with younger populations.

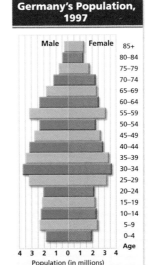

Source: U.S. Census Bureau

▲

Figure 15: Reading an age structure diagram

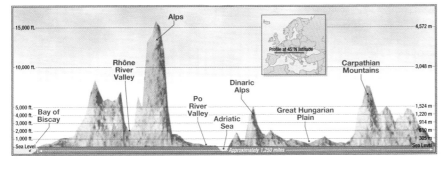

◄

Figure 16: Reading an elevation profile

Each unit atlas includes an elevation profile. See **Figure 16**. It is a side view, or profile, of a region along a line drawn between two points.

Vertical and horizontal distances are figured differently on elevation profiles. The vertical distance (the height of a mountain, for example) is exaggerated when compared to the horizontal distance between the two points. This technique is called vertical exaggeration. If the vertical scale were not exaggerated, even tall mountains would appear as small bumps on an elevation profile.

In each unit and chapter on the various regions of the world, you will find tables that provide basic information about the countries under study.

The countries of Spain and Portugal are listed on the left in the table in **Figure 17**. You can match statistical information on the right with the name of each country listed on the left. The categories of information are listed across the top of the table.

Graphic organizers can help you understand certain ideas and concepts. For example, the diagram in **Figure 18** helps you think about the uses of water. In this diagram, one water use goes in each oval. Graphic organizers can help you focus on key facts in your study of geography.

Time lines provide highlights of important events over a period of time. The time line in **Figure 19** begins at the left with 5000 B.C., when rice was first cultivated in present-day China. The time line highlights important events that have shaped the human and political geography of China.

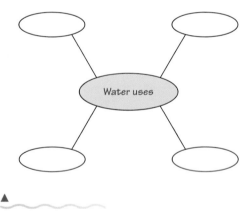

Spain and Portugal

COUNTRY	POPULATION/ GROWTH RATE	LIFE EXPECTANCY	LITERACY RATE	PER CAPITA GDP
Portugal	9,918,040 0.1%	73, male 79, female	90% (1995)	$14,600 (1998)
Spain	39,167,744 0.1%	74, male 82, female	97% (1995)	$16,500 (1998)
United States	272,639,608 0.9%	73, male 80, female	97% (1994)	$31,500 (1998)

Sources: Central Intelligence Agency, *The World Factbook 1999*; *The World Almanac and Book of Facts 1999*

▲
Figure 17: Reading a table

Water uses

▲
Figure 18: Graphic organizer

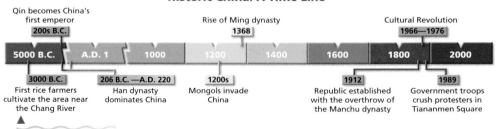

Historic China: A Time Line

Qin becomes China's first emperor
200s B.C.

Rise of Ming dynasty
1368

Cultural Revolution
1966—1976

| 5000 B.C. | A.D. 1 | 1000 | 1200 | 1400 | 1600 | 1800 | 2000 |

3000 B.C.
First rice farmers cultivate the area near the Chang River

206 B.C. —A.D. 220
Han dynasty dominates China

1200s
Mongols invade China

1912
Republic established with the overthrow of the Manchu dynasty

1989
Government troops crush protesters in Tiananmen Square

▲
Figure 19: Reading a time line

Corn: From Field to Consumer

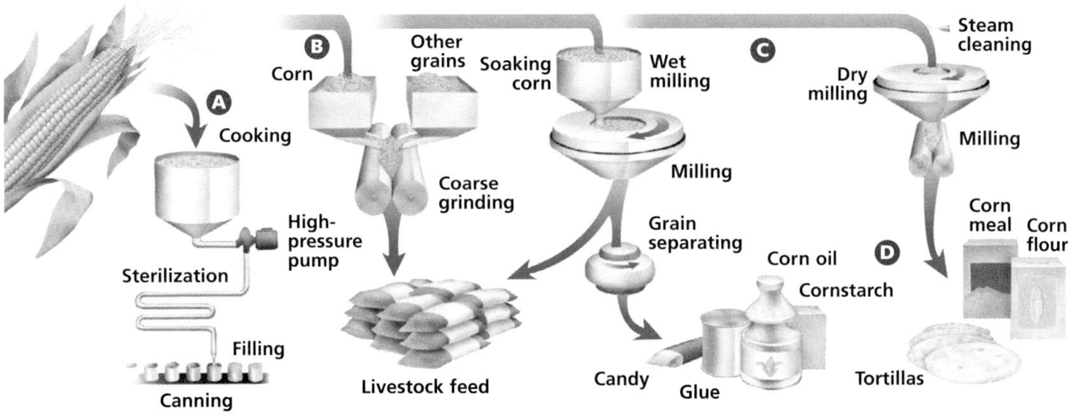

A. Corn can be processed in a variety of ways. Some corn is cooked and then canned.

B. Corn is ground and used for livestock feed.

C. Corn also might be wet-milled or dry-milled. Then grain parts

are used to make different products.

D. Corn by-products, such as cornstarch and corn syrup, are used to make breads, breakfast cereals, puddings, and snack foods. Corn oil is used for cooking.

Figure 20: Reading a flowchart

Flowcharts are visual guides that explain different processes. They lead the reader from one step to the next, sometimes providing both illustrations and text. The flowchart in **Figure 20** shows the different steps involved in harvesting corn and preparing it for use by consumers. The flowchart takes you through the steps of harvesting and processing corn. Captions guide you through flowcharts.

YOUR TURN

1. Look at the statistical table for Spain and Portugal in Figure 17. Which countries have the highest literacy rate?

2. Look at the China time line in Figure 19. Name two important events in China's history between 1200 and 1400.

3. Look at Figure 20. What are three corn products?

Your Turn

Answers

1. Spain and the United States

2. Mongols invade China, rise of Ming dynasty

3. Students should name three of the following products: livestock feed, candy, glue, corn oil, cornstarch, tortillas, cornmeal, corn flour

READING

A TIME-ZONE MAP

The sun is not directly overhead everywhere on Earth at the same time. Clocks are set to reflect the difference in the sun's position. Our planet rotates on its axis once every 24 hours. In other words, in one hour, it makes one twenty-fourth of a complete revolution. Since there are 360 degrees in a circle, we know that the planet turns 15 degrees of longitude each hour. (360° ÷ 24 = 15°) We also know that the planet turns in a west-to-east direction. Therefore, if a place on Earth has the sun directly overhead at this moment (noon), then a place 15 degrees to the west will have the sun directly overhead one hour from now. During that hour the planet will have rotated 15 degrees. As a result, Earth is divided into 24 time zones. Thus, time is an hour earlier for each 15 degrees you move westward on Earth. Time is an hour later for each 15 degrees you move eastward on Earth.

By international agreement, longitude is measured from the prime meridian. This meridian passes through the Royal Observatory in Greenwich, England. Time also is measured from Greenwich and is called

Greenwich mean time (GMT). For each time zone east of the prime meridian, clocks must be set one hour ahead of GMT. For each time zone west of Greenwich, clocks are set back

one hour from GMT. When it is noon in London, it is 1:00 P.M. in Oslo, Norway, one time zone east. However, it is 7 A.M. in New York City, five time zones west.

WORLD TIME ZONES

As you can see by looking at the map below, time zones do not follow meridians exactly. Political boundaries are often used to draw time-zone lines. In Europe and Africa, for example, time zones follow national boundaries. The mainland United States, meanwhile, is divided into four major time zones: Eastern, Central, Mountain, and Pacific. Alaska and Hawaii are in separate time zones to the west of the mainland.

Some countries have made changes in their time zones. For example, most of the United States has daylight savings time in the summer in order to have more evening hours of daylight.

The international date line is a north-south line that runs through the Pacific Ocean. It is located at 180°, although it sometimes varies from that meridian to avoid dividing countries.

At 180°, the time is 12 hours from Greenwich time. There is a time difference of 24 hours between the two sides of the 180° meridian. The 180° meridian is called the international date line because when you cross it, the date and day change. As you cross the date line from the west to the east, you gain a day. If you travel from east to west, you lose a day.

Answers
1. Answers will vary. If students are in the Eastern time zone, the time is the same as New York's. The time is one hour behind New York for each time zone moving westward.
2. five
3. five (includes Seychelles and Mauritius)
4. 9 A.M. The two locations are in the same time zone. The wide north-south distance between the locations makes no difference in time.

YOUR TURN

1. In which time zone do you live? Check your time now. What time is it in New York?
2. How many hours behind New York is Anchorage, Alaska?
3. How many time zones are there in Africa?
4. If it is 9 A.M. in the middle of Greenland, what time is it in São Paulo?

WRITING
ABOUT GEOGRAPHY

Writers have many different reasons for writing. In your study of geography, you might write to accomplish many different tasks. You might write a paragraph or short paper to express your own personal feelings or thoughts about a topic or event. You might also write a paper to tell your class about an event, person, place, or thing. Sometimes you may want to write in order to persuade or convince readers to agree with a certain statement or to act in a particular way.

You will find different kinds of questions at the end of each section, chapter, and unit throughout this textbook. Some questions will require in-depth answers. The following guidelines for writing will help you structure your answers so that they clearly express your thoughts.

Prewriting Prewriting is the process of thinking about and planning what to write. It includes gathering and organizing information into a clear plan. Writers use the prewriting stage to identify their audience and purpose for what is to be written.

The Writing Process

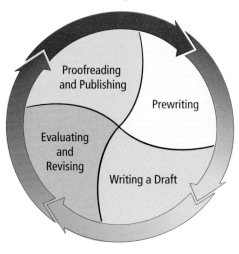

- Proofreading and Publishing
- Prewriting
- Evaluating and Revising
- Writing a Draft

Often, writers do research to get the information they need. Research can include finding primary and secondary sources. You will read about primary and secondary sources later in this handbook.

Writing a Draft After you have gathered and arranged your information, you are ready to begin writing. Many paragraphs are structured in the following way:

- **Topic Sentence:** The topic sentence states the main idea of the paragraph. Putting the main idea into the form of a topic sentence helps keep the paragraph focused.

- **Body:** The body of a paragraph develops and supports the main idea. Writers use a variety of information, including facts, opinions, and examples, to support the main idea.

- **Conclusion:** The conclusion summarizes the writer's main points or restates the main idea.

Evaluating and Revising Read over your paragraphs and make sure you have clearly expressed what you wanted to say. Sometimes it helps to read your paragraphs aloud or to ask someone else to read them. Such methods help you identify rough or unclear sentences and passages. Revise the parts of your paragraph that are not clear or that stray from your main idea. You might want to add, cut, reorder, or replace sentences to make your paragraph as clear as possible.

Proofreading and Publishing Before you write your final draft, read over your paragraphs and correct any errors in grammar, spelling, sentence structure, or punctuation. Common mistakes include misspelled place-names, incomplete sentences, and improper use of punctuation, such as commas. You should use a dictionary and standard grammar guides to help you proofread your work.

After you have revised and corrected your draft, neatly rewrite your paper. Make sure your final version is clean and free of mistakes. The appearance of your final version can affect how your audience perceives and understands your writing.

Practicing the Skill

1. What are the steps in the writing process?
2. How are most paragraphs formed?
3. Write a paragraph or short paper about your community for a visitor. When you have finished your draft, review it and then mark and correct any errors in grammar, spelling, sentence structure, or punctuation. At the bottom of your draft, list key resources—such as a dictionary— that you used to check and correct your work. Then write your final draft. When you are finished with your work, use pencils or pens of different colors to underline and identify the topic sentence, body, and conclusion of your paragraph.

DOING
RESEARCH

Research is at the heart of geographic inquiry. To complete a research project, you may need to use resources other than this textbook. For example, you may want to research specific places or issues not discussed in this textbook. You may also want to learn more about a certain topic that you have studied in a chapter. Following the guidelines below will help you plan and complete research projects for your class.

Planning The first step in approaching a research project is planning. Planning involves deciding on a topic and finding information about that topic.

- **Decide on a Topic.** Before starting any research project, you should decide on one topic. If you are working with a group, all group members should participate in choosing a topic. Sometimes a topic will be assigned to you, but at other times you may have to choose your own. Once you have settled on a topic, make sure you can find resources to help you research it.

- **Find Information.** In order to find a particular book, you need to know how libraries organize their materials. Libraries classify their books by assigning each book a call number that tells you its location. To find the call number, look in the library's card catalog. The card catalog lists books by author, by title, and by subject. Many libraries today have computerized card catalogs. Libraries often provide instructions on how to use their computerized card catalogs. If no instructions are available, ask a library staff member for help.

Practicing the Skill Answers
1. prewriting, writing a draft, evaluating and revising, proofreading and publishing
2. with a topic sentence, body, and conclusion
3. Paragraphs will vary but should use correct grammar, spelling, sentence structure, or punctuation. Students should include a list of resources.

Most libraries have encyclopedias, gazetteers, atlases, almanacs, and periodical indexes. Encyclopedias contain geographic, economic, and political data on individual countries, states, and cities. They also include discussions of historical events, religion, social and cultural issues, and much more. A gazetteer is a geographical dictionary that lists significant natural physical features and other places. An atlas contains maps and visual representations of geographic data. To find up-to-date facts, you can use almanacs, yearbooks, and periodical indexes.

References like *The World Almanac and Book of Facts* include historical information and a variety of statistics. Periodical indexes, particularly *The Reader's Guide to Periodical Literature*, can help you locate informative articles published in magazines. *The New York Times Index* catalogs the newspaper articles published in the *New York Times*.

You may also want to find information on the World Wide Web. The World Wide Web is the part of the Internet where people put files called Web sites for other people to access. To search the World Wide Web, you must use a search engine. A search engine will provide you with a list of Web sites that contain keywords relating to your topic. Search engines also provide Web directories, which allow you to browse Web sites by subject.

Organizing Organization is key to completing research projects of any size. If you are working with a group, every group member should have an assigned task in researching, writing, and completing your project. You and all the group members should keep track of the materials that you used to conduct your research. Then compile those sources into a bibliography and turn it in with your research project.

In addition, information collected during research should be organized in an efficient way. A common method of organizing research information is to use index cards. If you have used an outline to organize your research, you can code each index card with the appropriate main idea number and supporting detail letter from the outline. Then write the relevant information on that card. You might also use computer files in the same way. These methods will help you keep track of what information you have collected and what information you still need to gather.

Some projects will require you to conduct original research. This original research might require you to interview people, conduct surveys, collect unpublished information about your community, or draw a map of a local place. Before you do your original research, make sure you have all the necessary background information. Also, create a pre-research plan so that you can make sure all the necessary tools, such as research sources, are available.

Completing and Presenting Your Project Once you have completed your research project, you will need to present the information you have gathered in some fashion. Many times, you or your group will simply need to write a paper about your research. Research can also be presented in many other ways, however. For example, you could make an audiotape, a drawing, a poster board, a video, or a Web page to explain your research.

Practicing the Skill

1. What kinds of references would you need to research specific current events around the world?
2. Work with a group of four other students to plan, organize, and complete a research project on a topic of interest in your local community. For example, you might want to learn more about a particular individual or event that influenced your community's history. Other topics might include the economic features, physical features, and political features of your community.

ANALYZING
PRIMARY and SECONDARY SOURCES

When conducting research, it is important to use a variety of primary and secondary sources of information. There are many sources of first-hand geographical information, including diaries, letters, editorials, and legal documents such as land titles. All of these are primary sources. Newspaper articles are also considered primary sources, although they generally are written after the fact. Other primary sources include personal memoirs and autobiographies, which people usually write late in life. Paintings and photographs of particular events, persons, places, or things make up a visual record and are also considered primary sources. Because

they allow us to take a close-up look at a topic, primary sources are valuable geographic tools.

Secondary sources are descriptions or interpretations of events written after the events have occurred by persons who did not participate in the events they describe. Geography textbooks such as this one, as well as biographies, encyclopedias, and other reference works, are examples of secondary sources. Writers of secondary sources have the advantage of seeing what happened beyond the moment or place that is being studied. They can provide a perspective wider than that available to one person at a specific time.

How to Study Primary and Secondary Sources

1. **Study the Material Carefully.** Consider the nature of the material. Is it verbal or visual? Is it based on firsthand information or on the accounts of others? Note the major ideas and supporting details.

2. **Consider the Audience.** Ask yourself, "For whom was this message originally meant?" Whether a message was intended for the general public or for a specific private audience may have shaped its style or content.

3. **Check for Bias.** Watch for words or phrases that present a one-sided view of a person or situation.

4. **Compare Sources.** Study more than one source on a topic. Comparing sources gives you a more complete and balanced account of geographical events and their relationships to one another.

Practicing the Skill

1. What distinguishes secondary sources from primary sources?
2. What advantages do secondary sources have over primary sources?
3. Why should you consider the intended audience of a source?
4. Of the following, identify which are primary sources and which are secondary sources: a newspaper, a private journal, a biography, an editorial cartoon, a deed to property, a snapshot of a family vacation, a magazine article about the history of Thailand, an autobiography. How might some of these sources prove to be both primary and secondary sources?

CRITICAL
THINKING

The study of geography requires more than analyzing and understanding tools like graphs and maps. Throughout Holt People, Places, and Change, *you are asked to think critically about some of the information you are studying. Critical thinking is the reasoned judgment of information and ideas. The development of critical thinking skills is essential to learning more about the world around you. Helping you develop critical thinking skills is an important goal of Holt* People, Places, and Change. *The following critical thinking skills appear in the section reviews and chapter reviews of the textbook.*

Summarizing involves briefly restating information gathered from a larger body of information. Much of the writing in this textbook is summarizing. The geographical data in this textbook has been collected from many sources. Summarizing all the qualities of a region or country involves studying a large body of cultural, economic, geological, and historical information.

Finding the main idea is the ability to identify the main point in a set of information. This textbook is designed to help you focus on the main ideas in geography. The Read to Discover questions in each chapter help you identify the main ideas in each section. To find the main idea in any piece of writing, first read the title and introduction. These two elements may point to the main ideas covered in the text.

Also, write down questions about the subject that you think might be answered in the text. Having such questions in mind will focus your reading. Pay attention to any headings or subheadings, which may provide a basic outline of the major ideas. Finally, as you read, note sentences that provide additional details from the general statements that those details support. For example, a trail of facts may lead to a conclusion that expresses the main idea.

Comparing and contrasting involve examining events, points of view, situations, or styles to identify their similarities and differences. Comparing focuses on both the similarities and the differences. Contrasting focuses only on the differences. Studying similarities and differences between people and things can give you clues about the human and physical geography of a region.

Buddhist shrine, Myanmar   Stave church, Norway

Supporting a point of view involves identifying an issue, deciding what you think about it, and persuasively expressing your position. Your stand should be based on specific information. When taking a stand, state your position clearly and give reasons that support it.

Identifying points of view involves noting the factors that influence the opinions of an individual or group. A person's point of view includes beliefs and attitudes that are shaped by factors such as age, gender, race, and economic status. Identifying points of view helps us examine why people see things as they do. It also reinforces the realization that people's views may change over time or with a change in circumstances.

Identifying bias is an important critical thinking skill in the study of any subject. When a point of view is highly personal or based on unreasoned judgment, it is considered biased. Sometimes, a person's actions reflect bias. At its most extreme, bias can be expressed in violent actions against members of a particular culture or group. A less obvious form of bias is a stereotype, or a generalization about a group of people. Stereotypes tend to ignore differences within groups.

Political protest, India

Probably the hardest form of cultural bias to detect has to do with perspective, or point of view. When we use our own culture and experiences as a point of reference from which to make statements about other cultures, we are showing a form of bias called ethnocentrism.

Analyzing is the process of breaking something down into parts and examining the relationships between those parts. For example, to understand the processes behind forest loss, you might study issues involving economic development, the overuse of resources, and pollution.

Evaluating involves assessing the significance or overall importance of something. For example, you might evaluate the success of certain environmental protection laws or the effect of foreign trade on a society. You should base your evaluation on standards that others will understand and are likely to consider valid. For example, an evaluation of international relations after World War II might look at the political and economic tensions between the United States and the Soviet Union. Such an evaluation would also consider the ways those tensions affected other countries around the world.

Identifying cause and effect is part of interpreting the relationships between geographical events. A cause is any action that leads to an event; the outcome of that action is an effect. To explain geographical developments, geographers may point out multiple causes and effects. For example, geographers studying pollution in a region might note a number of causes.

Drought in West Texas　　　*Dallas, Texas*

Ecuador rain forest　　　*Cleared forest, Kenya*

Drawing inferences and drawing conclusions are two methods of critical thinking that require you to use evidence to explain events or information in a logical way. Inferences and conclusions are opinions, but these opinions are based on facts and reasonable deductions.

For example, suppose you know that people are moving in greater and greater numbers to cities in a particular country. You also know that poor weather has hurt farming in rural areas while industry has been expanding in cities. You might be able to understand from this information some of the reasons for the increased migration to cities. You could conclude that poor harvests have pushed people to leave rural areas. You might also conclude that the possibility of finding work in new industries may be pulling people to cities.

Making generalizations and making predictions are two critical thinking skills that require you to form specific ideas from a large body of information. When you are asked to generalize, you must take into account many different pieces of information. You then form a main concept that can be applied to all of the pieces of information. Many times making generalizations can help you see trends. Looking at trends can help you form a prediction. Making a prediction involves looking at trends in the past and present and making an educated guess about how these trends will affect the future.

Communications technology, rural Brazil

SKILLS

Like you, many people around the world have faced difficult problems and decisions. By using appropriate skills such as problem solving and decision making, you will be better able to choose a solution or make a decision on important issues. The following activities will help you develop and practice these skills.

Decision Making

Decision making involves choosing between two or more options. Listed below are guidelines to help you with making decisions.

1. **Identify a situation that requires a decision.** Think about your current situation. What issue are you faced with that requires you to take some sort of action?

2. **Gather information.** Think about the issue. Examine the causes of the issue or problem and consider how it affects you and others.

3. **Identify your options.** Consider the actions that you could take to address the issue. List these options so that you can compare them.

4. **Make predictions about consequences.** Predict the consequences of taking the actions listed for each of your options. Compare these possible consequences. Be sure the option you choose produces the results you want.

5. **Take action to implement a decision.** Choose a course of action from your available options, and put it into effect.

Problem Solving

Problem solving involves many of the steps of decision making. Listed below are guidelines to help you solve problems.

1. **Identify the problem.** Identify just what the problem or difficulty is that you are facing. Sometimes you face a difficult situation made up of several different problems. Each problem may require its own solution.

2. **Gather information.** Conduct research on any important issues related to the problem. Try to find the answers to questions like the following: What caused this problem? Who or what does it affect? When did it start?

3. **List and consider options.** Look at the problem and the answers to the questions you asked in Step 2. List and then think about all the possible ways in which the problem could be solved. These are your options—possible solutions to the problem.

4. **Examine advantages and disadvantages.** Consider the advantages and disadvantages of all the options that you have listed. Make sure that you consider the possible long-term effects of each possible solution. You should also determine what steps you will need to take to achieve each possible solution. Some suggestions may sound good at first but may turn out to be impractical or hard to achieve.

5. **Choose and implement a solution.** Select the best solution from your list and take the steps to achieve it.

6. **Evaluate the effectiveness of the solution.** When you have completed the steps needed to put your plan into action, evaluate its effectiveness. Is the problem solved? Were the results worth the effort required? Has the solution itself created any other problems?

Practicing the Skill

1. Chapter 24, Section 2: East Africa's History and Culture, describes the challenges of religious and ethnic conflict occurring in the region. Imagine that you are an ambassador to Rwanda. Use the decision-making guidelines to help you come up with a plan to help resolve the problems there. Be prepared to defend your decision.

2. Identify a similar problem discussed in another chapter and apply the problem-solving process to come up with a solution.

Becoming a Strategic Reader

by Dr. Judith Irvin

Everywhere you look, print is all around us. In fact, you would have a hard time stopping yourself from reading. In a normal day, you might read cereal boxes, movie posters, notes from friends, T-shirts, instructions for video games, song lyrics, catalogs, billboards, information on the Internet, magazines, the newspaper, and much, much more. Each form of print is read differently depending on your purpose for reading. You read a menu differently from poetry, and a motorcycle magazine is read differently than a letter from a friend. Good readers switch easily from one type of text to another. In fact, they probably do not even think about it, they just do it.

When you read, it is helpful to use a strategy to remember the most important ideas. You can use a strategy before you read to help connect information you already know to the new information you will encounter. Before you read, you can also predict what a text will be about by using a previewing strategy. During the reading you can use a strategy to help you focus on main ideas, and after reading you can use a strategy to help you organize what you learned so that you can remember it later. *Holt People, Places, and Change* was designed to help you more easily understand the ideas you read. Important reading strategies employed in *Holt People, Places, and Change* include:

A Tools to help you **preview and predict** what the text will be about

B Ways to help you use and analyze visual information

C Ideas to help you **organize the information** you have learned

A. Previewing and Predicting

How can I figure out what the text is about before I even start reading a section?

Previewing and **predicting** are good methods to help you understand the text. If you take the time to preview and predict before you read, the text will make more sense to you during your reading.

1 Usually, your teacher will set the purpose for reading. After reading some new information, you may be asked to write a summary, take a test, or complete some other type of activity.

"After reading about Spain and Portugal, you will work with a partner to present a history of the countries to a travel group..."

Previewing and Predicting

step 1 Identify your purpose for reading. Ask yourself what you will do with this information once you have finished reading.

▼

step 2 Ask yourself what is the main idea of the text and what are the key vocabulary words you need to know.

▼

step 3 Use signal words to help identify the structure of the text.

▼

step 4 Connect the information to what you already know.

2 As you preview the text, use **graphic signals** such as headings, subheadings, and boldface type to help you determine what is important in the text. Each section of *Holt People, Places, and Change* opens by giving you important clues to help you preview the material.

Looking at the section's **main heading** and subheadings can give you an idea of what is to come.

Read to Discover questions give you clues as to the section's main ideas.

Section 4 Spain and Portugal

Read to Discover
1. What were some major events in the history of Spain and Portugal?
2. What are the cultures of Spain and Portugal like?
3. What are Spain and Portugal like today?

Define
Moors
dialect
cork

Locate
Lisbon
Madrid
Barcelona

WHY IT MATTERS
Some Basque separatists have used violence to try to gain their independence from Spain. Use CNNfyi.com or other current events sources to find examples of this problem. Record your findings in your journal.

Paella, a popular dish in Spain

Define and **Locate** terms let you know the key vocabulary and places you will encounter in the section.

3 Other tools that can help you in previewing are **signal words**. These words prepare you to think in a certain way. For example, when you see words such as *similar to, same as,* or *different from,* you know that the text will probably compare and contrast two or more ideas. Signal words indicate how the ideas in the text relate to each other. Look at the list below for some of the most common signal words grouped by the type of text structures they include.

SIGNAL WORDS

Cause and Effect	Compare and Contrast	Description	Problem and Solution	Sequence or Chronological Order
because	different from	for instance	the question is	not long after
since	same as	for example	a solution	next
consequently	similar to	such as	one answer is	then
this led to...so	as opposed to	to illustrate		initially
if...then	instead of	in addition		before
nevertheless	although	most importantly		after
accordingly	however	another		finally
because of	compared with	furthermore		preceding
as a result of	as well as	first, second ...		following
in order to	either...or			on (date)
may be due to	but			over the years
for this reason	on the other hand			today
not only...but	unless			when

4 Learning something new requires that you connect it in some way with something you already know. This means you have to think before you read and while you read. You may want to use a chart like this one to remind yourself of the information already familiar to you and to come up with questions you want answered in your reading. The chart will also help you organize your ideas after you have finished reading.

What I know	What I want to know	What I learned

B. Use and Analyze Visual Information

How can all the pictures, maps, graphs, and time lines with the text help me be a stronger reader?

Using visual information can help you understand and remember the information presented in *Holt People, Places, and Change*. Good readers make a picture in their mind when they read. The pictures, charts, graphs, and diagrams that occur throughout *Holt People, Places, and Change* are placed strategically to increase your understanding.

1 You might ask yourself questions like these:

Why did the writer include this image with the text? What details about this image are mentioned in the text?

Analyzing Visual Information

step 1 As you preview the text, ask yourself how the visual information relates to the text.

▼

step 2 Generate questions based on the visual information.

▼

step 3 After reading the text, go back and review the visual information again.

▼

step 4 Connect the information to what you already know.

2 After you have read the text, see if you can answer your own questions.

→ *Why are windmills important?*

→ *What technology do windmills use to pump water?*

→ *How might environment affect the use of windmills?*

2 Maps, graphs, and charts help you organize information about a place. You might ask questions like these:

> *How does this map support what I have read in the text?*
>
> *What does the information in this bar graph add to the text discussion?*

→ *What is the purpose of this map?*

→ *What special features does the map show?*

→ *What do the colors, lines, and symbols on the map represent?*

Land Use and Resources

→ *What information is the writer trying to present with this graph?*

→ *Why did the writer use a bar graph to organize this information?*

Imports and Exports of Southern Europe

3 After reading the text, go back and review the visual information again.

4 Connect the information to what you already know.

C. Organize Information

Once I learn new information, how do I keep it all straight so that I will remember it?

To help you remember what you have read, you need to find a way of **organizing information**. Two good ways of doing this are by using graphic organizers and concept maps. **Graphic organizers** help you understand important relationships—such as cause and effect, compare/contrast, sequence of events, and problem/solution—within the text. **Concept maps** provide a useful tool to help you focus on the text's main ideas and organize supporting details.

Identifying Relationships

Using graphic organizers will help you recall important ideas from the section and give you a study tool you can use to prepare for a quiz or test or to help with a writing assignment. Some of the most common types of graphic organizers are shown below.

▶ Cause and Effect

Events in history cause people to react in a certain way. Cause-and-effect patterns show the relationship between results and the ideas or events that made the results occur. You may want to represent cause-and-effect relationships as one cause leading to multiple effects,

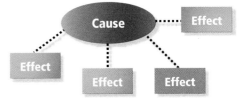

or as a chain of cause-and-effect relationships.

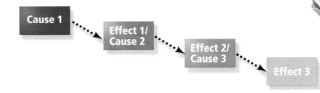

Constructing Graphic Organizers

step 1 Preview the text, looking for signal words and the main idea.

▼

step 2 Form a hypothesis as to which type of graphic organizer would work best to display the information presented.

▼

step 3 Work individually or with your classmates to create a visual representation of what you read.

▶ Comparing and Contrasting

Graphic organizers are often useful when you are comparing or contrasting information. Compare-and-contrast diagrams point out similarities and differences between two concepts or ideas.

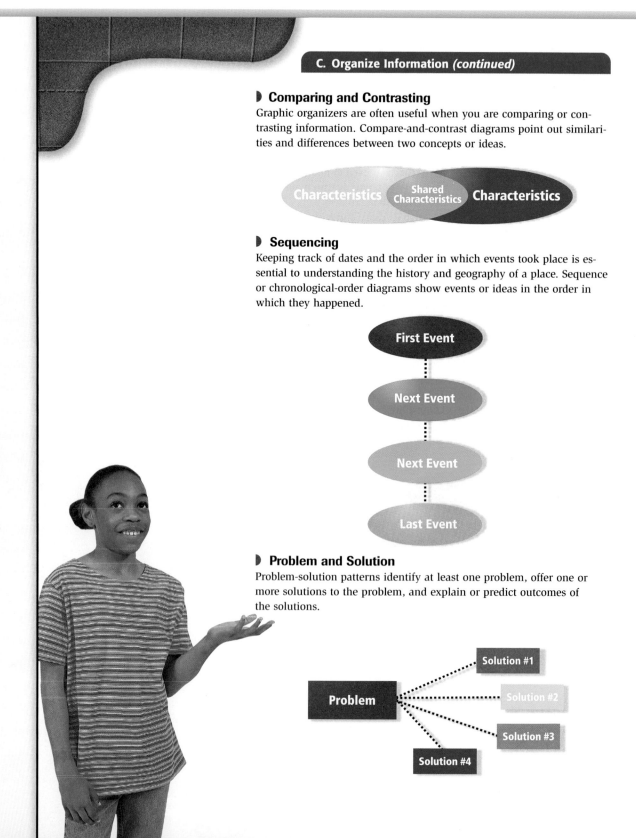

Characteristics   Shared Characteristics   Characteristics

▶ Sequencing

Keeping track of dates and the order in which events took place is essential to understanding the history and geography of a place. Sequence or chronological-order diagrams show events or ideas in the order in which they happened.

First Event

Next Event

Next Event

Last Event

▶ Problem and Solution

Problem-solution patterns identify at least one problem, offer one or more solutions to the problem, and explain or predict outcomes of the solutions.

Solution #1

Solution #2

Problem

Solution #3

Solution #4

Identifying Main Ideas and Supporting Details

One special type of graphic organizer is the concept map. A concept map allows you to zero in on the most important points of the text. The map is made up of lines, boxes, circles, and/or arrows. It can be as simple or as complex as you need it to be to accurately represent the text. Here are a few examples of concept maps you might use.

Constructing Concept Maps

step 1 Preview the text, looking for what type of structure might be appropriate to display as a concept map.

▼

step 2 Taking note of the headings, boldface type, and text structure, sketch a concept map you think could best illustrate the text.

▼

step 3 Using boxes, lines, arrows, circles, or any shapes you like, display the ideas of the text in the concept map.

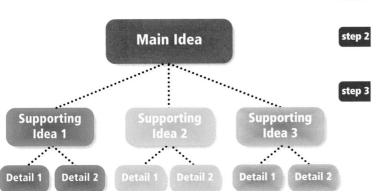

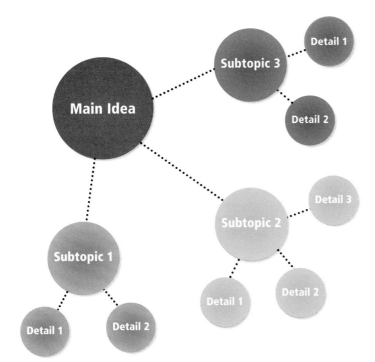

Standardized Test-Taking Strategies

A number of times throughout your school career, you may be asked to take standardized tests. These tests are designed to demonstrate the content and skills you have learned. It is important to keep in mind that in most cases the best way to prepare for these tests is to pay close attention in class and take every opportunity to improve your general social studies, reading, writing, and mathematical skills.

Tips for Taking the Test

1. Be sure that you are well rested.
2. Be on time, and be sure that you have the necessary materials.
3. Listen to the teacher's instructions.
4. Read directions and questions carefully.
5. **DON'T STRESS!** Just remember what you have learned in class, and you should do well.

Practice the strategies at go.hrw.com

Tackling Social Studies

The social studies portions of many standardized tests are designed to test your knowledge of the content and skills that you have been studying in one or more of your social studies classes. Specific objectives for the test vary, but some of the most common include the following:

1. Demonstrate an understanding of issues and events in history.
2. Demonstrate an understanding of geographic influences on historical issues and events.
3. Demonstrate an understanding of economic and social influences on historical issues and events.
4. Demonstrate an understanding of political influences on historical issues and events.
5. Use critical thinking skills to analyze social studies information.

Standardized tests usually contain multiple-choice and, sometimes, open-ended questions. The multiple-choice items will often be based on maps, tables, charts, graphs, pictures, cartoons, and/or reading passages and documents.

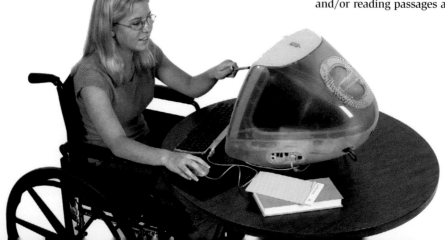

Tips for Answering Multiple-Choice Questions

1. If there is a written or visual piece accompanying the multiple-choice question, pay careful attention to the title, author, and date.

2. Then read through or glance over the content of the written or visual piece accompanying the question to familiarize yourself with it.

3. Next, read the multiple-choice question first for its general intent. Then reread it carefully, looking for words that give clues or can limit possible answers to the question. For example, words such as *most* or *best* tell you that there may be several correct answers to a question, but you should look for the most appropriate answer.

4. Read through the answer choices. Always read all of the possible answer choices even if the first one seems like the correct answer. There may be a better choice farther down in the list.

5. Reread the accompanying information (if any is included) carefully to determine the answer to the question. Again, note the title, author, and date of primary-source selections. The answer will rarely be stated exactly as it appears in the primary source, so you will need to use your critical thinking skills to read between the lines.

6. Think of what you already know about the time in history or person involved and use that to help limit the answer choices.

7. Finally, reread the question and selected answer to be sure that you made the best choice and that you marked it correctly on the answer sheet.

Strategies for Success

There are a variety of strategies you can prepare ahead of time to help you feel more confident about answering questions on social studies standardized tests. Here are a few suggestions:

1. Adopt an acronym—a word formed from the first letters of other words—that you will use for analyzing a document or visual piece that accompanies a question.

Helpful Acronyms

For a document, use **SOAPS**, which stands for

S Subject
O Overview
A Audience
P Purpose
S Speaker/author

For a picture, cartoon, map, or other visual piece of information, use **OPTIC**, which stands for

O Occasion (or time)
P Parts (labels or details of the visual)
T Title
I Interrelations (how the different parts of the visual work together)
C Conclusion (what the visual means)

2. Form visual images of maps and try to draw them from memory. Standardized tests will most likely include maps showing many features, such as states, countries, continents, and oceans. Those maps may also show patterns in settlement and the size and distribution of cities. For example, in studying the United States, be able to see in your mind's eye such things as where the states and major cities are located. Know major physical features, such as the Mississippi River, the Appalachian and Rocky Mountains, the Great Plains, and the various regions of the United States, and be able to place them on a map. Such features may help you understand patterns in the distribution of population and the size of settlements.

3. When you have finished studying a geographic region or period in history, try to think of who or what might be important enough for a standardized test. You may want to keep your ideas in a notebook to refer to when it is almost time for the test.

4. Standardized tests will likely test your understanding of the political, economic, and social processes that

shape a region's history, culture, and geography. Questions may also ask you to understand the impact of geographic factors on major events. For example, some may ask about the effects of migration and immigration on various societies and population change. In addition, questions may test your understanding of the ways humans interact with their environment.

5. For the skills area of the tests, practice putting major events and personalities in order in your mind. Sequencing people and events by dates can become a game you play with a friend who also has to take the test. Always ask yourself "why" this event is important.

6. Follow the tips under "Ready for Reading" below when you encounter a reading passage in social studies, but remember that what you have learned about history can help you in answering reading-comprehension questions.

Ready for Reading

The main goal of the reading sections of most standardized tests is to determine your understanding of different aspects of a piece of writing. Basically, if you can grasp the main idea and the writer's purpose and then pay attention to the details and vocabulary so that you are able to draw inferences and conclusions, you will do well on the test.

Tips for Answering Multiple-Choice Questions

1. Read the passage as if you were not taking a test.

2. Look at the big picture. Ask yourself questions like, "What is the title?", "What do the illustrations or pictures tell me?", and "What is the writer's purpose?"

3. Read the questions. This will help you know what information to look for.

4. Reread the passage, underlining information related to the questions.

Types of Multiple-Choice Questions

1. **Main Idea** This is the most important point of the passage. After reading the passage, locate and underline the main idea.

2. **Significant Details** You will often be asked to recall details from the passage. Read the question and underline the details as you read, but remember that the correct answers do not always match the wording of the passage precisely.

3. **Vocabulary** You will often need to define a word within the context of the passage. Read the answer choices and plug them into the sentence to see what fits best.

4. **Conclusion and Inference** There are often important ideas in the passage that the writer does not state directly. Sometimes you must consider multiple parts of the passage to answer the question. If answers refer to only one or two sentences or details in the passage, they are probably incorrect.

5. Go back to the questions and try to answer each one in your mind before looking at the answers.

6. Read all the answer choices and eliminate the ones that are obviously incorrect.

Tips for Answering Short-Answer Questions

1. Read the passage in its entirety, paying close attention to the main events and characters. Jot down information you think is important.

2. If you cannot answer a question, skip it and come back later.

3. Words such as *compare, contrast, interpret, discuss,* and *summarize* appear often in short-answer questions. Be sure you have a complete understanding of each of these words.

4. To help support your answer, return to the passage and skim the parts you underlined.

5. Organize your thoughts on a separate sheet of paper. Write a general statement with which to begin. This will be your topic statement.

6. When writing your answer, be precise but brief. Be sure to refer to details in the passage in your answer.

Targeting Writing

On many standardized tests, you will occasionally be asked to write an essay. In order to write a concise essay, you must learn to organize your thoughts before you begin writing the actual piece. This keeps you from straying too far from the essay's topic.

Tips for Answering Composition Questions

1. Read the question carefully.

2. Decide what kind of essay you are being asked to write. Essays usually fall into one of the following types: persuasive, classificatory, compare/contrast, or "how to." To determine the type of essay, ask yourself questions like, "Am I trying to persuade my audience?", "Am I comparing or contrasting ideas?", or "Am I trying to show the reader how to do something?"

3. Pay attention to key words, such as *compare, contrast, describe, advantages, disadvantages, classify,* or *speculate.* They will give you clues as to the structure that your essay should follow.

4. Organize your thoughts on a sheet of paper. You will want to come up with a general topic sentence that expresses your main idea. Make sure this sentence addresses the question. You should then create an outline or some type of graphic organizer to help you organize the points that support your topic sentence.

5. Write your composition using complete sentences. Also, be sure to use correct grammar, spelling, punctuation, and sentence structure.

6. Be sure to proofread your essay once you have finished writing.

Gearing Up for Math

On most standardized tests you will be asked to solve a variety of mathematical problems that draw on the skills and information you have learned in class. If math problems sometimes give you difficulty, have a look at the tips below to help you work through the problems.

Tips for Solving Math Problems

1. Decide what is the goal of the question. Read or study the problem carefully and determine what information must be found.

2. Locate the factual information. Decide what information represents key facts—the ones you must have to solve the problem. You may also find facts you do not need to reach your solution. In some cases, you may determine that more information is needed to solve the problem. If so, ask yourself, "What assumptions can I make about this problem?" or "Do I need a formula to help solve this problem?"

3. Decide what strategies you might use to solve the problem, how you might use them, and what form your solution will be in. For example, will you need to create a graph or chart? Will you need to solve an equation? Will your answer be in words or numbers? By knowing what type of solution you should reach, you may be able to eliminate some of the choices.

4. Apply your strategy to solve the problem and compare your answer to the choices.

5. If the answer is still not clear, read the problem again. If you had to make calculations to reach your answer, use estimation to see if your answer makes sense.

EXPLORING OUR WORLD

Direct students' attention to the photographs on these pages. Point out that in this unit students will learn there are two major branches of geography. One is physical geography, which deals with land, water, climate, and similar topics; the other is human geography, which involves people. Ask students which photos may relate more closely to physical geography *(rosettes, tropical landscape, iceberg)* and which relate more to human geography *(Carnival scene).*

Ask which photo shows a cold climate *(iceberg)*, and which shows a warm climate *(La Digue Island).*

On what familiar images or clues do we depend for the answers? *(Possible answers: ice, lush vegetation, palm trees)* Ask why the photo of the rosette plants does not give us much information about how warm or cold the climate may be. *(Because the plant is not familiar to most of us, we do not know where it grows.)*

You may want to invite students to speculate about the construction or meaning of the costumes in the Carnival photo. Lead a discussion about what kinds of parades are held in your community and what costumes the participants wear.

UNIT OBJECTIVES

1. Introduce geography as a field of study.
2. Explain Earth's position in space and the forces acting on Earth's land and water.
3. Analyze the interrelationships of wind, climate, and natural environments.
4. Identify major resources and how people use them.
5. Describe the development of cultures and the results of population expansion.
6. Learn to draw sketch maps and use them as geographic tools.

Your Classroom Time Line

To help you create a time line to display in your classroom, the most important dates and time periods discussed in each unit's chapters are compiled for you. Some additional dates have been inserted for clarity and continuity. Note that many dates, particularly those in the distant past, are approximate. In each unit, the lists begin in the sidebar on the page with the political map. You may want to have students use colored markers to differentiate among political, scientific, religious, and artistic events or achievements. You might also want to create your own categories.

Exploring Our World

CHAPTER 1
A Geographer's World

CHAPTER 2
Planet Earth

CHAPTER 3
Wind, Climate, and Natural Environments

CHAPTER 4
Earth's Resources

CHAPTER 5
The World's People

Iceberg in sea ice, Antarctica

Carnival parade in Valletta, Malta

A Physical Geographer in Mountain Environments

Professor Francisco Pérez studies tropical mountain environments. He is interested in the natural processes, plants, and environments of mountains. **WHAT DO YOU THINK?** *What faraway places would you like to study?*

I became attracted to mountains when I was a child. While crossing the Atlantic Ocean in a ship, I saw snow-capped Teide Peak in the Canary Islands rising from the water. It was an amazing sight.

As a physical geographer, I am interested in the unique environments of high mountain areas. This includes geological history, climate, and soils. The unusual conditions of high mountain environments have influenced plant evolution. Plants and animals that live on separate mountains sometimes end up looking similar. This happens because they react to their environments in similar ways. For example, several types of tall, weird-looking plants called giant rosettes grow in the Andes, Hawaii, East Africa, and the Canary Islands. Giant rosettes look like the top of a pineapple at the end of a tall stem.

I have found other strange plants, such as rolling mosses. Mosses normally grow on rocks. However, if a moss plant falls to the ground, ice crystals on the soil surface lift the moss. This allows it to "roll" downhill while it continues to grow in a ball shape!

I like doing research in mountains. They are some of the least explored regions of our planet. Like most geographers, I cannot resist the attraction of strange landscapes in remote places.

Rosette plants, Ecuador

La Digue Island, Seychelles

Understanding Primary Sources

1. What are three parts of the environment that Francisco Pérez studies?

2. Why do some plants that live on separate mountains look similar?

Sturgeonfish

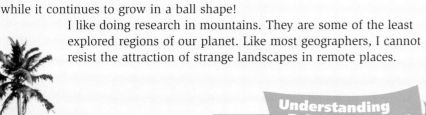

MORE FROM THE FIELD

Living things that are not related sometimes develop similar physical traits because they live in similar environments. This process is called convergent evolution. For example, tuna (fish) and dolphins (mammals) both have streamlined bodies and fins for living in the water.

For a land-based example, compare the serval of Africa, a cat, and the maned wolf of South America, a dog. Both have long necks, long legs, and large ears. They hunt small animals in grassy plains areas. Their long legs and necks elevate their ears above the grass. As a result, they can hear the slightest sound made by their prey.

Discussion: Refer students to the food web illustration in Section 3 of Chapter 3. Lead a discussion on how different plants or animals with similar characteristics might fit into the web.

Understanding Primary Sources
Answers

1. geological history, climate, soils

2. because they react to their environments in similar ways

CHAPTER 1

A Geographer's World
Chapter Resource Manager

Objectives	Pacing Guide	Reproducible Resources
SECTION 1 **Developing a Geographic Eye** (pp. 3–5) **1.** Identify the role perspective plays in the study of geography. **2.** Describe some of the issues or topics that geographers study. **3.** Name three levels geographers can use to view the world.	**Regular** .5 day **Block Scheduling** .5 day *Block Scheduling Handbook, Chapter 1*	**RS** Guided Reading Strategy 1.1 **SM** Geography for Life Activity 1
SECTION 2 **Themes and Essential Elements** (pp. 6–10) **1.** Identify tools geographers use to study the world. **2.** Identify what shapes Earth's features. **3.** Describe how humans shape the world. **4.** Describe how studying geography helps us understand the world.	**Regular** 1.5 days **Block Scheduling** .5 day *Block Scheduling Handbook, Chapter 1*	**RS** Guided Reading Strategy 1.2 **RS** Graphic Organizer 1 **E** Creative Strategies for Teaching World Geography, Lessons 1–3 **E** Lab Activities for Geography and Earth Science, Hands-On 2 **SM** Map Activity 1
SECTION 3 **The Branches of Geography** (pp. 11–13) **1.** Describe what is included in the study of human geography. **2.** Describe what is included in the study of physical geography. **3.** Identify the types of work geographers do.	**Regular** 1 day **Block Scheduling** .5 day *Block Scheduling Handbook, Chapter 1*	**RS** Guided Reading Strategy 1.3 **E** Lab Activities for Geography and Earth Science, Hands-On 1

Chapter Resource Key

RS Reading Support

IC Interdisciplinary Connections

E Enrichment

SM Skills Mastery

A Assessment

REV Review

ELL Reinforcement and English Language Learners

 Transparencies

 CD–ROM

 Music

 Video

 Internet

 Holt Presentation Maker Using Microsoft® Powerpoint®

 One-Stop Planner CD–ROM

See the *One-Stop Planner* for a complete list of additional resources for students and teachers.

One-Stop Planner CD–ROM

It's easy to plan lessons, select resources, and print out materials for your students when you use the *One-Stop Planner CD–ROM with Test Generator.*

Technology Resources

- One-Stop Planner CD–ROM, Lesson 1.1
- Global Skill Builder CD–ROM, Project 1
- Homework Practice Online
- HRW Go site

- One-Stop Planner CD–ROM, Lesson 1.2
- Homework Practice Online
- HRW Go site

- One-Stop Planner CD–ROM, Lesson 1.3
- Yourtown CD–ROM
- Homework Practice Online
- HRW Go site

Review, Reinforcement, and Assessment Resources

ELL	Main Idea Activity 1.1
REV	Section 1 Review, p. 5
A	Daily Quiz 1.1
ELL	English Audio Summary 1.1
ELL	Spanish Audio Summary 1.1

ELL	Main Idea Activity 1.2
REV	Section 2 Review, p. 10
A	Daily Quiz 1.2
ELL	English Audio Summary 1.2
ELL	Spanish Audio Summary 1.2

ELL	Main Idea Activity 1.3
REV	Section 3 Review, p. 13
A	Daily Quiz 1.3
ELL	English Audio Summary 1.3
ELL	Spanish Audio Summary 1.3

⚡ internet connect

HRW ONLINE RESOURCES

GO TO: go.hrw.com
Then type in a keyword.

TEACHER HOME PAGE
 KEYWORD: SJ3 TEACHER

CHAPTER INTERNET ACTIVITIES
 KEYWORD: SJ3 GT1

Choose an activity to:
- learn to use online maps.
- be a virtual geographer for a day.
- compare regions around the world.

CHAPTER ENRICHMENT LINKS
 KEYWORD: SJ3 CH1

CHAPTER MAPS
 KEYWORD: SJ3 MAPS1

ONLINE ASSESSMENT
Homework Practice
 KEYWORD: SJ3 HP1
 Standardized Test Prep Online
 KEYWORD: SJ3 STP1
 Rubrics
 KEYWORD: SS Rubrics

COUNTRY INFORMATION
 KEYWORD: SJ3 Almanac

CONTENT UPDATES
 KEYWORD: SS Content Updates

HOLT PRESENTATION MAKER
 KEYWORD: SJ3 PPT1

ONLINE READING SUPPORT
 KEYWORD: SS Strategies

CURRENT EVENTS
 KEYWORD: S3 Current Events

Chapter Review and Assessment

E	Readings in World Geography, History, and Culture 1, 2
SM	Critical Thinking Activity 1
REV	Chapter 1 Review, pp. 14–15
REV	Chapter 1 Tutorial for Students, Parents, Mentors, and Peers
ELL	Vocabulary Activity 1
A	Chapter 1 Test
	Chapter 1 Test Generator (on the One-Stop Planner)
	Audio CD Program, Chapter 1
A	Chapter 1 Test for English Language Learners and Special-Needs Students

Meeting Individual Needs

Ability Levels

Level 1 Basic-level activities designed for all students encountering new material

Level 2 Intermediate-level activities designed for average students

Level 3 Challenging activities designed for honors and gifted-and-talented students

English Language Learners Activities that address the needs of students with Limited English Proficiency

LAUNCH INTO LEARNING

Ask students to name a country they would like to visit and to give a reason why they want to travel to that particular country. *(Examples: France, for the food; China, to see the Great Wall)* Point out that their interests could probably be the subject of serious study by a geographer. *(Examples: A geographer may study patterns in food preferences among French people or the regional use of ingredients and cooking techniques. Another may use satellite technology to find forgotten sections of the Great Wall.)* Use several of the students' suggestions to show geography's wide range. Then ask students to create "geographic studies" based on their classmates' chosen destinations.

Section 1

Objectives

1. Explain the role perspective plays in the study of geography.

2. Describe some issues or topics that geographers study.

3. Identify the three levels geographers use to view the world.

LINKS TO OUR LIVES

You may want to share with your students the following reasons for gaining a basic understanding of geography as a field of study:

▶ Knowing the fundamentals of geography will help students learn more about all aspects of their world.

▶ Getting an overview from Chapter 1 will make it easier to grasp details in later chapters.

▶ Throughout the book, connections are made to the six essential elements of geography. These elements are explained fully in Chapter 1.

▶ Geography is an expanding field that includes a wide range of specializations. Students might want to consider geography as a career.

CHAPTER 1

A Geographer's World

Chart of the Mediterranean and Europe, 1559

Hand-held compass

GPS (global positioning satellite) receiver

LET'S GET STARTED

Select several photographs of scenes from around the world and display them in the classroom. Copy the following instructions on the chalkboard: *Choose one of the photographs and write down three questions you would like to ask about the place in the picture.* Call on students to read their questions aloud, and use their questions as the basis for a discussion about the issues that professional geographers study. Tell students that in Section 1 they will learn more about developing a geographic eye.

Building Vocabulary

Write the key terms on the chalkboard. Tell students that **perspective** is based on a word meaning "to look" and that **spatial** is based on a word meaning "space." Then, as a class, decide on a definition for **spatial perspective**. Compare this definition to the one in Section 1. Then, point out that **geography** is based on two Greek roots: *geō-*, which means "Earth," and *graphein*, which means "to write." Ask students to compare the meaning of the root words to the textbook's definition and explain the relationship. Finally, have students read the definitions for **urban** and **rural** and then provide examples of urban and rural areas in their region.

Section 1
Developing a Geographic Eye

Read to Discover

1. What role does perspective play in the study of geography?
2. What are some issues or topics that geographers study?
3. At what three levels can geographers view the world?

Define

perspective
spatial perspective
geography
urban
rural

WHY IT MATTERS

What factors would you consider if you were moving to a new town or city? You would probably want to know about its geography. Use **CNNfyi.com** or other **current events** sources to investigate a place you might like to live. Record your findings in your journal.

World map, 1598

Perspectives

People look at the world in different ways. Their experiences shape the way they understand the world. This personal understanding is called **perspective**. Your perspective is your point of view. A geographer's point of view looks at where something is and why it is there. This point of view is known as **spatial perspective**. Geographers apply this perspective when they study the arrangement of towns in a state. They might also use this perspective to examine the movement of cars and trucks on busy roads.

Geographers also work to understand how things are connected. Some connections are easy to see, like highways that link cities. Other connections are harder to see. For example, a dry winter in Colorado could mean that farms as far away as northern Mexico will not have enough water.

Geography is a science. It describes the physical and cultural features of Earth. Studying geography is important. Geographically informed people can see meaning in the arrangement of things on Earth. They know how people and places are related. Above all, they can apply a spatial perspective to real life. In other words, people familiar with geography can understand the world around them.

This fish-eye view of a large city shows highway patterns.
▼

✓ **READING CHECK:** *The World in Spatial Terms* What role does perspective play in the study of geography? Geographers use perspective when they study where something is and why it is there.

🌎 Teaching Objective 1

ALL LEVELS: (Suggested time: 10 min.) Discuss geographers' use of spatial perspective. Then have students examine the aerial photograph on the previous page and suggest why the highways are located where they are. **ENGLISH LANGUAGE LEARNERS**

🌎 Teaching Objectives 2–3

ALL LEVELS: (Suggested time: 20 min.) Copy the following graphic organizer onto the chalkboard, omitting the italicized answers. Use it to help students understand the issues geographers study and the level at which they view the world. **ENGLISH LANGUAGE LEARNERS**

STUDY OF GEOGRAPHY	
Issues/Topics	Levels
Earth's processes	local
relationships between people and environment	regional
governments	global
religion and food	local, regional
urban and rural areas	local, regional

Geographers study how people all around the world react to Earth's processes. Tristan da Cunha is one of a group of small islands in the South Atlantic Ocean about midway between Africa and South America. It is a British territory.

A volcano 6,760 feet (2,060 m) high dominates the island. Its peak is often shrouded in clouds. Lava flows have continually shaped the island's landscape. A volcanic eruption in 1961 forced the evacuation of the island's residents. After the danger passed, most of the Tristanians returned to their isolated island.

Critical Thinking: How have Earth's physical processes affected Tristanians?

Answer: They were forced to evacuate their homeland because of a volcano.

The movement of people is one issue that geographers study. For example, political and economic troubles led many Albanians to leave their country in 1991. Many packed onto freighters like this one for the trip. Geographers want to know how this movement affects the environment and other people.

Interpreting the Visual Record **How do you think Albania has been affected by so many people leaving the country?**

Geographic Issues

Issues geographers study include Earth's processes and their impact on people. Geographers study the relationship between people and environment in different places. For example, geographers study tornadoes to find ways to reduce loss of life and property damage. They ask how people prepare for tornadoes. Do they prepare differently in different places? When a tornado strikes, how do people react?

Geographers also study how governments change and how those changes affect people. Czechoslovakia, for example, split into Slovakia and the Czech Republic in 1993. These types of political events affect geographic boundaries. People react differently to these changes. Some people are forced to move. Others welcome the change.

Other issues geographers study include religions, diet (or food), **urban** areas, and **rural** areas. Urban areas contain cities. Rural areas contain open land that is often used for farming.

✓ **READING CHECK:** *The Uses of Geography* What issues or topics do geographers study? Earth's processes, the relationship between people and environment, changes of government, religions, diet, urban areas, and rural areas

Local, Regional, and Global Geographic Studies

With any topic, geographers must decide how large an area to study. They can focus their study at a local, regional, or global level.

Local Studying your community at the local, or close-up, level will help you learn geography. You know where homes and stores are located. You know how to find parks, ball fields, and other fun places. Over time, you see your community change. New buildings are constructed. People move in and out of your neighborhood. New stores open their doors, and others go out of business.

Visual Record Answer ▲

Answers will vary but students may mention possible political volatility of the area, a negative or fearful atmosphere, or fewer people for jobs.

internet connect

GO TO: go.hrw.com
KEYWORD: SJ3 CH1
FOR: Web sites about the geographer's world

CLOSE

Ask students to imagine that they are geographers from the planet Geog and they have landed on Earth. Have students list what human activities they would study first and what sources they would use in their research.

REVIEW AND ASSESS

Have students complete the Section Review. Then have students work in groups to create short quizzes based on the section's material. Have groups exchange quizzes and complete another group's quiz. Then have students complete Daily Quiz 1.1. **COOPERATIVE LEARNING**

RETEACH

Have students complete Main Idea Activity 1.1. Then have them illustrate one of the section's topics. Ask students to explain their illustrations. **ENGLISH LANGUAGE LEARNERS**

EXTEND

Have interested students conduct research on the history of the field of geography and its influence on society. They may want to concentrate on ancient Greek or Arabic achievements. Ask them to create illustrated charts showing their research. **BLOCK SCHEDULING**

▲

The southwest is a region within the United States. One well-known place that characterizes the landscape of the southwest is the Grand Canyon. The Grand Canyon is shown in the photo at left and in the satellite image at right.

Regional Regional geographers organize the world into convenient parts for study. For example, this book separates the world into big areas like Africa and Europe. Regional studies cover larger areas than local studies. Some regional studies might look at connections like highways and rivers. Others might examine the regional customs.

Global Geographers also work to understand global issues and the connections between events. For example, many countries depend on oil from Southwest Asia. If those oil supplies are threatened, some countries might rush to secure oil from other areas. Oil all over the world could then become much more expensive.

✓ **READING CHECK:** *The World in Spatial Terms* What levels do geographers use to focus their study of an issue or topic? local, regional, or global

Section Review 1

Define and explain: perspective, spatial perspective, geography, urban, rural

Reading for the Main Idea

1. *The Uses of Geography* How can a spatial perspective be used to study the world?

2. *The Uses of Geography* Why is it important to study geography?

Critical Thinking

3. **Drawing Inferences and Conclusions** How do threatening weather patterns affect people, and why do geographers study these patterns?

4. **Drawing Inferences and Conclusions** Why is it important to view geography on a global level?

Organizing What You Know

5. **Finding the Main Idea** Copy the following graphic organizer. Use it to examine the issues geographers study. Write a paragraph on one of these issues.

Section Review 1

Answers

Define For definitions, see: perspective, p. 3; spatial perspective, p. 3; geography, p. 3; urban, p. 4; rural, p. 4

Reading for the Main Idea

1. to understand how things are connected (NGS 3)

2. to see meaning in the arrangement of things on Earth and to understand the world (NGS 17, 18)

Critical Thinking

3. They can cause loss of life or property damage; to help people protect themselves from dangerous weather situations

4. Answers will vary but might include to gain an understanding of how events in one region can affect other regions.

Organizing What You Know

5. Answers will vary but should be issues geographers study.

Section 2

Objectives

1. Identify tools geographers use to study the world.
2. Identify what shapes Earth's features.
3. Examine how humans shape the world.
4. Explain how studying geography helps us understand the world.

Visual Record Answer ▶

near Cairo, near the pyramids

FOCUS

LET'S GET STARTED

Write the following question on the chalkboard: *Where is your favorite shopping mall or movie theater located?* Have students respond to the question. If a student names the actual address for the building, explain that he or she has provided its absolute location. Tell students that in Section 1 they will learn about the difference between absolute and relative location and other topics of geography.

Building Vocabulary

Write the key terms on the chalkboard. Ask what we mean when we say "It's all relative" and "Absolutely!" Ask students for suggestions on how those phrases could relate to **absolute location** and **relative location**. Have students look up the remaining key terms in the text or glossary and write sentences using them.

Section 2 Themes and Essential Elements

Read to Discover

1. What tools do geographers use to study the world?
2. What shapes Earth's features?
3. How do humans shape the world?
4. How does studying geography help us understand the world?

WHY IT MATTERS

Geographers often study the effect that new people have on a place. Use **CNNfyi.com** or other **current events** sources to find out how the arrival of new people has changed the United States or another country. Record your findings in your journal.

Define

absolute location
relative location
subregions
diffusion
levees

Tombs carved out of a mountain in Turkey

▲

The location of a place can be described in many ways.

Interpreting the Visual Record **Looking at the photo of this hotel in Giza, Egypt, and at the map, how would you describe Giza's location?**

Themes

The study of geography has long been organized according to five important themes, or topics of study. One theme, *location*, deals with the exact or relative spot of something on Earth. *Place* includes the physical and human features of a location. *Human-environment interaction* covers the ways people and environments affect each other. *Movement* involves how people change locations and how goods are traded, as well as the effects of these movements. *Region* organizes Earth into geographic areas with one or more shared characteristics.

✓ **READING CHECK:** *The Uses of Geography* What are the five themes of geography? location, place, human-environment interaction, movement, region

Six Essential Elements

Another way to look at geography is to study its essential elements, or most important parts. The six essential elements used to study geography are The World in Spatial Terms, Places and Regions, Physical Systems, Human Systems, Environment and Society, and The Uses of Geography. These six essential elements will be used throughout this textbook. They share many properties with the five themes of geography.

🌐 Teaching Objectives 1–2
ALL LEVELS: (Suggested time: 20 min.) Pair students and have each pair create a geography fact sheet for the school. Fact sheets should include the school's absolute and relative location as well as several of its physical or human characteristics. Ask volunteers to read their fact sheets to the class. Then discuss why pairs may have chosen different identifying features. **ENGLISH LANGUAGE LEARNERS, COOPERATIVE LEARNING**

🌐 Teaching Objectives 3–4
ALL LEVELS: (Suggested time: 30 min.) Have students draw maps of their neighborhoods, including homes, stores, streets, and other landmarks. Then have them locate their neighborhoods on a map of their town, city, or county. Tell students to label items on their maps that represent the relationship between environment and society. *(such as dams, recycling plants, airports, train stations, and highways)* Ask volunteers to share their maps with the class. **ENGLISH LANGUAGE LEARNERS**

Location Every place on Earth has a location. Location is defined by absolute and relative location.

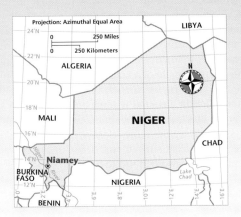

Absolute Location: the exact spot on Earth where something is found

Example: Niamey, the capital of Niger, is located at 13°31' north latitude and 2°07' east longitude.

Relative Location: the position of a place in relation to other places

Example: Yosemite National Park is north of Los Angeles, California, and east of San Francisco, California.

You Be the Geographer 1. Use an atlas to find the absolute location of your city or town.
2. Write a sentence describing the relative location of your home.

The World in Spatial Terms This element focuses on geography's spatial perspective. As you learned in Section 1, geographers apply spatial perspective when they look at the location of something and why it is there. The term *location* can be used in two ways. **Absolute location** defines an exact spot on Earth. For example, the address of the Smithsonian American Art Museum is an absolute location. The address is at 8th and G Streets, N.W., in Washington, D.C. City streets often form a grid. This system tells anyone looking for an address where to go. The grid formed by latitude and longitude lines also pinpoints absolute location. Suppose you asked a pilot to take you to 52° north latitude by 175° west longitude. You would land at a location on Alaska's Aleutian Islands.

 Relative location describes the position of a place in relation to another place. Measurements of direction, distance, or time can define relative location. For example, the following sentences give relative location. "The hospital is one mile north of our school." "Canada's border is about an hour's drive from Great Falls, Montana."

 A geographer must be able to use maps and other geographic tools and technologies to determine spatial perspective. A geographer must

▲
Places can be described by what they do not have. This photo shows the result of a long period without rain.

🧭 Linking Past to Present
Channeled Scablands Geographers think that large Ice Age floods originating in western Montana created the Channeled Scablands in eastern Washington state. A glacier blocked a river and created a glacial lake near Missoula in present-day Montana. When this ice dam broke, a wall of water perhaps 2,000 feet (610 m) high crashed through the region, carving out unusual landforms such as the Channeled Scablands—an area marked by channels, cliffs, and steep-sided canyons. Scientists suspect that water poured from the lake at 60 or more miles per hour and that the glacial lake near Missoula may have filled and emptied dozens of times.

Activity: Pair students and have them conduct research on the Channeled Scablands. Ask students to create a travel guide to the area that explains how the area was created and what tourists would see there. Guides should include photographs if possible.

You Be the Geographer Answers
1. Locations should be as accurate as the atlas used allows.
2. Sentences might refer to bodies of water, landforms, streets, or other features.

Teaching Objective 4

ALL LEVELS: (Suggested time: 30 min.) Copy the following graphic organizer onto the chalkboard, omitting the italicized answers. Use it to help students illustrate how regions and subregions help geographers understand our world. Using the United States as an example, tell students to identify regions *(such as the East or the Midwest)* and subregions *(such as their state)*. Remind students that regions and subregions may vary in size and can be categorized as cultural, economic, or political. Then have students identify subregions of one subregion and classify each into one of these three categories. **ENGLISH LANGUAGE LEARNERS**

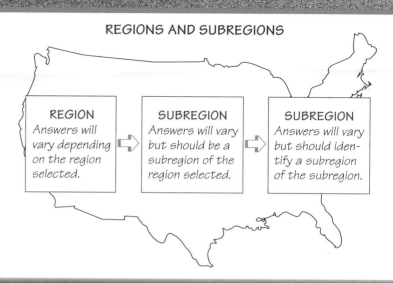

REGION
Answers will vary depending on the region selected.

⇨

SUBREGION
Answers will vary but should be a subregion of the region selected.

⇨

SUBREGION
Answers will vary but should identify a subregion of the subregion.

ENVIRONMENT AND SOCIETY

Recycling is another example of the relationship between society and the environment. Its success depends on developing new technology to reuse collected materials and protect the environment. Products made from recycled materials include boards made of sawdust mixed with scrap wood and shrinkwrap.

Recycled materials are even used in the manufacture of shoes and clothing. Three different recycled materials are used in some hiking boots. The soles are made from used tires, the innersole padding from white paper, and the fabric uppers from used plastic bottles. Plastic bottles are also recycled to create fabric for sweaters.

Critical Thinking: How do new recycling technologies affect the environment?

Answer: They help to protect it.

internet connect

GO TO: go.hrw.com
KEYWORD: SJ3 CH1
FOR: Web sites about recycling

Visual Record Answers ▲

the communication of ideas, the production of goods, trade, conflict, governments

▶

It represents a human adaptation to Egypt's environment.

People travel from place to place on miles of new roadway.
Interpreting the Visual Record What other forms of human systems are studied by geographers?

▼

Men in rural Egypt wear a long shirt called a *galabia*. This loose-fitting garment is ideal for people living in Egypt's hot desert climate. In addition, the galabia is made from cotton, an important agricultural product of Egypt.
Interpreting the Visual Record How does the *galabia* show how people have adapted to their environment?

▶

also know how to organize and analyze information about people, places, and environments using geographic tools.

Places and Regions Our world has a vast number of unique places and regions. Places can be described both by their physical location and by their physical and human features. Physical features include coastlines and landforms. They can also include lakes, rivers, or soil types. For example, Colorado is flat in the east but mountainous in the west. This is an example of a landform description of place. A place can also be described by its climate. For example, Greenland has long, cold winters. Florida has mild winters and hot, humid summers. Regions are areas of Earth's surface with one or more shared characteristics. To study a region more closely, geographers often divide it into smaller areas called **subregions**. Many of the characteristics that describe places can also be used to describe regions or subregions.

The Places and Regions element also deals with the human features of places and regions. Geographers want to know how people have created regions based on Earth's features and how culture and other factors affect how we see places and regions on Earth.

Physical Systems Physical systems shape Earth's features. Geographers study earthquakes, mountains, rivers, volcanoes, weather patterns, and similar topics and how these physical systems have affected Earth's characteristics. For example, geographers might study how volcanic eruptions in the Hawaiian Islands spread lava, causing landforms to change. They might note that southern California's shoreline changes yearly, as winter and summer waves move beach sand.

Geographers also study how plants and animals relate to these nonliving physical systems. For example, deserts are places with cactus and other plants as well as rattlesnakes and other reptiles, that can

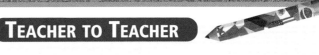

➤**ASSIGNMENT:** Have students recall the most beautiful, interesting, or exciting place they have ever visited. Then have them write words or phrases that describe that place in terms of landforms, climate, animal life, plant life, language spoken, common religion, history, customs, or other physical or human characteristics. Ask students to consult primary or secondary sources for additional information. Then have students write a description of their chosen locale's relative location and find its absolute location by calculating latitude and longitude.

▲
A satellite dish brings different images and ideas to people in a remote area of Brazil.
Interpreting the Visual Record **How might resources have affected the use of technology here?**

live in very dry conditions. Geographers also study how different types of plants, animals, and physical systems are distributed on Earth.

Human Systems People are central to geography. Our activities, movements, and settlements shape Earth's surface. Geographers study peoples' customs, history, languages, and religions. They study how people migrate, or move, and how ideas are communicated. When people move, they may go to live in other countries or move within a country. Geographers want to know how and why people move from place to place.

People move for many reasons. Some move to start a new job. Some move to attend special schools. Others might move to be closer to family. People move either when they are pushed out of a place or when they are pulled toward another place. In the Dust Bowl, for example, crop failures pushed people out of Oklahoma in the 1930s. Many were pulled to California by their belief that they would find work there. Geographers also want to know how ideas or behaviors move from one region to another. The movement of ideas occurs through communication. There are many ways to communicate. People visit with each other in person or on the phone. New technology allows people to communicate by e-mail. Ideas are also spread through films, magazines, newspapers, radio, and television. The movement of ideas or behaviors from one region to another is known as **diffusion**.

The things we produce and trade are also part of the study of human systems. Geographers study trading patterns and how countries depend on each other for certain goods. In addition, geographers look at the causes and results of conflicts between peoples. The study of governments we set up and the features of cities and other settlements we live in are also part of this study.

Environment and Society Geographers study how people and their surroundings affect each other. Their relationship can be examined in three ways. First, geographers study how humans depend on

▲
This woman at a railway station in Russian Siberia sells some goods that were once unavailable in her country.
Interpreting the Visual Record **Which essential element is illustrated in this photo?**

Display a picture of a well-known local landmark. Call on students to suggest how the six essential elements of geography relate to the landmark.

REVIEW AND ASSESS

Have students complete the Section Review. Then organize students into groups of four or five. Assign each group a city that appears on one of the Atlas maps in the textbook. Have students create a travel guide that describes the region in which the city is located. Then have students complete Daily Quiz 1.2. **COOPERATIVE LEARNING**

RETEACH

Have students complete Main Idea Activity 1.2. Then organize students into six groups and assign each group one of the six essential elements of geography. Ask members of each group to write a paragraph describing their element in relation to your school. **ENGLISH LANGUAGE LEARNERS**

EXTEND

Ask interested students to imagine that they have been hired to submit a building plan for a recreation center in their community. Tell them to use the six essential elements of geography to determine the center's location and construction features. Ask students to include a drawing of the building and a map showing its location within the community.
BLOCK SCHEDULING

Section Review 2

Answers

Define For definitions, see: absolute location, p. 7, relative location, p. 7, subregions, p. 8, diffusion, p. 9, levees, p. 10

Reading for the Main Idea

1. with maps and other geographic tools (NGS 1)

2. physical systems such as earthquakes, mountains, rivers, volcanoes, weather patterns (NGS 7)

Critical Thinking

3. through their activities, movements, settlements, modifications

4. provides clues to the past and helps geographers plan for the future

5. The World in Spatial Terms—using maps and other geographic tools to look at the world with a spatial perspective; Places and Regions—studying the physical and human features of a place; Physical Systems—systems that have shaped Earth's features; Human Systems—how people have shaped Earth's surface; Environment and Society—how people and their surroundings affect each other; The Uses of Geography—how geography helps us understand relationships among people, places, and the environment over time

▲
Open-air markets like this one in Mali provide opportunities for farmers to sell their goods.

their physical environment to survive. Human life requires certain living and non-living resources, such as freshwater and fertile soil for farming.

Geographers also study how humans change their behavior to be better suited to an environment. These changes or adaptations include the kinds of clothing, food, and shelter that people create. These changes help people live in harsh climates.

Finally, humans change the environment. For example, farmers who irrigate their fields can grow fruit in Arizona's dry climate. People in Louisiana have built **levees**, or large walls, to protect themselves when the Mississippi River floods.

The Uses of Geography Geography helps us understand the relationships among people, places, and the environment over time. Understanding how a relationship has developed can help in making plans for the future. For example, geographers can study how human use of the soil in a farming region has affected that region over time. Such knowledge can help them determine what changes have been made to the soil and whether any corrective measures need to be taken.

✓ **READING CHECK:** *The Uses of Geography* What are the six essential elements in studying geography? The World in Spatial Terms, Places and Regions, Physical Systems, Human Systems, Environment and Society, and The Uses of Geography

go.
hrw
.com
Homework Practice Online
Keyword: SJ3 HP1

Section Review 2

Define and explain: absolute location, relative location, subregions, diffusion, levees

Reading for the Main Idea

1. *The World in Spatial Terms* How do geographers study the world?

2. *Physical Systems* What shapes Earth's features? Give examples.

Critical Thinking

3. **Finding the Main Idea** How do humans shape the world in which they live?

4. **Analyzing Information** What benefits can studying geography provide?

Organizing What You Know

5. **Summarizing** Copy the following graphic organizer. Use it to identify and describe all aspects of each of the six essential elements.

Element	Description

Section 3

Objectives

1. Explain the study of human geography.
2. Describe the study of physical geography.
3. Investigate the types of work that geographers do.

Section 3 — The Branches of Geography

Read to Discover

1. What is included in the study of human geography?
2. What is included in the study of physical geography?
3. What types of work do geographers do?

Define

human geography
physical geography
cartography
meteorology
climatology

WHY IT MATTERS

Nearly every year, hurricanes hit the Atlantic or Gulf coasts of the United States. Predicting weather is one of the special fields of geography. Use **CNNfyi.com** or other **current events** sources to find out about hurricanes. Record your findings in your journal.

Map of an ancient fortress

Human Geography

The study of people, past or present, is the focus of **human geography**. People's location and distribution over Earth, their activities, and their differences are studied. For example, people living in different countries create different kinds of governments. Political geographers study those differences. Economic geographers study the exchange of goods and services across Earth. Cultural geography, population geography, and urban geography are some other examples of human geography. A professional geographer might specialize in any of these branches.

✓ **READING CHECK:** *Human Systems* How is human geography defined? as the study of people, past or present

A volunteer visits a poor area of Bangladesh. Geographers study economic conditions in regions to help them understand human geography.

Physical Geography

The study of Earth's natural landscapes and physical systems, including the atmosphere, is the focus of **physical geography**. The world is full of different landforms such as deserts, mountains, and plains. Climates affect these landscapes. Knowledge of physical systems helps geographers understand how a landscape developed and how it might change.

it to help students distinguish between the study of human geography and the study of physical geography to identify the types of work geographers do. **ENGLISH LANGUAGE LEARNERS**

Geography helps us understand the world.

Physical Geography
the study of
• *Earth's natural land-scapes and physical systems*
• *different landforms*

Types of work
include
• *cartography* • *meteorology* • *climatology*

Human Geography
the study of
• *people, past or present*
• *politics, economy, and culture*

Section Review 3

Answers

Define For definitions, see: human geography, p. 11; physical geography, p. 11; cartography, p. 13, meteorology, p. 13, climatology, p. 13

Reading for the Main Idea

1. Topics are the study of people, their location and distribution, their activities, and their differences. (NGS 9)

2. by tracking Earth's larger atmospheric systems (NGS 18)

Critical Thinking

3. to learn how a landscape developed and how it might change

4. issues such as population, pollution, endangered plants and animals, or decreased or increased economic activities

Organizing What You Know

5. meteorologist—tracks weather and atmospheric conditions; climatologist—tracks atmospheric systems

Connecting to Technology Answers

1. It can help planners build roads, dams, or other structures.

2. consequences, such as greater knowledge about population, profitable economic activities, or change to the environment

---

CONNECTING TO *Technology*

A mapmaker creates a digital map.

Maps are tools that can display a wide range of information. Traditionally, maps were drawn on paper and could not be changed to suit the user. However, computers have revolutionized the art of mapmaking.

Today, mapmakers use computers to create and modify maps for different uses. They do this by using a geographic information system, or GIS. A GIS is a computer system that combines maps and satellite photographs with other kinds of spatial data—information about places on the planet. This information might include soil types, population figures, or voting patterns.

Using a GIS, mapmakers can create maps that show geographic features and relationships. For example, a map showing rainfall patterns in a particular region might be combined with data

on soil types or human settlement to show areas of possible soil erosion.

The flexibility of a GIS allows people to seek answers to specific questions. Where should a new road be built to ease traffic congestion? How are changes in natural habitat affecting wildlife? These and many other questions can be answered with the help of computer mapping.

Computer Mapping

Understanding What You Read

1. How could a GIS help people change their environment?

2. What social, environmental, or economic consequences might future advances in GIS technology have?

Knowledge of physical and human geography will help you understand the world's different regions and peoples. In your study of the major world regions, you will see how physical and human geography connect to each other.

✓ **READING CHECK:** *Physical Systems* What is included in the study of physical geography? Earth's natural landscapes and physical systems, including the atmosphere

Working as a Geographer

Geography plays a role in almost every occupation. Wherever you live and work, you should know local geography. School board members know where children live. Taxi drivers are familiar with city streets. Grocery store managers know which foods sell well in certain areas.

CLOSE

Write the following statement on the chalkboard: *A cartographer's work is never done.* Ask students why might this be true. *(Possible answers: changes required by physical processes, new roads and suburbs, and political boundary changes)*

RETEACH

Have students complete Main Idea Activity 1.3. Then have them complete the following sentence: "I used my knowledge of geography today when I . . . " Ask volunteers to read their sentences to the class. **ENGLISH LANGUAGE LEARNERS**

REVIEW AND ASSESS

Have students complete the Section Review. Then pair students and have each pair locate newspaper or magazine articles that relate to some aspect of human or physical geography. Have pairs write a few sentences explaining the connection between the articles and human or physical geography. Then have students complete Daily Quiz 1.3. **COOPERATIVE LEARNING**

EXTEND

Have interested students conduct research on the history of cartography in an area that has been mapped since antiquity. Have them use copies of ancient maps to investigate how maps of that region have evolved over time and then present their findings to the class. **BLOCK SCHEDULING**

They also know where they can obtain these products throughout the year. Local newspaper reporters are familiar with town meetings and local politicians. Reporters also know how faraway places can affect their communities. Doctors must know if their towns have poisonous snakes or plants. City managers know whether nearby rivers might flood. Emergency workers in mountain towns check snow depth so they can give avalanche warnings. Local weather forecasters watch for powerful storms and track their routes on special maps.

Some specially trained geographers practice in the field of **cartography**. Cartography is the art and science of mapmaking. Today, most mapmakers do their work on computers. Geographers also work as weather forecasters. The field of forecasting and reporting rainfall, temperature, and other atmospheric conditions is called **meteorology**. A related field is **climatology**. These geographers, known as climatologists, track Earth's larger atmospheric systems. Climatologists want to know how these systems change over long periods of time. They also study how people might be affected by changes in climate.

Governments and a variety of organizations hire geographers to study the environment. These geographers might explore such topics as pollution, endangered plants and animals, or rain forests. Some geographers who are interested in education become teachers and writers. They help people of all ages learn more about the world. Modern technology allows people all over the world to communicate instantly. Therefore, it is more important than ever to be familiar with the geographer's world.

▲
Experts examine snow to help forecast avalanches. They study the type of snow, weather conditions, and landforms. For example, wet snow avalanches can occur because of the formation of a particular type of ice crystal, called depth hoar, near the ground.

✓ **READING CHECK:** *The Uses of Geography* What types of work do geographers perform? They make maps, work as weather forecasters, track atmospheric systems, or work as teachers or writers.

go. hrw .com Homework Practice Online
Keyword: SJ3 HP1

Section Review 3

Define and explain: human geography, physical geography, cartography, meteorology, climatology

Reading for the Main Idea

1. *Human Systems* What topics are included in the study of human geography?
2. *The Uses of Geography* How do people who study the weather use geography?

Critical Thinking

3. **Finding the Main Idea** Why is it important to study physical geography?

4. **Making Generalizations and Predictions** How might future discoveries in the field of geography affect societies, world economies, or the environment?

Organizing What You Know

5. **Categorizing** Copy the following graphic organizer. Use it to list geographers' professions and their job responsibilities.

Cartographer		
—makes maps		
—studies maps		

Building Vocabulary

For definitions, see: perspective, p. 3; spatial perspective, p. 3; geography, p. 3; urban, p. 4; rural, p. 4; absolute location, p. 7; relative location, p. 7; subregions, p. 8; diffusion, p. 9; levees, p. 10; human geography, p. 11; physical geography, p. 11; cartography, p. 13; metereology, p. 13; climatology, p. 13

Reviewing the Main Ideas

1. local, regional, and global; examples will vary but should reflect the appropriate level (NGS 3)
2. absolute location—directions would define an exact spot on Earth; relative location—directions would describe the position of a place in relation to another place (NGS 1)
3. the movement of ideas or behaviors from one cultural region to another; allows people to learn new things (NGS 9)
4. to study a region more closely (NGS 5)
5. because maps are useful; answers will vary but might include meteorology and climatology (NGS 1)

Have students complete the Chapter 1 Test.

RETEACH

Organize students into three groups—one group for each section. Have groups summarize their section in a paragraph. Call on volunteers to read their group's paragraph to the class. Ask members of the other groups to evaluate the paragraphs. **ENGLISH LANGUAGE LEARNERS, COOPERATIVE LEARNING**

PORTFOLIO EXTENSIONS

1. Different cultures use different methods for showing directions and locations. For example, long ago Polynesians developed shell maps to help them navigate in the vast Pacific Ocean. Have students work in groups to devise new ways to record information about a region familiar to them. Ask them to include a standard map of the area along with the new map. They should also write a legend or key for the new map.

2. Ask students to imagine that they are directing a documentary entitled *The Six Essential Elements of Geography and You.* Pair students and have each pair create six storyboards—one for each element. Storyboards should include a paragraph that describes how the scene represents the element.

CHAPTER 1 Review ANSWERS

Understanding Environment and Society

Information included in reports should be consistent with text material. Students should discuss the positive and negative aspects of the after-school program and of protecting endangered species. Use Rubric 12, Drawing Conclusions, to evaluate student work.

Thinking Critically

1. Answers will vary, but students might mention that a geographer identifies where things are so that connections can be made.

2. when—daily; how—answers will vary; examples might include building dams and irrigating fields

3. Students might mention the movement of people, trade networks, or the diffusion of ideas between groups.

4. both human and physical characteristics

5. by helping us see meaning in the arrangement of things on Earth

CHAPTER 1 Reviewing What You Know

Building Vocabulary

On a separate sheet of paper, write sentences to define each of the following words.

1. perspective
2. spatial perspective
3. geography
4. urban
5. rural
6. absolute location
7. relative location
8. levees
9. diffusion
10. subregions
11. human geography
12. physical geography
13. cartography
14. meteorology
15. climatology

Reviewing the Main Ideas

1. *(The World in Spatial Terms)* What are three ways to view geography? Give an example of when each type could be used.

2. *(The World in Spatial Terms)* What kind of directions would you give to indicate a place's absolute location? Its relative location?

3. *(Human Systems)* What is diffusion, and why is it important?

4. *(Places and Regions)* Why do geographers create subregions?

5. *(The World in Spatial Terms)* Why is cartography important? What types of jobs do geographers do?

Understanding Environment and Society

Land Use

You are on a committee that will decide whether to close a park near your school. One proposed use for the land is a building where after-school activities could be held. However, the park is the habitat of an endangered bird. Write a report describing consequences of the park closing. Then organize information from your report to create a proposal on what decision should be made.

Thinking Critically

1. **Analyzing Information** How can a geographer use spatial perspective to explain how things in our world are connected?

2. **Drawing Inferences and Conclusions** When and how do humans relate to the environment? Provide some examples of this relationship.

3. **Summarizing** How are patterns created by the movement of goods, ideas, and people?

4. **Finding the Main Idea** How are places and regions defined?

5. **Finding the Main Idea** How does studying geography help us understand the world?

FOOD FESTIVAL

Have students bring food items to class and use them to discuss human and physical geography. For example, a student may bring a can of green beans and note that certain soil, sunlight, and climate conditions must be present to grow the beans. Or a student may use the can to discuss how people in different parts of the country prepare green beans, or how farms and canneries affect local economies.

CHAPTER 1 REVIEW AND ASSESSMENT RESOURCES

Reproducible
- Readings in World Geography, History, and Culture 1, 2
- Critical Thinking Activity 1
- Vocabulary Activity 1

Technology
- Chapter 1 Test Generator (on the One-Stop Planner)

- Audio CD Program, Chapter 1 (English and Spanish)
- HRW Go site

Reinforcement, Review, and Assessment
- Chapter 1 Review, pp. 14–15

- Chapter 1 Tutorial for Students, Parents, Mentors, and Peers
- Chapter 1 Test
- Chapter 1 Test for English Language Learners and Special-Needs Students

Building Social Studies Skills

Map ACTIVITY

On a separate sheet of paper, match the letters on the map with their correct labels.

Africa	Europe
Antarctica	North America
Asia	South America
Australia	

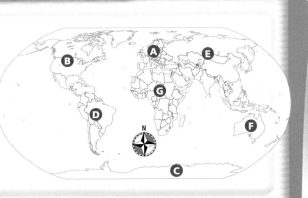

Mental Mapping Skills ACTIVITY

To help you understand the relationships between places, create a seating chart of your classroom. Then draw a sketch of the floor plan of your school. Discuss why certain areas are located in particular parts of the campus.

WRITING ACTIVITY

Write a letter persuading another student to enroll in a geography class. Include examples of professions that use geography and relate that information to the everyday life of a student. Be sure to use standard grammar, spelling, sentence structure, and punctuation.

Map Activity
A. Europe
B. North America
C. Antarctica
D. South America
E. Asia
F. Australia
G. Africa

Mental Mapping Skills Activity
Seating charts, sketches, and answers will vary but should be logical for your classroom and school.

Writing Activity
Letters will vary, but should include various professions that use geography. Letters should also relate the use of geography to the everyday life of a student. Use Rubric 25, Personal Letters, to evaluate student work.

Portfolio Activity
Answers will vary but should demonstrate the understanding that communities can be defined narrowly or broadly. Use Rubric 38, Writing to Classify, to evaluate student work. Call on students to share their sentences with the class.

Alternative Assessment

Portfolio ACTIVITY

Learning About Your Local Geography

Individual Project How do you define your community geographically? Is your community the area around your home or school? Write two or three sentences defining your community to share with the class.

internet connect

Internet Activity: go.hrw.com
KEYWORD: SJ3 GT1

Choose a topic to explore online:
- Learn to use online maps.
- Be a virtual geographer for a day.
- Compare regions around the world.

internet connect

GO TO: go.hrw.com
KEYWORD: SJ3 Teacher
FOR: a guide to using the Internet in your classroom

Planet Earth
Chapter Resource Manager

Objectives	Pacing Guide	Reproducible Resources
SECTION 1		
The Sun, Earth, and Moon (pp. 17–22) **1.** Name the objects that make up the solar system. **2.** Describe what causes the seasons. **3.** Identify the four parts of the Earth system.	**Regular** 1 day **Block Scheduling** .5 day *Block Scheduling Handbook, Chapter 2*	**RS** Guided Reading Strategy 2.1 **E** Environmental and Global Issues Activity 2 **SM** Geography for Life Activity 2 **IC** Interdisciplinary Activity for the Middle Grades 2 **E** Lab Activity for Geography and Earth Science, Demonstration 2
SECTION 2		
Water on Earth (pp. 23–27) **1.** Identify the processes that make up the water cycle and how they are connected. **2.** Describe how water is distributed on Earth. **3.** Explain how water affects people's lives.	**Regular** 1 day **Block Scheduling** .5 day *Block Scheduling Handbook, Chapter 2*	**RS** Guided Reading Strategy 2.2 **E** Interdisciplinary Activity for the Middle Grades 3 **E** Lab Activity for Geography and Earth Science, Hands-On 5 **E** Lab Activity for Geography and Earth Science, Demonstration 1
SECTION 3		
The Land (pp. 28–33) **1.** Describe primary landforms. **2.** Describe secondary landforms. **3.** Describe how humans interact with landforms.	**Regular** 1.5 days **Block Scheduling** .5 day *Block Scheduling Handbook, Chapter 2*	**RS** Guided Reading Strategy 2.3 **RS** Graphic Organizer 2 **E** Environmental and Global Issues Activity 3 **E** Lab Activities for Geography and Earth Science, Demonstrations 3, 4, 5, 6, 7, 9, and 10 **SM** Map Activity 2

Chapter Resource Key

RS Reading Support

IC Interdisciplinary Connections

E Enrichment

SM Skills Mastery

A Assessment

REV Review

ELL Reinforcement and English Language Learners

 Transparencies

 CD–ROM

 Music

 Video

 Internet

 Holt Presentation Maker Using Microsoft® Powerpoint®

 One-Stop Planner CD-ROM

See the *One-Stop Planner* for a complete list of additional resources for students and teachers.

One-Stop Planner CD–ROM

It's easy to plan lessons, select resources, and print out materials for your students when you use the *One-Stop Planner CD–ROM with Test Generator.*

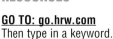

HRW ONLINE RESOURCES

GO TO: go.hrw.com
Then type in a keyword.

TEACHER HOME PAGE
 KEYWORD: SJ3 TEACHER

CHAPTER INTERNET ACTIVITIES
 KEYWORD: SJ3 GT2

Choose an activity to:
• learn more about Earth's seasons.
• discover facts about Earth's water.
• investigate earthquakes.

CHAPTER ENRICHMENT LINKS
 KEYWORD: SJ3 CH2

CHAPTER MAPS
 KEYWORD: SJ3 MAPS2

ONLINE ASSESSMENT
Homework Practice
 KEYWORD: SJ3 HP2
 Standardized Test Prep Online
 KEYWORD: SJ3 STP2
 Rubrics
 KEYWORD: SS Rubrics

COUNTRY INFORMATION
 KEYWORD: SJ3 Almanac

CONTENT UPDATES
 KEYWORD: SS Content Updates

HOLT PRESENTATION MAKER
 KEYWORD: SJ3 PPT2

ONLINE READING SUPPORT
 KEYWORD: SS Strategies

CURRENT EVENTS
 KEYWORD: S3 Current Events

Technology Resources

 One-Stop Planner CD–ROM, Lesson 2.1

 Geography and Cultures Visual Resources with Teaching Activity 1

Homework Practice Online

HRW Go site

Review, Reinforcement, and Assessment Resources

ELL Main Idea Activity 2.1
REV Section 1 Review, p. 22
 A Daily Quiz 2.1
ELL English Audio Summary 2.1
ELL Spanish Audio Summary 2.1

 One-Stop Planner CD–ROM, Lesson 2.2

 Geography and Cultures Visual Resources with Teaching Activity 7

 Earth: Forces and Formations CD–ROM/Seek and Tell/ Forces and Processes

Earth: Forces and Formations CD–ROM/Seek and Tell/The Earth's Surface

Our Environment CD–ROM/ Seek and Tell/Natural Resources

Homework Practice Online

HRW Go site

ELL Main Idea Activity 2.2
REV Section 2 Review, p. 27
 A Daily Quiz 2.2
ELL English Audio Summary 2.2
ELL Spanish Audio Summary 2.2

 One-Stop Planner CD–ROM, Lesson 2.3

Earth: Forces and Formations CD–ROM/Seek and Tell/Forces and Processes

Earth: Forces and Formations CD–ROM/Seek and Tell/The Earth's Surface

Homework Practice Online

HRW Go site

ELL ELL Main Idea Activity 2.3
REV Section 3 Review, p. 33
ELL English Audio Summary 2.3
ELL Spanish Audio Summary 2.3
 A Daily Quiz 2.3

Meeting Individual Needs

Ability Levels

Level 1 Basic-level activities designed for all students encountering new material

Level 2 Intermediate-level activities designed for average students

Level 3 Challenging activities designed for honors and gifted-and-talented students

English Language Learners Activities that address the needs of students with Limited English Proficiency

Chapter Review and Assessment

 E Readings in World Geography, History, and Culture 3 and 4

SM Critical Thinking Activity 2

REV Chapter 2 Review, pp. 34–35

REV Chapter 2 Tutorial for Students, Parents, Mentors, and Peers

ELL Vocabulary Activity 2

 A Chapter 2 Test

 Chapter 2 Test Generator (on the One-Stop Planner)

 Audio CD Program, Chapter 2

 A Chapter 2 Test for English Language Learners and Special-Needs Students

LAUNCH INTO LEARNING

Write *air, earth, fire,* and *water* on the chalkboard. Tell students that long ago, people thought everything was made up of these four elements. Ask students to identify different forms of these elements. Write their responses under the appropriate categories on the chalkboard. *(Examples: air—wind, tornadoes, ozone; earth—garden soil, landslides, mountains; fire—volcanoes, forest fires; water—rain, oceans, rivers, water from pipes and faucets)* Tell students that although we now know that air, earth, fire, and water are not elements, understanding their characteristics and relationships helps us understand geography and life on Earth.

Section 1

Objectives

1. Identify what objects make up the solar system.
2. Explain what causes the seasons.
3. Describe the four parts of the Earth system.

LINKS TO OUR LIVES

These are among the reasons why students should take an interest in this chapter's topics:

▶ It is easier to understand physical processes here on Earth if we first understand our planet's relationship to the solar system.

▶ New discoveries about the solar system are made almost every day. We need to have background knowledge if we are to understand those findings.

▶ To protect our supplies of freshwater and clean air, we should know more about these precious resources.

▶ By learning more about how land is formed and changed, we can save lives threatened by earthquakes, volcanoes, and other hazards.

CHAPTER 2

Planet Earth

Erupting volcano

Earth as seen from space

Fossilized shell

Galileo's telescope

Space observatory, Mauna Kea, Hawaii

LET'S GET STARTED

Write the following scenario on the chalkboard: *Imagine that you were born and raised in a dark cave. When you finally come out of the cave, you see the night sky, with the Moon and stars shining. How would you explain these bright objects?* Ask students to respond to the scenario. Have volunteers share their responses. *(Students will probably say that the Moon and the stars are close to or attached to a ceiling.)* Then ask students what they already know about the relationships involving Earth, the Sun, the Moon, the planets, and the stars. Tell students that in Section 1 they will learn more about the solar system.

Building Vocabulary

Write the key terms on the chalkboard in random order. Have students group the terms by using information they already know or from clues in the words themselves. *(Students are likely to group terms relating to the* **solar system**, *nouns that relate to Earth, and terms that have a common suffix.)* Ask students to find the definitions of the terms in Section 1 or the glossary. Then have them write sentences using the terms and explaining the connections between them.

Section 1
The Sun, Earth, and Moon

Read to Discover

1. What objects make up the solar system?
2. What causes the seasons?
3. What are the four parts of the Earth system?

Define

solar system	revolution	Tropic of
orbit	Arctic Circle	Capricorn
satellite	Antarctic Circle	equinoxes
axis	solstice	atmosphere
rotation	Tropic of Cancer	ozone

WHY IT MATTERS

In 2001 scientists labeled a rocky object beyond Pluto as the new largest minor planet. Use **CNNfyi.com** or other **current events** sources to discover more about this huge frozen rock, called 2001 KX76. Record your findings in your journal.

Mechanical model of the solar system

The Solar System

The **solar system** consists of the Sun and the objects that move around it. The most important of those objects are the planets, their moons, and relatively small rocky bodies called asteroids. Our Sun is a star at the center of our solar system. Every object in the system travels around the Sun in an **orbit**, or path. These orbits are usually elliptical, or oval shaped.

◀ **Visual Record Answer**

Neptune and Pluto

The Solar System

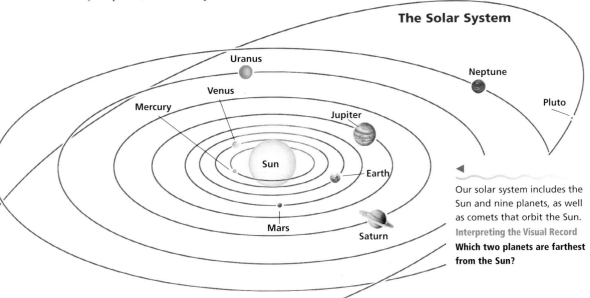

◀ Our solar system includes the Sun and nine planets, as well as comets that orbit the Sun.

Interpreting the Visual Record
Which two planets are farthest from the Sun?

17

Teaching Objective 1

LEVEL 1: (Suggested time: 15 min.) Organize students into groups. Have each group prepare a set of paper circles and label them with the names of the nine planets, the Sun, and Earth's moon. Have students arrange the elements of the solar system in the correct order on a desktop. Instruct groups to also create their own depiction of the asteroid belt and locate it correctly. **ENGLISH LANGUAGE LEARNERS, COOPERATIVE LEARNING**

LEVEL 2: (Suggested time: 40 min.) To show the relationships within our solar system, assign students to represent the Sun and each of the planets. Other students could represent major moons, including Earth's, or asteroids between Mars and Jupiter. Each student should wear a label identifying what he or she represents. Then take your class to the school's sports field for a simulation. Have the student representing the Sun stand at the goal line. Then ask students representing the planets to stand the correct distance from the "Sun," according to the chart on the following page. Later, lead a discussion on what students learned from the demonstration. *(Possible answer: The distances between outer planets are greater than between inner planets.)* **COOPERATIVE LEARNING**

National Geography Standard 15

Environment and Society

During a solar storm called a coronal mass ejection (CME), the Sun spews up to 10 billion tons (9 billion metric tons) of hot, electrically charged gas into space. The gas cloud travels at speeds reaching 1,250 miles per second (2,000 kilometers per second). If a really strong CME hits Earth, it can damage satellites, interrupt communication systems, and cause power blackouts. For example, in 1989 a CME left about 6 million people in Canada's Quebec province without electricity.

Technicians can prevent potential damage if they know when a CME is about to occur. Scientists know that periodic shifts in the Sun's magnetic field cause these ejections. They have also found that a gigantic "S" shape, seen in X-ray images taken of the Sun's surface, seems to indicate that the Sun may launch a CME within days. The "S" shapes, dubbed sigmoids, are areas about 50,000 miles (80,000 km) across, where the Sun's magnetic field has twisted back on itself.

Critical Thinking: How can scientists' knowledge about CMEs potentially affect the world?

Answer: By predicting CMEs, scientists can help people prepare for the consequences.

Mexico's Yucatán Peninsula has a crater that stretches 200 miles (322 km) wide. Scientists believe that it is the site where a giant asteroid crashed into Earth some 65 million years ago. They think the collision caused more than half of all species, including dinosaurs, to become extinct.

Tides are higher than normal when the gravitational pull of the Moon and the Sun combine. These tides, called spring tides, occur twice a month. Tides are lower than normal during neap tides, when the Sun and the Moon are at right angles. ▼

The planet nearest the Sun is Mercury, followed by Venus, Earth, and Mars. Located beyond the orbit of Mars is a belt of asteroids. Beyond this asteroid belt are the planets Jupiter and Saturn. Even farther from the Sun are the planets Uranus, Neptune, and Pluto.

The Moon Some of the planets in the solar system have more than one moon. Saturn, for example, has 18. Other planets have none. A moon is a **satellite**—a body that orbits a larger body. Earth has one moon, which is about one fourth the size of Earth. Our planet is also circled by artificial satellites that transmit signals for television, telephone, and computer communications. The Moon takes about 29½ days—roughly a month—to orbit Earth.

The Moon and Sun influence physical processes on Earth. This is because any two objects in space are affected by gravitational forces pulling them together. The gravitational effects of the Sun and the Moon cause tides in the oceans here on Earth.

The Sun Compared to some other stars, our Sun is small. It is huge, however, when compared to Earth. Its diameter is about 100 times the diameter of our planet. The Sun appears larger to us than other stars. This is because it is much closer to us than other stars. The Sun is about 93 million miles (150 million km) from Earth. The next nearest star is about 25 trillion miles (40 trillion km) away.

Scientists are trying to learn if other planets in our solar system could support life. Mars seems to offer the best possibility. It is not clear, however, if life can, or ever did, exist on Mars.

✓ **READING CHECK:** *Physical Systems* What are the main objects that make up the solar system? the Sun, the planets and their moons, asteroids

Effects of the Moon and Sun on Tides

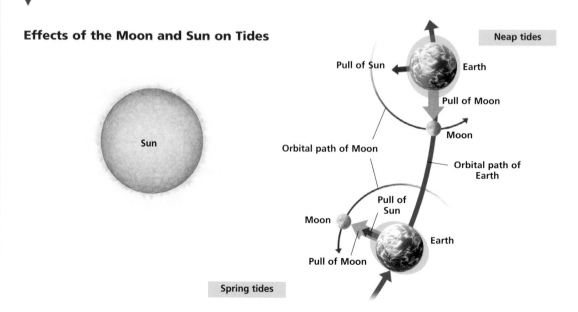

Planet	Distance from the Sun in AU*	Scaled distance in yards
Mercury	0.39	1
Venus	0.72	1.8
Earth	1	2.5
Mars	1.52	4
Jupiter	5.20	13
Saturn	9.54	24
Uranus	19.19	49
Neptune	30.06	76
Pluto	39.53	100

* AU stands for an astronomical unit, the average distance between Earth and the Sun.

LEVEL 3: Organize students into groups. Assign one planet to each group and have students conduct research on their planet's orbit. Students should find out how fast the planet travels and its distance from the Sun and the other planets as it travels. For example, Pluto's orbit swings inside Neptune's for part of its year. Have students use the information from their research to create a report. Make sure they create a bibliography with their report. Then have them set in motion the solar system from the Level 2 activity. Later, have each group present a brief summary of its research.
COOPERATIVE LEARNING

Earth

Geographers are interested in how different places on Earth receive different amounts of energy from the Sun. Differences in solar energy help explain why the tropics are warm, why the Arctic region is cold, and why day is warmer than night. To understand these differences, geographers study Earth's rotation, revolution, and the tilt of its **axis**. The axis is an imaginary line that runs from the North Pole through Earth's center to the South Pole. Rotation, revolution, and tilt control the amount of solar energy reaching Earth.

Rotation One complete spin of Earth on its axis is called a **rotation**. Each rotation takes 24 hours, or one day. Earth turns on its axis, but to us it appears that the Sun is moving. The Sun seems to "rise" in the east and "set" in the west. Before scientists learned that Earth revolves around the Sun, people thought that the Sun revolved around Earth. They thought Earth was at the center of the heavens.

Revolution It takes a year for Earth to orbit the Sun, or to complete one **revolution**. More precisely, it takes 365¼ days. To allow for this fraction of a day and keep the calendar accurate, every fourth year becomes a leap year. An extra day—February 29—is added to the calendar.

Tilt The amount of the Sun's energy reaching different parts of Earth varies. This is because Earth's axis is not straight up and down. It is actually tilted, or slanted, at an angle of 23.5° from vertical to the plane of Earth's orbit. Because of Earth's tilt, the angle at which the Sun's rays strike the planet is constantly changing as Earth revolves around the Sun. For this reason, the point where the vertical rays of

▲
Photographs taken from space can tell us about Earth.
Interpreting the Visual Record Where can you see the presence of water in this view of Earth?

internet connect
GO TO: go.hrw.com
KEYWORD: SJ3 CH2
FOR: Web sites about planet Earth

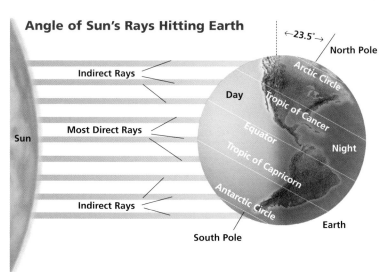

Angle of Sun's Rays Hitting Earth

←23.5°→
North Pole
Indirect Rays
Arctic Circle
Day
Tropic of Cancer
Most Direct Rays
Equator
Sun
Night
Tropic of Capricorn
Indirect Rays
Antarctic Circle
Earth
South Pole

◄
The tilt of Earth's axis and the position of the planet in its orbit determine where the Sun's rays will most directly strike the planet.
Interpreting the Visual Record Which areas of Earth receive only indirect rays from the Sun?

▲ **Visual Record Answers**

clouds, lakes, and oceans

◄

most areas except those closest to the equator

19

LEVELS 1 AND 2: (Suggested time: 15 min.) To help students understand how the seasons relate to the Sun's energy, ask two volunteers to perform this demonstration. Have one student act as Earth and give him or her a globe. Have the other student act as the Sun and give him or her a flashlight. Turn off the lights and have the volunteers sit on the floor. Ask the "Sun" to shine the flashlight on the globe, and ask "Earth" to slowly spin the globe. Have students notice which parts of the globe are most exposed to the Sun. Then have "Earth" make a complete revolution around the "Sun" while at the same time rotating the globe. As "Earth" revolves, have students identify the seasons in various parts of the world.

LEVEL 3: (Suggested time: 30 min.) On slips of paper write the names of cities around the world and various dates. Have each student select a city and find its location in the textbook's Atlas. Tell students to draw on a separate sheet of paper a diagram of Earth's position in relation to the Sun on that date. Then ask students to mark the location of their city and to write a sentence or two describing the season there on the assigned date.

COOPERATIVE LEARNING

Organize the class into small groups. Tell students that different cultures celebrate the beginning of the new year at different times. The starting point of a new year may be determined by seasonal occurrences such as the winter solstice or by a religious event. Ask each group to research the topic and prepare a presentation on the new year's observances of a particular culture or region. Tell them to be sure to include information on the historical origins of the observance.

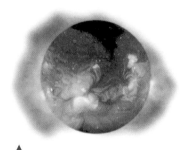

▲

The Sun's surface is always violently churning as heat flows outward from the interior.

As Earth revolves around the Sun, the tilt of the poles toward and away from the Sun causes the seasons to change.

Interpreting the Visual Record At what point is the North Pole tilted toward the Sun?

▼

the Sun strike Earth shifts north and south of the equator. These vertical rays provide more energy than rays that strike at an angle.

✓ **READING CHECK:** *Physical Systems* How do rotation, revolution, and tilt affect solar energy reaching Earth? They control the amount of solar energy because they determine the position of the Earth in relation to the Sun.

Solar Energy and Latitude

The angle at which the Sun's rays reach Earth affects temperature. In the tropics—areas in the low latitudes near the equator—the Sun's rays are nearly vertical throughout the year. In the polar regions—the areas near the North and South Poles—the Sun's rays are always at a low angle. As a result, the poles are generally the coldest places on Earth. The **Arctic Circle** is the line of latitude located 66.5° north of the equator. It circles the North Pole. The **Antarctic Circle** is the line of latitude located 66.5° south of the equator. It circles the South Pole.

The Seasons

Each year is divided into periods of time called seasons. Each season is known for a certain type of weather, based on temperature and amount of precipitation. Winter, spring, summer, and fall are examples of seasons that are described by their average temperature. "Wet" and "dry" seasons are described by their precipitation. The seasons change as Earth orbits the Sun. As this happens, the amount of solar energy received in any given location changes.

The Seasons

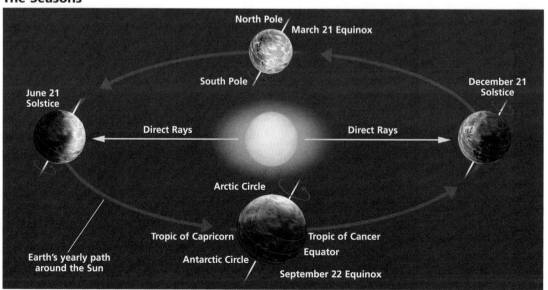

North Pole / March 21 Equinox

South Pole /

June 21 Solstice

December 21 Solstice

Direct Rays

Direct Rays

Arctic Circle

Tropic of Capricorn

Tropic of Cancer

Equator

Earth's yearly path around the Sun

Antarctic Circle

September 22 Equinox

Visual Record Answer ▶

June solstice

Teaching Objective 3

ALL LEVELS: (Suggested time: 20 min.) Copy the following graphic organizer onto the chalkboard, omitting the italicized answers. Use it to help students describe the four parts of the Earth system.

ENGLISH LANGUAGE LEARNERS

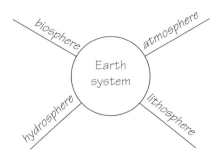

CLOSE

Call on students to suggest recent movies or television programs that depict the solar system in some way. Ask them to compare the portrayals of the scientific topics with what they have learned. Then ask them to discuss the accuracy of the filmed versions.

REVIEW AND ASSESS

Have students complete the Section Review. Then have them imagine that they are astronauts traveling through the solar system. Ask each student to write an entry in the ship's log that makes observations related to the key terms. Call on volunteers to read their entries to the class. Then have students complete Daily Quiz 2.1.

Solstice The day when the Sun's vertical rays are farthest from the equator is called a **solstice**. Solstices occur twice a year—about June 21 and about December 22. In the Northern Hemisphere the June solstice is known as the summer solstice. This is the longest day of the year and the beginning of summer. On this date the Sun's vertical rays strike Earth at the **Tropic of Cancer**. This is the line of latitude that is about 23.5° north of the equator. Six months later, about December 22, another solstice takes place. This is the winter solstice for the Northern Hemisphere. On this date the North Pole is pointed away from the Sun. The Northern Hemisphere experiences the shortest day of the year. On this date the Sun's rays strike Earth most directly at the **Tropic of Capricorn**. This line of latitude is about 23.5° south of the equator. In the Southern Hemisphere the seasons are reversed. June 21 is the winter solstice and December 22 is the summer solstice. The middle-latitude regions lie between the Tropic of Cancer and the Arctic Circle and between the Tropic of Capricorn and the Antarctic Circle.

Equinox Twice a year, halfway between summer and winter, Earth's poles are at right angles to the Sun. The Sun's rays strike the equator directly. On these days, called **equinoxes**, every place on Earth has 12 hours of day and 12 hours of night. Equinoxes mark the beginning of spring and fall. In the Northern Hemisphere the spring equinox occurs about March 21. The fall equinox occurs there about September 22. In the Southern Hemisphere the March equinox signals the beginning of fall, and the September equinox marks the beginning of spring.

Some regions on Earth, particularly in the tropics, have seasons tied to precipitation rather than temperature. Shifting wind patterns are one cause of seasonal change. For example, in January winds from the north bring dry air to India. By June the winds have shifted, coming from the southwest and bringing moisture from the Indian Ocean. These winds bring heavy rain to India. Some places in the United States also have seasons tied to moisture. East Coast states south of Virginia have a wet season in summer. These areas also have a hurricane season, lasting roughly from June to November. Some areas of the West Coast have a dry season in summer.

The seasons affect human activities. For example, in Minnesota, people shovel snow in winter to keep the walkways clear. Students waiting for a bus must wear warm clothes. The Sun rises late and sets early. As a result, people go to work and return home in darkness.

✓ **READING CHECK:** *Physical Systems* How do the seasons relate to the Sun's energy? *They are determined by the position of the Earth in its orbit around the Sun and how much solar energy is being received.*

This map shows Earth's temperatures on January 28, 1997.

Interpreting the Visual Record Which hemisphere is warmer in January?

▼

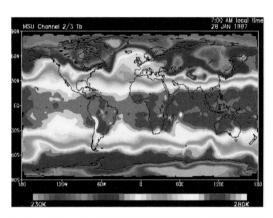

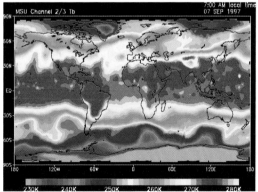

▲

This temperature map for September 7, 1997, shows summer in the Northern Hemisphere.

Across the Curriculum

SCIENCE
Life in the Hydrosphere

Deep in Earth's oceans, scientists have found organisms, such as bacteria and tube worms, that thrive in total darkness in superheated mineral-rich water. That organisms survive under these extreme conditions supports the theory that simple forms of life may exist on Europa, one of Jupiter's moons. Europa is covered by ice several miles thick. Images from NASA's *Galileo* spacecraft indicate that liquid water may have existed under the icy crust. This "ocean" may still exist, but scientists have no definitive proof that life-forms exist on Europa.

Activity: Have students conduct research on deep-sea vents on Earth called black smokers. Call on students to suggest plots for science-fiction stories based on their findings.

📶 **internet** connect

GO TO: go.hrw.com
KEYWORD: SJ3 CH2
FOR: Web sites about Europa

◀ **Visual Record Answer**

the Southern Hemisphere

Have students complete Main Idea Activity 2.1. Then organize students into three groups to create mobiles. One group's mobile should depict the objects of the solar system, the second the cause of seasons, and the third the relationships of the Earth system. **ENGLISH LANGUAGE LEARNERS**

Organize interested students into groups to research the discoveries of major figures in the history of astronomy, such as Nicolaus Copernicus, Galileo Galilei, Johannes Kepler, or Isaac Newton. Then have the groups work together to write a script for an imaginary conference call in which the scientists discuss their work. Ask the students to perform their scripts for the class. **BLOCK SCHEDULING**

Section Review 1

Answers

Define For definitions, see: solar system, p. 17; orbit, p. 17; satellite, p. 18; axis, p. 19; rotation, p. 19; revolution, p. 19; Arctic Circle, p. 20; Antarctic Circle, p. 20; solstice, p. 21; Tropic of Cancer, p. 21; Tropic of Capricorn, p. 21; equinoxes, p. 21; atmosphere, p. 22; ozone, p. 22

Reading for the Main Idea

1. the Sun, the planets, their moons, and asteroids (NGS 7)

2. atmosphere, lithosphere, hydrosphere, biosphere (NGS 3)

Critical Thinking

3. rotation, revolution, and tilt

4. Students might suggest that the seasons are reversed because the tilt of Earth's axis and Earth's rotation around the Sun affect the amount of solar energy received.

Organizing What You Know

5. Answers will vary but should include that a solstice occurs when the Sun's vertical rays are farthest from the equator and that an equinox occurs when the Sun's rays strike the equator directly.

Visual Record Answer ▶

the lake, rain, and snow

The Earth System

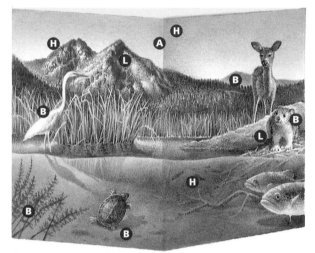

A Atmosphere   **B** Biosphere
L Lithosphere   **H** Hydrosphere

▲

The interactions of the atmosphere, lithosphere, hydrosphere, and biosphere make up the Earth system.

Interpreting the Visual Record Which items in this image are part of the hydrosphere?

Section Review 1

Define and explain: solar system, orbit, satellite, axis, rotation, revolution, Arctic Circle, Antarctic Circle, solstice, Tropic of Cancer, Tropic of Capricorn, equinoxes, atmosphere, ozone

Reading for the Main Idea

1. *Physical Systems* What are the major objects in the solar system?

2. *The World in Spatial Terms* What are the four parts of the Earth system?

The Earth System

Geographers need to be able to explain how and why places on Earth differ from each other. One way they do this is to study the interactions of forces and materials on the planet. Together, these forces and materials are known as the Earth system.

The Earth system has four parts: the **atmosphere**, the lithosphere, the hydrosphere, and the biosphere. The atmosphere is the layer of gases—the air—that surrounds Earth. These gases include nitrogen, oxygen, and carbon dioxide. The atmosphere also contains a form of oxygen called **ozone**. A layer of this gas helps protect Earth from harmful solar radiation. Another part of the Earth system is the lithosphere. The prefix *litho* means rock. The lithosphere is the solid, rocky outer layer of Earth, including the sea floor. The hydrosphere—*hydro* means water—consists of all of Earth's water, found in lakes, oceans, and glaciers. It also includes the moisture in the atmosphere. Finally, the biosphere—*bio* means life—is the part of the Earth system that includes all plant and animal life. It extends from high in the air to deep in the oceans.

By dividing Earth into these four spheres, geographers can better understand each part and how each affects the others. The different parts of the Earth system are constantly interacting in many ways. For example, a tree is part of the biosphere. However, to grow it needs to take in water, chemicals from the soil, and gases from the air.

✓ **READING CHECK:** *The World in Spatial Terms* What are the four parts of the Earth system? atmosphere, lithosphere, hydrosphere, biosphere

go.hrw.com
Homework Practice Online
Keyword: SJ3 HP2

Critical Thinking

3. **Summarizing** Which three things determine the amount of solar energy reaching places on Earth?

4. **Drawing Inferences and Conclusions** Why are the seasons reversed in the Northern and Southern Hemispheres?

Organizing What You Know

5. **Finding the Main Idea** Use this graphic organizer to explain solstice and equinox.

Solstice		Equinox
	⟨⊐⟨⊐⟩	

Section 2

Objectives

1. Examine the processes that make up the water cycle and how they are connected.
2. Identify how water is distributed on Earth.
3. Explain how water affects people's lives.

FOCUS

LET'S GET STARTED

Write the following instructions on the chalkboard: *Draw as many quick sketches as you can, in two minutes, of the ways you use water throughout the day.* Discuss completed drawings as a class. Then ask students to write a summary sentence beneath their drawings. *(Possible answers: We can't live without water. We use water every day in many ways.)* Display students' sketches around the classroom. Tell students that in Section 2 they will learn more about the role of water on Earth.

Building Vocabulary

Explain that adding *-tion* to a root verb usually turns it into a noun. Point out the verbs from which the terms *evaporation, condensation,* and *precipitation* are formed *(evaporate, condense, precipitate).* Have students use these key terms in sentences relating the terms to bodies of water in their region. Have students use the U.S. map in the textbook's Atlas for help. *(Example: Precipitation that falls in Illinois may end up in the Mississippi River.)*

Section 2 · Water on Earth

Read to Discover

1. Which processes make up the water cycle and how are they connected?
2. How is water distributed on Earth?
3. How does water affect people's lives?

Define

water vapor
water cycle
evaporation
condensation
precipitation
tributary
groundwater
continental shelf

WHY IT MATTERS

Scientists study other parts of our solar system to find out if water exists or might have existed elsewhere. Use **CNNfyi.com** or other **current events** sources to learn more about space agencies and their searches to detect water. Record your findings in your journal.

A limestone cavern

Characteristics of Water

Water has certain physical characteristics that influence Earth's geography. Water is the only substance on Earth that occurs naturally as a solid, a liquid, and a gas. We see water as a solid in snow and ice and as a liquid in lakes, oceans, and rivers. Water also occurs in the air as an invisible gas called **water vapor**.

Another characteristic of water is that it heats and cools slowly compared to land. Even on a very hot day, the ocean stays cool. A breeze blowing over the ocean brings cooler temperatures to shore. This keeps temperatures near the coast from getting as hot as they do farther inland. In winter the oceans cool more slowly than land. This generally keeps winters milder in coastal areas.

✔ **READING CHECK:** *Physical Systems* What are some important characteristics of water? occurs naturally as a solid, liquid, and gas; heats and cools slowly

The Water Cycle

The circulation of water from Earth's surface to the atmosphere and back is called the **water cycle**. The total amount of water on the planet does not change. Water, however, does change its form and its location.

Water rushes through the Stewart Mountain Dam in Arizona.

Section 2 · RESOURCES

Reproducible

◆ Guided Reading Strategy 2.2
◆ Interdisciplinary Activity for the Middle Grades 3
◆ Lab Activity for Geography and Earth Science, Hands-On 5
◆ Lab Activity for Geography and Earth Science, Demonstration 1

Technology

◆ One-Stop Planner CD–ROM, Lesson 2.2
◆ Homework Practice Online
◆ Geography and Cultures Visual Resources with Teaching Activity 7
◆ Earth: Forces and Formations, CD–ROM/Seek and Tell/Forces and Processes
◆ Earth: Forces and Formations, CD–ROM/Seek and Tell/The Earth's Surface
◆ Our Environment CD–ROM/Seek and Tell/Natural Resources
◆ HRW Go site

Reinforcement, Review, and Assessment

◆ Section 2 Review, p. 27
◆ Daily Quiz 2.2
◆ Main Idea Activity 2.2
◆ English Audio Summary 2.2
◆ Spanish Audio Summary 2.2

Teaching Objective 1

ALL LEVELS: (Suggested time: 20 min.) Pair students. Then, using the diagram of the water cycle as a guide, have each pair create a graphic organizer showing the role of evaporation, condensation, and precipitation in the cycle. **ENGLISH LANGUAGE LEARNERS, COOPERATIVE LEARNING**

Teaching Objective 2

LEVEL 1: (Suggested time: 15 min.) Pair students and tell each pair to create flash cards for the key terms that relate to how water is distributed on Earth. Students should write an appropriate key term on one side of a flash card and its definition on the reverse. Ask volunteers to share their flash cards with the class. **ENGLISH LANGUAGE LEARNERS**

LEVELS 2 AND 3: (Suggested time: 20 min.) Give each student a map of a river that flows through your region. Be sure that the map shows the river's full course. Have students label the river's headwaters, tributaries, and any reservoirs that have been created. To conclude, lead a discussion on how water is distributed on Earth.

PHYSICAL SYSTEMS

Each year as many as 40,000 icebergs break off from the glaciers of Greenland. One of Greenland's largest reported icebergs towered 550 feet (168 m) above the Atlantic Ocean's surface. In 1987 a huge tablelike berg broke away from an Antarctic ice sheet. This iceberg stretched some 100 miles (161 km). At high latitudes these gigantic icebergs can last as long as 10 years before melting.

Could we use the freshwater locked in icebergs? Experts have tried to design ways to tow icebergs to arid regions, but expense and other problems have prevented their success.

Discussion: Have students determine why towing icebergs has not been successful. Ask them to gather information about the subject and then brainstorm possible solutions to these obstacles. Ask them what the advantages and disadvantages of the solutions might be. Then have them choose a solution and demonstrate it with tubs of water and blocks of ice.

► The circulation of water from one part of the hydrosphere to another depends on energy from the Sun. Water evaporates, condenses, and falls to Earth as precipitation.
Interpreting the Visual Record How would a seasonal increase in the amount of the Sun's energy received by an area change the water cycle in that area?

This glacier is "calving"—a mass of ice is breaking off, forming an iceberg. Most of Earth's freshwater exists as ice.

▼

Visual Record Answer ►

It might increase evaporation.

The Water Cycle

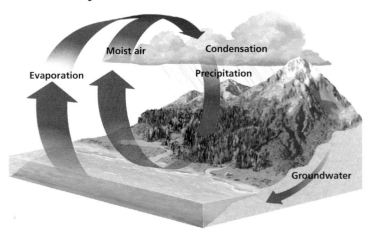

Moist air · Condensation · Evaporation · Precipitation · Groundwater

The Sun's energy drives the water cycle. **Evaporation** occurs when the Sun heats water on Earth's surface. The heated water evaporates, becoming water vapor and rising into the air. Energy from the Sun also causes winds to carry the water vapor to new locations. As the water vapor rises, it cools, causing **condensation**. This is the process by which water changes from a gas into tiny liquid droplets. These droplets join together to form clouds. If the droplets become heavy enough, **precipitation** occurs—that is, the water falls back to Earth. This water can be in the form of rain, hail, sleet, or snow. The entire cycle of evaporation, condensation, and precipitation repeats itself endlessly.

✓ **READING CHECK:** *Physical Systems* What is the water cycle? the circulation of water from Earth's surface to the atmosphere and back

Geographic Distribution of Water

The oceans contain about 97 percent of Earth's water. About another 2 percent is found in the ice sheets of Antarctica, the Arctic, Greenland, and mountain glaciers. Approximately 1 percent is found in lakes, streams, rivers, and under the ground.

Earth's freshwater resources are not evenly distributed. There are very dry places with no signs of water. Other places have many lakes and rivers. In the United States, for example, Minnesota is dotted with more than 11,000 lakes. Dry states such as Nevada have few natural lakes. In many dry places rivers have been dammed to create artificial lakes called reservoirs.

Teaching Objective 3

LEVEL 1: (Suggested time: 20 min.) Copy the following graphic organizer onto the chalkboard, omitting the italicized answers. Use it to help students explore how water affects people's lives. Have students copy the organizer into their notebooks and complete it.

ENGLISH LANGUAGE LEARNERS

WATER ISSUES

1. *floods*
2. *flood control/dams*
3. *availability of clean water*
4. *availability of adequate water supply*

Surface Water Water sometimes collects at high elevations, where rivers begin their flow down toward the lowlands and coasts. The first and smallest streams that form from this runoff are called headwaters. When these headwaters meet and join, they form larger streams. In turn, these streams join with other streams to form rivers. Any smaller stream or river that flows into a larger stream or river is a **tributary**. For example, the Missouri River is an important tributary of the Mississippi River, into which it flows near St. Louis, Missouri.

Lakes are usually formed when rivers flow into basins and fill them with water. Most lakes are freshwater, but some are salty. For example, the Great Salt Lake in Utah receives water from the Bear, Jordan, and Weber Rivers but has no outlet. Because the air is dry here, the rate of evaporation is very high. When the lake water evaporates, it leaves behind salts and minerals, making the lake salty.

Groundwater Not all surface water immediately returns to the atmosphere through evaporation. Some water from rainfall, rivers, lakes, and melting snow seeps into the ground. This **groundwater** seeps down until all the spaces between soil and grains of rock are filled. In some places, groundwater bubbles out of the ground as a spring. Many towns in the United States get their water from wells— deep holes dug down to reach the groundwater. Motorized pumps allow people to draw water from very deep underground.

Oceans Most of Earth's water is found in the oceans. The Pacific, Atlantic, Indian, and Arctic Oceans connect with each other. This giant body of water covers some 71 percent of Earth's surface. These oceans also include smaller regions called seas and gulfs. The Gulf of Mexico and the Gulf of Alaska are two examples of smaller ocean areas.

Surrounding each continent is a zone of shallow ocean water. This gently sloping underwater land—called the **continental shelf**—is important to marine life. Although the oceans are huge, marine life is concentrated in these shallow areas. Deeper ocean water is home to fewer organisms. Overall, the oceans average about 12,000 feet (about 3,700 m) in depth. The deepest place is the Mariana Trench in the Pacific Ocean, at about 36,000 feet (about 11,000 m) deep.

✓ **READING CHECK:** *The World in Spatial Terms* How is water distributed on Earth? *in the oceans; ice sheets of Antarctica, the Arctic, and Greenland; and in lakes, streams, rivers, and groundwater*

Groundwater

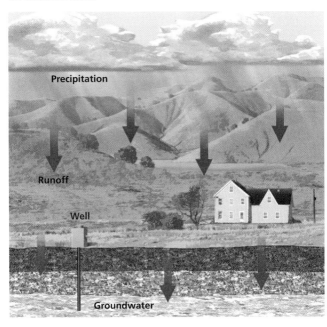

In some areas where rainfall is scarce, enough groundwater exists to support agriculture.

Interpreting the Visual Record How do people gain access to groundwater?

The Continental Shelf

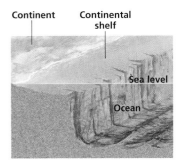

The continental shelf slopes gently away from the continents. The ocean floor drops steeply at the edge of the shelf.

EYE ON EARTH

We think of ocean waves as racing across miles of open water within hours. However, Rossby waves can take years to travel across the ocean. These large-scale waves may be only a few inches high, but they can be many miles long. Because they move slowly, Rossby waves can carry a "memory" of storms or other oceanic events that occurred years earlier. For example, oceanographers mapped a Rossby wave in 1994 that seemed to show evidence of a 1982–83 El Niño. Understanding Rossby waves helps meteorologists predict the weather because the waves may push powerful ocean currents away from their usual paths.

Activity: Have students conduct research on connections between ocean currents and weather prediction. Have students create graphic organizers of their findings.

▲ **Visual Record Answer**

by digging wells

➤**ASSIGNMENT:** Have students use their sketches from Let's Get Started to recall times they use water throughout the day. Have them write a list of those uses. Then have students propose how they would cope if the water supply were suddenly unavailable or polluted.

CLOSE

Tell students that in 1991, during the Persian Gulf War, a massive oil spill in the Persian Gulf threatened desalinization factories in the region. These plants remove salt from seawater and supply freshwater to countries in the region. Ask students to suggest how other natural and human-made disasters might affect water supplies.

National Geography Standard 8

Physical Systems Wetlands are vital to Earth's water supply. They are areas where the soil absorbs a great amount of water. Depending on the soil and plant life found in them, wetlands can also be called tidal flats, swamps, or bogs.

Wetlands are some of the most productive environmental systems on Earth. By absorbing a lot of moisture, they help prevent floods. The plants that grow in wetlands can remove pollution from sewage. Wetlands also serve as nurseries for fish, shrimp, and shellfish. About one third of the rare and endangered animals in the United States live in wetlands. Migrating birds depend on wetlands for rest and food. Coastal wetlands can help prevent erosion.

By the mid-1970s these vital elements in the biosphere were threatened. The continental United States had lost more than 50 percent of its original wetlands. Wetlands remain threatened worldwide.

Activity: Have students design a public-awareness campaign to focus attention on the importance of preserving the wetlands.

Visual Record Answer ▶

Answers will vary, but students might mention access to transportation, trade, and a water supply as reasons.

▲
This kelp forest is off the coast of southern California. The shallower parts of the oceans are home to many plants and animals.

Water Issues

Water plays an important part in our survival. As a result, water issues frequently show up in the news. Thunderstorms, particularly when accompanied by hail or tornadoes, can damage buildings and ruin crops. Droughts also can be deadly. In the mountains, heavy snowfalls sometimes cause deadly snow slides called avalanches. Heavy fog can make driving or flying dangerous. Geographers are concerned with these issues. They work on ways to better prepare for natural hazards.

Floods Water can both support and threaten life. Heavy rains can cause floods, which are the world's deadliest natural hazard. Floods kill four out of every ten people who die from natural disasters, including hurricanes, earthquakes, tornadoes, and thunderstorms.

Some floods occur in dry places when strong thunderstorms drop a large amount of rain very quickly. The water races along on the hard, dry surface instead of soaking into the ground. This water can quickly gather in low places. Creekbeds that are normally dry can suddenly surge with rushing water. People and livestock are sometimes caught in these flash floods.

Floods also happen in low-lying places next to rivers and on coastlines. Too much rain or snowmelt entering a river can cause the water to overflow the banks. Powerful storms, particularly hurricanes, can sometimes cause ocean waters to surge into coastal areas. Look at a

Davenport, Iowa, suffered severe flooding in 1993. Many cities are located along rivers even though there is a danger of floods.
Interpreting the Visual Record **Why are many cities located next to rivers?**

▶

Have students complete the Section Review. Then prepare a set of flash cards, each with a key term or definition on it. Hold up the cards and call on students to supply the correct term or definition. Then have students complete Daily Quiz 2.2.

Have interested students conduct research on a significant flood that occurred in the past 10 years. Tell students to provide a map of the areas hurt by the flood as well as information on how the flood affected the area's people, plants, animals, and economy. Students should also note the long-term effects of the flood. **BLOCK SCHEDULING**

Have students complete Main Idea Activity 2.2. Then call on students to relate each of the section's main concepts to specific locations on a map of your state. For example, students should locate tributaries, reservoirs, and areas that may be prone to flooding and explain why.
ENGLISH LANGUAGE LEARNERS

Section Review 2

map of the United States. You will see that most major cities are located either next to a river or along a coast. For this reason, floods can be a threat to lives and property.

Flood Control Dams are a means people use to control floods. The huge Hoover Dam on the Colorado River in the United States is one example. The Aswān Dam on the Nile River in Egypt and those on the Murray River in Australia are others. Dams help protect people from floods. They also store water for use during dry periods. However, dams prevent rivers from bringing soil nutrients to areas downstream. Sometimes farming is not as productive as it was before the dams were built.

Clean Water The availability of clean water is another issue affecting the world's people. Not every country is able to provide clean water for drinking and bathing. Pollution also threatens the health of the world's oceans—particularly in the shallower seabeds where many fish live and reproduce.

Water Supply Finding enough water to meet basic needs is a concern in regions that are naturally dry. The water supply is the amount of water available for use in a region. It limits the number of living things that can survive in a place. People in some countries have to struggle each day to find enough water.

✔ READING CHECK: *Environment and Society* What are some ways in which water affects people? through the weather, the availability of clean water for drinking and bathing

▲

The Atatürk Dam in Turkey is 604 feet (184 m) high.

Interpreting the Visual Record How have people used dams to change their physical environment?

Section Review 2

Define and explain: water vapor, water cycle, evaporation, condensation, precipitation, tributary, groundwater, continental shelf

Reading for the Main Idea

1. (*Physical Systems*) What are the steps in the water cycle?

2. (*Environment and Society*) What are some major issues related to water?

go. hrw .com

Homework Practice Online

Keyword: SJ3 HP2

Critical Thinking

3. **Analyzing Information** What are some important characteristics of water?

4. **Finding the Main Idea** Why is water supply a problem in some areas?

Organizing What You Know

5. **Sequencing** Copy the following graphic organizer. Use it to explain the water cycle.

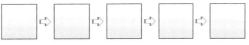

Section Review 2

Answers

Define For definitions, see: water vapor, p. 23; water cycle, p. 23; evaporation, p. 24; condensation, p. 24; precipitation, p. 24; tributary, p. 25; groundwater, p. 25; continental shelf, p. 25

Reading for the Main Idea

1. Evaporation leads to condensation, which leads to precipitation, and then the cycle repeats itself. (NGS 7)

2. floods, flood control, availability of clean water, and the water supply (NGS 15)

Critical Thinking

3. Water is the only substance on Earth that occurs naturally as a solid, a liquid, and a gas. Water also heats and cools slowly.

4. because these areas are naturally dry and do not have enough water to meet people's needs

Organizing What You Know

5. Answers will vary but should include evaporation, condensation, and precipitation as steps.

▲ **Visual Record Answer**

Dams provide people with the ability to control flooding and with water storage for dry periods. They can also provide electrical power.

27

Section 3

Objectives

1. Identify primary landforms.
2. Describe secondary landforms.
3. Explain how humans interact with landforms.

FOCUS

LET'S GET STARTED

Write the following question on the chalkboard: *What do we mean when we say "solid as a rock," "mountain of strength," or "older than dirt"?* Conduct a class discussion. Then ask students what the phrases suggest about Earth. *(Possible answers: that it is unchangeable, permanent)* Tell students that in Section 3 they will learn that Earth is actually in motion and can change dramatically.

Building Vocabulary

Write the key terms on slips of paper and have each student draw one from a hat. You may need to have duplicates of some terms. Ask students to find the definitions by skimming the text and read the definitions to the class. Then write *shapes on Earth's surface, slow movement,* and *fast change* on the chalkboard. Have the students determine the appropriate category for each term.

Section 3 RESOURCES

Reproducible

◆ Guided Reading Strategy 2.3
◆ Graphic Organizer 2
◆ Environmental and Global Issues Activity 3
◆ Lab Activities for Geography and Earth Science, Demonstrations 3, 4, 5, 6, 7, 9, and 10
◆ Map Activity 2

Technology

◆ One-Stop Planner CD–ROM, Lesson 2.3
◆ Homework Practice Online
◆ Earth: Forces and Formations CD–ROM/Seek and Tell/Forces and Processes
◆ Earth: Forces and Formations CD–ROM/Seek and Tell/The Earth's Surface
◆ HRW Go site

Reinforcement, Review, and Assessment

◆ Section 3 Review, p. 33
◆ Daily Quiz 2.3
◆ Main Idea Activity 2.3
◆ English Audio Summary 2.3
◆ Spanish Audio Summary 2.3

The Land

Read to Discover

1. What are primary landforms?
2. What are secondary landforms?
3. How do humans interact with landforms?

Define

landforms	core	subduction	alluvial fan
plain	mantle	earthquakes	floodplain
plateau	crust	fault	deltas
isthmus	magma	Pangaea	glaciers
peninsula	lava	weathering	
plate tectonics	continents	erosion	

WHY IT MATTERS

Earth's surface has been the focus of scientific research for centuries. Use CNN fyi.com logo or other **current events** sources to explore how scientists use maps to study the surface of Earth and other planets. Record your findings in your journal.

Mount Saint Helens, in southern Washington State

Landforms

Landforms are shapes on Earth's surface. One common landform is a **plain**—a nearly flat area. A **plateau** is an elevated flatland. An **isthmus** is a neck of land connecting two larger land areas. A **peninsula** is land bordered by water on three sides.

Primary Landforms

The theory of **plate tectonics** helps explain how forces raise, lower, and roughen Earth's surface. According to this theory, Earth's surface is divided into several large plates, or pieces. There are also many smaller plates. The plates slowly move. Some plates are colliding. Some are moving apart. Others are sliding by each other. Landforms created by tectonic processes are called primary landforms. These include masses of rock raised by volcanic eruptions and deep ocean trenches. The energy that moves the tectonic plates comes from inside Earth. The inner, solid **core** of the planet is surrounded by a liquid layer called the **mantle**. The outer, solid layer of Earth is called the **crust**. Currents of heat from the core travel outward through the mantle. When the currents reach the upper mantle, rocks can melt to form **magma**.

Earth's thin crust floats on top of the liquid mantle.

The Interior of Earth

Crust
(3–30 mi. or about 5–50 km)

Mantle
(1,800 mi. or about 2,900 km)

Outer core
(1,300 mi. or about 2,080 km)

Inner core
(860 mi. or about 1,390 km)

ATMOSPHERE

Teaching Objective 1

LEVEL 1: (Suggested time: 45 min.) Provide students with nature and tourism magazines. Have them find and cut out photographs that show various primary landforms. Have each student create a mural with his or her photos. Call on students to tell why they identified these landforms as primary landforms. **ENGLISH LANGUAGE LEARNERS**

LEVELS 2 AND 3: (Suggested time: 30 min.) Have each student create a three-panel brochure titled "When Plates Collide." Each panel should contain a description of a primary landform created when tectonic plates collide, a diagram of the process involved in creating the landform, and an example of a place in the world where that process is occurring. Ask volunteers to share their brochures with the class.

Plate Tectonics

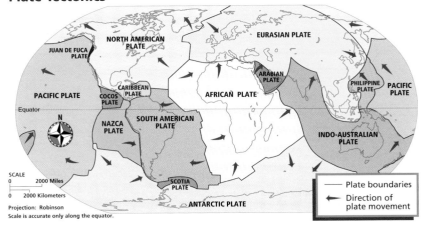

The plates that make up Earth's crust are moving, usually a few inches per year. This map shows the plates and the direction of their movement.

Magma sometimes breaks through the crust to form a volcano. After reaching Earth's surface, magma is called **lava**.

Plates cover Earth's entire surface, both the land and the ocean. In general, the plates under the oceans are made of dense rock. The plates on the **continents**—Earth's large landmasses—are made of lighter rock.

Plates Colliding When two plates collide, one plate can be pushed under another. When this occurs under the ocean, a very deep trench is sometimes created. This is happening near Japan, where the Pacific plate is slowly moving under the Eurasian and Philippine plates. Any time a heavier plate moves under a lighter one, trenches can form. This process is called **subduction**. **Earthquakes** are common in subduction zones. An earthquake is a sudden, violent movement along a fracture within Earth's crust. A series of shocks usually results from such a movement within the crust.

The borders of the Pacific plate move against neighboring plates. This causes volcanoes to erupt and earthquakes to strike in that area. The Pacific plate's edge has been called the Ring of Fire because it is rimmed by active volcanoes. Thousands of people have died and terrible destruction has resulted from the earthquakes and volcanoes there.

When a continental plate and an ocean plate collide, the lighter rocks of the continent do not sink. Instead, they crumple and form a mountain range. The Andes in South America were formed this way. When two continental plates collide, land is lifted, sometimes to great heights. The Himalayas, the world's highest mountain range, were created by the Indo-Australian plate pushing into the Eurasian plate.

Subduction and Spreading

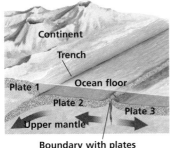

Where Plate 2 pushes under Plate 1, a deep trench forms. This process is called subduction. Where Plates 2 and 3 move apart, lava creates a mid-ocean ridge.

Plates Colliding

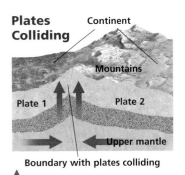

Where two continental plates collide, Earth's crust is pushed upward, forming a mountain range.

Cultural Kaleidoscope

Primary Landforms and Legends

According to Irish legend, a giant named Finn Mac Cool built the Giant's Causeway, a formation of about 40,000 stone columns on the coast of Northern Ireland. By one account, Mac Cool drove the columns into place so he could walk to Scotland to fight fellow giant Benandonner. After completing the causeway, Mac Cool returned to Ireland. When Benandonner saw the causeway, he walked across it to Ireland, where he saw Mac Cool asleep. Mac Cool's wife told Benandonner that the sleeping giant was her child. Thinking that if this was the child the father must surely be unbeatable, Benandonner fled back to Scotland, destroying the causeway as he went. Similar column formations exist across the North Channel where the "causeway" would have reached Scotland.

The basalt columns were actually formed 50–60 million years ago when lava flows cooled as they reached the sea. Pressure shaped the rock into columns with three to seven sides. The columns average 330 feet (100 m) high.

Activity: Have students identify an unusual landform in your state and find out how it was formed. Then have them write and act out legends that describe the feature's origin.

29

Primary Landforms	Secondary Landforms
created by tectonic processes	*landforms that result when primary landforms break down*
masses of rock raised by volcanic eruptions and deep ocean trenches	*caused by weathering and erosion*
include mountain ranges and mid-ocean ranges	*include floodplains and deltas*

Teaching Objective 2
LEVELS 2 AND 3: (Suggested time: 45 min.) Organize students into teams and have them explore the school grounds or a nearby park to find examples of erosion. Ask them to collect rocks and classify them according to which force seems to have eroded them. Also have students determine if and why any of the three forces of erosion are not present in their region. Have students present their findings to the class. *(Students might mention that glaciers are not a force of erosion in their regions.)*
COOPERATIVE LEARNING

EYE ON EARTH

Weathering and erosion are generally slow processes that lead to the creation of secondary landforms. Sometimes, however, the breaking down of primary landforms can be quite sudden.

For example, miners in Switzerland carved slate from the base of a mountain cliff, creating a huge overhang. Great cracks appeared in the cliff, and finally on September 11, 1881, millions of cubic yards of rock fell. The avalanche of rocks did not stop when it hit the valley floor. Instead, the broken rock continued up the valley's other side. More than 100 people were killed in this debris avalanche.

Critical Thinking: In what other ways do humans change their environment that might create conditions for debris avalanches?

Answer: cutting through mountains to build highways, constructing buildings on hillsides, cutting down trees, and other disturbances of steep terrains

Connecting to Technology Answers

1. could save lives and property
2. seismographs, tiltmeters, gravimeters, laser beams, and satellites

CONNECTING TO Technology

forecasting earthquakes

Since ancient times, people have tried to forecast earthquakes. A Chinese inventor even created a device to register earthquakes as early as A.D. 132.

The theory of plate tectonics gives modern-day scientists a better understanding of how and why earthquakes occur. Earthquake scientists, known as seismologists, have many tools to help them monitor movements in Earth's crust. They try to understand when and where earthquakes will occur.

The most common of these devices is the seismograph. It measures seismic waves—vibrations produced when two tectonic plates grind against each other. Scientists believe that an increase in seismic activity may signal a coming earthquake.

A scientist with a seismograph

Other devices show shifts in Earth's crust. Tiltmeters measure the rise of tectonic plates along a fault line. Gravimeters record changes in gravitational strength caused by rising or falling land. Laser beams can detect lateral movements along a fault line. Satellites can note the movement of entire tectonic plates.

Scientists have yet to learn how to forecast earthquakes with accuracy. Nevertheless, their ongoing work may one day provide important breakthroughs in the science of earthquake forecasting.

You Be the Geographer
1. What social and economic consequences might earthquake forecasting have?
2. What technology has helped with earthquake forecasting?

Plates Moving Apart When two plates move away from each other, hot lava emerges from the gap that has been formed. The lava builds a mid-ocean ridge—a landform that is similar to an underwater mountain range. This process is currently occurring in the Atlantic Ocean where the Eurasian plate and the North American plate are moving away from each other.

Plates Sliding Tectonic plates can also slide past each other. Earthquakes occur from sudden adjustments in Earth's crust. In California the Pacific plate is sliding northwestward along the edge of the North American plate. This has created the San Andreas Fault

30

Kay A. Knowles, of Montross, Virginia, suggests the following activity to help students identify landforms and predict where other landforms might be found. Display a wall map and have students point out examples of these landforms: continent, isthmus, peninsula, plain, and plateau. Also, ask students to identify places where they would expect to find these landforms: primary landform, secondary landform, alluvial fan, floodplain, and delta.

zone. A **fault** is a fractured surface in Earth's crust where a mass of rocks is in motion.

Tectonic plates move slowly—just inches a year. If, however, we could look back 200 million years, we would see that the continents have moved a long way. From their understanding of plate tectonics, scientists proposed the theory of continental drift. This theory states that the continents were once united in a single super-continent. They then separated and moved to the positions they are in today. Scientists call this original landmass **Pangaea** (pan-GEE-uh).

Continental Drift

About 200 million years ago it is believed there was only one continent, Pangaea, and one ocean, Panthalassa. The continental plates slowly drifted into their present-day positions.

✓ **READING CHECK:** **Physical Systems** How are primary landforms formed? *by the movement of the tectonic plates*

Secondary Landforms

The forces of plate tectonics build up primary landforms. At the same time, water, wind, and ice constantly break down rocks and cause rocky material to move. The landforms that result when primary landforms are broken down are called secondary landforms.

One process of breaking down and changing primary landforms into secondary landforms is called **weathering**. It is the process of breaking rocks into smaller pieces. Weathering occurs in several ways. Heat can cause rocks to crack. Water may get into cracks in rocks and freeze. This ice then expands with a force great enough to break the rock. Water can also work its way underground and slowly dissolve minerals such as limestone. This process sometimes creates caves. In

These steep peaks in Chile are part of the Andes.

Interpreting the Visual Record **How do these mountains show the effects of weathering and erosion?**

Have students study the map on p. 29 showing continental drift. Point out that the shapes of South America and Africa were an early clue in the development of the continental-drift theory. Ask students what other kinds of evidence might indicate that now-distant continents were once joined. *(Possible answers: corresponding fossils and mineral deposits)*

Have students complete the Section Review. Then pair students and instruct pairs to write sentences that illustrate the connection between two of the key terms. *(Example: A plateau is a plain that lies at a higher elevation.)* Call on volunteers to read their sentences to the class. Continue until all the major points have been covered. Then have students complete Daily Quiz 2.3.

Section Review 3

Answers

Define For definitions, see: landforms, p. 28; plain, p. 28; plateau, p. 28; isthmus, p. 28; peninsula, p. 28; plate tectonics, p. 28; core, p. 28; mantle, p. 28; crust, p. 28; magma, p. 28; lava, p. 29; continents, p. 29; subduction, p. 29; earthquakes, p. 29; fault, p. 31; Pangaea, p. 31; weathering, p. 31; erosion, p. 32; alluvial fan, p. 32; floodplain, p. 32; deltas, p. 32; glaciers, p. 33

Reading for the Main Idea

1. by tectonic processes (NGS 7)

2. weathering—heat, cold, water, and plants; erosion—water, glaciers, and wind (NGS 7)

Critical Thinking

3. Earth's surface is divided into seven plates. These plates move and create landforms.

4. People change landforms and adapt them as needed, building dams, canals, highways through mountains, etc.

Organizing What You Know

5. colliding—ocean trenches and mountain ranges created; earthquakes and volcanic eruptions common; moving apart—hot lava emerges and mid-ocean ridges are created; sliding—faults created

32

Elevation is the height of the land above sea level. An elevation profile is a cross-section used to show the elevation of a specific area.

Interpreting the Visual Record What is the range of elevation for Guadalcanal?

Elevation Profile: Guadalcanal

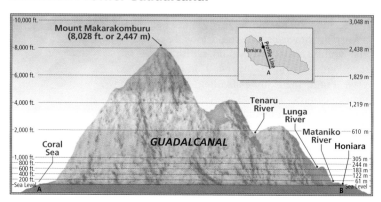

some areas small plants called lichens attach to bare rock. Chemicals in the lichens gradually break down the stone. Some places in the world experience large swings in temperature. In the Arctic the ground freezes and thaws, which tends to lift stones to the surface in unusual patterns. Regardless of which weathering process is at work, rocks eventually break down into sediment. These smaller pieces of rock are called gravel, sand, silt, or clay, depending on particle size. Once weathering has taken place, water, ice, or wind can move the material and create new landforms.

✓ **READING CHECK:** *Physical Systems* What is one way in which secondary landforms are created? **through weathering**

Erosion

Another process of changing primary landforms into secondary landforms is **erosion**. Erosion is the movement of rocky materials to another location. Moving water is the most common force that erodes and shapes the land.

River water, brown with sediment, enters the ocean.

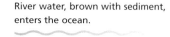

Water Flowing water carries sediment. This sediment forms different kinds of landforms depending on where it is deposited. For example, a river flowing from a mountain range onto a flat area, or plain, may deposit some of its sediment there. The sediment sometimes builds up into a fan-shaped form called an **alluvial fan**. A **floodplain** is created when rivers flood their banks and deposit sediment. A **delta** is formed when rivers carry some of their sediment all the way to the ocean. The sediment settles to the bottom where the river meets the ocean. The Nile and Mississippi Rivers have two of the world's largest deltas.

Waves in the ocean and in lakes also shape the land they touch. Waves can shape beaches into great dunes, such as on the shore of Long Island. The jagged coastline of Oregon also shows the erosive power of waves.

Have students complete Main Idea Activity 2.3. Then pair students and have them create brief descriptions of each main idea in the section. Have students consult atlases, globes, or encyclopedias to locate two specific examples of each feature or main idea. **ENGLISH LANGUAGE LEARNERS**

Have interested students conduct research on plate tectonics and prepare a report that focuses on one of the following topics: (1) how the theory was developed, and what evidence supports it; (2) the causes of plate movement and continental drift; and (3) practical applications of the theory, such as earthquake prediction and mineral exploration. Then, have students work together to create a script and several storyboards for a documentary film on the topic. Remind them to use standard grammar, spelling, sentence structure, and punctuation. **BLOCK SCHEDULING**

Glaciers In high mountain settings and in the coldest places on Earth are **glaciers**. These large, slow-moving rivers of ice have the power to move tons of rock.

Giant sheets of thick ice called continental glaciers cover Greenland and Antarctica. Over the past 2 million years Earth has experienced several ice ages—periods of extremely cold conditions. During each ice age continental glaciers covered most of Canada and the northern United States. The Great Lakes were carved out by the movement of a continental glacier.

Wind Wind also shapes the land. Strong winds can lift soils into the air and carry them across great distances. On beaches and in deserts wind can deposit large amounts of sand to form dunes.

Blowing sand can wear down rock. The sand acts like sandpaper to polish jagged edges. An example of rocks worn down by blowing sand can be seen in Utah's Canyonlands National Park.

✓ **READING CHECK:** *Physical Systems* What forces cause erosion? water, glaciers, wind

Waves 50 to 60 feet high, which the local people call Jaws, sometimes occur off the coast of Maui, Hawaii. They are caused by storms in the north Pacific and a high offshore ridge that focuses the waves' energy.

People and Landforms

Geographers study how people adapt their lives to different landforms. Deltas and floodplains, for example, are usually fertile places to grow food. People also change landforms. Engineers build dams to control river flooding. They drill tunnels through mountains instead of making roads over mountaintops. People have used modern technology to build structures that are better able to survive disasters like floods and earthquakes.

✓ **READING CHECK:** *Environment and Society* What are some examples of humans adjusting to and changing landforms? using fertile deltas for farming, drilling through mountains to make roads

Section Review 3

Homework Practice Online
Keyword: SJ3 HP2

Define and explain: landforms, plain, plateau, isthmus, peninsula, plate tectonics, core, mantle, crust, magma, lava, continents, earthquakes, fault, Pangaea, weathering, erosion, alluvial fan, floodplain, deltas, glaciers

Reading for the Main Idea

1. *Physical Systems* How are primary landforms created?

2. *Physical Systems* What forces cause weathering and erosion?

Critical Thinking

3. Summarizing What is plate tectonics?

4. Finding the Main Idea How do people affect landforms? Give examples.

Organizing What You Know

5. Identifying Cause and Effect Copy the following graphic organizer. Use it to describe the movement of plates and the movement's effects.

movement		resulting landforms and changes
	⇨	
	⇨	
	⇨	

Review
ANSWERS

Building Vocabulary
For definitions, see: solar system, p. 17; orbit, p. 17; solstice, p. 21; equinoxes, p. 21; atmosphere, p. 22; water cycle, p. 23; evaporation, p. 24; condensation, p. 24; precipitation, p. 24; landforms, p. 28; plate tectonics, p. 28; continents, p. 29; earthquakes, p. 29; weathering, p. 31; erosion, p. 32

Reviewing the Main Ideas

1. The rotation, revolution, and tilt of Earth limit the areas that sunlight can reach. (NGS 7)

2. The Sun's energy creates water vapor and causes winds, which carry water vapor to new locations. Water vapor rises, cools, condenses, forms clouds; precipitation occurs. (NGS 7)

3. Tectonic plates move and collide, raising, lowering, and roughening Earth's surface, and thereby shaping continents and primary landforms. (NGS 7)

4. A heavier plate is pushed under a lighter one, and sometimes a trench is formed. (NGS 7)

5. by weathering and erosion (NGS 7)

ASSESS

Have students complete the Chapter 2 Test.

RETEACH

Organize students into groups and have each group create one of three posters: a depiction of Earth's location in the solar system and how its tilt determines seasons, a cutaway view of Earth showing water distribution, or a cutaway view of Earth showing landforms and the processes that affect them, including ocean circulation. Compare the posters as a class to ensure that all major points have been covered.
ENGLISH LANGUAGE LEARNERS

PORTFOLIO EXTENSIONS

1. To demonstrate the importance of water, have students conduct research on the location of their school's drinking water. *(Possible water sources include rivers, lakes, reservoirs, and groundwater.)* Students should also find out how the school disposes of its wastewater. Have students create diagrams of their school's water-supply system.

2. Have students research the origin of local or nearby landforms. *(Possible origins include glacial action, volcanic action, and sedimentation.)* To present their findings, have students label landforms on a map according to how they were formed. You may want to have students design models that show the formation processes. Photograph the models and place the pictures in student portfolios.

Review
ANSWERS

Understanding Environment and Society

- Sources may include lakes, rivers, reservoirs, springs, and wells.
- Arid regions are more likely to have problems, but almost all areas have suffered occasional water shortages due to drought, broken pipes, flood, or other causes.
- Students' answers may include laws or ordinances regarding commercial and residential use.

Thinking Critically

1. because the atmosphere, lithosphere, hydrosphere, and biosphere are constantly interacting
2. Valleys, ravines, and canyons are cut by streams; deposits of river sediment form alluvial fans, floodplains, and deltas; glaciers move rock, carving landforms such as lakes; wind creates dunes and wears down rock.
3. These areas are usually good places for people to grow food.
4. (a) Land is lifted, creating mountains. (b) Lava emerges from the gap, creating a mid-ocean ridge.

Reviewing What You Know

Building Vocabulary

On a separate sheet of paper, write sentences to define each of the following words.

1. solar system
2. orbit
3. solstice
4. equinoxes
5. atmosphere
6. water cycle
7. evaporation
8. condensation
9. precipitation
10. landforms
11. plate tectonics
12. continents
13. earthquakes
14. weathering
15. erosion

Reviewing the Main Ideas

1. *(Physical Systems)* In what ways do Earth's rotation, revolution, and tilt help determine how much of the Sun's energy reaches Earth?
2. *(Physical Systems)* Explain how the Sun's energy drives the water cycle. Be sure to include a discussion of the three elements of the cycle.
3. *(Physical Systems)* How does plate tectonics relate to the continents and their landforms?
4. *(Physical Systems)* Describe how subduction zones are created.
5. *(Physical Systems)* Describe the different ways secondary landforms are created.

Understanding Environment and Society

Water Use
The availability and purity of water are important for everyone. Research the water supply in your own city or community and prepare a presentation on it. You may want to think about the following:

- Where your city or community gets its drinking water,
- Drought or water shortages that have happened in the past,
- Actions your community takes to protect the water supply.

Thinking Critically

1. **Drawing Inferences and Conclusions** How do the four parts of the Earth system help explain why places on Earth differ?
2. **Finding the Main Idea** Describe landforms that are shaped by water and wind.
3. **Drawing Inferences and Conclusions** Why do people continue to live in areas where floods are likely to occur?
4. **Identifying Cause and Effect** Describe the landforms that result (a) when two tectonic plates collide and (b) when two plates move away from each other.
5. **Summarizing** In what ways do people interact with landforms?
6. **Analyzing Information** How might mountains be primary and secondary landforms?

FOOD FESTIVAL

The Sun makes life on Earth possible. It can also be used to process foods. Have students make sun tea by filling a jar with water, adding tea bags, covering the jar, and then placing it in sunlight to steep. After a few hours, have students add ice to the tea and enjoy. You might also have students locate and bring to class sun-dried food items, such as beef jerky, raisins, or tomatoes. Challenge the class to research regions where fuel is scarce and devise ways people could use solar energy to cook food. Have them report their information to the class.

CHAPTER 2 — REVIEW AND ASSESSMENT RESOURCES

Reproducible
- Readings in World Geography, History, and Culture 3 and 4
- Critical Thinking Activity 2
- Vocabulary Activity 2

Technology
- Chapter 2 Test Generator (on the One-Stop Planner)

- Audio CD Program, Chapter 2
- HRW Go site

Reinforcement, Review, and Assessment
- Chapter 2 Review, pp. 34–35

- Chapter 2 Tutorial for Students, Parents, Mentors, and Peers
- Chapter 2 Test
- Chapter 2 Test for English Language Learners and Special-Needs Students

---

Building Social Studies Skills

Map ACTIVITY

On a separate sheet of paper, match the letters on the globe with their correct labels.

Tropic of Cancer
Tropic of Capricorn
equator

Arctic Circle
Antarctic Circle
North Pole
South Pole

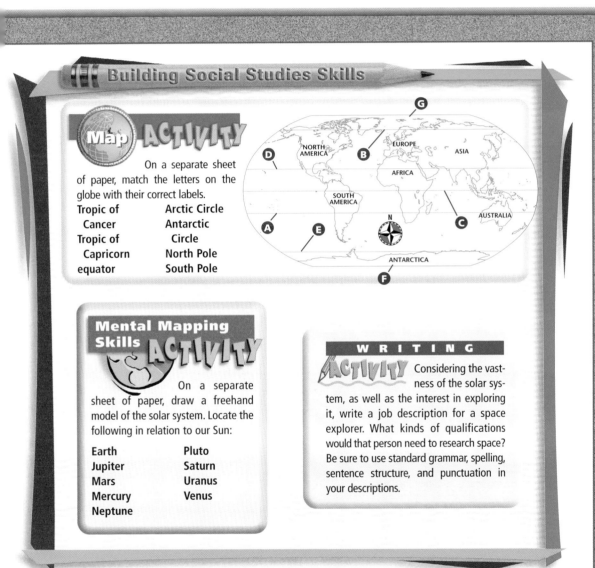

Mental Mapping Skills ACTIVITY

On a separate sheet of paper, draw a freehand model of the solar system. Locate the following in relation to our Sun:

Earth
Jupiter
Mars
Mercury
Neptune

Pluto
Saturn
Uranus
Venus

WRITING ACTIVITY

Considering the vastness of the solar system, as well as the interest in exploring it, write a job description for a space explorer. What kinds of qualifications would that person need to research space? Be sure to use standard grammar, spelling, sentence structure, and punctuation in your descriptions.

Alternative Assessment

Portfolio ACTIVITY

Learning About Your Local Geography

Individual Project Compare the latitude and longitude of your state's capital city with that of the capitals of three countries. How might the seasons be similar or different in each city?

internet connect
Internet Activity: go.hrw.com
KEYWORD: SJ3 GT2

Choose a topic to explore online:
- Learn more about Earth's seasons.
- Discover facts about Earth's water.
- Investigate earthquakes.

---

5. People can adapt to landforms, such as by growing food in river floodplains; change landforms to fit their needs, such as by building dams.
6. primary landform if created when two tectonic plates collide; ravines and contours might be created by erosion and weathering

Map Activity
A. Tropic of Capricorn
B. Arctic Circle
C. equator
D. Tropic of Cancer
E. Antarctic Circle
F. South Pole
G. North Pole

Mental Mapping Skills Activity
Use the diagram on p. 17 to check students' models.

Writing Activity
Answers will vary, but the information included should be consistent with text material. Use Rubric 31, Resumés, to evaluate student work.

Portfolio Activity
Cities in the Southern Hemisphere have winter while those in the Northern Hemisphere have summer.

internet connect
GO TO: go.hrw.com
KEYWORD: SJ3 Teacher
FOR: a guide to using the Internet in your classoom

CHAPTER 3

Wind, Climate, and Natural Environments

Chapter Resource Manager

Objectives	Pacing Guide	Reproducible Resources
SECTION 1		
Winds and Ocean Currents (pp. 37–41) 1. Describe how the Sun's energy changes the Earth. 2. Explain why wind and ocean currents are important.	**Regular** 1.5 days **Block Scheduling** .5 day *Block Scheduling Handbook, Chapter 3*	**RS** Guided Reading Strategy 3.1 **SM** Geography for Life Activity 3 **E** Lab Activities for Geography and Earth Science, Demonstrations 11 and 12
SECTION 2		
Earth's Climate and Vegetation (pp. 44–50) 1. Describe what is included in the study of weather. 2. Identify the major climate types, and the types of plants that live in each.	**Regular** 1.5 days **Block Scheduling** .5 day *Block Scheduling Handbook, Chapter 3*	**RS** Guided Reading Strategy 3.2 **RS** Graphic Organizer 3 **SM** Map Activity 3
SECTION 3		
Natural Environments (pp. 51–55) 1. Describe how environments affect life, and how they change. 2. Identify the substances that make up the different layers of soil.	**Regular** 1.5 days **Block Scheduling** .5 day *Block Scheduling Handbook, Chapter 3*	**RS** Guided Reading Strategy 3.3 **E** Environmental and Global Issues Activity 4 **E** Lab Activity for Geography and Earth Science, Hands-On 1

Chapter Resource Key

RS Reading Support
IC Interdisciplinary Connections
E Enrichment
SM Skills Mastery
A Assessment
REV Review

ELL Reinforcement and English Language Learners
 Transparencies
 CD–ROM
 Music

 Video
 go. hrw .com Internet
 Holt Presentation Maker Using Microsoft® Powerpoint®

One-Stop Planner CD–ROM

See the *One-Stop Planner* for a complete list of additional resources for students and teachers.

One-Stop Planner CD-ROM

It's easy to plan lessons, select resources, and print out materials for your students when you use the *One-Stop Planner CD–ROM with Test Generator.*

Technology Resources

One-Stop Planner CD–ROM, Lesson 3.1

Earth: Forces and Formations CD–ROM/Seek and Tell/Forces and Processes

Geography and Cultures Visual Resources with Teaching Activities 1–6 and 8

Homework Practice Online

HRW Go site

One-Stop Planner CD–ROM, Lesson 3.2

Earth: Forces and Formations CD–ROM/Seek and Tell/Forces and Processes

ARGWorld CD–ROM: Using Climagraphs to Interpret Seasons

Homework Practice Online

HRW Go site

One-Stop Planner CD–ROM, Lesson 3.3

Our Environment CD–ROM/ Seek and Tell/Natural Resources

Homework Practice Online

HRW Go site

Review, Reinforcement, and Assessment Resources

ELL	Main Idea Activity 3.1
REV	Section 1 Review, p. 41
A	Daily Quiz 3.1
ELL	English Audio Summary 3.1
ELL	Spanish Audio Summary 3.1

ELL	Main Idea Activity 3.2
REV	Section 2 Review, p. 50
A	Daily Quiz 3.2
ELL	English Audio Summary 3.2
ELL	Spanish Audio Summary 3.2

ELL	Main Idea Activity 3.3
REV	Section 3 Review, p. 55
A	Daily Quiz 3.3
ELL	English Audio Summary 3.3
ELL	Spanish Audio Summary 3.3

internet connect

HRW ONLINE RESOURCES

GO TO: go.hrw.com
Then type in a keyword.

TEACHER HOME PAGE
KEYWORD: SJ3 TEACHER

CHAPTER INTERNET ACTIVITIES
KEYWORD: SJ3 GT3

Choose an activity to:
• learn more about using weather maps.
• follow El Niño, an ocean phenomenon that affects weather.
• build a food web.

CHAPTER ENRICHMENT LINKS
KEYWORD: SJ3 CH3

CHAPTER MAPS
KEYWORD: SJ3 MAPS3

ONLINE ASSESSMENT
Homework Practice
KEYWORD: SJ3 HP3
Standardized Test Prep Online
KEYWORD: SJ3 STP3
Rubrics
KEYWORD: SS Rubrics

COUNTRY INFORMATION
KEYWORD: SJ3 Almanac

CONTENT UPDATES
KEYWORD: SS Content Updates

HOLT PRESENTATION MAKER
KEYWORD: SJ3 PPT3

ONLINE READING SUPPORT
KEYWORD: SS Strategies

CURRENT EVENTS
KEYWORD: S3 Current Events

Meeting Individual Needs

Ability Levels

Level 1 Basic-level activities designed for all students encountering new material

Level 2 Intermediate-level activities designed for average students

Level 3 Challenging activities designed for honors and gifted-and-talented students

English Language Learners Activities that address the needs of students with Limited English Proficiency

Chapter Review and Assessment

E	Readings in World Geography, History, and Culture 5 and 6	**A**	Chapter 3 Test
SM	Critical Thinking Activity 3		Chapter 3 Test Generator (on the One-Stop Planner)
REV	Chapter 3 Review, pp. 56–57		Audio CD Program, Chapter 3
REV	Chapter 3 Tutorial for Students, Parents, Mentors, and Peers	**A**	Chapter 3 Test for English Language Learners and Special-Needs Students
ELL	Vocabulary Activity 3		

LAUNCH INTO LEARNING

Give students two minutes to complete a quick sketch of a wild animal common in your region. Remind them that even crowded cities are home to wild animals. Have the students hold up their drawings. Then call on volunteers to suggest how the animal survives: what does it eat, how does it adapt to the heat or cold, where does it get water? Tell students that weather conditions and plants play a large role in determining where animals—and people—can live.

Section 1

Objectives

1. Analyze how the Sun's energy changes Earth.

2. Explain the importance of wind and ocean currents.

LINKS TO OUR LIVES

There are many reasons why studying climate and environmental issues is important. Here are some to share with your students:

▶ Ocean and wind currents can influence everything from vacation plans to our food supply.

▶ Every day, weather affects us all. We can predict it or accommodate it better if we understand it.

▶ We can prevent damage to the environment only if we understand the connections among climate, plants, animals, and humans.

▶ World events are often determined by environmental changes. We can predict events better if we recognize those cause-and-effect relationships.

CHAPTER 3

Wind, Climate, and Natural Environments

Diver, coral, and fish, Fiji Islands

Tornado in Saskatoon, Canada

Igloo at night, Alaska Range

36

LET'S GET STARTED

Copy the following instructions onto the chalkboard: *What would happen if, for one week, all the winds in the world were still? Write down your ideas.* Discuss student responses. *(Students might suggest that temperature differences would become extreme, that pollution would get worse, and that ocean waves would calm.)* Point out that although one cannot see it, smell it, or taste it, wind makes life as we know it possible on Earth. Tell students that in Section 1 they will learn more about winds and ocean currents.

Building Vocabulary

Write the key terms on the chalkboard and ask students if they are familiar with any of the words that make up the terms. Write students' responses beside the terms. *(Students might suggest that greenhouses are where flowers are grown, that **air pressure** measures the amount of air in tires, and that **currents** transmit electric power.)* Call on volunteers to find and read aloud the definitions in the text. Then ask volunteers to compare the text definitions to those already suggested. *(Example: Plants that would normally be killed by cold weather can survive inside a greenhouse because the glass traps heat. The **greenhouse effect** traps the Sun's energy.)*

Section 1 — Winds and Ocean Currents

Read to Discover

1. How does the Sun's energy change Earth?
2. Why are wind and ocean currents important?

Define

weather
climate
greenhouse effect
air pressure
front
currents

WHY IT MATTERS

Changes in ocean temperatures create currents that affect Earth's landmasses. Use **CNNfyi.com** or other **current events** sources to find examples of the effects of changes in ocean temperatures, such as the La Niña weather pattern. Record your findings in your journal.

Earth from space

The Sun's Energy

All planets in our solar system receive energy from the Sun. This energy has important effects. Among the most obvious effects we see on our planet are those on **weather** and **climate**. Weather is the condition of the atmosphere at a given place and time. Climate refers to the weather conditions in an area over a long period of time. How do you think the Sun's energy affects weather and climate?

Energy Balance Although Earth keeps receiving energy from the Sun, it also loses energy. Energy that is lost goes into space. As a result, Earth—as a whole—loses as much energy as it gets. Thus, Earth's overall temperature stays about the same.

As you learned in Chapter 2, the Sun does not warm Earth evenly. The part of Earth in daylight takes in more energy than it loses. Temperatures rise. However, the rest of Earth is in darkness. That part of Earth loses more energy than it gets. Temperatures drop. In addition, when the direct rays of the Sun strike Earth at the Tropic of Cancer, it is summer in the Northern Hemisphere. Temperatures are warm. The Southern Hemisphere, on the other hand, is having winter. Temperatures are lower. The seasons reverse when the Sun's direct rays move above the Tropic of Capricorn. We can see that in any one place temperatures vary from day to day. From year to year, however, they usually stay about the same.

Stored Energy Some of the heat energy that reaches Earth is stored. One place Earth stores heat is in the air. This keeps Earth's surface warmer than if there were no air around it. The process by which

Plants—like this fossil palm found in a coal bed—store the Sun's energy. When we burn coal—the product of long-dead plants—we release the energy the plants had stored.

internet connect

GO TO: go.hrw.com
KEYWORD: SJ3 CH3
FOR: Web sites about wind, climates, and environments

Section 1 RESOURCES

Reproducible
◆ Block Scheduling Handbook, Ch. 3
◆ Guided Reading Strategy 3.1
◆ Geography for Life Activity 3
◆ Lab Activities for Geography and Earth Science, Demonstrations 11 and 12

Technology
◆ One-Stop Planner CD–ROM, Lesson 3.1
◆ Homework Practice Online
◆ Earth: Forces and Formations CD–ROM/Seek and Tell/Forces and Processes
◆ Geography and Cultures Visual Resources with Teaching Activities 1–6 and 8
◆ HRW Go site

Reinforcement, Review, and Assessment
◆ Section 1 Review, p. 41
◆ Daily Quiz 3.1
◆ Main Idea Activity 3.1
◆ English Audio Summary 3.1
◆ Spanish Audio Summary 3.1

TEACH

Teaching Objectives 1–2

ALL LEVELS: (Suggested time: 10 min.) Copy the following graphic organizer onto the chalkboard, omitting the dashed arrows. Have each student complete the organizer by showing energy coming from the Sun *(large arrow)*, and being distributed by wind and ocean currents and by escaping into space *(small arrows)*. **ENGLISH LANGUAGE LEARNERS**

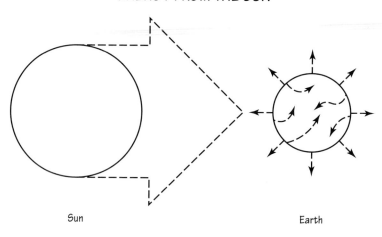

Sun   Earth

Linking Past to Present

The Little Ice Age Many scientists believe that the greenhouse effect is responsible for global temperature increases of recent decades. Temperature changes within recorded history have not always resulted in global warming, however.

During the Little Ice Age, which lasted roughly from the A.D. 1500s to the 1800s, the average global temperature dropped about 1°C.

Ice sheets advanced over farms and villages on Greenland. The Baltic Sea and Thames River, which now seldom freeze, then froze regularly. Crop failure, famine, and disease were common throughout Europe. Colonists in North America suffered through harsh winters also. Although personal accounts reveal that people endured hard times, they did not realize they were in the Little Ice Age. Climate changes are usually gradual and seen as normal within a person's lifetime.

Activity: Have students conduct research on the advance and retreat of glaciers since the Ice Age. Have them report their findings in a bar graph.

You Be the Geographer Answer ▶

Heat would build up and global average temperatures would rise.

The Greenhouse Effect

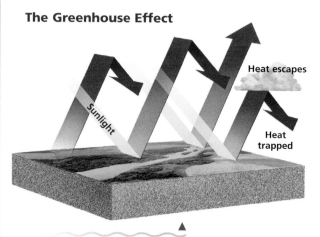

Light from the Sun passes through the atmosphere and heats Earth's surface. Most heat energy later escapes into space.

You Be the Geographer What would happen if too much heat energy remained trapped in the atmosphere?

This snow-covered waterfront town is located on Mackinac Island, Michigan.

Earth's atmosphere traps heat is called the **greenhouse effect**. In a greenhouse the Sun's energy passes through the glass and heats everything inside. The glass traps the heat, keeping the greenhouse warm.

Water and land store heat, too. As we learned in Chapter 2, water warms and cools slowly. This explains why in fall, long after temperatures have dropped, the ocean's water is only a little cooler than in summer. Land and buildings also store heat energy. For example, a brick building that has heated up all day stays warm after the Sun sets.

✔ **READING CHECK:** *Physical Systems* How does the Sun's energy affect Earth? It affects temperature and climate.

Wind and Currents

Air and water both store heat. When they move from place to place, they keep different parts of the world from becoming too hot or too cold. By moving air and water, winds and ocean currents move heat energy between warmer and cooler places. Different parts of the world are kept from becoming too hot or too cold.

When the wind is blowing, air is moving from one place to another. Everyone has experienced these local winds. Global winds also exist. They move air and heat energy around Earth. Ocean currents, which are caused by wind, also move heat energy.

Teaching Objective 2

ALL LEVELS: (Suggested time: 20 min.) Using the text atlas as a guide, have students draw a rough freehand map of the world and then add lines indicating the patterns of major ocean currents and wind currents described in the text. Students should label the continents and currents. **ENGLISH LANGUAGE LEARNERS**

LEVEL 2: (Suggested time: 40 min.) Have students write a children's book on wind and ocean currents. Topics discussed in the book should include air pressure and how the wind and ocean currents move energy around Earth. Students should also provide illustrations to further explain these subjects.

LEVEL 3: (Suggested time: 35 min.) Organize the students into groups of three. Assign each student air pressure, wind currents, or ocean currents as a topic. Have each student prepare a brief lesson on his or her topic to teach to the other group members. Have students teach the lesson. **COOPERATIVE LEARNING**

Air Pressure To understand why there are winds, we must understand **air pressure**. Air pressure is the weight of the air. Air is a mixture of gases. At sea level, a cubic foot of air weighs about 1.25 ounces (35 grams). We do not feel this weight because air pushes on us from all sides equally. The weight of air, however, changes with the weather. Cold air weighs more than warmer air. An instrument called a barometer measures air pressure.

When air warms, it gets lighter and rises. Colder air then moves in to replace the rising air. The result is wind. Wind travels from areas of high pressure to areas of low pressure. During the day land heats up faster than water. The air over the land heats up faster as well. Along the coast lower air pressure is located over land and higher air pressure is located over water. The air above land rises, and cool air flows in to shore to take its place. At night the land cools more quickly than the water. Air pressure over the land increases, and the wind changes direction.

Earth has several major areas where air pressure stays about the same throughout the year. Along the equator is an area of low air pressure. The pressure is low because the Sun is always warming this area.

▲

Wind shapes Earth and the life that thrives here. For example, this tree has grown in the direction blown by the area's prevailing winds.

Interpreting the Visual Record How do you think wind shapes Earth's landscape?

Reading a Weather Map

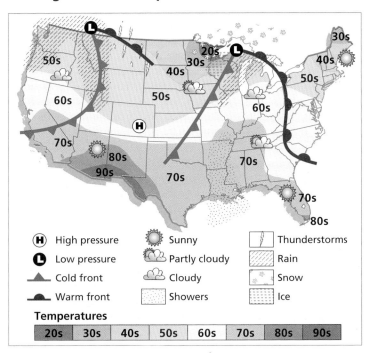

(H)	High pressure	☼	Sunny		Thunderstorms
(L)	Low pressure	☁	Partly cloudy		Rain
▲	Cold front	☁	Cloudy		Snow
●	Warm front		Showers		Ice

Temperatures

| 20s | 30s | 40s | 50s | 60s | 70s | 80s | 90s |

◄

Weather maps show atmospheric conditions as they currently exist or as they are forecast for a particular time period. Most weather maps have legends that explain what the symbols on the map mean. This map shows a cold front sweeping through the central United States. A low-pressure system is at the center of a storm bringing rain and snow to the Midwest. Notice that temperatures behind the cold front are considerably cooler than those ahead of the front.

You Be the Geographer What is the average temperature range for the southwestern states?

39

Tell students that during the 1700s sailors sometimes called the northern edge of the northeastern trade winds belt the "horse latitudes." Ask students to speculate how this region of wind current got its name. *(Ships caught in this calm region would often run low on supplies. To save water, the horses on board would be destroyed so the people could survive.)* Ask students to suggest nicknames for the other wind belts.

Have students complete the Section Review. Then pair students and assign one student the first Read to Discover question and assign the second student the other question. Have each student write three statements that address his or her assigned question. Have students within each pair exchange their statements. Then have students complete Daily Quiz 3.1.

COOPERATIVE LEARNING

EYE ON EARTH

The stormbeaches of southern England are known for their pebbles, which were washed up by rising sea levels and large waves some 10,000 years ago. The name of Chesil Beach comes from an old word for pebbles. Because of the prevailing winds and currents, the beach's pebbles vary in size. At the western end of this 16-mile (26 km) beach the pebbles are small, but they are larger farther east.

Fishermen coming ashore in a fog claim that they can determine their location along Chesil Beach by the size of the pebbles. Long ago, smugglers landing at night also checked the pebbles' size to determine their location.

Activity: Ask students to choose a coastal area of the United States to study. Then organize students into small groups and have each group conduct research on how winds and ocean currents have affected the dunes, sands, gravels, and other features of the area's beaches. Have each group present its findings to the class.

You Be the Geographer Answer ▶

An area of unstable weather forms.

Pressure and Wind Systems

▶

Winds between Earth's high- and low-pressure zones help regulate the globe's energy balance.

You Be the Geographer What happens when warm westerlies come into contact with cold polar winds?

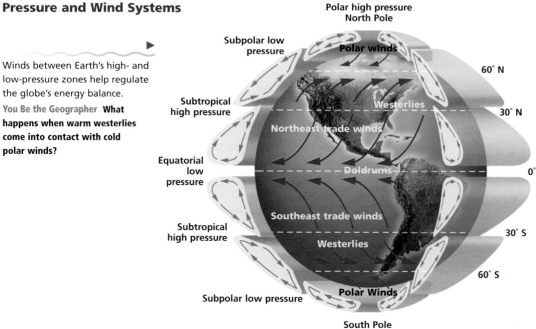

Polar high pressure
North Pole
Subpolar low pressure
Polar winds
60° N
Subtropical high pressure
Westerlies
30° N
Northeast trade winds
Equatorial low pressure
Doldrums
0°
Southeast trade winds
Subtropical high pressure
Westerlies
30° S
60° S
Subpolar low pressure
Polar Winds
South Pole
Polar high pressure

This ship uses wind to travel. As more wind catches the sails, the ship moves faster.

▼

This warm air rises along the equator and moves north and south. Some of the warm air cools and sinks once it reaches about 30° latitude on either side of the equator. This change in temperature causes areas of high air pressure in the subtropics. The pressure is also high at the North and South Poles because the air is so cold. This cold air flows away from the Poles. Because the cold air is heavier, it lifts the warmer air in its path. The subpolar regions have low air pressure.

When a large amount of warm air meets a large amount of cold air, an area of unstable weather forms. This unstable weather is called a **front**. When cold air from Arctic and Antarctic regions meets warmer air, a polar front forms. When this type of front moves through an area, it can cause storms.

Winds The major areas of air pressure create the "belts" of wind that move air around Earth. Winds that blow in the same direction over large areas of Earth are called prevailing winds.

The winds that blow in the subtropics are called the trade winds. In the Northern Hemisphere, the trade winds blow from the northeast. Before ships had engines, sailors used trade winds to sail from Europe to the Americas.

The westerlies are the winds that blow from the west in the middle latitudes. These are the most common winds in the United States. When you watch a weather forecast you can see how the westerlies push storms across the country from west to east.

Have students complete Main Idea Activity 3.1. Then organize the class into three groups and assign each group one of the Read to Discover questions. Have each group write, and then present to the class, three true-false statements related to its question. If the statement is false, call on a group member to explain why it is false. **ENGLISH LANGUAGE LEARNERS, COOPERATIVE LEARNING**

Have interested students conduct research on buildings designed to take advantage of stored energy. You might have them collect samples of materials from local building supply stores and organize a display. **BLOCK SCHEDULING**

Few winds blow near the equator. This area is called the doldrums. Here, warm air rises rather than blowing east or west.

Ocean Currents Winds make ocean water move in the same general directions as the air above it moves. Warm ocean water from the tropics moves in giant streams, or **currents**, to colder areas. Cold water moves in streams from the polar areas to the tropics. This moves energy between different places. Warm air and warm ocean currents raise temperatures. On the other hand, cold winds and cold ocean currents lower temperatures.

The Gulf Stream is an ocean current. It moves warm water north along the east coast of the United States. The Gulf Stream then moves across the Atlantic Ocean toward western Europe. The warm air that moves with it keeps winters mild. As a result, areas such as Ireland have warmer winters than areas in Canada that are just as far north.

Ocean currents also bring heat energy into the Arctic Ocean. In the winter, warm currents create openings in the ice. These openings give arctic whales a place to breathe. They also give people a place to catch fish. Cold water also flows out of the Arctic Ocean into warmer waters of the Pacific and Atlantic Oceans. The cold water sinks below warmer water, causing mixing. This mixing brings food to sea life.

▲

Plants and animals adapt to their particular environments on Earth.

Interpreting the Visual Record How do you think the bearded seal is able to live in Norway's cold environment?

✓ READING CHECK: *Physical Systems* How do winds and currents create patterns on Earth's surface? They move warm air and warm water from one place to another.

Section Review 1

Define and explain: weather, climate, greenhouse effect, air pressure, front, currents

Reading for the Main Idea

1. *Physical Systems* How does Earth's temperature stay balanced?

2. *Physical Systems* How does the greenhouse effect allow Earth to store the Sun's energy?

Critical Thinking

3. **Drawing Inferences and Conclusions** How would weather in western Europe be different if there were no Gulf Stream?

4. **Analyzing Information** Why were the trade winds important to early sailors?

Organizing What You Know

5. **Categorizing** Copy the following graphic organizer. Use it to show the names, locations, and directions of wind and air pressure belts.

Name	Location	Direction

go.hrw.com **Homework Practice Online**
Keyword: SJ3 HP3

Section Review 1

Answers

Define For definitions, see: weather, p. 37; climate, p. 37; greenhouse effect, p. 38; air pressure, p. 39; front, p. 40; currents, p. 41

Reading for the Main Idea

1. As Earth receives energy from the Sun, it also loses energy that escapes into space. (NGS 7)

2. Earth's atmosphere, water, and land all trap heat. (NGS 7)

Critical Thinking

3. Europe's winters would be colder.

4. Sailors used the trade winds to sail from Europe to the Americas.

Organizing What You Know

5. trade winds—subtropics, from the northeast; westerlies—middle latitudes, west to east; doldrums—equator, warm air rises but does not blow in any direction

◀ **Visual Record Answer**

The bearded seal has a layer of fat that insulates it from the cold.

41

Setting the Scene

Every year, hurricanes torment residents of the Caribbean islands, the coastlands bordering the Gulf of Mexico, and the East Coast of the United States. Because hurricanes get their strength from warm water, hurricane season lasts through summer and into fall. These storms carry tremendous energy. In one day an average hurricane releases at least 8,000 times the daily electrical power output of the United States. Severe hurricanes can cause billions of dollars of damage. Fewer lives are lost now than in years past, however, because early warning systems help predict the storms' paths and power. Coastal towns and cities evacuate people before the storms arrive. Satellites provide much of the information used to make storm predictions.

Building a Case

Have students read "Hurricane: Tracking a Natural Hazard" and follow the instructions in You Be the Geographer. Ask on what date the atmospheric pressure was lowest. *(10/27)* How did Mitch register on the Saffir-Simpson Scale that day? *(category 5)* Compare Hurricane Mitch with the hurricane that struck Galveston on September 8, 1900.

The storm headed for Galveston was first observed on August 30. The Weather Bureau placed Galveston under a storm warning on September 7. September 8 dawned rainy and gusty. Though the storm worsened, few residents left the city. At 6:30 P.M. a storm surge flooded the city. The lowest barometer reading was 27.91. Windspeed was estimated at more than 120 mph. By 10:00 P.M. much of the city was wrecked. As many as 8,000 city residents died.

HISTORICAL GEOGRAPHY

Hurricanes and typhoons—as these large storms are called when they occur in the Pacific Ocean—have changed history. Here are just three examples.

In 1281 the Mongol ruler Kublai Khan was ready to invade Japan, but a typhoon scattered his huge fleet of ships. A second storm, dubbed the Great Hurricane, ravaged the Caribbean in October 1780. It killed approximately 22,000 people and may be the deadliest hurricane on record. British and French fleets involved in the American Revolutionary War were both ravaged. Finally, in December 1944, during World War II, a sudden typhoon east of the Philippines caught the U.S. Third Fleet by surprise. Three destroyers, 146 aircraft, and several hundred men were lost.

Critical Thinking: How could early storm warning technology have changed world history?

Answer: Answers will vary but students might mention a successful invasion by Kublai Khan, fewer deaths in the Caribbean, and fewer ships and lives lost during the Revolutionary War and World War II.

➤ This Case Study feature addresses National Geography Standards 4, 15, and 17.

HURRICANE: TRACKING A NATURAL HAZARD

Hurricanes are large circulating storms that begin in tropical oceans. Hurricanes often move over land and into populated areas. When a hurricane approaches land, it brings strong winds, heavy rains, and large ocean waves.

The map below shows the path of Hurricane Fran in 1996. Notice how Fran moved to the west and became stronger until it reached land. It began as a tropical depression and became a powerful hurricane as it passed over warm ocean waters.

Scientists who study hurricanes try to predict where these storms will travel. They want to be able to warn people in the hurricane's path. Early warnings can help people be better prepared for the deadly winds and rain. It is a difficult job because hurricanes can change course suddenly. Hurricanes are one of the most dangerous natural hazards.

One way of determining a hurricane's strength is by measuring the atmospheric pressure inside it. The lower the pressure, the stronger the storm. Hurricanes are rated on a scale of one to five. Study Table 1 to see how wind speed and air pressure are used to help determine the strength of a hurricane.

Hurricane Mitch formed in October 1998. The National Weather Service (NWS) recorded Mitch's position and strength. They learned that Mitch's pressure was one of the lowest ever recorded. The NWS estimated that Mitch's maximum sustained surface winds reached 180 miles per hour.

Table 1: Saffir-Simpson Scale

HURRICANE TYPE	WIND SPEED MPH	AIR PRESSURE MB (INCHES)
Category 1	74–95	more than 980 (28.94)
Category 2	96–110	965–979 (28.50–28.91)
Category 3	111–130	945–964 (27.91–28.47)
Category 4	131–155	920–944 (27.17–27.88)
Category 5	more than 155	919 (27.16)

Source: Florida State University, <http://www.met.fsu.edu/explores/tropical.html>

Path of Hurricane Fran, 1996

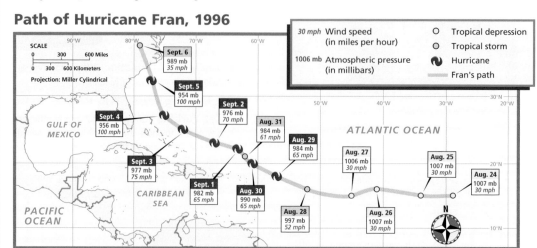

42

Drawing Conclusions

Lead a discussion comparing the two storms. According to wind speed and air pressure, what level storm was the Galveston hurricane? *(3)* Which was the stronger storm? *(Mitch)* Which hurricane lasted longer? *(Mitch)* Why did the 1900 storm kill so many people in such a short time? *(They had not evacuated the city.)* If there had been no warning system, how might Hurricane Mitch have affected the Caribbean region? *(It might have killed even more people.)*

What might have happened if Galveston had been warned earlier? Have students prepare and present an alternate newscast for the morning of September 9, 1900, based on this possibility.

Going Further: Thinking Critically

Locate detailed maps of the Gulf of Mexico or Atlantic coasts of the United States. Use maps of different areas or concentrate on one region. Have students work in groups to answer some or all of these questions:

- What cities and towns might be threatened by a hurricane? Can students estimate how many people live in the area?
- What routes could residents use to evacuate? What factors might slow evacuation? If they could travel about 30 mph (48 km/h), how far could people travel in one day? two days?
- What would happen if residents were warned just a few hours before a hurricane? What effect might an early warning system have on this region?

Table 2: Hurricane Mitch, 1998 Position and Strength

DATE	LATITUDE (DEGREES)	LONGITUDE (DEGREES)	WIND SPEED (MPH)	PRESSURE (MILLIBARS)	STORM TYPE
10/22	12 N	78 W	30	1002	Tropical depression
10/24	15 N	78 W	90	980	Category 2
10/26	16 N	81 W	130	923	Category 4
10/27	17 N	84 W	150	910	Category 5
10/31	15 N	88 W	40	1000	Tropical storm
11/01	15 N	90 W	30	1002	Tropical depression
11/03	20 N	91 W	40	997	Tropical storm
11/05	26 N	83 W	50	990	Tropical storm

Source: <http://www.met.fsu.edu/explores/tropical.html>

Hurricanes like Mitch cause very heavy rains in short periods of time. These heavy rains are particularly dangerous. The ground becomes saturated, and mud can flow almost like water. The flooding and mudslides caused by Mitch killed an estimated 10,000 people in four countries. Many people predicted that the region would not recover without help from other countries.

In the southeastern United States, many places have emergency preparedness units. The people assigned to these groups organize their communities. They provide food, shelter, and clothing for those who must evacuate their homes.

You Be the Geographer

1. Trace a map of the Caribbean. Be sure to include latitude and longitude lines.

2. Use the data about Hurricane Mitch in Table 2 to plot its path. Make a key with symbols to show Mitch's strength at each location.

3. What happened to Mitch when it reached land?

▲ This satellite image shows the intensity of Hurricane Mitch. With advanced technology, hurricane tracking is helping to save lives.

You Be the Geographer

1. Students may trace the map on the previous page. Or, provide an outline map to students.

2. On student maps, from its first position Hurricane Mitch should progress north-northwest toward Cuba, swing southwest toward Nicaragua, then back northeast across the Yucatán Peninsula on its way to the open Atlantic Ocean.

3. When it reached land, Mitch's strength weakened.

internet connect

GO TO: go.hrw.com
KEYWORD: SJ3 CH3
FOR: Web sites about hurricanes

43

Section 2

Objectives

1. Describe what is included in the study of weather.
2. Identify the major climate types, and describe the types of plants that live in each.

LET'S GET STARTED

Copy the following instructions onto the chalkboard: *Write down three words or phrases you use to describe the weather in our area. (Possible answers: muggy, bone-chilling, gully-washer, nor'easter, raining cats and dogs)* Point out that having so many terms for our weather indicates that weather is very important to us. Tell students that in Section 2 they will learn that climate and weather affect plants and practically all other forms of life.

Building Vocabulary

Write the key terms on the chalkboard. Call on volunteers to find and read aloud the definitions in the text or glossary. Point out that **monsoon** actually has two meanings. Originally, it referred to the wind systems that bring wet and dry seasons to tropical areas. Its more common usage refers to heavy rains brought in by the winds during summer. You may also want to introduce *orographic effect* as the term for what causes a **rain shadow**.

Section 2 RESOURCES

Reproducible
- Guided Reading Strategy 3.2
- Graphic Organizer 3
- Map Activity 3

Technology
- One-Stop Planner CD–ROM, Lesson 3.2
- Homework Practice Online
- Earth: Forces and Formations CD–ROM/Seek and Tell/Forces and Processes
- ARGWorld CD–ROM: Using Climagraphs to Interpret Seasons
- HRW Go site

Reinforcement, Review, and Assessment
- Section 2 Review, p. 50
- Daily Quiz 3.2
- Main Idea Activity 3.2
- English Audio Summary 3.2
- Spanish Audio Summary 3.2

Section 2   Earth's Climate and Vegetation

Read to Discover

1. What is included in the study of weather?
2. What are the major climate types, and what types of plants live in each?

Define

rain shadow
monsoon
arid
steppe climate

hurricanes
typhoons
tundra climate
permafrost

WHY IT MATTERS

Flooding often becomes a problem during severe weather. Use CNNfyi.com or other current events sources to find examples of the effects of flooding on nations around the world. Record your findings in your journal.

A barometer

Nature has many incredible sights. Lightning is one of the most spectacular occurrences. These flashes of light are produced by a discharge of atmospheric electricity.

▼

Weather

As you have read, the condition of the atmosphere in a local area for a short period of time is called weather. Weather is a very general term. It can describe temperature, amount of sunlight, air pressure, wind, humidity, clouds, and moisture.

When warm and cool air masses come together, they form a front. Cold air lifts the warm air mass along the front. The air that is moved to higher elevations is cooled. If moisture is present in the lifted air, this cooling causes clouds to form. The moisture may fall to Earth as rain, snow, sleet, or hail. Moisture that falls in any form is called precipitation.

Another type of lifting occurs when warm, moist air is blown up against a mountain and forced to rise. The air cools as it is lifted, just as when air masses collide. Clouds form, and precipitation falls. The side of the mountain facing the wind—the windward side—often gets heavy rain. By the time the air reaches the other side of the mountain—the leeward side—it has lost most of its moisture. This can create a dry area called a **rain shadow**.

✓ **READING CHECK:** *Physical Systems* What is included in the study of weather? temperature, amount of sunlight, air pressure, wind, humidity, clouds, moisture

Teaching Objective 1

LEVEL 1: (Suggested time: 30 min.) Have students read the section. Then pair students and have each pair create an informational brochure titled "The Study of Weather." In their brochures, students should identify and create a symbol for each of the elements of weather that can be studied. *(Brochures should identify temperature, amount of sunlight, air pressure, wind, humidity, clouds, and moisture.)* Ask volunteers to present and explain their brochures to the class.
ENGLISH LANGUAGE LEARNERS, COOPERATIVE LEARNING

LEVEL 2: (Suggested time: 30 min.) Prepare photocopies of the local weather map from your daily newspaper or the U.S. weather map in a national paper. Review the "Reading a Weather Map" diagram in Section 1. Have students work in pairs to circle and label the information on their map that illustrates the conditions that affect weather. *(See the Level 1 lesson for the correct elements of weather.)*
COOPERATIVE LEARNING

Landforms and Precipitation

Windward (wet)　　　Leeward (dry)

Snow

Rain　　Warming dry air

Cooling moist air

Rain Shadow

Ocean　　Inland

As moist air from the ocean moves up the windward side of a mountain, it cools. The water vapor in the air condenses and falls in the form of rain or snow. Descending, the drier air then moves down the leeward side of the mountain. This drier air brings very little precipitation to areas in the rain shadow.

Low-Latitude Climates

If you charted the average weather in your community over a long period of time, you would be describing the climate for your area. Geographers have devised ways of describing climates based mainly on temperature, precipitation, and natural vegetation. In general, an area's climate is related to its latitude. In the low latitudes—the region close to the equator—there are two main types of climates: humid tropical and tropical savanna.

Humid Tropical Climate A humid tropical climate is warm and rainy all year. People living in this climate do not see a change from summer to winter. This is because the Sun heats the region throughout the year. The heat in the tropics causes a great deal of evaporation and almost daily rainstorms. One of the most complex vegetation systems in the world—the tropical rain forest—exists in this climate. The great rain forests of Brazil, Indonesia, and Central Africa are in the equatorial zone.

Some regions at higher latitudes have warm temperatures all year but have strong wet and dry seasons. Bangladesh and coastal India, for example, have an extreme wet season during the summer. Warm, moist air from the Indian Ocean reaches land. The air rises and cools, causing heavy rains. The rains continue until the wind changes direction in the fall. This seasonal shift of air flow and rainfall is known as a **monsoon**. A monsoon may be wet or dry. The monsoon system is particularly important in Asia.

The people of Tamil Nadu, India, adjust to the monsoon season.

Interpreting the Visual Record
What problems might people face during the wet monsoon?

The monsoon is a seasonal wind that shifts direction twice a year. Summer monsoon winds bring heavy rains from the oceans to Asia. Beginning in October, the winter monsoon brings cool, dry air to India and China and rain to Indonesia and Australia.

The monsoon is vital to Asian agriculture and to the survival of billions of people, particularly those who rely on the success of a single harvest. The weather system formed by monsoon winds transfers heat to and from Asia and helps balance temperatures. Without the monsoons, regions that receive the Sun's direct rays would be scorched and other areas would be severely cold.

Critical Thinking: How might Asian farmers reduce their dependence on monsoons?

Answer: Farmers could dig deep wells or develop irrigation systems.

internet connect

GO TO: go.hrw.com
KEYWORD: SJ3 CH3
FOR: Web sites about sprites

◀ **Visual Record Answer**

flooded streets and buildings, contaminated water supplies, drowned livestock, and others

Teaching Objective 2

ALL LEVELS: (Suggested time: 10 min.) To help students understand the relationship between climate and latitude, copy the following graphic organizer onto the chalkboard, omitting the italicized answers. Pair students and have each pair complete the organizer by filling in the correct latitudinal groupings. Ask students what climate types are not represented on the organizer. *(Students should mention desert, steppe, and highland climates.)* Call on volunteers to describe those climates and their locations. Then have students describe the types of plants that live in each climate. **ENGLISH LANGUAGE LEARNERS, COOPERATIVE LEARNING**

high	*subarctic, tundra, and ice cap*
middle	*Mediterranean, humid subtropical, marine west coast, and humid continental*
low	*humid tropical and tropical savanna*
middle	*Mediterranean, humid subtropical, marine west coast, and humid continental*
high	*subarctic, tundra, and ice cap*

In terms of loss of life, the worst natural disaster in U.S. history is still the hurricane that hit Galveston, Texas, on September 8, 1900. As many as 8,000 people died in Galveston.

Palo Duro Canyon State Park in Texas attracts thousands of visitors each year.

Interpreting the Visual Record How can you tell that this canyon is located in a dry climate?

Visual Record Answer

lack of vegetation

Tropical Savanna Climate There is another type of tropical climate that has wet and dry seasons. However, this climate does not have the extreme shifts found in a monsoon climate. It is called a tropical savanna climate. The tropical savanna climate has a wet season soon after the warmest months. It has a dry season soon after the coolest months. Total rainfall, however, is fairly low. Vegetation in a tropical savanna climate is grass with scattered trees and shrubs.

✓ **READING CHECK:** *Physical Systems* Which climate regions are in the low latitudes? humid tropical and tropical savanna

Dry Climates

Temperature and precipitation are the most important parts of climate. Some regions, for example, experience strong hot and cold seasons. Other regions have wet and dry seasons. Some places are wet or dry all year long. If an area is **arid** (dry) it receives little rain. Arid regions usually have few streams and plants.

Desert Climate Most of the world's deserts lie near the tropics. The high air pressure and settling air keep these desert climate regions dry most of the time. Other deserts are located in the interiors of continents and in the rain shadows of mountains. Few plants can survive in the driest deserts, so there are many barren, rocky, or sandy areas. Dry air and clear skies permit hot daytime temperatures and rapid cooling at night.

Steppe Climate Another dry climate—the **steppe climate**—is found between desert and wet climate regions. A steppe receives more rainfall than a desert climate. However, the total amount of precipitation is still low. Grasses are the most common plants, but trees can grow along creeks and rivers. Farmers can grow crops but usually need to irrigate. Steppe climates occur in Africa, Australia, Central Asia, eastern Europe, in the Great Plains of the United States and Canada, and in South America.

✓ **READING CHECK:** *The World in Spatial Terms* What are the dry climates? desert and steppe

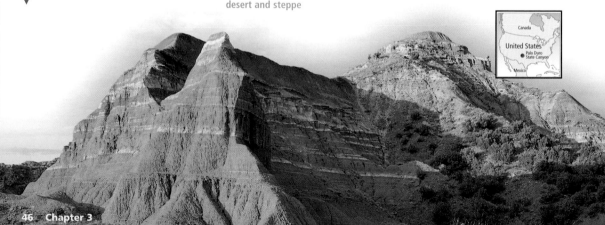

46

ALL LEVELS: (Suggested time: 10 min.) Tell students to use the world climate regions map on this page to locate the region where they live. Have them refer to the chart below the map to identify the state's climate region. Then tell them to read the description of the weather patterns for their climate region. Ask whether the description corresponds to their observations of weather patterns in the area. **ENGLISH LANGUAGE LEARNERS**

LEVEL 3: (Suggested time: 30 min.) Have students write trivia questions about each of the climate types. Tell students to refer to the world climate regions map on this page and the text atlas to help them write questions. *(Possible questions: What is the coldest temperature ever recorded at the South Pole? How hard can winds blow in the tropics? What is the hottest place in the middle latitudes?)* Then have students conduct research to find the answers to the questions.

World Climate Regions

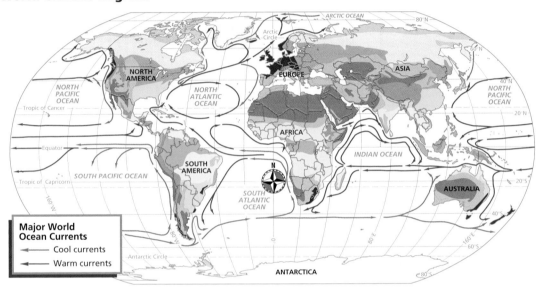

Major World Ocean Currents
- ◄─── Cool currents
- ◄─── Warm currents

	Climate		Geographic Distribution	Major Weather Patterns	Vegetation
Low Latitudes	HUMID TROPICAL		along the equator	warm and rainy year-round, with rain totaling anywhere from 65 to more than 450 in. (165–1,143 cm) a year	tropical rain forest
	TROPICAL SAVANNA		between the humid tropics and the deserts	warm all year; distinct rainy and dry seasons; at least 20 in. (51 cm) of rain during the summer	tropical grassland with scattered trees
Dry	DESERT		centered along 30° latitude; some middle-latitude deserts are in the interior of large continents and along their western coasts	arid; less than 10 in. (25 cm) of rain a year; sunny and hot in the tropics and sunny with wide temperature ranges during the day in middle latitudes	a few drought-resistant plants
	STEPPE		generally bordering deserts and interiors of large continents	semiarid; about 10–20 in. (25–51 cm) of precipitation a year; hot summers and cooler winters with wide temperature ranges during a day	grassland; few trees
Middle Latitudes	MEDITERRANEAN		west coasts in middle latitudes	dry, sunny, warm summers and mild, wetter winters; rain averages 15–20 in. (38–51 cm) a year	scrub woodland and grassland
	HUMID SUBTROPICAL		east coasts in the middle latitudes	hot, humid summers and mild, humid winters; rain year-round; coastal areas are in the paths of hurricanes and typhoons	mixed forest
	MARINE WEST COAST		west coasts in the upper-middle latitudes	cloudy, mild summers and cool, rainy winters; strong ocean influence; rain averages 20–60 in. (51–152 cm) a year	temperate evergreen forest
	HUMID CONTINENTAL		east coasts and interiors of upper-middle latitude continents	four distinct seasons; long, cold winters and short, warm summers; amounts of precipitation a year vary	mixed forest
High Latitudes	SUBARCTIC		higher latitudes of the interior and east coasts of continents	extremes of temperature; long, cold winters and short, warm summers; little precipitation all year	northern evergreen forest
	TUNDRA		high-latitude coasts	cold all year; very long, cold winters and very short, cool summers; little precipitation	moss, lichens, low shrubs; permafrost marshes
	ICE CAP		polar regions	freezing cold; snow and ice year-round; little precipitation	no vegetation
	HIGHLAND		high mountain regions	temperatures and amounts of precipitation vary greatly as elevation changes	forest to tundra vegetation, depending on elevation

Cultural Kaleidoscope

Religion and Weather Cultural and religious explanations of climate and weather have influenced people's understanding of the physical world. The Hopi's rain dance reveals that culture's attitude toward nature. By dancing and making offerings, the Hopi show respect to the kachinas, or spirits of ancestors, that they believe control the natural world. The Hopi believe the kachinas will return the favor by sending rain. Many other cultures have ceremonies designed to affect the weather.

Drought conditions in Texas during 1999 prompted community and church leaders there to pray for rain. The mayor of one city proclaimed a day of prayer for rain. One church pastor noted that although farmers can use the most modern equipment and seed, the success of a harvest is often beyond human control.

Activity: Have students conduct research on ways that weather has been incorporated into other cultures' religious beliefs or folklore and analyze the similarities and differences. Have students present their findings to the class.

►**ASSIGNMENT:** Prepare a list of 10 to 20 major world cities. Include cities in all climate zones (except for the ice cap zone). Give students copies of the list and have them locate the cities in the text atlas or on a classroom globe. Then have them write down each city's climate type and the temperatures, moisture, vegetation, sunlight levels, and storms that would be common there.

Maureen Dempsey of Spring Creek, Nevada, suggests the following activity to help students summarize climate types. Organize students into small groups. Have each group fold and staple 12 sheets of white paper together to create booklets titled *Climates of the World.* Students should describe each climate type on one page and illustrate it on the opposite page. They may draw pictures, cut pictures from magazines, or download images from the Internet. Each book should include a table of contents and a cover illustration.

National Geography Standard 14

Environment and Society.
Native to the Mediterranean region, the olive tree is important to the culture, history, and economy of that region. According to myth, the goddess Athena gave the ancient Greeks the olive tree to use for food, fuel, and healing wounds.

Wild olive trees are believed to have originated in Asia Minor and spread westward. The Romans then introduced the trees into regions they conquered. Later, Arabs began cultivating the olive. By the 1700s olive groves had been planted by the Spanish in places such as California, Mexico, and Chile.

Activity: Have students conduct research on other food plants that have been introduced into various geographical regions that share the same climate. Then have students create databases that show the means and results of the plants' introduction into different regions.

The Mediterranean coast of France is part of the Riviera. The Riviera is located in a Mediterranean climate region.

Interpreting the Visual Record
What features would make this area popular with tourists?

The Mediterranean has a variety of vegetation. This Mediterranean scrub forest is in Corfu, Greece.
Interpreting the Visual Record What kinds of vegetation do you see in this scrub forest?

Visual Record Answers ▲

Students might suggest hills, beaches, warm water, and the region's mild climate.

►

scrub bushes and trees

Middle-Latitude Climates

The middle latitudes are the two broad zones between Earth's polar circles (66.5° north and south latitudes) and the tropics (23.5° north and south latitudes). Most of the climates in the middle latitudes have cool or cold winters and warm or hot summers. Climates with wet and dry seasons are also found. Middle-latitude climates may include rain shadow deserts.

Mediterranean Climate Several climates have clear wet and dry seasons. One of these, the Mediterranean climate, takes its name from the Mediterranean region. This climate has hot, dry summers followed by cooler, wet winters. Much of southern Europe and coastal North Africa have a Mediterranean climate. Parts of California, Australia, South Africa, and Chile do as well. Vegetation includes scrub woodlands and grasslands.

Humid Subtropical Climate The southeastern United States is an example of the humid subtropical climate. Warm, moist air from the ocean makes this region hot and humid in the summer. Winters are mild, but snow falls occasionally. People in a humid subtropical climate experience **hurricanes** and **typhoons**. These are tropical storms that bring violent winds, heavy rain, and high seas. The humid subtropical climate supports areas of mixed forests where deciduous and coniferous forests blend. Deciduous trees lose their leaves during the fall each year. Coniferous trees have needle-shaped leaves that remain green year-round.

Marine West Coast Climate Some coastal areas of North America and much of western Europe have a marine west coast climate. Westerly winds carry moisture from the ocean across the land, causing winter rainfall. Evergreen forests can grow in these regions because of regular rain.

Point out to students that people live in almost every climate on Earth, including some of the most extreme. Have students describe those extreme climates and make suggestions on ways that people have adapted to them.

Have students complete the Section Review. Then have each student choose a climate type and write eight adjectives describing that climate. Have students read their descriptions to the class, and have other students guess which climate type is being described. Then have students complete Daily Quiz 3.2.

Humid Continental Climate Farther inland are regions with a humid continental climate. Winters in this region bring snowfall and cold temperatures, but there are some mild periods too. Summers are warm and sometimes hot. Most of the shifting weather in this climate region is the result of cold and warm air coming together along a polar front. Humid continental climates have four distinct seasons. Much of the midwestern and northeastern United States and southeastern Canada have a humid continental climate. This climate supports mixed forest vegetation.

✔ READING CHECK: *The World in Spatial Terms* What are the middle-latitude climates? Mediterranean, humid subtropical, marine west coast, humid continental

High-Latitude Climates

Closer to the poles we find another set of climates. They are the high-latitude climates. They have cold temperatures and little precipitation.

Subarctic Climate The subarctic climate has long, cold winters, short summers, and little rain. In the inland areas of North America, Europe, and Asia, far from the moderating influence of oceans, subarctic climates experience extreme temperatures. However, summers in these regions can be warm. In the Southern Hemisphere there is no land in the subarctic climate zone. As a result, boreal (BOHR-ee-uhl) forests are found only in the Northern Hemisphere. Trees in boreal forests are coniferous and cover vast areas in North America, Europe, and northern Asia.

Tundra Climate Farther north lies the **tundra climate**. Temperatures are cold, and rainfall is low. Usually just hardy plants, including mosses, lichens, and shrubs, survive here. Tundra summers are so short and cool that a layer of soil stays frozen all year. This frozen layer is called **permafrost**. It prevents water from draining into the soil. As a result, many ponds and marshes appear in summer.

Ice Cap Climate The polar regions of Earth have an ice cap climate. This climate

Wildlife eat the summer vegetation in the Alaskan tundra.

Interpreting the Visual Record Why is there snow on the mountain peaks during summer?

PHYSICAL SYSTEMS

Permafrost lies under some 20 to 25 percent of the world's land surface. It occurs in more than 50 percent of Russia and Canada and more than 80 percent of Alaska.

Permafrost gives scientists a window into the plant and animal life of the past. The various layers of permafrost contain plant and animal remains from different periods of Earth's history. Some of these layers are more than 30,000 years old.

Scientists also use permafrost to assess the rate of global warming. By studying ground temperatures and preserved plant and animal life, scientists can understand past climatic conditions and current temperature changes.

Critical Thinking: What might the presence of oak tree remains in a layer of permafrost indicate about past temperatures in a tundra region?

Answer: Students might suggest that the tundra region's temperatures were warmer in a previous era.

◄ **Visual Record Answer**

High elevations keep temperatures low.

Have students complete Main Idea Activity 3.2. Then have them choose a certain locale and create postcards they would send from there. On the picture side they should illustrate the climate for that area and on the other write a description of the weather on a typical day of their "vacation." Call on volunteers to read their postcards, until all the major climate types have been reviewed. Display the postcards around the classroom.

ENGLISH LANGUAGE LEARNERS

Have interested students perform a simple experiment to illustrate dew point—the temperature at which water vapor begins to condense. Put water into a coffee can and gradually add ice cubes while swirling a thermometer carefully in the icy water. Record the temperature at the moment when condensation forms on the outside of the coffee can. This temperature is the dew point. Have students conduct research on the relationship between relative humidity and dew point to explain the results of their experiment. Students may also want to investigate places in the world where very little rain falls and where dew sustains plants and animals. Have them briefly report their findings to the class.

BLOCK SCHEDULING

Section Review 2

Answers

Define For definitions, see: rain shadow, p. 44; monsoon, p. 45; arid, p. 46; steppe climate, p. 46; hurricanes, p. 48; typhoons, p. 48; tundra climate, p. 49; permafrost, p. 49

Reading for the Main Idea

1. Weather is the condition of the atmosphere at a given place and time. Climate refers to regional weather conditions over a long period of time. (NGS 7)

2. humid tropical, tropical savanna, desert, steppe, Mediterranean, humid subtropical, marine west coast, humid continental, subarctic, tundra, ice cap, and highland (NGS 3)

Critical Thinking

3. to be prepared for regular seasonal changes as well as potentially dangerous weather conditions

4. the closer a latitude is to the equator, the warmer the climate, generally

Organizing What You Know

5. For climate types, latitudes, and characteristics, see the table on p. 47.

Visual Record Answer ▶

The building is located underground and appears to be well insulated.

▲

The Amundsen-Scott South Pole Station is a research center in Antarctica.

Interpreting the Visual Record How does the design of this building show ways people have adapted to the ice cap climate?

is cold—the monthly average temperature is below freezing. Precipitation averages less than 10 inches (25 cm) annually. Animals adapted to the cold, like walruses, penguins, and whales, are found here. No vegetation grows in this climate.

✓ **READING CHECK:** *The World in Spatial Terms* How are tundra and ice cap climates different? Ice cap—snow and ice year-round, no vegetation; tundra—short, cool summer, hardy plants

Highland Climates

Mountains usually have several different climates in a small area. These climate types are known as highland climates. If you went from the base of a high mountain to the top, you might experience changes similar to going from the tropics to the Poles! The vegetation also changes with the elevation. It varies from thick forests or desert to tundra. Lower mountain elevations tend to be similar in temperature to the surrounding area. On the windward side, however, are zones of heavier rainfall or snowfall. As you go uphill, the temperatures drop. High mountains have a tundra zone and an icy summit.

✓ **READING CHECK:** *The World in Spatial Terms* What are highland climate regions? the climates around mountains; they differ greatly in a relatively small area

Homework Practice Online

Keyword: SJ3 HP3

Section Review 2

Define and explain: rain shadow, monsoon, arid, steppe climate, hurricanes, typhoons, tundra climate, permafrost

Reading for the Main Idea

1. *Physical Systems* How is weather different from climate?

2. *The World in Spatial Terms* What are the major climate regions of the world?

Critical Thinking

3. **Drawing Inferences and Conclusions** Why is it important to understand weather and climate patterns?

4. **Analyzing Information** How does latitude influence climate?

Organizing What You Know

5. **Categorizing** Copy the following graphic organizer. Use it to describe Earth's climate types.

Climate	Latitudes	Characteristics

Objectives

1. Explain how environments affect life, and describe how they change.
2. Identify the substances that make up the different layers of soil.

LET'S GET STARTED

Copy the following question onto the chalkboard: *What are three words or phrases that come to mind when you think of the plant life in this area?* If you are in a large city, you may want to specify a nearby park or wilderness area. *(Examples: forest, woods, prairie, cactus, scrub oak, wildflowers)* Write student responses on the chalkboard. As you teach the lesson, refer to the students' descriptions. Tell students that in Section 3 they will learn more about natural environments.

Building Vocabulary

Write **photosynthesis** on the chalkboard. Point out that *photo-* means "light." Have students find the definition in the text and relate the prefix's meaning to the term. What other words with this prefix do students know? *(Examples: photograph, photocopy)* Then have students create definitions for the Define terms that are compound words—**food chain, plant communities, plant succession**—based on what they know about the separate words. Check all definitions against the text. Have volunteers find and read aloud the remaining Define terms' definitions in the section.

Section 3 Natural Environments

Read to Discover

1. How do environments affect life, and how do they change?
2. What substances make up the different layers of soil?

Define

extinct
ecology
photosynthesis
food chain
nutrients

plant communities
ecosystem
plant succession
humus

WHY IT MATTERS

Plants and animals depend on their environment for survival. Use **CNN fyi.com** or other **current events** sources to find out how conservationists work to save animals whose environments are threatened. Record your findings in your journal.

Acorns, the nut of an oak tree

Reproducible
- Guided Reading Strategy 3.3
- Environmental and Global Issues Activity 4
- Lab Activity for Geography and Earth Science, Hands-On 1

Technology
- One-Stop Planner CD–ROM, Lesson 3.3
- Homework Practice Online
- Our Environment CD–ROM/Seek and Tell/Natural Resources
- HRW Go site

Reinforcement, Review, and Assessment
- Section 3 Review, p. 55
- Daily Quiz 3.3
- Main Idea Activity 3.3
- English Audio Summary 3.3
- Spanish Audio Summary 3.3

Environmental Change

Geographers examine the distribution of plants and animals and the environments they occupy. They also study how people change natural environments. Changes in an environment affect the plants and animals that live in that environment. If environmental changes are extreme, some types of plants and animals may become **extinct**. This means they die out completely.

Ecology and Plant Life The study of the connections among different forms of life is called **ecology**. One process connecting life forms is **photosynthesis**—the process by which plants convert sunlight into chemical energy. Roots take in minerals, water, and gases from the soil. Leaves take in sunlight and carbon dioxide from the air. Plant cells take in these elements and combine them to produce special chemical compounds. Plants use some of these chemicals to live and grow.

Plant growth is the basis for all the food that animals eat. Some animals, like deer, eat only plants. When deer eat plants, they store some of the plant food energy in their bodies. Other animals, like wolves, eat deer and indirectly get the plant food the deer ate. The plants, deer, and wolves together make up a **food chain**. A food chain is a series of organisms in which energy is passed along.

▲

Plants use a particular environment's sunlight, water, gases, and minerals to survive.

Teaching Objective 1

LEVELS 1 AND 2: (Suggested time: 30 min.) Pair students and have each pair write a paragraph to explain how environments affect life and how environments can change. *(Paragraphs should explain that different environments sustain different plants and animals and that changes to an ecosystem can result in the extinction of plants and animals. Paragraphs should also explain that environmental changes can be a result of natural disasters or human actions.)* Have volunteers read their paragraphs to the class. **COOPERATIVE LEARNING**

LEVEL 3: (Suggested time: 30 min.) Call on students to name a popular nearby natural area that includes a plant community, such as a park, forest, or riverbank. Ask students to identify potential natural and human events that could damage the ecosystem. *(Possible answers: fire, flood, drought, hurricane, clearing for new development)* Then have them gather information about the area and write a letter to the editor of the local newspaper listing the consequences of these damaging events. *(Possible answers: extinction of plants and animals)* Have students propose strategies in their letters for protecting the ecosystem. Then ask them to list the advantages and disadvantages of these options and choose one strategy to implement. Have them design a proposal for the city council that shows how to implement the solution and explains why it would be effective.

ENVIRONMENT AND SOCIETY

England's peppered moths are either light or dark. Scientist H. B. D. Kettlewell noticed that before 1848 there were few of the dark variety. Yet 50 years later, most were dark.

What had happened? Soot from the area's new factories had darkened the white birch trees, making it easier for moth-eating birds to spot the light moths that landed on them. As a result, fewer of the light moths survived to reproduce and pass on the gene for light coloring to later generations. The change in color was due to the process known as natural selection.

Discussion: What changes occurring in your area might give advantages or disadvantages to certain animals? *(Possible answers: Habitat destruction may favor animals that can live near humans. Drought may threaten species that require a steady supply of water.)*

Food chains rarely occur alone in nature. More common are food webs—interlocking networks of food chains. This food web includes tiny organisms, such as parasites and bacteria, as well as human beings and other large mammals.

Food Web

Not all predators are big. In Antarctica's dry areas a microscopic bacteria-eating worm that can survive years of being freeze-dried is at the top of the food chain.

Limits for Life Plants and animals cannot live everywhere. They are limited by environmental conditions. Any type of plant or animal tends to be most common in areas where it is best able to live, grow, and reproduce. A simple example shows why you do not find every kind of plant and animal in all regions on Earth. Trees do not survive in the tundra because it is too cold and there is not enough moisture. At the edge of the tundra, however, there are large boreal forests. Once in a while, wind carries tree seeds into the tundra. Some of the seeds sprout and grow. These young trees are generally small and weak. Eventually they die.

In this example life is limited by conditions of temperature and moisture. Other factors that can limit plant life include the amount of light, water, and soil **nutrients**. Nutrients are substances promoting growth. Some plant and animal life is limited by other plants and animals that compete for the same resources.

Teaching Objective 2

ALL LEVELS: (Suggested time: 10 min.) To help students understand the composition of the three soil layers, copy the following graphic organizer onto the chalkboard, omitting the italicized answers. Call on volunteers to supply the names of the three layers of soil and descriptions for each layer. Use students' answers to fill in the graphic organizer on the chalkboard. **ENGLISH LANGUAGE LEARNERS**

topsoil	*contains humus, insects, and plants*
subsoil	*contains deep roots*
broken rock	*contains rocks that eventually break down into more soil*

CONNECTING TO Science

Soil Factory

There are more than 2,000 species of harmless mushrooms.

The next time you see a fallen tree in the forest, do not think of it as a dead log. Think of it as a soil factory. As the tree decays and crumbles, it adds valuable nutrients to the forest soil. These nutrients enrich the soil. They also make it possible for new trees and plants to grow.

The fallen tree does not do its work alone, however. It is aided by many different living organisms that break down the wood and turn it into humus. Humus is a rich blend of organic material that mixes with the soil. A downed tree that lies on the forest floor is buzzing with the activity of hundreds of species of insects and plants that live and work inside it.

When a tree falls, insects, bacteria and other microorganisms invade the wood and start the process of decay. Insects like weevils, bark beetles, carpenter ants, and termites, among others, bore into the wood and break it down further. These insects, in turn, attract birds, spiders, lizards, and other predators who feed on insect life. Before long, the fallen tree is brimming with life. This happens even as the tree breaks apart on the forest floor.

Fallen trees provide as much as one third of the organic matter in forest soil. As forest ecologist Chris Maser notes, "Dead wood is no wasted resource. It is nature's reinvestment in biological capital."

You Be the Geographer

1. How do fallen trees decompose?
2. How is fertile soil produced by downed trees?

National Geography Standard 7

Physical Systems In the 1980s scientists noticed that spotted owls in the Pacific Northwest were declining. The old-growth forests where the owls lived were being cut down.

The owls became the subject of conflict between environmentalists who wanted to protect the owl's habitat and the timber industry and others who worried about the economic effects of ending lumbering. In the late 1980s the federal government put the owl on the Threatened Species List and set out to create a plan to conserve old-growth forests. Scientists are still working on a plan to track the owls to learn whether conservation plans have helped.

Activity: Have students create dioramas that show old-growth and second-growth forests. Students may need to do additional research.

Connecting to Science Answers

1. aided by different organisms that break down the wood
2. They add organic matter.

internet connect

GO TO: go.hrw.com
KEYWORD: SJ3 CH3
FOR: Web sites about forests

Plant Communities Groups of plants that live in the same area are called **plant communities**. In harsh environments, such as tundra, plant communities tend to be simple. They may be made up of just a few different types of plants. Regions that receive more rainfall and have more moderate temperatures tend to have more complex plant communities. The greatest variety of plants and animals can be found in the rain forests of the tropics. This environment is warm and moist all year.

Each plant community has plants adapted to the environment of the region. Near the Arctic, for example, the shortage of sunlight limits plant growth. Some of the flowers that grow here turn to follow the Sun as it moves across the sky. Thus the flowers collect as much sunlight as they can. In other regions there are plants adapted to survive in poor soils or with limited moisture. Some large trees, for example,

53

Have students look through the text's later chapters for photographs showing various plant communities or natural environments. Ask volunteers to choose a photograph and compare what they have learned in Section 3 with what they see in the photograph. Encourage the class to brainstorm how the plants and animals pictured fit into a food chain or ecosystem.

Have students complete the Section Review. Then have them create flash cards with important terms and phrases on one side and full descriptions on the back. Students should work in pairs to review the flash cards. Then have students complete Daily Quiz 3.3. **COOPERATIVE LEARNING**

Section Review 3

Answers

Define For definitions, see: extinct, p. 51; ecology, p. 51; photosynthesis, p. 51; food chain, p. 51; nutrients, p. 52; plant communities, p. 53; ecosystem, p. 54; plant succession, p. 54; humus, p. 55

Reading for the Main Idea

1. Plants and animals cannot adapt to every climate. A rain forest plant would not get enough moisture in a desert. (NGS 8)

2. layers of topsoil, subsoil, and broken rock; insects and bacteria live in the soil and produce humus (NGS 7)

Critical Thinking

3. the cold climate and predominance of ice

4. When natural or human forces disturb a plant community, the community may be replaced by a different group of plants suited to the new conditions.

Organizing What You Know

5. plants convert sunlight into energy through photosynthesis; an insect eats a plant; a fish eats the insect; a human eats the fish

You Be the Geographer Answer ▶

grasses and wildflowers

54

Forest Succession After a Fire

Ⓐ Forest fire in progress

Ⓑ Early plant growth

Ⓒ Middle stage

Ⓓ Forest recovered

Difficult conditions after a forest fire mean that the first plants to grow back must be very hardy.

You Be the Geographer In this series of photos, which plants are the first to grow back?

have deep roots that can reach water and soil nutrients far below the surface. Their spreading branches also collect large amounts of sunlight for photosynthesis.

Plants adapt to the sunlight, soil, and temperature of their region. They may also be suited to other plants found in their communities. For example, many ferns grow well in the shade found underneath trees. Some vines grow up tree trunks to reach sunlight.

All of the plants and animals in an area together with the nonliving parts of their environment—like climate and soil—form what is called an **ecosystem**. The size of an ecosystem varies, depending on how it is defined. A small pond, for example, can be considered an ecosystem. The entire Earth can also be considered an ecosystem.

Ecosystems can be affected by natural events like droughts, fires, floods, severe frosts, and windstorms. Human activities can also disturb ecosystems. This happens when land is cleared for development, new kinds of plants and animals are brought into an area, or pollution is released into air and water.

Plant Succession When natural or human forces disturb a plant community, the community may be replaced by a different group of plants suited to the new conditions. The gradual process by which one group of plants replaces another is called **plant succession**.

To better understand plant succession, imagine an area just after a forest fire. The first plants to return to the area need plenty of sunshine. These plants hold the soil in place. They also provide shade for the seeds of other plants. Gradually, seeds from small trees and shrubs grow under the protection of the first plants. These new plants grow taller and begin to take more and more of the sunlight. Many of the smaller plants die. Later, taller trees in the area replace the shorter trees and shrubs that grew at first.

It is important to remember that plant communities are not permanent. The conditions they experience change over time. Some changes affect a whole region. For example, a region's climate may gradually become colder, drier, warmer, or wetter. Additional changes may occur if new plants are introduced to the community.

✓ **READING CHECK:** (*Physical Systems*) How do environments affect life, and how do they change? They determine the plants and animals that live in the environment; natural or human forces can change them.

Soils

In any discussion of plants, plant communities, or plant succession, it is important to know about the soils that support plant life. All soils are not the same. The type of soil in an area can contribute to the kinds of plants that can be grown there. It can also affect how well a plant grows. Plants need soil with minerals, water, and small air spaces if they are to survive and grow.

Soils contain decayed plant and animal matter, called **humus**. Soils rich in humus are fertile. This means they can support an abundance of plant life. Humus is formed by insects and bacteria that live in soil. They break down dead plants and animals and make the nutrients available to plant roots. Insects also make small air spaces as they move through soils. These air pockets contain moisture and gases that plant roots need for growth.

The processes that break down rocks to form soil take hundreds or even thousands of years. Over this long period of time soil tends to form layers. If you dig a deep hole, you can see these layers. Soils typically have three layers. The thickness of each layer depends on the conditions in a specific location. The top layer is called topsoil. It includes humus, insects, and plants. The layer beneath the surface soil is called the subsoil. Only the deep roots of some plants, mostly trees, reach the subsoil. Underneath this layer is broken rock that eventually breaks down into more soil. As the rock breaks down it adds minerals to the soil.

Soils can lose their fertility in several ways. Erosion by water or wind can sweep topsoil away. Nutrients can also be removed from soils by leaching. This occurs when rainfall dissolves nutrients in topsoil and washes them down into lower soil layers, out of reach of most plant roots.

✓ **READING CHECK:** *Physical Systems* What are the physical processes that produce fertile soil? *Insects and bacteria live in the soil and form humus.*

Soil Layers

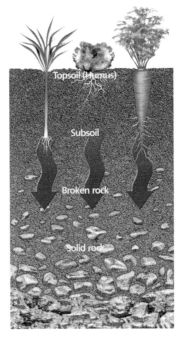

The three layers of soil are the topsoil, subsoil, and broken rock.
Interpreting the Visual Record What do you think has created the cracks in the rocky layer below the broken rock?

Section Review 3

Define and explain: extinct, ecology, photosynthesis, food chain, nutrients, plant communities, ecosystem, plant succession, humus

Reading for the Main Idea

1. *Physical Systems* Why cannot all plants and animals live everywhere? Provide an example to illustrate your answer.

2. *Physical Systems* What makes up soil, and what produces fertile soil?

Critical Thinking

3. **Analyzing Information** What keeps tundra plant communities simple?

4. **Summarizing** How does plant succession occur?

Organizing What You Know

5. **Sequencing** Copy the following graphic organizer. Use it to describe a food chain.

go.hrw.com Homework Practice Online
Keyword: SJ3 HP3

Have students complete the Chapter 3 Test.

RETEACH

Organize the class into three groups—one for each section. Have the groups discuss the text material among themselves, focusing on how factors discussed in the other sections affect, or are affected by, the concepts in their assigned sections. Then have them record these connections on a butcher-paper mural titled "Wind, Climate, and Natural Environments."
ENGLISH LANGUAGE LEARNERS, COOPERATIVE LEARNING

PORTFOLIO EXTENSIONS

1. Have students conduct research on ways sailors from the 1400s to 1700s used global wind currents to help them navigate. Students might compare the early sea voyages with the first nonstop balloon trip around the world in 1999. Using a world map students can study global wind currents and routes taken by adventurers. Have students write a paragraph on the results.

2. Have students obtain instructions for building a barometer, rain gauge, or home weather station. Ask them to build the device and then collect data for a certain length of time. Have them record their data in a chart and compare it with the official local weather reports. Take photographs of the measuring device(s), attach them to the reports, and place them with the reports in students' portfolios.

CHAPTER 3 Review
ANSWERS

Understanding Environment and Society
Outlines and presentations will vary. Students should present information on the formation of tornadoes as well as forecast changes and precautions. Use Rubric 29, Presentations, to evaluate student work.

Thinking Critically
1. Earth's atmosphere, water, and land trap heat.
2. They create warming and cooling currents that move according to the winds' directions.
3. to be prepared for regular seasonal changes as well as potentially dangerous weather conditions
4. to be able to correlate climates to geographic regions
5. to know which soils can support plant life and to understand how to maintain soil fertility

Map Activity
A. hurricanes on the East Coast
B. forest fires on the West Coast
C. hurricanes on the Gulf Coast
D. tornadoes in Texas, Oklahoma, Kansas, and Nebraska

CHAPTER 3 Reviewing What You Know

Building Vocabulary
On a separate sheet of paper, write sentences to define each of the following words.

1. weather
2. climate
3. greenhouse effect
4. air pressure
5. front
6. currents
7. rain shadow
8. monsoon
9. arid
10. permafrost
11. extinct
12. ecology
13. nutrients
14. ecosystem
15. humus

Reviewing the Main Ideas

1. *(The World in Spatial Terms)* What are the major wind belts? Describe each.
2. *(Physical Systems)* How is the area around mountains affected by precipitation?
3. *(The World in Spatial Terms)* Into what main divisions can climate be grouped?
4. *(Environment and Society)* What do people do to change ecosystems?
5. *(Physical Systems)* What elements make up soil? How is fertile soil created?

Understanding Environment and Society

Tornadoes
Prepare an outline for a presentation on tornadoes. As you prepare your presentation from the outline, think about the following:
• How tornadoes are formed.
• How experts are able to forecast their occurrence more accurately.
• The safety precautions taken by people in tornado-prone areas.

Thinking Critically

1. **Finding the Main Idea** How does Earth store the Sun's energy?
2. **Analyzing Information** What effect do wind patterns have on ocean currents?
3. **Drawing Inferences and Conclusions** Why is it important to study the weather?
4. **Drawing Inferences and Conclusions** Why is it important to understand the concept of latitude when learning about Earth's many climates?
5. **Drawing Inferences and Conclusions** Why is it important for scientists to study soils?

FOOD FESTIVAL

Fruits and vegetables in the supermarket come from a range of environments. Have students interview market managers to learn where produce items were grown. They could organize a display linking the fruits and vegetables with descriptions of the climate and soil conditions the plants require. Then wash, peel, and eat those items that do not require cooking.

Reproducible
- Readings in World Geography, History, and Culture 5 and 6
- Critical Thinking Activity 3
- Vocabulary Activity 3

Technology
- Chapter 3 Test Generator (on the One-Stop Planner)

- Audio CD Program, Chapter 3
- HRW Go site

Reinforcement, Review, and Assessment
- Chapter 3 Review, pp. 56–57

- Chapter 3 Tutorial for Students, Parents, Mentors, and Peers
- Chapter 3 Test
- Chapter 3 Test for English Language Learners and Special-Needs Students

---

Building Social Studies Skills

Map ACTIVITY

On a separate sheet of paper, match the letters on the map with their correct labels.

The following natural disasters are often experienced in the United States:

- **hurricanes on the East Coast**
- **hurricanes on the Gulf Coast**
- **forest fires on the West Coast**
- **tornadoes in Texas, Oklahoma, Kansas, and Nebraska**

Mental Mapping Skills ACTIVITY

Draw a freehand map of the globe. Draw lines to show the equator and low, middle, and high latitudes. Draw the continents in the appropriate areas.

WRITING ACTIVITY

After studying different climate regions, decide in which of the regions you would like to live. Write a journal entry describing what your life would be like in this particular area. Be sure to use standard grammar, spelling, sentence structure, and punctuation in your story.

Alternative Assessment

Portfolio ACTIVITY

Learning About Your Local Geography
Understanding Cause and Effect
Research your local weather patterns. Create a graph that shows how recent weather conditions have affected humans, plants, and animals in your part of the state.

internet connect

Internet Activity: go.hrw.com
KEYWORD: SJ3 GT3

Choose a topic to explore online:
- Learn more about using weather maps.
- Follow El Niño, an ocean phenomenon that affects weather.
- Build a food web.

Mental Mapping Skills Activity
Maps will vary; students should accurately indicate the locations of the equator and the low, middle, and high latitudes. The continents should be drawn in the appropriate areas.

Writing Activity
Journal entries will vary; each story should accurately describe the chosen region. Use Rubric 40, Writing to Describe, to evaluate student work.

Portfolio Activity
Flow charts will vary; however, they should show the relationship between recent conditions and humans, plants, and animals. Use Rubric 7, Charts, to evaluate student work.

internet connect

GO TO: go.hrw.com
KEYWORD: SJ3 Teacher
FOR: a guide to using the Internet in your classroom

CHAPTER 4

Earth's Resources
Chapter Resource Manager

Objectives	Pacing Guide	Reproducible Resources
SECTION 1 **Soil and Forests** (pp. 59–61) 1. Identify the processes that threaten soil fertility. 2. Describe why forests are valuable resources. 3. Identify the human activities that can help and hurt forests.	**Regular** .5 day **Block Scheduling** .5 day *Block Scheduling Handbook, Chapter 4*	**RS** Guided Reading Strategy 4.1 **SM** Geography for Life Activity 4 **SM** Map Activity 4
SECTION 2 **Water and Air** (pp. 62–64) 1. Describe why water is an important resource. 2. Identify what threatens our supply of freshwater and how we can protect these supplies. 3. Identify some problems caused by air pollution.	**Regular** .5 day **Block Scheduling** .5 day *Block Scheduling Handbook, Chapter 4*	**RS** Guided Reading Strategy 4.2 **E** Environmental and Global Issues Activities 1, 2, 5 **E** Lab Activities for Geography and Earth Science, Demonstration 8; Hands-On 3
SECTION 3 **Minerals** (pp. 65–67) 1. Explain what minerals are. 2. Identify the two types of minerals.	**Regular** .5 day **Block Scheduling** .5 day *Block Scheduling Handbook, Chapter 4*	**RS** Guided Reading Strategy 4.3
SECTION 4 **Energy Resources** (pp. 68–71) 1. Identify the three main fossil fuels. 2. Identify the four renewable energy sources. 3. Identify the issues that surround the use of nuclear power.	**Regular** .5 day **Block Scheduling** .5 day *Block Scheduling Handbook, Chapter 4*	**RS** Guided Reading Strategy 4.4 **RS** Graphic Organizer 4 **E** Environmental and Global Issues Activities 6, 7 **IC** Interdisciplinary Activity for the Middle Grades 4

Chapter Resource Key

RS Reading Support

IC Interdisciplinary Connections

E Enrichment

SM Skills Mastery

A Assessment

REV Review

ELL Reinforcement and English Language Learners

 Transparencies

 CD–ROM

 Music

 Video

 Internet

Holt Presentation Maker Using Microsoft® Powerpoint®

 One-Stop Planner CD–ROM

See the *One-Stop Planner* for a complete list of additional resources for students and teachers.

One-Stop Planner CD–ROM

It's easy to plan lessons, select resources, and print out materials for your students when you use the *One-Stop Planner CD–ROM with Test Generator.*

Technology Resources	Review, Reinforcement, and Assessment Resources	
One-Stop Planner CD–ROM, Lesson 4.1 Homework Practice Online HRW Go site	ELL	Main Idea Activity 4.1
	REV	Section 1 Review, p. 61
	A	Daily Quiz 4.1
	ELL	English Audio Summary 4.1
	ELL	Spanish Audio Summary 4.1
One-Stop Planner CD–ROM, Lesson 4.2 Homework Practice Online HRW Go site	ELL	Main Idea Activity 4.2
	REV	Section 2 Review, p. 64
	A	Daily Quiz 4.2
	ELL	English Audio Summary 4.2
	ELL	Spanish Audio Summary 4.2
One-Stop Planner CD–ROM, Lesson 4.3 Homework Practice Online HRW Go site	ELL	Main Idea Activity 4.3
	REV	Section 3 Review, p. 67
	A	Daily Quiz 4.3
	ELL	English Audio Summary 4.3
	ELL	Spanish Audio Summary 4.3
One-Stop Planner CD–ROM, Lesson 4.4 Homework Practice Online HRW Go site	ELL	Main Idea Activity 4.4
	REV	Section 4 Review, p. 71
	A	Daily Quiz 4.4
	ELL	English Audio Summary 4.4
	ELL	Spanish Audio Summary 4.4

internet connect

HRW ONLINE RESOURCES

GO TO: go.hrw.com
Then type in a keyword.

TEACHER HOME PAGE
KEYWORD: SJ3 TEACHER

CHAPTER INTERNET ACTIVITIES
KEYWORD: SJ3 GT4

Choose an activity to:
• trek through different kinds of forests.
• investigate global warming.
• make a recycling plan.

CHAPTER ENRICHMENT LINKS
KEYWORD: SJ3 CH4

CHAPTER MAPS
KEYWORD: SJ3 MAPS4

ONLINE ASSESSMENT
Homework Practice
KEYWORD: SJ3 HP4
Standardized Test Prep Online
KEYWORD: SJ3 STP4
Rubrics
KEYWORD: SS Rubrics

COUNTRY INFORMATION
KEYWORD: SJ3 Almanac

CONTENT UPDATES
KEYWORD: SS Content Updates

HOLT PRESENTATION MAKER
KEYWORD: SJ3 PPT4

ONLINE READING SUPPORT
KEYWORD: SS Strategies

CURRENT EVENTS
KEYWORD: S3 Current Events

Meeting Individual Needs

Ability Levels

Level 1 Basic-level activities designed for all students encountering new material

Level 2 Intermediate-level activities designed for average students

Level 3 Challenging activities designed for honors and gifted-and-talented students

English Language Learners Activities that address the needs of students with Limited English Proficiency

Chapter Review and Assessment

E	Readings in World Geography, History, and Culture 7 and 8
SM	Critical Thinking Activity 4
REV	Chapter 4 Review, pp. 72–73
REV	Chapter 4 Tutorial for Students, Parents, Mentors, and Peers
ELL	Vocabulary Activity 4
A	Chapter 4 Test
	Chapter 4 Test Generator (on the One-Stop Planner)
	Audio CD Program, Chapter 4
A	Chapter 4 Test for English Language Learners and Special-Needs Students

LAUNCH INTO LEARNING

Call on a student to select a classroom object at random. *(Examples: book, jacket, globe)* Ask students to name the various materials from which the object is made. *(Example: jacket—cotton, synthetic fibers, metal)* Challenge them to identify the origins of these materials *(cotton needs soil and water to grow, synthetic fibers from petroleum, metal zipper from ore)*. Tell students they are naming resources—supplies that can be used to create something else. Then tell students they will learn more about types and uses of resources in this chapter.

Section 1

Objectives

1. Describe the physical processes that produce fertile soil.
2. Describe the processes that threaten soil fertility.
3. Explain why forests are valuable resources and what efforts are being taken to preserve them.

LINKS TO OUR LIVES

You may want to emphasize the importance of studying the materials and substances on which we depend by discussing with your students the following points:

▶ Our quality of life—and life itself—can depend on how wisely we use our resources.

▶ Some resources may run out in the foreseeable future. We need to know how to make those resources last as long as possible.

▶ We make decisions about using resources every day, even if we do not realize it. We need to be well informed if we are to make good decisions.

▶ We must explore new sources of energy that can replace those, such as oil and natural gas, that cannot be renewed.

Earth's Resources

Precious gems

Wind turbines

Valley in the Andes of Ecuador

LET'S GET STARTED

Copy the following riddle onto the chalkboard and ask students to solve it.

It lies all around us. We don't give it a thought,
But it puts food on the table. What it gives can't be bought.

(*The answer is soil.*) Tell students that in Section 1 they will learn more about the importance of soil and forests.

Building Vocabulary

Write the key terms on the chalkboard. Point out that the prefix *re-* often means "again." So, **renewable resources** are those that can be new again. **Reforestation** indicates that a forest is growing again in the same place. To end a word in *-ation* indicates a process or action. Ask students to relate the suffix to the definitions of **crop rotation, desertification, deforestation,** and reforestation that they find in Section 1 or in the text's glossary. Then ask them to write sentences using each of the terms.

Section 1 — Soil and Forests

Read to Discover

1. What physical processes produce fertile soil?
2. What processes threaten soil fertility?
3. Why are forests valuable resources, and how are they being protected?

Define

renewable resources
crop rotation
terraces
desertification
deforestation
reforestation

WHY IT MATTERS

Many people work to balance concerns about saving the rain forests with local economic needs. Use **CNNfyi.com** or other **current events** sources to learn more about these efforts. Record your findings in your journal.

Corn, a major food crop

Section 1 RESOURCES

Reproducible
◆ Block Scheduling Handbook, Chapter 4
◆ Guided Reading Strategy 4.1
◆ Map Activity 4

Technology
◆ One-Stop Planner CD–ROM, Lesson 4.1
◆ Homework Practice Online
◆ HRW Go site

Reinforcement, Review, and Assessment
◆ Section 1 Review, p. 61
◆ Daily Quiz 4.1
◆ Main Idea Activity 4.1
◆ English Audio Summary 4.1
◆ Spanish Audio Summary 4.1

Soil

Soil is one of the most important **renewable resources** on Earth. Renewable resources are those that can be replaced by Earth's natural processes.

Soil types vary depending on geographic factors. As you learned in Chapter 3, soil contains rock particles and humus. It also contains water and gases. Because soil types vary, some are better able to support plant life than others.

Soil Fertility Soil conservation—protecting the soil's ability to nourish plants—is one challenge facing farmers. Plants must take up nutrients like calcium, nitrogen, phosphorus, and potassium in order to grow. These essential nutrients may become used up if fields are always planted with the same crops. Farmers can add these nutrients to soils in the form of fertilizers, sometimes called plant food. The first fertilizers used were manures. Later, chemical fertilizers were used to increase yields.

Some farmers choose not to use chemical fertilizers. Others cannot afford to use them. They rely on other ways to keep up the soil's ability to produce. One such method is **crop rotation**. This is a system of growing different crops on the same land over a period of years.

On this farm, fields are planted with corn and alfalfa. Alfalfa is a valued crop because it replaces nutrients in the soil.

Teaching Objective 1

ALL LEVELS: (Suggested time: 45 min.) Lead a discussion on the physical processes that produce fertile soil. Then have students find pictures in books and magazines that depict erosion, failing crops, desertification, or development onto farmland. Have them cut out or reproduce the pictures, paste them on posters, and label their posters with a paragraph that describes the problem illustrated and suggests a solution.
ENGLISH LANGUAGE LEARNERS

Teaching Objective 2

ALL LEVELS: (Suggested time: 25 min.) Copy the following graphic organizer onto the chalkboard, omitting the italicized answers. Use it to help students list ways people use forests. After discussing the completed organizer, point out that timber is an important natural resource. Ask them to describe how more timber can be produced. *(reforestation).*
ENGLISH LANGUAGE LEARNERS

other products — *recreational uses*
FORESTS
construction materials — *uses as fuel*

Linking Past to Present
Deforestation

Although they did not have access to modern technology, ancient peoples often dramatically affected natural landscapes. Some archaeologists believe that during the Neolithic era—beginning around 6000 B.C.—people may have deforested large areas of central and western Europe. They felled trees to clear land for farming and used the lumber for building and fuel.

Activity: Have students conduct research on deforestation that took place during ancient and modern times. Then have them compare the deforestation that took place during the two time periods in terms of causes, methods, and results. Ask students to draw conclusions about the long-term effects of deforestation.

Visual Record Answer ▶

reduce erosion by blocking wind

▶

A row of poplar trees divides farmland near Aix-en-Provence, France.
Interpreting the Visual Record **What effect will these trees have on erosion?**

Our Amazing Planet

About 47,000 aspen trees in Utah share a root system and a set of genes. These trees are actually a single organism. These aspens cover 106 acres (43 hectares) and weigh at least 13 million pounds (5.9 million kg). Together they are probably the world's heaviest living thing.

Salty Soil Salt buildup is another threat to soil fertility. In dry climates farmers must irrigate their crops. They use well water or water brought by canals and ditches. Much of this water evaporates in the dry air. When the water evaporates, it leaves behind small amounts of salt. If too much salt builds up in the soil, crops cannot grow.

Erosion The problem of soil erosion is faced by farmers all over the world. Farmers have to work to keep soil from being washed away by rainfall. Soil can also be blown away by strong winds. To prevent soil loss, some farmers plant rows of trees to block the wind. Others who farm on steep hillsides build **terraces** into the slope. Terraces are horizontal ridges like stair steps. By slowing water movement the terraces stop the soil from being washed away. They also provide more space for farming.

Loss of Farmland The loss of farmland is a serious problem in many parts of the world. In some places farming has worn out the soil, and it can no longer grow crops. Livestock may then eat what few plants remain. Without plants to hold the soil in place, it may blow away. The long-term process of losing soil fertility and plant life is called **desertification**. Once this process begins, the desert can expand as people move on to better soils. They often repeat the destructive practices, damaging ever-larger areas.

Farmland is also lost when cities and suburbs expand into rural areas. Nearby farmers sell their land. The land is then used for housing or businesses, rather than for agriculture. This is happening in many poorer countries. It also happens in richer nations like the United States.

✓ **READING CHECK:** **Environment and Society** What physical processes help soil fertility, and what threatens it? *fertilizers, crop rotation; salt buildup, erosion, too much farming, desertification, urbanization*

RETEACH

Have students complete Main Idea Activity 4.1. Then create a two-column chart on the chalkboard. Label the columns *Soil* and *Forests* and label the rows *Importance*, *Problems*, and *Solutions*. Fill in the chart as a class.
ENGLISH LANGUAGE LEARNERS

EXTEND

Assign groups of students a country. Have them identify and research a problem in their country's forests. Tell them to list and consider options to solve the problem and to discuss the merits of the options. Then have them draw up action plans for preserving their forests while providing for local needs. Invite the class to evaluate the plans. **COOPERATIVE LEARNING, BLOCK SCHEDULING**

Forests

Forests are renewable resources because new trees can be planted in a forest. If cared for properly, they will be available for future generations. Forests are important because they provide both people and wildlife with food and shelter. People depend on forests for a wide variety of products. Wood products include lumber, plywood, and shingles for building houses. Other wood-based manufactured products include cellophane, furniture, some plastics, and fibers such as rayon. Trees also supply fats, gums, medicines, nuts, oils, turpentine, waxes, and rubber. The forests are valuable not only for their products. People also use forests for recreational activities, such as camping and hiking.

Deforestation The destruction or loss of forest area is called **deforestation**. It is happening in the rain forests of Africa, Asia, and Central and South America. People clear the land and use it for farming, industry, and housing. Pollution also causes deforestation.

Protecting Forests Many countries, including the United States, are trying to balance their economic needs regarding forests with conservation efforts. Since the late 1800s Congress has passed laws to protect and manage forest and wilderness areas. In addition to protecting forests, people can also plant trees in places where forests have been cut down. This replanting is called **reforestation**. In some cases the newly planted trees can be "harvested" again in a few years.

✔ **READING CHECK:** *Environment and Society* Why is preserving forests important, and how can people contribute to it? They provide food, shelter, and materials for recreation, many products; through laws and reforestation.

▲
Villagers work on a reforestation project in Cameroon.

go.hrw.com **Homework Practice Online**
Keyword: SJ3 HP4

Section Review 1

Define and explain: renewable resources, crop rotation, terraces, desertification, deforestation, reforestation

Reading for the Main Idea

1. *Environment and Society* What physical processes produce fertile soil?
2. *Environment and Society* What do forests provide, and how can they be protected?

Critical Thinking

3. **Finding the Main Idea** Which two natural forces contribute to soil erosion? How can erosion be prevented?

4. **Drawing Inferences and Conclusions** Why is it important to manage forests?

Organizing What You Know

5. **Categorizing** Copy the following graphic organizer. Use it to discuss whether or not rain forests should be protected.

Reasons for . . .	Reasons against . . .

Section Review 1

Answers

Define For definitions, see: renewable resources, p. 59; crop rotation, p. 59; terraces, p. 60; desertification, p. 60; deforestation, p. 61; reforestation, p. 61

Reading for the Main Idea

1. using fertilizers and crop rotation, eliminating salt buildup, preventing erosion, preventing desertification (NGS 14, 16)

2. possible answers: lumber, plywood, shingles, cellophane, furniture, plastics, rayon, fats, gums, medicines, nuts, oil, turpentine, waxes, and rubber; protection through laws and reforestation (NGS 14, 16)

Critical Thinking

3. wind and water; through planting rows of trees, building terraces

4. possible answer: so they will continue to produce the goods people use, provide recreation, and shelter wildlife

Organizing What You Know

5. possible reasons for—products that forests supply, wildlife protection, oxygen supply; possible reasons against—land needed for farms, wood needed for industry and housing

Section 2

Objectives

1. Explain why water is an important resource.

2. Identify threats to our water supply, and describe ways to conserve it.

3. Analyze some of the problems caused by air pollution.

LET'S GET STARTED

Copy the following statement and question onto the chalkboard: *The average American uses 168 gallons (636 l) of water daily. What are some ways you and your family can save water?* (Students' answers might include turning off the tap while brushing teeth, quick showers, installing water-saving toilets, and running the dishwasher only when full.) Tell students that in Section 2 they will learn more about water and air.

Building Vocabulary

Write the key terms on the chalkboard. Call on volunteers to underline the parts of the words that resemble elements of familiar words *(semi-, aqu-)* and to tell what words they resemble. *(Possible answers: semiweekly, aquatic)* Have other students look up the key terms and infer how they relate to the familiar words. Ask students to find definitions for the remaining key terms and use the terms in sentences.

Section 2 RESOURCES

Reproducible

◆ Guided Reading Strategy 4.2

◆ Environmental and Global Issues Activities 1, 2, 5

◆ Lab Activities for Geography and Earth Science, Demonstration 8; Hands-On 3

Technology

◆ One-Stop Planner CD–ROM, Lesson 4.2

◆ Homework Practice Online

◆ HRW Go site

Reinforcement, Review, and Assessment

◆ Section 2 Review, p. 64

◆ Daily Quiz 4.2

◆ Main Idea Activity 4.2

◆ English Audio Summary 4.2

◆ Spanish Audio Summary 4.2

Section 2 Water and Air

Read to Discover

1. Why is water an important resource?

2. What threatens our supplies of freshwater, and how can we protect these supplies?

3. What are some problems caused by air pollution?

WHY IT MATTERS

Countries around the world have worked together to resolve problems regarding Earth's atmosphere. Use CNNfyi.com or other current events sources to learn about these efforts. Record your findings in your journal.

Define

semiarid	desalinization
aqueducts	acid rain
aquifers	global warming

A water tank

Water

Dry regions are found in many parts of the world, including the western United States. There are also **semiarid** regions—regions that receive a small amount of rain. Semiarid places are usually too dry for farming. However, these areas may be suitable for grazing animals.

Water Supply Many areas of high mountains receive heavy snowfall in the winter. When that snow melts, it forms rivers that flow from the mountains to neighboring regions. People in dry regions use various means to bring the water where it is needed for agricultural and other uses. They build canals, reservoirs, and **aqueducts**—artificial channels for carrying water.

Some places have water deep underground in **aquifers**. These are water-bearing layers of rock, sand, or gravel. Some are quite large. For example, the Ogallala Aquifer stretches across the Great Plains from Texas to South Dakota. People drill wells to reach the water in the aquifer.

People in dry coastal areas have access to plenty of salt water. However, they typically do not have enough freshwater. In Southwest Asia this situation is common. To create a supply of

As people make their homes in dry areas, they add to the demand for water.

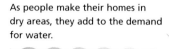

Teaching Objective 1

ALL LEVELS: (Suggested time: 30 min.) Have students design covers for a magazine titled *Water Weekly.* As part of their covers, have students include article titles that suggest why water is an important resource. Ask volunteers to present their covers to the class. Then display the covers around the classroom. **ENGLISH LANGUAGE LEARNERS**

Teaching Objectives 2–3

ALL LEVELS: (Suggested time: 30 min.) Copy the following graphic organizer onto the chalkboard, omitting the italicized answers, and have students copy it into their notebooks. Ask half of the class to fill in the water side and the other half to complete the air material. When they are finished, have students compare their charts and fill in the gaps.
ENGLISH LANGUAGE LEARNERS

	Water Pollution	Air Pollution
Causes	*chemical fertilizers, pesticides, industrial wastes*	*burning fuel*
Results	*reduced water supply, harm to other living things*	*threatens health, causes acid rain, may damage ozone layer*
Solutions	*limit pollutants*	*find cleaner energy sources*

freshwater, people in these places have built machines that take the salt out of seawater. This process, known as **desalinization**, is expensive and takes a lot of energy. However, in some places, it is necessary.

Water Conservation In recent decades people have developed new ways to save water. Many factories now recycle water. Farmers are able to irrigate their crops more efficiently. Cities build water treatment plants to purify water that might otherwise be wasted. Some people in dry climates are using desert plants instead of grass for landscaping. This means that they do not need to water as often.

It is important for people in all climates to conserve water. Wasting water in one location could mean that less water is available for use in other places.

Water Quality Industries and agriculture also affect the water supply. In many countries there are still places that cannot afford to build closed sewer systems. Some factories also operate without pollution controls because such controls would add a great deal of cost to operation.

Industrialized countries like the United States have water treatment plants and closed sewer systems. However, water can still be polluted when farmers use too much chemical fertilizer and pesticides. These chemicals can get into local streams. Waste from industries may also contain chemicals, metals, or oils that can pollute streams and rivers.

Rivers carry pollution to the oceans. The pollution can harm marine life such as fish and shellfish. Eating marine life from polluted waters can make people sick. So can drinking polluted water. Balancing industrial and agricultural needs with the need for clean water continues to be a challenge faced by many countries.

✓ READING CHECK: **Environment and Society** Why is water an important resource, and how does its availability affect people? People need it for drinking and farming; if water is polluted or wasted, people will not have enough for everyday life.

Air

Air is essential to life. Plants and animals need the gases in the air to live and grow.

Human activities can pollute the air and threaten the health of life on the planet. Burning fuels for heating, for transportation, and to power factories releases chemicals into the air. Particularly in large cities, these chemicals build up in the air. The chemicals create a mixture called smog.

Some cities have special problems with air pollution. Denver, Los Angeles, and Mexico City, for example, are located in bowl-shaped valleys that can trap air pollution. This pollution sometimes builds up to levels that are dangerous to people's health.

▲
Pivoting sprinklers irrigate these circular cornfields in Kansas.
Interpreting the Visual Record Why might irrigation be necessary in these fields?

Heavy smog clouds the Los Angeles skyline.
▼

People have been building dams for at least 5,000 years. More than 800,000 dams have been constructed around the world.

Dams and the reservoirs created by them can provide a steady water supply, hydroelectricity, protection from floods, and increased fish yields. However, they can also displace people, submerge farmland and cultural sites, disrupt animal migration patterns, and even increase the risk of certain diseases. Malaria, for example, is spread by mosquitoes that can breed in the standing water of reservoirs. For these and other reasons, dam construction has slowed in recent years.

Activity: Have students search news media for information on dams in your state. What controversies have arisen? What possible solutions have been suggested to address these controversies? Do you think that future scientific discoveries will lessen the negative effects of dams on the environment? Why or why not?

◄ **Visual Record Answer**

lack of adequate rainfall

Lead a discussion on water and air issues your community may face in coming years. *(Possible issues: reduced water supply from falling aquifer level, more pollution from increased traffic or industry, hurricanes threatening water supplies)* What might happen if new technology makes more water available in dry locations?

REVIEW AND ASSESS

Have students complete the Section Review. Then pair students. Have pairs find important terms in each subsection and take turns relating the terms to the main points. Then have students complete Daily Quiz 4.2.
COOPERATIVE LEARNING

RETEACH

Have students complete Main Idea Activity 4.2. Then have them design posters that illustrate the main points of the section. Ask volunteers to present their posters to the class. Then display the posters around the classroom. **ENGLISH LANGUAGE LEARNERS**

EXTEND

Have students conduct research on aqueducts built by the ancient Romans to supply freshwater to their cities. Students who like to work with their hands may want to construct a scale model. **BLOCK SCHEDULING**

Section Review 2

Answers

Define For definitions, see: semiarid, p. 62; aqueducts, p. 62; aquifers, p. 62; desalinization, p. 63; acid rain, p. 64; global warming, p. 64

Reading for the Main Idea
1. They build canals, reservoirs, and aqueducts.
   (NGS 14)
2. allow chemical fertilizers, pesticides, and industrial waste into water supply
   (NGS 14)
3. People burn fuel for heating, for transportation, and to power factories.
   (NGS 14)

Critical Thinking
4. The process of desalinization is expensive.

Organizing What You Know
5. possible answers— threatens health, creates acid rain, damages ozone layer, may cause global warming

This satellite image of the Southern Hemisphere shows a thinning in the ozone layer in October 1979.

▼

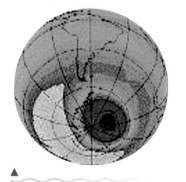

▲
By October 1992 the thinning area, shown in purple, had grown much larger.

Acid Rain When air pollution combines with moisture in the air, it can form a mild acid. This can be similar in strength to vinegar. When it falls to the ground, this moisture is called **acid rain**. It can damage or kill trees. Acid rain can also kill fish.

Many countries have laws to limit pollution. However, pollution is an international problem. Winds can blow away air pollution, but the wind is only moving the pollution to another place. Pollution can pass from one country to another. It can even pass from one continent to another. Countries that limit their own pollution can still be affected by pollution from other countries.

Pollution and Climate Change Smog and acid rain are short-term effects of air pollution. Air pollution may also have long-term effects by changing conditions in Earth's atmosphere. Certain kinds of pollution damage the ozone in the upper atmosphere. This ozone layer protects living things by absorbing harmful ultraviolet light from the Sun. Damage to Earth's ozone layer may cause health problems in people. For example, it could lead to an increase in skin cancer.

Another concern is **global warming**—a slow increase in Earth's average temperature. The Sun constantly warms Earth's surface. The gases and water vapor in the atmosphere trap some of this heat. This helps keep Earth warm. Without the atmosphere, this heat would return to space. Evidence suggests that pollution causes the atmosphere to trap more heat. Over time this would make Earth warmer.

Scientists agree that Earth's climate has warmed during the last century. However, they disagree about exactly why. Some scientists say that temperatures have warmed because of air pollution caused by human activities, particularly the burning of fossil fuels. Others think warmer temperatures have resulted from natural causes. Scientists also disagree about what has caused the thinning of the ozone layer.

✓ READING CHECK: *Environment and Society* What are the different points of view about air pollution and climate change? Some scientists think that air pollution has created warmer temperatures; others think that warmer temperatures are the result of natural changes.

go. hrw .com **Homework Practice Online**
Keyword: SJ3 HP4

Section Review 2

Define and explain: semiarid, aqueducts, aquifers, desalinization, acid rain, global warming

Reading for the Main Idea
1. *Environment and Society* How have people changed the environment to increase the water supply in drier areas?
2. *Environment and Society* How do people cause water pollution?
3. *Environment and Society* What kinds of human activities have polluted the air?

Critical Thinking
4. **Drawing Inferences and Conclusions** Why is desalinization rarely practiced?

Organizing What You Know
5. **Identifying Cause and Effect** Use this graphic organizer to explain air pollution.

(Causes:)———(Air Pollution)———(Effects:)

FOCUS

LET'S GET STARTED

Copy these questions onto the chalkboard: *Have you played the guessing game that begins by asking if something is animal, vegetable, or mineral? What does mineral mean in this context?* Students might suggest that a mineral is something that is not alive, rocks, or materials dug from the ground. Call on volunteers to list minerals with which they may already be familiar *(aluminum, building stone, precious and semiprecious stones, salt, and so on).* Tell students that in Section 3 they will learn more about minerals.

Building Vocabulary

Write the key terms on the chalkboard. Point out that the suffix *-ic* means "of or relating to" and that the prefix *non-* means "not." Have students relate these meanings to the word "metal" to arrive at definitions for **metallic** and **nonmetallic minerals**. Then ask students to use the glossary to define **minerals**. Call on a volunteer to use the definition of "renewable" in Section 1 to define **nonrenewable resources**.

Section 3

Minerals

Read to Discover

1. What are minerals?
2. What are the two types of minerals?

Define

nonrenewable resources

minerals

metallic minerals

nonmetallic minerals

WHY IT MATTERS

Most minerals are dug from deep in the ground by miners. Use **CNNfyi.com** or other **current events** sources to learn about life as a miner. Record your findings in your journal.

Quartz crystals

Section 3   RESOURCES

Reproducible
◆ Guided Reading Strategy 4.3

Technology
◆ One-Stop Planner CD–ROM, Lesson 4.3
◆ Homework Practice Online
◆ HRW Go site

Reinforcement, Review, and Assessment
◆ Section 3 Review, p. 67
◆ Daily Quiz 4.3
◆ Main Idea Activity 4.3
◆ English Audio Summary 4.3
◆ Spanish Audio Summary 4.3

Minerals

You have learned that renewable resources such as trees are always being produced. **Nonrenewable resources** are those that cannot be replaced by natural processes or are replaced very slowly.

Earth's crust is made up of substances called **minerals**. Minerals are an example of a nonrenewable resource. They provide us with many of the materials we need. More than 3,000 minerals have been identified, but fewer than 20 are common. Around 20 minerals make up most of Earth's crust. Minerals have four basic properties. First, they are inorganic. Inorganic substances are not made from living things or the remains of living things. Second, they occur naturally, rather than being manufactured like steel or brass. Third, minerals are solids in crystalline form, unlike petroleum or natural gas. Finally, minerals have a definite chemical composition or combination of elements. Although all minerals share these four properties, they can be very different from one another. Minerals are divided into two basic types: metallic and nonmetallic.

Metallic Minerals Metals, or **metallic minerals**, are shiny and can conduct heat and electricity. Metals are solids at normal room temperature. An exception is mercury, a metal that is liquid at room temperature.

Gold is one of the heaviest of all metals and is easily worked. For thousands of years people have highly valued gold. Precious metals are commonly made into jewelry and coins. Silver and platinum are other precious metals.

◀ **Chart Answer**

oxygen

The Most-Common Elements in Earth's Crust

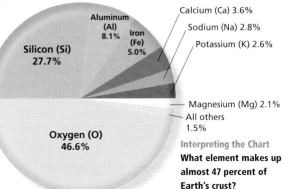

- Silicon (Si) 27.7%
- Aluminum (Al) 8.1%
- Iron (Fe) 5.0%
- Calcium (Ca) 3.6%
- Sodium (Na) 2.8%
- Potassium (K) 2.6%
- Magnesium (Mg) 2.1%
- All others 1.5%
- Oxygen (O) 46.6%

Interpreting the Chart

What element makes up almost 47 percent of Earth's crust?

Teaching Objectives 1–2

ALL LEVELS: (Suggested time: 30 min.) Copy the following graphic organizer onto the chalkboard, omitting the italicized answers. Use it to help students learn more about minerals.

ENGLISH LANGUAGE LEARNERS

MINERALS

Definition: *substances that make up Earth's crust*		
Types	*Metallic*	*Nonmetallic*
Characteristics	*shiny, good conductors of heat and electricity*	*varied appearance, not good conductors of heat and electricity*
Examples	*aluminum, iron, gold, silver, and other precious metals*	*quartz, talc, diamonds, gemstones, salt, and sulfur*

DAILY LIFE

One of the most widely used nonmetallic minerals is salt. Salt is plentiful, inexpensive, and essential to the health of people and animals. We are most familiar with salt as a cooking ingredient, but it is also used in many other products and processes.

Salt removes unwanted minerals from the water supply. Farm animals and poultry consume salt. Food processors, such as pickle makers, use tons of the mineral. Salt makes dyes colorfast in fabric. Film companies use salt in chemical solutions. Health spas offer salt baths and rubs. Salt producers have claimed that the precious mineral has 14,000 uses.

Activity: Have students examine labels of common household products and foods to find salt as an ingredient. You might also have them conduct research on the industrial uses for salt.

Connecting to Art
Answers
1. They use recycled objects to create art.
2. less-developed countries

internet connect
GO TO: go.hrw.com
KEYWORD: SJ3 CH4
FOR: Web sites about mining

CONNECTING TO Art

Art made from recycled products

Many Americans now make a habit of recycling. They put their cans, bottles, and newspapers by the curb for pickup or take them to recycling centers. However, some people do their recycling in a different way. They use their junk to create art.

Much recycled art is folk art. These objects have a practical purpose but are made with creativity and a sense of style. The making of folk art objects from junk is common in the world's poorer countries, where resources are scarce.

Some examples of recycled folk art include dust pans made from license plates (Mexico), jugs made from old tires (Morocco), briefcases made from flattened tin cans (Senegal), and a toy helicopter made from plastic containers and film canisters (Haiti). As scientist Stephen J. Gould has written, "In our world of material wealth, where so many broken items are thrown away rather than mended . . . we forget that most of the world fixes everything and discards nothing."

Americans do have a tradition of making recycled art, however. The Amish make quilts from old scraps of cloth. Other folk artists build whimsical figures out of bottle caps and wire. Some modern artists create sculptures from "found objects" like machine parts, bicycle wheels, and old signs. Junk art can even be fashion. One movie costume designer went to the Academy Awards ceremony wearing a dress made of credit cards!

Understanding What You Read
1. How do some societies turn junk into art?
2. Where is much recycled folk art made?

▲
Gold is a metallic mineral.

Iron is the cheapest metal. Iron can be combined with certain other minerals to make steel. Aluminum is another common metal. This lightweight metal is used in such items as soft drink cans and airplanes. We handle copper every time we pick up a penny.

Nonmetallic Minerals Minerals that lack the characteristics of a metal are called **nonmetallic minerals**. These vary in their appearance. Quartz, a mineral often found in sand, looks glassy. Talc has a pearly appearance. Most nonmetallic minerals have a dull surface and are poor conductors of heat or electricity.

Diamonds are minerals made of pure carbon. They are the hardest naturally occurring substance. The brilliant look of diamonds has made them popular gems. Their hardness makes them valuable for industrial use. Other gemstones, like rubies, sapphires, and emeralds, are also nonmetallic minerals.

Ask students to write letters to your local newspaper urging the community to begin, continue, or expand a recycling program.

Have students complete Main Idea Activity 4.3. Then prepare flash cards of the section's Define terms and names of minerals. As you display a card, call on a student to find the sentences that discuss the concept and explain the term to the class. **ENGLISH LANGUAGE LEARNERS**

REVIEW AND ASSESS

Have students complete the Section Review. Then call on a volunteer to name a common mineral. Instruct the student to call on another to suggest a use for that mineral. Continue until all the minerals discussed in the section have been reviewed. Then have students complete Daily Quiz 4.3.

EXTEND

Interested students may want to conduct research on the many ways that cultures have used precious and semiprecious stones throughout history. *(Examples: in paint, as medicine, as ritual objects, to decorate bookcovers)* Have students use their research to create entries for a pamphlet to accompany a museum exhibit displaying these uses. **BLOCK SCHEDULING**

Mineral and Energy Resources in the United States

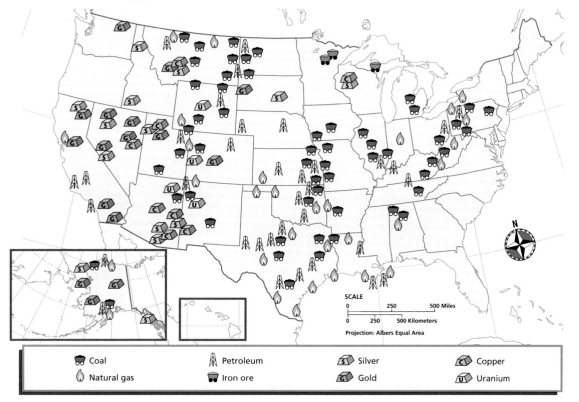

SCALE
0   250   500 Miles
0   250   500 Kilometers
Projection: Albers Equal Area

| Coal | Petroleum | Silver | Copper |
| Natural gas | Iron ore | Gold | Uranium |

Other mineral substances also have important uses. For example, people need salt to stay healthy. Sulfur is used in many ways, from making batteries to bleaching dried fruits. Graphite, another form of carbon, is used in making pencils.

✓ **READING CHECK:** *Physical Systems* What are the two types of minerals?
metallic and nonmetallic

Section Review 3

Define and explain: nonrenewable resources, minerals, metallic minerals, nonmetallic minerals

Reading for the Main Idea
1. *Physical Systems* What is a nonrenewable resource?
2. *Environment and Society* Why are minerals important? What is the difference between a metallic mineral and a nonmetallic mineral?

go.hrw.com **Homework Practice Online**
Keyword: SJ3 HP4

Critical Thinking
3. **Summarizing** How can minerals be used?
4. **Drawing Inferences and Conclusions** What makes gold such an important mineral?

Organizing What You Know
5. **Contrasting** Copy the following graphic organizer. Use it to contrast types of minerals.

Metallic minerals	Nonmetallic minerals

Section Review 3

Answers

Define For definitions, see: nonrenewable resources, p. 65; minerals, p. 65; metallic minerals, p. 65; nonmetallic minerals, p. 66

Reading for the Main Idea
1. cannot be replaced by natural processes or are replaced very slowly (NGS 7)
2. metallic—shiny, can conduct heat and electricity, most are solid at room temperature; nonmetallic—varied in appearance, poor conductors of heat and electricity (NGS 15)

Critical Thinking
3. Answer may include as jewelry and coins, to make steel, in soft drink cans and airplanes, for industrial use, to stay healthy, to make batteries, to bleach dried fruits, or in making pencils.
4. can be used as money and made into jewelry

Organizing What You Know
5. metallic—shiny, can conduct heat and electricity, most are solid at room temperature; nonmetallic—varied in appearance, poor conductors of heat and electricity

Section 4

Objectives

1. Describe the three main fossil fuels.

2. Identify four renewable energy sources.

3. Analyze the issues surrounding nuclear power.

FOCUS

LET'S GET STARTED

Copy the following instructions onto the chalkboard: *What are three ways you use energy? What are the sources of that energy?* Students may mention driving, cooking, heating, air conditioning, or operating appliances as ways to use energy. Possible answers for sources of energy might be natural gas, heating oil, gasoline, or electricity. Discuss responses. Then ask students how they would be affected if those energy sources were no longer available. Tell students that in Section 4 they will learn more about Earth's energy resources.

Building Vocabulary

Write the key terms on the chalkboard and underline *petro, hydro, geo,* and *sol.* Tell students that these Greek and Latin word fragments mean "rock," "water," "earth," and "sun" respectively. Have students infer the definitions of the Define terms that contain the word fragments and check the glossary to confirm. Ask students what other words they know that contain those prefixes. *(Possible answers: petrochemical, hydraulic, geology, solstice)* Call on volunteers to read the definitions for **fossil fuels** and **refineries**.

Section 4 RESOURCES

Reproducible
◆ Guided Reading Strategy 4.4
◆ Graphic Organizer 4
◆ Environmental and Global Issues Activities 6, 7
◆ Interdisciplinary Activity for the Middle Grades 4

Technology
◆ One-Stop Planner CD–ROM, Lesson 4.4
◆ Homework Practice Online
◆ HRW Go site

Reinforcement, Review, and Assessment
◆ Section 4 Review, p. 71
◆ Daily Quiz 4.4
◆ Main Idea Activity 4.4
◆ English Audio Summary 4.4
◆ Spanish Audio Summary 4.4

Section 4 Energy Resources

Read to Discover

1. What are the three main fossil fuels?
2. What are the four renewable energy sources?
3. What issues surround the use of nuclear power?

Define

fossil fuels
petroleum
refineries

hydroelectric power
geothermal energy
solar energy

WHY IT MATTERS

Automobile manufacturers are interested in building cars that use sources of power other than gasoline. Use **CNNfyi.com** or other current events sources to learn more about these changes in automobile design. Record your findings in your journal.

An oil tanker

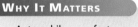

A miner digs coal in a narrow tunnel.

Nonrenewable Energy Resources

Most of the energy we use comes from the three main **fossil fuels**: coal, petroleum, and natural gas. Fossil fuels were formed from the remains of ancient plants and animals. These remains gradually decayed and were covered with sediment. Over long periods of time, pressure and heat changed these materials into fossil fuels. All fossil fuels are nonrenewable resources.

Coal Until the 1900s people mostly used wood and coal as sources of energy. Coal was used to make steel in giant furnaces and to run factories. However, the burning of coal polluted the air. The way coal is burned today has improved so that it releases less pollution. This new technology is more expensive, however. Some of the largest coal deposits are located in Australia, China, Russia, India, and the United States.

Petroleum Industrialized societies now use **petroleum**—an oily liquid—for a variety of purposes. When it is first pumped out of the ground, petroleum is called crude oil. It is then shipped or piped to **refineries**—factories where crude oil is processed, or refined. Petroleum is made into gasoline, diesel and jet fuels, and heating oil.

68

ALL LEVELS: (Suggested time: 30 min.) Copy the following graphic organizer onto the chalkboard, omitting the italicized answers. Pair students and have the pairs complete the organizer. Discuss any gaps in the students' charts. **ENGLISH LANGUAGE LEARNERS, COOPERATIVE LEARNING**

FUELS		Advantage	Disadvantage
F O S S I L	Coal	*plentiful*	*pollutes, nonrenewable*
	Petroleum	*plentiful*	*pollutes, not evenly distributed, nonrenewable*
	Natural gas	*plentiful, burns clean*	*hard to transport, nonrenewable*
R E N E W A B L E	Hydroelectric Power	*clean, renewable*	*affects habitats, drowns farms and forests*
	Wind Power	*clean, renewable*	*needs steady wind*
	Geothermal Energy	*clean, renewable*	*only in limited areas*
	Solar Energy	*clean, renewable*	*expensive*
	Nuclear Energy	*reduces need to burn fossil fuels*	*possible accidents, waste*

Who Has the Oil?

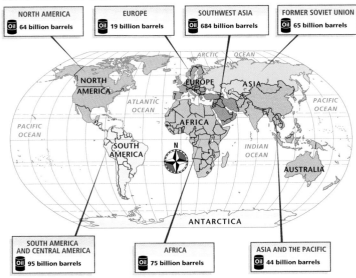

NORTH AMERICA	EUROPE	SOUTHWEST ASIA	FORMER SOVIET UNION
Oil 64 billion barrels	Oil 19 billion barrels	Oil 684 billion barrels	Oil 65 billion barrels

SOUTH AMERICA AND CENTRAL AMERICA	AFRICA	ASIA AND THE PACIFIC
Oil 95 billion barrels	Oil 75 billion barrels	Oil 44 billion barrels

Source: *BP Amoco Statistical Review of World Energy 2001*

◄ More than half of all known oil deposits are located in Southwest Asia. New deposits are still being found, but Southwest Asia will probably continue to hold the largest share.

Petroleum is not evenly distributed on Earth. Of the oil reserves that have been discovered, more than 65 percent are found in Southwest Asia. Most of that is in Saudi Arabia. North America has around 7 percent of the world's known oil, while South America has around 10 percent. Other regions have oil in small amounts.

Natural Gas The use of natural gas is growing rapidly. Natural gas is gas that comes from Earth's crust through natural openings or drilled wells. Large natural gas fields are found in Russia and Southwest Asia. Northern Canada also has large amounts of natural gas. However, the fields are located in the far north. Frozen seas and low temperatures make it difficult to pump and ship the gas safely.

Natural gas is the cleanest-burning fossil fuel. It produces much less air pollution than gasoline or diesel fuel. It is usually transported by pipeline, making it most useful for factories and electrical plants. It is also used for heating and cooking. Vehicles that run on natural gas have to carry the fuel in special, bulky containers. However, some large cities now have buses and taxis that use natural gas to help cut down on air pollution.

✓ **READING CHECK:** *Physical Systems*
What physical processes produced fossil fuels? The remains of ancient plants and animals decayed and were covered with sediment. Pressure and heat changed the materials to fossil fuels.

The Alaska pipeline carries oil from north to south across Alaska.

Interpreting the Visual Record **Why do you think this pipeline is above ground?**

SCIENCE

Methane Hydrates A methane hydrate crystal consists of a natural gas molecule surrounded by water molecules. Methane hydrates resemble regular ice, but unlike ice they can burn! The compounds are abundant in nature, particularly on the ocean floor, and might provide a new source of energy. In fact, the amount of carbon that is found in the global stores of hydrates may be twice as large as the total amount of carbon found in all other fossil fuels.

However, because they are formed under pressure, hydrates can disintegrate when removed from the ocean. Rockslides could make their removal from the ocean floor hazardous. Furthermore, in terms of its potential impact on global warming, methane as a greenhouse gas is 10 times stronger than carbon dioxide, which is released in the burning of traditional fossil fuels.

Discussion: Have students make predictions about the economic and environmental consequences of a new technology that would eliminate the hazards of using methane hydrates.

◄ **Visual Record Answer**

for easier access for repairs

69

Teaching Objective 1

LEVEL 2: (Suggested time: 20 min.) Have students create a flowchart showing how fossil fuels are formed. Invite volunteers to present their flowcharts to the class. **ENGLISH LANGUAGE LEARNERS**

➤**ASSIGNMENT:** Have students design yellow-pages ads for companies that harness and sell hydroelectric power, wind energy, geothermal energy, and solar energy. Challenge students to review previous chapters for hints about other potential sources of renewable energy *(tidal energy, wave energy)* and to design ads for them as well. **ENGLISH LANGUAGE LEARNERS**

TEACHER TO TEACHER

Jane Palmer of Sanford, Florida, suggests the following activity. Organize the class into six groups. Have two groups represent countries with many resources, another two groups represent countries with some resources, and the final two groups represent countries with few resources. Distribute building supplies (craft sticks, wood blocks, glue, and so on) to each group to represent that country's resources. For example, the countries that receive many materials would have more resources. Instruct each group to construct a small building using their supplies. Then ask students how they felt about using limited resources. Discuss challenges they faced and what could have been done to meet those challenges.

Section Review 4

Answers

Define For definitions, see: fossil fuels, p. 68; petroleum, p. 68; refineries, p. 68; hydroelectric power, p. 70; geothermal energy, p. 70; solar energy, p. 71

Reading for the Main Idea

1. coal, petroleum, natural gas; plant and animal remains decayed, were covered with sediment, and became fuel through heat and pressure. (NGS 7)

2. water, wind, geothermal energy, Sun; water—hydroelectric power, wind—wind turbines, geothermal energy—captures Earth's heat, Sun—solar panels absorb heat (NGS 7)

Critical Thinking

3. affect fish and wildlife habitats

4. nuclear accidents and waste

Organizing What You Know

5. renewable—can be replaced by Earth's natural processes; nonrenewable—cannot

---

Visual Record Answer ▶

People have taken advantage of a windy location through using turbines for power.

☑ **internet** connect

GO TO: go.hrw.com
KEYWORD: SJ3 CH4
FOR: Web sites about Earth's resources

Renewable Energy Resources

People have also learned to use several renewable energy resources. They include water power, wind power, heat from within Earth, and the Sun's energy.

Water The most commonly used renewable energy source is **hydroelectric power**. Hydroelectric power is the production of electricity by waterpower. Dams harness the energy of falling water to power generators. These generators produce electricity. Dams produce about 9 percent of the electricity used in the United States. Other countries producing hydroelectric power include Brazil, Canada, China, Egypt, New Zealand, Norway, and Russia.

Although hydroelectric power does not pollute the air, it does affect the environment. Fish and wildlife habitats can be affected when reservoirs, or artificial lakes, are created by dams. Reservoirs may also cover farmland and forests.

Wind For thousands of years, wind has powered sailing ships and boats. People have also long used wind power to turn windmills. Such mills were used to pump water out of wells or to grind grain into flour. In some places windmills are still used for these purposes.

Today, wind has a new use. It can create electricity by turning a system of fan blades called a turbine. "Wind farms" with hundreds of wind turbines have been built in some windy places.

Geothermal Energy The heat of Earth's interior—**geothermal energy**—can also be used to generate electricity. This internal heat escapes through hot springs and steam vents on Earth's surface. This geothermal energy can be captured to power electric generators.

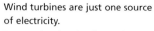

Wind turbines are just one source of electricity.

Interpreting the Visual Record How does this photo show human adaptation to the environment?

CLOSE

Have students imagine that they live in a community that is considering building a nuclear power plant. Ask each student to jot down two points—either for or against nuclear energy—that he or she would include in a speech at a public information meeting. Call on volunteers to make their points for both sides of the issue.

REVIEW AND ASSESS

Have students complete the Section Review. You might also have them prepare an outline of the section. Then have students complete Daily Quiz 4.4.

RETEACH

Have students complete Main Idea Activity 4.4. Then have them create crossword puzzles using the section's key terms and concepts. Ask students to exchange and solve each others' puzzles.
ENGLISH LANGUAGE LEARNERS

EXTEND

Have interested students conduct research on the Alaska pipeline—the controversies surrounding its construction, how those controversies were addressed, and the impact the pipeline has had on Alaska's economy.
BLOCK SCHEDULING

The Sun The Sun's heat and light are known as **solar energy**. This energy can be used to heat water or homes. Special solar panels also absorb solar energy to make electricity. In the past, converting solar energy into electricity was expensive. New research, however, may make solar energy cheaper in the future.

✓ **READING CHECK:** *Physical Systems* What are four renewable energy sources? water, wind, geothermal energy, the Sun

▲ Experimental cars like this one run on solar energy.

Nuclear Energy

In the late 1930s scientists discovered nuclear energy. They learned that energy could be released by changes in the nucleus, or core, of atoms. Nuclear energy was first used to make powerful bombs. Scientists found out that nuclear energy can also be used to produce electricity. Lithuania, France, Belgium, Ukraine, and Sweden rely on nuclear energy for much of their power. This reduces their dependence on imported oil.

There are serious concerns about the use of nuclear power. Several nuclear power plants have had accidents. In 1986 an accident at a nuclear reactor in Chernobyl in Ukraine killed dozens of people. It also caused cancer in thousands more. Homes and farms in the area of the reactor had to be abandoned.

Nuclear energy produces waste that remains dangerous for thousands of years. People do not agree on how nuclear waste can be stored or transported safely. Because of this some countries are reducing their dependence on nuclear energy. Denmark and New Zealand avoid it altogether. The United States does not intend to expand its use of nuclear energy. As a result, scientists continue to search for other renewable energy sources.

✓ **READING CHECK:** *Environment and Society* What are the concerns associated with nuclear power? accidents and nuclear waste

Section Review 4

Define and explain: fossil fuels, petroleum, refineries, hydroelectric power, geothermal energy, solar energy

Reading for the Main Idea

1. *Physical Systems* What are the three main fossil fuels, and how were they formed?

2. *Physical Systems* What are the four most common renewable energy sources? Describe how each of these harnesses energy.

go.hrw.com
Homework Practice Online
Keyword: SJ3 HP4

Critical Thinking

3. **Finding the Main Idea** How do dams affect the environment?

4. **Analyzing Information** What are some of the problems associated with nuclear energy?

Organizing What You Know

5. **Contrasting** Copy the following graphic organizer. Use it to contrast the characteristics of renewable and nonrenewable resources.

Renewable resources	↔	Nonrenewable resources

RETEACH

Organize the class into nine groups—for water, air, soil, forests, metallic minerals, nonmetallic minerals, nonrenewable energy sources, renewable energy sources, and nuclear energy. Have each group create an idea web of the material related to its topic. Compare, discuss, and display the webs around the classroom. **ENGLISH LANGUAGE LEARNERS, COOPERATIVE LEARNING**

PORTFOLIO EXTENSIONS

1. Have students collect useful minerals, label the samples, and construct flowcharts that trace the raw minerals through processing to their final uses. They may want to construct models of expensive or unavailable minerals. Photograph the project results for placement in student portfolios.

2. Ask students to name natural phenomena that link countries without respect to boundaries. *(winds, rain, waterways, wildlife migration)* Then have them explain ways that pollution created in one state or country can contaminate the air, water, or land of another. Have students create maps showing the spread of contaminants. Then have them explain why pollution control might be a source of conflict among countries. *(economic reasons)*

Review
ANSWERS

Understanding Environment and Society

Answers will vary, but the information included should be consistent with text material. Verify that students have compared environmental conditions and included a description of preventive measures. Use Rubric 29, Presentations, to evaluate student work.

Thinking Critically

1. Some people want to clear the land for farms, livestock, or industry. Others believe the rain forests should be protected and preserved. The rain forests are being deforested.

2. water—chemical fertilizers, pesticides, and industrial wastes in water supply; air—burning fuel for heating, transportation, and to power factories; pollution threatens health of people and animals, damages crops and forests, causes thinning in ozone layer, may cause global warming

3. provide materials we want and need

4. metallic—shiny, can conduct heat and electricity, most are solid at room temperature; nonmetallic—varied in

Reviewing What You Know

Building Vocabulary

On a separate sheet of paper, write sentences to define each of the following words.

1. renewable resources
2. desertification
3. deforestation
4. reforestation
5. aquifers
6. acid rain
7. nonrenewable resources
8. fossil fuels
9. hydroelectric power
10. solar energy

Reviewing the Main Ideas

1. *(Environment and Society)* Why is fertile soil important, and how is it created?

2. *(Environment and Society)* What actions have people taken to increase their supply of water?

3. *(Environment and Society)* What are minerals, and why are they important to us?

4. *(Environment and Society)* For what are fossil fuels used?

5. *(Environment and Society)* What are the four most common renewable energy sources, and how is each used?

Understanding Environment and Society

Chernobyl

On April 26, 1986, a nuclear reactor exploded in Chernobyl. Radiation caused serious problems in Ukraine and Belarus, as well as in Eastern and Western Europe. Create an outline for a presentation based on information about the following:

• Conditions of the environment then and now.

• Preventive measures taken to prevent future accidents.

Thinking Critically

1. **Drawing Inferences and Conclusions** Why does the issue of use of rain forests cause such disagreement? What is happening to Earth's rain forests?

2. **Summarizing** What are the major causes of water and air pollution, and how does pollution affect life on this planet?

3. **Drawing Inferences and Conclusions** Why are minerals valued by society?

4. **Contrasting** How are metallic minerals different from nonmetallic minerals?

5. **Drawing Inferences and Conclusions** Why does the use of nuclear energy continue to be a debated topic?

FOOD FESTIVAL

In countries where farmland is scarce, farmers are increasingly using hydroponic agriculture. Plants grown hydroponically are raised in a water-based nutrient solution. Ask students to visit a local grocery and to locate items in the produce department that are grown hydroponically. Students may need to ask workers at the grocery for help. The class may also want to try to grow plants, such as tomatoes, hydroponically.

CHAPTER 4
REVIEW AND ASSESSMENT RESOURCES

Reproducible
◆ Readings in World Geography, History, and Culture 7, 8
◆ Critical Thinking Activity 4
◆ Vocabulary Activity 4

Technology
◆ Chapter 4 Test Generator (on the One-Stop Planner)

◆ Audio CD Program, Chapter 4
◆ HRW Go site

Reinforcement, Review, and Assessment
◆ Chapter 4 Review, pp. 72–73

◆ Chapter 4 Tutorial for Students, Parents, Mentors, and Peers
◆ Chapter 4 Test
◆ Chapter 4 Test for English Language Learners and Special-Needs Students

Building Social Studies Skills

Map ACTIVITY

On a separate sheet of paper, match the letters on the map with their correct region. Then write the amount of oil known to be located in each region.

Africa	South and
Europe	Central
former Soviet	America
Union	Southwest Asia

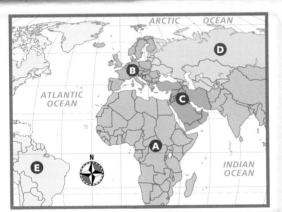

Mental Mapping Skills ACTIVITY

On a separate sheet of paper, draw a freehand map of the United States. Label areas that have copper. Compare this map with a physical map of the United States. What, if any, physical features are located in the same regions as copper deposits?

WRITING ACTIVITY

Write a short paper explaining which mineral is most important to you. Justify your selection with facts you have learned. Be sure to use standard grammar, spelling, sentence structure, and punctuation.

Alternative Assessment

Portfolio ACTIVITY

Learning About Your Local Geography

Environmental Issues Study your local environment. What issues of preservation and use are important to the people of your community? How does your government handle these issues?

internet connect

Internet Activity: go.hrw.com
KEYWORD: SJ3 GT4

Choose a topic to explore Earth's resources:
• Trek through different kinds of forests.
• Investigate global warming.
• Make a recycling plan.

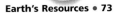

appearance, are poor conductors of heat and electricity

5. Nuclear energy offers great benefits but can create serious problems.

Map Activity
A. Africa
B. Europe
C. Southwest Asia
D. former Soviet Union
E. South and Central America

Mental Mapping Skills Activity
Maps will vary, but should show areas that have copper.

Writing Activity
The information included should be consistent with text material. Use Rubric 38, Writing to Classify, to evaluate student work.

Portfolio Activity
Answers will vary. Verify that students discuss both the nature of the local issues and how the local government has dealt with them. Use Rubric 30, Research, to evaluate student work.

internet connect

GO TO: go.hrw.com
KEYWORD: SJ3 Teacher
FOR: a guide to using the Internet in your classroom

Objectives	Pacing Guide	Reproducible Resources
SECTION 1		
Culture (pp. 75–80) **1.** Explain the term *culture*. **2.** Identify the influences that help cultures develop. **3.** Describe how agriculture has affected the development of culture.	**Regular** 2.5 days **Block Scheduling** 1 day *Block Scheduling Handbook, Chapter 5*	**RS** Guided Reading Strategy 5.1 **E** Creative Strategies for Teaching World Geography, Lessons 4, 5 **SM** Geography for Life Activity 5 **E** Lab Activity for Geography and Earth Science, Hands-On 4 **SM** Map Activity 5
SECTION 2		
Population, Economy, and Government (pp. 81–86) **1.** Explain why population density varies, and identify how the world's population has changed. **2.** Explain how geographers describe and measure economics. **3.** Identify the connections between economics and politics. **4.** Explain how governments differ.	**Regular** 2.5 days **Block Scheduling** .5 day *Block Scheduling Handbook, Chapter 5*	**RS** Guided Reading Strategy 5.2 **RS** Graphic Organizer 5
SECTION 3		
Population Growth Issues (pp. 87–89) **1.** Identify the problems associated with high and low population growth rates. **2.** Identify two different views of population growth and resources.	**Regular** .5 day **Block Scheduling** .5 day *Block Scheduling Handbook, Chapter 5*	**RS** Guided Reading Strategy 5.3 **E** Environmental and Global Issues Activity 8

Chapter Resource Key

RS Reading Support
IC Interdisciplinary Connections
E Enrichment
SM Skills Mastery
A Assessment
REV Review

ELL Reinforcement and English Language Learners
 Transparencies
 CD–ROM
 Music

 Video
 Internet
 Holt Presentation Maker Using Microsoft® Powerpoint®

  One-Stop Planner CD–ROM

See the *One-Stop Planner* for a complete list of additional resources for students and teachers.

One-Stop Planner CD–ROM

It's easy to plan lessons, select resources, and print out materials for your students when you use the *One-Stop Planner CD–ROM with Test Generator.*

Technology Resources

Review, Reinforcement, and Assessment Resources

One-Stop Planner CD–ROM, Lesson 5.1

Geography and Cultures Visual Resources with Teaching Activities 9 and 10

Homework Practice Online

HRW Go site

ELL	Main Idea Activity 5.1
REV	Section 1 Review, p. 80
A	Daily Quiz 5.1
ELL	English Audio Summary 5.1
ELL	Spanish Audio Summary 5.1

One-Stop Planner CD–ROM, Lesson 5.2

ARGWorld CD–ROM: Mapping Urban Growth

ARGWorld CD–ROM: Strategic Straits and Choke Points

ARGWorld CD–ROM: Technological Change and Factory Locations

ARGWorld CD–ROM: Representation in the United Nations

Homework Practice Online

HRW Go site

ELL	Main Idea Activity 5.2
REV	Section 2 Review, p. 86
A	Daily Quiz 5.2
ELL	English Audio Summary 5.2
ELL	Spanish Audio Summary 5.2

One-Stop Planner CD–ROM, Lesson 5.3

Yourtown CD–ROM

Homework Practice Online

HRW Go site

ELL	Main Idea Activity 5.3
REV	Section 3 Review, p. 89
A	Daily Quiz 5.3
ELL	English Audio Summary 5.3
ELL	Spanish Audio Summary 5.3

Meeting Individual Needs

Ability Levels

Level 1 Basic-level activities designed for all students encountering new material

Level 2 Intermediate-level activities designed for average students

Level 3 Challenging activities designed for honors and gifted-and-talented students

English Language Learners Activities that address the needs of students with Limited English Proficiency

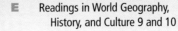

Chapter Review and Assessment

E	Readings in World Geography, History, and Culture 9 and 10		Chapter 5 Test Generator (on the One-Stop Planner)
SM	Critical Thinking Activity 5		Audio CD Program, Chapter 5
REV	Chapter 5 Review, pp. 90–91	A	Chapter 5 Test for English Language Learners and Special-Needs Students
REV	Chapter 5 Tutorial for Students, Parents, Mentors, and Peers		
ELL	Vocabulary Activity 5	A	Unit 1 Test for English Language Learners and Special-Needs Students
A	Chapter 5 Test		
A	Unit 1 Test		

LAUNCH INTO LEARNING

Display a news-story photo that shows elements of physical and human geography. *(Examples: natural disaster, new medicine discovered in rain forest)* Ask students to list the physical and human elements shown or implied in the photo. *(For example, for a hurricane: physical—rain, mud slides, beach erosion, high tides; human—people forced from homes, businesses hurt)* Point out to students that their interpretation of the photo would be incomplete if they left out either the physical or human issues. Similarly, the study of geography must include the human element. Tell students they will learn more about the human element of geography in this chapter.

Section 1

Objectives

1. Define culture and culture region.
2. Identify the importance of cultural symbols.
3. Trace how cultures develop.
4. Explain how agriculture affected culture.

LINKS TO OUR LIVES

You may want to emphasize the importance of learning the basics of human geography by sharing these points with your students:

▶ In order to understand individual cultures, we need to understand basic concepts of culture.

▶ Cultural differences affect world events and our daily lives. We should understand why cultures are different.

▶ We will participate more effectively in our own culture if we know how it developed.

▶ Population issues affect food production, international conflicts, and environmental questions—all of which can affect Americans directly.

▶ We must interpret population issues to make wise decisions about how we use resources.

CHAPTER 5

The World's People

The Colosseum, Rome, Italy

1998 Olympic opening ceremony, Nagano, Japan

Easter Island, Chile

Copy the following instructions onto the chalkboard: *What are some organized events that occur regularly in our community? Work with a partner to list as many as you can.* Have students write down their answers. Discuss responses. *(Possible answers: parades, festivals, garage sales, sports tournaments, concerts)* Point out to students that the events they listed are part of the community's culture. You may want to ask if students know of similar events elsewhere. If so, compare those events with the students' lists to note ways in which other communities have different cultures. Tell students that in Section 1 they will learn more about culture.

Building Vocabulary

Write the Define terms on the chalkboard. Underline the prefixes *multi-* and *sub-*. Explain that they mean, respectively, "many" and "below" or "almost." Ask students to find the terms' definitions in the text or glossary and to relate the prefix meanings to **multicultural** and **subsistence agriculture**. Call on students to read the remaining definitions. Ask each student to choose a Define term and write a sentence to define it. Call on students to read their sentences. Continue until all of the Define terms have been covered.

Section 1
Culture

Read to Discover

1. What is culture?
2. Why are cultural symbols important?
3. What influences how cultures develop?
4. How did agriculture affect the development of culture?

Define

culture	acculturation
culture region	symbol
culture traits	domestication
ethnic groups	subsistence agriculture
multicultural	commercial agriculture
race	civilization

WHY IT MATTERS

Throughout history, culture has both brought people together and created conflict among groups. Use **CNNfyi.com** or other **current events** sources to learn about cultural conflicts around the globe. Record your findings in your journal.

Flags at the South Pole

Section 1 RESOURCES

Reproducible
- ◆ Block Scheduling Handbook, Chapter 5
- ◆ Guided Reading Strategy 5.1
- ◆ Creative Strategies for Teaching World Geography, Lessons 4, 5
- ◆ Map Activity 5
- ◆ Geography for Life Activity 5
- ◆ Lab Activity for Geography and Earth Sciences, Hands-On 4

Technology
- ◆ One-Stop Planner CD–ROM, Lesson 5.1
- ◆ Homework Practice Online
- ◆ Geography and Cultures Visual Resources with Teaching Activities 9 and 10
- ◆ HRW Go site

Reinforcement, Review, and Assessment
- ◆ Section 1 Review, p. 80
- ◆ Daily Quiz 5.1
- ◆ Main Idea Activity 5.1
- ◆ English Audio Summary 5.1
- ◆ Spanish Audio Summary 5.1

Aspects of Culture

The people of the world's approximately 200 countries speak hundreds of different languages. They may dress in different ways and eat different foods. However, all societies share certain basic institutions, including a government, an educational system, an economic system, and religious institutions. These vary from society to society and are often based on that society's **culture**. Culture is a learned system of shared beliefs and ways of doing things that guides a person's daily behavior. Most people around the world have a national culture shared with people of their own country. They may also have religious practices, beliefs, and language in common with people from other countries. Sometimes a culture dominates a particular region. This is known as a **culture region**. In a culture region, people may share certain culture traits, or elements of culture, such as dress, food, or religious beliefs. West Africa is an example of a culture region. Culture can also be based on a person's job or age. People can belong to more than one culture and can choose which to emphasize.

Race and Ethnic Groups Cultural groups share beliefs and practices learned from parents, grandparents, and ancestors. These groups are sometimes called **ethnic groups**. An ethnic group's shared culture may include its religion, history, language, holiday traditions, and special foods.

When people from different cultures live in the same country, the country is described as **multicultural** or multiethnic. Many countries

Thousands of Czechs and Germans settled in Texas in the mid-1800s. Dancers from central Texas perform a traditional Czech dance.

▼

Teaching Objectives 1 and 3

ALL LEVELS: (Suggested time: 30 min.) Ask students to write on a slip of paper three of the cultures or culture regions to which they belong. Collect the suggestions. On the chalkboard, draw three columns. Label them *Culture, History,* and *Environment.* Choose a range of the student suggestions. As you record each group on the chart, lead a discussion on how it fulfills the definition of culture and how history and environment helped shape that culture. **ENGLISH LANGUAGE LEARNERS**

Teaching Objective 2

ALL LEVELS: (Suggested time: 45 min.) Point out that gestures are symbols and that they vary among cultures. Call on volunteers to demonstrate how they would show appreciation after a school play or choir concert. *(polite applause)* and after the home team's winning goal at a sports event. *(cheers, high fives).* Then have students conduct research on how people in other countries display approval in similar situations. Ask them to speculate how history and the environment have shaped the customs. Tell them to consider the effect of having a culture region divided by political boundaries. How might gestures or symbols in general change? You may need to extend this activity to another class period so that students can use other sources. **ENGLISH LANGUAGE LEARNERS**

DAILY LIFE

Peoples of many races and ethnic groups often enjoy the same entertainments. For example, a board game called *mancala* is popular in many parts of the world. *Mancala* is possibly the oldest board game in the world. Egyptians played this counting and strategy game before 1400 B.C. It is popular across Africa, among all ages and social classes.

Mancala is played on boards of different sizes and shapes or simply with shallow holes in the ground. Counters can be anything from seashells to seeds. Versions of *mancala* can be found thousands of miles from Africa, in Southeast Asia. The game's actual origin is not known, however.

Activity: Have students conduct research on the rules for playing *mancala.* Ask them to determine similarities and differences in rules from different countries. Then have them construct *mancala* boards from egg cartons. Hold a *mancala* tournament in your classroom.

⧉ internet connect

GO TO: go.hrw.com
KEYWORD: SJ3 CH5
FOR: Web sites about *mancala*

World Religions

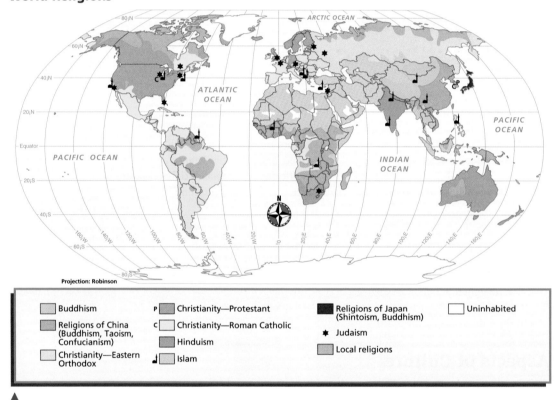

Buddhism	P ☐ Christianity—Protestant	■ Religions of Japan (Shintoism, Buddhism)	☐ Uninhabited
Religions of China (Buddhism, Taoism, Confucianism)	c ☐ Christianity—Roman Catholic	✶ Judaism	
Christianity—Eastern Orthodox	Hinduism	Local religions	
	Islam		

Projection: Robinson

▲

Religion is one aspect of culture.

A disc jockey sits at the control board of a Miami radio station that plays Cuban music.

▼

are multicultural. In some countries, such as Belgium, different ethnic groups have cooperated to form a united country. In other cases, such as in French-speaking Quebec, Canada, ethnic groups have frequently been in conflict. Sometimes, people from one ethnic group are spread over two or more countries. For example, Germans live in different European countries: Germany, Austria, and Czechoslovakia. The Kurds, who are a people with no country of their own, live mostly in Syria, Iran, Iraq, and Turkey.

Race is based on inherited physical or biological traits. It is sometimes confused with ethnic group. For example, the Hispanic ethnic group in the United States includes people who look quite different from each other. However, they share a common Spanish or Latin American heritage. As you know, people vary in physical appearance. Some of these differences have developed in response to climate factors like cold and sunlight. Because people have moved from region to region throughout history, these differences are not clear-cut. Each culture defines race in its own way, emphasizing particular biological and ethnic characteristics. An example can be seen in Rwanda, a country in East Africa. In this country, the Hutu and the Tutsi have carried on a bitter civil war. Although both are East African, each one

considers itself different from the other. Their definition of race involves height and facial features. Around the world, people tend to identify races based on obvious physical traits. However, these definitions of race are based primarily on attitudes, not actual biological differences.

Cultural Change Cultures change over time. Humans invent new ways of doing things and spread these new ways to others. The spread of one culture's ways or beliefs to another culture is called diffusion. Diffusion may occur when people move from one place to another. The English language was once confined to England and parts of Scotland. It is now one of the world's most widely spoken languages. English originally spread because people from England founded colonies in other regions. More recently, as communication among cultures has increased, English has spread through English-language films and television programs. English has also become an international language of science and technology.

People sometimes may borrow aspects of another culture as the result of long-term contact with another society. This process is called **acculturation**. For example, people in one culture may adopt the religion of another. As a result, they might change other cultural practices to conform to the new religion. For example, farmers who become Muslim may quit raising pigs because Islam forbids eating pork.

✓ **READING CHECK:** *Human Systems* What is the definition of culture? a learned system of shared beliefs and ways of doing things that guides a person's daily behavior

Cultural Differences

A **symbol** is a sign that stands for something else. A symbol can be a word, a shape, a color, or a flag. People learn symbols from their culture. The sets of sounds of a language are symbols. These symbols have meaning for the people who speak that language. The same sound may mean something different to people who speak another language. The word *bad* means "evil" in English, "cool" to teenagers, and "bath" in German.

If you traveled to another country, you might notice immediately that people behave differently. For instance, they may speak a different language or wear different clothes. They might celebrate different holidays or salute a different flag. Symbols reflect the artistic, literary, and religious expressions of a society or culture. They also reflect that society's belief systems. Language, clothing, holidays, and flags are all symbols. Symbols help people communicate with each other and create a sense of belonging to a group.

✓ **READING CHECK:** *Human Systems* How do symbols reflect differences among societies and cultures? They show the societies' particular cultural expressions or belief systems.

▲ Cultures mix in New York City's Chinatown.

Interpreting the Visual Record Why might immigrants to a new country settle in the same neighborhood?

internet connect

GO TO: go.hrw.com
KEYWORD: SJ3 CH5
FOR: Web sites about the world's people

Fans cheer for the U.S. Olympic soccer team.

Interpreting the Visual Record Why do you think symbols such as flags create strong emotions?

▼

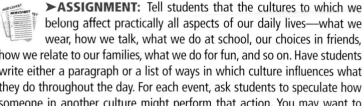

➤**ASSIGNMENT:** Tell students that the cultures to which we belong affect practically all aspects of our daily lives—what we wear, how we talk, what we do at school, our choices in friends, how we relate to our families, what we do for fun, and so on. Have students write either a paragraph or a list of ways in which culture influences what they do throughout the day. For each event, ask students to speculate how someone in another culture might perform that action. You may want to have students discuss what factors would affect that choice. For example, we may eat store-bought cereal for breakfast. In another culture, a Chinese teen might eat boiled rice instead because his or her family members raise rice on their farm.

Lois Jordan, of Nashville, Tennessee, suggests this research project to help students explore how minority groups can affect a country's development and culture. Organize the class into teams. Assign one country per team. Have each team use primary and secondary sources to research the country's minority groups and the geographical or historical reasons why the groups are there. Students should try to answer questions such as these: Are members of the minority groups spread throughout the country or concentrated in one area? In what ways are the minority cultures different from the majority culture? How do the majority and minority cultures relate to each other politically and socially? Have the students share their findings in a panel discussion.

Linking Past to Present
Nile River

Cultures Without the Nile River, the civilization whose people built the pyramids and the Sphinx could never have existed. The thousands of workers who labored on these and other monumental projects could be fed only because the river made intensive agriculture possible. The Nile Valley and Delta have been densely populated for thousands of years. Hunter-gatherers may have started to move into the Nile Valley by 12,000 B.C. By about 3,000 B.C. a great civilization flourished.

Alexandria, Egypt, was built next to the Nile Delta. During the 200s B.C. Alexandria may have been the largest metropolis in the world. Cleopatra's palace was there. During the 1980s, archaeologists began excavating the ruins of the city, which is now under water. They mapped the Royal Quarter, where Cleopatra had lived. Today, Alexandria is Egypt's second-largest city, with more than 4 million inhabitants.

Visual Record Answers

possible answers: embroidery, fancy headdress

possible answers: streets narrower and appear to be laid out in a random pattern, houses closer together and with flat roofs

▲ A couple prepares for a wedding ceremony in Kazakhstan.

Interpreting the Visual Record What aspects of these people's clothing indicate that they are dressed for a special event?

The layout of Marrakech, Morocco, is typical of many North African cities.

Interpreting the Visual Record How are the streets and houses of Marrakech different from those in your community?

Development of a Culture

All people have the same basic needs for food, water, clothing, and shelter. People everywhere live in families and mark important family changes together. They usually have rituals or traditions that go with the birth of a baby, the wedding of a couple, or the death of a grandparent. All human societies need to deal with natural disasters. They must also deal with people who break the rules of behavior. However, people in different places meet these needs in unique ways. They eat different foods, build different kinds of houses, and form families in different ways. They have different rules of behavior. Two important factors that influence the way people meet basic needs are their history and environment.

History Culture is shaped by history. A region's people may have been conquered by the same outsiders. They may have adopted the same religion. They may have come from the same area and may share a common language. However, historical events may have affected some parts of a region but not others. For example, in North America French colonists brought their culture to Louisiana and Canada. However, they did not have a major influence on the Middle Atlantic region of the United States.

Cultures also shape history by influencing the way people respond to the same historical forces. Nigeria, India, and Australia were all colonized by the British. Today each nation still uses elements of the British legal system, but with important differences.

Environment The environment of a region can influence the development of culture. For example, in Egypt the Nile River is central to people's lives. The ancient Egyptians saw the fertile soils brought by the flooding of the Nile as the work of the gods. Beliefs in mountain spirits were important in many mountainous regions of the world. These areas include Tibet, Japan, and the Andes of South America.

Call on students to suggest foods that are typical of a local culture. *(Possible answers: pierogi, quesadillas, dim sum, gumbo, grits)* Ask them to speculate how the popularity of that dish spread to the region. *(Possible answers: television ads, families moving to new regions, national restaurant franchises)* Point out how food is related to the main topics of the section.

Have students complete the Section Review. Then have them work in pairs to draw flow charts or other graphic organizers of the section material. Then have students complete Daily Quiz 5.1. **COOPERATIVE LEARNING**

Culture also determines how people use and shape their land-scape. For example, city plans are cultural. Cities in Spain and its former colonies are organized around a central plaza, or square, with a church and a courthouse. On the other hand, Chinese cities are oriented to the four compass points. American cities often follow a rectangular grid plan. Many French city streets radiate out from a central core.

✓ **READING CHECK:** *Human Systems* What are some ways in which culture traits spread? *through historical events such as conquest by outsiders or colonization*

Development of Agriculture

For most of human history people ate only wild plants and animals. When the food ran out in one place, they migrated, or moved to another place. Very few people still live this way today. Thousands of years ago, humans began to help their favorite wild plant foods to grow. They probably cleared the land around their campsites and dumped seeds or fruits in piles of refuse. Plants took root and grew. People may also have dug water holes to encourage wild cattle to come and drink. People began cultivating the largest plants and breeding the tamest animals. Gradually, the wild plants and animals changed. They became dependent on people. This process is called **domestication**. A domesticated species has changed its form and behavior so much that it depends on people to survive. Domestic sheep can no longer leap from rock to rock like their wild ancestors. However, the wool of domestic sheep is more useful to humans. It can be combed and twisted into yarn.

Domestication happened in many parts of the world. In Peru llamas and potatoes were domesticated. People in ancient Mexico and Central America domesticated corn, beans, squash, tomatoes, and hot peppers. None of these foods was grown in Europe, Asia, or Africa before the time of Christopher Columbus's voyages to the Americas. Meanwhile, Africans had domesticated sorghum and a kind of rice. Cattle, sheep, and goats were probably first raised in Southwest Asia. Wheat and rye were first domesticated in Central Asia. The horse was also domesticated there. These domesticated plants and animals were unknown in the Americas before the time of Columbus.

▲
This ancient Egyptian wall painting shows domesticated cattle.

Interpreting the Visual Record Can you name other kinds of domesticated animals?

◄ **Visual Record Answer**

possible answers: cats, chickens, dogs, ducks, geese, goats, horses, pigs, sheep, turkeys

Have students complete Main Idea Activity 5.1. Then have them work in groups to invent a previously unknown culture. Have them write sentences describing the features and development of their culture by using the Define terms. Discuss the invented cultures to check on students' understanding of key concepts. **ENGLISH LANGUAGE LEARNERS, COOPERATIVE LEARNING**

Have interested students use primary and secondary sources to research how archaeology has shed light on when and where various plants and animals were domesticated for use as food. Then challenge them to choose a certain time period and region and write recipes appropriate to the available foods. Students may want to prepare an "ancient" meal for the class. Substitutions will be necessary. **BLOCK SCHEDULING**

Section Review 1

Answers

Define For definitions, see: culture, p. 75; culture region, p. 75; culture trait, p. 75; ethnic groups, p. 75; multicultural, p. 75; race, p. 76; acculturation, p. 77; symbol, p. 77; domestication, p. 79; subsistence agriculture, p. 80; commercial agriculture, p. 80; civilization, p. 80

Reading for the Main Idea

1. People may belong to cultures based on where they live, their job, religious practices, beliefs, or age. (NGS 10)

2. Government, education, an economic system, religious institutions are basic to all. (NGS 10)

Critical Thinking

3. history—conquered by the same outsiders, have same religion or language; environment—affects religion, land use and planning; French colonists brought their culture to Louisiana; Nile River's influence on Egyptians

4. People built permanent settlements; surplus of food developed; population grew; civilizations formed.

Organizing What You Know

5. possible answers: religion, age, job, race, where we live, language and other cultural symbols

Thousands of years ago, domesticated dogs came with humans across the Bering Strait into North America. A breed called the Carolina dog may be descended almost unchanged from those dogs. The reddish yellow, short-haired breed also appears to be closely related to Australian dingoes.

Agriculture and Environment Agriculture changed the landscape. To make room for growing food, people cut down forests. They also built fences, dug irrigation canals, and terraced hillsides. Governments were created to direct the labor needed for these large projects. Governments also defended against outsiders and helped people resolve problems. People could now grow enough food for a whole year. Therefore, they stopped migrating and built permanent settlements.

Types of Agriculture Some farmers grow just enough food to provide for themselves and their own families. This type of farming is called **subsistence agriculture**. In the wealthier countries of the world, a small number of farmers can produce food for everyone. Each farm is large and may grow only one product. This type of farming is called **commercial agriculture**. In this system companies rather than individuals or families may own the farms.

Agriculture and Civilization Agriculture enabled farmers to produce a surplus of food—more than they could eat themselves. A few people could make things like pottery jars instead of farming. They traded or sold their products for food. With more food a family could feed more children. As a result, populations began to grow. More people became involved in trading and manufacturing. Traders and craftspeople began to live in central market towns. Some towns grew into cities, where many people lived and carried out even more specialized tasks. For example, cities often supported priests and religious officials. They were responsible for organizing and carrying out religious ceremonies. When a culture becomes highly complex, we sometimes call it a **civilization**.

✔ **READING CHECK:** *Environment and Society* In what ways did agriculture affect culture? Permanent settlements developed; a surplus of food developed; the population grew; civilizations developed.

Homework Practice Online
Keyword: SJ3 HP5

Section Review 1

Define and explain: culture, culture region, culture trait, ethnic groups, multicultural, race, acculturation, symbol, domestication, subsistence agriculture, commercial agriculture, civilization

Reading for the Main Idea

1. *Human Systems* How can an individual belong to more than one cultural group?

2. *Human Systems* What institutions are basic to all societies?

Critical Thinking

3. **Drawing Inferences and Conclusions** In what ways do history and environment influence or shape a culture? What examples can you find in the text that explain this relationship?

4. **Analyzing Information** What is the relationship between the development of agriculture and culture?

Organizing What You Know

5. **Summarizing** Copy the following graphic organizer. Use it to describe culture by listing shared beliefs and practices.

Culture

Section 2

Objectives

1. Explain why population density varies, and describe how the world's population has changed.
2. Identify ways that geographers describe and measure economies.
3. Identify the different economic systems.
4. Explain how governments differ.

FOCUS

LET'S GET STARTED

Write these instructions on the chalkboard: *Everyone except those on the first row should move to the back third of the room. I will explain soon.* Then tell students the crowded area represents a densely populated country, such as India, and the front of the room a thinly populated one, such as Mongolia. Discuss with students the advantages and disadvantages they would experience as citizens of these countries, based on their population density. Tell students that in Section 2 they will learn more about population issues.

Building Vocabulary

Write the Define terms on the chalkboard. Call on volunteers to read the definitions aloud. Then label some of them according to these categories: Economic Activities (**primary, secondary, tertiary, quaternary industries**); Ways to Measure Economic Development (**gross national product, gross domestic product, developed countries,** and **developing countries**); Economic Systems (**free enterprise; market, command, traditional economies**).

Section 2 — Population, Economy, and Government

Read to Discover

1. Why does population density vary, and how has the world's population changed?
2. How do geographers describe and measure economies?
3. What are the different types of economic systems?
4. How do governments differ?

Define

primary industries
secondary industries
tertiary industries
quaternary industries
gross national product
gross domestic product
developed countries
developing countries
free enterprise
factors of production
entrepreneurs
market economy
command economy
traditional economy
democracy
unlimited government
limited government

WHY IT MATTERS

Human population has increased dramatically in the past 200 years. Use **CNNfyi.com** or other **current events** sources to find current projections for global or U.S. populations. Record your findings in your journal.

Newborn baby

Section 2 RESOURCES

Reproducible
- Guided Reading Strategy 5.2
- Graphic Organizer 5

Technology
- One-Stop Planner CD–ROM, Lesson 5.2
- *ARGWorld* CD–ROM: Mapping Urban Growth
- *ARGWorld* CD–ROM: Strategic Straits and Choke Points
- *ARGWorld* CD–ROM: Technological Change and Factory Locations
- *ARGWorld* CD–ROM: Representation in the United Nations
- Homework Practice Online
- HRW Go site

Reinforcement, Review, and Assessment
- Section 2 Review, p. 86
- Daily Quiz 5.2
- Main Idea Activity 5.2
- English Audio Summary 5.2
- Spanish Audio Summary 5.2

Calculating Population Density

The branch of geography that studies human populations is called demography. Geographers who study it are called demographers. They look at such things as population size, density, and age trends. Some countries are very crowded. Others are only thinly populated. Demographers measure population density by dividing a country's population by its area. The area is stated in either square miles or square kilometers. For example, the United States has 74 people for every square mile (29/sq km). Australia has just 6 people per square mile (2.3/sq km). Japan has 869 people per square mile (336/sq km), and Argentina has 35 people per square mile (14/sq km).

These densities include all of the land in a country. However, people may not be able to live on some land. Rugged mountains, deserts, frozen lands, and other similar places usually have very few people. Instead, people tend to live in areas where the land can be farmed. Major cities tend to be located in these same regions of dense population.

READING CHECK: *Human Systems* What is population density?
country's population divided by its area

Shoppers crowd a street in Tokyo, Japan.

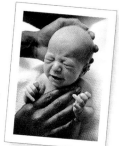

Differences in Population Density

Looking at this book's maps of world population densities allows us to make generalizations. Much of eastern and southern Asia is very

Teaching Objective 1

LEVEL 1: (Suggested time: 15 min.) Lead a class discussion on why people live where they do in your state. Encourage students to consider issues such as climate, availability of jobs, cultural and educational opportunities, and proximity to transportation facilities. Then call on volunteers to suggest how state population patterns can be compared to entire countries. *(People have the same basic needs and wants everywhere. More people live where their wants can be fulfilled most easily.)*

ENGLISH LANGUAGE LEARNERS

LEVELS 2 AND 3: (Suggested time: 30 min.) In advance, prepare names of countries on slips of paper. Include densely populated nations *(United Kingdom, Bangladesh, Japan)* and sparsely populated ones *(Mongolia, Australia, Canada)*. Have students pick a country and find population data on that country in the text or in other secondary sources. They should also look for information on the country's resources and climate. Then have students write a paragraph describing the influences on that country's population density. Ask students to use the Fast Facts features or a world almanac to calculate the population density for their chosen country.

National Geography Standard 9

Human Systems Our government maintains records of population densities in the United States. Every 10 years, the U.S. Census Bureau counts the people who live here and gathers information about them. Following are some interesting facts about the 2000 census form:

▶ For the first time, there was a question about grandparents as caregivers.

▶ Respondents could mark "one or more" for their racial group.

▶ The mail-in census form was printed in English, Spanish, Chinese, Tagalog, Vietnamese, and Korean. Language Assistance Guides were available in 44 other languages.

▶ Census workers counted homeless people at campgrounds, shelters, soup kitchens, and other places.

Discussion: Lead a discussion on what these facts about the census form can tell us about changes in American society.

Chart Answer ▶

1800

densely populated. There are dense populations in Western Europe and in eastern areas of North America. There are also places with very low population densities. Canada, Australia, and Siberia have large areas where few people live. The same is true for the Sahara Desert. Parts of Asia, South America, and Africa also have low population densities.

Heavily populated areas attract large numbers of people for different reasons. Some places have been densely populated for thousands of years. Examples include the Nile River valley in Egypt and the Huang River valley in China. These places have fertile soil, a steady source of water, and a good growing climate. These factors allow people to farm successfully. People are also drawn to cities. The movement of people from farms to cities is called urbanization. In many countries people move to cities when they cannot find work in rural areas. In recent years this movement has helped create huge cities. Mexico City, Mexico; São Paulo, Brazil; and Lagos, Nigeria, are among these giant cities. Some European cities experienced similar rapid growth when they industrialized in the 1800s. During that period people left farms to come to the cities to work in factories.

✓ **READING CHECK:** *Human Systems* Why does population density vary? because of what draws people to certain areas—fertile soil, water, good climate for agriculture, employment

World Population Growth

Source: U.S. Census

Interpreting the Chart Approximately when did world population growth begin to increase significantly?

▲

Population Growth

Researchers estimate that about 10,000 years ago the world's entire human population was less than 10 million. The annual number of births was roughly the same as the annual number of deaths.

After people made the shift from hunting and gathering to farming, more food was available. People began to live longer and have more children. The world's population grew. About 2,000 years ago, the world had some 200 million people. By A.D. 1650 the world's population had grown to about 500 million. By 1850 there were some 1 billion people. Better health care and food supplies helped more babies survive into adulthood and have children. By 1930 there were 2 billion people on Earth. Just 45 years later that number had doubled to 4 billion. In 1999 Earth's population reached 6 billion. By the year 2025 the world's population could grow to about 9 billion.

Births add to a country's population. Deaths subtract from it. The number of births per 1,000 people in a year is called the birthrate. Similarly, the death rate is the annual number of deaths per 1,000 people. The birthrate minus the death rate equals the rate of natural increase. This number is expressed as a percentage. A country's population changes when people enter or leave the country.

✓ **READING CHECK:** *Human Systems* What is the rate of natural increase? a country's birthrate minus its death rate

![globe icon] **Teaching Objective 2**
ALL LEVELS: (Suggested time: 20 min.) Have students create a time line of the world's population growth, using the years stated in the text as milestones. If some students complete their time lines early, you may want to have them compile their work to create one time line, enlarge it, and illustrate it on a butcher-paper mural.
ENGLISH LANGUAGE LEARNERS

![globe icon] **Teaching Objective 3**
LEVEL 1: (Suggested time: 30 min.) Organize the class into groups. Have each group examine one common object *(dish, book, sock, eyeglasses)* and discuss among themselves what role primary, secondary, tertiary, and quaternary industries have played in the production of the object. Then have each group create a flow chart to display their ideas.
ENGLISH LANGUAGE LEARNERS, COOPERATIVE LEARNING

LEVELS 2 AND 3: (Suggested time: 20 min.) Organize the class into two groups. Using the characteristics noted in the section, have one half write a what-I-did-today journal entry for a youngster in an imaginary developed country and the other half do the same for an imaginary developing country. Call on volunteers to read their entries. Lead a discussion on how the journal entries compare to the characteristics mentioned in the section.
COOPERATIVE LEARNING

Economic Activity

All of the activities that people do to earn a living are part of a system called the economy. This includes people going to work, making things, selling things, buying things, and trading services. Economics is the study of the production, distribution, and use of goods and services.

Types of Economic Activities Geographers divide economic activities into primary, secondary, tertiary, and quaternary industries. **Primary industries** are activities that directly involve natural resources or raw materials. These industries include farming, mining, and cutting trees.

The products of primary industries often have to go through several stages before people can use them. **Secondary industries** change the raw materials created by primary activities into finished products. For example, the sawmill that turns a tree into lumber is a secondary industry.

Tertiary industries handle goods that are ready to be sold to consumers. The stores that sell products are included in this group. The trucks and trains that move products to stores are part of this group. Banks, insurance companies, and government agencies are also considered tertiary industries.

The fourth part of the economy is known as **quaternary industries**. People in these industries have specialized skills. They work mostly with information instead of goods. Researchers, managers, and administrators fall into this category.

Economic Indicators A common means of measuring a country's economy is the **gross national product** (GNP). The GNP is the value of all goods and services that a country produces in one year. It includes goods and services made by factories owned by that country's citizens but located in foreign countries. Most geographers use **gross domestic product** (GDP) instead of GNP. GDP includes only those goods and services produced within a country. GDP divided by the country's population is called per capita GDP. This figure shows individual purchasing power and is useful for comparing levels of economic development.

ⓐ Primary industry: A dairy farmer feeds his cows.
ⓑ Secondary industry: Cheese is prepared in a factory.
ⓒ Tertiary industry: A grocer is selling cheese to a consumer.
ⓓ Quaternary industry: A technician inspects dairy products in a lab.

✔ **READING CHECK:** (*Human Systems*) What are primary, secondary, tertiary, and quaternary industries? primary—involve natural resources or raw materials; secondary—change raw materials into finished products; tertiary—handle goods for consumers; quaternary—skilled workforce working with information instead of goods

GLOBAL PERSPECTIVES

Beliefs and events can influence population issues. For example, in 1967 Iran became one of the first developing countries to institute a policy to slow population growth. The government declared family planning a human right and promoted it as a policy.

After Iran's 1979 Islamic Revolution, earlier efforts to limit family size were criticized as pro-West. The population growth rate climbed to more than 3 percent per year. However, a long war with Iraq badly damaged Iran. The leaders decided that it would be hard to rebuild their country while the population continued to expand so rapidly. As a result, the policy changed back to favoring small families.

Discussion: Lead a discussion on the relationship among religion, philosophical ideas, and Iranian culture with regards to population growth.

Teaching Objectives 3 and 4

ALL LEVELS: (Suggested time: 15 min.) Copy the following graphic organizer onto the chalkboard, omitting the italicized answers. Use it to help students understand the connection between economics and politics. Fill in the chart as a class. Point out that many countries do not follow the chart's model exactly but are a mixture of economic and political systems. Then have students discuss the concepts of limited and unlimited government. Have them work in groups to find examples of each type of government. **ENGLISH LANGUAGE LEARNERS, COOPERATIVE LEARNING**

LEVELS 2 AND 3: (Suggested time: 20 min.) Have students formulate and write hypotheses to explain why developed countries are usually based on free enterprise and democracy and why the economies of developing countries are often controlled by the government. Tell them to be sure to describe the benefits of the U.S. free enterprise system.

ECONOMICS AND POLITICS

	Developed Countries	Developing Countries
Economy	*free enterprise*	*government control, communism*
Political System	*democracy*	*communism*

Across the Curriculum

MATH

Per Capita GDP A country's gross domestic product (GDP) indicates the total size of its economy. However, it may not be a clear indication of an average citizen's wealth. For that purpose per capita GDP is probably more useful.

For example, China's GDP for 1998 was estimated at $4.4 trillion, making China a major economic power. However, China's per capita GDP was only $3,600, because the GDP figure was divided among China's immense population of more than 1 billion. In contrast, Chile's GDP the same year was about $184.6 billion. With a population of about 15 million, Chile's per capita GDP was $12,500. This comparison indicates that an average citizen in Chile is probably better off financially than an average citizen in China.

Activity: Use an almanac to provide students with GDP and population figures for several countries. Have each student choose a country and divide its gross GDP by its population to calculate its per capita GDP. Ask them to write a brief paragraph describing the information they found.

Chart Answer ▶

France and Poland

Economic Development

Geographers use various measures including GNP, GDP, per capita GDP, life expectancy, and literacy to divide the world into two groups. Industrialized countries like the United States, Canada, Japan, and most European countries are wealthier. They are called **developed countries**. These countries have strong secondary, tertiary, and quaternary industries. They have good health care systems. Developed countries have good systems of education and a high literacy rate. The literacy rate is the percentage of people who can read and write. Most people in developed countries live in cities and have access to telecommunications systems—systems that allow long-distance communication. Geographers study the number of telephones, televisions, or computers in a country. They sometimes use these figures to estimate the country's level of technology.

Developing countries make up the second group. They are in different stages of moving toward development. About two thirds of the world's people live in developing countries. These countries are poorer. Their citizens often work in farming or other primary industries and earn low wages. Cities are often crowded with poorly educated people hoping to find work. They usually have little access to health care or telecommunications. Some developing countries have made economic progress in recent decades. South Korea and Mexico are good examples. These countries are experiencing strong growth in manufacturing and trade. However, some of the world's poorest countries are developing slowly or not at all.

✓ **READING CHECK:** *Human Systems* How do geographers distinguish between developed and developing countries? by examining GNP, GDP, per capita GDP, life expectancy, and literacy

Comparing Developed and Developing Countries

COUNTRY	POPULATION	RATE OF NATURAL INCREASE	PER CAPITA GDP	LIFE EXPECTANCY	LITERACY RATE	TELEPHONE LINES
United States	281.4 million	0.9%	$ 36,200	77	97%	194 million
France	59.5 million	0.4%	$ 24,400	79	99%	35 million
South Korea	47.9 million	0.9%	$ 16,100	75	98%	24 million
Mexico	101.8 million	1.5%	$ 9,100	72	90%	9.6 million
Poland	38.6 million	−0.03%	$ 8,500	73	99%	8 million
Brazil	174.4 million	0.9%	$ 6,500	63	83%	17 million
Egypt	69.5 million	1.7%	$ 3,600	64	51%	3.9 million
Myanmar	42 million	0.6%	$ 1,500	55	83%	250,000
Mali	11 million	3.0%	$ 850	47	31%	23,000

Sources: Central Intelligence Agency, *The World Factbook 2001*

Interpreting the Visual Record **Which countries have the highest literacy rate?**

CLOSE

To illustrate the meaning of 6 billion, the approximate total world population, challenge students to calculate how long it would take for 6 billion seconds to go by *(approximately 190 years)*.

REVIEW AND ASSESS

Have students complete the Section Review. Then refer them to the list of Define terms. Call on students to create a sentence that relates the first term to the second, then the second to the third, and so on. Then have students complete Daily Quiz 5.2.

RETEACH

Have students complete Main Idea Activity 5.2. Then have students work in pairs to create web diagrams to illustrate the section's main points.
ENGLISH LANGUAGE LEARNERS, COOPERATIVE LEARNING

Economic Systems

Countries organize their economies in different ways. Most developed countries organize the production and distribution of goods and services in a system called **free enterprise**. The United States operates under a free enterprise system. There are many benefits to this system. Companies are free to make whatever goods they wish. Employees can seek the highest wages for their work. People, rather than the government, control the **factors of production**. Factors of production are the things that determine what goods are produced in an economy. They include the natural resources that are available for making goods for sale. They also include the capital, or money, needed to pay for production and the labor needed to manufacture goods. The work of **entrepreneurs** (ahn-truh-pruh-NUHRS) makes up a fourth factor of production. Entrepreneurs are people who start businesses in a free enterprise system. Business owners in a free enterprise system sell their goods in a **market economy**. In such an economy, business owners and customers make decisions about what to make, sell, and buy.

In contrast, the governments of some countries control the factors of production. The government decides what, and how much, will be produced. It also sets the prices of goods to be sold. This is called a **command economy**. Some countries with a command economy are governed by an economic and political system called communism. Under communism, the government owns almost all the factors of production. Very few countries today are communist. Cuba is an example of a communist nation.

Finally, there are some societies around the world that operate within a **traditional economy**. A traditional economy is one that is based on custom and tradition. Economic activities are based on laws, rituals, religious beliefs, or habits developed by the society's ancestors. The Mbuti people of the Democratic Republic of the Congo practice a traditional economy, for example.

✓ **READING CHECK:** *Human Systems* What are the three types of economies?
market, command, and traditional

◄

Large shopping malls, such as this one in New York City's Trump Tower, are common in countries that have market economies and a free enterprise system. Shoppers here can find a wide range of stores and goods concentrated in one area.

COOPERATIVE LEARNING

Organize the class into small groups. Tell students to imagine that an automobile company wants to build a factory in a developing country.

Have some members of each group play the role of automobile executives, and have others be officials of the developing country's government. Assign one member of each group to conduct research on (or invent) characteristics that describe the country (location, labor force, resources, environment, type of government). Ask students to address these questions and others while they negotiate the business deal: How much control will government officials have over the factory? Who will decide which designs to use? What will happen if the factory pollutes the air or water? What will happen if car buyers have complaints? Then set negotiations into motion. Challenge students to draw conclusions about the relationships between economics and politics.

EXTEND

No longer does one have to live in a large city to be employed by a large company. Have interested students conduct research on the effects of telecommuting—using an electronic linkup with a central office to work out of one's home—on population densities in the United States. Students may want to concentrate on one of your state's major cities and the towns nearby. Have students report their findings in a bar graph. You may want to have them interview a telecommuter for additional information.

BLOCK SCHEDULING

➤**ASSIGNMENT:** Have students discuss how different forms of government function in society. Then have them create editorial cartoons that express the different forms. Call on volunteers to present their cartoons to the class. Discuss and display all the cartoons.

Section Review 2

Answers

Define For definitions, see: population density, p. 81; primary industries, p. 83; secondary industries, p. 83; tertiary industries, p. 84; quaternary industries, p. 84; gross national product, p. 84; gross domestic product, p. 84; developed countries, p. 84; developing countries, p. 84; free enterprise, p. 85; factors of production, p. 85; market economy, p. 85; command economy, p. 85; traditional economy, p. 85; democracy, p. 86; communism, p. 86; unlimited governments, p. 86; limited government, p. 86

Reading for the Main Idea

1. good soil, water, jobs (NGS 15)

2. free enterprise, democracy, high per capita GDP, high life expectancy and literacy (NGS 14)

Critical Thinking

3. market—business owners and customers decide what to sell and buy; command—government makes decisions; traditional—based on custom (NGS 14)

4. in ancient Greece, results of American, French Revolutions; made governments accountable

Organizing What You Know

5. Answers will vary.

86

World Governments

Just as countries use different economic systems, they also have different ways of organizing their governments. Some countries are controlled by one ruler, such as a monarch or a dictator. For example, Saudi Arabia is ruled by a monarch. King Fahd bin Abd al-Aziz Al Saud is both the chief of state and the head of government.

In other countries, a relatively small group of people controls the government. Many countries—including the United States, New Zealand, and Germany—have democratic governments. In a **democracy**, voters elect leaders and rule by majority. Ideas about democratic government began in ancient Greece. The American and French Revolutions established the world's first modern democratic governments in the late 1700s. Today, most developed countries are democracies with free enterprise economies. However, some countries with democratic governments, such as Russia and India, struggle with economic issues.

How a government is organized determines whether it has limited or unlimited powers. **Unlimited governments**, such as the French monarchy before the French Revolution, have total control over their citizens. They also have no legal controls placed on their actions. In a **limited government**, government leaders are held accountable by citizens through their constitutions and the democratic process. These limitations help protect citizens from abuses of power. Today, many countries around the world, including the United States, have limited governments.

✓ READING CHECK: *Human Systems* What is the difference between limited and unlimited government? Unlimited governments have total control over their citizens, while limited governments hold leaders accountable to their citizens and follow a democratic process.

go.
hrw
.com
Homework Practice Online
Keyword: SJ3 HP5

Section Review 2

Define and explain: primary industries, secondary industries, tertiary industries, quaternary industries, gross national product, gross domestic product, developed countries, developing countries, free enterprise, factors of production, entrepreneurs, market economy, command economy, traditional economy, democracy, unlimited governments, limited government

Reading for the Main Idea

1. *Environment and Society* What geographic factors influence population density?

2. *Human Systems* What characteristics do developed countries share?

Critical Thinking

3. **Finding the Main Idea** What are the different economic systems? Describe each.

4. **Drawing Inferences and Conclusions** How did democracy develop, and why did it help create limited government?

Organizing What You Know

5. **Summarizing** Copy the following graphic organizer. Use it to study your local community and classify the businesses in your area.

Primary Industries	Secondary Industries	Tertiary Industries	Quaternary Industries
•	•	•	•
•	•	•	•

Section 3

Objectives

1. Identify problems associated with high and low population growth rates.
2. Compare two opposing views on population growth and resources.

LET'S GET STARTED

Copy the following instructions onto the chalkboard: *Look back through the Unit 1 chapters. Find a photo that you think shows a place with high population density. What made you choose that picture?* Allow students time to select a photo. Discuss student choices and what they see in the photos. Lead a discussion on how a high population density might affect daily life, as reflected in the photographs. Tell students that in Section 3 they will learn more about population growth and the issues associated with the topic.

Building Vocabulary

Write **carrying capacity** on the chalkboard. Have a student read the definition. Elicit a discussion comparing the carrying capacity of the entire Earth to that of a smaller environment. For example, have students consider the carrying capacity of your school. Ask students if they think the school could support a larger population or if it has reached its carrying capacity.

Section 3 — Population Growth Issues

Read to Discover

1. What problems are associated with high and low population growth rates?
2. What are two different views of population growth and resources?

Define

scarcity
carrying capacity

WHY IT MATTERS

Some people believe that the world's limited resources cannot support a rapidly increasing human population. Use **CNN fyi.com** or other **current events** sources to find out what issues are raised by world population growth. Record your findings in your journal.

Population sign for Anatone, Washington

Population Growth Rates

With the help of technology, humans can survive in a wide range of environments. People can build houses and wear clothing to survive in cold climates. Food can be grown in one place and shipped to another. For these reasons and others the human population has grown tremendously.

Population growth rates differ from place to place. Many developed countries have populations that are growing very slowly, holding steady, or even shrinking. However, the populations of most developing countries continue to grow rapidly.

Growth Rate Issues In general, a high population growth rate will hinder a country's economic development. Countries must provide jobs, education, and medical care for their citizens. A rapidly growing population can strain a country's resources and lead to **scarcity**—when demand is greater than supply. Many of the countries with the highest growth rates today are among the world's poorest.

However, a shortage of young people entering the workforce lowers a country's ability to produce goods. Young people are needed to replace older people who retire or die. Many countries with very low growth rates or shrinking overall populations must support a growing number of older people. These people may need more health care.

▲
An Inuit family in northern Canada uses a snowmobile to pull a sled.

🌐 **Teaching Objectives 1–2**

LEVEL 1: (Suggested time: 15 min.) Copy the following graphic organizer onto the chalkboard, omitting the italicized answers. Use it to help students understand population growth issues. Fill in the advantages and disadvantages boxes as a class. Then, using the chart, lead a discussion to summarize the two sides of the population and resources issue.

ENGLISH LANGUAGE LEARNERS

POPULATION GROWTH

	Advantages	Disadvantages
Low Population Growth Rate	*high standard of living, enough resources to go around*	*lowers country's ability to produce, large number of older people that may need financial support*
High Population Growth Rate	*increased political power, high productivity*	*hinders economic development, can strain resources*

Section Review 3

Answers

Define For definition, see: carrying capacity, p. 89

Reading for the Main Idea

1. Rapid population growth can hinder a country's economic development. (NGS 9)

2. Earth can easily support a much larger human population; Earth has reached its carrying capacity. (NGS 9)

Critical Thinking

3. possible answers: Humans can survive in a wide range of environments; people can build houses and wear clothing to survive in cold climates; food can be grown in one place and shipped to another.

4. through economic interdependence—trading—or through conflicts with other countries

Organizing What You Know

5. ability—new fertilizers, special seeds, new energy sources, better use of existing resources; inability—limited amount of land available for farming, shortage of freshwater, oil running out, pollution

Visual Record Answer ▶

small plots, animals used for power

Our Amazing Planet

About 1,000 years ago, people began carving out homes, churches, stables, and other "buildings" from the cone-shaped rock formations of Turkey's Cappadocia (ka-puh-DOH-shuh) region. Today, some of the larger spaces have been made into restaurants.

Many farmers in India still use traditional methods.

Interpreting the Visual Record What in this photo indicates a less developed agricultural society?

Uneven Resource Distribution Natural resources such as fresh water, minerals, and fertile land are not distributed evenly. A country's resources cannot always support its population. However, the country may be able to acquire needed resources by trading with other countries. In this system of economic interdependence, two countries can exchange resources or goods so that each gets what it needs. Japan, for example, has few energy resources but is a world leader in manufacturing. Japan sells manufactured goods to others, particularly the United States. Japan then uses the money to buy oil from Saudi Arabia.

The need for scarce resources usually leads countries to trade peacefully. However, it can also lead to military conflict. One country might try to take over a resource-rich area of a neighboring country. If we hope to avoid future wars over resources, the world's people must share resources more equally.

✔ **READING CHECK:** _Human Systems_ How does scarcity of resources affect international trade and economic interdependence? It causes countries to trade peacefully but can also lead to military conflict.

World Population and Resources

Some people think Earth can easily support a much larger human population. They base this view partly on history. For example, new fertilizers and special seeds mean more food can be grown today than ever before. They also think that scientists will probably discover new energy resources. The Sun, for example, is a vast energy source. Today we can only use a fraction of its power, however. Another way to support more people would be to make better use of existing resources. For example, we can recycle materials and reduce the amount of waste we generate.

Ask students how decisions they make now and in the future may affect population and resource distribution issues.

REVIEW AND ASSESS

Have students complete the Section Review. Then have students complete Daily Quiz 5.3.

RETEACH

Have students complete Main Idea Activity 5.3. Then use colored toothpicks to symbolize resources: water (blue), farmland (green), food (yellow), and mineral and energy resources (red). Each student is to represent a country.

Give many toothpicks of all colors to a few students, just a few toothpicks to more students, and just one or two to the remainder of the students. Explain the colors' meaning. Ask students to decide among themselves how to distribute their resources. After the exercise, elicit a discussion based on these questions: What happened when countries with few resources tried to increase their wealth? What might have happened if there were twice as many students in the class, but the same number of resources?
ENGLISH LANGUAGE LEARNERS

EXTEND

Have students conduct research on the role of population expansion in an era of invasion or conquest in world history, such as European colonialism. Ask them to create a cause-and-effect chart.
BLOCK SCHEDULING

A buildup of salt has ruined this field in southern Iraq. This land is now useless for growing crops, thus contributing to the lack of land suitable for farming.

Other people hold the opposite opinion. They argue that the world is already showing signs of reaching its **carrying capacity**. Carrying capacity is the maximum number of a species that can be supported by an area's scarce resources. The amount of land available for farming is shrinking. Many areas are experiencing a shortage of fresh water. Oil, a nonrenewable resource, will eventually run out. Pollution is damaging the atmosphere and the oceans. The rich nations of the world are not always willing to share with poorer nations. In the future, these people think food and water supplies will run short in many countries. People without enough to eat will become ill more easily, leading to widespread disease. Such problems may lead a country to invade its neighbors to capture resources.

These are challenging issues that reach into all areas of life. They will become even more important in the future. A better understanding of geography will help you understand and deal with these issues.

✓ **READING CHECK:** *Environment and Society* What are two different views on population growth and resources? Some people think Earth can support a much larger population; others think Earth is already reaching its carrying capacity.

go.
hrw
.com **Homework Practice Online**
Keyword: SJ3 HP5

Section Review 3

Define and explain: scarcity, carrying capacity

Reading for the Main Idea

1. (*Human Systems*) What problems are associated with rapid population growth?
2. (*Human Systems*) What are two viewpoints about future population growth?

Critical Thinking

3. **Analyzing Information** How has technology helped the worldwide human population grow?

4. **Finding the Main Idea** How do countries deal with the uneven distribution of resources?

Organizing What You Know

5. **Contrasting** Copy the following graphic organizer. Use it to discuss two arguments about population growth.

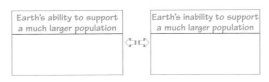

Earth's ability to support a much larger population		Earth's inability to support a much larger population

Building Vocabulary
For definitions, see: culture, p. 75; culture region, p. 75; culture traits, p. 75; ethnic groups, p. 75; multicultural, p. 75; acculturation, p. 77; symbol, p. 77; domestication, p. 79; subsistence agriculture, p. 80; civilization, p. 80; entrepreneurs, p. 85; limited government, p. 86; command economy, p. 85; market economy, p. 85; factors of production, p. 85; free enterprise, p. 85; carrying capacity, p. 89

Reviewing the Main Ideas

1. learned system of shared beliefs and ways of doing things that guide a person's daily behavior; to learn more about the world and its people (NGS 10)

2. subsistence—farmers growing just enough food for their own families; commercial—farmers growing food for consumers (NGS 14)

3. Some governments are controlled by one ruler, such as a monarch. In others, groups of people control the government. (NGS 9)

4. useful to indicate the country's overall level of technology (NGS 11)

5. A rapid population growth rate can strain a country's resources. (NGS 16)

89

The World's People • 89

ASSESS

Have students complete the Chapter 5 Test.

RETEACH

Organize the class into three groups—one for each section. Have students work in groups or alone to concentrate on topics within the sections. Tell them to create posters with captions to summarize their assigned section's main points. Display the posters. **COOPERATIVE LEARNING**

PORTFOLIO EXTENSIONS

1. Have students collect symbols and logos—on signs, clothing, and food and beverage labels, for example—that are part of American culture. They may need to photograph or draw some symbols. Ask students how they would interpret the images if they lived in another culture. Have them write brief explanations of the symbols' meaning. Include descriptions in portfolios.

2. Discuss what changes ancient peoples would have experienced as the population growth rate increased and changed the relative supply of the factors of production. Explain to students that writing did not develop until after the world's population had increased dramatically. Then have students create "cave paintings" on butcher paper to express those changes caused by the increased population growth rate.

Review
ANSWERS

Understanding Environment and Society
Answers will vary, but the information included should be consistent with text material. Presentations should address the points listed. Use Rubric 29, Presentations, to evaluate student work.

Thinking Critically
1. Possible answer: because differences are not clear-cut; people define races differently
2. doubled four times; by 2025; possible answers: development hindered, crowding, pollution, resource depletion
3. primary, secondary, tertiary, quaternary; farming, saw mill, bank, researcher
4. traditional, market, command; allows for economic development and individual freedom
5. Answers will vary according to students' opinions.
6. unlimited—totalitarian, undemocratic, no legal controls placed on their actions, French monarchy; limited—held accountable by their citizens, United States

Reviewing What You Know

Building Vocabulary

On a separate sheet of paper, write sentences to define each of the following words.

1. culture
2. ethnic groups
3. multicultural
4. acculturation
5. symbol
6. domestication
7. subsistence agriculture
8. civilization
9. limited government
10. entrepreneurs
11. command economy
12. market economy
13. factors of production
14. free enterprise
15. carrying capacity

Reviewing the Main Ideas

1. (*Human Systems*) What is culture, and why should people study it?
2. (*Environment and Society*) What is the difference between subsistence and commercial agriculture?
3. (*Human Systems*) What are some of the different ways countries organize governments?
4. (*Human Systems*) Why are telecommunications devices useful as economic indicators?
5. (*Environment and Society*) How do population growth rates affect resources?

Understanding Environment and Society

Domestication
Do you know how, when, and why people first domesticated dogs, cats, pigs, and hawks? What about oranges? Pick a domesticated plant or animal to research. As you prepare your presentation consider the following:
- Where the crop or animal was first domesticated.
- Differences between it and its wild ancestors.
- Humans spreading it to new areas.

Thinking Critically

1. **Drawing Inferences and Conclusions** Why are ethnic groups sometimes confused with races?
2. **Making Generalizations and Predictions** Over the past 2,000 years, how many times has world population doubled? By when is it projected to double again? What might be the effect of this increase?
3. **Finding the Main Idea** What are the four basic divisions of industry? Give examples.
4. **Summarizing** What are the three economic systems, and what are the benefits of the U.S. free-enterprise system?
5. **Making Generalizations and Predictions** Do you think Earth has a carrying capacity for its human population? Why or why not?
6. **Summarizing** Explain unlimited and limited government. Give examples of each.

FOOD FESTIVAL

Ask students to bring foods to class that reflect their families' culture. If most of the students are from similar backgrounds, have them concentrate on family variations of a common dish. For example, there are innumerable ways to make salsa, a spicy condiment for Mexican food. Jewish students of Eastern European heritage might compare how their families make kugel, a baked pudding of noodles or potatoes.

Reproducible
- Readings in World Geography, History, and Culture 9 and 10
- Critical Thinking Activity 5
- Vocabulary Activity 5

Technology
- Chapter 5 Test Generator (on the One-Stop Planner)
- HRW Go site

- Audio CD Program, Chapter 5

Reinforcement, Review, and Assessment
- Chapter 5 Review, pp. 90–91
- Chapter 5 Tutorial for Students, Parents, Mentors, and Peers

- Chapter 5 Test
- Chapter 5 Test for English Language Learners and Special-Needs Students
- Unit 1 Test
- Unit 1 Test for English Language Learners and Special-Needs Students

Building Social Studies Skills

Map ACTIVITY

On a separate sheet of paper, match the letters on the map with their correct labels.

Buddhism	**Christianity— Roman Catholic**
Christianity— Eastern Orthodox	
	Hinduism
Christianity— Protestant	**Islam**

Mental Mapping Skills ACTIVITY

Draw a map of the world and label Japan, Australia, the United Kingdom, and Argentina. Based on your knowledge of climates, population density, and resources, which of these countries probably depend on imported food? Which ones probably export food? Express this information on your map.

WRITING ACTIVITY

Study the economy of your local community. Has the local economy grown or declined since 1985? Why? Predict how your local area could change economically during the next 10 years. Be sure to use standard grammar, spelling, sentence structure, and punctuation.

Map Activity

A. Christianity—Eastern Orthodox
B. Hinduism
C. Islam
D. Christianity—Roman Catholic
E. Buddhism
F. Christianity—Protestant

Mental Mapping Skills Activity

Maps will vary, but listed places should be labeled in their approximate locations. Students should support their answers.

Writing Activity

Answers will vary, but the information included should be consistent with text material. Students' predictions should be supported by logical arguments. Use Rubric 37, Writing Assignments, to evaluate student work.

Portfolio Activity

Responses will vary, but students should support their arguments logically. Use Rubric 28, Posters, to evaluate student work.

Alternative Assessment

Portfolio ACTIVITY

Learning About Your Local Geography

Factors of Production Recall the discussion of the factors of production. Create a model showing how these factors influence the economy of your community.

internet connect

Internet Activity: go.hrw.com
KEYWORD: SJ3 GT5

Choose a topic to explore online:
- Visit famous buildings and monuments around the world.
- Compare facts about life in different countries.
- Examine world population growth.

internet connect

GO TO: go.hrw.com
KEYWORD: SJ3 Teacher
FOR: a guide to using the Internet in your classroom

Identifying Local Regions

Ask students to write down regions in which they live, go to school, shop, and pursue other daily activities. Call on volunteers for their responses. Record them on the chalkboard, placing the largest regions (continent, country) at the top and placing smaller regions (neighborhoods, boroughs, blocks) at the bottom. Note at what point descriptions of regions where students live differ from each other (probably at the town or neighborhood level). Draw a line separating these smaller divisions from the larger ones. Have students work in small groups to draw large maps showing these local regions and their relationships.

GEOGRAPHY SIDELIGHT

Physical, political, economic, and cultural changes can affect the way we define regions and draw their boundaries.

Physical changes are often slow to occur, but they are important. Beaches erode. Rivers run dry. Forests grow up on abandoned farmland. Political changes also bring regional redefinitions. For example, one large country may break up into several smaller ones, or small countries may unite to form a larger country.

Economic changes occur as different resources or economic activities decline or increase in importance. For example, a community once known for its manufacturing industries may now be part of an area that focuses on services. Finally, changing cultural characteristics can reshape regions as people of various language, religious, or ethnic groups move across regional boundaries.

Critical Thinking: What factors affect how we define regions?

Answer: physical, political, economic, and cultural changes

➤ This Focus On Regions feature addresses National Geography Standards 5 and 6.

What is a Region?

Think about where you live, where you go to school, and where you shop. These places are all part of your neighborhood. In geographic terms, your neighborhood is a region. A region is an area that has common features that make it different from surrounding areas.

What regions do you live in? You live on a continent, in a country, and in a state. These are all regions that can be mapped.

Regions can be divided into smaller regions called subregions. For example, Africa is a major world region. Africa's subregions include North Africa, West Africa, East Africa, central Africa, and southern Africa. Each subregion can be divided into even smaller subregions.

Regional Characteristics Regions can be based on physical, political, economic, or cultural characteristics. Physical regions are based on Earth's natural features, such as continents, landforms, and climates. Political regions are based on countries and their subregions, such as states, provinces, and cities. Economic regions are based on money-making activities such as agriculture or industries. Cultural regions are based on features such as language, religion, or ethnicity.

▲ East Africa is a subregion of Africa. It is an area of plateaus, rolling hills, and savanna grasslands.

Going Further: Thinking Critically

Have students compare their maps. Note areas where regions overlap and ask students how they might describe these areas *(transition zones)*. Also have them look at how maps of the same regions have different boundaries. Point out that these are perceived regions. The word *perceived* has to do with getting information from one's senses. Therefore, perceived regions differ according to how one "sees" the region. Ask students how they decided on the boundaries that they drew. Use their responses to illustrate the differences in perceived regions. Finally, review formal and functional regions and ask whether any regions shown on the maps fulfill the definitions.

Major World Regions

THE UNITED STATES AND CANADA

EUROPE

RUSSIA AND NORTHERN EURASIA

SOUTHWEST ASIA

SOUTH ASIA

EAST AND SOUTHEAST ASIA

AFRICA

MIDDLE AND SOUTH AMERICA

PACIFIC WORLD AND ANTARCTICA

PACIFIC WORLD AND ANTARCTICA

▲
The international border between Kenya and Tanzania is a clearly defined regional boundary.

Regional Boundaries All regions have boundaries, or borders. Boundaries are where the features of one region meet the features of a different region. Some boundaries, such as coastlines or country borders, can be shown as lines on a map. Other regional boundaries are less clear.

Transition zones are areas where the features of one region change gradually to the features of a different region. For example, when a city's suburbs expand into rural areas, a transition zone forms. In the transition zone, it may be hard to find the boundary between rural and urban areas.

Types of Regions There are three basic types of regions. The first is a formal region. Formal regions are based on one or more common features. For example, Japan is a formal region. Its people share a common government, language, and culture.

The second type of region is a functional region. Functional regions are based on movement and activities that connect different places. For example, Paris, France, is a functional region. It is based on the goods, services, and people that move throughout the city. A shopping center or an airport might also be a functional region.

The third type of region is a perceived region. Perceived regions are based on people's shared feelings and beliefs. For example, the neighborhood where you live may be a perceived region.

The three basic types of regions overlap to form complex world regions. In this textbook, the world is divided into nine major world regions (see map above). Each has general features that make it different from the other major world regions. These differences include physical, cultural, economic, historical, and political features.

Understanding What You Read

1. Regions can be based on what types of characteristics?

2. What are the three basic types of regions?

Understanding What You Read

Answers
1. Regions can be based on physical, political, economic, or cultural characteristics.

2. The three basic types of regions are: formal, functional, and perceived.

93

Going Further: Thinking Critically

Prepare two sets of words on slips of paper. One set should consist of action words, such as *go, fly, run, skip, dig, walk, swim, climb,* and *drive*. The other set should consist of nouns having to do with places, such as *tree, house, cliff, beach, road, river, mountain, church, rock, cave, highway,* and *ocean*. Students will draw at least one word of each type from a hat. Using their chosen words at least once, students should write travel or adventure stories set in familiar places. You may want to require that the stories be a certain length or that the students mention a certain number of places in their stories. Tell students that when they finish they should be able to draw maps of the settings through which the characters travel. When all the students have completed their stories, have them exchange papers to draw maps of each other's story settings.

PRACTICING

THE SKILL

1. The prime meridian extends through western Europe (England, western France, northeastern Spain) and western Africa (Algeria, Mali, Burkina Faso, Togo, Ghana).

2. Students' sketch maps should show the equator, Tropic of Cancer, Tropic of Capricorn, prime meridian, and continents in their approximate locations.

3. Answers will vary. Students might notice that the international date line does not cross any major landmasses, the Southern Hemisphere has much more water than land, South America extends farther south than Africa, or other similar facts.

➤ **This GeoSkills feature addresses National Geography Standards 1, 2, and 3.**

Building Skills for Life: Drawing Mental Maps

We create maps in our heads of all kinds of places—our homes, schools, communities, country, and the world. Some of these places we know well. Others we have only heard about. These images we carry in our heads are shaped by what we see and experience. They are also influenced by what we learn from news reports or other sources. Geographers call the maps that we carry around in our heads mental maps.

We use mental maps to organize spatial information about people and places. For example, our mental maps help us move from classroom to classroom at school or get to a friend's home. A mental map of the United States helps us list the states we would pass through driving from New York City to Miami.

We use our mental maps of places when we draw sketch maps. A sketch map showing the relationship between places and the relative size of places can be drawn using very simple shapes. For example, triangles and rectangles

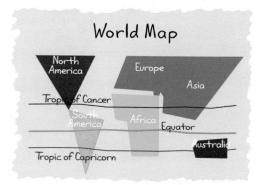

could be used to sketch a map of the world. This quickly drawn map would show the relative size and position of the continents.

Think about some simple ways we could make our map of the world more detailed. Adding the equator, Tropic of Cancer, and Tropic of Capricorn would be one way. Look at a map of the world in your textbook's Atlas. Note that the bulge in the continent of Africa is north of the equator. Also note that all of Asia is north of the equator. Next note that the Indian subcontinent extends south from the Tropic of Cancer. About half of Australia is located north of the Tropic of Capricorn. As your knowledge of the world increases, your mental map will become even more detailed.

THE SKILL

1. Look at the maps in your textbook's Atlas. Where does the prime meridian fall in relation to the continents?

2. On a separate sheet of paper, sketch a simple map of the world from memory. First draw the equator, Tropic of Cancer, Tropic of Capricorn, and prime meridian. Then sketch in the continents. You can use circles, rectangles, and triangles.

3. Draw a second map of the world from memory. This time, draw the international date line in the center of your map. Add the equator, Tropic of Cancer, and Tropic of Capricorn. Now sketch in the continents. What do you notice?

HANDS *on* GEOGRAPHY

Mental maps are personal. They change as we learn more about the world and the places in it. For example, they can include details about places that are of interest only to you.

What is your mental map of your neighborhood like? Sketch your mental map of your neighborhood. Include the features that you think are important and that help you find your way around. These guidelines will help you get started.

1. Decide what your map will show. Choose boundaries so that you do not sketch more than you need to.

2. Determine how much space you will need for your map. Things that are the same size in reality should be about the same size on your map.

3. Decide on and note the orientation of your map. Most maps use a directional indicator. On most maps, north is at the top.

4. Label reference points so that others who look at your map can quickly and easily figure out what they are looking at. For example, a major street or your school might be a reference point.

5. Decide how much detail your map will show. The larger the area you want to represent, the less detail you will need.

6. Use circles, rectangles, and triangles if you do not know the exact shape of an area.

7. As you think of them, fill in more details, such as names of places or major land features.

Lab Report

1. What are the most important features on your map? Why did you include them?

2. Compare your sketch map to a published map of the area. How does it differ?

3. At the bottom, list three ways that you could make your sketch map more complete.

GAINING A HISTORICAL PERSPECTIVE

Direct students' attention first to the large photograph and then to the smaller photographs on this page and the next. Ask students how they think all of the photographs are connected. *(The large photograph shows people digging at a site where there are ancient ruins and objects. The smaller photographs are some of the artifacts that have been found there.)* Have students read the captions and identify the objects. Point out that people must work very carefully at a site where there are building ruins and objects. Ask students why this is necessary *(in order not to destroy or damage anything; to preserve the historical record).* Also ask what kinds of tools students think people use at an archaeological site. *(Possible answers: hands, brushes, small picks, rakes, shovels.)*

Tell students that once an object is dug up, it is cataloged—numbered and listed in a book. Information about the type of object, where it was found, the date, and other appropriate information is recorded. Ask students why they think this might be important *(to keep track of objects; to let others who study the objects know about them).*

UNIT OBJECTIVES

1. Analyze life in prehistoric times and the development of early civilizations.
2. Examine changes in the political, social, and religious systems of the ancient world.
3. Identify the significance of the Middle Ages, Renaissance, Reformation, and Scientific and Industrial Revolutions.
4. Interpret the effects of European exploration, expansion, and colonization on other parts of the world.
5. Describe the political, social, religious, and military events that gave birth to the modern world.
6. List the political, social, religious, and military events that shaped the modern world we live in today.

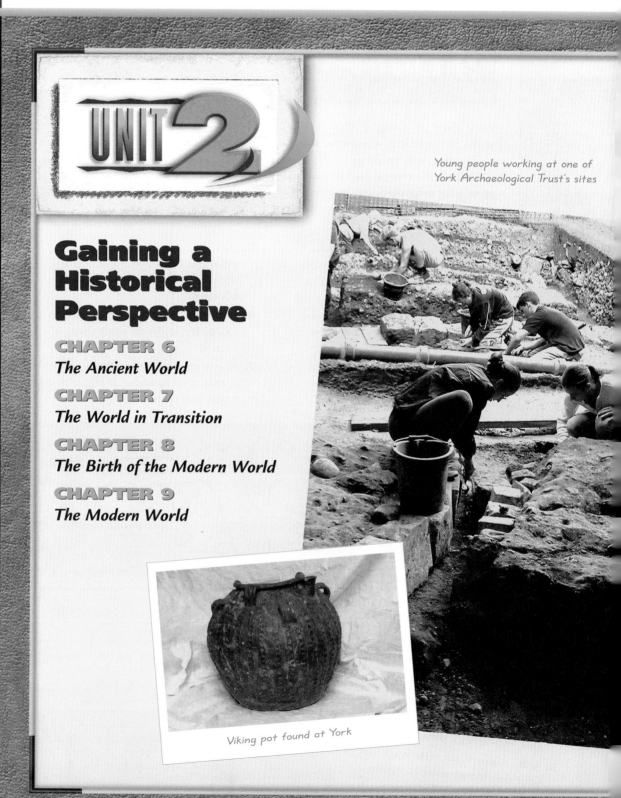

Gaining a Historical Perspective

Young people working at one of York Archaeological Trust's sites

Viking pot found at York

An Archaeologist at Work

Dr. Ailsa Mainman is an archaeologist. She studies the remains and ruins of past cultures. She lives in York, in the northeastern part of England. She works for the York Archaeological Trust. Here she describes her work.
WHAT DO YOU THINK? *What part of Dr. Mainman's work would you enjoy the most?*

I chose archaeology because I loved history but I wanted to be in touch with real things, not just books. It still gives me a real thrill to hold something that was found in one of our digs. The object might be hundreds, sometimes thousands of years old. I like to think about what the people and their lives were like.

We have schoolchildren who come in the summer to help out on the digs, including some from the United States. They wash, sort, and draw these finds (valuable discoveries). Then the children try to work out what the objects are. Even broken bits of pot or bone have a lot to tell us about how people used to live.

Everyone thinks Vikings were just fierce warriors and raiders. They were, of course, but they were so much more. We have recreated the Viking city of Jorvik, which thrived in York beginning in A.D. 866. You can travel through the streets and houses and see the artisans and craftsmen. The Vikings were skilled at many crafts. They worked with gold, silver, iron, antler, bone, glass, and many other materials.

The Vikings were also tremendous shipbuilders. They explored and then settled in Iceland, Greenland, and Canada. Their trade routes linked them with Eastern Europe and even China!

Leather boot found at York

Viking artifacts found at York

Understanding Primary Sources

1. What has Dr. Mainman and others learned about the Vikings from the finds at their archaeological digs?

2. How does Dr. Mainman describe the Vikings?

All images courtesy of the York Archaeological Trust, York, England

Comb found at York

Understanding Primary Sources
Answers
1. what the Vikings were like, how they lived, tools and other objects they made and used

2. fierce warriors and raiders, skilled craftworkers, shipbuilders, explorers, traders

In this unit, students will explore the story of human life from earliest times to the present in order to gain a historical perspective of world cultures. This story encompasses politics and government, science and technology, religion and ideas, the arts, and daily life. It describes events and patters that have influenced the course of world history from the Stone Age to the age of computers and space exploration, from the glories of ancient Rome and Greece to the brutality of the two world wars.

In prehistoric times, humans moved from hunting and gathering to using tools and practicing agriculture. Ancient civilizations developed complex societies and produced great and lasting works of art. Over centuries, curiosity and exploration led human beings to every corner of the world and new ways of thinking led to innovative governments with more individual freedoms. In the modern world—as in ancient times—people continue to wage wars, but they have tempered these violent periods with quests for peace. Students will end the unit by considering what contributions their own generation might make to the world's ongoing history.

Your Classroom Time Line

These are the major dates and time periods for this unit. You may want to have students watch for them as they progress through the unit.

c.* **3,700,000** B.C. Wandering hominids leave footprints in volcanic ash.

2,500,000 B.C. The first known stone tools are made.

c. **1,800,000** B.C. Hominids migrate from Africa to Asia.

c. **400,000** B.C.–**100,000** B.C. The first *Homo sapiens* appear.

c. **33,000** B.C. Cro-Magnon people create cave paintings.

c. **15,000** B.C. Humans now inhabit Africa, Europe, Asia, North America, and Australia.

c. **8000** B.C. Agricultural societies are developed in Mesopotamia.

c. **3200** B.C. Upper and Lower Egypt are united.

c. **2500** B.C. The Harappan civilization appears in the Indus River valley.

c. **2300** B.C. Indus River valley people trade with people of the Tigris and Euphrates River valleys.

c. **Late 1000s** B.C. The Zhou dynasty begins in China.

*c. stands for *circa* and means "about."

Chapter 6
3,700,000 B.C–A.D. 476
The Ancient World

Gold funeral mask of Pharaoh Tutankhamen

Greek vase showing potters at work

Ancient Chinese art

c. 400,000 B.C.– 100,000 B.C.
Global Events
The first *Homo sapiens* appear.

c. 3200 B.C.
Politics
Upper and Lower Egypt are united.

c. 800s B.C.–700s B.C.
Politics
Sparta and Athens develop into powerful city-states.

| 2,500,000 B.C. | 500,000 B.C. | 8000 B.C. | 4000 B.C. | 1 B.C. | A.D. 500 |

2,500,000 B.C.
Science and Technology
The first stone tools appear.

c. 8000 B.C.
Science and Technology
Agricultural societies develop in Mesopotamia.

c. Late 1000s B.C.
Politics
The Zhou dynasty begins in China.

c. 2500 B.C.
Global Events
The Harappan civilization appears in the Indus River valley.

A.D. 476
Global Events
The Western Roman Empire falls.

The Acropolis, Athens

98

Direct students' attention to the image of the Egyptian funeral mask on the opposite page and ask what they think the function of a funeral mask might have been *(Possible answer: showed how the dead person had looked in life)*. Tell students to examine the picture of the Acropolis. Call on volunteers to suggest how the structure is still standing after more than 2,000 years *(Possible answers: sound structure, good engineering)*. What does this building suggest about the ancient Greeks? *(Possible answer: advanced society that appreciated beauty and symmetry)* Ask students what kind of ship might have borne a carving like the one on the facing page.

(Possible answer: a war ship that sought to frighten enemies)

Direct students' attention to the picture of the crusaders on this page and have them find the first Crusade on the time line. Ask if they know who the crusaders were. Have students read the caption that accompanies the Gutenberg Bible and ask them to explain why they think the invention of movable type was an important historical development. Finally, ask students to compare the two portraits of people on this page and speculate as to why they were important. *(Catherine politically, Galileo as a scientist)*

Chapter 7
A.D. 432–1800
The World in Transition

Viking carving of a lion's head

Crusaders at the gates of Jerusalem

800
Politics
Charlemagne is crowned Emperor of the Romans by Pope Leo III.

800–900s
Politics
The Vikings invade Western Europe.

1347–1351
Global Events
The Black Death sweeps through Europe.

1492
Global Events
Christopher Columbus makes his first voyage to America.

1517
Daily Life
Martin Luther posts his 95 theses.

Catherine the Great, empress of Russia

| 1000 | 1200 | 1400 | 1600 | 1800 |

1096
Global Events
The first Crusade begins.

1271
Global Events
Marco Polo begins his trip to China.

c. 1450
Science and Technology
Johannes Gutenberg invents the movable type printing press.

1762
Politics
Catherine the Great becomes Empress of Russia.

1632
Science and Technology
Galileo proves that Earth revolves around the Sun.

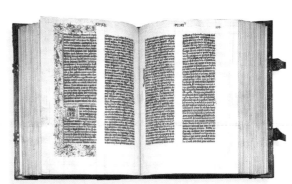

A Gutenberg Bible

Galileo Galilei

c. **800s** B.C.–**700s** B.C. Sparta and Athens develop into powerful city-states.

27 B.C. The Roman Republic becomes the Roman Empire.

A.D. **476** The Western Roman Empire falls.

800 Charlemagne is crowned Emperor of the Romans by Pope Leo III.

800–900s The Vikings invade Western Europe.

1096 The first Crusade begins.

1271 Marco Polo begins his trip to China.

c. **1345** Notre Dame Cathedral in Paris is completed.

1347–1351 The Black Death sweeps through Europe.

c. **1450** Johannes Gutenberg invents the movable type printing press.

1492 Christopher Columbus makes his first voyage to America.

1503–1506 Leonardo da Vinci paints the *Mona Lisa*.

c. **1514** Portuguese traders reach China.

1517 Martin Luther posts his 95 theses.

1558 Elizabeth I becomes queen of England.

1563 The first printing press is built in Europe.

99

USING THE ILLUSTRATIONS

Ask students to examine the picture of the new Globe Theatre. Explain that the theater was built to resemble the one where Shakespeare's plays were originally performed. Ask students to speculate why this theatre was built and what could be inferred about Shakespeare's importance in Britain *(built to honor Shakespeare; a very great writer)*.

Point out that the plate celebrating the coronation of William and Mary also honors figures in British history. Why might people have celebrated this corona-

tion? *(popular monarchs)* Tell students that the coronation was celebrated in England in 1691 after the Glorious Revolution.

Direct students' attention to the picture of women marching on Versailles. Explain that Versailles was the palace of the French king and queen before the French Revolution. Ask students what the women are carrying and why *(weapons, to attack the king and queen during the Revolution)*.

Your Classroom Time Line (continued)

1594–1595 William Shakespeare writes *Romeo and Juliet.*

1600 The British East India Company is created to control trade with Asia.

1607 Jamestown, Virginia, the first permanent English settlement in America, is founded.

1616 William Shakespeare dies.

1632 Galileo discovers that Earth revolves around the Sun.

1687 Isaac Newton publishes his most famous work, *Principia.*

1688 The Glorious Revolution occurs in England.

1690 John Locke publishes his ideas about government.

1708 British traders dominate trade between Europe and India.

1737 Samuel F.B. Morse invents the telegraph.

1747 The Black Death sweeps through Europe.

1762 Catherine the Great becomes Empress of Russia.

1763 The Seven Years' War ends.

1769 James Watt builds the first steam engine.

1776 The American colonies declare independence from Great Britain.

UNIT 2

Chapters 8 & 9
1550–Present

The Modern World

Plate celebrating the coronation of William III and Mary II

Women marching on Versailles during the French Revolution

1594-95
The Arts
William Shakespeare writes *Romeo and Juliet.*

1688
Politics
The Glorious Revolution occurs in England.

1763
Global Events
The Seven Years' War ends.

1789
Politics
The United States Constitution is ratified.

1789
Politics
The French Revolution begins.

1550 — **1650** — **1700** — **1750** — **1800**

1558
Politics
Elizabeth I becomes queen of England.

1687
Science and Technology
Isaac Newton publishes his most famous work, *Principia.*

1737
Science and Technology
Samuel F. B. Morse invents the telegraph.

1776
Politics
The American colonies declare independence from Great Britain.

1769
Science and Technology
James Watt builds the first steam engine.

Re-creation of Shakespeare's Globe Theatre

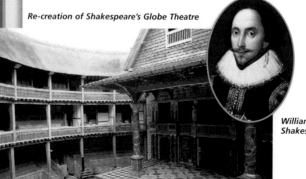

William Shakespeare

Early steam locomotive

Direct students' attention to the photograph of the Wright Brothers' plane. Have them contrast this plane with modern jets. Then encourage students to think about how the invention of the airplane changed the world.

Have students find the image on this page that shows the destruction caused by World War I. What was once in the place shown in the photograph? How can students tell? *(remains of buildings, paving stones on road)* Ask students to contemplate what it might feel like to be a person in a town destroyed by war.

Direct students' attention to the photo of Boris Yeltsin on this page and ask if they know he is. Ask students to describe Yeltsin's attitude in the picture *(Possible answers: happy, victorious)* Tell students that Yeltsin has been a major figure in recent Russian history and that they will learn about his accomplishments in this unit.

Finally, ask students to focus on the photograph of the September 11 interfaith memorial. Call on volunteers to speculate what the people in the photo were feeling and why have they lit candles.

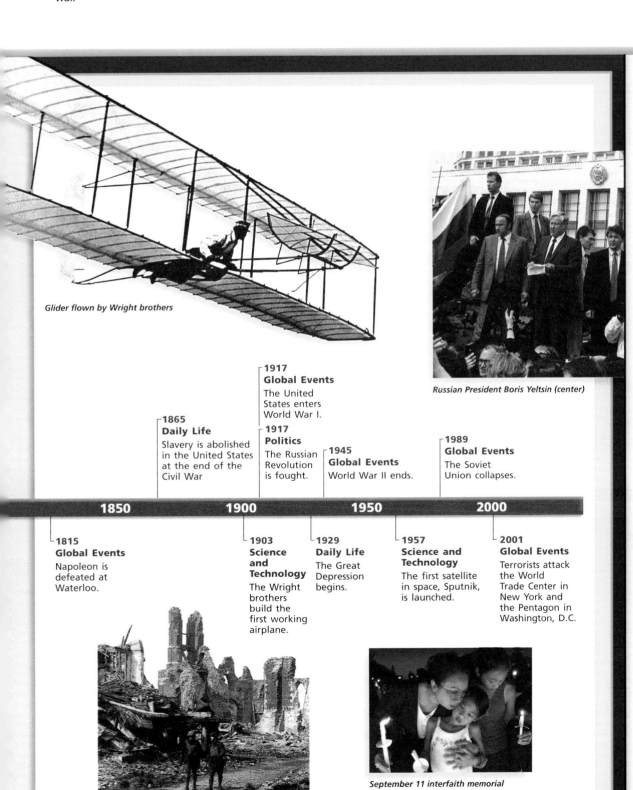

Glider flown by Wright brothers

Russian President Boris Yeltsin (center)

1865
Daily Life
Slavery is abolished in the United States at the end of the Civil War

1917
Global Events
The United States enters World War I.

1917
Politics
The Russian Revolution is fought.

1945
Global Events
World War II ends.

1989
Global Events
The Soviet Union collapses.

| 1850 | 1900 | 1950 | 2000 |

1815
Global Events
Napoleon is defeated at Waterloo.

1903
Science and Technology
The Wright brothers build the first working airplane.

1929
Daily Life
The Great Depression begins.

1957
Science and Technology
The first satellite in space, Sputnik, is launched.

2001
Global Events
Terrorists attack the World Trade Center in New York and the Pentagon in Washington, D.C.

World War I

September 11 interfaith memorial

Your Classroom Time Line (continued)

1780s The first factories open in England.

1789 The French Revolution begins.

1789 The United States Constitution is ratified.

1815 Napoleon is defeated at Waterloo.

1848 Revolutions break out across Europe.

1865 Slavery is abolished in the United States at the end of the Civil War.

1876 Alexander Graham Bell invents the telephone.

1903 The Wright brothers build the first working airplane.

1914 The Panama Canal opens.

1917 The Russian Revolution is fought.

1917 The United States enters World War I.

1918 World War I ends.

1929 The Great Depression begins.

1945 World War II ends.

1957 The first satellite in space, Sputnik, is launched.

1973 American forces pull out of the Vietnam War.

1989 The Soviet Union collapses.

2001 Terrorists attack the World Trade Center in New York and the Pentagon in Washington, D.C.

The Ancient World
Chapter Resource Manager

Objectives	Pacing Guide	Reproducible Resources
SECTION 1		
The Birth of Civilization (pp. 103–108)	**Regular** 1 day	**RS** Guided Reading Strategy 6.1
1. Describe the discoveries scientists have made about life in prehistoric times	**Block Scheduling** .5 day	**RS** Graphic Organizer 6
2. Identify the four characteristics of civilization.	*Block Scheduling Handbook, Chapter 6*	
3. Identify the locations of the first civilizations.		
SECTION 2		
Early Civilization in the Americas (pp. 109–13)	**Regular** 1 day	**RS** Guided Reading Strategy 6.2
1. Explain how the first people arrived in the Americas.	**Block Scheduling** .5 day	
2. Examine how geography and climate affected the cultures of North America.	*Block Scheduling Handbook, Chapter 6*	
3. Describe the main civilizations that developed in Central and South America.		
SECTION 3		
Greece and Rome (pp. 114–21)	**Regular** 2 days	**RS** Guided Reading Strategy 6.3
1. Describe how ancient Greek civilization developed.	**Block Scheduling** .5 day	**SM** Map Activity 6
2. Discuss the events that led to the birth of the Roman Empire.	*Block Scheduling Handbook, Chapter 6*	
3. Explain how Christianity began.		

Chapter Resource Key

RS Reading Support

IC Interdisciplinary Connections

E Enrichment

SM Skills Mastery

A Assessment

REV Review

ELL Reinforcement and English Language Learners

 Transparencies

 CD–ROM

 Music

 Video

 Internet

 Holt Presentation Maker Using Microsoft® Powerpoint®

 One-Stop Planner CD–ROM

See the *One-Stop Planner* for a complete list of additional resources for students and teachers.

One-Stop Planner CD–ROM

It's easy to plan lessons, select resources, and print out materials for your students when you use the *One-Stop Planner CD–ROM with Test Generator.*

Technology Resources

 One-Stop Planner CD–ROM, Lesson 6.1

 ARGWorld CD–ROM

 Homework Practice Online
HRW Go site

 One-Stop Planner CD–ROM, Lesson 6.2

 ARGWorld CD–ROM

Homework Practice Online
HRW Go site

 One-Stop Planner CD–ROM, Lesson 6.3

 ARGWorld CD–ROM

 Homework Practice Online
HRW Go site

Review, Reinforcement, and Assessment Resources

ELL	Main Idea Activity 6.1
REV	Section 1 Review, p. 108
A	Daily Quiz 6.1
ELL	English Audio Summary 6.1
ELL	Spanish Audio Summary 6.1

ELL	Main Idea Activity 6.2
REV	Section 2 Review, p. 113
A	Daily Quiz 6.2
ELL	English Audio Summary 6.2
ELL	Spanish Audio Summary 6.2

ELL	Main Idea Activity 6.3
REV	Section 3 Review, p. 121
A	Daily Quiz 6.3
ELL	English Audio Summary 6.3
ELL	Spanish Audio Summary 6.3

internet connect

HRW ONLINE RESOURCES

GO TO: go.hrw.com
Then type in a keyword.

TEACHER HOME PAGE
KEYWORD: SJ3 TEACHER

CHAPTER INTERNET ACTIVITIES
KEYWORD: SJ3 GT6

Choose an activity to:
- list different divisions of labor in early civilizations.
- report on Minoan civilization.
- create a newspaper article about ancient Athens.

CHAPTER ENRICHMENT LINKS
KEYWORD: SJ3 CH6

CHAPTER MAPS
KEYWORD: SJ3 MAPS6

ONLINE ASSESSMENT
Homework Practice
KEYWORD: SJ3 HP6
Standardized Test Prep Online
KEYWORD: SJ3 STP6
Rubrics
KEYWORD: SS Rubrics

COUNTRY INFORMATION
KEYWORD: SJ3 Almanac

CONTENT UPDATES
KEYWORD: SS Content Updates

HOLT PRESENTATION MAKER
KEYWORD: SJ3 PPT6

ONLINE READING SUPPORT
KEYWORD: SS Strategies

CURRENT EVENTS
KEYWORD: S3 Current Events

Meeting Individual Needs

Ability Levels

Level 1 Basic-level activities designed for all students encountering new material

Level 2 Intermediate-level activities designed for average students

Level 3 Challenging activities designed for honors and gifted-and-talented students

English Language Learners Activities that address the needs of students with Limited English Proficiency

Chapter Review and Assessment

E	Readings in World Geography, History, and Culture 29
SM	Critical Thinking Activity 6
REV	Chapter 6 Review, pp. 122–23
REV	Chapter 6 Tutorial for Students, Parents, Mentors, and Peers
ELL	Vocabulary Activity 6
A	Chapter 6 Test
A	Unit 2 Test
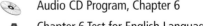	Chapter 6 Test Generator (on the One-Stop Planner)
	Audio CD Program, Chapter 6
A	Chapter 6 Test for English Language Learners and Special-Needs Students
A	Unit 2 Test for English Language Learners and Special-Needs Students

CHAPTER 6

Ask students what they think an archaeologist does and what a historian does. Discuss the differences. *(An archaeologist studies objects such as pottery, clothing, jewelry, and tools, while a historian studies written records such as documents, letters, and journals.)* Point out that the word *history* generally refers to anything that happened in the past. However, history specifically refers to events since people developed writing, about 5,000 years ago. Civilizations thrived long before people invented writing. Events that occurred before writing was developed make up the period referred to as prehistory. Tell students that they will learn about many cultures, both prehistoric and historic, in this chapter.

Section 1

Objectives

1. Examine the discoveries scientists have made about life in prehistoric times.
2. Describe the four characteristics of civilization.
3. Identify the locations of the first civilizations.

LINKS TO OUR LIVES

You might want to share with your students the following reasons for learning about early cultures:

▶ The study of ancient remains and artifacts affects our understanding of humans and of our place on Earth.

▶ Civilizations today share many characteristics with the first civilizations.

▶ The development of agriculture was key to the development of civilization. Agriculture is still an important issue, since the world still needs to be fed.

▶ Some parts of the world whose roots go back thousands of years are facing conflicts today.

▶ Members of some ethnic groups are proud to trace their heritage back to ancient times.

▶ Much of history is based on the exchange of ideas, goods, and technology, all of which are still important issues today.

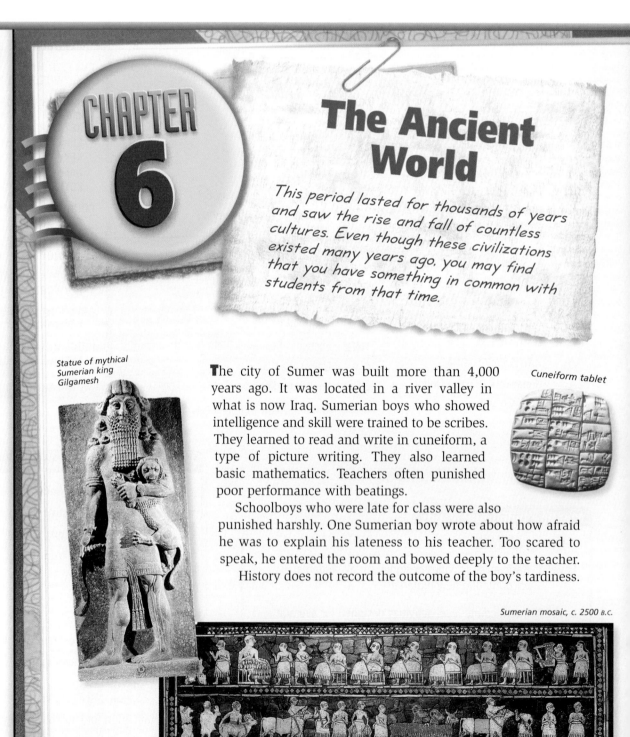

CHAPTER 6

The Ancient World

This period lasted for thousands of years and saw the rise and fall of countless cultures. Even though these civilizations existed many years ago, you may find that you have something in common with students from that time.

Statue of mythical Sumerian king Gilgamesh

Cuneiform tablet

The city of Sumer was built more than 4,000 years ago. It was located in a river valley in what is now Iraq. Sumerian boys who showed intelligence and skill were trained to be scribes. They learned to read and write in cuneiform, a type of picture writing. They also learned basic mathematics. Teachers often punished poor performance with beatings.

Schoolboys who were late for class were also punished harshly. One Sumerian boy wrote about how afraid he was to explain his lateness to his teacher. Too scared to speak, he entered the room and bowed deeply to the teacher.

History does not record the outcome of the boy's tardiness.

Sumerian mosaic, c. 2500 B.C.

Copy the following questions onto the chalkboard. *How would you survive if you were stranded on a remote island? How would you get food? What would you use for tools and shelter?* Discuss responses. *(Possible answers: food—hunt fish and small animals, gather seeds, nuts, berries; tools—sticks, rocks; shelter—branches, caves)* Point out to students that early humans had to survive in much the same way. Tell students that in Section 1 they will learn about early humans and how they lived.

Building Vocabulary

Write the key terms on the chalkboard. Remind students that the prefix *pre-* means "before." Ask students to infer the meaning of **prehistoric** *("before history")*. Ask students to identify other words that begin with the prefix *pre-*. Point out that the suffix *-tion* at the end of a word indicates a process or an action. Have students infer the meanings of the words **civilization** and **irrigation** using what they know about each base word and the suffix. Call on volunteers to read the definitions for all the terms from the text.

Section 1 — The Birth of Civilization

Read to Discover

1. What discoveries have scientists made about life in prehistoric times?
2. What are the four characteristics of civilization?
3. Where were the first civilizations located?

Define

hominid
prehistory
nomads
land bridges
irrigation

division of labor
history

Hominid skeleton from about 3 million years ago

WHY IT MATTERS

Scientists continue to uncover clues about how ancient people lived. Use **CNNfyi.com** or other **current events** sources to learn about recent discoveries. Record your findings in your journal.

Section 1 RESOURCES

Reproducible
◆ Block Scheduling Handbook, Chapter 6
◆ Guided Reading Strategy 6.1
◆ Graphic Organizer 6

Technology
◆ One-Stop Planner CD–ROM, Lesson 6.1
◆ Homework Practice Online
◆ HRW Go site

Reinforcement, Review, and Assessment
◆ Main Idea Activity 6.1
◆ Section 1 Review, p. 108
◆ Daily Quiz 6.1
◆ English Audio Summary 6.1
◆ Spanish Audio Summary 6.1

Prehistory

In the late 1970s in Tanzania, a country in East Africa, scientist Mary Leakey discovered parts of a skeleton dating back millions of years. She believed the bones were those of a **hominid**, an early human-like creature. Scientists use the remains of bodies and other objects they have found to make educated guesses about hominid life. For example, scientists can tell that hominids stood upright and used primitive tools made of stone.

The period during which hominids and even early humans lived is called **prehistory**. This means that no written records were made for historians to examine. The period of prehistory in which stone tools were used is called the Stone Age. It began about 2.5 million years ago and lasted for more than 2 million years.

Scientists carefully unearth objects in "digs" such as this one.
Interpreting the Visual Record What kinds of objects might these scientists be looking for?

◀ **Visual Record Answer**

skeletal remains and stone tools

103

Teaching Objective 1

ALL LEVELS: (Suggested time: 15 min.) Ask students to look in the text for examples of the types of information scientists have gathered about prehistoric times. Then ask students to provide specific examples for each one. *(Example: type of information—how early humans got food; specific examples—gathered plants, hunted, developed agriculture)*

LEVEL 1: (Suggested time: 30 min.) Have students complete the All Levels activity. Then copy the following graphic organizer onto the chalkboard, omitting the italicized answers. Use the organizer to help students examine the kinds of resources scientists believe early people used to survive. Have students work in pairs to complete it.
ENGLISH LANGUAGE LEARNERS, COOPERATIVE LEARNING

Across the Curriculum
TECHNOLOGY

Fire Archaeologists have discovered hearths, or fireplaces, along with the remains of hominids who lived approximately 500,000 years ago. With the ability to control fire, they apparently started to cook their meat, making it easier to chew. As a result, the need for powerful jaws and large teeth eventually declined. The mastery of fire also allowed people to live in colder regions of the world.

Critical Thinking: Ask students how hominids may have learned to master fire. *(Answer: Lightning may have started a fire, and hominids learned how to keep it burning or how to use the fire to start a new fire.)*

internet connect

GO TO: go.hrw.com
KEYWORD: SJ3 CH6
FOR: Web sites about archaeology

Visual Record Answer

because other materials were not available in sufficient quantities

This cave painting in Lascaux, France, was made before the invention of writing. People may have drawn on cave walls to express ideas.

internet connect

GO TO: go.hrw.com
KEYWORD: SJ3 CH6
FOR: Web sites about the ancient world

In northeastern Europe, early humans used the bones of giant mammoths to build shelters similar to this museum model.
Interpreting the Visual Record Why did people use bones to build a shelter instead of materials such as wood or stone?

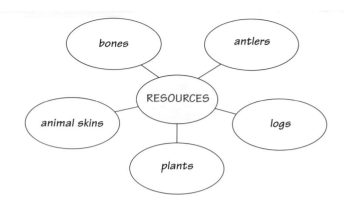

The First Humans Early humans that looked like modern people probably appeared during the Stone Age between 100,000 and 400,000 years ago. These first humans, called *Homo sapiens*, may have first lived in Africa. They were **nomads** who moved from place to place in search of food. They lived on seeds, fruits, nuts, and other plants that they gathered. In time they also began to hunt small animals.

Migration Within the last 1.7 million years, Earth has gone through several periods of very cold weather. Together these periods are known as the Ice Age. During each period, large parts of Earth's surface were covered with ice. Sea levels dropped, leaving strips of dry land called **land bridges** between continents. One such land bridge connected the eastern part of Asia with what is now Alaska. Scientists think that early humans and animals migrated from Africa, into Asia, and across the land bridge onto the North American continent. Over time, humans spread to all parts of the world.

Later Developments In time *Homo sapiens* began to make more advanced tools. They were able to hunt larger animals with spears. They made clothes from animal skins. They also learned how to control fire and how to use it for warmth and cooking. Between 37,000 and 27,000 years ago, people began to create art to express their ideas. Carved ivory figures show that some groups had time for activities besides hunting and tool making. Beautiful paintings found on the walls of caves in France and Spain show graceful, elegant animals such as bison, bulls, and horses.

LEVEL 2: (Suggested time: 30 min.) Have students work in pairs to create time lines that show what scientists believe to be the progress of humans from the appearance of *Homo sapiens* between 400,000 and 410,000 years ago up to the time that the first cities appeared about 5,000 years ago. On their time lines, have students list the developments and achievements of early humans. *(Example: 10,000 to 5,500 years ago.—made better tools)*

LEVEL 3: (Suggested time: 40 min.) Have students complete the Level 2 activity. Then have them combine their time lines into one and reproduce it on a long piece of butcher paper. Ask students to illustrate the time line as if it were a cave painting. Refer them to the illustration of the cave paintings of Lascaux on the previous page. You may want to provide additional resources and allocate extra time for this activity.

➤**ASSIGNMENT:** Tell students that agriculture, one of the most important achievements of early humans, affects all aspects of our daily lives. Have each student write a paragraph that tells how agriculture affects him or her personally each day. *(Example: Our breakfast cereal is made from grains grown by farmers. Many of the clothes we wear are made of cotton.)*

Later, between 10,000 and 5,500 years ago, people learned to make sharper tools by grinding and polishing stone. With better tools, people developed better methods of hunting. They made bows and arrows, which made hunting easier. They shaped fishhooks and harpoons from bones and antlers. People hollowed out logs to make canoes to fish in deep water and to cross rivers. Also around this time, people tamed the dog. Dogs helped people hunt. They may also have warned people if wild animals or strangers were approaching.

In the late Stone Age people learned to practice agriculture. We do not know why people made the change from gathering grains and other plants to growing them, but life changed drastically when they did. Instead of moving from place to place to hunt animals and gather food that grew wild, people began to stay in one place. They became farmers. People also domesticated animals such as cattle and sheep. That means people tamed animals that had been living wild.

The Importance of Agriculture Agriculture changed the ways in which people interacted with their environment. To grow food, people had to find ways to control and change their environment. They cleared forested areas to make room for fields. They invented **irrigation** systems, digging ditches and canals to move water from rivers to fields where crops grew.

Agriculture also changed the ways in which people interacted with each other. Because people who farmed stayed in one place, they began to live in larger groups and form societies. By about 9000 B.C., people began to live in permanent settlements and villages. Because farming made food more plentiful, populations increased. Small villages eventually grew into cities. In towns and cities, people shared new ideas and methods of doing things. Historians think that the first cities may have been founded more than 10,000 years ago. Jericho, the world's oldest known city, was founded at that time on the west bank of the Jordan River.

The dog was the first animal that humans tamed. Dogs and people have been companions for thousands of years. Some Stone Age people were even buried with their dogs.

Before people developed irrigation they depended on yearly floods of rivers such as the Nile to water fields.

▼

LEVEL 1: (Suggested time: 30 min) Have students list the four main characteristics shared by civilizations *(organized society, produce extra food, live in towns and cities with governments, practice division of labor)*. Ask each student to write a short paragraph explaining how each characteristic relates to the development of civilization. *(Answers will vary, but should note that organized society allows for cooperation and the development of government, laws, and customs; surplus food allows time for activities besides survival; government helps ensure order; and division of labor allows people to do different jobs.)*

LEVELS 2 AND 3: (Suggested time: 10 min) Have students complete the Level 1 activity. Then ask students to identify two more accomplishments characteristic of civilizations *(a calendar and some form of writing)*. Have students work in pairs to draw two-column charts with one of these characteristics at the top of each column. Then ask students to list the reasons why each characteristic contributes to a civilization. *(Examples: A calendar told people when to plant crops and when to expect rain; writing allowed people to keep records and communicate more easily.)* Lead a discussion about the significance of calendars and writing in our civilization.
COOPERATIVE LEARNING

Across the Curriculum

SCIENCE

Inca Farming Practices

Agriculture in South America began long ago. Scientific examination of organic material in the soil shows that people farmed the Andes region of Peru intensively from about 2000 B.C. to A.D. 100. When the region's climate cooled, farming declined, and soil erosion increased.

About 1230 A.D., the climate warmed, and the Inca took over the area. Archaeological evidence suggests that the Inca may have used sophisticated soil conservation techniques. They built a long canal to irrigate fields, terraced the hillsides to grow crops, and planted alder trees that helped curb soil erosion. After the Spanish conquest of Peru, the terraces deteriorated. Today, the alder trees are found only in a few remote ravines.

Activity: Have students conduct research on ancient farming practices used at various sites around the world and report their findings to the class.

Interpreting the Map Answer ▶

in parts of Southwest Asia and China

Emergence of Agriculture

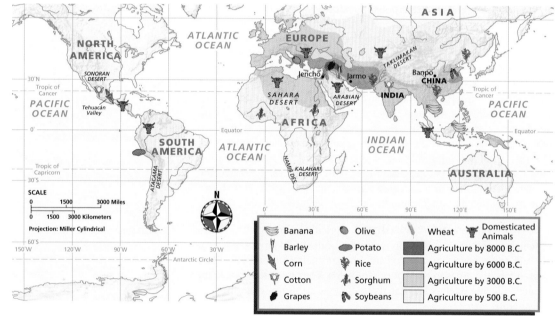

▲
The practice of agriculture spread over a period of thousands of years.
Interpreting the Map Where was agriculture first developed?

▲
People made stone and ivory tools during the Stone Age.

Because of the ways it changed people's lives, the development of agriculture was enormously important. In fact, learning how to grow food prepared the way for a new chapter in the story of human life—the story of civilization.

✓ **READING CHECK:** **Summarizing** How did agriculture change the ways in which people lived? *As farmers they didn't have to migrate to hunt for food.*

The Beginnings of Civilization

Historians describe civilization as having four basic characteristics. First, a civilization is made up of people who live in an organized society, not simply as a loosely connected group. Second, people are able to produce more food than they need to survive. Third, they live in towns or cities with some form of government. And fourth, they practice **division of labor**. This means that each person performs a specific job.

Agriculture and Civilization How did the development of agriculture affect the growth of civilization? Before agriculture, people spent almost all of their time simply finding food. When people were able to grow their own food, they could produce more than they needed to survive. This meant that some people did not have to grow food at all. They had time to develop other skills, such as making pottery, cloth, and other goods. These people could trade the goods they produced and the services they offered for food or other needs.

Teaching Objective 3

ALL LEVELS: (Suggested time: 20 min) Ask students to imagine that they are members of a farming community searching for a new home. Ask students what two features they would require in a new location *(rich, fertile soil and water for their crops)*. Then call on volunteers to explain briefly why the earliest civilizations developed in river valleys. Have them locate the four river valleys where civilization began on the world map in the textbook's atlas. You may want to extend the activity by leading a discussion about what hazards these river valleys may have posed, in contrast to their advantages. *(Possible answers: exposure to invasion, floods)*

CLOSE

Call on volunteers to discuss how the illustrations in this section reflect the development and achievements of early humans.

REVIEW AND ASSESS

Have students complete the Section Review. Then have pairs of students create a set of 10 question-and-answer flash cards about early humans and the birth of civilization. Pairs of students can then quiz each other with their cards. Then have students complete Daily Quiz 6.1.

ENGLISH LANGUAGE LEARNERS, COOPERATIVE LEARNING

Trade Once people began to trade, they had to deal with each other in more complex ways than before. Disagreements arose, creating a need for laws. Governments and priesthoods developed to fill that need. Governments made laws and saw that they were obeyed. Religion taught people what they should and should not do.

When people traded, they traveled to places where their goods were wanted and where they could get the things they wanted and needed. Some places where people exchanged goods grew into cities. In these cities, people traded not only goods but also ideas. Over time people built palaces, temples, and other public buildings in their cities.

The Development of Writing Trade, like business today, required people to keep records. Written languages may have developed from this need. The invention of written language began about 3,000 B.C. Farmers also needed a method to keep track of seasonal cycles. They had to know when it was time to plant new crops and when they could expect rain. Over time, they developed calendars.

Once they had writing and calendars, people began to keep written records of events. **History**, which is the written record of human civilization, had begun.

✓ **READING CHECK:** *Identifying Cause and Effect* How did trade lead to the development of writing? Written records were needed to produce and exchange goods.

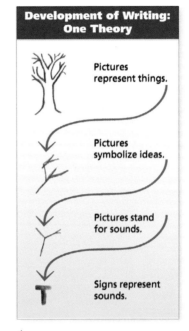

Development of Writing: One Theory

Pictures represent things.

Pictures symbolize ideas.

Pictures stand for sounds.

Signs represent sounds.

▲ The invention of the alphabet may have begun from pictures. This flowchart shows the possible development of the letter T.

Linking Past to Present

Trade and Travel in Egypt The development of travel by water has stimulated the growth of trade. Throughout history, water travel has usually been cheaper and faster than travel overland. As a result, goods transported by water are usually cheaper than those that must come by land from far away.

Living along the Nile River, the ancient Egyptians were some of the earliest developers of water transportation. An image on a pot dated from about 3200 B.C. shows that Egyptians were already using sails to travel on the Nile. Early boats floated north downriver toward the Mediterranean Sea. Then their captains could raise sails and be carried back upstream by the wind, which most of the time blows from the north. Rafts and barges carried goods up and down the Nile, and ferries traveled across the river. Egypt's rulers, the pharaohs, also used Nile River boats to send messages throughout the land.

River Valley Civilizations

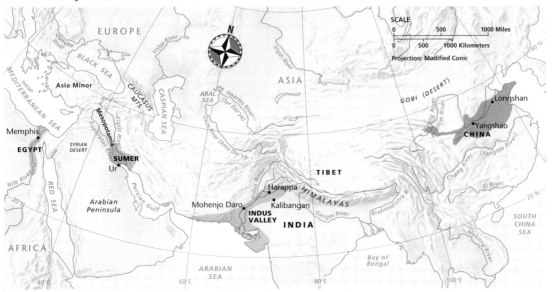

Have students complete Main Idea Activity English 6.1. Then have students copy the headings and subheadings in the section. Ask them to write sentences stating the main ideas for each heading and subheading.
ENGLISH LANGUAGE LEARNERS

Invite interested students to conduct research about scientists who study ancient hominid and human remains, the methods they use, and the information they gather. Students might research "Lucy"—the female hominid that may have lived 3 million years ago—whose skeletal remains were discovered by Donald Johanson in Ethiopia in 1974. Others might conduct research on the findings of Mary and Louis Leakey in Tanzania in the late 1970s. Encourage students to supplement their reports with graphics and illustrations. Have students present their findings to the class.
BLOCK SCHEDULING

Section Review 1

Answers

Define For definitions, see: hominid, p. 103; prehistory, p. 103; nomads, p. 104; land bridges, p. 104; irrigation, p. 105; division of labor, p.106; history, p. 107

Reading for the Main Idea

1. Possible answers: Hominids stood upright 2.5 million years ago; humans created art 37,000 years ago. (NGS 17)

2. People traded the goods they made and the services they offered for food or other goods they needed. (NGS 11)

3. Water and fertile soil were necessary for farming. (NGS 5)

Critical Thinking

4. Disagreements arising from trade created the need for law and order and for teachings about right and wrong behavior.

Organizing What You Know

5. Outer ovals—people live in an organized society; able to produce more food than they need to survive; live in towns or cities with a government; practice division of labor

Scientists discovered the remains of mud-brick houses at Catalhüyük (chah-TUHL-hoo-YOOKH) in Turkey. This was one of the world's first cities.

The First Civilizations

The world's first civilizations developed around four great river valleys. The earliest arose along the valley of the Tigris and Euphrates Rivers in Southwest Asia. Called Mesopotamia, this area was located in what is now Iraq. The ancient Egyptian civilization grew up around the valley of the Nile River. The first civilizations of India were centered on the Indus Valley. Early Chinese civilization began in the valley of the Huang, or Yellow River. (See the map on the previous page.)

These four great river valleys provided fertile soil for growing crops and rich water sources to irrigate crops. The people who lived in these valleys developed advanced civilizations. They learned how to make tools and weapons out of metal, first bronze and then iron. This was the end of the Stone Age. The civilizations that formed in each of these areas developed and declined for different reasons. All of them created written records of their cultures and societies. Thus, they mark the beginning of human history.

✓ **READING CHECK:** *Drawing Inferences* How did geography influence the beginnings of history? River valleys promoted irrigation and fertile soil for growing food.

Homework Practice Online Keyword: SJ3 HP6

Section Review 1

Define and explain: hominid, prehistory, nomads, history, irrigation, division of labor.

Reading for the Main Idea

1. *Human Systems* Describe some important discoveries which scientists have made about life in prehistoric times.

2. *Human Systems* How did the division of labor lead to the development of trade?

3. *Environment and Society* Why were early civilizations located around river valleys?

Critical Thinking

4. **Identifying Cause and Effect** How did the needs of early civilizations lead to the development of government and priesthoods?

Organizing What You Know

5. **Summarizing** Copy the following graphic organizer. Use it to describe the traits of civilization.

What Is Civilization?

Section 2

Objectives

1. Explain how the first people arrived in the Americas.
2. Examine how geography and climate affected the early cultures of North America.
3. Describe the main civilizations that developed in Central and South America.

FOCUS

(((•))) LETS GET STARTED

Copy the following questions and instructions onto the chalkboard: *What do you already know about the peoples who lived in the Americas before Europeans arrived? Write down three things about who these people were or where and how they lived.* As students report their responses, jot a summary on the chalkboard. Tell students that in this section they will find out if they are correct and that they will learn more about early civilizations in the Americas.

Building Vocabulary

Write the key terms on the chalkboard. Point out that **tepee** is from a word of the Dakota people. Then underline the root word *glyph* in **hieroglyphic**. Tell students that a glyph is a pictograph or a symbolic character. The prefix *hiero-* means "sacred" or "holy." Point out that hieroglyphics are sometimes found on the walls of ancient temples. Note that **quipu** is from Quechua, the language of the Inca. Then have students find all the terms in the text and read the definitions aloud.

Section 2 — Early Civilization in the Americas

Read to Discover

1. How did the first people arrive in the Americas?
2. How did geography and climate affect the cultures of North America?
3. What were the main civilizations that developed in Central and South America?

Define

migrated
tepees
hieroglyphics
quipu

Locate

Mexico
Yucatan Peninsula
Andes Mountains
Peru

WHY IT MATTERS

Early American cultures are part of our heritage. Use **CNNfyi.com** to find out about Native American issues in the news today. Record your findings in your journal.

The ancient Maya built this observatory to study the stars and planets.

Section 2 RESOURCES

Reproducible
◆ Guided Reading Strategy 6.2

Technology
◆ One-Stop Planner CD–ROM, Lesson 6.2
◆ Homework Practice Online
◆ HRW Go site

Reinforcement, Review, and Assessment
◆ Section 2 Review, p. 113
◆ Daily Quiz 6.2
◆ Main Idea Activity 6.2
◆ English Audio Summary 6.2
◆ Spanish Audio Summary 6.2

Peoples of North America, 2550 B.C.–A.D. 1550

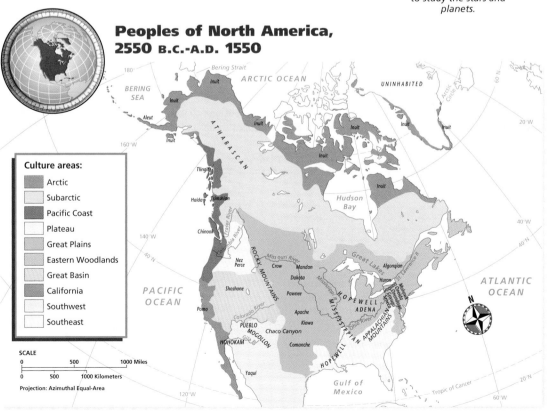

Culture areas:
- Arctic
- Subarctic
- Pacific Coast
- Plateau
- Great Plains
- Eastern Woodlands
- Great Basin
- California
- Southwest
- Southeast

SCALE
0   500   1000 Miles
0   500   1000 Kilometers
Projection: Azimuthal Equal-Area

Teaching Objective 1

ALL LEVELS: (Suggested time: 15 min.) Have students turn to the physical map of North America in the textbook atlas. Ask them to find the Bering Strait and to point out where land once connected Asia to North America. Then have students write a short description of how and why people may have crossed into North America and speculate about what they found when they arrived there. **ENGLISH LANGUAGE LEARNERS**

Teaching Objective 2

LEVEL 1: (Suggested time: 20 min.) Copy the following graphic organizer onto the chalkboard, omitting the italicized answers. Have students work in small groups to fill in the blanks, using information from the text. Ask students to use the geographical information to speculate about dwellings of the Eastern Woodlands peoples.
ENGLISH LANGUAGE LEARNERS, COOPERATIVE LEARNING

National Geography Standard 10

Human Systems Scientists who study human languages and how they are related are known as linguists. Linguists have found that in the past about 300 Native American languages were spoken in North America. Though most of these languages were not written, and some have been lost, the diversity that linguists find among the surviving languages is far greater than exists among modern European languages.

Critical Thinking: What can we infer about people whose languages are closely related? What can we infer from the fact that Native American languages differ from each other much more than do European languages?

Answer: If two languages are closely related, the people who speak them probably had close contact throughout history. Languages showing no similarities indicate widely separated development of cultures. We might infer that ancient Native American peoples from different regions had little contact with one another.

Visual Record Answer ▶

They were skilled builders and lived close to one another.

▲
Before people developed writing, they shared stories and ideas through art objects like this totem pole.

Many early dwellings in the Southwest were built along the sides of canyons or in cliffs.

Interpreting the Visual Record What can you tell about the early people who built these dwellings?

▼

The First Americans

The first people to settle in the Americas probably **migrated**, or moved, to North America from Asia. At that time, a strip of dry land connected Asia and North America across what is now the Bering Strait. Over time, however, huge glaciers melted and covered that strip of land. Many of these early people were nomads who depended on hunting, fishing, and collecting plants to live. Scientists think people probably followed migrating herds that moved into the Americas. Over time, some of these people migrated to the middle and eastern areas of North America. Others moved through Mexico and Central America.

The Americas stretch more than 11,000 miles from Greenland to the southern tip of South America. Almost every type of climate and type of land can be found somewhere in the Americas. Because environment affects the way people live, cultures developed differently in different places.

✓ **READING CHECK:** *Identifying Cause and Effect* How did early people move into the Americas? They migrated from Asia across the Bering Strait when it was a strip of dry land.

Early Cultures of North America

The Northwest Wherever early people settled, they learned to use what they found to build shelters and to feed themselves. (For example, people who settled along the Pacific coast lived close to the sea. Thus they largely depended on fish for food. Tall forests in the Northwest provided them with a rich supply of lumber.) Some people became expert woodworkers. They made canoes from the bark of trees. They carved great totem poles as symbols of their communities.

The Great Plains Most early peoples of the Great Plains were hunters. Huge herds of buffalo roamed over the wide-open spaces of the grassy plains. Plains people hunted buffalo. Because there were no horses in North America at this time, people had to hunt on foot. Sometimes they scared entire herds of buffalo into stampedes over high cliffs. At other times, they built corrals into which they drove the buffalo. These hunters used nearly every part of the buffalo in some way. They ate the meat and made tools from the bones. Buffalo hides were used for clothing and for tents called **tepees**. The buffalo was so important that it was sometimes considered sacred. Hunters held ceremonies before a hunt and gave thanks afterwards. However, not all people of the Great Plains followed the herds. Some became farmers who raised beans, corn, and squash.

	Northwest	Great Plains	Southwest	Eastern Woodlands
Characteristics of Areas Environment	near sea, tall forests	buffalo herds	dry and hot, few trees	forests
Type of Shelter	wooden dwellings	tepees made of buffalo hide	dwellings made of adobe, or clay brick	probably wooden dwellings
Main Type of Food	fish	buffalo meat	crops such as beans and corn	fruits, nuts, seeds, animals they hunted

LEVELS 2 and 3: (Suggested time: 10 min.) Ask students to find in the text at least one example of how geography affected the way early people lived in each area of North America. Have students write sentences about the examples they find. Discuss responses. *(Examples: Because there were many tall trees in the Northwest, people used wood to build dwellings, totem poles, and canoes; because there were few trees in the Southwest, people built homes out of clay bricks; because there were buffalo herds in the Great Plains, people ate buffalo meat and made clothing and dwellings out of buffalo hide.)* You may want to extend the activity by discussing what students already know about the effect of geography on ways of life in those same regions today.

The Southwest The southwestern part of North America has a hot and dry climate. People who moved there had to learn how to live in a desert environment. They had to find new ways to feed and clothe themselves and to water their crops. For example, the Hohokam developed irrigation systems to bring water to their fields. Even in very dry soil, they were able to grow crops such as beans, corn, and cotton. The Hohokam arrived in the American Southwest sometime between 300 B.C. and A.D 500. They remained in the area until the 1300s and 1400s. Few trees grow in the region, so the early people had to use other material to build their homes. One group, the Pueblo, found a way to build houses with adobe—sun-dried bricks made from clay and straw.

The Eastern Woodlands Unlike the Southwest, the Eastern Woodlands were covered with forests. The people who settled in these forests lived on the fruits, plants, seeds, and nuts they gathered. They also hunted. Over time, they learned to farm the land. Once people were able to produce more food than they needed, some group members had time to practice other activities. Some groups became known for their crafts.

The Hopewell were a group of early people who lived in the Ohio Valley region between 300 B.C. and 200 B.C. They are famous for the large earth mounds they built. From the air, some of these mounds look like animals. They may have been used as burial sites or for ceremonies. The jewelry, tools, and weapons found in the mounds show that the Hopewell were talented artists.

The Mississippians were another group who lived in the Eastern Woodlands. They were excellent farmers who developed a complex culture. Like the Hopewell, the Mississippians built huge earth mounds in the centers of their settlements.

✓ **READING CHECK:** *Comparing and Contrasting* How were the problems of survival in the Southwest different from those of other regions? Living in the desert required irrigation to grow food and unique ways to build shelter.

Empires of Mexico, Central America, and South America

While early cultures were developing in North America, other people were building great civilizations to the south. As early as 1500 B.C., people lived in villages in the areas known today as Mexico, Central America, and South America. Like most cultures, they learned to grow food and tame animals. In time these villages began to trade with each other. As trading centers were set up, cities grew.

Civilizations in Central and South America, c. 200 B.C.-A.D. 1535

Olmecs, c. 200
Maya, c. 600
Toltecs, c. 1100
Aztecs, c. 1478
Incas, c. 1500

SCALE
0   500   1000 Miles
0   500   1000 Kilometers
Projection: Miller Cylindrical

Linking Past to Present

La Venta The ancient Olmec settlement now called La Venta dates from between 800 B.C. and 400 B.C. It is located near the border of Tabasco and Veracruz, states of present-day Mexico. The orientations of the major structures have led scientists to propose that the buildings were aligned with a certain star or constellation. The site also contains a 100-foot clay cone-shaped mound, a ceremonial enclosure containing a number of tombs, and mosaic pavements of jaguar mask designs.

Activity: Have students conduct research to find pictures of structures left behind by the Olmec, particularly those found at La Venta. Then have some students create replicas of Olmec sculptures and display them for the class. Ask other students to write tourist guides for La Venta.

LEVEL 1: (Suggested time: 20 min.) Organize the class into pairs and have each pair choose a civilization of Central or South America. Have each pair create a graphic organizer to describe the civilization, using other organizers in this book as models. Students should make two copies of the organizer—one with the answers and one without. Have pairs exchange organizers, complete them, and then check the answers.
ENGLISH LANGUAGE LEARNERS, COOPERATIVE LEARNING

LEVELS 2 AND 3: (Suggested time: 40 min.) Have students create annotated and illustrated time lines of the major civilizations. Provide additional resources. Then ask each student to write a paragraph describing and comparing the length of time that the civilizations survived.

▶**ASSIGNMENT:** To help students learn more about the civilizations of Central and South America, have them conduct research on some of the ancient ruins still standing there today. Ask students to write postcards describing the ruins to friends or family members.

Across the Curriculum
Mathematics
The Maya Numbering System The Maya did not use digits as we do in our numbering system. Instead they used a system of dots and bars. One dot stood for the number 1, and one bar stood for the number 5. Other numbers were created by a combination of dots and bars. For example, one bar with two dots above it stood for the number 7, and three bars with three dots above them stood for 18.

The Mayan system also included a symbol for zero. It looked like the outline of a football. The zero is a useful mathematical concept because it works as a placeholder. For example, in the number 409 the zero shows that there are no tens.

Activity: Ask students to write their telephone numbers or their age using the Mayan numbering system.

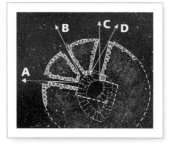

▲
Interpreting the Visual Record
This diagram shows the floor plan of Maya observatory. What do the four openings, A, B, C, and D, represent on this plan?

This Aztec Calendar Stone was used in ceremonies honoring the sun god. The god's face can be seen in the center of the stone.

▼

The Olmec The Olmec developed the earliest known civilization in Mexico. It appeared by about 1200 B.C. and lasted for some 800 years. Olmec society was made up of a large class of farmers and a small upper class. The upper class included the political, military, and religious leaders.

Historians have found objects that suggest the Olmec were an advanced society. For example, the Olmec carved gigantic heads out of stone. These heads, which can still be seen today, weigh as much as 40 tons. The stone from which they were carved came from a quarry nearly 50 miles away! Historians think that only a highly advanced society could find a way to move such large stones.

The Maya As hundreds of years passed, civilizations in the Americas became more advanced. Some historians believe the Maya were the most advanced early civilization in the Americas. Their civilization, which first appeared around 1500 B.C., lasted for hundreds of years. Maya culture reached its height between A.D. 300 and A.D. 900.

Ruins of beautiful Maya cities can still be seen in the forests of Guatemala, Honduras, and the Yucatán peninsula. These ruins show that the Maya were skilled architects and engineers. They built huge pyramids and special buildings from which they could study the skies. Their observations helped the Maya develop a 365-day calendar. The Maya also created the only complete writing system in the early Americas. This system was based on a type of picture writing called **hieroglyphics**. The Maya also developed a counting system that included the number zero. At some time after A.D. 800, the Maya civilization began to fall apart. Historians think that crops may have failed or that wars might have destroyed Maya cities.

The Aztec Some early peoples learned new skills from the civilizations they conquered. For example, in about A.D. 1200, a group of warriors called the Aztec invaded central Mexico. As they moved into the region, the Aztec learned many skills such as metalworking, pottery, and weaving from the peoples they conquered. Calendars and mathematics became part of Aztec life. By A.D. 1400 the Aztec had built their capital at the present site of Mexico City.

Aztec society was dominated by the military. Warriors gained glory, power, and wealth. The fierce Aztec quickly conquered many groups who lived in surrounding areas. They heavily taxed these peoples. Many of them came to resent the Aztec, and unrest grew. This weakened the Aztec Empire.

The Inca When the Aztec civilization was at its height, another civilization was growing in the Andes Mountains along the west coast of South America. These people, called the Inca, lived in the lands that make up the present-day countries of Peru, Ecuador, Chile, and Bolivia.

CLOSE

Challenge students with the task of recording information about your school—number of students, number of faculty, sports mascot, school colors, and so on—on a quipu. Ask them how they would differentiate among the categories of information.

REVIEW AND ASSESS

Have students complete the Section Review. Organize students into groups and assign one of the early cultures of the Americas to each group. Have groups compose two questions about their assigned cultures. Groups can take turns quizzing other groups with their questions. Then have students complete Daily Quiz 6.2. **ENGLISH LANGUAGE LEARNERS**

RETEACH

Have students complete Main Idea Activity 6.2. Then have students work in pairs to create posters with captions to summarize the section's main points. Display the posters in the classroom.
ENGLISH LANGUAGE LEARNERS, COOPERATIVE LEARNING

EXTEND

Have interested students conduct research on Aztec crafts such as weaving and pottery. Encourage them to use what they have learned to draw their own designs for objects similar to those the Aztec created. Invite students to present their findings and designs to the class. **BLOCK SCHEDULING**

The Inca emperors had complete authority over the people, but they used their power to improve the empire. For example, they had large storehouses built to hold extra food to be used when crops failed. The Inca built fortresses and irrigation systems. They also created a system of paved roads and bridges. The road system allowed relay runners to deliver messages to all parts of the empire.

Since tall mountains separated groups of people, the Inca grew into an empire of hundreds of groups who spoke different languages. To bring the people together, Inca rulers developed schools. Students, mainly from the wealthy classes, learned the Inca language. They also studied Inca law, religion, and history.

Although the Inca never developed a system of writing, they had a method of keeping records. They used a series of knots in strings called **quipu**. Officials kept track of harvests, population numbers, and important dates on quipus. The Inca people also produced ceramics, cloth goods, and metal objects.

European Contact In the early 1500s, Spanish explorers led by Hernán Cortés overthrew the Aztec empire. In the 1530s, Spaniards under Francisco Pizarro conquered the Inca and claimed their land for Spain. The Europeans had superior weapons and armor. They also carried diseases such as smallpox and measles that wiped out much of the local population.

By the end of the 1500s, Europeans controlled nearly all of Central and South America. In less than 100 years, millions of Indians had died. Some were killed in battle. Others died working as slaves in European mines. Most, however, were killed by the diseases carried by European soldiers. In some cases, whole cultures disappeared in just a few years.

✔ READING CHECK: *Analyzing* What were some achievements of the early civilizations of Mexico, Central America, and South America? Highly advanced monolithic sculptures and pyramids; 365-day calendar; hieroglyphic writing system; the concept of zero

Inca surgeons knew about pain-killing medicines called anesthetics. They could even perform operations on the brain.

Homework Practice Online
Keyword: SJ3 HP6

Section Review 2

Define and explain: migrated, tepees, hieroglyphics, quipu

Reading for the Main Idea

1. *Environment and Society* Why did the early peoples of North America develop different ways of living?

2. *Human Systems* Describe how the first people arrived in the Americas.

Critical Thinking

3. **Drawing Inferences** How can a culture's art give clues about its beliefs and practices?

4. **Identifying Cause and Effect** What are some things a civilization needs in order to grow?

Organizing What You Know

5. **Summarizing** Copy the following graphic organizer. Use it to list the resources that early peoples in North America used for survival.

Region	Resource(s)
Northwest	
Great Plains	
Eastern Woodlands	
Southwest	

Section Review 2

Answers

Define For definitions, see: migrate, p. 110; tepees, p. 110; hieroglyphic, p. 112; quipu, p. 113

Reading for the Main Idea

1. Varying climates and environments necessitated different methods of survival. (NGS 4)

2. Asian nomads probably followed migrating herds as they moved into the Americas. (NGS 9)

Critical Thinking

3. Art reveals what is important to a people in their daily lives and sometimes their views about death.

4. Possible answers: mathematics, calendars, arts and crafts, laws, and protection from disease

Organizing What You Know

5. Northwest—fish, lumber; Great Plains—buffalo, beans, corn, squash; Eastern Woodlands—forests, fruits, plants, seeds, nuts; Southwest—beans, corn, cotton, clay, straw

Section 3

Objectives

1. Describe how ancient Greek civilization developed.
2. Discuss the events that led to the birth and decline of the Roman Empire.
3. Explain how Christianity began.

Section 3 RESOURCES

Reproducible
◆ Guided Reading Strategy 6.3
◆ Map Activity 6

Technology
◆ One-Stop Planner CD–ROM, Lesson 6.3
◆ Homework Practice Online
◆ HRW Go site

Reinforcement, Review, and Assessment
◆ Section 3 Review, p. 121
◆ Daily Quiz 6.3
◆ Main Idea Activity 6.3
◆ English Audio Summary 6.3
◆ Spanish Audio Summary 6.3

FOCUS

LET'S GET STARTED

Copy the following instructions onto the chalkboard: *Examine the pictures in the section titled "The Greek and Roman Worlds." Write a sentence about what you can tell about the Greek and Roman civilizations by looking at these pictures. Discuss responses. (Possible answer: These civilizations had powerful leaders, beautiful art, and impressive buildings and other structures.)* Tell students that in this section they will learn about these great ancient civilizations.

Building Vocabulary

Write the key terms on the chalkboard. Call on individual students to locate the definitions in the text and to read them aloud. Then challenge students to write sentences that connect two of the terms. *(Example: A **direct democracy** and a **republic** are two forms of government.)*

Section 3 — The Greek and Roman Worlds

Read to Discover

1. How did ancient Greek civilization develop?
2. What events led to the birth and decline of the Roman Empire?
3. How did Christianity begin?

Define

city-states
direct democracy
republic
aqueduct

Locate

Aegean Sea
Greece
Crete
Athens
Italy
Rome

Adriatic Sea
Mediterranean Sea
Alps
Jerusalem

WHY IT MATTERS

Both the Greek and Roman governments were based on the idea that people can govern themselves. Use **CNNfyi.com** or other **current events** sources to find out about how citizens of the United States and other countries take part in government.

Greek coins, c. 500s B.C.

Aegean Civilization, c. 1450 B.C.-700 B.C.

BLACK SEA
Bosporus
MACEDONIA
THRACE
SEA OF MARMARA
Mt. Olympus
40°N
Dardanelles (Hellespont)
Troy
THESSALY
Asia Minor
GREECE
AEGEAN SEA
IONIA
MYCENAEANS
ATTICA
Mycenae
Athens
Tiryns
Peloponnesus
Pylos
Delos
Kos
IONIAN SEA
N
Thera
Rhodes
36°N
MINOANS
SCALE
Knossos
0   75   150 Miles
Crete
0   75   150 Kilometers
MEDITERRANEAN SEA
Projection: Lambert Conformal Conic
28°E

The Early Greeks

By 2000 B.C. civilizations were developing in the Nile River valley and the Fertile Crescent. At the same time another civilization was forming near the Balkan peninsula and the Aegean Sea. The people who settled this area later became known as the Greeks. From the Greeks came many of the ideas that formed the foundation for modern western civilization.

Geography and Greek Civilization

Geography has much to do with the way the early Greeks lived. Greece is a rugged country that is made up of many peninsulas and islands separated by narrow waters. The land is covered with high mountains. They separated groups of people who lived in the valleys. These landforms contributed to the development of separate communities rather than one large and united kingdom. Because it was difficult to travel through the mountains, some Greeks preferred to travel by sea. Many became fighters, sailors, and traders.

Teaching Objective 1

LEVEL 1: (Suggested time: 20 min.) Copy the following graphic organizer onto the chalkboard, omitting the italicized answers. Have students use it to compare and contrast the cultures of the Minoans and the Mycenaeans. Then ask students to create their own graphic organizers to compare and contrast the cultures of Sparta and Athens. *(Organizer designs will vary. Sparta—strong government, little personal freedom, powerful army, little development of arts and sciences; Athens—direct democracy, center of learning, literature and the arts, philosophy and science)*

ENGLISH LANGUAGE LEARNERS

	Minoans	Mycenaeans
Location	*island of Crete*	*Greek mainland*
Achievements	*palaces, writing system, beautiful art*	*fortlike walled cities*
Reasons for decline	*conquered by Mycenaeans*	*earthquakes, wars*

CONNECTING TO Art

The Toreador fresco

Frescoes Today, on the island of Crete, visitors can see the ruins of the palace of King Minos. When the palace was first built around 1500 B.C., the walls were covered with beautiful, colorful paintings called frescoes. The paintings were damaged over the years, but some of them have been carefully restored so that people can tell how they first looked. Many of the paintings show scenes from nature. Some are of birds, fish, dolphins, and other animals.

The largest of these paintings is called the *Toreador Fresco*. A toreador is a bullfighter. It shows ancient Minoan athletes jumping over a bull. The bull jumper at the right is a woman.

Understanding What You Read

1. Why did the paintings have to be restored?
2. What do the paintings tell us about life on the island of Crete in ancient times?

The Minoans About 100 years ago on the island of Crete, scientists found the remains of the earliest Greek civilization. By 2000 B.C., the Minoan people had developed a great civilization. From the evidence scientists found, we know that the Minoans built cities and grand palaces that even had running water. We also know that they developed a system of writing and that Minoan artists carved beautiful statues from gold, ivory, and stone. Because Crete's soil was very poor, farming was not very productive. Many people became sailors and fishers. By 1400 B.C., the Minoan civilization began to decline. It was conquered by a group that lived on the Greek mainland, the Mycenaeans (my-suh-NEE-unhz).

The Mycenaeans The Mycenaeans controlled the Greek mainland from about 1600 B.C. to about 1200 B.C. They were a warlike people who lived in tribes. Each tribe had its own chief. The Mycenaean tribes built fort-like cities surrounded by stone walls. They carried out raids on other peoples throughout the eastern Mediterranean. Once they conquered the Minoans, they adopted many aspects of their civilization. For example, they used the Minoan system of writing. By 1200 B.C. earthquakes and wars had destroyed most of the Mycenaean cities. A later Greek poet named Homer wrote a long poem called the *Iliad*, which pulls together about 400 years of historical events, legends, and folk tales. It tells the story of the Trojan War. The Mycenaeans were the Greeks Homer wrote about in that story.

READING CHECK: *Analyzing* How did the geography of the land affect the way the Minoans lived? *did not become farmers because the land was poor; earned living as fishers and sailors because of location close to large bodies of water.*

Across the Curriculum

ART

Architecture Although little is known about the earliest palaces, the later palaces of the Minoans are distinctive for the quality of their workmanship and the sophistication of their designs. The carefully planned palaces had wide staircases and multiple stories, vistas, and storage rooms. Artisans painted the interior walls with scenes of daily life. Even built-in bathtubs and running water could be found inside the palaces.

Critical Thinking: What do the remains of the palaces show about the society that built them?

Answer: The palaces show that the Minoans were quite advanced and that they were concerned with practical function as well as beauty.

Connecting to Art Answers

1. The paintings became damaged over the years.
2. The paintings show that the Minoans saw beauty in nature, that they liked sports, and that women participated in sports.

LEVEL 2: (Suggested time: 30 min.) Have students complete the Level 1 activity. Then use the text to review the physical features of Greece's geography. Ask students to write a few sentences about ways in which the development of Greek civilization was influenced by these features. *(Possible answers: many islands and high mountains—separate communities instead of one united kingdom; rugged land and nearness of sea—fighters, sailors, traders rather than farmers)*

LEVEL 3: (Suggested time: 30 min.) Point out to the class that the Greeks wrote many plays—some funny, some sad—that are still performed today. Organize the class into groups. Ask students to imagine that they are Greek playwrights. Have each group determine what events in Greek history the members would highlight in a play titled The Greatness of Greece and report the selections to the class. Then ask each group to outline a script for a brief scene from its play. You may want to extend the activity by having students perform their scenes. **COOPERATIVE LEARNING**

DAILY LIFE

In ancient Athens, men held all legal rights, and a woman who wished to be thought of as respectable would not even leave her house. Marriage was largely a financial agreement. Women played such a minor role in Greek society that they were often not named publicly until their death. As a result, historians have very little evidence about Athenian women.

Critical Thinking: Ask students to suggest ways that modern scholars have learned about the Greeks' attitudes toward women.

Answer: Student should suggest that scholars have probably learned about ancient Greek attitudes from written accounts of the time.

internet connect

GO TO: go.hrw.com
KEYWORD: SJ3 CH6
FOR: Web sites about Ancient Greece

Visual Record Answers ▶

It shows that Spartan soldiers were strong and wore elaborate helmets.

(Answer to second question) They valued both wisdom and military skill.

▶ Sparta's men were expected to serve in the military until they were 60 years old.
Interpreting the Visual Record What does this bronze figure show about Spartan soldiers?

This photo shows part of a temple in honor of Athena, the Greek goddess of warfare and wisdom. It was built between about 421 B.C. and 415 B.C. Athens was named for Athena.
Interpreting the Visual Record
What does this tell you about Athenian values?

The Greek City-States

Historians do not know much about how the Greeks lived for the next 400 years. However, by the 800s B.C. the Greeks lived in separate **city-states**. A city-state was a city or town that had its own government and laws and controls the land surrounding it. Each city-state had its own calendar, money, and system of weights and measures. Two of the largest city-states, Sparta and Athens, developed in different ways.

Sparta Spartans were very loyal to their city-state. It had a strong government that allowed little personal freedom. Sparta also had a powerful army. At the age of seven, boys left home to be trained as soldiers. When men grew older and left the army, they were expected to work for the public good. Girls received strict training at home. Both boys and girls were educated and encouraged to study music. However, Spartan culture left little time for the development of the arts, literature, philosophy, and science.

Athens The Athenians developed a form of government called **direct democracy**. In a direct democracy, citizens take part in making all decisions. Athenians also had courts where cases were decided by juries. Jurors were chosen by lot. The jury voted on each case by secret ballot. While the Athenians had much more freedom than the Spartans, Athenian women could not participate in government. Slavery was also permitted.

Athenians made great contributions to the arts, literature, philosophy, and science. Builders, artists, and sculptors created beautiful temples and other works. Athenian writers produced great works of literature. They were the first people to write drama, or plays. Scientists discovered new laws of mathematics and developed a system to classify animals and plants. The Greek physician Hippocrates is considered by many to be the father of medical science. Philosophers such as Socrates, Plato, and Aristotle studied questions of reality and human existence. Their thoughts formed the basis for many later ideas at the heart of Western culture. The 400s B.C. are known as the Golden Age of Athens.

✓ **READING CHECK:**
  Comparing and Contrasting
  How did Athenian way of life differ from the Spartan way of life? Spartans: a strict military life; Athenians: developed in the arts, in education, and in the sciences.

Teaching Objective 2

LEVEL 1: (Suggested time: 20 min.) Copy the following graphic organizer onto the chalkboard, omitting the italicized answers. Have pairs of students work together to complete and to understand cause-and-effect relationships as they read about Rome and its eventual decline. Then lead a discussion about the events in the chart, asking students to fill in more details. **ENGLISH LANGUAGE LEARNERS, COOPERATIVE LEARNING**

Cause	Effect
Romans build bridges and roads.	The Roman Republic grows.
Julius Caesar comes to power.	*Romans conquer more territory.*
Roman Empire controls vast lands.	Rome can no longer be ruled by one person.
Dishonest leaders neglect the empire.	*Invaders threaten Rome's borders, Rome is drained of its resources, civil wars begin, and rising prices and taxes make daily life hard.*

Alexander the Great

While Athens and other city-states fought, weakened, and declined, Macedonia, a kingdom to the north, gained strength. Macedonia's king, Philip II, took control of Greece and united it under his rule. When he was killed in 336 B.C., his son, who would become known as Alexander the Great, took his place. Alexander conquered much of the known world—including the Greek city-states—in a short period of time. He built cities that grew to be great centers of learning. Alexander admired the Greeks. Travel and trade along the routes that linked Alexander's empire to Asia spread Greek culture and ideas throughout the world. After Alexander died in 323 B.C., his empire was split into smaller kingdoms. In time, the Romans conquered Alexander's empire.

✓ **READING CHECK:** *Drawing Conclusions* How did Alexander help spread Greek culture throughout his empire? had widespread trade routes built; encouraged trade and travel; respected Greek ideas and culture.

The Early Romans

In about 750 B.C., while the city-states were growing in Greece, people called Latins were settling in villages along the Tiber River in Italy. In time, they united to form the city of Rome. During the late 600s B.C. they were conquered by a people called the Etruscans from the north. The Etruscans brought written language to Rome. Skilled and clever craftspeople, they built paved roads and sewers. Rome grew into a large and successful city. In time, the Greeks also settled in Italy. Their ideas and culture strongly influenced the Romans. Even the Roman religion was partly based on Greek beliefs.

Geography and Roman Civilization In some ways, the geography of Italy helped a great civilization to develop there. Italy is protected by the Alps to the north. The Mediterranean Sea to the west and the Adriatic Sea to the east of this boot-shaped peninsula made trade and travel easy. Not everything about Rome's location, however, was good. Geography also posed some problems. Passages through the Alps left Italy open to invasions. The peninsula's long coastlines left it open to attack from the sea.

✓ **READING CHECK:**
Identifying Cause and Effect
How did the geography of Italy help the Roman civilization grow?
location by the sea made trade and travel easy; protected by the Alps against northern invaders

Alexander's education prepared him to be a great leader. He got his military training from his father and his education from the great Greek philosopher Aristotle.

Linking Past to Present

Alexander's Cities Many legends surround Alexander the Great. One of the best-known involves his horse, Bucephalus. According to legend, Bucephalus was a gift to Philip, but no one could ride him. The horse shied and reared whenever someone tried to mount him. Alexander noticed that the horse was merely afraid of its own shadow. He turned the horse to face the sun and mounted him easily. Acknowledging his son's wisdom, Philip gave Bucephalus to Alexander, who rode him for many years. When the horse died on a campaign, Alexander founded a city at the spot and named it Alexandria Bucephala.

Activity: Have students conduct research to find maps showing Alexander's empire and the cities he founded, including Alexandria Bucephala. Then ask students to investigate the origins of the other cities whose names begin with Alexandria.

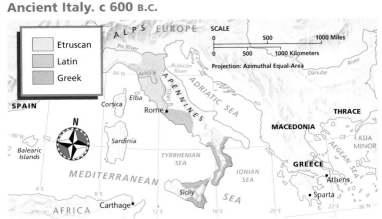

Ancient Italy. c 600 B.C.

LEVEL 2: (Suggested time: 15 minutes) Lead a class discussion comparing and contrasting the government of Greece—a direct democracy—with the government of Rome—a republic. Students should note that the main difference between the two is that in a direct democracy the people themselves vote on new laws, whereas in a republic the people elect officials to make laws for them. Ask students which type of government is more like the one we have in the United States today. *(The government of the United States is more like the government of Rome because we elect our lawmakers.)* Then have each student develop a time line of Rome's growth and decline. Compare and discuss time lines.

LEVEL 3: Suggested time: 20 min.) Ask students how Italy's location and geographic features created both advantages and disadvantages for Rome. *(The Alps protected Rome from invasion, and the sea made travel easy. Both the sea and the natural passes through the Alps left Rome open to invasion.)* Have students work in pairs to create ways to present this information graphically. Ask students to relate these factors to Rome's growth and decline. *(Students might create charts with two columns headed by plus and minus signs, listing advantages and disadvantages in the appropriate columns. They might draw maps with labeled arrows pointing to the appropriate features.)* **COOPERATIVE LEARNING**

COOPERATIVE LEARNING

Throughout history, communities have erected statues of people who contributed to the public good. Because many Romans viewed Julius Caesar as this kind of person, numerous statues of Caesar stood in Rome or in other parts of the Roman Empire. Today many of those statues are in museums.

Organize students into groups. Have each group select a person in your community who students believe deserves to be honored by a statue. Have each group decide on the statue's materials, size, and location. Then ask students to create a plaque that might appear at the base of the statue describing the person and his or her contributions to the community.

▲
Two of the men who killed Caesar were his friends.

Interpreting the Visual Record Why do you think some Roman leaders feared Caesar?

The Roman forum was the center of Rome's government.

▼

The Roman Republic

In 509 B.C. a group of wealthy Romans overthrew the Etruscan king. They replaced the Etruscan rule with a **republic**. A republic is a government in which voters elect leaders to run the state. For the next 200 years, the Romans fought many wars. Their well-trained armies built bridges and roads. The republic grew as the Romans gained new territories. In time, however, controlling such a vast area became a problem for Rome's leaders.

Julius Caesar By 60 B.C. a popular public speaker named Julius Caesar began to win support among Rome's poor. Caesar soon became a powerful general who conquered more territory for the republic. He extended the rule of Rome to present-day France. He even marched into Egypt. He put Cleopatra, a daughter of the ruling family, on the throne as a Roman ally. Victorius, Caesar returned to Rome in 46 B.C. Two years later Roman officials made him ruler for life. However, there were some Roman leaders who feared Caesar's power. On March 15, 44 B.C. Caesar was assassinated.

✓ **READING CHECK:** *Analyzing* What events led to Caesar's rise to power? developed a following among the poor; became a great general and conquered lands so that the Roman empire greatly expanded

Visual Record Answer ▶

Some Roman leaders may have felt he had too much power.

ALL LEVELS: (Suggested time: 40 min.) Organize students into small groups. Have each group create a chart with two headings—"Rome's Greatness" and "Rome's Decline." As students read about the Roman Empire, have them record details that contributed to Rome's greatness and details that contributed to Rome's decline on the chart. *(Examples: Rome's Greatness—well-trained armies, excellent roads and bridges, the Roman empire grew; Rome's Decline—dishonest leaders, invasions, civil wars, high taxes and prices)* **ENGLISH LANGUAGE LEARNERS, COOPERATIVE LEARNING**

Teaching Objective

ALL LEVELS: (Suggested time: 30 min.) Organize students into groups and ask each group to create a children's picture book to be titled The Rise of Christianity. Have each group create an outline for its book. You may want to have each student create text and illustrations for one or more pages. The book should summarize the key points that students have read in the text. You may want to provide additional resources.
COOPERATIVE LEARNING

The Roman Empire

After Caesar's death, his grandnephew Octavian became the leader of the Roman world. The Romans called Octavian *Augustus*, which meant the "honored one." He became known as Augustus Caesar. Historians refer to Augustus as the first emperor of Rome. Augustus, however, never actually used the title of emperor.

Under Augustus, the Roman republic became a great empire. Augustus sent his armies out to conquer new lands. Soon the empire extended from Spain to Syria and from Egypt to the Sahara Desert. In the north, the empire reached all the way to the Rhine and Danube Rivers.

Pax Romana The reign of Augustus began a 200-year period called the Pax Romana, which means "Roman Peace." During this time Rome was a stable and peaceful empire. Laws became more fair. Widespread trade created a strong economy. In order to trade, people had to travel, so the Roman army built roads and bridges that helped unite the vast empire. The Roman army also helped keep the peace by defending Rome's borders against outside invaders.

✓ **READING CHECK:** *Summarizing* What was the Roman Empire like during the Pax Romana? a stable and peaceful empire with fair laws; feeling of security against foreign invasion

The Roman army built many great roads and bridges. If you travel in Europe today, you may use the same roads the Romans built some 2000 years ago.

Trade in the Roman Empire A.D. **117**

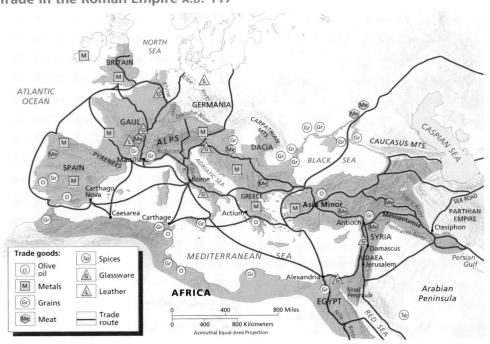

Trade goods:
- ⊙ Olive oil
- Ⓜ Metals
- Ⓖ Grains
- Ⓜ Meat
- Ⓢ Spices
- Ⓐ Glassware
- ⚠ Leather
- Trade route

0   400   800 Miles
0   400   800 Kilometers
Azimuthal Equal-Area Projection

Linking Past to Present

Latin The Latin alphabet was derived from the Etruscan alphabet. The earliest inscription in Latin was found on a cloak pin dating back to the 700s B.C. Today the Latin alphabet is the most widely used alphabetic system in the world. It is the alphabet of the English language and of most European languages.

Although today few people actually speak Latin, it has influenced many modern languages, including English. Numerous words from the medical, scientific, and legal professions are made up of Latin root words, prefixes, and suffixes. For example, the word *legal* comes from the Latin root, *legis*, meaning "law." Other words such as legislature and legitimate are also derived from this same root. The scientific classification system for plant and animal names is based on Latin words. For example, *felis domesticus* is the scientific term for "house cat."

Activity: Ask students to use a dictionary to find familiar words that are based on Latin words or roots.

TEACHER TO TEACHER

Peggy Altoff of Baltimore, Maryland, suggests the following activity to help students learn about the influence of Greece and Rome on the modern world. Have students compare the photographs of Greek and Roman buildings in this section and other sources to public buildings in your community. Have them describe the similarities and differences.

CLOSE

Lead a discussion on how the histories of Greece and Rome are different. *(Examples: Greece consisted of separate city-states, while Rome was more unified; Greece did not became an empire until Philip II, a Macedonian, took control; Greece contained direct democracies, while Rome was a republic; Roman rulers governed huge territories, but Greek rulers did not; Greece was conquered by the Roman Empire, while Rome fell to invading tribes.)* You may want to extend the activity by asking students in which civilization they would rather have lived and why.

Section Review 3

Answers

Define For definitions, see: city-states, p. 116; direct democracy, p. 116; republic, p. 118; aqueduct, p. 120

Reading for the Main Idea

1. Mountains separated settlements of people, allowing the growth of separate states. (NGS 5)

2. Augustus expanded the empire by conquering new lands. (NGS 10)

3. Jews believed Jesus to be their savior. His teachings became the foundation of Christianity. (NGS 6)

Critical Thinking

4. Direct democracy involves all citizens in making decisions. In the United States, elected leaders make decisions.

Organizing What You Know

5. Ancient Greece: direct democracy; Mycenaeans, Minoans; rugged mountains, many islands; Ancient Rome: Republic, then empire; Latins, Etruscans; boot-shaped peninsula, long coastlines

Visual Record Answer ▶

Jesus may have been referring to children's faithfulness, trust, and innocence.

120

▲
Aqueducts were stone canals that carried water from mountains to the city.

▲
This modern stained glass window shows Jesus surrounded by children. According to the Gospel of Matthew, Jesus said that people should change to become more like children.

Interpreting the Visual Record What do you think Jesus meant by this?

Rome's Achievements

The Romans made many important advances in science and engineering. They built temples, palaces, arenas, bridges, and roads. They figured out a way to transport water from one place to another by a system of aqueducts. An **aqueduct** is a sloped bridge-like structure that carries water. Skilled Roman architects designed buildings with domes and arches.

Some Romans were also very talented writers. The verses of ancient Roman poets such as Virgil, Horace, and Ovid are still read today. The biographer Plutarch wrote of the lives of famous Greeks and Romans. The Roman language, Latin, was used for many centuries. It became the basis for many modern languages including French, Spanish, Italian, Portuguese, and Romanian.

✓ **READING CHECK:** *Summarizing* What did Roman engineers contribute to their society? built great temples with domes and arches; created aqueducts to transport water.

The Rise of Christianity

In A.D. 6, the land called Judea came under Roman control. Judea was the home of the Jews. Because the Jews wanted a separate state, they rebelled against Roman rule. After a series of uprisings, the Romans sacked the city of Jerusalem. They destroyed all but the western wall of the city. In A.D. 135, the Roman emperor Hadrian defeated the Jews and banned them from the city of Jerusalem. The Jews continued to build communities outside of the city. They continued to practice their faith.

The Teachings of Jesus Judea was also the birthplace of Christianity. It was here that Jesus had begun to teach in about A.D. 27. His teachings were grounded in Jewish tradition. He taught his followers to believe in only one true God and to love others as they loved themselves. He was said to have performed miracles of healing. He defended the poor. People came from all over to hear Jesus speak. In time, the Romans began to fear that Jesus would lead an uprising.

Eventually, the Romans arrested Jesus. Soon afterward, he was nailed to a cross. His followers believed that Jesus rose from the dead and lived on Earth for 40 days. They believed that he then rose to heaven. Jesus' followers accepted that he was the Messiah, or the savior of the Jews. Soon, his disciples began to spread this message to other people. Christianity had begun to take root.

At first the Romans outlawed Christianity, but their efforts failed to prevent the new religion from spreading. Over the next 300 years, the Christian church became very large. Finally, in A.D. 312, the Roman emperor Constantine declared his support for Christianity. By the end of the century, Christianity had become the official religion of Rome.

✓ **READING CHECK:** *Identifying Cause and Effect* What are some of the events that led to the death of Jesus? developed a great following; Roman fear that he would lead an uprising

Have students complete the Section Review. Then have each student write a question about the content under each subhead. Call on students to read their questions and on others to provide answers. Continue until all questions have been answered correctly. Then have students complete Daily Quiz 6.3.

Ask interested students to conduct research about daily life in ancient Rome. Have students create displays with diagrams and illustrations depicting the food people ate, the homes they lived in, the kinds of work they did, and the entertainment they enjoyed . **BLOCK SCHEDULING**

RETEACH

Have students complete Main Idea Activity 6.3. Then organize students into three groups—one for each Read to Discover question. Have group members write three true-false statements that relate to their question. Ask class members to decide if statements are true or false and then change false statements to true statements. **ENGLISH LANGUAGE LEARNERS**

The Decline of the Roman Empire

In the early a.d. 200s, Rome faced troubled times. Ambitious generals frequently decided to seize power for themselves. Some of them even assassinated emperors and took their places. Over time, the army lost its loyalty to Rome. Soldiers became more interested in becoming wealthy than in defending the empire. Dishonest leaders fought for power. Their neglect of the empire made it possible for invaders to threaten the borders. Invasions were costly and drained the empire of its resources. Inside the empire, civil wars had begun. Daily life became more difficult as taxes and the cost of goods rose higher.

A Split in the Empire By A.D. 284 the Roman Empire could no longer be ruled well by one person. The emperor Diocletian selected a co-emperor to help rule. About 20 years later, the emperor Constantine, who had accepted Christianity, took over the eastern part of the empire. Constantine was a strong ruler. The empire in the east fared much better than the crumbling and weakened empire in the west.

The Fall of Rome As the years passed, invasions from the north continued. Groups such as the Vandals, Visigoths, and Huns set up tribal kingdoms within the empire. In 476, the last emperor of Rome was overthrown by invaders. This marked the end of the empire in the west. The empire of the east was able to fight off invaders. This part of the empire became known as the Byzantine Empire. It lasted until 1453 when it fell to the Ottoman Turks.

✔ **READING CHECK:** *Finding the Main Idea* How did weak leadership lead to the fall of Rome? neglected the empire; allowed invaders to set up tribal kingdoms and overthrow the last emperor in the west

Section Review 3

Define and explain: city-states, direct democracy, republic, aqueduct

Reading for the Main Idea

1. (*Environment and Society*) How did the geography of Greece lead to the development of its city-states?

2. (*Human Systems*) What events under the rule of Augustus helped the Roman republic become the Roman empire?

3. (*Human Systems*) How did the problems in Judea lead to the rise of Christianity?

go.hrw.com Homework Practice Online
Keyword: SJ3 HP6

Critical Thinking

4. **Comparing and Contrasting** How did the direct democracy of the Athenians differ from government in the United States today?

Organizing What You Know

5. Copy the following graphic organizer. Use it to compare the governments, people, and geography of ancient Greece and ancient Rome.

	Ancient Greece	Ancient Rome
Government		
Early People		
Geography		

Review ANSWERS

Building Vocabulary

For definitions, see: hominid, p. 103; prehistory, p. 103; nomads, p. 104; land bridges, p. 104; irrigation, p. 105; civilization, p. 106; division of labor, 106; history, p. 107; migrated, 110; hieroglyphics, p. 112; quipu, p. 113; city-states, p. 116; direct democracy, p. 116; republic, p. 118; aqueduct, p. 120

Reviewing the Main Ideas

1. People live in an organized society, produce more food than they need, live in towns or cities with governments, and practice division of labor. (NGS 12)

2. When people began to farm, they stayed in one place instead of living as wandering nomads. (NGS 15)

3. Possible answers: lumber, buffalo, fish, animals, plants, clay, stone (NGS 11)

4. City-states were separated by rugged landforms. They developed their own governments, laws, and cultures. (NGS 4)

5. Possible answers: architecture, literature, engineering, government (NGS 12)

ASSESS

Have students complete the Chapter 6 Test.

RETEACH

Organize the class into six teams. Have each time write ten short-answer questions about the content of the chapter. Be sure that students' questions cover the content of all sections. Then pair up the teams and have them quiz each other with their questions. **ENGLISH LANGUAGE LEARNERS, COOPERATIVE LEARNING**

PORTFOLIO EXTENSIONS

1. Ask student to imagine they are archaeologists excavating a site once inhabited by one of the ancient civilizations discussed in this chapter. Have them create a list of objects they think they might find that would indicate what life was like in that civilization. Then have students write explanations or draw pictures of each item.

2. Have students conduct research on the Olympic Games in ancient Greece. Students should include information on why the games were held, what events took place at the games, and who was allowed to participate and view the games. Then have students design a commemorative stamp to celebrate the ancient games, or design an original medal that may be awarded to winning participants.

CHAPTER 6 Review ANSWERS

Understanding History and Society

Presentations will vary but should include the effects of tool-making, agriculture, and trade on human civilization. Use Rubric 29, Presentations, to evaluate student work.

Thinking Critically

1. use of tools marked the beginning of technology necessary for development of civilization

2. All of these allowed people to keep records.

3. Possible answers: people will develop more advanced technology, further explore the universe, live peacefully

4. Similarities should include religion and achievements in the arts. Differences should include Rome's imperial government and conquests, and Greece's ideas and lack of political unity.

5. Possible answers: weak or corrupt rulers, war, economic problems, being conquered by other peoples

CHAPTER 6 Reviewing What You Know

Building Vocabulary

On a separate sheet of paper, write sentences to define each of the following words.

1. hominid
2. prehistory
3. nomads
4. land bridges
5. irrigation
6. civilization
7. division of labor
8. history
9. migrated
10. hieroglyphics
11. quipu
12. city-states
13. direct democracy
14. republic
15. aqueduct

Reviewing the Main Ideas

1. (*Human Systems*) What are the four characteristics of civilization?

2. (*Environment and Society*) How did the development of agriculture lead to the growth of villages and towns?

3. (*Environment and Society*) What are some of the resources early people in the Americas depended on for survival?

4. (*Environment and Society*) Why did the city-states in ancient Greece develop differently from each other?

5. (*Human Systems*) What were some of the ancient Romans' achievements?

Understanding History and Society

The Road to Civilization

People of the early civilizations lived very differently from the first humans. Make a flow chart that shows the development of civilization from nomadic hunters and gatherers to city dwellers. As you prepare your presentation about human development, consider the following:

• How people made better tools.
• How agriculture changed society.
• How trade affected the way people lived.

Thinking Critically

1. **Drawing Conclusions** Why is the ability to make and use tools an important step in human development?

2. **Drawing Inferences** Why are the developments of a calendar, a system of counting, and a system of writing so important to a civilization?

3. **Predicting** How do you think civilization might change in the future?

4. **Comparing and Contrasting** How were the cultures of the ancient Greeks and the ancient Romans similar? How were they different?

5. **Identifying Cause and Effect** Give at least two reasons why civilizations decline.

FOOD FESTIVAL

Barley is one of the world's oldest grain crops. It was grown in ancient times and was a staple in both Roman and Greek diets. Today, pearl barley, which has had its hull and outer bran removed is popular in soups and porridges and for making flour for flat breads. Have students find recipes for barley soup. Ask students to bring their soups to class to sample. You may wish to provide figs, dates, and raisins such as those that would have been found at a Roman banquet.

Reproducible
◆ Readings in World Geography, History, and Culture 29
◆ Critical Thinking Activity 6
◆ Vocabulary Activity 6

Technology
◆ Chapter 6 Test Generator (on the One-Stop Planner)

◆ Audio CD Program, Chapter 6
◆ HRW Go site

Reinforcement, Review, and Assessment
◆ Chapter 6 Review, pp. 122–123

◆ Chapter 6 Tutorial for Students, Parents, Mentors, and Peers
◆ Chapter 6 Test
◆ Chapter 6 Test for English Language Learners and Special-Needs Studentsy

Building Social Studies Skills

Map ACTIVITY

On a separate sheet of paper, match the letters on the map with their correct labels.

Macedon   Asia Minor
Ionian Sea   Athens
Aegean Sea   Crete
Knossos

Mental Mapping Skills ACTIVITY

On a separate sheet of paper, draw a freehand map of North America and label the following culture regions:

Eastern   Northwest
Woodlands   Southwest
Great Plains   Mexico

WRITING ACTIVITY

The *Iliad*, a long poem by the ancient Greek poet Homer, tells the story of a great war. Write a short poem that tells the story of another event from ancient history. Your poem does not have to rhyme.

Map Activity
A. Crete   **E.** Asia Minor
B. Knossos   **F.** Ionian Sea
C. Aegean Sea   **G.** Macedon
D. Athens

Mental Mapping Skills Activity
Maps will vary, but listed places should be labeled in their approximate locations.

Writing Activity
Poems will vary, but should accurately describe a historical event. Use Rubric 29, Poems and Songs, to evaluate student work.

Portfolio Activity
Posters will vary, but students should identify the first people who settled in your town or community. Use Rubric 28, Posters, to evaluate student work.

Alternative Assessment

Portfolio ACTIVITY

Learning About Your Local History

Early Settlers Research how your town or city was founded. On poster board, create a display that shows who the first people were to settle in your town.

internet connect

Internet Activity: go.hrw.com
KEYWORD: SJ3 GT6

Choose a topic to explore about the ancient world:
• List different divisions of labor in early civilization.
• Report on Minoan civilization.
• Create a newspaper article about ancient Athens.

internet connect

GO TO: go.hrw.com
KEYWORD: SJ3 Teacher
FOR: a guide to using the Internet in your classroom

CHAPTER 7

The World in Transition
Chapter Resource Manager

Objectives	Pacing Guide	Reproducible Resources
SECTION 1		
The Middle Ages (pp. 125–30) **1.** Define the Middle Ages. **2.** Describe society during the Middle Ages. **3.** Explain how the Middle Ages came to an end.	**Regular** 1 day **Block Scheduling** .5 day *Block Scheduling Handbook, Chapter 7*	**RS** Guided Reading Strategy 7.1 **RS** Graphic Organizer 7 **SM** Map Activity 7
SECTION 2		
Renaissance and Reformation (pp. 131–36) **1.** Identify the main interests of Renaissance scholars. **2.** Explain how people's lives changed during the Renaissance. **3.** Describe changes that took place during the Reformation and Counter-Reformation.	**Regular** 1 day **Block Scheduling** .5 day *Block Scheduling Handbook, Chapter 7*	**RS** Guided Reading Strategy 7.2
SECTION 3		
The Age of Exploration and Conquest (pp. 137–43) **1.** Describe the Scientific Revolution. **2.** Identify factors that triggered the Age of Exploration. **3.** Explain how the English monarchy differed from others in Europe. **4.** Describe the relationship between England and its American colonies.	**Regular** 1.5 days **Block Scheduling** .5 day *Block Scheduling Handbook, Chapter 7*	**RS** Guided Reading Strategy 7.3

Chapter Resource Key

RS Reading Support

IC Interdisciplinary Connections

E Enrichment

SM Skills Mastery

A Assessment

REV Review

ELL Reinforcement and English Language Learners

 Transparencies

 CD–ROM

 Music

 Video

 Internet

 Holt Presentation Maker Using Microsoft® Powerpoint®

 One-Stop Planner CD–ROM

See the *One-Stop Planner* for a complete list of additional resources for students and teachers.

One-Stop Planner CD–ROM

It's easy to plan lessons, select resources, and print out materials for your students when you use the *One-Stop Planner CD–ROM with Test Generator.*

Technology Resources

- One-Stop Planner CD–ROM, Lesson 7.1
- *ARGWorld* CD–ROM
- Homework Practice Online
- HRW Go site

- One-Stop Planner CD–ROM, Lesson 7.2
- *ARGWorld* CD–ROM
- Homework Practice Online
- HRW Go site

- One-Stop Planner CD–ROM, Lesson 7.3
- *ARGWorld* CD–ROM
- Homework Practice Online
- HRW Go site

Review, Reinforcement, and Assessment Resources

ELL	Main Idea Activity 7.1
REV	Section 1 Review, p. 130
A	Daily Quiz 7.1
ELL	English Audio Summary 7.1
ELL	Spanish Audio Summary 7.1

ELL	Main Idea Activity 7.2
REV	Section 2 Review, p. 136
A	Daily Quiz 7.2
ELL	English Audio Summary 7.2
ELL	Spanish Audio Summary 7.2

ELL	Main Idea Activity 7.3
REV	Section 3 Review, p. 143
A	Daily Quiz 7.3
ELL	English Audio Summary 7.3
ELL	Spanish Audio Summary 7.3

internet connect

HRW ONLINE RESOURCES

GO TO: go.hrw.com
Then type in a keyword.

TEACHER HOME PAGE
KEYWORD: SJ3 TEACHER

CHAPTER INTERNET ACTIVITIES
KEYWORD: SJ3 GT7

Choose an activity to:
- write a report on daily life in the Middle Ages.
- create a biography of a Renaissance artist or writer.
- learn more about an explorer described in this chapter.

CHAPTER ENRICHMENT LINKS
KEYWORD: SJ3 CH7

CHAPTER MAPS
KEYWORD: SJ3 MAPS7

ONLINE ASSESSMENT
Homework Practice
KEYWORD: SJ3 HP7
Standardized Test Prep Online
KEYWORD: SJ3 STP7
Rubrics
KEYWORD: SS Rubrics

COUNTRY INFORMATION
KEYWORD: SJ3 Almanac

CONTENT UPDATES
KEYWORD: SS Content Updates

HOLT PRESENTATION MAKER
KEYWORD: SJ3 PPT7

ONLINE READING SUPPORT
KEYWORD: SS Strategies

CURRENT EVENTS
KEYWORD: S3 Current Events

Meeting Individual Needs

Ability Levels

Level 1 Basic-level activities designed for all students encountering new material

Level 2 Intermediate-level activities designed for average students

Level 3 Challenging activities designed for honors and gifted-and-talented students

English Language Learners Activities that address the needs of students with Limited English Proficiency

Chapter Review and Assessment

SM	Critical Thinking Activity 7
REV	Chapter 7 Review, pp. 144–45
REV	Chapter 7 Tutorial for Students, Parents, Mentors, and Peers
ELL	Vocabulary Activity 7
A	Chapter 7 Test
A	Unit 2 Test
	Chapter 7 Test Generator (on the One-Stop Planner)
	Audio CD Program, Chapter 7
A	Chapter 7 Test for English Language Learners and Special-Needs Students
A	Unit 2 Test for English Language Learners and Special-Needs Students

LAUNCH INTO LEARNING

Ask students to suggest reasons why people travel today. *(Possible answers: for pleasure, for business, for religious purposes, to explore a new place)* Tell students that throughout history people have had similar reasons for travel. Some people were involved in the business of trading goods. Others went on pilgrimages to religious shrines and holy places. Still others went to search for and explore new lands in order to gain political and economic power. Point out that while all of these people were experiencing change away from home, the people, places, ideas, and institutions of the time were also changing. Tell students that they will learn about several periods of great change in Chapter 7.

Section 1

Objectives
1. Define the Middle Ages.
2. Describe society in the Middle Ages.
3. Explain how the Middle Ages came to an end.

LINKS TO OUR LIVES

You might want to share with your students the following reasons for learning about these periods of change:

▶ Some challenges facing governments today are similar to those faced in the Middle Ages and the Renaissance.

▶ Conflicts that began during this period are still flaring up in parts of the world today.

▶ Certain patterns of change that began in this period, such as the shift of population from rural areas to urban, have continued to the present.

▶ The works of many Renaissance artists are still admired today.

▶ In some parts of the world, the national boundaries recognized today were drawn by colonizing powers during this period.

The World in Transition

The next 2000 years in human history were a time of great change. In this chapter you will read about the growth of new empires, the development of new political systems, and the search for new ideas.

During the Middle Ages about 1,000 years ago, society was separated into distinct classes. Certain young women of the time who were born into noble families had to undergo training on how to behave.

Young girls from the families of lesser nobles often went to live in the households of higher-ranking noblewomen. There they would be trained in the skills and responsibilities that were expected of women in their rank.

Generally a young noblewoman was taught to sew, to weave, to cook, to play musical instruments, and to sing. She also learned the social conduct that was proper for women of the nobility. In some cases girls and young women were also instructed in the skills of household supervision.

Young noblewoman from the Middle Ages

Knight with female admirers

Papal palace in Avignon, France

((o)) LET'S GET STARTED

Copy the following question onto the board: *Why do you think the period of European history following the fall of the Roman Empire is called the Middle Ages?* Call on volunteers to share their answers. Point out that the name stems from the position of the period between the ancient and modern worlds. Tell students that they will learn more about the Middle Ages in this section.

Building Vocabulary

Write the key terms on the board. Call on volunteers to find the definitions of the terms in the glossary and to read them to the class. Then ask students write sentences that link pairs of related words. *(For example,* **vassals** *received* **fiefs** *from their lords.* **Knights** *practiced* **chivalry.** **Clergy** *and* **cathedrals** *were both parts of the Catholic Church.)*

Section 1

The Middle Ages

Read to Discover

1. What were the Middle Ages?
2. What was society like during the Middle Ages?
3. How did the Middle Ages come to an end?

Define

feudalism
nobles
fief
vassals
knight
chivalry

manors
serfs
clergy
cathedrals
Crusades

middle class
vernacular

Locate

France
England
Normandy
Norway
Denmark
Sweden

Jerusalem
Spain

A medieval knight

WHY IT MATTERS

The desire to control Jerusalem, which led to the Crusades, is still a cause of conflict. Use **CNNfyi.com** or other **current events** sources to learn what is happening in Jerusalem today. Record your findings in your journal.

The Rise of the Middle Ages

In the year A.D. 476, the last of the western Roman emperors was defeated by invading Germanic tribes. These invaders from the north brought new ideas and traditions that gradually developed into new ways of life for people in Europe. Historians see the years between the last of the Roman emperors and the beginnings of the modern world in about 1500 as a period of change. Because it falls between the ancient and modern worlds, this time in history is called the Middle Ages or the medieval period. *Medieval* comes from the Latin for "middle age."

The time from the 400s to around 1000 is known as the Early Middle Ages. As this period began, the Roman system of laws and government had broken down. Western Europe was in a state of disorder. It was divided into many kingdoms ruled by kings who had little authority. For example, Britain was largely controlled by two Germanic tribes, the Angles and the Saxons. These groups had established several independent kingdoms.

Nobles in the Middle Ages built their own castles for protection.

Teaching Objective 1

ALL LEVELS: (Suggested time: 10 min.) Have each student write a brief, dictionary-style definition of the term *Middle Ages*. Remind students that their definitions should note both the appropriate time period and the geographic area covered by the Middle Ages. When they have finished, have students compare their definitions with the definition in a classroom dictionary. Then call on volunteers to name characteristics that define the European Middle Ages. **ENGLISH LANGUAGE LEARNERS**

Scott Whitlow of Round Rock, Texas, suggests the following activity to help students better understand the Middle Ages. Ask students to name some images that come to mind when they think of the Middle Ages. Copy their list of images onto the chalkboard. *(Students will likely mention castles, knights in armor, princesses, swords, and so on.)* Refer to this list as you go through the main ideas in this section. At the end of the lesson, ask students if and how their ideas about the Middle Ages have changed.

Across the Curriculum
TECHNOLOGY

Weaponry The Franks were famous for their swords, which were in great demand across Europe. These swords were known for their balance and strength. Frankish foot soldiers used short swords and carried spears, bows and arrows, and shields. Soldiers who rode on horseback carried both long and short swords, spears, and round shields. Charlemagne's mounted soldiers also wore simple helmets and other basic forms of armor, representing what may be the early stages of the armored knight of later centuries.

Activity: Have students conduct research on the weapons of the Middle Ages. Ask them to make sketches of some of the weapons used during the period.

▲
Charlemagne (742-814) united most of the Christian lands in western Europe. He built libraries and supported the collection and copying of Roman books. His rule became a model for later kings in medieval Europe.

Interpreting the Visual Record

What qualities made Charlemagne a good king?

internet connect

GO TO: go.hrw.com
KEYWORD: SJ3 CH7
FOR: Web sites about the world in transition

▶

Between A.D. 600 and 1000, many people invaded western Europe.

Interpreting the Map What group of invaders were most active in the Mediterranean area?

The Franks The Franks were one of the Germanic tribes that moved into western Europe. In the 490s, Clovis, the king of the Frankish tribes, became a Christian and gained the support of the church. He conquered other Frankish tribes and won control of the territory of Gaul. Today this area is called France after the Franks. In 732 a Frankish army under King Charles Martel held off an army of Spanish Moors who had invaded his kingdom. This conflict is called the Battle of Tours. This defeat drove the Muslim Moors back into Spain, creating a border between the Christian and Muslim worlds.

The greatest Frankish king, Charlemagne (SHAR–luh–mayn), ruled from 768 to 814. A strong, smart leader, Charlemagne established schools and encouraged people to learn to read and write. His greatest accomplishment was to unite most of western Europe under his rule. Charlemagne's empire included most of the old Roman Empire plus some new additional territory. It later became known as the Holy Roman Empire because the pope declared Charlemagne "Emperor of the Romans."

After Charlemagne died, his grandsons weakened the empire by dividing it among themselves. Muslims invaded from the south. Slavs invaded from the east. From the north came the dreaded Vikings.

The Vikings During the 800s and 900s the Vikings were feared throughout western Europe. They came from what are now the countries of Denmark, Norway, and Sweden. The Vikings were not only farmers but also skilled sailors and fierce warriors who raided towns along the coasts of Europe. Eventually these invaders settled in England, Ireland, and other parts of Europe. A large Viking settlement in northwestern France gave that region its name. It is called Normandy, from the French word for "Northmen." The Vikings there came to be known as the Normans.

Visual Record Answer ▶

strength, intelligence, dedication

Map Answer ▶

the Vikings

Teaching Objective 2

LEVEL 1: (Suggested time: 30 min.) Copy the following graphic organizer onto the board, omitting the italicized answers. Organize students into groups and have them fill in the organizer with information about features of life in the Middle Ages. Ask volunteers to share their organizers with the class. **ENGLISH LANGUAGE LEARNERS, COOPERATIVE LEARNING**

Political System *based on feudalism; nobles held fiefs; lords very powerful*

Men and Women *men inherited property; women had few rights and little power*

LIFE IN THE MIDDLE AGES

Economy *limited trade; manorial system; serfs worked land*

Church *powerful and wealthy; great cathedrals*

Life in the Middle Ages

Feudalism Within 100 years of Charlemagne's death, the organized central government he had put in place was gone. By the 900s most of Europe was governed by local leaders under a system known as **feudalism**. It was a way of organizing and governing people based on land and service. In most feudal societies, the king, who owned all the land in his kingdom, granted some lands to **nobles**—people who were born into wealthy, powerful families. The grant of land was called a **fief**. Nobles had complete power over their land—power to collect taxes, enforce laws, and maintain armies. In return for land, they became **vassals** of the king. This means that they promised to serve the king, especially in battle. A noble could, in turn, grant fiefs to lesser nobles. In so doing, that noble would become a lord, and the lesser nobles would become his vassals. A vassal owed service—especially military service—to his lord.

Feudalism was a very complex system. Its rules varied from kingdom to kingdom. Feudal relationships in France, for example, were not the same as those in Germany. The relationships between kings and nobles in England were very different from those in either France or Germany. In addition, the nature of feudal relationships were constantly changing. Laws that governed a king's or a vassal's behavior one year might not apply just a few years later. It was sometimes very difficult, even during the Middle Ages, for people to keep track of their feudal obligations.

Nevertheless, powerful lords were the ruling class in Europe for more than 400 years. Some lords were so powerful that the king remained on the throne only with their support. Over time it became the custom that the owner of a fief would pass his land on to his son. By about 1100 the custom was that the eldest son inherited his father's land. Women had few rights when it came to owning property. If a woman who owned land married, her husband gained control of her land.

Knights The most common type of nobleman was the **knight**, or warrior, who received land from a lord in return for military service. Knights lived by a code of behavior called **chivalry**. This code said that a knight had to be brave, fight fairly, be loyal, and keep his word. He had to treat defeated enemies with respect and be polite to women. In battle, a knight wore heavy metal armor and a metal helmet. He carried a sword, a shield, a lance, and other weapons. Knights had plenty of opportunities to fight. In addition to large-scale wars that occurred during the Middle Ages, frequent smaller battles took place between lords who tried to seize each other's lands.

This stained-glass window shows a lord and his vassals.

Interpreting the Visual Record

What details in the picture show that the lord is more powerful than the vassals?

To become a knight, a boy usually had to come from a noble family. Boys began their training at the age of seven.

Interpreting the Visual Record **How do the knights in this picture look, and how do you think they feel?**

Across the Curriculum
LITERATURE
King Arthur

Legends King Arthur is a mythical figure from the Middle Ages. Legend tells that Arthur was the son of a king named Uther but was raised by one of the king's vassals, Sir Ector, who never told the boy of his ancestry. Arthur became king by pulling a magic sword out of a stone. At his favorite castle, Camelot, he presided over his loyal knights of the Round Table.

Sir Thomas Malory wrote *Le Morte d'Arthur (The Death of Arthur)* around 1469. Since then, many poems and stories have been written about the adventures of King Arthur and his knights. These include T.H. White's *The Once and Future King,* the movie *The Sword in the Stone,* and the Broadway musical *Camelot.*

Discussion: Ask students to share stories about King Arthur with which they are familiar. Lead a discussion about historical truths about the Middle Ages that can be seen in these tales.

◄ **Visual Record Answer**

king sitting on a throne and wearing a crown; vassals at a lower level

◄ **Visual Record Answer**

look strong and brave; possibly feel proud, excited, a little scared

127

LEVELS 2 AND 3: (Suggested time: 45 min.) Organize the class into groups. Ask each group to write a skit in which group members play the parts of figures in medieval society, such as kings, nobles, knights, and serfs. Tell students that their skits should demonstrate the ways in which various members of society interacted with one another. Call on volunteers to perform their skits for the class. **COOPERATIVE LEARNING**

Teaching Objective 3
ALL LEVELS: (Suggested time: 25 min.) Call on students to find in the text some reasons the Middle Ages ended. Ask volunteers to come forward and write these reasons on the board. *(Reasons include the growth of cities, increased interest in education and trade, a rise in the power of kings, and the weakening of the Church.)* Then lead a class discussion about how each of the given reasons contributed to the decline of the Middle Ages. **ENGLISH LANGUAGE LEARNERS**

HUMAN SYSTEMS

The Crusades brought Europeans and their customs to the Holy Land. After the capture of Jerusalem, the crusaders set up four small states. They introduced feudalism and subdivided the land among feudal lords. Trade between Europe and the Holy Land sprang up. Italian ships carried goods back and forth.

The European occupiers were affected as well. Christians and Muslims lived alongside each other and grew to respect each other. Many Europeans adopted Eastern customs and began to wear Eastern clothes and eat Eastern foods.

Activity: Ask students to imagine that they are Europeans living in one of the Crusader States. Have them write letters to family members back home comparing and contrasting life in the Holy Land with life in Europe.

This picture, made during the 1400s, shows peasants at work during a harvest.
Interpreting the Visual Record What do you think farming was like in the Middle Ages?

The Manorial System Trade declined after the end of the Roman Empire. Most people took up farming for a living. The large farm estates which some nobles developed were called **manors**. Such manors included large houses, farmed lands, wooded land, pastures, fields, and villages. The lord of a manor ruled over peasants called **serfs** who lived on his land. Serfs were poor and had no rights. They had to work the lord's land and give him part of their crops. They could not leave the manor without the lord's permission. A manor was usually self-sufficient. Almost everything people needed, including food and clothing, was produced right there.

The Church One of the largest and wealthiest landowners during the Middle Ages was the Catholic Church. Headed by the pope, the church was enormously powerful, with its own laws and courts.

Officials of the church were known as the **clergy**. Beneath the pope were bishops and priests. Other members of the clergy were monks, who lived in monasteries, and nuns, who lived in convents. While most ordinary people could not read or write, many members of the clergy were educated. In monasteries, monks prayed, studied, and copied ancient books.

Eventually, huge churches, called **cathedrals**, were built. Cathedrals cost a great deal of money and were beautifully decorated. Some of the most common decorations were elaborate stained-glass windows.

The Crusades In the late 1000s, the pope asked the lords of Europe to join in a great war against the Turks, who had gained control of Palestine, which the Christians called the Holy Land. This war turned into a long series of battles called the **Crusades**. The First Crusade lasted from 1095 to 1099. Crusaders captured Jerusalem and killed many of the Muslims and Jews who lived there. However, over the next 100 years, the Turks won back the land they had lost. Three more major Crusades were launched. Although the Holy Land was not recaptured, the Crusades led to important changes in Europe.

✓ **READING CHECK:** *Summarizing* How did feudalism make the nobles and their vassals depend on each other? Lords needed soldiers; vassals needed a means of support.

Visual Record Answer ▶

hard work, simple tools, no machinery

▶ Knights in the Middle Ages often fought from horseback.

Teaching Objectives 1–3

ALL LEVELS: (Suggested time: 35 min.) Organize students into groups and have each group construct a time line that illustrates one aspect of the Middle Ages. For example, one group might examine political developments while another looks at military history. Encourage group members to conduct research to find additional material for their time lines. Display the completed time lines in the classroom.

ENGLISH LANGUAGE LEARNERS, COOPERATIVE LEARNING

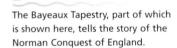

▲
The Bayeaux Tapestry, part of which is shown here, tells the story of the Norman Conquest of England.

The High Middle Ages

The Crusades brought about major economic and political changes in Europe. The period following the Crusades to about 1300 is known as the High Middle Ages.

Stronger Nations Many lords sold their lands to raise money in order to join the Crusades. Without land, they had no power. In addition, many lords died in the Crusades. With fewer powerful lords, kings grew stronger. By the end of the Middle Ages, England, France, and Spain had become powerful nations. Strong central governments and the decline of the nobility's power helped to bring about the end of feudalism in Europe.

In 1066 William, Duke of Normandy in northwestern France, claimed the English throne. He landed in England, defeated the Anglo-Saxon army, and was crowned King William I of England. He became known as William the Conqueror. William built a strong central government in England. When William's great-grandson John took the throne, however, he pushed the nobles too far by raising taxes. In 1215 a group of nobles forced King John to sign Magna Carta, one of the most important documents in European history.

Magna Carta stated that the king could not collect new taxes without the consent of the Great Council, a body of nobles and church leaders. The king could not take property without paying for it. Any person accused of a crime had the right to a trial by jury. The most important provision of Magna Carta was that the law, not the king, is the supreme power in England. The king had to obey the law. The Great Council was the forerunner of England's Parliament, which governs Great Britain today.

The Growth of Trade The Crusades increased Europeans' demand for Asian dyes, medicines, silks, and spices. People also began buying lemons, apricots, melons, rice, and sugar from Asia. In exchange, Europeans traded timber, leather, wine, glassware, and woolen cloth.

Increased trade led to the growth of manufacturing and banking. A **middle class** of merchants and craftsmen arose between the nobility and the peasants. The late medieval economy formed the basis for our modern economic system.

▲
People exchanged goods at trade fairs in the Middle Ages.

Across the Curriculum
MATH

Trade Fairs Fairs in the Middle Ages were colorful events. They were also the main method of carrying out trade. Because of fairs, easier ways to trade developed.

For example, some goods were sold by length or weight. Because people came from all over Europe to trade fairs, a standard system of weights and measures was needed. The troy weight was set to weigh gold and silver. This weight, named for the town of Troyes, France, is still used today.

Trade fairs also helped develop the bill of exchange. This note was a written promise to pay a sum of money at a later time.

Activity: Have students imagine that they are cloth traders at a fair in the Middle Ages. One meter of cloth is worth 3 ducats (a currency of the time). The spice trader nearby sells pepper at 12 ducats per ounce. How much cloth would they need to sell to buy 4 ounces of pepper? *(16 meters)*

REVIEW AND ASSESS

Have students complete the Section Review. Then have students work in pairs to review the section material. Have each student write one question about each of the subjects discussed in the text to ask his or her partner. Then have students complete Daily Quiz 7.1.

RETEACH

Have students complete Main Idea Activity 7.1. Then organize the class into groups. Tell each group to create a two-column chart with the headings "Early Middle Ages" and "High Middle Ages." Have students fill in the space beneath each heading with information about each period. Ask volunteers to write their charts on the board.
ENGLISH LANGUAGE LEARNERS, COOPERATIVE LEARNING

EXTEND

Have interested students conduct research on the art, music, or literature of the Middle Ages. Encourage students to choose a specific topic to research, such as stained glass windows or medieval instruments. Have each student present his or her findings in a report or poster. **BLOCK SCHEDULING**

Section Review 1

Answers

Define For definitions see: feudalism, 127; nobles, 127; fief, 127; vassals, 127; knight, 127; chivalry, 127; manor, 128; serf, 128; clergy, 128; cathedrals, 128; Crusades, 128; middle class, 129; vernacular, 130

Reading for the Main idea

1. refers to time between classical age and modern world (NGS 10)

2. feudalism; powerful lords sometimes more powerful than kings; women had few rights; little travel or trade; manorialism; lords owned self-sufficient manors; serfs worked land; most people not educated; church very powerful (NGS 11)

3. breakdown of feudalism and the manorial system, weakening of church power; growth of trade, growth of cities; growth of universities; stronger kings and nations (NGS 13)

Critical Thinking

4. took supreme power away from a king, and made the law the supreme power; idea has influenced national governments in modern times

Organizing What You Know

5. noble: collect taxes, enforce laws, maintain army
vassal: serve king or lord, especially in battle

Between 1347 and 1351 the Black Death killed about one third of Europe's entire population.

A lecturer teaches at a university during the Middle Ages. Medieval universities taught religion, the liberal arts, medicine, and law.

The Growth of Cities As Europe's economy got stronger, cities grew. Centers for trade and industry, cities attracted merchants and craftsmen as well as peasants, who hoped to find opportunities for better lives and more freedom. Both the manorial system and the feudal system began to fall apart.

During the Middle Ages, cities were crowded and dirty. When disease struck, it spread rapidly. In 1347 a deadly disease called the Black Death swept through Europe. Even this disastrous plague, however, had some positive effects. With the decrease in population came a shortage of labor. This meant that people could begin to demand higher wages for their work.

Education and Literature Most people could not read or write. As cities grew and trade increased, so did the demand and need for education. Between the late 1000s and the late 1200s, four important universities developed in England, France, and Italy. By the end of the 1400s, many more universities had opened throughout Europe.

Most people during the Middle Ages did not speak, read, or write Latin, the language of the Church. They spoke **vernacular** languages—everyday speech that varied from place to place. Writers such as Dante Alighieri in Italy and Geoffrey Chaucer in England began writing literature in vernacular languages. Dante is best known for *The Divine Comedy*. Chaucer's most famous work is *The Canterbury Tales*.

The End of the Middle Ages The decline of feudalism and the manorial system, the growth of stronger central governments, the growth of cities, and a renewed interest in education and trade brought an end to the Middle Ages. In addition, stronger kings challenged the power of the Catholic church. By the end of the 1400s, a new age had begun.

✓ **READING CHECK:** *Finding the Main Idea* Why did kings become more powerful after the Crusades? Lords sold land to raise money for the Crusades; many died in battle.

go.hrw.com Homework Practice Online
Keyword: SJ3 HP7

Section Review 1

Define and explain: feudalism, nobles, fief, vassals, knight, chivalry, manors, serfs, clergy, cathedrals, Crusades, middle class, vernacular

Reading for the Main Idea

1. *Human Systems* Why is the time between the A.D. 400s and about 1500 called the Middle Ages?

2. *Human Systems* What was life like in the Middle Ages?

3. *Human Systems* What led to the end of the Middle Ages?

Critical Thinking

4. **Evaluating** Why is Magna Carta considered one of the most important documents in European history?

Organizing What You Know

5. **Analyzing** Copy the following graphic organizer. Use the right-hand column to describe briefly each person's responsibility.

Person	Responsibility
noble	
vassal	

Section 2

Objectives

1. Identify the main interests of Renaissance scholars.

2. Explain how people's lives changed during the Renaissance.

3. Describe changes that occurred during the Reformation and Counter-Reformation.

FOCUS

LET'S GET STARTED

Copy the following instructions onto the chalkboard: *Look at the paintings from the Middle Ages in Section 1. Then look at the artwork from the Renaissance in Section 2. How are they different?* Discuss responses. Explain that art in the Renaissance became more sophisticated and realistic. Tell students that they will learn more about the Renaissance in this section.

Building Vocabulary

Write the key terms on the chalkboard. Underline *human* in **humanists**. Point out that humanist thinkers in the Renaissance stressed the importance of human reason. Then underline *protest* and *reform* in **Protestants** and **Reformation**. Explain that the Protestants of the Reformation protested against certain teachings and wanted to reform the Catholic Church. Point out that *counter-* in **Counter-Reformation** means "against." Tell students that this movement was staged against the effects of the Reformation. Call on a volunteer to read the definitions of all the key terms.

Section 2 — The Renaissance and Reformation

Read to Discover

1. What were the main interests of Renaissance scholars?

2. How did people's lives change during the Renaissance?

3. What changes took place during the Reformation and Counter-Reformation?

Define

Renaissance
humanists
Reformation
Protestants
Counter-Reformation

Locate

Rome
Florence

WHY IT MATTERS

Works of Renaissance art have remained popular for hundreds of years. Use **CNNfyi.com** or other current events sources to learn about Renaissance paintings displayed in museums around the world today. Record your findings in your journal.

Leonardo da Vinci's sketch of a flying machine

Section 2 RESOURCES

Reproducible
◆ Guided Reading Strategy 7.2

Technology
◆ One-Stop Planner CD–ROM, Lesson 7.2
◆ Homework Practice Online
◆ HRW Go site

Reinforcement, Review, and Assessment
◆ Main Idea Activity 7.2
◆ Section 2 Review, p. 136
◆ Daily Quiz 7.2
◆ English Audio Summary 7.2
◆ Spanish Audio Summary 7.2

New Interests and New Ideas

The Crusades and trade in distant lands caused great changes in Europe. During their travels, traders and Crusaders discovered scholars who had studied and preserved Greek and Roman learning. While trading in Southwest Asia and Africa, people learned about achievements in science and medicine. Such discoveries encouraged more curiosity. During the 1300s, this new creative spirit developed and sparked a movement known as the **Renaissance** (re-nuh-SAHNS). This term comes from the French word for "rebirth." The Renaissance brought fresh interest in exploring the achievements of the ancient world, its ideas, and its art.

Beginning of the Renaissance The Renaissance started in Italy. Italian cities such as Florence and Venice had become rich through industry and trade. Among the population was a powerful middle class. Many members of this class were wealthy and well-educated. They had many interests beside their work. Many studied ancient history, the arts, and education. They used their fortunes to support painters, sculptors, and architects, and to encourage learning. Scholars revived the learning of ancient Greece and Rome. Enthusiasm for art and literature increased. Over time the ideas of the Renaissance spread from Italy into other parts of Europe.

The Humanities As a result of increased interest in ancient Greece and Rome, scholars encouraged the study of subjects that had been taught in ancient Greek and Roman schools. These subjects, including history, poetry, and grammar, are called the humanities.

▲ The powerful Medici family ruled Florence for most of the Renaissance. Banker Cosimo de' Medici, seen here, was a great supporter of the arts.

Teaching Objective 1

LEVELS 1 AND 2: (Suggested time: 15 min.) Write the following questions on the chalkboard: *What events and discoveries sparked the Renaissance? What subjects were scholars interested in during this period?* Tell students to write down their responses. Then call on volunteers to share their answers and discuss them in class. Lead a class discussion about the ideas and beliefs that inspired Renaissance artists.
ENGLISH LANGUAGE LEARNERS

LEVEL 3: (Suggested time: 30 minutes) Have pairs of students work together to write fictional interviews with Renaissance figures like Michelangelo or Leonardo da Vinci. Tell students that their interviews should note the ideas that inspired their subjects to create the works for which they are famous. Call on volunteers to present their interviews in class and to discuss how their subjects represented the Renaissance spirit.
COOPERATIVE LEARNING

USING ILLUSTRATIONS

Have students examine the painting by Peter Breughel on this page. Ask them why the subject matter of the painting is unusual. *(Possible answer: It is unusual for a painting to focus on children at play.)* Also ask how the painting reflects Renaissance ideas. *(It shows normal people in their daily lives; it is realistic; it shows people enjoying themselves.)*

Activity: Have students find other examples of paintings by Breughel or other Renaissance artists. Ask them to write short descriptions of the works and how they reflect Renaissance ideas. Have students display the paintings and read their descriptions in class.

▲

Isabella d'Este was a very intelligent and powerful member of a wealthy Italian noble family. She was educated in languages and poetry. She supported the arts and hired noted architects to design parts of her palace.

Humanists, the people who studied these subjects, were practical. They wanted to learn more about the world and how things worked. Reading ancient texts helped them recover knowledge that had been forgotten or even lost. They believed that people should support the arts. They also thought that education was the only way to become a well-rounded person. People were urged to focus on what they could achieve in this life.

✓ **READING CHECK:** (*Cause and Effect*) How did the Renaissance begin?
Trade and Crusades bring people into contact with ideas; curiosity about the world awakened.

The Creative Spirit

During the Renaissance, interest in painting, sculpture, architecture, and writing was renewed. Inspired by Greek and Roman works, artists produced some of the world's greatest masterpieces for private buyers as well as for churches and other public places.

Art Leonardo da Vinci and Michelangelo truly represented the Renaissance. Leonardo achieved the Renaissance ideal of excelling in many things. He was not only a painter but also an architect, engineer, sculptor, and scientist. He sketched plants and animals. He made detailed drawings of a flying machine and a submarine. He used mathematics to organize space in his paintings and knowledge about the human body to make figures more realistic. Michelangelo was not only a brilliant sculptor, but also an accomplished painter, musician, poet, and architect.

Northern European merchants carried Italian paintings home, and painters went from northern Europe to study with Italian masters. In time, Renaissance ideas spread into northern and western Europe.

Visual Record Answer ▶

Possible answers: somersaults, tag, hoop spinning, tug-of-war, hanging from a crossbar, leap-frog

This painting by Pieter Brueghel shows children's games in the 1500s. Many Renaissance painters chose to focus their attention on daily life.
Interpreting the Visual Record **What activities do you see that children still do today?**

Teaching Objective 2

LEVEL 1: (Suggested time: 20 min.) Organize the class into groups and assign each group one aspect of life in the Renaissance. *(Possible topics include cities, foods, trade, and so on.)* Have each group write a paragraph that explains how their assigned aspect changed during the period. Call on a volunteer from each group to share its findings with the class. **ENGLISH LANGUAGE LEARNERS, COOPERATIVE LEARNING**

LEVEL 2: (Suggested time: 25 min.) Ask students to think about how daily life changed during the Renaissance. Then have each student write two journal entries. One entry should describe a typical day in the life of a teenager in the Middle Ages, while the other should describe a teenager's life in the Renaissance. Call on volunteers to read their entries and discuss how life changed during the Renaissance.
ENGLISH LANGUAGE LEARNERS

CONNECTING TO Technology

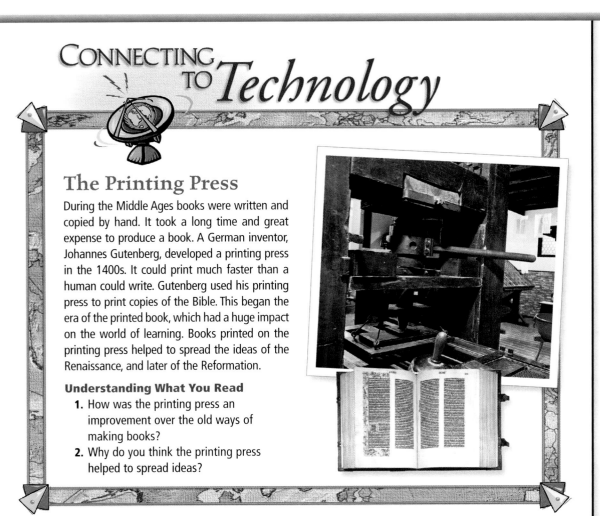

The Printing Press

During the Middle Ages books were written and copied by hand. It took a long time and great expense to produce a book. A German inventor, Johannes Gutenberg, developed a printing press in the 1400s. It could print much faster than a human could write. Gutenberg used his printing press to print copies of the Bible. This began the era of the printed book, which had a huge impact on the world of learning. Books printed on the printing press helped to spread the ideas of the Renaissance, and later of the Reformation.

Understanding What You Read
1. How was the printing press an improvement over the old ways of making books?
2. Why do you think the printing press helped to spread ideas?

Writing Writers of the time expressed the attitudes of the Renaissance. Popular literature was written in the vernacular, the people's language, instead of in Latin. Dutch writer Desiderius Erasmus criticized ignorance and superstition is his work *In Praise of Folly*. In *Gargantua*, French writer François Rabelais promoted the study of the arts and sciences. Spanish writer Miguel de Cervantes wrote *Don Quixote* in which he mocked the ideals of the Middle Ages. Italian writers such as Machiavelli and Baldassare Castiglione wrote handbooks of proper behavior for rulers and nobles.

Of all Renaissance writers, William Shakespeare is probably the most widely known. He was talented at turning popular stories into great drama. His plays and poetry show a great understanding of human nature. He used the popular English language of his time to skillfully express the thoughts and human feelings of his characters. Many of Shakespeare's subjects and ideas are still important to people today.

✓ **READING CHECK:** *Summarizing* What was the Renaissance attitude?
Curiosity about world; focus on achievement in this world; involvement in the arts; ed cation important.

William Shakespeare wrote such famous plays as *Romeo and Juliet*, *Hamlet*, and *Macbeth*.

▼

Across the Curriculum
LITERATURE
Thomas More's *Utopia* One important Renaissance figure was the English writer Thomas More. In 1516 More published *Utopia*, a book that became very popular throughout Europe. In the book, More criticized corrupt and harsh governments in Europe and the inequality in European society. He imagined a world in which all male citizens were equal and everyone worked to support society. As a result of More's book, the word *utopia* has come to mean "an ideal place or society."

Activity: Have students conduct research on great writers or thinkers of the Renaissance. Have each student create a collector's card (much like a baseball card) with a picture, short biography and list of the person's achievements.

Connecting to Technology Answers
1. Copying by hand took a long time and made books expensive. Printing was much quicker and reduced the costs of books.
2. Printed material can be passed around a wider area and more quickly than word of mouth.

During the Renaissance, more people learned to read and write. This painting shows a couple working together in their banking business.
Interpreting the Visual Record What does this painting suggest about the changing role of women during the Renaissance?

The city of Florence was the center of the early Renaissance. The large domed building is the Duomo, the cathedral of Florence

The Renaissance and Daily Life

The Renaissance was not only a time of learning, art, and invention. It was also a time of change in people's daily lives. As the manorial system of the Middle Ages fell apart, many peasants left the manors on which they had lived. Because there were fewer people to work the land, many of these peasants could now demand wages for their labor. For the first time, they had money to spend. As Europe's population began to increase again after the Black Death, however, prices rose very quickly. Only wealthy people could afford many goods.

Although they now had some money, most peasants were still poor. Some migrated to cities in search of work. Instead of raising their food, they bought it in shops. In the 1500s traders brought to Europe new vegetables such as beans, lettuce, melons, spinach, and tomatoes. Traders also brought new luxury items such as coffee and tea. As the idea of the printing press caught on in Europe, books became more common. More and more people learned how to read. Gradually a new way of life developed, and the quality of life slowly began to change.

✓ **READING CHECK:** *Comparing and Contrasting* In what ways did life in Europe change during the Renaissance? Migration to cities; less dependence on farming as a means of support; work for wages; new foods, more varied diets.

134

The Reformation

As humanism became more popular, people began to question their religious beliefs. Northern humanists thought the Roman Catholic Church had become too powerful and too worldly. They thought that it was too rich and owned too much land and that it had lost the true message of Jesus. Some people began to question the pope's authority. The humanists' claims sparked a movement that split the church in western Europe during the 1500s. This movement is called the **Reformation**.

Martin Luther A German monk named Martin Luther disagreed with the Catholic Church about how people should act. The Church taught that the way to heaven lay in attending church, giving money to the church, and doing good deeds. Luther said that the way to heaven was simply to have faith in God. He argued that the Bible was the only authority for Christians. The printing press helped Luther's ideas spread. He gained followers who became known as **Protestants** because they protested against the Catholic Church's teachings and practices. Luther eventually broke with the Church and founded the Lutheran Church.

John Calvin Another important thinker of the Reformation was John Calvin. Many of his ideas are similar to those of Martin Luther. Like Luther, Calvin taught that the Bible was the most important element of Christianity. Priests and other clergy were not necessary. Unlike Luther, however, Calvin believed that God had already decided who was going to go to heaven, even before these people were born. He encouraged his followers to dedicate themselves completely to God and to live lives of self-restraint.

Calvin's teachings were very popular, particularly in Switzerland. In 1536 he and his followers took over the city of Geneva. There they passed laws requiring that everyone live according to Calvinist teachings.

Henry VIII of England Henry brought major religious change to England. At first he was a great defender of the Roman Catholic Church, but this changed after a conflict with the pope. Henry wanted a son to inherit his throne, but his wife could not have more children after their daughter was born. Henry asked the pope for permission to divorce her, but the pope refused. Henry then claimed that the pope did not have authority over the powerful English monarchy. He broke away from the Roman Catholic Church and had laws passed that created the Church of England. The new Church granted Henry VIII a divorce.

▲
Martin Luther believed that God viewed all people of faith equally.

▲
Renaissance painter Hans Holbein the Younger created this famous portrait of King Henry VIII.

Have students complete the Section Review. Then hand out index cards and have students write questions for a quiz game about the Renaissance, the Reformation, and the Counter-Reformation. Have students use their cards to quiz each other. Then have students complete Daily Quiz 7.2.

Have interested students conduct research on the Renaissance in northern Europe, where art developed quite differently than it did in Italy. Encourage students to select one figure or work from the Northern Renaissance and compare it to the art of southern Europe in the same time period.
BLOCK SCHEDULING

RETEACH

Have students complete Main Idea Activity 7.2. Then have students create charts with three column headings—"Renaissance," "Reformation," and "Counter-Reformation"—and two row labels, "Origins" and "Features." Have them fill out the chart with the appropriate information.
ENGLISH LANGUAGE LEARNERS

Section Review 2

Answers

Define For definitions, see: Renaissance, p. 131; humanists, p. 131; Reformation, p. 135; Protestants, p. 135; Counter-Reformation, p. 136

Reading for the Main Idea

1. Travel brought people into contact with ideas; curiosity about the world awakened. (NGS 6)

2. developing cities; new foods; new sources of income; more freedom; work for hire (NGS 15)

3. many different churches; more emphasis on education; increased power of national governments and monarchs (NGS 12)

Critical Thinking

4. The Pope and the Church during the Middle Ages had tremendous power over nations and monarchs. The less powerful the Church became, the more powerful nations and governments became.

Organizing What You Know

5. d'Este: patron of the arts; Luther: started Protestantism; Gutenberg: developed a printing press; Leonardo: great artist and inventor; Shakespeare: popular plays and poems

Visual Record Answer ►

education, literacy

▲
Many schools, including the Dutch University of Leiden shown here, were established during the Reformation.
Interpreting the Visual Record What goal of the Renaissance humanists was shared by both Catholics and Protestants?

The Counter-Reformation In response to the rise of Protestantism, the Catholic Church attempted to reform itself. This movement is called the **Counter-Reformation**. Church leaders began to focus more on spiritual matters and on making Church teachings easier for people to understand. They also attempted to stop the spread of Protestantism. Since about 1478, Spanish leaders had put on trial and severely punished people who questioned Catholic teachings. Leaders of what was called the Spanish Inquisition saw their fierce methods as a way to protect the Catholic Church from its enemies. During the Counter-Reformation, the pope brought the Spanish Inquisition to Rome.

Results of Religious Struggle Terrible religious wars broke out in France, Germany, the Netherlands, and Switzerland after the Reformation. By the time these wars ended, important social and political changes had occurred in Europe. Many different churches arose in Europe.

A stronger interest in education arose. Catholics saw education as a tool to strengthen people's belief in the teachings of the Church. Protestants believed that people could find their own way to Christian faith by studying the Bible. Although both Catholics and Protestants placed importance on literacy, the ability to read, education did not make people more tolerant. Both Catholic and Protestant leaders opposed views that differed from their own.

As Protestantism became more popular, the Catholic Church lost some of its power. It was no longer the only church in Europe. As a result, it lost some of the tremendous political power it had held there. As the power of the Church and the pope decreased, the power of monarchs and national governments increased.

✓ **READING CHECK:** *Identifying Cause and Effect* How did the religious conflicts of the 1500s change life in Europe? Many different churches appeared; education more important; national governments gained power.

go. hrw .com
Homework Practice Online
Keyword: SJ3 HP7

Section Review 2

Define and explain: Renaissance, humanists, Reformation, Protestants, Counter-Reformation

Reading for the Main Idea

1. (*Human Systems*) What brought about the Renaissance?

2. (*Human Systems*) What were some important changes in daily life during the Renaissance?

3. (*Human Systems*) After the Reformation and Counter-Reformation, how was life in Europe different?

Critical Thinking

4. **Drawing Inferences** Why did national governments gain strength as the power of the Catholic Church declined?

Organizing What You Know

5. **Categorizing** Copy and complete the following graphic organizer with the achievements of some key people of the Renaissance and the Reformation.

Person	Achievement

Section 3

Objectives

1. Describe the Scientific Revolution.
2. Identify factors that triggered the Age of Exploration.
3. Explain how the English monarchy differed from others in Europe.
4. Describe the relationship between England and its American colonies.

FOCUS

LET'S GET STARTED

Copy the following question onto the chalkboard: *How has science changed the world? Write down your ideas.* Discuss student responses. Point out that many of the scientific advances of the modern age have their roots in the Scientific Revolution. Tell students that they will learn more about scientific advances and other developments of the 1500s and 1600s in this section.

Building Vocabulary

Write the key terms on the board. Ask students to read the definitions from the glossary aloud. Have students discuss what kind of government comes to mind when they hear **monarchy**. Point out that in addition to a monarchy, the United Kingdom has a **Parliament** that is responsible for making laws. This makes their government a **limited monarchy**. Have students read the definitions for the other key terms from the glossary.

Section 3 — Exploration and Conquest

Read to Discover

1. What was the Scientific Revolution?
2. What started the Age of Exploration?
3. How was the English monarchy different from others in Europe?
4. What was the relationship between England and its American colonies?

Define

Scientific Revolution
Age of Exploration
colony
mercantilism
absolute authority
monarchy
limited monarchy
Parliament
Puritan
constitution
Restoration

Locate

India
China
Spain
Portugal
Mexico
Bahamas
France
Russia
Austria
England

Section 3 RESOURCES

Reproducible
◆ Guided Reading Strategy 7.3

Technology
◆ One-Stop Planner CD–ROM, Lesson 7.3
◆ Homework Practice Online
◆ HRW Go site

Reinforcement, Review, and Assessment
◆ Main Idea Activity 7.3
◆ Section 3 Review, p. 143
◆ Daily Quiz 7.3
◆ English Audio Summary 7.3
◆ Spanish Audio Summary 7.3

WHY IT MATTERS

Countries often try to increase their power through conquest. Use CNNfyi.com or other **current events** sources to find examples of conquest going on today. Record your findings in your journal.

Galileo's telescope

The Scientific Revolution

You have read that Renaissance humanists encouraged learning, curiosity, and discovery. The spirit of the Renaissance paved the way for a development during the 1500s and 1600s known as the **Scientific Revolution**. During this period, Europeans began looking at the world in a different way. Using new instruments such as the microscope and the telescope, they made more accurate observations than were possible before. They set up scientific experiments and used mathematics to learn about the natural world.

This scientific approach produced new knowledge in the fields of astronomy, physics, and biology. For example, in 1609 Galileo Galilei built a telescope and observed the sky. He eventually proved that an earlier scientist, Copernicus, had been correct in saying that the planets circle the sun. Earlier, people had believed that the planets moved around the Earth. In 1687 Sir Isaac Newton explained the law of gravity. In the 1620s William Harvey discovered the circulation of blood.

Other discoveries and advances such as better ships, improved maps, compasses and other sailing equipment allowed explorers to venture farther over the seas than before. These discoveries paved the way for the Age of Exploration.

✓ **READING CHECK:** *Identifying Cause and Effect* What brought about the Scientific Revolution? Renaissance attitude of curiosity, a desire to learn and discover.

The model pictured below is of an English explorer's ship. Although they appear tiny and fragile by today's standards, ships like this carried European explorers to new lands around the world.

▼

Teaching Objective 1

LEVEL 1: (Suggested time: 20 min.) Ask students to name some of the defining characteristics of the Scientific Revolution. Then have each student write a paragraph summarizing those aspects. Tell them that their paragraphs should describe the Scientific Revolution by answering the five "W" questions: Who? What? When? Where? Why?
ENGLISH LANGUAGE LEARNERS

LEVELS 2 AND 3: (Suggested time: 45 min.) Provide students with encyclopedias or other reference books and have them conduct research about key figures of the Scientific Revolution. Then have each student prepare a brief report about his or her chosen person's life and achievements. Challenge students to evaluate the historic significance of the scientist's work and its influence on the modern world.

Across the Curriculum

SCIENCE

Early Mapmaking The interest of Renaissance scholars in the work of ancient geographers like Ptolemy led to improvements in mapmaking. Most scholars of the time knew the world was round. During the Renaissance, new information about Africa and Asia was added to ancient maps, but the Americas were not yet known.

With the Age of Exploration, mapmaking made great advances. The discovery of new lands and the use of navigational tools like the compass improved the accuracy of maps. The most famous mapmaker of the era was Gerardus Mercator, who published his first map of the world in 1538. Later he developed a technique, called the Mercator Projection, which allowed him to represent the curved surface of Earth on a flat map.

Discussion: Show students examples of old maps from reference books or the Internet. Lead a class discussion comparing and contrasting those maps with recent ones.

This map shows the routes taken by Portuguese, Spanish, French, English, and Dutch explorers. They sailed both east and west to discover new lands.

Interpreting the Map Find Magellan's course on the map. Describe the route he found from the Atlantic Ocean to the Pacific Ocean.
►

The Dutch artist Jan Vermeer, who painted during the 1600s, often showed his subjects in the midst of work activities. Vermeer's *The Astronomer* shows research and work connected to mapmaking.

▼

Map Answer ►

sailed around the tip of South America through what is now called the Strait of Magellan

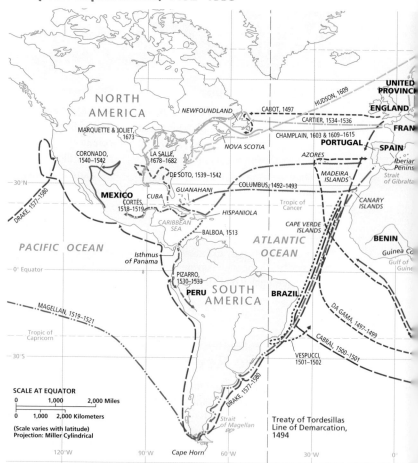

European Explorations, 1492–1535

SCALE AT EQUATOR
0   1,000   2,000 Miles
0   1,000   2,000 Kilometers
(Scale varies with latitude)
Projection: Miller Cylindrical

Treaty of Tordesillas Line of Demarcation, 1494

The Age of Exploration

Europeans were eager to find new and shorter sea routes so that they could trade with India and China for spices, silks, and jewels. The combination of curiosity, technology, and the demand for new and highly valued products launched a period known as the **Age of Exploration**.

A member of the Portuguese royal family named Prince Henry encouraged Portugal to become a leader in exploration. He wanted to find a route to the rich spice trade of India. Portuguese explorers did eventually succeed in finding a way. They reached India by sailing around Africa.

Hoping to find another route to India, Spain sponsored the voyage of the Italian navigator Christopher Columbus. He hoped to find a direct route to India by sailing westward across the Atlantic Ocean. In

LEVELS 1 AND 2: (Suggested time: 30 min.) Copy the following graphic organizer onto the board, omitting the italicized answers. Ask students to copy it into their notebooks. Then pair students and have each pair fill in the chart with factors that led to the beginning of the Age of Exploration. Call on volunteers to share their answers with the class. Discuss the information in students' charts.
ENGLISH LANGUAGE LEARNERS, COOPERATIVE LEARNING

Scientific factors	Economic factors	Early events
• *curiosity about the world* • *better ships, maps, and equipment*	• *desire for trade with Asia* • *search for shorter trade routes by sea*	• *Portuguese discovered sea route around Africa* • *Columbus landed in Americas*

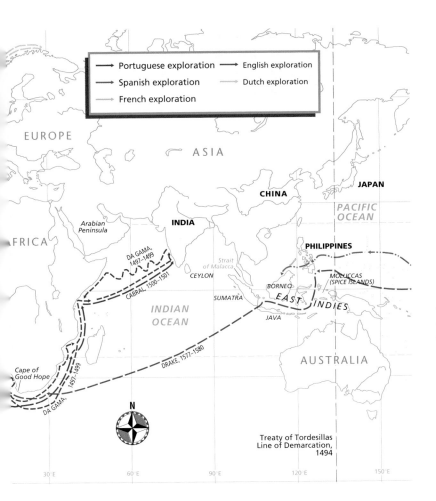

▲

Sailors of the 1500s had many new tools such as this astrolabe to hold a course and measure their progress. A sailor would sight a star along the bar of the astrolabe. By lining the bar with markings on the disk, he could figure out the latitude of the ship's position.

GLOBAL PERSPECTIVES

One way that colonial powers kept economic control over their colonies was to erect trade barriers, such as tariffs. These barriers helped prevent colonies from trading with other countries. Similar barriers still exist in varying degrees in modern nations, mainly as a way to protect local industries from foreign competition.

Today, however, many countries support free trade policies that lower barriers to foreign trade. Free trade agreements like the North American Free Trade Agreement (NAFTA) between the United States, Canada, and Mexico have reduced tariffs and stimulated foreign trade. Many economists believe that free trade benefits all nations in the long run. However, critics of free trade note that it can also cause job losses and result in the movement of factories from high-wage countries to low-wage countries.

Discussion: Lead a class discussion about the benefits and drawbacks of free trade. Ask students to evaluate the arguments, pro and con, and decide whether free trade is a good idea or not.

1492 Columbus reached an island in what is now called the Bahamas. Because he had no idea that the Americas lay between Europe and Asia, Columbus believed he had reached the east coast of India.

Later Spanish explorers who knew of the Americas were motivated more by the promise of conquest and riches than by curiosity and the opening of trade routes. The chart on the following page provides an overview of the major explorers from the late 1400s through the 1500s, their voyages, and their accomplishments.

Conquest and Colonization Over time the Spanish, French, English, Dutch, and others established American colonies. A colony is a territory controlled by people from a foreign land. As they expanded overseas, Europeans developed an economic theory called **mercantilism**. This theory said that a government should do everything it could to increase its wealth. One way it could do so was by

Teaching Objective 3

LEVEL 1: (Suggested time:15 min.) Copy the graphic organizer on the next page onto the board, omitting the italicized answers. Have students copy it into their notebooks and fill it in with descriptions of continental European monarchies and the British monarchy. Tell students that the space where the circles overlap should be filled in with features common to both types of monarchies. When students have finished, ask them how the English monarchy was different from others in Europe. *(Its power was limited by laws.)* **ENGLISH LANGUAGE LEARNERS**

National Geography Standard 10

Human Systems The conquest and colonization of the Americas brought a diffusion of culture and products between the Old and New Worlds. One result was the transfer of foods and animals, known as the Columbian Exchange.

The effects of this exchange are still felt today. For example, diets on both sides of the Atlantic changed. Potatoes from the Andes now feed millions of people in Europe. Corn, first grown by Native Americans, is a staple around the world. Wheat was not grown in the Americas until the Europeans arrived. The tomato first grew not in Italy, but in the Americas. Horses, cattle, goats, sheep, and other animals came from Europe, while North America contributed the turkey, the gray squirrel, and the muskrat.

Activity: Have students conduct research on the Columbian Exchange and create charts or posters detailing its features and effects.

Interpreting the Chart Which explorer do you think made the most important discovery? Give reasons for your answer.

European explorers sailed in search of goods like gold and cinnamon. In addition, some found new foods, like tomatoes, to bring back to Europe.

European Explorers

Name	Sponsoring Country	Date	Accomplishment
Christopher Columbus	Spain	1492–1504	Discovered islands in the Americas, claimed them for Spain
Amerigo Vespucci	Spain, Portugal	1497–1504	Reached America, realized it was not part of Asia
Vasco Núñez de Balboa	Spain	1513	Reached Pacific Ocean, proved that the Americas were not part of Asia
Ferdinand Magellan	Spain	1519	Made the first round-the-world voyage, first to reach Asia by water
Hernán Cortés	Spain	1519	Conquered Aztec civilization, brought smallpox to Central America
Francisco Pizarro	Spain	1530	Conquered Inca empire, claimed the land from present-day Ecuador to Chile for Spain
Jacques Cartier	France	1535	Claimed the Quebec region for France
Sir Francis Drake	England	1577–1580	Sailed around the World, claimed the California coast for England

selling more than it bought from other countries. A country that could get natural resources from colonies would not have to import resources from competing countries. The desire to win overseas sources of materials helped fuel the race for colonies. The Age of Exploration changed both Europe and the lands it colonized. Colonized lands did benefit from these changes. However, in general, Europe gained the most. During this time, goods, plants, animals, and even diseases were exchanged between Europe and the Americas.

The Slave Trade A tragic result of exploration and colonization was the spread of slavery. During the 1500s Europeans began to use enslaved Africans to work in their colonies overseas. In exchange for slaves, European merchants shipped cotton goods, weapons, and liquor to Africa. These slaves were sent across the Atlantic to the Americas, where they were traded for goods such as sugar and cotton. These goods were then sent to Europe in exchange for manufactured products to be sold in the Americas. Conditions aboard the slave ships were horrific, and slaves were treated brutally. Many died crossing the Atlantic.

✓ **READING CHECK:** (*Identifying Cause and Effect*) What were two results of European exploration? increased trade and the slave trade

Chart Answer

Answers will vary.

Features of European Monarchies

European Monarchies — Most Monarchies—absolute authority; monarchs' authority grew during period;

Shared features—headed by a king or queen;

British Monarchy — England—power limited by law; Parliament makes laws

LEVELS 2 AND 3: (Suggested time: 30 min.) Organize students into groups and have each group create a chart showing how Parliament exercised power over English rulers in the 1500s and 1600s. Charts should describe Parliament and its functions and trace its influence through events like the English Civil War, the Restoration, and the Glorious Revolution. Have students discuss their charts in class. **COOPERATIVE LEARNING**

Monarchies in Europe

Wealth flowed into European nations from their colonies. At the same time the Church's power over rulers and governments lost strength. The power of monarchs increased. In France, Russia, and Central Europe, monarchs ruled with **absolute authority**, meaning they alone had the power to make all the decisions about governing their nation. This situation would not change much until the 1700s.

France was ruled by a royal family called the Bourbons. Its most powerful member was Louis XIV, who ruled France from 1643 to 1715. Like many European monarchs, Louis believed that he had been chosen by God to rule. He had absolute control of the government and made all important decisions himself. Under Louis, France became a very powerful nation. In Russia, the Romanov dynasty came to power in the early 1600s. The most powerful of the Romanov czars was Peter the Great, who took the throne in 1682. He wanted to make Russia more like countries in Western Europe. Like Louis XIV, Peter the Great was an absolute monarch who strengthened his country. In Central Europe, two great families competed for power. The Habsburgs ruled the Austrian Empire, while the Hohenzollerns controlled Prussia to the north.

England's situation was different. When King John signed Magna Carta during the Middle Ages, he set a change in motion for England's government. England became a **limited monarchy**. This meant that the powers of the king were limited by law. By the 1500s, **Parliament**, an assembly made up of nobles, clergy, and common people, had gained the power to pass laws and make sure they were upheld.

English Civil War English monarchs such as Henry VIII and Elizabeth I had to work with or around Parliament to achieve their political goals. Later English monarchs fought with Parliament for power. Some even went to war over this issue. The struggle between king and Parliament reached its peak in the mid-1600s. Armies of Parliament supporters under Oliver Cromwell defeated King Charles I, ended the monarchy, and proclaimed England a commonwealth, a nation in which the people held most of the authority.

A special court tried Charles I for crimes against the people. Oliver Cromwell, a **Puritan**, took control of England. Puritans were a group of Protestants who thought that the Church of England was too much like the Catholic Church. The Puritans were a powerful group in Parliament at the time, and Cromwell was their leader.

Empress Maria Theresa of Austria was a member of the Habsburg family.

Oliver Cromwell led the Puritan forces that overthrew the English monarchy. He ruled England from 1653 to 1658.

ALL LEVELS: (Suggested time: 15 min.) Call on a volunteer to identify one policy passed by the British government with regard to its American colonies. Have the student write his or her answer on the chalkboard. Then call on another student to identify the response of the American colonists to this policy. Have that student write the response next to the first answer on the board. Repeat this process until students have names several policies that defined the relationship between Great Britain and the American colonies. Then lead a class discussion about the nature of this relationship. **ENGLISH LANGUAGE LEARNERS**

▶**ASSIGNMENT:** Have each student design a movie poster for a film set in or about the Scientific Revolution, the Age of Exploration, or another aspect of the 1500s and 1600s. Direct students to create names for their movies and a few characters that might appear in them. Students may also wish to suggest contemporary actors who could appear in the film. Call on volunteers to share their posters with the class. **COOPERATIVE LEARNING**

Section Review 3

Answers

Define For definitions, see: Scientific Revolution, p. 137; Age of Exploration, p. 138; colony, p. 139; mercantilism, p. 140; absolute authority, p. 141; monarchy, p. 141; limited monarchy, p. 141; Parliament, p. 141; Puritan, p. 141; constitution, p. 142; Restoration, p. 142

Reading for the Main Idea

1. Improvements in science and technology allowed sailors to explore distant lands. (NGS 12)

2. curiosity, a desire for new products, and desire for wealth (NGS 12)

3. others absolute; England became limited monarchy (NGS 13)

Critical Thinking

4. Policy of mercantilism antagonized the colonists by disregarding their rights.

Organizing What You Know

5. Europe to Africa—cotton goods, weapons, liquor; Americas to Europe—sugar and cotton; Europe to the Americas—manufactured goods; Africa to the Americas—slaves

Visual Record Answer ▶

to make a public example and public statement; a warning to any monarch who might try to gain power

The death warrant of Charles I was signed and sealed by members of Parliament. Parliament chose to behead King Charles I in public.

Interpreting the Visual Record
Why might Parliament have decided to have Charles I beheaded where everyone could see?

Charles II, shown here as a boy, became king following the fall of Cromwell's commonwealth in 1660.

Cromwell's Commonwealth Cromwell controlled England for about five years. He used harsh methods to create a government that represented the people. Twice he tried to establish a **constitution**, a document that outlined the country's basic laws, but his policies were unpopular. Discontent became widespread. In 1660, two years after Cromwell died, Parliament invited the son of Charles I to rule England. Thus the English monarchy was restored under Charles II. This period of English history was called the **Restoration**.

Last Change in Government Cheering crowds greeted Charles II when he reached London. One observer recalled that great celebrations were held in the streets, which were decorated with flowers and tapestries. People hoped that the Restoration would bring peace and progress to England.

Although England had a king again, the Civil War and Cromwell's commonwealth had made lasting changes in the government. Parliament strictly limited the king's power.

The Glorious Revolution When Charles II died, his brother became King James II. James's belief in absolute rule angered Parliament. They demanded that he give up the throne and invited his daughter, Mary, and her Dutch husband, William of Orange, to replace him. This transfer of power, which was accomplished without bloodshed, was called the Glorious Revolution. The day before William and Mary took the throne in 1689, they had to agree to a document called the Declaration of Rights. It stated that Parliament would choose who ruled the country. It also said that the ruler could not make laws, impose taxes, or maintain an army without Parliament's approval. By 1700 Parliament had replaced the monarchy as the major source of political power in England.

✓ READING CHECK: *Drawing Inferences* How did the English Civil War and events that followed affect the English government? Parliament strictly limited monarch's power.

REVIEW AND ASSESS

Have students complete the Section Review. Then ask each student to write a question about a person or event discussed in this section. Have students read their questions to the class and allow the class to answer. Then have students complete Daily Quiz 7.3.

RETEACH

Have students complete Main Idea Activity 7.3. Then organize students into groups and have each group write a skit based on the information in Section 3. Each skit should cover one of the three main topics discussed in this section. Call on volunteer groups to perform their skits for the class.
ENGLISH LANGUAGE LEARNERS, COOPERATIVE LEARNING

EXTEND

Have students conduct additional research on one of the explorers named on the chart in this section. Direct students to find information about the explorer's life and accomplishments. Then have each student prepare a brief biography of his or her chosen explorer and present it to the class.
BLOCK SCHEDULING

English Colonial Expansion

During the 1600s, English explorers began claiming and conquering lands overseas. In 1607 the British established Jamestown in what is now the state of Virginia. Jamestown was the first permanent English settlement in North America. In 1620, settlers founded Plymouth in what is now Massachusetts.

Mercantilism and the British Colonies The British government, with its policy of mercantilism, thought that the colonies should exist only for the benefit of England. Parliament passed laws that required colonists to sell certain products only to Britain, even if another country would pay a higher price. Other trade laws imposed taxes on sugar and other goods that the colonies bought from non-British colonies.

Resistance in the Colonies The American colonists saw these trade laws as a threat to their liberties. They found many ways to break the laws. For example, they avoided paying taxes whenever and however they could. Parliament, however, continued to impose new taxes. With each new tax, colonial resistance increased. Relations between England and the colonies grew steadily worse. The stage was set for revolution.

✓ **READING CHECK:** *Finding the Main Idea* How did England regard the American colonies? *as existing only for the benefit of England*

The settlers at Jamestown settled close by the James River.

Interpreting the Visual Record
Why do you think the colonists built their settlement in the manner shown here?

▼

The Granger Collection, New York

CHAPTER 7

Review
ANSWERS

Building Vocabulary
For definitions, see: feudalism, p. 127; vassals, p. 127; chivalry, p. 127; manors, p. 128; serfs, p. 128; clergy, p. 128; vernacular, p. 130; Renaissance, p. 131; humanists, p. 132; Reformation, p. 135; Protestants, p.134; Scientific Revolution, p. 137; colony, p. 138; mercantilism, p. 138; constitution, p. 141

Reviewing the Main Ideas
1. land (NGS 44)
2. philosophy that focused on life here and now rather than life after death, placed importance on individual achievement (NGS 10)
3. new churches, interest in education, reforms in Catholic Church; governments gained power; church's power diminished (NGS 12)
4. by fostering curiosity and through scientific and technological improvements (NGS 15)
5. idea that a country should do everything it can to increase its wealth (NGS 11)

◄ **Visual Record Answer**
for defense against attack

Section Review 3

Homework Practice Online
Keyword: SJ3 HP7

Define and explain: Scientific Revolution, Age of Exploration, colony, mercantilism, absolute authority, monarchy, limited monarchy, Parliament, Puritan, constitution, Restoration

Reading for the Main Idea
1. (*Human Systems*) How did the Scientific Revolution aid European exploration?
2. (*Places and Regions*) What prompted Europeans to explore and colonize land overseas?
3. (*Human Systems*) How was England different from other monarchies in Europe?

Critical Thinking
4. **Drawing Inferences** How did England's treatment of the American colonies set the stage for revolution?

Organizing What You Know
5. **Summarizing** Copy the following graphic organizer. Use it to show how the slave trade worked between Europe, Africa, and the Americas. Alongside each arrow, list the items that were traded along that route.

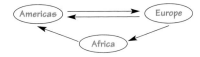

ASSESS

Have students complete the Chapter 7 Test.

RETEACH

Pair students and assign each pair one of the chapter's topics. Have each pair write the main points of the assigned topic on a transparency. Then have pairs teach the material to the class, using the transparency and an overhead projector. Students may write the main points on the chalkboard if no overhead projector is available. **ENGLISH LANGUAGE LEARNERS, COOPERATIVE LEARNING**

PORTFOLIO EXTENSIONS

1. Castles were a notable feature of the Middle Ages. Have students research medieval castles and create a guidebook for tourists exploring a castle. The guidebook should explain the different parts of the castle and what they were used for. Students should include illustrations to accompany their explanations.

2. Ask students to imagine that they are explorers of the early 1500s who wish to make a voyage to the New World. Have students work in groups to write a skit in which they petition the monarch of their country to finance the trip. In the skit, the explorer should provide reasons to convince the monarch that the trip will be worth while. Have each group perform its skit. Place scripts in student portfolios.

CHAPTER 7 Review ANSWERS

Understanding History and Society

Presentations will vary, but information included should be consistent with text. Students should demonstrate understanding that land was granted in exchange for military service.

Thinking Critically

1. Answers will vary. Students might say that it limited the king's power and stated that the law was the supreme power. It changed the source of political strength.

2. People who live in big cities and make their living through trade need to communicate with each other and keep written records.

3. Catholic Church lost much political power; monarchs and national governments gained power.

4. increased Great Britain's wealth, but pushed the colonists into rebellion

5. Parliament replaced the monarchy as the major political power in Great Britain.

CHAPTER 7 Reviewing What You Know

Building Vocabulary

On a separate sheet of paper, write a sentence to define each of the following words:

1. feudalism
2. vassals
3. chivalry
4. manors
5. serfs
6. clergy
7. vernacular
8. Renaissance
9. humanists
10. Reformation
11. Protestants
12. Scientific Revolution
13. colony
14. mercantilism
15. constitution

Reviewing the Main Ideas

1. *(Human Systems)* What did vassals receive in exchange for serving kings and nobles?
2. *(Human Systems)* What was humanism?
3. *(Human Systems)* What were the results of the Reformation and Counter-Reformation?
4. *(Environment and Society)* How did the Scientific Revolution pave the way for European exploration?
5. *(Human Systems)* What was mercantilism?

Understanding History and Society

The Feudal System
Throughout Europe in the Middle Ages, the feudal system was the most important political structure. Create a chart to explain the relationships among kings, lords, vassals, and knights. Consider the following:
• Who granted lands and who received the grants.
• What a king or lord expected from his vassals.

Thinking Critically

1. **Evaluating** Why was Magna Carta one of the most important documents in European history?

2. **Identifying Cause and Effect** Why did the growth of cities and trade during the High Middle Ages increase the need for education?

3. **Contrasting** How did political power in Europe change after the Counter-Reformation?

4. **Supporting a Point of View** How did mercantilism both help and harm Great Britain as a colonial power?

5. **Identifying Cause and Effect** How did the English Civil War change politics in Great Britain?

FOOD FESTIVAL

Many food combinations we take for granted today exist only because of the exchange of plants between the Americas and Europe. Salsa is a good example. Peppers and tomatoes came from the Americas, while onions and coriander—also called cilantro—were known in Europe. To make salsa combine 3 cups of chopped tomatoes, a half cup of chopped onion, 2 tablespoons of chopped coriander, and 1 or 2 chopped jalapeño peppers. Serve with tortilla chips. They are made from corn—another plant from the Americas.

CHAPTER 7 REVIEW AND ASSESSMENT RESOURCES

Reproducible
- Critical Thinking Activity 7
- Vocabulary Activity 7

Technology
- Chapter 7 Test Generator (on the One-Stop Planner)
- HRW Go site

- Audio CD Program, Chapter 7

Reinforcement, Review, and Assessment
- Chapter 7 Review, pp. 144–145

- Chapter 7 Tutorial for Students, Parents, Mentors, and Peers
- Chapter 7 Test
- Chapter 7 Test for English Language Learners and Special-Needs Students

Building Social Studies Skills

Map ACTIVITY

On a separate sheet of paper, match the letters on the map with their correct labels.

Spain	**South America**
Portugal	**Europe**
North America	
Mexico	

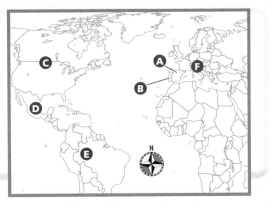

Mental Mapping Skills ACTIVITY

On a separate sheet of paper, draw a freehand map like the one in the Map Activity. On your map, sketch and label each of the following explorers' routes:

Columbus	**Drake**
Magellan	

WRITING ACTIVITY

Imagine that you are a peasant during the Middle Ages. Write a short dialogue in which you talk with another person about your lives on a manor. In your dialogue, discuss both positive and negative aspects of that life. Be sure to use standard grammar, spelling, sentence structure, and punctuation.

Alternative Assessment

Portfolio ACTIVITY

Learning About Your Local History

Local Artists and Writers The Renaissance produced many great artists and writers. Conduct research on the Internet or in your local library to find out about artists and writers in your state or region. Write a brief report on one of these artists or writers. Keep a list of sources you used to get your information.

internet connect

Internet Activity: go.hrw.com
KEYWORD: SJ3 GT7

Choose a topic to explore about the world in transition:
- Write a report on daily life in the Middle Ages.
- Create a biography of a Renaissance artist or writer.
- Learn more about an explorer described in this chapter.

Map Activity
A. Spain	E. South America
B. Portugal	
C. North America	F. Europe
D. Mexico	

Mental Mapping Skills Activity
Maps will vary, but should show the approximate routes taken by each explorer.

Writing Activity
Students' dialogues will vary but should accurately reflect aspects of the manorial system. Use Rubric 40, Writing to Describe, to evaluate student work.

Portfolio Activity
Reports will vary, but should accurately describe a local artist or writer. Use Rubric 30, Research, to evaluate student work.

internet connect

GO TO: go.hrw.com
KEYWORD: SJ3 Teacher
FOR: a guide to using the Internet in your classroom

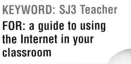

The Birth of the Modern World
Chapter Resource Manager

Objectives	Pacing Guide	Reproducible Resources
SECTION 1		
The Enlightenment (pp. 147–50) **1.** Define the Enlightenment **2.** Describe the ideas of government suggested by Enlightenment thinkers. **3.** Explain how the Enlightenment led to changes in society.	**Regular** 1 day **Block Scheduling** .5 day *Block Scheduling Handbook, Chapter 8*	**RS** Guided Reading Strategy 8.1 **RS** Graphic Organizer 8
SECTION 2		
The Age of Revolution (pp. 151–58) **1.** Describe the start and results of the American Revolution. **2.** Explain how the French Revolution changed France. **3.** Explain how Europe changed during and after the Napoléonic era.	**Regular** 1.5 days **Block Scheduling** .5 day *Block Scheduling Handbook, Chapter 8*	**RS** Guided Reading Strategy 8.2 **SM** Map Activity 8
SECTION 3		
The Industrial Revolution (pp. 159–63) **1.** Describe the beginnings of the Industrial Revolution. **2.** Explain how developments in transportation and communications spread industrial development. **3.** Identify business features that affected life in the Industrial Age.	**Regular** 1 day **Block Scheduling** .5 day *Block Scheduling Handbook, Chapter 8*	**RS** Guided Reading Strategy 8.3
SECTION 4		
Expansion and Reform (pp. 164–69) **1.** Describe changes that occurred in Europe and the United States after 1850. **2.** Identify factors that led to reform in the later 1800s. **3.** Explain how nationalism changed the map of Europe in the mid-1800s.	**Regular** 1 day **Block Scheduling** .5 day *Block Scheduling Handbook, Chapter 8*	**RS** Guided Reading Strategy 8.4

Chapter Resource Key

RS Reading Support
IC Interdisciplinary Connections
E Enrichment
SM Skills Mastery
A Assessment
REV Review

ELL Reinforcement and English Language Learners
 Transparencies
 CD–ROM
 Music

 Video
 Internet
 Holt Presentation Maker Using Microsoft® Powerpoint®

 One-Stop Planner CD–ROM

See the ***One-Stop Planner*** for a complete list of additional resources for students and teachers.

One-Stop Planner CD–ROM

It's easy to plan lessons, select resources, and print out materials for your students when you use the *One-Stop Planner CD–ROM with Test Generator.*

Technology Resources

 One-Stop Planner CD–ROM, Lesson 8.1

 ARGWorld CD–ROM

 Homework Practice Online

 HRW Go site

Review, Reinforcement, and Assessment Resources

ELL	Main Idea Activity 8.1	
REV	Section 1 Review, p. 150	
A	Daily Quiz 8.1	
ELL	English Audio Summary 8.1	
ELL	Spanish Audio Summary 8.1	

 One-Stop Planner CD–ROM, Lesson 8.2

 ARGWorld CD–ROM

 Homework Practice Online

HRW Go site

ELL	Main Idea Activity 8.2
REV	Section 2 Review, p. 158
A	Daily Quiz 8.2
ELL	English Audio Summary 8.2
ELL	Spanish Audio Summary 8.2

 One-Stop Planner CD–ROM, Lesson 8.3

 ARGWorld CD–ROM

 Homework Practice Online

 HRW Go site

ELL	Main Idea Activity 8.3
REV	Section 3 Review, p. 163
A	Daily Quiz 8.3
ELL	English Audio Summary 8.3
ELL	Spanish Audio Summary 8.3

 One-Stop Planner CD–ROM, Lesson 8.4

 ARGWorld CD–ROM

 Homework Practice Online

 HRW Go site

ELL	Main Idea Activity 8.4
REV	Section 3 Review, p. 169
A	Daily Quiz 8.4
ELL	English Audio Summary 8.4
ELL	Spanish Audio Summary 8.4

Meeting Individual Needs

Ability Levels

Level 1 Basic-level activities designed for all students encountering new material

Level 2 Intermediate-level activities designed for average students

Level 3 Challenging activities designed for honors and gifted-and-talented students

English Language Learners Activities that address the needs of students with Limited English Proficiency

Chapter Review and Assessment

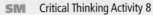

SM	Critical Thinking Activity 8		Chapter 8 Test Generator (on the One-Stop Planner)
REV	Chapter 8 Review, pp. 170–71		Audio CD Program, Chapter 8
REV	Chapter 8 Tutorial for Students, Parents, Mentors, and Peers	A	Chapter 8 Test for English Language Learners and Special-Needs Students
ELL	Vocabulary Activity 8		
A	Chapter 8 Test	A	Unit 2 Test for English Language Learners and Special-Needs Students
A	Unit 2 Test		

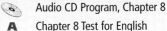

LAUNCH INTO LEARNING

Challenge students to write down the century in which the following concepts were introduced or became important: 1. Religious institutions should not control governments. 2. People have the right to decide on their country's form of government and to participate in it. 3. People are born equal and remain equal before the law. 4. Machines can produce more goods more quickly than can individual craftworkers. 5. A government should protect the rights of its citizens. Then point out that all of these now-common ideas either emerged or became widely accepted during the 1700s. Tell students they will learn more about how our modern world developed in this chapter.

Section 1

Objectives

1. Define the Enlightenment.
2. Describe the ideas about government suggested by Enlightenment thinkers.
3. Explain how the Enlightenment led to changes in society.

LINKS TO OUR LIVES

You might want to share with your students the following reasons for learning more about the 1700s and 1800s:

▶ Debates involving science and religion are still in the news and illustrate deep divisions between some segments of society.

▶ Monarchies still exist in the world. They differ widely in how well they meet the needs of their people.

▶ Revolutions that change governments still take place today. Some depose tyrants, while others bring tyrants to power.

▶ In much of the world, economies are based on industry and technology.

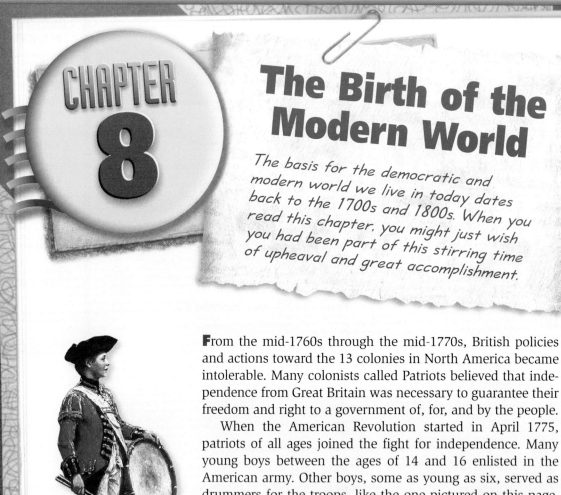

CHAPTER 8

The Birth of the Modern World

The basis for the democratic and modern world we live in today dates back to the 1700s and 1800s. When you read this chapter, you might just wish you had been part of this stirring time of upheaval and great accomplishment.

From the mid-1760s through the mid-1770s, British policies and actions toward the 13 colonies in North America became intolerable. Many colonists called Patriots believed that independence from Great Britain was necessary to guarantee their freedom and right to a government of, for, and by the people.

When the American Revolution started in April 1775, patriots of all ages joined the fight for independence. Many young boys between the ages of 14 and 16 enlisted in the American army. Other boys, some as young as six, served as drummers for the troops, like the one pictured on this page. Their job was to signal commands, which sometimes put them in the midst of battle.

The American navy had its share of young sailors as well. Small boys served as deckhands or "powder monkeys." They carried ammunition to the gunners during battle.

Drummer boy from the American Revolution

Revolutionary War cannon

Harbor at Charleston, South Carolina

LET'S GET STARTED
Write the word *Enlightenment* on the chalkboard. Ask students to copy it and then circle a familiar five-letter word they see within the word (light). Next, ask students what it means when a cartoonist shows a light bulb shining over a character's head. Discuss responses. *(Possible answer: The person has had a bright idea or "seen the light.")* Tell students that in Section 1 they will read about thinkers who had bright new ideas about government and society.

Building Vocabulary
Call on volunteers to locate the definitions of the words **secularism** and **individualism** in the text and read them aloud. Ask students what the words have in common. *(Possible answer: They are both about ideas.)* Next have students identify the two terms that are compound words (**popular sovereignty** *and* **social contract**). Call on volunteers to discuss the meanings of the individual words that make up each term. Then have students locate and read the definitions for the remaining terms. Ask students to write sentences for the terms.

Section 1 — The Enlightenment

Read to Discover
1. What was the Enlightenment?
2. What ideas about government did Enlightenment thinkers suggest?
3. How did the Enlightenment lead to changes in society?

Define
Enlightenment
reason
secularism
individualism

popular sovereignty
social contract

WHY IT MATTERS
Our American government and way of life are largely based on the ideas of the Enlightenment. Use **CNNfyi.com** or other current events sources to find out how these ideas continue to cause changes today all around the world. Record your findings in your journal.

Monticello, home of American Enlightenment thinker, Thomas Jefferson

Section 1 RESOURCES

Reproducible
◆ Block Scheduling Handbook, Chapter 8
◆ Guided Reading Strategy 8.1
◆ Graphic Organizer 8

Technology
◆ One-Stop Planner CD–ROM, Lesson 8.1
◆ Homework Practice Online
◆ HRW Go site

Reinforcement, Review, and Assessment
◆ Section 1 Review, p.150
◆ Daily Quiz 8.1
◆ Main Idea Activity 8.1
◆ English Audio Summary 8.1
◆ Spanish Audio Summary 8.1

The Birth of a New Age

You have read about how greatly European society changed during the Middle Ages, the Renaissance, the Reformation, and the Scientific Revolution. You have also learned how advances in science and technology paved the way for exploration and expansion overseas. As time went on, many changes in the world continued to take place, especially in people's thinking about their relationship with their nation's government.

From the mid 1600s through the 1700s, most European countries were ruled by monarchs. These rulers increasingly wanted absolute control over their governments and their subjects. Many, although not all, literary people, scientists, and philosophers, or thinkers, in these countries saw the need for change. They believed it was necessary to combat the political and social injustices people suffered every day. These critics claimed that the monarchs and nobility took too much from the common people and gave back too little. They wanted a new social order that was fairer for all people. These ideas were published in books, pamphlets, plays, and newspapers. Historians call this era of new ideas the **Enlightenment**.

✓ **READING CHECK:** *Identifying Cause and Effect* What about the political and social order of European nations caused many people to want change? monarchs' desire for absolute control over their governments and their subjects; people suffering from inequality and injustice in daily life.

▲
John Locke (1632-1704) was an important English philosopher. He is considered the founder of the Enlightenment in England.

TEACH

Teaching Objectives 1–2

LEVELS 1 AND 2: (Suggested time: 10 min.) Have students list some of the Enlightenment's beliefs about human beings and nature. *(Possible responses: The laws of nature govern all creatures; people can discover natural laws through reason; following natural law improves people's lives.)* Ask students what bigger idea these beliefs led to during the Enlightenment. *(Possible responses: They all led to the idea that people should have a part in governing themselves; people should have freedom.)*

LEVEL 3: (Suggested time: 15 min.) Pair students and tell each pair to create flash cards for the key terms *reason, secularism,* and *individualism.* Students should write each key term on one side of a flash card and its definition and details about how it affected the Enlightenment on the other side. **ENGLISH LANGUAGE LEARNERS, COOPERATIVE LEARNING**

Teaching Objective 3

ALL LEVELS: (Suggested time: 30 min.) Copy the graphic organizer on the following page onto the chalkboard, omitting the italicized answers. Have students work in pairs to complete the chart by listing three problems Enlightenment thinkers identified and their solutions to these problems. **COOPERATIVE LEARNING**

DAILY LIFE

The revolutionary ideas of Enlightenment thinkers did not spread rapidly to the general population of England and France. During the 1700s, most people in these countries could not read well, if they could read at all. Few common people had money or time to buy and read books. So in general, only privileged members of society met to discuss Enlightenment ideas. The daily lives of these wealthy and cultured people had little in common with the daily lives of most people. Nevertheless, the writings and conversations of Enlightenment thinkers gradually spread the new ideas far and wide.

Critical Thinking: How does requiring everyone to go to school and providing free education affect society?

Answer: If everyone can read, everyone has access to new ideas about government, and new ideas often lead to positive changes.

Mary Wollstonecraft was a British writer of the Enlightenment. She argued that women should have the same rights as men, including the right to an equal education.

🔲 internet connect

GO TO: go.hrw.com
KEYWORD: SJ3 CH8
FOR: Web sites about the birth of the modern world

Enlightenment Thinking

The Enlightenment is also called the Age of Reason. At this time, scientists began to use **reason**, or logical thinking, to discover the laws of nature. They believed that the laws of nature governed the universe and all its creatures. Some also thought there was a natural law that governed society and human behavior. They tried to use their powers of reasoning to discover this natural law. By following natural law, they hoped to solve society's problems and improve people's lives.

While religion was important to some thinkers, other thinkers played down its importance. Playing down the importance of religion became known as **secularism**. The ideas of secularism and **individualism**—a belief in the political and economic independence of individuals—would later influence some ideas about the separation of church and state in government. These ideas led to more rights for all people, individual freedoms, and government by the people.

The Enlightenment in England The English philosopher John Locke believed that natural law gave individuals the right to govern themselves. Locke wrote that freedom was people's natural state. He thought individuals possessed natural rights to life, liberty, and property. Locke also claimed people should have equality under the law.

Much of Locke's writing focused on government. Locke argued that government should be based on an agreement between the people and their leaders. According to Locke, people give their rulers the power to rule. If the ruler does not work for the public good, the people have the right to change the government. Locke's writings greatly influenced other thinkers of the Enlightenment. They also influenced the Americans who shaped and wrote the Declaration of Independence and the Constitution.

The Enlightenment in France In France, the thinkers of the Enlightenment believed that science and reason could work together to improve people's lives. They spoke out strongly for individual rights, such as freedom of speech and freedom of worship.

The Encyclopedia, published by Enlightenment philosophers, became the most famous publication of the period.

ENLIGHTENMENT IDEAS ABOUT SOCIETY	
Problems with Government or Society	Enlightenment Solutions to These Problems
Everyone has to belong to the same official faith.	People should have freedom of speech.
People should have freedom of religion.	People feel as though they have no political freedom.
People are being jailed for criticizing the king and nobility.	People must choose own government.

CLOSE

Ask students to study the illustrations and read the captions in Section 1. For each one, ask: *What does this illustration and caption tell you about the Enlightenment?* Discuss responses.

REVIEW AND ASSESS

Have students complete the Section Review. Then ask each student to choose a person, event, or idea from the section and write a few sentences describing the topic. Ask volunteers to read their sentences aloud while other students identify who or what is being described. Continue until all the major topics have been discussed. Then have students complete Daily Quiz 8.1.

Voltaire The French writer Voltaire was a leading voice of the Enlightenment. As a young man, Voltaire became a famous poet and playwright. He used his wit to criticize the French monarchy, the nobility, and the religious controls of the church. His criticisms got him into trouble. He eventually went to England after being imprisoned twice.

In England, Voltaire was delighted by the freedom of speech he found. In defense of this freedom, he wrote, "I may disapprove of what you say, but I will defend to the death your right to say it."

Voltaire also studied the writings of John Locke. When Voltaire returned to France, he published many essays and tales. These writings explored Enlightenment ideas, such as justice, good government, and human rights.

Rousseau Jean Jacques Rousseau (ROO-SOH) was another French thinker of the Enlightenment. He believed that people could only preserve their freedom if they chose their own government, and that good government must be controlled by the people. This belief is called **popular sovereignty**.

Rousseau's most famous book, *The Social Contract*, published in 1762, expressed his views. "Man was born free, and everywhere he is in chains," Rousseau wrote. He meant that people in society lose the freedom they have in nature. Like Locke, Rousseau believed that government should be based on an agreement made by the people. He called this agreement the **social contract**.

The Encyclopedia *The Encyclopedia* was the most famous publication of the Enlightenment. It brought together the writings of Voltaire, Rousseau, and other philosophers. The articles in *The Encyclopedia* covered science, religion, government, and the arts. Many articles criticized the French government and the Catholic church. Some philosophers went to jail for writing these articles. Nevertheless, the *Encyclopedia* helped spread Enlightenment ideas.

✔ **READING CHECK:** *Summarizing* What did the thinkers of the Enlightenment believe? A natural law governed human life and society; truth about natural law could be learned; human problems solved through reason.

In 1717, when Voltaire was 23, he spent eleven months in prison for making fun of the government. During that time, he wrote his first play. Its success made him the greatest playwright in France.

▲ Jean-Jacques Rousseau

◄

In France, writers and artists gathered each week at meetings like the one shown in this painting. Their purpose was to discuss the new ideas of the Enlightenment.
Interpreting the Visual Record
How might the group pictured here encourage the free sharing of ideas?

🎨 **Linking Past to Present**
The Enlightenment and the U.S. Constitution Point out the illustration of the U.S. Constitution on page 149 and have students read the caption. Ask them how Enlightenment ideas affected the kind of government we have in the United States today. *(The U.S. government is divided into three branches, as the philosopher Baron de Montesquieu thought governments should be.)*

Activity: Organize the class into three groups. Ask each group to conduct research on one of the three branches of the U.S. government—the executive, the legislative, or the judicial. Have the groups investigate the powers of its branch and the individuals and groups that are part of it. After each group has done its research, have students meet together to draw a chart that presents the three branches and their powers. Students might also create a chart that shows how the branches interact with each other.

◄ **Visual Record Answer**

Sitting in a semicircle in comfortable surroundings, people would feel free to exchange ideas and information.

Have students complete Main Idea Activity 8.1. Then organize the class into two groups. Ask the first group to pretend they are common people living in England or France before the Enlightenment. Have them list some of the problems they have with their society and government. Ask the second group to pretend they are typical Enlightenment thinkers. Ask them to explain the changes they would make in government and society and how these changes would improve people's lives. Have the groups compile their ideas on a Before and After chart. **ENGLISH LANGUAGE LEARNERS, COOPERATIVE LEARNING**

Section Review 1

Answers

Define For definitions, See: Enlightenment, p. 147; reason, p. 148; secularism, p. 148; individualism, p. 148; popular sovereignty p. 149; social contract, p. 149

Reading for the Main Idea

1. Thinkers tried to use reason to discover natural laws that governed society. (NGS 13)

2. equality under the law, basic rights, popular sovereignty. (NGS 13)

Critical Thinking

3. It included criticism of the government and the church.

4. Ideas such as equality, rights to life and liberty, and popular sovereignty appear in the Declaration of Independence. Three branches of government are in the U.S. Constitution.

Organizing What You Know

5. Locke—England; natural rights; equality; government based on agreement between people and leaders; Voltaire—France; freedom of speech; government good and just; Rousseau—France; social contract.

The philosopher Baron de Montesquieu (MOHN-tes-kyoo) thought governments should be divided into three branches. His ideas helped the writers of the U.S. Constitution form our government.

The Enlightenment and Society

When the philosophers began to publish their ideas, there was little freedom of expression in Europe. Most countries were ruled by absolute monarchs. Few people dared to criticize the court or the nobility. Most nations had official religions, and there was little toleration of other faiths.

As time passed, Enlightenment ideas about freedom, equality, and government became more influential. Eventually, they inspired the American and French revolutions. In that way, the Enlightenment led to more freedom for individuals and to government by the people.

go.hrw.com **Homework Practice Online**
Keyword: SJ3 HP8

Section Review 1

Define and explain Enlightenment, reason, secularism, individualism, popular sovereignty, social contract

Reading for the Main Idea

1. (*Human Systems*) Why was the Enlightenment also called the Age of Reason?

2. (*Human Systems*) What important ideas about government came from Enlightenment thinkers?

Critical Thinking

3. **Drawing Conclusions** Why might the French nobility and the church dislike *The Encyclopedia*?

4. **Analyzing** In what ways did John Locke and other philosophers of the Enlightenment help pave the way for democracy in the United States?

Organizing What You Know

5. **Categorizing** Copy the following graphic organizer. Use details from the chapter to fill it in. Then write a title for the chart.

Writer	Country	Important Ideas
Locke		
Voltaire		
Rousseau		

Section 2

Objectives

1. Describe the start and results of the American Revolution.
2. Examine how the French Revolution changed France.
3. Explain how Europe changed during and after the Napoléonic era.

FOCUS

LETS GET STARTED

Copy the following question onto the chalkboard: *What are some ways in which you might challenge governmental rules or laws that seem unjust?* Discuss responses and write a summary of students' suggestions on the chalkboard. Tell students that in Section 2 they will learn how both American and French citizens responded to their governments' injustices.

Building Vocabulary

Write the key terms on the chalkboard. Ask students to identify base words within some of the terms. *(Examples:* **Loyalists**—*loyal,* **alliance**—*ally,* **oppression**—*oppress,* **reactionaries**—*react)* Have students use what they know about the base words to speculate on the meanings of these key terms. Then have students locate and read the definitions for all the key terms, checking the inferences they have made about the meanings.

Section 2 The Age of Revolution

Read to Discover

1. What started the American Revolution and what were its results?
2. How did the French Revolution change France?
3. How did Europe change during and after the Napoléonic Era?

Define

Patriots
Loyalists
alliance
oppression

Reign of Terror
balance of power
reactionaries

WHY IT MATTERS

The revolutions of the 1700s in the United States and France gave many new rights and freedoms to ordinary citizens. Use **CNNfyi.com** or other current events sources to find examples of recent revolutions that have occurred in countries around the world.

American teapot with anti-Stamp Act slogan

Section 2 RESOURCES

Reproducible
◆ Guided Reading Strategy 8.2
◆ Map Activity 8

Technology
◆ One-Stop Planner CD–ROM, Lesson 8.2
◆ Homework Practice Online
◆ HRW Go site

Reinforcement, Review, and Assessment
◆ Section 2 Review, p. 158
◆ Daily Quiz 8.2
◆ Main Idea Activity 8.2
◆ English Audio Summary 8.2
◆ Spanish Audio Summary 8.2

The American Revolution

Enlightenment philosophers' ideas about freedom, equality, and government were not confined to Europe in the 1700s. By the 1750s, the British had established 13 colonies along the Atlantic Coast in North America. These British colonists had developed a new way of life and a new relationship with their home country. The colonists held their own elections and made their own laws. However, the colonists were still British subjects, and they had no representation in the British Parliament.

The Growing Conflict While the British had colonies along the Atlantic Coast in North America, the French colonies—New France—lay to the north and west. As British colonists pushed westward into French-controlled territory, tensions mounted.

France and Great Britain had long been enemies in Europe. In 1754, their conflict spilled over into North America, sparking the French and Indian War. In Europe this war was called the Seven Years' War. It began in 1756 and ended in 1763. As the victor in this war, the British gained control of most of North America.

To help pay for the war, the British taxed goods that their colonists in North America needed. Many colonists thought these new taxes were unfair, since they had no representatives in Parliament to express their views. Americans resisted the new taxes by refusing to buy British goods.

▲

The Stamp Act required Americans to purchase stamps like this one and to place them on many types of public documents.

Teaching Objective 1

LEVEL 1: (Suggested time: 30 min.) Copy the following graphic organizer onto the chalkboard, omitting the italicized answers. Pair students and have the pairs record causes of the American Revolution in the first column and results in the second column. Discuss the charts.
ENGLISH LANGUAGE LEARNERS, COOPERATIVE LEARNING

CAUSES AND RESULTS OF THE AMERICAN REVOLUTION

Causes of the American Revolution	Results of the American Revolution
Americans were influenced by Enlightenment ideas about freedom and popular sovereignty.	*The British recognized the independence of the United States.*
Americans were angered by taxation without representation.	*All land east of the Mississippi belonged to the United States.*
Colonies united against Great Britain by setting up and sending delegates to the Continental Congresses.	*Americans wrote the Articles of Confederation and the later the Constitution and set up a democracy.*

National Geography Standard 6

Places and Regions The 13 original colonies are usually grouped according to region. The Northern, or New England, Colonies were Connecticut, Massachusetts, New Hampshire, and Rhode Island. The Middle Colonies were Delaware, New Jersey, New York, and Pennsylvania. The Southern Colonies were Georgia, Maryland, North Carolina, South Carolina, and Virginia.

Activity: Have students label the 13 colonies on maps of the United States. Ask them to color code the colonies by region. You may also want students to conduct research on battles of the American Revolution and mark their locations on their maps.

The Declaration of Independence declared the American colonies free from British control. It was adopted on July 4, 1776—now celebrated as Independence Day.

The first battle of the American Revolution was fought in Lexington, Massachusetts, on April 19, 1775.

As their unhappiness increased, the colonies united against the British. In 1774, 12 colonies sent representatives to the First Continental Congress. The Congress pledged to stop trade with Britain until the colonies had representation in Parliament.

Some American colonists believed the best way to guarantee their rights was to break away from British rule. Colonists called **Patriots** wanted independence. They made up one third of the population. Another third, the **Loyalists**, wanted to remain loyal to Great Britain. The rest of the colonists were undecided.

The Declaration of Independence In 1776 the Continental Congress adopted the Declaration of Independence. Thomas Jefferson was the Declaration's main author. The Declaration clearly showed the influence of Enlightenment thinkers, especially Locke and Rousseau. The Declaration stated that that "all men are created equal" and have the right to "life, liberty, and the pursuit of happiness." The ideal of individual liberty was only applied in a limited way. Women and slaves were not included. Nevertheless, the Declaration was still a great step forward toward equality and justice.

Locke's and Rousseau's ideas about popular sovereignty were clearly seen in the Declaration. It stated that all powers of government come from the people. It said that no government can exist without the consent of its citizens and that government is created to protect individual rights. In addition, it stated that if a government fails to protect these rights, the people may change it and set up a new government.

War and Peace By the time the Declaration of Independence was written, the colonies were already at war with Great Britain. At first, the British seemed unbeatable. Then in late 1777, France formed an **alliance** with the Americans. An alliance is an agreement formed to help both sides. By helping the Americans, France hoped to weaken the British Empire.

In 1781 the American forces—commanded by George Washington—and their French allies defeated the main British army in Virginia. The Americans had won the Revolutionary War. The final peace terms were settled in the Treaty of Paris in 1783. The British recognized the independence of the United States. All land east of the Mississippi now belonged to the new country.

✓ **READING CHECK:** *Finding the Main Idea* How did the ideas of the Enlightenment influence the American Revolution? The ideas liberty, equality, and popular sovereignty inspired the move toward self-government.

LEVELS 2 AND 3: (Suggested time: 30 min.) Organize the students into small groups representing Patriots and Loyalists. Ask each group to give an account of the events leading up to the Revolution, as seen from its particular point of view. Similarly, the two groups can give their views of the Declaration of Independence and the events of the war and postwar period. You may want to have students present their views to the class in the form of skits or short plays. **COOPERATIVE LEARNING**

►**ASSIGNMENT:** Have students write newspaper editorials dated July 4, 1776, about the signing of the Declaration of Independence. Tell students to write the editorial from the point of view of a Patriot newspaper editor. In addition to giving factual details about this important document, the editorial should include a vision of the country that might one day be formed on the principles in the Declaration. You may want to extend the assignment by asking students what arguments Loyalists would raise against the Patriot editorials.

Effects of American Independence

In 1777 the Americans adopted a plan of government called the Articles of Confederation. The Articles set up a central government, but it was purposely weak. Many Americans did not trust that a central government would always protect the individual rights and liberties they had fought for in the Revolution. Thus Congress could not levy taxes or coin money. It could not regulate trade. Within 10 years, however, it became clear that a weak central government was not helping the country to work as a whole.

In May 1787 delegates from all the states met at a convention in Philadelphia to revise the Articles. The delegates soon realized that making changes in the Articles would not be enough. They decided instead to write a new constitution.

After choosing George Washington to preside over the convention, the delegates went to work. They wanted a strong central government. They also wanted some powers kept for the states. As a result, the new Constitution they wrote set up a federal system of government. This is a system of government in which power is divided between a central government and individual states. The central government was given several important powers. It could declare war, raise armies, and make treaties. It could coin money and regulate trade with foreign countries. The states and the people kept all other powers. The Constitution was approved in 1789. The federal government had three branches. Each branch acted as a check on the power of the others. The executive branch enforced the laws. The legislative branch made the laws. The judicial branch interpreted the laws.

The American Revolution and the writing of the U.S. Constitution were major events in world history. Enlightenment ideas were finally put into practice. The success of the American democracy also encouraged people around the world. They realized they could fight for political freedoms, too.

Of course, American democracy in 1789 was not perfect. Women had few rights, and slaves had no rights at all. Still, the world now had a democratic country that inspired the loyalty of most of its citizens.

✓ **READING CHECK:** *Comparing and Contrasting* How was the new Constitution different from the old Articles of Confederation? The Constitution gave the central government important powers, but allowed states to retain most of their powers to govern themselves.

The United States in 1783

▲ The Treaty of Paris doubled the size of the United States.

Interpreting the Map What nation controlled the region to the south of the United States? To the north? To the west?

▲ George Washington (1732–1799) led the American troops to victory in the Revolution and was elected the first president of the United States.

Teaching Objective 2

LEVEL 1: (Suggested time: 15 min.) Ask students to use the following as the main headings of an outline: Causes of the French Revolution, The Outbreak of the French Revolution,. The End of the Monarchy, and The French Republic. Beneath each main heading, ask students to enter at least two or three subheadings that provide important ideas about the heading.

LEVEL 2: (Suggested time: 30 min.) Have pairs of students turn the following headings and subheadings in the text into questions: Growing Discontent, The Outbreak of Revolution, The End of the Monarchy, The

French Republic. For example, the heading Growing Discontent might be rephrased as the question, *"Why was there growing discontent in France in the 1770s"?* Have students work together to write questions and then answer them. **COOPERATIVE LEARNING**

LEVEL 3: (Suggested time: 30 min.) Ask students to pretend they are Parisian citizens in July of 1789. Have them write diary entries about the events leading up to the storming of the Bastille. Each student's entry should include a description of the suffering that had taken place in Paris as well as an explanation for why the Bastille was chosen as a target.

USING ILLUSTRATIONS

Have students examine the painting of the Bastille on this page. Ask students why freeing the prisoners was a major accomplishment for the mob of angry citizens. *(Possible answer: The prison was a symbol of the king's power.)* Ask students to note other details about this historic event based on the painting. *(Possible answers: The mob seems to have set fire to part of the prison; the mob was armed with at least one cannon; some soldiers in uniform seem to have taken part in the attack.)* Invite students to draw conclusions about how the storming of the Bastille might have affected the revolutionaries. *(Possible answer: They would have been emboldened by this success to take further steps against the king and nobility.)*

You may also want to have students conduct further research on the storming of the Bastille and compare these facts with what they see in the illustration.

The nobles of France seemed to care little for the suffering of ordinary people. When Marie-Antoinette, the wife of King Louis XVI, was told that many peasants had no bread to eat, she is said to have replied, "Let them eat cake."

The French Revolution

For over 100 years France had been the largest and most powerful nation in Europe. For all of this time, a monarch with absolute power had ruled France. Yet within months of the beginning of the French Revolution in 1789, the king lost all power.

Growing Discontent As the United States won its independence, the French people struggled against **oppression**. Oppression is the cruel and unjust use of power against others. By the 1770s discontent with the nobility was widespread. Food shortages and rising prices led to widespread hunger. To make matter worse, the nobles, who owned most of the land, raised rents. Taxes were also raised on the peasants and middle classes while the nobles and the clergy paid no taxes. Some French people took to the streets, rioting against high prices and taxes.

At the same time, the French monarchy was losing authority and respect. Due to the king's expensive habits and spending on foreign wars, France was in deep debt. To pay the debts, King Louis XVI tried to tax the nobles and the clergy. When they refused to pay the taxes, France faced financial collapse.

The French peasants and middle classes had different complaints against the king. They did, however, share certain Enlightenment ideas. For example, they spoke of liberty and equality as their natural rights. These ideas united them against the king and nobles.

The Outbreak of Revolution

In 1789 a group representing the majority of the people declared itself to be the National Assembly. It was determined to change the existing government. This action marked the beginning of the French Revolution.

When Louis XVI moved troops into Paris and Versailles, there was fear that the soldiers would drive out the National Assembly by force. In Paris, the people took action. Angry city dwellers destroyed the Bastille prison, which they called a symbol of royal oppression. The violence spread as peasants attacked manor houses and monasteries throughout France.

On July 14, 1789, a crowd destroyed the Bastille, freeing its prisoners. Bastille Day marked the spread of the Revolution and is celebrated in France every July 14.

CAUSES OF THE REVOLUTIONS

Causes of the French Revolution
Widespread discontent with nobility
Food shortages and widespread hunger

Increasing belief in Enlightenment ideas of liberty and equality
Discontent over unfair taxes

Causes of the American Revolution
Anger over lack of representation in Parliament
The growing desire for independence/self-rule

The National Assembly quickly took away the privileges of the clergy and nobles. Feudalism was ended and peasants were freed from their old duties. The National Assembly also adopted the *Declaration of the Rights of Man and Citizen*. This document stated that men are born equal and remain equal under the law. It guaranteed basic rights and also defined the principles of the French Revolution—liberty, equality, and fraternity.

The End of the Monarchy In 1791 the National Assembly completed a constitution for France. This constitution allowed for the king to be the head of the government, but limited his authority. The constitution divided the government into three branches—executive, legislative, and judicial. Louis XVI pretended to agree to this new government. In secret, he tried to overthrow it. When he and his family tried to escape France in 1791, they were arrested and sent back to Paris.

The French Republic In 1792 a new group of people, the National Convention, gathered and declared France a republic. The National Convention also put Louis XVI on trial as an enemy of the state. Urged on by a lawyer named Maximilien Robespierre, the members found the king guilty. In 1793, the king was sent to the guillotine, a machine that dropped a huge blade to cut off a person's head.

Robespierre was the leader of a political group known as the Jacobins. Many of the Jacobins wanted to bring about sweeping reforms that would benefit all classes of French society. As the French Revolution went on, the Jacobins gained more and more power. By the time Louis XVI was executed, Robespierre was probably the most powerful man in France. He and his allies controlled the actions of the National Convention.

▲

This poster summarizes the main goals of the French Revolution: liberty, equality, and fraternity. Fraternity means "brotherhood."

The frequent use of the guillotine shocked people in France, Europe, and the United States. As a result, the French Revolution lost many supporters.

▼

Linking Past to Present

Louis XVI's lavish lifestyle contributed to his unpopularity. Maintaining his immense palace, Versailles, was a drain on the treasury. The palace was also a striking example of extravagance at the taxpayers' expense. While Louis XVI ruled, 10,000 people lived or worked in his household. There were special rooms for a clock, the king's throne, and the royal billiard table. The grounds at Versailles covered 2,000 acres and included miles of wide pathways and thousands of exotic plants. Versailles was sacked during the Revolution. Only a portion of its grandeur has been restored.

Activity: In 1999 a powerful storm swept through France and downed some 10,000 of the trees at Versailles. Have some students conduct research on the restoration of the palace's gardens. Ask a second group to investigate ongoing restoration of Versailles' interior. Have a third group research Versailles' role as a major tourist attraction.

Teaching Objective 3

LEVEL 1: (Suggested time: 30 minutes) Have pairs of students play the roles of ordinary French citizens who have lived through the Napoléonic era and the Congress of Vienna. Ask students to create conversations that the French citizens might have had about the accomplishments of their emperor and his later defeat. Students can also share their feelings about the events at the Congress of Vienna. **COOPERATIVE LEARNING**

LEVELS 2 AND 3: (Suggested time: 30 minutes) Have students work in groups to write newspaper headlines that tell about the accomplishments of Napoléon and the work of the Congress of Vienna. Ask each student to choose one of the group's headlines and write the opening, or lead, paragraph for an article about the event. You may want to provide additional resources. Have students read their paragraphs aloud to the class. You may want to extend the activity by having students complete the newspaper, with comics, an advice column, classified ads, and other features appropriate to the place and time. **COOPERATIVE LEARNING**

Across the Curriculum
MUSIC

National Anthems Play a recording of the French national anthem, *La Marseillaise*, for the class. Ask students to share their reactions to the music. Tell students that the anthem was written in 1792 by a captain in the French army, Rouget de Lisle. Originally a marching song, the words were sung by soldiers from Marseilles, a city in southern France, who were going to defend their homeland against invaders from Austria. Play the music again and ask students why a marching song is appropriate for a national anthem. *(Possible answer: The stirring rhythm inspires patriotism and feelings of pride among citizens.)*

Activity: Have students research the history of the American national anthem. Then have students compare and contrast *The Star-Spangled Banner* and *The Marseillaise.*

▲ A group called the Jacobins controlled the National Convention. Robespierre was a powerful leader of the Jacobins.

▲ Napoléon had a way of getting the public's attention. He was very popular with the French people.

Under Robespierre, the National Convention looked for other enemies. Anyone who had supported the king or criticized the revolution was a suspect. Thousands of people—nobles and peasants alike—died at the guillotine. This period, called the **Reign of Terror**, ended in 1794 when Robespierre himself was put to death. Despite the terror, the revolutionaries did achieve some goals. They replaced the monarchy with a republic. They also gave peasants and workers new political rights. They opened new schools and supported the idea of universal elementary education. They established wage and price controls in an effort to stop inflation. They abolished slavery in France's colonies. They encouraged religious tolerance.

Between 1795 and 1799, a government called the Directory tried to govern France. A new two-house legislature was created to make laws. This legislature also elected five officials called directors to run the government. The people selected to be directors, however, could not agree on many issues. They were corrupt and quarreled about many issues. They quickly became unpopular with the French people. In addition, by 1799 enemy armies were again threatening France. Food shortages were causing panic in the cities. Many French people concluded their country needed one strong leader to restore law and order.

✔ **READING CHECK:** *Summarizing* Why did many French peasants and poor workers support the Revolution? believed in liberty and equality, tired of being oppressed

The Napoléonic Era

In 1799 a young general named Napoléon Bonaparte overthrew the Directory and took control of the French government. Most people in France accepted Napoléon. In turn he supported the changes brought about by the Revolution. In 1804 France was declared an empire, and Napoléon was crowned emperor.

Napoléon as Emperor In France Napoléon used his unlimited power to restore order. He organized French law into one system—the Napoléonic Code. He set up the Bank of France to run the country's finances. Influenced by the Enlightenment, he built schools and universities.

A brilliant general, Napoléon won many land battles in Europe. By 1809 he ruled the Netherlands and Spain. He forced Austria and Prussia to be France's allies. He abolished the Holy Roman Empire. He also unified the northern Italian states into the Kingdom of Italy, under his control. Within five years of becoming emperor, Napoléon had reorganized and dominated Europe. Because of the important role that he played, the wars that France fought from 1796 until 1815 are called the Napoléonic Wars.

Napoléon also made changes in the lands he controlled. He put the Napoléonic Code into effect in the countries he conquered and abolished feudalism and serfdom. He also introduced new military techniques throughout Europe. Without intending to, the French increased feelings of loyalty and patriotism among the people Napoléon had conquered. In some places this increased opposition to French rule. Over time, the armies of Napoléon's enemies grew stronger.

In 1812 Napoléon invaded Russia with more than 500,000 soldiers. The invasion was a disaster. The cold Russian winter, hunger, and disease claimed the lives of most of the French soldiers. Napoléon finally ordered his soldiers to retreat.

Napoléon's Defeat The monarchs of Europe took advantage of Napoléon's weakened state. Prussia, Austria, and Great Britain joined together to invade France. These allies captured Paris in 1814. Napoléon gave up the throne and went into exile on the island of Elba, near Italy. Louis XVIII, the brother of the executed king, was made the new king of France.

The following year Napoléon made a short-lived attempt to retake his empire. This period is known as the Hundred Days. Between March and June of 1815, Napoléon regained control of France. The king fled into exile. Soon, however, Napoléon's enemies sent armies against him. The other European nations defeated him at Waterloo in Belgium. Napoléon was sent to St. Helena, a small island in the South Atlantic. He lived there under guard and died in 1821. In 1840 the British allowed the French to bring Napoléon's remains back to Paris, where they lie to this day.

Europe After Napoléon During the years of Napoléon's rule, France had become bigger and stronger than the other countries in Europe. After Napoléon's defeat, delegates from all over Europe met at the Congress of Vienna. Their goal was to bring back the **balance of power** in Europe. Having a balance of power is a way to keep peace by making sure no one nation or group of nations becomes too powerful.

Napoléon had not always upheld the ideals of the French Revolution, but he did extend their influence throughout Europe. This led other governments to fear that rebellions against monarchy might spread. Having defeated Napoléon, the major European powers wanted to restore order, keep the peace, and suppress the ideas of the revolution.

▲
This painting captures the glory of Napoléon as a military leader.

Governments of France, 1774-1814	
1774	Louis XVI becomes king.
1789	Third Estate, as the National Assembly, assumes power.
1791	Legislative Assembly, with Louis XVI as constitutional monarch, begins rule.
1799	Napoléon establishes himself as First Consul.
1804	Napoléon is crowned emperor.
1814	Napoléon is defeated and the monarchy is restored.

After being ruled as a republic and then an empire, France became a monarchy once again in 1814.

Play the *1812 Overture*, by Pyotr Ilich Tchaikovsky, for the class. This famous symphony commemorates Napoléon's ultimate defeat by the brutal Russian winter. Challenge students to listen for the Russian church music and folk music that the composer worked into the score. Tchaikovsky even incorporated the French national anthem into the music.

REVIEW AND ASSESS

Have students complete the Section Review. Then call on volunteers to read aloud terms, phrases, dates, and names from the section. Ask other students to explain the words' significance. Have students complete Daily Quiz 8.2.

RETEACH

Have students complete Main Idea Activity 8.2. Then have pairs of students portray TV news reporters and anchorpersons and report on events of the American or French Revolutions. Have pairs deliver their oral reports of the events in chronological order. **ENGLISH LANGUAGE LEARNERS**

EXTEND

Have interested students conduct research on one of the following historical American gatherings: the First Continental Congress (1775), the Second Continental Congress (1776), the Constitutional Convention (1787). Or, students can research these important French gatherings: the National Assembly (1789); the National Convention (1792), the Committee of Public Safety (1793). **BLOCK SCHEDULING**

Section Review 2

Answers

Define For definitions, see Patriots, p. 152; Loyalists, p. 152; alliance, p. 152; oppression, p. 154; Reign of Terror, p. 156; balance of power, p. 157; reactionaries, p. 158

Reading for the Main Idea

1. Colonists thought it was unfair that they had no representation in Parliament. (NGS 13)

2. Causes—rising costs of food and taxes; resentment toward the nobility and clergy; Effects—new ruling bodies, end of the monarchy; improvements in some areas of life and new political rights (NGS 11)

3. restored order to France, reorganized and dominated Europe, established reforms (NGS 13)

Critical Thinking

4. It was not until 1789 that the Constitution went into effect and the modern American political system was created.

Organizing What You Know

5. 1776—Declaration of Independence; 1783—Treaty of Paris; 1789—U.S. Constitution approved; 1789—Mob storms Bastille; 1799—Napoléon seizes control of France; 1812—Napoléon invades Russia; 1814—Congress of Vienna; 1815—Napoléon defeated

Many of Europe's royal families came to Vienna during the winter of 1814–15. They attended balls while diplomats and rulers discussed the situation of Europe after Napoléon.

Many delegates to the Congress of Vienna were **reactionaries**. Reactionaries not only oppose change. They would like to actually undo certain changes. In this case they wanted to return to an earlier political system. These delegates were not comfortable with the ideals of the French Revolution, such as liberty and equality. They worried that these ideals would overturn the monarchies in their own countries.

One of the most influential leaders at the Congress of Vienna was Prince Metternich of Austria. To protect his absolute power in Austria, Metternich suppressed ideas such as freedom of speech and of the press. He encouraged other leaders to censor newspapers and to spy on individuals they suspected of revolutionary activity.

The Congress of Vienna redrew the map of Europe. Lands that Napoléon had conquered were taken away from France. In the end, France's boundaries were returned to where they had been in 1790. Small countries around France were combined into bigger, stronger ones. This was done to prevent France from ever again threatening the peace of Europe. France also had to pay other countries for the damages it had caused. Ruling families were returned to their thrones in Spain, Portugal, and parts of Italy. Switzerland alone kept its constitutional government but had to promise to remain neutral in European wars.

The Congress of Vienna also led to an alliance between Great Britain, Russia, Prussia, and Austria. The governments of these countries agreed to work together to keep order in Europe. For 30 years the alliance successfully prevented new revolutions in Europe.

✓ **READING CHECK:** *Drawing Conclusions* Why did the other countries of Europe want to defeat Napoléon? to restore a balance of power in Europe

go.
hrw
.com
Homework Practice Online
Keyword: SJ3 HP8

Section Review 2

Define and explain Patriots, Loyalists, alliance, oppression, Reign of Terror, balance of power, reactionaries

Reading for the Main Idea

1. *(Human Systems)* Why did the 13 American colonies rebel against Great Britain?

2. *(Human Systems)* What were the causes and effects of the French Revolution?

3. *(Human Systems)* How did Napoléon change France and the rest of Europe?

Critical Thinking

4. **Supporting a Point of View** Many people argue that the United States was not really created until 1789. Why do you think this is so? Explain your answer.

Organizing What You Know

5. **Identifying Time Order** Copy the following time line. Use it to list some important events of both the American Revolution and the French Revolution.

1775   1780   1785   1790   1795   1800

Section 3

Objectives

1. Describe the beginnings of the Industrial Revolution.
2. Explain how developments in transportation and communications spread industrial development.
3. Identify business features that affected life in the Industrial Age.

FOCUS

LET'S GET STARTED

Write the word *revolution* on the chalkboard and ask students to think about its various meanings. Remind students that the previous section described political revolutions brought about by military means. Point out that a second definition for revolution is "a complete change." Have students name some major changes in society that have been or might be called revolutions. Tell students that they will learn about one of these, the Industrial Revolution, in this section.

Building Vocabulary

Write the key terms on the chalkboard. Circle the word *production* in **factors of production** and **mass production**. Ask students to provide a definition for *production (the act of making things)*. Based on this definition, have students infer what the key terms might mean *(mass production— making large numbers of things; factors of production— items needed to produce things)*. Then have students locate and read the definitions of the other key terms.

Section 3 The Industrial Revolution

Read to Discover

1. How did the Industrial Revolution begin?
2. What developments in transportation and communications helped spread industrial development?
3. What features of business affected life in the Industrial Age?

Define

Industrial Revolution
factors of production
capital
factories
capitalism
mass production

WHY IT MATTERS

The high standards of living that most American enjoy today were made possible by the Industrial Revolution. Use **CNNfyi.com** or other **current events** sources to find out more about industry and industrialized nations.

An early steam locomotive

Section 3 RESOURCES

Reproducible
◆ Guided Reading Strategy 8.3

Technology
◆ One-Stop Planner CD–ROM, Lesson 8.3
◆ Homework Practice Online
◆ HRW Go site

Reinforcement, Review, and Assessment
◆ Main Idea Activity 8.3
◆ Section 3 Review, p. 163
◆ Daily Quiz 8.3
◆ English Audio Summary 8.3
◆ Spanish Audio Summary 8.3

The Origins of the Industrial Revolution

In the early 1700s inventors began putting the ideas of the Scientific Revolution to work by creating many new machines. Advances in industry, business, transportation, and communications changed people's lives around the world in almost every way. This period, which lasted through the 1700s and 1800s, was called the **Industrial Revolution**.

New Needs in Agriculture The first stages of the Industrial Revolution took place in agricultural communities in Great Britain. Ways of dividing, managing, and using the land had changed greatly since the Middle Ages. People had begun to think about land in new ways. Wealthy farmers began to buy more land to increase the size of their farms. Small farmers, unable to compete with these large operations, sometimes lost their land. At the same time, Europe's population continued to grow, which meant that the demand for food grew as well. Farmers recognized the need to improve farming methods and increase production.

One such farmer was Jethro Tull. He invented a new farm machine, called a seed drill, for planting seeds in straight rows. More inventors soon followed with other new farm machines. The machinery made farms more productive, and farmers were able to grow more food with fewer workers. As a result, many farm workers lost their jobs. Many of these people moved to cities to look for other kinds of work.

▲ This painting shows the original McCormick reaper, used to cut grain. It was invented by Cyrus H. McCormick in 1831.

Teaching Objective 1

ALL LEVELS: (Suggested time: 30 min.) Copy the following graphic organizer onto the chalkboard, omitting the italicized answers. Ask students to copy it into their notebooks. Pair students and have each complete the chart with descriptions of how Great Britain was able to begin to industrialize. **ENGLISH LANGUAGE LEARNERS**

THE BEGINNINGS OF THE INDUSTRIAL REVOLUTION	
Factor of Production	How It Was Supplied in Britain
Land	*Britain had rich deposits of coal and iron ore and fast-moving rivers for water power.*
Labor	*Many workers left farms due to changes in agriculture. They looked for jobs in factories.*
Capital	*British had money and tools to buy and outfit factories.*

Linking Past to Present

The Textile Industry Changes in the factors of production, especially labor, have played an important role in the American textile industry over the last 200 years. In the early 1800s most textile mills were established in New England near fast-moving rivers that provided power. Displaced farm workers and immigrants provided much of the labor.

By the early 1900s the old textile mills of New England were shutting down as capitalists moved their operations to the South to take advantage of lower labor costs. Then, in the late 1900s, many southern mills were closed as textile companies again looked for cheaper labor in foreign countries.

Activity: Have students conduct research to create time lines illustrating key events in the history of the American textile industry.

Our Amazing Planet

One early water-powered machine in an English mill was said to spin more than 300 million yards of silk thread every day!

Factors of Production The Industrial Revolution began in Great Britain because the country had the right **factors of production**. These are items necessary for industry to grow. They include land, natural resources, workers, and **capital**. Capital refers to the money, and tools needed to make a product.

Great Britain had rich deposits of coal and iron ore. It also had many rivers to provide water power for **factories** and transportation. Money was available, since many British people had grown wealthy during the 1700s. They were willing to invest their money in new businesses. The British government allowed people to start businesses and protected their property. Labor was available since many ex-farm workers needed jobs.

 READING CHECK: (*Summarizing*) What factors of production helped Great Britain to develop early industries? land; natural resources including coal, iron ore, and many rivers; investment capital; many available workers; and political stability

The Growth of Industry

As mentioned earlier, agricultural needs led to new machines and methods for farming. People in other industries began to wonder how machines could help them as well. For example, before the early 1700s, British people had spun thread and woven their own cloth at home on simple spinning wheels and looms. It was a slow process, and the demand for cloth was always greater than the supply.

The Textile Industry To speed up cloth making in the early 1700s, English inventors built new types of spinning machines and looms. In 1769 Richard Arkwright invented a water-powered spinning machine. He eventually set up his spinning machines in mills and hired workers to run them. Workers earned a fixed rate of pay for a set number of hours of work. Arkwright brought his workers and machinery together in a large building called a factory. Arkwright's arrangement with his workers was the beginning of the factory system.

In 1785 Edmund Cartwright built a water-powered loom. It could weave cloth much faster than could a hand loom. In fact, one worker with a powered loom could produce as much cloth as several people with traditional ones. Each new invention that improved the spinning and weaving process led to more inventions and improvements.

▲ This painting shows one artist's view of a factory. By 1800, textiles made in English factories were shipped all over the world.

Teaching Objective 2

LEVEL 1: (Suggested time: 25 min.) Provide students with blank index cards. Have each student make a flash card for each of the following items: *water-powered spinning machines and looms, factories, steam engine, canals, paved roads, locomotive, telegraph.* On the back of each card, have students write a sentence or two that describes how each item helped industry develop and spread. **ENGLISH LANGUAGE LEARNERS, COOPERATIVE LEARNING**

LEVELS 2 AND 3: (Suggested time: 45 min.) Tell students to imagine they are the inventors of one of the machines described in this section. Ask each student to write a short speech in which he or she describes his invention and how it helped the Industrial Revolution to begin or spread. Encourage students to conduct additional research to find more information for their speeches. Call on volunteers to deliver their speeches.

The factory system soon spread to other industries. Machines were invented to make shoes, clothing, furniture, and other goods. Machines were also used for printing, papermaking, lumber and food processing, and for making other machines. More and more British people went to work in factories and mills.

The Steam Engine Early machines in factories were driven by water power. This system, however, had drawbacks. It meant that a factory had to be located on a stream or river, preferably next to a waterfall or dam. In many cases these streams and rivers were far from raw materials and overland transportation routes. The water flow in rivers can change from season to season, and sometimes rivers run dry. People recognized that a lighter, movable, and more dependable power source was needed. Many inventors thought using steam power to run machines was the answer.

Steam engines boil water and use the steam to do work. Early steam engines were not efficient though. In 1769 James Watt, a Scottish inventor, built a modern steam engine that did work well. With Watt's invention, steam power largely replaced water power. This meant that factories could be built anywhere.

The factory system changed the lives of workers. In the past, workers had taken years to learn their trades. In a factory, however, a worker could learn to run a machine in just a day or two. Factory owners hired unskilled workers—often young men, women, and children—and paid them as little as possible. As a result, the older skilled workers were often out of work.

✔ **READING CHECK:** *Cause and Effect* How was the textile industry created in Great Britain? the invention of machines to spin and weave produced enough cloth to sell at home and abroad

▲
New industries needed much steel for machinery. The Bessemer converter, invented in the 1850s, was a cheaper, better way to make steel.

HUMAN SYSTEMS

During the early decades of the Industrial Revolution, many serious problems arose in Great Britain. As more and more people migrated to industrial cities, severe overcrowding occurred. Unsanitary conditions led to outbreaks of disease. Workers, even young children, labored 10 to 14 hours a day, six days a week. Their factory jobs were monotonous and often dangerous. Wages were kept as low as possible, and workers often went hungry. Conditions were so bad that factory employees occasionally rioted, destroying the machines that they blamed for their misery.

Critical Thinking: Why do you think the conditions of industrial workers gradually improved during the 1800s?

Answer: Politicians began to act in the interests of the working class; laws were passed to keep workers safer and healthier.

The Spread of the Industrial Revolution

Great Britain quickly became the world's leading industrial power. British laws encouraged people to use capital to set up factories. Great Britain's stable government was good for industry too.

The rest of Europe did not develop industry as quickly. For one thing, the French Revolution and Napoléon's wars had disrupted Europe's economies. That made it difficult to put the factors of production to work. Many countries also lacked the resources needed to industrialize.

Teaching Objective 3

LEVEL 1: (Suggested time: 15 min.) Lead a class discussion on inventions and business practices that made mass production possible. *(Possible answers: Capitalists divided manufacturing into a series of steps; interchangeable parts were used to make products; assembly lines carried parts to workers.)* Then have students describe some of the effects of mass production for society in general. *(Possible answer: The cost of goods was lowered; more people could buy goods and enjoy a higher standard of living.)*

LEVELS 2 AND 3: (Suggested time: 20 min.) Ask students to imagine they have just gotten jobs in one of the large new factories of the late 1800s. In a letter to a relative or friend, have each student describe the factory in which he or she works and describe how his or her work is organized. Direct students to note in their letters which aspects of their jobs they like and which they dislike. Remind students that their letters should mention what type of product is made in the factory.

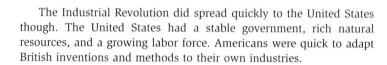

These steamboats from the 1850s carried people and goods on the Mississippi River.

The telegraph revolutionized communications in the 1850s. This device is a telegraph receiver.

The Industrial Revolution did spread quickly to the United States though. The United States had a stable government, rich natural resources, and a growing labor force. Americans were quick to adapt British inventions and methods to their own industries.

Transportation Since the Middle Ages, horse-drawn wagons had been the main form of transportation in Europe. Factory owners needed better transportation to get raw materials and send goods to market. To move goods faster, stone-topped roads were built in Europe and the United States. Canals were dug to link rivers. The steam engine was also put to work in transportation. In 1808 American inventor Robert Fulton built the first steamboat. Within a few decades, steamships were crossing the Atlantic.

Steam also powered the first railroads. An English engineer, George Stephenson, perfected a steam locomotive that ran on rails. By the 1830s, railways were being built across Great Britain, mainland Europe, and the United States.

Communication Even before 1800, scientists had known that electricity and magnetism were related. American inventor Samuel F. B. Morse put this knowledge to practical use. Morse sent an electrical current through a wire. The current made a machine at the other end click. Morse also invented a code of clicking dots and dashes to send messages this way.

Morse's inventions—the telegraph and the Morse code—brought about a major change in communications. Telegraph wires soon stretched across continents and under oceans. Suddenly information and ideas could travel at the speed of electricity.

Life in the Industrial Age

The 1800s are sometimes called the Industrial Age. This was an age of new inventions. It was a time when businesses found new ways to produce and distribute goods. The owners of factories in the Industrial Age often became very wealthy. Low factory wages, however, meant many workers faced poverty.

CLOSE

Call on students to summarize some of the ways in which the Industrial Revolution changed everyday life and work. Then ask students if they would have preferred to live before, during, or after the Industrial Revolution.

REVIEW AND ASSESS

Have students complete the Section Review. Call on students to choose one of the key terms from this section and summarize its significance with regard to the Industrial Revolution. Then have students complete Daily Quiz 8.3.

RETEACH

Have students complete Main Idea Activity 8.3. Ask students to imagine that they are a group of American capitalists from the late 1800s trying to decide what type of factory to build and where to build it. Have students discuss how the factors of production, as well as transportation and communication, might affect their decision. **ENGLISH LANGUAGE LEARNERS**

EXTEND

Invite students to conduct research on changes in the factory system since the Industrial Revolution. Encourage students to focus their attention on laws and policies that were passed to protect the rights and safety of workers. Call on volunteers to share their findings with the class. **BLOCK SCHEDULING**

The Rise of Capitalism In the late 1800s European and American individuals owned and operated factories. This economic system is called **capitalism**. In a capitalist system, individuals or companies, not the government, control the factors of production.

The early capitalists wanted to make as much profit as possible from their factories. They divided each manufacturing process into a series of steps. Each worker performed just one of the steps, over and over again. This division of labor meant workers could produce more goods in less time.

Factory owners used machines to make the parts for their products. These parts were identical and interchangeable. To speed up production, the parts were carried to the workers in the factory. Each worker added one part, and the product moved on to the next worker. This method of production is called an assembly line.

Mass Production The division of labor, interchangeable parts, and the assembly line made mass production possible. **Mass production** is a system of producing large numbers of identical items. Mass production lowered the cost of clothing, furniture, and other goods. It allowed more people to buy manufactured products and to enjoy a higher standard of living.

▲

In the early 1900s, Henry Ford used an assembly line to build cars.

Interpreting the Visual Record How do you think the assembly line might have made work easier for these people?

✓ **READING CHECK:** *Finding the Main Idea* What is capitalism and how did it affect the Industrial Revolution? private ownership of business; more capital available to invest in factories and inventions

Homework Practice Online
Keyword: SJ3 HP8

Section Review 3

Define and explain Industrial Revolution, factors of production, capital, factories, capitalism, mass production

Reading for the Main Idea

1. (*Environmental and Society*) Where did the Industrial Revolution begin and why?

2. (*Human Systems*) What advances in transportation and communications helped to spread the Industrial Revolution?

3. (*Human Systems*) How did capitalism and mass production affect people's standard of living in the late 1800s?

Critical Thinking

4. **Drawing Conclusions** Why do you think the steam engine was such an important invention of the Industrial Revolution?

Organizing What You Know

5. **Categorizing** Copy the following graphic organizer. Use it to describe some important inventions of the Industrial Revolution.

Invention	Inventor	Importance
seed drill		
spinning machine		
water-powered loom		
steam engine		
steam locomotive		
telegraph		

Section Review 3

Answers

Define For definitions, see: Industrial Revolution, p. 159; factors of production, p. 160; capital, p. 160; factory system, p. 160; capitalism, p. 163; mass production, p. 163

Reading for the Main Idea

1. Great Britain; sufficient land, resources, capital, labor; stable government (NGS 16)

2. roads, canals, steamship, locomotive, telegraph (NGS 11)

3. lowered the cost of goods, which improved standard of living (NGS 11)

Critical Thinking

4. allowed the development of mechanized industries and transportation systems

Organizing What You Know

5. seed drill—Jethro Tull, mechanized farming; spinning machine—Richard Arkwright, made thread easier to spin; water-powered loom—Edmund Cartwright, produced cloth faster; steam engine—James Watt, new source of power; steam locomotive—George Stephenson, transport people and goods quickly; telegraph—Samuel F.B. Morse, allowed long-distance communication

◄ **Visual Record Answer**

They only had to learn how to perform one task and did not have to move around much.

Section 4

Objectives

1. Describe changes that occurred in Europe and the United States after 1850.

2. Identify factors that led to reform in the later 1800s.

3. Explain how nationalism changed the map of Europe in the mid-1800s.

FOCUS

LET'S GET STARTED

Copy the following question onto the chalkboard: *What are some forms of new technology that affect your lives every day?* Discuss responses. *(Possible answers: cell phones, e-mail, personal computers, video games)* Ask students to name some older inventions that have made this newer technology possible. *(Possible answers: electricity, telephones, television)* Tell students that they will learn about the development of some of these inventions and how they affected peoples lives in Section 4.

Building Vocabulary

Write the key terms on the chalkboard. Explain that a number of these terms have Latin roots. **Literacy** comes from the Latin word *littera*, meaning "letter." **Emigrate** is derived from the verb *migrare*, meaning "to wander," and the prefix *e-*, meaning "out." **Suburbs** is a combination of the word *urbs*, meaning "city," and the prefix *sub-*, "below." Ask students to think of other words based on the same roots. *(Possible answers: literature, migration, urban, submarine)* Then have students find and read the definitions of all the key terms.

Section 4 RESOURCES

Reproducible

◆ Guided Reading Strategy 8.4

Technology

◆ One-Stop Planner CD–ROM, Lesson 8.4

◆ Homework Practice Online

◆ HRW Go site

Reinforcement, Review, and Assessment

◆ Main Idea Activity 8.4

◆ Section 3 Review, p. 169

◆ Daily Quiz 8.4

◆ English Audio Summary 8.4

◆ Spanish Audio Summary 8.4

Section 4 — Expansion and Reform

Read to Discover

1. How did life in Europe and America change after 1850?

2. What led to reforms in the later 1800s?

3. How did nationalism change the map of Europe in the mid-1800s?

Define

working class
literacy
emigrate
suburbs

reform
suffragettes
nationalism

WHY IT MATTERS

By the later 1800s, new ideas and technology began to improve city life. Use **CNNfyi.com** or other **current events** sources to find out about solutions to today's urban problems.

Thomas Edison's electric light bulb

The Rise of the Middle Class

The Industrial Revolution changed how people in Europe and America worked and lived. Industries and cities grew, and new inventions made life easier.

During the later 1800s many people became better educated. Some became wealthy. This group included bankers, doctors, lawyers, professors, engineers, factory owners, and merchants. Also in this group were the managers who helped keep industries running. Together these people and their families were known as the middle class. Membership in the middle class was based upon economic standing rather than upon birth.

The ideas of the middle class influenced many areas of life in Western Europe and in the United States. Over time, the middle class's wealth, social position, lifestyle, and political power grew. Government leaders began turning to some middle-class individuals for advice, particularly about business and industry.

Many middle-class families had enough money that women did not need to work outside the home. They cleaned, cooked, and took care of the children, often with hired help. In the mid-1800s, however, many middle-class women started to express a desire for roles outside the home.

Doctors were one of the groups who made up the middle class of the late 1800s. Medical advances made during this period made their jobs safer and more efficient.

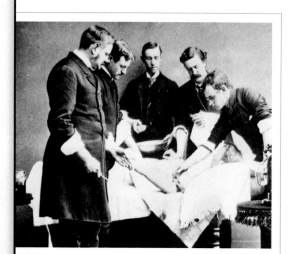

Teaching Objective 1

ALL LEVELS: (Suggested time: 30 minutes) Copy the following graphic organizer onto the chalkboard, omitting the italicized answers. Ask students to copy it into their notebooks. Pair students and have each pair fill in the chart, identifying one or more effects for each listed cause. Then invite students to add additional examples of causes and effects that they read about in this section. Call on volunteers to share their answers with the class. **ENGLISH LANGUAGE LEARNERS, COOPERATIVE LEARNING**

CHANGES IN THE LATE 1800S	
Cause:	Scientists learned more about food, health, and disease.
Effect:	*People were able to live longer and healthier lives.*
Cause:	Governments required all children to go to school.
Effect:	*Literacy became widespread. More books and magazines were published. People became more informed.*
Cause:	City dwellers had more time and money for entertainment.
Effect:	*Theaters and libraries opened. Sports teams became more organized. Parks were built*

For some women, doing something outside the home meant independence. It was also a way to earn a living. During the late 1800s more jobs opened up to women. They became nurses, secretaries, telephone operators, and teachers.

✓ **READING CHECK:** *Finding the Main Idea* What role did the middle class play in the society of the late 1800s? became the professional and management people of society; gave advice to the government about business and industry

The Growth of Society

The middle class was not the only group of people to enjoy the benefits of the Industrial Age. By the 1870s life was improving in some ways for both the middle class and the **working class,** people who worked in factories and mines.

Technology and Communication In the 1870s a tremendous new power source was developed. That power source was electricity. This led to a new wave of inventions in Western Europe and the United States. The electric generator produced the power needed to run all kinds of machines and engines. Thomas Edison's electric light bulb created a new way of lighting rooms, streets, and cities. Alexander Graham Bell's telephone made it possible to transmit the human voice over long distance.

In the late 1800s the first successful gasoline-driven automobile was built. In 1908 American inventor Henry Ford produced the Model T. This was the first automobile to become popular with American buyers.

Other Advances Advances in science and medicine also transformed people's lives. Scientists discovered more about the connection between food and health. This new knowledge, plus new information about diseases, made it possible for people to live healthier, longer lives.

Scientists of this time also made great advances in the fields of chemistry and physics. For example, they formed new theories about the structure of the atom and organized all known elements into the periodic table. It was also at this time that X-rays were discovered and first used in medicine. Scientists like Max Planck and Albert Einstein developed new ideas that changed the study of physics. Their ideas were the basis for the work of many later experiments.

Several new fields of study, together called the social sciences, gained popularity during this period. Scholars saw these fields as a way to study people as members of society. The social sciences include such fields as economics, politics, anthropology, and psychology. The study of history also changed. Historians searched for evidence of the past in documents, diaries, letters, and other written sources. As a result, new views of history began to emerge from their research.

▲ After 1870, more and more women went to high school. For the first time they began to study the same subjects as men did.

◄ Scientists of the 1800s used microscopes like this one to study cells.

Across the Curriculum

SCIENCE

The Fight Against Disease
In the struggle against disease, the discoveries of Louis Pasteur, Joseph Lister, and Robert Koch mark an important turning point. The French chemist Louis Pasteur identified bacteria and explained how they reproduce and cause disease. Joseph Lister, an English surgeon, used Pasteur's findings to develop *antisepsis*—the use of chemicals to kill disease-causing bacteria. Robert Koch, a German doctor, further confirmed Pasteur's findings by identifying the germs that caused tuberculosis and cholera.

Activity: Ask interested students to conduct research on one of these three scientists and to describe the experiments he used to make his discoveries.

Teaching Objective 2

LEVEL 1: (Suggested time: 15 min.) Have students draw three columns in their notebooks labeled *Reforms in Great Britain, Reforms in France*, and *Reforms in the United States*. Then have students look through the text to find examples of reforms to list under each heading. Call on volunteers to write their answers on the chalkboard.
ENGLISH LANGUAGE LEARNERS

LEVELS 2 AND 3: (Suggested time: 30 min.) Organize the students into small groups and have the members of each group imagine that they are living in Great Britain, France, or the United States in the mid-1800s. Have the groups discuss the reforms that are occurring in their respective countries and how these reforms might affect them. Then have each student write a letter to a member of another group describing the reforms and his or her thoughts about them. **COOPERATIVE LEARNING**

Linking Past to Present

Colleges for Women

By the end of the 1800s many countries offered elementary education for girls, but secondary education was limited. Some people argued that many subjects were not necessary or proper for women. In the United States, Great Britain, and France, secondary education for girls focused on languages, literature, and home economics—not the sciences or mathematics. Some people objected to these differences.

A British woman named Emily Davies urged her government to prepare women to attend universities. However, few colleges admitted women as students during the 1800s. Thus, colleges just for women began to appear in Great Britain and the United States. Some colleges today still admit only women.

Discussion: Lead a discussion based on the following questions: Is there still a need for women's colleges? Most colleges that once admitted only men are now co-educational. Should all colleges be co-educational?

Visual Record Answer ▶

Possible answers: People dressed up to play sport. Sports were popular with people who had money.

As greater numbers of people learned to read, newspapers competed for their attention. They published eye-catching stories and cartoons.

This is a painting of a croquet game in a public park. It reflects the increased participation in free-time activities during the late 1800s.

Interpreting the Visual Record What does this painting suggest about sports in the later 1800s?

Public Education After 1870, governments in Europe and the United States required all children to attend school. The spread of education had many benefits. As **literacy**—the ability to read—became widespread, more books and magazines were published. Newspapers that carried stories from all over the world also became very popular. By reading them, citizens became more informed about their governments.

Arts and Entertainment City dwellers of the late 1800s found themselves with more time and money for entertainment. Theaters opened to meet a demand for concerts, plays, and vaudeville shows. Art collections were made available to the public by displaying them in museums. Free public libraries opened in many cities. Sports became more organized, and cities began to sponsor teams with official rules and national competitions. Many cities began to construct public parks. These parks allowed people who lived in the cities to enjoy outdoor activities. By the end of the 1800s many of these parks had begun to include playgrounds for children.

A Growing Population One of the greatest changes of the later 1800s was the rapid growth of cities in the United States. Faced with crowded, dirty cities and seeking new opportunities, many Europeans chose to emigrate to the United States. To **emigrate** means to leave one country to live in another. The United States was not the only destination for these emigrants. Many also chose to seek new lives in South America, Africa, Australia, and New Zealand.

Between 1870 and 1900 more than 10 million people left Europe for the United States. These newcomers hoped to find economic opportunities. Some sought political and religious freedom as well.

Madeleine Schmitt of St. Louis, Missouri, suggests the following activity to help students understand the rise of realism in music and art. Have students listen to a piece of music written by a composer of the romantic period such as Brahms, Liszt, or Schubert. Ask students to name elements of the music that would define it as "romantic." Then have them compare how the "romantic" mood of such music differed from the real lives of most people. Lead a discussion about how this difference led to the rise of the realism movement.

Teaching Objective 3
ALL LEVELS: (Suggested time: 30 min.) Organize the class into six groups and assign each group one of the following topics: Italy before unification, Italy after unification, Germany before unification, Germany after unification, Russia before Alexander II, or Russia after Alexander II. Have each group design a poster that describes its assigned country during the appropriate time period. Encourage students to conduct additional research to find a map of the country to include on the poster. Display the posters in pairs grouped by the countries they describe.
ENGLISH LANGUAGE LEARNERS, COOPERATIVE LEARNING

CONNECTING TO *Art*

A New Art Period

Works of art not only show the values of an artist but also the values of the society in which the artist lives. The American and French Revolutions, for example, changed society deeply. Many artists were inspired to paint stirring scenes from history and nature. These artists were called *romantics*. Their scenes showed life to be more exciting and satisfying than it normally is.

By the mid-1800s, however, many artists had rejected romanticism. These artists instead wanted to portray life as it really was. A style called *realism* developed. The realists painted ordinary living conditions and familiar settings. They tried to re-create what they saw around them, accurately and honestly.

Honore Daumier was a French realist. He painted *The Washerwoman* during the Industrial Revolution in France, a time when many city workers were struggling to survive. The subject in the painting is one of these workers.

Understanding What You Read
1. How was realism different from romanticism?
2. What types of subjects and scenes might a realist paint?

DAILY LIFE

The development of public transportation encouraged city dwellers to move to outlying suburbs. The first suburbs tended to grow along train and trolley lines and close to the city center. However, as time passed and land costs increased, suburbs were built farther from cities and public transportation. The automobile eventually provided more mobility for people who lived in the suburbs.

Activity: Invite students to conduct research on how the areas surrounding a local city were suburbanized. Ask students how the locations of public transportation routes influenced the development of suburbs.

Connecting to Art
Answers
1. realism—ordinary scenes from life just as they are; romanticism—showed life to be more exciting than it really is
2. familiar settings; crowded slums, factories, nightclubs

City Improvements Faced with rapidly growing populations and changes in society, many cities needed civic improvements. Local governments began to provide water and sewer service to city dwellers. Many streets were also paved. In 1829 London organized a police force. Many other cities soon had police forces, too.

Cities around the world grew rapidly in the 1800s. By the early 1900s, more people lived in cities than in the country. Many cities—including New York, London, Paris, and Berlin—had populations of more than 1 million.

Cities also created public transportation systems. Horse-drawn streetcars and buses were used mainly within cities. Trains, however, could take people far outside a city. As a result, some people began to move outside cities to areas called **suburbs**. These people usually took trains into cities each morning and returned to their homes in the suburbs at night.

✓ **READING CHECK:** *Summarizing* What were some advances of the later 1800s? Some possible answers: electricity, electric light, telephone, new knowledge in science and medicine, automobile, new city transportation, public education

Our Amazing Planet

Between 1865 and 1900 most American cities doubled or tripled in size. Much of this growth was due to immigration. Many immigrants to the United States moved to New York City. By 1900 the city's population was nearly five times larger than it had been in 1850.

Ask students to name and discuss some of the changes that occurred during the second half of the 1800s that were important to people's well-being. As students name reforms, write them on the chalkboard. Then have the class vote on which reforms they consider to be the most important. Call on volunteers to explain their reasoning.

Have students complete the Section Review. Then name a person, event, or object described in this section. Ask students to explain its significance using the "W" questions: *who* or *what* was the person, event or object; *where* and *when* did it happen; and *why* is it significant? Then have students complete Daily Quiz 8.4.

Section Review 4

Answers

Define For definitions, see: working class, p. 165; literacy, p. 166; suburbs, p. 167; emigrate, p.167; reform, p. 168; suffragettes, p. 168; nationalism, p. 169

Reading for the Main Idea

1. advances in technology, communication, and other areas; water and sewer systems; streets paved and lit; police forces established (NGS 13)

2. improved working conditions and benefits; abolished slavery; men got the right to vote (NGS 13)

3. The love of country stirred some people to fight to unify the independent states in Italy and in Germany. (NGS 13)

Critical Thinking

4. rapid growth of cities; many civic improvements

Organizing What You Know

5. Giuseppe Garibaldi—Italy; led an army to unify the states of Italy; Otto von Bismarck—Prussia; Waged war with Austria and France to free German states under their control; united German states into the German Empire; Alexander II—Russia; freed the serfs and tried to modernize Russia

Emmeline Pankhurst (1858–1928) led many demonstrations and marches on behalf of the women's suffrage movement in Great Britain.

RÉPUBLIQUE FRANÇAISE.
Combat du peuple parisien dans les journées des 22, 23 et 24 Février 1848.

In 1848, French citizens overthrew of the king and set up the Second Republic.

Political and Social Reform

In addition to great advances in technology, communications, science, and medicine, the mid-1800s and early 1900s saw many political and social reforms. To **reform** something is to remove its faults. Around the world, citizens worked to improve their governments and societies.

Great Britain In Great Britain reformers passed laws that allowed male factory workers in cities to vote. Laws were also passed to improve conditions in factories, and slavery was abolished. New laws were passed to provide health insurance, unemployment insurance, and money for the elderly.

Beginning in the late 1800s, many women in Great Britain became **suffragettes**. These women campaigned for their right to vote. They were led by outspoken women like Emmeline Pankhurst. British women gained this right in 1928.

France A revolution in France in 1848 forced the king from the throne. A new government called the Second Republic was established. It guaranteed free speech and gave the vote to all men. In 1875, a new French constitution established the Third Republic. This constitution lasted for nearly 70 years.

United States The issue of slavery divided the United States in the mid-1800s. In late 1860 and early 1861 several southern states broke from the Union to form the Confederate States of America. The Civil War followed and raged until 1865 when the Confederacy surrendered. Congress then amended the Constitution to abolish slavery and grant citizenship to former slaves. The vote was given to all men, regardless of race or color.

Many reforms were the results of efforts by women such as Elizabeth Cady Stanton and Lucretia Mott. As early as 1848 they had campaigned for the abolition of slavery, equality for women, and the right to vote. In the 1890s and early 1900s, the movement for women's right to vote grew stronger. The Nineteenth Amendment to the Constitution, ratified in 1920, finally gave women this right.

✓ **READING CHECK:** *Comparing and Contrasting* How were reforms in Great Britain, France, and the United States in the mid-1800s and early 1900s similar? They all gave men the right to vote. Great Britain and the United States abolished slavery.

Have students complete Main Idea Activity 8.4. Then tell them to imagine that, like Rip Van Winkle, they had fallen asleep in 1840 and not woken up again until 1880. Ask: What changes might they spot walking around the streets of a major American city? What differences might they notice traveling through countries in Europe, such as England, France, Italy, Germany, and Russia? Call on students to share their answers.
ENGLISH LANGUAGE LEARNERS

Have interested students conduct research on the contributions of immigrants to American society. Students might wish to select one individual to study or may choose to look at a broader subject. Have students design collages that illustrate their findings. **BLOCK SCHEDULING**

Nationalism in Europe

Nationalism is the love of one's country more than the love of one's native region or state. In the 1800s, nationalism led to the unification of Italy and of Germany. It was also a driving force for change in Russia.

In the early 1800s the Congress of Vienna had divided Italy into several states, some of which were ruled by Austria. In the 1850s and 1860s, a nationalist named Giuseppe Garibaldi led a movement to unify these states. He and his army defeated the Austrians and their French allies and drove them out of Italy. Largely because of his efforts, most of present-day Italy had been unified by 1861. In that year Victor Emmanuel II was made the king of Italy.

Germany in the mid-1800s was a patchwork of 39 independent states. The largest was Prussia, ruled by William I. In 1862 he appointed Otto von Bismarck one of his advisers. Both men wanted to make Germany into a powerful unified country. Bismarck convinced the other German states to join in this effort and to declare war first on Austria, Prussia's chief rival, and then on France. After the war, the German states were joined together into the German Empire. William I became the first kaiser, or emperor.

In the 1800s Russia had more territory and people than any other country in Europe. Its economy, however, was not as developed as those of other countries. People from Russia's many ethnic groups felt very little unity with each other. In the 1850s Czar Alexander II tried to introduce major reforms. He freed all the serfs in Russia and introduced political changes. Later czars, however, tried to undo these reforms. Censorship and discrimination against minorities became widespread. This repression created an explosive situation in Russia. In 1905, a group of revolutionaries tried to overthrow the czar but failed.

▲
Otto von Bismarck (1815–1898) was known for his strong will and determination.

✓ **READING CHECK:** *Drawing Inferences* How did nationalism help reshape nations? led to the unification of Italy and Germany; led to social and political changes that brought more unity and equality to people in Russia

Section Review 4

Define and explain working class, literacy, emigrate, suburbs, reform, suffragettes, nationalism

Reading for the Main Idea

1. *Human Systems* What allowed people's lives to improve during the last half of the 1800s?

2. *Human Systems* How did reforms of the later 1800s and early 1900s affect people's lives?

3. *Human Systems* How did nationalism lead to the unification of Italy and Germany in the mid-1800s?

go.hrw.com **Homework Practice Online**
Keyword: SJ3 HP8

Critical Thinking

4. Analyzing What effect did immigration of the later 1800s have on the United States?

Organizing What You Know

5. Categorizing Copy the following chart. List the home country of each leader. Then give details about his accomplishments.

Leader	Country	Accomplishments
Giuseppe Garibaldi		
Otto von Bismarck		
Czar Alexander II		

CHAPTER 8

Review ANSWERS

Building Vocabulary
For definitions, see:
Enlightenment, p.147; reason, p.148; individualism, p.148; popular sovereignty, p.149; Patriots, p. 152; alliance, p. 152; oppression, p. 154; factors of production, p.160; capital, p.160; capitalism, p163; mass production, p.163; literacy, p. 166; emigrate, p. 167; reform, p.168; nationalism, p. 169

Reviewing the Main Ideas
1. a natural law that would help them solve society's problems and improve people's lives (NGS 10)

2. alliance with France (NGS 13)

3. land, natural resources, workers, and capital (NGS 14)

4. Governments required all children to go to school, so more public schools were built; spread of literacy led to more books, magazines, and newspapers read, so citizens became more informed (NGS 16)

5. led to unification of Italy and Germany and inspired reforms (NGS 10)

RETEACH

Organize students into groups and assign each group one of the following time spans: 1700–1750, 1751–1800, 1801–1850, 1851–1900. Have the members of each group list and describe some events and developments that occurred during its assigned time span. Working in chronological order, have members of each group present these events to the whole class, explaining why they are significant. **ENGLISH LANGUAGE LEARNERS, COOPERATIVE LEARNING**

 PORTFOLIO EXTENSIONS

1. Organize the class into small groups and ask each group to write a short script for a historical skit about one of the events described in this chapter. After students write and rehearse their scenes, ask them to perform their skits for the class. Take pictures of the performances and include these with the scripts in student portfolios.

2. Have students build dioramas that show memorable events from the American or French Revolution or scenes from everyday life during the Industrial Revolution or Age of Reform. Ask each student to write a short description of the scene he or she has depicted. Place these descriptions and photos of the dioramas in student portfolios.

Review
ANSWERS

Understanding History and Society
Presentations will vary, but should mention such things as sanitation, a police force, and public transportation. Use Rubric 7, Charts, to evaluate student work.

Thinking Critically
1. Enlightenment ideas of equality and popular sovereignty were the basis for the Declaration of Independence and Constitution.

2. to achieve a basic change in their political and economic situations

3. restored order to France; brought legal, financial, and educational changes to much of Europe; reorganized Europe's boundaries and governments

4. Answers will vary.

5. Members of the middle class owned most businesses and factories. Their wealth, social position, lifestyle, and political power affected all areas of society.

 Reviewing What You Know

Building Vocabulary

On a separate sheet of paper, write sentences to define each of the following.

1. Enlightenment
2. reason
3. individualism
4. popular sovereignty
5. Patriots
6. alliance
7. oppression
8. factors of production
9. capital
10. capitalism
11. mass production
12. literacy
13. emigrate
14. reform
15. nationalism

Reviewing the Main Ideas

1. (*Human Systems*) What did Enlightenment thinkers hope to discover?

2. (*Human Systems*) What led to the defeat of the British in the American Revolution?

3. (*Environment and Society*) What conditions in Great Britain gave rise to the Industrial Revolution?

4. (*Environment and Society*) How did education change greatly in Europe and the United States during the later 1800s and what were the benefits?

5. (*Human Systems*) What were the effects of nationalism on European nations in the late 1800s?

Understanding History and Society

Plans for Reform
Imagine you are the mayor of a large European city in the mid-1800s. Using a chart, make a presentation that lists and describes the reforms and changes you think should be made in your city. Consider the following:
- Healthy and safety issues.
- Communication and transportation needs.
- Education needs.

Thinking Critically

1. **Analyzing** How did Enlightenment ideas influence American democracy?

2. **Drawing Conclusions** Why did revolutions break out in America and France in the late 1700s?

3. **Analyzing** How did Napoléon Bonaparte change France and the rest of Europe in the early 1800s?

4. **Supporting a Point of View** Which three advances in the later 1800s do you think changed people's lives the most? Explain your answer.

5. **Evaluating** Why was the middle class so important to the Industrial Revolution?

FOOD FESTIVAL

American eating habits and diets have undergone a revolution since colonial days. Encourage students to conduct research on foods that were popular among people of the New England, Middle, and Southern colonies and to prepare a menu that might have been used in a colonial inn around 1750. If possible, invite students to prepare samples of some of these colonial dishes.

CHAPTER 8
REVIEW AND ASSESSMENT RESOURCES

Reproducible
◆ Readings in World Geography, History, and Culture 25, 26, and 27
◆ Critical Thinking Activity 8
◆ Vocabulary Activity 8

Technology
◆ Chapter 8 Test Generator (on the One-Stop Planner)

◆ Audio CD Program, Chapter 8
◆ HRW Go site

Reinforcement, Review, and Assessment
◆ Chapter 8 Review, pp. 170–171
◆ Chapter 8 Tutorial for

Students, Parents, Mentors, and Peers
◆ Chapter 8 Test
◆ Chapter 8 Test for English Language Learners and Special-Needs Students
◆ Unit 2 Test
◆ Unit 2 Test for English Language Learners and Special-Needs Students

Map ACTIVITY
On a separate sheet of paper, match the letters on the map with their correct labels.

Maine	Pennsylvania
New Hampshire	Maryland
Massachusetts	Virginia
Connecticut	North Carolina
New York	Georgia

Mental Mapping Skills ACTIVITY
On a separate sheet of paper, draw a freehand map of Europe in the late 1800s. Make a key for your map and label the following:

Great Britain	Italy
France	Russia
Germany	Austria

WRITING ACTIVITY
Imagine you are a news reporter in France in 1789, just as the French Revolution is breaking out. Write a news story about why the French people are in revolt and what they hope to achieve. Be sure to use standard grammar, spelling, sentence structure, and punctuation.

Map Activity
A. Maine
B. New Hampshire
C. Massachusetts
D. Connecticut
E. New York
F. Pennsylvania
G. Maryland
H. Virginia
I. North Carolina
J. Georgia

Mental Mapping Skills Activity
Maps will vary but listed places should be labeled in their approximate locations.

Writing Activity
News stories will vary but should include information about the French Revolution that is consistent with details in the text. Use Rubric 23, Newspapers, to evaluate student work.

Portfolio Activity
Answers will vary but should accurately describe local industries. Use Rubrics 7, Charts, and 30, Research, to evaluate student work.

Alternative Assessment

Portfolio ACTIVITY

Learning About Your Local History

Industries in Your Town Research one or two of the main industries of your town, region, or state. Create a chart to explain when, where, why, and how the industry or industries were developed.

🖉 **internet** connect

Internet Activity: go.hrw.com
KEYWORD: SJ3 GT8

Choose a topic to explore about the birth of the modern world:
• Explore the ideas of the Enlightenment.
• Investigate the causes of the French Revolution.
• Understand capitalism.

🖉 **internet** connect

GO TO: go.hrw.com
KEYWORD: SJ3 Teacher
FOR: a guide to using the Internet in your classroom

The Modern World
Chapter Resource Manager

Objectives	Pacing Guide	Reproducible Resources	
SECTION 1			
World War I (pp. 173–177) 1. Identify the causes of World War I. 2. Explain how science and technology made this war different from earlier wars. 3. Investigate how the world changed because of World War I.	**Regular** 1 day **Block Scheduling** .5 day *Block Scheduling Handbook, Chapter 9*	**RS** **RS**	Guided Reading Strategy 9.1 Graphic Organizer 9
SECTION 2			
The Great Depression and the Rise of Dictators (pp. 178–181) 1. Identify the causes of the Great Depression. 2. Describe a dictatorship. 3. Analyze how the Great Depression helped dictators come to power in Europe.	**Regular** 1 day **Block Scheduling** .5 day *Block Scheduling Handbook, Chapter 9*	**RS**	Guided Reading Strategy 9.2
SECTION 3			
Nationalism in Latin America (pp. 182–185) 1. Describe the European colonies in Latin America and how they won their independence. 2. Explain how dictatorships replaced some democracies in Latin America. 3. Describe how relations between the United States and Latin American countries have changed over time.	**Regular** .5 day **Block Scheduling** .5 day *Block Scheduling Handbook, Chapter 9*	**RS**	Guided Reading Strategy 9.3
SECTION 4			
World War II (pp. 186–191) 1. Identify the causes of World War II. 2. Describe the Holocaust. 3. Analyze how World War II came to an end.	**Regular** 1 day **Block Scheduling** .5 day *Block Scheduling Handbook, Chapter 9*	**RS**	Guided Reading Strategy 9.4
SECTION 5			
The World Since 1945 (pp. 192–199) 1. Describe the Cold War. 2. Identify scenes of conflict in the world since World War II. 3. Investigate important events that happened at the end of the 1900s.	**Regular** 1 day **Block Scheduling** .5 day *Block Scheduling Handbook, Chapter 9*	**RS** **E** **SM**	Guided Reading Strategy 9.5 Creative Strategies for Teaching World Geography, Lesson 10 Map Activity 9

Chapter Resource Key

RS Reading Support

IC Interdisciplinary Connections

E Enrichment

SM Skills Mastery

A Assessment

REV Review

ELL Reinforcement and English Language Learners

 Transparencies

 CD–ROM

 Music

 Video

 Internet

 Holt Presentation Maker Using Microsoft® Powerpoint®

  One-Stop Planner CD–ROM

See the *One-Stop Planner* for a complete list of additional resources for students and teachers.

One-Stop Planner CD–ROM

It's easy to plan lessons, select resources, and print out materials for your students when you use the *One-Stop Planner CD–ROM with Test Generator.*

Technology Resources	Review, Reinforcement, and Assessment Resources	
One-Stop Planner CD–ROM, Lesson 9.1	**ELL**	Main Idea Activity 9.1
ARGWorld CD–ROM	**REV**	Section 1 Review, p. 177
Homework Practice Online	**A**	Daily Quiz 9.1
HRW Go site	**ELL**	English Audio Summary 9.1
	ELL	Spanish Audio Summary 9.1
One-Stop Planner CD–ROM, Lesson 9.2	**ELL**	Main Idea Activity 9.2
ARGWorld CD–ROM	**REV**	Section 4 Review, p. 181
Homework Practice Online	**A**	Daily Quiz 9.2
HRW Go site	**ELL**	English Audio Summary 9.2
	ELL	Spanish Audio Summary 9.2
One-Stop Planner CD–ROM, Lesson 9.3	**ELL**	Main Idea Activity 9.3
ARGWorld CD–ROM	**REV**	Section 3 Review, p. 186
Homework Practice Online	**A**	Daily Quiz 19.3
HRW Go site	**ELL**	English Audio Summary 9.3
	ELL	Spanish Audio Summary 9.3
One-Stop Planner CD–ROM, Lesson 9.4	**ELL**	Main Idea Activity 9.4
ARGWorld CD–ROM	**REV**	Section 4 Review, p. 191
Homework Practice Online	**A**	Daily Quiz 9.4
HRW Go site	**ELL**	English Audio Summary 9.4
	ELL	Spanish Audio Summary 9.4
One-Stop Planner CD–ROM, Lesson 9.5	**ELL**	Main Idea Activity 9.5
ARGWorld CD–ROM	**REV**	Section 5 Review, p. 199
Homework Practice Online	**A**	Daily Quiz 9.5
HRW Go site	**ELL**	English Audio Summary 9.5
	ELL	Spanish Audio Summary 9.5

internet connect

HRW ONLINE RESOURCES

GO TO: go.hrw.com
Then type in a keyword.

TEACHER HOME PAGE
 KEYWORD: SJ3 TEACHER

CHAPTER INTERNET ACTIVITIES
 KEYWORD: SJ3 GT9

Choose an activity to:
• learn about the effects of the Treaty of Versailles.
• write a report about Anne Frank.
• create a poster about the causes and effects of global warming.

CHAPTER ENRICHMENT LINKS
 KEYWORD: SJ3 CH9

CHAPTER MAPS
 KEYWORD: SJ3 MAPS9

ONLINE ASSESSMENT
Homework Practice
 KEYWORD: SJ3 HP9
 Standardized Test Prep Online
 KEYWORD: SJ3 STP9
 Rubrics
 KEYWORD: SS Rubrics

COUNTRY INFORMATION
 KEYWORD: SJ3 Almanac

CONTENT UPDATES
 KEYWORD: SS Content Updates

HOLT PRESENTATION MAKER
 KEYWORD: SJ3 PPT9

ONLINE READING SUPPORT
 KEYWORD: SS Strategies

CURRENT EVENTS
 KEYWORD: S3 Current Events

Meeting Individual Needs

Ability Levels

Level 1 Basic-level activities designed for all students encountering new material

Level 2 Intermediate-level activities designed for average students

Level 3 Challenging activities designed for honors and gifted-and-talented students

English Language Learners Activities that address the needs of students with Limited English Proficiency

Chapter Review and Assessment

E	Readings in World Geography, History, and Culture 15, 23, 32, 34, 38, 74	**A**	Unit 2 Test
SM	Critical Thinking Activity 9		Chapter 9 Test Generator (on the One-Stop Planner)
REV	Chapter 9 Review, pp. 200–201		Audio CD Program, Chapter 9
REV	Chapter 9 Tutorial for Students, Parents, Mentors, and Peers	**A**	Chapter 9 Test for English Language Learners and Special-Needs Students
ELL	Vocabulary Activity 9	**A**	Unit 2 Test for English Language Learners and Special-Needs Students
A	Chapter 9 Test		

LAUNCH INTO LEARNING

Ask students to think about some things that have been invented or improved during their lifetimes. Then ask if any students have heard stories about their parents' lives when they were in school. Ask if any students know how their grandparents or great-grandparents lived when they were young. Encourage students to share these stories with the class. If no students have any family accounts, ask if they have seen pictures or read stories about life in the early to mid-1900s. Point out that nearly every aspect of society has undergone drastic changes since 1900. Tell students that they will learn about these changes—and the major events that caused them—in this chapter.

Section 1

Objectives
1. Identify the causes of World War I.
2. Examine how science and technology made this war different from earlier wars.
3. Investigate how the world changed because of World War I.

LINKS TO OUR LIVES

You might want to share with your students the following reasons for learning about the people and events of the 1900s:

▶ World War I and World War II led to the redrawing of national boundaries around the world. Some of the boundaries created after these wars are still a source of conflict.

▶ Countries that underwent great changes in the 1900s, such as Russia and other previously communist countries, are still affected by these changes.

▶ Organizations like the United Nations and NATO founded after World War II remain active in political affairs today.

▶ Science and technology continue to change our lives.

▶ Entertainment media developed in the 1900s—radio, movies, and television—are enjoyed by billions of people around the world.

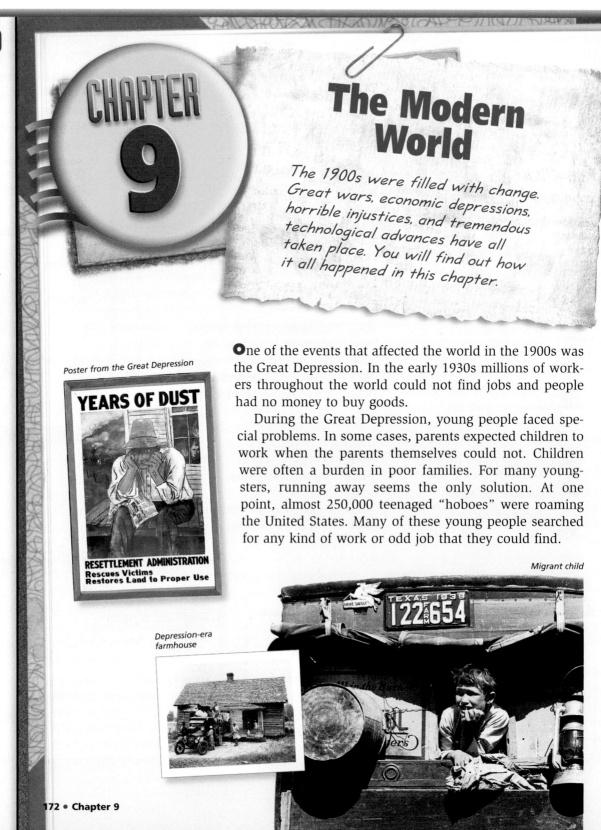

CHAPTER 9

The Modern World

The 1900s were filled with change. Great wars, economic depressions, horrible injustices, and tremendous technological advances have all taken place. You will find out how it all happened in this chapter.

Poster from the Great Depression

One of the events that affected the world in the 1900s was the Great Depression. In the early 1930s millions of workers throughout the world could not find jobs and people had no money to buy goods.

During the Great Depression, young people faced special problems. In some cases, parents expected children to work when the parents themselves could not. Children were often a burden in poor families. For many youngsters, running away seems the only solution. At one point, almost 250,000 teenaged "hoboes" were roaming the United States. Many of these young people searched for any kind of work or odd job that they could find.

Migrant child

Depression-era farmhouse

172

Copy this question and instructions onto the chalkboard: *What do you think the expression "world war" means?* Discuss student responses. *(Possible answer: a war involving many countries or affecting a large area)* Call on students to suggest some reasons a small or local conflict between two countries might turn into a world war. *(Students might say that a country could have many enemies or that other countries could take sides in an existing conflict.)* Tell students that in Section 1 they will learn about the causes and results of the First World War.

Building Vocabulary

Write the key terms on the chalkboard. Call on volunteers to find the definitions of the terms in the glossary and read them to the class. Point out the words **nationalism** and **militarism**, and circle the suffix *-ism*. Explain that this suffix indicates an idea, concept, or behavior. Ask students to use the suffix to explain the meaning of these two words. *(Possible answer: Nationalism is the idea that one should honor one's nation; militarists act in a way that glorifies the military.)* Ask students to name some other words ending in this suffix. *(Possible answers: terrorism, heroism, athleticism)*

Section 1 — World War I

Read to Discover

1. What were the causes of World War I?
2. How did science and technology make this war different from earlier wars?
3. How was the world changed because of World War I?

Define

militarism
U-boats
armistice

Locate

England
France
Germany
Russia
Austria-Hungary
Ottoman Empire
Serbia
Sarajevo

WHY IT MATTERS

World War I started in the Balkans—a region that is still the scene of much conflict. Use CNNfyi.com or other **current events** sources to learn what is happening in this region today. Record your findings in your journal.

German Poster from World War I

Section 1 RESOURCES

Reproducible
- Block Scheduling Handbook, Chapter 9
- Guided Reading Strategy 9.1
- Graphic Organizer 9

Technology
- One-Stop Planner CD–ROM, Lesson 9.1
- Homework Practice Online
- HRW Go site

Reinforcement, Review, and Assessment
- Main Idea Activity 9.1
- Section 1 Review, p. 177
- Daily Quiz 9.1
- English Audio Summary 9.1
- Spanish Audio Summary 9.1

Beginning of World War I

By the early 1900s, countries across Europe were competing for power. They built up strong armies to protect themselves and their interests. Powerful nations feared each other. Tensions were high. The stage was set for war.

The spirit of nationalism was still strong in Europe in the early 1900s. Nationalism is a fierce pride in one's country. Many European countries wanted more power and more land. They built strong armies, and threatened to use force to get what they wanted. The use of strong armies and the threat of force to gain power is called **militarism**.

Europe's leaders did not trust one another. To protect their nations against strong enemies, they formed alliances. An alliance is an agreement between countries. If a country is attacked, its allies—the members of the alliance—help it fight.

By 1907 Europe was divided into two opposing sides. Germany, Austria-Hungary, and Italy had formed one alliance. England, France, and Russia had formed another.

The attention of both alliances was soon drawn to the Balkans, a region in southeastern Europe. In 1878, Serbia, part of this region had become an independent country. Serbian nationalists now wanted control of Bosnia and Herzogovina, which belonged to Austria-Hungary.

On June 28, 1914, a Serbian nationalist shot and killed the heir to the Austro-Hungarian throne, Archduke Francis Ferdinand. As a result, Austria-Hungary declared war on Serbia. Russia supported Serbia; Germany supported Austria-Hungary. With Russia and its allies on one side and Germany and its allies on the other, conflict quickly spread.

▲

This drawing of the killing of Archduke Francis Ferdinand in Sarajevo was published in French newspapers.

Teaching Objective 1

ALL LEVELS: (Suggested time: 20 min.) Copy the following graphic organizer onto the board, omitting the italicized answers. Have students fill in the chart wit h descriptions and examples of the factors that led to World War I. After students have completed their charts, lead a class discussion about how each of the listed factors contributed to the outbreak of war.

THE CAUSES OF WORLD WAR I	
Nationalism	**Militarism**
People began to feel fierce pride in their countries. *Countries wanted more land.*	*Countries built strong armies.* *Strong countries used the threat of force to gain power.* *Countries formed alliances to maintain the balance of power.*

DAILY LIFE

In addition to working at important jobs and serving in the armed forces in Europe, American women contributed to the Allies' war effort by saving food. At the time of World War I, most homemakers were women. Some 20 million homemakers signed pledges promising not to serve meat on Mondays or bread on Wednesdays and to grow their own vegetables in "Victory Gardens."

In return, they received stickers to display in the windows of their homes, showing that they were helping with the war effort. Their voluntary rationing efforts helped the United States double its shipments of food to the Allies in Europe, who badly needed it.

Discussion: During times of crisis, why are people willing to make sacrifices that they might ordinarily complain about?

Possible answers: They feel patriotic; they want to help others; they want to join in with what others are doing.

internet connect

GO TO: go.hrw.com
KEYWORD: SJ3 CH9
FOR: Web sites about the modern world

The Allied Powers and the Central Powers divided Europe into two opposing sides.
▼

In August 1914 Germany declared war on Russia. Russia was allied with France, so Germany declared war on France, too. England declared war on Germany. Japan also declared war on Germany. England, France, Russia, and Japan became known as the Allied Powers. The alliance of Germany, Austria-Hungary, the Ottoman Empire, and Bulgaria was called the Central Powers. Later in the war, Italy left the Central Powers and joined the Allied Powers. Eventually, the Allied Powers included 32 countries.

✓ **READING CHECK:** *Identifying Cause and Effect* How did militarism and alliances help set the stage for war in Europe? Militarism: countries had standing armies ready to go to war. alliances: countries had to go to war to help their allies.

A New Kind of War

New weapons played a major role in World War I. Germany introduced submarines, which were called **U-boats**. This name is short for "underwater boats." Germany also introduced poison gas, which was later used by both sides and caused great loss of life. Other new weapons included large, long-range cannons and the machine gun. Machine guns could kill hundreds of people in a few minutes.

World War I was also the first war to use the airplane. At first airplanes were used mainly to observe enemy troops. Later, machine guns were placed on airplanes, so they could fire on troops and shoot at each other in the sky. England also introduced the tank during the war. This huge, heavy vehicle could not be easily stopped. With machine guns mounted on them, tanks could kill large numbers of soldiers.

Europe at the Beginning of World War I

(Map legend:)
Allied Powers
Central Powers
Neutral Countries
★ National capital

NORWAY, SWEDEN, FINLAND, Petrograd, RUSSIA, ATLANTIC OCEAN, NORTH SEA, DENMARK, BALTIC SEA, GREAT BRITAIN, NETHERLANDS, London, GERMAN EMPIRE, Berlin, English Channel, BELGIUM, LUXEMBOURG, Paris, ALSACE-LORRAINE, FRANCE, SWITZERLAND, Vienna, AUSTRIA-HUNGARY, Bay of Biscay, BOSNIA AND HERZEGOVINA, Sarajevo, SERBIA, ROMANIA, BLACK SEA, PORTUGAL, SPAIN, Corsica (FRENCH), Rome, ITALY, MONTENEGRO, Sardinia (ITALIAN), ALBANIA, Balkan Peninsula, BULGARIA, OTTOMAN EMPIRE, Danube R., Dnieper R., Rhine R., Vistula R., Tigris R., ADRIATIC SEA, CASPIAN SEA

SCALE
0   200   400 Miles
0   200 400 Kilometers
Projection: Azimuthal Equal-Area

Teaching Objective 2

ALL LEVELS: (Suggested time: 30 min.) Organize the class into four groups. Have each group prepare a chart, diagram, or report that shows how new weapons and technologies affected the way World War I was fought. Tell students that their projects should clearly demonstrate the differences between World War I and early wars in which the United States was involved. Encourage interested students to note developments from World War I that are still employed in wars today.
ENGLISH LANGUAGE LEARNERS, COOPERATIVE LEARNING

Teaching Objective 3

LEVELS 1 AND 2: (Suggested time: 10 min.) Ask students to find examples in the text of ways in which society changed after World War I. As students name changes, write them on the chalkboard. Then ask students to name the common theme that links all these things. (*People thought the world no longer made sense and felt the need to experiment with new ideas.*)

LEVEL 3: (Suggested time: 30 min.) Obtain samples of art and music from the era immediately following World War I. Play or show these materials to the class. Lead a discussion of how this music and art represent changes brought about by the war.

CONNECTING TO *Science*

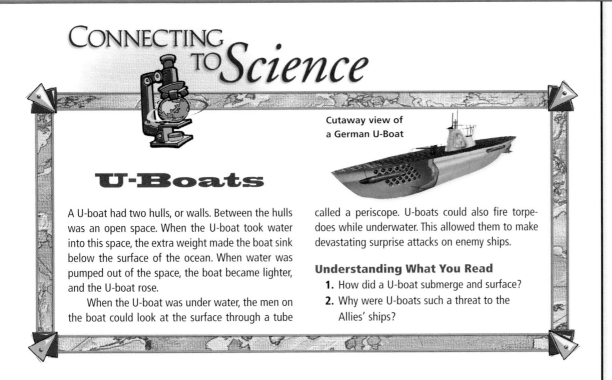

U-Boats

Cutaway view of a German U-Boat

A U-boat had two hulls, or walls. Between the hulls was an open space. When the U-boat took water into this space, the extra weight made the boat sink below the surface of the ocean. When water was pumped out of the space, the boat became lighter, and the U-boat rose.

When the U-boat was under water, the men on the boat could look at the surface through a tube called a periscope. U-boats could also fire torpedoes while underwater. This allowed them to make devastating surprise attacks on enemy ships.

Understanding What You Read

1. How did a U-boat submerge and surface?
2. Why were U-boats such a threat to the Allies' ships?

GLOBAL PERSPECTIVES

When Germany attacked France in 1914, the German army got as far as the Marne River, near Paris. But then their advance was stopped by the French and British armies. Both sides dug long deep trenches to protect their troops. These trenches extended across France and Belgium, from Switzerland to the North Sea. This broad band become known as the western front. The troops stayed in the trenches and attacked each other's positions, but neither side could gain much ground. Territory that was captured in one attack was quickly lost, sometimes in the very next attack. This kind of fighting became known as trench warfare.

Discussion: Lead a discussion on the following question: What do you think daily life was like for the soldiers in the trenches?

Understanding What You Read
Answers
1. by controlling the amount of water contained between its hulls
2. Possible answer: They could attack ships from underwater. Enemy ships could not see them approaching.

The Early Years of the War Early in the war, Germany attacked France. The German army almost reached Paris, the French capital. However, Russia attacked Germany and Austria-Hungary, forcing Germany's attention east. At sea, England used its powerful navy to stop supplies from reaching Germany by ship. Germany used its deadly U-boats to sink ships carrying supplies to Great Britain.

At first, both sides thought they would win a quick victory. They were wrong. Armies dug in for a long and costly fight. World War I would go on for four years.

The United States and World War I At first, the United States stayed out of World War I. In 1917, however, Germany tried to persuade Mexico to join the Central Powers. The Germans promised to help Mexico retake Arizona, New Mexico, and Texas from the United States after the war. This angered many Americans.

At the same time, German U-boats were attacking American ships carrying supplies to the Allies. Many ships were sunk and many Americans died.

The United States had another motivation for joining the war. The major Allied countries had moved toward democracy, but the Central Powers had not. President Woodrow Wilson told Congress that "the world must be made safe for democracy." On April 6, 1917, the United States declared war on Germany.

▲ American soldiers march through Paris during World War I.

✔ **READING CHECK:** *Summarizing* Why did the United States enter World War I? *Central Powers tried to get Mexico to join them; German U-boats attacked U.S. ships; many Americans died; the U.S. wanted to make the world safe for democracy.*

Point out to students that the term *World War I* was not actually used during the war itself. People who lived through the war often referred to it as the "war to end all wars." Ask students what they think people meant by this name. (*Possible answer: The war was so long and destructive that people thought it would discourage countries from ever going to war again.*) Then ask students why we no longer use this name. (*World War I did not, obviously, end warfare. World War II was fought later.*)

Have students complete the Section Review. Then organize students into teams. Give each team an index card. Have the members of each team choose an event or person discussed in the chapter and write a question about that event or person on their cards. Pass the cards from team to team. Ask each team to copy each question into their notebooks and to answer them. Continue until all each team has seen every question. Then have students complete Daily Quiz 9.1. **COOPERATIVE LEARNING**

HUMAN SYSTEMS

President Woodrow Wilson felt that World War I was a European affair. He wanted the United States to stay out of the war, and many Americans agreed. Even before entering the war on the Allied side in 1917, however, the United States sent food and weapons to both the Allies and the Central Powers.

Critical Thinking: What part did geography play in making Americans think that the war was a European affair?

Answer: The Atlantic Ocean separated the United States from Europe; the vast distance made Americans feel they were not involved in European matters.

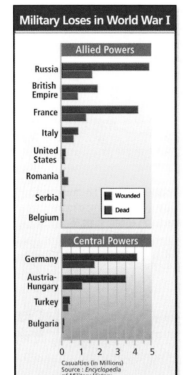

Interpreting the Graph Which of the Allied Powers had the highest number of total casualties? Which of the Central Powers had the highest casualties?

Graph Answer ►

Russia; Germany

The War Ends

During World War I, Russian citizens held protests and demonstrations because they did not have enough food and because so many Russians were dying in the war. The Russian army joined the people in their protests. In March 1917 the czar, or king, was overthrown and put in prison.

A new government was set up. Political groups called soviets, or councils, were also formed. The most powerful soviet leader was Vladimir Lenin. He offered the Russian people peace, food, and land. Lenin's ideas were part of an economic and political system known as communism. On November 7, 1917, Lenin's followers took control of Russia.

Lenin's government signed a peace treaty with the Central Powers, and Russia withdrew from the war. In 1918 Lenin's followers established a communist party. Some Russians wanted the czar to return, and Civil War broke out. The Communists won. In 1922 they renamed their country the Union of Soviet Socialist Republics, or the Soviet Union.

With Russia out of the war, the tide began to turn in favor of Germany. The German army advanced on Paris. However, when the United States entered the war, the German army was pushed back to its own border. Germany's allies began to surrender. At last, Germany itself surrendered. An **armistice** was signed. An armistice is an agreement to stop fighting.

The fighting stopped on November 11, 1918. More than 8.5 million soldiers had been killed, and 21 million more wounded. Millions who did not fight died from starvation, disease, and bombs.

Making Peace In January 1919 the Allied nations met near Paris to decide what would happen now that the war was over. This meeting came to be known as the Paris Peace Conference.

President Wilson wanted fair peace terms to end the war. He felt that harsh terms might lead to future wars. His ideas were called the Fourteen Points. These ideas called for no secret treaties, freedom of the seas for everyone, and the establishment of an association of nations to promote peace and international cooperation. That association, the League of Nations, was formed later but the United States never joined.

Other Allied leaders wanted to punish Germany. They felt that Germany had started the war and should pay for it. They believed that the way to prevent future wars was to make sure that Germany could never become powerful again.

The agreement these leaders finally reached became known as the Treaty of Versailles. Germany was forced to admit it had started the war and to pay money to the Allies. Germany also lost territory. The treaty stated that Germany could not make tanks, military planes, large weapons, or submarines. The United States never agreed to the Treaty of Versailles. It eventually signed a separate peace treaty with Germany.

▲ Allied Leaders at the Paris Peace Conference. President Woodrow Wilson is at the far right.

Have students complete Main Idea Activity 9.1. Then organize the class into four groups. Assign each group one of the topics discussed in this section. Have each group write a list of facts about its topic. Encourage students to illustrate their lists. Then have each group present and explain its list to the class. **ENGLISH LANGUAGE LEARNERS, COOPERATIVE LEARNING**

Have interested students conduct research on important battles or other major events of World War I. Ask them to prepare brief presentations on their findings. Have students give their presentations to the class, explaining the significance of the events they have researched. Encourage students to prepare maps or other visual aids to accompany their presentations. **BLOCK SCHEDULING**

A New Europe World War I changed the map of Europe. France and Belgium gained territory that had belonged to Germany. Austria, and Hungary became separate countries. Poland and Czechoslovakia gained their independence. Bosnia and Herzegovina, Croatia, Montenegro, Serbia, and Slovenia were united as Yugoslavia. Finland, Estonia, Latvia, and Lithuania, all of which had been part of Russia, also became independent nations. Bulgaria and the Ottoman Empire likewise lost territory.

✔ **READING CHECK:** *Comparing and Contrasting* How did the peace terms Woodrow Wilson wanted compare to those in the Treaty of Versailles? Wilson : fair peace terms that would not contribute to future wars. Treaty had the terms Allied leaders wanted: punish Germany and make sure it could not become powerful again.

A New World

After World War I, the world was very different. New ideas, new art, new music, and new kinds of books reflected the feeling that the world no longer made sense. Some writers called the people who had been through the war "the lost generation." Composers wrote music that sounded different from the music people were used to hearing. Many people thought it didn't sound pretty. Artists like Pablo Picasso and Salvador Dali created paintings that looked more like scenes from dreams than from the real world. People were tired of war. They wanted to have fun, and not worry so much about what might happen tomorrow. Jazz music, which gave musicians more freedom, became popular. Women wanted more freedom. They began to wear their hair and their skirts short. In the United States, women demanded and won the right to vote.

✔ **READING CHECK:** *Drawing Inferences* Why were there so many new ideas and new kinds of art after World War I? World no longer seemed to make sense; old ideas didn't work anymore; people wanted new ideas, more freedom, more fun, and not to worry about the future.

▲ Women vote in the United States, c. 1920.

go.hrw.com **Homework Practice Online**
Keyword: SJ3 HP9

Section Review 1

Define and explain: militarism, U-boats, armistice

Reading for the Main Idea

1. (*Human Systems*) What was Europe like just before World War I?

2. (*Human Systems*) What role did science and technology play in the war?

3. (*Human Systems*) How did World War I change the world?

Critical Thinking

4. **Drawing Inferences and Conclusions** How might World War I have been different if there had not been alliances in Europe?

Organizing What You Know

5. **Categorizing** Copy the following chart. Use it to list the members of each alliance.

Allied Powers	Central Powers

Section Review 1

Answers

Define Define For definitions see: nationalism, p.173; militarism, p.173; U-boats, p. 174; armistice, p.176

Reading for the Main idea

1. Militarism encouraged nations to build strong armies; nations did not trust one another; alliances meant countries had to go to war to help their allies. (NGS 13)

2. New weapons were a big reason for the war's high death toll. (NGS 14)

3. New weapons led to higher death tolls in future wars; the map of Europe changed; new nations created; League of Nations created; new ideas and new kinds of art, music, and books developed. (NGS 10)

Critical Thinking

4. The war would not have been as widespread; it might have been a war between only Serbia and Austria-Hungary.

Organizing What You Know

5. Allied Powers—England, France, Russia, Japan, Italy, United States; Central Powers—Germany, Austria-Hungary, Ottoman Empire, Bulgaria

Objectives

1. Identify the causes of the Great Depression.
2. Describe a dictatorship.
3. Analyze how the Great Depression helped dictators come to power in Europe.

LET'S GET STARTED

Copy the following question onto the board: *What do you think would happen if, suddenly, most people had no money?* Discuss student responses. (*Possible answers: People might go hungry; businesses might close as people stopped spending; many might panic or come to depend on government assistance.*) Point out that these things happened in the 1930s during the Great Depression. Tell students that in Section 2 they will learn more about the Great Depression and how it affected the world.

Building Vocabulary

Write the key terms on the board. Call on volunteers to find the definitions of the terms in the textbook, and read them to the class. Point out the suffix *–or* in the word **dictator,** meaning "one who." Ask students if they are familiar with the word *dictate.* Call on a volunteer to explain how its meaning is reflected in the definition of *dictator.* (*A dictator's rules dictate the actions of a country's people.*) Then ask what *collective* means (*working together*). Ask how this meaning relates to the definition of **collective farms.**

Reproducible
◆ Guided Reading Strategy 9.2

Technology
◆ One-Stop Planner CD–ROM, Lesson 9.2
◆ Homework Practice Online
◆ HRW Go site

Reinforcement, Review, and Assessment
◆ Main Idea Activity 9.2
◆ Section 2 Review, p. 181
◆ Daily Quiz 9.2
◆ English Audio Summary 9.2
◆ Spanish Audio Summary 9.2

Section 2 The Great Depression and the Rise of Dictators

Read to Discover

1. What led to the Great Depression?
2. What is a dictatorship?
3. How did the Great Depression help dictators come to power in Europe?

Define

stock market
bankrupt
The Great Depression
New Deal

dictator
fascism
communism
police state
collective farms

Locate

New York

WHY IT MATTERS

Some countries in today's world are ruled by dictators. Use **CNNfyi.com** or other **current events** sources to find examples of dictators and learn how they came to power. Record your findings in your journal.

Dorothea Lange's photograph Migrant Mother

Library of Congress

▲
During the stock market crash, people rushed to get money out of banks.

The Great Depression

During the 1920s industrialized countries such as the United States and Great Britain experienced great economic growth. However, less than 10 years later, many of these countries struggled with high rates of unemployment and poverty.

Causes of the Depression During World War I, much farmland in Europe was destroyed. Farmers all over the world planted more crops to sell food to European countries. Many American farmers borrowed money to buy farm machinery and more land. When the war ended, there was less demand for food in Europe. Prices went down. Farmers could not pay back the money they had borrowed. Many lost their land.

During the 1920s, the **stock market** did very well. The stock market is an organization through which shares of stock in companies are bought and sold. People who buy stock are buying shares in a company. If people sell their stock when the price per share has risen above the original price, they make a profit.

In the 1920s, stock prices rose very high and many people invested their money in the stock market. People thought stock prices would stay high, so they borrowed money to buy more stocks. Unfortunately, stock prices fell. Low stock prices made people rush to sell their shares before they lost any more money. So many people all selling stock at once drove prices down even more quickly. Finally the stock marked crashed, or hit bottom, on October 29, 1929.

Teaching Objective 1

ALL LEVELS: (Suggested time: 20 min.) Copy the following graphic organizer onto the board, omitting the italicized answers. Have each student complete the organizer with factors that led to the Great Depression. Ask volunteers to share their answers with the class. Use their completed organizers to lead a discussion about the beginnings of the Great Depression. **ENGLISH LANGUAGE LEARNERS**

People who had borrowed to buy stocks suddenly had to pay back the money. They rushed to the banks to take money out. But the banks did not have enough money to give everyone their savings all at once. In a very short time, banks, factories, farms, and people went **bankrupt**. This meant they had no more money.

◄
This is a breadline in New York City, during the Great Depression. People who could not find jobs stood in these lines to receive free food from the government.

This was the beginning of the **Great Depression**. All over the world, prices and wages fell, banks closed, business slowed or stopped, and people could not find jobs. Many people were poor. Many did not even have enough money to buy food. Some people sold apples on street corners to make a little money.

Governments around the world tried to lessen the effects of the Great Depression. Some limited the number of imports they allowed into their countries. They thought this would encourage citizens to buy products made by businesses in their own countries. This plan did not work. In fact, the loss of foreign markets for their products drove many countries even further into debt.

The New Deal In 1932 Franklin D. Roosevelt became president of the United States. He created a program to help end the Great Depression. This program was called the **New Deal**. The federal government gave money to each state to help people. The government also created jobs. It hired people to construct buildings and roads and work on other projects.

Laws were passed to regulate banks and stock exchanges better. In 1938, Congress passed the Fair Labor Standards Act. It established the lowest amount of money a worker could be paid to keep a healthy standard of living. Many people today refer to it as the minimum wage law. Congress also guaranteed workers the right to form unions so they could demand better pay and better working conditions. The Social Security Act, passed in 1935, created benefits for people who were unemployed or elderly.

Under the New Deal, the United States became deeply involved in the well-being of its citizens. For the first time, the government created large-scale social programs to better the lives of American citizens. The New Deal did not, however, completely end the Great Depression in the United States. Government programs helped the economy grow somewhat, but they were not enough to solve the economic crisis completely. The Great Depression would not end in the United States until World War II.

✓ **READING CHECK:** *Identifying Cause and Effect* How did the Great Depression start? Stock prices dropped; then rush to sell before the loss was great; selling panic drove stock values down more; many people lost all their money.

The Great Depression affected the entire world. By 1932, more than 30 million people throughout the world could not find jobs.

USING ILLUSTRATIONS

Direct students' attention to the picture of the New York City breadline on this page. Ask students what details they see in the image that reflect life during the Great Depression. List students' responses on the board and discuss each one. (*Possible answers: the line is long; many people needed free food; many people in the line are well dressed, showing that even people who once had money needed free food; all types of people needed free food; people seem to be getting only bread, but are willing to stand in line for it; people seem to be orderly; breadlines were common.*)

HUMAN SYSTEMS

When Mussolini's Fascist party first tried to gain power, it failed. The Fascists then turned to violence. They fought with communists in the streets of Italy's cities. They beat up people who tried to organize unions.

Discussion: Ask students how they would feel if a political party in the United States tried to gain power by using the same methods the Fascists used in Italy. Lead a discussion on how political groups in the United States today gain support for their ideas.

▲
When he became dictator of Italy, Benito Mussolini took the title *il Duce* (il DOO-chay), Italian for "the leader."

▲
Hitler was a very powerful speaker. He often twisted the truth in his speeches. He claimed that the bigger a lie was, the more likely people would be to believe it.

The Rise of Dictators in Europe

As the Great Depression continued in Europe, life got harder. Many people became unhappy with their governments, which were not able to help them. In some countries, people were willing to give up democracy to have strong leaders who promised them more money and better lives. It was easy for dictators to take control of these governments. A **dictator** is an absolute, or total, ruler. A government ruled by a dictator is called a dictatorship. Powerful dictators seized control of Italy and Germany.

Italy Becomes a Dictatorship Benito Mussolini told the Italians that he had the answers to their problems. Mussolini called his ideas **fascism** and started the Fascist Party. Fascism was a political movement that put the needs of the nation above the needs of the individual. The nation's leader was supposed to represent the will of the nation. The leader had total control over the people and the economy.

Fascist nations became strong through militarism. Their leaders feared communism for a good reason. **Communism** promised a society in which property would be shared by everyone. Mussolini promised that he would not let communists take over Italy. He also promised he would bring Italy out of the Great Depression and return Italy to the glory of the Roman Empire.

In 1924, the Fascist Party won Italy's national election. Mussolini took control of the government and became a dictator. He turned Italy into a police state. A **police state** is a country in which the government has total control over people and uses secret police to find and punish people who rebel or protest.

Germany After World War I many Germans felt that their government had betrayed them by signing the Treaty of Versailles. Many also blamed the German government for the unemployment and inflation brought by the Great Depression. Several groups attempted to overthrow and replace the old government. Eventually a new party called the Nazi Party gained power. It too was a fascist party. Adolf Hitler was its leader.

Hitler promised to break the treaty of Versailles. He said he would restore Germany's economy, rebuild Germany's military power, and take back territory that Germany had lost after the war. Hitler told the Germans that they were superior to other people. Many Germans eagerly listened to Hitler's message. They thought he would restore Germany to its former power.

CLOSE

Ask students to suggest images they think best represent the Great Depression in the United States. Discuss student responses.

REVIEW AND ASSESS

Have students complete the Section Review. Then organize students into four groups and have each group prepare five questions about one topic from this section. Have the class use these questions to play a quiz game. Then have students complete Daily Quiz 9.2.
ENGLISH LANGUAGE LEARNERS, COOPERATIVE LEARNING

RETEACH

Have students complete Main Idea Activity 9.2. Then organize the class into four groups. Assign each group one of the following countries: United States; Italy; Germany; Soviet Union. Have each group create a list of newspaper headlines that might have appeared in their country during the 1930s. Then have each share its headlines with the class.
ENGLISH LANGUAGE LEARNERS, COOPERATIVE LEARNING

EXTEND

Have interested students conduct research on programs of the New Deal. Have each of them write a brief report on one New Deal program and how it helped the country to recover from the Depression.
BLOCK SCHEDULING

The Nazis quickly gained power in Germany. In 1933 Hitler took control of the German government. He made himself dictator and used the title *der Führer* (FYOOR-ur), which is German for "the leader." He turned Germany into a police state. Newspapers and political parties that opposed the Nazis were outlawed. Groups of people that Hitler claimed were inferior, especially Jews, lost their civil liberties.

Hitler began to secretly rebuild Germany's army and navy. He was going to make Germany a mighty nation again. He called his rule the Third Reich. *Reich* is the German word for *empire*. In 1936 Hitler formed a partnership with Mussolini called the Rome-Berlin Axis.

The Soviet Union Russia had suffered terribly during World War I. Lenin's Communist government had promised an ideal society in which people would share things and live well. However, most Russians remained poor.

After Lenin died, Joseph Stalin gained control of the Communist Party. Stalin's government took land from farmers and forced farmers to work on large **collective farms** owned and controlled by the central government. Stalin also tried to industrialize the Soviet Union. However, for ordinary Russians Food and manufactured goods remained scarce.

Religious worship was forbidden. Artists were even told what kind of pictures to make. Secret police spied on people. If people did not obey Stalin's policies they were arrested and put in jail or killed. Scholars think that by 1939 more than 5 million people had been arrested, deported, sent to forced labor camps, or killed.

▲

For many years, many Soviet people thought Joseph Stalin was a great hero. Later, people became more aware of his responsibility for the deaths of millions of Russians for "crimes against the state."

✓ **READING CHECK:** *Analyzing* How did Hitler use the Treaty of Versailles to help him gain power in Germany? *Germans' problems blamed on Treaty of Versaille; Hitler promised to ignore the treaty and rebuild Germany's economy and military power.*

go.hrw.com
Homework Practice Online
Keyword: SJ3 HP9

Section Review 2

Define and explain: bankrupt, The Great Depression, New Deal, dictator, fascism, communism, police state, collective farms

Reading for the Main Idea

1. (*Human Systems*) What happened during the Great Depression?

2. (*Human Systems*) What is life like in a dictatorship?

3. (*Human Systems*) How did European dictators take advantage of the Great Depression to gain power?

Critical Thinking

4. **Making Inferences and Conclusions** How were Woodrow Wilson's concerns about the Treaty of Versailles proven correct by the rise of Adolf Hitler?

Organizing What You Know

5. **Identifying Cause and Effect** Copy the following graphic organizer. Fill it in to summarize what happened in the Great Depression.

Cause	Effect
Europe needs food during the war.	
The war ends and crop prices fall.	
Stock prices rise very high.	
People rush to sell their stocks.	
People rush to take money out of the banks.	

Section Review 2

Answers

Define For definitions, see: stock market, p 178; bankrupt, p.179; the Great Depression, p. 179; New Deal, p. 179; dictator, p.180; fascism, p.180; police state, p. 180 collective farms, p. 181

Reading for the Main Idea

1. prices and wages fell; banks closed; business slowed or stopped; people could not find jobs; many people were poor; dictators came to power in some countries (NGS 11)

2. no freedom of speech; no freedom of the press; no opposing political parties; no free elections; police state; secret police (NGS 13)

3. Great Depression caused economic problems; dictators said they would solve the problems (NGS 13)

Critical Thinking

4. Wilson felt harsh peace terms might lead to another war; Hitler used German anger about the treaty to gain power.

Organizing What You Know

5. Farmers plant more crops, borrow money for land and machinery; farmers cannot pay debts, lose their land; people borrow money to buy stocks; stock prices fall quickly; banks do not have enough money, people cannot get money.

Objectives

1. Describe the European colonies in Latin America and how they won their independence.

2. Explain how dictatorships replaced some democracies in Latin America.

3. Describe how relations between the United States and Latin American countries have changed over time.

FOCUS

LET'S GET STARTED

Copy the following questions onto the board: *What does it mean to be a good neighbor?* Discuss student responses. Then point out the countries of Latin America on a wall map. Tell students that in the early to mid-1900s, the leaders of the United States wanted to become a good neighbor to these countries. Tell students that in this section they will learn more about the countries of Latin America in the 1900s.

Building Vocabulary

Write **economic nationalism** and **revolutionaries** on the chalkboard. Ask students to recall definitions for the words *nationalism* and *revolution*. Then ask students how they think nationalism might be applied to economic affairs. Point out the suffix *–aries,* which means "people who belong to." Call on a volunteer to suggest a definition for *revolutionaries.* Then have a student read the definitions of all the key terms.

Section 3 RESOURCES

Reproducible
◆ Guided Reading Strategy 9.3

Technology
◆ One-Stop Planner CD–ROM, Lesson 9.3
◆ Homework Practice Online
◆ HRW Go site

Reinforcement, Review, and Assessment
◆ Main Idea Activity 9.3
◆ Section 3 Review, p. 186
◆ Daily Quiz 9.3
◆ English Audio Summary 9.3
◆ Spanish Audio Summary 9.3

Section 3 — Nationalism in Latin America

Read to Discover

1. What were European colonies in Latin America like, and how did they win their independence?

2. Why did dictatorships replace some democracies in Latin America?

3. How have relations between the United States and Latin American countries changed over time?

Define
revolutionaries
economic nationalism

Locate
Haiti
Cuba
Mexico
Central America
Panama Canal

WHY IT MATTERS

The United States is still very involved in Latin American affairs. Use CNNfyi.com or other **current events** sources to find examples of U.S. involvement in Latin America. Record your findings in your journal.

Red and green coffee beans still on the branch

▲

Explorer Hernán Cortés conquered the mighty Aztec civilization and claimed Mexico for Spain. Cortés is pictured here capturing the Aztec ruler Moctezuma II.

Interpreting the Visual Record
According to this picture, what advantages did the Spanish have over the Aztec?

Visual Record Answer ▶

better armor

The Changing Face of Latin America

Life in Latin America changed greatly in the 1500s after conquerors claimed the land for European countries. The Portuguese controlled Brazil. The much larger Spanish territory stretched from what is now Kansas all the way to the southern tip of South America.

Spanish colonists created a new civilization in the Americas. They forced many Indian people there to work their large farms and mine for silver and gold. Rich supplies of precious metals as well as agricultural products were shipped to Europe. Most of the workers had no land rights and received none of the wealth that resulted from their labor.

For more than three hundred years, Spain ruled its colonies strictly. Very little changed until the early 1800s. By that time, wars in Europe had weakened Spain. The people of the United States and France had overthrown their old governments. Colonial leaders throughout Latin America were inspired to break away from Spanish rule.

Moves for Independence

Over time, colonists in Latin America no longer thought of themselves as Europeans. Each colony had developed its own way of life, and wanted to control its own government. Colonists did not want to pay taxes to a country that was taking rich resources without giving much back. Nationalist movements began to gather strength.

Teaching Objective 1

LEVEL 1: (Suggested time: 10 min.) Copy the following graphic organizer onto the chalkboard, omitting the italicized answers. Ask students to copy it into their notebooks. Pair students and have them fill in each circle with the names of Latin American countries that once belonged to each European country. Then lead a discussion about the process by which these colonies won their independence. **ENGLISH LANGUAGE LEARNERS, COOPERATIVE LEARNING**

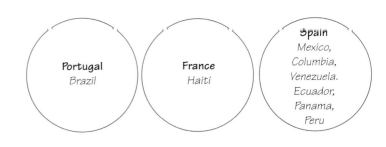

Portugal
Brazil

France
Haiti

Spain
Mexico, Columbia, Venezuela. Ecuador, Panama, Peru

◀ Mexican artist Diego Rivera painted pictures that showed the historical, social, and economic problems of his country. Other Latin American countries faced many of the same problems as Mexico.

 National Geography Standard 16

Environment and Society
In the 1800s a small group of people owned most of the land in northwestern Mexico, the area that is now California. These people were called *rancheros*. *Ranchero* families lived on large ranches where they raised sheep and horses. They also grew grain and grapes. The grapes were used to make wine. Because they owned so much land, *rancheros* had great economic and political power. It is estimated that at one time more than 8 million acres of land were controlled by only 800 *rancheros*—an average of 10,000 acres each!

Discussion: Ask students how the *ranchero* system would make it hard for poorer people to acquire land. Ask them to evaluate the *ranchero* system from the point of view of the *rancheros* and from the point of view of a worker who owns no land.

Hispaniola Columbus had claimed the Caribbean island of Hispaniola for Spain in the 1400s. France had later taken over the western end of the island. The French Revolution in Europe inspired the slaves. Led by Toussaint L'Ouverture, they rebelled in 1791 and eventually gained their freedom. In 1804 L'Ouverture took control of the newly independent country of Haiti. The former Spanish colony on Hispaniola broke away from Haiti in the mid-1800s and became known as the Dominican Republic.

Mexico In 1810 the Mexican people revolted against Spanish rule. After a struggle that lasted until 1821, Mexico finally won its independence. American and European investments helped Mexico grow economically, but most Mexicans remained poor.

Although Mexico was independent, political unrest continued over the next 100 years. During the 1820s and early 1830s, many Americans moved from the United States to a part of northern Mexico now known as Texas. Eventually, the Texans broke away from Mexico. In 1845 Texas became part of the United States. This event helped trigger a war between Mexico and the United States.

Unrest in Mexico increased in the early 1900s. The government was under the dictatorship of Porfirio Díaz. Once again, the rich people became richer and controlled much of the country's land. Most people, however remained poor and had no land of their own. In 1910, they began to revolt. The Mexican Revolution continued for many years and involved various leaders and groups.

Land reform was one of the main goals of the revolution. Over time, large farms were broken up and given to villages. At the same time, the Mexican government became more involved in the national economy than it had been before the revolution. Many foreign-owned businesses were forced out of Mexico.

Pancho Villa (VEE-yah) was a Mexican revolutionary and leader.

▼

183

LEVEL 2 AND 3: (Suggested time: 30 min.) Have each student write on an index card one reason that people in Latin American countries wanted to achieve independence. (*Possible answers: tired of forced labor, poverty, upset by strict rule by foreign country, forced to pay taxes to foreign country*) Call on volunteers to share their reasons with the class. As each reason is read, call on another student to explain the logic behind it. Finally, call on students to summarize the processes by which the countries of Latin America became independent.

 Teaching Objective 2

ALL LEVELS: (Suggested time: 30 min.) Call on a volunteer to recall the events that led to the rise of dictators in Europe during the Great Depression. (*People were poor. Future dictators promised to help people overcome their difficulties and to restore the former glory of their countries.*) Ask students how the rise of dictators in Latin America was similar (*largely caused by financial issues*) and how it was different (*were not trying to restore former glory*). Then have students design flow charts that explain the rise of dictators in Latin America. **COOPERATIVE LEARNING**

Cultural Kaleidoscope

Among the first serious rebellions against Spain's rule was an Indian uprising in the Viceroyalty of Peru in 1780. Led by Tupac Amarú, who claimed descent from an Inca emperor of the same name, a poorly armed force of more than 10,000 Indians overran the Peruvian highlands and attacked the city of Cuzco. Although Tupac was captured and executed by the Spanish in 1781, the revolt continued and spread to Bolivia, where La Paz was attacked twice before the Indians were finally defeated.

Critical Thinking: Tell students that the Spanish were exceptionally brutal in putting down this revolt. Ask why they reacted in this way.

Answer: Students may suggest that the European colonists were afraid of the threat of outside attack.

▲
Simón Bolívar (1783-1830) was called "the Liberator."

This painting depicts Theodore Roosevelt leading the Rough Riders into battle during the Spanish-American War.
▼

Central America Early Spanish settlers in Central America had established colonies with towns and large farm estates. As in Mexico, a few rich people owned the land, and most workers had no land rights. Although these colonies won independence from Spain, not much changed after Spanish officials left. Foreign countries such as the United States and Great Britain built railroads and developed big businesses. Most local people continued to be poor.

South America The demand for self-rule spread to other parts of the Americas. One of the most famous Latin American **revolutionaries** was Simón Bolívar. Called "the Liberator," Bolívar led a revolt against Spain that lasted 10 years. In 1821 Bolívar became president of a nation that eventually included the present-day countries of Colombia, Venezuela, Ecuador, and Panama. A few years later the Spanish were driven out of Peru. In 1825 the new country of Bolivia was named in honor of Bolívar. By the early 1830s, almost all the colonies in Latin America ruled themselves.

✓ **READING CHECK:** *Identifying Cause and Effect* How did feelings of nationalism affect Latin America in the 1800s? Political unrest grew as colonists no longer thought of themselves as Spaniards.

Challenges

The new nations of Latin America were now free to govern themselves. However, they still faced the challenge of improving people's lives. Revolutions changed the governments in these new countries, but they usually replaced one group of powerful families with another one. Life for most poor farmers, plantation workers, and city people hardly changed.

The United States and Latin America
During the late 1800s and early 1900s, the United States was a powerful political and economic force in Latin America. In 1898 the United States went to war with Spain to help Cuba win its independence. Many Americans felt that the Spanish were too harsh in their treatment of the Cubans. At the end of the war, Cuba and Puerto Rico came under U.S. control, as did the Philippines in the Pacific Ocean. Cuba became fully independent in 1902. In 1903 the United States helped Panama revolt against Colombia. Panama became an independent nation. The next year Panama allowed the United States to begin construction on the Panama Canal. The canal opened in 1914, connecting the Atlantic Ocean with the Pacific Ocean. The United States controlled the canal and the territory around it.

TEACH

Teaching Objective 3

ALL LEVELS: (Suggested time: 40 min.) On the board, draw a two-column chart with the headings Before and After. Have students provide details for the chart that describe relations between the United States and Latin America before and after President Franklin Roosevelt introduced the Good Neighbor Policy in the 1930s. Then lead a class discussion about how these relations have changed over time. You might point out that the United States no longer controls Cuba or the Panama Canal Zone, but that Puerto Rico is still a U.S. territory. **ENGLISH LANGUAGE LEARNERS, COOPERATIVE LEARNING**

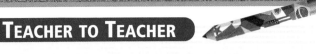

TEACHER TO TEACHER

Paul Horne of Columbia, South Carolina, suggests the following activity to help students understand the independence movement in Latin America. Have students write paragraphs comparing and contrasting the revolutions in Latin America with the American Revolution. Call on volunteers to read their essays to the class. Then lead a discussion about the ideas presented in the paragraphs.

In 1904 President Theodore Roosevelt announced that if the independence of any country in the Western Hemisphere were in danger, the United States would act to help. He also said that the United States would make sure that Latin American countries repaid loans. This policy angered many Latin Americans. They thought the plan was just an excuse for the United States to interfere in Latin American affairs. They did not think the United States had the right to take on this power.

✓ **READING CHECK:** *Evaluating* How did many Latin Americans feel about the relationship between the United States and Latin America? They felt that the United States was belittling them and interfering in their lives, and were angry.

The Rise of Dictators

Before World War I, Latin America was mainly a farming region. After the war, the region's economy started to grow and change. Oil became a major export. Mining grew rapidly. There was rapid growth of electric and hydroelectric power generation. The energy allowed many Latin American countries to begin to industrialize. A stronger economy led to changes in other areas of Latin American life. Cities grew. There were more jobs and opportunities for education. To supply labor for growing industries, some countries encouraged Europeans to immigrate.

Then the Great Depression struck. It affected Latin American nations in the same ways as other nations. Prices fell for many of Latin America's exports. Some countries could not pay their debts. People lost their jobs and could not find new jobs. More and more people became unhappy with their governments.

As in Europe, economic problems led to political problems. Some constitutional governments were overthrown, and dictators took over. In many Latin American countries, the military influenced or controlled the new governments.

In the past the United States had often stepped in to influence events in Latin America. This had led to tension between many Latin American countries and the United States. During the 1930s, President Franklin D. Roosevelt tried to improve relations between the United States and Latin America. He began a program he called the Good Neighbor Policy. The United States would cooperate with Latin American countries but would not interfere with their governments.

✓ **READING CHECK:** *Drawing Inferences* Why did President Roosevelt think the Good Neighbor Policy would improve relations between the United States and Latin America? United States interference had inadvertently caused political tensions in those countries, leading to the overthrow of democracies.

The Panama Canal is about 51 miles long. Before it was built, a ship sailing from New York City to San Francisco had to travel about 15,000 miles around South America. After the canal was built, the same trip was about 6,000 miles.

Like most of Latin America, Mexico City improved its economy in the 1920s

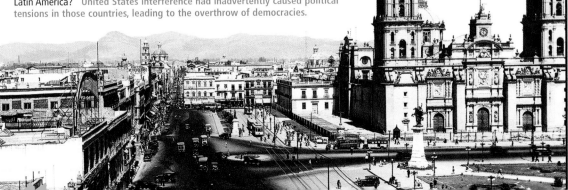

Section Review 3

Answers

Define For definitions, see: revolutionaries, p 184; economic nationalism, p.186

Reading for the Main Idea

1. The United States has exercised influence and control in Latin America; some Latin Americans have come to resent this. (NGS 13)

2. Economic nationalism gave Latin American countries more control of their industries and economies. (NGS 11)

Critical Thinking

3. Theodore Roosevelt's policy saw the United States actively involved in Latin American affairs; many felt the United States was interfering in their lives. Franklin Roosevelt's policy said the United States would cooperate with Latin American countries but would not interfere with their governments.

Organizing What You Know

4. 1791–1831—rebellion, political independence; 1898–1918—the United States was a powerful political and economic force; 1919–1933—economic growth, the Great Depression; 1934–1938—economic nationalism

Economic Independence

During the 1930s, many Latin American leaders wanted economic independence from the United States and Europe. Because of the global depression, international markets for Latin American goods were weak. Imported goods, however, were costly. Thus Latin American countries had no choice but to develop their own industries to make goods. This gave rise to **economic nationalism**. This means putting the economic interests of one's own country above the interests of other countries. Before this time other countries, such as the United States and Great Britain, had basically controlled the Latin American economy.

One example of economic nationalism can be seen in Mexico in 1938. Mexican workers wanted better pay from American- and British-owned oil companies, but the oil companies would not increase their wages. Mexican president Lázaro Cárdenas nationalized the oil industry. That means that the government seized ownership and control of all the oil companies in Mexico.

The Mexican president's actions angered Great Britain, but the Mexican and British governments eventually worked out an agreement. The Mexican people were happy that Mexico was gaining more control of its own economy. Today many Mexicans think of March 13, 1938—the day the oil companies were nationalized—as the beginning of Mexico's economic independence.

Mexican president Lázaro Cárdenas (center) nationalized Mexico's oil industry. Cárdenas, who came from a poor family, promised the Mexican people economic reform.

✓ **READING CHECK:** *Analyzing* Why did economic nationalism become important to the leaders of Latin America? As their countries became industrialized, it became necessary to take back the means of production from foreigners.

go.hrw.com
Homework Practice Online
Keyword: SJ3 HP9

Section Review 3

Define and explain: revolutionaries, economic nationalism

Reading for the Main Idea

1. (*Human Systems*) How would you describe the relationship between Latin America and the United States?

2. (*Human Systems*) How was Latin America changed by economic nationalism?

Critical Thinking

3. **Comparing/Contrasting** How was President Theodore Roosevelt's 1904 policy toward Latin America different from President Franklin D. Roosevelt's Good Neighbor Policy?

Organizing What You Know

4. **Summarizing** Copy the following graphic organizer. Fill it in, listing the movements, ideas, and policies that affected Latin America in each period.

1791-1831	1898-1918	1919-1933	1934-1938

Section 4

Objectives

1. Identify the causes of World War II.
2. Describe the Holocaust.
3. Analyze how World War II came to an end.

FOCUS

LET'S GET STARTED

Copy these instructions onto the board: *Think about television programs, movies, and books with which you are familiar. What images do you associate with World War II?* Discuss student responses. (*Students might mention Nazis, tanks, fighter planes, aircraft carriers, the Holocaust, or similar images.*) Tell students that they will learn more about the history of World War II in this section.

Building Vocabulary

Write the key terms on the chalkboard. Call on volunteers to locate and read the definitions in the glossary or text. Point out the prefix *anti-* in **anti-Semitism** means "against." *Semitism* comes from *Semite,* a word that refers to several peoples from Southwest Asia, including the Hebrews. **Genocide** is derived from the Latin words *gens,* which means "race" or "people," and *caedere,* meaning "to kill." Have students write a sentence using the word *aggression.*

Section 4 — World War II

Read to Discover

1. What were the causes of World War II?
2. What was the Holocaust?
3. How did World War II end?

Define

aggression
anti-Semitism
genocide
Holocaust

Locate

Japan
Italy
Germany
Hawaii
Ethiopia

Poland
Sicily
Pearl Harbor
Normandy

WHY IT MATTERS

Acts of aggression still happen. Use CNNfyi.com or other **current events** sources to find examples of aggression in today's world. Record your findings in your journal.

National Iwo Jima Memorial Monument

Section 4 RESOURCES

Reproducible
◆ Guided Reading Strategy 9.4

Technology
◆ One-Stop Planner CD–ROM, Lesson 9.4
◆ Homework Practice Online
◆ HRW Go site

Reinforcement, Review, and Assessment
◆ Main Idea Activity 9.4
◆ Section 2 Review, p. 191
◆ Daily Quiz 9.4
◆ English Audio Summary 9.4
◆ Spanish Audio Summary 9.4

Threats to World Peace

During the 1930s, Japan, Italy, and Germany committed acts of aggression against other countries. **Aggression** is warlike action, such as an invasion or an attack. At first, little was done to stop them. Eventually, their actions led to a full-scale war that involved much of the world.

In 1931, Japanese forces took control of Manchuria, a part of China. The League of Nations protested, but took no military action to stop Japan. Continuing its aggressive actions, Japan succeeded in controlling about one fourth of China by 1939. At about the same time, Italy invaded Ethiopia, a country in East Africa. Many countries protested, but they did not want to go to war again. Like Japan, Italy saw that the rest of the world would not try hard to stop its aggression.

Pablo Picasso painted *Guernica* after the Spanish town of the same name was bombed.

Interpreting the Visual Record **What human feelings about war does Picasso express?**

◀ **Visual Record Answer**

Picasso's broken figures show war victims' fear, suffering, horror, and despair.

Teaching Objective 1

ALL LEVELS: (Suggested time: 20 min.) Copy the following graphic organizer onto the board, omitting the italicized answers. Call on students to fill in the boxes on the chart with actions taken by each country that led to the outbreak of World War II. Then lead a discussion about the underlying causes that led to these actions.

ENGLISH LANGUAGE LEARNERS

ACTIONS AND EVENTS THAT LED TO WORLD WAR II		
Japan	1. *took control of Manchuria*	
	2. *conquered more of China*	
Italy	1. *invaded Ethiopia*	
Germany	1. *made Austria part of Germany*	
	2. *took over the Sudetenland*	
	3. *conquered Czechoslovakia*	
	4. *invaded Poland*	

DAILY LIFE

Even before World War II began, the leaders of Great Britain began making plans to protect its people from German air attacks. The British government sent thousands of children away from London to the countryside, where German bombers were less likely to attack. Moving children separated families, but it saved many lives.

When the war began, London was bombed night and day by German planes. People took shelter in subways and the basements of buildings. These bombing raids became known as the Blitz. The British people were famous for bravely carrying on with their normal lives as much as possible.

Discussion: Lead a discussion about what life was like during the Blitz. Call on students to share their ideas and impressions.

Women cry as they give the Nazi salute to German troops in Sudentenland.

Interpreting the Visual Record Why do you think these women are showing strong feelings?

Visual Record Answer ▶

Possible answer: They are unhappy that the Germans have occupied their homeland.

Spanish Civil War In 1936, civil war broke out in Spain. On one side were fascists led by General Francisco Franco. Both Italy and Germany sent troops and supplies to help Franco's forces. On the other side were Loyalists, people loyal to the elected Spanish government. The Soviet Union sent aid to the Loyalists. Volunteers from France, Great Britain, and the United States also fought on their side, but their help was not enough. In 1939, the fascists defeated the Loyalists.

Franco set up a dictatorship. He ended free elections and most civil rights. By the end of the 1930s, it was clear that fascism was growing in Europe.

Hitler's Aggressions In the late 1930s many Germans lived in Austria, Czechoslovakia, and Poland. Hitler wanted to unite these countries to bring all Germans together. In 1938 German soldiers marched into Austria, and Hitler declared Austria to be part of the Third Reich. Great Britain and France protested but did not attack Germany. Later that year Hitler took over the Sudentenland, a region of western Czechoslovakia. Other European countries were worried, but they still did not want a war. Hitler soon conquered the rest of Czechoslovakia.

Eventually Britain and France realized they could not ignore Hitler. They asked the Soviet Union to be their ally in a war against Germany. However, Soviet leader Joseph Stalin had made a secret plan with Hitler. They decided that their countries would never attack each other. This deal was called the German-Soviet nonaggression pact.

In September 1939 Hitler invaded Poland. Two days later, Great Britain and France declared war on Germany. World War II had begun. On one side were Germany, Italy, and Japan. They called themselves the Axis Powers, or the Axis. Great Britain, France, and other countries that fought against the Axis called themselves the Allies.

✓ **READING CHECK:** *Drawing Inferences* How might British and French leaders have prevented World War II? stopped aggression of Japan, Italy, and Germany; stopped Hitler before he became too powerful

Joseph Stalin (second from right) made an agreement that the Soviet Union and Germany would not attack each other.

Teaching Objective 2
ALL LEVELS: (Suggested time: 40 min.) Organize the class into groups and have each group design a museum exhibit about the Holocaust. Tell students that their designs should list the causes of the Holocaust and describe events associated with it. Call on students to share their ideas with the class. **ENGLISH LANGUAGE LEARNERS, COOPERATIVE LEARNING**

Teaching Objective 3
ALL LEVELS: (Suggested time: 30 min.) Have students work in pairs to create time lines of the last years of World War II. Direct students to fill their time lines with events from the text. Next to each entry on their time lines, have students note how each event contributed to ending the war. Display the completed time lines around the classroom. **ENGLISH LANGUAGE LEARNERS, COOPERATIVE LEARNING**

War

At the beginning of the war, the Germans won many victories. Poland fell in one month. In 1940 Germany conquered Denmark, Norway, the Netherlands, Belgium, and Luxembourg. In June 1940 Germany invaded and quickly defeated France. In less than one year, Hitler had gained control of almost all of western Europe. Next he sent German planes to bomb Great Britain. The British fought back with their own air force. This struggle became known as the Battle of Britain.

In June 1941, Hitler turned on his ally and invaded the Soviet Union. As winter set in, however, the Germans found themselves vulnerable to Soviet attacks. Without enough supplies, Hitler's troops were defeated by a combination of the freezing Russian winter and the Soviet Red Army. For the first time in the war, German soldiers were forced to retreat.

The United States Many people in the United States did not want their country to go to war. The United States sent supplies, food, and weapons to the British but did not actually enter the war until 1941. In that year Japan was taking control of Southeast Asia and the Pacific. Seeing the United States as a possible enemy, Japanese military leaders attempted to destroy the U.S. naval fleet in the Pacific. On December 7, 1941, Japan launched a surprise air attack on the naval base at Pearl Harbor, Hawaii. The attack sank or damaged U.S. battleships and killed more than 2,300 American soldiers. The next day President Franklin D. Roosevelt announced that the United States was at war with Japan. Great Britain also declared war on Japan. Three days later, Germany and Italy—both allies of Japan—declared war on the United States. In response, Congress declared war on both countries.

✔ **READING CHECK:** *Evaluating* How were Hitler's invasion of the Soviet Union and Japan's attack on Pearl Harbor turning points in the war? invasion of Soviet Union led to Germany's retreat; attack on Pearl Harbor brought U.S. into the war

◄ Carrying whatever belongings they can, people in northern France try to escape from attacks.

After France was forced to sign a peace agreement with Germany and Italy in 1940, some French groups escaped to North Africa or to Great Britain and formed the Free French army. Others remained in France and formed an "underground" movement to resist the Germans secretly.

▲ Newspapers around the country ran headlines similar to this one after the Japanese attack on Pearl Harbor.
Interpreting the Visual Record How do you think Americans felt when they saw headlines like this one?

National Geography Standard 9
Human Systems
In many occupied cities, the Nazis crowded Jews into small areas called *ghettos.* The ghetto in Warsaw, Poland, was surrounded by a wall topped with barbed wire and held 500,000 people in a small, cramped area. As the Germans began to deport Jews from the Warsaw ghetto to the death camps, German troops tried to enter the ghetto, only to be driven out by Jewish resistance forces. Starting in April 1943, about 600 to 1,000 Jews armed with little more than pistols battled 2,000 to 3,000 German soldiers and tanks for 27 days.

Discussion: Lead a class discussion about the meaning of the word *ghetto* today. Ask students how this meaning is related to the historical one.

◄ **Visual Record Answer**

Americans probably felt angry, sad, upset, frightened, and betrayed. Many probably wanted to fight back.

➤ASSIGNMENT: Tell students to imagine that they are American newspaper editors in 1941 immediately following the bombing of Pearl Harbor. Have each student write an editorial for his or her newspaper arguing whether or not the United States should enter the war.

Alfred J. Hamel, of Worcester, Massachusetts, suggests the following activity to help students understand the war in the Pacific. The American-led strategy called island hopping was eventually successful in defeating the Japanese. Have students conduct research on the Pacific war and draw maps that show which Pacific islands were taken by the Allies. Ask students to explain how the taking of these islands helped lead to the defeat of the Japanese.

National Geography Standard 17

The Uses of Geography
The war against Japan stretched across the Pacific Ocean. Therefore ships played a very important role in this part of World War II. U.S. submarines attacked Japanese shipping in an effort to cut off Japan's oil supply. Aircraft carriers transported planes to scenes of battle.

Early in the war, Japan advanced eastward across the Pacific Ocean by capturing many Pacific islands. To take back this territory, Allied forces used a strategy called "island hopping." They attacked only certain Japanese-held islands, skipping others, but leaving them without supplies. The plan was successful in stopping Japan and pushing its forces back across the Pacific.

Discussion: Lead a class discussion on this topic: How was the war in the Pacific different from the war in Europe?

▲
Anne Frank (1930-1945) was a Jewish teenager. During the Holocaust her family hid in an attic for two years to escape the Nazis. Anne kept a diary in which she wrote her thoughts and feelings.

Peace Memorial Park marks the spot where the first atomic bomb was dropped August 6, 1945, in Hiroshima, Japan

▼

The Holocaust

Hitler believed that Germans were a superior people, and planned to destroy or enslave people whom he believed were inferior. Hitler hated many peoples, but he particularly hated the Jews. Hatred of Jews is called **anti-Semitism**. The Nazis rounded up Europe's Jews and imprisoned them in concentration camps.

Death Camps In 1941, Hitler ordered the destruction of Europe's entire Jewish population. The Nazis built death camps in Poland to carry out this plan. People who could work were forced into slave labor. Those who could not work were sent to gas chambers where they were killed. Some Jews were shot in large groups. Thousands of other people died from conditions in the camps. The dead were buried in mass graves or burned in large ovens.

By the time the Nazi government fell, its leaders and followers had murdered an estimated 6 million European Jews. The Nazi **genocide**, the planned killing of a race of people, is called the **Holocaust**. Millions of non-Jews were also killed.

Resisting the Nazis Some Jews tried to fight back. Others hid. Most, however, were unable to escape. Many Europeans ignored what was happening to the Jews, but some tried to save people from the Holocaust. The Danes helped about 7,000 Jews escape to Sweden. In Poland and Czechoslovakia, the German businessman Oskar Schindler saved many Jews by employing them in his factories.

✓ **READING CHECK:** *Summarizing* What were Nazi concentration camps like? places for slave labor and for death in the gas chambers, or by being shot, or through starvation and disease

The End of the War

In 1942 the Germans tried to capture the Soviet city of Stalingrad. The battle lasted six months, but the Soviet defenders held out. The Germans were never able to take the city. This was a major blow to the Germans, who never fully recovered from this defeat. At the same time, American and British forces defeated the Germans in Africa. The war began to turn in favor of the Allies. That same year in the Pacific Japan lost several important battles. Led by the United States, Allied forces—including troops from Australia and New Zealand—began a campaign to regain some of the Pacific islands Japan had taken. Slowly, the Allies pushed the Japanese forces back across the Pacific Ocean.

In the summer of 1943 the Allies captured the island of Sicily in Italy. Italians forced Mussolini to resign, and Italy's new leader dissolved the Fascist Party. In September, Italy agreed to stop fighting the Allies.

Victory in Europe On June 6, 1944, Allied forces landed on the beaches of Normandy in northern France. This was the D-Day invasion. The invasion was a success. In August, Allied troops entered Paris. By September they were at Germany's western border. With the

Call on students to name individuals who played major roles in World War II. As people are named, call on other students to describe the actions for which they are known.

Have students complete Main Idea Activity 9.3. Then copy onto the chalkboard each of the major topics discussed in this section. Call on students to provide facts that apply to each topic.
ENGLISH LANGUAGE LEARNERS, COOPERATIVE LEARNING

REVIEW AND ASSESS

Have students complete the Section Review. Then ask students to write one sentence about the role each of the following countries played in World War II: Italy, Japan, Germany, England, Soviet Union, United States. Then have students complete Daily Quiz 9.4. **ENGLISH LANGUAGE LEARNERS**

EXTEND

Have interested students conduct research on the American home front during World War II. Direct students to pay particular attention to the roles of women during the war. Call on volunteers to share their findings with the class. **BLOCK SCHEDULING**

Soviets attacking Germany from the east, the Nazis' defenses fell apart. On April 30 Hitler killed himself, and within a week, Germany surrendered.

Victory over Japan Fighting continued in the Pacific. The Allies bombed Japan, but the Japanese would not surrender. Finally President Harry Truman decided to use the atomic bomb against Japan. On August 6, 1945, the most powerful weapon the world had ever seen was dropped on the city of Hiroshima. The bomb reduced the city to ashes and destroyed the surrounding area. About 130,000 people were killed and many more were injured. Countless more people died later. On August 9 another atomic bomb was dropped on the Japanese city of Nagasaki. Five days later Japan surrendered.

A New Age World War II resulted in more destruction than any other war in history. More than 50 million people were killed, and millions more were wounded. Unlike in most earlier wars, many of the people killed were civilians. **Civilians** are people who are not in the military. Millions were killed in the Holocaust. Thousands were killed by bombs dropped on cities in Europe and Japan. Thousands more died in prison camps in Japan and the Soviet Union. In time, people began to question how such cruel acts against human life and human rights were allowed to happen, and how they could be prevented in the future.

The American use of the atomic bomb began the atomic age. With it, came many questions and fears. How would this new weapon be used? What effect would it have on future wars? After World War II, world leaders would struggle with these questions.

✓ **READING CHECK:** *Analyzing* How was World War II unlike any war that came before it? most destructive war in history; human rights ignored in new ways; atomic bomb introduced

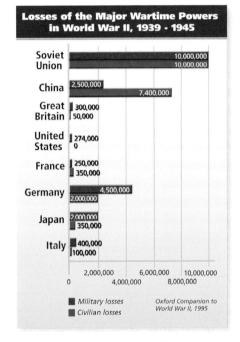

Losses of the Major Wartime Powers in World War II, 1939 - 1945

Soviet Union: 10,000,000 / 10,000,000
China: 2,500,000 / 7,400,000
Great Britain: 300,000 / 50,000
United States: 274,000 / 0
France: 250,000 / 350,000
Germany: 4,500,000 / 2,000,000
Japan: 2,000,000 / 350,000
Italy: 400,000 / 100,000

■ Military losses
■ Civilian losses

Oxford Companion to World War II, 1995

Interpreting the Graph What three countries had the highest civilian losses? What do you think caused these losses?

go.hrw.com
Homework Practice Online
Keyword: SJ3 HP9

Section Review 4

Define and explain: aggression, anti-Semitism, genocide, Holocaust

Reading for the Main Idea

1. (*Human Systems*) What events led to World War II?
2. (*Human Systems*) What happened during the Holocaust?
3. (*Human Systems*) What were the results of World War II?

Critical Thinking

4. **Cause and Effect** How did the rise of fascism in Europe lead to World War II?

Organizing What You Know

5. **Drawing Inferences and Conclusions** Copy the following graphic organizer. Fill it in, telling why each event was important in the war.

Event	Why Important?
Hitler invades Poland.	
Hitler gains control of western Europe.	
Germany invades the Soviet Union.	
Japan attacks Pearl Harbor.	
The Allies invade Europe on D-Day.	
The United States drops the atomic bomb on Japan.	

Section Review 4

Answers

Define Define For definitions see: aggression, p.187; anti-Semitism, p.189; genocide, p.190

Reading for the Main idea

1. acts of aggression by Japan, Italy, and Germany (NGS 13)
2. Jews were put in concentration camps where millions were murdered; millions of others who the Nazis thought were not pure were murdered. (NGS 9)
3. Axis defeated; millions of soldiers and civilians killed; enormous destruction; atomic age began (NGS 13)

Critical Thinking

4. Fascist nations became strong through militarism; Mussolini and Hitler set out to restore their nations' glory and power by invading other countries.

Organizing What You Know

5. World War II begins; Germany begins bombing England; Germans retreat for first time in war; United States enters war; Allies begin to retake Europe; Japan surrenders, war ends.

◄ **Graph Answer**

Soviet Union, China, Germany; locations of much of the fighting

Section 5

Objectives

1. Describe the Cold War.
2. Identify scenes of conflict in the world since World War II.
3. Investigate important events that happened at the end of the 1900s.

FOCUS

LET'S GET STARTED

Copy the following instructions onto the board: *What are some recent events that have affected the entire world? Write down a few of these events.* Allow students time to write their answers. Discuss the responses and list them on the board. Point out that our world is constantly changing. Tell students that in Section 5 they will learn some changes that have happened since 1945 and how they have shaped our world.

Building Vocabulary

Write **globalization** and **arms race** on the chalkboard. Underline the word *global* in *globalization*. Ask students to suggest possible meanings for the word based on the definition of the root word. Then point out that the word *arms* in *arms race* refers to weapons. Call on a volunteer to suggest a definition for the phrase. Then have a student locate and read the definitions for all the key terms.

Section 5 RESOURCES

Reproducible
- Guided Reading Strategy 9.5
- Creative Strategies for Teaching World Geography, Lesson 10
- Map Activity 9

Technology
- One-Stop Planner CD–ROM, Lesson 9.5
- Homework Practice Online
- HRW Go site

Reinforcement, Review, and Assessment
- Main Idea Activity 9.5
- Section 5 Review, p. 199
- Daily Quiz 9.5
- English Audio Summary 9.5
- Spanish Audio Summary 9.5

Section 5 The World Since 1945

Read to Discover

1. What was the Cold War?
2. Where in the world has conflict arisen since World War II?
3. What are some important events that happened at the end of the 1900s?

Define
bloc
arms race
partition
globalization

Locate
India
Israel
China
Taiwan
Korea
Vietnam
Cuba

WHY IT MATTERS

Created in 1945, the United Nations continues to promote international cooperation and peace. Use **CNNfyi.com** or other **current events** sources to find examples of UN action to stop violence. Record your findings in your journal.

compact discs

The Cold War

Although the Soviet Union and the United States were allies in World War II, the alliance fell apart after the war. The former allies clashed over ideas about freedom, government, and economics. Because this struggle did not turn into a shooting or "hot" war, it is known as the Cold War.

A Struggle of Ideas The struggle that started the Cold War was between two ideas—communism and capitalism. Communism is an economic system in which a central authority controls the government and the economy. Capitalism is a system in which businesses are privately owned. During the Cold War, the Soviet government functioned as a dictatorship which controlled the economy. However, the United States and other democratic nations practiced some form of capitalism.

In June 1948 the Soviets set up a blockade along the East German border to prevent supplies from getting into West Berlin. The people of West Berlin faced starvation. The United States and Great Britain organized an airlift to supply West Berlin. Food and supplies were flown in daily to the people.

192

Teaching Objective 1

ALL LEVELS: (Suggested time: 20 min.) Copy the following graphic organizer onto the board, omitting the italicized answers. Ask students to reread the discussion of the Cold War in their textbooks and to provide a definition for the term. Remind students that the Cold War was largely a conflict between the United States and its allies and the Soviet Union and its allies. Have students complete the graphic organizer with descriptions of these two sides. Lead a discussion about the completed organizer. **ENGLISH LANGUAGE LEARNERS, COOPERATIVE LEARNING**

Western Bloc	Eastern Bloc
•Strongest Country *United States*	•Strongest Country *Soviet Union*
•Economic System *capitalism*	•Economic System *communism*
•Political System *democracy*	•Political System *not democratic*
•Alliances *NATO*	•Alliances *Warsaw Pact*
•Economy *rapid economic growth*	•Economy *shortages of goods, food, money*

Europe Divided After World War II Joseph Stalin, leader of the Soviet Union, brought most countries in Eastern Europe under communist control. The Soviet Union and those communist-controlled countries were known as the Eastern bloc. A **bloc** is a group of nations united under a common idea or for a common purpose. The United States and the democracies in Western Europe were known as the Western bloc. While Western countries experienced periods of great economic growth, industries in most communist countries did not develop. People in these countries suffered from shortages of goods, food, and money.

Two Germanies After World War II the Allies divided Germany into four zones to keep it from becoming powerful again. Britain, France, the United States, and the Soviet Union each controlled a zone. Germany could no longer have an army, and the Nazi Party was outlawed. By 1948, the Western Allies were ready to unite their zones, but the Soviets did not want Germany united as a democratic nation. The next year the American, British, and French zones became the Federal Republic of Germany, or West Germany. The Soviets established the German Democratic Republic, or East Germany. The city of Berlin, although part of East Germany, was divided into East and West Berlin, with West Berlin under Allied control. The Berlin Wall, which became a famous symbol of the Cold War, separated the two parts of the city.

The Soviet Union After Stalin Joseph Stalin, who had led the Soviet Union through World War II, died in 1953. The next Soviet leader was Nikita Khrushchev. He criticized Stalin's policies and reduced the government's control over the economy.

During the 1950s and 1960s some Eastern bloc nations tried to break free of communism. In East Germany, Czechoslovakia, and Hungary, for example, people rebelled against Soviet control, but the Soviets crushed these revolts.

▲

The Berlin Wall did not stop people from trying to reach West Berlin. More than 130 people died trying to escape over the heavily guarded wall.

Interpreting the Visual Record Why do you think people risked death to escape communism?

ALL LEVELS: (Suggested time: 30 min.) Organize the class into six groups and assign each group one of the following areas of conflict: Southwest Asia, Korea, Cuba, Vietnam, Northern Ireland, and Yugoslavia. Have the members of each group discuss the conflicts that have arisen in their assigned area and prepare a chart that includes the approximate date of the conflict, reason for conflict, and the outcome. Have each group present its chart to the class and lead a discussion about the information in the chart. **ENGLISH LANGUAGE LEARNERS, COOPERATIVE LEARNING**

➤**ASSIGNMENT:** Have each student consult newspapers, television news programs, Internet news sites, or other sources to find information about a conflict raging in the world today. Have each student write a brief paragraph that tell something about this conflict. Students might note the cause of their chosen conflicts or efforts that have been undertaken to resolve them.

GLOBAL PERSPECTIVES

The United Nations is composed of six main bodies. The General Assembly investigates disputes and recommends action to resolve them. The Security Council these actions and approves all military actions. The International Court of Justice decides questions of international law. The Economic and Social Council sponsors trade and human rights organizations. The Trusteeship Council controls territories that are under UN supervision. Finally, the Secretariat runs the UN itself.

Critical Thinking: What are some things the UN can do to settle disputes?

Answer: encourage international communication, take military action, sponsor economic actions

▲
In this picture U.S. President Harry Truman signs the North Atlantic Pact, which created NATO. Shown above is the NATO emblem.

Interpreting the Visual Record How would you explain the NATO emblem?

▲
Mikhail Gorbachev, with his wife, Raisa. In 1990, Gorbachev won the Nobel Peace Prize for his reform work in the Soviet Union.

Visual Record Answer ▶

The flags are the flags of the NATO member nations; the compass stands for the four corners of the globe. The emblem means that NATO will protect its members wherever they are.

The United Nations World leaders did not want the Cold War to turn "hot." Although the League of Nations had failed to prevent World War II, people still wanted an international organization that could settle problems peacefully. In April, 1945, the United Nations (UN) was created. Its purpose was to solve economic and social problems as well as to promote international cooperation and maintain peace. Representatives of 50 countries formed the original United Nations. Today there are nearly 200 member nations. The six official languages of the United Nations are Arabic, Chinese, English, French, Russian, and Spanish. The headquarters of the United Nations are in New York City. It also has offices in Geneva, Switzerland, and Vienna, Austria.

New Alliances Fearing war but hoping to preserve peace, nations around the world formed new alliances. In 1949, 12 Western nations, including the United States, created the North Atlantic Treaty Organization (NATO). In 1954 the Southeast Asia Treaty Organization (SEATO) was created in an attempt to halt the spread of communism in Southeast Asia. Many Eastern bloc countries, including the Soviet Union, signed the Warsaw Pact in 1955. The Warsaw Pact countries had more total troops than the NATO members. This difference in the number of troops encouraged the Western powers to rely on nuclear weapons to establish a balance of power.

The End of the Cold War Throughout the Cold War, the Soviet Union and the United States had been a waging an **arms race**. The countries competed to create more advanced weapons and to have more nuclear missiles than each other. The arms race was expensive and took its toll on the already shaky Soviet economy.

In 1985, Mikhail Gorbachev became head of the Soviet Union. He reduced government control of the economy and increased individual liberties, such as freedom of speech and the press. He also improved relations with the United States.

These reforms in the Soviet Union encouraged democratic movements in Eastern bloc countries. In 1989, Poland and Czechoslovakia threw off communist rule. In November, the Berlin Wall came down. In October 1990, East and West Germany became one democratic nation. Soviet republics also began to seek freedom and independence. By the end of 1991, the Soviet Union no longer existed. The Cold War was over. The arms race could stop.

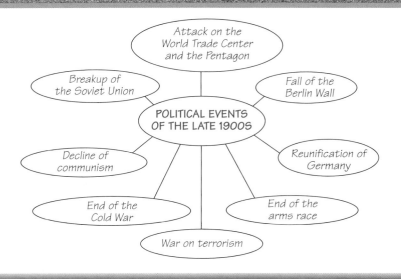

Teaching Objective 3

ALL LEVELS: (Suggested time: 20 min.) Copy the following graphic organizer onto the board, omitting the italicized answers. Ask students to copy it into their notebooks. Have student fill in the organizer with important political events that occurred at the end of the 1900s. Ask volunteers to present their completed charts to the class. Lead a class discussion about the events that appear on the graphic organizers.
ENGLISH LANGUAGE LEARNERS, COOPERATIVE LEARNING

In October 1990 young people in Berlin wave German flags to celebrate the reunification of Germany.

The breakup of the Soviet Union created several independent countries. Russia was the largest of these new nations. Its new leader was Boris Yeltsin. Under Yeltsin, Russia moved toward democracy. Yeltsin also improved Russia's relations with the West. In 2000, Vladimir Putin became leader of Russia. Under Putin, relations with the United States improved further.

Some tension arose, however, between Russia and the former Soviet republics. For example, Russia and the Ukraine clashed over military issues in the 1990s.

✓ **READING CHECK:** *Evaluating* How did the end of the Soviet Union affect the world? ended Cold War; stopped arms race; new countries created; improved relations between United States and Russia

Other World Conflicts

Since the end of World War II, many conflicts have shaken the world. Some have been resolved, while others continue to threaten world peace. The lessons of two world wars and the threat of mass destruction that would result from a nuclear war has kept these conflicts contained. World War III has not occurred, and the hope of people everywhere is that it never will.

Southwest Asia After World War I Britain said it would help create a Jewish homeland in Palestine, a region of Southwest Asia. Many Arab nations, however, wanted an Arab state in Palestine. In 1947 the UN voted to **partition**, or divide, Palestine, creating both a Jewish state and an Arab state. While the Arabs rejected this plan, Jewish leaders in Palestine accepted it. In May 1948, Israel was established as a Jewish state.

Israel's Knesset, or parliament, meets here. The Knesset is the supreme power in Israel.

LEVEL 1: (Suggested time: 10 min.) Call on students to name some of the advancements made at the end of the 1900s. As students name development, write them on the chalkboard. Discuss student responses. After the discussion, have students vote on which of the listed achievements they think will have the greatest influence on the future.
ENGLISH LANGUAGE LEARNERS

LEVEL 2: (Suggested time: 30 min.) Tell students to imagine that they are historians from the year 2250 who have been asked to write descriptions of the major events of the late 1990s and early 2000s for middle school students. Have each student write a short description of one event that explains its significance to the world. Encourage students to share their writing with the class.

GLOBAL PERSPECTIVES

In 2000, 50 years after the beginning of the Korean War, the leaders of North and South Korea met for a historic three-day summit at P'yŏngyang. The leaders agreed to reunited families separated by political boundaries since the Korean War and to promote economic development in both countries. They also pledged to work toward reunification through increased cultural, athletic, medical, and environmental cooperation and exchanges.

Part of the agreement went into effect immediately. Two hundred families were brought together later that year. In addition, members of the North and South Korean Olympic teams marched together at the 2000 Olympics.

Activity: Have students use the Internet, almanacs, and other current events sources to find recent information about reunification efforts.

At the end of the Korean War, the two sides set up a neutral area, called the demilitarized zone, or DMZ. It is a buffer zone, and no military forces from either side may enter the area.

▼

The establishment of Israel enraged many Palestinian Arabs and Arab nations. Attacked by neighboring Arab countries, Israel fought back. By early 1949 a cease-fire was reached. Israel survived, but Palestinian Arabs had no homeland. In 1967, tensions between Israel and its Arab neighbors exploded into war again. In what became known as the Six-Day War, Israel captured territory from Egypt, Syria, and Jordan. After the Six-Day War, the Palestine Liberation Organization (PLO), led by Yasir Arafat, launched many attacks on Israel.

The many attempts to bring peace to the Middle East have failed. The Israelis and the Arabs do not trust each other. Both sides make demands that the other side will not meet. When one side commits acts of violence, the other side strikes back. The failure to achieve peace in the Middle East continues to be one of the most disturbing issues facing the world.

Korea At the end of world War II the Soviet Union controlled northern Korea. U.S. troops controlled southern Korea. A Communist government took power, and in 1950, North Korea invaded South Korea. The United Nations sent troops to stop the invasion. The Korean War lasted until 1953, when a cease-fire was signed. Korea remains divided.

Cuba In 1959 Fidel Castro established a Communist government in Cuba. In 1961 President John F. Kennedy approved an invasion of Cuba by anti-Castro forces. The invasion failed, and Castro turned to the Soviet Union for support. The Soviet Union, which by this time possessed nuclear weapons, sent nuclear missiles to Cuba. Kennedy demanded that the missiles be withdrawn, but Khruschev refused. NATO and Warsaw Pact military forces prepared for combat. For several days the Cuban missile crisis held the world on the brink of nuclear war. Finally, the Soviet Union agreed to remove its missiles, and the United States promised it would not invade Cuba.

Vietnam Vietnam had been a colony of France for more than 60 years. In 1945 Ho Chi Minh, a Communist leader, declared Vietnam independent. In 1954 Vietnam became divided. North Vietnam was communist. South Vietnam was not. In the late 1950s, when North Vietnam invaded South Vietnam, the United States sent troops to South Vietnam to fight the communists. Many Americans were unhappy that the country had become involved in this war, and American troops pulled out of Vietnam in 1973. South Vietnam surrendered in 1975. Nearly 1.7 million Vietnamese and about 58,000 Americans lost their lives in the Vietnam War. In 1976 Vietnam was united under a Communist government.

Northern Ireland When the Republic of Ireland gained independence from Great Britain in 1922, the territory of Northern Ireland remained part of Britain. The Protestant majority in Northern Ireland controlled both the government and the economy. This caused resentment among Northern Irish Catholics. During the late 1960s Catholic protests began to turn violent. The British have tried to resolve the conflict both through political means and military force. While the situation has improved, a permanent peaceful solution has not yet been found.

The Breakup of Yugoslavia Yugoslavia was created after World War I by uniting several formerly independent countries. These included Bosnia, Croatia, Slovenia, and Serbia. After the fall of communism in eastern Europe, Eastern Orthodox Serbs tried to dominate parts of Yugoslavia where people were mainly Roman Catholic or Muslim. Fighting broke out between Serbia and Croatia, which was mainly Roman Catholic. Yugoslavia was once again divided into several countries in the early 1990s, but this did not end the violence. In 1992 Bosnian Serbs began a campaign of terror and murder intended to drive the Muslims out of Bosnia. Finally, NATO bombed Serbian targets in 1995, and the fighting stopped. Several Serb leaders were tried as war criminals.

The War on Terrorism On September 11, 2001, terrorists attacked the World Trade Center in New York City and the Pentagon in Washington, D.C. Following these attacks, the United States asked for support of nations around the world. Many nations, including Russia, China, Cuba, Pakistan, and Saudi Arabia, supported the United States in what President George W. Bush called a "war on terrorism." That show of support indicated that nations of the world might be beginning to leave behind some of the struggles of the last century.

✓ **READING CHECK:** _Evaluating_ What has prevented conflicts in different parts of the world from becoming world wars? fear of another world war; fear of the destruction from a nuclear war

After the Vietnam War more than a million people fled Vietnam by boat.

Interpreting the Visual Record What kinds of conditions might make people willing to leave their homes?

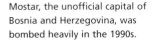
Mostar, the unofficial capital of Bosnia and Herzegovina, was bombed heavily in the 1990s.

Ask students to list the three most significant challenges they think the world faces today. Lead a brief discussion about these issues and steps that have been taken to address them.

Have students complete the Section Review. Then have each student write four multiple choice questions about one of the topics discussed in this section. Collect the questions and redistribute them to the class so that each student receives questions prepared by another student to answer. Then have students complete Daily Quiz 9.5. **ENGLISH LANGUAGE LEARNERS**

Section Review 5

Answers

Define For definitions, see: bloc, p. 192; arms race, p. 194; partition, p. 195; globalization, p. 199

Reading for the Main Idea

1. struggle between communism and capitalism; arms race; new alliances among countries (NGS 13)

2. Israeli Jews and Palestinian Arabs disagree over who should control Palestine. Many Arabs want to eliminate the country of Israel. (NGS 13)

3. space travel; computers; the Internet; genetics; cloning; global cooperation (NGS 10)

Critical Thinking

4. Answers will vary, but should be supported.

Organizing What You Know

5. Stalin—brought eastern Europe under communist control; Gorbachev—began reform in Soviet Union, improved relations with United States; Yeltsin—led Russia toward democracy; Bush—brought nations together to fight terrorism; Arafat—created PLO

Visual Record Answer ▶

courage, generosity, willingness to take risks in order to help others

▲

After the terrorist attacks, members of New York City's fire and police departments risked their lives to help people.

Interpreting the Visual Record **What qualities can a terrible disaster bring out in people?**

▲

The launch of the *Sputnik* satellite, shown above, began a space race between the U.S. and the Soviet Union to see who would reach the moon first.

Progress and Problems

During the late 1900s enormous advances of all kinds occurred in science, medicine, space travel, communication, and technologies. Also during this period, however, problems developed that must be addressed in the new century.

Space On October 4, 1957, the Soviet Union launched *Sputnik*, the world's first space satellite. The space age had begun. In 1961 Soviet Yury Gagarin became the first person to travel into space. In 1969 Neil Armstrong, an American, became the first person to walk on the moon. Beginning in the late 1990s, the United States, Russia, and 14 other nations worked together to build an International Space Station (ISS) The first crew members reached the ISS late in 2000. Researchers in space have conducted many useful experiments in geography, engineering, medicine, and other fields of study.

Genetics Genes are small units in the body that determine all our physical characteristics. Knowledge of genetics is leading to new cures for diseases. It has also given scientists the ability to clone, or make an identical copy of an animal. In 1997, for example, scientists cloned a sheep. Although cloning has raised some ethical debates that have not been resolved, it is still a remarkable scientific advance.

The Computer Age The first modern computer, called ENIAC, was built in 1946. It was so large that it filled an entire room. By the end of the 1950s, however, the manufacturing of computers was a growing industry. Over time, computers became smaller and faster. Now many people around the world use computers every day at home, at work, and at school. They are important parts of many products, from cars to medical equipment to rockets. New technologies like the Internet and the World Wide Web have made computers even more useful. Today computers can send messages around the world in just a few seconds.

Have students complete Main Idea Activity 9.5. Then pair students and have each pair write newspaper headlines that describe five of the events described in this section. Encourage the pairs to choose the five events they think are most significant in world affairs. Call on volunteers to share their headlines with the class. Display students' headlines around the classroom. **ENGLISH LANGUAGE LEARNERS, COOPERATIVE LEARNING**

Have interested students conduct research on an environmental issue facing the United States and prepare a presentation on this issue. Presentations should include a description of the issue addressed as well as recommendations for solving problems related to the issue. Have students give their presentations to the class. **BLOCK SCHEDULING**

CHAPTER 9

Review

Globalization Faster ways of traveling and communicating have brought countries and cultures around the world closer together. As a result, people's lifestyles are becoming more similar. For example, people in different parts of the world now eat the same kinds food, wear the same types of jeans, and listen to the same music. This process, in which connections around the world increase and cultures share similar practices, is called **globalization**. One example of globalism is the European Union (EU). The EU includes twelve European nations that agreed to drop trade barriers and to adopt a common currency—the Euro.

An Endangered Environment Modern technology has improved the world in many ways, but it has also caused problems. For example, burning certain fuels causes pollution. As rain falls through the polluted air, it becomes acid rain, which kills trees and plants. Burning these same fuels also releases excess carbon dioxide into the atmosphere. Some scientists say this causes a greenhouse effect, trapping heat and causing temperatures on Earth to rise. Some scientists also say that Earth's ozone layer is thinning out. Ozone is a gas that shields our planet from some of the sun's rays which are harmful to plants and animals.

Looking Ahead The world faces many challenges, but the prospect for the future is hopeful if we can meet them. If we can live in peace, preserve the environment, and use science and technology for the benefit of all, the world of tomorrow can be a truly wonderful place.

✓ **READING CHECK:** *Summarizing* How have some modern scientific advances changed the world? space travel, computers, the Internet, knowledge of genetics, identifying the greenhouse effect

These coins and bills are part of the European Union's currency system, the Euro.

▼

ANSWERS

Building Vocabulary
For definitions, see: militarism, p. 173; armistice, p. 176; stock market, p. 178; bankrupt, p. 179; dictator, p. 180; fascism, p. 180; police state, p. 180; revolutionaries, p. 184; economic nationalism, p. 186; aggression, p. 187; anti-Semitism, p. 189; genocide, p. 190; bloc, p. 192; arms race, p. 194; partition, p. 195

Reviewing the Main Ideas
1. nationalism , militarism; killing of Archduke Francis Ferdinand in Sarajevo; Austria-Hungary declaring war on Serbia (NGS 13)

2. promised to solve economic woes of the Great Depression and to restore countries' former glory (NGS 11)

3. encouraged Latin American nation's fight for political and economic independence (NGS 11)

4. Japan's attack on Pearl Harbor (NGS 13)

5. new alliances; world divided into Western bloc and Eastern bloc; Germany divided into communist and noncommunist sections; West's attempts to stop spread of communism; arms race; space race (NGS 13)

Section Review 5

Define and explain: bloc, arms race, partition, globalization

Reading for the Main Idea

1. (*Human Systems*) What happened during the Cold War?

2. (*Human Systems*) What are the reasons for conflict in Southwest Asia?

3. (*Human Systems*) What were some important advances made in 1900s?

go.hrw.com Homework Practice Online
Keyword: SJ3 HP9

Critical Thinking

4. **Supporting a Point of View** Do you think most countries believe it is in their best interest to keep peace? Why?

Organizing What You Know

5. **Lead in to come?** Copy the following graphic organizer. Fill it in, telling how each person shaped events in the late 20th century.

Joseph Stalin	
Eleanor Roosevelt	
Mao Zedong	
Mikhail Gorbachev	
George W. Bush	

199

RETEACH

Organize the class into five groups and assign each group one of the sections in this chapter. Have each group create the front page of a newspaper that describes the events discussed in their assigned sections. The page should include a few short articles with headlines addressing the major topics in the section. Discuss the information in each newspaper.
ENGLISH LANGUAGE LEARNERS, COOPERATIVE LEARNING

PORTFOLIO EXTENSIONS

1. Organize the class into groups and have each group create a short play that depicts a major event from Chapter 9. Have each group present its play to the class. Remind students that they should create programs to accompany their plays, listing the cast, describing the event depicted, and explaining the plot of the play. Place these programs in student portfolios.

2. Have students write paragraphs or create models that represent what they consider to be the most significant development of the late 1900s. Encourage students to be creative in their work. Place the paragraphs or photographs of the models in student portfolios.

Review
ANSWERS

Understanding History and Society
The Changing World Map
Charts will vary but should note countries created after World War I, after World War II, and the breakup of the Soviet Union. Use Rubric 7, Charts, to evaluate student work.

Thinking Critically
1. Germans were angry about Treaty of Versailles; Hitler used this anger to gain support and power.
2. gave money to states to help people, created jobs
3. The United States has become less directly involved in Latin American affairs. This changed in response to Latin American anger toward American interference
4. Aggression helped Germany, Italy, and Japan gain power and territory; war was finally declared to stop the aggression
5. Possible answers: Students might mention the lessons of two world wars, UN intervention, or the threat of mass destruction posed by a nuclear war.

Reviewing What You Know

Building Vocabulary

On a separate sheet of paper, write sentences to define each of the following words.

1. militarism
2. armistice
3. stock market
4. bankrupt
5. dictator
6. fascism
7. police state
8. revolutionaries
9. economic nationalism
10. aggression
11. anti-Semitism
12. genocide
13. bloc
14. arms race
15. globalization

Reviewing the Main Ideas

1. (Human Systems) What ideas, actions, and incidents were major causes of World War I?
2. (Human Systems) How did dictators come to power in Europe during the 1930s?
3. (Human Systems) How did nationalism change Latin America?
4. (Human Systems) What single event finally caused the United States to enter World War II?
5. (Human Systems) How did the Cold War shape the second half of the 1900s?

Understanding Environment and Society

The Changing World Map
The world map has changed a great deal during the 1900s. Create a presentation that shows some of these changes. Your presentation should include a chart that describes the changes and shows where and when they happened. As you prepare your chart, consider the changes that occurred:
• After World War I.
• After World War II.
• After the Cold War.

Thinking Critically

1. **Drawing Conclusions** How did the Treaty of Versailles help set the stage for World War II?
2. **Evaluating** How did President Roosevelt's program, the New Deal, help workers?
3. **Comparing and Contrasting** How has U.S. policy in Latin America changed over the years? Why did it change?
4. **Analyzing** What role did aggression, and the response of world leaders to aggression, play in World War II?
5. **Evaluating** What prevented conflicts from becoming major world wars in the late 1900s?

FOOD FESTIVAL

The expression "An army marches on its stomach" suggests how important food is to soldiers. In World War II, the U.S. military developed K rations so soldiers would have something to eat when no other food was available. Soldiers carried K rations in their packs. K rations contained biscuits, canned meat, coffee, sugar, chewing gum, chocolate, and other nonperishable foods. Sometimes K rations were the only food soldiers had to eat.

CHAPTER 9 REVIEW AND ASSESSMENT RESOURCES

Reproducible
◆ Readings in World Geography, History, and Culture 15, 23, 32, 34, 38, 74
◆ Critical Thinking Activity 9
◆ Vocabulary Activity 9

Technology
◆ Chapter 9 Test Generator (on the One-Stop Planner)

◆ HRW Go site
◆ Audio CD Program, Chapter 9

Reinforcement, Review, and Assessment
◆ Chapter 9 Review, pp. 200–201
◆ Chapter 9 Tutorial for

Students, Parents, Mentors, and Peers
◆ Chapter 9 Test
◆ Chapter 9 Test for English Language Learners and Special-Needs Students
◆ Unit 2 Test
◆ Unit 2 Test for English Language Learners and Special-Needs Students

Building Social Studies Skills

Map ACTIVITY

On a separate sheet of paper, match the letters on the map with their correct labels.

Austria
France
Germany
Great Britain
Italy
Poland
Russia

Mental Mapping Skills ACTIVITY

On a separate sheet of paper, draw a freehand map of Western Europe. Label Germany, Italy, France, and Great Britain. Shade the countries that sided with the Allies in World War II one color. Shade the countries that were Axis Powers a different color. Make a key for your map.

WRITING ACTIVITY

Imagine that you are making a film about a family in the United States during the Great Depression. Write a summary of the film you would like to make. List characters who will appear in the film, events that will occur, problems that will arise, and how the characters will solve them. Be sure to use standard grammar, spelling, sentence structure, and punctuation.

Alternative Assessment

Portfolio ACTIVITY

Learning About Your Local History

Historical People and Places
Research how your town is connected to an event from the 1900s. This link might be a person, place, building, or statue. Write a report on this person, place, or object and explain its historical importance. Include photographs or drawings in your report.

⧉ internet connect

Internet Activity: go.hrw.com
KEYWORD: SJ3 GT9

Choose a topic about the modern world.
• Learn about the effects of the Treaty of Versailles.
• Write a report about Anne Frank.
• Create a poster about the causes and effects of global warming.

Map Activity
A. Great Britain   **E.** Austria
B. France   **F.** Italy
C. Germany   **G.** Poland
D. Russia

Mental Mapping Skills Activity
Maps should show correct locations of Germany, Italy, France, and Great Britain. They should show that Great Britain and France were on the side of the Allies and that Germany and Italy were on the side of the Axis.

Writing Activity
Summaries will vary but they should be consistent with text material. Students should mention events such as the stock market crash, poverty, and the run on the banks. Use Rubric 40, Describe, to evaluate student work.

Portfolio Activity
Reports will vary. Students should identify a local person, place, or thing that is connected to a local historical event and write about this connection. Use Rubric 42, Writing to Inform, to evaluate student work.

⧉ internet connect

GO TO: go.hrw.com
KEYWORD: SJ3 Teacher
FOR: a guide to using the Internet in your classroom

Studying Government Agencies

Tell students that many different agencies of the U.S. government are responsible for guaranteeing the safety and security of American citizens from terrorists and terrorism. These include the Federal Bureau of Investigation (FBI), the Central Intelligence Agency (CIA), the Immigration and Naturalization Service (INS), the Federal Aviation Agency (FAA), and the recently formed Office of Homeland Security. Organize the class into groups and assign one of the federal agencies to each group. Ask each group to conduct research on its assigned agency. Direct the groups to create charts about these agencies. In one column of the chart, have students list the agency's main responsibilities and tasks. In the other, have them list ways in which the agency might combat terrorism either at home or around the world. Have each group present its chart to the class. Then stage a roundtable discussion about these federal agencies and the fight against terrorism.

GEOGRAPHY SIDELIGHT

One reason for the initial success of the war on terrorism was that Afghanistan had become a friendless regime. Isolated geographically, Afghanistan had cut cultural ties with the rest of the world. Few countries recognized the Taliban as the legitimate government.

The Taliban, an Islamic fundamentalist group, rose to power in the late 1980s. By the late 1990s they controlled more than 90 percent of the country. The United States imposed sanctions against Afghanistan in 1999 when the Taliban refused to turn over Osama bin Laden, who had ordered the bombing of American embassies in Kenya and Tanzania the previous year. In early 2001, the Taliban angered countries around the world when it destroyed two ancient Buddha statues in Bamiyan, near Kabul.

Critical Thinking: How might the war on terrorism have been different had Afghanistan had more allies?

Answer: The coalition might have had to fight Afghanistan's allies and trading partners.

➤ **This Focus On Government feature addresses National Geography Standard 13.**

FOCUS ON GOVERNMENT

Combating Global Terrorism

Early in the morning of September 11, 2001, two passenger jets crashed into the two towers of the World Trade Center. Another jet hit the Pentagon outside of Washington, D.C., and a fourth crashed in southern Pennsylvania. All of these planes were piloted by terrorists, people who use violence to achieve political goals. An investigation by the U.S. government led officials to conclude that al Qaeda, a terrorist organization led by Osama bin Laden, was responsible for these attacks. At the time of the attacks, bin Laden was living in Afghanistan, a landlocked country in Southwest Asia.

The War on Terror The terrorist attacks of September 11, 2001, were not the first on American soil or American interests. Terrorists tried to destroy the World Trade Center in New

The New York City sky filled with smoke after the attack.

▼

▲

President George W. Bush addressed a joint session of Congress on September 20. During the session he pledged to use the full might of the nation in a war on international terrorism

York in 1993. American citizens and embassies abroad have also been targeted. U.S. allies have also been victims of terrorists. For years, the U.S. government and its allies went to great lengths to combat terrorism. The terrible attacks of September 11, however, demanded the strongest possible retaliation.

Calling the attacks an act of war, President George W. Bush declared war on terrorism. "Either you are with us or you are with the terrorists," President Bush told the world.

Forming an International Coalition

The United States wanted to combat terrorists on a global scale. To do that, President Bush formed an international coalition. Great Britain was an early partner in the coalition, and Prime Minister Tony Blair enlisted the support of many nations. A new era in global relations began as China, Russia, and more than 50 other countries joined the coalition.

Going Further: Thinking Critically

Ask students to discuss why it is essential for the different agencies of the United States to share information and cooperate among themselves to fight terrorism. Have students speculate on how the Office of Homeland Security might make this kind of cooperation possible. Ask students to use the results of their research in the previous activity to suggest kinds of information that each agency might be able to provide to other organizations. Finally, using what they have learned, have students work together to design a diagram that shows how the various government agencies they have studied work together to combat terrorism.

Asia Political

▲

Afghanistan is a landlocked country. It shares borders with Pakistan, Iran, Turkmenistan, Uzbekistan, and Tajikistan.

▲

Much of Afghanistan is covered by rugged mountainous terrain. This landscape provided hiding places for al Qaeda terrorist camps and cells.

Some countries supplied military support for the fight against terrorists in Afghanistan. Others, including Pakistan and Uzbekistan, allowed coalition forces to use their air bases near Afghanistan. The coalition included many Muslim and Arab nations.

Coalition forces achieved victory quickly in Afghanistan. By the end of 2001, the Taliban, Afghanistan's ruling government, was forced from power and many al Qaeda terrorists were captured or killed. Despite this success, the war on terrorism is not over. Terrorist cells are still active in other countries. President Bush warned that new battles will have to be fought to eliminate them.

A New Government for Afghanistan

After the fall of the Taliban, representatives from all over Afghanistan formed a new government. Their country faced difficult problems. Years of war had left the country in ruins. As part of the war on terrorism, coalition forces provided food and medical care and also promised to rebuild Afghanistan.

Homeland Security While fighting terrorism in Afghanistan, the U. S. government also acted to block future terrorist attacks at home. The nation's borders and coastline are patrolled more carefully. New procedures make air travel more secure.

President Bush also created the Office of Homeland Security. Its job is to coordinate different security and intelligence agencies. As a result, officials can better identify people with links to terrorists.

Understanding What You Read

1. How are the United States and other nations fighting global terrorism?

2. In what ways are Americans defending their homeland against terror?

Going Further: Thinking Critically

Explain that the development of photography has contributed a great deal to what we know about the past leaders of our country. Gather images of painted portraits of early presidents, such as George Washington and Thomas Jefferson, and show them to the class. Point out that we have no photographs of these men. Next show a few photographs of Abraham Lincoln. Explain to students that these photographs were taken at a time when the camera was still a relatively new invention. Finally, show some candid photographs of recent presidents, such as John F. Kennedy playing with his children or Bill Clinton playing golf. Explain that our modern presidents usually face dozens of photographers almost every day.

After students study the visual evidence you have presented, ask them to speculate about how photography may have changed our attitudes and ideas about our leaders and other public figures. How might general attitudes toward leaders been different in the past before photography and video? *(Students may note that images and posed photographs present a more formal view of individuals, while candid shots make even powerful figures seem more approachable and human.)*

PRACTICING
THE SKILL

1. Students should identify the approximate year or decade when each photograph was taken. Depending on the photos, students might list details that describe clothing and home furnishing styles, buildings, automobiles, toys, tools, and other everyday objects.

2. Suggest that students cover the caption before they begin to study the photo. In creating the "story" that a photo tells, ask students to think about the values that the photographer or subjects in the photo seem to advance.

3. Ask students to choose photos that tell about different areas and aspects of American life. These areas, for example, might include business, sports, technology, art and culture, education, and so on.

➤ This GeoSkills feature addresses National Geography Standards 2, 4, 6, 15, and 18.

▲
Pieter Brueghel the Younger painted this scene in the 1500s.

Building Skills for Life: Using Visual Evidence

We learn about the past from many different sources. One important source of information is visual evidence. Paintings, statues, drawings, and photographs are all examples of visual evidence. Pictures and paintings provide many clues about life in the past. Sometimes they show the clothing people wore, the houses they lived in, or the games they played. Often they document, or prove, that certain events were an important part of a culture.

As with all historical sources, you need to study visual evidence carefully. Often, artists and photographers only show what they want you to see. To use visual evidence effectively, follow these steps:

1. **Identify the visual evidence.** Study a picture or painting carefully. Pay attention to details. Ask yourself who the subjects of the picture are and what they might be doing. Look at the painting on this page. How would you describe the people you see? What are they doing?

2. **Evaluate the visual evidence.** Remember, a picture or photo does not always show the whole story. The artist or photographer might have left out details on purpose. Thinking about the artist's purpose will help your decide whether visual evidence is reliable. What purpose do you think Brueghel had for painting his picture? Do you think it is a reliable record of how some people lived?

3. **Learn from the visual evidence.** After studying a picture carefully, use the details that you note to draw conclusions. Your conclusions should help you understand how some people lived at a particular time and place in the past. What conclusions can you draw from the details in Brueghel's painting?

PRACTICING THE SKILL

1. Study an old family album or library book that contains photographs taken long ago in your city or state. What do the pictures tell you about life in the time in which they were taken?

2. Study a picture in a newspaper or magazine without reading the caption. What story does the photo tell about our culture? Why do you think the photographer has chosen to document this event in this way?

3. What visual evidence would best capture our way of life today? Choose five pictures. Explain why each one might help people of the future understand American life today.

HANDS on GEOGRAPHY

Paintings, drawings, statues, sculpture, and photographs are all types of visual evidence. Historians study the details in visual evidence to draw conclusions about the past.

Look at the two examples of visual evidence below. The first is a sculpture. It shows a battle that took place in ancient Greece about 2,500 years ago. The second is a painting from 1862 of a British railway station. What details do you notice in each example? What information do they help you to know?

▶

This relief shows the Battle of Marathon that took place in 490 B.C.

◀ Artist William Frith painted this scene in 1862.

Lab Report

1. Look at the sculpture of the Battle of Marathon. What conclusions can you draw about warfare in ancient Greece?

2. Look at the painting. What do you think the artist's purpose was painting this scene?

UNIT 3

UNITED STATES AND CANADA

UNIT OBJECTIVES

1. Describe the landforms, climates, and resources found in the United States and Canada.
2. Examine the geography, history, and cultures of the United States.
3. Identify the similarities and differences that exist between various regions of the United States.
4. Examine the historical and cultural geography of Canada.
5. Interpret special-purpose maps, graphs, and charts to better understand the interrelationships of each country's physical and human geography.

USING THE ILLUSTRATIONS

Direct students' attention to the photographs on these pages. Ask them to cover the captions and to determine which of the smaller photos was most likely taken in the United States. *(people standing near the American flag and Statue of Liberty replica).* How can they tell? *(Possible answer: because the flag and Statue of Liberty are symbols of the United States)* Remind students that a symbol is a sign that stands for something else. Ask what the Statue of Liberty stands for. *(Possible answers: freedom, a new life for immigrants)* Ask which other images on the pages may serve as symbols *(CN Tower, buffalo)* and what they may symbolize. *(Possible answers: modern Canada, the American West)* You may want to point out that although the buffalo is sometimes used as a symbol for the United States, the largest unrestricted buffalo herd is in Canada.

Then point out that the mountain and lake scene, although in Canada, looks like many places in the United States because the Rocky Mountains extend through both countries. in fact, the two countries share several landforms, such as the Great Plains. The United States and Canada share all but one of the Great Lakes.

UNIT 3

United States and Canada

CHAPTER 10 & 11
The United States

CHAPTER 12
Canada

Canadian Rocky Mountains

CN Tower and SkyDome, Toronto, Canada

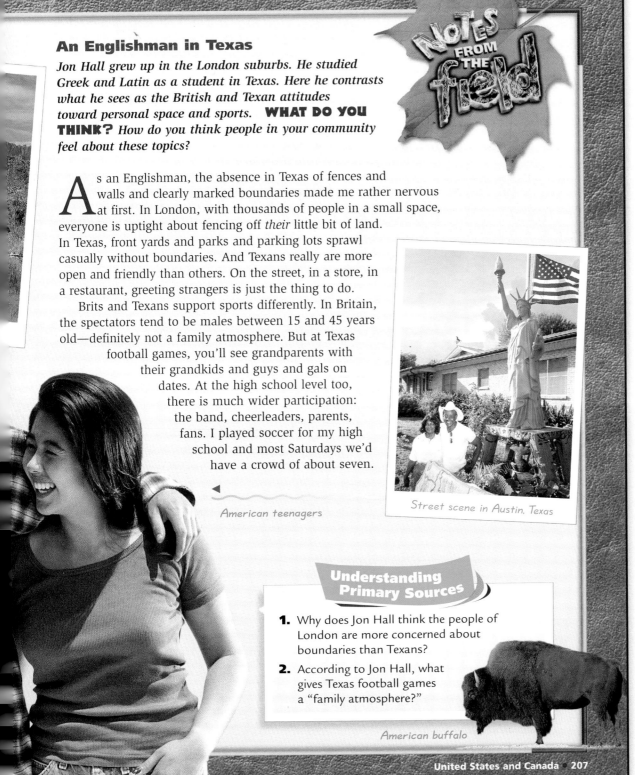

An Englishman in Texas

Jon Hall grew up in the London suburbs. He studied Greek and Latin as a student in Texas. Here he contrasts what he sees as the British and Texan attitudes toward personal space and sports. **WHAT DO YOU THINK?** *How do you think people in your community feel about these topics?*

As an Englishman, the absence in Texas of fences and walls and clearly marked boundaries made me rather nervous at first. In London, with thousands of people in a small space, everyone is uptight about fencing off *their* little bit of land. In Texas, front yards and parks and parking lots sprawl casually without boundaries. And Texans really are more open and friendly than others. On the street, in a store, in a restaurant, greeting strangers is just the thing to do.

Brits and Texans support sports differently. In Britain, the spectators tend to be males between 15 and 45 years old—definitely not a family atmosphere. But at Texas football games, you'll see grandparents with their grandkids and guys and gals on dates. At the high school level too, there is much wider participation: the band, cheerleaders, parents, fans. I played soccer for my high school and most Saturdays we'd have a crowd of about seven.

American teenagers

Street scene in Austin, Texas

Understanding Primary Sources

1. Why does Jon Hall think the people of London are more concerned about boundaries than Texans?

2. According to Jon Hall, what gives Texas football games a "family atmosphere?"

American buffalo

MORE FROM THE FIELD

Although the Texas plains may appear to be wide-open, fences now divide the land.

Early settlers in eastern Texas used ditches, mud fences, and hedges of the thorny bois d'arc tree, also known as the Osage orange, to fence off their property. Bois d'arc fences were said to be "pig tight, horse high, and bull strong." Dry western Texas had fewer fencing materials available, however, and westward expansion slowed as a result. In the 1870s barbed wire, consisting of thornlike barbs wrapped around smooth wires, was introduced. The invention was reportedly advertised as being "light as air, stronger than whiskey, and cheap as dirt."

Activity: Ask students to imagine that they are the first ranchers or farmers to settle in their area. What materials would they use if they wanted to build fences or mark their property?

Understanding Primary Sources
Answers

1. because so many people are crowded into a small space

2. families, people on dates, wider participation, better attendance

207

OVERVIEW

In this unit, students will learn about the tremendous geographical differences within the United States and Canada.

The landscape of the United States is immensely varied, from high mountains to flat coastal plains. There are also many climates represented. The diverse population reflects the many cultures that have played a role in the country's history. Although agriculture in the United States is highly productive, few people live on farms. Most of the population lives in cities and towns. Workers in the United States are part of the most technologically advanced economy in the world. The country's wealth is based on its many natural resources, free enterprise economy, and democratic government.

Many landforms extend from the United States into Canada, the second-largest country in the world. Our northern neighbor has a colder climate and a much smaller population. Most Canadians also enjoy a high standard of living. France and Great Britain influenced Canada's culture. Asians form another important immigrant group. The Inuit have recently gained greater control over their traditional homelands.

PEOPLE IN THE PROFILE

Note that the elevation profile crosses southeastern Pennsylvania. This farming region is known for its large Amish population.

The Amish, a conservative Christian group related to the Mennonites, were one of the many groups that left Europe for the Americas for religious freedom. The first large group of Amish to settle in Lancaster County, Pennsylvania, arrived in the early 1700s. Amish clothing styles echo the same era—dark colors, hooks and eyes instead of buttons, long skirts and bonnets for women, wide-brimmed black hats for men. The Amish do not use telephones, electric lights, or cars. They use horse-drawn farm machinery and buggies.

There are special customs for Amish weddings, which are held in November and December. Instead of an engagement ring, the young man gives his fiancée china or a clock. The bride's simple blue wedding dress will later serve as her Sunday church attire. On the wedding day, several hundred guests gather at the bride's home. After a three-hour service everyone sits down to a big dinner.

Activity: Have students research Amish culture and then lead a class discussion on some of the characteristics of Amish culture.

The United States and Canada

Elevation Profile

15,000 ft.	4,572 m
10,000 ft.	3,048 m
5,000 ft.	1,524 m
4,000 ft.	1,220 m
3,000 ft.	914 m
2,000 ft.	610 m
1,000 ft.	305 m
Sea Level	Sea Level

Rocky Mountains
Sierra Nevada
Coast Range
Pacific Ocean
Interior Plains
Allegheny Plateau
Appalachian Mountains
Atlantic Coastal Plain
Atlantic Ocean

Profile at 40° N latitude

Approximately 2,600 miles

The United States and Canada:
Comparing Sizes

GEOSTATS:

Highest point in North America: Mount McKinley, Alaska—20,320 ft. (6,194 m)

Lowest point in North America: Death Valley, California—282 ft. below sea level (86 m below sea level)

Largest lake in North America: Lake Superior—31,800 sq. mi. (82,362 sq km)

World's highest tides: Bay of Fundy, Nova Scotia, Canada—as high as 70 ft. (21 m)

Focus students' attention on the **physical map** on this page. Have each student write a sentence that describes a physical feature in terms of its relative location. *(Examples: The Labrador Peninsula is in eastern Canada, between Hudson Bay and the* Labrador Sea. The Rocky Mountains are in western North America. The Great Plains are east of the Rocky Mountains.) Call on volunteers to read their sentences. Continue until all the major physical features have been located.

United States and Canada: Physical

UNIT 3 ATLAS

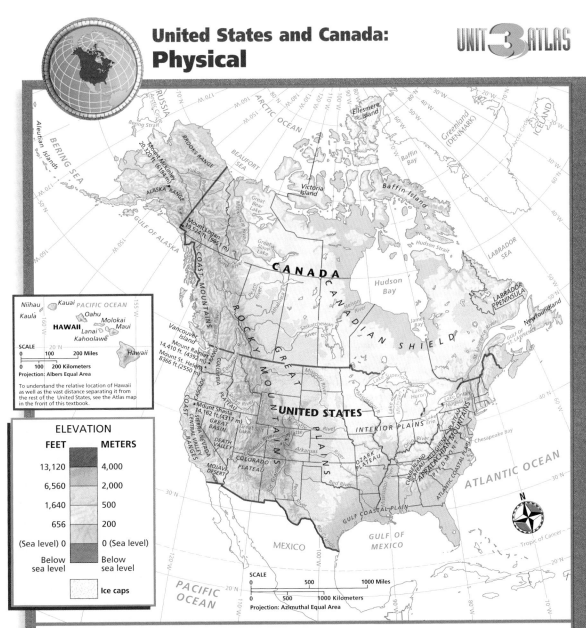

ELEVATION

FEET	METERS
13,120	4,000
6,560	2,000
1,640	500
656	200
(Sea level) 0	0 (Sea level)
Below sea level	Below sea level
Ice caps	

1. (Places and Regions) Where is the highest point in North America? Which southwestern desert includes an area below sea level?

2. (Places and Regions) Which two large lakes are entirely within Canada?

3. (Places and Regions) What are four tributaries of the Mississippi River?

Critical Thinking

4. **Drawing Inferences and Conclusions** Compare this map to the **population map** of the region. What is one physical feature that many of the largest cities have in common? What is the connection between the cities' locations and these physical features?

Physical Map
Answers
1. Mount McKinley in Alaska; Mojave Desert

2. Great Bear Lake, Great Slave Lake

3. Red River, Arkansas River, Missouri River, Ohio River

Critical Thinking
4. located on bodies of water; aids trade, transportation, industry, communication

USING THE POLITICAL MAP

 Have students use the **political map** on this page to locate their home state. Ask them to compare it to the **physical map** to identify their state's landforms, rivers, lakes, and seacoasts.

Ask students to name the capitals of the United States *(Washington, D.C.)* and Canada *(Ottawa).*

internet connect

ONLINE ATLAS
GO TO: go.hrw.com
KEYWORD: SJ3 MapsU3

Your Classroom Time Line

These are the major dates and time periods for this unit. Have students enter them on the time line. You may want to watch for these dates as students progress through the unit.

c.* 18,000 B.C. People first cross into North America.

c. A.D. 700 Anasazi develop irrigation system.

c. 1000 The first Europeans to attempt settlement in Canada arrive.

late 1400s European exploration of the North Atlantic coast of North America resumes.

1500s Europeans begin settling in North America.

*c. stands for *circa* and means "about."

Political Map

Answers

1. Rio Grande

Critical Thinking

2. Canadian; because much of the Mexican border runs through a desert

3. Nunavut; Quebec; because Nunavut includes many islands

United States and Canada: Political

UNIT 3 ATLAS

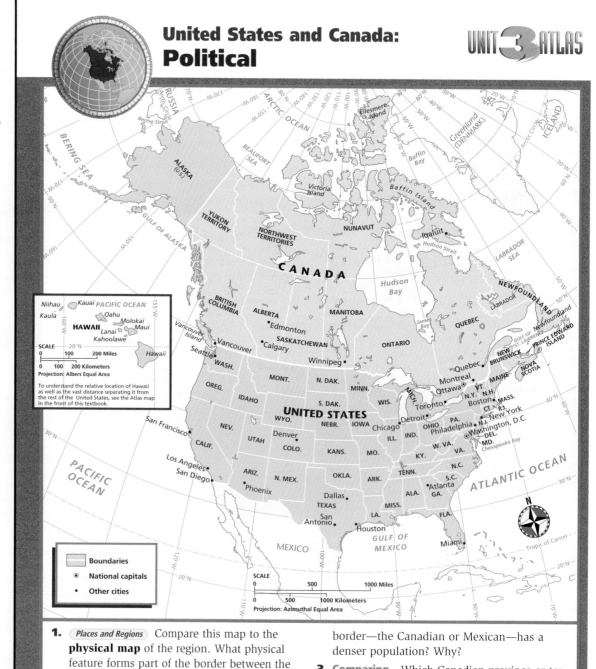

1. _Places and Regions_ Compare this map to the **physical map** of the region. What physical feature forms part of the border between the United States and Mexico?

Critical Thinking

2. Comparing Compare this map to the **climate** and **population maps**. Which U.S. border—the Canadian or Mexican—has a denser population? Why?

3. Comparing Which Canadian province or territory appears to be the largest? the second largest? Why is it hard to compare their sizes?

USING THE CLIMATE MAP

 Direct students' attention to the climate map of the United States and Canada on this page. Ask students what climate type covers most of Canada's territory *(subarctic)* and what climate regions of the United States are not found in Canada *(humid subtropical, Mediterranean, tropical savanna, desert, humid tropical).*

Have students compare this map to the **population map** on the next page. Call on volunteers to identify major cities and the climate regions in which they are located. *(Examples: Dallas—humid subtropical; Los Angeles—Mediterranean; Phoenix—desert; Toronto—humid continental; Vancouver—marine west coast)*

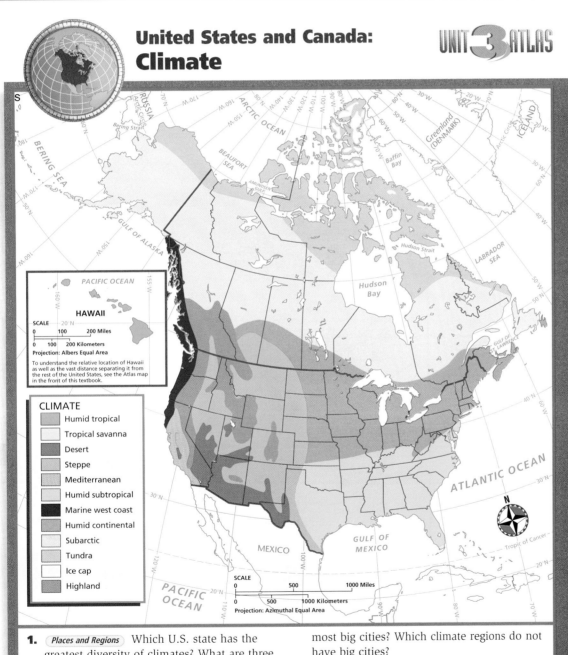

United States and Canada: Climate

UNIT 3 ATLAS

CLIMATE
- Humid tropical
- Tropical savanna
- Desert
- Steppe
- Mediterranean
- Humid subtropical
- Marine west coast
- Humid continental
- Subarctic
- Tundra
- Ice cap
- Highland

HAWAII

SCALE
0   100   200 Miles
0   100   200 Kilometers
Projection: Albers Equal Area

To understand the relative location of Hawaii as well as the vast distance separating it from the rest of the United States, see the Atlas map in the front of this textbook.

SCALE
0   500   1000 Miles
0   500   1000 Kilometers
Projection: Azimuthal Equal Area

1. (*Places and Regions*) Which U.S. state has the greatest diversity of climates? What are three states that have only one climate type?

2. (*Places and Regions*) Compare this map to the **physical** and **population maps** of the region. Which climate region contains the most big cities? Which climate regions do not have big cities?

Critical Thinking

3. Comparing Compare this map to the **physical map**. What may have limited settlement of the U.S. Great Plains?

Your Classroom Time Line (continued)

1600s The first enslaved Africans are brought to English colonies in North America.

1608 Quebec City is founded.

1756–63 The Seven Years' War, known in North America as the French and Indian War, is fought.

1776 The 13 American colonies declare independence from Great Britain.

1867 The United States buys Alaska from Russia.

1867 The British Parliament creates the Dominion of Canada.

1870 Manitoba becomes a province.

1885 The Canadian Pacific Railroad is completed.

Climate Map

Answers

1. California; any of the states that are one solid color

2. humid continental; ice cap, tundra, subarctic

Critical Thinking

3. lack of water

USING THE POPULATION MAP

While students examine the **population map** on this page, ask them to locate cities they have lived in, visited, or heard about from movies, television, or sports events. You might ask them to tell what they know about those cities.

Ask students which country has the largest area where practically no one lives *(Canada)*. Have students compare this map to the **political map** to determine which of the U.S. states appear to have the lowest overall population densities *(Alaska and Nevada)*.

internet connect

ONLINE ATLAS
GO TO: go.hrw.com
KEYWORD: SJ3 MapsU3

Your Classroom Time Line (continued)

1898 The United States annexes Hawaii.

1905 Alberta and Saskatchewan become provinces.

1917–18 The United States fights in World War I.

1941 Japan bombs Pearl Harbor, Hawaii. The United States enters World War II.

1945 World War II ends.

1949 Newfoundland becomes Canada's 10th province.

1990 The Cold War ends.

Population Map
Answers

1. United States

2. Toronto, Montreal

3. northeast coast; because it has the densest population

Critical Thinking

4. Charts, graphs, databases, and models should accurately reflect population figures for cities on the map.

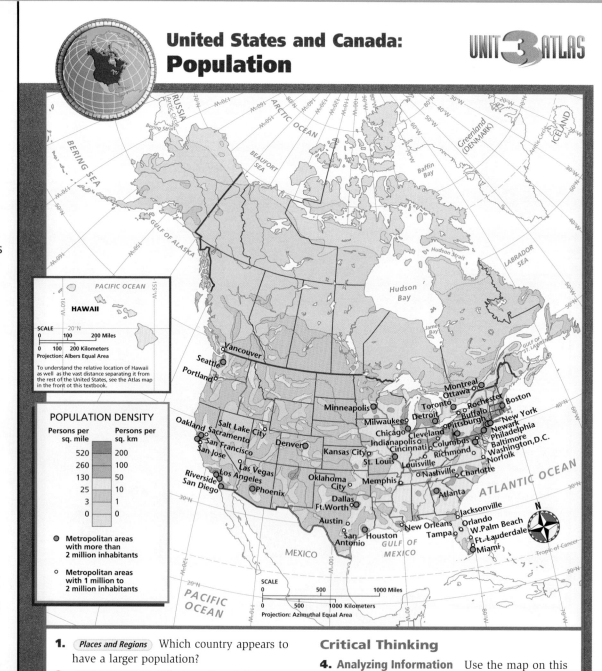

United States and Canada: Population

UNIT 3 ATLAS

POPULATION DENSITY

Persons per sq. mile	Persons per sq. km
520	200
260	100
130	50
25	10
3	1
0	0

⊙ Metropolitan areas with more than 2 million inhabitants

○ Metropolitan areas with 1 million to 2 million inhabitants

1. *Places and Regions* Which country appears to have a larger population?

2. *Places and Regions* What are Canada's two largest cities?

3. *Places and Regions* Based on the map, which area of the United States do you think was settled first? Why?

Critical Thinking

4. Analyzing Information Use the map on this page to create a chart, graph, database, or model of population centers in the United States and Canada.

USING THE LAND USE AND RESOURCES MAP

Have students locate the state in which they live on the **land use and resources map**. Tell them to identify the economic activities and resources that are found in their state. Ask students to use the map to identify economic patterns in the United States and Canada. Have each student write a sentence to summarize these patterns. Have students use other maps to help them write their sentences. *(Examples: There are more nuclear power plants in the eastern United States than in the western part of the country. Large parts of Canada have limited economic activity. Coal mining is common in the Appalachian Mountains.)*

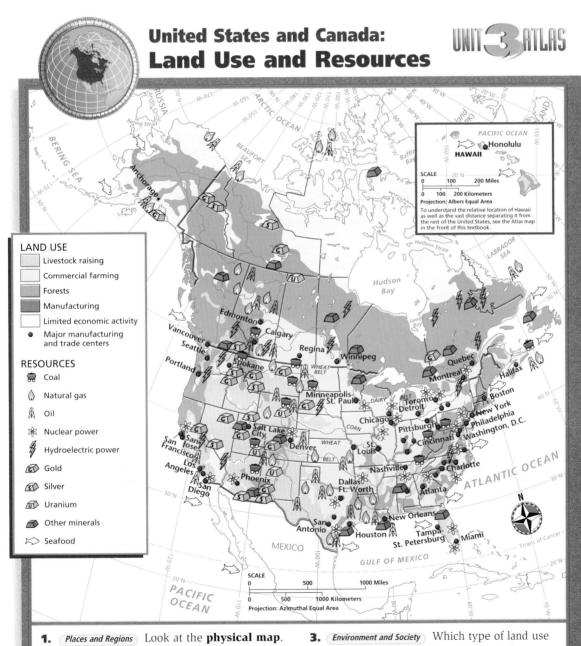

United States and Canada: Land Use and Resources

UNIT 3 ATLAS

LAND USE
- Livestock raising
- Commercial farming
- Forests
- Manufacturing
- Limited economic activity
- ● Major manufacturing and trade centers

RESOURCES
- Coal
- Natural gas
- Oil
- Nuclear power
- Hydroelectric power
- Gold
- Silver
- Uranium
- Other minerals
- Seafood

HAWAII — Honolulu — PACIFIC OCEAN

SCALE
0   100   200 Miles
0   100   200 Kilometers
Projection: Albers Equal Area

To understand the relative location of Hawaii as well as the vast distance separating it from the rest of the United States, see the Atlas map in the front of this textbook.

SCALE
0   500   1000 Miles
0   500   1000 Kilometers
Projection: Azimuthal Equal Area

1. *(Places and Regions)* Look at the **physical map**. In which area of North America are gold, silver, and uranium found?

2. *(Places and Regions)* Which two states on the Gulf of Mexico produce large amounts of oil and natural gas?

3. *(Environment and Society)* Which type of land use is most common throughout Canada?

Critical Thinking

4. **Analyzing Information** Create a chart, graph, database, or model of economic activities in the United States and Canada.

Land Use and Resources Map

Answers

1. in the west, concentrated in the Rocky Mountains

2. Texas and Louisiana

3. forests

Critical Thinking

4. Charts, graphs, databases, and models should accurately reflect infromation from the map.

UNITED STATES AND CANADA

FAST FACTS ACTIVITIES

LEVEL 1: (Suggested time: 30 min.) Call students' attention to the population figures for the United States and Canada. Have them calculate how many times larger the U.S. population is than Canada's. Round off to the nearest whole number. *(Divide the U.S. population total by Canada's population total. The result is about 8.8, which should be rounded up to 9.)*

Tell students the United States has about 74 people per square mile and Canada about 8 people per square mile. Then review the climate map at the beginning of this unit with students. Have them identify the main climate types in Canada *(subarctic and tundra)*. Ask students why Canada has a lower population density than the United States. *(Areas with subarctic and tundra climates are unsuitable for large permanent settlements.)*

LEVEL 2: (Suggested time: 30 min.) Organize the class into small groups. Draw students' attention to the nicknames listed under each state. Ask the groups to identify three states whose nicknames refer to historical events *(such as Alaska, Connecticut, Delaware)*, to physical characteristics *(such as Illinois, Michigan, South Dakota)*, and to climatic and economic conditions *(such as California, Florida, Maine)*.

The United States

UNITED STATES
CAPITAL: Washington, D.C.
AREA: 3,717,792 sq. mi. (9,629,091 sq km)
POPULATION: 281,421,906
MONEY: U.S. dollar (US$)
LANGUAGES: English, Spanish (spoken by a large minority)

Alabama
CAPITAL: Montgomery
NICKNAME: Heart of Dixie, Camellia State

Alaska
CAPITAL: Juneau
NICKNAME: The Last Frontier (unofficial)

Arizona
CAPITAL: Phoenix
NICKNAME: Grand Canyon State

Arkansas
CAPITAL: Little Rock
NICKNAME: Natural State, Razorback State

California
CAPITAL: Sacramento
NICKNAME: Golden State

Colorado
CAPITAL: Denver
NICKNAME: Centennial State

Connecticut
CAPITAL: Hartford
NICKNAME: Constitution State, Nutmeg State

Delaware
CAPITAL: Dover
NICKNAME: First State, Diamond State

Florida
CAPITAL: Tallahassee
NICKNAME: Sunshine State

Georgia
CAPITAL: Atlanta
NICKNAME: Peach State, Empire State of the South

Hawaii
CAPITAL: Honolulu
NICKNAME: Aloha State

Idaho
CAPITAL: Boise
NICKNAME: Gem State

Illinois
CAPITAL: Springfield
NICKNAME: Prairie State

Indiana
CAPITAL: Indianapolis
NICKNAME: Hoosier State

Iowa
CAPITAL: Des Moines
NICKNAME: Hawkeye State

Kansas
CAPITAL: Topeka
NICKNAME: Sunflower State

Kentucky
CAPITAL: Frankfort
NICKNAME: Bluegrass State

Louisiana
CAPITAL: Baton Rouge
NICKNAME: Pelican State

Maine
CAPITAL: Augusta
NICKNAME: Pine Tree State

Maryland
CAPITAL: Annapolis
NICKNAME: Old Line State, Free State

Massachusetts
CAPITAL: Boston
NICKNAME: Bay State, Old Colony

Michigan
CAPITAL: Lansing
NICKNAME: Great Lakes State, Wolverine State

Minnesota
CAPITAL: St. Paul
NICKNAME: Gopher State, North Star State

Mississippi
CAPITAL: Jackson
NICKNAME: Magnolia State

Missouri
CAPITAL: Jefferson City
NICKNAME: Show Me State

Montana
CAPITAL: Helena
NICKNAME: Treasure State

Nebraska
CAPITAL: Lincoln
NICKNAME: Cornhusker State

Nevada
CAPITAL: Carson City
NICKNAME: Sagebrush State, Battle Born State, Silver State

New Hampshire
CAPITAL: Concord
NICKNAME: Granite State

New Jersey
CAPITAL: Trenton
NICKNAME: Garden State

New Mexico
CAPITAL: Santa Fe
NICKNAME: Land of Enchantment

New York
CAPITAL: Albany
NICKNAME: Empire State

Countries, states, provinces, and territories not drawn to scale.

214

Challenge each group to create nicknames for the Canadian provinces. Nicknames should be based on characteristics similar to those that inspired the nicknames of the U.S. states. Have students use this unit's atlas for help. *(Example: British Columbia could be nicknamed The Pacific Province.)*

LEVEL 3: (Suggested time: 45 min.) Have students imagine that they work for the Canadian travel bureau. It is their job to create materials that tell travelers about the provinces and entice tourists to vacation there.

Assign each student a province or territory. Have each student create an illustrated travel poster for their province. Ask them to highlight the capital city and include a nickname for the province. Have students use this unit's atlas for information. *(Example: British Columbia—the Pacific Province. Come climb our mountains, camp our forests, and sail along the lovely Fraser River. Explore Vancouver, our capital city!)*

UNIT 3 ATLAS

North Carolina
CAPITAL: Raleigh
NICKNAME:
Tar Heel State, Old North State

Pennsylvania
CAPITAL: Harrisburg
NICKNAME:
Keystone State

Texas
CAPITAL: Austin
NICKNAME:
Lone Star State

North Dakota
CAPITAL: Bismarck
NICKNAME:
Peace Garden State

Rhode Island
CAPITAL: Providence
NICKNAME:
Little Rhody, Ocean State

Utah
CAPITAL: Salt Lake City
NICKNAME:
Beehive State

West Virginia
CAPITAL: Charleston
NICKNAME:
Mountain State

Ohio
CAPITAL: Columbus
NICKNAME:
Buckeye State

South Carolina
CAPITAL: Columbia
NICKNAME:
Palmetto State

Vermont
CAPITAL: Montpelier
NICKNAME:
Green Mountain State

Wisconsin
CAPITAL: Madison
NICKNAME:
Badger State

Oklahoma
CAPITAL: Oklahoma City
NICKNAME:
Sooner State

South Dakota
CAPITAL: Pierre
NICKNAME:
Coyote State, Mount Rushmore State

Virginia
CAPITAL: Richmond
NICKNAME:
Old Dominion

Wyoming
CAPITAL: Cheyenne
NICKNAME:
Equality State, Cowboy State

Oregon
CAPITAL: Salem
NICKNAME:
Beaver State

Tennessee
CAPITAL: Nashville
NICKNAME:
Volunteer State

Washington
CAPITAL: Olympia
NICKNAME:
Evergreen State

Canada

CANADA
CAPITAL: Ottawa
AREA:
3,890,695 sq. mi.
(9,976,140 sq km)
POPULATION: 31,592,805
MONEY:
Canadian dollar (Can$)
LANGUAGES:
English (official), French (official)

Nova Scotia
CAPITAL:
Halifax

Saskatchewan
CAPITAL:
Regina

Nunavut
CAPITAL:
Iqaluit

Yukon Territory
CAPITAL:
Whitehorse

Alberta
CAPITAL:
Edmonton

New Brunswick
CAPITAL:
Fredericton

Ontario
CAPITAL:
Toronto

British Columbia
CAPITAL:
Victoria

Newfoundland
CAPITAL:
St. John's

Prince Edward Island
CAPITAL:
Charlottetown

Manitoba
CAPITAL:
Winnipeg

Northwest Territories
CAPITAL:
Yellowknife

Quebec
CAPITAL:
Quebec

Sources: Central Intelligence Agency, *The World Factbook 1999; The World Almanac and Book of Facts 1999*

CHAPTER 10

The Geography and History of the United States

Chapter Resource Manager

Objectives	Pacing Guide	Reproducible Resources
SECTION 1		
Physical Geography (pp. 216–21)	**Regular** 1.5 days	**RS** Guided Reading Strategy 10.1
1. Describe the major physical features of the United States.		**RS** Graphic Organizer 10
2. Examine the climate regions found in the United States.	**Block Scheduling** .5 day	**SM** Geography for Life Activities 5, 6
3. Identify the natural resources of the United States.	*Block Scheduling Handbook, Chapter 10*	**IC** Interdisciplinary Activity for Middle Grades 10
SECTION 2		
The History of the United States (pp. 222–29)	**Regular** 2 days	**RS** Guided Reading Strategy 10.2
1. Explain how the early colonists from Europe changed North America.		**E** Creative Strategies for Teaching World Geography, Lesson 6
2. Describe ways the United States changed during the late 1700s and early 1800s.	**Block Scheduling** 1 day	**SM** Map Activity 10
3. Analyze how the United States has changed in the last 50 years.	*Block Scheduling Handbook, Chapter 10*	

Chapter Resource Key

RS Reading Support

IC Interdisciplinary Connections

E Enrichment

SM Skills Mastery

A Assessment

REV Review

ELL Reinforcement and English Language Learners

 Transparencies

 CD–ROM

 Music

 Video

 Internet

 Holt Presentation Maker Using Microsoft® Powerpoint®

 One-Stop Planner CD–ROM

See the *One-Stop Planner* for a complete list of additional resources for students and teachers.

One-Stop Planner CD–ROM

It's easy to plan lessons, select resources, and print out materials for your students when you use the *One-Stop Planner CD–ROM with Test Generator.*

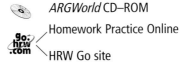

Technology Resources

- One-Stop Planner CD–ROM, Lesson 10.1
- *ARGWorld* CD–ROM
- Homework Practice Online
- HRW Go site

Review, Reinforcement, and Assessment Resources

ELL	Main Idea Activity 10.1
REV	Section 1 Review, p. 221
A	Daily Quiz 10.1
ELL	English Audio Summary 10.1
ELL	Spanish Audio Summary 10.1

- One-Stop Planner CD–ROM, Lesson 10.2
- *ARGWorld* CD–ROM
- Homework Practice Online
- HRW Go site

ELL	Main Idea Activity 10.2
REV	Section 2 Review, p. 229
A	Daily Quiz 10.2
ELL	English Audio Summary 10.2
ELL	Spanish Audio Summary 10.2

internet connect

HRW ONLINE RESOURCES

GO TO: go.hrw.com
Then type in a keyword.

TEACHER HOME PAGE
KEYWORD: SJ3 TEACHER

CHAPTER INTERNET ACTIVITIES
KEYWORD: SJ3 GT10

Choose an activity to:
- Experience life in the early colonies.
- Search letters and journal entries written by soldiers in the Civil War.
- Explore the advance of civil rights in the United States.

CHAPTER ENRICHMENT LINKS
KEYWORD: SJ3 CH10

CHAPTER MAPS
KEYWORD: SJ3 MAPS10

ONLINE ASSESSMENT
Homework Practice
KEYWORD: SJ3 HP10
Standardized Test Prep Online
KEYWORD: SJ3 STP10
Rubrics
KEYWORD: SS Rubrics

COUNTRY INFORMATION
KEYWORD: SJ3 Almanac

CONTENT UPDATES
KEYWORD: SS Content Updates

HOLT PRESENTATION MAKER
KEYWORD: SJ3 PPT10

ONLINE READING SUPPORT
KEYWORD: SS Strategies

CURRENT EVENTS
KEYWORD: S3 Current Events

Meeting Individual Needs

Ability Levels

Level 1 Basic-level activities designed for all students encountering new material

Level 2 Intermediate-level activities designed for average students

Level 3 Challenging activities designed for honors and gifted-and-talented students

English Language Learners Activities that address the needs of students with Limited English Proficiency

Chapter Review and Assessment

E	Readings in World Geography, History, and Culture 11, 12
SM	Critical Thinking Activity 10
REV	Chapter 10 Review, pp. 232–33
REV	Chapter 10 Tutorial for Students, Parents, Mentors, and Peers
ELL	Vocabulary Activity 10
A	Chapter 10 Test
A	Unit 3 Test
	Chapter 10 Test Generator (on the One-Stop Planner)
	Audio CD Program, Chapter 10
A	Chapter 10 Test for English Language Learners and Special-Needs Students
A	Unit 3 Test for English Language Learners and Special-Needs Students

LAUNCH INTO LEARNING

Provide a small group of volunteers with the lyrics to "America the Beautiful," and ask them to sing the first verse. As students sing, write key phrases such as *amber waves of grain, purple mountain majesties, fruited plain* on the chalkboard. Ask students what these phrases mean and what they tell us about our country. *(Possible answer: The United States has fertile farmlands and high mountains.)* Tell students they will learn more about our country's physical features and history in this chapter.

Section 1

Objectives

1. Describe the major physical features of the United States.
2. Examine the climate regions found in the United States.
3. Identify the natural resources of the United States.

LINKS TO OUR LIVES

You might want to share with your students the following reasons for learning about the geography and history of the United States:

▶ Economic development and population movement are both affected by geographical features.

▶ Some of the issues that led to the founding of this country, such as the proper role of government, are still being debated today.

▶ Many of the trends and patterns that have appeared throughout the history of the United States are still evident in our country today.

▶ We can be more effective citizens if we understand the physical and human geography of our country.

CHAPTER 10

The Geography and History of the United States

Many different groups have settled in the United States. Here you will read about one of the first of these groups, the Pilgrims.

Puritan couple dressed for church

In November 1620 a small ship sailing from England was blown off course into a harbor near what is now Provincetown, Massachusetts. It was the *Mayflower*, which had been bound for Virginia. The weary voyagers "fell upon their knees and blessed the God of Heaven, who had brought them over the vast and furious ocean," wrote William Bradford, a leader of the group.

The voyagers had indeed been over a "furious ocean." For more than two months, the *Mayflower* had been blown about by violent storms. The passengers were sick and weak, and they had landed far north of their destination at the wrong time of year. The "howling wilderness" terrified many of them, and the weather was cold and getting colder. While their shipmates huddled on board, some of the passengers searched for a place to settle. By mid-December they had found a site—Plymouth, named after the English port from which they had sailed.

Pilgrims landing at Plymouth Rock

Early American colonial settlement

216

LET'S GET STARTED

Copy the following questions onto the chalkboard: *How would you describe the land and climate of your hometown? Is it flat, hilly, or mountainous? Is the climate cold, hot, dry, or rainy?* Ask students to respond in writing and discuss their responses. Tell students that in Section 1 they will learn about the physical geography of the United States.

Building Vocabulary

Write the key terms on the chalkboard. Have students find definitions for them in Section 1 or in the glossary. Call on students to read the definitions aloud to the class. Then ask volunteers to use a wall map of the United States as a visual aid to explain the meaning of **contiguous** and **Continental Divide**. Have other volunteers find examples of **basins**.

Using the Physical-Political Map

Have students examine the map on this page. Ask volunteers to identify the principal landforms and bodies of water. Have students locate the areas where they live and name other areas they have visited. Then have students compare the physical and political features of the areas mentioned.

Section 1 — Physical Geography

Read to Discover

1. What are the major physical features of the United States?
2. What climate regions are found in the United States?
3. What natural resources does the United States have?

Define

contiguous
Continental Divide
basins

Locate

Coastal Plain
Appalachian Mountains
Interior Plains
Rocky Mountains
Great Lakes
Mississippi River
Great Plains
Columbia River
Great Basin
Colorado Plateau
Sierra Nevada
Cascade Range
Aleutian Islands

WHY IT MATTERS

Some of the U.S. states are at risk for earthquakes. Use **CNNfyi.com** or other **current events** sources to find out about earthquakes in the United States. Record your findings in your journal.

Bald eagle

Section 1 RESOURCES

Reproducible
- Block Scheduling Handbook, Chapter 10
- Guided Reading Strategy 10.1
- Geography for Life Activities 5, 10
- Interdisciplinary Activity for Middle Grades 6

Technology
- One-Stop Planner CD–ROM, Lesson 10.1
- Homework Practice Online
- Geography and Cultures Visual Resources with Teaching Activities 11–17
- HRW Go site

Reinforcement, Review, and Assessment
- Section 1 Review, p. 221
- Daily Quiz 10.1
- Main Idea Activity 10.1
- English Audio Summary 10.1
- Spanish Audio Summary 10.1

The United States: Physical-Political

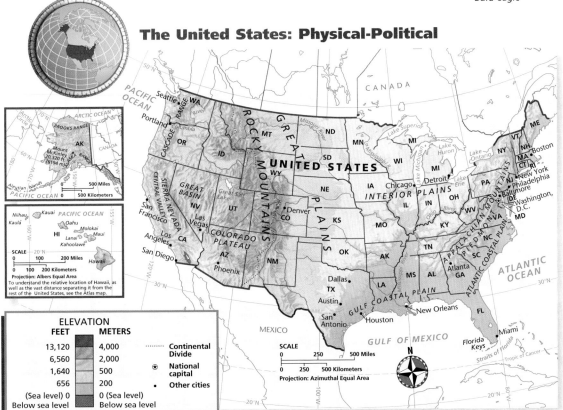

ELEVATION

FEET	METERS
13,120	4,000
6,560	2,000
1,640	500
656	200
(Sea level) 0	0 (Sea level)
Below sea level	Below sea level

········ Continental Divide
⊛ National capital
• Other cities

SCALE
0   250   500 Miles
0   250   500 Kilometers
Projection: Azimuthal Equal Area

TEACH

Teaching Objective 1

LEVEL 1: (Suggested time: 15 min.) Ask students to identify types of landforms. *(Possible answers: mountains, plains, and rivers)* Write these types on the chalkboard and then have volunteers list—under each landform category—specific examples of these physical features that can be found in the United States. Tell students to write a paragraph summarizing the information in their notebooks.
ENGLISH LANGUAGE LEARNERS

LEVELS 2 AND 3: (Suggested time: 30 min.) Pair students and have each pair sketch an elevation profile of the United States from New York to San Francisco. You may want to have some students sketch other elevation profiles such as along specific lines of latitude. Students should label the mountain ranges, rivers, lakes, plains, and other major physical features on their elevation profiles. **COOPERATIVE LEARNING**

National Geography Standard 15

Environment and Society
Although most people associate earthquakes with California, several significant earthquakes have shaken the Interior Plains. On December 16, 1811, the first of three magnitude 8 earthquakes to rock the central Mississippi River valley awakened the residents of New Madrid, in the Missouri Territory. No recorded earthquake has ever exceeded a value of 9 on the Richter scale. Thousands of aftershocks followed during the winter of 1811–12. The greatest quake was felt as far away as Chicago, New Orleans, Boston, and even parts of Canada. Although the land heaved, buckled, and cracked, few people were killed because the region was sparsely populated.

Today, that region is home to millions of people. Some scientists worry that the next big earthquake may occur there.

internet connect

GO TO: go.hrw.com
KEYWORD: SJ3 CH10
FOR: Web sites about
earthquakes

Visual Record Answer ▶

Students might note the smooth, gentle slopes, which appear to be heavily eroded.

218

internet connect

GO TO: go.hrw.com
KEYWORD: SJ3 CH10
FOR: Web sites about the
United States

The Great Smoky Mountains of Tennessee are a range of the Appalachian Mountains.

Interpreting the Visual Record
What clues from this photograph might tell you that these mountains are very old?

▼

Physical Features

The 48 **contiguous** American states and the District of Columbia lie between the Atlantic and Pacific Oceans. Contiguous states are those that border each other. Two states are not contiguous: Alaska, to the northwest of Canada, and Hawaii, in the Pacific Ocean. The United States also has territories in the Pacific Ocean and the Caribbean Sea. We will now look at the physical features of the 50 states. You can use the map on the next page to follow along.

The East The eastern United States rises from the Coastal Plain to the Appalachian Mountains. The Coastal Plain is a low region that lies close to sea level. It rises gradually inland. The Coastal Plain stretches from New York to Mexico along the Atlantic and Gulf of Mexico coasts.

The Appalachians include mountain ranges and river valleys from Maine to Alabama. The mountains are very old. Erosion has lowered and smoothed the peaks for more than 300 million years. The highest mountain in the Appalachians rises to just 6,684 feet (2,037 m).

At the foot of the Appalachians, between the mountains and the Coastal Plain, is the Piedmont. The Piedmont is a region of rolling plains. It begins in New Jersey and extends as far south as Alabama.

The Interior Plains Vast plains make up most of the United States west of the Appalachians. This region is called the Interior Plains. The Interior Plains stretch westward to the Rocky Mountains.

After the last ice age, glaciers shrank. They left rolling hills, lakes, and major river systems in the northern Interior Plains. The Great Lakes were created by these retreating ice sheets. From west to east, the Great Lakes are Lake Superior, Lake Michigan, Lake Huron, Lake Erie, and Lake Ontario.

The Mississippi River and its tributaries drain the Interior Plains. Along the way, they deposit rich soils that produce fertile farmlands. A tributary is a stream or river that flows into a larger stream or river. Tributaries of the Mississippi include the Missouri and Ohio Rivers.

Teaching Objective 2

LEVELS 1 AND 2: (Suggested time: 20 min.) Copy the following graphic organizer onto the chalkboard, omitting the italicized answers. Use it to help students examine climate regions in the United States. Have students supply words and phrases describing the climate regions and use their descriptions to complete the organizer. Students should copy the chart into their notebooks. **ENGLISH LANGUAGE LEARNERS**

LEVEL 3: (Suggested time: 30 min.) Pair students and tell each pair to prepare a weather forecast for a television station in one of the following cities: Buffalo, New York; Chicago, Illinois; Denver, Colorado; Fairbanks, Alaska; Los Angeles, California; Miami, Florida; Phoenix, Arizona; or Seattle, Washington. Have pairs present their forecasts to the class, using props, maps, and graphics. **COOPERATIVE LEARNING**

Climate Regions in the United States				
The East	**The Interior**	**The West**	**Hawaii**	**Alaska**
humid continental, humid subtropical	*humid continental, humid subtropical, steppe*	*steppe, highland, desert, marine west coast, Mediterranean*	*tropical, tropical savanna*	*subarctic, tundra*

Physical Regions of the United States and Canada

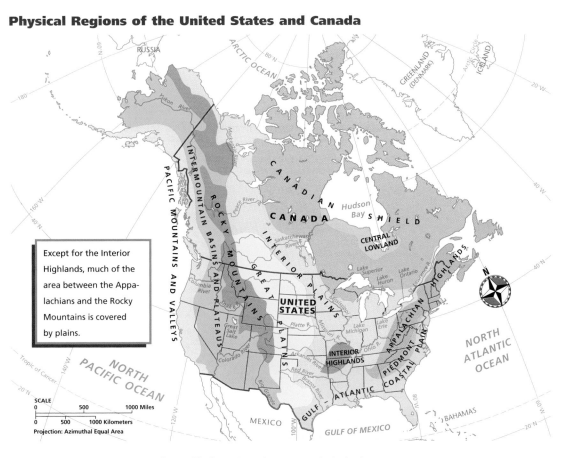

Except for the Interior Highlands, much of the area between the Appalachians and the Rocky Mountains is covered by plains.

In some places on its way to the Gulf of Mexico, the great Mississippi River is 1.5 miles (2.4 km) wide!

The flattest part of the Interior Plains is the Great Plains region. It lies closest to the Rocky Mountains. The region has a higher elevation than the rest of the Interior Plains. The Great Plains extend from Mexico in the south into Canada in the north.

The West As we continue westward, we reach the Rocky Mountains. The Rockies include a series of mountain ranges separated by high plains and valleys. They extend from Mexico to the cold Arctic. Many Rocky Mountain peaks reach more than 14,000 feet (4,267 m).

Running along the crest of the Rockies is the **Continental Divide**. It divides the flow of North America's rivers. Rivers east of the divide, such as the Missouri, flow eastward. Rivers west of the divide, such as the Columbia, flow westward.

West of the Rockies is a region of plateaus and **basins**. A basin is a region surrounded by higher land, such as mountains. The Great Basin in Nevada and Utah, for example, is surrounded by high mountains. Southeast of the Great Basin is

The Rio Grande cuts through Big Bend National Park along the Texas-Mexico border.

Interpreting the Visual Record What landforms are found in this area?

Teaching Objective 3

ALL LEVELS: (Suggested time: 30 min.) Tell students to imagine that they are economists who are representing the United States at a foreign trade conference. They must give a presentation on the natural resources of the United States. *(Presentations should discuss farmlands and their products, coal and other minerals, forests, and natural beauty.)* Students' presentations should also discuss how these natural resources contribute to the U.S. economy. Reintroduce the concepts of primary, secondary, tertiary, and quaternary industries to students and ask them to refer to these concepts and give examples of them in their presentations.

ENGLISH LANGUAGE LEARNERS

CLOSE

Have a volunteer point to random places on a wall map of the United States. Ask the class to describe the physical features, climate(s), and natural resources of those locations. Ask students how these factors would affect the daily lives and economic activities of people living there.

EYE ON EARTH

One of the most unique physical features of the United States is the Everglades. The Everglades, a marshland, once covered almost 9 million acres (3.6 million hectares) in southern Florida. During the last 100 years much of the land has been drained and used for residential development and farming. The Everglades National Park includes about one sixth of the original marshland.

The Everglades ecosystem depends on a huge supply of freshwater. Excess rainwater used to overflow the southern rim of Lake Okeechobee in south-central Florida and then travel in a broad, shallow sheet of water through the Everglades to the Gulf of Mexico. Much of the water is now redirected to the large farms and population centers of South Florida. As a result, the ecosystem has suffered. Of the species that live in the Everglades, more than 50 are endangered or threatened. These include the American crocodile and the Florida panther. A plan proposed in 1999 to restore a supply of freshwater to the Everglades may take decades to implement and cost more than $7 billion.

Do you remember what you learned about tectonic activity? See Chapter 2 to review.

Ⓐ Winter storm in Massachusetts
Ⓑ Spring meadow in California
Ⓒ Summer in Florida
Ⓓ Fall in Colorado

the Colorado Plateau. The Colorado Plateau lies in Utah, Colorado, New Mexico, and Arizona. West of the Great Basin lie the Sierra Nevada, the Coast Ranges, and many valleys. The Cascade Range in the Pacific Northwest was formed by volcanic eruptions. Volcanic eruptions and earthquakes are a danger in the Pacific states. In this area, the North American plate is colliding with the Pacific plate.

Hawaii and Alaska Farther west, Hawaii and Alaska also experience tectonic activity. The Hawaiian Islands were formed by large volcanoes that have risen from the floor of the Pacific Ocean. Alaska's Aleutian (uh-LOO-shuhn) Islands also have volcanic origins.

Southeastern and south-central Alaska are very mountainous. The highest mountain in North America is Mount McKinley, in the Alaska Range. The American Indian name for Mount McKinley is Denali. It soars to a height of 20,320 feet (6,194 m).

✔ **READING CHECK:** *Places and Regions* What are the major physical features of the United States? plains, mountain ranges, rivers, river valleys, lakes, Continental Divide, plateaus, basins

Climate

The climates of the United States are varied. The country has 11 climate types—the greatest variety of climates of any country.

The East Most of the eastern United States is divided into two climate regions. In the north is a humid continental climate with snowy winters and warm, humid summers. Southerners experience the milder winters and warm, humid summers of a humid subtropical climate. Coastal areas often experience tropical storms. Southern Florida, with a tropical savanna climate, is warm all year.

The Interior Plains Humid continental, humid subtropical, and steppe climates meet in the interior states. People there sometimes experience violent weather, such as hail and tornadoes. The steppe climate of the Great Plains supports wide grasslands. Summers are hot, and droughts can be a problem. Blinding snowstorms called blizzards sometimes occur during winters, which can be very cold.

The West Climates in the West are mostly dry. Steppe and varied highland climates dominate much of the Rocky Mountain region. Temperatures and the amount of rain and snow vary.

Have students complete the Section Review. Then tell each student to create a crossword puzzle using 10 terms. The terms and their clues should cover the main ideas of the section. Pair students and have them solve each other's puzzles. Then have students complete Daily Quiz 10.1. **COOPERATIVE LEARNING**

RETEACH

Have students complete Main Idea Activity 10.1. Then organize students into three groups. Assign each group one of the following aspects of the United States: physical features, climates, or natural resources. Have groups produce skits or infomercials on their topics. Groups should write a script and provide some visual resources. Have the groups present their skits or infomercials to the class. **ENGLISH LANGUAGE LEARNERS, COOPERATIVE LEARNING**

EXTEND

Have interested students conduct research on the landforms, rivers, climates, and unique attractions of a national park. Ask students to imagine that they are park rangers and have them create a brief orientation on the park's attractions to present to a group of tourists. **BLOCK SCHEDULING**

Much of the Southwest has a desert climate. The West Coast has two climate types. The forested north has a wet and mild marine west coast climate. A drier Mediterranean climate is found in the south.

Except in the southeast, most of Alaska has very cold subarctic and tundra climates. Hawaii is the only state within the tropics. Northeasterly trade winds bring rain to eastern sides of the islands. Hawaii's western slopes have a drier tropical savanna climate.

✓ **READING CHECK:** *The World in Spatial Terms* What climate regions are found in the United States? humid continental, humid subtropical, tropical savanna, steppe, varied highland, desert, marine west coast, Mediterranean, subarctic, tundra, tropical, tropical savanna

Natural Resources

The United States has many resources. Some of the most productive farmlands in the world are found in the Interior Plains. Ranches and farms produce beef, wheat, corn, and soybeans. California, Florida, Texas, and other areas grow fruit, vegetables, and cotton.

Alaska, California, Texas, and other states supply oil and natural gas. Coal and other minerals are found in Appalachian and western states. Gold and silver mines also operate in some western states.

Forests, especially in the Northwest and the Southeast, are important sources of lumber. The Atlantic Ocean, Gulf of Mexico, and Pacific Ocean are rich sources of fish and other seafood. The natural beauty of the country is also a valuable resource for tourism. All these resources help support industry and other economic activities.

✓ **READING CHECK:** *Places and Regions* What natural resources does the United States have? productive farm and ranch land, oil and natural gas, minerals, precious metals, forests, seafood

Our Amazing Planet

Redwoods, the tallest trees in the world, grow in California and Oregon. They can grow well over 300 feet (90 m) high with trunks 20 feet (6 m) in diameter. Some redwood trees are more than 1,500 years old!

Section Review 1

Define and explain: contiguous, Continental Divide, basins

Working with Sketch Maps On a map of the United States that you draw or that your teacher provides, label the following: Coastal Plain, Appalachian Mountains, Interior Plains, Rocky Mountains, Great Lakes, Mississippi River, Great Plains, Columbia River, Great Basin, Colorado Plateau, Sierra Nevada, Cascade Range, and Aleutian Islands.

Reading for the Main Idea

1. *Places and Regions* What are the two major mountain regions in the United States? What major landform region lies betweeen them?

2. *Places and Regions* What resources are found in the country? Why are they important for the economy?

3. *Places and Regions* What parts of the United States have particularly rich farmlands?

Critical Thinking

4. **Drawing Inferences and Conclusions** The Rockies are higher than the Appalachians. How is this fact a clue to the relative age of the two mountain systems?

Organizing What You Know

5. **Categorizing** Copy the following graphic organizer. Use it to categorize the major physical features and climates of the East, Interior Plains, and West.

	Physical Features	Climates
East		
Interior Plains		
West (including Alaska, and Hawaii)		

go. hrw .com **Homework Practice Online**

Keyword: SJ3 HP10

Section Review 1

Answers

Define For definitions, see: contiguous, p. 218; Continental Divide, p. 219; basins, p. 219

Working with Sketch Maps Maps will vary, but listed places should be labeled in their approximate locations.

Reading for the Main Idea

1. the Rocky and Appalachian Mountains; the Interior Plains (NGS 4)

2. farmland, oil, natural gas, coal, minerals, forests, seafood, natural beauty; pump money into the U.S. economy (NGS 4)

3. the Interior Plains (NGS 4)

Critical Thinking

4. The Rockies are younger because they have not been eroded as much as the Appalachians.

Organizing What You Know

5. East—mountain ranges, river valleys, and rolling plains; humid continental, humid subtropical, and tropical savanna; Interior Plains—plains, hills, lakes, and major river systems; humid continental, humid subtropical, and steppe; West—mountain ranges, plateaus, basins, valleys, plains, and hills; steppe, varied highlands, desert, marine west coast, Mediterranean, subarctic, tundra, and tropical

Section 2

Objectives

1. Explain how the early colonists from Europe changed North America.

2. Describe ways the United States changed during the late 1700s and early 1800s.

3. Analyze how the United States has changed in the last 50 years.

FOCUS

LET'S GET STARTED

Write the following on the board: *Name five events that you think were important in United States history.* Discuss student responses. As students name events, write them on the chalkboard. Keep a list of these events to refer to at the end of the section. Tell students that in Section 2 they will learn more about the people and events that helped shape the United States we live in today.

Building Vocabulary

Write the key terms on the board. Call on volunteers to locate the definitions in the text and read them aloud. Then ask students to write sentences that connect two of the terms. *(Example: **Abolitionists** were probably pleased by the **Emancipation Proclamation**.)* Call on students to share their sentences with the class.

Section 2 RESOURCES

Reproducible

◆ Guided Reading Strategy 10.2
◆ Creative Strategies for Teaching World Geography, Lesson 6
◆ Map Activity 10

Technology

◆ One-Stop Planner CD-ROM, Lesson 10.2
◆ Homework Practice Online
◆ HRW Go site

Reinforcement, Review, and Assessment

◆ Main Idea Activity 10.2
◆ Section 2 Review, p. 229
◆ Daily Quiz 10.2
◆ English Audio Summary 10.2
◆ Spanish Audio Summary 10.2

Section 2 The History of the United States

Read to Discover

1. How did early colonists from Europe change North America?

2. How did the United States change during the late 1700s and early 1800s?

3. How has the United States changed in the last 50 years?

Define

plantations
frontier
ratified
pioneers
abolitionist
secede
Emancipation
 Proclamation
emigrating

WHY IT MATTERS

The idea that people have right to decide their own government is still at work today. Use **CNNfyi.com** or other current events sources to find out about how people in the United States take part in the government. Record your findings in your journal.

An American quilt from the 1800s

Colonizing North America

You know that the history of the United States is connected to the history of Europe and many other countries in the world. You have already learned some early U.S. history. However, there is much more to know about how our land became the great country it is today. Our history has been made by people from many different cultures. It is filled with earth-shattering events and ordinary, everyday happenings. It is a story of great discoveries, courageous journeys, sweeping political movements, and constant change.

The Indians of North America The first people who shaped the land we live on were various American Indian tribes. They had lived in what is now the United States for thousands of years before Europeans arrived. From the Atlantic coast to the Pacific coast, American Indians developed ways of life that were strongly influenced by their environments. For examples, the Iroquois hunted in the Woodlands of the Northeast. The Natchez raised crops in the rich soil of the Southeast. The Lakota Sioux and other Plains people were both buffalo hunters and farmers. In the Southwest, the Navajo and Apache mostly hunted and herded animals while the Hopi and the Zuni (zoo-nee) mostly farmed. The Tlingit (TLING-kut) of Alaska fished.

▲
Plains Indians hunted buffalo for food, and used the skins to make clothing and cover their teepees.

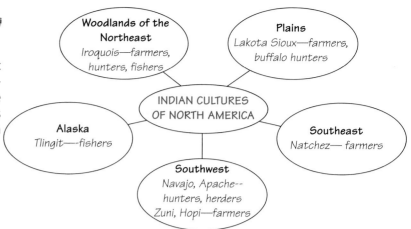

Woodlands of the Northeast
Iroquois—farmers, hunters, fishers

Plains
Lakota Sioux—farmers, buffalo hunters

INDIAN CULTURES OF NORTH AMERICA

Alaska
Tlingit—-fishers

Southeast
Natchez— farmers

Southwest
Navajo, Apache-- hunters, herders Zuni, Hopi—farmers

The New Colonists In the 1500s Europeans from Spain, France, the Netherlands, and other countries began to travel to the Americas. During the 1600s the English also began claiming lands and establishing colonies. In 1607 they established Jamestown in what is now Virginia. It was the first permanent English settlement in America. In 1620, settlers founded Plymouth in what is now Massachusetts. Sometimes the settlers made friends with the Indians living there. More often Indians and settlers fought.

Overland travel was difficult in the early colonies. For a long time water transportation was the colonists' main link to the outside world. In fact nearly all early colonial settlements were ports located on natural harbors or navigable rivers. New settlers migrated by sea to the growing coastal towns and inland trading posts on rivers.

Enslaved Africans In the early 1600s thousands of enslaved Africans were brought to North America. Most were brought to the southern colonies to work on **plantations**. A plantation is a large farm that grows mainly one crop to sell. Plantations were common in the southern colonies because of the area's rich soils and mild climates. Most colonial plantations produced large crops of cotton or tobacco.

A much larger group of enslaved Africans were brought to the islands of the West Indies in the Caribbean Sea and to South America. They were forced to work on large sugarcane plantations. Thousands of Africans died during the passage across the Atlantic.

▲ Thousands of enslaved Africans were brought to colonial Jamestown in Dutch ships like the one shown in this painting.

◄ This painting shows a southern plantation as a small community.
Interpreting the Visual Record How do you think products from this plantation were sent to markets?

▲ **Visual Record Answer**

by ship or boat

LEVEL 2: (Suggested time: 45 min.) Organize the class into pairs. Have each pair choose one of the original 13 colonies and prepare a short report that describes it. These reports should note the American Indians who first lived in the area, the Europeans who first settled the colony, and what daily life was like there. Encourage students to describe how life changed in the region over time. **COOPERATIVE LEARNING**

LEVEL 3: Have each student write a short magazine article that describes an event from colonial America. Tell students that they should also make notes about what kinds of pictures they would include with their articles. Have students combine their articles into a class magazine about colonial history.

Cultural Kaleidoscope
Sybil Ludington's

Ride Many students are probably familiar with the story of Paul Revere's ride. Less well-known than Revere but equally courageous was a 16-year-old girl named Sybil Ludington, daughter of Colonel Henry Ludington of the New York Militia. On the night of April 26, 1777, Sybil rode 40 miles through the besieged countryside around Carmel, New York, to awaken her father's regiment. By dawn the soldiers had assembled and were on their way to halt a British raid on nearby Danbury, Connecticut.

Activity: Have students use a road atlas to locate the villages of Carmel and Ludingtonville—named for Sybil's family—in what is now Putnam County, New York. Have them also locate Danbury, Connecticut, and trace the route the militia may have taken from Ludingtonville to Danbury.

Visual Record Answer ▶

became angry, united against the British

▲

This painting shows a scene of the Boston Massacre, March 5, 1770. British troops in Boston opened fire on an angry crowd, killing several.

Interpreting the Visual Record How do you think the people of Boston reacted to the Boston Massacre?

The 13 Colonies

Life in the Colonies By the mid-1700s many British citizens had settled in 13 colonies along the Atlantic coast. Some came in search of wealth. Others came seeking religious freedom. Settlers had also come from Scotland, Ireland, Germany, France, Sweden, and other countries. They contributed greatly to the growth of the colonies. Many individuals owned and worked small farms, but the climate and soils in the southern colonies were ideal for a large-scale agricultural economy. The colonies in the north—the New England and Middle colonies—became centers for trade, shipbuilding, and fishing. Cities such as Boston and New York became major seaports.

Farming, trading, and fishing were only a few ways the colonists earned a living. There were also craftsworkers, such as carpenters and blacksmiths, who produced a variety of goods. As colonial economies grew, colonists did not need to import as many goods from England.

Most of the people who came to America settled along the coast. As these areas became more crowded people began to move to the **frontier**. The frontier referred to land to the west of the colonies that was not already settled by the Europeans. These unsettled lands, however, were not empty. They were Indian lands.

The British were not alone in taking over Indian lands. The French and Spanish also claimed lands that had long been home to American Indians. The French had established a fur-trading business with some Indians, and many Indian tribes remained allies of the French for many years.

Trouble Brewing At the same time, British colonists were moving onto lands claimed by the French. The British and French had already been fighting wars in Europe for years. In 1754 the conflict spilled over to North America in what was called the French and Indian War. The British colonists fought not only the French but also their Indian allies.

By 1763 Great Britain had won the war. However, it had been a very expensive war. To raise money, the British Parliament began to impose new restrictions on the colonists and establish new taxes. The colonists said that Parliament had no right to tax them without the consent of their own colonial assemblies. They called this "taxation without representation." Some refused to buy British goods.

A worried Parliament sent British soldiers to Boston to control the protests. Each new British act only made the colonies work more closely together. Slowly the 13 colonies were becoming united.

✓ **READING CHECK:** *Summarizing* What was colonial life like during the 1600s and early 1700s? difficult to travel; needed to work hard; work reflected environment; no interference by British Parliament until taxes imposed unfairly on the colonists; colonists eventually rebelled, declared independence

Teaching Objective 2

ALL LEVELS: (Suggested time: 20 min.) Organize the class into small groups. Have each group create an illustrated time line that shows the most important events and developments in the history of the United States from the late 1700s through the early 1800s. Have groups compare their completed time lines and discuss any discrepancies in the events that appear on them. Display the completed time lines around the classroom. **ENGLISH LANGUAGE LEARNERS, COOPERATIVE LEARNING**

LEVEL 1: (Suggested time: 20 min.) Ask each student to write a few sentences that explain how each of the following might have felt about the election of Abraham Lincoln as President: a northern abolitionist, a plantation owner in South Carolina, an enslaved African American in Virginia. Then lead a class discussion on the events following Lincoln's election that led to the Civil War. **ENGLISH LANGUAGE LEARNERS**

American Independence

In April 1775, British troops near Boston tried to capture gunpowder and weapons stored by the colonists. Fighting broke out near the towns of Lexington and Concord. The American Revolution had begun.

The Declaration of Independence was signed on July 4, 1776, but the American Revolution dragged on for five years. Eventually, however, George Washington was able to defeat British general Cornwallis and get him to surrender in 1781. This battle, the Battle of Yorktown, was the last major battle of the Revolution.

In 1787 the Americans wrote a new Constitution that set up the federal system of government with three main branches—executive, legislative, and judicial. Each had distinct powers and acted as a check on the others so that no one branch had all the power. This is the system of government we have today. The Constitution was **ratified**, or approved, by the states in 1788 and took effect in 1789.

The signing of the U.S. Constitution took place in Philadelphia on September 17, 1787.

Westward Expansion

In the 1790s and 1800s many Americans and immigrants migrated west of the Appalachians in search of more and better farmland. These new settlers were called **pioneers**. They were people who were leading the way into new areas. The boundaries of the United States shifted as settlement spread westward. By the 1820s, pioneers had crossed the Mississippi River and settled as far south as Texas. By the mid-1800s the country stretched from the Atlantic to the Pacific coast.

Albert Bierstadt, *Emigrants Crossing the Plains*, 1867, oil on canvas, AO1.IT: The National Cowboy Hall of Fame and Western Heritage Center, Oklahoma City, OK

Linking Past to Present
The Dust Bowl

As settlers moved west, they faced many new challenges. The prairie farmland was so thick with tangled grass roots that it broke most plows. In 1837, a steel plow that could slice through the heavy sod was invented. Soon the vast grasslands of the plains turned into farmlands.

In the 1930s, however, the area was hit by years of drought. Without the thick grass roots to hold it in place, the rich soil of the Great Plains dried up and blew away. Heavy winds swept giant dust storms across the region, which became known as the "Dust Bowl." The drought ended in the 1930s and early 1940s, and farmers once again began to grow grain. Now farmers practice new farming techniques to prevent the soil from blowing away.

Activity: Have students locate the general area of the Dust Bowl on a wall map.

◄

This painting by American artist Albert Bierstadt (1830–1902) shows a group of pioneers heading west across the Great Plains.

Interpreting the Visual Record What does Bierstadt's painting suggest about American ideas of the West during the 1800s?

◄ **Visual Record Answer**

Possible answers: rugged, grand, majestic, unspoiled, inviting

225

LEVEL 2: (Suggested time: 20 min.) Organize the class into small groups and have each group write a short play set in the United States shortly after the Revolutionary War. Direct students to describe in their plays some of the challenges facing the new country as well as some of the opportunities enjoyed by its citizens. Call on volunteer groups to perform their plays for the class. **COOPERATIVE LEARNING**

LEVEL 3: (Suggested time: 40 min.) Organize the class into two groups. Tell one group to imagine that they are American Indians living on the Plains in 1850. Tell the other group to imagine they are homesteaders who hope to own farms in the western United States. Have group members discuss their thoughts about land ownership and land as a resource. Then stage a debate between the two groups in which each side is allowed to express its thoughts and ideas. **COOPERATIVE LEARNING**

DAILY LIFE

What would it be like to homestead on the frontier in the 1880s? Three modern American families got the chance to find out in 2001 by spending five months in an isolated area of Montana with only the tools and supplies a homesteading family would have had. They had no electricity or running water, and the nearest store was seven hours away by foot. The project was an experiment conducted by a public television station. Unlike other "reality" programming, this was not a contest and there were no prizes.

Discussion: What do you think would be the hardest part of living as these families did? What do you think you might learn from such an experience?

internet connect

GO TO: go.hrw.com
KEYWORD: SJ3 CH10
FOR: Web sites about
Westward Expansion

Map Answer ▶

through treaties, purchases, and annexations; the Louisiana Purchase

Visual Record Answer ▶

many tents, few permanent buildings

▶

Within 100 years of independence, the borders of the United States had expanded to the Pacific Ocean. New land was added through treaties, purchases, and wars.

Interpreting the Map How did the United States gain most of its territory after 1783? Which was the largest single addition?

Most of the people in mining camps were young, unmarried men. The camps were rather wild, with a lot of crime and fighting.

Interpreting the Visual Record What parts of this scene suggest that the town has grown rapidly?

▼

Territorial Expansion of the United States

The government sold land cheaply or gave it away to encourage people to settle new areas. Many people arrived on the Pacific coast after gold was discovered in California in 1848. However, few people settled in the deserts and mountains of the western United States or in the Great Plains. They called this area the Great American Desert. Most people believed it was too dry to support farming.

As pioneers moved westward they began to have bitter conflicts with American Indians. Many of the Indians did not own land as individuals. Instead, they considered land a shared resource. As settlers occupied and divided up land, they pushed American Indians farther west and onto reservations. Many died from warfare or from diseases carried by settlers.

✓ **READING CHECK:** *Summarizing* What problem developed between the settlers and the American Indians as the United States expanded? settlers moved onto Indian lands and pushed Indians out; many Indians died from warfare or from diseases

The Granger Collection, New York

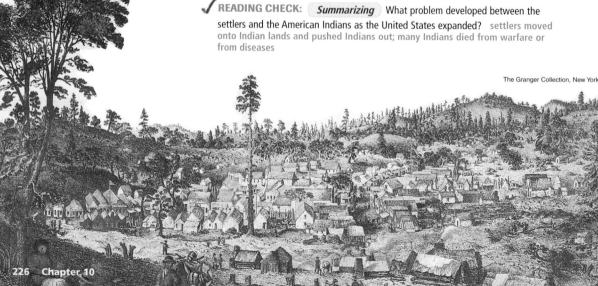

226 Chapter 10

226

Teaching Objectives 1–2

ALL LEVELS: (Suggested time: 25 min.) Copy the following graphic organizer onto the chalkboard, omitting the italicized answers. Have students complete the chart with descriptions of the colonies and states that developed in New England and the Middle Atlantic States and those that developed in the South. Once students have completed their charts, lead a class discussion about differences between these regions. Ask students to suggest how the differences they have noted contributed to the outbreak of the Civil War. **ENGLISH LANGUAGE LEARNERS, COOPERATIVE LEARNING**

	New England and Middle States	Southern States
type(s) of farms	*mostly small farms*	*small farms and large plantations*
industrialization	*industry and railroads, factories*	*little industrialization*
growth of cities	*large cities, major seaports*	*few large cities*
workers	*general population*	*dependence on slave labor*

The Civil War

By 1830 the northeastern United States was industrializing. Industries and railroads spread. In the South, however, the economy was based on export crops like tobacco and cotton. As you have read, farmers grew those crops on plantations with the labor of enslaved Africans. The North, with its factories and large cities, had less need for slave labor. Economic differences between the North and South, and the South's insistence on maintaining slavery, eventually led to war.

The Slavery Issue The Missouri Compromise of 1820 let Missouri enter the Union as a slave state and divided the rest of U.S. territory into slave and free areas. Many people, however, wanted to abolish slavery altogether. As new states entered the Union, bitter arguments raged over whether each state would allow slavery. As the election of 1860 approached, the issue of slavery threatened to rip apart the country.

In 1860 Abraham Lincoln was elected President. He was an **abolitionist**, someone who wanted to end slavery. In response, South Carolina **seceded** from, or left, the United States. Other southern states soon followed. These states formed the Confederate States of America.

Lincoln argued that states had no right under the Constitution to secede. He said that the government must put down the rebellion. The long, bloody Civil War began in 1861. It caused much suffering, as families were divided and property was destroyed. The war finally ended in 1865 when the Confederacy surrendered. Bitter feelings about the war, however, lasted into the next century.

During the war, Lincoln issued the **Emancipation Proclamation**. This document freed slaves in the states that were in rebellion. When the war ended, Congress passed three amendments ending slavery and guaranteeing the former slaves rights. However, the laws were often ignored. Life did not get much better for many former slaves.

President Abraham Lincoln

More than 600,000 Americans died in the Civil War. This was more than in any other war the United States has fought before or since.

During the Civil War soldiers on each side were often ordered to attack well-defended positions. This led to many casualties.

Interpreting the Visual Record Whose point of view do you think the artist is trying to portray? Why?

Brian Callahan of Patterson, New York, suggests the following activity to help students learn about the issues that brought about the Civil War. Have students conduct research and hold a mock trial of abolitionist John Brown. Assign students to the roles of Brown, defense attorneys, and prosecutors, while others will be the jury. Still others can act as reporters. They can write newspaper stories and editorials. As in a real trial, all evidence discovered by the prosecution must be shared with the defense team. Ask a teacher from another class to act as judge and hold the trial.

Teaching Objective 3

ALL LEVELS: (Suggested time: 20 min.) Have students choose an event or discovery of the 1900s they think most changed people's lives in the United States. Have them draw pictures to illustrate the event. Call on volunteers to share and explain their choices.

➤**ASSIGNMENT:** Have each student interview a parent, grandparent, or other adult to learn which event the adult thinks was the most significant event of the 1900s. Tell students they should ask why the adult feels this way. Have students share their findings with the class.

National Geography Standard 9

Human Systems In 1875, the federal government built facilities for "processing" immigrants on Ellis Island, a tiny island in New York harbor named for Sam Ellis, a merchant who lived during the American Revolution. The first immigrant to enter the United States through Ellis Island was fifteen-year-old Annie Moore of County Cork, Ireland. She arrived on January 1, 1892.

From that day until it closed in 1954, Ellis Island was the entry point for more than 12 million immigrants. Some 40 percent of all U.S. citizens can trace their ancestry to someone who came through Ellis Island. Today the buildings are a museum where visitors can search an immigration database to find their ancestors.

Activity: Encourage students to conduct research to learn when and where their ancestors arrived in the United States.

internet connect

GO TO: go.hrw.com
KEYWORD: SJ3 CH10
FOR: Web sites about Ellis Island

The development of steamships during the second half of the 1800s allowed a great increase in immigration to the United States and Canada. In this image from the late 1800s, immigrants in search of new opportunities come ashore at Ellis Island, New York.

Elizabeth Cady Stanton (1815–1902) campaigned for women's rights. Her efforts helped start the women's suffrage movement, to get women the right to vote.

Growth and Expansion After the War Most of the Civil War had been fought in the South. It took a long time for the South to recover and rebuild. In the meantime, the rest of the country changed rapidly during the later 1800s. More railroads and industry expanded, settlers pushed west, and more and more people were immigrating, or moving, to the United States.

The first transcontinental railroad was completed in 1869. It linked the eastern United States with California. This railroad made it much easier to move goods and people across the country. Railroads also allowed major cities to develop far from navigable waterways.

With new agricultural machinery, farms could produce more food using fewer people than ever before. Irrigation and better plows allowed farmers to grow crops in the Great American Desert. As a result, people's views of the region changed and they began to settle the area.

The development of industry attracted more people to the country's growing cities. Some came from rural areas. However, many were immigrants, mostly from Europe. Many European immigrants settled in the industrial cities of the Northeast. Millions of immigrants poured into the United States during the later 1800s and early 1900s, making the nation the most diverse in the world. By 1920 more Americans lived in cities than rural areas.

✓ **READING CHECK:** *Finding the Main Idea* Why did many Americans and immigrants move westward during the 1800s? Industry expanded. Railroads made travel easier and allowed cities to grow away from waterways. Better agricultural machines led to more farming.

On May 10, 1869, the Union Pacific and Central Pacific railroads met at Promontory Point, Utah, joining East Coast to West.

REVIEW AND ASSESS

Have students complete the Section Review. Then have each student list on separate slips of paper significant events discussed in the section. Collect the slips and have students take turns drawing two of the slips of paper and telling which event occurred first. Then have students complete Daily Quiz 10.2.

RETEACH

Have students complete Main Idea Activity 10.2. Then write the following headings on the chalkboard: *The Indians, The New Colonists, Enslaved Africans, Life in the Colonies, Trouble Brewing,* and so on. Call on students to supply a one sentence summary of the material under each heading.
ENGLISH LANGUAGE LEARNERS

EXTEND

Have interested students conduct further research into life in a particular decade of American history. Direct students to focus their attention on changes or developments that occurred during their chosen decades. Have students create posters or dioramas to present their findings.
BLOCK SCHEDULING

The 1900s and Beyond

In the 1900s the United States experienced major social, economic, and technological changes. The country fought in World War I in 1917–18 and suffered through the Great Depression in the 1930s. U.S. forces also fought in World War II from 1941 to 1945. Since the mid-1900s the United States has been one of the richest and most powerful countries in the world.

After World War II the United States and the Soviet Union became rivals in the Cold War. Both countries built huge military forces and developed nuclear weapons. The two countries never formally went to war against each other. However, they supported different sides in small wars around the world. Since the collapse of the Soviet Union in 1991, the United States and Russia have had better relations.

The 1950s and 1960s saw the rise of the Civil Rights Movement. Presidents John F. Kennedy and Lyndon B. Johnson introduced reforms which ended many forms of segregation and discrimination. Hispanics, Native Americans, and women also fought for and won gains in the fight for equal treatment and equal opportunity.

The last decade has seen rapid advances in computer technology. The development of the microchip shrank the size and cost of computers and made them accessible to many Americans. The Internet, a global network of computers, helps link people around the world. Other communication tools such as cellular phones and fax machines have made communication easier as well. All of this technology has sparked an information revolution that has changed our lives forever. More than ever, the future of the United States holds promise for all Americans.

▲
Dr. Martin Luther King, Jr., was a key leader of the Civil Rights Movement. He delivered his famous "I Have a Dream" speech at the Lincoln Memorial in Washington, D.C., in August 1963.

✓ **READING CHECK:** *Finding the Main Idea* How have computer technology and the Internet changed Americans' lives? provide information and communication between people all over the world

go.hrw.com **Homework Practice Online**
Keyword: SJ3 HP10

Section Review 2

Define and explain: plantations, frontier, ratified, pioneers, abolitionist, secede, Emancipation Proclamation

Reading for the Main Idea

1. *Environment and Society* How did North America change after the early colonists arrived?

2. *Human Systems* What major events changed in the United States between the late 1700s and the mid-1800s?

3. *Human Systems* How has the United States changed since the mid-1900s?

Critical Thinking

4. **Drawing Conclusions** During which period in U.S. history do you think our country changed the most? Why?

Organizing What You Know

5. **Identifying Time Order** Copy the time line below into your notebook. Use it to list some important dates, periods, and events that occurred during the 1800s.

|———————————————————————|
1800   1900

Section Review 2

Answers

Define For definitions, see: plantations, p. 223; frontier, p. 224; ratified, p. 225; pioneers, p. 225; abolitionist, p. 227; secede, p. 227; Emancipation Proclamation, p. 227

Reading for the Main Idea

1. first inhabited only by Indians, then populated by Europeans; growth of colonies; society connected by trade and transportation routes (NGS 14)

2. American Revolution, Declaration of Independence, U.S. Constitution, westward expansion, gold rush, Civil War (NGS 12)

3. became rich and powerful; gains in civil rights and equal opportunity for minorities and women; advances in technology and communication (NGS 13)

Critical Thinking

4. Answers will vary but should be supported.

Organizing What You Know

5. Possible answers: westward expansion, gold rush, Civil War, Emancipation Proclamation, completion of the transcontinental railroad, immigration.

LEVEL 1: Organize the class into small groups. To help students understand how the United States was attacked on September 11, 2001, have students take notes as you recall the events of that day. *(8:45 AM–north tower struck; 9:00 AM–south tower struck; 9:40 AM–Pentagon struck; 10:00 AM—south tower collapses; 10:30 AM—north tower collapses, United Airlines Flight 93 crashes.)* Then ask students to create a time line of these events. **ENGLISH LANGUAGE LEARNERS**

LEVEL 2: Tell students that the United States responded to the attacks with military, diplomatic, and humanitarian measures. Assign each student a country mentioned in the text as supporting the efforts to eliminate global terrorism. Have students conduct research to obtain information about that country. Then ask students to make a paper link. On that link, students should identify the leader of the country and examples of the assistance given by that country to fight terrorism. Once the links are completed, connect them to form a chain to show how countries are linked to fight terrorism. **BLOCK SCHEDULING**

GLOBAL PERSPECTIVES

The Taliban movement began in the schools of Afghanistan and western Pakistan where a strict version of Islam was taught. In fact, Taliban means "the students" in both Pashto and Farsi, two languages commonly spoken in Afghanistan. It is derived from an Arabic root.

In 1995 a civil war broke out in Afghanistan, and Mullah Muhammad Omar formed a militia made up of his students. This militia became known as the Taliban. The Taliban took control of Afghanistan in 1996 and imposed its interpretation of Islam on Afghan citizens. Taliban laws regarding women were especially strict. Women had to cover their faces, they could not attend school or go to work, and they were forbidden to talk to a man not related to them. It has been reported that those who violated the laws were brutally tortured, beaten, and even executed.

Activity: Have students research and report on the government, society, and economy of Afghanistan today.

CONNECTING TO *History*

An ATTACK on America

On the morning of September 11, 2001, the United States experienced one of the greatest tragedies in its history. Terrorists—individuals who use violence to achieve political goals—hijacked four commercial airliners flying out of airports in Boston, Massachusetts; Newark, New Jersey; and Washington, D.C. The terrorists crashed two of the planes into the Twin Towers of New York City's World Trade Center, causing the buildings to collapse. The third plane hit the Pentagon, and the fourth crashed in rural Pennsylvania. Thousands of people were killed and injured in the attacks.

Smoke rises from the site of the World Trade Center towers.

People around the nation watched in shock and disbelief as the crisis unfolded on television. The administration of President George W. Bush and members of the U.S. Congress declared these attacks acts of war against the United States. The president tried to reassure the nation.

President George W. Bush, addressing Congress

❝ Terrorist attacks can shake the foundations of our biggest buildings, but they cannot touch the foundation of America. These acts shattered steel, but they cannot dent the steel of American resolve. ❞
—President George W. Bush,
Speech on September 11, 2001

World leaders expressed their concern for the people of the United States. Referring to World Wars I and II, European Commission president Romano Prodi offered support. "In the darkest days of European history, America stood close by us and today we stand close by America." British prime minister Tony Blair stated that the attacks were "an attack on the free and democratic world everywhere." People around the world gathered to remember the victims of the attacks, holding candlelight services and moments of silence.

Within only a few days of the attacks, a massive federal investigation uncovered many details about the hijackers and their plans. The president and other U.S. officials worked quickly to build an inter-

LEVEL 3: Bring newspapers to class that contain various stories about the attacks on September 11, 2001. Then have students use their textbook and the newspapers to conduct research on the steps U.S. leaders took to find those responsible and bring them to justice. Ask students to write two or three paragraphs outlining what steps were taken to find those responsible for the attacks.

ALL LEVELS: Lead a classroom discussion on what students think ought to be done with the World Trade Center site. Ask: Should the towers be rebuilt? Should a memorial be built? Encourage students to conduct research to find out about ideas that have been suggested by various people and groups. Ask students to share their findings with the rest of the class.

CHAPTER 10

Review
ANSWERS

national coalition, or alliance, to fight terrorism. Included in the coalition were members of the North Atlantic Treaty Organization. Russia also offered diplomatic and other support, as did Japan, Jordan, Egypt, and several other nations.

Multifaith memorial service for the victims

The investigation led to terrorist Osama bin Laden and his network, al Qaeda, based in Afghanistan. Bin Laden and his network were also suspects in previous overseas attacks against the United States. U.S. leaders demanded that the ruling government of Afghanistan, the Taliban, turn bin Laden over to U.S. authorities, but Taliban leaders refused. With international support, the United

Firefighters played a key role in the search and recovery efforts at the attack sites.

States and Great Britain launched attacks on Taliban and al Qaeda targets. As Americans tightened security around the nation, President Bush warned Americans that the fight against terrorism would be a long one.

Understanding What You Read

1. What is terrorism?
2. Why is international support needed to fight terrorism?

U.S. aircraft carrier departing as part of a battlegroup launched in response to the September 11 attacks

Building Vocabulary

For definitions, see: contiguous, p. 108; Continental Divide, p. 109; basins, p. 109; plantations, p. 113; frontier, p. 224; ratified, p. 225; pioneers, p. 225; abolitionist, p. 227; secede, p. 227; Emancipation Proclamation, p. 227

Reviewing the Main Ideas

1. Rocky Mountains, Appalachian Mountains (NGS 4)

2. in the West, particularly the Southwest (NGS 4)

3. colonies established; native Indians forced off their lands; changes in land use; growth of population and cities (NGS 14)

4. people moved westward to the Pacific; death of many Indians; Civil War; ending of slavery; people traveled more; tremendous immigration; more people moved to cities (NGS 12)

5. became a world power; Civil Rights movement; great advances in technology; information revolution (NGS 10)

231

ASSESS

Have students complete the Chapter 10 Test.

RETEACH

Organize students into pairs and assign one section of this chapter to each pair. Have each pair of students create an illustrated poster that shows the main ideas of its assigned section. Direct students to write captions to identify any images they include on their posters. Then have students use the posters to teach the material to rest of the class.
ENGLISH LANGUAGE LEARNERS, COOPERATIVE LEARNING

PORTFOLIO EXTENSIONS

1. Have students make relief maps of the United States from papier maché, clay, or flour-salt dough. Models should show the gradual rise in elevation of the U.S. from east to west, the Great Plains, the Rocky Mountains and the Appalachians. Students can paint their maps with different colors to show varying elevations and provide a key. Have students write a paragraph explaining what they did, and include a photograph of their finished maps for their portfolios.

2. Have students conduct research about the peak years of immigration to the United States, from about 1845 to 1920. Have them organize the information into bar graphs to show how the immigrants' countries of origin changed through the years.

CHAPTER 10 Review ANSWERS

Understanding History and Society

Answers will vary, but students should mention that the Transcontinental Railroad provided the first efficient transportation link between the eastern and western United States. Use Rubric 29, Presentations, to evaluate student work.

Thinking Critically

1. The Appalachians are less steep and rugged than the Rocky Mountains. This is because the Appalachians are older and have been eroded more.

2. The Mississippi River deposits rich soils that produce fertile farmlands.

3. Indians—believed the land was to be shared; Pioneers—felt they had a right to claim and own land

4. Most of the Civil War had been fought in the South, and there was a great deal of damage and destruction of property.

5. The Internet links people in the United States with people around the world. It allows people to access information easily and almost instantaneously.

CHAPTER 10 Reviewing What You Know

Building Vocabulary

On a separate sheet of paper, write sentences to define each of the following words.

1. contiguous
2. Continental Divide
3. basins
4. plantations
5. frontier
6. ratified
7. pioneers
8. abolitionist
9. secede
10. Emancipation Proclamation

Reviewing the Main Ideas

1. (Places and Regions) What are the two major mountain regions of the United States?

2. (Places and Regions) Where are the dry climate regions located in the United States?

3. (Environment and Society) How did North America change in the 1600s and 1700s when European colonists arrived?

4. (Human Systems) How did people's lives change in the United States between 1830 and 1900?

5. (Environment and Society) What changes occurred in the United States after World War II?

Understanding History and Society

The Transcontinental Railroad
The completion of the first transcontinental railroad linked the West Coast of the United States with the East Coast. Prepare a presentation about the building of this railroad, considering the following:
- When and where the building of the railroad lines began, and when and where it was completed.
- Problems railroad workers faced.
- The effect the Transcontinental Railroad had on American life.

Thinking Critically

1. **Contrasting** How is the physical geography of the Appalachians different from that of the Rocky Mountains? Why is this true?

2. **Drawing Inferences and Conclusions** Why do you think the Mississippi River is important to agriculture in the South?

3. **Contrasting** Contrast the pioneers' and American Indians' viewpoints about the land in North America and its use.

4. **Finding the Main Idea** Why did it take the South a long time to recover and rebuild after the Civil War?

5. **Finding the Main Idea** Why is the Internet such an important technological advance in the United States?

232

FOOD FESTIVAL

Have students interview parents and other adults to find how food fashions and cooking methods have changed over time. *(Examples: convenience foods, TV dinners, "fast" food, the fondue craze, and so on.)* Invite students to bring samples to share with the class. Interested students may also conduct research on how kitchens have changed (electric and gas stoves replacing wood burners, refrigerators replacing ice boxes). Challenge students to identify connections between changes in eating and cooking habits and changes in society. *(Example: Convenience foods became popular when women began to work outside of the home.)*

CHAPTER 10 REVIEW AND ASSESSMENT RESOURCES

Reproducible
- Critical Thinking Activity 10

Technology
- Chapter 10 Test Generator (on the One-Stop Planner)
- Audio CD Program, Chapter 10

Reinforcement, Review, and Assessment
- Chapter 10 Review, pp. 232–233
- Chapter 10 Tutorial for Students, Parents, Mentors, and Peers
- Vocabulary Activity 10

- Chapter 10 Test
- Chapter 10 Test for English Language Learners and Special-Needs Students

---

Building Social Studies Skills

Map ACTIVITY

On a separate sheet of paper, match the letters on the map with their correct labels.

Atlantic Ocean	Great Lakes
Pacific Ocean	New York City
Alaska	Los Angeles
Hawaii	Miami
Gulf of Mexico	Chicago

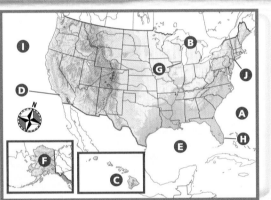

Mental Mapping Skills ACTIVITY

On a separate sheet of paper, draw a freehand outline map of the United States. Label the following territories that were purchased in the 1800s:

Louisiana Purchase	Mexican Cession
Texas Annexation	Gadsden Purchase
Oregon Territory	

WRITING ACTIVITY

You might be familiar with the song "America the Beautiful." It reminds the listener of the physical beauty of the United States—"from sea to shining sea." Write a short poem, song, or rap describing the physical geography of the eastern, interior, or western United States.

Alternative Assessment

Portfolio ACTIVITY

Learning About Your Local History

Historical Buildings Research the buildings in your town to find out which are the oldest or which have historical significance. Take a photograph or make a drawing of one building. Write a paragraph to explain the part the building played in your town's history.

internet connect

Internet Activity: go.hrw.com
KEYWORD: SJ3 GT10

Choose a topic to explore about the United States:
- Experience life in the early colonies.
- Search letters and journal entries written by soldiers in the Civil War.
- Explore the advance of civil rights in the United States.

Map Activity

A. Atlantic Ocean	**F.** Alaska
B. Great Lakes	**G.** Chicago
C. Hawaii	**H.** Miami
D. Los Angeles	**I.** Pacific
E. Gulf of Mexico	**J.** New York City

Mental Mapping Skills Activity

Maps will vary, but listed areas should be labeled in their approximate locations.

Writing Activity

Answers will vary, but the information included should be consistent with text material. Use Rubric 26, Poems and Songs, to evaluate student work.

Portfolio Activity

Paragraphs will vary, but students should identify a building in the town of historical significance. Use Rubric 42, Writing to Inform, to evaluate student work.

internet connect

GO TO: go.hrw.com
KEYWORD: SJ3 Teacher
FOR: a guide to using the Internet in your classroom

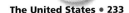

CHAPTER 11

The Regions of the United States
Chapter Resource Manager

Objectives	Pacing Guide	Reproducible Resources
SECTION 1		
The Northeast (pp. 235–39) 1. Describe the features of the Northeast. 2. Describe the landforms and climates found in New England and the Middle Atlantic states. 3. Explain how physical features affect the economies of New England and the Middle Atlantic states.	**Regular** 1 day **Block Scheduling** .5 day *Block Scheduling Handbook, Chapter 11*	**RS** Guided Reading Strategy 11.1 **RS** Graphic Organizer 11 **E** Creative Strategies for Teaching World Geography, Lesson 7 **E** Cultures of the World Activity 1 **IC** Interdisciplinary Activity for the Middle Grades 7, 8
SECTION 2		
The South (pp. 240–43) 1. Describe the landforms found in the South. 2. Describe the climates of the South. 3. Identify the resources and industries important to the economy of the South.	**Regular** .5 day **Block Scheduling** .5 day *Block Scheduling Handbook, Chapter 11*	**RS** Guided Reading Strategy 11.2
SECTION 3		
The Midwest (pp. 244–47) 1. Describe the landform regions of the Midwest. 2. Identify some agricultural products of the Midwest. 3. Describe the main industries of the Midwest.	**Regular** .5 day **Block Scheduling** .5 day *Block Scheduling Handbook, Chapter 11*	**RS** Guided Reading Strategy 11.3
SECTION 4		
The Interior West (pp. 248–52) 1. Identify major landform regions of the Interior West. 2. Describe the climates of the Interior West. 3. List some economic activities of the Interior West.	**Regular** 1 day **Block Scheduling** .5 day *Block Scheduling Handbook, Chapter 11*	**RS** Guided Reading Strategy 11.4
SECTION 5		
The Pacific States (pp. 253–57) 1. Identify the landforms found in the Pacific states. 2. Explain how the climates of the Pacific states differ. 3. Explain how the Pacific states contribute to the economy of the United States.	**Regular** 1 day **Block Scheduling** .5 day *Block Scheduling Handbook, Chapter 11*	**RS** Guided Reading Strategy 11.5 **SM** Map Activity 11

Chapter Resource Key

RS Reading Support

IC Interdisciplinary Connections

E Enrichment

SM Skills Mastery

A Assessment

REV Review

ELL Reinforcement and English Language Learners

 Transparencies

 CD–ROM

 Music

 Video

 Internet

 Holt Presentation Maker Using Microsoft® Powerpoint®

  One-Stop Planner CD–ROM

See the *One-Stop Planner* for a complete list of additional resources for students and teachers.

One-Stop Planner CD-ROM

It's easy to plan lessons, select resources, and print out materials for your students when you use the *One-Stop Planner CD-ROM with Test Generator.*

Technology Resources	Review, Reinforcement, and Assessment Resources	
● One-Stop Planner CD–ROM, Lesson 11.1	**ELL**	Main Idea Activity 11.1
● *ARGWorld* CD–ROM	**REV**	Section 1 Review, p. 239
● Homework Practice Online	**A**	Daily Quiz 11.1
go.hrw.com	**ELL**	English Audio Summary 11.1
HRW Go site	**ELL**	Spanish Audio Summary 11.1
● One-Stop Planner CD–ROM, Lesson 11.2	**ELL**	Main Idea Activity 11.2
● *ARGWorld* CD–ROM	**REV**	Section 2 Review, p. 243
● Homework Practice Online	**A**	Daily Quiz 11.2
go.hrw.com	**ELL**	English Audio Summary 11.2
HRW Go site	**ELL**	Spanish Audio Summary 11.2
● One-Stop Planner CD–ROM, Lesson 11.3	**ELL**	Main Idea Activity 11.3
● *ARGWorld* CD–ROM	**REV**	Section 3 Review, p. 247
● Homework Practice Online	**A**	Daily Quiz 11.3
go.hrw.com	**ELL**	English Audio Summary 11.3
HRW Go site	**ELL**	Spanish Audio Summary 11.3
● One-Stop Planner CD–ROM, Lesson 11.4	**ELL**	Main Idea Activity 11.4
● *ARGWorld* CD–ROM	**REV**	Section 4 Review, p. 252
● Homework Practice Online	**A**	Daily Quiz 11.4
go.hrw.com	**ELL**	English Audio Summary 11.4
HRW Go site	**ELL**	Spanish Audio Summary 11.4
● One-Stop Planner CD–ROM, Lesson 11.5	**ELL**	Main Idea Activity 11.5
● *ARGWorld* CD–ROM	**REV**	Section 4 Review, p. 257
● Homework Practice Online	**A**	Daily Quiz 11.5
go.hrw.com	**ELL**	English Audio Summary 11.5
HRW Go site	**ELL**	Spanish Audio Summary 11.5

internet connect

HRW ONLINE RESOURCES

GO TO: go.hrw.com
Then type in a keyword.

TEACHER HOME PAGE
KEYWORD: SJ3 TEACHER

CHAPTER INTERNET ACTIVITIES
KEYWORD: SJ3 GT11

Choose an activity to:
• examine growth and development along the Great Lakes and St. Lawrence River.
• create a brochure on Hawaii's environment.
• create a newspaper article on the Gulf Coast of Texas.

CHAPTER ENRICHMENT LINKS
KEYWORD: SJ3 CH11

CHAPTER MAPS
KEYWORD: SJ3 MAPS11

ONLINE ASSESSMENT
Homework Practice
KEYWORD: SJ3 HP11
Standardized Test Prep Online
KEYWORD: SJ3 STP11
Rubrics
KEYWORD: SS Rubrics

COUNTRY INFORMATION
KEYWORD: SJ3 Almanac

CONTENT UPDATES
KEYWORD: SS Content Updates

HOLT PRESENTATION MAKER
KEYWORD: SJ3 PPT11

ONLINE READING SUPPORT
KEYWORD: SS Strategies

CURRENT EVENTS
KEYWORD: S3 Current Events

Meeting Individual Needs

Ability Levels

Level 1 Basic-level activities designed for all students encountering new material

Level 2 Intermediate-level activities designed for average students

Level 3 Challenging activities designed for honors and gifted-and-talented students

English Language Learners Activities that address the needs of students with Limited English Proficiency

Chapter Review and Assessment

SM	Critical Thinking Activity 11	●	Chapter 11 Test Generator (on the One-Stop Planner)
REV	Chapter 11 Review, pp. 258–59	●	Audio CD Program, Chapter 11
REV	Chapter 11 Tutorial for Students, Parents, Mentors, and Peers	**A**	Chapter 11 Test for English Language Learners and Special-Needs Students
ELL	Vocabulary Activity 11		
A	Chapter 11 Test	**A**	Unit 3 Test for English Language Learners and Special-Needs Students
A	Unit 3 Test		

LAUNCH INTO LEARNING

Ask students the following question: *How many different ways can you think of to identify the part of the United States in which we live?* Discuss student responses. Point out that people often use many different names to refer to a single part of the country. For example, the phrases *New England* and *the Northeast* can be used to identify the same area. Texas is sometimes considers part of the South, sometimes part of the West, and sometimes part of the Southwest. Tell students that they will learn about one system for dividing the United States into regions in this chapter.

Section 1

Objectives

1. Describe the features of the Northeast.

2. Describe the landforms and climates found in New England and the Middle Atlantic states.

3. Explain how physical features affect the economies of New England and the Middle Atlantic states.

LINKS TO OUR LIVES

You might want to share with your students the following reasons for learning more about the regions of the United States:

▶ Each region shares some features with the other regions but also has some unique aspects. We can better understand our country by understanding the similarities and differences among the regions.

▶ Each region has a unique history that contributes to the history of our country as a whole.

▶ The United States is a wealthy and beautiful country. We need to learn how to appreciate and protect the many advantages our country enjoys.

▶ Learning about the geography of the United States might help us decide where we would like to visit or live someday.

CHAPTER 11

The Regions of the United States

The different regions that make up our country are similar in some ways and quite unique in others. What do you already know about the different regions?

In this chapter you'll learn about the land, climate, and economy of the different regions in the United States. In the rest of the book, you will meet people from all over the world. Each of them has cultural traditions—beliefs and behaviors—that are different from yours. However, each has many things in common.

To learn about American culture, we are going to start by meeting an American: you. What beliefs do you have? What do you like to eat? What language do you speak at home? What kind of music do you like? Which holidays do you celebrate? These things and others reflect something about our American culture as well as our diversity.

LET'S GET STARTED

Copy the following instructions onto the board: *What is life like in a big city? Write down your ideas.* Discuss student responses. Point out that most people in the Northeast live in cities and that many cities in the region are very large. Identify some of the major cities of the Northeast on a wall map, including Baltimore, Boston, New York, Philadelphia, and Washington, D.C. Tell students that they will learn more about life in this largely urban area in Section 1.

Using the Physical-Political Map

Have students examine the map on this page. Ask them to describe the region's physical geography *(mostly plains along the coast,* with higher elevations to the west*)*. Tell students that many cities in the Northeast are ports. Ask them to identify some of the bodies of water on which these cities lie (*Atlantic Ocean, Penobscot Bay, Gulf of Maine, Massachusetts Bay, Cape Cod Bay, Delaware Bay, Chesapeake Bay, Lake Erie, Lake Ontario*).

Building Vocabulary

Write **megalopolis**, **moraine**, and **estuary** on the chalkboard and have students read their definitions. Point out that the prefix *megalo-* means "large" or "of giant size" and that *-polis* means "city." Thus, the literal meaning of *megalopolis* is "large city." Ask students what *moraine* and *estuary* have in common. *(They both describe physical features.)*

Section 1 The Northeast

Read to Discover

1. What are some features of the Northeast?
2. What kinds of landforms and climates are found in New England and the Middle Atlantic states?
3. How do physical features affect the economies of New England and the Middle Atlantic states?

Define

megalopolis
moraine
second-growth forests
estuary

Locate

White Mountains
Green Mountains
Longfellow Mountains
Berkshire Hills
Cape Cod
Nantucket

Martha's Vineyard
Chesapeake Bay
Susquehanna River
Washington, D.C.
Philadelphia
Baltimore

WHY IT MATTERS

Millions of tourists visit the Northeast every year. Use **CNNfyi.com** or other **current events** sources to learn about a place you might like to visit there. Record your findings in your journal.

The Statue of Liberty in New York City

Section 1 RESOURCES

Reproducible
- Block Scheduling Handbook, Chapter 11
- Guided Reading Strategy 11.1
- Graphic Organizer 11
- Creative Strategies for Teaching World Geography, Lesson 7
- Cultures of the World Activity 1
- Interdisciplinary Activities for Middle Grades 7, 8

Technology
- One-Stop Planner CD–ROM, Lesson 11.1
- Homework Practice Online
- HRW Go site

Reinforcement, Review, and Assessment
- Main Idea Activity 11.1
- Section 1 Review, p.239
- Daily Quiz 11.1
- English Audio Summary 11.1
- Spanish Audio Summary 11.1

Northeastern United States: Physical-Political

ELEVATION

FEET	METERS
13,120	4,000
6,560	2,000
1,640	500
656	200
(Sea level) 0	0 (Sea level)
Below sea level	Below sea level

⊛ National capital
★ State capitals
• Other cities

CANADA
QUEBEC
St. Lawrence River
Georgian Bay
ONTARIO
Aroostook Valley
NEW BRUNSWICK
St. John River
LONGFELLOW MTS.
MAINE
NOVA SCOTIA
Lake Champlain
Burlington
WHITE MTS.
Mt. Washington (6288 ft. 1917 m)
Augusta
ADIRONDACK MOUNTAINS
VERMONT
Montpelier
NEW HAMPSHIRE
Concord
Portland
Penobscot Bay
GULF OF MAINE
Lake Ontario
Oneida Lake
Lake George
Manchester
Nashua
Rochester
Syracuse
Albany
GREEN MTS.
MASSACHUSETTS
Boston
Massachusetts Bay
Niagara Falls
Buffalo
Erie Canal
Mohawk River
Finger Lakes
BERKSHIRE HILLS
Worcester
Cape Cod Bay
Cape Cod
ATLANTIC OCEAN
Lake Erie
Erie
NEW YORK
CATSKILL MTS.
CONNECTICUT
Hartford
Providence
Plymouth
Nantucket
Newport
Martha's Vineyard
OHIO
Allegheny River
MOUNTAINS
Bridgeport
New Haven
Long I. Sound
RHODE ISLAND
PENNSYLVANIA
POCONO MTS.
Jersey City
New York
Long Island
Pittsburgh
Allentown
Newark
NEW JERSEY
Harrisburg
PIEDMONT
Trenton
Ohio River
Gettysburg
PLAIN
Philadelphia
Camden
INDIANA
Monongahela River
Wilmington
COASTAL
PINE BARRENS
Atlantic City
WEST VIRGINIA
Baltimore
Dover
DELAWARE
Washington, D.C.
Annapolis
Delaware Bay
Huntington
Charleston
Kanawha River
APPALACHIAN
FALL LINE
MARYLAND
Delmarva Peninsula
KENTUCKY
VIRGINIA
ATLANTIC
Chesapeake Bay

SCALE
0   100   200 Miles
0   100   200 Kilometers
Projection: Two-Point Equidistant

N

Teaching Objective 1

LEVEL 1: (Suggested time: 35 min.) Provide art supplies and old magazines to the class and have students create collages that illustrate some features of the Northeast. Direct students to draw images and to cut out photographs from the magazines to use in their collages. Ask volunteers to present their collages to the class and describe the features they depict. **ENGLISH LANGUAGE LEARNERS**

LEVELS 2 AND 3: (Suggested time: 30 min.) Pair students and provide them with an assortment of newspapers. Have students look through the newspapers for articles with datelines from the cities of the megalopolis or other cities from the Northeast. Have each pair select one article and explain to the class what it says about the city in which it was written. Lead a class discussion about the cities of the Northeast and their significance to the rest of the country. **COOPERATIVE LEARNING**

COOPERATIVE LEARNING

Organize the class into small groups and assign each group one of the cities of the megalopolis. Tell students that each group represents the chamber of commerce of its assigned city and that each group should create a brochure designed to attract businesses to its city.

Each group will need to discuss its assigned city's attributes, particularly features that make it a desirable location for business. The brochures should describe established industries and businesses, transportation systems, educational facilities, cultural activities, and so on. Remind students that their brochures should also include pictures and have them create drawings or cut pictures from magazines to illustrate their work. Call on groups to share their brochures with the class.

The Northeast

NEW ENGLAND STATES	MIDDLE ATLANTIC STATES
Connecticut	Delaware
Maine	Maryland
Massachusetts	New Jersey
New Hampshire	New York
Rhode Island	Pennsylvania
Vermont	West Virginia

internet connect

GO TO: go.hrw.com
KEYWORD: SJ3 CH11
FOR: Web sites about regions of the United States

New York City is an important seaport.

Interpreting the Visual Record Why do you suppose shipping helped the major cities of the Northeast grow?

A View of the Northeast

Although the Northeast is the smallest region in the United States, it is the most heavily populated. It has the country's greatest concentration of cities, factories, banks, universities, and transportation centers. The region is home to New York City, the country's largest city, and Washington, D.C., its capital.

The states of the Northeast are grouped into two subregions. The New England states are located in the northeastern portion of the country. The Middle Atlantic states stretch from the Appalachians to the Atlantic coast. There are six states in each subregion.

Megalopolis Overlapping the two subregions of the Northeast is a huge densely populated urban area. This area is called a **megalopolis**. A megalopolis is a string of cities that have grown together. This urban area stretches along the Atlantic coast from Boston to Washington, D.C. Major cities that are part of the megalopolis include New York City, Philadelphia, and Baltimore. At least 40 million people live in this area.

Except for Washington, D.C., all of these cities were founded during the colonial era. They grew because they were important seaports. Today they are major industrial and financial centers as well. These cities are connected by roads, railroads, and airline routes. Many of the world's largest companies have home offices in these cities. Washington, D.C. is the country's capital and the center of government offices. D.C. stands for District of Columbia.

✓ **READING CHECK:** *Places and Regions* Into what areas is the Northeastern region divided? two sub-regions, each with six states and overlapping string of cities known as megalopolis

Visual Record Answer

Shipping industries brought much income and attracted many new businesses to the cities.

 Teaching Objective 2

LEVEL 1: (Suggested time: 15 min.) Have each student write three questions about the landforms and climates of New England and the Middle Atlantic states. *(Examples: What kind of landform dominates much of northern New England? What is a northeaster? What process shaped many of the landforms of the Northeast?)* Direct students to write each question on one side of an index card and to write its answer on the other side. Then have volunteers ask their questions in front of the class and have classmates answer them. **ENGLISH LANGUAGE LEARNERS**

LEVELS 2 AND 3: (Suggested time: 20 min.) Have pairs of students consult the text, map, and images in this sections to prepare an itinerary for a scenic tour of New England. Tell students to identify towns, states, and major landforms along their planned routes. Ask them to choose a season for travel and to describe the typical weather conditions they might encounter at that time of year. **COOPERATIVE LEARNING**

New England

The New England states are famous for their scenic beauty and varied landforms. The subregion has also been the site of many important events in American history. The Pilgrims landed here in 1620 and were soon followed by other colonists. In the late 1700s, New England was the center of the American Revolution.

Landforms The Appalachian Mountains cross much of northern New England. The Appalachian system is actually made up of many small ranges. In New England, these include New Hampshire's White Mountains, Vermont's Green Mountains, and Maine's Longfellow Mountains. At 6,288 feet (1,917 m), Mount Washington in the White Mountains is the highest peak. Because of glacial erosion, northern New England has thousands of lakes.

Southern New England mainly is a hilly region. As glaciers pushed to the Atlantic coast, they left rock material in the form of hundreds of hills. The Berkshires of western Massachusetts are the highest hilly region. The only plains in New England are along the Atlantic coast and in the Connecticut River valley.

New England's coastline varies greatly from north to south. In the north, the scenic coast of Maine is made of granite. Granite is a hard, speckled rock. Maine's coast is rugged and rocky with many narrow inlets, bays, peninsulas, and islands.

Glaciers are responsible for the features of the southern New England coast. Cape Cod and the islands of Nantucket and Martha's Vineyard are glacial **moraine** materials. A moraine is a ridge of rocks, gravel, and sand piled up by the huge ice sheets.

Climate New England has a humid continental climate. Each autumn the region's brightly colored leaves attract tourists. Winter sports fans like its snowy winters. In the summer, fog is common along the coast. Very rarely during the summer, a hurricane may strike the coast of southern New England. More common are winter storms from the North Atlantic called northeasters. These storms bring cold, snowy weather with strong winds and high ocean waves.

Economy Dairy farming is the region's most important agricultural activity. Other important crops are cranberries and potatoes. However, a short growing season and rocky terrain limit farming in New England. As a result, most farms in this region are small. Some are now used as second homes or retirement retreats.

Cool, shallow waters off the coast are good fishing areas. Cod and shellfish are the most valuable seafood. The U.S. government has set rules to prevent overfishing in some areas.

▲
Small towns and beautiful scenery can be found throughout Vermont and the rest of New England.

Do you remember what you learned about glaciers? See Chapter 2 to review.

PLACES AND REGIONS

In designing license plates for automobiles, state governments typically select an aspect of the state's geography, economy, history, or tourist attractions to highlight. The lobster appearing on Maine's license plates, for example, refers to the state's physical and economic geography.

Activity: Invite students to create license-plate designs for the other New England states. Have students suggest symbols representing the physical geography of the states or tourist attractions associated with landforms and climate. Encourage students to consult almanacs and encyclopedias for additional information about each state.

Teaching Objective 3

ALL LEVELS: (Suggested time: 30 min.) Copy the following graphic organizer onto the board, omitting the italicized answers. Have students complete the chart to describe the relationships between the physical features and the economies of New England and the Middle Atlantic states. Call on volunteers to share their organizers with classmates. Then lead a class discussion about the economic activities of the Northeast.
ENGLISH LANGUAGE LEARNERS

Region	Physical Features	Economic Activities
New England	• *rocky terrain* • *coastal waters* • *seaport* • *scenic natural sites*	• *little farming* • *seafood industry* • *shipping industry* • *tourist industry*
Middle Atlantic States	• *good soil* • *coal* • *seaport*	• *good farming* • *mining industry* • *shipping industry* • *tourist industry*

Across the Curriculum
MATH

The Appalachian Trail The Appalachian Trail is a footpath that extends from Maine to northern Georgia. The 2,167 mile (3,488 km) trail closely follows the ridge line of the Appalachian Mountains. It passes through 14 states and eight national forests. More than 4 million people use some part of the trail annually. Most are short-term hikers, but each years several thousand people try to hike the entire trail in one continuous journey. Most of these so-called thru-hikers start in the South in early spring and hike the entire trail in five to six months.

Activity: Have students calculate how far hikers must travel each day in order to complete the entire trail in five months *(approximately 14 miles a day).*

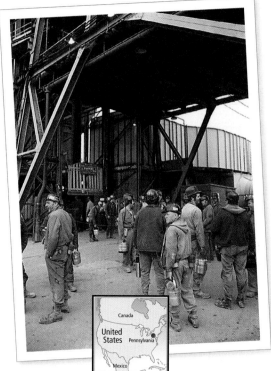

▲
These Pennsylvania miners are waiting to enter one of the many coal mines in the Middle Atlantic states.

Shipbuilding was a major industry in colonial New England. Builders used wood from the region's forests. Nearly all of New England's forests today are **second-growth forests**. These are the trees that cover an area after the original forest has been cut.

New England was the country's first industrial area. Textile mills and shoe factories were built along rivers and swift streams. The water was a handy power source for factories. Today, many banks, investment houses, and insurance companies are based in the region. In addition, the area has many respected colleges and universities. These schools include Harvard, Yale, the Massachusetts Institute of Technology, and many others.

✔ **READING CHECK:** *Places and Regions* Why was New England better suited for industrial growth than for agriculture? agriculture limited by terrain and climate; rivers and streams as close and available power source

Middle Atlantic States

About one fifth of the U.S. population lives in the Middle Atlantic states subregion. It is one of the world's most industrialized and urbanized areas.

Landforms Three landform regions cross the Middle Atlantic states. They are the Coastal Plain, the Piedmont, and the Appalachian Mountains. The Coastal Plain stretches across all of the Middle Atlantic states except West Virginia. This flat plain does not rise much above sea level. Long Island and Chesapeake Bay are both part of the Coastal Plain.

Chesapeake Bay is fed by the Susquehanna River. It is the largest **estuary** on the Atlantic coast. An estuary is a body of water where salty seawater and freshwater mix. As Ice Age glaciers melted along this coast, the sea level rose. The rising sea filled low river valleys and formed the bay.

Inland from the Coastal Plain is the Piedmont. This region slopes down from the Appalachians to the Coastal Plain. Where these regions meet, rivers plunge over rapids and waterfalls. These rivers supplied waterpower for the early towns and cities.

The northern part of the Appalachian Mountains crosses the Middle Atlantic states. Several major rivers cut through the Appalachians here on their way to the Atlantic Ocean. These include the Potomac, Susquehanna, Delaware, and Hudson rivers.

Climates The Middle Atlantic states have two major climate types. The north has a humid continental climate, while the south has a humid subtropical climate. Summers in both climate regions can be

CLOSE

Ask students to name factors that make the Northeast important to the country as a whole. *(Possible answers: its industrial and financial centers, government offices, universities, factories, natural resources, scenic beauty, historic sites)*

REVIEW AND ASSESS

Have students complete the Section Review. Then pair students. Have each pair create an outline based on the material discussed in this section. Have partners take turns using the outline to give an oral summary of the information. Then have students complete Daily Quiz 11.1.
COOPERATIVE LEARNING

RETEACH

Have students complete Main Idea Activity 11.1. Then arrange students in three groups to create storyboards for documentary films about the Northeast. One group should focus on the megalopolis, one on landforms and climates, and the third on economic activities.
ENGLISH LANGUAGE LEARNERS, COOPERATIVE LEARNING

EXTEND

Have interested students conduct research on the use of coal in the United States. Ask them to draw a line graph showing its use over the past 50 years and to draw conclusions from their graphs. **BLOCK SCHEDULING**

very hot and humid. Winds passing over the warm Gulf Stream and the Gulf of Mexico bring hot humid air inland.

During winter, arctic air from the north sometimes enters the region. Winters are colder away from the coast, particularly in the north. Snowfall can be heavy in some areas, especially in the Appalachians and upstate New York. Like the New England coast, the Middle Atlantic coast may experience hurricanes and northeasters.

Economy Soils are better for farming in the Middle Atlantic states than in New England. However, the region's expanding cities have taken over most of the available farmland.

Major coal-mining areas are found in the Appalachians, particularly in West Virginia and Pennsylvania. Coal is used in steelmaking. The steel industry helped make Pittsburgh, in western Pennsylvania, the largest industrial city in the Appalachians.

Today nearly every kind of manufacturing and service industry can be found in the Middle Atlantic states. Major seaports allow farmers and companies to ship their products to markets around the world. In addition, tourists visit natural and historical sites. These include Niagara Falls between New York and Canada and Gettysburg, a major Civil War battleground in Pennsylvania.

✓ **READING CHECK:** *Physical Systems* How are the climates of the Middle Atlantic states similar to those of New England? humid, continental climates

▲
Developments like Camden Yards baseball stadium and a new harbor have strengthened the economy of Baltimore, Maryland.

Homework Practice Online
Keyword: SJ3 HP11

Section Review 1

Define and explain: megalopolis, moraine, second-growth forest, estuary

Working with Sketch Maps On a map of the United States that you draw or that your teacher provides, label the following: White Mountains; Green Mountains; Longfellow Mountains; Berkshire Hills; Cape Cod; Nantucket; Martha's Vineyard; Chesapeake Bay; Susquehanna River; Washington, D.C.; Philadelphia; Baltimore.

Reading for the Main Idea

1. *Human Systems* What are some characteristics of the Northeast?

2. *Human Systems* What types of landforms and climates are found in New England and in the Middle Atlantic states?

3. *Human Systems* How do landforms and climate affect the economies of New England and the Middle Atlantic states?

Critical Thinking

4. **Drawing Conclusions** What kinds of environmental challenges might threaten the cities of the megalopolis? Why do you think this?

Organizing What You Know

5. **Categorizing** Categorizing Copy the following graphic organizer. Use it to categorize the major landforms and climates of the Northeast.

	Landforms	Climates
New England		
Middle Atlantic states		

Section Review 1

Answers

Define For definitions, see: megalopolis, p. 236; moraine, p. 237; estuary, p. 238

Working with Sketch Maps Maps will vary, but listed places should be labeled in their approximate locations.

Reading for the Main Idea

1. heavily populated; many cities, factories, universities, banks, transportation centers (NGS 4)

2. hills, mountains, valleys, peninsulas, islands, coastal plain; cold, snowy winters; warm or hot summers; northeasters (NGS 4)

3. good farmland in Middle Atlantic; good fishing areas and seaports; coal deposits; tourist sites (NGS 4)

Critical Thinking

4. Possible answers: overcrowding, pollution

Organizing What You Know

5. New England— Appalachians, glacial hills, plains, granite coastline, moraine islands; humid continental; Middle Atlantic— Coastal Plain, Piedmont, Appalachians; humid continental, humid subtropical

Section 2

Objectives

1. Describe the landform regions found in the South.
2. Describe the climates of the South.
3. Identify the resources and industries important to the economy of the South.

Section 2 The South

Read to Discover

1. What landform regions are found in the South?
2. What are the characteristics of climates in the South?
3. What resources and industries are important to the economy of the South?

Define

barrier islands
wetlands
sediment
diversify

Locate

Everglades
Okefenokee Swamp
Mississippi Delta
Blue Ridge Mountains
Great Smoky Mountains
Cumberland Plateau
Ozark Plateau

High Plains
Atlanta
Houston
New Orleans
Miami
Dallas

WHY IT MATTERS

Hurricanes have caused great damage to the southern states. Use **CNNfyi.com** or other **current events** sources to learn about hurricanes that have hit this region in the past. Record your findings in your journal.

Plantation house in Louisiana

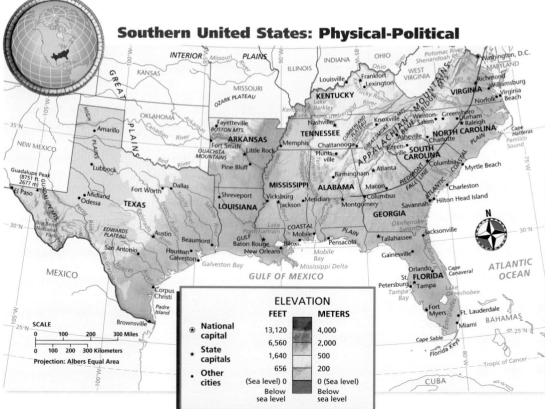

Southern United States: Physical-Political

ALL LEVELS: (Suggested time: 20 min.) Copy the following graphic organizer onto the chalkboard, omitting the italicized answers. Use it to help students identify and describe the various landform regions of the South. Pair students and have each pair find material in the text to complete the chart. Then lead a class discussion about the physical geography of the South. **ENGLISH LANGUAGE LEARNERS, COOPERATIVE LEARNING**

Landform Region	Coastal Plain	Piedmont	Ozark Plateau	Interior Plains
Description	*low, wetlands, barrier islands*	*upland region, rolling hills*	*rugged, hilly, Arkansas River*	*some places low and hilly; mountains, basins*
Location	*along Atlantic Ocean and Gulf of Mexico*	*Virginia, Carolinas, Georgia; inland from Coastal Plain*	*mostly in Arkansas*	*east and west of Ozarks; Kentucky, Tennessee, Texas*

Landforms

The South stretches in a great arc from Virginia to Texas. The 12 states of this region are Virginia, North Carolina, South Carolina, Georgia, Florida, Alabama, Mississippi, Tennessee, Kentucky, Arkansas, Louisiana, and Texas. Texas, the largest state in the region, shares a long border with Mexico.

The Coastal Plain is the major landform region of the South. It stretches inland from the Atlantic Ocean and the Gulf of Mexico. **Barrier islands** line most of the shore along this low area. A barrier island is a long, narrow, sandy island separated from the mainland. Many coastal areas of the South are covered by **wetlands**. Wetlands are land areas that are flooded for at least part of the year.

Two of the largest wetlands areas found along the Coastal Plain are the Everglades in Florida and the Okefenokee Swamp in Georgia. In Louisiana, hundreds of coastal marsh areas can be seen on the Mississippi River Delta. The Mississippi is one of the largest rivers in the world. It carries small bits of mud, sand, or gravel called **sediment** that collects at the river's mouth. This sediment has built up to form the large Mississippi Delta—a low, swampy area cut by small streams.

Inland from the Coastal Plain lies the Piedmont, an upland region. Its rolling hills cover much of Virginia, the Carolinas, and Georgia. Farther from the coast are the southern Appalachian Mountains. Landforms in this region include the Blue Ridge and Great Smoky Mountains and the Cumberland Plateau. The Appalachians cross parts of Virginia, the Carolinas, Kentucky, Tennessee, Georgia, and Alabama.

Another landform of the South is the Ozark Plateau. This ancient plateau lies mainly in Arkansas but extends into Oklahoma and Missouri as well. It is a rugged, hilly region. The Arkansas River runs between the Ozark Plateau and another upland area, the Ouachita (WASH-i-taw) Mountains. Although they are called mountains, the Ouachitas are really rugged hills.

To the east and west of the Ozarks are the rolling hills of the Interior Plains. These plains cover most of central Kentucky and Tennessee. Eastern Texas is also part of the Interior Plains region. Most of central and western Texas is part of the Great Plains. In some places the land rises between 2,000 and 5,000 feet (600 and 1,500 m). While eastern Texas is low and hilly, southwest Texas is a region of mountain ranges and desert basins.

✔ **READING CHECK:** *Places and Regions* What landform regions are found in the South? Coastal Plain, Piedmont, Appalachian Mountains, Ozark Plateau, Interior Plains.

▲ The Florida Everglades is a rich habitat for plants and wildlife.
Interpreting the Visual Record What happens to natural habitats when developers drain wetlands?

Linking Past to Present

The Galveston Hurricane The worst natural disaster in U.S. history was a hurricane that struck Galveston Island in Texas on September 8, 1900. Storm warnings had been issued, but fallen wires kept the news from spreading. Many people ignored the warnings that did reach the island. Consequently, few people had evacuated when all bridges to the mainland collapsed.

Wind gusts estimated at 130 miles per hour (209 kmh) drove huge waves onto the shore. Seawater poured into the city of Galveston, flooding some streets to a depth of 15 feet (4.6 m). As many as 8,000 people died in the city, and perhaps 4,000 more on the rest of the island. Contemporary estimates of property damage range as high as $30 million. This figure is equivalent to a much higher number today.

◄ **Visual Record Answer**

Possible answer: Plants and animals disappear as their habitats are destroyed.

ALL LEVELS: (Suggested time: 30 min.) Direct students to create a climate map of the South. Ask them to create symbols for the different types of weather common to this region and to draw the symbols in appropriate locations on a map. Remind students to include keys with their maps. **ENGLISH LANGUAGE LEARNERS**

Teaching Objective 3

ALL LEVELS: (Suggested time: 45 min.) Organize the class into five groups: citrus farmers, oil workers, fishers and shrimpers, textile manufacturers, and high-tech industries. Have each group write a script for a television commercial that highlights its industry's importance to the South and to the rest of United States. Have groups present their commercials to the class. **COOPERATIVE LEARNING**

GLOBAL PERSPECTIVES

In addition to its many aerospace-related industries, the South hosts some of the major U.S. space exploration and launching facilities, including the John F. Kennedy Space Center in Cape Canaveral, Florida, and the Lyndon B. Johnson Space Center in Houston, Texas. The space shuttles are launched from Cape Canaveral. During flight, however, they are controlled by teams in Houston.

The shuttles put satellites into place and maintain them, thus playing an important role in global communications and in monitoring Earth's weather and environment. The shuttles also carry experiments and projects for other countries.

Discussion: Lead a discussion based on the following question: What benefits might international cooperation in the exploration of space bring to participating nations?

Visual Record Answer ▶

Warm tropical weather and mild winters attract tourists to the region.

242

Millions of tourists visit the South every year.

Interpreting the Visual Record
How does the region's climate help attract visitors?

▼

Climate

Most people who live in the South, particularly along the Coastal Plain, are used to a humid subtropical climate. Winters are mild, and summers are long, hot, and humid. Not all of the South, however, has a subtropical climate. In higher areas such as the Appalachians and Ozarks, temperatures are cooler. Because it is so large, Texas experiences several different climates. These include humid, subtropical, desert, and highland climates.

Most of the South receives between 40 and 60 inches (100 and 150 cm) of rainfall per year. Thunderstorms, which often create dangerous lightning and tornadoes, bring much of this rain to the region. Some areas of the Appalachians may experience occasional winter snowstorms. During late summer and early fall, hurricanes may strike coastal areas.

✔ **READING CHECK:** *Places and Regions* What types of storms are common in the South? Thunderstorms, tornadoes, snowstorms, hurricanes.

Economy

Historically, the South was mainly rural and agricultural. In recent decades, however, the region has attracted many new industries. As the economy has grown, so have many cities. Houston is now the fourth-largest city in the country. Dallas-Fort Worth, Atlanta, Miami, and New Orleans have also grown to be major cities and commercial centers.

Agriculture Although agriculture is no longer the South's major economic activity, many small towns and farming areas stretch across the region. The region has long been a major producer of cotton, tobacco, and citrus fruit. In recent years, some farmers have decided to **diversify**. This means that farmers are producing a variety of crops instead of just one.

Resources The coastal waters of the Gulf of Mexico and the Atlantic are rich in ocean life. Though all the southern coastal states have fishing industries, Louisiana and Texas are the leaders. The shallow waters around the Mississippi Delta produce great quantities of oysters, shrimp, and other seafood. The major mineral and energy resources of the region include coal, sulfur, salt, phosphates, oil, and natural gas. Oil is found in Texas and Louisiana. These two states also lead the country in the production of sulfur. Florida has the country's largest phosphate-mining industry. Phosphates are used to make fertilizer. Much of it is exported to Japan and other Asian nations.

Industry Today many textile factories operate in the Piedmont areas of Georgia, the Carolinas, and Virginia. The Texas Gulf Coast and the lower Mississippi River area have huge oil refineries. Houston, New Orleans, and other major seaports ship the oil. Some cities such as Austin, Texas, also have computer, software, and publishing companies.

Ask students to name cities or other locations in the South that they would like to visit. Call on volunteers to explain their choices to the class.

REVIEW AND ASSESS

Have students complete the Section Review. Have students work in pairs to create simple graphic organizers about one of the topics discussed in this section. Then have students complete Daily Quiz 11.2.
COOPERATIVE LEARNING

RETEACH

Have students complete Main Idea Activity 11.2. Then organize students into three groups and assign each group one of the following topics: landforms, climates, or resources and industries. Have each group create a poster related to its assigned topic. Display and discuss the posters.
ENGLISH LANGUAGE LEARNERS, COOPERATIVE LEARNING

EXTEND

Tell students that the South has been the birthplace of many musical styles. Have interested students conduct research on southern music. Styles include bluegrass, *conjunto*, country western, jazz, Cajun, zydeco. Students may want to play for the class brief examples of the music.
BLOCK SCHEDULING

A Louisiana shrimper clears the day's catch. The Gulf Coast is the richest shrimp-producing region in the United States.

Interpreting the Visual Record How do you think oil spills could affect the local economy here?

Warm weather and beautiful beaches draw many vacationers to resorts in the South. People vacation in eastern Virginia, Florida, and the coastal islands of the Carolinas and Texas. Tourist attractions in cities like New Orleans, San Antonio, and Nashville also attract many visitors.

Many cities in the South have important links with countries in Central and South America. Miami is an important travel connection with Caribbean countries, Mexico, and South America. Atlanta, Houston, and Dallas also are major transportation centers.

✓ **READING CHECK:** *Places and Regions* How does geography affect the economic resources of the South? Supports agriculture, industry, and the transportation of goods.

Florida produces about 75 percent of the oranges and grapefruits grown in the United States.

go.
hrw
.com
Homework Practice Online
Keyword: SJ3 HP11

Section Review 2

Define and explain: barrier islands, wetlands, sediment, diversify

Working with Sketch Maps On the map that you created in Section 1, label the 12 states of the South. Then label the following: Everglades, Okefenokee Swamp, Mississippi Delta, Blue Ridge Mountains, Great Smoky Mountains, Cumberland Plateau, Ozark Plateau, High Plains, Atlanta, Houston, New Orleans, Miami, Dallas.

Reading for the Main Idea

1. *Places and Regions* What areas of the South are wetlands?

2. *Places and Regions* Which southern state has many climates? What are these climates?

3. *Places and Regions* How has the economy of the South changed in recent years?

Critical Thinking

4. **Supporting a Point of View** Use what you learned in this section, and what you know to support this statement: Industry should continue to grow in the South.

Organizing What You Know

5. **Summarizing** Copy the following graphic organizer. Use it to describe the resources and industries that contribute to the economy of the South.

Economic Activities in the South

Section Review 2

Answers

Define For definitions, see: barrier islands, p. 241; wetlands, p. 241; sediment, p. 241; diversify, p. 243

Working with Sketch Maps Maps will vary, but listed places should be labeled in their approximate locations.

Reading for the Main Idea

1. Everglades, Okefenokee Swamp, Mississippi Delta, coastal marsh areas (NGS 4)

2. Texas; humid subtropical, desert, highland (NGS 4)

3. It has become more urban and industrial. (NGS 12)

Critical Thinking

4. Answers will vary, but students should note that the South's climate and geography attract business and workers.

Organizing What You Know

5. farming, fishing, textile workers, tourism, oil refineries, computer/software companies, publishing companies

◄ **Visual Record Answer**

pollute beaches that attract vacationers; tourism, the main business, dries up

Section 3

Objectives

1. Describe the landform regions of the Midwest.
2. Identify some agricultural products of the Midwest.
3. Describe the industries of the Midwest.

Section 3 RESOURCES

Reproducible
◆ Guided Reading Strategy 11.3

Technology
◆ One-Stop Planner CD–ROM, Lesson 11.3
◆ Homework Practice Online
◆ HRW Go site

Reinforcement, Review, and Assessment
◆ Main Idea Activity 11.3
◆ Section 3 Review, p. 247
◆ Daily Quiz 11.3
◆ English Audio Summary 11.3
◆ Spanish Audio Summary 11.3

FOCUS

LET'S GET STARTED

Copy the following instructions and questions onto the chalkboard: *Find the Great Lakes on the section map. Which lake is largest? Which is smallest? Which border Canada?* Discuss responses. Challenge students to make a word using the first letter in the name of each lake. Point out that many people use the word *HOMES* to help them remember the lakes' names. Then tell students that in Section 3 they will learn more about the importance of the Great Lakes to the Midwest.

Building Vocabulary

Write the key terms on the board and have students locate and read the definitions in the text or in the glossary. Ask students to think of problems that might be caused by **droughts**. *(Possible answers: Farmers could not grow crops; rivers might dry up; fires could start in dry areas.)* Then call on students to point out on a wall map the locations of the **Corn Belt** and the **Dairy Belt**.

Section 3 — The Midwest

Read to Discover

1. What types of landforms and climates are found in the Midwest?
2. What are some agricultural products grown in the region?
3. What are the main industries of the Midwest?

WHY IT MATTERS

Managing pollution is a challenge for the U.S. government. Use CNNfyi.com or other current events sources to find out pollution problems and clean-up efforts in the Midwest. Record your findings in your journal.

Define
droughts
Corn Belt
Dairy Belt

Locate
Chicago
Detroit

Gateway Arch, St. Louis, Missouri

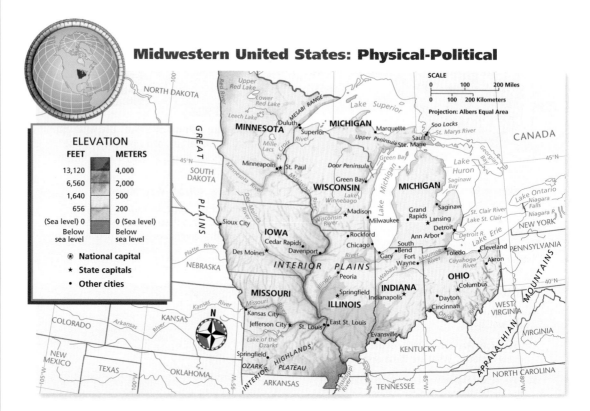

Midwestern United States: Physical-Political

ELEVATION

FEET	METERS
13,120	4,000
6,560	2,000
1,640	500
656	200
(Sea level) 0	0 (Sea level)
Below sea level	Below sea level

⊛ National capital
★ State capitals
• Other cities

Teaching Objective 1

ALL LEVELS: (Suggested time: 30 min.) Copy the following graphic organizer onto the board, omitting the italicized words and check marks. Ask students to copy the table into their notebooks. Pair students and have each pair fill in the names of the Midwestern states in the first column. Then have them identify the physical features of the states by placing check marks in the boxes below features located in or lakes that border each state. **ENGLISH LANGUAGE LEARNERS, COOPERATIVE LEARNING**

PHYSICAL FEATURES OF THE MIDWEST					
State	Interior Plains	Lake Superior	Lake Michigan	Lake Huron	Lake Erie
Ohio	√				√
Michigan	√	√	√	√	√
Indiana	√		√		
Illinois	√		√		
Wisconsin	√	√	√		
Minnesota	√	√			
Iowa	√				
Missouri	√	√			

Landforms and Climate

The Midwest includes eight states: Ohio, Michigan, Indiana, Illinois, Wisconsin, Minnesota, Iowa, and Missouri. All but Iowa and Missouri have shorelines on the Great Lakes, the largest freshwater lake system in the world.

Most of the Midwest lies within the Interior Plains, a region of flat plains and low hills. The most rugged areas are in the Ozark Plateau, which stretches into southern Missouri. During the Ice Age the northern part of the region was eroded by glaciers. These glaciers created the Great Lakes. They also left behind thin soils and many thousands of smaller lakes.

The entire Midwest has a humid continental climate with four distinct seasons. The whole region experiences cold arctic air and snow in winter. The region is subject to thunderstorms, tornadoes, and occasional summer **droughts**. Droughts are periods when little rain falls and crops are damaged.

✓ **READING CHECK:** *Places and Regions* In which landform and climate regions are the Midwest located? Interior Plains; humid continental

Economy

Agriculture Good soils and flat land have helped make the Midwest one of the world's great farming regions. Farmers produce corn, dairy products, and soybeans and raise cattle. The core of the Midwest's corn-growing region is the **Corn Belt**. It stretches from central Ohio to central Nebraska. Much of the corn is used to feed livestock, such as beef cattle and hogs.

States in the **Dairy Belt** are major producers of milk, cheese, and other dairy products. Most of the Dairy Belt is pasture. Located north of the Corn Belt, the area includes Wisconsin and most of Minnesota and Michigan.

Corn: From Field to Consumer

The northern part of the Midwest has thousands of lakes. Minnesota alone has more than 10,000 lakes.

A Corn can be processed in a variety of ways. Some corn is cooked and then canned.

B Corn is ground and used for livestock feed.

C Corn also might be wet-milled or dry-milled. Then grain parts are used to make different products.

D Corn by-products, such as cornstarch and corn syrup, are used to make breads, breakfast cereals, and snack foods.

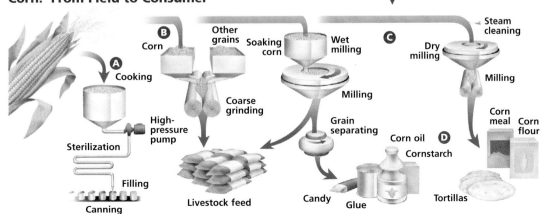

ENVIRONMENT AND SOCIETY

The Corn Belt could easily be called the Soy Belt instead. The production of soybeans has been an economic boon to American farmers. It has also helped to improved soil quality. The plant was introduced into the region at the beginning of the 1800s. A visitor to China smuggled a few soybean seeds out of that country, where the plant had been domesticated for about 5,000 years, and brought them to the United States.

Farmers often rotate soybeans with corn, a process that reduced insect infestations and disease. In addition, soybeans, which are legumes, fix nitrogen in the soil. Corn needs nitrogen-rich soil to thrive. The beans themselves are consumed by humans and livestock and are made into numerous products.

Critical Thinking: Why might farmers in the Midwest want to plant soybeans?

Answer: The plants improve soil quality, allowing other crops to be grown. They can also be sold.

ALL LEVELS: (Suggested time: 45 min.) Organize students into four or five groups. Have the groups meet to discuss where they would choose to live and what jobs they would want to pursue if they lived in the Midwest. Ask students if they would prefer to live in the Corn Belt or the Dairy Belt and be involved in agriculture, or live in one of the Midwest's big cities, such as Chicago, St. Louis, or Detroit, where they might work in business or manufacturing. Ask students to be specific in their answers. For example, they should name the crops they might grow as farmers or the industry in which they might work in a city. Have the groups create balance sheets on which they record the advantages and disadvantages of each lifestyle. Allow groups to compare their charts.

ENGLISH LANGUAGE LEARNERS, COOPERATIVE LEARNING

➤ASSIGNMENT: Provide students with a copy of Carl Sandburg's poem "Chicago." Ask them to read it and then write a poem about another important Midwestern city in a style that, like Sandburg's, describes the city as if it were a person. Ask volunteers to read their poems to the class.

EYE ON EARTH

Long before ships sailed on the Great Lakes, the Algonquin and Iroquois knew that crossing the lakes in the fall was dangerous. Legends told of a Mighty Serpent of the Lakes whose wrath caused great gusts of wind and walls of water to drown travelers. Sailors today speak of the Witch of November, who is said to stir up gale-force winds and gigantic waves. There is no witch, of course, but November storms on the lakes can be fierce.

Meteorologists suggest that the jet stream in the area pulls down cold Arctic air. When it meets warm, humid air from the Gulf of Mexico, violent storms can result.

Activity: Have students explore the basis of legends related to other locations. Have them report their findings in paragraphs.

Connecting to Technology
Answers
1. Duluth, Sault Ste. Marie, Green Bay, Milwaukee, Chicago, Gary, Detroit, Toledo, Cleveland, Erie, Buffalo; were near water and easy shipping
2. Lake Superior; Lake Ontario

CONNECTING TO *Technology*

The Great Lakes and St. Lawrence Seaway

The St. Lawrence Seaway permits ships to go between the Great Lakes and the Atlantic Ocean. This waterway's canals and locks allow ships to move from one water level to another. The difference in water levels is significant. For example, Lake Erie is about 570 feet (174 m) above sea level. Farther to the east, Montreal is about 100 feet (30 m) above sea level. Ships moving from one water level to another enter a lock. The water level inside the lock is raised or lowered. It is changed to match the water level in the waterway ahead of the ship. From there, the ship can move on to the next lock.

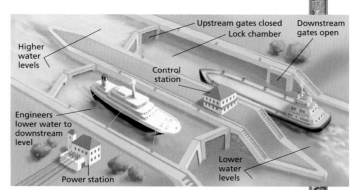

Moving through a canal lock

Understanding What You Read
1. What U.S. cities are located along the shores of the Great Lakes? Why might they have been established here?

2. Which of the Great Lakes lies at the highest elevation? Which lies at the lowest elevation?

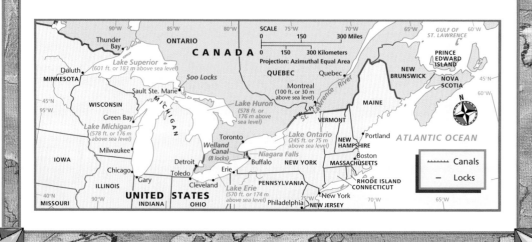

CLOSE

Have students take turns naming Midwestern states. Have each student point out the state on a wall map of the United States and relate a fact about its landforms, agricultural products, or industries.

REVIEW AND ASSESS

Have students complete the Section Review. Then have each student create a crossword puzzle based on the material in Section 3. Tell students to include clues for 10 terms. Pair students and invite them to solve each other's puzzles. Then ask students to complete Daily Quiz 11.3.
COOPERATIVE LEARNING

RETEACH

Have students complete Main Idea Activity 11.3. Organize the class into four groups to plan state pavilions for a Midwest Fair. Assign two states to each group and have members design displays to educate visitors about the landforms, agriculture, and industry of their states.
ENGLISH LANGUAGE LEARNERS

EXTEND

Have interested students choose a Midwestern state and compile a fact sheet on that state. Students should include information such as population, area, state capital, state bird, state flower, nickname, and state motto. They may provide additional facts such as how the state got its name and famous people born in the state.
BLOCK SCHEDULING

Many of the Midwest's products are shipped to markets by water. One route is along the Mississippi River to the Gulf of Mexico. The other is through the Great Lakes.

Industry Chicago, Illinois, is one of the busiest shipping ports on the Great Lakes and has one of the world's busiest airports. Chicago is linked to the rest of the region by highways and railroads. In the late 1800s, its industries attracted many immigrants. They worked in steel mills, meat-packing plants, and other businesses. Today, Chicago is the third-largest city in the United States.

Other Midwest cities such as Cleveland, Detroit, and Milwaukee were also founded on important transportation routes, either on the Great Lakes or on major rivers. The locations of these cities gave industries access to nearby farm products, coal, and iron ore. Those resources have supported thriving industries such as food processing, iron, steel, machinery, and automobile manufacturing. Detroit, Michigan, has been the nation's leading automobile producer since the early 1900s.

On the upper Mississippi River, Minneapolis and St. Paul are major distribution centers for agricultural and industrial products from the upper Midwest. On the western bank of the Mississippi River is St. Louis. About two hundred years ago, it became the center for pioneers heading west and the nation's leading riverboat port.

The Midwest's traditional industries declined in the late 1900s. In addition, industrial pollution threatened the Great Lakes and surrounding areas. In response, companies have modernized their plants and factories. The region has also attracted new industries. Many produce high-technology products. The Midwest is again a prosperous region. Stricter pollution laws have made many rivers and the Great Lakes much cleaner.

✔ **READING CHECK:** *Places and Regions* What are the major economic activities of the Midwest? *agriculture, industry, transportation*

▲
The Sears Tower (at left) in downtown Chicago is one of the world's tallest buildings.

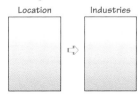
go. **Homework**
hrw **Practice**
.com **Online**
Keyword: SJ3 HP11

Section Review 3

Define and explain: droughts, Corn Belt, Dairy Belt

Working with Sketch Maps On the map that you created in Section 2, label the eight states of the Midwest. Then label Chicago, Detroit, and St. Louis.

Reading for the Main Idea

1. *Places and Regions* Which Midwestern states are in the Corn Belt? Which are in the Dairy Belt?

2. *Places and Regions* Why are Chicago and Detroit important Midwestern cities?

3. *Human Systems* How do farm and industrial products from the Midwest get to markets elsewhere?

Critical Thinking

4. **Drawing Inferences and Conclusions** How might the physical geography of the Midwest be different if the Ice Age had not occurred?

Organizing What You Know

5. **Cause and Effect** Copy the following graphic organizer. Use it to show how location on important transportation routes helped the economy of Midwest cities to grow.

Location	Industries
	⇨

Section Review 3

Answers

Define For definitions, see: droughts, p. 245; Corn Belt, p. 245; Dairy Belt, p. 245

Working with Sketch Maps Maps will vary, but listed places should be labeled in their approximate locations.

Reading for the Main Idea

1. Corn Belt—Ohio, Indiana, Illinois, Iowa, Missouri; Dairy Belt—Wisconsin, Minnesota, Michigan (NGS 4)

2. Chicago—busy transportation center with airports, highways, railroads, Great Lakes port; Detroit—nation's leading auto producer (NGS 12)

3. many shipped by water on the Mississippi River or through the Great Lakes (NGS 11)

Critical Thinking

4. fewer lakes and rivers; soils less fertile

Organizing What You Know

5. Location—ports on Great Lakes, major rivers; Industries— food processing, iron, steel, machinery, automobiles, high-technology products

Section 4

Objectives

1. Identify the major landform regions of the Interior West.
2. Describe the climates of the Interior West.
3. List some economic activities of the Interior West.

FOCUS

LET'S GET STARTED

Copy the following question onto the chalkboard: *What images come to mind when you think of the Great Plains?* Have students write down their answers. Discuss responses. *(Many students will probably mention large, flat areas covered by prairie grasses.* Point out that the region also has a variety of other landforms, including hills, mountains, and sand dunes. Tell students they will learn more about the Great Plains and other features of the Interior West in Section 4.

Building Vocabulary

Write the key terms on the chalkboard and have students read their definitions from the glossary or text. Ask students which words relate to landforms (**badlands**), climate (**chinooks**), agriculture (**Wheat Belt, center-pivot irrigation**), mining (**strip mining**), and tourism (**national parks**).

Section 4 RESOURCES

Reproducible
- Guided Reading Strategy 11.4

Technology
- One-Stop Planner CD–ROM, Lesson 11.4
- Homework Practice Online
- HRW Go site

Reinforcement, Review, and Assessment
- Main Idea Activity 11.4
- Section 4 Review, p. 252
- Daily Quiz 11.4
- English Audio Summary 11.4
- Spanish Audio Summary 11.4

Section 4 — The Interior West

Read to Discover

1. What are the major landform regions of the Interior West?
2. What are the characteristics of the Interior West's climates?
3. What economic activities are important in the Interior West?

Define
badlands
chinooks
Wheat Belt
center-pivot irrigation
strip mining
national parks

Locate
Phoenix
Las Vegas
Denver

Hopi Kachina

WHY IT MATTERS

Parts of the Interior West have experienced droughts. Use CNNfyi.com or other **current events** sources to find out about droughts in this region. Record your findings in your journal.

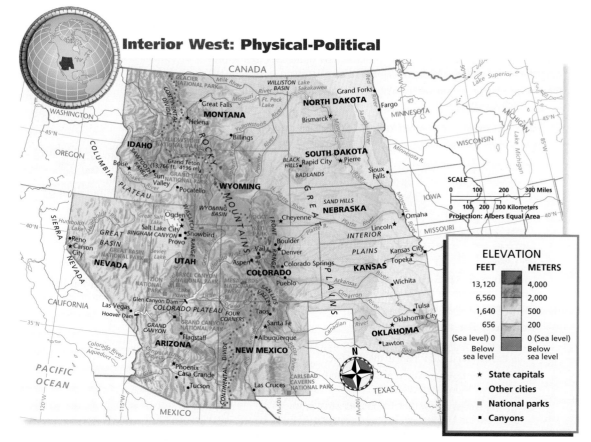

Interior West: Physical-Political

ELEVATION

FEET	METERS
13,120	4,000
6,560	2,000
1,640	500
656	200
(Sea level) 0	0 (Sea level)
Below sea level	Below sea level

★ State capitals
• Other cities
■ National parks
▪ Canyons

Teaching Objective 1

LEVEL 1: (Suggested time: 40 min.) Have students imagine they are landscape artists who recently visited the Interior West. Ask them to sketch four of the landforms they saw there. Display the drawings. Then lead a class discussion about the physical features of the Interior West. **ENGLISH LANGUAGE LEARNERS**

LEVELS 2 AND 3: (Suggested time: 30 min.) Have each student write a list of six terms that are related to the landforms of the Interior West. Then have each student exchange lists with a partner and write a paragraph or two describing the landforms of the Interior West. Tell students that they should use all of the terms they have received in their descriptions. Call on volunteers to share their work with the class. **COOPERATIVE LEARNING**

Landforms

The Interior West includes the states of North Dakota, South Dakota, Nebraska, Kansas, Oklahoma, Montana, Wyoming, Colorado, Idaho, Utah, Nevada, New Mexico, and Arizona. These states occupy three landform regions: the Great Plains, the Rocky Mountains, and the Intermountain West.

The Great Plains were formed over millions of years by the depositing of sediment from mountains. Rivers carried this sediment onto the plains, slowly increasing their elevation. The plains are known for their flat, sweeping horizons. A few areas, however, feature more varied landforms. The Sand Hills of Nebraska are ancient, grass-covered sand dunes. In the Dakotas, rugged areas of soft rock called **badlands** are found. Badlands are areas that have been eroded by wind and water into small gullies. They have little vegetation or soil.

West of the Great Plains are the Rocky Mountains. They stretch from the Arctic through Idaho, Montana, Wyoming, Utah, Colorado, and New Mexico. The Rockies are actually a series of mountain ranges, passes, and valleys.

Two other major landforms, the Great Basin and the Colorado Plateau, are located west of the Rockies. They are both part of the Intermountain West region. The rivers of the Great Basin do not reach the sea. Instead they flow into low basins and dry up. There they leave behind dry lake beds or salt flats. The Colorado Plateau is known for its deep canyons. The largest of these is the Grand Canyon.

✓ **READING CHECK:** *Places and Regions* What kinds of landforms are found in the Interior West? *flat plains, sand dunes, mountain ranges, passes, valleys, basins, canyons*

▲ Storms like this one are common on the Great Plains of Wyoming.

◄ The Rocky Mountains create beautiful views in much of Colorado. The highest mountain peaks reach elevations where trees cannot grow.

One of the most recognizable symbols of the United States is Mount Rushmore in Keystone, South Dakota. This monument is a symbol of the birth and development of the country, personified in the 60-foot (18 m) busts of Presidents George Washington, Thomas Jefferson, Abraham Lincoln, and Theodore Roosevelt. Between 1927 and 1941, artist Gutzon Borglum and 400 workers carved the likenesses of these presidents into the Black Hills.

Discussion: Lead a class discussion about why these four presidents where chosen for Mount Rushmore. *(Possible answers: Washington represents the country's founding. Jefferson embodies its political philosophy. Lincoln signifies preservation, and Roosevelt stands for expansion and conservation.)*

Teaching Objective 2

ALL LEVELS: (Suggested time: 20 min.) Copy the following graphic organizer onto the chalkboard, omitting the italicized answers. Ask students to copy the organizer into their notebooks. Have students complete the chart with words that describe the climates of the three Interior West subregions. Discuss the completed charts.

ENGLISH LANGUAGE LEARNERS

CLIMATES OF THE INTERIOR WEST		
Great Plains	**Rocky Mountains**	**Intermountain West**
• *steppe climate* • *dust storms* • *droughts* • *chinooks*	• *highland climate* • *semiarid grass-lands* • *snowy winter slopes*	• *desert climate* • *steppe climate* • *almost no rain in some areas*

COOPERATIVE LEARNING

Tell students that the Grand Canyon is among North America's most spectacular physical features. Rock layers hundreds of millions of years old are visible in the colorful sides of the canyon. Rocks at the bottom of the canyon are several billion years old. Point out that this scenic area is a national park.

Organize the class into groups of four or five. Direct each group to conduct research and write a brief report on Grand Canyon National Park. The report should note when the park was established and what attractions it offers. Students should also note steps the government is taking to correct problems caused by overcrowding in the park. Have students share their reports.

internet connect

GO TO: go.hrw.com
KEYWORD: SJ3 CH11
FOR: Web sites about Grand
Canyon National Park

▲

These cacti are native to the desert climate of Arizona. The short plant is a teddy bear cholla (CHOY-yuh). The tall cactus is a saguaro (suh-WAHR-uh).

Do you remember what you learned about the steppe climate? See Chapter 3 to review.

▶

The Colorado River flows through Marble Canyon in Arizona.
Interpreting the Visual How do you think this canyon was formed?

Visual Record Answer ▶

erosion caused by flowing water

Climate

Most of the Great Plains region has a steppe climate. They are semiarid and become drier toward the west. Temperatures can vary greatly. Winter temperatures in some areas can drop to -40°F (-40°C), while summer temperatures might rise above 110°F (43°C). Droughts are a major climate hazard in this region. During drought periods, dust storms may cover huge areas. In addition, strong, dry winds blow from the Rocky Mountains onto the Great Plains. These winds are called **chinooks**.

The highland climates of the Rocky Mountain region vary, depending on elevation. Semiarid grasslands usually are found at the foot of the mountains. Most of the slopes, on the other hand, are forested. The forests capture the winter snowfall and form the source for the rivers that flow across the Great Plains. Climates in the Great Basin and the Colorado Plateau also vary. Arizona, New Mexico, and Nevada have mostly desert climates. Utah and southern Idaho have mainly steppe climates. Some parts of this region lie in the rain shadow created by the Rocky Mountains. These areas receive almost no rain at all. This is particularly true in low-lying areas of the Great Basin.

Because the region's climate is so varied, many types of vegetation grow in the Interior West. The Great Plains were once covered with grasses, shrubs, and sagebrush. Most of the native vegetation, however, has been cut to allow farming and ranching. The drier desert areas of the Southwest have less vegetation. Bushes and cacti are the most common plants in the region.

✓ **READING CHECK:** *Places and Regions* Why do the Rockies have several different types of climate? different elevations

Teaching Objective 3

LEVEL 1: (Suggesting time: 20 min.) Organize the class into three groups representing ranchers, farmers, and miners. Have each student write down one or two facts from the textbook about his or her assigned economic group. Call on students in each group to state a different fact until all the information has been covered. Then lead a discussion about the economies of the states of the Interior West. **COOPERATIVE LEARNING**

LEVELS 2 AND 3: (Suggesting time: 45 min.) Arrange the class into two groups to prepare a debate. Ask one group to prepare arguments supporting agriculture as the most important economic feature of the Interior West. The other group should argue that mining is more important to the region. Encourage students to conduct additional research for more information. Stage the debate. **COOPERATIVE LEARNING**

Economy

Ranching became important in the Interior West in the 1800s. Great herds of cattle and flocks of sheep roamed the Great Plains. Today, both ranching and wheat farming are common. The greatest wheat-growing area is known as the **Wheat Belt**. It stretches across the Dakotas, Montana, Nebraska, Kansas, Oklahoma, Colorado, and Texas.

Irrigation Because this region receives relatively little rainfall, farmers depend on irrigation to water their crops. Much of the farmland in the Interior West must be irrigated. One method of irrigation uses long sprinkler systems mounted on wheels. The wheels rotate slowly. In this way, the sprinkler irrigates the area within a circle. This is called **center-pivot irrigation**.

Historically most of the water used to irrigate fields in the Interior West was drawn from underground aquifers. Overuse of this water, however, has drained much of the water from these aquifers. As a result, some farmers have begun to seek new sources for water. They want to preserve some of the region's valuable groundwater.

Wheat from the Great Plains is shipped to other countries.

◄ Workers have dug this pit to mine coal near Sheridan, Wyoming. Wyoming has the largest coal deposits in the Interior West.

Linking Past to Present

The Comstock Lode In the late 1850s prospectors discovered gold in western Nevada. Miners near the small community of Virginia City complained that the small deposits of gold they found had to be dug out of a sticky blue-gray mud that clung to their picks and shovels. This troublesome mud, however, was soon identified as the richest deposit of silver ore ever found in the United States. It became known as the Comstock Lode.

Silver mining was the most important activity in Nevada for many years, but the metal began to decline in value in the 1870s. The Comstock Lode, stripped of its gold and silver, was abandoned by 1900. Tourism has become Virginia City's primary industry.

Critical Thinking: How did the Comstock Lode help settlements in Nevada grow?

Answer: Many prospectors were lured to the area by the potential to get rich.

251

Section Review 4

Answers

Define For definitions, see: badlands, p. 249; chinooks, p. 250; Wheat Belt, p. 251; center-pivot irrigation, p. 251; strip mining, p. 251; national parks, p. 252

Working with Sketch Maps Maps will vary, but listed places should be labeled in their approximate locations.

Reading for the Main Idea

1. Great Plains, Rocky Mountains, Great Basin, Colorado Plateau; steppe, highland, desert (NGS 4)

2. has plains suitable for ranching and wheat farming, deposits of minerals for mining, and scenic areas for tourism (NGS 11)

Critical Thinking

3. possible answer: to preserve natural beauty, environment

4. probably not; without it people might not tolerate hot climate

Organizing What You Know

5. Badlands—gullies, no plants or soil; wind and water erosion; Great Basin—dry lake beds, salt flats; rivers flowing into low basins and drying up; Grand Canyon—deep gorges; carved by Colorado River

Our Amazing Planet

Yellowstone is the country's oldest national park, dating back to 1872. Yellowstone National Park has about 200 geysers—including the famous Old Faithful—and 10,000 hot springs.

Mining and Industry Mining is a key economic activity in the Rocky Mountains. Early prospectors struck large veins of gold and silver there. Today Arizona, New Mexico, and Utah are leading copper-producing states. Nevada is a leading gold-mining state. Lead and many other ores are also found in the Interior West.

However, mining can cause problems. For example, coal miners in parts of the Great Plains strip away soil and rock. This process is called **strip mining**. This kind of mining leads to soil erosion and other problems. Today, laws require miners to restore damaged areas.

One of the Interior West's greatest resources is its natural beauty. The U.S. government has set aside large scenic areas known as **national parks**. Among these are Yellowstone, Grand Teton, Rocky Mountain, and Glacier National Parks. Tourists are also drawn to other areas of the Rocky Mountains. Ski resorts like Aspen and Vail in Colorado and Taos in New Mexico attract many people to the region. Another popular attraction is Mount Rushmore. In the early 1900s, sculptors carved the faces of four presidents into one of the Black Hills of South Dakota. Las Vegas, Nevada, is likewise one of the country's most popular tourist destinations.

Interior West cities like Phoenix, Arizona, and Denver, Colorado, are growing rapidly. The population of Phoenix, for example, has nearly doubled since 1980. Many retirement communities have sprung up in the desert. After World War II, the federal government built dams, military bases, and major highways in the area. The widespread use of air conditioning has made the Phoenix area even more attractive.

✓ **READING CHECK:** *Places and Regions* What makes agriculture possible in the climate of the Great Plains? *irrigation*

Homework Practice Online
Keyword: SJ3 HP11

Section Review 4

Define and explain: badlands, chinooks, Wheat Belt, center-pivot irrigation, strip mining, national parks

Working with Sketch Maps On the map that you created for Section 3, label the 13 Interior West states. Then label Phoenix and Las Vegas.

Reading for the Main Idea

1. *Places and Regions* What landform regions and climates are found in the Interior West?

2. *Human Systems* How does geography contribute to the economy of the Interior West?

Critical Thinking

3. **Drawing Inferences** Why has the U.S. government created national parks in this region?

4. **Drawing Inferences and Conclusions** Would cities such as Phoenix be experiencing such rapid growth if air conditioning had not been invented? Why?

Organizing What You Know

5. **Identifying Cause and Effect** Copy the following graphic organizer. Use it to describe landforms of the Interior West and what created them.

Landform	Description	Formation
Badlands		
Great Basin		
Grand Canyon		

Section 5

Objectives

1. Identify the types of landforms found in the Pacific states.
2. Explain how the climates of the Pacific states differ.
3. Explain how the Pacific states contribute to the economy of the United States.

Section 5 The Pacific States

Read to Discover

1. **What types of landforms do the different Pacific states share?**
2. **How do the climates of the Pacific states differ?**
3. **How do the Pacific states contribute to the economy of the United States?**

Define
caldera

Locate
Coast Ranges
Sierra Nevada
Death Valley
Willamette Valley
Cascades
Los Angeles
Seattle

WHY IT MATTERS

Several computer and software companies are located in the Pacific states. Use CNN fyi.com or other **current events** sources to find out what changes have taken place in these industries. Record your findings in your journal.

Computer microchip

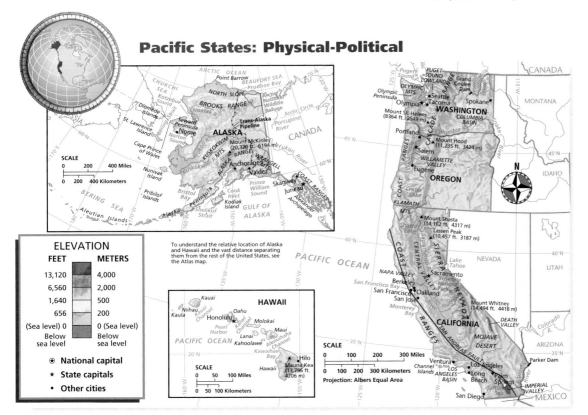

Pacific States: Physical-Political

To understand the relative location of Alaska and Hawaii and the vast distance separating them from the rest of the United States, see the Atlas map.

ELEVATION

FEET	METERS
13,120	4,000
6,560	2,000
1,640	500
656	200
(Sea level) 0	0 (Sea level)
Below sea level	Below sea level

⊛ National capital
★ State capitals
• Other cities

Do you remember what you learned about plate tectonics? See Chapter 2 to review.

The San Andreas Fault caused a severe earthquake in Los Angles in 1994. Scientists warn that future severe earthquakes threaten California.

A View of the Pacific States

If you looked at a map, you might wonder why California, Oregon, Washington, Alaska, and Hawaii are grouped together as a region. Alaska and Hawaii are separated from each other and from the 48 contiguous states. Contiguous states are those that border each other. Yet these states share a physical environment characterized by mountains, volcanoes, and earthquakes. They share cultural, political, and economic similarities as well. In addition, each of the states is working to protect fragile wilderness areas, fertile agricultural lands, and valuable natural resources. In many cases, these features are what first attracted people to the region.

Landforms

California California can be divided into four major landform areas. They are the Coast Ranges, the Sierra Nevada, the Central Valley, and the desert basins and ranges. The Coast Ranges form a rugged coastline along the Pacific.

The Sierra Nevada range lies east of the Coast Ranges. It is one of the longest and highest mountain ranges in the United States. Mount Whitney, the highest peak in the 48 contiguous states, rises above the Sierra Nevada.

Between the Sierra Nevada and the Coast Ranges is a narrow plain known as the Central Valley. This plain stretches more than 400 miles (644 km). The Central Valley is irrigated by rivers that flow from the Sierra Nevada.

To the east of the Sierra Nevada are desert basins and mountain ranges. Included in this region is Death Valley, the lowest point in all of North America.

Earthquakes are common in California due to the San Andreas Fault system. This fault was formed where the Pacific and North American plates meet. The Pacific plate is slowly moving northward along the San Andreas Fault, past the North American plate. The shock waves caused by this activity can create severe earthquakes.

Oregon and Washington Four landform regions dominate Oregon and Washington. As in California the Coast Ranges form the scenic Pacific Northwest coastline. Just east of these mountains are two lowlands areas. The Puget Sound Lowland is in Washington, and the Willamette Valley lies in Oregon. These rich farmlands contain most of the population of each state.

LEVEL 3: (Suggested time: 15 min.) List the following Pacific states on the chalkboard: Oregon, Washington, Alaska, Hawaii. Ask students to identify the physical features of these states that are related to water in some way. *(Possible answers: Oregon—Willamette Valley, Crater Lake; Washington—Puget Sound, Columbia Basin; Alaska—Aleutian Islands, lakes, ice fields, glaciers; Hawaii—Hawaiian Islands, coral reefs)* Then ask students to name some landforms created by tectonic activity. *(Possible answers: calderas, Hawaiian islands)* Call on volunteers to explain how these landforms are similar to those found in California.

Patricia Britt of Durant, Oklahoma, suggests the following activity to help students understand the five Pacific states. After students have read this section, have each student plan a trip to one of the states. Provide students with itinerary forms, atlases, and travel guides. Students should plan where they will stay, what they will see, and about how much the trip might cost. They should also write journal entries describing the landforms, bodies of water, climates, and vegetation they might see on their trip. Students might also discuss regional foods they might enjoy and with which they are familiar.

Crater Lake is what remains of a volcanic mountain that erupted more than 6,000 years ago.

The Cascades are a volcanic mountain range stretching across both states into northern California. The range includes Oregon's Crater Lake, the deepest lake in the United States. Crater Lake fills a huge **caldera**. A caldera is a large depression formed after a major eruption and collapse of a volcanic mountain.

East of the Cascades is a region of dry basins and mountains. Much of this area is known as the Columbia Basin. It is drained by the Columbia River.

Alaska and Hawaii Alaska occupies a huge peninsula that juts out from the northwestern part of North America. The volcanic Aleutian Islands form an arc to the southwest. The mountain ranges of Alaska are some of the most rugged in the world. The state has more than 3 million lakes. Ice fields and glaciers cover 4 percent of the state.

The Hawaiian Islands are a chain of eight major islands and more than 100 smaller islands. Although the islands are volcanic in origin, only one has active volcanoes. This is Hawaii, the largest of the islands. The Hawaiian Islands have scenic coasts with coral reefs that erode into fine, white beach sand.

✓ READING CHECK:
Identifying Cause and Effect
What causes severe earthquakes in California? shock waves created by the movement of the Pacific plate northward along the San Andreas Fault, past the North American plate

Sunlight reflects off the rugged cliffs of Kalau Valley on the Hawaiian island of Kauai.
Interpreting the Visual Why do you suppose few roads have been built in some parts of Kauai?

Across the Curriculum
MUSIC
Slack Key Guitar Spanish and Mexican vaqueros, or cowboys, probably brought the first guitars to Hawaii. In the 1830s, King Kamehameha III hired vaqueros to control the wild cattle population on the islands. Some Hawaiians began playing these guitars and adapting them to suit Hawaiian musical rhythms. By loosening the instrument's strings, the musicians created a rich, full sound. They also used traditional drum and dance rhythms. The Hawaiian guitar style is called *ki ho'alu*, or "slack key."

Activity: Play a recording of Hawaiian slack key guitar music and a recording of American folk guitar music. Ask students to listen for and to describe the differences between the two recordings. Have students discuss why they think this style became popular.

◄ **Visual Record Answer**

The steep, rugged cliffs would make road building impossible.

255

Teaching Objective 2

ALL LEVELS: (Suggested time: 30 min.) Organize the class into several groups and provide each group with a large piece of butcher paper. Have each group draw a large freehand map of the Pacific West states and mark various climates found in the Pacific states.

ENGLISH LANGUAGE LEARNERS

Teaching Objective 3

ALL LEVELS: (Suggested time: 15 min.) Copy the following graphic organizer onto the chalkboard, omitting the italicized answers. Have student fill in the space below each state's name with some of the economic activities found there.

ECONOMIC CHARACTERISTICS OF THE PACIFIC WEST STATES				
California	**Oregon**	**Washington**	**Alaska**	**Hawaii**
agriculture	*forests*	*forests*	*oil*	*tourism*
aerospace	*fish*	*fish*	*forests*	*agriculture*
construction		*software*	*fish*	
computers				
software				
entertainment				
tourism				

Section Review 5

Answers

Define For definitions, see: contiguous, p. 254; caldera, p. 255

Working with Sketch Maps Maps will vary, but listed places should be labeled in their approximate locations.

Reading for the Main Idea

1. mountains, valleys, basins, lowland, islands (NGS 4)

2. marine west coast, Mediterranean, desert, steppe, subarctic, tundra, humid tropical (NGS 4)

3. forests, fish, oil; nuts, fruit, vegetables, fruit; software, tourism, aerospace (NGS 11)

Critical Thinking

4. might make natural areas less inviting and hurt tourism

Organizing What You Know

5. California—Coast Ranges, Sierra Nevada, Central Valley, desert basins; Oregon—Coast Ranges, Willamette Valley, Cascades, Columbia Basin; Washington—Coast Ranges, Puget Sound Lowland, Cascades; Alaska—mountains, ice fields, glaciers; Hawaii—mountains, islands, beaches

▲ Only animals that can survive cold harsh climates, like the polar bear, live in northern Alaska.

Our Amazing Planet

Part of the island of Kauai in Hawaii is considered the wettest place in the world.

Climate

The climates of the Pacific states vary. Seven different climates can be found within this region. They range from tropical to tundra climates.

California The northern coast of California has a marine west coast climate. Along the southern coast and in central California, the climate is Mediterranean. Temperatures are warm all year, even in winter. Summers are dry with hot winds. The basins of eastern California experience desert and steppe climates. Summer temperatures in Death Valley often reach 120° F (49° C).

Oregon and Washington The Cascades divide Oregon and Washington into two climate regions. To the west, the climate is marine west coast. Temperatures are mild year-round, with cloudy, rainy winters and warm, sunny summers. To the east of the Cascades are drier desert and steppe climates.

Alaska and Hawaii A marine west coast climate is found along the southeast coast of Alaska. Most of the state, especially the interior, has a subarctic climate. Summers are short, while winters are long and severe. Most precipitation is in the form of snow. The northern area of Alaska along the Arctic Ocean has a tundra climate.

Hawaii has little daily or seasonal temperature change. During Honolulu's coldest month, the temperature averages 72° F (22° C). It averages 81° F (27° C) during the warmest month. The climate of the eastern slopes of the islands is humid tropical with heavy rainfall.

✓ **READING CHECK:** *Supporting a Point of View* Which climate of the Pacific states would you prefer? Why? Answers will vary, but should be supported by details about a particular climate found in the Pacific states.

► The Trans-Alaska Pipeline snakes across mountain ranges and tundra.

Have students complete the Section Review. Then begin an outline on the chalkboard titled *Geography of the Pacific States*. Write three headings under the title: *Landforms, Climates,* and *Economy.* Have students work in pairs to complete the outline. Then have students complete Daily Quiz 11.5.
COOPERATIVE LEARNING

Have students complete Main Idea Activity 11.5. Then have them work in groups to make cards. Give each group ten index cards. On each of five cards, have students write the name of a Pacific state. On each remaining card, have them name or draw an important physical feature from each state. Have groups match each other's cards.
ENGLISH LANGUAGE LEARNERS

Have students conduct research on the oil resources of Alaska. Ask them to focus their research on the following questions: How important is oil to the economy of Alaska and our country? What debates have taken place over Alaska's oil resources?
BLOCK SCHEDULING

Economy

Each of the Pacific states contributes to the economy of the United States. California is the leading agricultural producer and leading industrial state. Crops include cotton, nuts, vegetables, and fruit. Aerospace, construction, entertainment, computers, software, and tourism are important industries in the state.

Forests and fish are two of the most important resources in Oregon and Washington. Seattle, Washington, is home to many important industries, including a major computer software company.

Alaska's economy is largely based on oil, forests, and fish. Hawaii's natural beauty, mild climate, and fertile soils are its most important resources. Hawaii's volcanic soils and climate are ideal for growing sugarcane, pineapples, and coffee. Millions of tourists visit the islands each year. Both states lack diverse agriculture and industry. As a result most goods must be imported from other states.

 READING CHECK: *Drawing Conclusions* What can you conclude about how Hawaii contributes to the U.S. economy as a result of having a mild climate and fertile soils? Answers will vary, but should include that Hawaii contributes as an agricultural producer.

The cutting down of many trees in Olympic National Forest of Washington has left irregular patterns in the forest.

Interpreting the Visual Record
Why do you think the lumber industry has been the focus of environmental debates?

go.hrw.com **Homework Practice Online**
Keyword: SJ3 HP11

Section Review 5

Define and explain: caldera

Working with Sketch Maps On the map that you created for Section 4, label the five Pacific states. Then label the following places: Coast Ranges, Sierra Nevada, Death Valley, Willamette Valley, Cascades, Los Angeles, Seattle.

Reading for the Main Idea

1. *Places and Regions* What are five physical features of the Pacific states?

2. *Places and Regions* What are the seven climates of the Pacific states?

3. *Human Systems* What resources, agricultural products, and industries are important to the economy of the Pacific states?

Critical Thinking

4. **Analyzing** How might severe hurricane damage affect Hawaii's economy?

Organizing What You Know

5. **Categorizing** Copy the following graphic organizer. Use it to categorize the different landform areas in each of the Pacific states.

State	Landform Areas
California	
Oregon	
Washington	
Alaska	
Hawaii	

Building Vocabulary
For definitions see: megalopolis, p. 236; moraine, p. 237; estuary, p. 238; barrier islands, p. 241; wetlands, p. 241; sediment, p. 241; diversify, p. 243; droughts, p. 245; Corn Belt, p. 245; Dairy Belt, p. 245; badlands, p. 245; chinooks, p. 250; Wheat Belt, p. 251; strip mining, p. 251; caldera, p. 255

Reviewing the Main Ideas

1. The cities are important seaports and are connected by roads, railroads, and airline routes. (NGS 5)

2. The land is low and the water near and around them rises. (NGS 4)

3. good soils, flat land (NGS 4)

4. soil and rock being stripped away by mining; damaged areas are being restored (NGS 5)

5. share a physical environment characterized by mountains, volcanoes, and earthquakes, also have some resources in common (NGS 5)

◀ **Visual Record Answer**

Environmentalists want to protect the forests.

257

ASSESS

Have students complete the Chapter 11 Test.

RETEACH

Organize the class into five groups. Assign each group one region of the United States. Ask students to review the landforms, climates, resources, industries, and cities of their assigned region. Provide butcher paper and divide it into five frames. Have each group create a mural scene in one of the frames that illustrate these aspects of their region. Display the murals.
ENGLISH LANGUAGE LEARNERS, COOPERATIVE LEARNING

PORTFOLIO EXTENSIONS

1. Literature that uses local color incorporates characteristics—such as speech patterns, specific locations, or regional occupations—of a particular place. Have each student write a short story using local color about their own area or another within the United States. Ask them to underline words that identify the characteristics of the location.

2. Have students use the library and other community resources to find out about a conflict in or near their area between resource development and environmental protection. Students should gather information on both sides of the debate. Ask students to present their findings as a written interview or news report.

CHAPTER 11 Review ANSWERS

Understanding Environment and Society

Answers will vary, but students should mention that coal mining is important in Middle Atlantic and Interior West states and that strip mining causes soil erosion. Use Rubric 9, Presentations, to evaluate student work.

Thinking Critically

1. No, regions are defined also by common landforms, climates, and economies.
2. Possible answers: glaciers; waves and tides; river rapids; waterfalls; earthquakes; volcanoes; wind; storms
3. Tourism brings money into the regions and contributes to their economies.
4. Answers will vary.
5. They are great source of income for a region. Students most likely will predict that these industries will grow even larger because technology keeps advancing.

CHAPTER 11 Reviewing What You Know

Building Vocabulary

On a separate sheet of paper, write sentences to define each of the following words.

1. megalopolis
2. moraine
3. estuary
4. barrier islands
5. wetlands
6. sediment
7. diversify
8. droughts
9. Corn Belt
10. Dairy Belt
11. badlands
12. chinooks
13. Wheat Belt
14. strip mining
15. caldera

Reviewing the Main Ideas

1. (Places and Regions) What has helped the megalopolis in the Northeast become a major industrial and financial center?
2. (Places and Regions) Why are many coastal areas of the South covered by wetlands?
3. (Places and Regions) What factors have helped to make the Midwest a great farming region?
4. (Places and Regions) What mining problems has the Interior West faced and how are they being corrected?
5. (Places and Regions) Why are California, Oregon, Washington, Alaska, and Hawaii grouped together as a region?

Understanding History and Society

Resource Use

Coal is an important natural resource in parts of the United States. The coal mining industry has created many jobs. As you prepare an oral report about this industry, consider the following:

• In which states coal mining is important.
• How coal is mined.
• How coal mining affects the environment.

Thinking Critically

1. **Evaluating** Are regions defined solely on the basis of location? Explain your answer.
2. **Analyzing** What natural forces have contributed to shaping the landforms in the regions of the United States?
3. **Drawing Conclusions** Why is tourism important to many regions?
4. **Supporting a Point of View** In which region or state in particular would you most like to live in the future? Explain your answer.
5. **Making Generalizations and Predictions** Why have most regions developed high-technology industries? What do you think will happen in the future to these industries?

FOOD FESTIVAL

Each region of the United States contributes to the country's food supply. Have each student create a dinner menu that includes foods from each region. Tell students to choose items that are specialties of the regions. Students may wish to look in cookbooks to learn ways that regional foods are prepared. Invite students to decorate the their menus and, if possible, to bring samples of some recipes to class.

CHAPTER 11
REVIEW AND ASSESSMENT RESOURCES

Reproducible
◆ Readings in World Geography, History, and Culture 11 and 12
◆ Critical Thinking Activity 11
◆ Vocabulary Activity 11

Technology
◆ Chapter 11 Test Generator (on the One-Stop Planner)

◆ Audio CD Program, Chapter 11
◆ HRW Go site
Reinforcement, Review, and Assessment
◆ Chapter 11 Review, pp. 258–259
◆ Chapter 11 Tutorial for Students, Parents,

Mentors, and Peers
◆ Chapter 11 Test
◆ Chapter 11 Test for English Language Learners and Special-Needs Students
◆ Unit 3 Test
◆ Unit 3 Test for English Language Learners and Special-Needs Students

---

Building Social Studies Skills

Map ACTIVITY

On a separate sheet of paper, match the letters on the map with their correct labels.

Aleutian Islands	Detroit
Arctic Ocean	Hawaii
Cape Cod	New Orleans
Chesapeake Bay	Phoenix
Crater Lake	Seattle

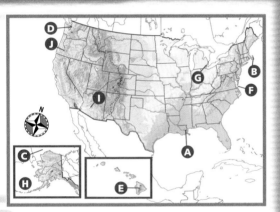

Mental Mapping Skills ACTIVITY

On a separate sheet of paper, draw a freehand map of the United States. Make a key for your map and label the following:

Everglades	Ozark Plateau
Grand Canyon	San Andreas Fault
Mississippi Delta	Lake Erie
White Mountains	Sierra Nevada

WRITING ACTIVITY

Each region of the United States welcomes tourists. Pleasant climates, natural sites, and historic sites attract visitors. Write a short travel ad for a magazine describing the attractions of one region. Be sure to use standard grammar, spelling, sentence structure, and punctuation.

Map Activity
A. New Orleans	G. Detroit
B. Cape Cod	H. Aleutian Islands
C. Arctic Ocean	
D. Seattle	I. Phoenix
E. Island of Hawaii	J. Crater Lake
F. Chesapeake Bay	

Mental Mapping Skills Activity
Maps will vary but listed places should be labeled in their approximate locations.

Writing Activity
Answers will vary, but the information included should be consistent with text material. Use Rubric 2, Advertisements, and Rubric 40, Writing to Describe, to evaluate student work.

Portfolio Activity
Charts will vary, but students should identify examples of their state's diversity. Use Rubric 20, Map Creation, and Rubric 28, Posters, to evaluate student work.

Alternative Assessment

Portfolio ACTIVITY

Learning About Your Local Geography
Research Project Research the climate of your state. Find out the typical weather conditions for the area in which you live. Create a chart that shows the average temperature, rainfall, and other precipitation for each season.

☑ internet connect
Internet Activity: go.hrw.com
KEYWORD: SJ3 GT11

Choose a topic to explore about the regions of the United States:
• Examine growth and development along the Great Lakes and St. Lawrence River.
• Create a brochure on Hawaii's environment.
• Create a newspaper article on the Gulf Coast of Texas.

☑ internet connect

GO TO: go.hrw.com
KEYWORD: SJ3 Teacher
FOR: a guide to using the Internet in your classroom

Canada
Chapter Resource Manager

Objectives	Pacing Guide	Reproducible Resources
SECTION 1 **Physical Geography** (pp. 261–63) 1. Identify Canada's major landforms, rivers, and lakes. 2. Identify the major climate types and natural resources of Canada.	**Regular** 1 day **Block Scheduling** .5 day *Block Scheduling Handbook, Chapter 12*	**RS** Guided Reading Strategy 12.1 **IC** Interdisciplinary Activity for the Middle Grades 6
SECTION 2 **History and Culture** (pp. 264–68) 1. Describe how France and Britain affected Canada's history. 2. Explain how immigrants have influenced Canadian culture.	**Regular** 1.5 days **Block Scheduling** .5 day *Block Scheduling Handbook, Chapter 12*	**RS** Guided Reading Strategy 12.2 **RS** Graphic Organizer 12 **E** Cultures of the World Activity 1
SECTION 3 **Canada Today** (pp. 269–73) 1. Describe how regionalism has affected Canada. 2. Identify the major areas and provinces into which Canada is divided.	**Regular** .5 day **Block Scheduling** .5 day *Block Scheduling Handbook, Chapter 12*	**RS** Guided Reading Strategy 12.3 **SM** Map Activity 12 **SM** Geography for Life Activity 12

Chapter Resource Key

RS Reading Support

IC Interdisciplinary Connections

E Enrichment

SM Skills Mastery

A Assessment

REV Review

ELL Reinforcement and English Language Learners

 Transparencies

 CD–ROM

 Music

 Video

 Internet

 Holt Presentation Maker Using Microsoft® Powerpoint®

 One-Stop Planner CD–ROM

See the *One-Stop Planner* for a complete list of additional resources for students and teachers.

One-Stop Planner CD–ROM

It's easy to plan lessons, select resources, and print out materials for your students when you use the *One-Stop Planner CD–ROM with Test Generator.*

Technology Resources

 One-Stop Planner CD–ROM, Lesson 12.1

 Geography and Cultures Visual Resources with Teaching Activities 11–16

 Homework Practice Online

HRW Go site

 One-Stop Planner CD–ROM, Lesson 12.2

 Homework Practice Online

HRW Go site

 One-Stop Planner CD–ROM, Lesson 12.3

Homework Practice Online

HRW Go site

Review, Reinforcement, and Assessment Resources

ELL	Main Idea Activity 12.1
REV	Section 1 Review, p. 263
A	Daily Quiz 12.1
ELL	English Audio Summary 12.1
ELL	Spanish Audio Summary 12.1

ELL	Main Idea Activity 12.2
REV	Section 2 Review, p. 268
A	Daily Quiz 12.2
ELL	English Audio Summary 12.2
ELL	Spanish Audio Summary 12.2

ELL	Main Idea Activity 12.3
REV	Section 3 Review, p. 273
A	Daily Quiz 12.3
ELL	English Audio Summary 12.3
ELL	Spanish Audio Summary 12.3

🖅 internet connect

HRW ONLINE RESOURCES

GO TO: go.hrw.com
Then type in a keyword.

TEACHER HOME PAGE
KEYWORD: SJ3 TEACHER

CHAPTER INTERNET ACTIVITIES
KEYWORD: SJ3 GT12

Choose an activity to:
• take a trip to the Yukon Territory.
• learn about Canada's First Nations.
• meet the people of Quebec.

CHAPTER ENRICHMENT LINKS
KEYWORD: SJ3 CH12

CHAPTER MAPS
KEYWORD: SJ3 MAPS12

ONLINE ASSESSMENT
Homework Practice
KEYWORD: SJ3 HP12
Standardized Test Prep Online
KEYWORD: SJ3 STP12
Rubrics
KEYWORD: SS Rubrics

COUNTRY INFORMATION
KEYWORD: SJ3 Almanac

CONTENT UPDATES
KEYWORD: SS Content Updates

HOLT PRESENTATION MAKER
KEYWORD: SJ3 PPT12

ONLINE READING SUPPORT
KEYWORD: SS Strategies

CURRENT EVENTS
KEYWORD: S3 Current Events

Meeting Individual Needs

Ability Levels

Level 1 Basic-level activities designed for all students encountering new material

Level 2 Intermediate-level activities designed for average students

Level 3 Challenging activities designed for honors and gifted-and-talented students

English Language Learners Activities that address the needs of students with Limited English Proficiency

Chapter Review and Assessment

E	Readings in World Geography, History, and Culture 13 and 14
SM	Critical Thinking Activity 12
REV	Chapter 12 Review, pp. 274–75
REV	Chapter 12 Tutorial for Students, Parents, Mentors, and Peers
ELL	Vocabulary Activity 12
A	Chapter 12 Test
A	Unit 3 Test
	Chapter 12 Test Generator (on the One-Stop Planner)
	Audio CD Program, Chapter 12
A	Chapter 12 Test for English Language Learners and Special-Needs Students
A	Unit 3 Test for English Language Learners and Special-Needs Students

LAUNCH INTO LEARNING

Ask students what images or impressions of Canada and Canadians they have received from movies and television. *(Students may mention hockey, snow, lumberjacks, Royal Canadian Mounted Police, or other images.)* Ask which they think are true, which are false, and which are exaggerations. Tell them they will learn to distinguish between some facts and fiction about Canada in this chapter. You may want to ask students what impressions they think Canadians may have about citizens of the United States. Which of those do students think are true and which false?

Section 1

Objectives

1. Describe Canada's major land-forms, rivers, and lakes.
2. Identify Canada's major climate types and natural resources.

LINKS TO OUR LIVES

You may want to emphasize the importance of knowing about Canada's geography by sharing these reasons with your students:

▶ Canada and the United States share the longest unguarded boundary in the world. We are also allies.

▶ Each country is the other's most important trading part-ner. Changes in either country's government can affect that relationship.

▶ We share a language, some aspects of history, and many cultural traditions.

▶ Canada is a beautiful country that offers spectacular scenery, fascinating historical sites, and multicultural entertainment for the visitor.

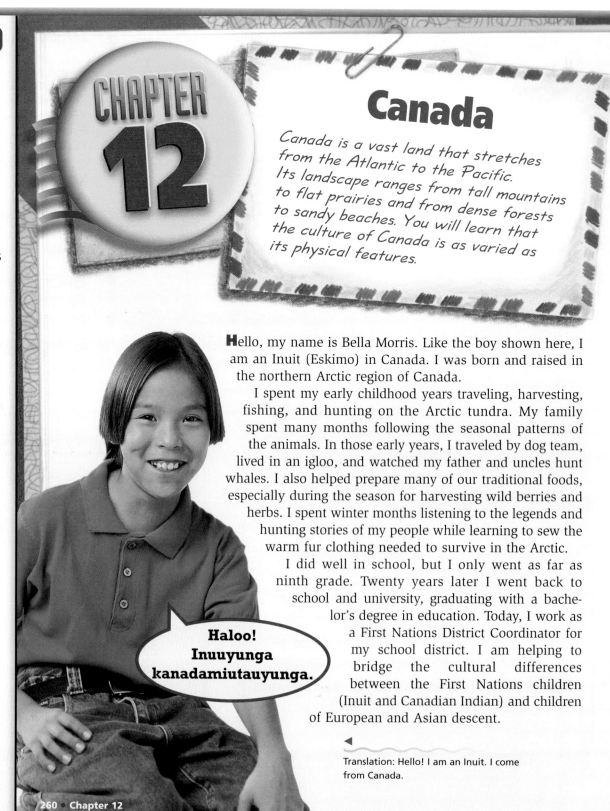

Canada

Canada is a vast land that stretches from the Atlantic to the Pacific. Its landscape ranges from tall mountains to flat prairies and from dense forests to sandy beaches. You will learn that the culture of Canada is as varied as its physical features.

Hello, my name is Bella Morris. Like the boy shown here, I am an Inuit (Eskimo) in Canada. I was born and raised in the northern Arctic region of Canada.

I spent my early childhood years traveling, harvesting, fishing, and hunting on the Arctic tundra. My family spent many months following the seasonal patterns of the animals. In those early years, I traveled by dog team, lived in an igloo, and watched my father and uncles hunt whales. I also helped prepare many of our traditional foods, especially during the season for harvesting wild berries and herbs. I spent winter months listening to the legends and hunting stories of my people while learning to sew the warm fur clothing needed to survive in the Arctic.

I did well in school, but I only went as far as ninth grade. Twenty years later I went back to school and university, graduating with a bache-lor's degree in education. Today, I work as a First Nations District Coordinator for my school district. I am helping to bridge the cultural differences between the First Nations children (Inuit and Canadian Indian) and children of European and Asian descent.

Haloo! Inuuyunga kanadamiutauyunga.

◀

Translation: Hello! I am an Inuit. I come from Canada.

LET'S GET STARTED
Copy the following instructions onto the chalkboard: *During frontier days in the United States, people went west to find wealth and adventure. Why do you think Canadians might go north today? Discuss the question with a partner.* Allow students time to talk with their partners. Discuss student responses. *(Northern Canada is undeveloped and sparsely populated, but its natural resources and beauty offer great opportunities.)* Tell students they will learn more about Canada's physical features and resources in Section 1.

Using the Physical-Political Map
Have students examine the map on this page. Call on volunteers to name the bodies of water that surround Canada *(Pacific Ocean, Atlantic Ocean, Arctic Ocean, Baffin Bay, and Labrador Sea)* and to describe the land boundary that separates it from the United States *(straight line for most of its length).* Ask students to compare the countries' sizes and to find the largest and smallest of Canada's provinces and territories. Ask how their sizes appear to compare to the largest and smallest states in the United States.

Section 1

Physical Geography

Read to Discover

1. What are Canada's major landforms, rivers, and lakes?
2. What are the major climate types and natural resources of Canada?

Define

potash
pulp
newsprint

Locate

Rocky Mountains
Appalachian Mountains
Canadian Shield
Hudson Bay
Great Lakes
St. Lawrence River
Great Bear Lake
Great Slave Lake

WHY IT MATTERS

The United States and Canada share a 5,526-mile-long border. Use **CNNfyi.com** or other current events sources to find out what political and environmental issues surround this border. Record your findings in your journal.

A Canadian coin

Section 1 RESOURCES

Reproducible
◆ Block Scheduling Handbook, Chapter 12
◆ Guided Reading Strategy 12.1
◆ Interdisciplinary Activity for the Middle Grades 6

Technology
◆ One-Stop Planner CD–ROM, Lesson 12.1
◆ Homework Practice Online
◆ Geography and Cultures Visual Resources with Teaching Activities 11–16
◆ HRW Go site

Reinforcement, Review, and Assessment
◆ Section 1 Review, p. 263
◆ Daily Quiz 12.1
◆ Main Idea Activity 12.1
◆ English Audio Summary 12.1
◆ Spanish Audio Summary 12.1

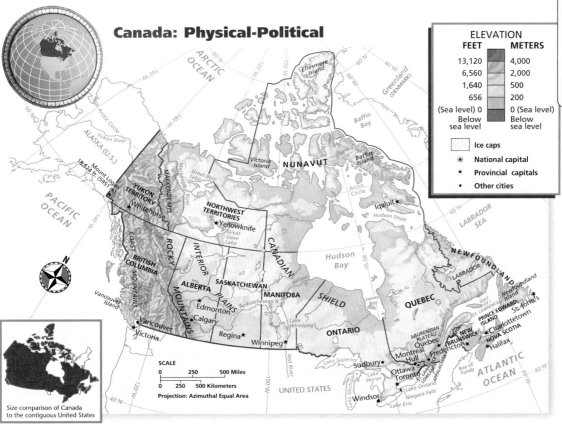

Canada: Physical-Political

ELEVATION

FEET		METERS
13,120		4,000
6,560		2,000
1,640		500
656		200
(Sea level) 0		0 (Sea level)
Below sea level		Below sea level

Ice caps
⊛ National capital
★ Provincial capitals
• Other cities

SCALE
0   250   500 Miles
0   250   500 Kilometers
Projection: Azimuthal Equal Area

Size comparison of Canada to the contiguous United States

Teaching Objectives 1–2

ALL LEVELS: (Suggested time: 30 min.) Copy the following graphic organizer onto the chalkboard, omitting the italicized answers. Use it to help students describe Canada's major landforms, rivers and lakes, climates, and resources. Have students copy the chart into their notebooks and complete it. **ENGLISH LANGUAGE LEARNERS**

Landforms	Lakes and Rivers	Climates	Resources
• *Coast Mountains and Rocky Mountains* • *Canadian Shield* • *fertile farmland in St. Lawrence River Valley and Great Lakes region*	• *Great Lakes* • *thousands of lakes and rivers, many carved by glaciers* • *St. Lawrence River links Great Lakes to Atlantic Ocean*	• *central and eastern— humid continental* • *southwest— marine west coast* • *central and north—subarctic* • *far north— tundra*	• *rich fishing areas* • *tourism* • *fertile soil* • *minerals— nickel, zinc, uranium, lead, copper, gold, silver, coal, potash* • *oil and natural gas* • *forests*

internet connect

GO TO: go.hrw.com
KEYWORD: SJ3 CH12
FOR: Web sites about Canada

EYE ON EARTH

The harsh climate of Canada's western mountains makes life there difficult but not impossible.

Nunataks are small, isolated mountaintops that poke through ice fields, such as those in the Yukon's St. Elias Mountains. These microenvironments, most no bigger than a few acres, shelter tiny flower meadows, lichens, mosses, and rare insects and spiders.

Small rabbitlike animals called pikas are the only year-round mammal residents of the nunataks. They supplement their diet of plants with the bodies of dead birds that wander too far into the ice fields. How the pikas find enough mates to maintain their populations, since they usually venture only a few yards in any direction, is still unknown.

Activity: Have students conduct research on nunataks and other microenvironments around the world. Ask them to construct scale models to show the interrelationships of the microenvironment's plants and animals.

BUILD on WHAT You Know

Do you remember what you learned about glaciers? See Chapter 2 to review.

Banff National Park is Canada's oldest and most famous national park. The park's many visitors are treated to spectacular views of Canada's Rocky Mountains.

Interpreting the Visual Record What types of climate would you expect to find here?

Physical Features

The physical geography of Canada has much in common with that of the United States. Both countries share some major physical regions. For example, the Coast Mountains and the Rocky Mountains extend into western Canada. Broad plains stretch across the interiors of both countries. Also, the Appalachian Mountains extend into southeastern Canada.

The Canadian Shield, a region of rocky uplands, borders Hudson Bay. To the south are some of Canada's most fertile soils, in the St. Lawrence River Valley and the Great Lakes region. The Great Lakes are Lake Erie, Lake Huron, Lake Michigan, Lake Ontario, and Lake Superior.

Canada has thousands of lakes and rivers. Many of Canada's lakes were carved out by Ice Age glaciers. The Great Bear and Great Slave are two of Canada's larger lakes. The most important river is the St. Lawrence. The St. Lawrence River links the Great Lakes to the Atlantic Ocean.

✔ **READING CHECK:** **Places and Regions** What are the major physical features of Canada? *mountains, Canadian Shield, fertile soil in the St. Lawrence River Valley and Great Lakes region, Great Lakes, Great Bear, Great Slave, and St. Lawrence River*

Climate

The central and eastern parts of southern Canada have a humid continental climate. The mildest area of Canada is in the southwest. This region has a marine west coast climate. Here, winters are rainy and heavy snow falls in the mountains. Much of central and northern Canada has a subarctic climate. The far north has tundra and ice-cap climates. Permafrost underlies about half of Canada.

✔ **READING CHECK:** **Places and Regions** What are Canada's climates? *humid continental, marine west coast, subarctic, tundra, and ice cap*

Visual Record Answer

highland, subarctic

CLOSE

Tell students these details about Canada's great outdoors: Wood Buffalo National Park, in Alberta and the Northwest Territories, contains the largest herd of free-roaming bison in the world. The greatest one-day snowfall in Canada's history occurred in January 1974, when more than 46 inches (118 cm) fell at Lakelse Lake, British Columbia.

REVIEW AND ASSESS

Have students complete the Section Review. Then assign each student landforms, rivers and lakes, climate, or resources. Ask each student to write a fill-in-the-blank question, with answer, on his or her topic. Then collect students' questions and use them to quiz the class. Then have students complete Daily Quiz 12.1.

RETEACH

Have students complete Main Idea Activity 12.1. Then focus students' attention on the Section 1 illustrations. Have the students write new captions for them that include information on the section's main topics. **ENGLISH LANGUAGE LEARNERS**

EXTEND

Have students investigate acid rain's effect on Canada. Some members of Canada's government believe that pollution from U.S. industries causes much of the acid rain in the country. Ask students to write mock correspondence between Canadian and U.S. environmental officials that reflects the students' findings. **BLOCK SCHEDULING**

Section Review 1

Answers

Define For definitions, see: potash, p. 263; pulp, p. 263; newsprint, p. 263

Working with Sketch Maps Maps will vary, but listed places should be labeled in their approximate locations.

Reading for the Main Idea

1. Rocky Mountains, Coast Mountains, interior plains, Appalachian Mountains, Canadian Shield, and St. Lawrence River Valley (NGS 4)

2. St. Lawrence River (NGS 4)

Critical Thinking

3. glacial action, humid climate

4. southern part of Pacific coast; marine west coast climate is influenced by nearby Pacific Ocean.

Organizing What You Know

5. Webs will vary but should include the following: coastal waters, lakes, and streams—fish, tourism; fertile soil—farms, ranching; minerals—nickel, zinc, uranium, lead, copper, gold, silver, coal; potash—fertilizer; oil and gas—fuel; trees—lumber, pulp.

Resources

Canada's Atlantic and Pacific coastal waters are among the world's richest fishing areas. Canada's many lakes and streams provide freshwater fish and attract tourists as well. Wheat farmers and cattle producers benefit from Canada's fertile soil.

Minerals are the most valuable of Canada's natural resources. The Canadian Shield contains many mineral deposits. Canada is a leading source of the world's nickel, zinc, and uranium. Lead, copper, gold, silver, and coal are also present. Saskatchewan has the world's largest deposits of **potash**, a mineral used to make fertilizer. Alberta produces most of Canada's oil and natural gas.

A belt of coniferous forests stretches across Canada from Labrador to the Pacific coast. These trees provide lumber and **pulp**. Pulp—softened wood fibers—is used to make paper. The United States, the United Kingdom, and Japan get much of their **newsprint** from Canada. Newsprint is cheap paper used mainly for newspapers.

✓ **READING CHECK:** *Environment and Society* How do Canada's major resources affect its economy? Coastal waters, lakes, streams provide fish, attract tourists; fertile soil is good for farming, ranching; minerals, oil, natural gas are most valuable economic resources; forests provide lumber, pulp.

▲ Canada's resources can be shipped to markets through Vancouver in British Columbia.

▲ Logs are floated down British Columbia's rivers and lakes to sawmills. Timber is an important resource in Canada.

go.hrw.com **Homework Practice Online** Keyword: SJ3 HP12

Section Review 1

Define and explain: potash, pulp, newsprint

Working with Sketch Maps On a map of Canada that you draw or that your teacher provides, label the following: Rocky Mountains, Appalachian Mountains, Canadian Shield, Hudson Bay, Great Lakes, St. Lawrence River, Great Bear Lake, and Great Slave Lake.

Reading for the Main Idea

1. *Places and Regions* What are Canada's major landforms?

2. *Places and Regions* What river links the Great Lakes to the Atlantic Ocean?

Critical Thinking

3. **Drawing Inferences and Conclusions** Why do you think there are so many lakes in Canada?

4. **Drawing Inferences and Conclusions** Where would you expect to find Canada's mildest climate? Why?

Organizing What You Know

5. **Finding the Main Idea** Use the following graphic organizer to identify Canadian resources and their economic benefits.

Natural Resources

Section 2

Objectives

1. Discuss the effect France and Britain had on Canada's history.
2. Describe how immigrants have influenced Canadian culture.

FOCUS

LET'S GET STARTED

Copy the following instructions onto the chalkboard: *Keep in mind what you have learned about Canada's physical geography. What natural resources do you think would draw Europeans to Canada?* Have students write down their answers. Discuss student responses. *(possible answers: fish, furs from the forests, farmland, minerals)* Tell students that in Section 2 they will learn more about the original inhabitants and settlers who have influenced Canada's culture.

Building Vocabulary

Have students find the Define terms' definitions in the text or glossary. All of the terms come from Latin words. *Provincia* means the same as *province.* Add an *s* for the plural form, **provinces.** *Dominus,* from which we get **dominion,** means "master." You may want to introduce and relate the word *dominate* if it is unfamiliar to students. **Métis** comes from *mixticius,* meaning "mixed."

Section 2 RESOURCES

Reproducible
◆ Guided Reading Strategy 12.2
◆ Cultures of the World Activity 1

Technology
◆ One-Stop Planner CD–ROM, Lesson 12.2
◆ Homework Practice Online
◆ HRW Go site

Reinforcement, Review, and Assessment
◆ Section 2 Review, p. 268
◆ Daily Quiz 12.2
◆ Main Idea Activity 12.2
◆ English Audio Summary 12.2
◆ Spanish Audio Summary 12.2

Section 2 History and Culture

Read to Discover
• How did France and Britain affect Canada's history?
• How have immigrants influenced Canadian culture?

Define
provinces
dominion
Métis

Locate
Quebec
Ontario
Nova Scotia
New Brunswick
Newfoundland

Prince Edward Island
British Columbia
Manitoba
Alberta
Saskatchewan

WHY IT MATTERS

European contact forever changed Native Americans who lived in areas that are now the United States and Canada. Use CNNfyi.com or other current events sources to find out what life is like for Native Americans in Canada today.

Totem pole from British Columbia

▲
A French artist shows an early expedition in Canada led by French explorer Jacques Cartier. Cartier explored the St. Lawrence River area up to present-day Montreal in the 1500s.

History

Cree. Déné. Inuit. Mohawk. Ojibwa. These are among the names of Canada's First Nations. Native Canadians lived in all parts of what is now Canada before Europeans arrived.

European Settlement The first Europeans to settle in Canada were probably the Vikings, or Norse. They arrived around A.D. 1000. Norse settlement of North America either failed or was abandoned. European exploration resumed at the end of the 1400s. At that time explorers and fishers from several areas of western Europe began crossing the Atlantic.

These newcomers affected the native inhabitants in many different ways. For example, European diseases such as smallpox killed many people. It is difficult to estimate how many people died. There were also other effects, including the mixing of cultures. Europeans greatly valued the furs that Native Canadians hunted and trapped. The Canadians valued European metal goods like kettles and axes. The two groups began to adopt aspects of each other's culture, including foods, clothing, and means of transportation.

Teaching Objectives 1–2

ALL LEVELS: (Suggested time: 20 min.) Copy the following graphic organizer onto the chalkboard, omitting the italicized answers. Use it to help students trace major events in Canadian history as the country steps into the 2000s. Lead a discussion about the most important aspects of each step and the effect that France and Britain have had on Canada's history. **ENGLISH LANGUAGE LEARNERS**

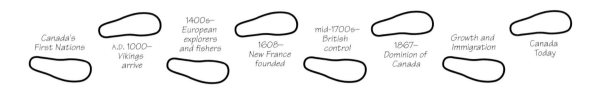

Canada's First Nations — *A.D. 1000—Vikings arrive* — *1400s—European explorers and fishers* — *1608—New France founded* — *mid-1700s—British control* — *1867—Dominion of Canada* — *Growth and Immigration* — Canada Today

The French built the Fortress of Louisbourg in Nova Scotia in the early 1700s. It was an important city and trading port in New France. The British captured the fortress in 1758.

Interpreting the Visual Record What materials were used to build the fortress? What does that tell you about Nova Scotia's resources?

New France France was the first European country to successfully settle parts of what would become Canada. Quebec City was founded in 1608. The French called their new territories in North America New France. At its height, New France included much of eastern Canada and the central United States. New France was important for several reasons. It was part of the French Empire. It was a base that could be used to spread France's religion and culture. It was also an important commercial area for France's empire.

France and Britain were rival colonial powers. Part of their competition included building and defending their empires around the world. The French built trade and diplomatic relations with American Indians. The French did this to increase their power and influence on the continent. Furs, fish, and other products were traded between New France and other parts of the French Empire. Manufactured goods from France and other countries in Europe became the main imports to New France. French missionaries tried to convert native people to Christianity. Some did become Christians, while others held to their traditional beliefs.

New France lasted a century and a half before it was conquered by the British. During that time, it shaped the geography of Canada in important ways. The descendants of French settlers form one of Canada's major ethnic groups today. Almost a quarter of present-day Canadians are of French ancestry. This has deeply affected Canada's culture and politics.

British Conquest and American Revolution The Seven Years' War (1756–63) was mainly fought in Europe. This same period of conflict in the American colonies was called the French and Indian

PLACES AND REGIONS

For visitors to the city of Quebec there are vivid reminders of French settlement.

The city overlooks the St. Lawrence River. Most of the old part of the city of Quebec (Vieux Québec in French) sits at the top of a cliff and is surrounded by a wall. In fact, Vieux Québec is the only walled city in the Americas north of Mexico. The walled city and its environs are sometimes referred to as Haute-Ville, or Upper Town. The Citadelle is part of the city's fortifications. Construction of the star-shaped fort began in 1820 and required more than 30 years. Military troops are still stationed at the Citadelle.

Tourists can descend from the Haute-Ville by taking stairs, including wooden steps called the "Breakneck Stairs," or by riding a modern tram. At the base of the cliff is Lower Town, or Basse-Ville. Here is the Place Royale, a district of narrow streets where Samuel de Champlain founded New France in 1608.

Activity: Have students conduct additional research on major Canadian cities and prepare maps for walking sightseeing tours.

◀ **Visual Record Answer**

wood, stone; plentiful trees and stones

265

Teaching Objective 2

LEVEL 2: (Suggested time: 30 min.) Pair students. Give each pair a yard of adding-machine tape or several pages of accordion-folded computer paper. Point out that Canada developed like a ribbon unrolling from east to west along its southern border. Instruct each pair to arrange and label its paper with East Coast on the right end and West Coast on the left. Have them place Viking explorers, fur traders, railroad builders, ranchers, farmers, and other groups on the paper according to where and when they settled. **COOPERATIVE LEARNING**

LEVEL 3: (Suggested time: 45 min.) Have students conduct a debate on which group of immigrants has most influenced Canadian culture. Ask them to use outside resources. You may need one class period for students to prepare and another for the debate itself.

➤ASSIGNMENT: In advance, write the names of places in Canada on slips of paper. Be sure the places you choose appear on either the chapter or unit map. Have students draw the slips from a hat. Then have them imagine that they are exchange students living in the places they drew. Ask them to write letters home to their families, describing their physical surroundings and the regional history. You may want to have students research and report on a local festival that reflects the heritage of their Canadian home away from home. Students might also illustrate their letters.

Cultural Kaleidoscope

Acadians and Cajuns The term *Acadian* refers to French immigrants who settled in Nova Scotia and nearby areas in the 1600s. Great Britain won these lands from France in 1713. In 1755 British authorities ordered the Acadians to leave. Some fled to nearby Quebec, while others were scattered far and wide. Many families were separated. This episode inspired Henry Wadsworth Longfellow's poem "Evangeline." Some Acadians settled in Louisiana, which was then a French colony, where they became known as Cajuns.

Chart Answer ▶

people of British Isles origin

internet connect

GO TO: go.hrw.com
KEYWORD: SJ3 CH12
FOR: Web sites about Vancouver

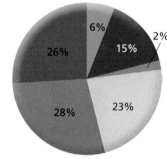

Canadian Ethnic Groups

6%
2%
15%
26%
28%
23%

- British Isles origin
- French origin
- Other European
- Canadian Indian
- Mixed background
- Other, mostly Asian, African, Arab

Source: Central Intelligence Agency, *The World Factbook 2001*

Interpreting the Chart **What is the largest ethnic group in Canada?**

The British flag flies alongside the provincial flag at government buildings in Victoria, British Columbia. Canada is a member of the British Commonwealth of Nations. In fact, the British monarch also is Canada's.

War. As a result of that war the British took control of New France. A small number of French went back to France. However, the great majority of the *habitants* (inhabitants) stayed. For most of them, few changes occurred in their daily activities. They farmed the same land, prayed in the same churches, and continued to speak French. Few English-speaking settlers came to what is now called Quebec.

The American Revolution pushed new groups of people into other areas of British North America. Many United Empire Loyalists—Americans who remained loyal to the king of England—fled north. Most Loyalists either could not stay in the new United States or were too afraid to remain. For some, their political views may have led them to move to British-controlled territories. For many others, it was the desire to get free land. This movement of people was part of a larger westward migration of pioneers.

After the American Revolution, the borders of British North America were redrawn. Quebec was divided into two colonies. Lower Canada was mostly French-speaking, and Upper Canada was mostly English-speaking. The boundary between Upper and Lower Canada forms part of the border between the **provinces** of Quebec and Ontario today. Provinces are administrative divisions of a country. To the east, Nova Scotia (noh-vuh SKOH-shuh) was also divided. A new province called New Brunswick was created where many of the British Loyalists lived.

Creation of Canada For two generations these colonies developed separately. The British also maintained colonies in Newfoundland, Prince Edward Island, the western plains, and the Pacific coast. The colonies viewed themselves as different from other parts of the British Empire. Therefore, the British Parliament created the Dominion of Canada in 1867. A **dominion** is a territory or area of influence. For Canadians, the creation of the Dominion was a statement of independence. They saw a future separate from that of the

United States. The motto of the new Dominion was "from sea to sea." However, it included only New Brunswick, Nova Scotia, and the southern parts of Ontario and Quebec.

How would Canadians create a nation from sea to sea? With railroads. The Canadian heartland in Ontario and Quebec was already well served by railroads. From the heartland the Intercolonial Railway would run east to the Atlantic Ocean. The Canadian Pacific Railroad would be the first of three transcontinental railroads running west to the Pacific Ocean. It was completed in 1885. Both British Columbia and Prince Edward Island were quickly added to the Dominion.

Canada also acquired vast lands between the original provinces and British Columbia. It also expanded to the north. Much of this land was bought from the Hudson's Bay Company, a British fur-trading business. Most of the people in this area were Canadian Indians and **Métis** (may-TEES). Métis, people of mixed European and native ancestry, considered themselves a separate group. With the building of the railroads and the signing of treaties with Native Canadians, the way was opened for settlement of the area. Manitoba became a province in 1870. Alberta and Saskatchewan followed in 1905.

Government Canada is a federation today. It has a central government led by a prime minister. Its 10 provincial governments are each led by a premier. Canada's central government is similar to our federal government. Its provincial governments are much like our state governments. A federal system lets people keep their feelings of loyalty to their own province. At the same time they remain part of a larger national identity.

✓ READING CHECK: **Human Systems** How is Canada's government similar to that of the United States? It has a central government like our federal government and provincial governments like our state governments.

Culture

A history of colonial rule and waves of immigration have shaped Canada today. The country is home to a variety of ethnic groups and cultures. They have combined to form a single country and Canadian identity.

Immigration During the late 1800s and early 1900s, many immigrants from Europe came to Canada. Many farmed, but others worked in mines, forests, and factories. British Columbia became the first Canadian province to have a substantial Asian minority. Many Chinese Canadians helped build the railroads.

These immigrants played an important part in the economic boom that Canada experienced in the early 1900s. Quebec, New Brunswick,

This train follows the Bow River in Banff National Park in Alberta.
Interpreting the Visual Record How did railroad technology help people change their environment?

▼

Canada				
COUNTRY	POPULATION/ GROWTH RATE	LIFE EXPECTANCY	LITERACY RATE	PER CAPITA GDP
Canada	31,592,805 0.99%	76, male 83, female	97%	$24,800
United States	281,421,906 0.9%	74, male 80, female	97%	$36,200

Sources: Central Intelligence Agency, *The World Factbook 2001;* U.S. Census Bureau

Interpreting the Chart Which country has the greater growth rate?

Linking Past to Present
Lighthouses

European immigrants to Canada faced a difficult voyage across the North Atlantic Ocean. Even when they were within sight of their goal, the dangers were not over, for ships could easily be wrecked on the rocky coast.

Newfoundland lighthouses helped sailors find their way to safety. The region's first lighthouse was set up in 1813 to mark the narrow channel that connects St. John's harbor to the ocean.

In the 1800s and early 1900s, many European ships bound for North America set their course for the Cape Race Lighthouse on the southeastern tip of Newfoundland Island. Millions of European immigrants first set foot in North America at Cape Race.

Modern technology has made the old lighthouses obsolete. Most have been automated and no longer require human operators.

Activity: Have students write a lighthouse-keeper's log for several days in the 1800s. Ask them to consider the adventures and problems the keeper might have had.

▲ **Visual Record Answer**

It allowed them to settle more of Canada and reach more remote locations.

◄ **Chart Answer**

Canada

details with webs or other graphics. Then have the teams meet to fill in any gaps and display connections among the facts they have recorded. *(For example, add the Vikings before the French arrival. Connect Asian immigrants with French and British influences by means of business, foods, and festivals.)* Then have students complete Daily Quiz 12.2.
COOPERATIVE LEARNING

RETEACH

Have students complete Main Idea Activity 12.2. Then organize the class into pairs. Have each student write a list of five facts he or she considers the most important ones from Section 2. Then have the students exchange papers with their partners and supply an explanation, cause, result, or other

detail for each of the facts. Discuss the statements and details until all major points have been covered. **ENGLISH LANGUAGE LEARNERS, COOPERATIVE LEARNING**

EXTEND

Canada's national anthem is "O Canada." Have students explain the lyrics to the anthem and then write a new national anthem for Canada, using information in the text as a basis. You may want to let students use a tune with which they are already familiar and simply write new lyrics.
BLOCK SCHEDULING

Section Review 2

Answers

Define For definitions, see: provinces, p. 266; dominion, p. 266; Métis, p. 267

Working with Sketch Maps Maps will vary, but listed places should be labeled in their approximate locations. All of the provinces and territories except Ontario, New Brunswick, Prince Edward Island, and Nova Scotia use lines of latitude for boundaries.

Reading for the Main Idea

1. Colonization influenced language, government, customs, other aspects of culture. (NGS 9)

2. through their work—on farms and railroads, in mines, forests, and in factories—contributed to economic boom of the early 1900s (NGS 9)

Critical Thinking

3. brought metal goods, disease, traded with Native Canadians

4. created by British Parliament as a dominion because the colonies viewed themselves as different from other parts of British Empire

Organizing What You Know

5. Subsequent boxes should be labeled Manitoba, Saskatchewan, Alberta, and British Columbia.

Visitors enjoy the sights and sounds of a children's festival in Vancouver. Vancouver and other Canadian cities are home to large and increasingly diverse populations.

▼

and Ontario produced wheat, pulp, and paper. British Columbia and Ontario supplied minerals and hydroelectricity. By the 1940s Canadians enjoyed one of the highest standards of living in the world.

Movement to Cities In recent years Canadians have moved from farms to the cities. Some settlements in Newfoundland—which became Canada's 10th province in 1949—and rural Saskatchewan disappeared because the people left. Many Canadians have moved to Ontario to find jobs. Others moved to British Columbia for its mild climate. Resources such as oil, gas, potash, and uranium have changed the economies of the western provinces. The economic center of power remains in the cities of southern Ontario and southwestern Quebec. Toronto is now Canada's largest city. Many Canadian businesses have their main offices there.

After World War II, another wave of immigrants from Europe came to Canada. They were joined by other people from Africa, the Caribbean, Latin America, and particularly Asia. Asian businesspeople have brought a great deal of wealth to Canada's economy. Most immigrants have settled in Canada's large cities. Toronto has become one of the most culturally diverse cities in the world. Many Canadians now enjoy Thai, Vietnamese, and other Asian foods. Chinese New Year parades and other colorful festivals attract tourists.

✓ **READING CHECK:** *Human Systems* How has immigration changed Canada? It has given Canada a significant Asian population, helped create an economic boom, and added to the wealth of Canada's economy and to the cultural diversity of the country.

go. hrw .com **Homework Practice Online**
Keyword: SJ3 HP12

Section Review 2

Define and explain: provinces, dominion, Métis

Working with Sketch Maps On the map you created in Section 1, label Quebec, Ontario, Nova Scotia, New Brunswick, Newfoundland, Prince Edward Island, British Columbia, Manitoba, Alberta, and Saskatchewan. Which provinces seem to use lines of latitude as boundaries?

Reading for the Main Idea

1. *Human Systems* How did French and British colonization influence Canada's history?

2. *Human Systems* How did immigrants contribute to Canada?

Critical Thinking

3. **Identifying Cause and Effect** Why did Europeans come to Canada? What was the effect of the arrival of Europeans on the First Nations?

4. **Summarizing** How and why was the country of Canada created?

Organizing What You Know

5. **Sequencing** Copy the following graphic organizer. Use it to explain how the Dominion of Canada developed. Add boxes as needed.

Quebec ⇨ Ontario ⇨ [ ]

Section 3

Objectives

1. Analyze how regionalism has affected Canada.
2. Identify Canada's major areas and provinces.

Section 3 Canada Today

Read to Discover

1. How has regionalism affected Canada?
2. Into what major areas and provinces is Canada divided?

Define

regionalism
maritime
Inuit

Locate

Gulf of St. Lawrence	Montreal	Vancouver
Labrador	Toronto	Yukon Territory
Halifax	Ottawa	Northwest Territories
Windsor	Edmonton	
Quebec City	Calgary	Nunavut
	Winnipeg	

WHY IT MATTERS

French- and English-speaking Canadians continue to have cultural and political differences. Use **CNNfyi.com** or other **current events** sources to find out what conflicts have taken place between these groups. Record your findings in your journal.

The maple leaf, a symbol of Canada

Regionalism

English is the main language in most of Canada. In Quebec, however, French is the dominant language. The cultural differences between English and French Canada have created problems. When Canadians from different regions discuss important issues, they are often influenced by **regionalism**. Regionalism refers to the strong connection that people feel toward their region. Sometimes, this connection is stronger than their connection to their country as a whole.

Regionalism was very important in America during the 1800s. At that time the United States split into the North and the South as its citizens fought the Civil War. The country was divided over issues such as slavery. People supported whichever group shared their beliefs. In Canada many residents of Quebec, or Quebecois (kay-buh-KWAH), believe their province should be given a special status. Quebecois argue that this status would recognize the cultural differences between their province and the rest of Canada. Some even want Quebec to become independent.

On the other hand, many English-speaking Canadians think Quebec already has too many privileges. Others, especially in western Canada, want the provinces to have more freedom from national control. Most Canadians, however, still support a united Canada. Strong feelings of regionalism will continue to be an important issue in Canada's future.

✓ **READING CHECK:** *Human Systems* Why do some people in Quebec want independence from the rest of Canada?

▲ Demonstrators say *"oui"* (yes) to independence for Quebec.

Teaching Objective 1

ALL LEVELS: (Suggested time: 20 min.) Copy the following graphic organizer onto the chalkboard, omitting the italicized answers. Use it to help students understand how regionalism has caused conflict in Canada. Call on students to suggest phrases to fill in the ovals. You may want to extend the activity by challenging students to draw editorial cartoons to illustrate the issues. **ENGLISH LANGUAGE LEARNERS**

Quebecois believe Quebec should have special status.

Some Quebecois want independence for Quebec.

Regionalism in Canada

English-speaking Canadians believe there are too many privileges already for Quebec.

Other provinces, particularly in western Canada, want more freedom from national control.

GLOBAL PERSPECTIVES

Most residents of Canada's eastern provinces and the country's heartland live close to the United States, both physically and culturally.

Modern technology has increased communications between the two countries. U.S. companies control much of Canada's film, video, recording, and book publishing industries. As a result, Canadians are bombarded with American music, movies, and other examples of U.S. popular culture. This disturbs Canadians who want to maintain their own cultural identity.

In response, the Canadian government has passed rules that require that a certain percentage of all television and musical radio programming be written or performed by Canadians. The government has made these rules to help protect Canadian culture from U.S. influence.

Discussion: Ask how students would feel if only Canadian performers appeared on MTV or if their local theater showed only Canadian movies. Then ask if they support the Canadian government's efforts to protect Canadian popular culture. Why or why not?

Visual Record Answer ▶

because the many fish in the waters surrounding the peninsula are the basis of the region's economy

▲
The harbor at Avalon Peninsula is home to many fishing boats.
Interpreting the Visual Record Why do you think most people in the Maritime Provinces live in coastal cities?

Our Amazing Planet

Canada's Bay of Fundy has some of the highest tides in the world — up to 70 feet (21 m). The tide brings water from the North Atlantic into the narrow bay. The bore, or leading wave of the incoming water, can roar like a big truck as the tide rushes in.

The Eastern Provinces

New Brunswick, Nova Scotia, and Prince Edward Island are often called the Maritime Provinces. **Maritime** means "on or near the sea." Each of these provinces is located near the ocean. Prince Edward Island is a small island, and Nova Scotia occupies a peninsula. New Brunswick has coasts on the Gulf of St. Lawrence and on the Bay of Fundy. Newfoundland is usually not considered one of the Maritime Provinces. It includes the island of Newfoundland and a large region of the mainland called Labrador.

A short growing season and poor soils make farming difficult in the eastern provinces. Most of the region's economy is related to forestry and fishing.

Many people in the eastern provinces are descendants of families that emigrated from the British Isles. In addition, many French-speaking families have moved from Quebec to New Brunswick. Most of the region's people live in coastal cities. The cities have industrial plants and serve as fishing and shipping ports. Halifax, Nova Scotia, is the region's largest city.

✓ **READING CHECK:** *Places and Regions* Why are the eastern provinces called the Maritime Provinces? because they are located near the sea

The Heartland

Inland from the eastern provinces are Quebec and Ontario. More than half of all Canadians live in these heartland provinces. In fact, the chain of cities that extends from Windsor, Ontario, to the city of Quebec is the country's most urbanized region.

LEVEL 1: (Suggested time: 45 min.) Explain that Canada's independence day is called Canada Day and occurs on July 1. Organize students into 13 groups or pairs—one for each province or territory. Have each group create a banner for a Canada Day parade to represent its part of the country. You may also want to have the groups write slogans for their provinces or territories. Have students use this unit's atlas and material in Sections 1 and 2 when creating their banners.
ENGLISH LANGUAGE LEARNERS, COOPERATIVE LEARNING

LEVELS 2 AND 3: (Suggested time: 30 min.) Have students work in pairs to create acronym sentences to help them remember Canada's provinces and territories in clockwise order, starting with Newfoundland. Ask students to write one or two sentences comprised of words that begin with *n, p, n, n, q, o, m, s, a, b, y, n,* and *n.* Sentences may be humorous or serious. *(Example: Noisy penguins never need quiet old motorcycles. Strong astronauts become your nearest neighbors.)* Ask a student to read the sentence(s) while his or her partner tells for which province or territory each word stands. Then ask other students to provide a fact for each province or territory. **COOPERATIVE LEARNING**

CONNECTING TO *Literature*

Anne of Green Gables
by Lucy Maud Montgomery

Canada's smallest province, Prince Edward Island, was the birthplace (in 1874) of one of the country's best-loved writers. Lucy Maud Montgomery based her Anne of Green Gables *series on the island she loved. She created characters and situations that lived in the minds of readers. Since its publication in 1908,* Anne of Green Gables *put the tiny island on the map, inspiring tours and festivals. It has even drawn tourists from as far away as Japan. They come to capture the spirit of the brave orphan. In this passage, young Anne is being driven to Green Gables by her new guardian, Matthew. He is kind but hardly imaginative!*

Anne says, "When I don't like the name of a place or a person I always imagine a new one and always think of them so. There was a girl at the asylum [orphanage] whose name was Hepzibah Jenkins, but I always imagined her as Rosalia DeVere." . . .

They had driven over the crest of a hill. Below them was a pond. . . . A bridge spanned it midway and from there to its lower end, where an amber-hued belt of sand hills shut it in from the dark-blue gulf beyond, the water was a glory of many shifting hues. . . .

"That's Barry's pond," said Matthew.

"Oh, I don't like that name, either. I shall call it—let me see—the Lake of Shining Waters. Yes, that is the right name for it. I know because of the thrill. When I hit on a name that suits exactly it gives me a thrill. Do things ever give you a thrill?"

Matthew ruminated.[1] "Well now, yes. It always kind of gives me a thrill to see them ugly white grubs[2] that spade up in the cucumber beds. I hate the look of them."

"Oh, I don't think that can be exactly the same kind of thrill. Do you think it can? There doesn't seem to be much connection between grubs and lakes of shining waters, does there? But why do other people call it Barry's pond?"

"I reckon because Mr. Barry lives up there in that house."

Analyzing Primary Sources
1. Why is *Anne of Green Gables* popular outside of Prince Edward Island?
2. How does imagination affect a person's view of the world?

Definitions [1]ruminate: to think over in the mind slowly [2]grubs: wormlike insect larvae

---

HUMAN SYSTEMS

Lucy Maud Montgomery's *Anne of Green Gables* series is popular around the world, particularly in Japan. Japanese tourists flock to Prince Edward Island to see the places associated with Anne. Every year several Japanese couples marry in the same room where the author was wed.

Many reasons have been proposed for the connection Japanese readers feel with the spunky orphan. When the book first appeared in Japan in 1952, there were many orphans in the war-torn country. Some of them read the Anne books in school and were inspired by her example. Another reason may be Anne's devotion to her studies and her efforts to win a scholarship, since education is highly valued in Japan. Also, the descriptions of Prince Edward Island's beauty seem to appeal to many Japanese readers.

Activity: Have students read various selections from the *Anne of Green Gables* series and discuss how people of different cultures might respond to the passages.

Connecting to Literature Answers
1. Answers will vary but should mention Anne's bravery in the face of hardship.
2. possible answer: makes one's view of the world more interesting or fun

271

Lead a class discussion on the advantages and disadvantages Canadians may face because they live in a huge country with a relatively small population. *(possible answers for advantages: many opportunities for economic growth, minimal crowding, fewer problems with pollution or traffic; possible answers for disadvantages: inconveniently long distances for pleasure or business travel, lack of unity, many people cut off from educational or cultural opportunities)*

Have students complete the Section Review. Then call on a volunteer to describe a place in Canada by starting a sentence with "I am thinking of a place where...." The student should list physical, economic, or cultural details without revealing the location. That student then calls on another to tell which place was described. Continue until all the provinces, territories, and major cities have been discussed. Then have students complete Daily Quiz 12.3.

Section Review 3

Answers

Define For definitions, see: regionalism, p. 269; maritime, p. 270; Inuit, p. 273

Working with Sketch Maps Places should be labeled in approximate locations; most in southeast or on Pacific coast; both regions have waterways for trade, travel

Reading for the Main Idea

1. divisions between English, French speakers and between western, eastern provinces (NGS 6)

2. eastern; heartland; western (prairie and British Columbia); north (NGS 3)

Critical Thinking

3. financial, industrial, governmental, educational, and cultural center

4. created for native Inuit; very small population, harsh environment

Organizing What You Know

5. eastern—New Brunswick, Nova Scotia, Prince Edward Island, Newfoundland; heartland—Quebec, Ontario; western—Manitoba, Saskatchewan, Alberta, British Columbia; north—Yukon and Northwest Territories, Nunavut

Visual Record Answer ▶

flat, with fertile soil

▲
Day draws to a close in Toronto, Ontario. The city is located at the site of a trading post from the 1600s. Today Toronto is one of North America's major cities.

Canada's Prairie Provinces are home to productive farms like this one near Brandon, Manitoba.

Interpreting the Visual Record What can you tell about the physical geography of this region of Canada?

Quebec The city of Quebec is the capital of the province. The city's older section has narrow streets, stone walls, and French-style architecture. Montreal is Canada's second-largest city and one of the largest French-speaking cities in the world. About 3.3 million people live in the Montreal metropolitan area. It is the financial and industrial center of the province. Winters in Montreal are very cold. People in the city center use underground passages and overhead tunnels to move between buildings.

Ontario Ontario is Canada's leading manufacturing province. It is also Canada's most populous province. About 4.7 million people live in the metropolitan area of Toronto, Ontario's capital. Toronto is a major center for industry, finance, education, and culture. Toronto's residents have come from many different regions, including China, Europe, and India. Between Toronto and Montreal lies Ottawa, Canada's capital. In Ottawa many people speak both English and French. It has grand government buildings, parks, and several universities.

✓ **READING CHECK:** **Places and Regions** What are the major cities of Canada's heartland? Quebec, Montreal, Toronto, Ottawa

The Western Provinces

Farther to the west are the major farming regions of Manitoba, Saskatchewan, and Alberta. These three provinces are called the Prairie Provinces. Along the Pacific coast is British Columbia.

Have students complete Main Idea Activity 12.3. Then have students work in pairs to create a third column for the graphic organizer in the Section 3 Review. Title the third column "What I should remember about this province or territory." Discuss student responses.
ENGLISH LANGUAGE LEARNERS

Have students report on the life and achievements of Adrienne Clarkson, who in 1999 became the first Asian Canadian to be appointed governor general, the personal representative of Queen Elizabeth II to Canada. Madame Clarkson, born Adrienne Poy, came to Canada in 1942 from Hong Kong. **BLOCK SCHEDULING**

The Prairie Provinces More people live in Quebec than in all of the Prairie Provinces combined. The southern grasslands of these provinces are part of a rich wheat belt. Farms here produce far more wheat than Canadians need. The extra wheat is exported. Oil and natural gas production also are important in Alberta. Rocky Mountain resorts in western Alberta attract many tourists. The major cities of the Prairie Provinces are Edmonton, Calgary, and Winnipeg.

British Columbia British Columbia is Canada's westernmost province. This mountainous province has rich natural resources, including forests, salmon, and important minerals. Almost 4 million people live in British Columbia. Nearly half of them are in the coastal city of Vancouver. Vancouver is a multicultural city with large Chinese and Indian populations. It also is a major trade center.

✓ **READING CHECK:** *Environment and Society* How does geography affect the location of economic activities in the Prairie Provinces? grasslands good for growing wheat; oil and natural gas reserves also add to economy

The Canadian North

Canada's vast northern lands include the Yukon Territory, the Northwest Territories, and Nunavut (NOO-nah-vuht). Nunavut is a new territory created for the **Inuit** (Eskimos) who live there. *Nunavut* means "Our Land" in the Inuit language. Nunavut is part of Canada, but the people have their own local government. The three territories cover more than one third of Canada but are home to only about 100,000 people. Boreal forests, tundra, and frozen Arctic ocean waters separate isolated towns and villages.

✓ **READING CHECK:** Who lives in Nunavut? the Inuit

In the St. Elias Mountains in the Yukon Territory is the world's largest nonpolar ice field. The field covers an area of 15,822 square miles (40,570 sq km) and stretches into Alaska.

The Inuit have long hunted beluga whales for meat, blubber, and skin.

go.hrw.com
Homework Practice Online
Keyword: SJ3 HP12

Section Review 3

Define and explain: regionalism, maritime, Inuit

Working with Sketch Maps On the map that you created in Section 2, locate and label the major cities of Canada's provinces and territories. Where are most of these major cities located? What may have led to their growth?

Reading for the Main Idea

1. *Places and Regions* How does regionalism affect Canada's culture?
2. *The World in Spatial Terms* Into which provincial groups is Canada divided?

Critical Thinking

3. **Drawing Inferences and Conclusions** What makes the heartland a good area in which to settle?
4. **Finding the Main Idea** Why was Nunavut created?

Organizing What You Know

5. **Categorizing** Use the following graphic organizer to identify Canada's regions and provinces.

Region	Provinces

CHAPTER 12

Review
ANSWERS

Building Vocabulary
For definitions, see: potash, p. 263; pulp, p. 263; newsprint, p. 263; provinces, p. 266; dominion, p. 266; Métis, p. 267; regionalism, p. 269; maritime, p. 270; Inuit, p. 273

Reviewing the Main Ideas

1. second-largest (NGS 4)
2. fish, fertile soil, forests, minerals, oil, and natural gas; southern grasslands of the Prairie Provinces (NGS 4)
3. Native Canadians, Vikings, British, French, Asians, Africans, Caribbean people, Latin Americans, and others (NGS 9)
4. by speaking French and asking that Quebec have special status or be independent (NGS 10)
5. possible answers: on the St. Lawrence River, northeast of Ottawa, southeast of Quebec; Canada's capital and several universities located there (NGS 4)

273

ASSESS

Have students complete the Chapter 12 Test.

RETEACH

Have students meet in small groups to discuss how the essential elements of geography relate to Canada. Assign one student to lead the discussion and another to take notes. When the groups have finished their discussions, have a volunteer from each group use the notes to summarize the discussion for the class. **ENGLISH LANGUAGE LEARNERS, COOPERATIVE LEARNING**

PORTFOLIO EXTENSIONS

1. Organize the class into small groups. Each group will serve as a planning committee to host an international geographers' tour of Canada. Each committee should prepare a detailed, illustrated itinerary that includes as much variety as possible. Have students place copies of the tour package in their portfolios, along with a description of their own contributions to the project.

2. Have students locate Canadian newspapers or magazines. Ask them to compare how an event in the United States is discussed in Canadian and U.S. media and explain how the coverage differs. Instruct students to determine if each country's interests affect the article's main points and style. Have students place both articles and their comparisons in their portfolios.

Review
ANSWERS

Understanding Environment and Society

Presentations will vary, but information included should be consistent with text material. Students should support their arguments and conclude that resources will be severely reduced or totally depleted if no steps are taken. Use Rubric 29, Presentations, to evaluate student work.

Thinking Critically

1. federation; central and provincial governments similar to central U.S. government and states

2. possible answer: has created ethnic and cultural diversity

3. 150 years of French colonization created a strong French-speaking population that currently is in conflict with the English-speaking population.

4. possible answers: shipping, shipbuilding, seafood processing, lumber processing, furniture manufacturing

5. Maritime Provinces—good fishing waters; heartland provinces—urban centers of finance and industry; Prairie Provinces—good farmland; British Columbia—prime location for Pacific trade

Reviewing What You Know

Building Vocabulary

On a separate sheet of paper, write sentences to define each of the following words.

1. potash
2. pulp
3. newsprint
4. provinces
5. dominion
6. Métis
7. regionalism
8. maritime
9. Inuit

Reviewing the Main Ideas

1. (*Places and Regions*) How large is Canada compared to the rest of the world's countries?

2. (*Places and Regions*) What types of natural resources can be found in Canada? Where is Canada's major wheat-farming area?

3. (*Human Systems*) What groups of people have made Canada their home?

4. (*Human Systems*) How do the people of Quebec show their regionalism?

5. (*Places and Regions*) What is Ottawa's relative location? Why is the city important?

Understanding Environment and Society

Resource Use

The Canadian economy depends on industries such as fishing, lumber, mining, and fossil fuels. As you create a presentation for your class, you may want to think about:

• Actions the Canadian government might take to conserve Canada's resources.

• What might happen if nothing is done to conserve Canada's resources.

Thinking Critically

1. **Comparing** How is Canada's government similar to that of the United States?

2. **Finding the Main Idea** How has immigration changed Canada?

3. **Drawing Inferences and Conclusions** How have past events shaped current conflicts in Canada?

4. **Drawing Inferences and Conclusions** If fishing and lumber are the primary industries of the Maritime Provinces, what are some likely secondary industries? List three industries.

5. **Summarizing** What geographic factors are responsible for economic activities in the Canadian Provinces?

FOOD FESTIVAL

Many Canadians enjoy afternoon tea—snacks or a light meal with a cup of hot tea. Have students research and bring teatime foods that represent the many peoples who have influenced Canadian culture. For example, cucumber sandwiches and small, thick pancakes called crumpets could represent the British. Crusty bread and cheese could stand for French immigrants. Students representing Asians could bring steamed dumplings or egg rolls. Be sure to include Canadian Indians and Inuit.

CHAPTER 12 REVIEW AND ASSESSMENT RESOURCES

Reproducible
◆ Reading in World Geography, History, and Culture 13 and 14
◆ Critical Thinking Activity 12
◆ Vocabulary Activity 12

Technology
◆ Chapter 12 Test Generator (on the One-Stop Planner)

◆ Audio CD Program, Chapter 12
◆ HRW Go site

Reinforcement, Review, and Assessment
◆ Chapter 12 Review, pp. 274–275

◆ Chapter 12 Tutorial for Students, Parents, Mentors, and Peers
◆ Chapter 12 Test
◆ Chapter 12 Test for English Language Learners and Special-Needs Students
◆ Unit 3 Test
◆ Unit 3 Test for English Language Learners and Special-Needs Students

Building Social Studies Skills

Map ACTIVITY

On a separate sheet of paper, match the letters on the map with their correct labels.

Great Slave Lake
Great Bear Lake
Newfoundland
Prince Edward Island
Manitoba

Alberta
Toronto
Calgary
Vancouver
Nunavut

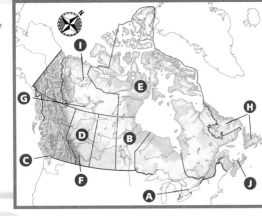

Mental Mapping Skills ACTIVITY

On a separate sheet of paper, draw a map of Canada and label the following:

British Columbia
Great Lakes
Hudson Bay
Ontario

Ottawa
Quebec
Saskatchewan
Yukon Territory

WRITING ACTIVITY

Write three paragraphs explaining where in Canada you might like to live and why. Name a province and describe the geographic, climatic, political, and cultural features that make it attractive to you. Be sure to use standard grammar, sentence structure, spelling, and punctuation.

Alternative Assessment

Portfolio ACTIVITY

Learning About Your Local Geography

Individual Project Canada's cold climate helped make winter sports popular there. What sports are popular in your region? How do they compare with Canada's sports? Create a thematic map of your region that shows the location of different sporting activities.

◢ **internet** connect

Internet Activity: go.hrw.com
KEYWORD: SJ3 GT12

Choose a topic to explore about Canada:
• Take a trip to the Yukon Territory.
• Learn about Canada's First Nations.
• Meet the people of Quebec.

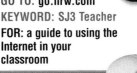

Map Activity
A. Toronto
B. Manitoba
C. Vancouver
D. Alberta
E. Nunavut
F. Calgary
G. Great Slave Lake
H. Newfoundland
I. Great Bear Lake

Mental Mapping Skills Activity
Maps will vary, but the listed places should be labeled in their approximate locations.

Writing Activity
Answers will vary, but information included should be consistent with text material. Check to see that students have written about geographic, climatic, political, and cultural features. Use Rubric 37, Writing Assignments, to evaluate each student's work.

Portfolio Activity
Students should make a clear connection between their region's climate and the sports popular there. Use Rubric 37, Writing Assignments, to evaluate student work.

◢ **internet** connect

GO TO: go.hrw.com
KEYWORD: SJ3 Teacher
FOR: a guide to using the Internet in your classroom

Detecting Cultural Bias

Recognizing cultural bias is an important skill in the study of human geography. Tell students that the word *bias* means "prejudice." Then ask them to create a definition for the term *cultural bias. (Cultural bias is the influence that a person's culture has on his or her view of other cultures and people.)* Tell students that cultural bias can cause a person to base decisions only on his or her own culture or point of view. All people view the world from the context of their own culture, but to make good decisions that do not harm others, we should try to take other cultures and their viewpoints into account.

Organize the class into three groups and assign each group a participating country in NAFTA—Canada, the United States, or Mexico. Have each group conduct research on the popularity of NAFTA in their assigned country. Ask students to evaluate how cultural bias may have played a role in the decisions made regarding NAFTA.

NAFTA is not the first effort made by the United States and Mexico to benefit their economic relationship. In the 1960s the Mexican government began a program to develop its northern border region. Under the Border Industrialization Program, the government encouraged foreign companies to build factories in Mexico to manufacture goods for export. Many of these factories, or *maquiladoras*, are owned by U.S. companies.

Despite economic growth along the border, problems developed. Because Mexico's environmental regulations were not rigorously enforced, industrial waste poured into the air and water of both countries. Many Mexicans feared that Mexico was becoming "Americanized." Some workers in the United States feared that Mexico's economic growth was taking place at their expense. These problems are related to concerns some people have had about NAFTA.

Activity: Have students research *maquiladoras* and their impact on Mexican society. Ask students to present their research in an oral report.

➤ **This Focus On Economy feature addresses National Geography Standards 5, 11, and 16.**

NAFTA

Borders and Barriers Canada, Mexico, and the United States are the three largest countries in North America. They have very large and productive economies. Borders separate each country and create barriers between them. For example, one barrier is the use of different types of currency in each country. In Canada, people use Canadian dollars. In Mexico, they use Mexican pesos. In the United States, people use U.S. dollars. You cannot buy a cheeseburger in Canada with Mexican pesos because of this barrier. Another barrier is the different laws that each country has. This creates a barrier to law enforcement.

The NAFTA agreement was signed by the leaders of Mexico, the United States, and Canada in 1992 and took effect on January 1, 1994. NAFTA created the largest free trade area in the world.

▼

Some barriers between Canada, the United States, and Mexico are being removed. For example, barriers to trade are changing. The North American Free Trade Agreement (NAFTA) is causing these changes. With NAFTA, Canada, the United States, and Mexico have agreed to follow some common rules about trading. These new rules make it easier to trade across borders.

Beginnings of NAFTA NAFTA began on January 1, 1994. The purpose of NAFTA is to improve nearly all aspects of trade between the three countries. It promotes free trade by lowering or removing taxes on goods traded between Canada, the United States, and Mexico. For example, NAFTA eliminates tariffs between them. A tariff is a kind of tax. Tariffs are paid on goods that are exported to other countries. For example, when apples are exported from the United States to Japan, Japan collects a tariff.

Because of NAFTA, many agricultural and industrial products are not taxed when they are imported from the other member countries. Therefore, these goods can be easily traded from one country to another. NAFTA also eases barriers for corporations. For example, a company might be based in the United States. That company is now able to set up factories in Mexico without the problem of special taxes or tariffs.

Going Further: Thinking Critically

Have students recall the observations made in the previous activity. Note that each country has had both benefits and problems as a result of NAFTA. Recall these with the class and record them on the chalkboard. Then have students return to their groups to create a proposal to improve NAFTA that capitalizes on the benefits and proposes solutions to the problems.

NAFTA Trade Flows

▶ This map shows trade among the United States and its NAFTA partners in billions of U.S. dollars. Numbers in parentheses show the percentage of each country's total exports. Trade among all three countries has greatly increased since NAFTA began in 1994.

CANADA
234.2
(86%)

178.5
(23%)

UNITED STATES

108.6
(14%)

ATLANTIC
OCEAN

PACIFIC
OCEAN

123.6
(73.6%)

MEXICO

Removing Barriers Some people in both the United States and Canada have criticized the NAFTA agreement. They believe NAFTA has caused many Americans and Canadians to lose their jobs. Some corporations have closed factories in the United States and Canada. These corporations then opened new factories in Mexico. People are also worried that NAFTA might harm the environment. Under NAFTA, fewer environmental laws and regulations must be followed. This may cause environmental problems such as increased pollution or pesticide use.

Other people support NAFTA. They point out that trade among Canada, the United States, and Mexico has increased since NAFTA began. There are now more jobs in all three countries. For example, the United States exports more agricultural products to Mexico now. NAFTA also helps all three countries by increasing their total production of goods. For example, T-shirts are made from cotton. With NAFTA, cotton grown in Mexico, the United States, or Canada is cheaper than cotton from other countries. This helps cotton growers in all NAFTA countries.

The borders between Canada, the United States, and Mexico are slowly becoming less important. NAFTA is removing trade barriers between these countries and increasing their economic connections.

Understanding What You Read

1. What is the purpose of NAFTA?
2. What are some arguments in favor of NAFTA? What are some arguments against NAFTA?

Understanding What You Read

Answers

1. The purpose of NAFTA is to promote free trade by lowering or removing taxes on goods traded between Canada, the United States, and Mexico.

2. Some arguments against NAFTA are that NAFTA has caused many people in the United States and Canada to lose their jobs and that NAFTA will result in harm to the environment. Some arguments in favor of NAFTA are that trade among Canada, the United States, and Mexico has increased since NAFTA began, which has resulted in an increase in the number of jobs in all three countries. Supporters also feel that NAFTA benefits all three countries by increasing their total production of goods.

Going Further: Thinking Critically

Have students collect old photos of buildings, roads, parks, factories, farms, monuments, and other features of their community or a nearby city. The public library or a local historical society would be a good resource. You will need to make arrangements to duplicate the photos. For some places there are published books that contain historic photos.

Then have students enlarge a published map or draw a large map of the community on butcher paper or posterboard. When the class has collected enough photos, have the students attach them at the edges of the map. Then have students draw arrows that indicate the past or present location of the pictured features. Below each photo, have students attach a snapshot that shows what that place looks like today.

Have students work individually or in pairs to study the photo-map collage and to write paragraphs to describe how the community has changed. Here are some questions to help them get started:

- Have many old buildings been torn down? What has replaced them?
- Have old buildings been restored or modernized?
- How are changes in transportation methods shown in the photos?
- How much open space remains? Have vacant lots and park land been filled in with buildings?
- Are the businesses in the old photos still open? Have they expanded or declined?
- Has the community become more or less crowded?

PRACTICING THE SKILL

1. Students might observe changes in landscapes—new parks, road construction, or housing developments; changes in architecture—new designs for homes and other buildings; or changes in the kinds of new businesses being established—daycare centers, megaplex cinemas, outlet malls, and so on. Students should note whether they think the changes are big or small.

2. Students should identify the person from whom they gathered information about the community. Informants might mention physical changes in architecture or landscape, or cultural changes, as shown by the types of entertainment and recreation people pursue. Informants may mention places that no longer exist.

3. Students might notice differences in style, color, and design as well as content. Students should identify out-of-date information if possible.

➤ This GeoSkills feature addresses National Geography Standards 1, 3, 4, 14, and 17.

Building Skills for Life: Analyzing Changing Landscapes

▲

This map from the mid-1600s shows Quebec when it was a French colony.

The world and its people are always changing. Sometimes these changes happen slowly. They may even be so slow that they are hard to notice. For example, some mountain ranges have built up over millions of years. These ranges are now slowly eroding. Because mountains look almost the same every day, it may be hard to notice changes.

Other times, the world changes quickly. Mudslides can bury villages so fast that people do not have time to escape. In less than 30 seconds, houses and crops might be completely destroyed.

Geographers want to know what the world is like today. However, they also need to know what it was like in the past. Analyzing how places have changed over time can help us better understand the world. For example, suppose you took a trip to Quebec. You would notice that most people there speak French. If you knew that Quebec used to be a French colony, it would help you understand why French is the main language today.

Understanding how places have changed in the past can also help us predict future changes. This is an important goal for people such as city planners, government officials, and transportation engineers.

As you study geography, try to remember that the world is always changing. Think about how places have changed and why these changes are important. Ask yourself how places might change in the future. Be aware that maps, photographs, newspapers, and other tools of geography can become outdated. After all, the world does not wait for geography!

THE SKILL

1. Find an area in your community that you think is changing quickly. Observe the changes. Are new businesses, houses, or roads being built? Are the changes big or small?

2. Talk to a parent, grandparent, or neighbor. Ask how your community, town, or city has changed during the last 20 to 30 years. What are some of the major changes? Does this person remember some old places that are no longer there?

3. Look for an old book, magazine, or map. Try to find one that is at least 30 years old. How is it different from a current book, magazine, or map?

HANDS on GEOGRAPHY

You can analyze changing landscapes by comparing old photographs of a place to new ones. Important changes might be easy to see. For example, new buildings and new roads often stand out on the more recent photographs.

Look at the two photographs below. They show a small section of Las Vegas, Nevada. The photograph on the left was taken in 1963. The photograph on the right was taken 35 years later in 1998. Now compare the two photographs. What do you notice?

◄ Las Vegas, 1963

▲ Las Vegas, 1998

Lab Report

1. How do you know these two photographs show the same place? What evidence do you see?

2. How did this section of Las Vegas change between 1963 and 1998? Do you think these are big changes or small ones?

3. Find a street map of Las Vegas and try to figure out exactly what part of the city the photographs show. Then compare the map with the 1998 photograph. What information does the map give about this section of the city?

Lab Report

Answers

1. The same street layouts are visible in each photograph.

2. Students should notice an increase in the number of buildings and a decrease in the amount of open space. They may also note changes such as new roads and parking lots. Students should conclude that these are big changes.

3. Students might suggest names of streets and highways, elevations, names of public buildings, or other features.

MIDDLE AND SOUTH AMERICA

Direct students' attention to the photographs on these pages. Remind students that animals have adapted in many ways so they can survive in the ecosystems where they live. Focus attention on the frog and butterfly. Point out that the frog's skin secretes a deadly poison. Ask how the creatures' distinctive appearance may help them survive. *(Possible answers: Attackers may not see the butterfly because they can see through its wings. The frog's color may attract mates or warn attackers.)*

Tell students that the buildings in the photos are separated by about 1,660 miles (2,671 km) and that one was built hundreds of years before the other. Yet the two building sites have at least two qualities in common. Ask what they might be. *(Possible answers: apparently built of durable materials, use many straight lines)*

Finally, focus attention on the dancer in the center of the page. You may want to have students try to estimate how many yards of fabric and ribbon were required to make her costume.

UNIT OBJECTIVES

1. Describe the landforms, bodies of water, climates, and resources of Middle and South America.
2. Analyze the influence of ancient American civilizations, European colonialism, and recent events on societies in Middle and South America.
3. Identify the political, economic, social, cultural, and environmental challenges facing the region's countries.
4. Interpret special-purpose diagrams to better understand human-environment interaction in Middle and South America.

Middle and South America

Congress Building, Brasília, Brazil

Dancer, Puerto Vallarta, Mexico

Machu Picchu, Peru, and llama

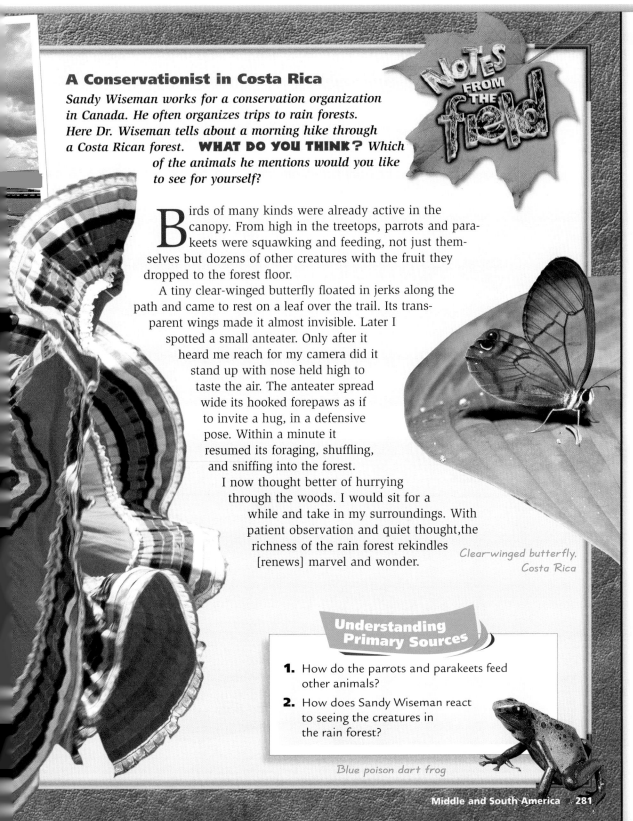

A Conservationist in Costa Rica

Sandy Wiseman works for a conservation organization in Canada. He often organizes trips to rain forests. Here Dr. Wiseman tells about a morning hike through a Costa Rican forest. **WHAT DO YOU THINK?** *Which of the animals he mentions would you like to see for yourself?*

Birds of many kinds were already active in the canopy. From high in the treetops, parrots and parakeets were squawking and feeding, not just themselves but dozens of other creatures with the fruit they dropped to the forest floor.

A tiny clear-winged butterfly floated in jerks along the path and came to rest on a leaf over the trail. Its transparent wings made it almost invisible. Later I spotted a small anteater. Only after it heard me reach for my camera did it stand up with nose held high to taste the air. The anteater spread wide its hooked forepaws as if to invite a hug, in a defensive pose. Within a minute it resumed its foraging, shuffling, and sniffing into the forest.

I now thought better of hurrying through the woods. I would sit for a while and take in my surroundings. With patient observation and quiet thought, the richness of the rain forest rekindles [renews] marvel and wonder.

Clear-winged butterfly, Costa Rica

Understanding Primary Sources

1. How do the parrots and parakeets feed other animals?

2. How does Sandy Wiseman react to seeing the creatures in the rain forest?

Blue poison dart frog

MORE FROM THE FIELD

Costa Rica, with its incredible biodiversity, is an ecotourism success story. A recent study found that more than 66 percent of all tourists traveling to Costa Rica had visited a natural protected area.

Nature preserves can protect animals while welcoming tourists. For example, many visitors to Costa Rica's Monteverde Cloud Forest Preserve want to see magnificent birds called quetzals. While the female quetzals are building nests, trails in their nesting areas are closed. The trails open again while the birds incubate their eggs and tolerate visitors more easily.

Activity: Have students conduct research on one of the animals mentioned in Notes from the Field. Ask them to speculate on the accommodations visitors to the animal's habitat may need to make.

Understanding Primary Sources
Answers

1. by dropping fruit to the forest floor

2. sits down and takes in his surroundings

OVERVIEW

In this unit, students will learn about the physical geography and cultures of Middle and South America.

Much of the region is within the tropics; tropical and savanna climates predominate. Mountainous areas have highland climates. Although subsistence and commercial agriculture are widespread, large areas are undeveloped and almost completely uninhabited. Cities are concentrated along coasts and in highland areas. Industrial growth draws people to the rapidly expanding cities.

The region's cultures reflect its history of colonization by Spain, Portugal, and other European countries. In most countries people of mixed heritage are in the majority, with European and indigenous peoples in the minority.

The countries of Middle and South America face many challenges, such as improving standards of living, increasing personal freedoms, and protecting fragile ecosystems.

PEOPLE IN THE PROFILE

Note that the elevation profile crosses the Brazilian Highlands, which includes the state of Bahia. Today, Salvador, the state capital, is perhaps the largest center of African culture in the Americas.

Many Africans were brought to Bahia as slaves. Their influence is readily seen in spicy foods, music, crafts, and a unique folk dance called *capoeira*.

There are several theories about the origins of *capoeira*, which evolved from an Angolan martial art into a dance. The most widely accepted theory suggests that *capoeira* was created in the 1600s by escaped slaves who had fled to the forests and formed independent villages called *quilombos*. These people defended their villages with *capoeira*, a fighting style that they had developed in secret.

In 1974 *capoeira* was recognized as the national sport of Brazil. As practiced today, *capoeira* combines dancing, fighting, acrobatics, and music.

Critical Thinking: How does *capoeira* reflect Africans' resistance to slavery?

Answer: It was developed as a secret fighting style by escaped slaves to protect their villages.

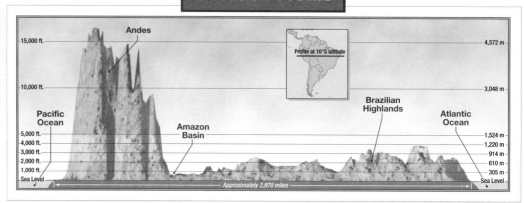

UNIT 4 ATLAS **The World in Spatial Terms**

Middle and South America

Elevation Profile

Andes
15,000 ft. — 4,572 m
Profile at 10°S latitude
10,000 ft. — 3,048 m
Pacific Ocean
Brazilian Highlands
Atlantic Ocean
Amazon Basin
5,000 ft. — 1,524 m
4,000 ft. — 1,220 m
3,000 ft. — 914 m
2,000 ft. — 610 m
1,000 ft. — 305 m
Sea Level — Sea Level
Approximately 2,970 miles

The United States and Middle and South America:
Comparing Sizes

GEOSTATS:

World's highest waterfall: Angel Falls, Venezuela—3,212 ft. (979 m)

World's largest river system: Amazon River drains 2,053,318 square miles (5,318,094 sq km)

World's highest capital city: La Paz, Bolivia—12,001 ft. (3,658 m)

World's largest tropical rain forest: Amazon rain forest

Highest mountain in South America: Mount Aconcagua, Argentina—22,834 ft. (6,960 m)

USING THE PHYSICAL MAP

As students examine the map on this page, call their attention to the physical features they will study in this unit. Ask them to name the major landforms of South America *(Amazon Basin, Andes)*. You may want to review the meaning of *basin* and relate the word to the elevation patterns of Brazil. Ask students where the mountains are concentrated *(along the western*

edge of the landmass). Then ask students to name other countries whose highest mountains are in the west *(United States, Canada)*.

Call on volunteers to identify the major rivers of South America and the countries through which they flow. Point out that Middle America has few major rivers.

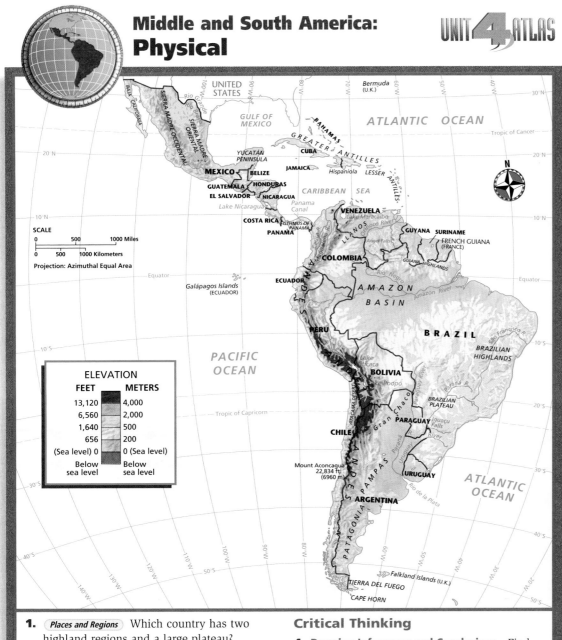

Middle and South America: Physical

UNIT 4 ATLAS

internet connect

ONLINE ATLAS
GO TO: go.hrw.com
KEYWORD: SJ3 MapsU4
FOR: Web links to online maps of the region

ELEVATION

FEET	METERS
13,120	4,000
6,560	2,000
1,640	500
656	200
(Sea level) 0	0 (Sea level)
Below sea level	Below sea level

1. **(Places and Regions)** Which country has two highland regions and a large plateau?

2. **(Places and Regions)** Which country has two large peninsulas that extend into different oceans?

3. **(Places and Regions)** Which island groups separate the Caribbean Sea from the Gulf of Mexico and the Atlantic Ocean?

Critical Thinking

4. **Drawing Inferences and Conclusions** Find central Mexico on the map. Why do you think east-west travel might be difficult in this area? What physical feature do you think would make travel easier in northern Brazil?

Physical Map

Answers

1. Brazil

2. Mexico

3. Greater Antilles and Lesser Antilles

Critical Thinking

4. presence of two mountain ranges—the Sierra Madre Occidental and the Sierra Madre Oriental; Amazon River

283

USING THE POLITICAL MAP

Focus students' attention on the capital cities on the **political map** of Middle and South America on this page. Ask students to name some of the capital cities that are located on or very near a seacoast *(Havana, Panama City, Caracas, Lima, Buenos Aires, Montevideo, Georgetown, Paramaribo)* and to name one reason why so many capitals are located on seacoasts *(for trade and communication with other regions).* Then ask which capitals are located closer to the countries' center *(Mexico City, Bogotá, Quito, Santiago, Brasília)* and what might be the advantage of these more centrally located capitals *(more accessible to all parts of the country).* Finally, ask which country has two capital cities *(Bolivia).*

Your Classroom Time Line

These are the major dates and time periods for this unit. Have students enter them on the time line you created earlier. You may want to watch for these dates as students progress through the unit.

c.* 10,000 B.C. The first people arrive in Mesoamerica.

c. 1500 B.C. Small farming villages are established in Mesoamerica.

c. 900 B.C. Peru's first advanced civilization reaches its height.

c. A.D. 800 Maya civilization begins to collapse.

c. 1200 The Aztec start to move into central Mexico.

c. 1300 The Aztec establish Tenochtitlán.

1492 Christopher Columbus sails into the Caribbean.

1500s Spain and Portugal establish colonies in the Americas.

*c. stands for *circa* and means "about."

Political Map

Answers

1. Panama

2. Mexico; Brazil

Critical Thinking

3. Río Negro, Orinoco, Paraguay, Paraná, and Uruguay Rivers

4. located in dense rain forest

Middle and South America: Political

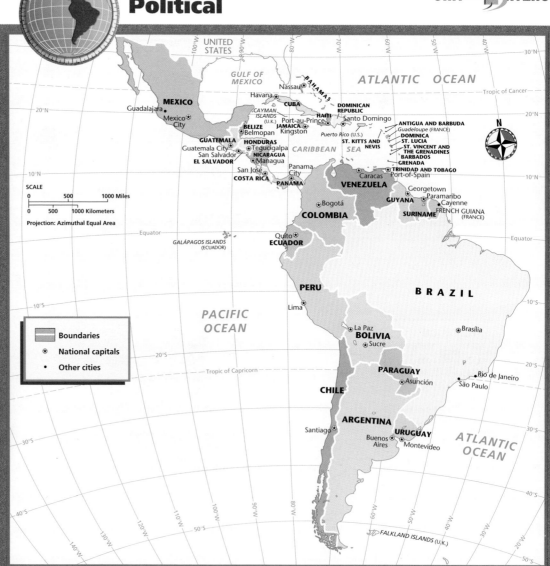

1. **Places and Regions** Which country connects Middle America and South America?

2. **Places and Regions** What is the largest country in Middle America? in South America?

Critical Thinking

3. **Comparing** Compare this map to the **physical map** of the region. What are some rivers that mark the borders between different countries in South America?

4. **Analyzing Information** Compare this map to the **land use and resources** and **physical maps** of the region. Why would it be difficult to mark the boundaries where Colombia, Brazil, and Peru meet?

Direct students' attention to the **climate map** of Middle and South America on this page. Point out the wide variety of climate types in the region. Ask students to compare this map to the **physical map** of the region. You might have them write sentences to describe the relative locations of the main climate regions of South America. *(Examples: The highland climate runs the length of the Andes. A humid tropical climate exists in the Amazon Basin.)* Ask students to study the map legend and to choose the climate that is not found on the mainland of the region *(subarctic).*

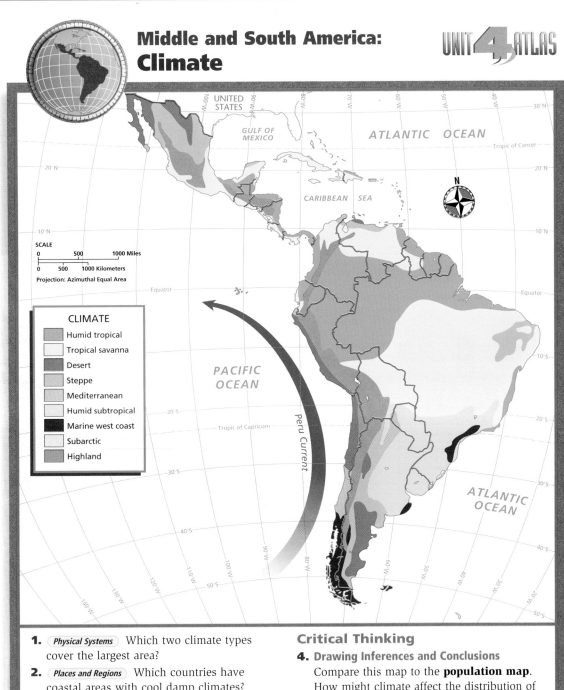

Middle and South America: Climate

UNIT 4 ATLAS

CLIMATE

- Humid tropical
- Tropical savanna
- Desert
- Steppe
- Mediterranean
- Humid subtropical
- Marine west coast
- Subarctic
- Highland

SCALE
0 500 1000 Miles
0 500 1000 Kilometers
Projection: Azimuthal Equal Area

1. *(Physical Systems)* Which two climate types cover the largest area?

2. *(Places and Regions)* Which countries have coastal areas with cool damp climates?

3. *(Places and Regions)* Compare this map to the **physical map** of the region. What type of climate does most of the Amazon Basin have?

Critical Thinking

4. Drawing Inferences and Conclusions
Compare this map to the **population map**. How might climate affect the distribution of Brazil's population?

Your Classroom Time Line *(continued)*

1521 Hernán Cortés conquers the Aztec.

1535 Inca Empire ends.

1600s–1700s The English, French, Dutch, and Danish establish Caribbean colonies.

1800s Many Central and South American countries gain their independence.

1804 Haiti wins its independence from France.

1810 Miguel Hidalgo y Costilla begins the Mexican revolt against the Spanish.

1830 Venezuela becomes independent.

1846 The Mexican War begins.

1872 José Hernández publishes his epic poem *The Gaucho Martin Fierro.*

Climate Map

Answers
1. humid tropical and tropical savanna
2. Chile, Argentina, Brazil
3. humid tropical

Critical Thinking
4. population distribution apparently affected by humid tropical climate

285

USING THE POPULATION MAP

Have students examine the **population map** on this page. Ask them to describe the general pattern of population distribution in South America. (*Possible answers: Most people live near the coast or in highland regions; the interior is very sparsely populated.*)

Then ask students to suggest reasons why the population is distributed in this way. (*Possible answers: The coasts are more accessible than the interior. The highlands have a more temperate climate than lower elevations.*)

Your Classroom Time Line (continued)

1898 The United States takes Cuba from Spain in the Spanish-American War.

1902 Cuba gains its independence.

1903 Panama becomes independent.

1910 The Mexican Revolution begins.

1914 The United States finishes the Panama Canal.

1920 The Mexican Revolution ends.

1959 Fidel Castro seizes power in Cuba.

1960s The United States begins ban on trade with Cuba and restrictions on travel by U.S. citizens to Cuba.

1973–74 Juan Perón serves his second term as president of Argentina.

1979 The Sandinistas overthrow a dictator in Nicaragua.

Population Map

Answers
1. French Guiana
2. Cuba, Hispaniola

Critical Thinking
3. Charts, graphs, databases, and models should accurately reflect population figures on the map.

Middle and South America: Population

UNIT 4 ATLAS

POPULATION DENSITY

Persons per sq. mile	Persons per sq km
520	200
260	100
130	50
25	10
3	1
0	0

● Metropolitan areas with more than 2 million inhabitants

○ Metropolitan areas with 1 million to 2 million inhabitants

SCALE
0   500   1000 Miles
0   500   1000 Kilometers
Projection: Azimuthal Equal Area

1. *Places and Regions* Which area in South America has no place with more than three people per square mile?

2. *Places and Regions* Compare this map to the **physical map** of the region. Which two islands have cities with populations of more than 2 million?

Critical Thinking

3. **Analyzing Information** Use the map on this page to create a chart, graph, database, or model of population centers in Middle and South America.

Have students examine the map on this page. Call on volunteers to name the countries that have oil or natural gas reserves *(Mexico, Venezuela, Colombia, Ecuador, Bolivia, Chile, Argentina, Brazil)*. Ask students to identify areas in South America where hydroelectric power has been developed *(on rivers in the northeast and southeast)*. Then ask how the availability of oil, natural gas, or hydroelectric power might affect the industrial development of individual countries. *(Possible answer: Because factories use large amounts of power, industrial development could probably increase if these energy resources were available.)*

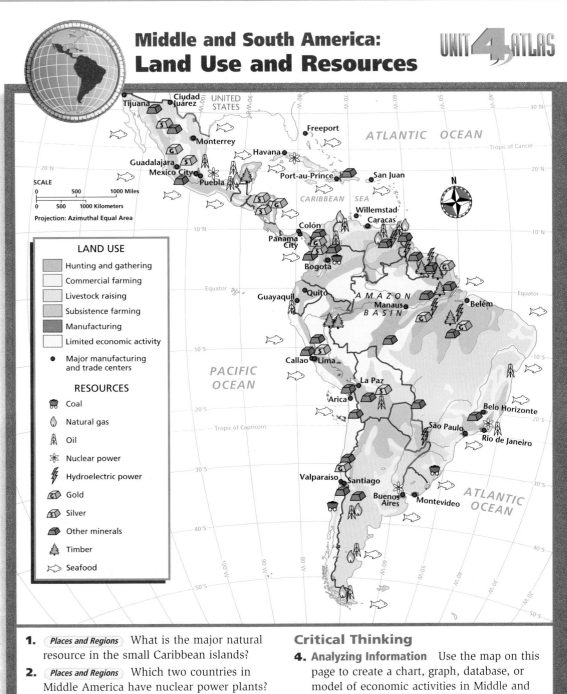

Middle and South America: Land Use and Resources

UNIT 4 ATLAS

LAND USE

- Hunting and gathering
- Commercial farming
- Livestock raising
- Subsistence farming
- Manufacturing
- Limited economic activity
- ● Major manufacturing and trade centers

RESOURCES

- Coal
- Natural gas
- Oil
- Nuclear power
- Hydroelectric power
- Gold
- Silver
- Other minerals
- Timber
- Seafood

1. *Places and Regions* What is the major natural resource in the small Caribbean islands?

2. *Places and Regions* Which two countries in Middle America have nuclear power plants? Which South American countries have them?

3. *Places and Regions* What types of agriculture are important in the Buenos Aires region?

Critical Thinking

4. Analyzing Information Use the map on this page to create a chart, graph, database, or model of economic activities in Middle and South America.

Your Classroom Time Line (continued)

1980s Civil war in El Salvador begins.

1983 Argentina's defeat in the Falklands War leads to governmental reforms.

1990s Volcanic eruptions cause great damage on the island of Montserrat.

1990s The collapse of the Soviet Union hurts Cuba's economy.

1990s The Children's Movement of Colombia is nominated for the Nobel Peace Prize.

1990s Chile becomes a democracy.

1990s Civil wars in Central America decline.

1994 NAFTA takes effect.

1999 The United States gives up control of the Panama Canal to Panama.

Land Use and Resources Map

Answers

1. seafood
2. Mexico, Cuba; Brazil, Argentina
3. livestock raising and commercial farming

Critical Thinking

4. Charts, graphs, databases, and models should accurately reflect information from the map.

287

MIDDLE AND SOUTH AMERICA

LEVEL 1: (Suggested time: 30 min.) Have students examine the figures that describe the number of children born per woman. Students may notice that many of the figures contain decimals. Invite students to suggest how to interpret "less than one" child. (*The decimals are the result of calculating an average. For example, consider a group of 10 women. If one woman had one child, another had seven, five had three each, and the rest had none, the average number of children per woman would be 2.3. If the group of women were considered a country, that country would report that an average of 2.3 children were born per woman.*)

Have students list the six countries with the highest number of children per woman (*Belize, Bolivia, Guatemala, Haiti, Honduras, and Paraguay*). Then have students identify the six countries with the lowest number of children per woman (*Barbados, Cuba, Dominica, Puerto Rico, and St. Vincent and the Grenadines, Trinidad and Tobago*).

LEVEL 2: (Suggested time: 45 min.) Have students create a chart with four columns labeled Country, Children Born per Woman, Literacy Rate, and Percent of Population below the Poverty Level. Ask them to fill in their charts with the six countries with the highest number of children born per woman and the

---

UNITED STATES OF AMERICA

CAPITAL:
Washington, D.C.

AREA:
3,717,792 sq. mi.
(9,629,091 sq km)

POPULATION:
281,421,906

MONEY:
U.S. dollar (USD)

LANGUAGES:
English, Spanish (spoken by a large minority)

CHILDREN BORN/WOMAN:
2.06

Middle and South America

ANTIGUA AND BARBUDA

CAPITAL:
Saint John's

AREA:
171 sq. mi. (442 sq km)

POPULATION:
66,970

MONEY:
East Caribbean dollar (XCD)

LANGUAGES:
English

CHILDREN BORN/WOMAN:
2.3

BELIZE

CAPITAL:
Belmopan

AREA:
8,867 sq. mi. (22,966 sq km)

POPULATION:
256,062

MONEY:
Belizean dollar (BZD)

LANGUAGES:
English (official), Spanish, Mayan, Garifuna (Carib)

CHILDREN BORN/WOMAN:
4.1

ARGENTINA

CAPITAL:
Buenos Aires

AREA:
1,068,296 sq. mi.
(2,766,890 sq km)

POPULATION:
37,384,816

MONEY:
Argentine peso (ARS)

LANGUAGES:
Spanish (official), English, Italian, German, French

CHILDREN BORN/WOMAN:
2.4

BOLIVIA

CAPITAL:
La Paz, Sucre

AREA:
424,162 sq. mi.
(1,098,580 sq km)

POPULATION:
8,300,463

MONEY:
boliviano (BOB)

LANGUAGES:
Spanish, Quechua, Aymara (all official)

CHILDREN BORN/WOMAN:
3.5

BAHAMAS

CAPITAL:
Nassau

AREA:
5,382 sq. mi. (13,940 sq km)

POPULATION:
297,852

MONEY:
Bahamian dollar (BSD)

LANGUAGES:
English, Creole (among Haitian immigrants)

CHILDREN BORN/WOMAN:
2.3

BRAZIL

CAPITAL:
Brasília

AREA:
3,286,470 sq. mi.
(8,511,965 sq km)

POPULATION:
174,468,575

MONEY:
real (BRL)

LANGUAGES:
Portuguese (official), Spanish, English, French

CHILDREN BORN/WOMAN:
2.1

BARBADOS

CAPITAL:
Bridgetown

AREA:
166 sq. mi. (430 sq km)

POPULATION:
275,330

MONEY:
Barbadian dollar (BBD)

LANGUAGES:
English

CHILDREN BORN/WOMAN:
1.6

CHILE

CAPITAL:
Santiago

AREA:
292,258 sq. mi.
(756,950 sq km)

POPULATION:
15,328,467

MONEY:
Chilean peso (CLP)

LANGUAGES:
Spanish

CHILDREN BORN/WOMAN:
2.2

Countries not drawn to scale.

countries with the lowest number. Countries should be listed from the highest to the lowest figure.

Provide a copy of the table on the next pages to students. Ask them to suggest how literacy rates and poverty relate to the children per woman figures. *(Answers will vary.)* Have students add the relevant literacy and poverty figures to their charts.

Explain that high birthrates are generally associated with low education and low income, while low birthrates are generally linked with high education and high income. Have students refer to their charts to determine whether the figures they have gathered tend to support this statement. Then have students write a paragraph explaining the possible reasoning behind this statement, and whether they feel the data support it. *(Countries with the lowest children per woman figures also had high literacy rates. No conclusion can be drawn regarding poverty level as these figures were not available for all the countries. Every country with a large number of children born per woman, with the exception of Paraguay, had a low literacy rate and a large percentage of its population living below poverty level. Because there are countries with high literacy rates that also have a large number of children born per woman—Grenada, for example—this statement should only be taken as a generalization.)*

UNIT 4 **ASSESSMENT RESOURCES**

Reproducible
◆ Unit 4 Test
◆ Unit 4 Test for English Language Learners and Special-Needs Students

internet connect go.hrw.com

COUNTRY STATISTICS
GO TO: go.hrw.com
KEYWORD: SJ3 Facts U4

Highlights of Country Statistics
• *CIA World Factbook*
• Library of Congress country studies
• Flags of the world

UNIT 4 ATLAS

COLOMBIA
CAPITAL: Bogotá
AREA: 439,733 sq. mi. (1,138,910 sq km)
POPULATION: 40,349,388
MONEY: Colombian peso (COP)
LANGUAGES: Spanish
CHILDREN BORN/WOMAN: 2.7

COSTA RICA
CAPITAL: San José
AREA: 19,730 sq. mi. (51,100 sq km)
POPULATION: 3,773,057
MONEY: Costa Rican colon (CRC)
LANGUAGES: Spanish (official), English
CHILDREN BORN/WOMAN: 2.5

CUBA
CAPITAL: Havana
AREA: 42,803 sq. mi. (110,860 sq km)
POPULATION: 11,184,023
MONEY: Cuban peso (CUP)
LANGUAGES: Spanish
CHILDREN BORN/WOMAN: 1.6

DOMINICA
CAPITAL: Roseau
AREA: 291 sq. mi. (754 sq km)
POPULATION: 70,786
MONEY: East Caribbean dollar (XCD)
LANGUAGES: English (official), French patois
CHILDREN BORN/WOMAN: 2.0

DOMINICAN REPUBLIC
CAPITAL: Santo Domingo
AREA: 18,815 sq. mi. (48,730 sq km)
POPULATION: 8,581,477
MONEY: Dominican peso (DOP)
LANGUAGES: Spanish
CHILDREN BORN/WOMAN: 3.0

ECUADOR
CAPITAL: Quito
AREA: 109,483 sq. mi. (283,560 sq km)
POPULATION: 13,183,978
MONEY: U.S. dollar (USD)
LANGUAGES: Spanish (official), ethnic languages including Quechua
CHILDREN BORN/WOMAN: 3.1

EL SALVADOR
CAPITAL: San Salvador
AREA: 8,124 sq. mi. (21,040 sq km)
POPULATION: 6,237,662
MONEY: Salvadoran colon (SVC)
LANGUAGES: Spanish, Nahua
CHILDREN BORN/WOMAN: 3.3

FRENCH GUIANA (Overseas department of France)
CAPITAL: Cayenne
AREA: 35,135 sq. mi. (91,000 sq km)
POPULATION: 177,562
MONEY: French franc (FRF); euro (EUR)
LANGUAGES: French
CHILDREN BORN/WOMAN: 3.2

GRENADA
CAPITAL: St. George's
AREA: 131 sq. mi. (340 sq km)
POPULATION: 89,227
MONEY: East Caribbean dollar (XCD)
LANGUAGES: English (official), French patois
CHILDREN BORN/WOMAN: 2.5

Sources: Central Intelligence Agency, *The World Factbook 2001; The World Almanac and Book of Facts 2001;* pop. figures are 2001 estimates.

Literacy and Poverty in Middle and South America		
Country	Literacy Rate (Percent)	People below Poverty Level (Percent)
Antigua and Barbuda	89	*
Argentina	96	37
Bahamas	98	*
Barbados	97	*
Belize	70	33
Bolivia	83	70
Brazil	83	17.4
Chile	95	22
Colombia	91	55
Costa Rica	95	20.6
Cuba	96	*
Dominica	94	*
Dominican Republic	82	25
Ecuador	90	50

Literacy and Poverty in Middle and South America		
Country	Literacy Rate (Percent)	People below Poverty Level (Percent)
El Salvador	72	48
French Guiana	83	*
Grenada	98	*
Guatemala	64	60
Guyana	98	*
Haiti	45	80
Honduras	73	53
Jamaica	85	34.2
Mexico	90	27
Nicaragua	66	50
Panama	91	*
Paraguay	92	36
Peru	89	49
Puerto Rico	89	*

*data not available

GUATEMALA

CAPITAL: Guatemala City

AREA: 42,042 sq. mi. (108,890 sq km)

POPULATION: 12,974,361

MONEY: quetzal (GTQ)

LANGUAGES: Spanish, Quiche, Cakchiquel, Kekchi

CHILDREN BORN/WOMAN: 4.6

MEXICO

CAPITAL: Mexico City

AREA: 761,602 sq. mi. (1,972,550 sq km)

POPULATION: 101,879,171

MONEY: Mexican peso (MXN)

LANGUAGES: Spanish, Mayan, Nahuatl, ethnic languages

CHILDREN BORN/WOMAN: 2.6

GUYANA

CAPITAL: Georgetown

AREA: 83,000 sq. mi. (214,970 sq km)

POPULATION: 697,181

MONEY: Guyanese dollar (GYD)

LANGUAGES: English, ethnic dialects

CHILDREN BORN/WOMAN: 2.1

NICARAGUA

CAPITAL: Managua

AREA: 49,998 sq. mi. (129,494 sq km)

POPULATION: 4,918,393

MONEY: gold cordoba (NIO)

LANGUAGES: Spanish (official), English, ethnic languages

CHILDREN BORN/WOMAN: 3.2

HAITI

CAPITAL: Port-au-Prince

AREA: 10,714 sq. mi. (27,750 sq km)

POPULATION: 6,964,549

MONEY: gourde (HTG)

LANGUAGES: French (official), Creole

CHILDREN BORN/WOMAN: 4.4

PANAMA

CAPITAL: Panama City

AREA: 30,193 sq. mi. (78,200 sq km)

POPULATION: 2,845,647

MONEY: balboa (PAB)

LANGUAGES: Spanish (official), English

CHILDREN BORN/WOMAN: 2.3

HONDURAS

CAPITAL: Tegucigalpa

AREA: 43,278 sq. mi. (112,090 sq km)

POPULATION: 6,406,052

MONEY: lempira (HNL)

LANGUAGES: Spanish, ethnic languages

CHILDREN BORN/WOMAN: 4.2

PARAGUAY

CAPITAL: Asunción

AREA: 157,046 sq. mi. (406,750 sq km)

POPULATION: 5,734,139

MONEY: guarani (PYG)

LANGUAGES: Spanish (official), Guarani

CHILDREN BORN/WOMAN: 4.1

JAMAICA

CAPITAL: Kingston

AREA: 4,243 sq. mi. (10,990 sq km)

POPULATION: 2,665,636

MONEY: Jamaican dollar (JMD)

LANGUAGES: English, Creole

CHILDREN BORN/WOMAN: 2.1

PERU
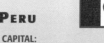

CAPITAL: Lima

AREA: 496,223 sq. mi. (1,285,220 sq km)

POPULATION: 27,483,864

MONEY: nuevo sol (PEN)

LANGUAGES: Spanish (official), Quechua (official), Aymara

CHILDREN BORN/WOMAN: 3.0

Countries not drawn to scale.

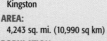

Literacy and Poverty in Middle and South America		
Country	Literacy Rate (Percent)	People below Poverty Level (Percent)
St. Kitts and Nevis	97	*
St. Lucia	67	*
St. Vincent and the Grenadines	96	*
Suriname	93	*
Trinidad and Tobago	98	21
Uruguay	97	*
Venezuela	91	67
United States	97	12.7

*data not available

LEVEL 3: (Suggested time: 40 min.) Have students compose an essay discussing some of the issues that adults in the United States take into account when making decisions about family size. In addition, have students tell how the issues may differ in a Middle or South American country. Students should refer to this unit's Fast Facts feature to support the statements they make in their essay. *(Essays should address economic, social, and educational issues.)*

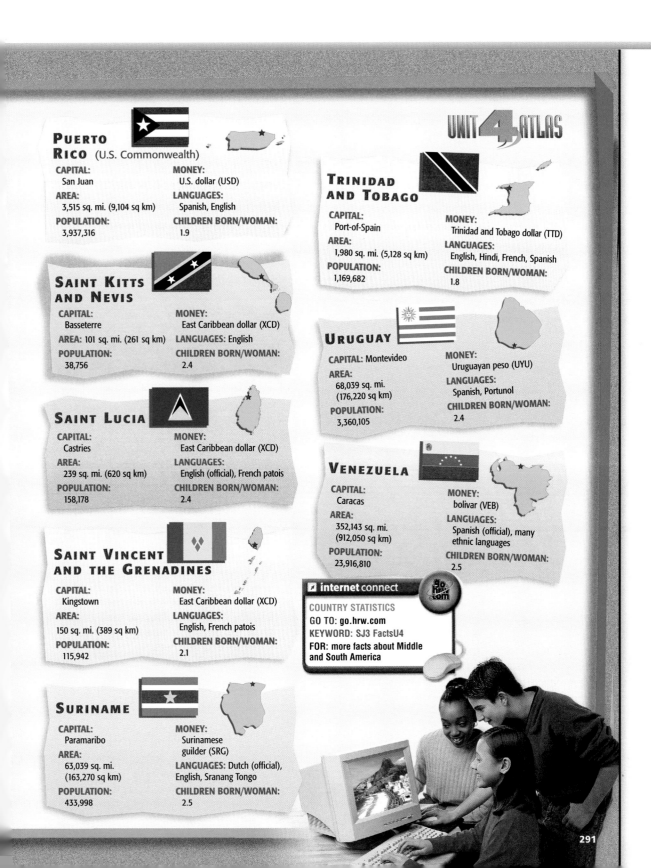

UNIT 4 ATLAS

PUERTO RICO (U.S. Commonwealth)

CAPITAL:
San Juan

AREA:
3,515 sq. mi. (9,104 sq km)

POPULATION:
3,937,316

MONEY:
U.S. dollar (USD)

LANGUAGES:
Spanish, English

CHILDREN BORN/WOMAN:
1.9

SAINT KITTS AND NEVIS

CAPITAL:
Basseterre

AREA: 101 sq. mi. (261 sq km)

POPULATION:
38,756

MONEY:
East Caribbean dollar (XCD)

LANGUAGES: English

CHILDREN BORN/WOMAN:
2.4

SAINT LUCIA

CAPITAL:
Castries

AREA:
239 sq. mi. (620 sq km)

POPULATION:
158,178

MONEY:
East Caribbean dollar (XCD)

LANGUAGES:
English (official), French patois

CHILDREN BORN/WOMAN:
2.4

SAINT VINCENT AND THE GRENADINES

CAPITAL:
Kingstown

AREA:
150 sq. mi. (389 sq km)

POPULATION:
115,942

MONEY:
East Caribbean dollar (XCD)

LANGUAGES:
English, French patois

CHILDREN BORN/WOMAN:
2.1

SURINAME

CAPITAL:
Paramaribo

AREA:
63,039 sq. mi. (163,270 sq km)

POPULATION:
433,998

MONEY:
Surinamese guilder (SRG)

LANGUAGES: Dutch (official), English, Sranang Tongo

CHILDREN BORN/WOMAN:
2.5

TRINIDAD AND TOBAGO

CAPITAL:
Port-of-Spain

AREA:
1,980 sq. mi. (5,128 sq km)

POPULATION:
1,169,682

MONEY:
Trinidad and Tobago dollar (TTD)

LANGUAGES:
English, Hindi, French, Spanish

CHILDREN BORN/WOMAN:
1.8

URUGUAY

CAPITAL: Montevideo

AREA:
68,039 sq. mi. (176,220 sq km)

POPULATION:
3,360,105

MONEY:
Uruguayan peso (UYU)

LANGUAGES:
Spanish, Portunol

CHILDREN BORN/WOMAN:
2.4

VENEZUELA

CAPITAL:
Caracas

AREA:
352,143 sq. mi. (912,050 sq km)

POPULATION:
23,916,810

MONEY:
bolivar (VEB)

LANGUAGES:
Spanish (official), many ethnic languages

CHILDREN BORN/WOMAN:
2.5

internet connect

COUNTRY STATISTICS
GO TO: go.hrw.com
KEYWORD: SJ3 FactsU4
FOR: more facts about Middle and South America

Objectives	Pacing Guide	Reproducible Resources
SECTION 1 **Physical Geography** (pp. 293–96) 1. Identify the main physical features of Mexico. 2. Identify the climate types, plants, and animals found in Mexico. 3. Identify Mexico's main natural resources.	**Regular** 1 day **Block Scheduling** .5 day *Block Scheduling Handbook, Chapter 13*	**RS** Guided Reading Strategy 13.1
SECTION 2 **History and Culture** (pp. 297–301) 1. Identify some early cultures that developed in Mexico. 2. Describe what Mexico was like under Spanish rule and after independence. 3. Identify some important features of Mexican culture.	**Regular** 1 day **Block Scheduling** 1 day *Block Scheduling Handbook, Chapter 13*	**RS** Guided Reading Strategy 13.2 **RS** Graphic Organizer 13 **SM** Geography for Life Activity 13 **SM** Map Activity 13 **E** Cultures of the World Activity 2 **IC** Interdisciplinary Activity for the Middle Grades 12
SECTION 3 **Mexico Today** (pp. 304–07) 1. Describe the kind of government and economy Mexico has today. 2. Identify the important features of Mexico's six culture regions.	**Regular** 1 day **Block Scheduling** .5 day *Block Scheduling Handbook, Chapter 13*	**RS** Guided Reading Strategy 13.3 **E** Creative Strategies for Teaching World Geography, Lesson 9 **E** Environmental and Global Issues Activity 2

Chapter Resource Key

RS Reading Support

IC Interdisciplinary Connections

E Enrichment

SM Skills Mastery

A Assessment

REV Review

ELL Reinforcement and English Language Learners

 Transparencies

 CD–ROM

 Music

 Video

 Internet

 Holt Presentation Maker Using Microsoft® Powerpoint®

 One-Stop Planner CD–ROM

See the *One-Stop Planner* for a complete list of additional resources for students and teachers.

One-Stop Planner CD–ROM

It's easy to plan lessons, select resources, and print out materials for your students when you use the *One-Stop Planner CD–ROM with Test Generator.*

■ internet connect

HRW ONLINE RESOURCES

GO TO: go.hrw.com
Then type in a keyword.

TEACHER HOME PAGE
KEYWORD: SJ3 TEACHER

CHAPTER INTERNET ACTIVITIES
KEYWORD: SJ3 GT13

Choose an activity to:
• travel along Mexico's coastlines.
• see the arts and crafts of Mexico.
• use ancient Maya hieroglyphs.

CHAPTER ENRICHMENT LINKS
KEYWORD: SJ3 CH13

CHAPTER MAPS
KEYWORD: SJ3 MAPS13

ONLINE ASSESSMENT
Homework Practice
KEYWORD: SJ3 HP13
Standardized Test Prep Online
KEYWORD: SJ3 STP13
Rubrics
KEYWORD: SS Rubrics

COUNTRY INFORMATION
KEYWORD: SJ3 Almanac

CONTENT UPDATES
KEYWORD: SS Content Updates

HOLT PRESENTATION MAKER
KEYWORD: SJ3 PPT13

ONLINE READING SUPPORT
KEYWORD: SS Strategies

CURRENT EVENTS
KEYWORD: S3 Current Events

Technology Resources

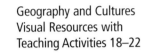

- One-Stop Planner CD–ROM, Lesson 13.1
- Geography and Cultures Visual Resources with Teaching Activities 18–22
- Homework Practice Online
- HRW Go site

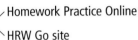

- One-Stop Planner CD–ROM, Lesson 13.2
- Homework Practice Online
- HRW Go site

- One-Stop Planner CD–ROM, Lesson 13.3
- Homework Practice Online
- HRW Go site

Review, Reinforcement, and Assessment Resources

ELL	Main Idea Activity 13.1
REV	Section 1 Review, p. 296
A	Daily Quiz 13.1
ELL	English Audio Summary 13.1
ELL	Spanish Audio Summary 13.1

ELL	Main Idea Activity 13.2
REV	Section 2 Review, p. 301
A	Daily Quiz 13.2
ELL	English Audio Summary 13.2
ELL	Spanish Audio Summary 13.2

ELL	Main Idea Activity 13.3
REV	Section 3 Review, p. 307
A	Daily Quiz 13.3
ELL	English Audio Summary 13.3
ELL	Spanish Audio Summary 13.3

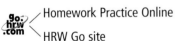

Meeting Individual Needs

Ability Levels

Level 1 Basic-level activities designed for all students encountering new material

Level 2 Intermediate-level activities designed for average students

Level 3 Challenging activities designed for honors and gifted-and-talented students

English Language Learners Activities that address the needs of students with Limited English Proficiency

Chapter Review and Assessment

E	Readings in World Geography, History, and Culture 15 and 16
SM	Critical Thinking Activity 13
REV	Chapter 13 Review, pp. 308–09
REV	Chapter 13 Tutorial for Students, Parents, Mentors, and Peers
ELL	Vocabulary Activity 13
A	Chapter 13 Test
	Chapter 13 Test Generator (on the One-Stop Planner)
	Audio CD Program, Chapter 13
A	Chapter 13 Test for English Language Learners and Special-Needs Students

LAUNCH INTO LEARNING

Display a picture, poster, or reproduction of the Mexican flag. Explain to students that the illustrations on flags often tell us what a nation's citizens value about their country. Point out the eagle holding the snake in its beak and the cactus on which it rests. Ask students what these designs may tell us about how the Mexican people feel about their country. *(Possible answers: They are proud and free, like the eagle. They have survived in a sometimes harsh environment.)* Tell students they will learn more about the physical and human geography of Mexico in this chapter.

Section 1

Objectives

1. Describe the main physical features of Mexico.
2. Name the climate types, plants, and animals that are found in Mexico.
3. Identify Mexico's main natural resources.

LINKS TO OUR LIVES

Consider using the following points to explain the importance of learning about Mexico:

▶ Mexico is one of only two countries that have land borders with the United States.

▶ More immigrants come to the United States from Mexico than from any other country.

▶ Students have used or consumed many products, particularly vegetables and fruits, from Mexico. In fact, 88.6 percent of Mexico's exports go to the United States.

▶ Mexico is the second largest importer of goods produced in the United States.

▶ Millions of Americans, whatever their background, enjoy the foods, music, festivals, and other traditions of Mexico.

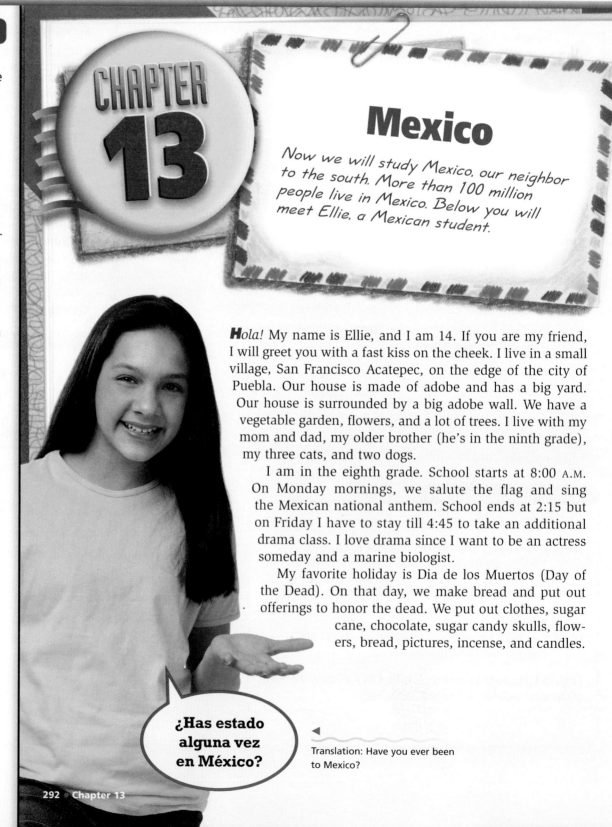

CHAPTER 13

Mexico

Now we will study Mexico, our neighbor to the south. More than 100 million people live in Mexico. Below you will meet Ellie, a Mexican student.

Hola! My name is Ellie, and I am 14. If you are my friend, I will greet you with a fast kiss on the cheek. I live in a small village, San Francisco Acatepec, on the edge of the city of Puebla. Our house is made of adobe and has a big yard. Our house is surrounded by a big adobe wall. We have a vegetable garden, flowers, and a lot of trees. I live with my mom and dad, my older brother (he's in the ninth grade), my three cats, and two dogs.

I am in the eighth grade. School starts at 8:00 A.M. On Monday mornings, we salute the flag and sing the Mexican national anthem. School ends at 2:15 but on Friday I have to stay till 4:45 to take an additional drama class. I love drama since I want to be an actress someday and a marine biologist.

My favorite holiday is Dia de los Muertos (Day of the Dead). On that day, we make bread and put out offerings to honor the dead. We put out clothes, sugar cane, chocolate, sugar candy skulls, flowers, bread, pictures, incense, and candles.

¿Has estado alguna vez en México?

◀ Translation: Have you ever been to Mexico?

LET'S GET STARTED

Copy the following instructions onto the chalkboard: *Look at the map on this page. What kind of climate do you think the Mexican Plateau has? Why do you think so?* Have students explain their predictions. *(Possible answers: It is fairly dry, because the mountains along Mexico's eastern and western edges seem to create a rain shadow. Or, the fact that there are few rivers may indicate that there is little surface water.)* Tell students that in Section 1 they will learn more about Mexico's physical geography.

Using the Physical-Political Map

Have students examine the map on this page. Ask them to describe the relative locations of Mexico, the Mexican Plateau, the Yucatán Peninsula, the Isthmus of Tehuantepec, the Río Bravo, the Gulf of California, and Baja California. You may want to explain that *baja* means "lower."

Building Vocabulary

Write the key term **sinkholes** on the chalkboard. Have students divide the word into its two parts and describe what each one means. Call on a volunteer to read the definition of the word in the text. You may want to ask if students know of sinkholes in your state or region. *(Missouri, Kentucky, and Florida are three states where sinkholes are common.)*

Section 1

Physical Geography

Read to Discover

1. What are the main physical features of Mexico?
2. What climate types, plants, and animals are found in Mexico?
3. What are Mexico's main natural resources?

Define

sinkholes

Locate

Gulf of Mexico
Baja California
Gulf of California
Río Bravo (Rio Grande)
Mexican Plateau
Sierra Madre Oriental
Sierra Madre Occidental
Mount Orizaba
Yucatán Peninsula

Scarlet macaw

WHY IT MATTERS

The Río Bravo (Rio Grande) marks the border between the United States and Mexico. Use **CNNfyi.com** or other **current events** sources to find information on this major river. Record your findings in your journal.

Mexico: Physical-Political

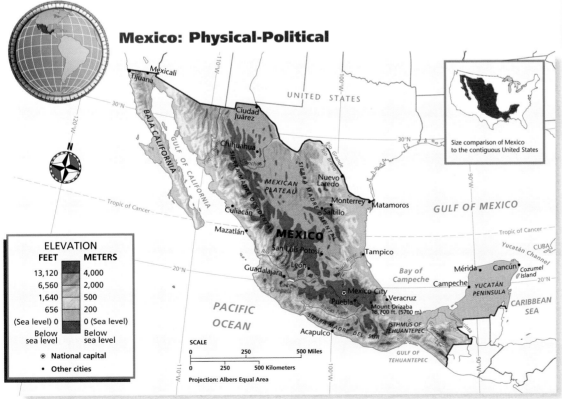

Size comparison of Mexico to the contiguous United States

ELEVATION

FEET	METERS
13,120	4,000
6,560	2,000
1,640	500
656	200
(Sea level) 0	0 (Sea level)
Below sea level	Below sea level

⊛ National capital
• Other cities

SCALE
0   250   500 Miles
0   250   500 Kilometers
Projection: Albers Equal Area

TEACH

Teaching Objectives 1–3

LEVELS 1 AND 2: (Suggested time: 20 min.) Copy the following graphic organizer onto the chalkboard, omitting the italicized answers. Use it to help students learn about the physical features of Mexico. Organize the class into four groups and assign each group one of the organizer topics. Have groups use their textbook to find as much information about their topic as possible. Ask students to come to the chalkboard to complete the organizer with information from their group.
ENGLISH LANGUAGE LEARNERS, COOPERATIVE LEARNING

MEXICO			
Landforms	**Bodies of Water**	**Climates/ Animals/Plants**	**Resources**
Isthmus of Tehuantepec, Baja California, Mexican Plateau, Sierra Madre Occidental, Sierra Madre Oriental, Yucatán Peninsula, Valley of Mexico	*Río Bravo, Gulf of Mexico, Caribbean Sea, Pacific Ocean, Gulf of California*	*deserts, steppe, savanna, humid tropical; cougars, coyotes, deer, anteaters, jaguars, monkeys, parrots; desert scrub, rain forests, dry grasslands*	*oil, gold, silver, copper, lead, zinc*

EYE ON EARTH

Mexico's Popocatépetl volcano is located just 45 miles (72 km) east of Mexico City. It is North America's second-highest volcano. Farmers have settled near the volcano for centuries because the area's rich soil, sunshine, and reliable rainfall ensure good crops.

The most recent significant volcanic activity at Popocatépetl dates to the 1920s. Both U.S. and Mexican scientists extensively monitor the volcano, because more than 30 million people live within view of the crater. Hundreds of thousands live close enough to be threatened by mudflows, rock showers, and clouds of hot gases and ash.

Activity: Have students locate Popocatépetl on a map and determine the relative location of nearby major population centers.

internet connect

GO TO: go.hrw.com
KEYWORD: SJ3 CH13
FOR: Web sites about Popocatépetl

Visual Record Answer ▶

Climate would vary according to elevation, from mild to freezing.

Volcanic Mount Orizaba rises southeast of Mexico City.
Interpreting the Visual Record What kinds of climates do you think people living in this area might experience?

internet connect

GO TO: go.hrw.com
KEYWORD: SJ3 CH13
FOR: Web sites about Mexico

▲ Thousands of monarch butterflies migrate south to the mountains of central Mexico for the winter.

Physical Features

Mexico is a large country. It has a long coast on the Pacific Ocean. It has a shorter one on the Gulf of Mexico. In far southern Mexico, the two bodies of water are just 137 miles (220 km) apart. This part of Mexico is called the Isthmus of Tehuantepec (tay-WAHN-tah-pek). The Caribbean Sea washes the country's sunny southeastern beaches. Beautiful Caribbean and Pacific coastal areas attract many tourists.

Baja (BAH-hah) California is a long, narrow peninsula. It extends south from Mexico's northwestern border with the United States. The peninsula separates the Gulf of California from the Pacific Ocean. One of Mexico's few major rivers, the Río Bravo, forms the Mexico-Texas border. In the United States this river is called the Rio Grande.

Plateaus and Mountains Much of Mexico consists of a rugged central region called the Mexican Plateau. The plateau's wide plains range from 3,700 feet (1,128 m) to 9,000 feet (2,743 m). Isolated mountain ridges rise much higher. Two mountain ranges form the edges of the Mexican Plateau. The Sierra Madre Oriental rise to the east. The Sierra Madre Occidental lie in the west.

At the southern end of the plateau lies the Valley of Mexico. Mexico City, the capital, is located there. The mountains south of Mexico City include towering, snowcapped volcanoes. Volcanic eruptions and earthquakes are a threat in this area. The highest peak, Mount Orizaba (oh-ree-SAH-buh), rises to 18,700 feet (5,700 m).

The Yucatán The Yucatán (yoo-kah-TAHN) Peninsula is generally flat. Limestone underlies much of the area, and erosion has created numerous caves and **sinkholes**. A sinkhole is a steep-sided depression formed when the roof of a cave collapses. The climate in the northern part of the peninsula is hot and dry. Scrub forest is the main vegetation. Farther south, rainfall becomes much heavier. Tropical rain forests cover much of the southern Yucatán.

✔ **READING CHECK:** *Places and Regions* What are Mexico's major physical features? coasts on the Pacific Ocean and Gulf of Mexico; Caribbean Sea; the Río Bravo; Mexican Plateau; Sierra Madre; Oriental and Sierra Madre Occidental; Valley of Mexico; Yucatán Peninsula

LEVEL 3: (Suggested time: 45 min.) Have students complete the Levels 1 and 2 lesson. Then ask students to use the information from the graphic organizer to design a Web site on Mexico. The Web site should include a sample home page as well as sample pages for landforms, bodies of water, climates, animals, plants, and natural resources. Remind students to include other elements common to Web sites, such as a frequently asked questions page or a multimedia map page.

Marcia J. Clevenger, of Charleston, West Virginia, suggests this activity to help students summarize their knowledge about Mexico. Have students search the Internet, newsmagazines, and newspapers for articles about Mexico, such as material on politics, economic challenges, and social issues. They should then create summaries, using your classroom's multimedia equipment to the greatest extent available to them. For example, students might simply combine their information onto a paper collage. Other students might use a tape recorder and an overhead projector. If your school has the necessary equipment, have students scan their information into a computer and create a multimedia presentation.

The States of Mexico

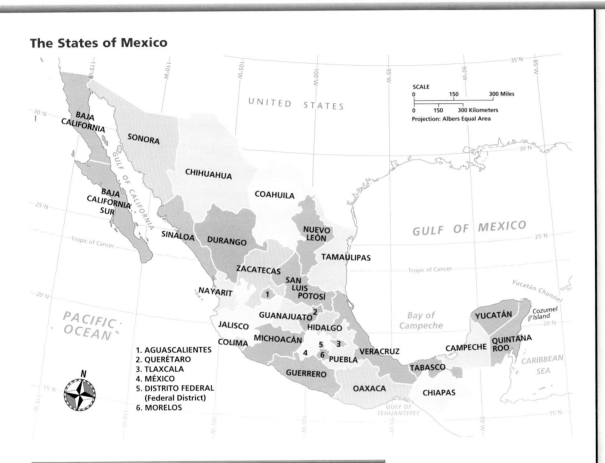

Across the Curriculum
SCIENCE

From Mexico to Mars

Mexico is home to a wide variety of plants and animals, including a unique group of species that dwells underground. Cueva de Villa Luz, a sulfur-spring cave in the Mexican state of Tabasco, contains organisms that use sulfur compounds rather than photosynthesis to make their food.

Other microbes consume the sulfur-eating bacteria. Tiny invertebrates eat the microbes. They, in turn, are eaten by spiders, worms, and water bugs. Tiny fish feast on this food supply. Local people, descendants of the Maya, catch the fish by sprinkling a natural pesticide in the water. The fish come to the surface gasping for air.

Some scientists believe that Mars, a sulfur-rich planet, may contain similar places. They suspect that such places could provide habitats where life could exist beneath the hostile surface of that planet.

Activity: Have students draw the food chain that exists in the sulfur-spring cave.

Climate, Vegetation, and Wildlife

Mexico's climate varies by region. Its mountains, deserts, and forests also support a variety of plants and animals.

A Tropical Area Mexico extends from the middle latitudes into the tropics. It has desert, steppe, savanna, and humid tropical climates. Most of northern Mexico is dry. There, Baja California's Sonoran Desert meets the Chihuahuan (chee-WAH-wahn) Desert of the plateau. Desert scrub vegetation and dry grasslands are common. Cougars, coyotes, and deer can be found in some areas of the north.

The forested plains along Mexico's southeastern coast are hot and humid much of the year. Summer is the rainy season. Forests cover about 20 percent of Mexico's land area. Tropical rain forests provide a home for anteaters, jaguars, monkeys, parrots, and other animals.

Many varieties of cactus thrive in the Sonoran Desert.

Interpreting the Visual Record Why do you think large trees are not found in the Sonoran Desert?

◀ **Visual Record Answer**

not enough water to support them

Point out on a map the major landforms and climate regions of Mexico. Have students describe how these physical features might affect the way in which people live.

REVIEW AND ASSESS

Have students complete the Section Review. Then distribute index cards to students. Have students develop questions and answers for a quiz game using these categories: physical features, climate types, vegetation, animals, and resources. Have students use their cards to quiz each other. Then have students complete Daily Quiz 13.1.

RETEACH

Have students complete Main Idea Activity 8.1. Display a blank wall map of Mexico and have each student create symbols for one physical feature, one resource, and one climate type. Then have students glue their symbols to the map and lead a discussion about the map.
ENGLISH LANGUAGE LEARNERS

EXTEND

Have students use primary and secondary sources to conduct research on a place in Mexico and create an illustrated brochure that addresses the six essential elements of geography. Students can also research the customs of one of the ethnic groups in Mexico and present an oral presentation based on their information. **BLOCK SCHEDULING**

Section Review 1

Answers

Define For definition, see: sinkholes, p. 294

Working with Sketch Maps Maps will vary, but listed places should be labeled in their approximate locations. The Yucatán is flat and has caves and sinkholes. It is hot and dry with scrub forest in the north. Farther south are tropical rain forests and heavier rainfall.

Reading for the Main Idea
1. Baja California, Río Bravo, Mexican Plateau, Sierra Madre Oriental, Sierra Madre Occidental, Valley of Mexico, Mount Orizaba, Yucatán Peninsula (NGS 4)
2. desert, steppe, savanna, humid tropical (NGS 4)

Critical Thinking
3. possible answers: coastal areas, Valley of Mexico, along rivers
4. possible answer: Mountains and lack of rivers could limit movement and communication.

Organizing What You Know
5. landforms—mountains, plateaus, caves, sinkholes; bodies of water—Caribbean Sea, Gulf of Mexico, Río Bravo, Gulf of California, Pacific Ocean; climate—desert, steppe, savanna, humid tropical; resources—oil, metals, limited water

So much water has been pumped from under Mexico City that the ground is sinking, or subsiding. Parts of the central city have subsided about 25 feet (7.6 m) in the last 100 years. This has damaged buildings, pipes, sewers, and subway tunnels.

Climate Variations In some areas changes in elevation cause climates to vary widely within a short distance. Many people have settled in the mild environment of the mountain valleys. The valleys along Mexico's southern coastal areas also have pleasant climates.

The areas of high elevation on the Mexican Plateau experience surprisingly cool temperatures. Freezing temperatures sometimes reach as far south as Mexico City.

✔ **READING CHECK:** *Places and Regions* What are Mexico's climate zones?
desert, steppe, savanna, humid tropical

Resources

Petroleum is Mexico's most important energy resource. Oil reserves lie primarily under southern and Gulf coastal plains as well as offshore in the Gulf of Mexico. In 1997 Mexico had the world's eighth-largest crude oil reserves.

Mining is also important in Mexico. Some gold and silver mines begun centuries ago are still in operation. Fresnillo has been a silver-mining center since 1569. New mines have been developed in Mexico's northern and southern mountains. Silver is the most valuable part of Mexico's mining industry. In 1997 one silver mine in the state of Zacatecas produced 706 tons (641,037 kg) of silver. Mexico also produces large amounts of copper, gold, lead, and zinc.

Water is a limited resource in parts of Mexico. Water scarcity, particularly in the dry north, is a serious issue.

✔ **READING CHECK:** *Environment and Society* What problems might water scarcity cause for Mexican citizens? not enough clean water to drink, farm with, use for power

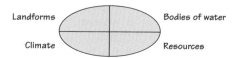

Homework Practice Online
Keyword: SJ3 HP13

Section Review 1

Define and explain: sinkholes

Working with Sketch Maps On a map of Mexico that you draw or that your teacher provides, label the following: Gulf of Mexico, Baja California, Gulf of California, Río Bravo (Rio Grande), Mexican Plateau, Sierra Madre Oriental, Sierra Madre Occidental, Mount Orizaba, and Yucatán Peninsula. In a caption, describe the Yucatán.

Reading for the Main Idea
1. *Places and Regions* What are Mexico's major physical features?
2. *Places and Regions* What are the major climate zones of Mexico?

Critical Thinking
3. **Drawing Inferences and Conclusions** Based on physical features and climate, which areas of Mexico provide the best conditions for people to live?
4. **Drawing Inferences and Conclusions** How do you think Mexico's geography affects movement and communication between different parts of the country?

Organizing What You Know
5. **Summarizing** Copy the following graphic organizer. Use it to describe Mexico's physical geography.

Landforms   Bodies of water
Climate   Resources

Section 2

Objectives

1. Analyze the early cultures that developed in Mexico.

2. Describe what Mexico was like under Spanish rule and after independence.

3. Identify some important features of Mexican culture.

FOCUS

LET'S GET STARTED

Copy the following statement and questions onto the chalkboard: *Mexico and the United States both started out as European colonies. How are they different today? What might be the cause for these differences?* Discuss students' answers and tell them that in Section 2 they will learn more about Mexico's colonial period and independence.

Building Vocabulary

Write the key terms on the chalkboard. Pronounce the terms and have students repeat them aloud. Call on volunteers to find and read aloud the definitions of the terms. Ask: What English words does the word **conquistadores** resemble? *(conquer, conquest)* Call on students to make connections among these words. Ask: Which of the other key terms refer to groups of people? *(mestizos, mulattoes)* Which refer to farming or farmland? *(chinampas, ejidos, haciendas)* Call on volunteers to compose sentences using **epidemic**, **empire**, or **missions**.

Section 2 — History and Culture

Read to Discover

1. What early cultures developed in Mexico?

2. What was Mexico like under Spanish rule and after independence?

3. What are some important features of Mexican culture?

Define

chinampas
conquistadores
epidemic
empire
mestizos

mulattoes
missions
ejidos
haciendas

WHY IT MATTERS

There are many aspects of Spanish influence still found in present-day Mexico. Use CNN**fyi**.com or other **current events** sources to find information about Mexican history. Record your findings in your journal.

An Aztec serpent pendant

Section 2 RESOURCES

Reproducible
◆ Guided Reading Strategy 13.2
◆ Graphic Organizer 13
◆ Geography for Life Activity 13
◆ Cultures of the World Activity 2
◆ Interdisciplinary Activity for the Middle Grades 12
◆ Map Activity 13

Technology
◆ One-Stop Planner CD–ROM, Lesson 13.2
◆ Homework Practice Online
◆ HRW Go site

Reinforcement, Review, and Assessment
◆ Section 2 Review, p. 301
◆ Daily Quiz 13.2
◆ Main Idea Activity 13.2
◆ English Audio Summary 13.2
◆ Spanish Audio Summary 13.2

Early Cultures

Mesoamerica (*meso* means "in the middle") is the cultural area including Mexico and much of Central America. Many scientists think the first people to live in Mesoamerica arrived from the north about 12,000 years ago. By about 5,000 years ago, people in Mesoamerica were growing beans, peppers, and squash. They also domesticated an early form of corn. It eventually developed into the corn that we see today.

By about 1500 B.C. many people throughout the region were living in small farming villages. Along the humid southern coast of the Gulf of Mexico lived the Olmec people. The Olmec built temples, pyramids, and huge statues. They traded carved jade and obsidian, a volcanic stone, throughout eastern Mexico.

By about A.D. 200, other complex cultures were developing in what is now Mexico. Many of these civilizations had large city centers. Those centers had apartments, great avenues, open plazas, and pyramid-shaped temples. Some temple areas throughout Mesoamerica had stone ball courts. Players on those courts competed in a game somewhat like basketball. However, this ball game was not simply a sport. It had deep religious importance. Players who lost might be sacrificed to the gods.

Maya ruins include stone carvings and pyramid-type structures. Figures such as this Chac Mool are thought to represent the rain god.

TEACH

Teaching Objective 1

LEVEL 1: (Suggested time: 30 min.) Organize students into four groups. Have each group create a time line from 1500 B.C. to A.D. 1519 that includes information on the early cultures that developed in Mexico. Tell them to use the text to find information for the time line and to draw pictures illustrating the items. Have students present their time lines to the class. **ENGLISH LANGUAGE LEARNERS, COOPERATIVE LEARNING**

LEVELS 2 AND 3: (Suggested time: 20 min.) Have students imagine that they are participating in an archaeological dig at a site in Mexico. At this site they are uncovering artifacts of early Mexican civilizations. Have students write a report detailing the artifacts that they have found. The reports should include information identifying the civilization and the location of the site.

DAILY LIFE

Before the arrival of the Spanish in the Americas, Mexican Indians ate protein-rich moth larvae, grasshoppers, locusts, ant larvae, and live worms. Now these and other exotic preconquest foods are popular in some Mexican restaurants.

Among the foods eaten by the Aztec was a grayish fungus called *cuitlacoche* (weet-lah-KOH-chay), known in the United States as corn smut. American farmers once considered the fungus a problem. Now, however, some cultivate it because the fungus has become a popular gourmet item. *Cuitlacoche* is sold canned or frozen in gourmet markets and can be used much like cooked mushrooms.

Critical Thinking: Why might farmers switch from fighting the fungus to cultivating it?

Answer: The gourmet item might bring a higher price than corn, resist drought better, or have other advantages.

This copy of an original Aztec tax record shows items that were to be collected as tribute.
Interpreting the Visual Record What items besides tunics and headgear can you identify?

▼

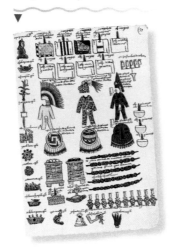

The Maya ruins of Tulum, an ancient Maya ceremonial and religious center, lie on the Caribbean coast.
Interpreting the Visual Record Why do you think the Maya built a large wall around this city?

▼

The Maya Maya civilization developed in the tropical rain forest of southeastern Mexico, Guatemala, Belize, and Honduras. Maya city-states were at their peak between about A.D. 250 and 800. The Maya made accurate astronomical calculations and had a detailed calendar. Modern scholars can now read some Maya writing. This has helped us understand their civilization.

The Maya grew crops on terraced hillsides and on raised fields in swampy areas. They dug canals, piling the rich bottom mud onto the fields alongside the canals. This practice enriched the soil. Using this productive method of farming, the Maya supported an extremely dense population.

Sometime after A.D. 800, Maya civilization collapsed. The cities were abandoned. This decline may have been caused by famine, disease, warfare, or some combination of factors. However, the Maya did not die out. Millions of people of Maya descent still live in Mexico and Central America today.

The Aztec A people called the Aztec began moving into central Mexico from the north about A.D. 1200. They later established their capital on an island in a lake in the Valley of Mexico. Known as Tenochtitlán (tay-nawch-teet-LAHN), this capital grew into a splendid city. Its population in 1519 is estimated to have been at least 200,000. It was one of the largest cities in the world at that time. The Aztec conquered other Indian peoples around them. They forced these peoples to pay taxes and provide captives for sacrifice to Aztec gods.

The Aztec practiced a version of raised-field agriculture in the swampy lakes of central Mexico. The Aztec called these raised fields **chinampas** (chuh-NAM-puhs). There they grew the corn, beans, and peppers that most people ate. Only rich people in this society ate meat. On special occasions common people sometimes ate dog meat. The upper classes also drank chocolate.

✓ **READING CHECK:** What were the main features of Mexico's early civilizations?
Maya—lived in city-states, made astronomical calculations, kept a calendar, farmed; Aztec—built Tenochtitlán, conquered other peoples, farmed

Visual Record Answers ▲

Answers may vary, but students should note objects from the picture.

▶

as protection from enemies, hurricane-driven waves, or other threats

298

298   Chapter 13

Colonial Mexico — *Spanish, American Indians, and Africans mixed cultures. Missions established to spread religion. Ejidos taken away, replaced by haciendas.*

→ **1821 Independence** →

After Independence — *Texas revolts, later joins the United States; in war over U.S.-Mexico border, Mexico loses land to United States; 1920 revolution changes government and restores ejidos back to peasants.*

Connecting to History

TENOCHTITLÁN

Tenochtitlán—now Mexico City—was founded by the Aztec in the early 1300s. According to their legends, they saw an eagle sitting atop a cactus on a swampy island in Lake Texcoco. The eagle held a snake in its mouth. A prophecy had instructed them to build a city where they saw such an eagle.

Within 200 years this village had become an imperial capital and the largest city in the Americas. It was a city of pyramids, palaces, markets, and gardens. Canals and streets ran through the city. Stone causeways connected the island to the mainland.

Bernal Díaz, a Spanish soldier, described his first view of the Aztec capital in 1519.

❝When we saw so many cities and villages built in the water and other great towns on dry land and that straight and level Causeway going towards [Tenochtitlán], we were amazed and said that it was like the enchantments they tell of in the legend of Amadis, on account of the great towers and temples and buildings rising from the water, and all built of masonry. And some of our soldiers asked whether the things that we saw were not a dream.❞

The Spaniards went on to conquer the Aztec and destroy Tenochtitlán. On the ruins they built Mexico City. They also drained Lake Texcoco to allow the city to expand.

Understanding What You Read

1. Where and when was Tenochtitlán first built?
2. What was the Spaniards' reaction to Tenochtitlán?

A sketch map of the Aztec capital of Tenochtitlán

Colonial Mexico and Independence

Spanish Conquest Hernán Cortés, a Spanish soldier, arrived in Mexico in 1519 with about 600 men. These **conquistadores** (kahn-kees-tuh-DAWR-ez), or conquerors, had both muskets and horses. These were unknown in the Americas at that time. However, the most important factor in the conquest was disease. The native people of the Americas had no resistance to European diseases. The first **epidemic**, or widespread outbreak, of smallpox struck central Mexico in 1520. The death toll from disease greatly weakened the power of the Aztec. In 1521 Cortés completed his conquest of the Aztec and the other American Indian peoples of southern Mexico. They named the territory New Spain.

Colonial Mexico During this period, Spain ruled an **empire**. An empire is a system in which a central power controls a number of

299

Teaching Objective 3

LEVEL 1: (Suggested time: 25 min.) Pair students and have them design and write a children's book on the culture and customs of Mexico. Remind students that a children's book should include a cover, be easy to read, and contain illustrations or pictures. Bind the books with string or yarn and display them around the classroom.
ENGLISH LANGUAGE LEARNERS, COOPERATIVE LEARNING

LEVELS 2 AND 3: (Suggested time: 25 min.) Ask students to imagine that they are traveling through Mexico. Have students prepare several journal entries about their trip. Entries should include descriptions of cultural events and customs as well as differences in languages that they might find on the trip.

►**ASSIGNMENT:** Ask students to imagine that they are Mexican Indians who have recently returned to Mexico during the time of Spanish exploration and colonization. Tell students that they had been away from Mexico for 10 years and therefore were not there for the arrival of the Spanish. Have students write an account of how Mexico has changed during their years away.

CLOSE

Have students take turns using one of the key terms in a sentence that describes an aspect of Mexican history or culture.

DAILY LIFE

One of the many cultural traditions and customs from Mexico is the *quinceañera* (KEEN-se-ahn-ye-rah). Many girls throughout Latin America and parts of the United States spend up to a year planning a *quinceañera*. The *quinceañera* celebration marks a girl's 15th birthday. It begins with a thanksgiving mass. The girl, wearing a pastel gown and a matching tiara or headpiece, is seated at the foot of the altar during the mass. She can have up to seven maids of honor and as many male attendants. A big fiesta follows the mass. The party can rival a wedding in terms of its preparation, food, and costs.

The *quinceañera* tradition may have its roots in the Aztec practice of marking girls' transition into womanhood. The Aztec ceremony stressed the importance of following society's rules.

Critical Thinking: Why might *quinceañeras* be important in Hispanic communities?

Answer: Students might say it is a historical celebration that strengthens community, religious, and family ties.

The Santa Prisa Cathedral rises above the rest of Taxco in southern Mexico. The Spaniards built many beautiful cathedrals in Mexico.

territories. New Spain was just one of Spain's colonies in the Americas. Mainland New Spain extended north from Mexico to what is now northern California and south to Panama.

After the initial conquest, Spanish and American Indian peoples and cultures mixed. This mix formed a new Mexican identity. The Spaniards called people of mixed European and Indian ancestry **mestizos** (me-STEE-zohs). When enslaved Africans were brought to Spanish America, they added to this blend of cultures. The Spaniards called people of mixed European and African ancestry **mulattoes** (muh-LA-tohs). Africans and Indians also intermarried.

Large areas of northern Mexico were left to the Roman Catholic Church to explore and to rule. Church outposts known as **missions** were scattered throughout the area. Mission priests often learned Indian languages and tried to convert the Indians to Catholicism.

Colonial Economy At first, the Spaniards were mainly interested in mining gold and silver in Mexico. Gradually, agriculture also became an important part of the colonial economy. Indians did most of the hard physical labor on farms and in mines. Many of them died from disease and overwork. Therefore, Spaniards began to bring enslaved Africans to the Americas as another source of labor.

Before the arrival of the Spaniards, Indian communities owned and worked land in groups. The lands they worked in common are called *ejidos* (e-HEE-thohs). After the conquest, the Spanish monarch granted **haciendas** (hah-see-EN-duhs), or huge expanses of farmlands, to favored people. Peasants, usually Indians, lived and worked on these haciendas. Cattle ranching operated according to a similar system.

Independence and After In 1810 a Catholic priest named Miguel Hidalgo y Costilla began a revolt against Spanish rule. Hidalgo was killed in 1811. However, fighting continued until independence was won in 1821.

Fifteen years later, Texas broke away from Mexico. Texas became part of the United States in 1845. Shortly after, the United States and Mexico argued over the location of their common border. This conflict led to a war in which Mexico lost about half its territory. Americans know this conflict as the Mexican War. Mexicans usually refer to it as the War of the North American Invasion.

In the early 1900s many Mexicans grew unhappy with the government of military leader Porfirio Díaz. As a result, in 1910 the Mexican Revolution broke out. Fighting between various leaders and groups lasted until 1920.

One of the major results of the revolution was land reform. The new government took land from the haciendas. This land was given back to peasant villages according to the old *ejido* system. Today *ejidos* make up about half of Mexico's farmland.

✔ **READING CHECK:** *Human Systems* How did Mexico gain independence from Spain? Mexican priest Miguel Hidalgo y Costilla began a revolt in 1810; fighting continued until independence was won in 1821.

◄ Mexicans light candles for relatives on the Day of the Dead.
Interpreting the Visual Record What celebrations in your community reflect special customs?

Culture and Customs

About 89 percent of Mexico's people are Roman Catholic. The Day of the Dead is an example of how cultures mix in Mexico. This holiday takes place on November 1 and 2. These are the same dates the Catholic Church celebrates All Saints' Day and All Souls' Day. The Day of the Dead honors dead ancestors. Families often place different foods on the graves of dead relatives. This recalls the Indian belief that the dead need material things just as the living do.

In Mexico ethnic identity is associated more with culture than with ancestry. One major indicator of a person's ethnic identity is language. Speaking one of the American Indian languages identifies a person as Indian. People who speak only Spanish are usually not considered Indian. This may be true even if they have Indian ancestors.

✓ **READING CHECK:** *Human Systems* What is the significance of the Day of the Dead? honors dead ancestors—recalls both Indian belief that the dead need material things and the religious holidays of All Saints' Day and All Souls' Day

Section Review 2

Define and explain: *chinampas*, conquistadores, epidemic, empire, mestizos, mulattoes, missions, *ejidos*, haciendas

Working with Sketch Maps On the map you created in Section 1, shade the areas once controlled by the Maya and Aztec. Which civilization occupied the Yucatán Peninsula?

Reading for the Main Idea
1. *Human Systems* What were some notable achievements of the Maya and Aztec civilizations?
2. *Human Systems* What were the effects of Spanish rule on Mexico?

go.hrw.com Homework Practice Online
Keyword: SJ3 HP13

Critical Thinking
3. **Analyzing Information** What advantages did Cortés have in his conquest of the Aztec?
4. **Finding the Main Idea** What is a sign of a person's ethnic identity in Mexico?

Organizing What You Know
5. **Identifying Cause and Effect** Copy the following graphic organizer. Use it to develop a cause-and-effect chart to identify events related to the Mexican War.

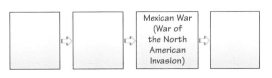
Mexican War (War of the North American Invasion)

Section Review 2

Answers

Define For definitions, see: *chinampas*, p. 298; conquistadores, p. 299; epidemic, p. 299; empire, p. 299; mestizos, p. 300; mulattoes, p. 300; missions, p. 300; *ejidos*, p. 300; haciendas, p. 300

Working with Sketch Maps Maya—southeastern Mexico; Aztec—central Mexico

Reading for the Main Idea
1. Maya—astronomical calculations, calendar, writing system, complex agricultural systems; Aztec—built Tenochtitlán, grew crops on *chinampas* (NGS 10)
2. blend of cultures, spread of Roman Catholicism, mining of gold and silver, the division of land into haciendas (NGS 13)

Critical Thinking
3. muskets, horses
4. language

Organizing What You Know
5. first box—Texas breaks away from Mexico and later becomes U.S. state; second box—Mexico and the United States dispute border; fourth box—Mexico loses territory

▲ **Visual Record Answer**

Students might mention holidays and community activities.

301

Setting the Scene

Globalization has given rise to complex manufacturing, shipping, and marketing networks. Companies headquartered in one country can finance, construct, and manage factories in other countries. Additionally, parts manufactured in one country can be assembled in another country into finished goods. These goods can then be shipped to and sold in a third country. *Maquiladoras* are a concrete example of this system in operation and illustrate the importance of cheap labor in locating manufacturing enterprises. The North American Free Trade Agreement (NAFTA) removed trade barriers among Canada, Mexico, and the United States. Since the agreement was implemented in 1994, Mexican trade with the United States and Canada has nearly doubled. The number of *maquiladoras* has increased.

Building a Case

Have students read " *Maquiladoras* Along the U.S.–Mexico Border." Then ask the students the following questions: What are *maquiladoras? (factories that can be foreign-owned along the U.S.–Mexico border)* When were the first *maquiladoras* built? *(mid-1960s)* What are some of the cities where *maquiladoras* are found? *(Tijuana, Ciudad Juárez, Nuevo Laredo, Monterrey)* What are some of the goods manufactured in *maquiladoras*? *(vacuum cleaners, automobile parts, electronics)* Present students with the following facts about the lower standard of living in Mexico: High wages in Mexico are limited to a small portion of the workforce—the top 20 percent of income earners receive 55 percent of the total wages paid. In 1998, 27 percent of the population was estimated to live in poverty. These economic realities enable U.S.-based companies to pay lower wages to Mexican workers than to American workers.

HISTORICAL GEOGRAPHY

Although the scale and speed of global trade is unique to the modern age, far-flung trading networks have existed throughout history. The colonial plantation system of the New World is one example. First developed in the 1400s by the Portuguese, plantation agriculture was the dominant economic system in the Americas during the colonial era. Specialized crops, such as cotton, tobacco, indigo, sugarcane, and rice, were grown on large estates that relied on poorly paid workers or slave labor. After harvest, the crops were shipped mostly to European markets for sale.

Activity: Have students conduct research on the trading routes important to the colonial plantation system. Suggest that students find out which crops were raised for export, which modern countries had extensive plantation economies, and the sources of plantation labor.

➤ **This Case Study feature addresses National Geography Standards 1 and 11.**

Visual Record Answer ▶

agricultural land use on U.S. side and dense urban and industrial use on Mexico's side of the border

MAQUILADORAS ALONG THE U.S.-MEXICO BORDER

Geographers are often interested in borders. Borders are more than just lines on a map. They are places where different countries, cultures, and ways of life meet. For example, the U.S.-Mexico border separates a developing country from a very rich country. This fact has had an important influence on the location of industries in the region.

On the Mexican side of the border there are many *maquiladoras* (mah-kee-lah-DOHR-ahs)—factories that can be owned by foreign companies. The first *maquiladoras* were built in the mid-1960s. Since then, their numbers have increased dramatically. There are now more than 2,000 *maquiladoras* in Mexico. They employ hundreds of thousands of workers.

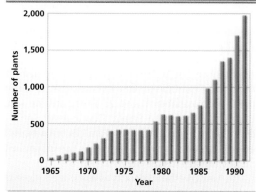

Growth in *Maquiladora* Plants

Source: Instituto Nacional de Estadística Geografía e Informática, Mexico

▲
The U.S.-Mexico border separates Mexicali, Mexico, from Calexico, California. The large buildings on the left are *maquiladoras*.

Interpreting the Visual Record How has the border influenced land use in this area?

Many *maquiladoras* are owned by American companies. They produce goods such as vacuum cleaners, automobile parts, and electronics. These goods are usually exported to the United States. Most *maquiladoras* are located within about 20 miles (32 km) of the U.S.-Mexico border. Cities such as Tijuana, Ciudad Juárez, and Nuevo Laredo have many *maquiladoras*.

Why have so many American companies chosen to build factories across the border in Mexico? The location allows companies to take advantage of the difference in wealth between the two countries. Mexico has a much lower standard of living than the United States. Workers there get paid less. Therefore, factories in Mexico can produce goods more cheaply. The United States has a higher standard of living. People there have more money to buy goods. They buy the products that have been

Drawing Conclusions

Lead students in a discussion of the global flow of goods and money. Discuss how the producers and consumers of goods do not always live in the same place. What criteria do manufacturers use to decide where to locate new factories? *(low wages, transportation costs, tax breaks, or fewer environmental protections)* Look at a road atlas of North America. Locate the Mexican cities where *maquiladoras* are concentrated. *(Tijuana, Ciudad Juarez, Nuevo Laredo, Monterrey)* What are the important interstate highways that might carry finished goods from these areas to consumers in the United States? *(Tijuana: I 5 and I 8; Ciudad Juarez: I 25 and I 10; Nuevo Laredo and Monterrey: I 35)* Have students trace a route from one of these cities to their hometown. Challenge them to look for items in their homes that were manufactured in Mexico.

Going Further: Thinking Critically

Maquiladoras are only one example of the global manufacturing network. Multinational manufacturing enterprises have also been active in Asia. China, Taiwan, South Korea, and India all have large workforces willing to work for relatively low wages. Have students discuss the following issues:

- Many of the Asian export-manufacturing areas are near port cities. Why is it important for factories to be located there?
- Some countries do not have strong regulations governing working conditions. Pollution controls, child labor restrictions, minimum-wage laws, and workweek standards are often not as strict as those in the United States and Western Europe. How might this affect where American and European companies locate their factories? Challenge students to monitor news stories about foreign working conditions.

U.S.-Mexico Border Region

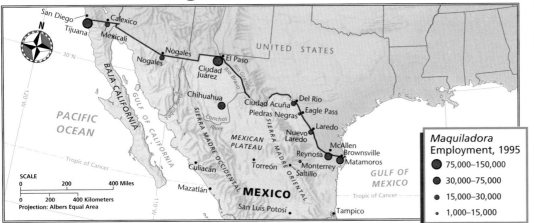

Maquiladora Employment, 1995
- ● 75,000–150,000
- ● 30,000–75,000
- ● 15,000–30,000
- · 1,000–15,000

Understanding What You Read

1. Companies want to employ low-wage workers while remaining close to wealthy markets. They have therefore built *maquiladoras* close to the U.S.-Mexico border.

2. Goods produced include household appliances, like vacuum cleaners. Automobile parts and electronics are also manufactured in Mexico.

assembled in Mexico and shipped to American stores.

American companies have other reasons for building factories in Mexico. The Mexican government has given tax breaks to foreign companies that own and operate *maquiladoras*. It has encouraged the growth of *maquiladoras*. They create jobs for Mexican citizens and bring money into the country. Also, environmental laws in Mexico are not as strict as in the United States. Therefore, companies in Mexico do not have to spend as much money to control pollution.

Maquiladoras have become an important part of the border economy. They are also a good example of how borders can influence the landscape. The U.S.-Mexico border has attracted *maquiladoras* because it is more than just a line on a map. It separates two countries with different cultures, economies, and governments. Some American companies have built factories on the Mexican side of the border that export goods to the United States. These companies benefit from the different standards of living in the two countries and the border that separates them.

◄ Trucks carrying goods from Mexico to the United States are inspected by U.S. Customs officials in Laredo, Texas.

Understanding What You Read

1. Why have some American companies built factories in Mexico?

2. What are some of the goods that are produced in *maquiladoras*?

Section 3

Objectives

1. Describe the government and economy of Mexico today.
2. Identify the important features of Mexico's six culture regions.

FOCUS

LET'S GET STARTED

Copy the following question onto the chalkboard: *What kinds of factors distinguish one region from another?* Have students discuss their answers. *(Possible answers: economic, political, and social)* Tell students that in Section 3 they will learn about the six culture regions of Mexico as well as about its government and economy.

Building Vocabulary

Write the key terms on the chalkboard. Ask volunteers to find and read aloud the definitions of **cash crops**, **smog**, and **slash-and-burn agriculture**. Relate **inflation** to blowing up a balloon. Pronounce *maquiladoras* for them and have them repeat it and give its meaning. Then have students volunteer to use two of the terms in a sentence.

Section 3 RESOURCES

Reproducible
◆ Guided Reading Strategy 13.3
◆ Creative Strategies for Teaching World Geography, Lesson 9
◆ Environmental and Global Issues Activity 2

Technology
◆ One-Stop Planner CD–ROM, Lesson 13.3
◆ Homework Practice Online
◆ HRW Go site

Reinforcement, Review, and Assessment
◆ Section 3 Review, p. 307
◆ Daily Quiz 13.3
◆ Main Idea Activity 13.3

Chart Answer ▶

that many Mexican citizens live in poverty

Section 3 — Mexico Today

Read to Discover

1. What kind of government and economy does Mexico have today?
2. What are the important features of Mexico's six culture regions?

Define
inflation
cash crops
smog
maquiladoras
slash-and-burn agriculture

Locate
Tijuana
Ciudad Juárez
Acapulco
Mazatlán
Cancún
Mexico City
Guadalajara
Tampico
Campeche
Monterrey

WHY IT MATTERS

The North American Free Trade Agreement (NAFTA), signed by the United States, Mexico, and Canada, went into effect in 1994. Use **CNNfyi.com** or other **current events** sources to find out more about NAFTA. Record your findings in your journal.

Bean stew in a tortilla bowl

COUNTRY	POPULATION/ GROWTH RATE	LIFE EXPECTANCY	LITERACY RATE	PER CAPITA GDP
Mexico	101,879,171 1.5%	69, male 75, female	90%	$9,100
United States	281,421,906 0.9%	74, male 80, female	97%	$36,200

Mexico

Sources: Central Intelligence Agency, *The World Factbook 2001;* U.S. Census Bureau

Interpreting the Chart What does the per capita GDP in Mexico indicate about its economic development?

Government and Economy

Like that of the United States, Mexico's government includes an elected president and a congress. However, in Mexico one political party—the Partido Revolucionario Institucional (PRI)—controlled the Mexican government for 71 years. This control ended in 2000 when Vicente Fox of the National Action Party was sworn in as president.

For many years, Mexico operated more on the principles of a command economy in which the government controlled economic activity. In recent decades, however, Mexico has worked to reduce government control of the economy. Mexico's economy is growing and modernizing. However, living standards are much lower than in other nations.

Mexico began to export oil in 1911 and is a leading oil exporter today. However, problems began when the price of oil fell in the 1980s. Since then, Mexico has wrestled with debts to foreign banks, high unemployment, and **inflation**. Inflation is the rise in prices that occurs when currency loses its buying power.

The North American Free Trade Agreement (NAFTA) has helped Mexico's economy. NAFTA took effect in 1994. It made trade between Mexico, the United States, and Canada easier. Mexico's leaders hope increased trade will create more jobs in their country.

Teaching Objective 1

ALL LEVELS: (Suggested time: 20 min.) Copy the following graphic organizer onto the chalkboard, omitting the italicized answers. Organize the class into five groups. Assign each group one of the organizer topics and have students use their textbooks to find information. Ask volunteers from each group to write their information onto the organizer. **ENGLISH LANGUAGE LEARNERS, COOPERATIVE LEARNING**

GOVERNMENT
elected president and a congress, Partido Revolucionario Institucional controlled government for 71 years

ECONOMY—Challenges
foreign debts, poverty, high unemployment, inflation

MEXICO

ECONOMY—Agriculture
farming, coffee, sugarcane, livestock ranching, cash crops

ECONOMY—Industry
oil, mining, manufacturing many foreign companies build factories in Mexico

ECONOMY—Tourism
visit old colonial sites, Maya and Aztec monuments, coastal resorts

Agriculture Agriculture has long been an important part of the Mexican economy. In fact, farming is the traditional focus of life in the country. This is true even though just 12 percent of the land can grow crops. Rainfall supports agriculture in the southern part of the Mexican Plateau and in southern valleys. Plantations in coastal lowlands and on mountain slopes along the Gulf of Mexico also produce important crops. Those crops include coffee and sugarcane. Northern drylands are important for livestock ranching.

High demand in the United States has encouraged a shift to the growing of **cash crops**. A cash crop is produced primarily to sell, rather than for the farmer to eat. Trucks bring Mexican vegetables, fruits, and other cash crops to the United States.

Industry Many Mexicans work in primary and secondary industries. These include the oil industry, mining, and manufacturing. The country's fastest-growing industrial centers lie along the U.S. border. Tijuana (tee-HWAH-nah) and Ciudad Juárez (syoo-thahth HWAHR-es) are two of these major industrial centers.

Many U.S. and other foreign companies have built factories in Mexico. This is because wages are lower there. Mexican workers in these factories assemble products for export to other countries. (See the Case Study in this chapter.)

Tourism Tourism and other service industries are also important to Mexico's economy. Many tourists visit old colonial cities and Maya and Aztec monuments. Popular coastal cities and resorts include Acapulco, Mazatlán, and Cancún. Acapulco and Mazatlán are located on Mexico's Pacific coast. Cancún is on the Yucatán Peninsula.

✔ **READING CHECK:** *Environment and Society* How does geography affect the location of economic activities in Mexico? Good rainfall produces land that can support agriculture; oil and mineral reserves support mining and drilling; good beaches support tourism.

▲

Gourds—hard-shelled ornamental fruits—are just some of Mexico's many agricultural products.

Puerto Vallarta—a popular beach resort—is on the Pacific coast.
Interpreting the Visual Record How do you think the growth of tourism has changed settlement patterns?

▼

Across the Curriculum
ART

Murals and Mosaics One cultural feature of Mexico that attracts tourists is the beautiful murals. Murals decorate the walls of many public buildings in Mexico. One of the most famous mural artists was Diego Rivera, who painted frescoes and designed mosaics showing scenes from Mexican history and daily life. Architect Juan O'Gorman designed the huge mosaic mural on the library at the National Autonomous University of Mexico. These murals along with old colonial architecture combine to form a distinct style found in many Mexican cities.

Discussion: Lead a discussion on the relationship between Mexican history and murals and mosaics.

🖳 internet connect

GO TO: go.hrw.com
KEYWORD: SJ3 CH13
FOR: Web sites about murals

◄ **Visual Record Answer**

Students might suggest that the new jobs generated by increased tourism would probably draw more people to live in those areas.

LEVEL 1: (Suggested time: 30 min.) Organize students into six groups and assign each group a culture region. Have students design and draw on a posterboard a commemorative stamp that highlights distinctive characteristics of their assigned culture region. **ENGLISH LANGUAGE LEARNERS, COOPERATIVE LEARNING**

LEVEL 2: (Suggested time: 30 min.) Organize the class into six groups. Assign each group one of the culture regions. Give each group six index cards and have the group write one question about its region on each card. Answers for each question should be written on the back of the card. When all groups have completed their questions, collect all the cards and quiz the class. Keep score to see which group answers the most questions correctly. **COOPERATIVE LEARNING**

LEVEL 3: (Suggested time: 20 min.) Have students design their own graphic organizer to identify the characteristics of the six culture regions of Mexico. Organizers may be in a variety of formats, such as charts or idea webs. Have volunteers copy their graphic organizer onto the chalkboard. Then lead a class discussion on the culture regions.

CLOSE

Ask volunteers to give examples of the relationship between environment and society in each of the six culture regions of Mexico.

Section Review 3

Answers

Define For definitions, see: inflation, p. 304; cash crops, p. 305; smog, p. 306; *maquiladoras*, p. 307; slash-and-burn agriculture, p. 307

Working with Sketch Maps Maps will vary, but listed places should be labeled in their approximate locations. Mexico City is the national capital. Acapulco, Tampico, Cancún, Tijuana, Campeche, and Mazatlán are near sea level. Guadalajara, Mexico City, Ciudad Juárez, and Monterrey are at higher elevations.

Reading for the Main Idea

1. debts to foreign banks, high unemployment, inflation (NGS 12)

2. Greater Mexico City—most developed and crowded; Central Interior—fertile valleys; Oil Coast—forested plain with growing population; Southern Mexico—poorest region; Northern Mexico—prosperous and modern; Yucatán—tourism, sparsely populated (NGS 10)

Critical Thinking

3. hurt the country's economy because Mexico is a major exporter of oil

4. Farmers buy the food with the cash they earn from selling their crops.

Visual Record Answer ▶

smoke and fog

Do you remember what you learned about population density? See Chapter 5 to review.

Mexico City was 250 miles (400 km) away from the center of the 1985 earthquake on Mexico's Pacific coast. However, the city suffered heavy damage and thousands of deaths. The city sits on a former lake bed. The loose soil under the city made the damage worse.

Smog covers Mexico City. The city continues to work toward solutions to challenges facing it as one of the world's largest cities.

Interpreting the Visual Record What two words do you think the word *smog* comes from?

▶

Mexico's Culture Regions

Mexico's 31 states and one federal district can be grouped into six culture regions. These regions are highly diverse in their resources, climate, population, and other features.

Greater Mexico City Greater Mexico City is Mexico's most developed and crowded region. This area includes Mexico City and about 50 smaller cities. More than 20 million people live there. It is one of the most heavily populated urban areas in the world.

Mexico City is also one of the world's most polluted cities. Thousands of factories and millions of automobiles release exhaust and other pollutants into the air. Surrounding mountains trap the resulting **smog**—a mixture of smoke, chemicals, and fog. Smog can cause health problems like eye irritation and breathing difficulties.

Wealth and poverty exist side by side in Mexico City. The city has very poor slums. It also has busy highways, modern office buildings, high-rise apartments, museums, universities, and old colonial cathedrals.

Central Interior Mexico's central interior region lies north of the capital. It extends toward both coasts. Many cities here began as mining or ranching centers during the colonial period. Small towns with a central square and a colonial-style church are common.

The region has many fertile valleys and small family farms. In recent years the central interior has attracted new industries from overcrowded Mexico City. As a result, cities like Guadalajara are growing rapidly.

Oil Coast The forested coastal plains between Tampico and Campeche (kahm-PAY-chay) were once lightly settled. However, the population has grown as oil production in this region has increased. In addition, large forest areas are being cleared for farming and ranching.

REVIEW AND ASSESS

Have students complete the Section Review. Then organize students into six groups and assign each group a culture region. Have each group use the text and map to prepare five questions, with answers, about the economy and geography of the culture regions of Mexico for a quiz game. When questions are completed, have the class play the quiz game. Then have students complete Daily Quiz 13.3. **COOPERATIVE LEARNING**

RETEACH

Have students complete Main Idea Activity 13.3. Organize students into two groups and have them create a wall map of Mexico. Have one group label the cultural regions of Mexico and the other group draw or glue symbols depicting the economic activities of the country. Display and discuss the maps. **ENGLISH LANGUAGE LEARNERS, COOPERATIVE LEARNING**

EXTEND

Have interested students find out more about Mexico's tourist industry by contacting a travel agent. Ask students to plan a vacation itinerary based on their research. **BLOCK SCHEDULING**

Southern Mexico Many people in southern Mexico speak Indian languages. They live in the country's poorest region. It has few cities and little industry. Subsistence farming is common. Poverty and corrupt local governments have led to unrest. In the 1990s people in the state of Chiapas staged an antigovernment uprising.

Northern Mexico Northern Mexico has become one of the country's most prosperous and modern areas. NAFTA has helped the region's economy grow. Monterrey and Tijuana are important cities here. Factories called ***maquiladoras*** (mah-kee-lah-DORH-ahs) are located along the northern border and are often foreign-owned.

American music, television, and other forms of entertainment are popular near the border. Many Mexicans cross the border to shop, work, or live in the United States. In recent decades, the U.S. government has increased its efforts to stop illegal immigration across the border.

The Yucatán Most of the Yucatán Peninsula is sparsely populated. Mérida is this region's major city. As in other parts of Mexico, some farmers in the region practice **slash-and-burn agriculture**, in which an area of forest is burned to clear it for planting. The ashes enrich the soil. After a few years of planting crops, the soil is exhausted. The farmer then moves on to a new area of forest. Farmers can return to previously farmed areas years later.

Maya ruins and sunny beaches have made tourism a major industry in this area. The popular resort of Cancún and the island of Cozumel are located here.

✓ **READING CHECK:** *Human Systems* What are the six culture regions of Mexico? Greater Mexico City, Central Interior, Oil Coast, Southern Mexico, Northern Mexico, the Yucatán

▲
Celebrations, such as Danza de Los Viejitos—Dance of the Old Men— are popular throughout Mexico.

go.hrw.com **Homework Practice Online** Keyword: SJ3 HP13

Section Review 3

Define and explain: inflation, cash crops, smog, *maquiladoras*, slash-and-burn agriculture

Working with Sketch Maps On the map you created in Section 2, label Tijuana, Ciudad Juárez, Acapulco, Mazatlán, Cancún, Mexico City, Guadalajara, Tampico, Campeche, and Monterrey. Which city is the national capital? Which cities are located at or near sea level? Which are located at higher elevations?

Reading for the Main Idea

1. *Human Systems* What are three economic problems faced by Mexico in recent decades?

2. *Human Systems* What are Mexico's six culture regions? Describe a feature of each.

Critical Thinking

3. Analyzing Information How have changes in the price of oil affected Mexico?

4. Drawing Inferences and Conclusions Why would farmers in Mexico grow only cash crops?

Organizing What You Know

5. Categorizing Copy the following graphic organizer. Use it to list key facts about the population and economy of each culture region in Mexico.

	Population	Economy
Greater Mexico City		
Central interior		
Oil coast		
Southern Mexico		
Northern Mexico		
Yucatán		

Organizing What You Know

5. Greater Mexico City—densely populated, business and industry; Central Interior—small towns, farming and new industries; Oil Coast—growing population, oil and ranching; Southern Mexico—poorest people, farming; Northern Mexico—prosperous and modern, economic links to the United States; Yucatán—sparsely populated, farming and tourism

CHAPTER **13**

Review ANSWERS

Building Vocabulary

For definitions, see: conquistadores, p. 299; epidemic, p. 299; empire, p. 299; mestizos, p. 300; missions, p. 300; haciendas, p. 300; inflation, p. 304; cash crops, p. 305; smog, p. 306; *maquiladoras*, p. 307

ASSESS

Have students complete the Chapter 13 Test.

RETEACH

Organize students into five groups and assign each group one of the following topics: vegetation, wildlife, mining and petroleum resources, culture and civilizations, or agriculture and industry. Have each group create a map, design map-key illustrations for its topic, and place the illustrations on the map in the appropriate areas. Display and discuss the completed map.

PORTFOLIO EXTENSIONS

1. Have a Mexican fiesta! Organize the class into groups to research and plan various aspects of the party—costumes, dances, decorations, games, music, refreshments—so that it is as authentic as possible. Students should write summaries of their findings and take photographs of their contributions. Place summaries and photos in portfolios.

2. Have students create mosaics depicting Mexican history and culture using construction paper cut into small pieces. You may want to assign specific topics to individual students or allow them to choose topics. Pin mosaics onto a bulletin board to form a collage. Later, place mosaics in portfolios.

Review
ANSWERS

Reviewing the Main Ideas

1. rugged central plains region (NGS 4)

2. because oil production is located there (NGS 4)

3. high death toll weakened the Aztec Empire (NGS 12)

4. Spaniards, Indians, mestizos, mulattoes (NGS 10)

5. Mexican agricultural products are exported to the United States. Many *maquiladoras* are located along the border. Trade is easier because of NAFTA. (NGS 11)

Understanding Environment and Society

Students may mention beaches and ancient ruins and note that Mexico has developed tourist resorts along the coasts. Benefits may include an economic boost; problems may include a strain on environment and resources. Use Rubric 29, Presentations, to evaluate student work.

Thinking Critically

1. differences in elevation

2. terraced hillsides, dug canals and used bottom mud to enrich soil

3. Spanish conquistadores were Roman Catholic. Spanish missions spread the religion.

Reviewing What You Know

Building Vocabulary

On a separate sheet of paper, write sentences to define each of the following

1. conquistadores
2. epidemic
3. empire
4. mestizos
5. missions
6. haciendas
7. inflation
8. cash crops
9. smog
10. *maquiladoras*

Reviewing the Main Ideas

1. *(Places and Regions)* What is the Mexican Plateau?

2. *(Places and Regions)* Why are the coastal plains now more heavily settled?

3. *(Human Systems)* How did European diseases affect the Indians in Mexico?

4. *(Human Systems)* What were the main ethnic divisions in New Spain?

5. *(Human Systems)* How are agriculture and industry in Mexico and the United States related?

Understanding Environment and Society

Encouraging Tourism

Mexico's pleasant coastal climate has helped its tourist industry grow. Create a graph and presentation on the growth of tourism in Mexico. As you prepare your presentation you may want to think about the following:

• Mexico's popular tourist attractions
• Mexico's promotion of tourism
• Benefits and problems of tourism

Thinking Critically

1. **Drawing Inferences and Conclusions** Why do climates in southern Mexico vary so widely?

2. **Finding the Main Idea** How did the Maya and Aztec modify their environment to suit their needs?

3. **Analyzing Information** Why is Roman Catholicism the most common religion in Mexico?

4. **Drawing Inferences and Conclusions** Mexican workers are paid much less than U.S. workers. Why do you think this is so?

5. **Finding the Main Idea** Think about what you have learned about agriculture in Mexico. What is the major drawback of slash-and-burn agriculture?

FOOD FESTIVAL

Have students prepare guacamole as it might have been made 2,000 years ago in Oaxaca. Boil 4 cups of water in a saucepan. Add 3 large husked, washed tomatillos. Turn off heat and allow tomatillos to soften—about five minutes. Remove from water and puree in blender. Chill. Combine the flesh of 2 large avocados and ½ minced serrano or jalapeño pepper in a bowl. Add tomatillo puree and mash until combined. Serve with baked tortilla chips.

CHAPTER 13 REVIEW AND ASSESSMENT RESOURCES

Reproducible
◆Readings in World Geography, History, and Culture 15 and 16
◆Critical Thinking Activity 13
◆Vocabulary Activity 13

Technology
◆Chapter 13 Test Generator (on the One-Stop Planner)

◆Audio CD Program, Chapter 13
◆HRW Go site

Reinforcement, Review, and Assessment
◆Chapter 13 Review, pp. 308–09

◆Chapter 13 Tutorial for Students, Teachers, Mentors, and Peers
◆Chapter 13 Test
◆Chapter 13 Test for English Language Learners and Special-Needs Students

---

Building Social Studies Skills

Map ACTIVITY

On a separate sheet of paper, match the letters on the map with their correct labels.

Río Bravo (Rio Grande)
Mexico City
Mount Orizaba
Yucatán Peninsula

Tijuana
Ciudad Juárez
Acapulco
Cancún
Guadalajara

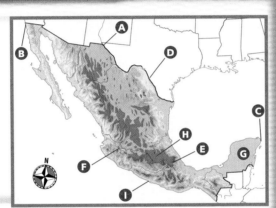

Mental Mapping Skills ACTIVITY

On a separate sheet of paper, draw a freehand map of Mexico. Make a key for your map and label the following:

Baja California
Gulf of California
Gulf of Mexico
Mexican Plateau
Sierra Madre Occidental
Sierra Madre Oriental

WRITING ACTIVITY

Imagine that you are a Spanish soldier reporting back to Spain on the situation in Mexico. Research and write a short report on the effects of disease on the Aztec. Include a map, chart, graph, model, or database illustrating the information in your report. Be sure to use standard grammar, spelling, sentence structure, and punctuation.

4. possible answers: prices lower in Mexico, lower standard of living
5. exhausts the soil

Map Activity
A. Ciudad Juárez
B. Tijuana
C. Cancún
D. Río Bravo
E. Mount Orizaba
H. Mexico City
I. Acapulco
F. Guadalajara
G. Yucatán Peninsula

Mental Mapping Skills Activity
Maps will vary, but listed places should be labeled in their approximate locations.

Writing Activity
Reports should contain information about disease and the Aztec. Use Rubric 37, Writing Assignments, to evaluate student work.

Portfolio Activity
Data should include as many factors as possible, including pollutants from factories and vehicles. Use Rubric 30, Research, to evaluate student work.

Alternative Assessment

Portfolio ACTIVITY

Learning About Your Local Geography
Research Project Contact your local air quality board for information on the air quality of your area. If possible, collect data going back several years. What factors account for the patterns you see over time?

internet connect

Internet Activity: go.hrw.com
KEYWORD: SJ3 GT13

Choose a topic to explore about Mexico:
• Travel along Mexico's coastlines.
• See the arts and crafts of Mexico.
• Use ancient Maya hieroglyphs

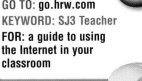

internet connect

GO TO: go.hrw.com
KEYWORD: SJ3 Teacher
FOR: a guide to using the Internet in your classroom

CHAPTER 14

Central America and the Caribbean Islands

Chapter Resource Manager

Objectives	Pacing Guide	Reproducible Resources
SECTION 1 **Physical Geography** (pp. 311–13) 1. Identify the physical features of Central America and the Caribbean islands. 2. Identify the climates found in the region. 3. Identify the natural resources in the region.	**Regular** .5 day **Block Scheduling** .5 day *Block Scheduling Handbook, Chapter 14*	**RS** Guided Reading Strategy 14.1 **E** Lab Activities for Geography and Earth Science, Demonstrations 5, 6, 7
SECTION 2 **Central America** (pp. 314–18) 1. Describe what Central America's early history was like. 2. Describe how the region's history is reflected in its people today. 3. Describe what the countries of Central America are like today.	**Regular** 1.5 days **Block Scheduling** .5 day *Block Scheduling Handbook, Chapter 14*	**RS** Guided Reading Strategy 14.2 **RS** Graphic Organizer 14 **E** Cultures of the World Activity 2 **IC** Interdisciplinary Activities for the Middle Grades 9, 11, 12
SECTION 3 **The Caribbean Islands** (pp. 319–23) 1. Describe the Caribbean's history. 2. Describe how the region's history is reflected in its people today. 3. Describe what the countries of the Caribbean are like today.	**Regular** .5 day **Block Scheduling** .5 day *Block Scheduling Handbook, Chapter 14*	**RS** Guided Reading Strategy 14.3 **SM** Geography for Life Activity 14 **SM** Map Activity 14

Chapter Resource Key

RS Reading Support

IC Interdisciplinary Connections

E Enrichment

SM Skills Mastery

A Assessment

REV Review

ELL Reinforcement and English Language Learners

 Transparencies

 CD–ROM

 Music

 Video

 Internet

 Holt Presentation Maker Using Microsoft® Powerpoint®

 One-Stop Planner CD–ROM

See the *One-Stop Planner* for a complete list of additional resources for students and teachers.

One-Stop Planner CD–ROM

It's easy to plan lessons, select resources, and print out materials for your students when you use the *One-Stop Planner CD–ROM with Test Generator.*

internet connect

HRW ONLINE RESOURCES

<u>GO TO: go.hrw.com</u>
Then type in a keyword.

TEACHER HOME PAGE
KEYWORD: SJ3 TEACHER

CHAPTER INTERNET ACTIVITIES
KEYWORD: SJ3 GT14

Choose an activity to:
• plan an ecotour of Central America.
• learn more about the Panama Canal.
• hunt for hurricanes.

CHAPTER ENRICHMENT LINKS
KEYWORD: SJ3 CH14

CHAPTER MAPS
KEYWORD: SJ3 MAPS14

ONLINE ASSESSMENT
Homework Practice
KEYWORD: SJ3 HP14
Standardized Test Prep Online
KEYWORD: SJ3 STP14
Rubrics
KEYWORD: SS Rubrics

COUNTRY INFORMATION
KEYWORD: SJ3 Almanac

CONTENT UPDATES
KEYWORD: SS Content Updates

HOLT PRESENTATION MAKER
KEYWORD: SJ3 PPT14

ONLINE READING SUPPORT
KEYWORD: SS Strategies

CURRENT EVENTS
KEYWORD: S3 Current Events

Technology Resources

 One-Stop Planner CD–ROM, Lesson 14.1

 Geography and Cultures Visual Resources with Teaching Activities 18–22

 Homework Practice Online

HRW Go site

 One-Stop Planner CD–ROM, Lesson 14.2

Homework Practice Online

HRW Go site

 One-Stop Planner CD–ROM, Lesson 14.3

 ARGWorld CD–ROM: Beach Resorts and Ecotourism Around the Caribbean

Homework Practice Online

HRW Go site

Review, Reinforcement, and Assessment Resources

ELL	Main Idea Activity 14.1
REV	Section 1 Review, p. 313
A	Daily Quiz 14.1
ELL	English Audio Summary 14.1
ELL	Spanish Audio Summary 14.1

ELL	Main Idea Activity 14.2
REV	Section 2 Review, p. 318
A	Daily Quiz 14.2
ELL	English Audio Summary 14.2
ELL	Spanish Audio Summary 14.2

ELL	Main Idea Activity 14.3
REV	Section 3 Review, p. 323
A	Daily Quiz 14.3
ELL	English Audio Summary 14.3
ELL	Spanish Audio Summary 14.3

Meeting Individual Needs

Ability Levels

Level 1 Basic-level activities designed for all students encountering new material

Level 2 Intermediate-level activities designed for average students

Level 3 Challenging activities designed for honors and gifted-and-talented students

English Language Learners Activities that address the needs of students with Limited English Proficiency

Chapter Review and Assessment

E	Creative Strategies for Teaching World Geography, Lesson 9
E	Readings in World Geography, History, and Culture 17, 18, and 19
SM	Critical Thinking Activity 14
REV	Chapter 14 Review, pp. 324–25
REV	Chapter 14 Tutorial for Students, Parents, Mentors, and Peers
ELL	Vocabulary Activity 9
A	Chapter 14 Test
	Chapter 14 Test Generator (on the One-Stop Planner)
	Audio CD Program, Chapter 14
A	Chapter 14 Test for English Language Learners and Special-Needs Students

LAUNCH INTO LEARNING

Have students look at the photographs throughout the chapter. Ask them to use the photographs to identify opportunities and challenges this region might have. Discuss responses. *(Possible answers: The region's many ruins and natural beauty may offer opportunities for tourism. The varied environment and the isolation of the region's islands might cause economic challenges.)* Explain that although the region has a fascinating history and widespread natural beauty, it also faces many challenges. Tell students they will learn more about the region in this chapter.

Section 1

Objectives

1. Describe the physical features of Central America and the Caribbean islands.
2. Identify the climates found in the region.
3. Describe the region's natural resources.

LINKS TO OUR LIVES

You may want to reinforce interest in Central America and the Caribbean islands by pointing out the following facts:

▶ Many Central American and Caribbean immigrants have come to the United States to escape political violence and economic problems.

▶ The United States has a long history of military and economic involvement in the region.

▶ Some forms of music that students might enjoy, such as reggae, originated in the Caribbean region.

▶ Historic sites, mild weather, and sunny beaches attract tourists from around the world to Central America and the Caribbean islands.

Central America and the Caribbean Islands

Now we will study Central America and the islands of the Caribbean. First we will meet a student from Jamaica.

I am Aldayne and I am 11 years old. I live with my mother in Kingston, Jamaica. My house is gray—the door is green and the windows are blue. It is all on one floor, with a yard where we grow roses.

I am in the sixth grade. I am studying social studies, math, English, and history. We have three school terms a year. We go to school for three months and then have one month off. My favorite vacation month is in summer. That's when I can go to the country with my father. I catch birds, go fishing, run boats, and have cookouts.

When school is over at 2:00 P.M., I go downtown to my mother's shop. She is a dressmaker. I do my homework and sometimes do errands for her. By the time I get home, all my homework is done. While my mother makes dinner, I clean my shoes, get my clothes together, and get ready for the next day. My favorite dinners are rice and peas with spring chicken, or crispy fried chicken with chips (french fries).

Hello from Jamaica!

LET'S GET STARTED

Copy the following question and saying onto the chalkboard: *What do you think this ancient Maya saying suggests about the Maya's view of nature? "Who cuts the trees as he pleases cuts short his own life."* Allow students to write their responses. *(possible answer: indicates they realized human life depends on the environment)* Lead a discussion based on students' answers. You may want to compare the Maya belief to what students know about American Indians. Tell students that in Section 1 they will learn more about the region's environment.

Using the Physical-Political Map

Have students examine the map on this page. Discuss the meaning of *isthmus (a narrow piece of land that connects two other landmasses like a bridge).* Point out that Central America is an isthmus. Have students identify the region's main bodies of water and physical features. Point out that many goods pass through the Panama Canal, which plays a key economic role in the hemisphere.

Building Vocabulary

Write the key terms on the chalkboard and call on volunteers to locate and read the definitions aloud. Ask students to match **archipelago**, **cloud forest**, and **bauxite** with these categories of physical geography: resources, landforms, and vegetation. *(archipelago—landforms; cloud forest—vegetation; and bauxite—resources)*

Section 1 — Physical Geography

Read to Discover

1. What are the physical features of Central America and the Caribbean islands?
2. What climates are found in the region?
3. What natural resources does the region have?

Define

archipelago
cloud forest
bauxite

Locate

Central America
Caribbean Sea
Cuba
Jamaica
Greater Antilles
Hispaniola
Puerto Rico
Lesser Antilles
Virgin Islands
Trinidad and Tobago
Bahamas

WHY IT MATTERS

Hurricanes are frequent dangers in the Caribbean region. Use **CNNfyi.com** or other **current events** sources to find out about recent hurricanes. Record your findings in your journal.

A conch shell

Section 1 RESOURCES

Reproducible
◆ Block Scheduling Handbook, Chapter 14
◆ Guided Reading Strategy 14.1
◆ Lab Activities for Geography and Earth Science, Demonstrations 5, 6, 7

Technology
◆ One-Stop Planner CD–ROM, Lesson 14.1
◆ Homework Practice Online
◆ Geography and Cultures Visual Resources with Teaching Activities 18–22
◆ HRW Go site

Reinforcement, Review, and Assessment
◆ Section 1 Review, p. 313
◆ Daily Quiz 14.1
◆ Main Idea Activity 14.1
◆ English Audio Summary 14.1
◆ Spanish Audio Summary 14.1

Central America and the Caribbean Islands: Physical-Political

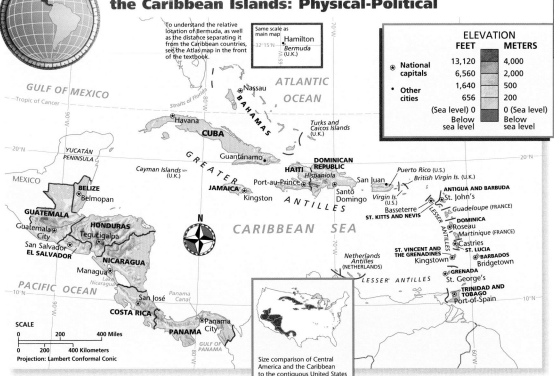

To understand the relative location of Bermuda, as well as the distance separating it from the Caribbean countries, see the Atlas map in the front of the textbook.

Same scale as main map
Hamilton
Bermuda (U.K.)
32°15'N
65°W

ELEVATION

	FEET	METERS
⊛ National capitals	13,120	4,000
• Other cities	6,560	2,000
	1,640	500
	656	200
	(Sea level) 0	0 (Sea level)
	Below sea level	Below sea level

GULF OF MEXICO
Tropic of Cancer
Straits of Florida
ATLANTIC OCEAN
Nassau
BAHAMAS
Havana
CUBA
Turks and Caicos Islands (U.K.)
20°N
YUCATÁN PENINSULA
Guantánamo
Cayman Islands (U.K.)
GREATER ANTILLES
Guatánamo
HAITI
DOMINICAN REPUBLIC
Hispaniola
Puerto Rico (U.S.)
British Virgin Is. (U.K.)
20°N
MEXICO
BELIZE
Belmopan
Port-au-Prince
JAMAICA
Kingston
San Juan
Santo Domingo
Virgin Is. (U.S.)
ANTIGUA AND BARBUDA
St. John's
GUATEMALA
Guatemala City
HONDURAS
Tegucigalpa
San Salvador
EL SALVADOR
NICARAGUA
Managua
Lake Nicaragua
Basseterre
ST. KITTS AND NEVIS
Guadeloupe (FRANCE)
DOMINICA
Roseau
Martinique (FRANCE)
Castries
ST. LUCIA
Netherlands Antilles (NETHERLANDS)
ST. VINCENT AND THE GRENADINES
Kingstown
BARBADOS
Bridgetown
CARIBBEAN SEA
LESSER ANTILLES
GRENADA
St. George's
PACIFIC OCEAN
10°N
San José
Panama Canal
COSTA RICA
PANAMA
Panama City
GULF OF PANAMA
TRINIDAD AND TOBAGO
Port-of-Spain
10°N

SCALE
0   200   400 Miles
0   200   400 Kilometers
Projection: Lambert Conformal Conic

Size comparison of Central America and the Caribbean to the contiguous United States

311

Teaching Objectives 1–3

LEVEL 1: (Suggested time: 20 min.) Copy the graphic organizer at right onto the chalkboard, omitting the italicized answers. Call on students to provide words and phrases to describe the landforms, bodies of water, climates, and resources of the region. **ENGLISH LANGUAGE LEARNERS**

LEVELS 2 AND 3: (Suggested time: 25 min.) Have students use the graphic organizer in the Level 1 activity as a basis for this activity. Ask them to write a diary entry for a Central American or a Caribbean islander who works as a volcanologist (scientist who studies volcanoes), forest ranger, hurricane tracker, or miner. Encourage students to use additional resources.

Physical Geography of Central America and the Caribbean Islands

Landforms	Bodies of Water
natural land bridge, archipelago, Greater Antilles, Lesser Antilles, Cocos plate, Caribbean plate	*Caribbean Sea, Pacific Ocean, Lake Nicaragua, Atlantic Ocean, Gulf of Mexico*
Climates	**Resources**
mild highland, humid tropical, tropical savanna	*land for agriculture, timber, bauxite, copper*

PHYSICAL SYSTEMS

Montserrat, a small British territory in the Caribbean Sea, had just recovered from the devastating Hurricane Hugo of 1989 when its Soufrière Hills volcano erupted after lying dormant for hundreds of years. The eruption in 1997 set fire to the capital city and set fires in an oil plant. Rain mixed with volcanic ash created acid rain, damaging Montserrat's environmental systems.

Most of Montserrat is now considered a danger zone, and many residents have moved elsewhere. Decreased tourism and increased business closings have weakened its economy. On the positive side, however, farmers note that the ash has provided nutrients for its farmland.

Activity: Have students investigate the effect of volcanoes on Central America and the Caribbean islands. Ask them to prepare a line graph to show the dates of the events and the estimated costs of the damages they caused.

Visual Record Answer

humid, rainy

312

internet connect

GO TO: go.hrw.com
KEYWORD: SJ3 CH14
FOR: Web sites about Central America and the Caribbean islands

Our Amazing Planet

Dominica, in the far southeast of the region, receives up to 250 inches (835 cm) of rain per year! In contrast, Miami, Florida, receives only about 56 inches (140 cm).

Forested mountains rise above Roseau, capital of the island country of Dominica. Dominica is in the Lesser Antilles.

Interpreting the Visual Record What might the lush forest tell us about Dominica's climate?

Physical Features

Central America and the Caribbean Sea are home to 20 countries and a number of island territories. Central America includes Guatemala, Belize, Honduras, El Salvador, Nicaragua, Costa Rica, and Panama. The Caribbean islands include Cuba, Jamaica, Haiti, and the Dominican Republic. There also are nine smaller island countries.

Central America forms a bridge between North and South America. No place on this isthmus is more than 125 miles (200 km) from the sea. Mountains separate the Caribbean and Pacific coastal plains.

The Caribbean islands form an **archipelago** (ahr-kuh-PE-luh-goh), or large group of islands. The Caribbean archipelago is arranged in a long curve. It stretches from south of Florida to South America.

There are two main island groups in the Caribbean archipelago. The four large islands of the Greater Antilles (an-TI-leez) are Cuba, Jamaica, Hispaniola, and Puerto Rico. The small islands of the Lesser Antilles stretch from the Virgin Islands to Trinidad and Tobago. Another island group, the Bahamas, lies outside the Caribbean, east of Florida. It includes nearly 700 islands and thousands of reefs.

Earthquakes and volcanic eruptions are frequent in this region. Colliding plates cause this tectonic activity. The Cocos plate collides with and dives under the Caribbean plate off Central America's western coast. Another plate boundary lies to the east. The Caribbean plate borders the North American plate there. Volcanic eruptions can cause great damage.

✓ **READING CHECK:** (*Places and Regions*) What are the physical features of Central America and the Caribbean islands? *Central America—land bridge, mountains, plains; Caribbean islands—archipelago*

Climate and Vegetation

Along the Caribbean coast of Central America are humid tropical plains. The area also has dense rain forests. Inland mountains rise into mild highland climates. Most of the people live there because the temperatures are more moderate. Much of the original savanna vegetation inland

CLOSE

Have students list as many products as they can that are made from aluminum, which comes from bauxite. *(possible answers: soda cans, foil, house siding, pots, and pans)*

REVIEW AND ASSESS

Have students complete the Section Review. Then pair students. Ask each student to create a five-question crossword puzzle based on the section's main ideas. Have students exchange puzzles with their partners and solve them. Then have students complete Daily Quiz 14.1.
COOPERATIVE LEARNING

RETEACH

Have students complete Main Idea Activity 14.1. Then have students create a chart with three columns labeled Physical Features, Climates, and Natural Resources. Have students write three facts in each column.
ENGLISH LANGUAGE LEARNERS

EXTEND

Have interested students conduct research on how Panama's physical geography affected both the Panama Canal's design and the methods used to build it. Ask students to prepare diagrams of the construction process and the canal's lock system. Have students discuss any technological innovations used during construction. **BLOCK SCHEDULING**

has been cleared. It has been replaced by plantations and ranches. The Pacific coast has a warm and sunny tropical savanna climate.

Some of Central America's mountain areas are covered by **cloud forest**. This is a high-elevation, very wet tropical forest where low clouds are common. It is home to numerous plant and animal species.

The islands of the Caribbean have pleasant humid tropical and tropical savanna climates. Winters are usually drier than summers. The islands receive 40 to 60 inches (102 to 152 cm) of rainfall each year. On some islands the bedrock is mostly limestone. Water drains quickly. As a result, drought conditions are common.

Hurricanes are a danger in the region. Hurricanes are tropical storms that bring violent winds, heavy rain, and high seas. Most occur between June and November. They can cause great destruction and loss of life.

✓ **READING CHECK:** *Environment and Society* What effect do hurricanes have on people in the region? They cause great destruction and loss of life.

Resources

Agriculture in the region can be profitable where volcanic ash has enriched the soil. Coffee, bananas, sugarcane, and cotton are major crops. Timber is exported from the rain forests of Belize and Honduras. Tourism is the most important industry, particularly in the Caribbean islands.

The region has few mineral resources. However, Jamaica has large reserves of **bauxite**, the most important aluminum ore. There are huge copper deposits in Panama. Energy resources are limited. This makes the region dependent on energy imports and limits economic development.

✓ **READING CHECK:** *Environment and Society* Why do you think tourism is the most important industry in the region? because of the region's beautiful beaches and waters

BUILD on WHAT You Know

Do you remember what you learned about tropical forests? See Chapter 3 to review.

go.hrw.com
Homework Practice Online
Keyword: SJ3 HP14

Section Review 1

Define and explain: archipelago, cloud forest, bauxite

Working with Sketch Maps On a map of the region that you draw or that your teacher provides, label the following: Central America, Caribbean Sea, Cuba, Jamaica, Greater Antilles, Puerto Rico, Hispaniola, Lesser Antilles, Virgin Islands, Trinidad and Tobago, and the Bahamas. Identify continents Central America links.

Reading for the Main Idea

1. *Places and Regions* What two major island groups make up the Caribbean archipelago?

2. *Places and Regions* In what climate region do most Central Americans live and why?

3. *Places and Regions* What are some of the region's crops?

Critical Thinking

4. **Making Generalizations and Predictions** What natural hazards do you think will continue to be a problem for the region? Why?

Organizing What You Know

5. **Categorizing** Copy the following graphic organizer. Use it to describe the climates, vegetation, and resources found in the region.

Climates	Vegetation	Resources

Section Review 1

Answers

Define For definitions, see: archipelago, p. 312; cloud forest, p. 313; bauxite, p. 313

Working with Sketch Maps Maps will vary, but listed places should be labeled in their approximate locations. North America and South America are linked by Central America.

Reading for the Main Idea

1. the Greater Antilles and the Lesser Antilles (NGS 4)

2. in the highland climates; because they are more moderate (NGS 4)

3. coffee, bananas, sugarcane, and cotton (NGS 4)

Critical Thinking

4. possible answers: hurricanes because of the region's climate and location; earthquakes and volcanic eruptions because of the region's continuing tectonic plate activity

Organizing What You Know

5. climates—humid tropical plains, mild highland, warm and sunny tropical savanna; vegetation—dense rain forests, plants ranging from oak trees to orchids; resources—bauxite, coffee, bananas, sugarcane, cotton, timber, and copper

Section 2 Central America

Read to Discover

1. What was Central America's early history like?
2. How is the region's history reflected in its people today?
3. What are the countries of Central America like today?

Define

cacao
dictators
cardamom
civil war
ecotourism

Locate

Guatemala City
Lake Nicaragua
San José
Panama City
Panama Canal

WHY IT MATTERS

Civil wars caused great damage and disruption in Central America during the 1980s and early 1990s. Use **CNN fyi.com** or other **current events** sources to find out about conditions in Central American countries today. Record your findings in your journal.

Ancient jaguar pendant from Panama

▲ Maya ruins can be found here in Belize and in other parts of Central America and Mexico.

History

More than 35 million people live in the countries of Central America. These countries have a shared history.

Early History The early peoples of Central America developed different cultures and societies. The Maya, for example, built large cities with pyramids and temples. People of Maya descent still live in Guatemala and parts of Mexico. Many of their ancient customs and traditions still influence modern life.

In the early 1500s European countries began establishing colonies in the region. Most of Central America came under the control of Spain. In the 1600s the British established the colony of British Honduras, which is now Belize. The British also occupied the Caribbean coast of Nicaragua.

European colonists established large plantations. They grew crops like tobacco and sugarcane. They forced the Central American Indians to work on the plantations. Some Indians were sent to work in gold mines elsewhere in the Americas. In addition, many Africans were brought to the region as slaves.

Teaching Objective 1

LEVEL 1: (Suggested time: 30 min.) In advance, prepare slips of paper—one for each student—with descriptions of different people who played a role in Central America's history. (examples: a Spanish colonist, a British colonist, a Central American Indian, or an African) Repeat roles if necessary. Ask each student to draw a slip of paper and to write a few sentences describing how the person he or she chose influenced the region's early history. To conclude, lead a discussion about important dates and events in Central America's early history.
ENGLISH LANGUAGE LEARNERS

LEVELS 2 AND 3: (Suggested time: 40 min.) Organize the class into groups. In advance, locate and duplicate brief articles—one article per group—about some aspect of the early history of Central America. Give members of each group copies of the same article. Give students time to read it. Then review the six essential elements of geography with students. Have them discuss among themselves how the information in their article illustrates one, some, or all of the elements. Then lead a class discussion about important dates and events in the region's early history.
COOPERATIVE LEARNING

Independence Costa Rica, El Salvador, Guatemala, Honduras, and Nicaragua declared independence from Spain in 1821. They formed the United Provinces of Central America, but separated from each other in 1838–1839. Panama, once part of Colombia, became independent in 1903. The British left Nicaragua in the late 1800s. British Honduras gained independence as Belize in 1981.

✓ **READING CHECK:** *Human Systems* What role did Europeans play in the region's history? They colonized the region.

Culture

Central America's colonial history is reflected in its culture today. In what ways do you think this is true?

People, Languages, and Religion The region's largest ethnic group is mestizo. Mestizos are people of mixed European and Indian ancestry. People of African ancestry make up a significant minority. Various Indian peoples also live in the region.

Spanish is the official language in most countries. However, many people speak Indian languages. In the former British colony of Belize, English is the official language.

Many Central Americans practice religions brought to the region by Europeans. Most are Roman Catholics. Spanish missionaries converted many Indians to Catholicism. However, Indian religions have influenced Catholicism in the region. Protestant Christians are a large minority in some countries, particularly Belize.

Lake Nicaragua was probably once part of an ocean bay. It is a freshwater lake, but it also has some oceanic animal life. This animal life includes the Lake Nicaragua shark. It grows to a length of about 8 to 10 feet (2.4 to 3 m).

Guatemalans celebrate the Christian holiday *Semana Santa*, or Holy Week. The street is covered with colored sand and flowers.
Interpreting the Visual Record How important do Christian holidays appear to be in the region?

National Geography Standard 13

Human Systems The United Provinces of Central America was formed in 1823, shortly after the member countries had declared their independence from Mexico and two years since their break with Spain. Each country retained an independent government and constitution. The countries hoped that unification would give them more power to deal with larger countries and would strengthen the members' already similar cultures.

However, there was division within the union early in its history. Conservatives preferred keeping control in the hands of individual states and the Roman Catholic leadership. Liberals supported a strong central government with limited church involvement. Civil war erupted in 1826 and continued off and on through 1840, by which time the union had disbanded. As a result, Central America remained a collection of small, relatively weak countries.

◀ **Visual Record Answer**

Judging from the number of participants and their clothing, it is very important.

Teaching Objective 2

LEVEL 1: (Suggested time: 10 min.) Copy the graphic organizer at right onto the chalkboard, omitting the italicized answers. Use it to help students understand the region's culture. Ask students to copy the organizer into their notebooks and fill it in to show how the region's people, languages, religion, food, and festivals reflect its history. The first answer is filled in for them. **ENGLISH LANGUAGE LEARNERS**

LEVELS 2 AND 3: (Suggested time: 20 min.) Have students write a paragraph to support this thesis statement: "The European conquest of Central America has affected almost every aspect of life in the region." *(Students should include the effects on the Central American Indian population and the introduction of Roman Catholicism in their discussions.)* Call on volunteers to read their paragraphs to the class.

Aspects of Central American Culture

Before European Conquest	After European Conquest
• Central American Indian peoples	• mestizos, people of African ancestry
• Indian languages	• *Spanish and English languages*
• Indian religions	• *Christianity*
• native food crops	• *corn, sweet potatoes, hot peppers, tomatoes, and cacao*
	• *saints' feast days*

Linking Past to Present

Cacao Cacao trees are native to Central America. Chocolate is made from cacao. Spaniards brought chocolate to Europe after learning about its uses from the Aztecs.

In the 1980s Costa Rica's cacao industry was destroyed by a fungus called *Monilia* pod rot. Today, scientists are reintroducing the cacao tree to Costa Rica. Instead of clear-cutting the forest to create farmland, researchers have selected specific areas to cut. Once cleared, the land is replanted with cacao trees.

This process has many benefits. Cacao is the perfect crop for small farmers. Employment has increased and insecticide use is down.

Activity: Have students conduct research on other foods known to the early peoples of Central America. Ask them to find out if or how those foods are still cultivated in the region.

Connecting to Math
Answers
1. religion, mathematics, and astronomy
2. They were partially based on religion.

CONNECTING TO *Math*

Stone Maya calendar

The Maya created one of the most advanced civilizations in the Americas. They built impressive cities of stone and invented a complex writing system. They also created calendars that allowed them to plan for the future.

The Maya based their calendar system on religion, mathematics, and astronomy. They believed that the days, months, and years were represented by different gods. Those gods had powers that helped determine the course of events. The Maya also understood the mathematical concepts of zero and numerical positioning. Numerical positioning is the idea that numbers can have different values depending on their position in a sequence, such as the 1 in 10 and 100. The Maya also calculated the length of the solar year. They predicted eclipses and lunar cycles.

MAYA CALENDAR

Based on this knowledge, the Maya created two calendars. One was a religious calendar of 260 days. The other was a solar calendar of 365 days. These two calendars were used together to record historical events and to determine future actions. Using calendars, the Maya chose the best times to plant crops, hold festivals, and crown new rulers. They even believed that their calendars would help them predict the future.

Understanding What You Read
1. What were the foundations of the calendar system developed by the Maya?
2. How did the Maya's belief systems affect their use of calendars?

Food and Festivals Central America shares many traditional foods with Mexico and South America. These foods include corn and sweet potatoes. The region is also home to tomatoes, hot peppers, and **cacao** (kuh-KOW). Cacao is a small tree on which cocoa beans grow.

As in other Catholic countries, each town and country celebrates special saints' feast days. Images of the saints are paraded through the streets. A community might sponsor a fair and have dancing or plays.

Government From time to time, Central American countries have been ruled by **dictators**. A dictator is a person who rules a country with complete authority. The opponents of dictators often are arrested or even killed. Such abuses are one reason for limiting the power of government. Today the region's countries have elected governments.

✓ **READING CHECK:** *Human Systems* Why should government powers be limited? **to prevent abuses of power**

ALL LEVELS: (Suggested time: 30 min.) Tell students to imagine that they are tourists in Central America. Then have each student write a postcard to a family member or friend that describes pertinent features of two Central American countries. *(See the Section Review answers on the following page for the correct features.)* Ask volunteers to read their postcards to the class.

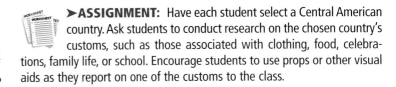

➤**ASSIGNMENT:** Have each student select a Central American country. Ask students to conduct research on the chosen country's customs, such as those associated with clothing, food, celebrations, family life, or school. Encourage students to use props or other visual aids as they report on one of the customs to the class.

CLOSE

Remind students that in the 1800s several Central American countries formed the United Provinces of Central America. Ask the class to imagine that the same countries have decided to form the United States of Central America. Lead a discussion about how such a union might affect trade and political relationships with the United States of America.

Central America Today

Now we will take a closer look at each of the countries of Central America today.

Guatemala Guatemala is the most populous country in Central America. More than 12 million people live there. Nearly half of the country's people are Central American Indians. Many speak Maya languages. A majority of Guatemala's population are mestizos.

Most Maya live in isolated villages in the country's highlands. Fighting between rebels and government forces has killed more than 100,000 Guatemalans since the 1960s. Guatemalans hope that recent peace agreements will end this conflict.

Coffee is Guatemala's most important crop. The country also is a major producer of **cardamom**, a spice used in Asian foods.

Belize Belize is located on the Caribbean coast of the Yucatán Peninsula. It has the smallest population in Central America. Only about 236,000 people live there. The country's Maya ruins, coral reefs, and coastal resorts attract many tourists.

Honduras Honduras is a country of rugged mountains. Most people live in mountain valleys and along the northern coast. Transportation is difficult in the rugged terrain. Only 15 percent of the land is suitable for growing crops. Fruit is an important export.

El Salvador El Salvador lies on the Pacific side of Central America. Volcanic ash has made the country's soils the most fertile in the region. Important crops include coffee and sugarcane.

Most Salvadorans live in poverty. A few powerful families own much of the best land. These conditions were a major reason behind a long **civil war** in the 1980s. A civil war is a conflict between two or more groups within a country. The war killed many Salvadorans and slowed economic progress. Salvadorans have been rebuilding their country since the war ended in 1992.

Nicaragua Nicaragua is the largest Central American country. It has coasts on both the Caribbean Sea and the Pacific Ocean. Lake Nicaragua, near the Pacific coast, is the largest lake in Central America.

Nicaragua also has been rebuilding since the end of a long civil war. Its civil war ended in 1990. Free elections that year ended the rule of the Sandinistas. The Sandinistas had overthrown a dictator in 1979. They then ruled Nicaragua without elections. Today, the country is a democracy with many political voices.

▲

A woman in the Guatemalan highlands weaves traditional fabrics.

Interpreting the Visual Record
How have available resources affected this woman's use of technology for weaving?

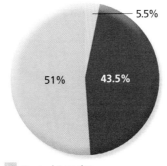

U.S. Banana Imports

5.5%

51% 43.5%

▢ **Central America**
◼ **South America**
▢ **Rest of the world**

Source: U.S. Bureau of the Census Trade Data

▲

Central America's warm, wet climate is ideal for growing bananas.
Interpreting the Chart **From where do most of the U.S. banana imports come?**

EYE ON EARTH

Belize is a scuba diver's paradise. The clear Caribbean waters near Belize offer a splendid view of three bodies called atolls. An atoll is a ring- or horseshoe-shaped formation of coral that is often as big as an island and surrounded by open sea. Atolls also enclose a shallow lagoon. The Belize atolls began forming some 70 million years ago. They are home to more than 60 species of stony coral and 200 species of colorful fish.

Most atolls are found in the Pacific and Indian Oceans. Only four atolls exist in the Western Hemisphere. The Belize atolls, the barrier reef to their west, and the vegetation native to the atolls and reefs together constitute a rich and fragile ecosystem.

Activity: Have students conduct research on the atolls and create illustrations of the food chains that exist in an atoll. Hold a class discussion on how these fragile environments might be protected.

▲ **Visual Record Answer**

She is using sticks and other natural materials for her frame.

◀ **Chart Answer**

from Central and South America

317

Have students complete the Section Review. Then have each student make a flash card for each Central American country. Instruct students to write the country's name on one side and a descriptive statement about that country on the other. Have students use their flash cards to quiz each other. Then have students complete Daily Quiz 14.2.

RETEACH

Have students complete Main Idea Activity 14.2. Organize the class into seven groups, one for each Central American country, and give each group a large sheet of paper. Have students draw and label pictures to illustrate important facts about their assigned country. Discuss and display students' pictures. **ENGLISH LANGUAGE LEARNERS, COOPERATIVE LEARNING**

CHALLENGE AND EXTEND

Have interested students read accounts by the early explorers, historians, and archaeologists who encountered the ancient ruins of Central America. Ask them to compare the early accounts with what it is like to visit those places today and share their impressions with the class. **BLOCK SCHEDULING**

Section Review 2

Answers

Define For definitions, see: cacao, p. 316; dictators, p. 316; cardamom, p. 317; civil war, p. 317; ecotourism, p. 318

Working with Sketch Maps Places should be labeled in their approximate locations.

Reading for the Main Idea

1. Spain and Great Britain (NGS 12)
2. holding parades, fairs, dances, and plays (NGS 10)

Critical Thinking

3. mix of peoples, languages, religions provides evidence of groups who settled there
4. stable democratic government and progress in reducing poverty

Organizing What You Know

5. Guatemala—largest population; Belize—smallest population; Honduras—mountainous; El Salvador—civil war ended in 1992; Nicaragua—largest; Costa Rica—democratic; Panama—Panama Canal

Chart Answer ▲

Costa Rica, Panama; perhaps stability of Costa Rica's government and the Panama Canal

COUNTRY	POPULATION/ GROWTH RATE	LIFE EXPECTANCY	LITERACY RATE	PER CAPITA GDP
Belize	256,062 2.7%	69, male 74, female	70%	$3,200
Costa Rica	3,773,057 1.7%	74, male 79, female	95%	$6,700
El Salvador	6,237,662 1.9%	66, male 74, female	72%	$4,000
Guatemala	12,974,361 2.6%	64, male 69, female	64%	$3,700
Honduras	6,406,052 2.4%	68, male 71, female	73%	$2,700
Nicaragua	4,918,393 2.2%	67, male 71, female	66%	$2,700
Panama	2,845,647 1.3%	73, male 79, female	91%	$6,000
United States	281,421,906 0.9%	74, male 80, female	97%	$36,200

Sources: Central Intelligence Agency, *The World Factbook 2001*; U.S. Census Bureau

Interpreting the Chart Which two Central American countries have the highest per capita GDP? Why might this be?

Costa Rica Costa Rica has a long history of stable, democratic government. In the last half of the 1900s, the country remained at peace. During that time, many of its neighbors were torn by civil wars. Costa Rica has also made important progress in reducing poverty.

Costa Rica's capital, San José, is located in the central highlands. Many coffee farms are also located in the highlands. Coffee and bananas are important Costa Rican crops.

Many travelers are attracted to Costa Rica's rich tropical rain forests and national parks. **Ecotourism**—the practice of using an area's natural environment to attract tourists—is an important part of Central American and Caribbean economies.

Panama Panama lies between Costa Rica and Colombia. Most Panamanians live in areas near the Panama Canal. Canal fees and industries make the canal area the country's most prosperous region.

The Panama Canal links the Pacific Ocean to the Caribbean Sea and Atlantic Ocean. The United States finished the canal in 1914. The canal played an important role in U.S. economic and foreign policies. It also allowed the United States to control territory and extend its influence in the area. The United States controlled the canal until 1999. Then, as called for by a 1978 treaty, Panama took it over.

✔ **READING CHECK:** *The Uses of Geography* Why might Panama want control of the canal? The canal is an important transportation corridor for trade; fees and industries near it help the surrounding area prosper.

go.hrw.com Homework Practice Online Keyword: SJ3 HP14

Section Review 2

Define and explain: cacao, dictators, cardamom, civil war, ecotourism

Working with Sketch Maps On the map you created in Section 1, label the Central American countries, Guatemala City, Lake Nicaragua, San José, Panama City, and the Panama.

Reading for the Main Idea

1. *Human Systems* What two European powers had Central American colonies by the late 1600s?
2. *Human Systems* How do many Central American communities honor certain Roman Catholic saints?

Critical Thinking

3. **Finding the Main Idea** How do Central America's people, languages, and religions reflect the region's history?
4. **Contrasting** How has Costa Rica's history differed from that of its Central American neighbors?

Organizing What You Know

5. **Categorizing** Copy the following graphic organizer. Use it to write at least one important fact about each Central American country today. Add as many rows as needed to list all the countries.

Guatemala	

Section 3

Objectives

1. Analyze the history of the Caribbean.
2. Discuss how the region's history is reflected in its people today.
3. Describe the Caribbean countries of today.

FOCUS

LET'S GET STARTED

Copy the following instructions and questions onto the chalkboard: *Look at the photos in Section 3 of your textbook. Which of the places pictured would you like to visit? Why?* Have students write their answers. Discuss student responses. Ask students to keep what they have written at hand so they can compare their original impressions with what they learn. Tell students that in Section 3 they will learn more about the history and culture of the Caribbean islands.

Building Vocabulary

Write the key terms on the chalkboard. Invite students to suggest meanings of these words, encouraging them to use their knowledge of word parts to derive the meanings. For example, *refuge*, the base of **refugees**, means "a safe place." Thus, a refugee is someone who seeks a safe place. Call on students to suggest how *cooperate* and **cooperatives** might be connected and then to check their suggestions against the text. Tell students that **guerrilla** is related to the Spanish word for war, *guerra*. Have students find the remainder of the definitions in the chapter text.

Section 3
The Caribbean Islands

Read to Discover

1. What was the Caribbean's history like?
2. How is the region's history reflected in its people today?
3. What are the countries of the Caribbean like today?

Define

Santería
calypso
reggae
merengue
guerrilla
refugees
cooperatives
plantains
commonwealth

Locate

Havana
Port-au-Prince
Santo Domingo

WHY IT MATTERS

Cuba and the United States have had a difficult relationship for many years. Use CNN**fyi**.com or other current events sources to learn about problems that trouble U.S.–Cuban relations. Record your findings in your journal.

A guiro, a musical instrument

Section 3 RESOURCES

Reproducible
◆ Guided Reading Strategy 14.3
◆ Map Activity 14
◆ Geography for Life Activity 14

Technology
◆ One-Stop Planner CD–ROM, Lesson 14.3
◆ *ARGWorld* CD–ROM: Beach Resorts and Ecotourism Around the Caribbean
◆ HRW Go site

Reinforcement, Review, and Assessment
◆ Section 3 Review, p. 323
◆ Daily Quiz 14.3
◆ Main Idea Activity 14.3
◆ English Audio Summary 14.3
◆ Spanish Audio Summary 14.3

History

The Caribbean islands include 13 independent countries. All are former European colonies.

Early History Christopher Columbus first sailed into the Caribbean Sea for Spain in 1492. He thought he had reached the Indies, or the islands near India. He called the islands the West Indies and the people who lived there Indians. Spain established colonies there. Many Caribbean Indians died from disease or war.

In the 1600s and 1700s, the English, French, Dutch, and Danish also established Caribbean colonies. They built large plantations on the islands. Crops included sugarcane, tobacco, and cotton. Europeans brought Africans to work as slaves.

Independence A slave revolt won Haiti its independence from France in 1804. By the mid-1800s the Dominican Republic had also won independence. The United States took Cuba from Spain in the Spanish-American War in 1898. Cuba gained independence in 1902. Other Caribbean countries did not gain independence until the last half of the 1900s.

Tourists still visit Christophe's Citadel in Haiti. The fortress was built in the early 1800s after Haiti won independence from France.

✓ **READING CHECK:** *Human Systems* What was the early history of the Caribbean islands? *Christopher Columbus arrived in 1492; Spain, England, France, Holland, and Denmark established colonies there, built plantations, and brought enslaved Africans.*

319

Central America and the Caribbean Islands • 319

Teaching Objectives 1–2

ALL LEVELS: (Suggested time: 45 min.) Organize students into small groups. Have each group write a short poem to describe the early history of the Caribbean. Ask students to mention the various peoples, events, and dates that pertain to the region's history. *(Students should mention Columbus sailing into the region in 1492; European colonization in the 1600s and 1700s; the enslavement of Africans; Haiti's independence from France in 1804; the U.S. seizure of Cuba in the Spanish-American War in 1898; Cuba's independence in 1902; other countries' independence in the late 1900s.)* **COOPERATIVE LEARNING**

Teaching Objective 3

ALL LEVELS: (Suggested time: 25 min.) Organize the class into groups, one for each Caribbean island country. Give students 15 minutes to read the information about their island. Then read aloud a fact about any of the Caribbean islands today. When group members think the fact is about their island, they are to raise their hands. Ask the class for a consensus on the correct answer, confirm the answer, and continue to the next fact. **COOPERATIVE LEARNING**

Across the Curriculum
MUSIC

Calypso and Reggae
Calypso is a type of folk music originally sung by African slaves on the plantations of Trinidad. Forbidden to speak while they worked, the enslaved people conveyed messages to each other in song. Therefore, the words of a calypso song are at least as important as the tune. "Day-O," a calypso tune that became popular in the United States, warns banana pickers that the tarantula spider can lurk in a bunch of bananas. The worker also expresses his weariness in the song.

Critical Thinking: How did slavery influence the development of calypso?

Answer: Slaves sang warnings to one another and expressed their feelings about their lives and work in their songs.

internet connect

GO TO: go.hrw.com
KEYWORD: SJ3 CH14
FOR: Web sites about
Caribbean music

Chart Answer ▶

Students might mention better standard of living and access to quality health care in some nations.

Caribbean Islands				
COUNTRY	**POPULATION/ GROWTH RATE**	**LIFE EXPECTANCY**	**LITERACY RATE**	**PER CAPITA GDP**
Antigua and Barbuda	66,970 0.7%	69, male 73, female	89%	$8,200
Bahamas	297,852 .93%	67, male 74, female	98%	$15,000
Barbados	275,330 0.5%	71, male 76, female	97%	$14,500
Cuba	11,184,023 0.4%	74, male 79, female	96%	$1,700
Dominica	70,786 -.98%	71, male 77, female	94%	$4,000
Dominican Republic	8,581,447 1.6%	71, male 76, female	82%	$5,700
Grenada	89,227 -0.06%	63, male 66, female	98%	$4,400
Haiti	6,964,549 1.4%	47, male 51, female	45%	$1,800
Jamaica	2,665,636 0.5%	74, male 78, female	85%	$3,700
St. Kitts and Nevis	38,756 -0.11%	68, male 74, female	97%	$7,000
St. Lucia	158,178 1.23%	69, male 76, female	67%	$4,500
St. Vincent and the Grenadines	115,942 0.4%	71, male 74, female	96%	$2,800
Trinidad and Tobago	1,169,682 -0.51%	66, male 71, female	98%	$9,500
United States	281,421,906 0.9%	74, male 80, female	97%	$36,200

Sources: Central Intelligence Agency, *The World Factbook 2001;* U.S. Census Bureau

Interpreting the Chart Why might life expectancy be greater in some Caribbean island countries than in others?

Culture

Today, nearly every Caribbean island shows the signs of past colonialism and slavery. These signs can be seen in the region's culture.

People, Languages, and Religion
Most islanders are of African or European descent or are a mixture of the two. Much smaller numbers of Asians also live there. Chinese and other Asians came to work on the plantations after slavery ended in the region.

English, French, and mixtures of European and African languages are spoken on many islands. For example, Haitians speak French and Creole. Creole is a Haitian dialect of French. Spanish is spoken in Cuba, the Dominican Republic, Puerto Rico, and some small islands. Dutch is the main language on several territories of the Netherlands.

Another sign of the region's past are the religions practiced there. Protestant Christians are most numerous on islands that were British territories. Former French and Spanish territories have large numbers of Roman Catholics. On all the islands, some people practice a combination of Catholicism and traditional African religions. One of these religions is **Santería**. Santería began in Cuba and spread to nearby islands and parts of the United States. It has roots in West African religions and traditions.

Food, Festivals, and Music Caribbean cooking today relies on fresh fruits, vegetables, and fish or meat. Milk or preserved foods like cheese or pickled fish are seldom used. Cooking has been influenced by foods brought from Africa, Asia, and elsewhere. For example, the samosa—a spicy, deep-fried pastry—has its origins in India. Other popular foods include mangoes, rice, yams, and okra.

People on each Caribbean island celebrate a variety of holidays. One of the biggest and most widespread is Carnival. Carnival is a time of feasts and parties before the Christian season of Lent. It is celebrated with big parades and beautiful costumes.

The islands' musical styles are popular far beyond the Caribbean. Trinidad and Tobago is the home of steel-drum and **calypso** music. Jamaica is famous as the birthplace of **reggae** music. **Merengue** is the national music and dance of the Dominican Republic.

Caribbean musical styles have many fans in the United States. However, the United States has influenced Caribbean culture as well. For example, baseball has become a popular sport in the region. It is

➤**ASSIGNMENT:** Have students conduct research on a specific aspect of Caribbean culture, such as food, visual arts, literature, religion, or music. Have each student create an illustrated time line to show how the chosen cultural trait has changed through time. *(example of time line title: Caribbean Music through the Ages)* Have students present their time lines to the class.

Teaching Objectives 1–3

ALL LEVELS: (Suggested time: 20 min.) Copy the graphic organizer at right onto the chalkboard, omitting the italicized answers. Use it to help students describe the Caribbean islands. Call on students to fill in the chart with basic facts. Then lead a discussion on the traits the island cultures have in common.

The Caribbean Islands

	History	Government and Economics	People and Culture
Cuba	• *taken from Spain by the United States in 1898* • *gained independence in 1902*	• *communist government* • *trade with United States restricted* • *farms organized into cooperatives* • *sugarcane, tourism*	• *most populous country in region* • *origin of Santería*
Haiti	• *won independence from France in 1804*	• *poorest country in Americas* • *many corrupt governments* • *coffee, sugarcane, plantains*	• *densely populated* • *many refugees in United States*
Dominican Republic	• *former Spanish colony*	• *more developed than Haiti* • *agriculture, tourism*	• *education, health care, and housing are improving*
Puerto Rico	• *former Spanish colony*	• *commonwealth of United States* • *most developed island in region*	• *debate over its future*

particularly popular in the Dominican Republic and Cuba. A number of successful professional baseball players in the United States come from Caribbean countries.

✓ **READING CHECK:** *Human Systems* How do the cultures of the Caribbean islands reflect historical events? *They reflect the islands' colonial history and slavery.*

The Caribbean Islands Today

Now we will look at the largest island countries. We also will examine the island territory of Puerto Rico.

Cuba Cuba is the largest and the most populous country in the Caribbean. It is about the size of Tennessee but has more than twice the population. It is located just 90 miles (145 km) south of Florida. Havana, the capital, is the country's largest and most important city.

Cuba has had a Communist government since Fidel Castro seized power in 1959. Cuba has supported Communist **guerrilla** movements trying to overthrow other governments. A guerrilla takes part in irregular warfare, such as raids.

Many Cubans who oppose Castro have become **refugees** in the United States. A refugee is someone who flees to another country, usually for economic or political reasons. Many Cuban refugees and their families have become U.S. citizens. Most live in Florida.

The U.S. government has banned trade with Cuba. It also has restricted travel by U.S. citizens to the island since the 1960s. For years Cuba received economic aid and energy supplies from the Soviet Union. The collapse of the Soviet Union in the early 1990s has hurt Cuba's economy.

Today, private businesses remain limited in Cuba. Most farmland is organized into **cooperatives** and government-owned sugarcane

▲ A girl joins the Carnival celebration in Trinidad and Tobago.

Interpreting the Visual Record **What does the elaborate costume imply about the popularity of Carnival?**

A worker cuts sugarcane in central Cuba. Sugarcane is Cuba's most important cash crop.

▼

DAILY LIFE

Since the colonial era the inhabitants of the West Indies have celebrated Carnival. The holiday falls just before Lent, a holy time for Roman Catholics.

The celebration is rooted in pre-Christian European festivals that allowed general merrymaking and temporary equality for all social classes. In 1264 the pope declared it a Christian holiday. The idea of social equality of ancient festivals is echoed in Carnival. With everyone dressed in wild, colorful costumes, no one can tell who is rich and who is poor.

Carnival's blend of Catholic and pre-Christian characteristics typifies Caribbean culture's unique combination of European and pre-colonial traditions. Caribbean Carnivals include grand parades with music, floats, and parties for all.

Discussion: Have students discuss how modern-day Carnivals are influenced by the region's history.

internet connect

GO TO: go.hrw.com
KEYWORD: SJ3 CH14
FOR: Web sites about Carnival

▲ **Visual Record Answer**

Possible answer: implies that Carnival is so popular that people will spend a great deal of time or money on such a costume

321

Frank Thomas of Austin, Texas, suggests this activity to introduce students to the foods and cultures of Central America and the Caribbean islands. Organize students into groups of three to five students. Provide each group with an unfamiliar fruit from the region. Examples include ugli fruit, kiwano, star fruit, plumcot, horned fruit, plantain, and mango. These fruits are available in many large supermarkets or ethnic markets, or may be ordered from suppliers. If possible, supply more than one piece of fruit per group. Do not divulge the fruits' names at this point. Ask the students to describe the exterior of the fruit in detail: texture, color, firmness, scent, and so on. Ask them to speculate what the fruits look like inside and what they taste like. Then cut open and slice the fruit. Ask students to describe the interior: texture, color, seeds or pits, juice, and taste. Ask students to create their own name for the fruit. Then reveal the fruits' names. Supply reference materials for each group. Ask them to research their fruit and report on how it is used in the region's cuisine. You may want students to create new recipes using the fruit, other ingredients from Central America and the Caribbean islands, and regional cooking techniques.

Section Review 3

Answers

Define For definitions, see: Santería, p. 320; calypso, p. 320; reggae, p. 320; merengue, p. 320; guerrilla, p. 321; refugees, p. 321; cooperatives, p. 321; plantains, p. 322; commonwealth, p. 323

Working with Sketch Maps Havana, Port-au-Prince, and Santo Domingo should be located accurately.

Reading for the Main Idea

1. in 1492 and continuing into the 1600s and 1700s (NGS 12)

2. Most of the population is of African and/or European descent. (NGS 10)

3. relief from corrupt politics and unstable economies (NGS 13)

Critical Thinking

4. colonization, revolution, civil war, and eventual freedom; the remnants of colonialism still present a problem in modern-day life in many of these nations

Organizing What You Know

5. 1492—Columbus sails into Caribbean Sea; 1600–1700—Europeans establish Caribbean colonies; 1804—Haiti wins independence; 1902—Cuba gains independence; 1959—Castro seizes power in Cuba

plantations. A cooperative is an organization owned by its members and operated for their mutual benefit.

Sugarcane remains Cuba's most important crop and export. Tourism has also become an important part of the economy. There has been debate in the United States over ending the ban on trade and travel to Cuba.

Haiti Haiti occupies the mountainous western third of the island of Hispaniola. It is the poorest country in the Americas. It is also one of the most densely populated. Its people have suffered under many corrupt governments during the last two centuries. Many Haitian refugees have come to the United States to escape poverty and political violence.

Port-au-Prince (pohr-toh-PRINS) is the national capital and center of industry. Coffee and sugarcane are two of the country's most important crops. Most Haitians farm small plots. Many grow **plantains**. Plantains are a type of banana used in cooking.

Dominican Republic The Dominican Republic occupies the eastern part of Hispaniola. It is a former Spanish colony. The capital is Santo Domingo. Santo Domingo was the first permanent European settlement in the Western Hemisphere.

Other Island Countries

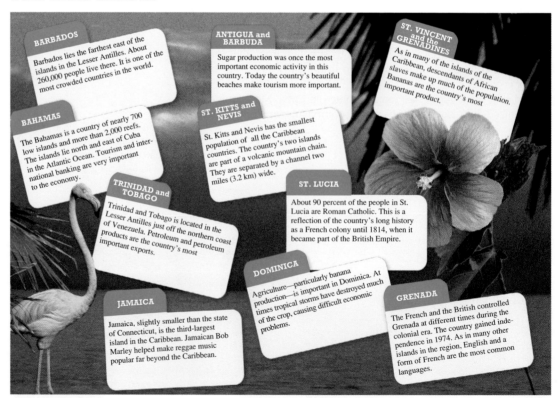

BARBADOS Barbados lies the farthest east of the islands in the Lesser Antilles. About 260,000 people live there. It is one of the most crowded countries in the world.

ANTIGUA and BARBUDA Sugar production was once the most important economic activity in this country. Today the country's beautiful beaches make tourism more important.

ST. VINCENT and the GRENADINES As in many of the islands of the Caribbean, descendants of African slaves make up much of the population. Bananas are the country's most important product.

BAHAMAS The Bahamas is a country of nearly 700 low islands and more than 2,000 reefs. The islands lie north and east of Cuba in the Atlantic Ocean. Tourism and international banking are very important to the economy.

ST. KITTS and NEVIS St. Kitts and Nevis has the smallest population of all the Caribbean countries. The country's two islands are part of a volcanic mountain chain. They are separated by a channel two miles (3.2 km) wide.

ST. LUCIA About 90 percent of the people in St. Lucia are Roman Catholic. This is a reflection of the country's long history as a French colony until 1814, when it became part of the British Empire.

TRINIDAD and TOBAGO Trinidad and Tobago is located in the Lesser Antilles just off the northern coast of Venezuela. Petroleum and petroleum products are the country's most important exports.

DOMINICA Agriculture—particularly banana production—is important in Dominica. At times tropical storms have destroyed much of the crop, causing difficult economic problems.

JAMAICA Jamaica, slightly smaller than the state of Connecticut, is the third-largest island in the Caribbean. Jamaican Bob Marley helped make reggae music popular far beyond the Caribbean.

GRENADA The French and the British controlled Grenada at different times during the colonial era. The country gained independence in 1974. As in many other islands in the region, English and a form of French are the most common languages.

Call on volunteers to read aloud the summaries about Other Island Countries in this section. Ask other students to point out their locations on a wall map.

Have students complete Main Idea Activity 14.3. Assign each student one country in the Caribbean islands, and have students produce a brief news report about the country's history, culture, or current conditions.
ENGLISH LANGUAGE LEARNERS

REVIEW AND ASSESS

Have students complete the Section Review. Ask each student to create three multiple-choice questions, with answers, covering the information presented in this section. Have students trade and check each other's quizzes. Then have students complete Daily Quiz 14.3.

EXTEND

Have students research the history of the Taino, an early people who inhabited Puerto Rico before the Spanish conquest. You may want to have other students investigate Puerto Rican culture in major U.S. cities or the issue of Puerto Rican statehood. Have students present their findings to the class.
BLOCK SCHEDULING

The Dominican Republic is not a rich country. However, its economy, education, health care, and housing are more developed than Haiti's. Agriculture and tourism are important parts of the Dominican Republic's economy.

Puerto Rico Puerto Rico is the easternmost of the Greater Antilles. Once a Spanish colony, today it is a U.S. **commonwealth**. A commonwealth is a self-governing territory associated with another country. Puerto Ricans are U.S. citizens. However, they have no voting representation in the U.S. Congress.

Unemployment is higher and wages are lower in Puerto Rico than in the United States. Still, American aid and investment has helped make the economy of Puerto Rico more developed than those of other Caribbean islands. Puerto Ricans continue to debate whether their island should remain a commonwealth. Some want it to become an American state or an independent country.

Other Islands Jamaica, in the Greater Antilles, is the largest of the remaining Caribbean countries. The smallest country is St. Kitts and Nevis in the Lesser Antilles. The smallest U.S. state, Rhode Island, is nearly 12 times larger! For more about other Caribbean countries and the Bahamas, see the Other Island Countries illustration.

A number of Caribbean and nearby islands are territories of other countries. These territories include the U.S. and British Virgin Islands. The Netherlands and France also have Caribbean territories. Bermuda, an Atlantic island northwest of the Caribbean, is a British territory.

✔ **READING CHECK:** *Human Systems* How are Puerto Rican citizens' political rights different from those of other U.S. citizens? They have no vote in Congress.

▲

Ocho Rios, Jamaica, and other beautiful Caribbean resorts and beaches attract many tourists.

Interpreting the Visual Record Why do you think cruise ships like the one in the photo are a popular way to travel in the Caribbean?

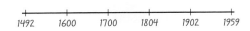

go.hrw.com
Homework Practice Online
Keyword: SJ3 HP14

Section Review 3

Define and explain: Santería, calypso, reggae, merengue, guerrilla, refugees, cooperatives, plantains, commonwealth

Working with Sketch Maps On the map you created in Section 2, label the Caribbean countries, Havana, Port-au-Prince, and Santo Domingo.

Reading for the Main Idea
1. *Human Systems* When did European powers establish colonies in the Caribbean islands?
2. *Human Systems* What ethnic groups make up the region's population today?

3. *Human Systems* Why have Cubans and Haitians come to the United States as refugees?

Critical Thinking
4. **Comparing/Contrasting** What do the histories of Caribbean countries have in common with the history of the United States? How are they different?

Organizing What You Know
5. **Sequencing** Copy the following graphic organizer. Use it to show important events and periods in the history of the Caribbean islands since 1492.

```
+      +      +      +      +      +
1492  1600  1700  1804  1902  1959
```

Building Vocabulary
For definitions, see: archipelago, p. 312; cloud forest, p. 313; bauxite, p. 313; cacao, p. 316; dictators, p. 316; cardamom, p. 317; civil war, p. 317; ecotourism, p. 318; Santería, p. 320; calypso, p. 320; guerrilla, p. 321; refugees, p. 321; cooperatives, p. 321; plantains, p. 322; commonwealth, p. 323

Reviewing the Main Ideas
1. Greater Antilles and Lesser Antilles (NGS 4)
2. 1492; colonial control, migration, plantation-based economies, slavery, ethnic diversity, cultural blending (NGS 12)
3. Guatemala; Cuba (NGS 4)
4. Examples include calypso, reggae, and merengue; they have won fans and spread around the world. (NGS 10)
5. It is communist. (NGS 13)

▲ **Visual Record Answer**

Possible answer: They allow tourists the opportunity to travel to the many islands of the Caribbean.

323

PORTFOLIO EXTENSIONS

1. Organize students into small groups. Have each group select one country from the region. Be sure that as many countries as possible are selected. Each group should design storyboards for a commercial about the country, including information on the country's history, people, and resources. Have students present their commercials to the class. Place copies of storyboards in portfolios.

2. Have students conduct research on an archaeological site in Central America or the Caribbean islands. Ask them to find or draw a picture of the site. Have each student create a model of an artifact from the site, such as a pottery jar, jewelry, tool, or figurine. Ask students to share their pictures and models with the class. Photograph student work for portfolios.

Review
ANSWERS

Understanding Environment and Society

Students may say that Belize and Costa Rica are among the most popular destinations in the region. Activities include exploring ancient ruins, national parks, and rain forests; visiting coastal resorts; and participating in festivals. People may alter the land by taking or changing things as they visit. The sheer volume of people in sensitive areas may be harmful. Use Rubric 29, Presentations, to evaluate student work.

Thinking Critically

1. resulted in diverse mixture of European, African, and Indian peoples

2. governments elected by the people

3. seeking relief from corrupt government and unstable economies

4. coffee, sugarcane, fruit; good soil, economic dependence on farming

5. earthquakes, volcanoes, and hurricanes; by moving to safer locations and/or by adjusting their lives and activities

Reviewing What You Know

Building Vocabulary

On a separate sheet of paper, write sentences to define each of the following words.

1. archipelago
2. cloud forest
3. bauxite
4. cacao
5. dictators
6. cardamom
7. civil war
8. ecotourism
9. Santería
10. calypso
11. guerrilla
12. refugees
13. cooperatives
14. plantains
15. commonwealth

Reviewing the Main Ideas

1. (*Places and Regions*) What two main island groups make up the Caribbean archipelago?

2. (*Human Systems*) When did European powers begin establishing colonies in Central America and the Caribbean islands? What were some effects on the region?

3. (*Places and Regions*) Which is the most populous country in Central America? in the Caribbean?

4. (*Human Systems*) What are some musical styles with origins in the Caribbean? What influence have they had abroad?

5. (*Human Systems*) How is the government of Cuba organized?

Understanding Environment and Society

Resource Use

Prepare a presentation and thematic map on ecotourism in the region. As you prepare your report, you may want to consider the following points:

• Which countries are among the most popular tourist destinations, and why.

• Which activities are most popular among tourists.

• How humans threaten the natural environment of the region.

Thinking Critically

1. **Drawing Inferences and Conclusions** How did slavery affect the ethnic diversity of the region?

2. **Comparing** What do the governments of all Central American countries have in common?

3. **Finding the Main Idea** Why have refugees from Cuba and Haiti come to the United States?

4. **Summarizing** What agricultural products are grown in many Central American and Caribbean countries? Why?

5. **Analyzing Information** What are some of the natural and environmental hazards in Central America and the Caribbean islands? How do you think people may have coped with these hazards?

FOOD FESTIVAL

Postre de mangoes is a delicious dessert from the Caribbean islands. It is easy to make. Mangoes are available in most supermarkets. Place a pound of sliced mangoes, 15 ounces sweetened condensed milk, and the juice of a lemon in a blender and blend until smooth. Chill mixture in dessert dishes for two to three hours. Serve topped with fresh fruit and berries that have been marinated in orange juice. Sprinkle with pine nuts if desired. Note that this recipe serves six. Be prepared to increase proportions to serve your entire class.

Building Social Studies Skills

On a separate sheet of paper, match the letters on the map with their correct labels.

Guatemala	Haiti
Nicaragua	Dominican Republic
Panama	Puerto Rico
Cuba	Bahamas
Jamaica	Havana

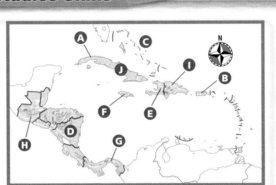

Mental Mapping Skills ACTIVITY

On a separate sheet of paper, draw a freehand map of Central America and the Caribbean. Make a key for your map and label the following:

Caribbean Sea	Lesser Antilles
Central America	Panama Canal
Greater Antilles	Santo Domingo

WRITING ACTIVITY

Imagine that you have been asked to write a paragraph for a travel brochure about Central America and the Caribbean. Your paragraph should include information that tourists would want to know. This includes general information about the region's climates, vegetation, food, festivals, and other special celebrations. Be sure to use standard grammar, sentence structure, spelling, and punctuation.

Alternative Assessment

Portfolio ACTIVITY

Learning About Your Local Geography
Research Project Are Caribbean musical styles popular in your community? Design a poster advertising a festival of music that is popular where you live.

internet connect

Internet Activity: go.hrw.com
KEYWORD: SJ3 GT14

Choose a topic to explore about Central America and the Caribbean islands:
- Plan an ecotour of Central America.
- Learn more about the Panama Canal.
- Hunt for hurricanes.

Map Activity
A. Havana
B. Puerto Rico
C. Bahamas
D. Nicaragua
E. Haiti
F. Jamaica
G. Panama
H. Guatemala
I. Dominican Republic
J. Cuba

Mental Mapping Skills Activity
Maps will vary. However, the listed places should be labeled in their approximate locations.

Writing Activity
Answers will vary, but the information included should be consistent with text materials. Check to see that students have included accurate information about the region's climates, vegetation, food, and festivals. Use Rubric 40, Writing to Describe, to evaluate student work.

Portfolio Activity
Answers will vary, but students may describe such musical styles as rock, rap, techno, jazz, and pop. Use Rubric 2, Advertisements, to evaluate festival posters.

internet connect

GO TO: go.hrw.com
KEYWORD: SJ3 Teacher
FOR: a guide to using the Internet in your classroom

CHAPTER 15

Caribbean South America
Chapter Resource Manager

Objectives	Pacing Guide	Reproducible Resources
SECTION 1		
Physical Geography (pp. 327–29) 1. Identify the major landforms and rivers of Caribbean South America. 2. Identify the climate and vegetation types found in the region. 3. Identify the natural resources of this region.	**Regular** .5 day **Block Scheduling** .5 day *Block Scheduling Handbook, Chapter 15*	**RS** Guided Reading Strategy 15.1 **RS** Graphic Organizer 15 **SM** Map Activity 15
SECTION 2		
Colombia (pp. 330–33) 1. Identify the main periods of Colombia's history. 2. Describe what Colombia is like today.	**Regular** .5 day **Block Scheduling** .5 day *Block Scheduling Handbook, Chapter 15*	**RS** Guided Reading Strategy 15.2 **E** Cultures of the World Activity 2
SECTION 3		
Venezuela (pp. 334–36) 1. Describe how the Spanish contributed to Venezuela's history. 2. Identify some characteristics of Venezuela's culture.	**Regular** .5 day **Block Scheduling** .5 day *Block Scheduling Handbook, Chapter 15*	**RS** Guided Reading Strategy 15.3 **E** Cultures of the World Activity 2 **SM** Geography for Life Activity 15
SECTION 4		
The Guianas (pp. 337–39) 1. Identify the countries that influenced the early history of the Guianas. 2. Describe how Guyana, Suriname, and French Guiana are similar today.	**Regular** .5 day **Block Scheduling** .5 day *Block Scheduling Handbook, Chapter 15*	**RS** Guided Reading Strategy 15.4 **E** Cultures of the World Activity 2

Chapter Resource Key

RS Reading Support

IC Interdisciplinary Connections

E Enrichment

SM Skills Mastery

A Assessment

REV Review

ELL Reinforcement and English Language Learners

 Transparencies

 CD–ROM

 Music

 Video

 Internet

Holt Presentation Maker Using Microsoft® Powerpoint®

 One-Stop Planner CD–ROM

See the *One-Stop Planner* for a complete list of additional resources for students and teachers.

One-Stop Planner CD–ROM

It's easy to plan lessons, select resources, and print out materials for your students when you use the *One-Stop Planner CD–ROM with Test Generator.*

Technology Resources	Review, Reinforcement, and Assessment Resources	
One-Stop Planner CD–ROM, Lesson 15.1	**ELL**	Main Idea Activity 15.1
Geography and Cultures Visual Resources with Teaching Activities 18–22	**REV**	Section 1 Review, p. 329
Earth: Forces and Formations CD–ROM/Seek and Tell/Forces and Processes	**A**	Daily Quiz 15.1
Homework Practice Online	**ELL**	English Audio Summary 15.1
HRW Go site	**ELL**	Spanish Audio Summary 15.1
One-Stop Planner CD–ROM, Lesson 15.2	**ELL**	Main Idea Activity 15.2
Homework Practice Online	**REV**	Section 2 Review, p. 333
HRW Go site	**A**	Daily Quiz 15.2
	ELL	English Audio Summary 15.2
	ELL	Spanish Audio Summary 15.2
One-Stop Planner CD–ROM, Lesson 15.3	**ELL**	Main Idea Activity 15.3
Homework Practice Online	**REV**	Section 3 Review, p. 336
HRW Go site	**A**	Daily Quiz 15.3
	ELL	English Audio Summary 15.3
	ELL	Spanish Audio Summary 15.3
One-Stop Planner CD–ROM, Lesson 15.4	**ELL**	Main Idea Activity 15.4
Homework Practice Online	**REV**	Section 4 Review, p. 339
HRW Go site	**A**	Daily Quiz 15.4
	ELL	English Audio Summary 15.4
	ELL	Spanish Audio Summary 15.4

internet connect

HRW ONLINE RESOURCES

GO TO: go.hrw.com
Then type in a keyword.

TEACHER HOME PAGE
KEYWORD: SJ3 TEACHER

CHAPTER INTERNET ACTIVITIES
KEYWORD: SJ3 GT15

Choose an activity to:
• trek through the Guiana Highlands.
• search for the treasures of El Dorado.
• ride with the *llaneros* of Venezuela.

CHAPTER ENRICHMENT LINKS
KEYWORD: SJ3 CH15

CHAPTER MAPS
KEYWORD: SJ3 MAPS15

ONLINE ASSESSMENT
Homework Practice
KEYWORD: SJ3 HP15
Standardized Test Prep Online
KEYWORD: SJ3 STP15
Rubrics
KEYWORD: SS Rubrics

COUNTRY INFORMATION
KEYWORD: SJ3 Almanac

CONTENT UPDATES
KEYWORD: SS Content Updates

HOLT PRESENTATION MAKER
KEYWORD: SJ3 PPT15

ONLINE READING SUPPORT
KEYWORD: SS Strategies

CURRENT EVENTS
KEYWORD: S3 Current Events

Meeting Individual Needs

Ability Levels

Level 1 Basic-level activities designed for all students encountering new material

Level 2 Intermediate-level activities designed for average students

Level 3 Challenging activities designed for honors and gifted-and-talented students

English Language Learners Activities that address the needs of students with Limited English Proficiency

Chapter Review and Assessment

E	Creative Strategies for Teaching World Geography, Lesson 9
IC	Interdisciplinary Activity for the Middle Grades 12
E	Readings in World Geography, History, and Culture 20 and 21
SM	Critical Thinking Activity 15
REV	Chapter 15 Review, pp. 340–41
REV	Chapter 15 Tutorial for Students, Parents, Mentors, and Peers
ELL	Vocabulary Activity 15
A	Chapter 15 Test
	Chapter 15 Test Generator (on the One-Stop Planner)
	Audio CD Program, Chapter 15
A	Chapter 15 Test for English Language Learners and Special-Needs Students

LAUNCH INTO LEARNING

Have each student compile a list of all the different climates and landforms that exist in the United States. Then ask each student to compile a list of the various regions of the world from which immigrants to the United States have come. Ask volunteers to share their lists with the class. Tell students that, like the United States, Caribbean South America has a variety of landforms and climates as well as an ethnically diverse population. Tell students that in this chapter they will learn more about the physical geography, history, and important natural resources of Caribbean South America.

Section 1

Objectives

1. Identify and describe the major landforms and rivers of Caribbean South America.
2. Describe the climate and vegetation types found in the region.
3. List the natural resources of the region.

LINKS TO OUR LIVES

You might point out to the students that there are many reasons why we should know more about Caribbean South America, these among them:

▶ Caribbean South America is home to tropical rain forests. These rain forests support many species of plants and animals as well as human inhabitants. Destruction of rain forests threatens plant, animal, and human life and could contribute to significant global warming.

▶ Gas prices in the United States are affected by the supply of foreign oil. Because Venezuela is one of the leading suppliers of oil to the United States, developments in the Venezuelan economy could affect the U.S. economy and U.S. consumers.

▶ Caribbean Latin America has rich and diverse cultural traditions that might be of interest to people in the United States.

Caribbean South America

Caribbean South America is a region of varied landscapes. The fertile valleys in the Andes and the rich land near the Caribbean shore were important to Spain's empire.

My name is Jorge and I live in Armenia in northwestern Colombia. Armenia is a big city about 8 hours from the capital, Bogotá. I live in a big house in the northern part of the city with my three younger brothers, my mother and father, my grandmother, and our big black dog, Rocca. Our house has two floors around a courtyard with flowers—it is pink with yellow shutters.

My father is a merchant and a farm-owner. The farm, or hacienda, is in the country about 45 minutes away by car. About 60 people work for my father there, growing coffee, plantains, yucca, and fruits like strawberries and oranges. We also raise chickens and pigs. My dad has a fleet of five trucks to carry our produce into the city for sale to groceries and restaurants. My mom has three people to help with the cooking at the farm.

¡Hola! ¿Cómo estas?

▲

Translation: Hello! How are you?

LET'S GET STARTED

Copy the following instructions onto the chalkboard: *Study the map of Caribbean South America. Find a geographic characteristic that Caribbean South America shares with the United States.* Ask volunteers to share their answers with the class. *(Students might mention that in addition to rivers and mountains, Caribbean South America stretches from the Atlantic Ocean to the Pacific Ocean.)* Tell students that in Section 1 they will learn more about the physical geography of the region.

Using the Physical-Political Map

Ask students to examine the map on this page and name the landforms and rivers of Caribbean South America. Point out that the Andes in Colombia separate into three parallel ranges, or cordilleras, like the frayed end of a rope. Have students predict which parts of Caribbean South America are the most heavily populated and the most agriculturally productive.

Teaching Objectives 1–3

ALL LEVELS: (Suggested time: 25 min.) Copy the graphic organizer on the following page onto the chalkboard, omitting the italicized answers. Use it to help students learn about the landforms, rivers, climates, vegetation, and natural resources of Caribbean South America. Pair students and ask each pair to complete the organizer. Ask volunteers to share their answers with the class. **ENGLISH LANGUAGE LEARNERS, COOPERATIVE LEARNING**

Section 1 — Physical Geography

Read to Discover

1. What are the major landforms and rivers of Caribbean South America?
2. What climate and vegetation types are found in the region?
3. What are the natural resources of this region?

Define

cordillera
tepuís
Llanos

Locate

Andes
Guiana Highlands
Orinoco River

WHY IT MATTERS

The varied physical geography of Caribbean South America has contributed to the cultural diversity of the region. Use **CNNfyi.com** or other **current events** sources to learn more about the geography of Caribbean South America. Record your findings in your journal.

Spectacled caiman—a crocodile

Section 1 — RESOURCES

Reproducible
- Block Scheduling Handbook, Chapter 15
- Guided Reading Strategy 15.1
- Graphic Organizer 15
- Map Activity 15

Technology
- One-Stop Planner CD–ROM, Lesson 15.1
- Homework Practice Online
- Geography and Cultures Visual Resources with Teaching Activities 18–22
- Earth: Forces and Formations CD–ROM/Seek and Tell/Forces and Processes
- HRW Go site

Reinforcement, Review, and Assessment
- Section 1 Review, p. 329
- Daily Quiz 15.1
- Main Idea Activity 15.1
- English Audio Summary 15.1
- Spanish Audio Summary 15.1

Caribbean South America: Physical-Political

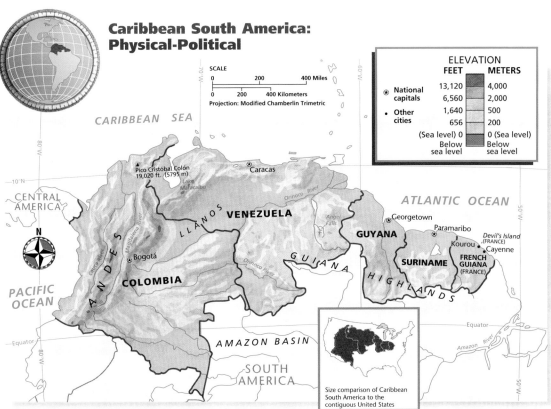

SCALE
0   200   400 Miles
0   200   400 Kilometers
Projection: Modified Chamberlin Trimetric

ELEVATION		
	FEET	METERS
⊛ National capitals	13,120	4,000
	6,560	2,000
• Other cities	1,640	500
	656	200
	(Sea level) 0	0 (Sea level)
	Below sea level	Below sea level

CARIBBEAN SEA

Pico Cristóbal Colón 19,020 ft. (5795 m)
Caracas
Lake Maracaibo
Orinoco River
CENTRAL AMERICA
LLANOS
VENEZUELA
Angel Falls
ATLANTIC OCEAN
Georgetown
GUYANA
Paramaribo
Devil's Island (FRANCE)
Kourou
Cayenne
SURINAME
FRENCH GUIANA (FRANCE)
ANDES
Magdalena River
Bogotá
COLOMBIA
Orinoco River
GUIANA HIGHLANDS
PACIFIC OCEAN
Equator
AMAZON BASIN
SOUTH AMERICA
Equator
Amazon River

Size comparison of Caribbean South America to the contiguous United States

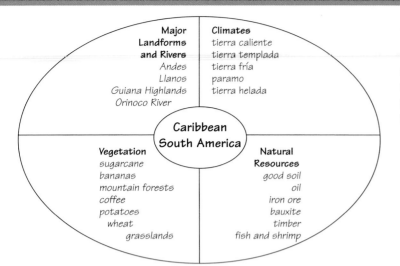

Major Landforms and Rivers
- Andes
- Llanos
- Guiana Highlands
- Orinoco River

Climates
- tierra caliente
- tierra templada
- tierra fría
- paramo
- tierra helada

Caribbean South America

Vegetation
- sugarcane
- bananas
- mountain forests
- coffee
- potatoes
- wheat
- grasslands

Natural Resources
- good soil
- oil
- iron ore
- bauxite
- timber
- fish and shrimp

►**ASSIGNMENT:** Tell students to imagine that it is the early 1500s and that they are Spanish explorers who are journeying through Caribbean South America for the first time. Ask each student to write journal entries that describe the landforms, rivers, climate, vegetation, and natural resources of the region. Journal entries should be descriptive, and students should feel free to invent vocabulary to name what they encounter. Ask volunteers to share their entries with the class.

PLACES AND REGIONS

There are more than 100 *tepuís* in Venezuela, many of them unexplored. Years of isolation have resulted in the development of unique plant and animal life on the summits of these formations. One such plant is an orchid whose blossoms are no bigger than a pinhead.

Tepuís provide a standard by which to judge the effect that humans' activities have had on the environment. For example, of the plants found in the British Isles, 30 percent are nonnative, having been introduced accidentally or deliberately. In contrast, 100 percent of the species on the summits of the *tepuís* are native.

Critical Thinking: Why might we expect nonnative plants to be introduced to the *tepuís* at some point?

Answer: Students might suggest that as more people explore the *tepuís*, someone will inadvertently introduce a nonnative species.

Visual Record Answer ▶

tierra caliente, tierra templada, tierra fría and *paramo*

☑ internet connect

GO TO: go.hrw.com
KEYWORD: SJ3 CH15
FOR: Web sites about Caribbean South America

Environments in the Andes change with elevation. Five different elevation zones are commonly recognized.

Interpreting the Visual Record

In which elevation zones is it possible to grow crops?

▼

Physical Features

A rugged landscape and dense forests have often separated peoples and cultures in this region. In the west, the Andes (AN-deez) rise above 18,000 feet (5,486 m). Here the mountain range forms a three-pronged **cordillera** (kawr-duhl-YER-uh). A cordillera is a mountain system made up of parallel ranges. Many active volcanoes and earthquakes shake these mountains.

In the east the Guiana Highlands have been eroding for millions of years. However, some of the steep-sided plateaus are capped by sandstone layers that have resisted erosion. These unusual formations are called **tepuís** (tay-PWEEZ). They can reach approximately 3,000 to 6,000 feet (914 to 1,829 m) above the surrounding plains.

Between these two upland areas are the vast plains of the Orinoco (OHR-ee-NOH-koh) River basin. These plains are the **Llanos** (LAH-nohs) of eastern Colombia and western Venezuela. The northeastern edge of the Guiana Highlands slopes down to a fertile coastal plain in Guyana, Suriname, and French Guiana.

Of the region's many rivers, the Orinoco is the longest. It flows for about 1,281 miles (2,061 km) through the region on its way to the Atlantic Ocean. Large oceangoing ships can travel upriver on the Orinoco for about 225 miles (362 km).

Some remarkable animals live in and around the Orinoco. They include aggressive meat-eating fish called piranhas (puh-RAH-nuhz), 200-pound (90-kg) catfish, and crocodiles as long as 20 feet (6 m). More than 1,000 bird species live in the Orinoco River basin.

✔ **READING CHECK:** *Places and Regions* What are this region's major landforms? *the Andes, the Guiana Highlands, tepuís, Orinoco River and river basin, Llanos*

Elevation Zones in the Andes

Tierra helada

Paramo

Tree Line

Tierra fría

Tierra templada

Tierra caliente

Sea Level

☐ **Tierra helada** Above 16,000 feet (4,877 m) Permanently covered with snow	☐ **Tierra fría** 6,000 to 10,000 feet (1,829 to 3,048 m) Potatoes, wheat, oats, barley, beans, corn, rye	☐ **Tierra caliente** Sea Level to 3,000 feet (914 m) Bananas, cacao, rice, sugarcane
☐ **Paramo** 10,000 to 16,000 feet (3,048 to 4,877 m) Potatoes, grasslands and hardy shrubs, grazing	☐ **Tierra templada** 3,000 to 6,000 feet (914 to 1,829 m) Coffee, corn, wheat, cotton, potatoes, sugarcane, tobacco	

Tell students that volcanoes can melt snow very quickly. This happened in Colombia in November 1985. Heat from a volcano melted snow and ice on its steep peak. As the water ran down the mountain, it gathered mud and debris. By the time the mud and debris hit the town of Armero, it was a wall 130 feet (40 m) tall. About 25,000 people were killed. Ask students what benefits volcanoes might have. *(They create fertile soil.)*

Have students complete Main Idea Activity 15.1. Then trace a wall map of the region. Have students create one symbol for each of the following features: landforms; wildlife and vegetation; climate types; or natural resources. Have students draw or glue the symbols onto the map. **ENGLISH LANGUAGE LEARNERS**

Have students complete the Section Review. Then have students create eight matching quiz questions that relate a country to its landforms, climates, or natural resources. Ask students to exchange and complete their quizzes. Then have students complete Daily Quiz 10.1.

Organize interested students into small groups. Have each group conduct research on the animals that live in and around the Orinoco River and create a short multimedia presentation. Have groups deliver their presentations to the class. **BLOCK SCHEDULING**

Climate and Vegetation

The climates of the Andes are divided by elevation into five zones. The *tierra caliente* (tee-E-ruh kal-ee-EN-tee), or "hot country," refers to the hot and humid lower elevations near sea level. There is little difference between summer and winter temperatures in this region. Crops such as sugarcane and bananas are grown here.

Higher up the mountains the air becomes cooler. Moist climates with mountain forests are typical here. This zone of pleasant climates is called *tierra templada* (tem-PLAH-duh), or "temperate country." Coffee is a typical crop grown in this area. The next zone is the *tierra fría* (FREE-uh), or "cold country." The *tierra fría* has forests and grasslands. Farmers can grow potatoes and wheat. Bogotá, Colombia's capital, lies in the *tierra fría*. Above the tree line is a zone called the *paramo* (PAH-rah-moh). Grasslands and hardy shrubs are the usual vegetation. Frost may occur on any night of the year in this zone. The *tierra helada* (el-AH-dah), or "frozen country," is the zone of highest elevation. It is always covered with snow.

✓ **READING CHECK:** *Places and Regions* What are the region's climate elevation zones? *tierra caliente, tierra templada, tierra fría, paramo, tierra helada*

Resources

Good soil and moderate climates help make the region a rich agricultural area. The region has other valuable resources, including oil, iron ore, and bauxite. Lowland forests provide timber. Coastal areas yield fish and shrimp. Some rivers in the region are used to produce hydroelectric power.

✓ **READING CHECK:** *Environment and Society* How do geographic factors affect economic activities in this region? *Good soil and moderate climate—farming; minerals and petroleum; mining and drilling; forests—timber harvesting; coast—fishing*

South America's capybara is the world's largest rodent. Also called water hogs, capybaras live along lakes and rivers in wet tropical climates.

Section Review 1

Define and explain: cordillera, *tepuís,* Llanos

Working with Sketch Maps On a map of Caribbean South America that you draw or that your teacher provides, label the following: Andes, Guiana Highlands, and Orinoco River.

Reading for the Main Idea

1. *Physical Systems* What effect do the Andes have on the region's climate?

2. *Environment and Society* Why is the Orinoco River important?

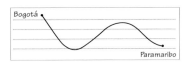
go.hrw.com **Homework Practice Online**
Keyword: SJ3 HP15

Critical Thinking

3. **Drawing Inferences and Conclusions** How might Colombia's location affect its trade?

4. **Analyzing Information** Which physical features make farming easier in the region? Why?

Organizing What You Know

5. **Analyzing Information** Use the graphic organizer to describe the climate and vegetation found in traveling from Bogotá to Paramaribo, Suriname.

Bogotá

Paramaribo

Section Review 1

Answers

Define For definitions, see: cordillera, p. 328; *tepuís,* p. 328; Llanos, p. 328

Working with Sketch Maps The Andes, Guiana Highlands, and Orinoco River should be located accurately.

Reading for the Main Idea

1. create a range of climates (NGS 7)

2. It creates a corridor to the Atlantic and supports a diverse animal population. (NGS 15)

Critical Thinking

3. allows easy access to Central American trade as well as South American trade and sea trade

4. Varied elevations enable many different crops to be grown.

Organizing What You Know

5. *tierra fría*—cooler mountainous region with forests and grasslands; *tierra templada*—moist, pleasant climates with mountain forests; *tierra caliente*—hot, humid climates with tropical vegetation and crops

Section 2

Objectives
1. Discuss the main periods of Colombia's history.
2. Describe Colombia today.

LET'S GET STARTED

Copy the following instruction onto the chalkboard: *Compile a list of places in the Americas that were named in honor of Christopher Columbus.* Point out that at one time many people believed that the Americas should have been named Columbia, after Columbus. Ask volunteers to share their answers with the class. *(Students might mention Columbus, Ohio; the District of Columbia; the Columbia River; and the country Colombia).* Tell students that in Section 2 they will learn more the history and culture of Colombia.

Building Vocabulary

Write the terms **El Dorado** and **cassava** on the chalkboard. Ask students to guess which term has European origins and which term has American Indian origins. *(Cassava has Taino Indian origins, and El Dorado has Spanish origins.)* Ask volunteers to find the terms in the glossary and to read the definitions aloud. Then have them use the terms in a sentence.

Section 2 RESOURCES

Reproducible
◆ Guided Reading Strategy 15.2
◆ Cultures of the World Activity 2

Technology
◆ One-Stop Planner CD–ROM, Lesson 15.2
◆ Homework Practice Online
◆ HRW Go site

Reinforcement, Review, and Assessment
◆ Section 2 Review, p. 333
◆ Daily Quiz 15.2
◆ Main Idea Activity 15.2
◆ English Audio Summary 15.2
◆ Spanish Audio Summary 15.2

Section 2
Colombia

Read to Discover
1. What are the main periods of Colombia's history?
2. What is Colombia like today?

Define
El Dorado
cassava

Locate
Colombia
Bogotá
Cauca River
Magdalena River

WHY IT MATTERS

Colombia remains a country troubled by violence and conflict. Use CNNfyi.com or other **current events** sources to learn more about problems in Colombia. Record your findings in your journal.

Pre-Columbian gold armor

Giant stone figures near the headwaters of the Magdalena River are part of the San Agustín culture.

Interpreting the Visual Record How might the location of these figures in a remote area have helped to preserve them?

Visual Record Answer

The figures' isolation protected them from theft or vandalism.

Early History

Advanced cultures have lived in Colombia for centuries. Some giant mounds of earth, stone statues, and tombs found in Colombia are more than 1,500 years old.

The Chibcha In western Colombia, the Chibcha people had a well-developed civilization. The Chibcha practiced pottery making, weaving, and metalworking. Their gold objects were among the finest in ancient America.

The Chibcha had an interesting custom. New rulers were covered with gold dust and then taken to a lake to wash the gold off. Gold and emerald objects were thrown into the water as the new ruler washed. This custom inspired the legend of **El Dorado** (el duh-RAH-doh), or "the Golden One." The old legend of El Dorado describes a marvelous, rich land.

Spanish Conquest Spanish explorers arrived on the Caribbean coast of South America about 1500. They were helping to expand Spain's new empire. The Spanish conquered the Chibcha and seized much of their treasure. Spaniards and their descendants set up large estates. Powerful Spanish landlords forced South American Indians and enslaved Africans to work the land.

🌐 **Teaching Objective 1**

ALL LEVELS: (Suggested time: 30 min.) Organize students into groups and tell them to imagine that they are editing a book about Colombia's history. Have each group write a dust-jacket summary that briefly discusses the three historical periods prior to Colombia's modern era. *(Summaries should mention the correct historical eras.)* Ask volunteers to read their dust-jacket summaries to the class.

ENGLISH LANGUAGE LEARNERS, COOPERATIVE LEARNING

Marcia Caldwell of Austin, Texas, suggests the following activity to help students understand Colombia's history. Pair students and have each pair create a mini-scrapbook about Colombia's past. They should use text and illustrations to present information about the different periods of the history of Colombia. The pages of the book can be attached with staples, string, or brads. *(Scrapbooks should mention early history, including the Chibcha civilization and Spanish conquest; Spanish rule, including the estate system and slavery; and independence, including Gran Colombia, the establishment of Colombia, and church-state conflicts.)*

Independence In the late 1700s people in Central and South America began struggling for independence from Spain. After independence was achieved, the Republic of Gran Colombia was created. It included Colombia, Ecuador, Panama, and Venezuela. In 1830 the republic dissolved, and New Granada, now Colombia, was created. Present-day Panama was once part of New Granada.

After independence, debate raged in Colombia. People argued over how much power the central government and the Roman Catholic Church should have. Part of the problem had to do with the country's rugged geography. The different regions of Colombia had little contact with each other. They developed separate economies and identities. Uniting these different groups into one country was hard. Outbreaks of violence throughout the 1800s and 1900s killed thousands of people.

COUNTRY	POPULATION/ GROWTH RATE	LIFE EXPECTANCY	LITERACY RATE	PER CAPITA GDP
Colombia	40,349,388 1.6%	67, male 75, female	91%	$6,200
United States	281,421,906 0.9%	74, male 80, female	97%	$36,200

Colombia

Sources: Central Intelligence Agency, *The World Factbook 2001;* U.S. Census Bureau

Interpreting the Chart How much larger is the U.S. population than that of Colombia?

✔ **READING CHECK:** *Environment and Society* What geographic factors influenced Colombia's ability to control its territory? *rugged geography that separated the different regions of the country*

Colombia Today

Colombia is Caribbean South America's most populous country. The national capital is Bogotá, a city located high in the eastern Andes. Most Colombians live in the fertile valleys and basins among the mountain ranges because those areas are moderate in climate and good for farming. Rivers, such as the Cauca and Magdalena, flow down from the Andes to the Caribbean. They help connect settlements between the mountains and the coast. Cattle ranches are common in the Llanos. Few people live in the tropical rain forest regions in the south.

The guard tower of an old Spanish fort stands in contrast with the modern buildings of Cartagena, Colombia.

Interpreting the Visual Record
Why might the Spanish have chosen this location for a fort?

▼

Teaching Objective 2

LEVEL 1: (Suggested time: 15 min.). Copy the following graphic organizer onto the chalkboard, omitting the italicized answers. Use it to help students learn about life in present-day Colombia. Have each student complete the organizer. Ask volunteers to share their answers with the class. **ENGLISH LANGUAGE LEARNERS**

LEVELS 2 AND 3: (Suggested time: 30 min.) Tell students to imagine that they are Colombian businesspeople who are trying to attract foreign investors and businesses to Colombia. Pair students and have each pair create an illustrated brochure that describes the economic resources and cultural traditions of Colombia. *(See the graphic organizer from the Level 1 lesson for the economic resources and cultural traditions.)* Ask volunteers to share their brochures with the class. **COOPERATIVE LEARNING**

Aspects of the Economy

> *coffee*
> *bananas*
> *cassava*
> *flowers*
> *oil*
> *iron ore, gold,*
> *coal, tin, emeralds*

Colombia Today

Aspects of Culture

> *regional isolation*
> *cultural diversity*
> *traditional cultures*
> *Roman Catholicism*
> *soccer*
> *tejo*

ENVIRONMENT AND SOCIETY

Cassava, which originated in Central America, is now grown in areas of Africa, Asia, and the Philippines as well as the Americas. Cassava was introduced to other continents through what is known as the Columbian Exchange. This exchange began with Christopher Columbus, who transported American food plants across the Atlantic.

Cassava is now the main food of approximately 500 million people. It tolerates drought, poor soil, and harsh climate conditions. Moreover, the cassava plant is versatile. Tapioca can be processed from the cassava root, and the leaves can be eaten as a vegetable or used as livestock feed.

Activity: Have students conduct research on the food plants introduced from the Americas to other parts of the world or introduced to the Americas through the Columbian Exchange.

Connecting to Science
Answers

1. Because it treated malaria, it became the subject of power struggles between nations. It remains a treatment for heart disease.

2. When the Axis Powers seized the Netherlands, they cut off the supply to the Allies, leading to the development of a black market for the drug.

Fighting Malaria

Botanical print of a gray cinchona

Malaria is a disease usually transmitted by mosquitoes. It is common in the tropics. For centuries malaria was also widespread in Europe, but Europeans had no remedy. Native peoples in the South American rain forest did have a treatment, though. They used the powdered bark of the cinchona tree, which contains the drug quinine. The history of quinine and the struggle to obtain it is a story of great adventure.

The Spanish first discovered cinchona in the 1500s, when they conquered Peru. Shipments of the bark were soon arriving in Europe, where quinine was produced. Later, some countries tried to control the supply of bark. However, the Dutch smuggled cinchona seeds out of South America. They set up their own plantations in the East Indies. Before long, the Netherlands controlled most of the world's supply of quinine.

During World War II, the Axis Powers seized the Netherlands. As a result, the Allies lost their source of quinine. A crisis was prevented when some quinine was smuggled out of Germany and sold on the black market. Since then, scientists have developed synthetic drugs for the treatment of malaria. However, quinine remains an important drug. It is used to treat heart disease and is a key ingredient in tonic water.

Understanding What You Read

1. How did native peoples' use of the cinchona tree change the world?
2. How did political decisions during World War II affect the use of quinine?

Do you remember what you learned about renewable and nonrenewable resources? See Chapter 4 to review.

Economy Colombia's economy relies on several valuable resources. Rich soil produces world-famous Colombian coffee. Only Brazil produces more coffee. Other major export crops include bananas, corn, rice, and sugarcane. **Cassava** (kuh-SAH-vuh), a tropical plant with starchy roots, is an important food crop. Colombian farms also produce flowers that are exported around the world. In fact, only the Netherlands exports more cut flowers than Colombia.

In recent years oil has become Colombia's leading export. Oil is found mainly in eastern Colombia. Other natural resources include iron ore, gold, coal, and tin. Most of the world's emeralds also come from Colombia.

Even with these rich resources, many Colombians have low incomes. Colombia faces the same types of problems as other countries in Central and South America. For example, urban poverty and rapid population growth remain a challenge in Colombia.

CLOSE

Have each student attempt to describe Colombia as completely as possible using only five adjectives or descriptive phrases. Ask volunteers to share their descriptions with the class. Conclude by supplying any adjectives or descriptions that students may have omitted.

REVIEW AND ASSESS

Have students complete the Section Review. Then pair students and have each pair write a short essay question about Colombia. Have each pair exchange its test with another pair and complete the test it receives. Then have students complete Daily Quiz 15.2. **COOPERATIVE LEARNING**

RETEACH

Have students complete Main Idea Activity 15.2. Then pair students and have each pair create a collage to depict various aspects of Colombia's history, culture, and economy. Ask volunteers to present and explain their collages to the class. **ENGLISH LANGUAGE LEARNERS, COOPERATIVE LEARNING**

EXTEND

Have interested students conduct research on the legend of El Dorado. Then have each student write a short poem on the Chibcha custom that inspired the legend or the later search for a mythical golden city. Ask volunteers to read their poems to the class. **BLOCK SCHEDULING**

◄ Folk dancers get ready to perform in Neiva, Colombia.

Interpreting the Visual Record What aspects of culture can you see in this photograph?

Cultural Life The physical geography of Colombia has isolated its regions from one another. This is one reason why the people of Colombia are often known by the area in which they live. African traditions have influenced the songs and dances of the Caribbean coast. Traditional music can be heard in some remote areas. In addition to music, many Colombians enjoy soccer. They also play a Chibcha sport called *tejo*, a type of ringtoss game. Roman Catholicism is the country's main religion.

Conflict is a serious problem in Colombia today. Border conflicts with Venezuela have gone on for many years. Many different groups have waged war with each other and with Colombia's government. These groups have controlled large areas of the country. Many farmers have been forced off of their land, and the economy has been damaged. Because of this instability, the future of Colombia is uncertain.

✔ **READING CHECK:** *Human Systems* How have historical events affected life in Colombia today? *African traditions originating from enslaved Africans influence contemporary song and dance; border conflicts cause violence and damage the economy.*

Section Review 2

Define and explain: El Dorado, cassava

Working with Sketch Maps On the map you created in Section 1, label Colombia, Bogotá, and the Cauca and Magdalena Rivers.

Reading for the Main Idea

1. (*Human Systems*) Why did Spanish explorers come to Colombia?

2. (*Human Systems*) What are some characteristics of Colombia's culture?

Homework Practice Online
Keyword: SJ3 HP15

Critical Thinking

3. **Finding the Main Idea** How have Colombia's varied landscapes affected its history?

4. **Summarizing** How have conflicts in Colombia affected its economy?

Organizing What You Know

5. **Sequencing** Copy the following graphic organizer. Use it to describe Colombia's historical periods.

Early history	Spanish period	Independence	Colombia today

Section Review 2

Answers

Define For definitions, see: El Dorado, p. 330; cassava, p. 332

Working with Sketch Maps Places should be labeled appropriately.

Reading for the Main Idea

1. They were looking for a route to the Pacific Ocean. (NGS 9)

2. isolated regions with different cultural traditions, Spanish and South American Indian cultural influences (NGS 9)

Critical Thinking

3. hindered unity

4. damaged it

Organizing What You Know

5. early history—advanced cultures, Chibcha civilization; Spanish period—forced labor on large estates; independence—establishment of Gran Colombia and New Granada, disputes between central government and Roman Catholic Church, violence; Colombia today—reliance on oil and other primary products, diverse cultural traditions, low incomes, internal and external conflicts

◄ **Visual Record Answer**

architecture, clothing, dance, hairstyles

333

Section 3
Venezuela

Read to Discover

1. How did the Spanish contribute to Venezuela's history?
2. What are some characteristics of Venezuela's culture?

Define
indigo
caudillos
llaneros
pardos

Locate
Venezuela
Caracas
Lake Maracaibo

WHY IT MATTERS

Changing oil prices can have a dramatic effect on the Venezuelan economy. Use CNNfyi.com or other **current events** sources to find information about current oil prices. Record your findings in your journal.

Empanadas

History of Venezuela

There were many small tribes of South American Indians living in Venezuela before the Spanish arrived. Most were led by chiefs and survived by a combination of hunting and farming.

Spanish Conquest Christopher Columbus landed on the Venezuelan coast in 1498. By the early 1500s the Spanish were exploring the area further. They forced South American Indians to dive for pearls and pan for gold. There was little gold, however. The settlers had to turn to agriculture. They grew **indigo** (IN-di-goh) and other crops. Indigo is a plant used to make a deep blue dye. South American Indians were forced to work the fields. When many of them died, plantation owners brought in enslaved Africans to take their place. Some slaves were able to escape. They settled in remote areas and governed themselves.

Indigo was an important crop in Venezuela during colonial times.

Margarita Island in Venezuela was the site of a Spanish fort in the 1500s.

Teaching Objective 1

ALL LEVELS: (Suggested time: 25 min.) Organize students into small groups and have each group draw pictures depicting how the Spanish influenced Venezuela's history. *(Pictures should include Columbus's landing in 1498; the creation of a plantation economy with indigo as a primary crop; and the pressing of Indians into forced labor and Africans into slavery.)* Have groups present their pictures to the class.
ENGLISH LANGUAGE LEARNERS, COOPERATIVE LEARNING

Teaching Objective 2

ALL LEVELS: (Suggested time: 20 min.) To help students learn about Venezuela's culture, copy the following graphic organizer onto the chalkboard, omitting the italicized answers. Have each student complete the organizer, and ask volunteers to share their answers with the class.
ENGLISH LANGUAGE LEARNERS

Ancestry
- *Pardos constitute a majority of the population.*
- *Venezuelan Indians are a small minority.*

Language
- *Spanish is the official language.*
- *More than 25 other languages are spoken.*

Venezuelan Culture

Religion
- *Most Venezuelans are Roman Catholic.*
- *Venezuelan Indian religions are practiced.*

Sports and Leisure
- *Joropo is the national dance.*
- *Baseball, soccer, and toros coleados are popular sports.*

Independence Partly because the colony was so poor, some people in Venezuela revolted against Spain. Simón Bolívar led the fight against the Spanish armies. Bolívar is considered a hero in many South American countries because he led wars of independence throughout the region. The struggle for independence finally ended in 1830, when Venezuela became an independent country.

Throughout the 1800s Venezuelans suffered from dictatorships and civil wars. The country's military leaders were called **caudillos** (kow-THEE-yohs). After oil was discovered, some caudillos kept the country's oil money for themselves. In 1958 the last dictator was forced out of power.

Oil Wealth By the 1970s Venezuela was earning huge sums of money from oil. This wealth allowed part of the population to buy luxuries. However, about 80 percent of the population still lived in poverty. Many of these people moved to the cities to find work. Some settled on the outskirts in shacks that had no running water, sewers, or electricity.

Venezuela's wealth drew many immigrants from Europe and from other South American countries. However, in the 1980s oil prices dropped sharply. Because Venezuela relied on oil for most of its income, the country suffered when prices decreased.

✓ **READING CHECK:** **Human Systems** How did the Spanish contribute to Venezuela's history? They explored and settled there, farmed, and brought enslaved Africans to work fields after many of the South American Indians died as a result of forced labor.

Our Amazing Planet

Venezuela is home to the anaconda—the longest snake in the world. Adult anacondas are more than 15 feet (4.6 m) long.

Venezuela Today

Most Venezuelans live along the Caribbean coast and in the valleys of the nearby mountains. About 85 percent live in cities and towns. Caracas (kuh-RAHK-uhs), the capital, is the center of Venezuelan culture. It is a large city with a modern subway system, busy expressways, and tall office buildings. However, slums circle the city. Poverty in rural areas is also widespread. Still, Venezuela is one of South America's wealthiest countries. It is developing rapidly.

Economy Venezuela's economy is based on oil production. Lake Maracaibo (mah-rah-KY-boh) is a bay of the Caribbean Sea. The rocks under the lake are particularly rich in oil. However, the country is trying to reduce its dependence on oil income.

The Guiana Highlands in the southeast are rich in other minerals, such as iron ore for making steel. Dams on tributaries of the Orinoco River produce hydroelectricity.

Venezuela				
COUNTRY	POPULATION/ GROWTH RATE	LIFE EXPECTANCY	LITERACY RATE	PER CAPITA GDP
Venezuela	23,916,810 1.6%	70, male 77, female	91%	$6,200
United States	281,421,906 0.9%	74, male 80, female	97%	$36,200

Sources: Central Intelligence Agency, *The World Factbook 2001;* U.S. Census Bureau

Interpreting the Chart What is the average life expectancy for someone from Venezuela?

◄ **Chart Answer**

73 years

CLOSE

Tell students that there are three ways to get to Venezuela's spectacular Angel Falls. You can fly over them, but clouds may keep you from seeing them. You can hike for four or five days through the jungle. Or you can take a dugout canoe trip, travel by jeep around the rapids, and hike for an hour to the falls. Poll the class to see which methods of travel students would prefer.

REVIEW AND ASSESS

Have students complete the Section Review. Then pair students and have each pair create a flash card for each of the Define and Locate terms in this section. Have students within the pair quiz each other. Then have students complete Daily Quiz 15.3. **COOPERATIVE LEARNING**

RETEACH

Have students complete Main Idea Activity 15.3. Then have students design travel posters to attract tourists to Venezuela. Display posters around the classroom. **ENGLISH LANGUAGE LEARNERS**

EXTEND

Have interested students conduct research on Simón Bolívar. Then have each student write a short eulogy for Bólivar. Ask volunteers to read their eulogies to the class. **BLOCK SCHEDULING**

Section Review 3

Answers

Define For definitions, see: indigo, p. 334; caudillos, p. 335; llaneros, p. 336; pardos, p. 336

Working with Sketch Maps Venezuela, Caracas, and Lake Maracaibo should be located accurately. Its elevation leads to a temperate climate.

Reading for the Main Idea

1. Settlers found little of the gold they had sought. (NGS 15)
2. Simón Bolívar (NGS 13)

Critical Thinking

3. Its economy would be less likely to suffer from hardships resulting from lower oil prices.
4. The Caribbean coast has a greater population density.

Organizing What You Know

5. history—some caudillos kept oil money; urban poverty—many poor people moved to cities, but oil wealth was not distributed evenly, and many continued to live in poverty; economy—reliant on oil production

Visual Record Answer ▶

Answers will vary but might include the llaneros' clothing and equipment.

Llaneros herd cattle on the large ranches of the Llanos.

Interpreting the Visual Record How do these *llaneros* look similar to cowboys in the United States?

Agriculture Northern Venezuela has small family farms and large commercial farms. **Llaneros** (lah-NE-rohs)—cowboys of the Venezuelan Llanos—herd cattle on the many ranches in this region. Few people live in the Guiana Highlands. Some small communities of South American Indians practice traditional slash-and-burn agriculture there.

Cultural Life More than two thirds of Venezuela's population are **pardos**. They are people of mixed African, European, and South American Indian ancestry. Native groups make up only about 2 percent of the population. They speak more than 25 different languages. Spanish is the official language. Most of the people are Roman Catholics. Some Venezuelan Indians follow the religious practices of their ancestors.

The *joropo*, a lively foot-stomping couples' dance, is Venezuela's national dance. *Toros coleados* is a local sport. In this rodeo event, the contestant pulls a bull down by grabbing its tail. Baseball and soccer are also popular in Venezuela.

✓ **READING CHECK:** *Human Systems* What are some aspects of Venezuela's culture? Two thirds of population are *pardos*; native groups make up only 2 percent of population; many languages spoken; most people Roman Catholic; local sports include *toros coleados*, baseball, soccer.

go. hrw .com **Homework Practice Online**
Keyword: SJ3 HP15

Section Review 3

Define and explain: indigo, caudillos, llaneros, pardos

Working with Sketch Maps On the map you created in Section 2, label Venezuela, Caracas, and Lake Maracaibo. How does elevation affect the climate of Caracas?

Reading for the Main Idea

1. (*Environment and Society*) Why did Spanish settlers in Venezuela have to turn to agriculture?
2. (*Human Systems*) Who led Venezuela's revolt against Spain?

Critical Thinking

3. **Drawing Inferences and Conclusions** Why might Venezuela try to reduce its dependence on oil exports?
4. **Comparing** Compare the population densities of the Caribbean coast and the Guiana Highlands.

Organizing What You Know

5. **Analyzing Information** Copy the following graphic organizer. Use it to explain how oil is related to the history, urban poverty, and economy of Venezuela.

	History	
Oil	Urban poverty	
	Economy	

336

Section 4

Objectives

1. Identify the countries that influenced the early history of the Guianas.
2. Compare Guyana, Suriname, and French Guiana today.

FOCUS

LET'S GET STARTED

Copy the following passage onto the chalkboard: *The same treaty that transferred New York City from the Dutch to the English transferred Suriname from the English to the Dutch. What if this trade had not taken place?* Discuss responses. *(Possible answers: New York might still have a Dutch culture. There might be a separatist movement there, like there is in Quebec.)* Tell students that in Section 4 they will learn more about the history of the Guianas, of which Suriname is a part.

Building Vocabulary

Ask a volunteer to read the definition of the term **indentured servants** from a dictionary. Point out that the root *indent* comes from the notch that was made in the copies of the contracts to make sure they matched. Indentured servants legally sold their labor for an extended period of time. Then ask a volunteer to read the definition of **pidgin languages** from the glossary.

Section 4

The Guianas

Read to Discover

1. Which countries influenced the early history of the Guianas?
2. How are Guyana, Suriname, and French Guiana similar today?

Define

indentured servants
pidgin languages

Locate

Guyana
Suriname
French Guiana
Georgetown

Paramaribo
Devil's Island
Kourou
Cayenne

WHY IT MATTERS

French Guiana is the site of a modern space center. Use **CNNfyi.com** or other **current events** sources to find space-related sites that talk about the European Space Agency. Record your findings in your journal.

Decorated Djuka canoe paddle

Section 4 RESOURCES

Reproducible
◆ Guided Reading Strategy 15.4
◆ Cultures of the World Activity 2

Technology
◆ One-Stop Planner CD–ROM, Lesson 15.4
◆ Homework Practice Online
◆ HRW Go site

Reinforcement, Review, and Assessment
◆ Section 4 Review, p. 339
◆ Daily Quiz 15.4
◆ Main Idea Activity 15.4
◆ English Audio Summary 15.4
◆ Spanish Audio Summary 15.4

Early History of the Guianas

Dense tropical rain forests cover much of the region east of Venezuela. Rugged highlands lie to the south. The physical environment of this region kept it somewhat isolated from the rest of South America. Thus, the three countries known as the Guianas (gee-AH-nuhz) have a history quite different from the rest of the continent.

European Settlement Spain was the first European country to claim the Guianas. The Spanish eventually lost the region to settlers from Great Britain, France, and the Netherlands. Sometimes a war fought in Europe determined which country held this corner of South America. The Europeans established coffee, tobacco, and cotton plantations. They brought Africans to work as slaves on these plantations. Sugarcane later became the main crop.

Asian Workers European countries made slavery illegal in the mid-1800s. Colonists in the Guianas needed a new source of labor for their plantations. They brought **indentured servants** from India, China, and Southeast Asia. Indentured servants agree to work for a certain period of time, often in exchange for travel expenses. As these people worked together, they developed **pidgin languages**. Pidgin languages are simple so that people who speak different languages can understand each other.

✓ **READING CHECK:** *Human Systems* What countries influenced the early history of the Guianas? Spain, Great Britain, France, the Netherlands

This Hindu temple is located in Cayenne, French Guiana.

Interpreting the Visual Record
Which architectural features resemble other buildings you have seen? Which features are different?
▼

◄ **Visual Record Answer**

Answers will vary.

Teaching Objective 1

ALL LEVELS: (Suggested time: 15 min.) Copy the following graphic organizer onto the chalkboard, omitting the italicized answers. Use it to help students learn about the countries that influenced the history of the Guianas. When students have completed the organizer, ask volunteers to share their answers with the class. **ENGLISH LANGUAGE LEARNERS**

Teaching Objective 2

ALL LEVELS: (Suggested time: 20 min.) Pair students and have each pair create a chart that depicts the similarities and differences between Guyana, Suriname, and French Guiana today. *(Charts should mention government, resources, and population.)* Ask volunteers to share their posters with the class. **ENGLISH LANGUAGE LEARNERS, COOPERATIVE LEARNING**

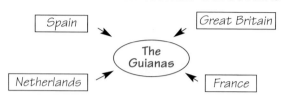

COUNTRIES THAT INFLUENCED THE GUIANAS

Spain → The Guianas ← Great Britain
Netherlands → The Guianas ← France

Section Review 4

Answers

Define For definitions, see: indentured servants, p. 337; pidgin languages, p. 337

Working with Sketch Maps Places should be labeled in their approximate locations.

Reading for the Main Idea

1. coffee, tobacco, cotton, and sugarcane (NGS 14)

2. Slavery had been outlawed by European countries, and settlers needed a new source of labor (NGS 9)

Critical Thinking

3. enjoy benefits of French citizenship

4. heavily forested

Organizing What You Know

5. Guyana—independent country, bauxite an important resource; Suriname—independent country, aluminum and forestry important to economy; French Guiana—forestry and fishing important to economy, depends on imports of food and energy, is part of France; common—diverse population, agriculture confined to coastal area

Chart Answer ▶

Answers will vary, but might mention disease, people moving to new locations for economic opportunities, or fewer young people having children.

The Guianas

Country	Population/ Growth Rate	Life Expectancy	Literacy Rate	Per Capita GDP
French Guiana	177,562 2.7%	73, male 80, female	83%	$6,000
Guyana	697,181 .07%	61, male 66, female	98%	$4,800
Suriname	433,998 0.6%	69, male 74, female	93%	$3,400
United States	281,421,906 0.9%	74, male 80, female	97%	$36,200

Sources: Central Intelligence Agency, *The World Factbook 2001*; U.S. Census Bureau

Interpreting the Chart Why might the populations of Guyana and Suriname be growing slowly?

Our Amazing Planet

The goliath bird-eating spider of northeastern South America is the largest spider in the world. The record holder had a leg span more than 11 inches (28 cm) across. That is as big as a dinner plate!

A woman cooks cassava cakes in Bigi Poika, Suriname.

The Guianas Today

The area formerly known as British Guiana gained its independence in 1966 and became Guyana. In 1975 Dutch Guiana broke away from the Netherlands to become Suriname. French Guiana remains a part of France.

Guyana *Guyana* (gy-AH-nuh) is a South American Indian word that means "land of waters." Nearly all of Guyana's agricultural lands are located along the narrow coastal plains. Guyana's most important agricultural products are rice and sugar. The country's major mineral resource is bauxite.

Guyana has a diverse population. About half of its people are of South Asian descent. Most of these people farm small plots of land or run small businesses. About one third of the population is descended from African slaves. These people control most of the large businesses and hold most of the government positions. More than one third of the country's population lives in Georgetown, the capital.

Suriname The resources and economy of Suriname (soohr-uh-NAH-muh) are similar to those of Guyana. Many farms in Suriname are found in coastal areas. Aluminum is a major export. Interior forests also supply lumber for export to other countries.

Like Guyana, Suriname has a diverse population. There are South Asians, Africans, Chinese, Indonesians, and people of mixed heritage. Muslim, Hindu, Roman Catholic, and Protestant houses of worship line the streets of the national capital, Paramaribo (pah-rah-MAH-ree-boh). Nearly half of the country's people live there.

Tell students to read the chart on the previous page that presents statistics about the populations of the Guianas and the United States. Ask students what they might conclude from this chart. *(Students might suggest that although many people in the Guianas are poor, they are relatively well educated.)*

REVIEW AND ASSESS

Have students complete the Section Review. Then pair students. Have each pair write a quiz consisting of eight multiple-choice questions about the Guianas. Have pairs exchange and complete their quizzes. Then have students complete Daily Quiz 15.4. **COOPERATIVE LEARNING**

RETEACH

Have students complete Main Idea Activity 15.4. Then have each student write three sentences for each of the Guianas. Sentences should relate to each country's culture, history, and natural resources. **ENGLISH LANGUAGE LEARNERS**

EXTEND

Have interested students conduct research on pidgin languages and where such languages have developed. Ask each student to present his or her research in a short written report. **BLOCK SCHEDULING**

CHAPTER 15

Review
ANSWERS

Carnival is a time for celebration in Cayenne, French Guiana.

French Guiana French Guiana (gee-A-nuh) has a status in France similar to that of a state in the United States. It sends representatives to the French Parliament in Paris. France used to send some of its criminals to Devil's Island. This island was a prison colony just off French Guiana's coast. Prisoners there suffered terribly from overwork. Devil's Island was closed in the early 1950s.

Today, forestry and shrimp fishing are the most important economic activities. Agriculture is limited to the coastal areas. The people of French Guiana depend heavily on imports for their food and energy. France developed the town of Kourou (koo-ROO) into a space center. The European Space Agency launches satellites from this town.

More than 160,000 people live in French Guiana, mostly in coastal areas. About two thirds of the people are descended from Africans. Other groups include Europeans, Asians, and South American Indians. The national capital is Cayenne (keye-EN).

✓ **READING CHECK:** ***Human Systems*** How is French Guiana different from Guyana and Suriname? *It is not an independent country.*

Building Vocabulary
For definitions, see: cordillera, p. 328; *tepuís*, p.328; Llanos, p. 328; El Dorado, p. 330; cassava, p. 332; indigo, p. 334; caudillos, p. 335; *llaneros*, p. 336; *pardos*, p. 336; indentured servants, p. 337; pidgin languages, p. 337

Reviewing the Main Ideas
1. *tierra caliente, tierra templada, tierra fría, paramo,* and *tierra helada* (NGS 4)
2. good soil, mineral deposits, and timber (NGS 4)
3. Geographic isolation led to the formation of separate economies and identities, making it difficult to create a unified national identity. (NGS 15)
4. Poverty led many to revolt against the colonial rulers. (NGS 13)
5. It is a French territory instead of an independent country. (NGS 13)

Understanding Environment and Society
Answers will vary, but the information included should be consistent with text material. Students may find that if the industry performs badly, the

Section Review 4

Define and explain: indentured servants, pidgin languages

Working with Sketch Maps On the map you created in Section 3, label Guyana, Suriname, French Guiana, Georgetown, Paramaribo, Devil's Island, Kourou, and Cayenne.

Reading for the Main Idea
1. ***Environment and Society*** What crops did early settlers raise in the Guianas?
2. ***Human Systems*** Why did colonists in the Guianas bring indentured servants from Asia?

Homework Practice Online
Keyword: SJ3 HP15

Critical Thinking
3. **Drawing Inferences and Conclusions** Why might the people of French Guiana prefer to remain a part of France?
4. **Drawing Inferences and Conclusions** Why do you think few people live in the interior of the Guianas?

Organizing What You Know
5. **Categorizing** Copy the following graphic organizer. Use it to describe the Guianas. Fill in the ovals with features of Guyana, Suriname, and French Guiana. In the center, list features they have in common.

ASSESS

Have students complete the Chapter 15 Test.

RETEACH

Tell students to imagine that they are foreign exchange students to Caribbean South America. Have each student create an illustrated letter that he or she would send home after traveling through and living in the region. **ENGLISH LANGUAGE LEARNERS**

PORTFOLIO EXTENSIONS

1. Have students conduct research on the flowers of Caribbean South America. Then have them re-create selected species with colored tissue paper, pipe cleaners, and markers. Attach a botanical label to the stem of each flower and decorate the classroom with them. Photograph the flowers for inclusion in student portfolios.

2. Pair students and have each pair conduct research on cultures that celebrate Carnival. Then have each pair design a Carnival costume and commemorative stamp. Students should write a short explanation of how their stamps and costumes relate to the traditions of Carnival.

Review
ANSWERS

effects on the country's economy will be devastating. They might suggest development of services related to the oil industry or the development of new export industries. Use Rubric 29, Presentations, to evaluate student work.

Thinking Critically
1. creates a corridor to the Atlantic; diverse animal population
2. If the population expands while resources remain constant or are not distributed evenly, poverty increases.
3. The region's colonizers practiced Roman Catholicism.
4. to live in a more temperate climate zone rather than a hot, heavily forested zone
5. They developed as a means of communication among people who spoke different languages.

Map Activity
A. Orinoco River
B. Lake Maracaibo
C. Andes
D. Llanos
E. Magdalena River
F. Devil's Island
G. Guiana Highlands

340

Reviewing What You Know

Building Vocabulary

On a separate sheet of paper, write sentences to define each of the following

1. cordillera
2. *tepuís*
3. Llanos
4. El Dorado
5. cassava
6. indigo
7. caudillos
8. *llaneros*
9. *pardos*
10. indentured servants
11. pidgin languages

Reviewing the Main Ideas

1. (*Places and Regions*) What are the five common elevation zones in the Andes region?
2. (*Places and Regions*) What are three resources found in Caribbean South America?
3. (*Environment and Society*) How did geography affect political upheavals throughout the history of Colombia?
4. (*Human Systems*) Why did Venezuela revolt against Spanish colonial rule?
5. (*Human Systems*) How does the political status of French Guiana differ from that of Guyana and Suriname?

Understanding Environment and Society

Resource Use
Venezuela and Colombia both depend on income from oil exports. Create a presentation in which you use a map, graph, chart, model, or database showing the problems this could cause. You may want to think about the following:
- Why it is risky for a country to rely heavily on a single industry.
- How a country might diversify its export products.

Thinking Critically

1. **Finding the Main Idea** Why is the Orinoco River important to Caribbean South America?
2. **Finding the Main Idea** How might rapid population growth contribute to poverty in the region?
3. **Analyzing Information** Why is Roman Catholicism the main religion in Caribbean South America?
4. **Drawing Inferences and Conclusions** Why do you think most Venezuelans live along the Caribbean coast and in the valleys of the nearby mountains?
5. **Drawing Inferences and Conclusions** How are pidgin languages in the Guianas an example of cultural cooperation?

FOOD FESTIVAL

Plantains are similar to bananas. Fried plantains are popular in Colombia. Speciality markets may have a thin, crisp variety similar to potato chips in the snack section or a thicker version in the frozen food case. To make them at home, peel four large plantains and cut into three or four pieces. Fry in hot vegetable oil until golden. Remove the plantains and pound flat. Return them to the oil and refry briefly. Remove and place on paper towels to drain. Sprinkle with salt.

CHAPTER 15 REVIEW AND ASSESSMENT RESOURCES

Reproducible
◆ Readings in World Geography, History, and Culture 20 and 21
◆ Critical Thinking Activity 15
◆ Vocabulary Activity 15

Technology
◆ Chapter 15 Test Generator (on the One-Stop Planner)

◆ Audio CD Program, Chapter 15
◆ HRW Go site

Reinforcement, Review, and Assessment
◆ Chapter 15 Review, pp. 340–41

◆ Chapter 15 Tutorial for Students, Parents, Mentors, and Peers
◆ Chapter 15 Test
◆ Chapter 15 Test for English Language Learners and Special-Needs Students

Building Social Studies Skills

Map ACTIVITY

On a separate sheet of paper, match the letters on the map with their correct labels.

Andes
Guiana Highlands
Orinoco River
Magdalena River

Lake Maracaibo
Devil's Island
Llanos

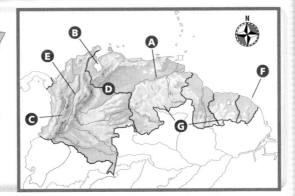

Mental Mapping Skills ACTIVITY

On a separate sheet of paper, draw a freehand map of Caribbean South America. Make a key for your map and label:

Atlantic Ocean
Caribbean Sea
Colombia
French Guiana

Guyana
Pacific Ocean
Suriname
Venezuela

WRITING ACTIVITY

Imagine that you are a teacher living in Caracas, Venezuela. Use the chapter map to write a quiz, with answers, for students in your geography class on the geography, economy, and people of Venezuela. Be sure to use standard grammar, sentence structure, spelling, and punctuation.

Alternative Assessment

Portfolio ACTIVITY

Learning About Your Local Geography

Research Project Many South American Indians build homes using local resources. Draw illustrations of some of the oldest buildings. Describe the materials used and their possible origin.

internet connect

Internet Activity: go.hrw.com
KEYWORD: SJ3 GT15

Choose a topic to explore about Caribbean South America:
• Trek through the Guiana Highlands.
• Search for the treasures of El Dorado.
• Ride with the *llaneros* of Venezuela.

Mental Mapping Skills Activity

Maps will vary. However, listed places should be labeled in their approximate locations.

Writing Activity

Quizzes will vary, but the information included should be consistent with text material. Check to see that students have written appropriate questions using the map. Use Rubric 40, Writing to Describe, to evaluate student work.

Portfolio Activity

Answers will vary, but students should locate and identify the oldest buildings in the community. Illustrations should be labeled with the correct information. Use Rubric 30, Research, to evaluate student work.

internet connect

GO TO: go.hrw.com
KEYWORD: SJ3 Teacher
FOR: a guide to using the Internet in your classroom

Atlantic South America
Chapter Resource Manager

Objectives	Pacing Guide	Reproducible Resources
SECTION 1 **Physical Geography** (pp. 343–46) **1.** Identify the landforms and rivers found in Atlantic South America. **2.** Describe the region's climates, vegetation, and wildlife. **3.** Name some of the region's important resources.	**Regular** .5 day **Block Scheduling** .5 day *Block Scheduling Handbook, Chapter 16*	**RS** Guided Reading Strategy 16.1 **RS** Graphic Organizer 16 **E** Creative Strategies for Teaching World Geography, Lesson 8 **IC** Interdisciplinary Activity for the Middle Grades 10
SECTION 2 **Brazil** (pp. 347–50) **1.** Describe the history of Brazil. **2.** Identify the important characteristics of Brazil's people and culture. **3.** Describe what Brazil's four major regions are like today.	**Regular** 1 day **Block Scheduling** 1 day *Block Scheduling Handbook, Chapter 16*	**RS** Guided Reading Strategy 16.2 **E** Cultures of the World Activity 2 **SM** Geography for Life Activity 16 **SM** Map Activity 16
SECTION 3 **Argentina** (pp. 351–54) **1.** Describe the history of Argentina. **2.** Identify the important characteristics of Argentina's people and culture. **3.** Describe what Argentina is like today.	**Regular** .5 day **Block Scheduling** .5 day *Block Scheduling Handbook, Chapter 16*	**RS** Guided Reading Strategy 16.3 **E** Cultures of the World Activity 2
SECTION 4 **Uruguay and Paraguay** (pp. 355–57) **1.** Describe what the people and economy of Uruguay are like today. **2.** Describe what the people and economy of Paraguay are like today.	**Regular** 1 day **Block Scheduling** .5 day *Block Scheduling Handbook, Chapter 16*	**RS** Guided Reading Strategy 16.4 **E** Cultures of the World Activity 2

Chapter Resource Key

RS Reading Support

IC Interdisciplinary Connections

E Enrichment

SM Skills Mastery

A Assessment

REV Review

ELL Reinforcement and English Language Learners

 Transparencies

 CD–ROM

 Music

 Video

 Internet

 Holt Presentation Maker Using Microsoft® Powerpoint®

 One-Stop Planner CD–ROM

See the *One-Stop Planner* for a complete list of additional resources for students and teachers.

One-Stop Planner CD–ROM

It's easy to plan lessons, select resources, and print out materials for your students when you use the *One-Stop Planner CD–ROM with Test Generator.*

Technology Resources	Review, Reinforcement, and Assessment Resources
One-Stop Planner CD–ROM, Lesson 16.1 Geography and Cultures Visual Resources with Teaching Activities 18–22 Homework Practice Online HRW Go site	**ELL** Main Idea Activity 16.1 **REV** Section 1 Review, p. 346 **A** Daily Quiz 16.1
One-Stop Planner CD–ROM, Lesson 16.2 Our Environment CD–ROM/ Seek and Tell/People Affecting Nature *ARGWorld* CD–ROM: Colonial History in South America Homework Practice Online HRW Go site	**ELL** Main Idea Activity 16.2 **REV** Section 2 Review, p. 350 **A** Daily Quiz 16.2
One-Stop Planner CD–ROM, Lesson 16.3 Homework Practice Online HRW Go site	**ELL** Main Idea Activity 16.3 **REV** Section 3 Review, p. 354 **A** Daily Quiz 16.3
One-Stop Planner CD–ROM, Lesson 16.4 Homework Practice Online HRW Go site	**ELL** Main Idea Activity 16.4 **REV** Section 4 Review, p. 357 **A** Daily Quiz 16.4

⧈ internet connect

HRW ONLINE RESOURCES

GO TO: go.hrw.com
Then type in a keyword.

TEACHER HOME PAGE
 KEYWORD: SJ3 TEACHER

CHAPTER INTERNET ACTIVITIES
 KEYWORD: SJ3 GT16

Choose an activity to:
• journey along the Amazon River.
• compare the people of Uruguay and Paraguay.
• celebrate the Brazilian Carnival.

CHAPTER ENRICHMENT LINKS
 KEYWORD: SJ3 CH16

CHAPTER MAPS
 KEYWORD: SJ3 MAPS16

ONLINE ASSESSMENT
Homework Practice
 KEYWORD: SJ3 HP16
 Standardized Test Prep Online
 KEYWORD: SJ3 STP16
 Rubrics
 KEYWORD: SS Rubrics

COUNTRY INFORMATION
 KEYWORD: SJ3 Almanac

CONTENT UPDATES
 KEYWORD: SS Content Updates

HOLT PRESENTATION MAKER
 KEYWORD: SJ3 PPT16

ONLINE READING SUPPORT
 KEYWORD: SS Strategies

CURRENT EVENTS
 KEYWORD: S3 Current Events

Meeting Individual Needs

Ability Levels

Level 1 Basic-level activities designed for all students encountering new material

Level 2 Intermediate-level activities designed for average students

Level 3 Challenging activities designed for honors and gifted-and-talented students

English Language Learners Activities that address the needs of students with Limited English Proficiency

Chapter Review and Assessment

E	Creative Strategies for Teaching World Geography, Lesson 9	**ELL**	Vocabulary Activity 16
IC	Interdisciplinary Activity for the Middle Grades 12	**A**	Chapter 16 Test
E	Readings in World Geography, History, and Culture 22, 23, and 24		Chapter 16 Test Generator (on the One-Stop Planner)
SM	Critical Thinking Activity 16		Audio CD Program, Chapter 16
REV	Chapter 16 Review, pp. 358–59	**A**	Chapter 16 Test for English Language Learners and Special-Needs Students
REV	Chapter 16 Tutorial for Students, Parents, Mentors, and Peers		

LAUNCH INTO LEARNING

Draw two columns on the chalkboard. Label one *Rainforest Animals*, the other *Rainforest Plants*. Call on volunteers to suggest items for both categories. *(Possible answers: jaguars, monkeys, sloths; mahogany trees, vines, orchids)* Then tell students that the purpose of studying this chapter will be to examine the importance of the rain forest in particular and to expand their knowledge of Atlantic South America in general. Add that these countries affect the economy and security of the United States and that the destruction of rain forests may be affecting the entire world. Tell students they will learn more about the physical and cultural geography of Atlantic South America in this chapter.

Section 1

Objectives

1. Investigate the landforms and rivers of Atlantic South America.
2. Analyze the region's climates, vegetation, and wildlife.
3. Identify some of the region's important resources.

LINKS TO OUR LIVES

You may want to point out the following reasons that make it important to study the countries of Atlantic South America:

▶ Consumers in the United States enjoy a wide range of agricultural products grown in the region.

▶ Brazil is so large that the health of its economy affects economies throughout Middle and South America. The U.S. economy can also be affected.

▶ The United States shares with the region some elements of history, such as periods of colonization and immigration, and culture, such as that of Argentina's gauchos.

▶ The destruction of tropical rain forests in the region may have severe consequences for Earth's climate and environment. However, conservation of the forests presents many economic and other challenges.

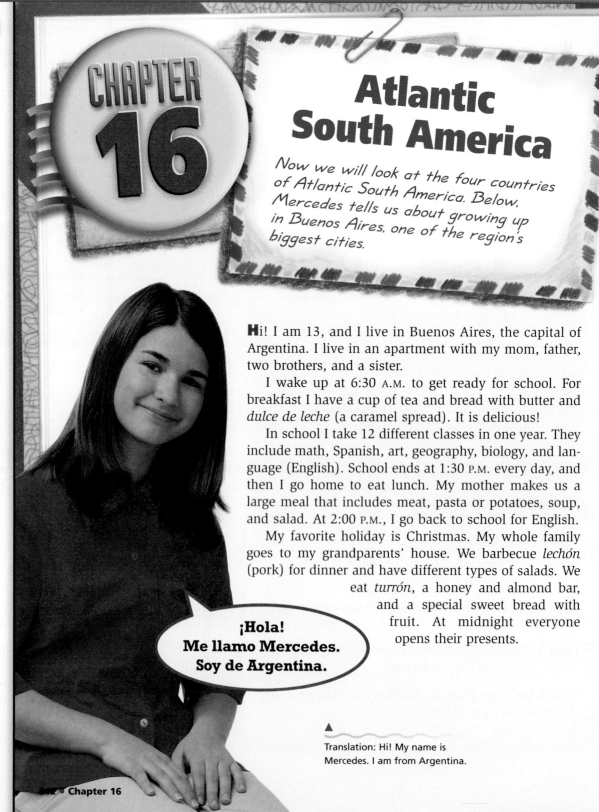

CHAPTER 16

Atlantic South America

Now we will look at the four countries of Atlantic South America. Below, Mercedes tells us about growing up in Buenos Aires, one of the region's biggest cities.

Hi! I am 13, and I live in Buenos Aires, the capital of Argentina. I live in an apartment with my mom, father, two brothers, and a sister.

I wake up at 6:30 A.M. to get ready for school. For breakfast I have a cup of tea and bread with butter and *dulce de leche* (a caramel spread). It is delicious!

In school I take 12 different classes in one year. They include math, Spanish, art, geography, biology, and language (English). School ends at 1:30 P.M. every day, and then I go home to eat lunch. My mother makes us a large meal that includes meat, pasta or potatoes, soup, and salad. At 2:00 P.M., I go back to school for English.

My favorite holiday is Christmas. My whole family goes to my grandparents' house. We barbecue *lechón* (pork) for dinner and have different types of salads. We eat *turrón*, a honey and almond bar, and a special sweet bread with fruit. At midnight everyone opens their presents.

¡Hola!
Me llamo Mercedes.
Soy de Argentina.

Translation: Hi! My name is Mercedes. I am from Argentina.

LET'S GET STARTED

Copy the following instructions onto the chalkboard: *Look at the map of Atlantic South America at the beginning of the chapter. Based on what you see on the map, what recreational activities might be available or not available to the residents of the countries there?* Allow students time to write their responses. Discuss their suggestions. *(Possible answers: Argentina—can go hiking in the mountains, Paraguay—cannot go sailing on the ocean, and so on)* Tell students that in Section 1 they will learn more about how the physical geography affects life in Atlantic South America.

Using the Physical-Political Map

Have students examine the map on this page. Call on volunteers to point out the major physical features of the area *(Amazon River, Andes, Brazilian Highlands)* Have students speculate what the climate and vegetation of each area might be. *(Possible answers: warm, humid, rain forest; cold, highland, shrubs and limited vegetation; temperate, mixed forests)*

Section 1 — Physical Geography

Read to Discover

1. What landforms and rivers are found in Atlantic South America?
2. What are the region's climates, vegetation, and wildlife like?
3. What are some of the region's important resources?

Define

Pampas
estuary
soil exhaustion

Locate

Amazon River
Brazilian Highlands
Brazilian Plateau
Gran Chaco
Patagonia

Tierra del Fuego
Andes
Paraná River
Paraguay River
Río de la Plata

WHY IT MATTERS

Itaipu Dam, on the Paraná River between Brazil and Paraguay, is an important source of hydroelectric power. Use **CNNfyi.com** or other **current events** sources to find out more about dam projects and their consequences. Record your findings in your journal.

Amazon rain forest monkey

Section 1 RESOURCES

Reproducible
◆ Block Scheduling Handbook, Chapter 16
◆ Guided Reading Strategy 16.1
◆ Graphic Organizer 16
◆ Creative Strategies for Teaching World Geography, Lesson 8
◆ Interdisciplinary Activity for the Middle Grades 10

Technology
◆ One-Stop Planner CD–ROM, Lesson 16.1
◆ Homework Practice Online
◆ Geography and Cultures Visual Resources with Teaching Activities 18–22
◆ HRW Go site

Reinforcement, Review, and Assessment
◆ Section 1 Review, p. 346
◆ Daily Quiz 16.1
◆ Main Idea Activity 16.1
◆ English Audio Summary 16.1
◆ Spanish Audio Summary 16.1

Atlantic South America: Physical-Political

SOUTH AMERICA

AMAZON BASIN
Manaus
Belém
Equator

BRAZIL

PACIFIC OCEAN

MATO GROSSO PLATEAU
Brasília
BRAZILIAN HIGHLANDS
Salvador

ATLANTIC OCEAN

BRAZILIAN PLATEAU

PARAGUAY
Asunción
Rio de Janeiro
São Paulo
Tropic of Capricorn

GRAN CHACO

Mount Aconcagua 22,834 ft. (6969 m)
Córdoba
Rosario
URUGUAY
Montevideo
PAMPAS
Buenos Aires
Río de la Plata

ARGENTINA

A N D E S

PATAGONIA

TIERRA DEL FUEGO
CAPE HORN

SCALE
0   500   1000 Miles
0   500   1000 Kilometers
Projection: Modified Chamberlin Trimetric

ELEVATION
FEET	METERS
13,120	4,000
6,560	2,000
1,640	500
656	200
(Sea level) 0	0 (Sea level)
Below sea level	Below sea level

⊛ National capitals
• Other cities

Size comparison of Atlantic South America to the contiguous United States

343

Teaching Objective 1

ALL LEVELS: (Suggested time: 30 min.) Ask each student to prepare a simple schematic drawing to show how the region's landforms relate to its rivers. Provide colored markers so students can differentiate among elevations and among rivers. **ENGLISH LANGUAGE LEARNERS**

Teaching Objective 2

ALL LEVELS: (Suggested time: 35 min.) Use the text to review the climates, vegetation, and wildlife of the region with the class. Then have students compare and contrast these characteristics of Atlantic South America with those of the United States. Then have each student write a paragraph describing the similarities and differences between Atlantic South America and the United States.

Teaching Objective 3

ALL LEVELS: (Suggested time: 30 min.) Copy the graphic organizer on the next page onto the chalkboard, omitting the italicized answers. Use it to help students identify the region's important resources. Have students work in pairs to fill in the blanks by using the land use and resources map in this unit's atlas and information in Section 1. **ENGLISH LANGUAGE LEARNERS, COOPERATIVE LEARNING**

ENVIRONMENT AND SOCIETY

Although climbers can ascend Mount Aconcagua without special equipment, its unpredictable weather threatens the unwary mountaineer. At the summit, climbers may be exposed to extremely low temperatures.

A particular danger is the *viento blanco*, or "white wind." This condition occurs when strong high-altitude winds blow from the west and create a cloud of microscopic ice crystals near the mountain's summit. Even when weather on the lower slopes is fine, a *viento blanco* can cause snow and high winds to occur farther up the mountain. Being caught inside the *viento blanco* cloud can be very dangerous.

Critical Thinking: What effect does the weather at Mount Aconcagua have on climbers?

Answer: It is dangerous for climbers because of the extreme cold and *viento blanco*.

internet connect

GO TO: go.hrw.com
KEYWORD: SJ3 CH16
FOR: Web sites about
Atlantic South America

Ranchers herd sheep on the Pampas of Argentina.

Interpreting the Visual Record What other economic activities are common in flat plains regions such as this?

▼

Physical Features

The region of Atlantic South America includes four countries: Brazil (bruh-ZIL), Argentina (ahr-juhn-TEE-nuh), Uruguay (oo-roo-GWY), and Paraguay (pah-rah-GWY). This vast region covers about two thirds of South America. Brazil alone occupies nearly half of the continent.

Plains and Plateaus This region's landforms include mainly plains and plateaus. The Amazon River basin in northern Brazil is a giant, flat flood plain. To the southeast are the Brazilian Highlands, a region of old, eroded mountains. Farther west is the Brazilian Plateau, an area of upland plains.

South of the Brazilian Plateau is a lower region known as the Gran Chaco (grahn CHAH-koh). The Gran Chaco stretches across parts of Paraguay, Bolivia, and northern Argentina. It is an area of flat, low plains covered with low trees, shrubs, and savannas.

In central Argentina are the wide, grassy plains of the **Pampas**. Patagonia, a desert region of dry plains and plateaus, is in southern Argentina. A cool, windswept island called Tierra del Fuego lies at the southern tip of the continent. Tierra del Fuego and nearby small islands are divided between Argentina and Chile.

Mountains The Andes, South America's highest mountains, extend north-south along Argentina's border with Chile. Here we find the Western Hemisphere's highest peak, Mount Aconcagua (ah-kohn-KAH-gwah). It rises to 22,834 feet (6,960 m).

River Systems The world's largest river system, the Amazon, flows eastward across northern Brazil. The Amazon River is about 4,000 miles (6,436 km) long. It extends from the Andes Mountains to the Atlantic Ocean. Hundreds of tributaries flow into it. Together they drain a vast area. This area includes parts of nearly all of the countries of northern and central South America.

Visual Record Answer ▶

possible answers: farming, oil and mineral exploration

Resources of Atlantic South America

Area of Atlantic South America	Resource
Amazon River basin	rain forest—for food, wood, rubber, medicinal plants; gold, other minerals
Brazilian Highlands	oil
Brazilian Plateau	hydroelectric power, minerals
Gran Chaco	hydroelectric power
Pampas	oil, natural gas, minerals
Patagonia	oil, natural gas

Teaching Objectives 1–3

ALL LEVELS: (Suggested time: 45 min.) Have students imagine that they are going to travel from Manaus to the top of Mount Aconcagua. Organize the class into groups. Have each group plan a trip route, make a list of supplies needed, and then write a journal describing the landforms, bodies of water, vegetation, animal life, and resources they would see on their journey. Ask groups to share their travelogues with the class. The activity may require more than one class period.
COOPERATIVE LEARNING

As a result, the Amazon also carries more water than any other river. About 20 percent of the water that runs off Earth's surface flows down the Amazon. This freshwater lowers the salt level of Atlantic waters for more than 100 miles (161 km) out.

The Paraná (pah-rah-NAH) River system drains much of the central part of the region. The Paraná River is 3,030 miles (4,875 km) long. It forms part of Paraguay's borders with Brazil and Argentina. Water from the Paraná flows into the Paraguay River. It continues on to the Río de la Plata (REE-oh day lah PLAH-tah) and the Atlantic Ocean beyond. The Río de la Plata is an **estuary**. An estuary is a partially enclosed body of water where salty seawater and freshwater mix.

✓ **READING CHECK:** *Places and Regions* What are the region's major land-forms and rivers? Amazon River and basin, Brazilian Highlands, Brazilian Plateau, Gran Chaco, Pampas, Andes, Paraná River, Rio de la Plata

Climate, Vegetation, and Wildlife

Tropical, moist climates are found in northern and coastal areas. They give way to cooler climates in southern and highland areas.

The Rain Forest The Amazon River basin's humid tropical climate supports the world's largest tropical rain forest. Rain falls almost every day in this region. The Amazon rain forest may contain the world's greatest variety of plant and animal life. Animals there include meat-eating fish called piranhas and predators such as jaguars and giant anacondas. The sloth, a mammal related to anteaters, moves slowly through the trees. It feeds on vegetation.

Plains and Plateaus Climates in the Brazilian Highlands vary widely. In the north, the coastal region is covered mostly with tropical rain forests and savannas. Inland from the coast, the highlands become drier and are covered with grasslands. The southeastern highlands have a mostly humid subtropical climate like the southeastern United States. These moist environments are major agricultural areas.

Southwest of the highlands, the Gran Chaco has a humid tropical climate. Because the Gran Chaco is flat, water drains slowly. Summer rains can turn areas of the region into marshlands. Wildlife there includes giant armadillos, pumas, red wolves, and at least 60 snake species.

The temperate grasslands of the Pampas stretch for almost 400 miles (644 km) southwest of the Río de la Plata. The rich soils and humid subtropical climate make the Pampas a major farming region. Farther south, warm summers and cold winters cause great annual temperature ranges in the Patagonia desert. The Andes block the Pacific Ocean's rain-bearing storms from reaching the area.

✓ **READING CHECK:** *Places and Regions* What are the climates of the region? humid tropical, highland, savanna, humid subtropical

South America's tropical rain forest blankets an area of about 2.3 million square miles (nearly 6 million sq. km). That is more than two thirds the area of the contiguous United States!

A sloth makes its way along tree branches in the Amazon rain forest.
Interpreting the Visual Record What do its long, sharp claws allow the sloth to do?

EYE ON EARTH

The Amazon River basin is home to many creatures. One of the most remarkable is the anaconda, a large South American snake of the boa family. Anacondas are the longest snakes in the Western Hemisphere and the heaviest snakes in the world.

The anaconda, also called the water boa, lies in wait by the water's edge. To kill, the anaconda pulls its prey beneath the water's surface until the prey drowns. Anacondas feed on birds, fish, large rodents, and small mammals.

Activity: Have students create a food web of plants and animals in the rain forest, including anacondas. Students may need to consult other sources.

internet connect

GO TO: go.hrw.com
KEYWORD: SJ3 CH16
FOR: Web sites about anacondas

◄ **Visual Record Answer**

possible answers: hang upside down from branches, rest safely for long periods

345

Tell the class about the Amazon ant, also known as a slave-making ant. They attack the nests of other ants and kidnap the helpless offspring. When the young become adults, they act as slaves to the Amazon ants. Amazon ants need the slave ants because their jaws are so long and curved that they cannot feed themselves or dig nests. The slave ants perform these tasks.

REVIEW AND ASSESS

Have students complete the Section Review. Then have each student create a five-question crossword puzzle based on the main ideas from each section. Pair students and have them complete each other's puzzles. Then have students complete Daily Quiz 16.1.

RETEACH

Have students complete Main Idea Activity 16.1. Then have students read the Read to Discover questions for each section and skim the section to find the answers. Ask volunteers to write the answers on the board. **ENGLISH LANGUAGE LEARNERS**

EXTEND

Have interested students conduct research on the Amazon River basin and create a diorama to show the physical features, vegetation, and animals of the area. **BLOCK SCHEDULING**

Section Review 1

Answers

Define For definitions, see: Pampas, p. 344; estuary, p. 345; soil exhaustion, p. 346

Working with Sketch Maps Maps will vary, but listed places should be labeled in their approximate locations.

Reading for the Main Idea
1. Brazilian Highlands, Gran Chaco, Pampas (NGS 4)
2. drains the largest area and carries more water than any other river (NGS 4)
3. The Andes prevent the Pacific Ocean's rain-bearing storms from reaching the area. (NGS 4)

Critical Thinking
4. As the soil is exhausted, people will clear new areas to plant.

Organizing What You Know
5. climates—tropical moist climates, some cooler climates; vegetation—rain forest, marshlands, grasslands; wildlife—piranhas, jaguars, sloths, armadillos, pumas, red wolves, snakes; resources—rain forest, minerals, hydroelectric power

Visual Record Answer ▶

possible answers: cheaper, easier, requires fewer tools, disposes of debris

Resources

One of the region's greatest resources is the Amazon rain forest. The rain forest provides food, wood, natural rubber, medicinal plants, and many other products. However, large areas of the rain forest are being cleared for mining, ranching, and farming.

Commercial agriculture is found throughout the region. In some areas, however, planting the same crop every year has caused **soil exhaustion**. Soil exhaustion means that the soil has lost nutrients needed by plants. Overgrazing is also a problem in some places.

The region's mineral wealth includes gold, silver, copper, and iron. There are oil deposits in the region, particularly in Brazil and Patagonia. Some of the region's large rivers provide hydroelectric power. One of the world's largest hydroelectric dams is the Itaipu Dam on the Paraná River. The dam lies between Brazil and Paraguay.

✓ READING CHECK: **Environment and Society** How have humans modified the region's environment? through logging, mining, ranching, farming, hydroelectric dams

◀

People are clearing large areas of the Amazon rain forest by burning and by cutting.

Interpreting the Visual Record Why do you think someone would choose to clear rain forest areas by burning rather than by cutting?

go. hrw .com Homework Practice Online
Keyword: SJ3 HP16

Section Review 1

Define and explain: Pampas, estuary, soil exhaustion

Working with Sketch Maps On an outline map of the region that you draw or that your teacher provides, label the following: Amazon River, Brazilian Highlands, Brazilian Plateau, Gran Chaco, Patagonia, Tierra del Fuego, Andes, Paraná River, Paraguay River, and Río de la Plata.

Reading for the Main Idea
1. *Places and Regions* What major landforms lie between the Amazon River basin and Patagonia?

2. *Places and Regions* What sets the Amazon River apart from all of the world's other rivers?

3. *Places and Regions* Why is the Patagonia desert dry?

Critical Thinking
4. **Drawing Inferences and Conclusions** How do you think soil exhaustion in Brazil could lead to more deforestation?

Organizing What You Know
5. **Categorizing** Copy the following graphic organizer. Use it to describe Atlantic South America.

Climates	Vegetation and wildlife	Resources

Section 2

Objectives

1. Examine the history of Brazil.
2. Identify important characteristics of Brazil's people and culture.
3. Describe what Brazil's four regions are like today.

FOCUS

LET'S GET STARTED

Copy the following instructions onto the chalkboard: *What celebrations are held regularly in our area? When do they occur? What do they celebrate?* Discuss responses. Tell students that in Brazil, Carnival is a pre-Lent festival celebrated with music, dancing, parades, and colorful costumes. Explain that Lent is more widely observed in mainly Catholic countries like Brazil, but people of many religions unite for Carnival celebrations. Tell students that in Section 2 they will learn more about the diversity of Brazil.

Building Vocabulary

Write **favelas** on the chalkboard and call on a volunteer to read the term's definition from a dictionary. *(huge slum areas located around Brazilian cities).* Explain that the term has Portuguese origins. Ask students to investigate the history of and current conditions in the favelas of Brazil. *(Possible answers: usually poor living conditions, high mortality rates, some choose to live there by necessity, others because of tradition)*

Section 2 Brazil

Read to Discover

1. What is the history of Brazil?
2. What are important characteristics of Brazil's people and culture?
3. What are Brazil's four major regions like today?

Define

favelas

Locate

Rio de Janeiro
São Paulo
Manaus
Belém

Salvador
São Francisco River
Mato Grosso Plateau
Brasília

WHY IT MATTERS

The Amazon rain forest is a tremendous resource for Brazil. Use CNNfyi.com or other **current events** sources to find out more about the Amazon region. Record your findings in your journal.

Brazil nuts

Section 2 RESOURCES

Reproducible

◆ Guided Reading Strategy 16.2
◆ Cultures of the World Activity 2
◆ Geography for Life Activity 16
◆ Map Activity 16

Technology

◆ One-Stop Planner CD–ROM, Lesson 16.2
◆ Homework Practice Online
◆ Our Environment CD–ROM/Seek and Tell/People Affecting Nature
◆ *ARGWorld* CD–ROM: Colonial History in South America
◆ HRW Go site

Reinforcement, Review, and Assessment

◆ Section 2 Review, p. 350
◆ Daily Quiz 16.2
◆ Main Idea Activity 16.2
◆ English Audio Summary 16.2
◆ Spanish Audio Summary 16.2

History

Most Brazilians are descended from three groups of immigrants. The first group included peoples who probably migrated to the Americas from northern Asia long ago. The second was made up of Portuguese and other Europeans who came after 1500. Africans brought as slaves made up the third group.

First Inhabitants Brazil's first human inhabitants arrived in the region thousands of years ago. They spread throughout the tropical rain forests and savannas.

These peoples developed a way of life based on hunting, fishing, and small-scale farming. They grew crops such as sweet potatoes, beans, and cassava. The root of the cassava plant is ground up and used as an ingredient in many foods in Brazil today. It also is used to make tapioca. Tapioca is a common food in grocery stores in the United States and other countries.

Europeans and Africans After 1500, Portuguese settlers began to move into Brazil. Favorable climates and soil helped make Brazil a large sugar-growing colony. Colonists brought slaves from Africa to work alongside Brazilian Indians on large sugar plantations. Sugar plantations eventually replaced forests along the Atlantic coast.

As elsewhere in the region, Brazilian Indians fought back against early European colonists. However, the Indians could not overcome powerful European forces.

▼

Teaching Objective 1

ALL LEVELS: (Suggested time: 30 min.) Pair students and have pairs create a time line listing important events in the history of Brazil. Ask them to write a brief statement by each event indicating its significance. Display and discuss the time lines. **COOPERATIVE LEARNING**

Teaching Objective 2

ALL LEVELS: (Suggested time: 20 min.) Lead a class discussion comparing and contrasting the cultures of Brazil and the United States. Include the topics of religion, diversity, language, and food.

➤ASSIGNMENT: Have students consider the advantages and disadvantages that Brazil brings with it into the 2000s. Ask students to list factors that will help Brazil in its future and obstacles that Brazil must overcome if its future is to be bright. Ask students to evaluate their lists and to write a brief prediction for Brazil's future.

DAILY LIFE

The samba is a dance style popular in Brazil. Samba schools, or neighborhood social clubs, exist throughout Rio de Janeiro's favelas. Every year, the samba schools compete to have the best entry in the Carnival parade. Members select a story or theme and then spend almost a year creating music, dance, and costumes to accompany it.

Activity: Ask a volunteer to learn the basic samba steps by consulting outside sources or a dance organization in your community. Then ask that student to teach the steps to the entire class. Organize the class into groups. Have each group prepare a samba dance entry for a samba contest. Have students perform their dances and judge their classmates' entries.

Visual Record Answer ▶

They are the major colors on Brazil's flag.

▲
Brazilian fans cheer on their country's soccer team. Soccer is very popular throughout Atlantic South America. Brazilian teams often compete in international tournaments, such as the World Cup.

Interpreting the Visual Record

Why do you think Brazilian fans are wearing the colors green and gold?

Farther inland, Portuguese settlers set up cattle ranches. These ranches provided hides and dried beef for world markets. In the late 1600s and early 1700s gold and precious gems were discovered in the southeast. A mining boom drew adventurers from around the world. The coastal city of Rio de Janeiro grew during this boom. In the late 1800s southeastern Brazil became a major coffee producer. This coffee boom promoted the growth of São Paulo.

Brazil gained independence from Portugal in 1822. An emperor ruled the country until 1889. Dictators and elected governments have ruled the country at various times since then. Today, Brazil has an elected president and legislature. Like the United States, Brazil provides its citizens the opportunity to participate in the political process through voting and other political activities.

✓ **READING CHECK:** **Human Systems** How have different peoples influenced Brazilian history? through farming, immigration, colonization

People and Culture

Nearly 40 percent of Brazil's more than 170 million people are of mixed African and European descent. More than half of Brazilians are ethnic European. They include descendants of Portuguese, Spaniards, Germans, Italians, and Poles. Portuguese is the official language. Some Brazilians also speak Spanish, English, French, Japanese, and Indian languages.

Religion Brazil has the world's largest population of Roman Catholics. About 70 percent of Brazilians are Catholic. Some Brazilians also practice Macumba (mah-KOOM-bah). Macumba combines African, Indian, and Catholic religious ideas and practices.

Carnival Other aspects of Brazilian life also reflect the country's mix of cultures. For example, Brazilians celebrate Carnival before the Christian season of Lent. However, the celebration mixes traditions from Africa, Brazil, and Europe. During Carnival, Brazilians dance the samba, which was adapted from an African dance.

Food Other examples of immigrant influences can be found in Brazilian foods. In parts of eastern Brazil, an African dish called *vatapá* (vah-tah-PAH) is popular. *Vatapá* mixes seafood, sauces, and red peppers. Many Brazilians also enjoy eating *feijoada* (fay-ZHWAH-da), a stew of black beans and meat. It is traditionally served on Saturday to large groups of people. *Feijoada* has many regional varieties.

✓ **READING CHECK:** **Human Systems** How has cultural borrowing affected Brazilian culture? Many languages are spoken; Macumba developed; Carnival is an important holiday celebration; samba is a popular dance; a variety of influences can be seen in the food.

Teaching Objective 3

LEVEL 1: (Suggested time: 10 min.) Copy the following graphic organizer onto the chalkboard, omitting the italicized answers. Use it to help students understand the characteristics of Brazil's regions. Point out that it is shaped roughly like Brazil. Have students copy the organizer into their notebooks and complete it. **ENGLISH LANGUAGE LEARNERS**

LEVELS 2 AND 3: (Suggested time: 30 min.) Organize the class into four groups. Assign each group one of the four regions of Brazil. Have students prepare a short presentation on why a certain new industry should operate from the region. Each presentation should include the economic, social, and environmental effects of the new industry. **COOPERATIVE LEARNING**

Brazil

The Amazon — *dense rain forest; isolated Brazilian Indian villages; Manaus—major city; Belém—Atlantic port; large mining district; tensions among Brazilian Indians, settlers, miners*

The Interior — *savannas and dry woodlands, could become agricultural area, Brasília, the national capital*

The Northeast — *old colonial cities like Salvador, suffers from drought, poorest region, favelas*

The Southeast — *coffee-growing area, Rio de Janeiro, richest region, most people, São Paulo*

▲ Day fades into night in Rio de Janeiro, Brazil's second-largest city. Sugarloaf Mountain stands near the entrance to Guanabara Bay. The Brazilian city is often referred to simply as Rio.

Brazil Today

Brazil is the largest and most populous country in South America. It ranks as the fifth-largest country in the world in both land area and population. Brazil also has the region's largest economy. Many Brazilians are poor, but the country has modern and prosperous areas. We will explore these and other areas by dividing the country into four regions. Those regions are the Amazon, the northeast, the southeast, and the interior. We will start in the Amazon and move southward.

The Amazon The Amazon region covers much of northern and western Brazil. Isolated Indian villages are scattered throughout the region's dense rain forest. Some Indians had little contact with out-siders until recently.

The major inland city in the region is Manaus. More than 1 million people live there. It is the Amazon's major river port and indus-trial city. South of the Atlantic port of Belém is a large mining district. New roads and mining projects are bringing more people and develop-ment to this region. However, development is destroying large areas of the rain forest. It also threatens the way of life of Brazilian Indians who live there. This development has created tensions among the Indians, new settlers, and miners.

The Northeast Northeastern Brazil includes many old colonial cities, such as Salvador. It is Brazil's poorest region. Many people there can-not read, and health care is poor. The region suffers from drought and has had trouble attracting industry. Cities in the region have huge slums called **favelas** (fah-VE-lahs).

Do you remember what you learned about deforestation? See Chapter 4 to review.

Brazil

COUNTRY	POPULATION/ GROWTH RATE	LIFE EXPECTANCY	LITERACY RATE	PER CAPITA GDP
Brazil	174,468,575 0.9%	59, male 68, female	83%	$6,500
United States	281,421,906 0.9%	74, male 80, female	97%	$36,200

Sources: Central Intelligence Agency, *The World Factbook 2001*; U.S. Census Bureau

Interpreting the Chart How much greater is the U.S. per capita GDP than the Brazilian per capita GDP?

◀ **Chart Answer**

$29,700

Have students pick the Brazilian region where they would most like to live. Ask students to explain why they would prefer that region and what occupation they would expect to pursue there.

Have students complete the Section Review. Then have students write five questions, along with the answers, using the information in the section. Collect the questions and answers. Organize the class into two teams and have the teams compete against each other. Then have students complete Daily Quiz 16.2.

Have students complete Main Idea Activity 16.2. Have class members imagine that they are in charge of planting a time capsule for Brazil that will be unearthed in 100 years. It should reflect Brazil's physical and human geography. Have the class decide what should be included in the time capsule.
ENGLISH LANGUAGE LEARNERS

Ask interested students to investigate how international organizations are working to save Brazil's rain forests and to present their findings to the class. Then have the class discuss ways they could help.
BLOCK SCHEDULING

Section Review 2

Answers

Define For definition, see: favelas, p. 349

Working with Sketch Maps Brasília was built to help develop Brazil's interior. Although it was designed for 500,000 people, the city now has about 1.5 million residents.

Reading for the Main Idea

1. Africans, Portuguese, Spaniards, Germans, Italians, and Poles (NGS 10)

2. northeast; southeast (NGS 4)

Critical Thinking

3. possible answers: conflicts involving control of land and wealth, fear of destroying Brazilian Indians' way of life

4. possible answer: one of the world's most ethnically diverse countries, resulting in a variety of languages, cultures, religions, foods, customs

Organizing What You Know

5. economic development, mining, road building, population growth, built new cities, hydroelectric development, industrial development

Chart Answer ►

Brazil

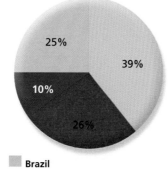

Major Producers of Coffee

39%
25%
10%
26%

- Brazil
- Colombia
- Rest of Middle and South America
- Rest of the world

Source: United Nations, Food and Agriculture Organization

Interpreting the Chart Which single South American country produces the most coffee?

The Southeast Large favelas are also found in other Brazilian cities, particularly in the southeast. Most of Brazil's people live in the southeast. However, in contrast to the northeast, the southeast is Brazil's richest region. It is rich in natural resources and has most of the country's industries and productive farms. The southeast is one of the most productive coffee-growing regions in the world.

The giant cities of São Paulo and Rio de Janeiro are located in the southeast. More than 17.7 million people live in and around São Paulo. It is the largest urban area in South America and the fourth largest in the world. The city is also Brazil's main industrial center.

Rio de Janeiro lies northeast of São Paulo. More than 10 million people live there. The city was Brazil's capital from 1822 until 1960. Today, Rio de Janeiro remains a major seaport and is popular with tourists.

The Interior The interior region is a frontier land of savannas and dry woodlands. It begins in the upper São Francisco River basin and extends to the Mato Grosso Plateau. The region's abundant land and mild climate could one day make it an important agricultural area.

Brasília, the national capital, is located in this region. Brazil's government built the city during the 1950s and 1960s. Government officials hoped the new city would help develop Brazil's interior. It has modern buildings and busy highways. Nearly 2 million people live in Brasília, although it was designed for only 500,000.

✓ **READING CHECK:** *Places and Regions* What are Brazil's four main regions?
the Amazon, the Northeast, the Southeast, the Interior

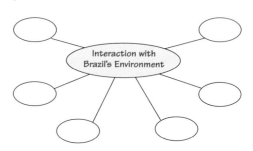

go.hrw.com **Homework Practice Online**
Keyword: SJ3 HP16

Section Review 2

Define and explain: favelas

Working with Sketch Maps On the map you created in Section 1, label Brazil, Rio de Janeiro, São Paulo, Manaus, Belém, Salvador, São Francisco River, Mato Grosso Plateau, and Brasília. In a margin box, write a caption explaining why Brazil's government built Brasília and how the city has grown over time.

Reading for the Main Idea

1. *Human Systems* From what major immigrant groups are many Brazilians descended?

2. *Places and Regions* What is Brazil's poorest region? What is its richest and most populated region?

Critical Thinking

3. **Drawing Inferences and Conclusions** Why do you think development in the Amazon has caused conflicts among miners, settlers, and Brazilian Indians?

4. **Finding the Main Idea** How did immigration influence Brazilian culture today?

Organizing What You Know

5. **Summarizing** Copy the following graphic organizer. Use it to list how early Brazilian Indians and European settlers and their descendants have used Brazil's natural resources and natural environment. Write each example in a circle radiating from the central circle. Create as many or as few circles as you need.

Interaction with Brazil's Environment

Objectives

1. Explore the history of Argentina.
2. Determine the important characteristics of Argentina's people and culture.
3. Explain what Argentina is like today.

FOCUS

LET'S GET STARTED

Copy the following question onto the chalkboard: *Have you heard about Eva Perón? What do you already know about her?* Allow students time to write their answers. *(Possible answer: She was the subject of a popular musical and movie.)* Discuss responses. Then tell students that Eva Perón was a real person and an important figure in Argentina's history. You may want to read a brief biography of Eva Perón to the class. Refer students to the photo of Perón in this section. Tell students that in Section 3 they will learn more about Argentina's history and culture.

Building Vocabulary

Write the key terms *encomienda* and **gauchos** on the chalkboard. Have volunteers find and read the definitions aloud. Explain that both of the terms have Spanish origins. Have students write a sentence using both terms. *(Example: Gauchos and the encomienda system both had an effect on Argentina's agriculture.)* Then write **Mercosur** on the chalkboard. Have a student read the definition. Give an example of other trade agreements or organizations, such as NAFTA.

Section 3
Argentina

Read to Discover

1. What is the history of Argentina?
2. What are important characteristics of Argentina's people and culture?
3. What is Argentina like today?

Define

encomienda
gauchos
Mercosur

Locate

Buenos Aires
Córdoba
Rosario

Reproducible
◆ Guided Reading Strategy 16.3
◆ Cultures of the World Activity 2

Technology
◆ One-Stop Planner CD–ROM, Lesson 16.3
◆ Homework Practice Online
◆ HRW Go site

Reinforcement, Review, and Assessment
◆ Section 3 Review, p. 354
◆ Daily Quiz 16.3
◆ Main Idea Activity 16.3
◆ English Audio Summary 16.3
◆ Spanish Audio Summary 16.3

WHY IT MATTERS

Argentina has carried out reforms to mend its economy and promote economic growth. Use CNNfyi.com or other **current events** sources to find out about economic reforms in Latin America or other parts of the world. Record your findings in your journal.

Argentine gaucho's belt buckle

History

Like most of South America, what is now Argentina was originally home to groups of Indians. Groups living in the Pampas hunted wild game. Farther north, Indians farmed and built irrigation systems.

Early Argentina In the 1500s Spanish conquerors spread into southern South America. They moved into the region in search of riches they believed they would find there. They called the region Argentina, meaning "land of silver" or "silvery one."

The first Spanish settlement in Argentina was established in the early 1500s. Spanish settlements were organized under the ***encomienda*** system. Under that system, the Spanish monarch gave land to colonists. These landowners were granted the right to force Indians living there to work the land.

The Pampas became an increasingly important agricultural region during the colonial era. Argentine cowboys, called **gauchos** (GOW-chohz), herded cattle and horses on the open grasslands. Colonists eventually fenced off their lands into huge ranches. They hired gauchos to tend their herds of livestock. Today the gaucho, like the American cowboy, is vanishing. Still, the gaucho lives on in Argentine literature and popular culture.

In 1816 Argentina gained independence. However, a long period of instability and violence followed. Many Indians, particularly in the Pampas, were killed in wars with the Argentine

Examples of Spanish-style architecture can be seen throughout Argentina. This Roman Catholic church was built in the 1600s in Córdoba.

Teaching Objective 1

LEVEL 1: (Suggested time: 15 min.) Ask students to write newspaper headlines describing events during Argentina's major historical periods. (*Example: New irrigation system built in north will help farmers.*)

LEVELS 2 AND 3: (Suggested time: 45 min.) Have students use other resources to conduct research on the era that is the subject of one of the headlines they wrote for the Level 1 activity. Ask them to write the newspaper article to match the headline.

Teaching Objective 2

ALL LEVELS:

(Suggested time: 10 min.) Copy the following graphic organizer onto the chalkboard. Use it to demonstrate the ethnic makeup of Argentina's population. Then ask students to summarize other features of Argentine culture.

Ethnic makeup of Argentina

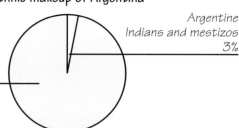

Descendants of Spanish, Italian, or other European settlers 97%

Argentine Indians and mestizos 3%

Argentina and Great Britain fought a war for control of a small group of islands that lies about 300 miles (483 km) east of the Strait of Magellan. The Argentines call the islands the Islas Malvinas and had long claimed ownership of them. Since 1833 the British have ruled what they call the Falkland Islands.

In April 1982, Argentine troops seized the Falklands. Argentina and Britain then fought air, sea, and land battles for control of the territory. Argentina surrendered in June 1982.

Discussion: Ask students to explain the significance of the Falkland Islands. Have students find the Falklands on a map and analyze why both Britain and Argentina would claim these remote islands.

Chart Answer ▶

People in the United States live slightly longer than those in Argentina.

Visual Record Answer ▶

People of all ages are participating.

Argentina

COUNTRY	POPULATION/ GROWTH RATE	LIFE EXPECTANCY	LITERACY RATE	PER CAPITA GDP
Argentina	37,384,816 1.2%	72, male 79, female	96%	$12,900
United States	281,421,906 0.9%	74, male 80, female	97%	$36,200

Sources: Central Intelligence Agency, *The World Factbook 2001*; U.S. Census Bureau

Interpreting the Chart How does life expectancy in the United States differ from that in Argentina?

Couples practice the Argentine tango on a Buenos Aires sidewalk. Varieties of the tango are popular in some other Middle and South American countries and in parts of Spain.

Interpreting the Visual Record How can you tell that the tango appeals to Argentines of all ages?

▶

government. As a result, Argentina has a small Indian population today. Most of these wars had ended by the late 1870s.

Modern Argentina New waves of European immigrants came to Argentina in the late 1800s. Immigrants included Italians, Germans, and Spaniards. Exports of meat and other farm products to Europe helped make the country richer.

However, throughout much of the 1900s, Argentina struggled under dictators and military governments. These unlimited governments abused human rights. Both the country's economy and its people suffered. In 1982 Argentina lost a brief war with the United Kingdom over the Falkland Islands. Shortly afterward, Argentina's last military government gave up power to an elected government.

✔ **READING CHECK:** *Human Systems* How was Argentina's government organized during much of the 1900s? **as a dictatorship or military government**

People and Culture

Argentina's culture has many European ties. Most of the more than 37 million Argentines are Roman Catholic. Most also are descended from Spanish, Italian, or other European settlers. Argentine Indians and mestizos make up only about 3 percent of the population. Spanish is the official language. English, Italian, German, and French are also spoken there.

Beef is an important agricultural product and a big part of the Argentine diet. A popular dish is *parrillada*. It includes sausage and steak served on a small grill. Supper generally is eaten after 9 P.M.

✔ **READING CHECK:** *Human Systems* Why are so many languages spoken in Argentina? **because of the many different cultures that make up Argentine society**

Teaching Objective 3

LEVEL 1: (Suggested time: 30 min.) Have students imagine that they are photographers who will take photos for an exhibit titled Daily Life in Argentina. Have students draw what they would photograph for the exhibit. Be sure they include information on people, places, cultural events, and holidays. Display the drawings. **ENGLISH LANGUAGE LEARNERS**

LEVELS 2 AND 3: (Suggested time: 30 min.) Explain that Argentina faces many challenges at home and abroad. Have the class brainstorm ideas through discussion and generate a list of challenges. Have students work with a partner to create a flowchart of causes, effects, and possible solutions.

Joann Sadler of Buffalo, New York, suggests this activity. Have students refer to the beginning of the chapter and the introductory letter by Mercedes. Have them note some of the similarities and differences between themselves and her. Ask them to write a letter to Mercedes telling her about themselves and their community. Ask students to include some information similar to what she provides. Then organize the students into pairs or triads and have them critique each other's letters. Call on volunteers to read their letters to the class.

CONNECTING TO *Literature*

Argentine gaucho

The Gaucho Martín Fierro

José Hernández was born in 1834. A sickly boy, he was sent to regain his health on Argentina's Pampas. The gauchos lived freely on the plains there, herding cattle. As an adult, Hernández took part in his country's political struggles. He fled to Brazil after a failed revolt. In 1872 he published his epic poem, The Gaucho Martín Fierro. *In this excerpt, Martín Fierro recalls his happier times as an Argentine cowboy.*

Ah, those times! . . . you felt proud
to see how a man could ride.
When a gaucho really knew his job,
even if the colt went over backwards,
not one of them wouldn't land on his feet
with the halter-rein in his hand. . . .

Even the poorest gaucho
had a string of matching horses;
he could always afford some amusement,

and people were ready for
anything. . . .
Looking out across the land
you'd see nothing but cattle
and sky.

When the branding-time came round
that was work to warm you up!
What a crowd! lassoing the running steers
and keen to hold and throw them. . . .
What a time that was! in those days surely
there were champions to be seen. . . .

And the games that would get going
when we were all of us together!
We were always ready for them,
as at times like those
a lot of neighbors would turn up
to help out the regular hands.

Analyzing Primary Sources

1. What does Martín Fierro remember about the "old days"?
2. In what ways is a gaucho's life similar to that of an American cowboy?

Across the Curriculum
LITERATURE
Latin American

Fiction Like José Hernández, Jorge Luis Borges (1899–1986) was an Argentine writer. Borges won acclaim for his distinctive works of prose and poetry.

Borges was born in Buenos Aires. During the 1920s he wrote about the city in his book of poems, *Fervor de Buenos Aires*. In many ways Borges's work set a standard for Latin American writers. Many of his works have an otherworldly quality that readers find entrancing.

Activity: Have students read some English translations of Borges's poems about Buenos Aires. Then have them conduct research on other cities in the region and write original poems about them.

Connecting to Literature
Answers

1. possible answers: gauchos were respected, times were better and happier, people worked together, work more plentiful
2. possible answers: tasks and skills similar, participated in competitions, popular, strong, worked as team

Argentina Today

Argentina has rich natural resources and a well-educated population. The Pampas are the most developed agricultural region. About 12 percent of Argentina's labor force works in agriculture. Large ranches and farms produce beef, wheat, and corn for export to other countries.

Much of Argentina's industry is located in and around Buenos Aires, the national capital. Buenos Aires is the second-largest urban area in South America. Its location on the coast and near the Pampas has contributed to its economic development. It is home to nearly a

CLOSE

Call on volunteers to describe how the illustrations in this section reflect the history and culture of Argentina.

REVIEW AND ASSESS

Have students complete the Section Review. Then have students work in pairs to write lyrics for a gaucho's song about Argentina's history, culture, or economy. Call on volunteers to read their lyrics aloud and explain their connection to Argentine society. Then have students complete Daily Quiz 16.3. **COOPERATIVE LEARNING**

RETEACH

Have students complete Main Idea Activity 16.3. Assign each of three groups one of the following topics: Argentina's history, Argentina's people and culture, or Argentina today. Have each group create a radio newscast on the assigned topic. **ENGLISH LANGUAGE LEARNERS, COOPERATIVE LEARNING**

EXTEND

Buenos Aires is a very cosmopolitan city. Have students conduct research on the cultural life of Buenos Aires and create a weekend entertainment supplement for the city's newspaper. **BLOCK SCHEDULING**

Section Review 3

Answers

Define For definitions, see: *encomienda*, p. 351; *gauchos*, p. 351; Mercosur, p. 354

Working with Sketch Maps Places should be labeled in their approximate locations.

Reading for the Main Idea

1. killed in wars with the Argentine government; few Indians today (NGS 12)

2. Buenos Aires area; because of its beneficial position in the country (NGS 4)

Critical Thinking

3. variety of languages spoken, foods like sausage and steaks, Roman Catholic religion

4. promotes economic cooperation among Argentina, Brazil, Paraguay, Uruguay, and to a lesser extent, Chile; provides for increased imports, exports, and job opportunities; the countries are neighbors and are logical economic partners

Organizing What You Know

5. 1500s: Spanish settle Argentina; 1816: Argentina gains its independence; about 1816–80: Indian–government wars; late 1800s: new European immigrants; 1983: return of democracy to Argentina

Wide European-style avenues stretch through Buenos Aires. Large advertisements compete for attention in the lively city.

third of all Argentines. Other large Argentine cities include the interior cities of Córdoba and Rosario.

Recent economic reforms in Argentina have been designed to help businesses grow. In addition, Argentina is a member of **Mercosur**. Mercosur is a trade organization that promotes economic cooperation among its members in southern and eastern South America. Brazil, Uruguay, and Paraguay also are members. Chile and Bolivia are associate members of Mercosur.

Argentina has made progress since the return of democracy in 1983. Today, Argentina has an elected president and legislature. The economy has been growing. The country still faces challenges, but Argentines now have more political freedom and economic opportunities.

✔ **READING CHECK:** *Places and Regions* What is Argentina like today? more democratic, well-educated population, large urban areas, growing economy

go.hrw.com
Homework Practice Online
Keyword: SJ3 HP16

Section Review 3

Define and explain: *encomienda*, *gauchos*, Mercosur

Working with Sketch Maps On the map you created in Section 2, label Argentina, Buenos Aires, Córdoba, and Rosario.

Reading for the Main Idea

1. *Human Systems* What happened to Argentine Indians in the Pampas in the 1800s? How did that affect Argentine society?

2. *Places and Regions* Where is much of Argentina's industry located? Why?

Critical Thinking

3. **Finding the Main Idea** In what ways does Argentine culture reflect European influences? What are some examples of these influences?

4. **Analyzing Information** How does Mercosur help Argentina's economy grow today? How does location play a part in that process?

Organizing What You Know

5. **Sequencing** Copy the following time line. Use it to identify important dates, events, and periods in the history of Argentina.

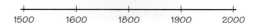

1500   1600   1800   1900   2000

Section 4

Objectives

1. Investigate the people and economy of Uruguay.
2. Describe the people and economy of Paraguay.

FOCUS

LET'S GET STARTED

Copy these instructions onto the chalkboard: *Look at the photos in this section of the chapter for clues about life in Paraguay and Uruguay. What conclusions can you draw?* Allow students time to complete their answers. *(Possible answers: modern and colonial-era buildings, handmade cloth, people dressed like cowboys)* Discuss responses. Have students predict what life is like in these countries based on their observations. Tell students that in Section 4 they will learn more about Uruguay and Paraguay.

Building Vocabulary

Write the key term **landlocked** on the chalkboard. Have students generate a list of descriptive characteristics implied by the term. Call on a volunteer to find the definition in the textbook and read it aloud. Explain that Paraguay is a landlocked country. Have volunteers use a world map to locate other landlocked countries. Discuss the advantages and disadvantages a landlocked country may have.

Section 4 Uruguay and Paraguay

Read to Discover

1. What are the people and economy of Uruguay like today?
2. What are the people and economy of Paraguay like today?

Define

landlocked

Locate

Montevideo
Asunción

WHY IT MATTERS

In the past, both Uruguay and Paraguay have had military rulers. Use **CNNfyi.com** or other **current events** sources to find out how military rule differs from democracy. Record your findings in your journal.

Handcrafted Paraguayan fabric

Section 4 RESOURCES

Reproducible
◆ Guided Reading Strategy 16.4
◆ Cultures of the World Activity 2

Technology
◆ One-Stop Planner CD–ROM, Lesson 16.4
◆ Homework Practice Online
◆ HRW Go site

Reinforcement, Review, and Assessment
◆ Section 4 Review, p. 357
◆ Daily Quiz 16.4
◆ Main Idea Activity 16.4
◆ English Audio Summary 16.4
◆ Spanish Audio Summary 16.4

Uruguay

Uruguay lies along the Río de la Plata. The Río de la Plata is the major estuary and waterway of southern South America. It stretches about 170 miles (274 km) inland from the Atlantic Ocean.

Uruguay's capital, Montevideo (mawn-tay-bee-THAY-oh), is located on the north shore of the Río de la Plata. The city is also the business center of the country.

Like its neighbors, Uruguay at times has been ruled by the military. However, in general the country has a strong democratic tradition of respect for political freedom.

Portugal claimed Uruguay during the colonial era. However, the Spanish took over the area by the 1770s. By that time, few Uruguayan Indians remained. Uruguay declared independence from Spain in 1825.

The People People of European descent make up about 88 percent of Uruguay's population. About 12 percent of the population is either mestizo, African, or Indian.

Roman Catholicism is the main religion in the country. Spanish is the official language, but many people also speak Portuguese.

More then 90 percent of Uruguayans live in urban areas. More than a third of the people live in and around Montevideo. The country has a high literacy

Government buildings and monuments surround Independence Square in Montevideo, the capital of Uruguay.

▼

◀ **Visual Record Answer**

warm, tropical climate

TEACH

Teaching Objectives 1–2

LEVEL 1: (Suggested time: 15 min.) Copy the following graphic organizer onto the chalkboard, omitting the italicized answers. Have volunteers fill in the organizer.

LEVELS 2 AND 3: (Suggested time: 30 min.) Have students create an advertisement for a business magazine enticing foreign investment in either Paraguay or Uruguay.

Uruguay
People
88% European; most Roman Catholic; Spanish, Portuguese; 90% urban; high literacy rate
Economy
led by Brazil and Argentina; agriculture, livestock; hydroelectric power

Paraguay
People
95% mestizo; Spanish, Guaraní, most Roman Catholic
Economy
controlled by few families and companies; surplus hydroelectricity; agriculture; not much industry

Section Review 4

Answers

Define For definition, see: landlocked, p. 356

Working with Sketch Maps Montevideo is the capital of Uruguay and center of government. Asunción is the capital of Paraguay.

Reading for the Main Idea

1. 1825 from Spain; 1811 from Spain (NGS 13)

2. east of Paraguay River (NGS 4)

3. erected hydroelectric projects on the Paraná River to create more power; surplus power sold to Brazil and Argentina (NGS 14)

Critical Thinking

4. no access to the ocean, restriction of trade, must maintain good relations with neighbors for imports and exports

Organizing What You Know

5. Uruguay—88 percent of population of European descent, 12 percent mestizo or of African descent, Catholic, Spanish and Portuguese spoken; Paraguay—95 percent of population mestizo, 5 percent Paraguayan Indian or of European descent, Catholic, Spanish and Guaraní spoken

Chart Answer ▶

much lower

356

The town of Colonia del Sacramento, Uruguay, was founded in 1680. Today it is an internationally recognized historical site.

rate. In addition, many Uruguayans have good jobs and can afford a wide range of consumer goods.

Economy Uruguay's economy is tied to the economies of Brazil and Argentina. In fact, more than half of Uruguay's foreign trade is with these two Mercosur partners. In addition, many Brazilians and Argentines vacation at beach resorts in Uruguay.

Uruguay's humid subtropical climate and rich soils have helped make agriculture an important part of the economy. As in Argentina, ranchers graze livestock on inland plains. Beef is an important export.

Uruguay has few mineral resources. An important source of energy is hydroelectric power. One big challenge is developing the poor rural areas in the interior, where resources are in short supply.

✔ **READING CHECK:** *Places and Regions* What geographic factors support agriculture in Uruguay? the humid subtropical climate and rich soils

Uruguay and Paraguay				
COUNTRY	POPULATION/ GROWTH RATE	LIFE EXPECTANCY	LITERACY RATE	PER CAPITA GDP
Paraguay	5,734,139 2.6%	71, male 77, female	92%	$4,750
Uruguay	3,360,105 0.7%	72, male 79, female	97%	$9,300
United States	281,421,906 0.9%	74, male 80, female	97%	$36,200

Sources: Central Intelligence Agency, *The World Factbook 2001;* U.S. Census Bureau

Interpreting the Chart How does Paraguay's per capita GDP compare to that of Uruguay?

Paraguay

Paraguay shares borders with Bolivia, Brazil, and Argentina. It is a **landlocked** country. Landlocked means it is completely surrounded by land, with no direct access to the ocean.

The Paraguay River divides the country into two regions. East of the river is the country's most productive agricultural land. The region west of the river is part of the Gran Chaco. This region has low trees and thorny shrubs. Ranchers graze livestock in some parts of western Paraguay.

Spanish settlers claimed Paraguay in the early 1500s. The country won independence from Spain in 1811. Paraguay was ruled off and on by dictators until 1989. Today, the country has an elected government.

Have students complete Main Idea Activity 16.4. Then organize the class into two groups. Ask students to imagine that one group is from Paraguay and the other group is from Uruguay. Hold a class debate over which country is a better place to live. Be sure to include debate questions about all aspects of the section. **ENGLISH LANGUAGE LEARNERS**

COOPERATIVE LEARNING

The People About 95 percent of Paraguayans are mestizos. European descendants and Paraguayan Indians make up the rest of the population. Spanish is the official language. Almost all people speak both Spanish and Guaraní (gwah-ruh-NEE), an Indian language. As in Uruguay, most people are Roman Catholic.

Asunción (ah-soon-SYOHN) is Paraguay's capital and largest city. It is located along the Paraguay River near the border with Argentina.

Economy Much of Paraguay's wealth is controlled by a few rich families and companies. These families and companies have great influence over the country's government.

Agriculture is an important part of Paraguay's economy. Much of the economy is traditional—many people are subsistence farmers. They grow just enough to feed themselves and their families. In fact, nearly half of the country's workers are farmers. They grow corn, cotton, soybeans, and sugarcane, some for profit. Paraguay also has a market economy, with thousands of small businesses but not much industry. Many Paraguayans have moved to neighboring countries to find work.

Paraguay's future may be promising as the country learns how to use its resources effectively. For example, the country has built hydroelectric dams on the Paraná River. These dams provide Paraguay with much more power than it needs. Paraguay sells the surplus electricity to Brazil and Argentina.

▲

These Paraguayan cowboys are drinking yerba maté (yer-buh MAH-tay). The herbal tea is popular in the region. It is made from the leaves and shoots of a South American shrub called the maté.

✔ **READING CHECK:** *Human Systems* How is Paraguay's economy organized?
partly traditional—subsistence farming—and partly market—small businesses

Homework Practice Online
Keyword: SJ3 HP16

Section Review 4

Define and explain: landlocked

Working with Sketch Maps On the map you created in Section 3, label Uruguay, Paraguay, Montevideo, and Asunción. In a box in the margin, briefly explain the significance of Montevideo and Asunción.

Reading for the Main Idea

1. *Human Systems* When and from what country did Uruguay win independence? What about Paraguay?

2. *Places and Regions* Where is Paraguay's most productive agricultural land?

3. *Environment and Society* How has Paraguay used its natural resources for economic development?

Critical Thinking

4. **Comparing** What disadvantages do you think a landlocked country might have when compared with countries that are not landlocked?

Organizing What You Know

5. **Comparing/Contrasting** Copy the following graphic organizer. Use it to compare and contrast the populations of Uruguay and Paraguay.

Uruguay	Paraguay

CHAPTER 16 Review ANSWERS

Building Vocabulary
For definitions, see: Pampas, p. 344; estuary, p. 345; soil exhaustion, p. 346; favelas, p. 349; *encomienda*, p. 351; gauchos, p. 351; Mercosur, p. 354; landlocked, p. 356

Reviewing the Main Ideas
1. Brazil; Spain (NGS 13)
2. the Amazon and Paraná River systems; transportation corridors, hydroelectric power (NGS 15)
3. southeast; northeast (NGS 4)
4. The Río de la Plata is an estuary between Uruguay and Argentina. (NGS 4)
5. Pampas (NGS 4)

Understanding Environment and Society
Answers will vary, but students should discuss government efforts to expand hydroelectric power through the use of dams. Use Rubric 29, Presentations, to evaluate student work.

Have students complete the Chapter 16 Test.

RETEACH

Have students review information about the history and present conditions in the Atlantic South American countries. Organize students into groups. Then have the groups develop storyboards for trailers for movies to be set in the region. The movies should focus on either the past or present conditions in the different countries assigned.
ENGLISH LANGUAGE LEARNERS

PORTFOLIO EXTENSIONS

1. Have students develop a bumper sticker representing a city, subregion, or country of the region of Atlantic South America. Encourage creativity and concise statements. The bumper stickers should be no larger than 4 inches by 11 inches. Display them when completed. Then place them in students' portfolios.

2. Have students conduct a debate between indigenous peoples who live in the rain forest and developers who want to turn it into grazing land. Following the debate, have each student write a position statement supporting one side or the other. Some students may prefer a compromise position.

Review
ANSWERS

Thinking Critically

1. All countries are primarily Roman Catholic. Primary languages are Portuguese and Spanish. Foods are a mixture of European and South American Indian influences. Political systems are modeled after European systems.

2. for economic development and for population growth

3. northern Brazil—tropical rain forest climate; southern Argentina—dry, cold climate

4. to promote economic cooperation among the countries

5. Paraguay—95 percent mestizo, 5 percent of European descent or Paraguayan Indian; Uruguay—88 percent of European descent, 12 percent Uruguayan Indian, of African descent or mestizo

Map Activity

A. Tierra del Fuego
B. Buenos Aires
C. Brasília
D. Paraná River
E. Brazilian Highlands
F. Patagonia
G. São Paulo
H. Manaus
I. Asunción
J. Montevideo

Reviewing What You Know

Building Vocabulary

On a separate sheet of paper, write sentences to define each of the following words.

1. Pampas
2. estuary
3. soil exhaustion
4. favelas
5. *encomienda*
6. gauchos
7. Mercosur
8. landlocked

Reviewing the Main Ideas

1. (*Human Systems*) Which country in Atlantic South America was a Portuguese colony? What colonial power controlled the other countries before they became independent?

2. (*Places and Regions*) What major river systems drain much of northern and central Atlantic South America? What economic roles do they play?

3. (*Places and Regions*) Which is the richest region in Brazil? Which is the poorest?

4. (*Places and Regions*) What and where is the Río de la Plata?

5. (*Places and Regions*) What is the most important agricultural area in Argentina?

Understanding Environment and Society

Resource Use

Paraguay and Uruguay both harness energy through hydroelectric dams. Create a presentation on the use of dams for hydroelectric power. Consider the following:

- Government decisions to expand the use of hydroelectric power.
- Benefits of hydroelectric power.
- How hydroelectric power is harnessed.

Thinking Critically

1. How are European influences reflected in the religions, languages, and other cultural characteristics of the countries of Atlantic South America?

2. Why are large areas of Brazil's tropical rain forest being cleared?

3. How do the climates of northern Brazil and southern Argentina differ?

4. How might Mercosur help the economies of the region's countries?

5. How are the ethnic populations of Uruguay and Paraguay different?

FOOD FESTIVAL

Tapioca is a starch taken from the root of the cassava, a tropical plant native to South America. Cassava, also called manioc, has been the chief starch eaten by most of the area's indigenous peoples. The roots are washed and then reduced to pulp, which is strained and dried. It forms small pellets we know as tapioca. Tapioca can be eaten plain or with flavoring. Brazilians favor sweetened tapioca with ginger and cinnamon.

Prepare tapioca according to package instructions. Serve with sugar, honey, syrup, or other flavors. Make larger quantities as needed.

CHAPTER 16 REVIEW AND ASSESSMENT RESOURCES

Reproducible
- Readings in World Geography, History, and Culture 22, 23, and 24
- Critical Thinking Activity 16
- Vocabulary Activity 16

Technology
- Chapter 16 Test Generator (on the One-Stop Planner)

- HRW Go site
- Audio CD Program, Chapter 16

Reinforcement, Review, and Assessment
- Chapter 16 Review, pp. 358–59

- Chapter 16 Tutorial for Students, Parents, Mentors, and Peers
- Chapter 16 Test
- Chapter 16 Test for English Language Learners and Special-Needs Students

Building Social Studies Skills

Map ACTIVITY

On a separate sheet of paper, match the letters on the map with their correct labels.

Brazilian Highlands	Manaus
Patagonia	Brasília
Tierra del Fuego	Buenos Aires
Paraná River	Montevideo
São Paulo	Asunción

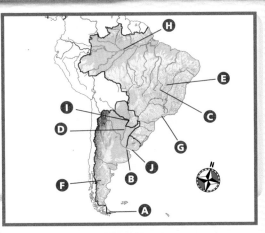

Mental Mapping Skills ACTIVITY

On a separate sheet of paper, draw a map of Atlantic South America and label the following:

Amazon River	Gran Chaco
Argentina	Pampas
Brazil	Paraguay
Brazilian Plateau	Uruguay

WRITING ACTIVITY

Imagine that you are a gaucho working on an Argentine ranch in the 1800s. Write a short song or poem that describes your daily life as a gaucho. Be sure to use standard grammar, sentence structure, spelling, and punctuation.

Mental Mapping Skills Activity
Maps will vary, but listed places should be labeled in their approximate locations.

Writing Activity
Answers will vary, but the information included should be consistent with text material. Use Rubric 26, Poems and Songs, to evaluate student work.

Portfolio Activity
Answers will vary, but the information included should be consistent with the text material. Use Rubric 29, Presentations, to evaluate student work.

Alternative Assessment

Portfolio ACTIVITY

Learning About Your Local Geography
Research Project Ranching is a major economic activity in Atlantic South America. Interview a member of a business organization. Then create a business profile of your community. Identify the role of entrepreneurs in your profile.

internet connect

Internet Activity: go.hrw.com
KEYWORD: SJ3 GT16

Choose a topic to explore about Atlantic South America:
- Journey along the Amazon River.
- Compare Uruguayans and Paraguayans.
- Celebrate the Brazilian Carnival.

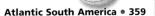

internet connect

GO TO: go.hrw.com
KEYWORD: SJ3 Teacher
FOR: a guide to using the Internet in your classroom

Pacific South America

Chapter Resource Manager

CHAPTER 17

Objectives	Pacing Guide	Reproducible Resources
SECTION 1 **Physical Geography** (pp. 361–64) 1. Identify the major physical features of the region. 2. Describe the climates and vegetation that exist in the region. 3. Identify the region's major resources.	**Regular** 1 day **Block Scheduling** .5 day *Block Scheduling Handbook, Chapter 17*	**RS** Guided Reading Strategy 17.1
SECTION 2 **History and Culture** (pp. 365–69) 1. Name some achievements of the region's early cultures. 2. Describe what the Inca Empire was like. 3. Describe the role that Spain played in the region's history. 4. Identify the governmental problems the region's people have faced.	**Regular** 1.5 days **Block Scheduling** .5 day *Block Scheduling Handbook, Chapter 17*	**RS** Guided Reading Strategy 17.2 **RS** Graphic Organizer 17 **E** Cultures of the World Activity 2 **E** Lab Activity for Geography and Earth Science, Hands-On 4 **SM** Map Activity 17
SECTION 3 **Pacific South America Today** (pp. 370–73) 1. Identify the three regions of Ecuador. 2. Describe how Bolivia might develop its economy. 3. Identify the features of Peru's regions. 4. Describe how Chile is different from its neighbors in Pacific South America.	**Regular** .5 day **Block Scheduling** .5 day *Block Scheduling Handbook, Chapter 17*	**RS** Guided Reading Strategy 17.3 **E** Creative Strategy for Teaching World Geography, Lesson 9 **SM** Geography for Life Activity 17

Chapter Resource Key

RS Reading Support

IC Interdisciplinary Connections

E Enrichment

SM Skills Mastery

A Assessment

REV Review

ELL Reinforcement and English Language Learners

 Transparencies

 CD–ROM

 Music

 Video

 Internet

Holt Presentation Maker Using Microsoft® Powerpoint®

 One-Stop Planner CD–ROM

See the *One-Stop Planner* for a complete list of additional resources for students and teachers.

One-Stop Planner CD-ROM

It's easy to plan lessons, select resources, and print out materials for your students when you use the *One-Stop Planner CD–ROM with Test Generator.*

Technology Resources

One-Stop Planner CD–ROM, Lesson 17.1

Geography and Cultures Visual Resources with Teaching Activities 18–23

HRW Go site

One-Stop Planner CD–ROM, Lesson 17.2

HRW Go site

One-Stop Planner CD–ROM, Lesson 17.3

ARGWorld CD–ROM: Mapping Population Change in Bolivia

HRW Go site

Review, Reinforcement, and Assessment Resources

ELL	Main Idea Activity 17.1
REV	Section 1 Review, p. 364
A	Daily Quiz 17.1

ELL	Main Idea Activity 17.2
REV	Section 2 Review, p. 369
A	Daily Quiz 17.2

ELL	Main Idea Activity 17.3
REV	Section 3 Review, p. 373
A	Daily Quiz 17.3

internet connect

HRW ONLINE RESOURCES

GO TO: go.hrw.com
Then type in a keyword.

TEACHER HOME PAGE
KEYWORD: SJ3 TEACHER

CHAPTER INTERNET ACTIVITIES
KEYWORD: SJ3 GT17

Choose an activity to:
• analyze Chile's climate.
• hike the Inca trail and visit Machu Picchu.
• learn about the languages of the Andes.

CHAPTER ENRICHMENT LINKS
KEYWORD: SJ3 CH17

CHAPTER MAPS
KEYWORD: SJ3 MAPS17

ONLINE ASSESSMENT
Homework Practice
KEYWORD: SJ3 HP17
Standardized Test Prep Online
KEYWORD: SJ3 STP17
Rubrics
KEYWORD: SS Rubrics

COUNTRY INFORMATION
KEYWORD: SJ3 Almanac

CONTENT UPDATES
KEYWORD: SS Content Updates

HOLT PRESENTATION MAKER
KEYWORD: SJ3 PPT17

ONLINE READING SUPPORT
KEYWORD: SS Strategies

CURRENT EVENTS
KEYWORD: S3 Current Events

Meeting Individual Needs

Ability Levels

Level 1 Basic-level activities designed for all students encountering new material

Level 2 Intermediate-level activities designed for average students

Level 3 Challenging activities designed for honors and gifted-and-talented students

English Language Learners Activities that address the needs of students with Limited English Proficiency

Chapter Review and Assessment

IC	Interdisciplinary Activity for the Middle Grades 17	**A**	Unit 34 Test
E	Readings in World Geography, History, and Culture 25, 26, and 27		Chapter 17 Test Generator (on the One-Stop Planner)
SM	Critical Thinking Activity 17		Audio CD Program, Chapter 17
REV	Chapter 17 Review, pp. 374–75	**A**	Chapter 17 Test for English Language Learners and Special-Needs Students
REV	Chapter 17 Tutorial for Students, Parents, Mentors, and Peers	**A**	Unit 4 Test for English Language Learners and Special-Needs Students
ELL	Vocabulary Activity 17		
A	Chapter 17 Test		

LAUNCH INTO LEARNING

Ask students to imagine that a glorious empire has arisen in your region and they are to design its new capital city. The city is to be laid out in the shape of an animal. Ask students which animal they would choose as appropriate and to sketch a basic city plan in the shape of that animal. Call on volunteers to sketch their city plans on the chalkboard. Then tell them that, according to tradition, Cuzco, the capital of the vast Inca Empire of the Pacific South America region, was built in the shape of a cougar. Ask what this might say about what the Inca people valued. *(possible answers: strength, power)* Tell students that in Chapter 17 they will learn more about the ancient and recent cultures of Pacific South America.

Section 1

Objectives

1. Identify the major physical features of the region.
2. Investigate the region's climates and vegetation.
3. Describe the region's major resources.

LINKS TO OUR LIVES

You might want to share with your students the following reasons for learning about Pacific South America:

▶ Studying the region's past civilizations, such as the Inca, can help us understand the cultures of the people who live there today.

▶ Chile is the world's top copper producer. Copper is used in electrical wire, industrial machinery, consumer products, and coinage.

▶ Peru and Bolivia are major producers of coca leaves—the source of cocaine. Cocaine use remains a big problem in the United States.

▶ The weather event known as El Niño affects both Pacific South America and the United States.

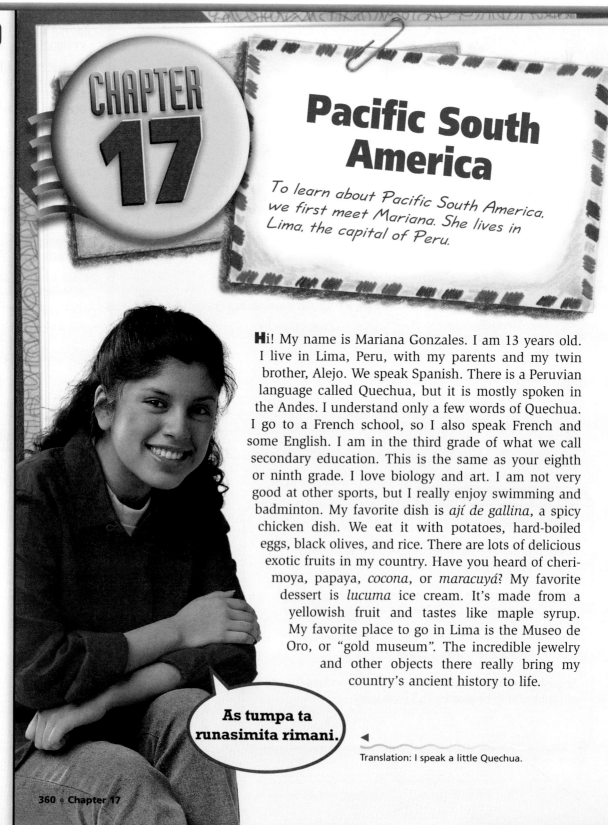

Pacific South America

To learn about Pacific South America, we first meet Mariana. She lives in Lima, the capital of Peru.

Hi! My name is Mariana Gonzales. I am 13 years old. I live in Lima, Peru, with my parents and my twin brother, Alejo. We speak Spanish. There is a Peruvian language called Quechua, but it is mostly spoken in the Andes. I understand only a few words of Quechua. I go to a French school, so I also speak French and some English. I am in the third grade of what we call secondary education. This is the same as your eighth or ninth grade. I love biology and art. I am not very good at other sports, but I really enjoy swimming and badminton. My favorite dish is *ají de gallina*, a spicy chicken dish. We eat it with potatoes, hard-boiled eggs, black olives, and rice. There are lots of delicious exotic fruits in my country. Have you heard of cherimoya, papaya, *cocona*, or *maracuyá*? My favorite dessert is *lucuma* ice cream. It's made from a yellowish fruit and tastes like maple syrup. My favorite place to go in Lima is the Museo de Oro, or "gold museum". The incredible jewelry and other objects there really bring my country's ancient history to life.

As tumpa ta runasimita rimani.

◀

Translation: I speak a little Quechua.

360

LET'S GET STARTED

Copy the following instructions onto the chalkboard: *Refer to the physical-political map at the beginning of the chapter. Locate Quito, the capital of Ecuador, and estimate the city's latitude. What do you think Quito's climate is like?* Allow students to record their answers. Discuss responses.*(Students should note that Quito is almost on the equator. Some students may suggest that the climate is hot.)* Point out that instead of a hot tropical climate, Quito has a moderate, springlike climate. Ask students what factor besides latitude affects the climate there *(elevation)*. Tell students that in Section 1 they will learn more about how the region's landforms affect climate.

Using the Physical-Political Map

Have students examine the map on this page. Call on a volunteer to list the names of the countries in this region. Point out that the Andes—the world's longest mountain chain on land—extend through the region and cross the equator. Ask students to suggest how the mountains influence these countries.

Building Vocabulary

Write the key terms on the chalkboard and call on volunteers to find the definitions of the terms in the textbook and to read them to the class. Ask students which terms are related to land *(selvas)* and which are related to water (**strait, Peru Current, El Niño**). Point out that El Niño affects the land also by causing heavy rainfall along the Pacific coast of South America. Have students locate the Strait of Magellan, the Peru Current, and the *selvas* on the physical-political map.

Section 1 Physical Geography

Read to Discover

1. What are the major physical features of the region?
2. What climates and vegetation exist in this region?
3. What are the region's major resources?

Define

strait
selvas
Peru Current
El Niño

Locate

Andes	Iquitos
Strait of Magellan	Altiplano
Tierra del Fuego	Lake Titicaca
Cape Horn	Lake Poopó
Amazon River	Atacama Desert

WHY IT MATTERS

The Atacama Desert offers a large variety of natural features including mountains, geysers, and the "Valley of the Moon." Use CNNfyi.com or other current events sources to discover other natural wonders of this area. Record your findings in your journal.

Peruvian hummingbird

Section 1 RESOURCES

Reproducible
◆ Block Scheduling Handbook, Chapter 17
◆ Guided Reading Strategy 17.1

Technology
◆ One-Stop Planner CD–ROM, Lesson 17.1
◆ Homework Practice Online
◆ Geography and Cultures Visual Resources with Teaching Activities 18–23
◆ HRW Go site

Reinforcement, Review, and Assessment
◆ Section 1 Review, p. 364
◆ Daily Quiz 17.1
◆ Main Idea Activity 17.1
◆ English Audio Summary 17.1
◆ Spanish Audio Summary 17.1

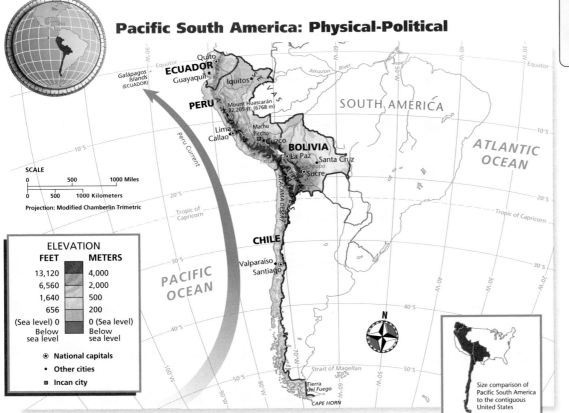

Pacific South America: Physical-Political

SCALE
0   500   1000 Miles
0   500   1000 Kilometers
Projection: Modified Chamberlin Trimetric

ELEVATION

FEET	METERS
13,120	4,000
6,560	2,000
1,640	500
656	200
(Sea level) 0	0 (Sea level)
Below sea level	Below sea level

⊛ National capitals
• Other cities
■ Incan city

Size comparison of Pacific South America to the contiguous United States

361

Teaching Objectives 1–3

LEVEL 1: (Suggested time: 20 min.) Copy the following graphic organizer onto the chalkboard, omitting the italicized answers. Tell students it represents a simplified elevation profile, or cutaway view, of Pacific South America near the region's widest point. Call on students to label the profile with the region's landforms, bodies of water, climates, vegetation, and resources. Encourage them to use the Unit Atlas as a resource.
ENGLISH LANGUAGE LEARNERS

Physical Geography of Pacific South America

Andes and Altiplano (glaciers, ice cap, Lake Titicaca, Lake Poopó, dry)

eastern plains (Amazon River rain forest, grass lands, oil, natural gas, silver, gold, other minerals)

Pacific coast (Atacama Desert, fog, minerals)

Pacific Ocean (Peru Current, fish)

PLACES AND REGIONS

Ecuador's physical features also include a cluster of islands far from the South American mainland. The Galápagos Islands lie about 600 miles (965 km) west of Ecuador's Pacific coast. Many of the animals and plants that live there are found nowhere else on Earth. Marine iguanas and giant tortoises that can live 150 years are among the island's most unusual inhabitants. The Galápagos Islands have become a popular tourist attraction. Increased tourism has damaged the islands' fragile environment. Now there are limits on the number of tourists and the type of tours allowed.

Activity: Have students conduct research on the Galápagos Islands and identify their unique plants and animals. Ask students to suggest how tourism might hurt specific species. Then ask students to check their predictions against ecologists' most-recent findings.

internet connect
GO TO: go.hrw.com
KEYWORD: SJ3 CH17
FOR: Web sites about the Galápagos Islands

Visual Record Answer ▶

Their ability to travel long distances without water allows them to survive in the region's dry climate.

internet connect
GO TO: go.hrw.com
KEYWORD: SJ3 CH17
FOR: Web sites about Pacific South America

BUILD on WHAT You Know

Do you remember what you learned about earthquakes? See Chapter 2 to review.

▶

Llamas are used to carry loads in the Altiplano region. Llamas are related to camels and can travel long distances without water.
Interpreting the Visual Record Why are llamas well suited to the environment of the Andes?

Physical Features

Shaped like a shoestring, Chile (CHEE-lay) stretches about 2,650 miles (4,264 km) from north to south. However, at its broadest point Chile is just 221 miles (356 km) wide. As its name indicates, Ecuador (E-kwuh-dawr) lies on the equator. Peru (puh-ROO) forms a long curve along the Pacific Ocean. Bolivia (buh-LIV-ee-uh) is landlocked.

Mountains The snowcapped Andes run through all four of the region's countries. Some ridges and volcanic peaks rise more than 20,000 feet (6,096 m) above sea level. Because two tectonic plates meet at the region's edge, earthquakes and volcanoes are constant threats. Sometimes these earthquakes disturb Andean glaciers, sending ice and mud rushing down the mountain slopes.

In Chile's rugged south, the peaks are covered by an ice cap. This ice cap is about 220 miles (354 km) long. Mountains extend to the continent's southern tip. There, the Strait of Magellan links the Atlantic and Pacific Oceans. A **strait** is a narrow passageway that connects two large bodies of water. The large island south of the strait is Tierra del Fuego, or "land of fire." It is divided between Chile and Argentina. At the southernmost tip of the continent, storms swirl around Chile's Cape Horn. Many other islands lie along Chile's southern coast.

Tributaries of the Amazon River begin high in the Andes. In fact, ships can travel upriver about 2,300 miles (3,700 km) on the Amazon. This allows ships to sail from the Atlantic Ocean to Iquitos, Peru.

Altiplano In Ecuador, the Andes range splits into two ridges. The ridges separate further in southern Peru and Bolivia. A broad, high plateau called the Altiplano lies between the ridges.

Rivers on the Altiplano have no outlet to the sea. Water collects in two large lakes, Lake Titicaca and salty Lake Poopó (poh-oh-POH). Lake Titicaca lies at 12,500 feet (3,810 m) above sea level. Large ships carry freight and passengers across it. The lake covers about 3,200 square miles (8,288 sq km).

✓ **READING CHECK:** *Places and Regions* What are the major physical features of the region? Andes, ice cap, Strait of Magellan, Tierra del Fuego, Amazon and its tributaries, Altiplano, Lake Titicaca, Lake Poopó

LEVELS 2 AND 3: (Suggested time: 30 min.) Organize the class into four groups. Point out that four aspects of Pacific South America's physical geography—El Niño, the Peru Current, the movement of tectonic plates, and the location of resources—can have dramatic effects on life in the region. Assign each group one of the four topics. Have each group create two graphic organizers—one that shows how its topic affects life in the area and another to show how local residents might adapt to that aspect of their physical geography. **COOPERATIVE LEARNING**

CLOSE

Go through Section 1 page by page, having students cover the photo captions. Ask them to speculate where the photographs were taken, based on what they learned in Section 1. Ask: *What in the photos indicates the location?* Have students read the captions to check their predictions.

REVIEW AND ASSESS

Have students complete the Section Review. Then ask each student to use the maps in the Unit Atlas to choose a place in Pacific South America to depict. Tell students to write a sentence describing that place in terms of its physical features. Have one volunteer read a sentence and have another student try to guess the location. Continue until all the major places and points have been discussed. Then have students complete Daily Quiz 17.1.

▲
Canoes are an important form of transportation along the Napo River in eastern Ecuador.
Interpreting the Visual Record What type of vegetation does this region have?

Climate and Vegetation

Climate and vegetation vary widely in Pacific South America. Some areas, such as Chile's central valley, have a mild Mediterranean climate. Other areas have dry, wet, or cold weather conditions.

Grasslands and Forests Mountain environments change with elevation. The Altiplano region between the mountain ridges is a grassland with few trees. Eastern Ecuador, eastern Peru, and northern Bolivia are part of the Amazon River basin. These areas have a humid tropical climate. South Americans call the thick tropical rain forests in this region *selvas*. In Bolivia the rain forest changes to grasslands in the southeast. Far to the south, southern Chile is covered with dense temperate rain forests. Cool, rainy weather is typical of this area of rain forests.

Deserts Northern Chile contains the Atacama Desert. This desert is about 600 miles (965 km) long. Rain is extremely rare, but fog and low clouds are common. They form when the cold **Peru Current** chills warmer air above the ocean's surface. Cloud cover keeps air near the ground from being warmed by the Sun. The area receives almost no sunshine for about six months of the year. Yet it seldom rains. As a result, coastal Chile is one of the cloudiest—and driest— places on Earth.

In Peru, rivers cut through the dry coastal region. They bring snowmelt down from the Andes. About 50 rivers cross Peru. The rivers have made settlement possible in these dry areas.

Our Amazing Planet

The Atacama Desert is one of the driest places on Earth. Some spots in the desert have not received any rain for more than 400 years. Average rainfall is less than 0.004 inches (0.01 cm) per year.

EYE ON EARTH

While in the Americas from 1799 to 1804, German scientist Alexander von Humboldt studied the Peru Current. He discovered how the current's cold water helps create the coastal desert. The current is sometimes called the Humboldt Current.

Cold water wells up from the depths of the Pacific Ocean, bringing nutrients closer to the surface. This makes the Peru Current area one of the world's richest fishing grounds. Anchovies and big fish such as tuna that feed on the smaller fish are particularly common. Seabirds that eat the anchovies drop guano on coastal islands, where it is harvested for fertilizer.

Activity: Have students create a cutaway view of the Peru Current and the ocean's surface. Drawings should show the relationship between nutrients, fish, and birds.

◄ **Visual Record Answer**

a tropical rain forest

Have students complete Main Idea Activity 17.1. Then organize the class into groups. Tell each group to trace a wall map of the region. Students should choose or create symbols, colors, or shading patterns to depict land-forms, bodies of water, climates, or resources. Encourage them to use the Unit Atlas for reference. Display maps around the classroom. **ENGLISH LANGUAGE LEARNERS, COOPERATIVE LEARNING**

Have interested students conduct research on the physical and human geography of Tierra del Fuego. Ask them to investigate how the residents have adapted to the harsh climate and isolation. Have students report their findings in a letter from a teenager in Tierra del Fuego, who is trying to per-suade a friend to visit. **BLOCK SCHEDULING**

Section Review 1

Answers

Define For definitions, see: strait, p. 362; *selvas*, p. 363; Peru Current, p. 231; El Niño, p. 364

Working with Sketch Maps Lake Titicaca is 12,500 feet (3,810 m) above sea level and covers about 3,200 square miles (8,288 sq km).

Reading for the Main Idea
1. the Andes (NGS 4)
2. weather pattern that causes heavy rains along the coast (NGS 7)

Critical Thinking
3. rain forest, mountains, deserts; changes in climate—warm and humid, cold and dry, cool and damp
4. lack of water; possible answers: collect and condense fog, bring snow down from Andes, dig deeper wells, divert more water from rivers, remove salt from seawater

Organizing What You Know
5. western—the Andes; high-land, desert; varied vegeta-tion depending on elevation, temperate rain forest in south; rivers from the Andes; central—Altiplano; dry; grass-land; Lake Titicaca and Lake Poopó; eastern—plains; humid tropical; *selvas*, grass-lands; Amazon River

▲
A Bolivian miner uses a jackhammer to mine for tin.

El Niño About every two to seven years, an ocean and weather pattern affects the dry Pacific coast. This weather pattern is called **El Niño**. Cool ocean water near the coast warms. As a result, fish leave what is normally a rich fish-ing area. Areas along the coast often suffer flooding from heavy rains. El Niño is caused by the buildup of warm water in the Pacific Ocean. Ocean and weather events around the world can be affected. Some scientists think that greenhouse gases made a long El Niño during the 1990s even worse.

✓ **READING CHECK:** *Physical Systems* How does El Niño affect Earth? Ocean water near the Pacific coast warms, driving away fish; coasts are hit by heavy rains; ocean and weather events around the world can be affected.

Resources

The countries of Pacific South America have many impor-tant natural resources. The coastal waters of the Pacific Ocean are rich in fish. Forests in southern Chile and east of the Andes in Peru and Ecuador provide lumber. In addition, the region has oil, natural gas, silver, gold, and other valuable mineral resources. Bolivia has large deposits of tin. It also has resources such as copper, lead, and zinc. Chile has large copper deposits. In fact, Chile is the world's leading producer and exporter of copper.

✓ **READING CHECK:** *Environment and Society* How has Chile's copper supply affected its economy? It has made Chile a leading producer and exporter of copper.

Homework Practice Online Keyword: SJ3 HP17

Section Review 1

Define strait, *selvas*, Peru Current, El Niño

Working with Sketch Maps On a map of South America that you draw or that your teacher pro-vides, label the following: Andes, Strait of Magellan, Tierra del Fuego, Cape Horn, Amazon River, Iquitos, Altiplano, Lake Titicaca, Lake Poopó, and Atacama Desert. Where is Lake Titicaca, and what is it like?

Reading for the Main Idea
1. (*Places and Regions*) What is the main landform region of Pacific South America?
2. (*Physical Systems*) What is El Niño? What effect does El Niño have on coastal flooding in Peru?

Critical Thinking
3. **Finding the Main Idea** What would you encounter on a journey from Iquitos, Peru, to Tierra del Fuego?
4. **Making Generalizations and Predictions** What problem would people living on the dry coast of Chile experience, and how could they solve it?

Organizing What You Know
5. **Categorizing** Copy the following graphic organ-izer. Use it to describe the landforms, climate, vegeta-tion, and sources of water of Pacific South America.

	Western	Central	Eastern
Landforms			
Climate			
Vegetation			
Water sources			

Section 2

Objectives

1. Explain some achievements of the region's early cultures.
2. Describe the Inca Empire.
3. Analyze the role Spain played in the region's history.
4. Identify governmental problems the region's people have faced.

FOCUS

LET'S GET STARTED

Copy the following question onto the chalkboard: *How might the Inca have communicated across their empire without wheeled vehicles and a written language?* Allow students to write down their answers. Summarize student responses on the chalkboard. Refer to the list as students read Section 2 and check off any suggestions that match the section's information. Tell students that in Section 2 they will learn more about the Inca and other times in the region's history.

Building Vocabulary

Write the key terms on the chalkboard. Point out that **quinoa** and **quipus** are from Quechua, a native language of the Andes region. **Viceroy** is based on a French word, *roi*, which means "king." **Creoles** is from *criollo*, a Spanish word for a person native to a certain place. French gives us **coup**, a short version of *coup d'état*, which means "stroke of state." Have students find the definitions in the textbook or glossary. Call on volunteers to relate them to the word origins.

Section 2 History and Culture

Read to Discover

1. What were some achievements of the region's early cultures?
2. What was the Inca Empire like?
3. What role did Spain play in the region's history?
4. What governmental problems have the region's people faced?

Define

quinoa
quipus
viceroy
creoles
coup

Locate

Cuzco
Machu Picchu

An ancient gold ornament from Peru

WHY IT MATTERS

Hiram Bingham of Yale University discovered the ruins of the "lost city" of Machu Picchu in 1911. Use **CNNfyi.com** or other **current events** sources to learn more about famous historical sites in Pacific South America. Record your findings in your journal.

Section 2 RESOURCES

Reproducible
- Guided Reading Strategy 17.2
- Graphic Organizer 17
- Cultures of the World Activity 2
- Lab Activity for Geography and Earth Science, Hands-On 4
- Map Activity 17

Technology
- One-Stop Planner CD–ROM, Lesson 17.2
- Homework Practice Online
- HRW Go site

Reinforcement, Review, and Assessment
- Section 2 Review, p. 369
- Daily Quiz 17.2
- Main Idea Activity 17.2
- English Audio Summary 17.2
- Spanish Audio Summary 17.2

Early Cultures

Thousands of years ago, agriculture became the basis of the region's economy. To raise crops in the steep Andes, early farmers cut terraces into the mountainsides. Peoples of the region developed crops that would be important for centuries to come. They domesticated **quinoa** (KEEN-wah), a native Andean plant that yields nutritious seeds. They also grew many varieties of potatoes. Domesticated animals included the llama (LAH-muh) and alpaca (al-PA-kuh). Both have thick wool and are related to camels. Early inhabitants raised and ate guinea pigs, which are related to mice. The people wove many fabrics from cotton and wool. These fabrics had complicated, beautiful patterns.

Peru's first advanced civilization reached its height in about 900 B.C. The main town was located in an Andean valley. This town contained large stone structures decorated with carved jaguars and other designs. Later, people in coastal areas used sophisticated irrigation systems to store water and control flooding. They also built pyramids about 100 feet (30 m) high. Huge stone carvings remain near the Bolivian shores of Lake Titicaca. They were carved by the people of the Tiahuanaco (TEE-uh-wuh-NAH-koh) culture. Another people scratched outlines of animals and other shapes into the surface of the Peruvian desert. These designs are hundreds of feet long.

✓ **READING CHECK:** *Human Systems* What were some achievements of the region's early cultures? domesticated plants and animals, wove fabrics with complicated patterns, carved stone structures, irrigated, scratched elaborate designs into the surface of the desert

This Peruvian woman is separating seeds from the quinoa plant.

▼

Teaching Objectives 1–2

LEVEL 1: (Suggested time: 20 min.) Copy the following graphic organizer onto the chalkboard, omitting the italicized answers. Pair students and have the pairs record details about the cultures' history and achievements on the chart. Students may want to add more bullets than are provided. Discuss the charts. **ENGLISH LANGUAGE LEARNERS, COOPERATIVE LEARNING**

Achievements and Characteristics of Pre-Inca and Inca Civilizations

Pre-Inca Cultures	Inca
• *in area for thousands of years* • *agriculture basis of economy* • *cut terraces into hillsides* • *domesticated quinoa, grew potatoes* • *raised llamas and alpacas* • *raised and ate guinea pigs*	• *conquered other Andean peoples by early 1500s* • *controlled area from southern Colombia to central Chile* • *skilled at stonework and metalwork* • *built irrigation projects, road system, suspension bridges* • *no wheeled vehicles, horses, or writing system* • *used quipus*

DAILY LIFE

People of the Andes domesticated the potato as early as 1,800 years ago. However, when potatoes were brought to Europe they were regarded with suspicion. In the 1500s, Sir Walter Raleigh helped prove potatoes could be eaten safely.

Today, the potato is one of the world's most important food crops. It has been a staple food for many people and contains essential nutrients, including vitamin C. In the United States people enjoy mashed potatoes, baked potatoes, French fries, potato chips, potato salad, hash browns, home fries, and other dishes using the humble tuber. Potatoes are popular in their native South America also. One may buy as many as 60 varieties of potatoes in a single Andean village market.

Activity: Have students conduct research on the spread of potatoes from South America to the rest of the globe. Students should create a world map that illustrates their findings.

Peru • Machu Picchu • Brazil • Bolivia • Chile

The ruins of Machu Picchu, an Inca city, were discovered in 1911.
Interpreting the Visual Record Why might the Inca have chosen this site for a settlement?

Quipus were used by the Inca to keep records.
Interpreting the Visual Record How might quipus have made it easier for the Inca to control their empire?

Visual Record Answer

easy to defend as a fortress city

▶

helped communicate important information

The Inca

By the early 1500s, one people ruled most of the Andes region. This group, the Inca, conquered the other cultures around them. They controlled an area reaching from what is now southern Colombia to central Chile. The Inca Empire stretched from the Pacific Coast inland to the *selvas* of the Amazon rain forest. Perhaps as many as 12 million people from dozens of different ethnic groups were included. The Inca called their empire Tawantinsuyu (tah-WAHN-tin-SOO-yoo), which means "land of the four quarters." Four highways that began in the Inca capital, Cuzco (KOO-skoh), divided the kingdom into four sections.

The Inca Empire The Inca adopted many of the skills of the people they ruled. They built structures out of large stone blocks fitted tightly together without cement. Buildings in the Andean city of Machu Picchu have survived earthquakes and the passing of centuries. Inca metalworkers created gold and silver objects, some decorated with emeralds. Artists made a garden of gold plants with silver stems. They even made gold ears for the corn plants.

Perhaps the Inca's greatest achievement was the organization of their empire. Huge irrigation projects turned deserts into rich farmland that produced food for the large population. A network of thousands of miles of stone-paved roads connected the empire. Along the highways were rest houses, temples, and storerooms. The Inca used storerooms to keep supplies of food. For example, they stored potatoes that had been freeze-dried in the cold mountain air.

To cross the steep Andean valleys, the Inca built suspension bridges of rope. The Inca had no wheeled vehicles or horses. Instead, teams of runners carried messages throughout the land. An important message could be moved up to 150 miles (241 km) in one day.

The runners did not carry letters, however, because the Inca did not have a written language. Instead, they used **quipus** (KEE-pooz). Quipus were complicated systems of knots tied on strings of various colors. Numerical information about important events, populations, animals, and grain supplies was recorded on quipus. Inca officials were trained to read the knots' meaning.

Civil War Although it was rich and efficient, the Inca Empire did not last long. When the Inca emperor died in 1525, a struggle began. Two of his sons fought over who would take his place. About seven years later, his son Atahualpa (ah-tah-WAHL-pah) won the civil war.

✓ **READING CHECK:** *Human Systems* What was the Inca's greatest achievement? the organization of the empire through irrigation and roads

LEVELS 2 AND 3: (Suggested time: 45 min.) Organize the class into groups. Provide each group with strings or cords of different colors. Have the groups create their own quipus as devices to help them remember major events in the history of Pacific South America. Students will need to assign colors to topics and devise a knotting system. Challenge them to work without writing or taking notes. After they have recorded the events on their quipus, have the students imagine that they are officials reporting to the Inca emperor. They should use their quipus to recite their histories to the class. **COOPERATIVE LEARNING**

➤**ASSIGNMENT:** Have students write a newspaper story based on an event or series of events in the history of Pacific South America. You may want to extend the activity by asking students to complete the newspaper page by adding appropriate pictures, recipes, cartoons, an advice column, advertisements, sports story, or other items ordinarily found in a newspaper.

CONNECTING TO Technology

Inca Roads

The road system was one of the greatest achievements of Inca civilization. The roads crossed high mountains, tropical rain forests, and deserts.

The main road, Capac-nan, or "royal road," connected the capitals of Cuzco and Quito. These cities were more than 1,500 miles (2,414 km) apart. This road crossed jungles, swamps, and mountains. It was straight for most of its length. Inca roads were built from precisely cut stones. One Spanish observer described how the builders worked.

❝ *The Indians who worked these stones used no mortar; they had no steel for cutting and working the stones, and no machines for transporting them; yet so skilled was their work that the joints between the stones were barely noticeable.* ❞

Another longer highway paralleled the coast and joined Capac-nan, creating a highly

Inca roads had stairways to cross steep peaks.

efficient network. Along these roads were rest houses. The Inca also built suspension bridges to span deep ravines. They built floating bridges to cross wide rivers.

When the Spanish conquered the region in the 1530s, they destroyed the road system. Today, only fragments of the Inca roads still exist.

Understanding What You Read

1. How might the road system have allowed the Inca to control their territory?
2. What were the main features of the Inca road system?

Across the Curriculum
TECHNOLOGY

Mummies and Medicine
Some of the famed Inca roads led deep into the Andes, where archaeologists have found hundreds of mummies. These mummies include the remains of children who had been buried alive as sacrifices during the Inca period. The region's dry air helped preserve the bodies.

Modern medical technology allows researchers to learn details about how these ancient people lived and died. DNA testing has proved that tuberculosis existed in the Americas 1,000 years before Europeans arrived.

Activity: Have students conduct research on the events and processes that led to the preservation of the bodies in the Andes. Then have students write short reports to present their findings.

Connecting to Technology Answers
1. They could more quickly and easily reach distant parts of the empire.
2. straight, used precisely cut stones, were connected to form a network, had rest houses, suspension bridges, and floating bridges

Spain in Pacific South America

While on his way to Cuzco to be crowned, Atahualpa met Spanish explorer Francisco Pizarro. Pizarro's small group of men and horses had recently landed on the continent's shore. Pizarro wanted Inca gold and silver.

Conquest and Revolt Pizarro captured Atahualpa, who ordered his people to fill a room with gold and silver. These riches were supposed to be a ransom for his freedom. However, Pizarro ordered the Inca emperor killed. The Spaniards continued to conquer Inca lands.

Teaching Objectives 3–4

LEVEL 1: (Suggested time: 20 min.) Have students create a time line of events in the history of Pacific South America, beginning in 1525 and ending in the present. **ENGLISH LANGUAGE LEARNERS**

LEVELS 2 AND 3: (Suggested time: 40 min.) Have students complete the Level 1 activity and ask them try to predict which events from the distant past may have contributed to governmental problems of the recent past. Have students draw arrows that connect events from the distant past to recent governmental problems. Have them include descriptions of how the event in the distant past may have contributed to unstable governments of the recent past. If some students finish their work early, have them check their predictions in resource materials and report to the class on their findings.

TEACHER TO TEACHER

Patricia Britt of Durant, Oklahoma, suggests the following activity to help students learn about the countries of Pacific South America: Organize students into groups and assign a country to each group. Provide almanacs and other resources. Have each group create a poster with the country's flag, a map, and illustrations of important aspects of the country, such as its struggle for independence, unique cultural features, and national heroes. When the groups present their information to the class, ask some students to bring samples of the country's food, find ways to dress for the occasion, or teach the class a song from the country.

National Geography Standard 17

The Uses of Geography

The location of natural resources affects history. In the late 1800s Chile wanted full use of the nitrate deposits in western Bolivia. At the time, Bolivia's territory stretched all the way to the Pacific Ocean. In 1879 Chilean forces invaded Bolivia, and the War of the Pacific began.

In 1884 Chile defeated Bolivia and Peru, which had aided Bolivia. As a result of its defeat, Bolivia had to give up Antofagasta, its Pacific port city, thus becoming landlocked. Relations between Bolivia and Chile remain strained today because Bolivia has not given up hope of regaining access to the Pacific Ocean.

Critical Thinking: Why are relations between Bolivia and Chile strained?

Answer: because in 1884 Bolivia had to give up Antofagasta to Chile, and it has not been returned

Visual Record Answer ▶

Students might suggest that the Spanish tried to replace Inca religious practices with Catholicism.

The stonework of an Inca temple in Cuzco now forms the foundation of this Catholic church.

Interpreting the Visual Record What does this photo suggest about how the Spanish viewed the religious practices of the Inca?

Simón Bolívar was a soldier and political leader in the early 1800s. He fought against the Spanish and helped win independence for several South American countries.

By 1535 the Inca Empire no longer existed. The last Inca rulers fled into the mountains. A tiny independent state in the foothills of the eastern Andes survived several decades more.

The new Spanish rulers often dealt harshly with the South American Indians. Many Indians had to work in gold and silver mines and on the Spaniards' plantations. Inca temples were replaced by Spanish-style Roman Catholic churches. A **viceroy**, or governor, appointed by the king of Spain enforced Spanish laws and customs.

From time to time, the people rebelled against their Spanish rulers. In 1780 and 1781 an Indian named Tupac Amarú II (too-PAHK ahm-AHR-oo) led a revolt. This revolt spread, but was put down quickly.

Independence By the early 1800s, the desire for independence had grown in South America. The people of Pacific South America began to break away from Spain. **Creoles**, American-born descendants of Europeans, were the main leaders of the revolts. Chile became an independent country in 1818. Ecuador achieved independence in 1822, and Peru became independent two years later. Bolivia became independent in 1825. In the 1880s Bolivia lost a war with Chile. As a result, Bolivia lost its strip of seacoast and became landlocked.

✓ **READING CHECK:** (**Human Systems**) What was Spain's role in the region's history? conquered Inca Empire, replaced temples with Catholic churches, ruled native population with Spanish laws; by the 1800s Spain had lost Pacific South America to independence movements

Government

Since gaining their independence, the countries of Pacific South America have had periods during which their governments were unstable. Often military leaders have taken control and limited citizens' rights. However, in recent decades the region's countries have moved toward more democratic forms of government. These governments now face the challenge of widespread poverty.

Acquire and play a selection of Andean flute music for the class. Ask students to comment on how Andean flute music sounds different from music they are familiar with.

REVIEW AND ASSESS

Have students complete the Section Review. Then call out terms, phrases, dates, names, and place names from Section 2. Ask students to identify if the word relates to the region's early history, the Spanish era, or its governments. Have another student describe the item's significance. Then have students complete Daily Quiz 17.2.

RETEACH

Have students complete Main Idea Activity 17.2. Then have students create outlines of the section material. **ENGLISH LANGUAGE LEARNERS**

EXTEND

Have interested students use primary and secondary sources to research a past Andean civilization such as the Chavin, Nazca, Moche, Huari, Tiahuanaco, or Inca cultures. Others may want to conduct research on some aspect of Spanish involvement in the region. Ask students to create a written report and an illustrative poster. **BLOCK SCHEDULING**

Ecuador is a democracy. Ecuador's government is working to improve housing, medical care, and literacy. Bolivia has suffered from a series of violent revolutions and military governments. It is now also a democracy. Bolivia has had one of the most stable elected governments in the region in recent years.

Peru's recent history has been particularly troubled. A terrorist group called the Sendero Luminoso, or "shining path," was active in the 1980s. The group carried out deadly guerrilla attacks. With the arrest of the group's leader in 1992, hopes for calm returned. Peru has an elected president and congress.

Chile has also recently ended a long violent period. In 1970, Chileans elected a president who had been influenced by communist ideas. A few years later he was overthrown and killed during a military **coup** (KOO). A coup is a sudden overthrow of a government by a small group of people. In the years after the coup, the military rulers tried to crush their political enemies. The military government was harsh and often violent. Thousands of people were imprisoned or killed. In the late 1980s the power of the rulers began to weaken. After more than 15 years of military rule, Chileans rejected the military dictatorship. A new, democratic government was created. Chileans now enjoy many new freedoms.

▲ Young Peruvians hold a political campaign rally.

✔ **READING CHECK:** *Human Systems* How has unlimited government been a problem for some of the nations in the region? Military dictatorships have taken over and crushed all opposition by killing and imprisoning people.

go.
hrw
.com
Homework Practice Online
Keyword: SJ3 HP17

Section Review 2

Define and explain: quinoa, quipus, viceroy, creoles, coup

Working with Sketch Maps On the map you created in Section 1, label Cuzco and Machu Picchu. Then shade in the area ruled by the Inca. What may have limited Inca expansion eastward?

Reading for the Main Idea

1. *Human Systems* How did the Inca communicate across great distances?

2. *Environment and Society* What attracted the Spanish conquerors to Pacific South America?

Critical Thinking

3. **Analyzing Information** What governmental problems have many of the region's nations experienced?

4. **Drawing Inferences and Conclusions** Why do you think many leaders of the independence movement were creoles rather than Spaniards?

Organizing What You Know

5. **Sequencing** Copy the following graphic organizer. Use it to show important events in the history of Pacific South America.

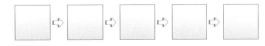

Section Review 2

Answers

Define For definitions, see: quinoa, p. 365; quipus, p. 366; viceroy, p. 368; creoles, p. 368; coup, p. 369

Working with Sketch Maps Cuzco and Machu Picchu should be labeled in their approximate locations. The shaded area should extend from southern Colombia to central Chile; the Amazon rain forest may have limited eastward expansion.

Reading for the Main Idea

1. by using swift runners, a good road system, and quipus (NGS 12)

2. gold and silver (NGS 15)

Critical Thinking

3. They have had unlimited governments with military dictatorships.

4. Creoles may have felt less loyalty to Spain, since they had not been born there.

Organizing What You Know

5. possible answer: early cultures, Inca Empire, conquest by Spain, independence movements, modern period of governmental instability

Section 3

Objectives

1. Identify the three regions of Ecuador.

2. Suggest how Bolivia might develop its economy.

3. Describe the features of Peru's regions.

4. Explain how Chile is different from its neighbors in Pacific South America.

FOCUS

LET'S GET STARTED

Copy the following question onto the chalk-board: *Is there a type of weather that you have never experienced?* Allow students to record their responses. *(Students living in southern or tropical regions might answer that they never have seen snow).* Discuss responses. Ask students if they can imagine never having seen rain. Tell them that many residents of the world's driest desert, the Atacama in Chile, have never seen rain. Tell students that in Section 3 they will learn more about Chile and its neighbors.

Building Vocabulary

Write the key terms on the chalkboard and have students read the definitions in Section 3 or the glossary. Remind students that they have already learned a word from **Quechua**, the language spoken by the Inca. What is it? *(quinoa)* Tell them that millions of people in Pacific South America speak Quechua and other native languages. Ask: How did a European language replace Quechua in much of the region? *(spread by the Spanish)* The Spanish language has given us **junta**.

Section 3 · Pacific South America Today

Read to Discover

1. **What are the three regions of Ecuador?**

2. **How might Bolivia develop its economy?**

3. **What are the features of Peru's regions?**

4. **How is Chile different from its neighbors in Pacific South America?**

Define

Quechua

junta

Locate

Guayaquil	Callao
Quito	Lima
La Paz	Santiago
Sucre	Valparaíso
Santa Cruz	

WHY IT MATTERS

Spanish explorers named the Galapagos Islands in the 1500s. Use **CNNfyi.com** or other **current events** sources to learn more about the Galapagos Islands and the environmental concerns of that area. Record your findings in your journal.

A spicy Andean stew

Ecuador Today

Many of Ecuador's people live in the coastal lowland. The country's largest city, Guayaquil (gwy-ah-KEEL), is located there. Guayaquil is Ecuador's major port and commercial center. The coastal lowland has valuable deposits of natural gas. It is also an important agricultural region. Rich fishing waters lie off the coast.

The Andean region in the heart of Ecuador is where Quito, the national capital, is located. Open-air markets and Spanish colonial buildings attract many tourists to Quito. Modern buildings surround the old city.

Large numbers of people are moving to the third region, the eastern lowlands. Here in the Amazon Basin are economically essential oil deposits.

Spanish is the official language of Ecuador. However, about 19 percent of the population speaks South American Indian languages such as **Quechua** (KE-chuh-wuh). Quechua was the language of the Inca. Ecuador's Indians are politically active and are represented in the parliament. However, many of them continue to live in terrible poverty.

Ecuador's Galápagos Islands are famous for animals, like these giant tortoises, found nowhere else in the world.

✓ **READING CHECK:** *Human Systems*
How do Ecuador's Indians participate in and influence the political process? politically active, represented in parliament

LEVEL 1: (Suggested time: 30 min.) Copy the following graphic organizer onto the chalkboard, omitting the italicized answers. Ask students to copy it into their notebooks. Pair students and have the pairs fill in the chart. Discuss any gaps in students' charts. Then lead a discussion on possible ways Bolivia might develop its economy.
ENGLISH LANGUAGE LEARNERS, COOPERATIVE LEARNING

Geographic and Economic Features of Ecuador, Bolivia, and Peru

	Ecuador	Bolivia	Peru
Regions	*coastal lowland, Andes, eastern lowlands*	*coastal plain, Altiplano, eastern plains*	*coastal plain, Andes, eastern lowlands*
Major Cities	*Guayaquil, Quito*	*La Paz, Sucre, Santa Cruz*	*Lima, Callao*
Economy	*natural gas, farming, fishing, oil*	*fertile soil, natural gas, metals*	*forests for lumber and fruit, minerals, hydroelectric power, tourism*

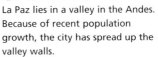

La Paz lies in a valley in the Andes. Because of recent population growth, the city has spread up the valley walls.
Interpreting the Visual Record How might the growth of La Paz affect the region's environment?

Bolivia Today

Bolivia has two capitals. La Paz is located in a valley of the Altiplano. At 12,001 feet (3,658 m), it is the highest capital in the world. It is also Bolivia's chief industrial city. Bolivia's congress meets in La Paz but the supreme court meets in Sucre (soo-kray), farther south. The country's fastest-growing region surrounds the city of Santa Cruz, east of the Andes.

In the plains of eastern Bolivia there are few roads and little money for investment. However, the region's fertile soil, adequate rainfall, and grazing land can help in its development. The country has other valuable resources, including natural gas and various metals, such as tin. However, coups and revolutions have slowed development. Bolivia remains a poor country.

Bolivia's population has the highest percentage of Indians of any South American country. Many Bolivian Indians follow customs and lifestyles that have existed for centuries. They often dress in traditional styles. Women wear full, sweeping skirts and derby hats. Men wear striped ponchos.

Bolivian music is bright and festive. Common instruments include flutes, drums, bronze gongs, and copper bells. The charango is a string instrument that resembles a small guitar. Its sound box is made from the shell of an armadillo.

Shoppers buy food at a vegetable market in La Paz.

READING CHECK: *Human Systems* What is Bolivia's music like? bright and festive with lots of instruments, including a local instrument called the charango

ENVIRONMENT AND SOCIETY

Ecotourism may provide more income for Bolivia. Madidi National Park, in the northwestern part of the country, will probably become a major tourist destination after accommodations are built.

The park covers about 4.7 million acres and it is slightly smaller than the state of New Jersey. Within its borders are glaciers, a rain forest, flat grassland, a dry forest, and a cloud forest. Because there are so many different habitats, the variety of plants and animals in the park is immense. An estimated 1,000 species of birds can be found there—about 300 more species than live in the United States and Canada combined.

Madidi's future is uncertain. The Bolivian government wants to build a dam at the park's southeastern border. Rising waters behind the dam would flood about 1,000 square miles. Bolivia does not need all the hydroelectric power the dam would generate, but it could sell the extra energy to Brazil. Construction of the dam could cost as much as $3 billion.

◀ **Visual Record Answer**

erosion of slopes and air pollution

371

LEVELS 2 AND 3: (Suggested time: 45 min.) Have students choose one of the four countries of Pacific South America. Ask them to imagine that they are running for election to the top office in their chosen country's government. Then have students write campaign speeches that demonstrate their understanding of the country's situation and that propose solutions for its remaining problems. Encourage students to consult outside reference materials for information. Call on volunteers to deliver their speeches.

Teaching Objective 4

ALL LEVELS: (Suggested time: 30 min.) Have students use the information from the previous lesson to compare and contrast Chile with its Pacific South America neighbors. Students should note political and economic differences. Have students present their information in a graphic organizer. Ask for volunteers to present their organizers to the class. **ENGLISH LANGUAGE LEARNERS**

CLOSE

Have students imagine that they have been hired by a rock band from Pacific South America to design the cover for the band's next CD. Tell students that the band's lyrics deal with political troubles, economic challenges, and cultural features specific to the region. You may want to have the students give the band a name, write song titles, or even lyrics for some of the songs.

Section Review 3

Answers

Define For definitions, see: Quechua, p. 370; junta, p. 373

Working with Sketch Maps Dense rain forest limits settlements.

Reading for the Main Idea
1. rich oil deposits (NGS 4)
2. to find jobs in industry and government (NGS 9)

Critical Thinking
3. If demand for copper declines or prices fall, there are few other ways for the country to make money.
4. Students might suggest that foreigners may have been nervous about investing.

Organizing What You Know
5. possible answers: Bolivia—two capitals, largest number of Indians, unstable governments; Chile—one of the most stable countries, advanced economy; Ecuador—population concentrated on coast, natural gas and oil deposits, fishing, Quechua, politically active Indians; Peru—political violence, important tourist industry, Quechua

Chart Answer ▶

because it has been politically unstable

Pacific South America

COUNTRY	POPULATION/ GROWTH RATE	LIFE EXPECTANCY	LITERACY RATE	PER CAPITA GDP
Bolivia	8,300,463 1.8%	62, male 67, female	83%	$2,600
Chile	15,328,467 1.1%	73, male 79, female	95%	$10,100
Ecuador	13,183,978 2%	69, male 74, female	90%	$2,900
Peru	27,483,864 1.7%	68, male 73, female	89%	$4,550
United States	281,421,906 0.9%	74, male 80, female	97%	$36,200

Sources: Central Intelligence Agency, *The World Factbook 2001;* U.S. Census Bureau

Interpreting the Chart Why is Bolivia's per capita GDP lower than those of other countries in the region?

Peru Today

Peru is making progress in its struggle against poverty and political violence. However, the government has been criticized for using harsh methods to solve political problems.

Peru, like Ecuador, has three major regions. The dense rain forests in eastern Peru provide lumber. Tropical fruit trees grow there. The Amazon River flows through this region.

The Andes highlands include the Altiplano and the heartland of what was the Inca Empire. Stone structures from the Inca period draw thousands of tourists to Machu Picchu and Cuzco. Potatoes and corn are among the crops grown in this region. Many of the people in the highlands are South American Indians. Millions of Peruvians speak Quechua.

Important mineral deposits are located near the Pacific coast. This is Peru's most modern and developed region. Hydroelectric projects on coastal rivers provide energy. The seaport city of Callao (kah-YAH-oh) serves Lima (LEE-mah), the capital, a few miles inland. Nearly one third of all Peruvians live in these two cities. Callao is Peru's leading fishing and trade center. Industry and government jobs draw many people from the countryside to Lima.

✓ **READING CHECK:** *Human Systems* What draws people to Callao and Lima? fishing and trading, jobs in industry and government

Tourists explore icy landscapes in southern Chile. ▼

Have students complete the Section Review. Then ask each student to write a "What Am I?" question based on the Section 3 material. *(Example: "I am the language spoken by the Inca. What am I?" Answer: Quechua)* Call on volunteers for their questions. Then have students complete Daily Quiz 17.3.

RETEACH

Have students complete Main Idea Activity 17.3. Then organize students into groups. Ask students to imagine that their group represents an American company interested in investing several million dollars in one country in Pacific South America. Each group should select the best country for its investment, choose business categories, and identify factors that may threaten the investment.
ENGLISH LANGUAGE LEARNERS, COOPERATIVE LEARNING

EXTEND

Have students conduct research on one of the nature preserves of Pacific South America, such as Manu National Park in Peru or Cayambe-Coca Ecological Reserve in Ecuador. Ask them to highlight how protecting the preserve or park fits into the country's plans for development. You might have students report their findings in the form of a proposal to the commerce or interior department of the country's government. **BLOCK SCHEDULING**

Chile Today

In the late 1980s Chile ended the rule of a **junta** (HOOHN-tuh) and became a democracy. A junta is a small group of military officers who rule a country after seizing power. Chile is now also one of the most stable countries in South America. Its economy is also one of the most advanced. Chile's prospects for the future seem bright.

Chile's economy is based on mining, fishing, forestry, and farming. Copper mining is especially important. It accounts for more than one third of the country's exports. In fact, Chile has the world's largest open-pit copper mine. It is located in the Atacama Desert near the town of Chuquicamata.

About one third of all Chileans live in Central Chile. It includes the capital, Santiago, and its nearby seaport, Valparaíso (bahl-pah-rah-EE-soh). The mild Mediterranean climate allows farmers to grow a wide range of crops. Grapes grow well there, and wine is exported around the world. Cool, mountainous southern Chile has forests, oil, and farms. Few people live there, however. Northern Chile includes the Atacama Desert. Croplands along valleys there are irrigated by streams flowing down from the Andes.

Although poverty remains a problem, Chile's economy is becoming stronger. Small businesses and factories are growing quickly. More Chileans are finding work, and wages are rising. Chile hopes that the price of its main export, copper, remains high. Chile's economy would suffer if the world price of copper fell.

Chile wants to expand its trade links with the United States. Some people have suggested that Chile join the North American Free Trade Agreement (NAFTA). This free-trade group currently includes Canada, the United States, and Mexico.

✓ **READING CHECK:** *Environment and Society* How does copper affect Chile's economy? *It supports it; it could also hurt it if the price for copper fell.*

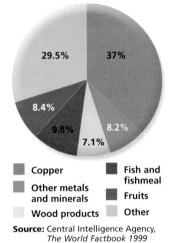

Chile's Export Products

29.5% • 37% • 8.4% • 9.8% • 7.1% • 8.2%

- ■ Copper
- ■ Other metals and minerals
- □ Wood products
- ■ Fish and fishmeal
- ■ Fruits
- □ Other

Source: Central Intelligence Agency, *The World Factbook 1999*

Interpreting the Chart **What percentage of Chile's export products is metals and minerals?**

go.hrw.com
Homework Practice Online
Keyword: SJ3 HP17

Section Review 3

Define and explain: Quechua, junta

Working with Sketch Maps On the map you created in Section 2, label Guayaquil, Quito, La Paz, Sucre, Santa Cruz, Callao, Lima, Santiago, and Valparaíso. Why are there so few cities in the eastern part of the region?

Reading for the Main Idea

1. *Places and Regions* What features are drawing people to Ecuador's eastern lowlands?

2. *Human Systems* Why do many people move from the countryside to Lima, Peru?

Critical Thinking

3. **Drawing Inferences and Conclusions** What might be a disadvantage of Chile depending so heavily on its copper industry?

4. **Analyzing Information** Why have changes in Bolivia's government slowed development?

Organizing What You Know

5. **Summarizing** Copy the following graphic organizer. Use it to write phrases that describe each country.

Bolivia	
Chile	
Ecuador	
Peru	

Review ANSWERS

Building Vocabulary
For definitions, see: strait, p. 362; *selvas*, p. 363; El Niño, p. 364; quinoa, p. 365; quipus, p. 366; viceroy, p. 368; creoles, p. 368; coup, p. 369; Quechua, p. 370; junta, p. 373

Reviewing the Main Ideas

1. broad, high-plateau plain between Andes ridges; grassland with few trees (NGS 4)

2. cool ocean water near the coast warms, fish leave, heavy rains, flooding (NGS 7)

3. irrigation projects, road system, suspension bridges (NGS 12)

4. captured and killed Atahualpa, conquered Inca lands (NGS 13)

5. unstable governments that limited citizens' rights (NGS 13)

6. northern—Atacama Desert, copper mining, irrigated farms; central—heartland, dense population, major cities, mild climate, range of crops; southern—forests, oil, farms, few people (NGS 5)

◄ **Chart Answer**

45.2

Pair students and assign each pair one of the chapter's topics. Have each pair write the main points of the assigned topic on a transparency. Then have pairs teach the material to the class, using the transparency and an overhead projector. Students may write the main points on the chalkboard if no overhead projector is available.

ENGLISH LANGUAGE LEARNERS, COOPERATIVE LEARNING

PORTFOLIO EXTENSIONS

1. Inca builders created some exquisite stone architecture. Have students create a guidebook for several of the best sites for viewing Inca stonework. The guidebook should tell what structures are at the site and what the buildings were used for, if known. Students should draw or duplicate illustrations of the sites. You may also want to require a map. Some students may want to further investigate Inca stonecutting and construction methods.

2. Have students build scale models of the Andes region from papier mâché or flour-salt dough. Models should show how the mountains extend to South America's tip at a lower altitude, forming the islands of southern Chile. Place photos of models in portfolios.

CHAPTER 17 Review ANSWERS

Understanding Environment and Society

Presentations will vary, but information included should be consistent with text. Students should conclude that any drop in tin production will cause a reduction in Potosí's population. Use Rubric 29, Presentations, to evaluate student work.

Thinking Critically

1. because the region was conquered by Spaniards in the 1500s

2. Students might suggest that the descendants of Spanish rulers, creoles, and Indians have different interests and loyalties. Also, people in different areas within the region have different interests because they have different ways of life.

3. might make it poorer by limiting trade possibilities because all trade would have to be overland

4. They are politically active and represented in the parliament; answers will vary but should mention the importance of voluntary civic participation.

CHAPTER 17 Reviewing What You Know

Building Vocabulary

On a separate sheet of paper, write sentences to define each of the following words.

1. strait
2. *selvas*
3. El Niño
4. quinoa
5. quipus
6. viceroy
7. creoles
8. coup
9. Quechua
10. junta

Reviewing the Main Ideas

1. *(Places and Regions)* What is the Altiplano, and what kind of vegetation is found there?

2. *(Physical Systems)* What are some of the effects of El Niño?

3. *(Human Systems)* What are three types of construction projects that helped the Inca organize their empire?

4. *(Human Systems)* What happened when the Spaniards arrived in Pacific South America?

5. *(Human Systems)* What trend have the countries of this region experienced in their forms of government?

6. *(Places and Regions)* How are northern, central, and southern Chile different?

Understanding Environment and Society

Population and Industry
Potosí, Bolivia, has long been a major mining center in the Americas. Create a presentation, including a map, graph, chart, model, or database, on aspects of Potosís mining industry. Consider the following:

- Where tin is mined in the area.
- How much tin Potosí produces each year.
- How a drop in tin production might affect Potosí.

Include a five-question quiz on your presentation to challenge fellow students.

Thinking Critically

1. **Analyzing Information** Why is Spanish the most widely spoken language in Pacific South America?

2. **Drawing Inferences and Conclusions** Why do you think the region's countries have had so many unstable governments since gaining their independence?

3. **Finding the Main Idea** How might Bolivia's landlocked position affect the country's economy?

4. **Analyzing Information** What role do Ecuador's Indians play in its political system? Identify and explain the importance of this role.

FOOD FESTIVAL

Quinoa is high in protein and low in unsaturated fats. It can be purchased at health food stores, some supermarkets, and from mail-order companies. Before using, the quinoa grains must be washed thoroughly in five changes of fresh cold water to remove their bitter coating. The grain can then be steamed or boiled. Quinoa can be substituted for rice in many recipes. Have students experiment with substituting quinoa for rice in familiar dishes. You might also challenge them to find quinoa recipes and prepare them for the class.

CHAPTER 17 REVIEW AND ASSESSMENT RESOURCES

Reproducible
◆ Readings in World Geography, History, and Culture 25, 26, and 27
◆ Critical Thinking Activity 17
◆ Vocabulary Activity 17

Technology
◆ Chapter 17 Test Generator (on the One-Stop Planner)

◆ Audio CD Program, Chapter 17
◆ HRW Go site

Reinforcement, Review, and Assessment
◆ Chapter 17 Review, pp. 374–75
◆ Chapter 17 Tutorial for Students, Parents, Mentors, and Peers
◆ Chapter 17 Test
◆ Chapter 17 Test for English Language Learners and Special-Needs Students
◆ Unit 4 Test
◆ Unit 4 Test for English Language Learners and Special-Needs Students

Building Social Studies Skills

Map ACTIVITY

On a separate sheet of paper, match the letters on the map with their correct labels.

Peru Current
Strait of Magellan
Tierra del Fuego
Amazon River
Lake Poopó
Atacama Desert
Quito
La Paz
Lima
Santiago

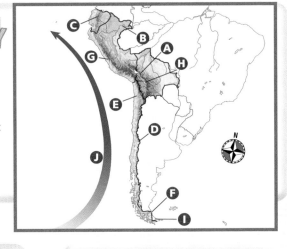

Mental Mapping Skills ACTIVITY

On a separate sheet of paper, draw a freehand map of Pacific South America. Make a key for your map and label the following:

Altiplano
Andes
Bolivia
Chile
Ecuador
Peru

WRITING ACTIVITY

Imagine that you are making a film about Ecuador, Bolivia, Peru, or Chile. Write a summary of the film you would like to make. List land features, industries, peoples, and customs that you want to cover. Explain why you think these topics would be interesting to audiences. Be sure to use standard grammar, spelling, sentence structure, and punctuation.

Map Activity

A. La Paz
B. Amazon River
C. Quito
D. Santiago
E. Atacama Desert
F. Strait of Magellan
G. Lima
H. Lake Poopó
I. Tierra del Fuego
J. Peru Current

Mental Mapping Skills Activity

Maps will vary, but listed places should be labeled in their approximate locations.

Writing Activity

Information included should be consistent with text material. Check that students discuss physical features, industries, peoples, and customs and explain why the chosen topics would be interesting to audiences. Use Rubric 43, Writing to Persuade, to evaluate student work.

Portfolio Activity

Answers will vary, but information included should be consistent with text material. Percentages on pie charts should total 100 percent. Use Rubric 7, Charts, to evaluate student work.

Alternative Assessment

Portfolio ACTIVITY

Learning About Your Local Geography

Languages Most people in Pacific South America speak Spanish. Research languages in your area. Make a pie chart showing the percentage of people who speak each language.

internet connect

Internet Activity: go.hrw.com
KEYWORD: SJ3 GT17

Choose a topic to explore Pacific South America:
• Analyze Chile's climate.
• Hike the Inca Trail and visit Machu Picchu.
• Learn about Andean languages.

internet connect

GO TO: go.hrw.com
KEYWORD: SJ3 Teacher
FOR: a guide to using the Internet in your classroom

Relating Geography and Agriculture

Have students conduct an experiment over a span of several days to determine the relative salt tolerance of common plants. Ask students to bring in samples of small house plants, weeds, grasses, vegetables, and other plants easily acquired in your area. All plants should be in pots with adequate soil. If possible, place them in pots that are a similar size. Have students create artificial seawater by adding a small amount of table salt to a large pitcher of water. Then have them water all the plants equally. Ask them to decide how much salt to add each day and to increase the salt concentration in the water by that much each day. Then have them give all the plants the same amount of water. Ask students to keep a log of which plants live the longest and therefore have the highest salt tolerance of the group. You may want to extend the activity by having students contact your state's agriculture department for information about any salt-tolerant crops in your region.

Civilization might have developed differently if glasswort had been available to the people of Sumer. Evidence shows that a buildup of salt in Mesopotamia's soils contributed to the civilization's decline.

Salinization is a result of careless irrigation. During its early periods, Sumer controlled salinization by limiting irrigation and allowing land to lie fallow every other year. As the growing population required more food, however, these practices ended.

At one point, the city of Umma blocked the canals that fed water to territory it disputed with the city of Girsu. In retaliation, Girsu's king dug a new canal. Although the canal brought in much-needed water, it caused the water table to rise, bringing accumulated salts into the soil.

Sumerian farmers switched from wheat to salt-tolerant barley. Crop yields dropped to one third of their earlier levels. The effect on the region's cities must have been devastating.

Critical Thinking: How did the Girsu king's canal affect the environment?

Answer: It caused the water table to rise, bringing salts into the soil.

➤ **This Focus On Environment feature addresses National Geography Standards 14, 15, and 16.**

Seawater Agriculture

Water is one of the most important resources on Earth. All life needs water to survive. As the world's population grows, more and more places are facing water shortages. For example, freshwater is scarce in Sonora and Baja California, Mexico. People have used up most of the groundwater in the region for agriculture. The water table is so low that many wells can no longer be used.

About 97 percent of Earth's water is in oceans. However, seawater is too salty for people or animals to drink. Most plants will die quickly if they are exposed to seawater. Desalinization machines can make seawater more useful. These machines take the salt out of seawater. However, this process is expensive. Many countries cannot afford it.

Growing Plants in the Desert Some people want to grow salt-tolerant plants in deserts and irrigate them with seawater. Finding crops that can grow in these conditions would allow people to grow more food. Such crops would also ease pressure on freshwater supplies.

Using desert land could also help the environment. Farming in deserts might keep people from clearing forests. The demand for farmland is so great in some places that using the desert could make a difference. In the coming years, new farmland will be needed to feed the world's growing population.

Saltbush, shown here in California's Owens Valley, is one type of salt-tolerant plant that can be used to feed livestock

▼

Going Further: Thinking Critically

Point out Saudi Arabia on a map and tell students that much of the country is desert. Add that in Saudi Arabia, the wealth from the nation's oil resources helps fund expensive desalinization plants along the Persian Gulf. These factories remove salt from seawater, turning it into freshwater. The process is generally very expensive, as desalinization plants require a great deal of energy. However, Saudi Arabia uses its own oil to run the plants. Have students compare and contrast desalinization and seawater agriculture as solutions to the growing freshwater shortage worldwide. In particular, students should conduct research on the use of these methods in Saudi Arabia and the Sonoran Desert of Mexico. Have students prepare a paragraph concluding which approach best suits each region. The paragraph should contain reasoning for the students' conclusions.

Many plants do grow along seacoasts and in deserts. Mangrove trees, for example, grow in shallow coastal waters. Some researchers have tried growing various salt-tolerant plants in the coastal deserts of Sonora, Mexico. They were looking for salt-tolerant plants that could be used to feed livestock or people. Three plants—glasswort (*Salicornia*), sea blite (*Suaeda*), and saltbush (*Atriplex*), proved useful for feeding livestock.

A Perfect Crop? Glasswort may be the most promising salt-tolerant plant. Fresh glasswort is a vegetable that can be eaten. Known as "sea asparagus," it is an expensive delicacy in Europe. Glasswort also produces seeds that are high in protein and oil. The oil can be used for cooking. It is similar to safflower oil and has a nutty taste. After the oil is removed, seed meal is left. Seed meal can be used as a protein supplement for livestock. Glasswort's usefulness could greatly help Mexico. For example, Mexico imports seeds used to make cooking oil. Using glasswort seeds grown in Mexico would help the economy and create jobs.

Glasswort is a productive plant. It can be as productive as soybeans or other crops. Growing glasswort does require more water than most crops grown with freshwater. However, irrigating a field with seawater in Sonora cost about one third what it would cost to use freshwater.

Glasswort requires frequent watering because it absorbs water slowly. The water that stays in the ground becomes much saltier than seawater. As a result, the plants must be watered frequently. In sandy desert soils the water quickly drains back to the sea. Therefore, salts do not build up in the root zone. Salt buildup in the soil affects some 20 percent of the world's fields irrigated with freshwater. For example, this has happened in California's Central Valley. When salt buildup occurs, drainage systems must be built. These systems are very expensive.

Growing glasswort can help the environment in another way. Glasswort fields can recycle wastewater from coastal shrimp farms. This

▲
Glasswort grows in shallow, salty water. Research has shown that glasswort can be irrigated with seawater and used to feed both people and animals.

wastewater is high in nutrients. If it is released directly into the ocean, it can cause algae blooms or disease. However, if this water is used to irrigate glasswort fields, the plants absorb many of the nutrients. The water can then be drained back into the ocean.

Researchers hope their work in Mexico will help people in other dry regions. However, it is still too early to tell whether it will be profitable to grow glasswort on a large scale.

Understanding What You Read

1. Why do people want to grow crops with seawater?

2. What are some uses of the glasswort plant?

Understanding What You Read

Answers

1. People want to grow crops with seawater because as the world's populations grow, more and more places are facing shortages of freshwater.

2. The glasswort plant can be eaten as a vegetable. It produces seeds that are high in protein and oil. The oil can be used for cooking. The seed meal that remains after the oil is removed can be used as a protein supplement for livestock.

Going Further: Thinking Critically

Prepare in advance slips of paper on which you have written the names of countries that have experienced heavy migration, either into, out of, or from one region to another within the country. Here are some possibilities: Canada, Ireland, Mauritania, Mexico, Brazil, Thailand, and Vietnam. You may want to omit the United States because the wide availability of information could put other groups at a disadvantage.

Organize the class into as many groups as you have countries and have a representative of each group draw a country name from a hat. Have the students work together to investigate their chosen country's migration patterns. Statistics are available from the *CIA World Factbook* online. Newspapers and newsmagazines can provide detail. Ask students to concentrate on the concept of push and pull—what social, environmental, or economic factors pull or push the residents into or out of the place.

When their research is complete, have students write and perform skits to illustrate some of the issues they have investigated. Their skits may take place in the country where new residents have arrived or the place they have left. After the performances, lead a class discussion on what students have learned from the activity.

PRACTICING THE SKILL

1. Students might suggest the following: push—wars, discrimination, unemployment, limited resources, natural disasters, famine, forced military service; pull—jobs or higher salary, greater political freedom, freedom from war, plentiful resources, or family members already living in the other place.

2. If possible, students should discuss their mother's and father's ancestors. Students may also research a friend's family. Students' ancestors may have come to the United States for any of the reasons listed in the answer to the previous question or for other reasons. Some students may mention that their ancestors were brought here against their will as slaves.

3. Students may suggest any place in the world and describe their emotions about migrating.

➤ **This GeoSkills feature addresses National Geography Standards 6, 9, 10, 12, and 15.**

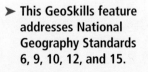

Many people from Japan migrated to the area around São Paulo, Brazil in the 1900s.

Building Skills for Life: Understanding Migration Patterns

People have always moved to different places. This movement is called migration. Understanding migration patterns is very important in geography. These patterns help explain why certain places are the way they are today. In South America, for example, there are people whose ancestors came from Africa, Asia, Europe, and North America. All of these people migrated to South America sometime in the past.

Why do people migrate? There are many different reasons. Sometimes, people do not want to move, but they are forced to. This is called forced migration. Other times, people migrate because they are looking for a better life. This is an example of voluntary migration.

Geographers who study migration patterns often use the words *push* and *pull*. They identify situations that push people out of places. For example, wars often push people away. They also identify situations that pull people to new places. A better job might pull someone to a

new place. Usually, people migrate for a combination of reasons. They might be pushed from a place because there is not enough farmland. They might also be pulled to a new place by good farmland.

Geographers are also interested in barriers to migration. Barriers make it harder for people to migrate. There are cultural barriers, economic barriers, physical barriers, and political barriers. For example, deserts, mountains, and oceans can make it harder for people to migrate. Unfamiliar languages and ways of life can also block migration.

Migration changes people and places. Both the places that people leave and the places where they arrive are changed. Can you think of some ways that migration has changed the world?

THE SKILL

1. List some factors that can push people out of a place or pull them to a new place.

2. Research the migration of your ancestors or a friend's ancestors. Where did they come from? When did they migrate? Why?

3. Imagine you had to migrate to a new place. Where would you go? Why would you pick this place? Do you think you would be scared, excited, or both?

HANDS on GEOGRAPHY

The following passage was written by a woman from Argentina who migrated to the United States. Read the passage and then answer the Lab Report questions.

My nephew and I came from a faraway country called Argentina. When I mention this country to others, they say, "Oh, Argentina! It is so beautiful!" However, as beautiful as Argentina is, life there is very difficult now.

In Argentina, there are three social classes: rich, middle class, and poor. The middle class, of which I am a member, is the largest. We are the workers and the businesspeople. Because of bad economic decisions and corrupt government, the middle class has almost disappeared. Factories have closed, and people have no work. The big companies move to other countries. The small businesses depend on the big companies and often have to close. Argentina has a very high rate of unemployment. Many people are hungry.

My husband and I had some friends who worked in Argentina. They advised us to sell everything we owned and move to Houston.

They told us that life was better there and they had relatives who would help us find a house to rent. They told us a lawyer would help us put our papers in order so we could work. I had a small sewing shop with some sewing machines, which I sold so I could travel.

▲
Poverty and unemployment are problems in some parts of Argentina.
Interpreting the Visual Record **What evidence can you see of poor living conditions?**

Lab Report

1. What pushed the author out of Argentina?
2. Why did she choose Houston as her destination?
3. According to the author, what do other people think Argentina is like? does she agree?

EUROPE

Direct students' attention to the photographs on these pages. Tell them that instead of streets Venice, Italy, has canals. The city's only paved streets are narrow and for pedestrians only. Venice was built on a group of islands in a lagoon. You may want to read the sidebar feature about Venice's gondolas in the chapter on southern Europe. Ask students to speculate about all the ways their lives would be different if the major thoroughfares of their community were canals instead of streets.

Remind students that various forces of erosion shape the land. Ask how the valley in the small photo may have been created *(by a glacier)*. Tell students that in the chapter on northern Europe they will learn more about fjords, like the one pictured on this page, and how they were formed.

Direct students' attention to the photograph of the swan. Point out that Europe is densely populated and that very few large wild animals live there except in nature preserves.

Finally, refer to the photo taken in Prague. The writer Franz Kafka lived on the street pictured. The house in the center is tiny, and visitors must stoop to enter it.

UNIT OBJECTIVES

1. Describe the major landforms, bodies of water, climates, and resources of Europe.

2. Explain the historical development of the European nations and the influence of European cultures on other parts of the world.

3. Identify different cultural groups in Europe and understand how they interact with one another.

4. Discuss Europe's influence in international affairs.

5. Interpret special-purpose maps to analyze relationships among climates, population patterns, and economic resources in Europe.

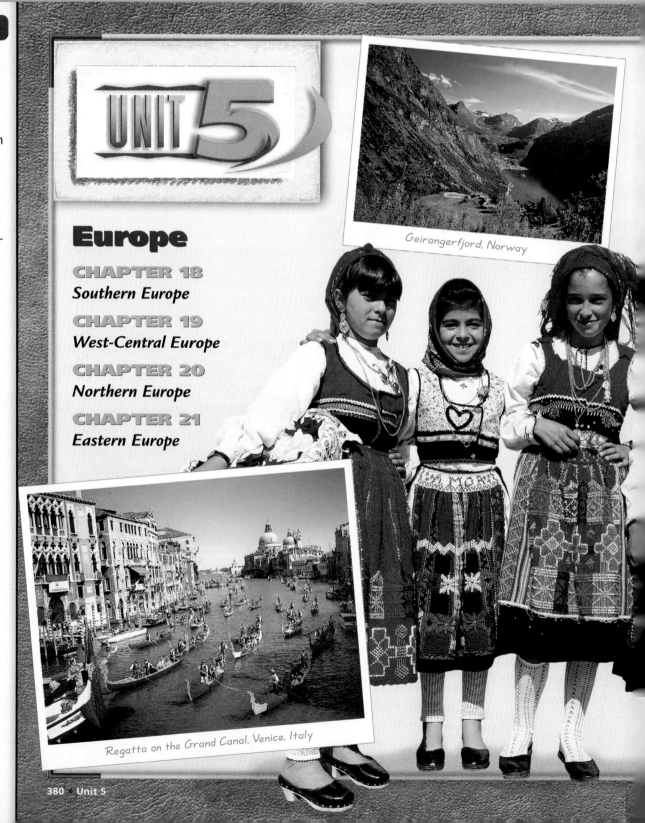

Europe

Geirangerfjord, Norway

Regatta on the Grand Canal, Venice, Italy

380

Notes from the field

A Professor in the Czech Republic

Meredith Walker teaches English at Clemson University. She also teaches English As a Second Language courses. Here she describes a visit to Prague, the capital of the Czech Republic. **WHAT DO YOU THINK?** *Does Prague sound like a city you would like to see?*

The castle in Prague sits on a hill high above the Vltava River. This river divides the city. A castle has stood on that hillside for 1,000 years. Rising from the inner courtyard of the castle is a huge medieval building, St. Vitus' Cathedral. Together, the castle and cathedral look almost magical, especially at night when spotlights shine on them. The castle is the most important symbol of the city. A great Czech writer, Franz Kafka, believed that the castle influenced everything and everyone in the city.

Another important landmark in Prague is Charles Bridge, one of eight bridges that cross the Vltava River. The bridge is part of the route that kings once traveled on their way to the castle to be crowned. Today, large crowds of tourists walk there. Many stop to photograph some of the 22 statues that line the bridge. No cars are allowed on the bridge today.

The first time I walked across the bridge, heading up toward the castle, it was at night. Fireworks were lighting the sky all around me. It was a colorful welcome.

Street scene in Prague, Czech Republic

Portuguese girls in native costume

Whooper swan

Understanding Primary Sources

1. How does the city's physical geography help make the castle a symbol of Prague?

2. How did the fireworks display affect Meredith Walker's feelings about Prague?

MORE FROM THE FIELD

Events in Prague have given the world an unusual word: *defenestration*, which means "a throwing of a person or thing out of a window." The word is taken from the Latin word *fenestra*, which means "window." In 1419, religious reformers defenestrated the entire town council. A second case was part of a Protestant rebellion. Three Catholic officials were thrown from a window in Prague's Hradčany Castle. The Second Defenestration of Prague, as the episode was called, led directly to the Thirty Years' War (1618–48).

In the aftermath of the Communist takeover of 1948, Jan Masaryk, the non-Communist foreign minister, committed suicide or was assassinated in a fall from his office window. Controversy arose after poet and screenwriter Bohumil Hrabal fell to his death from the window of his fifth-floor hospital room in 1997.

Activity: Have students conduct research on how certain buildings in Eastern Europe relate to the local or regional history of the area in which they are located.

Understanding Primary Sources

Answers

1. It is on a hill.

2. It made her feel more welcome.

OVERVIEW

In this unit students will learn about the landscape and cultures of one of the world's most densely settled continents.

A wide range of climates, from steppe to ice-cap, has allowed Europeans to use the land in many ways. In general, the farms are so productive that the majority of the population can live in cities and towns and work in manufacturing and service jobs.

Europe's past is dynamic and turbulent. Countless leaders have gone into battle to control its resources. Explorers, conquerors, and colonists spread out from Europe and have influenced most parts of the world. The continent was at the center of two world wars. Ethnic conflicts continue to trouble postwar Europe. Recently created republics struggle to find their place among the other countries.

Over the centuries, European art, architecture, music, literature, and drama have been valued by people around the world. The region's international influence has grown since many of the countries have joined the European Union.

PEOPLE IN THE PROFILE

Note that the elevation profile crosses the Great Hungarian Plain. The nation of Hungary dates back to when the nomadic Magyar people, whose ancestors came from central Russia, moved into the sparsely populated plain during the A.D. 890s.

The Magyars spent much of their time on horseback. They lived mainly on meat, mare's milk, and fish. The Magyars were organized into clans, which in turn banded together into tribes.

Until the mid-900s Magyars raided far into western Europe. They were skilled with their favorite weapon, the bow and arrow, and their horses were fast. The Magyars captured slaves and stole treasure until Otto the Great of Germany stopped them. According to tradition, they adopted Christianity on Christmas Day, 1000, when, with the pope's approval, the Magyar king, Stephen I, was crowned.

Today, most Hungarians still call themselves Magyars. There are large Magyar minorities in other countries, such as Romania. In fact, perhaps half of the world's Magyars live in countries other than Hungary.

Critical Thinking: Where did the Magyar people originate?

Answer: central Russia

Europe

Elevation Profile

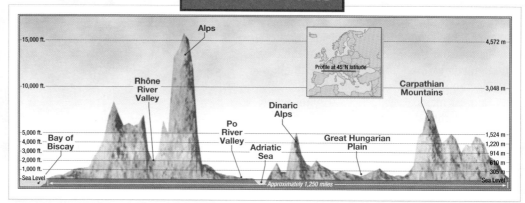

The United States and Europe: Comparing Sizes

GEOSTATS:

World's largest island: Greenland—839,999 sq. mi. (2,175,597 sq km)

World's smallest independent country: Vatican City—0.17 sq. mi. (0.44 sq km)

World's northernmost town: Longyearbyen, Spitsbergen Island, Norway—about 77.5°N

Direct students' attention to the **physical map** on this page. Ask a volunteer to name the highest physical feature on the map and its location *(Mont Blanc, on France's border with Italy and Switzerland)*. Then ask them to identify the area that is below sea level *(large area in the Netherlands)*. Have students compare this map to the **population map** and have them suggest a problem that people of the Netherlands had to solve before the country could support a large population *(how to keep the North Sea from flooding the land)*.

Ask students to suggest areas where no people live *(southeast Iceland, most of Greenland)*. Then ask how they could guess this without looking at the population map *(presence of ice caps)*.

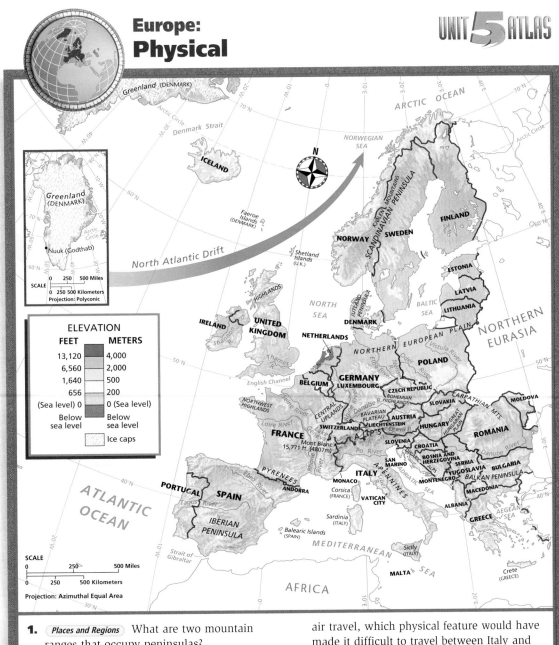

Europe: Physical

UNIT 5 ATLAS

1. **(Places and Regions)** What are two mountain ranges that occupy peninsulas?
2. **(Places and Regions)** What are the two major plains of Europe? Which is larger?

Critical Thinking

3. **Analyzing Information** In the days before air travel, which physical feature would have made it difficult to travel between Italy and the countries to its north?

4. **Analyzing Information** Which physical feature would have made travel between Greece, Italy, and Spain fairly easy?

Physical Map

Answers

1. Apennines and Kjølen Mountains

2. Northern European Plain and Great Hungarian Plain; Northern European Plain

Critical Thinking

3. Alps

4. Mediterranean Sea

383

USING THE POLITICAL MAP

Tell the class that Europe is often called a peninsula of peninsulas. On a wall map, trace the outline of Europe so that students can observe how the entire continent is a peninsula of the Eurasian land mass. Then ask them to look at the **political map** on this page and to identify countries or pairs of countries that are peninsulas. *(Examples: Spain and Portugal, Italy, Denmark, Norway and Sweden, Greece)*

Also ask students to identify the largest country in Europe and those that are so small that their area is not apparent on the map *(France; Andorra, Liechtenstein, Malta, Monaco, San Marino, Vatican City)*. Then ask which country would be the largest if territory under its control were included in its area *(Denmark, because it controls Greenland)*. Point out that most of the large islands and island groups in the Mediterranean Sea are not independent countries. Call on volunteers to identify the countries to which the islands belong *(Balearic Islands, Spain; Corsica, France; Sardinia and Sicily, Italy; Crete, Greece)*.

Your Classroom Time Line

These are the major dates and time periods for this unit. Have students enter them on the time line you created earlier. You may want to watch for these dates as students progress through the unit.

c.* 16,000 B.C. Early peoples create cave paintings at Altamira, Spain.

c. 2000 B.C. Complex civilization exists on Crete.

c. 800 B.C. City-states organize on Greek mainland.

c. 750 B.C. The Latins establish Rome.

c. 750–450 B.C. The Celts come to the British Isles.

c. 400s B.C. Construction of the Parthenon begins.

c. 300 B.C. Euclid writes *Elements*.

*c. stands for *circa* and means "about."

Political Map

Answers

1. Spain, France

2. Pyrenees

Critical Thinking

3. Bosnia and Herzegovina, Slovenia

4. It is separated from other countries by the English Channel, the Atlantic Ocean, and the North Sea.

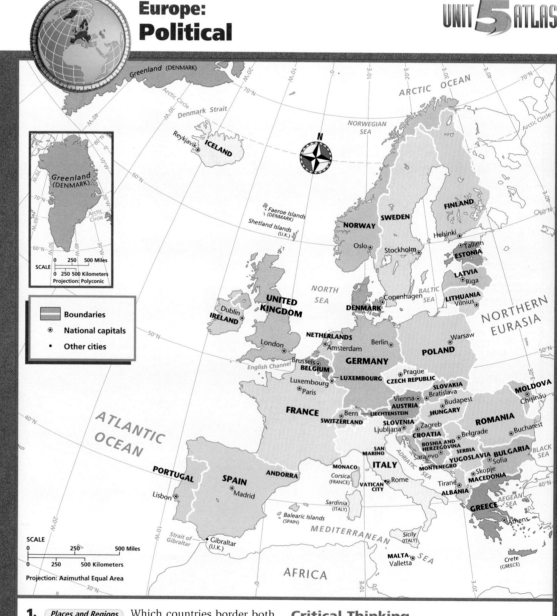

Europe: **Political**

1. **Places and Regions** Which countries border both the Atlantic and the Mediterranean?

2. **Places and Regions** Compare this map to the **physical map** of the region. Which physical feature helps form the boundary between Spain and France?

Critical Thinking

3. **Comparing** Which countries have the shortest coastlines on the Adriatic Sea?

4. **Drawing Inferences and Conclusions** The United Kingdom has not been invaded successfully since A.D. 1066. Why?

USING THE CLIMATE MAP

Direct students' attention to the **climate map** on this page. Ask one student to call out the names of the individual countries that make up Europe and have other students name the main climate types in each country. Have students compare this map to the **political map** to answer these questions: Which island countries have only a marine west coast climate? *(Ireland, United Kingdom)* Which Atlantic coast country has only a Mediterranean climate? *(Portugal)* Which large Eastern European country is divided about equally between marine west coast and humid continental climate regions? *(Poland)*

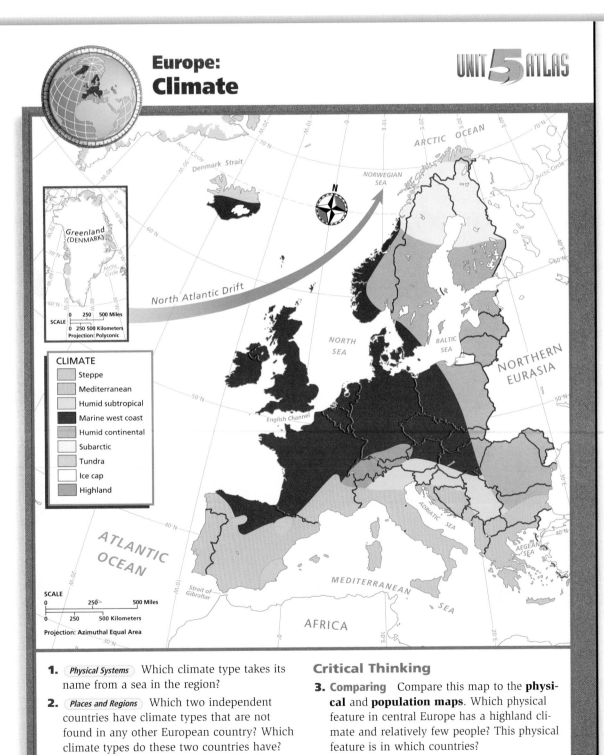

Europe: Climate

UNIT 5 ATLAS

CLIMATE
- Steppe
- Mediterranean
- Humid subtropical
- Marine west coast
- Humid continental
- Subarctic
- Tundra
- Ice cap
- Highland

Greenland (DENMARK)

SCALE
0   250   500 Miles
0   250 500 Kilometers
Projection: Polyconic

North Atlantic Drift

ATLANTIC OCEAN

SCALE
0   250   500 Miles
0   250   500 Kilometers
Projection: Azimuthal Equal Area

1. **Physical Systems** Which climate type takes its name from a sea in the region?

2. **Places and Regions** Which two independent countries have climate types that are not found in any other European country? Which climate types do these two countries have?

Critical Thinking

3. **Comparing** Compare this map to the **physical** and **population maps**. Which physical feature in central Europe has a highland climate and relatively few people? This physical feature is in which countries?

Your Classroom Time Line (continued)

200s–100s B.C. Rome begins to expand.

A.D. **100s** The Roman Empire is at its greatest extent.

early 300s Constantine adopts Christianity.

400s The Roman Empire is divided into two parts.

400s Angles and Saxons migrate to the British Isles.

400s Huns invade eastern Europe.

476 Rome falls to invaders.

early 700s Moors conquer the Iberian Peninsula.

800 Charlemagne is crowned Emperor of the Romans.

1066 The duke of Normandy invades the British Isles.

1100s England conquers Ireland.

1200s The Mongols invade Hungary.

1300s The Renaissance begins in Italy.

Climate Map

Answers

1. Mediterranean

2. Iceland and Spain; tundra, marine west coast, ice cap; marine west coast, Mediterranean, steppe

Critical Thinking

3. Alps; France, Italy, Germany, Austria, Switzerland

385

USING THE POPULATION MAP

As students examine the **population map** on this page, point out that many European cities are located on or near rivers. Have students compare this map to the **physical map** to identify some of these cities and rivers. Students may need to use the chapter maps to confirm detail. *(Possible answers: London—Thames, Paris—Seine, Rome—Tiber, Rotterdam—Rhine, Budapest—Danube)*

Ask students to identify the countries with the largest areas of high population density *(United Kingdom, Netherlands, Belgium, Germany, Italy).*

Your Classroom Time Line (continued)

1337–1453 England and France fight the Hundred Years' War.

1453 Ottoman Turks conquer Constantinople.

1492 King Ferdinand and Queen Isabella take Granada, Spain, from Moors. Christopher Columbus sails to America.

1500s Germany becomes the center of the Reformation.

1503–06 Leonardo da Vinci paints *Mona Lisa.*

1588 The English defeat the Spanish Armada.

1616 William Shakespeare dies.

1789 The French Revolution begins.

1815 Napoléon Bonaparte is defeated.

1840s The Irish potato famine occurs.

1867 The Austro-Hungarian Empire is created.

1871 Prussia unites Germany.

Population Map

Answers

1. cold subarctic, tundra climates

2. Latvia, Lithuania

Critical Thinking

3. Charts, graphs, databases, and models should accurately reflect population figures for cities on the maps.

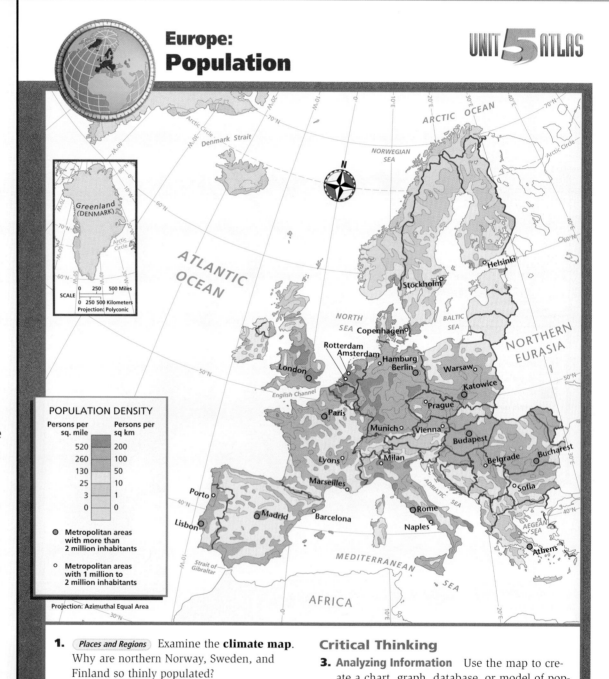

Europe:
Population

UNIT 5 ATLAS

POPULATION DENSITY

Persons per sq. mile	Persons per sq km
520	200
260	100
130	50
25	10
3	1
0	0

● Metropolitan areas with more than 2 million inhabitants

○ Metropolitan areas with 1 million to 2 million inhabitants

Projection: Azimuthal Equal Area

1. *Places and Regions* Examine the **climate map**. Why are northern Norway, Sweden, and Finland so thinly populated?

2. *Places and Regions* Compare this map to the **political map**. Which countries have between 25 and 130 persons per square mile in all areas?

Critical Thinking

3. Analyzing Information Use the map to create a chart, graph, database, or model of population centers in Europe.

Focus students' attention on the **land use and resources map** on this page. Ask students to name countries that do not have major manufacturing and trade centers *(Iceland, Estonia, Latvia, Lithuania, Yugoslavia, Albania, Macedonia, Croatia, Bosnia and Herzegovina)* and to identify other economic activi- ties that are important to these countries *(commercial farming, livestock raising, some mineral production)*. Then ask which countries appear to have the largest number of major manufacturing and trade centers. *(Germany and Italy)*

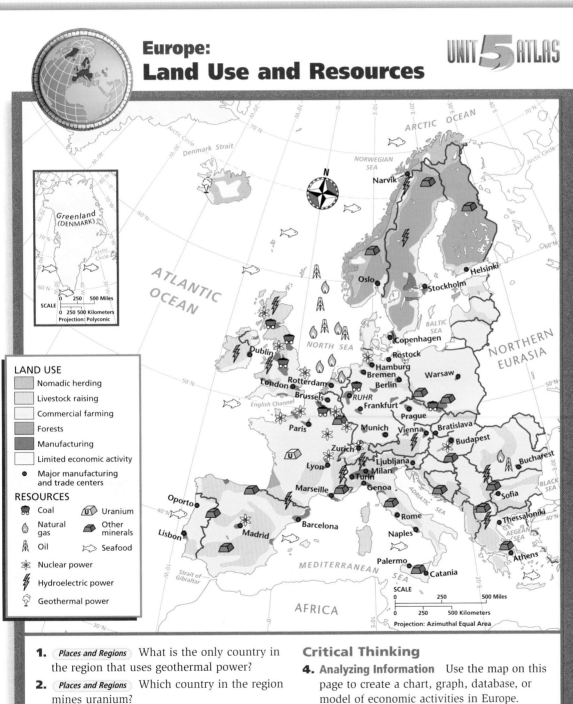

Europe: Land Use and Resources

UNIT 5 ATLAS

LAND USE

- Nomadic herding
- Livestock raising
- Commercial farming
- Forests
- Manufacturing
- Limited economic activity
- ● Major manufacturing and trade centers

RESOURCES

- Coal
- U Uranium
- Natural gas
- Other minerals
- Oil
- Seafood
- Nuclear power
- Hydroelectric power
- Geothermal power

SCALE
0 250 500 Miles
0 250 500 Kilometers
Projection: Polyconic

Greenland (DENMARK)

SCALE
0 250 500 Miles
0 250 500 Kilometers
Projection: Azimuthal Equal Area

1. (Places and Regions) What is the only country in the region that uses geothermal power?

2. (Places and Regions) Which country in the region mines uranium?

3. (Places and Regions) In which body of water is oil and gas production concentrated?

Critical Thinking

4. Analyzing Information Use the map on this page to create a chart, graph, database, or model of economic activities in Europe.

Your Classroom Time Line (continued)

1914–18 World War I is fought.

1921 Most of Ireland becomes independent.

1933 The Nazi Party seizes power in Germany.

1936–39 The Spanish Civil War is fought.

1939 World War II begins.

1944 Allies invade Normandy.

1945 Germany is defeated.

1956 Revolt in Hungary against the Soviet Union fails.

1961 The Berlin Wall is built.

1974 Greece becomes a republic.

1975 Francisco Franco's rule of Spain ends.

1989 Soviet control of Eastern Europe ends.

1989 The Berlin Wall is opened.

1993 Czechoslovakia divides into two countries.

1990s Civil war and fighting in former Yugoslavia begins.

Land Use and Resources Map

Answers

1. Iceland
2. France
3. North Sea

Critical Thinking

4. Charts, graphs, databases, and models should accurately reflect information from the map.

387

Fast FACTS

EUROPE

LEVEL 1: (Suggested time: 30 min.) Ask students to look at the population and area figures on these pages. Tell them that the United States has a population of more than 281.4 million people living in an area of 3,717,792 square miles. Have students work with a partner to compare the U.S. population per square mile with that of the European countries. Ask the students to decide if the population in the United States is more or less dense than the population of the European countries. *(Population density in Europe is higher than it is in the United States.)*

Then ask each pair of students to share their answer with another pair of students. Encourage the students to define the concept of population density and to demonstrate how they determined whether the United States or European countries tended to have a higher population density. *(Divide population by square miles. A larger number indicates a higher density—more people per square mile.)*

LEVEL 2: (Suggested time: 25 min.) Have students examine the figures for population and cars for the European countries. Mention that the U.S. population of approximately 281 million owns more than 130 million cars. Those figures translate roughly to one car for every two people in the United States, which

UNITED STATES OF AMERICA

CAPITAL:
Washington, D.C.
AREA:
3,717,792 sq. mi.
(9,629,091 sq km)
POPULATION:
281,421,906
MONEY:
U.S. dollar (US$)
LANGUAGES:
English, Spanish (spoken by a large minority)
CARS:
131,838,538

Fast FACTS Europe

ALBANIA
CAPITAL: Tiranë
AREA:
11,100 sq. mi. (28,748 sq km)
POPULATION: 3,510,484
MONEY: lek (ALL)
LANGUAGES:
Albanian, Greek
CARS: data not available

ANDORRA
CAPITAL:
Andorra la Vella
AREA:
181 sq. mi. (468 sq km)
POPULATION: 67,627
MONEY:
euro (€), 1-01-2002
LANGUAGES:
Catalan (official), French
Castilian
CARS: 35,358

AUSTRIA
CAPITAL: Vienna
AREA:
32,378 sq. mi.
(83,858 sq km)
POPULATION: 8,150,835
MONEY:
euro (€),
1-01-2002
LANGUAGES: German
CARS: 3,780,000

BELGIUM
CAPITAL:
Brussels
AREA:
11,780 sq. mi. (30,510 sq km)
POPULATION: 10,258,762
MONEY:
euro (€), 1-01-2002
LANGUAGES:
Dutch, French, German
CARS: 4,420,000

BOSNIA AND HERZEGOVINA
CAPITAL:
Sarajevo
AREA:
19,741 sq. mi. (51,129 sq km)
POPULATION: 3,922,205
MONEY:
marka (BAM)
LANGUAGES:
Croatian, Serbian, Bosnian
CARS: data not available

BULGARIA
CAPITAL: Sofia
AREA:
42,822 sq. mi.
(110,910 sq km)
POPULATION: 7,707,495
MONEY:
lev (BGL)
LANGUAGES:
Bulgarian
CARS: 1,650,000

CROATIA
CAPITAL:
Zagreb
AREA:
21,831 sq. mi.
(56,542 sq km)
POPULATION: 4,334,142
MONEY:
Croatian kuna (HRK)
LANGUAGES:
Croatian
CARS: 698,000

CZECH REPUBLIC
CAPITAL: Prague
AREA:
30,450 sq. mi.
(78,866 sq km)
POPULATION: 10,264,212
MONEY:
Czech koruna (CZK)
LANGUAGES:
Czech
CARS: 4,410,000

DENMARK
CAPITAL:
Copenhagen
AREA:
16,639 sq. mi.
(43,094 sq km)
POPULATION: 5,352,815
MONEY:
Danish krone (DKK)
LANGUAGES:
Danish, Faroese, Greenlandic
(an Inuit dialect), German
CARS: 1,790,000

ESTONIA
CAPITAL:
Tallinn
AREA:
17,462 sq. mi.
(45,226 sq km)
POPULATION: 1,423,316
MONEY:
Estonian kroon (EEK)
LANGUAGES:
Estonian (official), Russian,
Ukrainian, English, Finnish
CARS: 338,000

Countries not drawn to scale.

is far more than in most European countries. You may want to invite students to verify this statement by making rough estimates of the European figures.

Have the students refer to the accompanying table listing the miles of railroads in selected European countries. Point out the figure for the United States, and mention that the United States is larger than all of the European countries combined. (Students may verify this statement by totaling the areas of the European countries and comparing the result to the area of the United States.) Ask students to write a sentence or two explaining why they think Europe has fewer cars. *(Possible answers: Europe's*

population density may indicate that many people live in apartments and do not have a place to park a car. Because Europe has an extensive rail system, people do not need to rely as much on cars.)

Reproducible
◆ Unit 5 Test
◆ Unit 5 Test for English Language Learners and Special-Needs Students

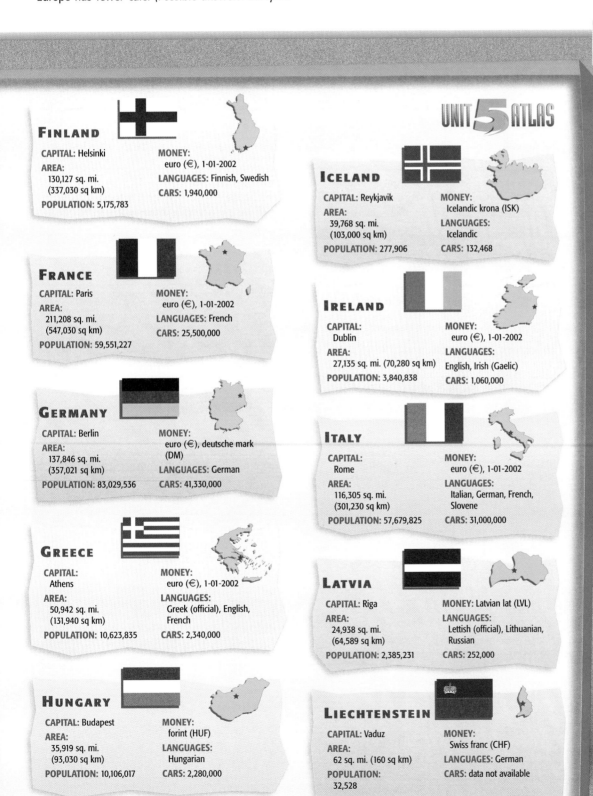

UNIT 5 ATLAS

FINLAND
CAPITAL: Helsinki
AREA:
130,127 sq. mi.
(337,030 sq km)
POPULATION: 5,175,783
MONEY:
euro (€), 1-01-2002
LANGUAGES: Finnish, Swedish
CARS: 1,940,000

FRANCE
CAPITAL: Paris
AREA:
211,208 sq. mi.
(547,030 sq km)
POPULATION: 59,551,227
MONEY:
euro (€), 1-01-2002
LANGUAGES: French
CARS: 25,500,000

GERMANY
CAPITAL: Berlin
AREA:
137,846 sq. mi.
(357,021 sq km)
POPULATION: 83,029,536
MONEY:
euro (€), deutsche mark (DM)
LANGUAGES: German
CARS: 41,330,000

GREECE
CAPITAL: Athens
AREA:
50,942 sq. mi.
(131,940 sq km)
POPULATION: 10,623,835
MONEY:
euro (€), 1-01-2002
LANGUAGES:
Greek (official), English, French
CARS: 2,340,000

HUNGARY
CAPITAL: Budapest
AREA:
35,919 sq. mi.
(93,030 sq km)
POPULATION: 10,106,017
MONEY:
forint (HUF)
LANGUAGES:
Hungarian
CARS: 2,280,000

ICELAND
CAPITAL: Reykjavik
AREA:
39,768 sq. mi.
(103,000 sq km)
POPULATION: 277,906
MONEY:
Icelandic krona (ISK)
LANGUAGES:
Icelandic
CARS: 132,468

IRELAND
CAPITAL:
Dublin
AREA:
27,135 sq. mi. (70,280 sq km)
POPULATION: 3,840,838
MONEY:
euro (€), 1-01-2002
LANGUAGES:
English, Irish (Gaelic)
CARS: 1,060,000

ITALY
CAPITAL:
Rome
AREA:
116,305 sq. mi.
(301,230 sq km)
POPULATION: 57,679,825
MONEY:
euro (€), 1-01-2002
LANGUAGES:
Italian, German, French, Slovene
CARS: 31,000,000

LATVIA
CAPITAL: Riga
AREA:
24,938 sq. mi.
(64,589 sq km)
POPULATION: 2,385,231
MONEY: Latvian lat (LVL)
LANGUAGES:
Lettish (official), Lithuanian, Russian
CARS: 252,000

LIECHTENSTEIN
CAPITAL: Vaduz
AREA:
62 sq. mi. (160 sq km)
POPULATION:
32,528
MONEY:
Swiss franc (CHF)
LANGUAGES: German
CARS: data not available

Sources: Central Intelligence Agency, *The World Factbook 2001; The World Almanac and Book of Facts 2001;* population figures are 2001 estimates.

 internet connect

COUNTRY STATISTICS
GO TO: **go.hrw.com**
KEYWORD: **SJ3 FactsU5**

Highlights of Country Statistics
• *CIA World Factbook*
• Library of Congress country studies
• Flags of the world

MILES OF RAILROAD IN SELECTED EUROPEAN COUNTRIES AND THE UNITED STATES

Austria	3,524	Norway	2,485
Belgium	2,093	Poland	14,904
Denmark	1,780	Romania	7,062
Finland	3,641	Slovakia	2,277
France	19,847	Spain	8,252
Germany	54,994	Sweden	6,756
Greece	1,537	Switzerland	3,132
Hungary	8,190	United Kingdom	23,518
Italy	9,944	United States	137,900
Netherlands	1,702		

Source: *The World Almanac and Book of Facts 2001*

LITHUANIA

CAPITAL: Vilnius
AREA:
25,174 sq. mi.
(65,200 sq km)
POPULATION:
3,610,535
MONEY:
litas (LTL)
LANGUAGES:
Lithuanian (official), Polish, Russian
CARS: 653,000

MONACO

CAPITAL:
Monaco
AREA:
0.75 sq. mi. (1.95 sq km)
POPULATION:
31,842
MONEY:
euro (€), 1-01-2002
LANGUAGES:
French (official), English, Italian, Monegasque
CARS: 17,000

LUXEMBOURG

CAPITAL:
Luxembourg
AREA:
998 sq. mi. (2,586 sq km)
POPULATION:
429,080
MONEY: euro (€),
Luxembourg franc (LuxF)
LANGUAGES:
Luxembourgish, German, French
CARS: 231,666

NETHERLANDS

CAPITAL: Amsterdam
AREA:
16,033 sq. mi.
(41,526 sq km)
POPULATION: 15,981,472
MONEY:
euro (€), 1-01-2002
LANGUAGES: Dutch
CARS: 5,810,000

MACEDONIA

CAPITAL: Skopje
AREA:
9,781 sq. mi.
(25,333 sq km)
POPULATION: 2,046,209
MONEY:
Macedonian denar (MKD)
LANGUAGES:
Macedonian, Albanian, Turkish, Serbo-Croatian
CARS: 263,000

NORWAY

CAPITAL: Oslo
AREA:
125,181 sq. mi.
(324,220 sq km)
POPULATION: 4,503,440
MONEY:
Norwegian krone (NOK)
LANGUAGES: Norwegian
CARS: 1,760,000

MALTA

CAPITAL:
Valletta
AREA:
122 sq. mi. (316 sq km)
POPULATION:
394,583
MONEY:
Maltese lira (MTL)
LANGUAGES:
Maltese (official), English (official)
CARS: 122,100

POLAND

CAPITAL: Warsaw
AREA:
120,728 sq. mi.
(312,685 sq km)
POPULATION: 38,633,912
MONEY:
zloty (PLN)
LANGUAGES:
Polish
CARS: 7,520,000

MOLDOVA

CAPITAL: Chişinău
AREA:
13,067 sq. mi.
(33,843 sq km)
POPULATION: 4,431,570
MONEY:
Moldovan leu (MDL)
LANGUAGES:
Moldovan (official), Russian, Gagauz
CARS: 169,000

PORTUGAL

CAPITAL: Lisbon
AREA:
35,672 sq. mi.
(92,391 sq km)
POPULATION: 10,066,253
MONEY:
euro (€), 1-01-2002
LANGUAGES: Portuguese
CARS: 2,950,000

Countries not drawn to scale.

LEVEL 3: (Suggested time: 30 min.) Tell students that Europe has an economic organization called the European Union (EU). It was formed to strengthen the economies of European nations. In 1991, countries of the EU began to discuss adopting a common currency, to be called the euro. Point out the euro symbol in the Fast Facts data. By May of 1998, nearly every interested EU country had met the requirements for membership, which include maintaining a low inflation rate, low budget deficit, and low interest rate. Soon after, the European National Bank was formed. Then, on January 1, 2002, the countries adopted the euro as their standard currency.

Ask students to consider the effect that changing to a common currency might have on a nation's culture, economy, and relationships with neighboring countries. Have them imagine that they live in Europe in 1991 and are weighing the pros and cons of changing to the euro. In one paragraph, students should give reasons why they support their country's adoption of a common currency. In another paragraph, they should defend a position against adopting the euro. Invite them to refer back to their arguments as they learn more about the culture, economies, and relationships among European countries.

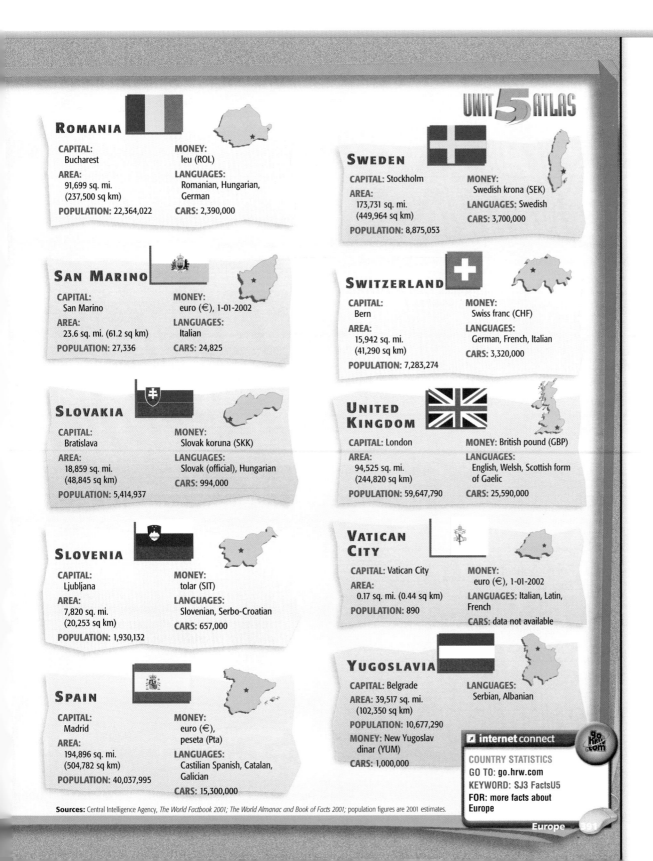

UNIT 5 ATLAS

ROMANIA

CAPITAL:
Bucharest

AREA:
91,699 sq. mi.
(237,500 sq km)

POPULATION: 22,364,022

MONEY:
leu (ROL)

LANGUAGES:
Romanian, Hungarian, German

CARS: 2,390,000

SAN MARINO

CAPITAL:
San Marino

AREA:
23.6 sq. mi. (61.2 sq km)

POPULATION: 27,336

MONEY:
euro (€), 1-01-2002

LANGUAGES:
Italian

CARS: 24,825

SLOVAKIA

CAPITAL:
Bratislava

AREA:
18,859 sq. mi.
(48,845 sq km)

POPULATION: 5,414,937

MONEY:
Slovak koruna (SKK)

LANGUAGES:
Slovak (official), Hungarian

CARS: 994,000

SLOVENIA

CAPITAL:
Ljubljana

AREA:
7,820 sq. mi.
(20,253 sq km)

POPULATION: 1,930,132

MONEY:
tolar (SIT)

LANGUAGES:
Slovenian, Serbo-Croatian

CARS: 657,000

SPAIN

CAPITAL:
Madrid

AREA:
194,896 sq. mi.
(504,782 sq km)

POPULATION: 40,037,995

MONEY:
euro (€),
peseta (Pta)

LANGUAGES:
Castilian Spanish, Catalan, Galician

CARS: 15,300,000

SWEDEN

CAPITAL: Stockholm

AREA:
173,731 sq. mi.
(449,964 sq km)

POPULATION: 8,875,053

MONEY:
Swedish krona (SEK)

LANGUAGES: Swedish

CARS: 3,700,000

SWITZERLAND

CAPITAL:
Bern

AREA:
15,942 sq. mi.
(41,290 sq km)

POPULATION: 7,283,274

MONEY:
Swiss franc (CHF)

LANGUAGES:
German, French, Italian

CARS: 3,320,000

UNITED KINGDOM

CAPITAL: London

AREA:
94,525 sq. mi.
(244,820 sq km)

POPULATION: 59,647,790

MONEY: British pound (GBP)

LANGUAGES:
English, Welsh, Scottish form of Gaelic

CARS: 25,590,000

VATICAN CITY

CAPITAL: Vatican City

AREA:
0.17 sq. mi. (0.44 sq km)

POPULATION: 890

MONEY:
euro (€), 1-01-2002

LANGUAGES: Italian, Latin, French

CARS: data not available

YUGOSLAVIA

CAPITAL: Belgrade

AREA: 39,517 sq. mi.
(102,350 sq km)

POPULATION: 10,677,290

MONEY: New Yugoslav dinar (YUM)

CARS: 1,000,000

LANGUAGES:
Serbian, Albanian

internet connect

COUNTRY STATISTICS
GO TO: go.hrw.com
KEYWORD: SJ3 FactsU5
FOR: more facts about Europe

Sources: Central Intelligence Agency, *The World Factbook 2001; The World Almanac and Book of Facts 2001;* population figures are 2001 estimates.

Southern Europe
Chapter Resource Manager

Objectives	Pacing Guide	Reproducible Resources
SECTION 1 **Physical Geography** (pp. 393–95) 1. Identify the major landforms and rivers of southern Europe. 2. Identify the major climate types and resources of this region.	**Regular** .5 day **Block Scheduling** .5 day *Block Scheduling Handbook, Chapter 18*	**RS** Guided Reading Strategy 18.1 **RS** Graphic Organizer 18 **E** Creative Strategies for Teaching World Geography, Lessons 10 and 11 **SM** Geography for Life Activity 18 **IC** Interdisciplinary Activity for Middle Grades 4
SECTION 2 **Greece** (pp. 396–99) 1. Identify some achievements of the ancient Greeks. 2. Identify two features of Greek culture. 3. Describe what Greece is like today.	**Regular** 1 day **Block Scheduling** .5 day *Block Scheduling Handbook, Chapter 18*	**RS** Guided Reading Strategy 18.2 **E** Cultures of the World Activity 3 **IC** Interdisciplinary Activity for Middle Grades 16
SECTION 3 **Italy** (pp. 400–03) 1. Describe the early history of Italy. 2. Describe how Italy has added to world culture. 3. Describe what Italy is like today.	**Regular** 1 day **Block Scheduling** .5 day *Block Scheduling Handbook, Chapter 18*	**RS** Guided Reading Strategy 18.3 **E** Cultures of the World Activity 3 **SM** Map Activity 18
SECTION 4 **Spain and Portugal** (pp. 404–07) 1. Identify some major events in the history of Spain and Portugal. 2. Describe the cultures of Spain and Portugal. 3. Describe what Spain and Portugal are like today.	**Regular** 1 day **Block Scheduling** .5 day *Block Scheduling Handbook, Chapter 18*	**RS** Guided Reading Strategy 18.4 **E** Cultures of the World Activity 3

Chapter Resource Key

RS	Reading Support	**ELL**	Reinforcement and English Language Learners
IC	Interdisciplinary Connections		
E	Enrichment		Transparencies
SM	Skills Mastery		CD–ROM
A	Assessment		Music
REV	Review		Video

 Internet

 Holt Presentation Maker Using Microsoft® Powerpoint®

 One-Stop Planner CD–ROM

See the *One-Stop Planner* for a complete list of additional resources for students and teachers.

One-Stop Planner CD-ROM

It's easy to plan lessons, select resources, and print out materials for your students when you use the *One-Stop Planner CD–ROM with Test Generator.*

| internet connect |
HRW ONLINE RESOURCES

GO TO: go.hrw.com
Then type in a keyword.

TEACHER HOME PAGE
KEYWORD: SJ3 TEACHER

CHAPTER INTERNET ACTIVITIES
KEYWORD: SJ3 GT18

Choose an activity to:
- explore the islands and peninsulas on the Mediterranean coast.
- take an online tour of ancient Greece.
- learn the story of pizza.

CHAPTER ENRICHMENT LINKS
KEYWORD: SJ3 CH18

CHAPTER MAPS
KEYWORD: SJ3 MAPS18

ONLINE ASSESSMENT
Homework Practice
KEYWORD: SJ3 HP18
Standardized Test Prep Online
KEYWORD: SJ3 STP18
Rubrics
KEYWORD: SS Rubrics

COUNTRY INFORMATION
KEYWORD: SJ3 Almanac

CONTENT UPDATES
KEYWORD: SS Content Updates

HOLT PRESENTATION MAKER
KEYWORD: SJ3 PPT18

ONLINE READING SUPPORT
KEYWORD: SS Strategies

CURRENT EVENTS
KEYWORD: S3 Current Events

Technology Resources

	Review, Reinforcement, and Assessment Resources

One-Stop Planner CD–ROM, Lesson 18.1

Geography and Cultures Visual Resources with Teaching Activities 24–29

Homework Practice Online

HRW Go site

- **ELL** Main Idea Activity 18.1
- **REV** Section 1 Review, p. 395
- **A** Daily Quiz 18.1
- **ELL** English Audio Summary 18.1
- **ELL** Spanish Audio Summary 18.1

One-Stop Planner CD–ROM, Lesson 18.2

Homework Practice Online

HRW Go site

- **ELL** Main Idea Activity 18.2
- **REV** Section 2 Review, p. 399
- **A** Daily Quiz 18.2
- **ELL** English Audio Summary 18.2
- **ELL** Spanish Audio Summary 18.2

One-Stop Planner CD–ROM, Lesson 18.3

ARGWorld CD–ROM: Conditions and Connections in Renaissance Europe

Homework Practice Online

HRW Go site

- **ELL** Main Idea Activity 18.3
- **REV** Section 3 Review, p. 403
- **A** Daily Quiz 18.3
- **ELL** English Audio Summary 18.3
- **ELL** Spanish Audio Summary 18.3

One-Stop Planner CD–ROM, Lesson 18.4

Homework Practice Online

HRW Go site

- **ELL** Main Idea Activity 18.4
- **REV** Section 4 Review, p. 407
- **A** Daily Quiz 18.4
- **ELL** English Audio Summary 18.4
- **ELL** Spanish Audio Summary 18.4

Meeting Individual Needs

Ability Levels

Level 1 Basic-level activities designed for all students encountering new material

Level 2 Intermediate-level activities designed for average students

Level 3 Challenging activities designed for honors and gifted-and-talented students

English Language Learners Activities that address the needs of students with Limited English Proficiency

Chapter Review and Assessment

- **IC** Interdisciplinary Activity for the Middle Grades 13, 14, 15
- **E** Readings in World Geography, History, and Culture 28, 29, and 30
- **SM** Critical Thinking Activity 18
- **REV** Chapter 18 Review, pp. 276–77
- **REV** Chapter 18 Tutorial for Students, Parents, Mentors, and Peers
- **ELL** Vocabulary Activity 18
- **A** Chapter 18 Test
- Chapter 18 Test Generator (on the One-Stop Planner)
- Audio CD Program, Chapter 18
- **A** Chapter 18 Test for English Language Learners and Special-Needs Students

LAUNCH INTO LEARNING

In a democracy such as the United States, people participate in the government in a variety of ways. Citizens vote, hold office, write letters, and demonstrate peacefully. Ask students where they think Americans got these ideas. Have them locate Greece and Italy on the map on the following page. Tell students that in the 400s B.C. citizens of the Greek city-state of Athens voted on their government's decisions. Romans began to elect people to represent them in their government at about the same time. Tell students they will learn more about these cultures, what came after them, and the nearby countries of Spain and Portugal in this chapter.

Section 1

Objectives

1. Identify the major landforms and rivers of southern Europe.
2. Examine the major climate types and resources of this region.

LINKS TO OUR LIVES

These are reasons why studying southern Europe might interest students:

▶ Many ideas that are basic to the Western world's way of life originated with the ancient Greeks and Romans.

▶ Spanish and Portuguese sailors opened the New World to European exploration and settlement.

▶ Artists, musicians, writers, scientists, and scholars from southern Europe have made significant contributions to the world's cultural heritage.

▶ The U.S. economy is closely linked with the economies of these countries.

CHAPTER 18

Southern Europe

Southern Europe's peninsulas, islands, mountains, and plateaus form a beautiful region. Tourists enjoy visiting this region to see its historical and cultural treasures.

Ciao. I am Paolo. I am 11 years old. I live in an apartment with my parents and my *nonna*, which is Italian for "grandma." Nonna is teaching me how to cook while she makes dinner for the family. I have no brothers or sisters, but on Sundays my aunts and uncles and three cousins all come for a big lunch. After lunch, the cousins play outside on the playground swings.

In the mornings, I walk to school with my mother before she goes to her job as a professor. My school is not too strict. We study English, Italian, and religion. After school I practice with my team at the swim club.

My favorite holiday is Christmas. On Christmas Eve we go to my other grandmother's house for a special dinner. We have smoked salmon, then grilled trout and spaghetti with mussels. Then there are special Christmas sweets: Pandoro—a cake sprinkled with sugar, and Torrone—a candy log with chocolate and nuts.

Abito a Roma, la capitale dell'Italia.

Translation: I live in Rome, the capital of Italy.

LET'S GET STARTED

Copy the following instructions onto the chalkboard: *Look through Section 1 and find a picture you like. What do you want to know about that picture? Write down a question.* Ask volunteers to share their questions with the class. Discuss some of the questions. Invite students to suggest answers to each other's questions. Tell students that in Section 1 they will learn more about the physical geography of southern Europe.

Using the Physical-Political Map

Have students examine the map on this page. Remind students that Europe is often called a peninsula of peninsulas. Point out the Iberian, Italian, and Greek peninsulas and call on students to name the bodies of water around them. Ask students how the countries' locations may have influenced the region's history and economy. *(Access to the sea would help trade and travel.)*

Section 1 — Physical Geography

Read to Discover

1. What are the major landforms and rivers of southern Europe?
2. What are the major climate types and resources of this region?

Define

mainland
sirocco

Locate

Mediterranean Sea
Strait of Gibraltar

Iberian Peninsula
Cantabrian Mountains
Pyrenees Mountains
Alps
Apennines
Aegean Sea
Peloponnesus

Ebro River
Douro River
Tagus River
Guadalquivir River
Po River
Tiber River

WHY IT MATTERS

Southern Europe's location on peninsulas encouraged its cultures to become great seafarers. Use **CNNfyi.com** or other current events sources to find examples of trade on the region's oceans today. Record your findings in your journal.

Cave painting from 12,000 B.C.

Section 1 RESOURCES

Reproducible
- Block Scheduling Handbook, Chapter 18
- Graphic Organizer 18
- Guided Reading Strategy 18.1
- Geography for Life Activity 18
- Interdisciplinary Activity for the Middle Grades 4
- Creative Strategies for Teaching World Geography, Lessons 10 and 11

Technology
- One-Stop Planner CD–ROM, Lesson 18.1
- Homework Practice Online
- Geography and Cultures Visual Resources with Teaching Activities 24–29
- HRW Go site

Reinforcement, Review, and Assessment
- Section 1 Review, p. 263
- Daily Quiz 18.1
- Main Idea Activity 18.1
- English Audio Summary 18.1
- Spanish Audio Summary 18.1

Southern Europe: Physical-Political

EUROPE

Bay of Biscay

ATLANTIC OCEAN

CANTABRIAN MTS.

Oporto
PORTUGAL
Lisbon

SPAIN
IBERIAN PENINSULA
MESETA
Madrid
Barcelona

Mont Blanc
15,771 ft. (4,807 m)
ALPS
Milan Venice
Turin Genoa
PO VALLEY
Florence
SAN MARINO
ANDORRA

APENNINES
ADRIATIC SEA

VATICAN CITY
ITALY
Rome
Naples

Sardinia (ITALY)
Mount Vesuvius
4,190 ft. (1,277 m)
Pompeii

Thessaloniki
Mount Olympus
9,570 ft. (2,917 m)
PINDUS MTS.
GREECE
Patras Athens
Olympia Piraeus
PELOPONNESUS Sparta

Seville Granada
Mount Mulhacen
11,407 ft. (3,477 m)
Málaga
GIBRALTAR (U.K.)

Strait of Gibraltar

Valencia

Balearic Islands (SPAIN)

MEDITERRANEAN SEA

Palermo
Sicily (ITALY)
Mount Etna
10,902 ft. (3,323 m)

IONIAN SEA

Crete (GREECE)

AFRICA

ELEVATION

	FEET	METERS
National capitals	13,120	4,000
Other cities	6,560	2,000
	1,640	500
Historic sites	656	200
	(Sea level) 0	0 (Sea level)
	Below sea level	Below sea level

SCALE
0 100 200 300 Miles
0 100 200 300 Kilometers
Projection: Azimuthal Equal Area

Size comparison of southern Europe to the contiguous United States

393

Teaching Objectives 1–2

ALL LEVELS: (Suggested time: 30 min.) Copy the following graphic organizer onto the chalkboard, omitting the italicized answers. Use it to help students understand the physical geography of southern Europe. Have students complete the organizer. **ENGLISH LANGUAGE LEARNERS**

The Physical Geography of SOUTHERN EUROPE

Shared Characteristics:
1. peninsulas 2. mountains 3. rivers 4. climate 5. resources

Spain and Portugal	Italy	Greece
1. *Iberian Peninsula*	1. *shaped like a boot*	1. *largest peninsula is Peloponnesus*
2. *Cantabrian and Pyrenees*	2. *southern Alps, Apennines*	2. *very mountainous*
3. *several east-west*	3. *Po and Tiber*	3. *most rivers are short*
4. *some semiarid climates; northern Spain is cool and humid*	4. *sirocco*	4. *warm and sunny*
5. *trade, fishing, iron ore, beaches*	5. *trade, marble*	5. *bauxite, chromium, lead, marble, and zinc*

EYE ON EARTH

Sicily's Mount Etna is the tallest and most active volcano in Europe. Its name comes from a Greek word meaning "I burn." The mountain has three ecological zones, each with its own vegetation. The lowest zone is fertile and rich in citrus fields, olive groves, and vineyards. Catania, a city of about 330,000 residents, is located on the mountain's lowest slopes. Forests are found farther up the mountain in the second ecological zone. Ash, sand, and lava fragments cover the mountain at heights more than 6,500 feet (1,980 m) in the final zone.

Geologists estimate that Mount Etna has been active for more than 2.5 million years. The mountain has had more than 100 serious eruptions in the past 2,500 years. The resulting lava flows have repeatedly destroyed villages, fields, and vineyards.

Discussion: Point out Catania and Mount Etna on a map of Italy. Ask students to identify any advantages that might counterbalance the city's dangerous location. *(possible answers: seaport, fertile fields from lava flows)*

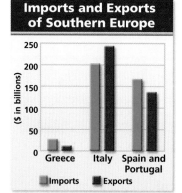

Imports and Exports of Southern Europe

Source: Central Intelligence Agency, *The World Factbook, 2001*

Interpreting the Graph Which country has the fewest imports and exports?

Greece, a land of mountains and sea, is home to the ancient city of Lindos on the island of Rhodes.

Graph Answer

Greece

Physical Features

Southern Europe is also known as Mediterranean Europe because most of its countries are on the sea's shores. The Mediterranean Sea stretches some 2,300 miles (3,700 km) from east to west. *Mediterranean* means "middle of the land" in Latin. In ancient times, the Mediterranean was considered the center of the Western world, since it is surrounded by Europe, Africa, and Asia. The narrow Strait of Gibraltar (juh-BRAWL-tuhr) links the Mediterranean to the Atlantic Ocean.

The Land Southern Europe is made up of three peninsulas. Portugal and Spain occupy one, Italy occupies another, and Greece is located on a third peninsula. Portugal and Spain are on the Iberian (eye-BIR-ee-uhn) Peninsula. Much of the peninsula is a high, rocky plateau. The Cantabrian (kan-TAY-bree-uhn) and the Pyrenees (PIR-uh-neez) Mountains form the plateau's northern edge. Italy's peninsula includes the southern Alps. A lower mountain range, the Apennines (A-puh-nynz), runs like a spine down the country's back. Islands in the central and western Mediterranean include Italy's Sicily and Sardinia (sahr-DI-nee-uh), as well as Spain's Balearic (ba-lee-AR-ik) Islands.

Greece's **mainland**, or the country's main landmass, extends into the Aegean (ee-JEE-uhn) Sea in many jagged little peninsulas. The largest one is the Peloponnesus (pe-luh-puh-NEE-suhs). Greece is mountainous and includes more than 2,000 islands. The largest island is Crete (KREET).

On all three peninsulas, coastal lowlands and river valleys provide excellent areas for growing crops and building cities. Soils on the region's uplands are thin and stony. They are also easily eroded. In this area of young mountains, earthquakes are common. They are particularly common in Greece and Italy.

The Rivers Several east-west rivers cut through the Iberian Peninsula. The Ebro River drains into the Mediterranean. The Douro, Tagus, and Guadalquivir (gwah-thahl-kee-VEER) Rivers, however, flow to the Atlantic Ocean. The Po (POH) is Italy's largest river. It creates a fertile agricultural region in northern Italy. Farther south, along the banks of the much smaller Tiber River, is the city of Rome.

✓ **READING CHECK:** *Physical Systems* What physical processes cause problems in parts of southern Europe? erosion, earthquakes

CLOSE

Have students review the questions they wrote for the Let's Get Started Activity. Ask students if they can answer more of them now. What else do they need to learn to find the answers?

REVIEW AND ASSESS

Have students complete the Section Review. Then organize students into teams. Give each team an index card with a Define/Locate term on one side. Have teams create a question about the word or term on their card. Pass the cards from team to team. Ask each team to read the question aloud and to answer it. Continue until all the questions have been answered. Then have students complete Daily Quiz 18.1.
COOPERATIVE LEARNING

RETEACH

Have students complete Main Idea Activity 18.1. Then have volunteers come to the chalkboard and write explanations of the main physical features discussed in Section 1. **ENGLISH LANGUAGE LEARNERS**

EXTEND

Have interested students conduct research on the connections between the physical geography of southern Europe and economic activities relating to agriculture. Have students consider climate, soil, and vegetation in their study. Ask them to report their findings in the form of a graph or a map.
BLOCK SCHEDULING

Climate and Resources

Much of southern Europe enjoys a warm, sunny climate. Most of the rain falls during the mild winter. Rainfall sometimes causes floods and mudslides due to erosion from overgrazing and deforestation. A hot, dry wind from North Africa called a **sirocco** (suh-RAH-koh) picks up some moisture over the Mediterranean Sea. It blows over Italy during spring and summer. The Po Valley is humid. Northern Italy's Alps have a highland climate. In Spain, semiarid climates are found in pockets. Northern Spain is cool and humid.

Southern Europeans have often looked to the sea for trade. Important Mediterranean ports include Barcelona, Genoa, Naples, Piraeus (py-REE-uhs)—the port of Athens—and Valencia. Lisbon, the capital of Portugal, is an important Atlantic port. The Atlantic Ocean supports Portugal's fishing industry. Although the Mediterranean suffers from pollution, it has a wealth of seafood.

The region's resources vary. Northern Spain has iron ore mines. Greece mines bauxite, chromium, lead, and zinc. Italy and Greece quarry marble. Falling water generates hydroelectricity throughout the region's uplands. Otherwise, resources are scarce.

The region's sunny climate and natural beauty have long attracted visitors. Millions of people explore castles, museums, ruins, and other cultural sites each year. Spain's beaches help make that country one of Europe's top tourist destinations.

▲ Workers prepare to separate a giant block of marble from a wall in a quarry in Carrara, Italy.

✓ **READING CHECK:** *Places and Regions* What climate types and natural resources are found in the region? *mostly warm, sunny; also highland, semiarid, cool, humid; seafood; iron ore, bauxite, chromium, lead, zinc, marble*

Section Review 1

Define and explain: mainland, sirocco

Working with Sketch Maps On a map of southern Europe that you draw or that your teacher provides, label Greece, Italy, Spain, and Portugal. Also label the Mediterranean Sea and the Strait of Gibraltar.

Reading for the Main Idea

1. *Places and Regions* Why is southern Europe known as Mediterranean Europe?

2. *Places and Regions* What countries occupy the region's three main peninsulas?

3. *Environment and Society* Why might people settle in river valleys and in coastal Southern Europe?

go.hrw.com **Homework Practice Online**
Keyword: SJ3 HP18

Critical Thinking

4. **Drawing Inferences and Conclusions** In what ways could the region's physical geography aid the development of trade?

Organizing What You Know

5. **Summarizing** Copy the following graphic organizer. Use it to list the region's physical features, climates, and resources.

	Physical Features	Climate	Resources
Spain and Portugal			
Italy			
Greece			

Section Review 1

Answers

Define For definitions, see: mainland, p. 394; sirocco, p. 395

Working with Sketch Maps Maps will vary, but listed places should be in their approximate locations.

Reading for the Main Idea
1. because most of its countries are on the sea's shores (NGS 4)
2. Portugal, Spain, Italy, and Greece (NGS 4)
3. access to water transportation, trade, natural resources like fertile soil, moderate climate (NGS 15)

Critical Thinking
4. access to the sea, many good harbors

Organizing What You Know
5. Spain and Portugal—on the Iberian Peninsula rocky plateau, Cantabrian and Pyrenees Mountains, Balearic Islands; east-west rivers; warm, sunny; seafood, iron ore, beaches; Italy—southern Alps, Apennines, islands of Sicily and Sardinia; Po and Tiber Rivers; warm, sunny, sirocco; seafood, marble; Greece—large and small peninsulas, mountains, Crete; warm, sunny; seafood, bauxite, chromium, lead, zinc, marble

395

Section 2

Objectives

1. Describe some of the achievements of the ancient Greeks.
2. Identify two features of Greek culture.
3. Examine what Greece is like today.

LET'S GET STARTED

Copy the following questions onto the chalkboard: *What is one way your life would be different if U.S. citizens had no say in the government? What if you did not know about atoms or the true size of Earth?* Ask students to respond to the questions in writing. Ask volunteers to read their answers. Tell students that our form of government, physics, and geography are part of what we have learned from the ancient Greeks. Tell students that they will learn more about Greece in Section 2.

Building Vocabulary

Write the key terms on the chalkboard. Have students suggest a meaning for **city-states**, based on the component words, and check their definitions against the text. Have students find the definition of **mosaic**. Tell them that the word comes from *muse*. The muses were nine Greek goddesses of the arts. Ask students how mosaic might relate to goddesses of the arts.

Section 2 RESOURCES

Reproducible
◆ Guided Reading Strategy 18.2
◆ Cultures of the World Activity 3
◆ Interdisciplinary Activity for the Middle Grades 16

Technology
◆ One-Stop Planner CD–ROM, Lesson 18.2
◆ Homework Practice Online
◆ HRW Go site

Reinforcement, Review, and Assessment
◆ Section 2 Review, p. 399
◆ Daily Quiz 18.2
◆ Main Idea Activity 18.2
◆ English Audio Summary 18.2
◆ Spanish Audio Summary 18.2

Visual Record Answer ▶

Students might suggest that it was considered a sacred place and was made of durable materials.

Greece

Read to Discover

1. What were some of the achievements of the ancient Greeks?
2. What are two features of Greek culture?
3. What is Greece like today?

Define
city-states
mosaics

Locate
Athens
Thessaloníki

WHY IT MATTERS

The Olympic Games first started in Greece. Today we continue the tradition and hold both Summer and Winter Games. Use **CNNfyi.com** or other current events sources to find out more about the modern Olympic Games. Record your findings in your journal.

Ancient coin of Alexander the Great

History

The Greek islands took an early lead in the development of trade and shipping between Asia, Africa, and Europe. By about 2000 B.C. large towns and a complex civilization existed on Crete.

Ancient Greece About 800 B.C. Greek civilization arose on the mainland. The mountainous landscape there favored small, independent **city-states**. Each Greek city-state, or *polis*, was made up of a city and the land around it. Each had its own gods, laws, and form of government. The government of the city-state of Athens was the first known democracy. Democracy is the form of government in which all citizens take part. Greek philosophers, artists, architects, and writers made important contributions to Western civilization. For example, the Greeks are credited with inventing theater. Students still study ancient Greek literature and plays.

Eventually, Greece was conquered by King Philip. Philip ruled Macedonia, an area north of Greece. About 330 B.C. Philip's son, Alexander the Great, conquered Asia Minor, Egypt, Persia, and part of India. His empire combined Greek culture with influences from Asia and Africa. In the 140s B.C. Greece and Macedonia were conquered by the Roman Empire.

The Greeks believed that the Temple of Delphi—shown below—was the center of the world.

Interpreting the Visual Record Why do you think remains of this temple have lasted for so many years?

Teaching Objective 1

ALL LEVELS: (Suggested time: 20 min.) Copy the following graphic organizer onto the chalkboard, omitting the italicized answers. Use it to help students understand the achievements of the ancient Greeks.
ENGLISH LANGUAGE LEARNERS

Ancient Greek Achievements

Athens—first known democracy

city-states

Eastern Orthodox Christianity

Ancient Greek Achievements

Byzantine Empire

The Byzantine Empire About A.D. 400 the Roman Empire was divided into two parts. The western half was ruled from Rome. It soon fell to Germanic peoples the Romans called barbarians. Barbarian means both *illiterate* and *wanderer*. The eastern half of the Roman Empire was known as the Byzantine Empire. It was ruled from Constantinople. Constantinople was located on the shore of the Bosporus in what is now Turkey. This city—today known as Istanbul— served as a gathering place for people from Europe and Asia. The Byzantine Empire carried on the traditions of the Roman Empire for another 1,000 years. Gradually, an eastern form of Christianity developed. It was influenced by Greek language and culture. It became known as Eastern Orthodox Christianity. It is the leading form of Christianity in Greece, parts of eastern Europe, and Russia.

Turkish Rule In 1453 Constantinople was conquered by the Ottoman Turks, a people from Central Asia. Greece and most of the rest of the region came under the rule of the Ottoman Empire. It remained part of this empire for nearly 400 years. In 1821 the Greeks revolted against the Turks, and in the early 1830s Greece became independent.

Government In World War II Greece was occupied by Germany. After the war Greek communists and those who wanted a king and constitution fought a civil war. When the communists lost, the military took control. Finally, in the 1970s the Greek people voted to make their country a republic. They adopted a new constitution that created a government with a president and a prime minister.

✓ **READING CHECK:** *Human Systems* What were some of the achievements of ancient Greece? democracy, contributions of Greek philosophers, artists, architects, and writers

Culture

Turkish influences on Greek art, food, and music can still be seen. However, Turkey and Greece disagree over control of the islands and shipping lanes of the Aegean Sea.

Religion Some 98 percent of Greeks are Eastern Orthodox Christians, commonly known as Greek Orthodox. Easter is a major holiday and cause for much celebration. The traditional Easter meal is eaten on Sunday—roasted lamb, various vegetables, Easter bread, and many desserts. Because the Greek Orthodox Church has its own calendar, Christmas and Easter are usually celebrated one to two weeks later than in the West.

▲
The Acropolis in Athens was built in the 400s B.C. The word *acropolis* is Greek for "city at the top."

Interpreting the Visual Record Why would it be important to build a city on a hill?

▲
The people of Karpathos and their religious leaders are participating in an Easter celebration.

Linking Past to Present
The Seafaring Greeks Greece has long been closely linked to the sea. Many Greek cities were established where ships could drop anchor. Naval power allowed Athens to dominate the Aegean area in the 400s B.C.

Modern Greece also depends on the sea. Shipping and tourism are its major sources of income. The Greek merchant fleet—more than 2,000 ships—is one of the world's largest.

Critical Thinking: Point out to students that ancient Romans developed their navy more slowly than did the Greeks. Have students consult the physical-political map in this chapter and then ask them why might this be so.

Answer: Italy has fewer harbors, and therefore travel by sea was more difficult there. Also, because Romans could easily cross the Apennines and trade inland, they did not depend as heavily on the sea.

▲ **Visual Record Answer**

Students might suggest that a city on a hill is easier to defend against attack.

397

Teaching Objective 2

ALL LEVELS: (Suggested time: 20 min.) Pair students and have one student in each pair write a few sentences about religion in Greece. Have the other student write about the arts. Then have pairs explain their topics to each other. **ENGLISH LANGUAGE LEARNERS, COOPERATIVE LEARNING**

Teaching Objective 3

ALL LEVELS: (Suggested time: 30 min.) Pair students and give each pair two sheets of heavy white paper, several pieces of colored paper, scissors, and glue. Have students cut the colored paper into small squares and then arrange the squares to create mosaics depicting some of the agricultural products of Greece. **ENGLISH LANGUAGE LEARNERS, COOPERATIVE LEARNING**

➤**ASSIGNMENT:** Have students write about some of the objects in their everyday lives that are related to ideas from ancient Greece. *(examples: Newspapers reflect freedom of speech, which is important to our democracy. Televisions broadcast dramas, which the Greeks developed.)*

CLOSE

Tell students that the first Olympic Games were held in Greece in 776 B.C. Stage a Class Olympics based on what students have learned about Greece. Events might be organized around physical features, history, or culture. Organize team or individual events.

ENVIRONMENT AND SOCIETY

Greece became an industrial nation in the 1970s. Since then heavy air and water pollution have made some people ill and eroded the marble of many ancient monuments and statues. Recent efforts have reduced some air pollution, but heavy traffic still dirties the air. In some places untreated sewage and industrial wastes have polluted the Mediterranean Sea.

Connecting to Math
Answers

1. using deduction in mathematical proofs, developing the Pythagorean Theorem, stating the basic principles of geometry, and calculating the value of *pi*

2. the creation of a model of the solar system, the estimation of the circumference of Earth, the treatment of medicine as a science, the gathering of information on plants and animals, and an understanding of the importance of observation and classification

CONNECTING TO *Math*

Greek postage stamp of the Pythagorean Theorem

Greek civilization made many contributions to world culture. We still admire Greek art and literature. Greek scholars also paved the way for modern mathematics and science.

More than 2,000 years ago Thales (THAY-leez), a philosopher, began the use of deduction in mathematical proofs. Pythagoras (puh-THAG-uh-ruhs) worked out an equation to calculate the dimensions of a right triangle. The equation became known as the Pythagorean Theorem. By 300 B.C. Euclid (YOO-kluhd) had stated the basic principles of geometry in his book *Elements*. Soon after, Archimedes (ahr-kuh-MEED-eez) calculated the value of *pi*. This value is used to measure circles and spheres. He also explained how and why the lever, a basic tool, works.

Aristarchus (ar-uh-STAHR-kuhs), an astronomer, worked out a model of the solar system. His model placed the Sun at the center of the universe. Eratosthenes (er-uh-TAHS-thuh-neez) estimated the circumference of Earth with great accuracy.

Two important figures in the life sciences were Hippocrates (hip-AHK-ruh-teez) and Aristotle (AR-uh-staht-uhl). Hippocrates was a doctor who treated medicine as a science. He understood that diseases have natural causes. Aristotle gathered information on a wide variety of plants and animals. He helped establish the importance of observation and classification in the study of nature. In many ways, the Greeks began the process of separating scientific fact from superstition.

Greek Math and Science

Understanding What You Read
1. How did Greeks further the study of mathematics?
2. What were some of the Greeks' scientific achievements?

This Greek vase shows a warrior holding a shield. Kleophrades is thought to have made this vase, which dates to 500 B.C.

This mosaic of a dog bears the inscription "Good Hunting."

The Arts The ancient Greeks produced beautiful buildings, sculpture, poetry, plays, pottery, and gold jewelry. They also made **mosaics** (moh-ZAY-iks)—pictures created from tiny pieces of colored stone—that were copied throughout Europe. The folk music of Greece shares many features with the music of Turkey and Southwest Asia. In 1963 the Greek writer George Seferis won the Nobel Prize in literature.

✓ **READING CHECK:** Why is there a Turkish influence in Greek culture? because the Ottoman Turks conquered Greece and controlled it for nearly 400 years

Have students complete the Section Review. Then have them reread the subsection on Greek history in Section 2. Pair students and tell one student to close his or her textbook and to summarize the main points. The other student in each pair should keep the book open and ask questions to encourage responses. Have students reverse the roles for the next subsection. Continue until the entire section is completed. Then have students complete Daily Quiz 18.2. **COOPERATIVE LEARNING**

RETEACH

Have students complete Main Idea Activity 18.2. Then tell students about the battle in 490 B.C. at Marathon, where Greek soldiers defeated the Persians. According to legend, a messenger ran more than 26 miles to report the news. Modern marathons commemorate the event. Hold a marathon in class. Organize students into teams, mark off 26 spaces on the floor, and allow "runners" to advance one "mile" for each fact from Section 2 that the team can report to the class.
ENGLISH LANGUAGE LEARNERS

EXTEND

Have interested students conduct research on Greek mythology and how it explained natural phenomena. Each student should focus on one myth. Then have students compare the Greek explanation with their culture's explanation. Ask volunteers to present their findings to the class.
BLOCK SCHEDULING

Greece Today

When people think of Greece now, they often recall the past. For example, many have seen pictures of the Parthenon, a temple built in the 400s B.C. in Athens. It is one of the world's most photographed buildings.

Economy Greece today lags behind other European nations in economic growth. More people work in agriculture than in any other industry. However, only about 19 percent of the land can be farmed because of the mountains. For this reason old methods of farming are used rather than modern equipment. Farmers raise cotton, tobacco, vegetables, wheat, lemons, olives, and raisins.

Service and manufacturing industries are growing in Greece. However, the lack of natural resources limits industry. Tourism and shipping are key to the Greek economy.

Cities About 20 percent of the Greek labor force works in agriculture. About 40 percent of Greeks live in rural areas. In the past few years, people have begun to move to the cities to find better jobs.

Athens, in central Greece, is the capital and by far the largest city. About one third of Greece's population lives in the area in and around Athens. Athens and its seaport, Piraeus, have attracted both people and industries. Most of the country's economic growth is centered there. However, the city suffers from air pollution, which causes health problems. Air pollution also damages historical sites, such as the Parthenon. Greece's second-largest city is Thessaloníki. It is the major seaport for northern Greece.

✓ **READING CHECK:** *Environment and Society* How does scarcity of natural resources affect Greece's economy? *It limits industry and forces Greece to rely on tourism and shipping.*

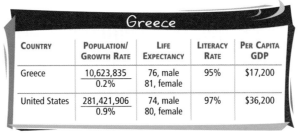

COUNTRY	POPULATION/ GROWTH RATE	LIFE EXPECTANCY	LITERACY RATE	PER CAPITA GDP
Greece	10,623,835 0.2%	76, male 81, female	95%	$17,200
United States	281,421,906 0.9%	74, male 80, female	97%	$36,200

Sources: Central Intelligence Agency, *The World Factbook 2001;* U.S. Census Bureau

Interpreting the Chart Why might life expectancy be higher in Greece than in the United States?

go.hrw.com Homework Practice Online
Keyword: SJ3 HP18

Section Review 2

Define and explain: city-states, mosaics

Working with Sketch Maps On the map you created in Section 1, label Athens and Thessaloníki. What physical features do these cities have in common? What economic activities might they share?

Reading for the Main Idea
1. (*Human Systems*) What groups influenced Greek culture?
2. (*Human Systems*) For what art forms is Greece famous?

Critical Thinking
3. **Drawing Inferences and Conclusions** How did the physical geography of this region influence the growth of major cities?
4. **Finding the Main Idea** On what does Greece rely to keep its economy strong?

Organizing What You Know
5. **Sequencing** Create a time line that documents the history of ancient Greece from 2000 B.C. to A.D. 1453.

2000 B.C. ———————————————— A.D. 1453

Objectives

1. Explore the early history of Italy.
2. Describe how Italy has added to world culture.
3. Explain what Italy is like today.

FOCUS

LET'S GET STARTED

Copy the following question onto the chalkboard: *What do you already know about Rome or the Roman Empire?* Discuss responses. *(Possible answers: chariot races, gladiators, the persecution of Christians)* Tell students that although Rome spread and maintained its power by force, it also made lasting contributions to literature, language, law, city planning, architecture, engineering, and other fields. Tell students that in Section 3 they will learn more about ancient Rome and modern Italy.

Building Vocabulary

Write the key terms on the chalkboard. Explain that **pope** comes from the Latin word *papa*, meaning "father," and that the pope is considered the father of the Catholic church. Tell students that the word **Renaissance** combines *re-*, which means "again," with a Latin word, *nasci*, which means "to be born." Renaissance means "to be born again" or "rebirth." Point out that the Renaissance was a rebirth of learning. Call on a volunteer to find and read the definition of **coalition governments**.

Reproducible

◆ Guided Reading Strategy 18.3
◆ Cultures of the World Activity 3
◆ Map Activity 18

Technology

◆ One-Stop Planner CD–ROM, Lesson 18.3
◆ Homework Practice Online
◆ *ARGWorld* CD–ROM: Conditions and Connections in Renaissance Italy
◆ HRW Go site

Reinforcement, Review, and Assessment

◆ Map Activity 18
◆ Section 3 Review, p. 403
◆ Daily Quiz 18.3
◆ Main Idea Activity 18.3
◆ English Audio Summary 18.3
◆ Spanish Audio Summary 18.3

Visual Record Answer ▶

Students might suggest that the contruction is of high quality.

Section 3
Italy

Read to Discover

1. What was the early history of Italy like?
2. How has Italy added to world culture?
3. What is Italy like today?

Define

pope
Renaissance
coalition governments

Locate

Rome
Genoa
Naples
Milan
Turin
Florence

WHY IT MATTERS

Leonardo da Vinci and Galileo are just two famous Italians who have made significant contributions to science and art. Use CNNfyi.com or other current events sources to find examples of recent Italian scientists and artists. Record your findings in your journal.

Artifact from a warrior's armor, 400s B.C.

▲
The Appian Way was a road from Rome to Brindisi. It was started in 312 B.C. by the emperor Claudius.

Interpreting the Visual Record How do you think this road has withstood more than 2,000 years of use?

History

About 750 B.C. a tribe known as the Latins established the city of Rome on the Tiber River. Over time, these Romans conquered the rest of Italy. They then began to expand their rule to lands outside Italy.

Roman Empire At its height about A.D. 100, the Roman Empire stretched westward to what is now Spain and Portugal and northward to England and Germany. The Balkans, Turkey, parts of Southwest Asia, and coastal North Africa were all part of the empire. Roman laws, roads, engineering, and the Latin language could be found throughout this huge area. The Roman army kept order, and people could travel safely throughout the empire. Trade prospered. The Romans made advances in engineering, including roads and aqueducts—canals that transported water. They also learned how to build domes and arches. Romans also produced great works of art and literature.

About A.D. 200, however, the Roman Empire began to weaken. The western part, with its capital in Rome, fell in A.D. 476. The eastern part, the Byzantine Empire, lasted until 1453.

Roman influences in the world can still be seen today. Latin developed into the modern languages of French, Italian, Portuguese, Romanian, and Spanish. Many English words have Latin origins as well. Roman laws and political ideas have influenced the governments and legal systems of many modern countries.

Teaching Objectives 1–2

ALL LEVELS: (Suggested time: 40 min.) Copy the following graphic organizer onto the chalkboard, omitting the italicized answers. Use it to help students explore the early history of Italy and how it has added to world culture. Complete the organizer as a class. Then pair students and have each pair write one sentence about each entry. **ENGLISH LANGUAGE LEARNERS, COOPERATIVE LEARNING**

The History and Culture of ITALY

History	Culture
• *750 B.C. Rome established*	• *Latin*
• *Roman Empire*	• *Roman Catholic Church*
• *Christianity*	• *Mediterranean diet*
• *Renaissance*	• *glassware*
• *coalition governments*	• *jewelry*
	• *painting*
	• *sculpture*

The Colosseum is a giant amphitheater. It was built in Rome between A.D. 70 and 80 and could seat 50,000 people.

Interpreting the Visual Record
For what events do you think the Colosseum was used? What type of modern buildings look like this?

Christianity began in the Roman province of Judaea (modern Israel and the West Bank). It then spread through the Roman Empire. Some early Christians were persecuted for refusing to worship the traditional Roman gods. However, in the early A.D. 300s the Roman emperor, Constantine, adopted Christianity. It quickly became the main religion of the empire. The **pope**—the bishop of Rome—is the head of the Roman Catholic Church.

The Renaissance Beginning in the 1300s a new era of learning began in Italy. It was known as the **Renaissance** (re-nuh-SAHNS). In French this word means "rebirth." During the Renaissance, Italians rediscovered the work of ancient Roman and Greek writers. Scholars applied reason and experimented to advance the sciences. Artists pioneered new techniques. Leonardo da Vinci, painter of the *Mona Lisa*, was also a sculptor, engineer, architect, and scientist. Another Italian, Galileo Galilei, perfected the telescope and experimented with gravity.

Christopher Columbus opened up the Americas to European colonization. Although Spain paid for his voyages, Columbus was an Italian from the city of Genoa. The name *America* comes from another Italian explorer, Amerigo Vespucci.

Government Italy was divided into many small states until the late 1800s. Today Italy's central government is a democracy with an elected parliament. Italy has had many changes in leadership in recent years. This has happened because no political party has won a majority of votes in Italian elections. As a result, political parties must form **coalition governments**. A coalition government is one in which several parties join together to run the country. Unfortunately, these coalitions usually do not last long.

✓ **READING CHECK:** *Human Systems* How is the Italian government different from that of the United States? *many changes in leadership, coalition governments*

Leonardo da Vinci painted the *Mona Lisa* about 1503–06.

Interpreting the Visual Record Why do you think Leonardo's painting became famous?

Across the Curriculum
TECHNOLOGY

Roman Engineering The Romans were the best civil engineers of the ancient world. They used arches, concrete, and tunnel-like vaults to create bridges and buildings.

The Romans also built more than 50,000 miles of roads, which crossed an area that now includes 30 countries. The roads were built in layers of rough stone, gravel, and sand. Smooth paving stones covered the surface, which was higher in the center so that water would drain to the sides.

The Romans were also famous for their aqueducts, or covered stone water channels. Aqueducts supplied the city of Rome with more than 200 million gallons of water each day.

Activity: Organize students into groups to research, build, and label models of aqueducts, baths, or arenas.

☑ internet connect

GO TO: go.hrw.com
KEYWORD: SJ3 CH18
FOR: Web sites about Rome

▲ **Visual Record Answer**

sporting events and public displays; stadiums

◄

Students might mention the subject's mysterious smile.

Teaching Objective 3

ALL LEVELS: (Suggested time: 20 min.) Organize students into triads. Have each triad write a brief magazine article titled Italy Today. Articles should focus on Italy's economy, its cities, and the differences between northern and southern Italy.
ENGLISH LANGUAGE LEARNERS, COOPERATIVE LEARNING

CLOSE

Have volunteers write on the chalkboard one fact about Roman and Italian contributions to culture. To prompt students, you might first write categories such as architecture, politics, religion, art, food, and fashion on the chalkboard.

REVIEW AND ASSESS

Have students complete the Section Review. Then organize the class into groups of four, to represent the four horses of a Roman chariot team. Have each group review the sections. Then stage a chariot race by clearing a space in the classroom, asking questions of each group, and allowing the groups to take one step forward with each correct answer. Then have students complete Daily Quiz 18.3.

RETEACH

Have students complete Main Idea Activity 18.3. Then create sets of index cards with the names of major historical periods and famous Italians written on them. Organize the class into groups. Give each group a set of cards.

USING ILLUSTRATIONS

Focus students' attention on the *Mona Lisa* on the previous page. Remind students that Italian artist Leonardo da Vinci painted the *Mona Lisa* about 500 years ago. Ask them to identify other places they may have seen the image. *(possible answers: T-shirts, cartoons, and advertisements)* Point out that images of the *Mona Lisa* have been popular for many years and that the painting is recognized around the world. How have modern media helped spread the image to people around the world? You may want to challenge students to search the media for more images that began as part of another country's culture but are now encountered worldwide.

This view of the Pantheon in Rome shows the oculus—the opening at the top. The Pantheon was built as a temple to all Roman gods.
Interpreting the Visual Record Why did the Romans need to design buildings that let in light?

▼

In Rome people attend mass in St. Peter's Square, Vatican City. Vatican City is an independent state within Rome.
Interpreting the Visual Record How does a plaza, or open area, help create a sense of community?

▼

Culture

People from other places have influenced Italian culture. During the Renaissance, many Jews who had been expelled from Spain moved to Italian cities. Jews often had to live in segregated areas called ghettoes. Today immigrants have arrived from former Italian colonies in Africa. Others have come from the eastern Mediterranean and the Balkans.

Religion and Food Some 98 percent of Italians belong to the Roman Catholic Church. The leadership of the church is still based in the Vatican in Rome. Christmas and Easter are major holidays in Italy. Italians also celebrate All Saints Day on November 1 by cleaning and decorating their relatives' graves.

Italians enjoy a range of regional foods. Recipes are influenced by the history and crops of each area. In the south, Italians eat a Mediterranean diet of olives, bread, and fish. Dishes are flavored with lemons from Greece and spices from Africa. Tomatoes, originally from the Americas, have become an important part of the diet. Some Italian foods, such as pizza, are popular in the United States. Modern pizza originated in Naples. Northern Italians eat more rice, butter, cheeses, and mushrooms than southern Italians.

The Arts The ancient Romans created beautiful glassware and jewelry as well as marble and bronze sculptures. During the Renaissance, Italy again became a center for art, particularly painting and sculpture. Italian artists discovered ways to make their paintings more lifelike. They did this by creating the illusion of three dimensions. Italian writers like Francesco Petrarch and Giovanni Boccaccio wrote some of the most important literature of the Renaissance. More recently, Italian composers have written great operas. Today, Italian designers, actors, and filmmakers are celebrated worldwide.

✓ **READING CHECK:** *Human Systems* What are some examples of Italian culture? Italian food, such as pizza; glassware, jewelry, painting, sculpture, literature, opera, film

Visual Record Answers ▲

because they had no electricity or gas to create artificial light

▶

by allowing people to come together

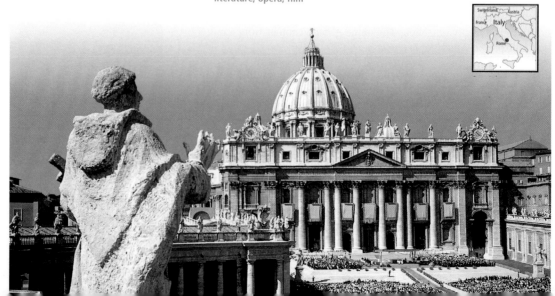

Have students place the cards in correct chronological order as quickly as possible. Have each group select one card and use the text to tell the class about the time period or person on that card.
ENGLISH LANGUAGE LEARNERS, COOPERATIVE LEARNING

EXTEND

Have interested students conduct research on the major languages derived from Latin—French, Italian, Portuguese, and Spanish. Challenge them to find and compare common words in those languages. Students might create web diagrams to share their findings with the class. **BLOCK SCHEDULING**

TEACHER TO TEACHER

Lois Jordan, of Nashville, Tennessee, suggests the following activity to help students learn more about Italian culture. Have students conduct research on what typical Italians might eat. Then have students compare the Italian diet with the diet in a region or country with different physical and climatic characteristics, such as northern Canada. Students might investigate how climate and natural resources affect diet, how much of the average person's income is spent on food, and whether most people eat their meals at home or elsewhere.

Italy Today

Italy is slightly smaller than Florida and Georgia combined, with a population of about 57 million. A shared language, the Roman Catholic Church, and strong family ties continue to bind Italians together.

Economy After its defeat in World War II, Italy rebuilt its industries in the north. Rich soil and plenty of water make the north Italy's "breadbasket," or wheat growing area. Italy's most valuable crop is grapes. Although grapes are grown throughout the country, northern Italy produces the best crops. These grapes help make Italy the world's largest producer of wine. Tourists are also important to Italy's economy. They visit northern and central Italy to see ancient ruins and Renaissance art. Southern Italy remains poorer with lower crop yields. Industrialization there also lags behind the north. Tourist resorts, however, are growing in the south and promise to help the economy.

Cities The northern cities of Milan, Turin, and Genoa are important industrial centers. Their location near the center of Europe helps companies sell products to foreign customers. Also in the north are two popular tourist sites. One is Venice, which is famous for its romantic canals and beautiful buildings. The other is Florence, a center of art and culture. Rome, the capital, is located in central Italy. Naples, the largest city in southern Italy, is a major manufacturing center and port.

✓ **READING CHECK:** *Environment and Society* What geographic factors influence Italy's economy? Rich soil and plenty of water are good for growing crops, particularly grapes.

Italy

COUNTRY	POPULATION/ GROWTH RATE	LIFE EXPECTANCY	LITERACY RATE	PER CAPITA GDP
Italy	57,679,825 .07%	76, male 83, female	98%	$22,100
United States	281,421,906 0.9%	74, male 80, female	97%	$36,200

Sources: Central Intelligence Agency, *The World Factbook 2001*; U.S. Census Bureau

Interpreting the Chart What is the difference in the growth rate of Italy and the United States?

Section Review 3

Define and explain: pope, Renaissance, coalition governments

Working with Sketch Maps On the map you created in Section 2, label Florence, Genoa, Milan, Naples, Rome, and Turin. Why are they important?

Reading for the Main Idea

1. *Human Systems* What were some of the important contributions of the Romans?

2. *Human Systems* What are some art forms for which Italy is well known?

Critical Thinking

3. **Finding the Main Idea** Which of Italy's physical features encourage trade? Which geographical features make trading difficult?

4. **Analyzing Information** Why is the northern part of Italy known as the country's "breadbasket"?

Organizing What You Know

5. **Finding the Main Idea** Copy the following graphic organizer. Use it to describe the movement of goods and ideas to and from Italy during the early days of trade and exploration.

[ ] ⇨ Italy ⇨ [ ]

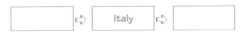
Homework Practice Online Keyword: SJ3 HP18

Section Review 3

Answers

Define For definitions, see: pope, p. 401; Renaissance, p. 401; coalition governments, p. 401

Working with Sketch Maps Maps will vary, but listed places should be labeled in their approximate locations. These cities serve as centers of culture, industry, and government.

Reading for the Main Idea

1. Latin, art, literature, roads, aqueducts, domes, arches, laws (NGS 10)

2. architecture, glassware, jewelry, sculpture, painting, literature, opera, drama, and film (NGS 10)

Critical Thinking

3. access to the sea and its location at the center of the Mediterranean; few major rivers or harbors and its central mountains

4. its plentiful wheat harvests

Organizing What You Know

5. to—Jews expelled from Spain, African and Balkan immigrants; from—laws, roads, engineering, language, art, literature, architecture, Christianity, Renaissance ideas

▲ **Chart Answer**

Italy's growth is declining, while that of the United States is increasing.

403

Objectives

1. Identify some major events in the history of Spain and Portugal.
2. Describe the cultures of Spain and Portugal.
3. Investigate what Spain and Portugal are like today.

FOCUS

LET'S GET STARTED

Copy the following instructions onto the chalkboard: *Use the Fast Facts features in your textbook to find countries other than Spain or Portugal where Spanish or Portuguese is spoken. How many can you list?* Discuss students' lists. Ask students how they think these languages spread so far. Tell them that in Section 4 they will learn more about the history and culture of Spain and Portugal.

Building Vocabulary

Write the key terms on the chalkboard. Write *moor* next to **Moors**. Explain that the capitalized term refers to Muslims from North Africa who conquered much of Spain in the A.D. 700s. Point out that **dialect** comes from Greek, Latin, and French roots that mean "between" or "over" (dia-) and "to speak" (-lect). Have students relate the roots to the term. Conclude by having a volunteer find and read aloud the definition of the term **cork**.

Section 4 RESOURCES

Reproducible
◆ Guided Reading Strategy 18.4
◆ Cultures of the World Activity 3

Technology
◆ One-Stop Planner CD–ROM, Lesson 18.4
◆ HRW Go site

Reinforcement, Review, and Assessment
◆ Section 4 Review, p. 407
◆ Daily Quiz 18.4
◆ Main Idea Activity 18.4
◆ English Audio Summary 18.4
◆ Spanish Audio Summary 18.4

Visual Record Answer ▶

Blades catch the wind and allow a mechanical system to draw water.

Section 4 Spain and Portugal

Read to Discover

1. What were some major events in the history of Spain and Portugal?
2. What are the cultures of Spain and Portugal like?
3. What are Spain and Portugal like today?

WHY IT MATTERS

Some Basque separatists have used violence to try to gain their independence from Spain. Use CNNfyi.com or other current events sources to find examples of this problem. Record your findings in your journal.

Define

Moors
dialect
cork

Locate

Lisbon
Madrid
Barcelona

Paella, a popular dish in Spain

History

Beautiful paintings of bison and other animals are found in caves in northern Spain. Some of the best known are at Altamira and were created as early as 16,000 B.C. Some cave paintings are much older. These paintings give us exciting clues about the early people who lived here.

Ancient Times Spain has been important to Mediterranean trade for several thousand years. First, the Greeks and then the Phoenicians, or Carthaginians, built towns on Spain's southern and eastern coasts. Then, about 200 B.C. Iberia became a part of the Roman Empire and adopted the Latin language.

These windmills in Consuegra, Spain, provided water for the people of the region.

Interpreting the Visual Record How do you think windmills pump water?

▼

Teaching Objective 1
ALL LEVELS: (Suggested time: 30 min.) Tell students to imagine that they are tour guides in charge of a trip to Spain and Portugal. Before they leave on the trip they must present a brief history of Spain and Portugal to their clients. Pair students and have each pair write a presentation. Ask volunteers to present their histories to the class.
ENGLISH LANGUAGE LEARNERS, COOPERATIVE LEARNING

Teaching Objectives 2–3
ALL LEVELS: (Suggested time: 20 min.) Copy the graphic organizer on the following page onto the chalkboard, omitting the italicized answers. Use it to help students compare the cultures of Spain and Portugal and what the countries are like today. Have each student complete it.
ENGLISH LANGUAGE LEARNERS

The Muslim North Africans, or **Moors**, conquered most of the Iberian Peninsula in the A.D. 700s. Graceful Moorish buildings, with their lacy patterns and archways, are still found in Spanish and Portuguese cities. This is particularly true in the old Moorish city of Granada in southern Spain.

Great Empires From the 1000s to the 1400s Christian rulers fought to take back the peninsula. In 1492 King Ferdinand and Queen Isabella conquered the kingdom of Granada, the last Moorish outpost in Spain. That same year, they sponsored the voyage of Christopher Columbus to the Americas. Spain soon established a large empire in the Americas.

The Portuguese also sent out explorers. Some of them sailed around Africa to India. Others crossed the Atlantic and claimed Brazil. In the 1490s the Roman Catholic pope drew a line to divide the world between Spain and Portugal. Western lands, except for Brazil, were given to Spain, and eastern lands to Portugal.

With gold and agricultural products from their American colonies, and spices and silks from Asia, Spain and Portugal grew rich. In 1588 Philip II, king of Spain and Portugal, sent a huge armada, or fleet, to invade England. The Spanish were defeated, and Spain's power began to decline. However, most Spanish colonies in the Americas did not win independence until the early 1800s.

Government In the 1930s the king of Spain lost power. Spain became a workers' republic. The new government tried to reduce the role of the church and to give the nobles' lands to farmers. However, conservative military leaders under General Francisco Franco resisted. A civil war was fought from 1936 to 1939 between those who supported Franco and those who wanted a democratic form of government. Franco's forces won the war and ruled Spain until 1975. Today Spain is a democracy, with a national assembly and prime minister. The king also plays a modest role as head of state.

Portugal, like Spain, was long ruled by a monarch. In the early 1900s the monarchy was overthrown. Portugal became a democracy. However, the army later overthrew the government, and a dictator took control. A revolution in the 1970s overthrew the dictatorship. For a few years disagreements between the new political parties brought violence. Portugal is now a democracy with a president and prime minister.

✓ READING CHECK: ***Human Systems*** How did Spain and Portugal move from unlimited to limited governments? both were ruled by monarchs and military governments, and now both have democratic governments

The interior of the Great Mosque in Córdoba, Spain, shows the lasting beauty of Moorish architecture. A cathedral was built within the mosque after Christians took back the city.

One of the world's most endangered wild cats is the Iberian lynx. About 50 survive in a preserve on the Atlantic coast of Spain.

Across the Curriculum
MUSIC
Flamenco The musical and dance performance known as flamenco developd in southern Spain hundreds of years ago. Flamenco reflects Gypsy, Andalusian, and Arabic influences, among others. Early flamenco combined *cante* (song) and *baile* (dance), accompanied by rhythmic handclapping.

The golden age of flamenco lasted from 1869 to 1910. It became popular in cafés, and many performers added *guitarra*, or guitar playing. In the 1900s jazz, salsa, and bossa nova influenced flamenco. Flamenco has gained popularity recently, and some of its spontaneity has been replaced by rehearsed routines.

Activity: Have interested students conduct research on the popularity of flamenco guitar music and the rhythmic handclapping, called *palmas*, that accompanies it. Play a recording of a flamenco song in class. Ask students who have researched *palmas* to lead the class in clapping.

internet connect

GO TO: go.hrw.com
KEYWORD: SJ3 CH18
FOR: Web sites about flamenco

◄ **Chart Answer (p. 406)**

They are almost the same.

Tell students about *castell* building—a sport popular in the Spanish region of Catalonia. *Castellers* build castles out of people. Dressed in traditional clothing, a group pushes together tightly to form the foundation of the castle. Other people press in against them. Lighter, barefoot participants then scramble over their backs to stand on their shoulders. Finally a child climbs to the top. In this way, Catalonians build human towers about 35 feet (10.6 m) tall. Ask students to speculate on how this custom began.

The Culture of SPAIN and PORTUGAL

Food and Festivals	The Arts	Today
• olives and olive oil • limes • wine • fish • wheat • foods from the Americas • Roman Catholic holidays • bullfights	• porcelain • fado singers • flamenco dancers • Picasso	• European Union • agricultural products: wine, fruit, olive oil, olives, and cork • clothing • timber products • cars and trucks • tourism

---

Section Review 4

Answers

Define For definitions see: Moors, p. 273; dialect, p. 274; cork, p. 275

Working with Sketch Maps Maps will vary, but should correctly depict the locations of the three cities. Lisbon and Barcelona are seaports while Madrid is located inland.

Reading for the Main Idea

1. Portuguese fado music and Spanish flamenco dancing reflect African influence. (NGS 10)

2. by joining the European Union, which provides free trade, travel, exchange of workers, and rapid growth (NGS 11)

Critical Thinking

3. Greeks, Phoenicians, Romans, and Moors

4. 1930s—king lost power and Spain became a workers' republic; 1936–1939—civil war, which led to conservative military rule under Francisco Franco until 1975; since 1975—democracy

Organizing What You Know

5. Graphic organizers will vary but should include the following information: A.D. 700s—Moorish advance on area; 1492—kingdom of Granada conquered and Columbus sails to the Americas; 1588—defeat of Spanish Armada.

Spain and Portugal

Country	Population/ Growth Rate	Life Expectancy	Literacy Rate	Per Capita GDP
Portugal	10,066,253 0.2%	72, male 80, female	87%	$15,800
Spain	40,037,995 0.1%	75, male 83, female	97%	$18,000
United States	281,421,906 0.9%	74, male 80, female	97%	$36,200

Sources: Central Intelligence Agency, *The World Factbook 2001;* U.S. Census Bureau

Interpreting the Chart How do the growth rates of these countries compare?

Flamenco dancers perform at a fair in Málaga, Spain.

Culture

The most widely understood Spanish **dialect** (DY-uh-lekt), or variation of a language, is Castilian. This is the form spoken in central Spain. Spanish and Portuguese are not the only languages spoken on the Iberian Peninsula, however. Catalan is spoken in northeastern Spain (Catalonia). Basque is spoken by an ethnic group living in the Pyrenees.

Spain faces a problem of unrest among the Basque people. The government has given the Basque area limited self-rule. However, a small group of Basque separatists continue to use violence to protest Spanish control.

Food and Festivals Spanish and Portuguese foods are typical of the Mediterranean region. Many recipes use olives and olive oil, lemons, wheat, wine, and fish. Foods the explorers brought back from the Americas—such as tomatoes and peppers—are also important.

Both Spain and Portugal remain strongly Roman Catholic. The two countries celebrate major Christian holidays like Christmas and Easter. As in Italy, each village has a patron saint whose special day is the occasion for a fiesta, or festival. A bull fight, or *corrida*, may take place during the festival.

The Arts Spanish and Portuguese art reflects the many peoples who have lived in the region. The decoration of Spanish porcelain recalls Islamic art from North Africa. The sad melodies of the Portuguese fado singers and the intense beat of Spanish flamenco dancing also show African influences. In the 1900s the Spanish painter Pablo Picasso boldly experimented with shape and perspective. He became one of the most famous artists of modern times.

✓ **READING CHECK:** How has the mixture of different ethnic groups created some conflict in Spanish society? Some Basque people want to be separate from Spain and are using violence to achieve that goal.

Have students complete the Section Review. Then organize the class into triads. Have each triad work together to create an outline of Section 4. Call on volunteers to write their outlines on the chalkboard. Compare the outlines and fill in any blanks. To conclude have students complete Daily Quiz 18.4. **COOPERATIVE LEARNING**

Have students complete Main Idea Activity 18.4. Then have them suggest words that describe Spain or Portugal. Have the class decide whether the word relates to the physical or the cultural geography of the region. **ENGLISH LANGUAGE LEARNERS**

Have interested students conduct research on a Spanish artist, such as El Greco, Goya, Murillo, or Velázquez. Ask students to prepare posters illustrating a work by their chosen artist. Have students include a statement about how the artist's work reflects Spanish cultural traditions. **BLOCK SCHEDULING**

Spain and Portugal Today

Like Greece and Italy, both Spain and Portugal belong to the European Union (EU). The EU allows free trade, travel, and exchange of workers among its members. The economies of Spain and Portugal have been growing rapidly. However, they remain poorer than the leading EU countries.

Agricultural products of Spain and Portugal include wine, fruit, olives, olive oil, and **cork**. Cork is the bark stripped from a certain type of oak tree. Spain exports oranges from the east, beef from the north, and lamb from ranches on the Meseta. Portugal also makes and exports clothing and timber products. Spain makes cars and trucks, and most of its industry is located in the north. Tourism is also an important part of the Spanish economy. This is particularly true along Spain's coasts and on the Balearic Islands.

Portugal's capital and largest city is Lisbon. It is located on the Atlantic coast at the mouth of the Tagus River. Madrid, Spain's capital and largest city, is located inland on the Meseta. Spain's second-largest city is the Mediterranean port of Barcelona.

✔ **READING CHECK:** *Places and Regions* What are Spain and Portugal like today? rapidly growing economies with agricultural exports and industry

Porto, Portugal, combines modern industry with the historical sea trade.

A worker uses an ax to strip the bark from a cork oak.

go.hrw.com
Homework Practice Online
Keyword: SJ3 HP18

Section Review 4

Define and explain: Moors, dialect, cork

Working with Sketch Maps On the map that you created in Section 3, label Lisbon, Madrid, and Barcelona. How are Lisbon and Barcelona different from Madrid?

Reading for the Main Idea

1. *Human Systems* How do the performing arts of this region reflect different cultures?

2. *Human Systems* How have Spain and Portugal worked to improve their economies?

Critical Thinking

3. **Analyzing Information** Which groups influenced the culture of Spain and Portugal?

4. **Summarizing** How was the government of Spain organized during the 1900s?

Organizing What You Know

5. **Sequencing** Copy the following graphic organizer. Use it to list important events in the history of Spain and Portugal from the 700s to the 1600s.

☐ ⇨ ☐ ⇨ ☐ ⇨ ☐ ⇨ ☐

CHAPTER **18**

Review
ANSWERS

Building Vocabulary
For definitions, see: mainland, p. 394; sirocco, p. 395; city-states, p. 396; mosaics, p. 398; pope, p. 401; Renaissance, p. 401; coalition governments, p. 401; Moors, p. 405; dialect, p. 406; cork, p. 407

Reviewing the Main Ideas

1. peninsulas (Iberian, Italian, Greek); mountains (Cantabrian, Pyrenees, Alps, Apennines); a volcano (Vesuvius); and thousands of islands (NGS 4)

2. warm, sunny climate; natural beauty; numerous historical and cultural sites (NGS 15)

3. northern—importance of wine industry, more industrialization, prosperous; southern—less industrialized, poorer (NGS 12)

4. Roman Catholicism; in the Vatican in Rome (NGS 10)

5. by acquiring and trading gold and agricultural products from their New World colonies, and spices and silks from Asia (NGS 11)

Have students complete the Chapter 18 Test.

RETEACH

Have students prepare segments for a television broadcast titled Focus on Southern Europe. Organize the class into five groups. Assign each group one of the following topics: physical features, major historical events, contributions to world culture, economies, and cities. Have groups write scripts and present their newscasts to the class.
ENGLISH LANGUAGE LEARNERS, COOPERATIVE LEARNING

PORTFOLIO EXTENSIONS

COOPERATIVE LEARNING

1. Organize students into five groups and have each group do research and create an illustrated time line showing the development of Greek contributions to one of the following topics: art, architecture, literature, science, and philosophy. Have the groups collaborate to make connections between the events on the time lines. *(Example: a scene from the Iliad might be depicted on artwork)*

2. Not all citizens of modern Spain consider themselves Spaniards. Many people living in the country's northeastern Basque region want an independent Basque nation. Have students conduct research on the Basque conflict and debate whether the region should be allowed to break away from Spain. Place research notes and debate outlines in student portfolios.

Review
ANSWERS

Understanding Environment and Society
Presentations will vary, but the information included should be consistent with text material. Use Rubric 40, Writing to Describe, to evaluate student work.

Thinking Critically

1. Answers will vary but should refer to the region's access to the Mediterranean Sea and the Atlantic Ocean.

2. art, food, and music; the proximity of the two countries and former Turkish rule of Greece

3. its location in the center of the Mediterranean world

4. Olives are used to make olive oil; grapes are used to make wine; and other staple crops such as lemons, mushrooms, tomatoes, and wheat are used in many dishes.

5. the region's cultural events, historical sites, and pleasant climate

Reviewing What You Know

Building Vocabulary

On a separate sheet of paper, write sentences to define each of the following words.

1. mainland
2. sirocco
3. city-states
4. mosaics
5. pope
6. Renaissance
7. coalition governments
8. Moors
9. dialect
10. cork

Reviewing the Main Ideas

1. *(Places and Regions)* What are the important physical features of southern Europe?

2. *(Environment and Society)* Why has southern Europe been a major tourist attraction for centuries?

3. *(Human Systems)* What are the major economic differences between northern and southern Italy?

4. *(Human Systems)* What is the main religion in Italy? Where is the leadership of this religion based?

5. *(Human Systems)* How did both Spain and Portugal become wealthy during the 1400s and 1500s?

Understanding Environment and Society

The Arts
Environment influences a culture's art. For example, grapes might be featured in a mosaic. Create a presentation about the influence of the environment on the arts. Write a description of each piece and explain how it might have been influenced by the artist's environment.

Thinking Critically

1. **Drawing Influences and Conclusions** In what ways do you think the geography of southern Europe made trade and exploration possible?

2. **Summarizing** What parts of Greek culture have been most strongly influenced by Turkish customs? Why is this the case?

3. **Drawing Inferences and Conclusions** Recall what you have learned about the Roman Empire. What about the Italian peninsula made it a good location for a Mediterranean empire?

4. **Drawing Inferences and Conclusions** How are agricultural products of southern Europe used in food?

5. **Finding the Main Idea** Why do many tourists continue to visit historical cities in southern Europe?

FOOD FESTIVAL

This is the perfect opportunity for a pizza party. You may want to have students bring different toppings for purchased crusts. Or, have students conduct research on the history of pizza and make one as historically accurate as possible. For example, in Italy, pizza is sometimes made without tomato sauce. To celebrate Spain, enjoy tapas. These appetizers can range from olives, cubes of cheese, or ham to fancier cold omelets, stuffed peppers, or small sandwiches.

CHAPTER 18

REVIEW AND ASSESSMENT RESOURCES

Reproducible
◆ Readings in World Geography, History, and Culture 28, 29, and 30
◆ Critical Thinking Activity 18
◆ Vocabulary Activity 18

Technology
◆ Chapter 18 Test Generator (on the One-Stop Planner)

◆ HRW Go site
◆ Audio CD Program, Chapter 18

Reinforcement, Review, and Assessment
◆ Chapter 18 Review, pp. 408–09

◆ Chapter 18 Tutorial for Students, Parents, Mentors, and Peers
◆ Chapter 18 Test
◆ Chapter 18 Test for English Language Learners and Special-Needs Students

Building Social Studies Skills

Map ACTIVITY

On a separate sheet of paper, match the letters on the map with their correct labels.

Alps	Po River
Apennines	Tiber River
Aegean Sea	Naples
Peloponnesus	Sicily
Ebro River	Meseta

Mental Mapping Skills ACTIVITY

On a separate sheet of paper, draw a freehand map of southern Europe. Make a key for your map and label the following:

Athens	Portugal
Greece	Rome
Italy	Spain
Mediterranean Sea	Strait of Gibraltar

WRITING ACTIVITY

Find a recording of Portuguese fado music. Then write a review that explains what the lyrics of the songs reveal about Portuguese culture. Be sure to use standard grammar, spelling, sentence structure, and punctuation in your review.

Map Activity

A. Sicily	F. Peloponnesus
B. Po River	G. Meseta
C. Tiber River	H. Naples
D. Apennines	I. Ebro River
E. Aegean Sea	J. Alps

Mental Mapping Skills Activity

Maps will vary but listed places should be labeled in their approximate locations.

Writing Activity

Reviews will vary but should include information about Portuguese culture. Use Rubric 37, Writing Assignments, to evaluate student work.

Portfolio Activity

Answers will vary, but the information included should be consistent with text material. Use Rubric 7, Charts, to evaluate student work.

Alternative Assessment

Portfolio ACTIVITY

Learning About Your Local Geography
Individual Project The cultures of southern European influence the United States in many ways. Investigate southern European influences in your community. List some of those influences in a chart.

internet connect
Internet Activity: go.hrw.com
KEYWORD: SJ3 GT18

Choose a topic to explore southern Europe:
- Explore the islands and peninsulas on the Mediterranean coast.
- Take an online tour of ancient Greece.
- Learn the story of pizza.

internet connect
GO TO: go.hrw.com
KEYWORD: SJ3 Teacher
FOR: a guide to using the Internet in your classroom

West–Central Europe
Chapter Resource Manager

Objectives	Pacing Guide	Reproducible Resources
SECTION 1		
Physical Geography (pp. 411–13) 1. Identify the area's major landform regions. 2. Describe the role rivers, canals, and harbors play in the region. 3. Identify west-central Europe's major resources.	**Regular** .5 day **Block Scheduling** .5 day *Block Scheduling Handbook, Chapter 19*	**RS** Guided Reading Strategy 19.1 **RS** Graphic Organizer 19
SECTION 2		
France (pp. 414–17) 1. Identify which foreign groups affected the historical development of France. 2. Identify the main features of French culture. 3. Identify the products France exports.	**Regular** 1 day **Block Scheduling** .5 day *Block Scheduling Handbook, Chapter 19*	**RS** Guided Reading Strategy 19.2 **E** Cultures of the World Activity 3 **E** Creative Strategies for Teaching World Geography, Lessons 10 and 11 **SM** Geography for Life Activity 19
SECTION 3		
Germany (pp. 418–21) 1. Describe the effects wars have had on Germany. 2. Identify Germany's major contributions to world culture. 3. Describe how the division of Germany affected its economy.	**Regular** 1 day **Block Scheduling** .5 day *Block Scheduling Handbook, Chapter 19*	**RS** Guided Reading Strategy 19.3 **E** Creative Strategies for Teaching World Geography, Lessons 10 and 11 **SM** Map Activity 19
SECTION 4		
The Benelux Countries (pp. 422–24) 1. Describe how larger countries influenced the Benelux countries. 2. Describe what this region's culture is like. 3. Describe what the Benelux countries are like today.	**Regular** 1 day **Block Scheduling** .5 day *Block Scheduling Handbook, Chapter 19*	**RS** Guided Reading Strategy 19.4 **E** Creative Strategies for Teaching World Geography, Lessons 10 and 11
SECTION 5		
The Alpine Countries (pp. 425–27) 1. Identify some of the major events in the history of the Alpine countries. 2. Identify some of the cultural features of the region. 3. Describe how the economies of Switzerland and Austria are similar.	**Regular** 1 day **Block Scheduling** .5 day *Block Scheduling Handbook, Chapter 19*	**RS** Guided Reading Strategy 19.5 **E** Creative Strategies for Teaching World Geography, Lessons 10 and 11

Chapter Resource Key

RS Reading Support
IC Interdisciplinary Connections
E Enrichment
SM Skills Mastery
A Assessment
REV Review

ELL Reinforcement and English Language Learners
 Transparencies
 CD–ROM
 Music

 Video
 Internet
 Holt Presentation Maker Using Microsoft® Powerpoint®

  One-Stop Planner CD–ROM

See the *One-Stop Planner* for a complete list of additional resources for students and teachers.

One-Stop Planner CD–ROM

It's easy to plan lessons, select resources, and print out materials for your students when you use the *One-Stop Planner CD–ROM with Test Generator.*

Technology Resources	Review, Reinforcement, and Assessment Resources
One-Stop Planner CD–ROM, Lesson 19.1 Geography and Cultures Visual Resources with Teaching Activities 24–30 Homework Practice Online HRW Go site	**ELL** Main Idea Activity 19.1 **REV** Section 1 Review, p. 413 **A** Daily Quiz 19.1 **ELL** English Audio Summary 19.1 **ELL** Spanish Audio Summary 19.1
One-Stop Planner CD–ROM, Lesson 19.2 Homework Practice Online HRW Go site	**ELL** Main Idea Activity 19.2 **REV** Section 2 Review, p. 417 **A** Daily Quiz 19.2 **ELL** English Audio Summary 19.2 **ELL** Spanish Audio Summary 19.2
One-Stop Planner CD–ROM, Lesson 19.3 Homework Practice Online HRW Go site	**ELL** Main Idea Activity 19.3 **REV** Section 3 Review, p. 421 **A** Daily Quiz 19.3 **ELL** English Audio Summary 19.3 **ELL** Spanish Audio Summary 19.3
One-Stop Planner CD–ROM, Lesson 19.4 Homework Practice Online HRW Go site	**ELL** Main Idea Activity 19.4 **REV** Section 4 Review, p. 424 **A** Daily Quiz 19.4 **ELL** English Audio Summary 19.4 **ELL** Spanish Audio Summary 19.4
One-Stop Planner CD–ROM, Lesson 19.5 *ARGWorld* CD–ROM: Ski Resorts in Switzerland Homework Practice Online HRW Go site	**ELL** Main Idea Activity 19.5 **REV** Section 5 Review, p. 427 **A** Daily Quiz 19.5 **ELL** English Audio Summary 19.5 **ELL** Spanish Audio Summary 19.5

internet connect

HRW ONLINE RESOURCES

GO TO: go.hrw.com
Then type in a keyword.

TEACHER HOME PAGE
KEYWORD: SJ3 TEACHER

CHAPTER INTERNET ACTIVITIES
KEYWORD: SJ3 GT19

Choose an activity to:
• tour the land and rivers of Europe.
• travel back in time to the Middle Ages.
• visit schools in Belgium and the Netherlands.

CHAPTER ENRICHMENT LINKS
KEYWORD: SJ3 CH19

CHAPTER MAPS
KEYWORD: SJ3 MAPS19

ONLINE ASSESSMENT
Homework Practice
KEYWORD: SJ3 HP19
Standardized Test Prep Online
KEYWORD: SJ3 STP19
Rubrics
KEYWORD: SS Rubrics

COUNTRY INFORMATION
KEYWORD: SJ3 Almanac

CONTENT UPDATES
KEYWORD: SS Content Updates

HOLT PRESENTATION MAKER
KEYWORD: SJ3 PPT19

ONLINE READING SUPPORT
KEYWORD: SS Strategies

CURRENT EVENTS
KEYWORD: S3 Current Events

Meeting Individual Needs

Ability Levels

Level 1 Basic-level activities designed for all students encountering new material

Level 2 Intermediate-level activities designed for average students

Level 3 Challenging activities designed for honors and gifted-and-talented students

English Language Learners Activities that address the needs of students with Limited English Proficiency

Chapter Review and Assessment

IC	Interdisciplinary Activities for the Middle Grades 13, 14, 15	**A** Chapter 19 Test
E	Readings in World Geography, History, and Culture 31–33	Chapter 19 Test Generator (on the One-Stop Planner)
SM	Critical Thinking Activity 14	Audio CD Program, Chapter 19
REV	Chapter 19 Review, pp. 428–29	**A** Chapter 19 Test for English Language Learners and Special-Needs Students
REV	Chapter 19 Tutorial for Students, Parents, Mentors, and Peers	
ELL	Vocabulary Activity 19	

CHAPTER 19

Write the names of the countries of west-central Europe on the chalkboard. Then ask students leading questions about the countries. *(Examples: What is the Eiffel Tower, and where is it located? What country used to be two separate countries but has been reunited? What kind of cheese is full of holes? What country do you associate with tulips and windmills?)* Students will probably know the answers to some of these questions. Encourage them to ask other questions based on yours. *(Examples: What do windmills actually do? Why are they important?)* Create a web of student responses and questions on the chalkboard. Have a volunteer copy the web into his or her notes to allow class members to compare their initial questions to what they learn in this chapter.

Section 1

Objectives

1. Describe the area's major landform regions.
2. Analyze the role that rivers, canals, and harbors play in the region.
3. Identify west-central Europe's major resources.

LINKS TO OUR LIVES

There are many reasons why American students should know more about the countries of west-central Europe. Here are a few of them:

▶ All of the countries of west-central Europe are friendly to the United States and have strong economic ties to our country.

▶ Although Germany was our enemy in both world wars, it is now a major ally. The United States has military bases in Germany.

▶ A large percentage of Americans trace their ancestry to the region.

▶ The region has produced some of the world's greatest composers, writers, and artists.

▶ Millions of American tourists enjoy the region's beauty, cuisine, culture, and sports.

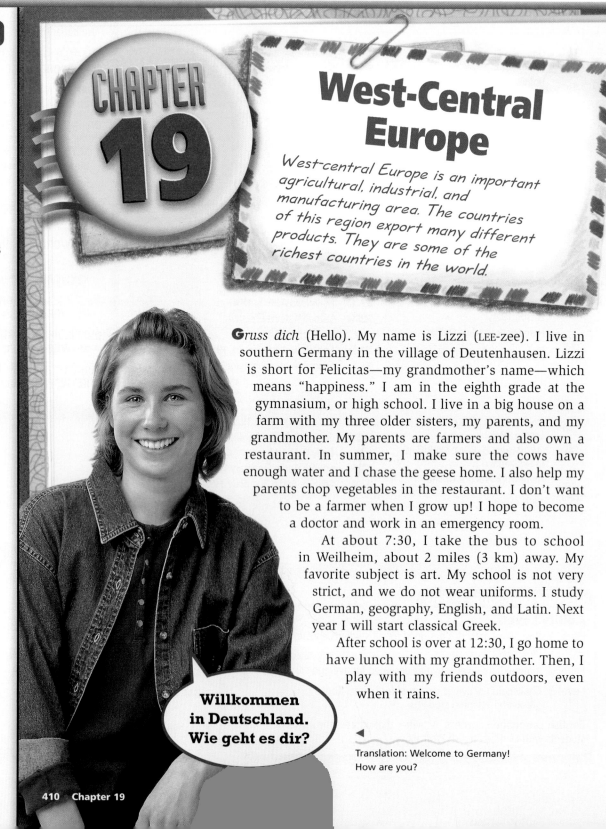

CHAPTER 19

West-Central Europe

West-central Europe is an important agricultural, industrial, and manufacturing area. The countries of this region export many different products. They are some of the richest countries in the world.

G*russ dich* (Hello). My name is Lizzi (LEE-zee). I live in southern Germany in the village of Deutenhausen. Lizzi is short for Felicitas—my grandmother's name—which means "happiness." I am in the eighth grade at the gymnasium, or high school. I live in a big house on a farm with my three older sisters, my parents, and my grandmother. My parents are farmers and also own a restaurant. In summer, I make sure the cows have enough water and I chase the geese home. I also help my parents chop vegetables in the restaurant. I don't want to be a farmer when I grow up! I hope to become a doctor and work in an emergency room.

At about 7:30, I take the bus to school in Weilheim, about 2 miles (3 km) away. My favorite subject is art. My school is not very strict, and we do not wear uniforms. I study German, geography, English, and Latin. Next year I will start classical Greek.

After school is over at 12:30, I go home to have lunch with my grandmother. Then, I play with my friends outdoors, even when it rains.

Willkommen in Deutschland. Wie geht es dir?

◀

Translation: Welcome to Germany! How are you?

Copy the following question onto the chalkboard: *What are some of the sports at which European athletes have excelled?* Discuss responses. *(Possible answers: bicycling, skiing, ice skating, sailing, or mountain climbing)* Then ask students to suggest what these sports might indicate about the landforms of west-central Europe. *(The region includes mountains, water, plains, cold and warm climates.)* Tell students that they will learn about the physical geography of west-central Europe in Section 1.

Using the Physical-Political Map

Have students examine the map on this page. Then have them name countries that fit these categories: countries on the North Sea *(Germany, Netherlands, Belgium, France)*, countries with elevations over 6,560 feet (1,999 m) *(France, Germany, Switzerland, Austria, Liechtenstein)*, landlocked countries *(Switzerland, Austria, Liechtenstein, Luxembourg)*, country on the Bay of Biscay and the English Channel *(France)*. Have students list the countries through which the Loire, Rhine, and Danube Rivers flow.

Section 1 — Physical Geography

Read to Discover

1. Where are the area's major landform regions?
2. What role do rivers, canals, and harbors play in the region?
3. What are west-central Europe's major resources?

Define

navigable
loess

Locate

Northern European Plain
Pyrenees
Alps
Seine River
Rhine River
Danube River
North Sea
Mediterranean Sea
English Channel
Bay of Biscay

WHY IT MATTERS

Nuclear power is important in west-central Europe. However, nuclear reactors can pose problems for the environment. Use **CNNfyi.com** or other current events sources to find out more about alternative sources of power. Record your findings in your journal.

Neuschwanstein Castle, Germany

Section 1 RESOURCES

Reproducible
- Block Scheduling Handbook, Chapter 19
- Graphic Organizer 19
- Guided Reading Strategy 19.1

Technology
- One-Stop Planner CD–ROM, Lesson 19.1
- Homework Practice Online
- Geography and Visual Resources with Teaching Activity 24–30
- HRW Go site

Reinforcement, Review, and Assessment
- Section 1 Review, p. 413
- Daily Quiz 19.1
- Main Idea Activity 19.1
- English Audio Summary 19.1
- Spanish Audio Summary 19.1

West-Central Europe: Physical-Political

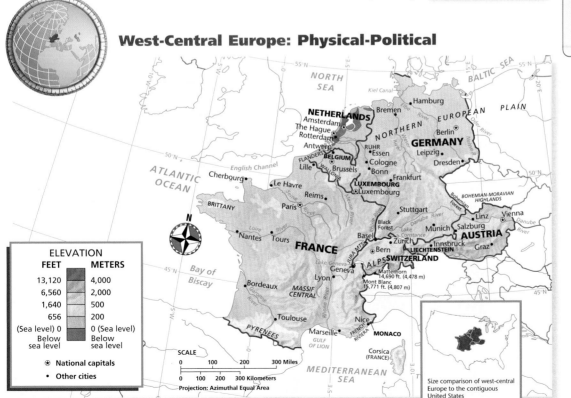

ELEVATION

FEET	METERS
13,120	4,000
6,560	2,000
1,640	500
656	200
(Sea level) 0	0 (Sea level)
Below sea level	Below sea level

⊛ National capitals
• Other cities

Size comparison of west-central Europe to the contiguous United States

🌐 **Teaching Objectives 1–3**

ALL LEVELS: (Suggested time: 20 min.) Copy the following graphic organizer onto the chalkboard, omitting the italicized answers. Have each student complete the organizer by filling in west-central Europe's major landform regions; the role of canals, rivers, and harbors in the region; and the region's major resources. Ask volunteers to share their answers with the class. **ENGLISH LANGUAGE LEARNERS**

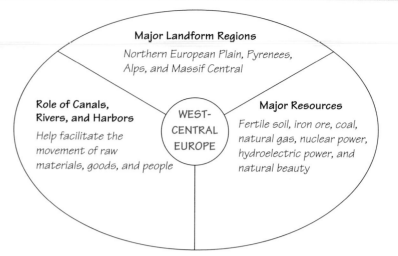

Major Landform Regions
Northern European Plain, Pyrenees, Alps, and Massif Central

Role of Canals, Rivers, and Harbors
Help facilitate the movement of raw materials, goods, and people

WEST-CENTRAL EUROPE

Major Resources
Fertile soil, iron ore, coal, natural gas, nuclear power, hydroelectric power, and natural beauty

---

EYE ON EARTH

During the 1880s commercial fishers would take about 250,000 Atlantic salmon each year from the Rhine River. However, in 1958 the last known salmon was pulled from the river. What had happened?

For decades, the mighty Rhine had been dredged and straightened. This process altered the water's flow, clarity, and temperature. Overfishing and dumping of toxic chemicals from factories depleted the salmon and other species. Then, in 1986, following a fire at a chemical plant, over 30 tons (27 metric tons) of dyes, herbicides, pesticides, fungicides, and mercury poured into the river. The Rhine seemed to be poisoned forever.

European governments and citizens responded by banding together to clean up the river. Finally, in 1990, an Atlantic salmon was pulled from the Sieg, a tributary of the Rhine. The "king of fish" had returned to live and breed.

Discussion: Lead a discussion on European citizens' working to clean up the Rhine River and the importance of civic participation in democratic societies.

Visual Record Answer ▶

Many industries might locate near the Rhine to facilitate the transportation of goods.

▲
The Rhine River has been an important transportation route since Roman times.

Interpreting the Visual Record How might the Rhine influence the location of German industry?

🖳 **internet** connect

GO TO: go.hrw.com
KEYWORD: SJ3 CH19
FOR: Web sites about west-central Europe

The Alps have many large glaciers, lakes, and valleys.
▼

Physical Features

West-central Europe includes France, Germany, Belgium, the Netherlands, Luxembourg, Switzerland, and Austria. Belgium, the Netherlands, and Luxembourg are called the Benelux countries. The word Benelux is a combination of the first letters of each country's name. They are also sometimes called the Low Countries. Large areas of Switzerland and Austria lie in the Alps mountain range. For this reason, they are called the Alpine countries.

Lowlands The main landform regions of west-central Europe are arranged like a fan. The outer edge of the fan is the Northern European Plain. Brittany, a peninsula jutting from northern France, rises slightly above the plain. In Belgium and the Netherlands, the Northern European Plain dips below sea level.

Uplands Toward the middle of the fan a wide band of uplands begins at the Pyrenees (PIR-uh-neez) Mountains. Another important uplands region is the Massif Central (ma-SEEF sahn-TRAHL) in France. Most of the southern two thirds of Germany is hilly. The Schwarzwald (SHFAHRTS-vahlt), or Black Forest, occupies the southwestern corner of Germany's uplands region.

Mountains At the center of the fan are the Alps, Europe's highest mountain range. Many peaks in the Alps reach heights of more than 14,000 feet (4,267 m). The highest peak, France's Mont Blanc (mawn BLAHN), reaches to 15,771 feet (4,807 m). Because of their high elevations, the Alps have large glaciers and frequent avalanches. During the Ice Age, glaciers scooped great chunks of rock out of the mountains, carving peaks such as the Matterhorn.

✓ **READING CHECK:** *Places and Regions* What are the area's major land-forms? Northern European Plain; uplands—Pyrenees Mountains and Massif Central, Black Forest; Alps

Climate and Waterways

West-central Europe's marine west coast climate makes the region a pleasant place to live. Winters can be cold and rainy, but summers are mild. However, areas that lie farther from the warming influence of the North Atlantic are colder. For example, central Germany receives more snow than western France. The Alps have a highland climate.

Snowmelt from the Alps feeds west-central Europe's many **navigable** rivers. Navigable rivers are deep enough and wide enough to be used by ships. France has four major rivers: the Seine (SEN), the Loire (LWAHR), the Garonne (gah-RAWN), and the

CLOSE

Call on volunteers to use the physical-political map to name the landforms and rivers they would encounter if they were to travel eastward from Cherbourg to Berlin or southward from Hamburg to Marseille.

REVIEW AND ASSESS

Have students complete the Section Review. Then call on a student to choose a place on the physical-political map of west-central Europe and on another student to describe its physical geography. Continue until all major features have been covered. Then have students complete Daily Quiz 19.1.

RETEACH

Have students complete Main Idea Activity 19.1. Then pair students and have each pair create a jumble puzzle of 10 words from the key terms or places on the physical-political map. Students should scramble the letters of each term and write a clue describing it.

ENGLISH LANGUAGE LEARNERS, COOPERATIVE LEARNING

EXTEND

Have interested students conduct research on the mistral—a dry, cold northerly wind that blows from the Alps through the Rhone Valley, or the foehn—a warm, dry wind that blows down from the Alps into Switzerland. Ask students to use their research to create a poster illustrating what creates these winds and how they affect daily life.

Rhone (ROHN). Germany has five major rivers: the Rhine (RYN), the Danube (DAN-yoob), the Elbe (EL-buh), the Oder (OH-duhr), and the Weser (VAY-zuhr). These rivers and the region's many canals are important for trade and travel. Many large harbor cities are located where rivers flow into the North Sea, Mediterranean Sea, English Channel, or Bay of Biscay. The region's heavily indented coastline has hundreds of excellent harbors.

✓ **READING CHECK:** *Environment and Society* What economic role do rivers, canals, and harbors play in west-central Europe? They are important for trade and travel.

Resources

Most of the forests that once covered west-central Europe were cut down centuries ago. The fields that remained are now some of the most productive in the world. Germany's plains are rich in **loess** (LES)—fine, wind-blown soil deposits. Germany and France produce grapes for some of the world's finest wines. Switzerland's Alpine pastures support dairy cattle.

The distribution of west-central Europe's mineral resources is uneven. Germany and France have deposits of iron ore but must import oil. Energy resources are generally in short supply in the region. However, there are deposits of coal in Germany and natural gas in the Netherlands. Nuclear power helps fill the need for energy, particularly in France and Belgium. Alpine rivers provide hydroelectric power in Switzerland and Austria. Natural beauty is perhaps the Alpine countries' most valuable natural resource, attracting millions of tourists every year.

✓ **READING CHECK:** *Environment and Society* What geographic factors contribute to the economy of the region? Rich soils are good for farming; Alpine pastures support dairy cattle; lack of energy resources leads to energy imports; natural beauty supports tourism.

The Grindelwald Valley in Switzerland has excellent pastures.

Do you remember what you learned about hydroelectric power? See Chapter 4 to review.

go. hrw .com **Homework Practice Online**
Keyword: SJ3 HP19

Section Review 1

Define and explain: navigable, loess

Working with Sketch Maps On a map of west-central Europe that you draw or that your teacher provides, label the following: the Northern European Plain, Alps, North Sea, Mediterranean Sea, English Channel, and Bay of Biscay.

Reading for the Main Idea

1. *Places and Regions* What are the landform regions of west-central Europe?

2. *Places and Regions* What type of climate dominates this region?

Critical Thinking

3. **Making Generalizations and Predictions** What might be the advantages of having many good harbors and navigable rivers?

4. **Drawing Inferences and Conclusions** How do you think an uneven distribution of resources has affected this region?

Organizing What You Know

5. **Categorizing** Copy the following graphic organizer. Use it to describe the major rivers of west-central Europe. Add rows as needed.

River	Country/Countries	Flows into. . .

Section Review 1

Answers

Define For definitions, see: navigable, p. 412; loess, p. 413

Working with Sketch Maps Maps will vary, but listed places should be labeled in their approximate locations.

Reading for the Main Idea

1. the Northern European Plain, the Pyrenees, the Alps, and the Massif Central (NGS 4)

2. marine west coast (NGS 4)

Critical Thinking

3. They would help move materials, goods, and people.

4. influenced types of goods and energy the region produces, what resources it must import

Organizing What You Know

5. Seine—France, English Channel; Loire—France, Bay of Biscay; Garonne—France, Bay of Biscay; Rhone—France, Mediterranean; Rhine—Switzerland, Germany, the Netherlands, North Sea; Danube—Germany and Austria; Elbe—Germany, North Sea; Oder—Germany, the Baltic; and Weser—Germany, North Sea

Section 2

Objectives

1. Discuss which foreign groups have affected France's history.
2. Describe the main features of French culture.
3. Identify the products that France exports.

Section 2 RESOURCES

Reproducible
◆ Guided Reading Strategy 19.2
◆ Cultures of the World Activity 3
◆ Geography for Life Activity 19
◆ Creative Strategies for Teaching World Geography, Lessons 10 and 11

Technology
◆ One-Stop Planner CD–ROM, Lesson 19.2
◆ Homework Practice Online
◆ HRW Go site

Reinforcement, Review, and Assessment
◆ Section 2 Review, p. 417
◆ Daily Quiz 19.2
◆ Main Idea Activity 19.2
◆ English Audio Summary 19.2
◆ Spanish Audio Summary 19.2

FOCUS

LET'S GET STARTED

Copy the following question onto the chalkboard: *What are three things that come to mind when you think of France?* Discuss responses. *(Possible answers: fashion, food, Eiffel Tower)* Then ask students where they might have received their impressions of France. *(movies, television commercials, news broadcasts)* You may want to ask what important aspects of the country are missing from the list. *(Possible answers: history, daily life, economy)* Tell students that in Section 2 they will learn more about France.

Building Vocabulary

Copy the key terms **medieval**, **NATO**, and **impressionism** onto the chalkboard. Tell students that *medieval* is one of the most frequently misspelled words in the English language. Have a volunteer read aloud the word's definition and history from the text. For **NATO**, introduce the concept of acronyms. Finally, tell students that **impressionism** comes from a painting by Claude Monet titled *Impression—Sunrise*. Ask students why Monet may have given his painting this title. *(It was his impression of the sunrise, not a faithful reproduction of every detail.)*

Section 2 France

Read to Discover

1. Which foreign groups affected the historical development of France?
2. What are the main features of French culture?
3. What products does France export?

WHY IT MATTERS

France is a key member of NATO, the North Atlantic Treaty Organization. Use **CNNfyi.com** or other **current events** sources to find out more about NATO. Record your findings in your journal.

Define

medieval
NATO
impressionism

Locate

Brittany
Normandy
Paris
Marseille
Nice

French croissants

▲ In this illustration messengers inform Charlemagne of a recent military victory.

History

France has been occupied by people from many other parts of Europe. In ancient times, France was part of a region known as Gaul. Thousands of years ago, people moved from eastern Europe into Gaul. These people spoke Celtic languages related to modern Welsh and Gaelic. Breton is a Celtic language still spoken in the region of Brittany.

Early History About 600 B.C. the Greeks set up colonies on Gaul's southern coast. Several centuries later, the Romans conquered Gaul. They introduced Roman law and government to the area. The Romans also established a Latin-based language that developed into French.

Roman rule lasted until the A.D. 400s. A group of Germanic people known as the Franks then conquered much of Gaul. It is from these people that France takes its name. Charlemagne was the Franks' greatest ruler. He dreamed of building a Christian empire that would be as great as the old Roman Empire. In honor of this, the pope crowned Charlemagne Emperor of the Romans in

Teaching Objective 1

ALL LEVELS: (Suggested time: 15 min.) Copy the following graphic organizer onto the chalkboard, omitting the italicized answers. Have each student complete the organizer by describing how Celts, Romans, Franks, and Normans affected France's historical development. Ask volunteers to share their answers with the class. **ENGLISH LANGUAGE LEARNERS**

INFLUENCES ON FRANCE'S EARLY HISTORY

Celts	Romans	Franks	Normans
• migrated from eastern Europe to Gaul	*• conquered Gaul*	*• conquered Gaul*	*• migrated from Normandy and conquered England*
• introduced Celtic languages, including Breton	*• introduced Roman law and government and established Latin-based language that developed into French*	*• Frankish emperor Charlemagne strengthened government and improved education*	*• Norman kings of England claimed throne of France, which led to Hundred Years' War*

A.D. 800. During his rule, Charlemagne did much to strengthen government and improve education and the arts in Europe.

The Franks divided Charlemagne's empire after his death. Invading groups attacked from many directions. The Norsemen, or Normans, were one of these groups. They came from northern Europe. The area of western France where the Normans settled is known today as Normandy.

The period from the collapse of the Roman Empire to about 1500 is called the Middle Ages, or **medieval** period. The word medieval comes from the Latin words *medium,* meaning "middle," and *aevum,* meaning "age." During much of this period kings in Europe were not very powerful. They depended on cooperation from nobles, some of whom were almost as powerful as kings.

In 1066 a noble, the duke of Normandy, conquered England, becoming its king. As a result, the kings of England also ruled part of France. In the 1300s the king of England tried to claim the throne of France. This led to the Hundred Years' War, which lasted from 1337 to 1453. Eventually, French armies drove the English out of France. The French kings then slowly increased their power over the French nobles.

During the Middle Ages the Roman Catholic Church created a sense of unity among many Europeans. Many tall, impressive cathedrals were built during this time. Perhaps the most famous is the Cathedral of Notre Dame in Paris. It took almost 200 years to build.

Revolution and Napoléon's Empire From the 1500s to the 1700s France built a global empire. The French established colonies in the Americas, Asia, and Africa. During this period most French people lived in poverty and had few rights. In 1789 the French Revolution began. The French overthrew their king and established an elected government. About 10 years later a brilliant general named Napoléon Bonaparte took power. As he gained control, he took the title of emperor. Eventually, Napoléon conquered most of Europe. Napoléon built new roads throughout France, reformed the French educational system, and established the metric system of measurement. In 1815 an alliance including Austria, Great Britain, Prussia, and Russia finally defeated Napoléon. The French king regained the throne.

World Wars During World War I (1914–18) the German army controlled parts of northern and eastern France. In the early years of World War II, Germany defeated France and occupied the northern and western parts of the country. In 1944, Allied armies including U.S., British, and Canadian soldiers landed in Normandy and drove the Germans out. However, after two wars in 30 years France was devastated. Cities, factories, bridges, railroad lines, and train stations had been destroyed. The North Atlantic Treaty Organization, or **NATO**, was formed in 1949 with France as a founding member. This military alliance was created to defend Western Europe against future attacks.

▲

French and English knights clash in this depiction of the Hundred Years' War.

▲

Napoléon Bonaparte became the ruler of France and conquered most of Europe.

Linking Past to Present

Links between England and France During part of the Middle Ages, England and France were closely linked politically. In 1066 William, duke of Normandy conquered. Later, King Henry II of England also ruled Normandy and western France because his wife had inherited a huge piece of France. Only after its defeat in the Hundred Years' War did England lose most of its French territory. For the next several centuries, England and France were often in conflict.

In 1994 a new era in British-French cooperation dawned with the completion of the "Chunnel." It consists of two railroad tunnels and one service tunnel that link France and England under the English Channel. These tunnels have helped stimulate trade and economic development.

Critical Thinking: How did substantial resources affect the creation of the Chunnel?

Answer: A project like the Chunnel would be very expensive and could only be completed by wealthy countries.

internet connect

GO TO: go.hrw.com
KEYWORD: SJ3 CH19
FOR: Web sites about the Chunnel

Teaching Objectives 2–3

LEVEL 1: (Suggested time: 30 min.) Pair students and have each pair create a collage depicting some of the main features of French culture as well as products that France exports. *(The collages' images and symbols should pertain to Roman Catholicism; predominance of the French language; traditions of wine and cheese making; various immigrant cultures; French literary, artistic, and philosophical traditions; and exports such as wheat, olives, cars, airplanes, shoes, clothing, machinery, and chemicals.)* Display collages around the classroom. **ENGLISH LANGUAGE LEARNERS, COOPERATIVE LEARNING**

LEVELS 2 AND 3: (Suggested time: 30 min.) Have each student write a short poem that describes aspects of France's culture and economy. Have volunteers recite their poems to the class.

TEACHER TO TEACHER

Rebecca Minnear of Las Vegas, Nevada, suggests the following activity to help students learn more about France: Have each student create two postcards to illustrate and describe aspects of France's economy, history, and culture. Students should create an illustration for one side of the postcard. On the reverse side of the card, students should write a note to a friend or family member describing the significance of the illustration. Have volunteers present and explain their postcards to the class.

DAILY LIFE

Have the class look up the word *baccalaureate* in a dictionary. *(The word refers to the bachelor's degree bestowed by a college or university, or to a religious service for graduates.)*

Explain that in France *baccalaureate* refers to a difficult examination students take when they have completed high school. Students who want to go to a university or find a good job must do well on *"le bac."* The emphasis placed on the baccalaureate exam reflects the high value the French place on education.

Activity: Have students acquire information about various placement examinations from a high school counselor. Then lead a discussion comparing the French examination system to systems found in the United States.

Visual Record Answers ▲

no need to convert currencies, eased trade

▶

modern vehicle, otherwise still traditional

Answers

Define For definitions, see: medieval, p. 415; NATO, p. 415; impressionism, p. 417

416

▲

The euro replaced the currencies of most of the individual EU countries.
Interpreting the Visual Record **What are the advantages of a shared currency?**

▲

Many French cheeses, such as Brie, Camembert, and Roquefort, are named after the places where they are made.

Workers harvest grapes at a vineyard in the Rhone Valley near Lyon.
Interpreting the Visual Record **Has modern technology changed the grape-growing process?**

▶

Government In the 1950s and 1960s most French colonies in Asia and Africa achieved independence. However, France still controls several small territories around the world. Today, France is a republic with a parliament and an elected president. France is also a founding member of the European Union (EU). France is gradually replacing its currency, the franc, with the EU currency, the euro.

✓ **READING CHECK:** **Human Systems** Which foreign groups have affected France's historical development? Greeks, Romans, Franks, Normans, English, Germans

Culture

About 90 percent of French people are Roman Catholic, and 3 percent practice Islam. Almost all French citizens speak French. However, small populations of Bretons in the northwest and Basques in the southwest speak other languages. In Provence-Alpes-Côte d'Azur and Languedoc-Roussillon in the south and on the island of Corsica, some people speak regional dialects along with French. Immigrants from former colonies in Africa, the Caribbean, and Southeast Asia also influence French culture through their own styles of food, clothing, music, and art.

Customs In southern France people eat Mediterranean foods like wheat, olives and olive oil, cheeses, and garlic. In the north food is more likely to be prepared with butter, herbs, and mushrooms. Wine is produced in many French regions, and France produces more than 400 different cheeses. French people celebrate many festivals, including Bastille Day on July 14. On this date in 1789 a mob stormed the Bastille, a royal prison in Paris. The French recognize this event as the beginning of the French Revolution.

Display impressionist paintings for students and lead a discussion about how they depict the effects of light.

REVIEW AND ASSESS

Have students complete the Section Review. Then teach the class the phrase *"Je sais"* (ZHUH SAY)—meaning "I know" in French. Call on volunteers to say *Je sais* and follow it with an important fact pertaining to Section 2. Then have students complete Daily Quiz 19.2.

RETEACH

Have students complete Main Idea Activity 19.2. Then organize students into small groups and have each group prepare an outline of Section 2.
ENGLISH LANGUAGE LEARNERS, COOPERATIVE LEARNING

EXTEND

Have interested students compile a "Who's Who" of famous French musicians, writers, political leaders, and other historical figures. Examples include Joan of Arc, Napoléon, Debussy, and Hugo, among others. Have each student choose two or three figures and write a short description of his or her chosen figures' significance. Arrange the descriptions in alphabetical order. Compile the biographies into a scrapbook, and use it as a classroom reference.

The Arts and Literature France has a respected tradition of poetry, philosophy, music, and the visual arts. In the late 1800s and early 1900s France was the center of an artistic movement called **impressionism**. Impressionist artists tried to capture the rippling of light rather than an exact, realistic image. Famous impressionists include Monet, Renoir, and Degas. French painters, like Cézanne and Matisse, influenced styles of modern painting. Today, France is a world leader in the arts and film industry.

France

COUNTRY	POPULATION/ GROWTH RATE	LIFE EXPECTANCY	LITERACY RATE	PER CAPITA GDP
France	59,551,227 0.4%	75, male 83, female	99%	$24,400
United States	281,421,906 0.9%	74, male 80, female	97%	$36,200

Sources: Central Intelligence Agency, *The World Factbook 2001*; U.S. Census Bureau

Interpreting the Chart How do France's life expectancy and literacy rate compare to those of the United States?

✓ **READING CHECK:** *Human Systems* How did French art affect the world?
French painters are respected around the world and influenced modern painting.

France Today

France is a major agricultural and industrial country. Its resources, labor force, and location in the heart of Europe have helped spur economic growth. France exports wheat, olives, wine, and cheeses as well as other dairy products. French factories produce cars, airplanes, shoes, clothing, machinery, and chemicals. France's largest city is Paris, which has nearly 10 million people in its metropolitan area. Other major cities include Marseille, Nice, Lyon, and Lille. France's major cities are linked by high-speed trains and excellent highways.

✓ **READING CHECK:** *Human Systems* What are some products that France exports? wheat, olives, wine, cheeses, cars, airplanes, shoes, clothing, machinery, chemicals

Homework Practice Online
Keyword: SJ3 HP19

Section Review 2

Define and explain: medieval, NATO, impressionism

Working with Sketch Maps On the map you created in Section 1, label Brittany, Normandy, Paris, Marseille, and Nice.

Reading for the Main Idea
1. *Human Systems* What were the main periods of French history?
2. *Human Systems* What are the main features of French culture?

Critical Thinking
3. **Finding the Main Idea** What were some long-lasting achievements of Charlemagne and Napoléon?
4. **Summarizing** What is the French economy like?

Organizing What You Know
5. **Identifying Cause and Effect** Copy the following graphic organizer. Use it to list the causes and effects of the Hundred Years' War.

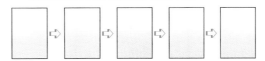

Working with Sketch Maps Maps will vary, but listed places should be labeled in their approximate locations.

Reading for the Main Idea
1. Greek colonization, Romans, Franks, medieval period, colonial empire, French Revolution, Napoleonic era, world wars (NGS 13)
2. mostly Roman Catholic; French language; immigrant cultures; popularity of wine and cheese; respected cultural traditions (NGS 10)

Critical Thinking
3. Charlemagne—strengthened government, helped to improve education and arts in Europe; Napoléon—built new roads, reformed French educational system, instituted metric system in France
4. major agricultural and industrial producer and exporter; resources, location

Organizing What You Know
5. duke of Normandy conquers England; Norman kings of England continue to rule part of France; English king tries to claim French throne; Hundred Years' War; French drive English out

▲ **Chart Answer**

They are higher.

Section 3

Objectives

1. Examine the effects that wars have had on Germany.

2. Describe Germany's major contributions to world culture.

3. Analyze how the division of Germany affected its economy.

LET'S GET STARTED

Copy the following passage onto the chalkboard: *Imagine that the American Civil War had resulted in the creation of two separate countries. Imagine that the two "Americas" were separated for 40 years before they were reunified. What is one problem that the country might face after reunification?* Discuss responses. Explain that after World War II Germany was divided into two countries that have since been reunited. Tell students that in Section 3 they will learn more about Germany.

Building Vocabulary

Locate a newspaper headline that includes the word *reform*. Discuss with students how the word is used in the headline and article. Compare modern political reform with the **Reformation** of the 1500s. Then tell students that the word **Holocaust** comes from the Greek *holokauston*, which means "that which is completely burnt." Tell students that the Holocaust refers to the mass killing, or genocide, of millions of Jews and other people by the Nazis during World War II. Finally, have a volunteer read aloud the definition of the word **chancellor** from the glossary.

Section 3 RESOURCES

Reproducible

◆ Guided Reading Strategy 19.3

◆ Map Activity 19

◆ Creative Strategies for Teaching World Geography, Lessons 10 and 11

Technology

◆ One-Stop Planner CD–ROM, Lesson 19.3

◆ Homework Practice Online

◆ HRW Go site

Reinforcement, Review, and Assessment

◆ Section 3 Review, p. 421

◆ Daily Quiz 19.3

◆ Main Idea Activity 19.3

◆ English Audio Summary 19.3

◆ Spanish Audio Summary 19.3

Visual Record Answer ▶

on a hill overlooking a river

Section 3
Germany

Read to Discover

1. What effects have wars had on Germany?

2. What are Germany's major contributions to world culture?

3. How did the division of Germany affect its economy?

Define

Reformation
Holocaust
chancellor

Locate

Berlin
Bonn
Essen
Frankfurt

Munich
Hamburg
Cologne

WHY IT MATTERS

The Berlin Wall had fallen by 1990, and East and West Germany were reunited as a single nation. What have been the effects of reunification? Use **CNNfyi.com** or other **current events** sources to find out more about Germany. Record your findings in your journal.

A VW Turbo beetle

This medieval castle overlooks a German town.

Interpreting the Visual Record What geographic features made this a good place to build a fortress?

▼

History

Many Germans are descendants of tribes that migrated from northern Europe in ancient times. The Romans conquered the western and southern fringes of the region. They called this land Germania, from the name of one of the tribes that lived there.

The Holy Roman Empire When the Roman Empire collapsed, the Franks became the most important tribe in Germany. The lands ruled by the Frankish king Charlemagne in the early 800s included most of what is now Germany. Charlemagne's empire was known as the Holy Roman Empire.

Reformation and Unification During the 1500s Germany was the center of the **Reformation**—a movement to reform Christianity. The reformers were called Protestants. Protestants rejected many practices of the Roman Catholic Church. At the time, Germany was made up of many small states. Each state was ruled by a prince who answered to the Holy Roman emperor. Many of the princes became Protestants. This angered the Holy Roman emperor, who was Catholic. He sent armies against the princes. Although the princes won the right to choose the religion of their states, conflict continued. This conflict eventually led to the Thirty Years' War (1618–48). This war was costly. Many towns were destroyed and nearly one third of the

Teaching Objective 1

ALL LEVELS: (Suggested time: 10 min.) Copy the following graphic organizer onto the chalkboard, omitting the italicized answers. Ask each student to complete the organizer by describing the effects of the Thirty Years' War and World Wars I and II on Germany. Ask volunteers to share their answers with the class. **ENGLISH LANGUAGE LEARNERS**

EFFECTS OF WARS ON GERMANY

Thirty Years' War	World War I	World War II
• *many towns destroyed* • *nearly one third of the population died*	• *lost territory and overseas colonies* • *paid heavy fines after the war*	• *Jewish population was nearly wiped out* • *divided into two countries*

Teaching Objective 1

LEVELS 2 AND 3: (Suggested time: 30 min.) Tell students to imagine that they are Germans who oppose their country's involvement in any military actions. Then ask each student to write an antiwar speech. Students' speeches should reference the effects of previous wars on Germany. Ask volunteers to deliver their speeches to the class.

population died. Germany remained divided for more than 200 years. In the late 1800s Prussia, the strongest state, united Germany.

World Wars In 1914 national rivalries and a conflict in the Balkans led to World War I. Austria, Germany, and the Ottoman Empire, later joined by Bulgaria, fought against Britain, France, and Russia, later joined by Italy and the United States. By 1918 Germany and its allies were defeated.

During the 1920s Austrian war veteran Adolf Hitler led a new political party in Germany called the Nazis. The Nazis took power in 1933. In the late 1930s Germany invaded Austria, Czechoslovakia, and finally Poland, beginning World War II. By 1942 Germany and Italy had conquered most of Europe. The Nazis forced many people from the occupied countries into concentration camps to be enslaved or killed. About 6 million Jews and millions of other people were murdered in a mass killing called the **Holocaust**.

To defeat Germany, several countries formed an alliance. These Allies included Britain, the Soviet Union, the United States, and many others. The Allies defeated Germany in 1945. Germany and its capital, Berlin, were divided into Soviet, French, British, and U.S. occupation zones. Britain, France, and the United States later combined their zones to create a democratic West Germany with its capital at Bonn. In its zone, the Soviet Union set up the Communist country of East Germany with an unlimited totalitarian government. Its capital became East Berlin; however, West Berlin became part of West Germany. In 1961 the East German government built the Berlin Wall across the city to stop East Germans from escaping to the West.

Reunification and Modern Government West Germany's roads, cities, railroads, and industries were rebuilt after the war with U.S. financial aid. East Germany was also rebuilt, but it was not as prosperous as West Germany. Unlike the West German government, the East German government allowed people very little freedom. Also, its command economy—managed by the government—was less productive than the free enterprise, market system of West Germany. In the late 1980s East Germans and people throughout Eastern Europe demanded democratic reform. In 1989 the Berlin Wall was torn down. In 1990 East and West Germany reunited. Germany's capital again became Berlin. Today, all Germans enjoy democratic rights. A parliament elects the president and prime minister, or **chancellor**. Germany is a member of the EU and NATO.

✓ **READING CHECK:** *Human Systems* How were the economies of East and West Germany organized following World War II? East Germany—command economy controlled by a communist government; West Germany—free enterprise, market-based economy

▲ German youth salute Adolf Hitler in Nürnberg in 1938.

▲ For nearly 30 years the Berlin Wall separated East and West Berlin. Many people in West Berlin protested by painting graffiti on the wall. **Interpreting the Visual Record** Why would the government make the wall solid instead of a barrier that would allow people visual access?

Cultural Kaleidoscope

The Jews of Berlin When Adolf Hitler came to power in 1933, more than 170,000 Jews lived in Berlin. By 1945 Berlin's Jewish population stood at some 5,000. The vast majority of Berlin's Jews had fled or been killed.

Now Berlin's Jewish community is growing again. Many of the city's new Jewish residents are emigrants from Russia. New Jewish shops, restaurants, schools, and even a chess team have been established. A new Jewish museum has opened. A synagogue that had been damaged by the Nazis and almost destroyed by Allied bombs has been restored.

Activity: Have each student conduct research on the history of Jews in Germany and create a time line to depict major events.

◄ **Visual Record Answer**

Possible answer: The government did not want people to communicate through the wall or see that conditions were better on the other side.

ALL LEVELS: (Suggested time: 30 min.) Tell students to imagine that they have been hired to design a Web page that highlights Germany's major contributions to world culture. Then have each student design a mock Web page that includes headlines and links. *(Web pages should mention the invention of movable metal type, the development of classical music, including the work of Bach and Beethoven as well as the operas of Wagner, and German traditions in literature and the arts.)* Ask volunteers to present their Web pages to the class. **ENGLISH LANGUAGE LEARNERS**

LEVEL 1: (Suggested time: 15 min.) Tell students to imagine that they are journalists who are writing about how the division of Germany affected the country's economy. Then pair students and have pairs write three headlines that describe the effects of the country's division on the economy. Have volunteers read their headlines to the class. **ENGLISH LANGUAGE LEARNERS, COOPERATIVE LEARNING**

LEVELS 2 AND 3: (Suggested time: 30 min.) Have each student create an editorial cartoon that might accompany an article printed below one of the headlines from the Level 1 activity.

Across the Curriculum
ART
Landscape Painting Some of the world's greatest landscape artists have been from west-central Europe. Among these are the Dutch painter Vincent van Gogh (1853–90), French artist Antoine Watteau (1684–1721), and German painter Caspar David Friedrich (1774–1840).

Activity: Have students choose a landscape artist of the 1700s or 1800s from west-central Europe. Ask them to compare how the artist portrayed the region's landscape with photographs they find in the text and other sources. Are the physical features painted realistically? Was the painting meant to be realistic? Have students write a paragraph or two comparing and contrasting their chosen painter's depiction of the landscape with the way the landscape appears in photographs.

internet connect

GO TO: go.hrw.com
KEYWORD: SJ3 CH19
FOR: Web sites about landscapes

Visual Record Answer ▶

It is a festive, celebratory time.

Chart Answer (p. 421) ▶

Answers will vary, but students might say that Germans value education.

Crowds gather in a German town for a Christmas market. Christmas markets have been popular in Germany for more than 400 years. From the beginning of Advent until Christmas, booths are set up on the market place in most cities. Here people can buy trees, decorations, and gifts.

Interpreting the Visual Record What does this photo suggest about the importance of Christmas in Germany?

▶

Culture

About 34 percent of Germans are Roman Catholic, and 38 percent are Protestant. Most other Germans have no religious association. Many of these people are from eastern Germany, where the communist government suppressed religion from 1945 to 1990.

Diversity About 90 percent of Germany's inhabitants are ethnic Germans. However, significant numbers of Turks, Poles, and Italians have come to Germany to live and work. These "guest workers" do not have German citizenship. Germany has also taken in thousands of refugees from Eastern Europe during the last 50 years.

Customs Traditional German food emphasizes the products of the forests, farms, and seacoasts. Each region produces its own varieties of sausage, cheese, wine, and beer. German celebrations include Oktoberfest; *Sangerfast,* a singing festival; and *Fastnacht,* a religious celebration. The major German festival season is Christmas. The Germans began the custom of bringing an evergreen tree indoors at Christmas and decorating it with candles.

The Arts and Literature Germany has a great tradition of literature, music, and the arts. The first European to print books using movable metal type was a German, Johannes Gutenberg. In the 1700s and 1800s, Germany led Europe in the development of classical music. World-famous German composers include Johann Sebastian Bach and Ludwig van Beethoven. The operas of Richard Wagner revived the folktales of ancient Germany.

✓ **READING CHECK:** *Human Systems* What technology and other contributions have Germans made to world culture? movable type, classical music

CLOSE

Play for the class a work by one of the great German composers, such as Johann Sebastian Bach, Ludwig van Beethoven, or Johannes Brahms—sometimes called the "Three Bs."

REVIEW AND ASSESS

Have students complete the Section Review. Then pair students and have each student write five questions based on Section 3 information. Have students exchange questions with their partners and answer them. Then have students complete Daily Quiz 19.3. **COOPERATIVE LEARNING**

RETEACH

Have students complete Main Idea Activity 19.3. Then organize students into small groups and have each group create a poster to show how one of the six essential elements of geography applies to a topic in Section 3. Have volunteers present their posters to the class. **ENGLISH LANGUAGE LEARNERS, COOPERATIVE LEARNING**

EXTEND

Have interested students conduct research on how German immigrants have influenced the customs, religion, politics, language, and food of a specific U.S. region. Then have each student write a short play to share his or her findings. **BLOCK SCHEDULING**

Germany Today

Germany has a population of 83 million, more people than any other European country. Germany also has Europe's largest economy. Nearly one fourth of all goods and services produced by the EU come from Germany.

Economy Ample resources, labor, and capital have made Germany one of the world's leading industrial countries. The nation exports a wide variety of products. You may be familiar with German automakers like Volkswagen, Mercedes-Benz, and BMW. The German government provides education, medical care, and pensions for its citizens, but Germans pay high taxes. Unemployment is high. Many immigrants work at low-wage jobs. These "guest workers" are not German citizens and cannot receive many government benefits. Since reunification, Germany has struggled to modernize the industries, housing, and other facilities of the former East Germany.

Cities Germany's capital city, Berlin, is a large city with wide boulevards and many parks. Berlin was isolated and economically restricted during the decades after World War II. However, Germans are now rebuilding their new capital to its former splendor.

Near the Rhine River and the coal fields of Western Germany is a huge cluster of cities, including Essen and Düsseldorf. They form Germany's largest industrial district, the Ruhr. Frankfurt is a city known for banking and finance. Munich is a manufacturing center. Other important cities include Hamburg, Bremen, Cologne, and Stuttgart.

✓ **READING CHECK:** *Human Systems* How did the division of Germany affect its economy? It delayed the modernization of East German industry, housing, and other facilities.

Germany

COUNTRY	POPULATION/ GROWTH RATE	LIFE EXPECTANCY	LITERACY RATE	PER CAPITA GDP
Germany	83,029,536 0.3%	74, male 81, female	99%	$23,400
United States	281,421,906 0.9%	74, male 80, female	97%	$36,200

Sources: Central Intelligence Agency, *The World Factbook 2001*; U.S. Census Bureau

Interpreting the Chart What might the literacy rate of Germany suggest about its culture?

Section Review 3

Define and explain: Reformation, Holocaust, chancellor

Working with Sketch Maps On the map you created in Section 2, label Berlin, Bonn, Essen, Frankfurt, Munich, Hamburg, and Cologne.

Reading for the Main Idea

1. *Human Systems* How did wars affect the development of Germany in the 1900s?

2. *Human Systems* What are some notable features of the German economy?

Critical Thinking

3. **Drawing Inferences and Conclusions** How has Germany's history influenced the religious makeup of the population?

4. **Summarizing** What have been some results of the unification of Germany in 1990?

Organizing What You Know

5. **Sequencing** Create a time line listing key events in the history of Germany from 1000 B.C. to 1990.

go.hrw.com **Homework Practice Online**
Keyword: SJ3 HP19

1000 B.C. ——————————————— A.D. 1990

Section Review 3

Answers

Define For definitions, see: Reformation, p. 418; Holocaust, p. 419; chancellor, p. 419

Working with Sketch Maps Maps will vary, but listed places should be labeled in their approximate locations.

Reading for the Main Idea

1. World War I—lost territory, had to pay heavy fines; World War II—Jewish population nearly wiped out, country divided (NGS 13)

2. leading industrial producer with many notable exports, including automobiles (NGS 11)

Critical Thinking

3. Reformation gave rise to large Protestant population; Holocaust killed all but small number of Jews.

4. capital was moved from Bonn to Berlin, and country struggling to modernize eastern zone

Organizing What You Know

5. 800s—Franks established Holy Roman Empire; 1500s—Reformation; 1618–48—Thirty Years' War; late 1800s—Germany united; 1914–18—World War I; 1933—Nazis seize power; late 1930s–1945—World War II; 1961—Berlin Wall built; 1989—wall torn down; 1990—Germany reunited

421

Objectives

1. Investigate how the Benelux countries were influenced by larger countries.

2. Describe what the region's culture is like.

3. Discuss what the Benelux countries are like today.

Section 4 The Benelux Countries

Read to Discover

1. How were the Benelux countries influenced by larger countries?
2. What is this region's culture like?
3. What are the Benelux countries like today?

WHY IT MATTERS

The Benelux countries are key members of the European Union (EU). Use **CNNfyi.com** or other current events sources to find out more about the membership, functions, and goals of the EU. Record your findings in your journal.

Define

cosmopolitan

Locate

Flanders Antwerp
Wallonia Brussels
Amsterdam

Dutch wooden shoes

▲
The Dutch city of Rotterdam is one of the world's busiest ports.

Interpreting the Visual Record Why might this city be an important transportation center?

History

Celtic and Germanic tribes once lived in this region, as in most of west-central Europe. They were conquered by the Romans. After the fall of the Roman Empire and the conquests of Charlemagne, the region was ruled alternately by French rulers and by the Holy Roman emperor.

In 1555 the Holy Roman emperor presented the Low Countries to his son, King Philip II of Spain. In the 1570s the Protestants of the Netherlands won their freedom from Spanish rule. Soon after, the Netherlands became a great naval and colonial power. Belgium had been ruled at times by France and the Netherlands. However, by 1830 Belgium had broken away to become an independent kingdom.

Both world wars scarred this region. Many of the major battles of World War I were fought in Belgium. Then in World War II Germany occupied the Low Countries. In 1949 Belgium, the Netherlands, and Luxembourg were founding members of NATO. Later they joined the EU. Today, each of the three countries is ruled by a parliament and a monarch. The monarchs' duties are mostly ceremonial. The Netherlands controls several Caribbean islands. However, its former colonies in Asia and South America are now independent.

✓ **READING CHECK:** *Human Systems* How are the governments of the Benelux countries organized? *Each is ruled by a parliament and a monarch, who has mostly ceremonial duties.*

TEACH

Teaching Objectives 1–3

ALL LEVELS: (Suggested time: 40 min.) Copy the graphic organizer at the right onto the chalkboard, omitting the italicized answers. Have each student complete it by filling in information about the influence of larger countries on the Benelux countries and the region's culture. Then lead a discussion based on how students might spend their day if they lived in the Netherlands, Belgium, or Luxembourg.

ENGLISH LANGUAGE LEARNERS

►**ASSIGNMENT:** Have each student write a paragraph that analyzes the economies of the Benelux region and proposes a reason why the Benelux countries might have become cosmopolitan.

Influences from Larger Countries	Culture of the Benelux Countries
• *Ruled by France and the Holy Roman Empire* • *Netherlands ruled by Spain* • *Belgium ruled by France and the Netherlands* • *WWI battles fought in Belgium* • *Low Countries occupied by Germany during WWII*	• *Belgians and Luxembourgers predominantly Roman Catholic* • *Dutch evenly divided between Catholics, Protestants, and nonreligious persons* • *Dutch spoken in Netherlands* • *Flemish and French spoken in Belgium*

CONNECTING TO Technology

Dutch Polders

A polder in the Netherlands

Much of the Netherlands lies below sea level and was once covered with water. For at least 2,000 years, the Dutch have been holding back the sea. First they lived on raised earthen mounds. Later they built walls or dikes to keep the water out. After building dikes, the Dutch installed windmills to pump the water out of reclaimed areas, called polders.

Using this system, the Dutch have reclaimed large amounts of land. Cities like Amsterdam and Rotterdam sit on reclaimed land. The dike and polder system has become highly sophisticated. Electric pumps have largely replaced windmills, and dikes now extend along much of the country's coastline. However, this system is difficult to maintain. It requires frequent and expensive repairs. Creating polders has also produced sinking lowlands and other environmental damage. As a result, the Dutch are considering changes to the system. These changes might include restoring some of the polders to wetlands and lakes.

Understanding What You Read

1. What are polders?
2. How did the Dutch use technology to live on land previously under water?

Culture

The people of Luxembourg and Belgium are mostly Roman Catholic. The Netherlands is more evenly divided among Catholic, Protestant, and those who have no religious ties.

Dutch is the language of the Netherlands. Flemish is a language related to Dutch that is spoken in Flanders, the northern part of Belgium. Belgium's coast and southern interior are called Wallonia. People in Wallonia speak mostly French and are called Walloons. In the past, cultural differences between Flemish and Walloons have produced conflict in Belgium. Today street signs and other notices are often printed in both Flemish and French. The Benelux countries are also home to immigrants from Asia and Africa.

These children in Brussels, Belgium, are wearing traditional clothing.

PLACES AND REGIONS

Antwerp, Belgium, is one of the world's four leading diamond-cutting centers. (The others are New York, Tel Aviv, and Mumbai.) According to legend, the first diamond was cut in Antwerp in 1476. Since the 1500s establishments that cut and deal in diamonds have flourished in a neighborhood near the central train station.

At Antwerp's Diamond Museum, employees demonstrate the art of cutting and polishing diamonds. Priceless jewelry sparkles in the museum's treasure chamber.

Activity: Have students conduct research on the diamond industry. Tell students to create a map showing sources, processing centers, and major markets.

Connecting to Technology Answers

1. areas reclaimed from the sea
2. They have used dikes, windmills, and electric pumps to reclaim land.

Section Review 4

Answers

Define For definition, see: cosmopolitan, p. 424

Working with Sketch Maps Maps will vary, but listed places should be labeled in their approximate locations.

423

Benelux Countries

COUNTRY	POPULATION/GROWTH RATE	LIFE EXPECTANCY	LITERACY RATE	PER CAPITA GDP
Belgium	10,258,762 0.2%	75, male 81, female	98%	$25,300
Luxembourg	442,972 1.3%	74, male 81, female	100%	$36,400
Netherlands	15,807,641 0.5%	75, male 81, female	99%	$24,400
United States	281,421,906 0.9%	74, male 80, female	97%	$36,200

Sources: Central Intelligence Agency, *The World Factbook 2001;* U.S. Census Bureau

Interpreting the Chart Which country's per capita GDP is closest to that of the United States?

The region's foods include dairy products, fish, and sausage. The Dutch spice trade led to dishes flavored with spices from Southeast Asia. The Belgians claim they invented french fries, which they eat with mayonnaise.

The Netherlands and Belgium have been world leaders in fine art. In the 1400s and 1500s, Flemish artists painted realistic portraits and landscapes. Dutch painters like Rembrandt and Jan Vermeer experimented with different qualities of light. In the 1800s Dutch painter Vincent van Gogh portrayed southern France with bold brush strokes and bright colors.

✔ READING CHECK: *Human Systems* What is the relationship between cultures in Belgium? many different cultures; in the past, Walloons and Flemish have had conflict but now cooperate

The Benelux Countries Today

The Netherlands is famous for its flowers, particularly tulips. Belgium and the Netherlands export cheeses, chocolate, and cocoa. Amsterdam and Antwerp, Belgium, are major diamond-cutting centers. The Netherlands also imports and refines oil. Luxembourg earns much of its income from services such as banking. The region also produces steel, chemicals, and machines. Its **cosmopolitan** cities are centers of international business and government. A cosmopolitan city is one that has many foreign influences. Brussels, Belgium, is the headquarters for many international organizations such as the EU and NATO.

✔ READING CHECK: *Places and Regions* What are the Benelux countries like today? They are major exporters and are very cosmopolitan.

go. hrw .com **Homework Practice Online**
Keyword: SJ3 HP19

Section Review 4

Define and explain: cosmopolitan

Working with Sketch Maps On the map you created in Section 3, label Flanders, Wallonia, Amsterdam, Antwerp, and Brussels.

Reading for the Main Idea

1. *Human Systems* What are the main cultural features of the Benelux countries?

2. *Human Systems* In what ways do the people of the Benelux countries differ?

Critical Thinking

3. **Drawing Inferences and Conclusions** Why might the economies of the Benelux countries be dependent on international trade?

4. **Analyzing Information** Why have groups in Belgium been in conflict?

Organizing What You Know

5. **Comparing** Copy the following graphic organizer. Use it to compare the Benelux countries' industries.

Belgium	Luxembourg	Netherlands

Section 5

Objectives

1. Identify some of the major events in the history of the Alpine countries.
2. Describe some of the cultural features of the region.
3. Compare the economies of Switzerland and Austria.

LET'S GET STARTED

Copy the following passage onto the chalkboard: *Look up* cacao *in the textbook's index. What is it? Where is it grown? Why might we think of Switzerland when we think of chocolate?* Discuss responses. Tell students that in 1876 a Swiss man added concentrated milk to chocolate for the first time, forming milk chocolate. Today Switzerland imports the raw materials for making chocolate and exports the finished product. Tell students that in Section 5 they will learn more about Switzerland and Austria.

Building Vocabulary

Copy the terms **cantons** and **nationalism** onto the chalkboard and call on volunteers to read the definitions aloud from the glossary. Ask students what other names they know that refer to political divisions of territory. *(Examples: state, province, territory, county)* Canton comes from a Latin word meaning "corner." So, students might think of a canton as a corner of Switzerland. Nationalism contains the word *nation*. People who want to form their own country or nation are nationalists.

Section 5 — The Alpine Countries

Read to Discover

1. What are some of the major events in the history of the Alpine countries?
2. What are some cultural features of this region?
3. How are the economies of Switzerland and Austria similar?

WHY IT MATTERS

When people think of chocolate they often think of Switzerland, one of the world's leading producers of chocolate. Use CNNfyi.com or other **current events** sources to find out more about chocolate. Record your findings in your journal.

Define

cantons
nationalism

Locate

Geneva
Salzburg
Vienna
Zurich
Basel
Bern

Swiss cuckoo clock

Reproducible
◆ Guided Reading Strategy 19.5
◆ Creative Strategies for Teaching World Geography, Lessons 10 and 11

Technology
◆ One-Stop Planner CD–ROM, Lesson 19.5
◆ Homework Practice Online
◆ *ARGWorld* CD–ROM: Ski Resorts in Switzerland
◆ HRW Go site

Reinforcement, Review, and Assessment
◆ Section 5 Review, p. 427
◆ Daily Quiz 19.5
◆ Main Idea Activity 19.5
◆ English Audio Summary 19.5
◆ Spanish Audio Summary 19.5

History

Austria and Switzerland share a history of Celtic occupation, Roman and Germanic invasions, and rule by the Holy Roman Empire.

Switzerland Swiss **cantons**, or districts, gradually broke away from the Holy Roman Empire, and in the 1600s Switzerland became independent. Today Switzerland is a confederation of 26 cantons. Each controls its own internal affairs, and the national government handles defense and international relations. Switzerland's location in the high Alps has allowed it to remain somewhat separate from the rest of Europe. It has remained neutral in the European wars of the last two centuries. Switzerland has not joined the United Nations, EU, or NATO. Because of this neutrality, the Swiss city of Geneva is home to many international organizations.

Austria During the Middle Ages, Austria was a border region of Germany. This region was the home of the Habsburgs, a powerful family of German nobles. From the 1400s onward the Holy Roman emperor was always a Habsburg. At the height of their power the Habsburgs ruled Spain and the Netherlands, as well as large areas of Germany, eastern Europe, and Italy. This empire included different

The International Red Cross in Geneva has offices in almost every country in the world.

Interpreting the Visual Record What symbol of the Red Cross is displayed on this building?

◀ **Visual Record Answer**

the flag

TEACH

Teaching Objectives 1–3

ALL LEVELS: (Suggested time: 30 min.) Copy the following graphic organizer onto the chalkboard, omitting the italicized answers. Pair students and have each pair complete the organizer with major events in the history of the Alpine countries, cultural features of each country, and economic features of each country. Ask volunteers to share their answers with the class. **ENGLISH LANGUAGE LEARNERS, COOPERATIVE LEARNING**

COMPARING AUSTRIA AND SWITZERLAND

	Austria	Switzerland
History	*Invasion by Celts, Romans, and Germanic tribes; ruled by the Holy Roman Empire; part of Habsburg, Austrian, and Austro-Hungarian Empires; became republic; annexed by Germany; became republic*	*Invasion by Celts, Romans, and Germanic tribes; ruled by the Holy Roman Empire; gained independence in the 1600s*
Culture	*Predominantly Roman Catholic and German-speaking with small minority of Slovenes and Croatians; known for classical music*	*Majority of Swiss are Roman Catholic or Protestant.*
Economy	*Dairy products, including cheese; Vienna is Austria's commercial and industrial center.*	*Dairy products, including cheese; manufacturer of watches, optical instruments, and other machinery; Zurich is the center of Swiss banking.*

Section Review 5

Answers

Define For definitions, see: cantons, p. 425; nationalism, p. 426

Working with Sketch Maps Maps will vary, but listed places should be labeled in their approximate locations.

Reading for the Main Idea
1. Switzerland—independence in 1600s, neutral; Austria—part of larger empires, republic, annexed by Germany, and republic (NGS 13)
2. Swiss—three languages; Austrian—mainly Roman Catholic, German-speaking, Slovene and Croatian minorities, known for music (NGS 10)

Critical Thinking
3. Switzerland's mountains helped the country remain neutral.
4. Napoléon's conquest led to end of Holy Roman Empire; Germany drew Austria into World War II.

Organizing What You Know
5. Students should mention the elements of culture, language, economics, and history in the section.

Visual Record Answers

It flows through many countries.

▲
The Danube River passes through Vienna, the capital of Austria.

Interpreting the Visual Record How does this river influence movement and trade?

▲
Austrians wearing carved wooden masks celebrate the return of spring and milder weather.

ethnic groups, each with its own language, government, and system of laws. The empire was united only in its allegiance to the emperor and in its defense of the Roman Catholic religion.

With the conquests of Napoléon after 1800, the Holy Roman Empire was formally eliminated. It was replaced with the Austrian Empire, which was also under Habsburg control. When Napoléon was defeated, the Austrian Empire became the dominant power in central Europe.

Through the 1800s the diverse peoples of the empire began to develop **nationalism**, or a demand for self-rule. In 1867 the Austrians agreed to share political power with the Hungarians. The Austrian Empire became the Austro-Hungarian Empire. After World War I the empire was dissolved. Austria and Hungary became separate countries. Shortly before World War II the Germans took over Austria and made it part of Germany. After the war, the Allies occupied Austria. Today Austria is an independent member of the EU.

 READING CHECK: *Human Systems* What were the major events in the history of the Alpine countries? Switzerland—breaking away from Holy Roman Empire; end of that empire, beginning of Austrian Empire, shared power with Hungarians, World Wars I and II

Culture

About 46 percent of the population in Switzerland is Roman Catholic, and 40 percent is Protestant. Austria's population is mainly Roman Catholic. Only about 5 percent of its people are Protestant, while 17 percent follow Islam or other religions.

Languages and Diversity About 64 percent of Swiss speak German, 19 percent speak French, and 8 percent speak Italian. Small groups in the southeast speak a language called Romansh. Other European languages are also spoken in Switzerland. Austria is almost entirely German-speaking, but contains small minorities of Slovenes and Croatians.

Customs Christmas is a major festival in both countries. People make special cakes and cookies at this time. In rural parts of Switzerland people take cattle up to the high mountains in late spring

Refer students to the comments in the chapter introduction. Have each student write a similar self-introduction for a boy or girl from Switzerland or Austria, using information from Section 5. Ask volunteers to share their introductions with the class.

Have students complete Main Idea Activity 19.5. Then have each student choose and illustrate a major concept from Section 5. Display and discuss the illustrations. **ENGLISH LANGUAGE LEARNERS, COOPERATIVE LEARNING**

REVIEW AND ASSESS

EXTEND

Have students complete the Section Review questions. Then have students work in pairs to create flash cards of Define and Locate terms and key concepts from the section. Ask them to quiz each other using the flash cards. Then have students complete Daily Quiz 19.5.

Have interested students conduct research on how Switzerland has been able to maintain neutrality since the 1500s. Students might focus on whether or how the country's physical geography has helped it to maintain neutrality.

---

and return in the fall. Their return is celebrated by decorating homes and cows' horns with flowers. A special feast is also prepared.

The Alpine region is particularly well known for its music. In the 1700s Mozart wrote symphonies and operas in the Austrian city of Salzburg. Every year a music festival is held there in his honor. Austria's capital, Vienna, is also known as a center of music and fine art.

✓ **READING CHECK:** *Human Systems* What role have the arts played in this region? *The region is well known for its music.*

The Alpine Countries

COUNTRY	POPULATION/ GROWTH RATE	LIFE EXPECTANCY	LITERACY RATE	PER CAPITA GDP
Austria	8,150,835 0.2%	75, male 81, female	98%	$25,500
Switzerland	7,283,274 0.3%	77, male 83, female	99%	$28,600
United States	281,421,906 0.9%	74, male 80, female	97%	$36,200

Sources: Central Intelligence Agency, *The World Factbook 2001;* U.S. Census Bureau

Interpreting the Chart How do the populations of the Alpine countries compare with that of the United States?

The Alpine Countries Today

Switzerland and Austria both produce dairy products, including many kinds of cheese. Switzerland is also famous for the manufacturing of watches, optical instruments, and other machinery. Swiss chemists discovered how to make chocolate bars. Switzerland is a major producer of chocolate, although it must import the cocoa beans.

Switzerland and Austria are linked to the rest of Europe by excellent highways, trains, and airports. Several long tunnels allow trains and cars to pass through mountains in the Swiss Alps. Both countries attract many tourists with their mountain scenery, lakes, and ski slopes.

Located on the Danube, Vienna is Austria's commercial and industrial center. Switzerland's two largest cities are both in the German-speaking north. Zurich is a banking center, while Basel is the starting point for travel down the Rhine to the North Sea. Switzerland's capital is Bern, and Geneva is located in the west.

✓ **READING CHECK:** *Human Systems* How are the economies of Switzerland and Austria similar? *They both produce dairy products and attract tourists.*

go. hrw .com **Homework Practice Online** Keyword: SJ3 HP19

Section Review 5

Define and explain: cantons, nationalism

Working with Sketch Maps On the map you created in Section 4, label Geneva, Salzburg, Vienna, Zurich, Basel, and Bern.

Reading for the Main Idea

1. *Human Systems* What were the main events in the history of the Alpine countries?

2. *Human Systems* What are some notable aspects of Swiss and Austrian culture?

Critical Thinking

3. **Drawing Inferences and Conclusions** How might geography have been a factor in Switzerland's historical neutrality?

4. **Drawing Inferences and Conclusions** How have foreign invasions of Austria shaped its history?

Organizing What You Know

5. **Comparing/Contrasting** Use this graphic organizer to compare and contrast the culture, language, economies, and history of Switzerland and Austria.

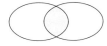

Building Vocabulary

For definitions, see: navigable, p. 412; loess, p. 413; medieval, p. 415; impressionism, p. 417; Reformation, p. 418; Holocaust, p. 419; chancellor, p. 419; cosmopolitan, p. 424; cantons, p. 425; nationalism, p. 426

Reviewing the Main Ideas

1. marine west coast climate with cold and rainy winters but mild summers; Alps have a highland climate (NGS 4)

2. led to the Thirty Years' War in which nearly one third of the population died (NGS 13)

3. coal, natural gas, nuclear power, and hydroelectric power (NGS 4)

4. Students should name five of the following: Monet, Renoir, Degas, Cézanne, Matisse, Rembrandt, Vermeer, van Gogh, Bach, Beethoven, Wagner, or Mozart. (NGS 10)

5. NATO and the EU (NGS 11, 13)

▲ **Chart Answer**

They are much smaller.

ASSESS

Have students complete the Chapter 19 Test

RETEACH

Copy the web created for the Launch into Learning activity. Erase or cover some of the labels and connections. Duplicate the altered web and give each student a copy. Have students work in pairs to fill in the missing information and provide additional links and labels. Display and discuss the webs. **ENGLISH LANGUAGE LEARNERS, COOPERATIVE LEARNING**

PORTFOLIO EXTENSIONS

1. Avalanches are a constant threat in the Austrian and Swiss Alps. Have students conduct research on Alpine avalanche control measures. They might build scale models to demonstrate the most common methods. Cotton batting can substitute for snow. Place photos of the models and written descriptions of the project in portfolios.

2. Have students choose one of the countries of west-central Europe for a virtual vacation. Allot a specific amount of money for students to spend on their trip. Have them plan transportation, accommodations, food, sightseeing, and souvenirs within their budget. Have students prepare an itinerary with the costs for each major expenditure. You may want to have students find prices in each country's currency and convert the prices to dollars.

CHAPTER 19
Review
ANSWERS

Understanding Environment and Society

Reports will vary, but the information included should address the topics suggested. Use Rubric 42, Writing to Inform, to evaluate student work.

Thinking Critically

1. Rivers have facilitated trade; the Alps have somewhat hindered travel and trade.

2. by building tunnels through the Alps

3. the Rhine River, Jura Mountains, Pyrenees, and Brittany; borders with Belgium, Luxembourg, and part of the border with Germany

4. The EU has its headquarters in Brussels.

5. low population growth, long life expectancy, high literacy rate, and high incomes; each of the nations is economically developed

CHAPTER 19
Reviewing What You Know

Building Vocabulary

On a separate sheet of paper, write sentences to define each of the following words.

1. navigable
2. loess
3. medieval
4. impressionism
5. Reformation
6. Holocaust
7. chancellor
8. cosmopolitan
9. cantons
10. nationalism

Reviewing the Main Ideas

1. (*Places and Regions*) What is the climate of west-central Europe like?

2. (*Human Systems*) What were the effects of religious conflicts on Germany?

3. (*Places and Regions*) What energy resources are available to the countries of this region?

4. (*Human Systems*) Name five artists from west-central Europe who have made notable contributions to culture.

5. (*Human Systems*) What international organizations have created ties between countries of west-central Europe since World War II?

Understanding Environment and Society

Cleaning up Pollution

Create a presentation on East German industries' pollution of the environment. Include a chart, graph, database, model, or map showing patterns of pollutants, as well as the German government's clean-up effort. Consider the following:

• Pollution in the former East Germany.

• Pollution today.

• Costs of stricter environmental laws.

Write a five-question quiz, with answers about your presentation to challenge fellow students.

Thinking Critically

1. **Drawing Inferences and Conclusions** What geographic features have encouraged travel and trade in west-central Europe? What geographic features have hindered travel and trade?

2. **Finding the Main Idea** How have the people of Switzerland altered their environment?

3. **Analyzing Information** What landform regions give France natural borders? Which French borders do not coincide with physical features?

4. **Drawing Inferences and Conclusions** Why might Brussels, Belgium, be called the capital of Europe?

5. **Comparing** What demographic factors are shared by all countries of west-central Europe today? How do they reflect levels of economic development?

FOOD FESTIVAL

The countries of west-central Europe are famous for their cheese. Point out that cheese comes in hundreds of varieties, based on these factors and others: the type of milk used to make the cheese (cow, goat, or sheep); fresh or ripened; soft, hard, semisoft, or semifirm; and the herbs, spices, mold spores or bacteria that have been added to the cheese. Students could search the supermarket for different types of cheese from west-central Europe, and the class could hold a cheese-tasting party. Serve French or German breads and a beverage with the cheese.

CHAPTER 19

REVIEW AND ASSESSMENT RESOURCES

Reproducible
◆ Readings in World Geography, History, and Culture 31, 32, and 33
◆ Vocabulary Activity 19

Technology
◆ Chapter 19 Test Generator (on the One-Stop Planner)

◆ Audio CD Program, Chapter 19

Reinforcement, Review, and Assessment
◆ Chapter 19 Review, pp. 428–29

◆ Chapter 19 Tutorial for Students, Parents, Mentors, and Peers
◆ Chapter 19 Test
◆ Chapter 19 Test for English Language Learners and Special-Needs Students

Building Social Studies Skills

Map ACTIVITY

On a separate sheet of paper, match the letters on the map with their correct labels.

Northern European Plain	Danube River
Pyrenees	North Sea
Alps	Mediterranean Sea
Seine River	Paris
Rhine River	Berlin

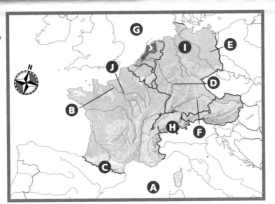

Mental Mapping Skills ACTIVITY

On a separate sheet of paper, draw a freehand map of west-central Europe. Make a key for your map and label the following:

Austria	Luxembourg
Belgium	the Netherlands
France	North Sea
Germany	Switzerland

WRITING ACTIVITY

Imagine you are taking a boat tour down the Rhine River. You will travel from Basel, Switzerland, to Rotterdam, the Netherlands. Keep a journal describing the places you see and the stops you make. Be sure to use standard grammar, spelling, sentence structure, and punctuation.

Map Activity

A. Mediterranean Sea
B. Seine River
C. Pyrenees
D. Rhine River
E. Berlin
F. Danube River
G. North Sea
H. Alps
I. Northern European Plain
J. Paris

Mental Mapping Skills Activity

Maps will vary, but listed places should be labeled in their approximate locations.

Writing Activity

Journals will vary, but the place descriptions should be consistent with text material. Use Rubric 15, Journals, to evaluate student work.

Portfolio Activity

Maps will vary, but they should identify important products and the areas in which they are grown. Maps should be accompanied by a discussion explaining why various products are grown in different regions. Use Rubric 20, Map Creation, to evaluate student work.

Alternative Assessment

Portfolio ACTIVITY

Learning About Your Local Geography Agricultural Products Some countries in west-central Europe are major exporters of agricultural products. Research the agricultural products of your state. Draw a map that shows important products and areas they are produced.

internet connect

Internet Activity: go.hrw.com
KEYWORD: SJ3 GT19

Choose a topic to explore about west-central Europe:
• Tour the land and rivers of Europe.
• Travel back in time to the Middle Ages.
• Visit Belgian and Dutch schools.

internet connect

GO TO: go.hrw.com
KEYWORD: SJ3 Teacher
FOR: a guide to using the Internet in your classroom

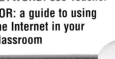

Northern Europe
Chapter Resource Manager

CHAPTER 20

Objectives	Pacing Guide	Reproducible Resources
SECTION 1		
Physical Geography (pp. 431–33) 1. Describe the region's major physical features. 2. Identify the region's most important natural resources. 3. Identify the climates that are found in northern Europe.	**Regular** .5 day **Block Scheduling** .5 day *Block Scheduling Handbook, Chapter 20*	**RS** Guided Reading Strategy 20.1 **RS** Graphic Organizer 20 **E** Creative Strategies for Teaching World Geography, Lessons 10 and 11 **E** Lab Activity for Geography and Earth Science, Demonstration 4
SECTION 2		
The United Kingdom (pp. 434–37) 1. Identify some important events in the history of the United Kingdom. 2. Describe what the people and culture of the country are like. 3. Describe the United Kingdom of today.	**Regular** 1 day **Block Scheduling** .5 day *Block Scheduling Handbook, Chapter 20*	**RS** Guided Reading Strategy 20.2 **E** Cultures of the World Activity 3 **SM** Map Activity 20
SECTION 3		
The Republic of Ireland (pp. 440–42) 1. Identify the key events in Ireland's history. 2. Describe the people and culture of Ireland. 3. Identify the kinds of economic changes Ireland has experienced in recent years.	**Regular** 1 day **Block Scheduling** .5 day *Block Scheduling Handbook, Chapter 20*	**RS** Guided Reading Strategy 20.3 **E** Cultures of the World Activity 3
SECTION 4		
Scandinavia (pp. 443–47) 1. Describe the people and culture of Scandinavia. 2. Identify some important features of each of the region's countries, plus Greenland and Lapland.	**Regular** 1.5 days **Block Scheduling** .5 day *Block Scheduling Handbook, Chapter 20*	**RS** Guided Reading Strategy 20.4 **E** Cultures of the World Activity 3 **SM** Geography for Life Activity 20

Chapter Resource Key

RS Reading Support

IC Interdisciplinary Connections

E Enrichment

SM Skills Mastery

A Assessment

REV Review

ELL Reinforcement and English Language Learners

 Transparencies

 CD–ROM

 Music

 Video

 Internet

 Holt Presentation Maker Using Microsoft® Powerpoint®

 One-Stop Planner CD–ROM

See the *One-Stop Planner* for a complete list of additional resources for students and teachers.

One-Stop Planner CD–ROM

It's easy to plan lessons, select resources, and print out materials for your students when you use the *One-Stop Planner CD–ROM with Test Generator.*

Technology Resources	Review, Reinforcement, and Assessment Resources	
One-Stop Planner CD–ROM, Lesson 20.1	ELL	Main Idea Activity 20.1
Geography and Cultures Visual Resources with Teaching Activities 24–29	REV	Section 1 Review, p. 433
Homework Practice Online	A	Daily Quiz 20.1
HRW Go site	ELL	English Audio Summary 20.1
	ELL	Spanish Audio Summary 20.1
One-Stop Planner CD–ROM, Lesson 20.2	ELL	Main Idea Activity 20.2
Homework Practice Online	REV	Section 2 Review, p. 437
HRW Go site	A	Daily Quiz 20.2
	ELL	English Audio Summary 20.2
	ELL	Spanish Audio Summary 20.2
One-Stop Planner CD–ROM, Lesson 20.3	ELL	Main Idea Activity 20.3
Homework Practice Online	REV	Section 3 Review, p. 442
HRW Go site	A	Daily Quiz 20.3
	ELL	English Audio Summary 20.3
	ELL	Spanish Audio Summary 20.3
One-Stop Planner CD–ROM, Lesson 20.4	ELL	Main Idea Activity 20.4
Homework Practice Online	REV	Section 4 Review, p. 447
HRW Go site	A	Daily Quiz 20.4
	ELL	English Audio Summary 20.4
	ELL	Spanish Audio Summary 20.4

internet connect

HRW ONLINE RESOURCES

GO TO: go.hrw.com
Then type in a keyword.

TEACHER HOME PAGE
KEYWORD: SJ3 TEACHER

CHAPTER INTERNET ACTIVITIES
KEYWORD: SJ3 GT20

Choose an activity to:
- explore the island and fjords on the Scandinavian coast.
- visit historic palaces in the United Kingdom.
- investigate the history of skiing.

CHAPTER ENRICHMENT LINKS
KEYWORD: SJ3 CH20

CHAPTER MAPS
KEYWORD: SJ3 MAPS20

ONLINE ASSESSMENT
Homework Practice
KEYWORD: SJ3 HP20
Standardized Test Prep Online
KEYWORD: SJ3 STP20
Rubrics
KEYWORD: SS Rubrics

COUNTRY INFORMATION
KEYWORD: SJ3 Almanac

CONTENT UPDATES
KEYWORD: SS Content Updates

HOLT PRESENTATION MAKER
KEYWORD: SJ3 PPT20

ONLINE READING SUPPORT
KEYWORD: SS Strategies

CURRENT EVENTS
KEYWORD: S3 Current Events

Meeting Individual Needs

Ability Levels

Level 1 Basic-level activities designed for all students encountering new material

Level 2 Intermediate-level activities designed for average students

Level 3 Challenging activities designed for honors and gifted-and-talented students

English Language Learners Activities that address the needs of students with Limited English Proficiency

Chapter Review and Assessment

IC	Interdisciplinary Activities for the Middle Grades 13, 14, 15
E	Readings in World Geography, History, and Culture 34, 35, 36
SM	Critical Thinking Activity 20
REV	Chapter 20 Review, pp. 448–49
REV	Chapter 20 Tutorial for Students, Parents, Mentors, and Peers
ELL	Vocabulary Activity 20
A	Chapter 20 Test
	Chapter 20 Test Generator (on the One-Stop Planner)
	Audio CD Program, Chapter 20
A	Chapter 20 Test for English Language Learners and Special-Needs Students

LAUNCH INTO LEARNING

Select a brief folktale from one of the countries of northern Europe. Two possibilities are "Three Billy Goats Gruff" from Norway or one of the King Arthur stories from Great Britain. Read the story aloud to the class. You may want to invite students to sketch illustrations for the story as you read. When you have finished the story, ask students if they have heard it before. Discuss responses. Explain that Americans have inherited many cultural elements from northern Europe, such as the stories that were told around the fire on long winter nights. Tell students they will learn more about the countries of northern Europe and the connections the United States has with those countries in Chapter 20.

Section 1

Objectives

1. Describe the major physical features of northern Europe.
2. Identify the region's most important natural resources.
3. Examine the climates found in northern Europe.

LINKS TO OUR LIVES

You may wish to point out to students the following reasons why we should study northern Europe:

▶ The United States has strong historical and economic ties to the countries of northern Europe.

▶ Denmark, Iceland, Norway, and the United Kingdom are members of the North Atlantic Treaty Organization (NATO). They are political allies of the United States.

▶ Millions of Americans have northern European ancestry.

▶ American culture has been influenced beyond measure by northern Europe. The language, law, commerce, government, literature, art, holidays, and food found in the United States—all these and more bear the imprint of northern Europe.

CHAPTER 20

Northern Europe

Now we will study the countries of northern Europe. First we meet Lars, a student in Norway. He lives in a place where the Sun does not rise during much of the winter.

Hi! My name is Lars. I am 13, and I live in Tromsø, one of the northernmost cities in Europe. I am in my seventh year at school. In school we study Norwegian, plus English, French or German, social studies, science, music, art, and cooking. If I do well in junior high, I will go to an academic high school and prepare for a university.

Usually I walk to school, which is about 3 km (1.9 miles) away. In the winter, everyone skis to school. The Sun never shines on many winter days because we live north of the Arctic Circle. On January 20, when the Sun appears again for just a few minutes, we celebrate Sun Day.

In the summer the Sun never sets. This is my favorite time of the year. It still can be cold then. Last summer the temperature was mostly around 6° or 7°C (about 43° or 44°F).

Jeg bor i midnattssolens land.

Translation: I live in the Land of the Midnight Sun.

LET'S GET STARTED

Copy the following instructions onto the chalkboard: *Look at the physical-political map of northern Europe. What outdoor sports might you enjoy if you lived in the lake region of Finland, in central Norway, in southern England, or on the western coast of Ireland? (Possible answers: Finland—ice skating, Norway—skiing, England—soccer, Ireland—sailing)* Give students time to write down their answers. Discuss responses. Point out that it is possible to participate in many sports in northern Europe because the region has a mix of landforms and climates. Tell students that in Section 1 they will learn more about the physical geography of northern Europe.

Using the Physical-Political Map

Have students examine the map on this page. Call on a volunteer to name the countries in the region. Have students classify the countries as islands or peninsulas. Ask students to name the seas that surround the region's countries and to speculate on the influence the sea has had on the economic activities of the people of northern Europe.

Section 1
Physical Geography

Read to Discover

1. What are the region's major physical features?
2. What are the region's most important natural resources?
3. What climates are found in northern Europe?

Define

fjords   lochs   North Atlantic Drift

Locate

British Isles
English Channel
North Sea
Great Britain

Ireland
Iceland
Greenland
Scandinavian
  Peninsula

Jutland Peninsula
Kjølen Mountains
Northwest Highlands
Shannon River
Baltic Sea

WHY IT MATTERS

An important physical feature of Northern Europe is the North Sea. Use CNN**fyi**.com or other **current events** sources to learn more about America's dependence on oil. Record your findings in your journal.

A Viking ship post

Section 1 RESOURCES

Reproducible
- Block Scheduling Handbook, Chapter 20
- Guided Reading Strategy 20.1
- Graphic Organizer 20
- Creative Strategies for Teaching World Geography, Lessons 10 and 11
- Lab Activity for Geography and Earth Science, Demonstration 4

Technology
- One-Stop Planner CD–ROM, Lesson 20.1
- Homework Practice Online
- Geography and Cultures Visual Resources with Teaching Activities 24–29
- HRW Go site

Reinforcement, Review, and Assessment
- Section 1 Review, p. 433
- Daily Quiz 20.1
- Main Idea Activity 20.1
- English Audio Summary 20.1
- Spanish Audio Summary 20.1

Northern Europe: Physical-Political

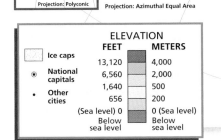

To understand the relative location of Greenland, as well as the vast distance separating it from Europe, see the Atlas map in the front of this textbook.

SCALE
0   250   500 Miles
0   250 500 Kilometers
Projection: Azimuthal Equal Area

GREENLAND
(DENMARK)
Nuuk (Godthab)

SCALE
0   250   500 Miles
0 250 500 Kilometers
Projection: Polyconic

ICELAND
Reykjavik

ATLANTIC OCEAN

NORWEGIAN SEA

LAPLAND

Faeroe Islands
(DEN.)

Shetland Islands
(U.K.)

Orkney Islands
(U.K.)

SCANDINAVIAN PENINSULA
KJØLEN MTS.
GULF OF BOTHNIA

FINLAND

NORWAY   SWEDEN

Bergen
Oslo
Stockholm
Helsinki
Göteborg

Denmark Strait

Arctic Circle

NORTHERN IRELAND (U.K.)
SCOTLAND
UNITED KINGDOM
Belfast
Glasgow

NORTH SEA

DENMARK
JUTLAND PENINSULA
Copenhagen

BALTIC SEA

IRELAND
Galway
Shannon River
Cork
Dublin
WALES
Cardiff
Manchester
Birmingham
London
ENGLAND

EUROPE

IRISH SEA

English Channel

ELEVATION

	FEET	METERS
Ice caps		
	13,120	4,000
National capitals	6,560	2,000
	1,640	500
Other cities	656	200
	(Sea level) 0	0 (Sea level)
	Below sea level	Below sea level

Size comparison of northern Europe to the contiguous United States

Teaching Objectives 1–3

ALL LEVELS: (Suggested time: 20 min.) Copy the graphic organizer at right onto the chalkboard, omitting the italicized answers. Use it to help students classify the landforms, lakes and rivers, natural resources, and climates of northern Europe. Call on volunteers to point out the listed physical features on a wall map. Then have students brainstorm occupations in northern Europe that depend on the listed physical features and resources. *(Example: oil—oil-rig worker)* **ENGLISH LANGUAGE LEARNERS**

Physical Geography of Northern Europe

Landforms		Climates
hills of Ireland highlands of Great Britain Kjölen Mountains lowlands of southeastern Great Britain and southern Scandinavia glaciers fjords	*marine west coast caused by North Atlantic Drift humid continental subarctic tundra*	

Lakes and rivers		Natural resources
lochs of Scotland Shannon River — longest in region	*North Sea, Baltic Sea fish, forests, soil oil and natural gas geothermal and hydroelectric power*	

internet connect
GO TO: go.hrw.com
KEYWORD: SJ3 CH20
FOR: Web sites about northern Europe

National Geography Standard 14

Environment and Society
Northern Europe has many lakes, rivers, and streams. Some of these waterways are polluted.

During the 1960s increased industrial production in Europe caused airborne pollutants to drift far over the continent. These pollutants combined with water vapor in the air and fell as acid rain. Forests in Scandinavia show the damaging effects.

In winter, snow carries the pollutants to the land. When the snow melts, the acid makes its way into lakes and streams. Fish have been killed off in many lakes of Norway and Sweden. Acid rain has also corroded many important buildings.

Activity: Have students conduct research on efforts by northern European governments and organizations to solve pollution problems. Allow them to summarize their findings in a written report or in an oral presentation to the class.

internet connect
GO TO: go.hrw.com
KEYWORD: SJ3 CH20
FOR: Web sites about pollution

Visual Record Answer ▶

carved by glaciers

Our Amazing Planet

Scotland's Loch Ness contains more fresh water than all the lakes in England and Wales combined. It is deeper, on average, than the nearby North Sea.

Fjords like this one shelter many harbors in Norway.

Interpreting the Visual Record
How are fjords created? ▼

Physical Features

Northern Europe includes several large islands and peninsulas. The British Isles lie across the English Channel and North Sea from the rest of Europe. They include the islands of Great Britain and Ireland and are divided between the United Kingdom and the Republic of Ireland. This region also includes the islands of Iceland and Greenland. Greenland is the world's largest island.

To the east are the Scandinavian and Jutland Peninsulas. Denmark occupies the Jutland Peninsula and nearby islands. The Scandinavian Peninsula is divided between Norway and Sweden. Finland lies farther east. These countries plus Iceland make up Scandinavia.

Landforms The rolling hills of Ireland, the highlands of Great Britain, and the Kjølen (CHUH-luhn) Mountains of Scandinavia are part of Europe's Northwest Highlands region. This is a region of very old, eroded hills and low mountains.

Southeastern Great Britain and southern Scandinavia are lowland regions. Much of Iceland is mountainous and volcanic. More than 10 percent of it is covered by glaciers. Greenland is mostly covered by a thick ice cap.

Coasts Northern Europe has long, jagged coastlines. The coastline of Norway includes many **fjords** (fee-AWRDS). Fjords are narrow, deep inlets of the sea set between high, rocky cliffs. Ice-age glaciers carved the fjords out of coastal mountains.

Lakes and Rivers Melting ice-age glaciers left behind thousands of lakes in the region. In Scotland, the lakes are called **lochs**. Lochs are found in valleys carved by glaciers long ago.

Northern Europe does not have long rivers like the Mississippi River in the United States. The longest river in the British Isles is the Shannon River in Ireland. It is just 240 miles (390 km) long.

✓ **READING CHECK:** *Places and Regions* What are the physical features of the region? large islands, peninsulas, hills, mountains, glaciers, ice cap, fjords, lakes

Natural Resources

Northern Europe has many resources. They include water, forests, and energy sources.

Water The ice-free North Sea is especially important for trade and fishing. Parts of the Baltic Sea freeze over during the winter months. Special ships break up the ice to keep sea lanes open to Sweden and Finland.

Forests and Soil Most of Europe's original forests were cleared centuries ago. However,

CLOSE

Challenge students to suggest ways that northern Europe's physical geography may have shaped its history. *(Examples: Long coastlines and ice-free ports make long-distance trade and conquest possible. The isolation of the British Isles may have discouraged some invaders.)*

REVIEW AND ASSESS

Have students complete the Section Review. Then have students write a description of a country or region in northern Europe without naming it. Call on volunteers to read their descriptions and on others to determine the place described. Then have students complete Daily Quiz 20.1.

RETEACH

Have students complete Main Idea Activity 20.1. Then organize them into groups to create a database of facts on the region. Each database should include physical features, resources, and climate types for each area.
ENGLISH LANGUAGE LEARNERS, COOPERATIVE LEARNING

EXTEND

Have interested students conduct research on Surtsey, a volcanic island that emerged off the coast of Iceland in 1963. Ask students to include information on how the island was formed, the plants that now grow there, and the animals that either live on or visit the island. Have students write reports of these findings. **BLOCK SCHEDULING**

Sweden and Finland still have large, coniferous forests that produce timber. The region's farmers grow many kinds of cool-climate crops.

Energy Beneath the North Sea are rich oil and natural gas reserves. Nearly all of the oil reserves are controlled by the nearby United Kingdom and Norway. However, these reserves cannot satisfy all of the region's needs. Most countries import oil and natural gas from southwest Asia, Africa, and Russia. Some, such as Iceland, use geothermal and hydroelectric power.

✓ **READING CHECK:** *Environment and Society* In what way has technology allowed people in the region to keep the North Sea open during the winter?
Special ships break up the ice.

Climate

Despite its northern location, much of the region has a marine west coast climate. Westerly winds blow over a warm ocean current called the **North Atlantic Drift**. These winds bring mild temperatures and rain to the British Isles and coastal areas. Atlantic storms often bring even more rain. Snow and frosts may occur in winter.

Central Sweden and southern Finland have a humid continental climate. This area has four true seasons. Far to the north are subarctic and tundra climates. In the forested subarctic regions, winters are long and cold with short days. Long days fill the short summers. In the tundra region it is cold all year. Only small plants such as grass and moss grow there.

✓ **READING CHECK:** *Places and Regions* What are the region's climates?
marine west coast, humid continental, subarctic, tundra

Do you remember what you learned about ocean currents? See Chapter 3 to review.

Homework Practice Online
Keyword: SJ3 HP20

Section Review 1

Define and explain: fjords, lochs, North Atlantic Drift

Working with Sketch Maps On an outline map that you draw or that your teacher provides, label the following: British Isles, English Channel, North Sea, Great Britain, Ireland, Iceland, Greenland, Scandinavian Peninsula, Jutland Peninsula, Kjølen Mountains, Northwest Highlands, Shannon River, and Baltic Sea. In the margin, write a short caption explaining how the North Atlantic Drift affects the region's climates.

Reading for the Main Idea

1. *Places and Regions* Which parts of northern Europe are highland regions? Which parts are lowland regions?

2. *Places and Regions* What major climate types are found in northern Europe?

Critical Thinking

3. **Finding the Main Idea** How has ice shaped the region's physical geography?

4. **Making Generalizations and Predictions** Think about what you have learned about global warming. How might warmer temperatures affect the climates and people of northern Europe?

Organizing What You Know

5. **Summarizing** Copy the following graphic organizer. Use it to describe the region's important natural resources.

Water	Forests and soil	Energy

Section Review 1

Answers

Define For definitions, see: fjords, p. 300; lochs, p. 432; North Atlantic Drift, p. 433

Working with Sketch Maps Places should be labeled in their approximate locations. The North Atlantic Drift brings mild temperatures and rain to the British Isles and coastal areas.

Reading for the Main Idea

1. hills of Ireland, highlands of Great Britain, Kjölen Mountains, much of Iceland; southeastern Great Britain, southern Scandinavia (NGS 4)

2. marine west coast, humid continental, subarctic, tundra (NGS 4)

Critical Thinking

3. carved fjords, left lakes behind

4. Possible answers: melt Iceland's glaciers and Greenland's ice cap, fill fjords, flood lowlands

Organizing What You Know

5. water—North Sea, Baltic Sea; forests and soil—coniferous forests in Sweden and Finland, soil for cool-climate crops; energy—oil, natural gas, geothermal and hydroelectric power

Section 2

Objectives

1. Identify some important events in the history of the United Kingdom.

2. Describe the people and culture of the country.

3. Explain what the United Kingdom is like today.

FOCUS

LET'S GET STARTED

Copy the following "formula" and question onto the chalkboard: *Celts + Romans + Angles + Saxons + Jutes + Vikings + Normans + Indians + Pakistanis + others = British people. What does this formula mean? (It refers to some of the groups of people who have come to live in what is now the United Kingdom.)* Have students write their answers. Discuss responses. Explain that the British Isles have been a "melting pot" for centuries and that its population continues to become more diverse. Tell students they will learn more about the United Kingdom in Section 2.

Building Vocabulary

Write the key terms on the chalkboard. Ask volunteers to read the definitions aloud and use each one in a sentence. Point out that the word **textile** is often used as an adjective, such as in "textile mill" or "textile market." Explain that in the United Kingdom Parliament passes laws. The United Kingdom also has a reigning king or queen; he or she is called the monarch. The combination of these two parts of the government creates a **contitutional monarchy**. Ask students if they know any place names or streets that include **glen**. The word has become common in the English language.

Section 2 RESOURCES

Reproducible
◆ Guided Reading Strategy 20.2
◆ Map Activity 20
◆ Cultures of the World Activity 3

Technology
◆ One-Stop Planner CD–ROM, Lesson 20.2
◆ Homework Practice Online
◆ HRW Go site

Reinforcement, Review, and Assessment
◆ Section 2 Review, p. 437
◆ Daily Quiz 20.2
◆ Main Idea Activity 20.2
◆ English Audio Summary 20.2
◆ Spanish Audio Summary 20.2

Section 2 — The United Kingdom

Read to Discover

1. What are some important events in the history of the United Kingdom?
2. What are the people and culture of the country like?
3. What is the United Kingdom like today?

Define
textiles
constitutional monarchy
glen

Locate
England
Scotland
Wales
Northern Ireland
Irish Sea
London
Birmingham
Manchester
Glasgow
Cardiff
Belfast

WHY IT MATTERS

The United States has been heavily influenced by the British people. Use CNNfyi.com or other current events sources to find out about present-day ties between the United States and Great Britain. Record your findings in your journal.

British crown

Early peoples of the British Isles built Stonehenge in stages from about 3100 B.C. to about 1800 B.C.

This beautiful Anglo-Saxon shoulder clasp from about A.D. 630 held together pieces of clothing.

History

Most of the British are descended from people who came to the British Isles long ago. The Celts (KELTS) are thought by some scholars to have come to the islands around 450 B.C. Mountain areas of Wales, Scotland, and Ireland have remained mostly Celtic.

Later, from the A.D. 400s to 1000s, new groups of people came. The Angles and Saxons came from northern Germany and Denmark. The Vikings came from Scandinavia. Last to arrive in Britain were the Normans from northern France. They conquered England in 1066. English as spoken today reflects these migrations. It combines elements from the Anglo-Saxon and Norman French languages.

A Global Power England became a world power in the late 1500s. Surrounded by water, the country developed a powerful navy that protected trade routes. In the 1600s the English began establishing colonies around the world. By the early 1800s they had also united England, Scotland, Wales, and Ireland into one kingdom. From London the United Kingdom built a vast British Empire. By 1900 the empire covered nearly one fourth of the world's land area.

Teaching Objective 1

LEVEL 1: (Suggested time: 45 min.) Organize students into groups. Ask each group to make an illustrated time line of a particular 100-year period in the history of England. Be sure all significant eras are covered. Ask students to include on their time lines significant historical events and achievements that have helped make Great Britain the nation it is today. Encourage students to consult library resources for information.
ENGLISH LANGUAGE LEARNERS, COOPERATIVE LEARNING

LEVELS 2 AND 3: (Suggested time: 45 min.) Have students complete the time line in the Level 1 activity. Then ask them to choose a historical figure who lived during their 100-year period. The person chosen may be famous, such as Queen Elizabeth I or William Shakespeare, or may be a representative of now anonymous citizens, such as a worker in a textile mill of the early Industrial Revolution. Have students write monologues in which their chosen figures reflect on how they fit into or played a role in British history.

The United Kingdom also became an economic power in the 1700s and 1800s. It was the cradle of the Industrial Revolution, which began in the last half of the 1700s. Large supplies of coal and iron and a large labor force helped industries grow. The country also developed a good transportation network of rivers, canals, and railroads. Three of the early industries were **textiles**, or cloth products, shipbuilding, iron, and later steel. Coal powered these industries. Birmingham, Manchester, and other cities grew up near Britain's coal fields.

Decline of Empire World wars and economic competition from other countries weakened the United Kingdom in the 1900s. All but parts of northern Ireland became independent in 1921. By the 1970s most British colonies also had gained independence. Most now make up the British Commonwealth of Nations. Members of the Commonwealth meet to discuss economic, scientific, and business matters.

The United Kingdom still plays an important role in world affairs. It is a leading member of the United Nations (UN), the European Union (EU), and the North Atlantic Treaty Organization (NATO).

The Government The United Kingdom's form of government is called a **constitutional monarchy**. That is, it has a monarch—a king or queen—but a parliament makes the country's laws. The monarch is the head of state but has largely ceremonial duties. Parliament chooses a prime minister to lead the national government.

In recent years the national government has given people in Scotland and Wales more control over local affairs. Some people think Scotland might one day seek independence.

✓ **READING CHECK:** *Human Systems* How are former British colonies linked today? Most of them make up the British Commonwealth of Nations.

▲
Queen Elizabeth I (1533–1603) ruled England as it became a world power in the late 1500s.

The British government is seated in London, the capital. The Tower Bridge over the River Thames [TEMZ] is one of the city's many famous historical sites.

Interpreting the Visual Record
Why do you think London became a large city?

▼

USING ILLUSTRATIONS

Direct students' attention to the image of Queen Elizabeth I on this page. Have them describe her clothing and jewelry. Have students contrast Elizabeth's clothing to today's styles. Point out that the queen set fashion trends of the day. Ask students who establishes today's fads and fashions.

Then focus students' attention on the photograph of London. Lead a discussion about how the London cityscape compares to cities with which the students are familiar. Your questions might include: Do U.S. cities have more or fewer skyscrapers than London? Do U.S. cities have more or less open space in their downtown areas?

🔲 **internet** connect

GO TO: go.hrw.com
KEYWORD: SJ3 CH20
FOR: Web sites about England

◄ **Visual Record Answer**

its location on a major river and near the sea

ALL LEVELS: (Suggested time: 10 min.) Copy the following graphic organizer onto the chalkboard, omitting the italicized answers. Call on volunteers to complete it. Add more boxes as needed. Then lead a discussion of how the United Kingdom's role in world affairs has changed over the years. Have students identify historical events and factors that have led to these changes. **ENGLISH LANGUAGE LEARNERS**

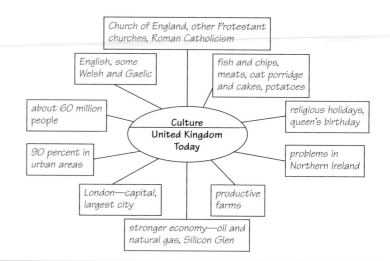

Graphic organizer:
- *Church of England, other Protestant churches, Roman Catholicism*
- *English, some Welsh and Gaelic*
- *fish and chips, meats, oat porridge and cakes, potatoes*
- *about 60 million people*
- **Culture United Kingdom Today**
- *religious holidays, queen's birthday*
- *90 percent in urban areas*
- *problems in Northern Ireland*
- *London—capital, largest city*
- *productive farms*
- *stronger economy—oil and natural gas, Silicon Glen*

Cultural Kaleidoscope

Pakistanis in Britain The population of Great Britain has become increasingly diverse over the years. People from around the world, particularly from countries that were once part of the British Empire, have immigrated to Great Britain. During the 1950s and 1960s many people immigrated to Britain from Pakistan. Today Britain's Pakistanis number some 500,000. Most live in large cities, where some have established restaurants and other small businesses.

Pakistanis living in Britain face many challenges, including discrimination. The unemployment rate among them is high partly because some recent immigrants speak little English. In addition, many Pakistani children in Britain receive a poor education. However, British-born Pakistanis have better job prospects. This segment of the population has grown up speaking English.

Activity: Have students research immigrant education in Britain and the United States and compare them.

Visual Record Answer ▲

Students might suggest that television exposes people around the world to each other's cultures.

Chart Answer ▶

It is higher.

▲

Millions of Americans watched the Beatles, a British rock band, perform on television in 1964. The Beatles and other British bands became popular around the world. **Interpreting the Visual Record How do you think television helps shape world cultures today?**

The United Kingdom				
COUNTRY	**POPULATION/ GROWTH RATE**	**LIFE EXPECTANCY**	**LITERACY RATE**	**PER CAPITA GDP**
United Kingdom	59,647,790 0.2%	75, male 81, female	99%	$22,800
United States	281,421,906 0.9%	74, male 80, female	97%	$36,200

Sources: Central Intelligence Agency, *The World Factbook 2001*; U.S. Census Bureau

Interpreting the Chart How does the literacy rate in the United Kingdom compare with that of the United States?

Culture

Nearly 60 million people live in the United Kingdom today. English is the official language. Some people in Wales and Scotland also speak the Celtic languages of Welsh and Gaelic [GAY-lik]. The Church of England is the country's official church. However, many Britons belong to other Protestant churches or are Roman Catholic.

Food and Festivals Living close to the sea, the British often eat fish. One popular meal is fish and chips—fried fish and potatoes. However, British food also includes different meats, oat porridge and cakes, and potatoes in many forms.

The British celebrate many religious holidays, such as Christmas. Other holidays include the Queen's official birthday celebration in June. In July many Protestants in Northern Ireland celebrate a battle in 1690 in which Protestants defeated Catholic forces. In recent years the day's parades have sometimes sparked protests and violence between Protestants and Catholics.

Art and Literature British literature, art, and music have been popular around the world. Perhaps the most famous British writer is William Shakespeare. He died in 1616, but his poetry and plays, such as *Romeo and Juliet*, remain popular. In the 1960s the Beatles helped make Britain a major center for modern popular music. More recently, British performers from Elton John to the Spice Girls have attracted many fans.

✓ **READING CHECK:** *Human Systems* What aspects of British culture have spread around the world? British literature, art, and music have been popular around the world.

The United Kingdom Today

Nearly 90 percent of Britons today live in urban areas. London, the capital of England, is the largest city. It is located in southeastern England. London is also the capital of the whole United Kingdom.

More than 7 million people live in London. The city is a world center for trade, industry, and services, particularly banking and insurance. London also has one of the world's busiest airports. Many tourists visit London to see its famous historical sites, theaters, and shops. Other important cities include Glasgow, Scotland; Cardiff, Wales; and Belfast, Northern Ireland.

Ask students to name rock bands, television programs, fashions, cars, and fads from the United Kingdom. Ask why these influences travel to the United States so easily. (Possible answer: A common language and many cultural ties link the countries.)

Have students complete Main Idea Activity 20.2. Then write the headings *History, Government, Culture,* and *Economy* on the chalkboard. Call on students to supply terms and phrases related to the United Kingdom for each heading. **ENGLISH LANGUAGE LEARNERS**

REVIEW AND ASSESS

EXTEND

Have students complete the Section Review. Then pair students. Ask each student to write three questions and answers related to the content of Section 2. Have students take turns reading an answer and challenging their partner to guess the corresponding question. Then have students complete Daily Quiz 20.2. **COOPERATIVE LEARNING**

Have interested students conduct research on differences between British and American word usage. Have students create a chart to present notable differences in usage. **BLOCK SCHEDULING**

The Economy Old British industries like mining and manufacturing declined after World War II. Today, however, the economy is stronger. North Sea reserves have made the country a major producer of oil and natural gas. Birmingham, Glasgow, and other cities are attracting new industries. One area of Scotland is called Silicon Glen. This is because it has many computer and electronics businesses. **Glen** is a Scottish term for a valley. Today many British work in service industries, including banking, insurance, education, and tourism.

Agriculture Britain's modern farms produce about 60 percent of the country's food. Still, only about 1 percent of the labor force works in agriculture. Important products include grains, potatoes, vegetables, and meat.

Northern Ireland One of the toughest problems facing the country has been violence in Northern Ireland. Sometimes Northern Ireland is called Ulster. The Protestant majority and the Roman Catholic minority there have bitterly fought each other. Violence on both sides has resulted in many deaths.

Many Catholics believe they have not been treated fairly by the Protestant majority. Therefore, many want Northern Ireland to join the mostly Roman Catholic Republic of Ireland. Protestants fear becoming a minority on the island. They want to remain part of the United Kingdom. Many people hope that recent agreements made by political leaders will lead to a lasting peace. In 1999, for example, Protestant and Roman Catholic parties agreed to share power in a new government. However, there have been problems putting that agreement into effect.

✔ **READING CHECK:** *Human Systems* What has been the cause of conflict in Northern Ireland? historical differences between the Protestant majority and the Roman Catholic minority

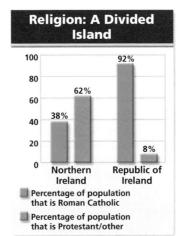

Religion: A Divided Island

Source: *The Statesman's Year Book*, 1998–99.

■ Percentage of population that is Roman Catholic
■ Percentage of population that is Protestant/other

Interpreting the Graph How does the number of Roman Catholics in the Republic of Ireland differ from that of Northern Ireland?

Section Review 2

Define and explain: textiles, constitutional monarchy, glen

Working with Sketch Maps On the map you created in Section 1, label the United Kingdom, England, Scotland, Wales, Northern Ireland, Irish Sea, London, Birmingham, Manchester, Glasgow, Cardiff, and Belfast.

Reading for the Main Idea

1. *Human Systems* What peoples came to the British Isles after the Celts? When did they come?
2. *Places and Regions* What was the British Empire?

Critical Thinking

3. **Contrasting** How is the British government different from the U.S. government?
4. **Drawing Inferences and Conclusions** Why do you think Protestants in Northern Ireland want to remain part of the United Kingdom?

Organizing What You Know

5. **Sequencing** Create a timeline that lists important events in the period.

800 B.C.   A.D. 2000

go.hrw.com
Homework Practice Online
Keyword: SJ3 HP20

Section Review 2

Answers

Define For definitions, see: textiles, p. 435; constitutional monarchy, p. 435; glen, p. 437

Working with Sketch Maps Places should be labeled in their approximate locations.

Reading for the Main Idea

1. Angles, Saxons, Vikings—A.D. 400s to 1000s; Normans—1066 (NGS 9)
2. colonies around the world established by England (NGS 4)

Critical Thinking

3. Britain—constitutional monarchy; U.S.—constitutional democracy
4. Possible answers: fear of losing influence or of change, desire to stay in majority

Organizing What You Know

5. 750–450 B.C.—Celts arrive; A.D. 400s–1000s—Angles, Saxons, Vikings, Normans arrive; late 1500s—England a world power; 1600s—colonies established; 1700s–1800s—Industrial Revolution; 1900—empire at its height; by 1921—United Kingdom weakened; 1970s—most colonies independent

▲ **Graph Answer**

Roman Catholics make up most of the population.

Setting the Scene

During the mid-1800s, London was a crowded, dirty place for many of its citizens. Health care was limited by misinformation. Sanitation was poor. These factors contributed to the spread of disease. Cholera was one of the worst. Cholera victims experienced symptoms such as diarrhea and vomiting. Dehydration ultimately killed them.

Building a Case

Have students read the Case Study feature. Then have students fill in this graphic organizer on the progression of events and the effect that Dr. Snow's map had on the control of cholera.

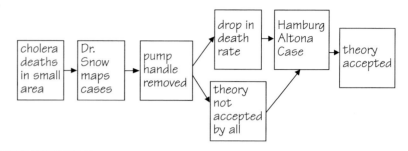

HISTORICAL GEOGRAPHY

The United States had its own epidemic of huge proportions early in the 1900s, but many people have never heard of it.

In March 1918, a soldier in Kansas went to an army base hospital complaining of a fever, sore throat, and headache. The soldier had influenza, more commonly called the flu. By the end of the week, the base hospital had treated about 500 ill people.

Soon, Americans were dying by the thousands. New York City reported 851 deaths in one day. San Francisco reported 5,000 new cases in December 1918. In all, more than half a million Americans died of the flu and related diseases during the epidemic.

Americans were also fighting World War I at the time U.S. soldiers accidentally spread the disease to Europe. By the time the epidemic ended in 1919, the flu had killed some 30 million people worldwide.

Activity: Have students conduct research on the spread of influenza during 1918–19 and record their findings on a world map. Can they draw any conclusions about the spread of the disease from the results?

➤ This Case Study feature addresses National Geography Standards 1, 3, 12, 15 and 18.

MAPPING THE SPREAD OF CHOLERA

Medical geographers want to discover why a disease occurs in a particular place. Does a disease occur in a certain type of environment? Is there a pattern to the way a disease spreads? Mapping is one tool medical geographers use to answer these questions. They first used maps in this way to fight cholera.

Cholera has existed in India for hundreds of years. It did not appear in Europe, however, until the 1800s. At that time better transportation systems helped spread cholera around the world. For example, in 1817 India experienced an unusually bad outbreak of the disease. India was then part of the British Empire. British soldiers and ships carried cholera to new places. By 1832 the disease had spread to the British Isles and to North America.

No one knew what caused cholera. In fact, no one knew about bacteria or how they caused disease. What was known was that sick people suffered from diarrhea and vomiting. They often died quickly. In just 10 days in 1854, more than 500 people in one London neighborhood died from cholera. Dr. John Snow thought he knew why.

Dr. Snow believed that people got cholera from dirty water. To test his theory, he mapped the location of some of London's public water pumps. (Houses at that time did not have running water.) Then he marked the location of each cholera death on his map. He found most of the deaths were scattered around the water pump on Broad Street. He persuaded officials to remove the pump's handle. After the pump was shut down, there were few new cases of cholera. Not everyone, however, believed Dr. Snow's evidence.

This illustration shows London's Regent Street in about 1850. An outbreak of cholera in this neighborhood killed hundreds of people in 1854.

Drawing Conclusions

Why was Dr. Snow's map instrumental in solving the cholera riddle? Are there other factors besides dirty water that might have caused the results recorded on the map? Challenge students to answer these questions by writing and acting out a skit of Dr. Snow's meeting with the London authorities. Conclude by asking students to suggest other issues that we might understand more fully if they were plotted on a map.

Going Further: Thinking Critically

Have students contact public health departments for information on allergies in the United States. Then have them answer these questions:

- Where are allergy cases common? Can any connections be made between the cases and natural features, such as types of vegetation?
- Do people in areas where air pollution is a problem suffer more from allergies? *(Students should provide evidence to support their conclusions.)*
- What actions should be taken next if mapping allergy sufferers and allergens (substances that cause allergic reactions) reveals clear connections? *(Students might provide small- or large-scale solutions to allergy outbreaks.)*

Cholera Deaths in London, 1854

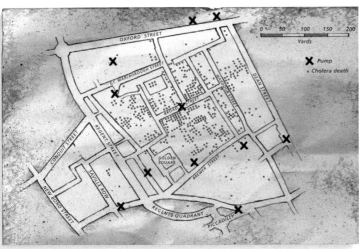

Dr. Snow plotted cholera deaths for the first 10 days of September 1854 on a map of the neighborhood. Dr. Snow believed that people were getting impure water from the Broad Street pump. A check of the pump showed that a leaking sewer had contaminated the pump's water.

Interpreting the Map What does the map tell you about the relationship between deaths and the Broad Street pump?

Bacteria was discovered in the 1880s. Yet many people still believed that bad air from river mud or swamps spread diseases like cholera. It took another epidemic to convince everyone that dirty water made people sick.

In 1892 an outbreak of cholera in the German port city of Hamburg ended the debate. Once again, the locations of people who became sick were mapped. This mapping was the key to proving that water could spread cholera.

At the time, a street separated Hamburg and another town, Altona. Both towns got their water from the Elbe River. There was, however, an important difference in the two towns' water supplies. Altona had a system for cleaning its water. Hamburg did not. People living on the Hamburg side of the street got sick. Those on the Altona side of the street did not. The air on both sides of the street was the same. Therefore, no one could argue that bad air caused the outbreak. Hamburg then moved quickly to install a system to clean its water.

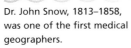

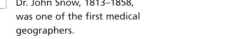

Dr. John Snow, 1813–1858, was one of the first medical geographers.

Understanding What You Read

1. Why did Dr. Snow make a map of a neighborhood's cholera deaths and the location of its water pumps?
2. Why was it important that people on the Hamburg side of the street got sick but people on the Altona side did not?

Understanding What You Read

1. Dr. Snow wanted to test his theory that people got cholera from impure water. By mapping the location of water pumps and cholera cases, he hoped to make a visual connection between the two factors.
2. It showed that bad air was not causing the cholera outbreak, since people on either side of the street breathed the same air. However, the two sides got their water from different sources. Altona had a water purification system, but Hamburg did not.

internet connect

GO TO: go.hrw.com
KEYWORD: SJ3 CH20
FOR: Web sites about cholera

◄ **Interpreting the Map**

Deaths were concentrated around the Broad Street pump.

Section 3

Objectives

1. Identify key events in Ireland's history.
2. Describe the people and culture of Ireland.
3. Explain the economic changes Ireland has experienced in recent years.

Visual Record Answer ▶

rainy temperate

FOCUS

LET'S GET STARTED

Copy the following questions onto the chalkboard: *Which Irish holiday is celebrated by many Americans? How is it celebrated? (Saint Patrick's Day, wearing green, parades)* Have students write their answers. Explain that Irish immigrants have contributed much to our culture. Point out that although it is now a fairly prosperous country, Ireland has had difficult times in its past. Because of those difficulties, many people left Ireland for the United States. Tell students that in Section 3 they will learn more about the Republic of Ireland.

Building Vocabulary

Write the key terms on the chalkboard and ask students to locate their definitions in the section text. Call on volunteers to read the definitions aloud. Ask students if they recall news about **famine** in other countries. You may want to ask if they have heard of the Irish potato famine. Explain that from 1845 to 1849, a disease killed so many potato plants in Ireland that hundreds of thousands of people starved. Potatoes had been a major food source, particularly among the poor. Explain that **bog** comes from the ancient language of Ireland, Gaelic. **Peat** is found in bogs.

Section 3 The Republic of Ireland

Read to Discover

1. What are the key events in Ireland's history?
2. What are the people and culture of Ireland like?
3. What kinds of economic changes has Ireland experienced in recent years?

Define
famine
bog
peat

Locate
Dublin
Cork
Galway

WHY IT MATTERS

Millions of Americans trace their heritage to Ireland. Use **CNNfyi.com** or other current events sources to learn about Ireland and its people today. Record your findings in your journal.

An Irish harp

History

The Irish are descendants of the Celts. Irish Gaelic, a Celtic language, and English are the official languages. Most people in Ireland speak English. Gaelic is spoken mostly in rural western areas.

English Conquest England conquered Ireland in the A.D. 1100s. By the late 1600s most of the Irish had become farmers on land owned by the British. This created problems between the two peoples. Religious differences added to these problems. Most British were Protestant, while most Irish were Roman Catholic. Then, in the 1840s, millions of Irish left for the United States and other countries because of a poor economy and a potato **famine**. A famine is a great shortage of food.

Independence The Irish rebelled against British rule. In 1916, for example, Irish rebels attacked British troops in the Easter Rising.

Stone fences divide the green fields of western Ireland, the Emerald Isle.
Interpreting the Visual Record What kind of climate would you expect to find in a country with rich, green fields such as these?

▼

Teaching Objectives 1–3

LEVEL 1: (Suggested time: 20 min.) Copy the following graphic organizer onto the chalkboard, omitting the italicized answers. Have students fill in important eras in Ireland's history. Then lead a discussion about how Ireland's history has influenced its culture and the recent economic changes in the country. **ENGLISH LANGUAGE LEARNERS**

LEVELS 2 AND 3: (Suggested time: 30 min.) Ask students to imagine that they are the prime minister of Ireland and that they are running for re-election. Have them write campaign speeches in which they display their knowledge of Ireland's past and take credit for the country's economic progress. Call on volunteers to deliver their speeches to the class.

| *Celtic Ireland* | → | *English conquest* | → | *Immigration to United States* | → | *Easter Rising* | → | *Independence* |

At the end of 1921, most of Ireland gained independence. Some counties in northern Ireland remained part of the United Kingdom. Ties between the Republic of Ireland and the British Empire were cut in 1949.

Government Ireland has an elected president and parliament. The president has mostly ceremonial duties. Irish voters in 1990 elected a woman as president for the first time.

The parliament makes the country's laws. The Irish parliament chooses a prime minister to lead the government.

✓ **READING CHECK:** (*Human Systems*) What are some important events in Ireland's history? English conquest, Easter Rising, independence for most of Ireland, end of ties between Republic of Ireland and the British Empire

Culture

Centuries of English rule have left their mark on Irish culture. For example, today nearly everyone in Ireland speaks English. Irish writers, such as George Bernard Shaw and James Joyce, have been among the world's great English-language writers.

A number of groups promote traditional Irish culture in the country today. The Gaelic League, for example, encourages the use of Irish Gaelic. Gaelic and English are taught in schools and used in official documents. Another group promotes Irish sports, such as hurling. Hurling is an outdoor game similar to field hockey and lacrosse.

Elements of Irish culture have also become popular outside the country. For example, traditional Irish folk dancing and music have attracted many fans. Music has long been important in Ireland. In fact, the Irish harp is a national symbol. Many musicians popular today are from Ireland, including members of the rock band U2.

More than 90 percent of the Irish today are Roman Catholic. St. Patrick's Day on March 17 is a national holiday. St. Patrick is believed to have brought Christianity to Ireland in the 400s.

✓ **READING CHECK:** (*Human Systems*) What is Irish culture like today? English is almost universal; groups promote Irish traditions, such as the use of Gaelic and Irish sports; music is important; 90 percent of Irish are Roman Catholic.

Ireland Today

Ireland used to be one of Europe's poorest countries. Today it is a modern, thriving country with a strong economy and growing cities.

▲ Many Irish enjoy a meal of lamb chops with mustard sauce, soda bread, carrots, and mashed potatoes.

Linking Past to Present

St. Patrick The young man who would later be called Saint Patrick was living in western England when he was captured by raiders and sold into slavery in Ireland. He is credited with converting the Irish people to Christianity in the A.D. 400s. He also introduced the Roman alphabet and Latin literature to Ireland.

According to legend, Patrick drove all the snakes in Ireland into the sea. However, evidence shows that a great freeze affected Ireland and much of the Northern Hemisphere until some 15,000 years ago. Any snakes living in Ireland would have been killed off. Because land snakes cannot migrate across water, there are no snakes living in Ireland today. The legend may refer to Patrick's eliminating ancient non-Christian beliefs from Ireland. Snakes were among many pre-Christian symbols.

Activity: Have students work in groups to conduct research on the historical background of the holidays celebrated in their communities. You may want to have them concentrate on their community's celebration of St. Patrick's Day. Have them summarize their findings in a written or oral report.

◄ **Chart Answer (p. 442)**

They are the same.

Play a recording of a Celtic Irish song, such as one by the Chieftains or DeDanaan, and one by a modern Irish rock band, such as U2. Ask students to compare the two songs and explain how they express characteristics of traditional and modern Irish society.

Have students complete Main Idea Activity 20.3. Then write *Ireland* vertically on the chalkboard. Have students suggest a phrase or sentence beginning with each letter on the board that describes Ireland. *(Example: I—Island; Ireland is an island country.)* **ENGLISH LANGUAGE LEARNERS**

REVIEW AND ASSESS

EXTEND

Have students complete the Section Review. Then organize the class into groups—one for each major topic in the section. Have each group write four original statements about its topic. Have students write their sentences on the chalkboard. Discuss each statement. Then have students complete Daily Quiz 20.3. **COOPERATIVE LEARNING**

Ask interested students to conduct research on popular sports in the Republic of Ireland. They may want to investigate which sports teens enjoy, which have professional status, how much income sports generate for the country, or the sports stars of today. Have students report their findings in a simulated television sports broadcast. **BLOCK SCHEDULING**

Section Review 3

Answers

Define For definitions, see: famine, p. 440; bog, p. 442; peat, p. 442

Working with Sketch Maps Dublin is the capital and largest city, with many factories. It is the country's center for education, banking, and shipping.

Reading for the Main Idea

1. poor economy, potato famine (NGS 9)

2. change from agricultural to industrial economy, investment by foreign companies (NGS 11, 12)

Critical Thinking

3. possible answer: desire to remain independent and retain their own language, religion

4. more demand for housing as people moved there to work, more people able to pay higher prices

Organizing What You Know

5. Ireland: problems arose between Brit. landowners, Irish farmers; many Irish emigrated; Irish rebelled, gained independence; most Irish speak English; Irish contributed to arts, literature; most Irish are Roman Catholic; economy improved; main cities—Dublin, Cork, Galway. See Section 2 for information about UK.

Ireland

COUNTRY	POPULATION/ GROWTH RATE	LIFE EXPECTANCY	LITERACY RATE	PER CAPITA GDP
Ireland	3,840,838 1.1%	74, male 80, female	98%	$21,600
United States	281,421,906 0.9%	74, male 80, female	97%	$36,200

Sources: Central Intelligence Agency, *The World Factbook 2001*; U.S. Census Bureau

Interpreting the Chart How does life expectancy in Ireland compare with that of the United States?

Economy Until recently, Ireland was mostly an agricultural country. This was true even though much of the country is either rocky or boggy. A **bog** is soft ground that is soaked with water. For centuries, **peat** dug from bogs has been used for fuel. Peat is made up of dead plants, usually mosses.

Today Ireland is an industrial country. Irish workers produce processed foods, textiles, chemicals, machinery, crystal, and computers. Finance, tourism, and other service industries are also important.

How did this change come about? Ireland's low taxes, well-educated workers, and membership in the European Union have attracted many foreign companies. Those foreign companies include many from the United States. These companies see Ireland as a door to millions of customers throughout the EU. In fact, goods from their Irish factories are exported to markets in the rest of Europe and countries in other regions.

Cities Many factories have been built around Dublin. Dublin is Ireland's capital and largest city. It is a center for education, banking, and shipping. Nearly 1 million people live there. Housing prices rapidly increased in the 1990s as people moved there for work.

Other cities lie mainly along the coast. These cities include the seaports of Cork and Galway. They have old castles, churches, and other historical sites that are popular among tourists.

✓ **READING CHECK:** *Human Systems* What important economic changes have occurred in Ireland and why? industrial country; low taxes, well-educated workers, and EU membership have attracted foreign companies

go.hrw.com
Homework Practice Online
Keyword: SJ3 HP20

Section Review 3

Define and explain: famine, bog, peat

Working with Sketch Maps On the map you created in Section 2, label Ireland, Dublin, Cork, and Galway. In the margin explain the importance of Dublin to the Republic of Ireland.

Reading for the Main Idea

1. *Human Systems* What were two of the reasons many Irish moved to the United States and other countries in the 1800s?

2. *Human Systems* What are some important reasons why the economy in Ireland has grown so much in recent years?

Critical Thinking

3. **Drawing Inferences and Conclusions** Why do you think the Irish fought against British rule?

4. **Drawing Inferences and Conclusions** Why do you suppose housing prices rapidly increased in Dublin in the 1990s?

Organizing What You Know

5. **Comparing/Contrasting** Copy the following graphic organizer. Use it to compare and contrast the history, culture, and governments of the Republic of Ireland and the United Kingdom.

Ireland	United Kingdom
Conquered by England in the 1100s	Created vast world empire

Section 4

Objectives

1. Describe the people and culture of Scandinavia.
2. Identify important features of the region's countries, Greenland, and Lapland.

Section 4

Scandinavia

Read to Discover

1. What are the people and culture of Scandinavia like?
2. What are some important features of each of the region's countries, plus Greenland, and Lapland?

Define

neutral
uninhabitable
geysers

Locate

Oslo
Bergen
Stockholm
Göteborg
Copenhagen
Nuuk (Godthab)

Reykjavik
Gulf of Bothnia
Gulf of Finland
Helsinki
Lapland

WHY IT MATTERS

Much of the fish Americans eat comes from the nations of Scandinavia. Use **CNNfyi.com** or other **current events** sources to learn more about the economic importance of these nations. Record your findings in your journal.

Smoked salmon, a popular food in Scandinavia

People and Culture

Scandinavia once was home to fierce, warlike Vikings. Today the countries of Norway, Sweden, Denmark, Iceland, and Finland are peaceful and prosperous.

The people of the region enjoy high standards of living. They have good health care and long life spans. Each government provides expensive social programs and services. These programs are paid for by high taxes.

The people and cultures in the countries of Scandinavia are similar in many ways. For example, the region's national languages, except for Finnish, are closely related. In addition, most people in Scandinavia are Lutheran Protestant. All of the Scandinavian countries have democratic governments.

✔ **READING CHECK:**
Human Systems How are the people and cultures of Scandinavia similar? Most languages closely related, most people are Lutheran Protestant, all governments are democratic.

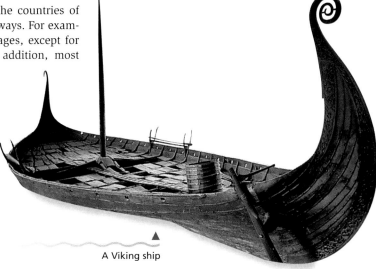

▲ A Viking ship

Teaching Objective 1

ALL LEVELS: (Suggested time: 10 min.) Copy the following graphic organizer onto the chalkboard, omitting the italicized answers. Call on students to fill in the lines with characteristics the Scandinavian countries share. **ENGLISH LANGUAGE LEARNERS**

peaceful and prosperous

high standards of living

good health care, long life spans

social programs and services government-sponsored

Scandinavia

high taxes

languages closely related (except Finnish)

Lutheran Protestants

democratic governments

ENVIRONMENT AND SOCIETY

Norway's long shape, severe winter climate, and sparsely populated countryside make it difficult and expensive to provide transportation services throughout the country. Winter weather limits car travel—the main road between Oslo and Bergen is closed for four to six months.

Because transportation costs are high, the Norwegian government provides or helps to pay for some forms of transportation. The government-owned railway loses money each year. There are simply too few train riders outside the large cities to make the service profitable.

Critical Thinking: Why might the Norwegian government continue to provide rail services even though it loses money?

Answer: Students might suggest that a national transportation system is vital for commerce, national defense purposes, or as a public good.

Chart Answer ▲

They are higher than those of the United States and Denmark.

Visual Record Answer ▶

thwart land-based attacks; create increased access to waterways for trade

Scandinavia

COUNTRY	POPULATION/ GROWTH RATE	LIFE EXPECTANCY	LITERACY RATE	PER CAPITA GDP
Denmark	5,352,815 0.3%	74, male 80, female	100%	$25,500
Finland	5,175,783 0.2%	74, male 81, female	100%	$22,900
Iceland	277,906 0.5%	77, male 82, female	100%	$24,800
Norway	4,503,440 0.5%	76, male 82, female	100%	$27,700
Sweden	8,875,053 .02%	77, male 83, female	99%	$22,200
United States	281,421,906 0.9%	74, male 80, female	97%	$36,200

Sources: Central Intelligence Agency, *The World Factbook 2001;* U.S. Census Bureau

Interpreting the Chart What is noteworthy about life expectancy in Iceland, Norway, and Sweden?

Riddarhomem—the Isle of Knights— is one of several islands on which the original city of Stockholm was built.

Interpreting the Visual Record Why might a group of islands be a good place to build a city?

▶

Norway

Norway is a long, narrow, and rugged country along the western coast of the Scandinavian Peninsula. Norway once was united with Denmark and then Sweden. In 1905 Norway became independent. Today Norway is a constitutional monarchy with an elected parliament.

About 75 percent of the people live in urban areas. The largest cities are the capital, Oslo, and Bergen on the Atlantic coast. Oslo is a modern city. It lies at the end of a wide fjord on the southern coast. The city is Norway's leading seaport, as well as its industrial and cultural center.

Norway has valuable resources, especially oil and natural gas. However, Norway's North Sea oil fields are expected to run dry over the next century. A long coastline and location on the North Sea have helped make Norway a major fishing and shipping country. Fjords shelter Norway's harbors and its fishing and shipping fleets.

Sweden

Sweden is Scandinavia's largest and most populous country. It is located between Norway and Finland. Most Swedes live in cities and towns. The largest cities are Stockholm, which is Sweden's capital, and Göteborg. Stockholm is located on the Baltic Sea coast. It is a beautiful city of islands and forests. Göteborg is a major seaport.

Like Norway, Sweden is a constitutional monarchy. The country has been at peace since the early 1800s. Sweden remained **neutral** during World Wars I and II. A neutral country is one that chooses not to take sides in an international conflict.

Sweden's main sources of wealth are forestry, farming, mining, and manufacturing. Wood, iron ore, automobiles, and wireless telephones are exports. Hydroelectricity is important.

Teaching Objective 2

LEVEL 1: (Suggested time: 30 min.) Organize the class into five groups. Assign one of the region's countries to each group. Have the groups design a new flag for their assigned country using colors and symbols that refer to the text material. Tell students that the flags should represent the important features of the region's countries as well as Lapland and Greenland. Ask each group to choose a spokesperson to explain its meaning to the class. Then have the spokespeople make their presentations. Finally, display students' flags around the classroom.

ENGLISH LANGUAGE LEARNERS, COOPERATIVE LEARNING

LEVELS 2 AND 3: (Suggested time: 45 min.) Have students create brochures to encourage people from other countries to invest in Scandinavian countries. Brochures should provide an introduction to the country's physical and human geography and summarize each country's economic and political characteristics as well as other important features. Instruct students to make a clear connection in their brochures between their assigned country's characteristics and its economic future. Ask students to add a catchy slogan for the country. (example: "Iceland—Cold Name, Warm Welcome") You may want to have students add drawings or pictures from magazines to add interest. Additional research may be necessary.

CONNECTING TO Art

A stave church in Norway

Stave Churches

In Norway you will find some beautiful wooden churches built during the Middle Ages. They are known as stave churches because of their corner posts, or staves. The staves provide the building's basic structure. Today stave churches are a reminder of the days when Viking and Christian beliefs began to merge in Norway.

As many as 800 stave churches were built in Norway during the 1000s and 1100s. Christianity was then beginning to spread throughout the country. It was replacing the old religious beliefs of the Viking people. Still, Viking culture is clearly seen in stave buildings.

Except for a stone foundation, stave churches are made entirely of wood. Workers used methods developed by Viking boat builders. For example, wood on Viking boats was coated with tar to keep it from rotting. Church builders did the same with the wood for their churches. They also decorated the churches with carvings of dragons and other creatures. The stave church at Urnes even has a small Viking ship decorated with nine candles.

When the plague, or Black Death, arrived in Norway about 1350, many communities were abandoned. Many stave churches fell apart. Others were replaced by larger stone buildings. Today only 28 of the original buildings remain. They have been preserved for their beauty and as reminders of an earlier culture.

Understanding What You Read
1. What are staves?
2. How did stave churches reflect new belief systems in Norway?

Across the Curriculum

LITERATURE

Hans Christian Andersen
Among the many contributions made to the arts by Scandinavians, some of the most popular are the stories written by Danish author Hans Christian Andersen. His fairy tales are among the most frequently translated works in the world. Andersen also wrote plays and novels.

The author published his first book of stories, *Tales, Told for Children,* in 1835. It included the classic "Princess and the Pea," which was the inspiration for the Broadway musical *Once upon a Mattress.* Andersen's story "The Little Mermaid" was the basis for a popular animated movie of the same name. A bronze statue of a mermaid was placed in Copenhagen's harbor to commemorate Andersen's heroine.

Critical Thinking: In what ways have Andersen's tales spread beyond Scandinavia?

Answer: They have become popular around the world and have been used for a musical and a movie.

Connecting to Art Answers
1. corner posts of wooden churches
2. They were built as Christianity was replacing Viking religious beliefs.

Denmark

Denmark is the smallest and most densely populated of the region's countries. Most of Denmark lies on the Jutland Peninsula. About 500 islands make up the rest of the country.

Denmark is also a constitutional monarchy. The capital and largest city is Copenhagen. It lies on an island between the Jutland Peninsula and Sweden. Some 1.4 million people—about 25 percent of Denmark's population—live there.

About 60 percent of Denmark's land is used for farming. Farm products, especially meat and dairy products, are important exports. Denmark also has a modern industrial economy. Industries include food processing, machinery, furniture, and electronics.

445

Kay A. Knowles of Montross, Virginia, suggests the following activity to show how cultures can differ within the same region. First, have students conduct research on these topics as they relate to the British Isles and Scandinavia: origins, religions, languages, ethnic groups, conflicts, and customs. Then have students create charts to organize and present their findings.

Lead a class discussion comparing what students have learned from the activity to other countries or regions they have studied. Challenge them to draw conclusions.

Lead a discussion on how a Viking might react if he or she visited Scandinavia today. Ask: Which aspects of the region might be familiar? Which would be unfamiliar?

Section Review 4

Answers

Define For definitions, see: neutral, p. 444; uninhabitable, p. 446; geysers, p. 446

Working with Sketch Maps They are descendants of hunters from northern Asia, speak languages related to Finnish, earn money from reindeer herding or tourism.

Reading for the Main Idea

1. closely related languages (except for Finnish), Lutheran Protestant religion, democratic governments (NGS 10)

2. by catching fish, using hot water from geysers for heat (NGS 15)

3. country's original settlers not Vikings; Finnish not related to other Scandinavian languages (NGS 10)

Critical Thinking

4. possible answer: They may cause the cultures to become more separate and lead to independence for Greenland.

Organizing What You Know

5. Answers will vary but should be consistent with text.

Visual Record Answer ▶

The interior is icy and uninhabitable.

▲
Greenland's capital lies on the island's southwestern shore.

Interpreting the Visual Record
Why do most people in Greenland live along the coast?

The Great Geysir in southwestern Iceland can spout water nearly 200 feet (61 m) into the air. Some geysers shoot steam and boiling water to a height of more than 1,600 feet (nearly 500 m)!

Greenland

The huge island of Greenland is part of North America, but it is a territory of Denmark. Greenland's 56,000 people have their own government. They call their island Kalaallit Nunaat. The capital is Nuuk, also called Godthab. Most of the island's people are Inuit (Eskimo). Fishing is the main economic activity. Some Inuit still hunt seals and small whales.

The island's icy interior is **uninhabitable**. An uninhabitable area is one that cannot support human settlement. Greenland's people live mostly along the southwestern coast in the tundra climate regions.

Iceland

Between Greenland and Scandinavia is the country of Iceland. This Atlantic island belonged to Denmark until 1944. Today it is an independent country. It has an elected president and parliament.

Unlike Greenland, Iceland is populated mostly by northern Europeans. The capital and largest city is Reykjavik (RAYK-yuh-veek). Nearly 40 percent of the country's people live there.

Icelanders make good use of their country's natural resources. For example, about 70 percent of the country's exports are fish. These fish come from the rich waters around the island. In addition, hot water from Iceland's **geysers** heats homes and greenhouses. The word *geyser* is an Icelandic term for hot springs that shoot hot water and steam into the air. Volcanic activity forces heated underground water to rise from the geyser.

Finland

Finland is the easternmost of the region's countries. It lies mostly between two arms of the Baltic Sea: the Gulf of Bothnia and the Gulf of Finland. The capital and largest city is Helsinki, which is located on the southern coast.

Have students complete the Section Review. Then have students outline the main points of Section 4. Ask them to choose one topic from their outlines and create a simple graphic organizer to illustrate it. Call on volunteers to draw their organizers on the chalkboard and explain their use. Then have students complete Daily Quiz 20.4.

RETEACH

Have students complete Main Idea Activity 20.4. Then organize students into groups and assign each group one of the Scandinavian countries. Have each group become "experts" on their country. Give them chalk for writing

on the board or materials for using the overhead projector, allow time for collaboration, and have them present a lesson on their country to the class.
ENGLISH LANGUAGE LEARNERS, COOPERATIVE LEARNING

EXTEND

Have interested students conduct research on the important role Finnish ski troops played during the Soviet invasion of their country in 1939–40. Have students act out their findings in a "news broadcast" from the "front."
BLOCK SCHEDULING

The original Finnish settlers probably came from northern Asia. Finnish belongs to a language family that includes Estonian and Hungarian. About 6 percent of Finns speak Swedish. Finland was part of Sweden from the 1100s to 1809. It then became part of Russia. Finland gained independence at the end of World War I.

As in the other countries of the region, trade is important to Finland. The country is a major producer of paper and other forest products as well as wireless telephones. Metal products, shipbuilding, and electronics are also important industries. Finland imports energy and many of the raw materials needed in manufacturing.

Lapland

Across northern Finland, Sweden, and Norway is a culture region known as Lapland. This region is populated by the Lapps, or Sami, as they call themselves.

The Sami are probably descended from hunters who moved to the region from northern Asia. The languages they speak are related to Finnish. The Sami have tried to keep their culture and traditions, such as reindeer herding. Many now earn a living from tourism.

✔ **READING CHECK:** (*Human Systems*) Around what activities are the economies of the countries and territories discussed in this section organized? *Some are modern industrial economies, while others are organized around fishing, farming, or tourism.*

▲

Young Sami couples here are dressed in traditional clothes for the Easter reindeer races in northern Norway.

Interpreting the Visual Record What climates would you find in the lands of the Sami?

go. hrw .com **Homework Practice Online**
Keyword: SJ3 HP20

Section Review 4

Define and explain: neutral, uninhabitable, geysers

Working with Sketch Maps On the map you created in Section 3, label the countries of Scandinavia, Oslo, Bergen, Stockholm, Göteborg, Copenhagen, Nuuk (Godthab), Reykjavik, Gulf of Bothnia, Gulf of Finland, Helsinki, and Lapland. In the margin describe the people of the Lapland region.

Reading for the Main Idea

1. (*Human Systems*) What are two of the cultural similarities among the peoples of Scandinavia?

2. (*Environment and Society*) In what ways have Icelanders adapted to their natural environment?

3. (*Human Systems*) How have the history and culture of Finland been different from that of other countries in Scandinavia?

Critical Thinking

4. **Making Generalizations and Predictions** How do you think the location of Greenland and the culture of its people will affect the island's future relationship with Denmark?

Organizing What You Know

5. **Summarizing** Copy the following graphic organizer. Label the center of the organizer "Scandinavia." In the ovals, write one characteristic of each country and region discussed in this section. Then do the same for the other countries, as well as for Greenland and Lapland.

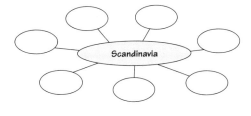
Scandinavia

Building Vocabulary
For definitions, see: fjords, p. 432; lochs, p. 432; North Atlantic Drift, p. 433; constitutional monarchy, p. 435; famine, p. 440; bog, p. 442; peat, p. 442; neutral, p. 444; uninhabitable, p. 446; geysers, p. 446

Reviewing the Main Ideas

1. Great Britain, Ireland; Atlantic Ocean, North Sea, English Channel, Irish Sea (NGS 4)

2. marine west coast, humid continental, subarctic, tundra (NGS 4)

3. Greenland (NGS 4)

4. Most colonies gained their independence. The empire became the British Commonwealth of Nations. (NGS 13)

5. Roman Catholicism and Protestantism are the two major religions there; violent Catholic and Protestant groups have fought over whether or not Northern Ireland should remain part of the United Kingdom. (NGS 10)

◄ **Visual Record Answer**

subarctic or tundra

447

Have students complete the Chapter 20 Test.

RETEACH

Organize the class into sections for the United Kingdom, Ireland, and Scandinavia. Ask students to work with a partner within their section to review the landforms, rivers, climates, resources, economy, history, governments, and cultural traits of their assigned areas. Then have the pairs meet within their section and compile lists of what they think are the 10 most important facts one should know about their region.

PORTFOLIO EXTENSIONS

1. Have students conduct research on John Cabot, Sir Francis Drake, Sir Walter Raleigh, or Captain James Cook and create presentations on the role the explorer played in the creation of the British Empire. Students' presentations should include an illustrated map and a version of the explorer's diary or ship's log.

2. For almost 100 years, Northern Ireland has had a tradition of political murals. Have students conduct research on the conflicts in Northern Ireland and draw murals advocating lasting peace on butcher paper. Photograph the murals and place the photograph in students' portfolios.

Review

ANSWERS

Understanding Environment and Society

The North Sea is an important location for the production of oil and gas. Natural gas was discovered off the Netherlands in 1959. Norway, Denmark, Germany, Netherlands, and the United Kingdom are the main countries that drill in the North Sea. Use Rubric 29, Presentations, to evaluate student work.

Thinking Critically

1. Answers will vary, but students should mention the themes in some of these art forms, such as Shakespearean literature or Beatles music, make them universally appealing.

2. saved transportation costs because coal, which fueled factories, was available nearby

3. history of English control; English is language of commerce, government, education

4. Possible: adapted building and clothing styles; used skis, snowshoes, sleds for travel; learned to make a living from what environment makes possible, such as reindeer

Reviewing What You Know

Building Vocabulary

On a separate sheet of paper, write sentences to define each of the following words.

1. fjords
2. lochs
3. North Atlantic Drift
4. constitutional monarchy
5. famine
6. bog
7. peat
8. neutral
9. uninhabitable
10. geysers

Reviewing the Main Ideas

1. (*Places and Regions*) What islands make up the British Isles? What bodies of water surround the British Isles?

2. (*Places and Regions*) What are the three main climates in northern Europe?

3. (*Places and Regions*) What large North Atlantic island is a territory of Denmark?

4. (*Human Systems*) What happened to the British Empire?

5. (*Human Systems*) What are the two major religions in Northern Ireland? What do these religions have to do with violence in Northern Ireland?

Understanding Environment and Society

Resource Use

Prepare a presentation, along with a map, graph, chart, model, or database, on the distribution of oil and natural gas in the North Sea. You may want to think about the following:

- How oil and natural gas are recovered there.
- How countries divided up the North Sea's oil and natural gas.

Write a five-question quiz, with answers, about your presentation to challenge fellow students.

Thinking Critically

1. **Drawing Inferences and Conclusions** Why do you think British literature, art, and music have been popular around the world?

2. **Drawing Inferences and Conclusions** Why do you think industrial cities like Birmingham and Manchester in Great Britain grew up near coal deposits?

3. **Drawing Inferences and Conclusions** Why do you think most Irish speak English rather than Gaelic?

4. **Making Generalizations and Predictions** How do you think Scandinavians have adapted to life in these very cold environments?

5. **Finding the Main Idea** Why are climates in the British Isles milder than in much of Scandinavia?

FOOD FESTIVAL

In Swedish, the word *smorgasbord* means "bread and butter table," but a smorgasbord is not just a table loaded with buttered bread. It is a complete buffet-style meal, with a variety of open-faced sandwiches, sliced meats, marinated or pickled fish, cheeses, hot or cold cooked vegetables, salads, and desserts. To create your own Swedish smorgasbord, have students bring as many of the listed food items as they can. Set the food on a large table and let students help themselves.

CHAPTER 20 REVIEW AND ASSESSMENT RESOURCES

Reproducible
- Readings in World Geography, History, and Culture 34, 35, and 36
- Vocabulary Activity 20

Technology
- Chapter 20 Test Generator (on the One-Stop Planner)

- Audio CD Program, Chapter 20 (English and Spanish)
- HRW Go site

Reinforcement, Review, and Assessment
- Chapter Review, pp. 448–49

- Chapter 20 Tutorial for Students, Parents, Mentors, and Peers
- Chapter 20 Test
- Chapter 20 Test for English Language Learners and Special-Needs Students

Building Social Studies Skills

Map ACTIVITY

On a separate sheet of paper, match the letters on the map with their correct labels.

London	Oslo
Manchester	Stockholm
Belfast	Copenhagen
Dublin	Reykjavik
Cork	Helsinki

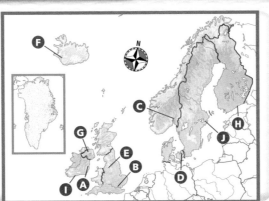

Mental Mapping Skills ACTIVITY

On a separate sheet of paper, draw a freehand map of northern Europe. Make a key for your map and label the following:

Baltic Sea	Ireland
Denmark	Norway
Finland	Scandinavian
Great Britain	Peninsula
Iceland	Sweden

WRITING ACTIVITY

Use print resources to find out more about the Vikings and how they lived in their cold climate. Write a short story set in a Viking village or on a Viking voyage. Describe daily life in the village or on the voyage. Include a bibliography showing references you used. Be sure to use standard grammar, spelling, sentence structure, and punctuation.

herding; developed diverse energy sources
5. Winds blowing over the North Atlantic Drift bring mild temperatures and rain.

Map Activity
A. Dublin	F. Reykjavik
B. London	G. Belfast
C. Oslo	H. Helsinki
D. Copenhagen	I. Cork
E. Manchester	J. Stockholm

Mental Mapping Skills Activity
Maps will vary, but listed places should be labeled in their approximate locations.

Writing Activity
Check stories to see that students have introduced several aspects of the Vikings' environment. Use Rubric 39, Writing to Create, to evaluate student work.

Portfolio Activity
Information included should be consistent with text material. Use Rubric 29, Presentations, to evaluate student work.

Alternative Assessment

Portfolio ACTIVITY

Learning About Your Local Geography
Research Project Northern Europe's social programs are supported by taxes. Research the taxes that people in your area must pay. Interview residents, asking their feelings about taxes.

internet connect

Internet Activity: go.hrw.com
KEYWORD: SJ3 GT20

Choose a topic to explore northern Europe:
- Explore the islands and fjords on the Scandinavian coast.
- Visit historic palaces in the United Kingdom.
- Investigate the history of skiing.

internet connect

GO TO: go.hrw.com
KEYWORD: SJ3 Teacher
FOR: a guide to using the Internet in your classroom

449

Eastern Europe

Chapter Resource Manager

Objectives	Pacing Guide	Reproducible Resources
SECTION 1		
Physical Geography (pp. 415–53) 1. Identify the major physical features of Eastern Europe. 2. Identify the climates and natural resources of the region.	**Regular** .5 day **Block Scheduling** .5 day *Block Scheduling Handbook, Chapter 21*	**RS** Guided Reading Strategy 21.1
SECTION 2		
The Countries of Northeastern Europe (pp. 454–59) 1. Identify the peoples who contributed to the early history of northeastern Europe. 2. Describe how northeastern Europe's culture was influenced by other cultures. 3. Describe how the political organization of the region has changed since World War II.	**Regular** 2.5 days **Block Scheduling** 1 day *Block Scheduling Handbook, Chapter 21*	**RS** Guided Reading Strategy 21.2 **RS** Graphic Organizer 21 **E** Cultures of the World Activity 3 **E** Creative Strategies for Teaching World Geography, Lesson 11 **SM** Geography for Life Activities 14 and 21 **SM** Map Activity 21
SECTION 3		
The Countries of Southeastern Europe (pp. 460–65) 1. Describe how southeastern Europe's early history helped to shape its modern societies. 2. Describe how culture affects the region. 3. Describe how the region's past has contributed to current conflicts.	**Regular** 2 days **Block Scheduling** .5 day *Block Scheduling Handbook, Chapter 21*	**RS** Guided Reading Strategy 21.3 **E** Cultures of the World Activity 3 **E** Creative Strategies for Teaching World Geography, Lesson 11 **SM** Geography for Life Activity 21

Chapter Resource Key

RS Reading Support

IC Interdisciplinary Connections

E Enrichment

SM Skills Mastery

A Assessment

REV Review

ELL Reinforcement and English Language Learners

 Transparencies

 CD–ROM

 Music

 Video

 Internet

 Holt Presentation Maker Using Microsoft® Powerpoint®

 One-Stop Planner CD–ROM

See the ***One-Stop Planner*** for a complete list of additional resources for students and teachers.

One-Stop Planner CD–ROM

It's easy to plan lessons, select resources, and print out materials for your students when you use the *One-Stop Planner CD–ROM with Test Generator.*

Technology Resources

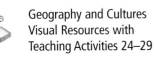

One-Stop Planner CD–ROM, Lesson 21.1

Geography and Cultures Visual Resources with Teaching Activities 24–29

Homework Practice Online

HRW Go site

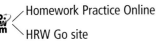

One-Stop Planner CD–ROM, Lesson 21.2

Homework Practice Online

HRW Go site

One-Stop Planner CD–ROM, Lesson 21.3

Homework Practice Online

HRW Go site

Review, Reinforcement, and Assessment Resources

ELL Main Idea Activity 21.1
REV Section 1 Review, p. 453
A Daily Quiz 21.1
ELL English Audio Summary 21.1
ELL Spanish Audio Summary 21.1

ELL Main Idea Activity 21.2
REV Section 2 Review, p. 459
ELL English Audio Summary 21.2
ELL Spanish Audio Summary 21.2
A Daily Quiz 21.2

ELL Main Idea Activity 21.3
REV Section 3 Review, p. 465
A Daily Quiz 21.3
ELL English Audio Summary 21.3
ELL Spanish Audio Summary 21.3

internet connect

HRW ONLINE RESOURCES

GO TO: go.hrw.com
Then type in a keyword.

TEACHER HOME PAGE
KEYWORD: SJ3 TEACHER

CHAPTER INTERNET ACTIVITIES
KEYWORD: SJ3 GT21

Choose an activity to:
• investigate the conflicts in the Balkans.
• take a virtual tour of Eastern Europe.
• learn about Baltic amber.

CHAPTER ENRICHMENT LINKS
KEYWORD: SJ3 CH21

CHAPTER MAPS
KEYWORD: SJ3 MAPS21

ONLINE ASSESSMENT
Homework Practice
KEYWORD: SJ3 HP21
Standardized Test Prep Online
KEYWORD: SJ3 STP21
Rubrics
KEYWORD: SS Rubrics

COUNTRY INFORMATION
KEYWORD: SJ3 Almanac

CONTENT UPDATES
KEYWORD: SS Content Updates

HOLT PRESENTATION MAKER
KEYWORD: SJ3 PPT21

ONLINE READING SUPPORT
KEYWORD: SS Strategies

CURRENT EVENTS
KEYWORD: S3 Current Events

Meeting Individual Needs

Ability Levels

Level 1 Basic-level activities designed for all students encountering new material

Level 2 Intermediate-level activities designed for average students

Level 3 Challenging activities designed for honors and gifted-and-talented students

English Language Learners Activities that address the needs of students with Limited English Proficiency

Chapter Review and Assessment

IC Interdisciplinary Activities for the Middle Grades 13, 14, 15

E Readings in World Geography, History, and Culture 37, 38, and 39

SM Critical Thinking Activity 21

REV Chapter 21 Review, pp. 466–67

REV Chapter 21 Tutorial for Students, Parents, Mentors, and Peers

ELL Vocabulary Activity 21

A Chapter 21 Test

A Unit 5 Test

Chapter 21 Test Generator (on the One-Stop Planner)

Audio CD Program, Chapter 21

A Chapter 21 Test for English Language Learners and Special-Needs Students

A Unit 5 Test for English Language Learners and Special-Needs Students

LAUNCH INTO LEARNING

Focus students' attention on the map on the following page. Call on volunteers to locate several familiar place names on the map and tell where they have heard the names before. *(Possible answers: Transylvanian Alps—Dracula movies, Bosnia or Kosovo—news broadcasts, Poland—history lessons on World War II)* Discuss what these associations indicate about the region's history. *(rich history, involved in wars)* Tell students that cultural conflicts are part, but not all, of the region's complex history. Tell students that they will learn more about the history and cultures of Eastern Europe in this chapter.

Section 1

Objectives

1. Identify the major physical features of Eastern Europe.
2. Describe the climates and natural resources of the region.

LINKS TO OUR LIVES

You may wish to point out the following reasons why we should know more about Eastern Europe:

▶ Ethnic conflicts in Yugoslavia have resulted in military involvement by the United States, NATO, and the UN.

▶ Many countries in the region are establishing democratic governments after decades of communism.

▶ Eastern Europe is a cultural and economic crossroads between Western Europe and Asia.

▶ Many Americans can trace their ancestry to Eastern Europe.

▶ The region's foods, festivals, literature, music, and other traditions can be enjoyed by all.

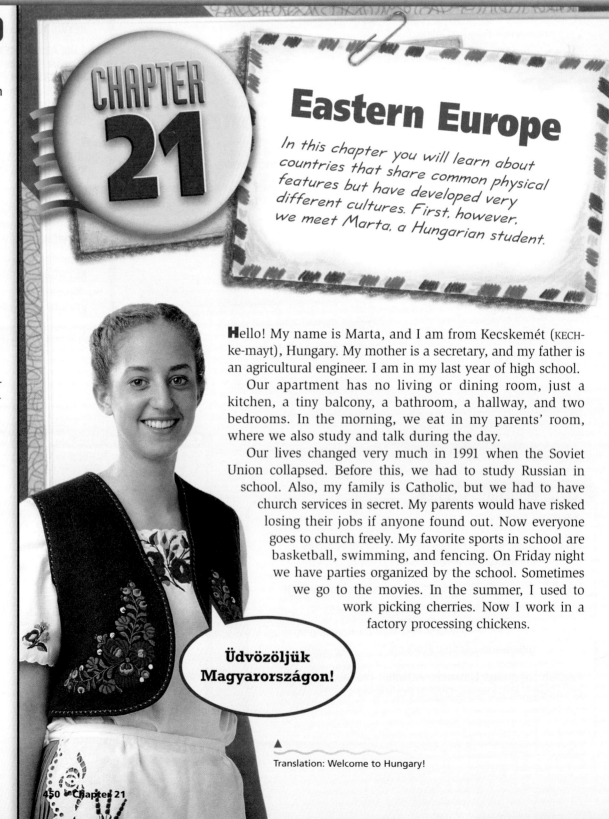

CHAPTER 21

Eastern Europe

In this chapter you will learn about countries that share common physical features but have developed very different cultures. First, however, we meet Marta, a Hungarian student.

Hello! My name is Marta, and I am from Kecskemét (KECH-ke-mayt), Hungary. My mother is a secretary, and my father is an agricultural engineer. I am in my last year of high school.

Our apartment has no living or dining room, just a kitchen, a tiny balcony, a bathroom, a hallway, and two bedrooms. In the morning, we eat in my parents' room, where we also study and talk during the day.

Our lives changed very much in 1991 when the Soviet Union collapsed. Before this, we had to study Russian in school. Also, my family is Catholic, but we had to have church services in secret. My parents would have risked losing their jobs if anyone found out. Now everyone goes to church freely. My favorite sports in school are basketball, swimming, and fencing. On Friday night we have parties organized by the school. Sometimes we go to the movies. In the summer, I used to work picking cherries. Now I work in a factory processing chickens.

Üdvözöljük Magyarországon!

▲ Translation: Welcome to Hungary!

LET'S GET STARTED

Copy the following instructions onto the chalkboard: *Look at the map and photo in Section 1. Imagine that they illustrate magazine articles about Eastern Europe's physical geography. What might be the titles for the articles? Write down your ideas for one of the illustrations. (Example: for the photo of the Danube River—"Danube: the Lifeblood Flowing through Romania's Heart")* Discuss student responses. Tell students that in Section 1 they will learn more about the physical geography of Eastern Europe.

Using the Physical-Political Map

Tell the class that Eastern Europe has been invaded from different directions many times over the centuries. Have students examine the map on this page. Ask them to speculate how Eastern Europe's physical geography might have contributed to the region being a crossroads. *(center of the continent, many rivers for navigation, few barriers to invasion)*

Building Vocabulary

Write the key terms on the chalkboard. Have students find and read the definitions for **oil shale**, **lignite**, and **amber** in the Section 1 text or glossary. Ask: What do these materials have in common? *(They are natural resources found in Eastern Europe.)*

Section 1
Physical Geography

Read to Discover

1. What are the major physical features of Eastern Europe?
2. What climates and natural resources does this region have?

Define

oil shale
lignite
amber

Locate

Baltic Sea
Adriatic Sea
Black Sea
Danube River
Dinaric Alps
Balkan Mountains
Carpathian Mountains

WHY IT MATTERS

The Danube and its tributaries are important transportation links in Eastern Europe. The rivers also help spread pollution. Use **CNNfyi.com** or other **current events** sources to find examples of pollution problems facing the region's rivers. Record your findings in your journal.

Amber

Section 1 RESOURCES

Reproducible
◆ Block Scheduling Handbook, Chapter 21
◆ Guided Reading Strategy 21.1

Technology
◆ One-Stop Planner CD–ROM, Lesson 21.1
◆ Homework Practice Online
◆ Geography and Cultures Visual Resources with Teaching Activities 24–29
◆ HRW Go site

Reinforcement, Review, and Assessment
◆ Section 1 Review, p. 4531
◆ Daily Quiz 21.1
◆ Main Idea Activity 21.1
◆ English Audio Summary 21.1
◆ Spanish Audio Summary 21.1

Eastern Europe: Physical-Political

ELEVATION

FEET	METERS
13,120	4,000
6,560	2,000
1,640	500
656	200
(Sea level) 0	0 (Sea level)
Below sea level	Below sea level

⊛ National capitals
• Other cities

Size comparison of Eastern Europe to the contiguous United States

Projection: Azimuthal Equal Area

ALL LEVELS: (Suggested time: 20 min.) Copy the following graphic organizer onto the chalkboard, omitting the italicized answers. Point out that it shows the general location of Eastern Europe's main regions. Call on students to provide words and phrases to summarize the physical features, climates, and resources of each region. Encourage students to use the unit's atlas as an additional resource. Then have students speculate how the location and availability of resources might affect economic development.

LEVELS 2 AND 3: (Suggested time: 20 min.) Have students complete the All Levels lesson. Then have each student select an Eastern European country and write a paragraph that describes its landforms, climates, rivers, and resources. Again, encourage students to use the unit's atlas as a resource. Call on volunteers to read their paragraphs to the class. Then have students speculate what effect pollution might have on the country they selected. *(Possible answer: harm rivers and air)*

The Physical Geography of Eastern Europe

Baltic Countries	Heartland	Balkan Countries
• *plains*	• *plains, mountains*	• *mountains*
• *cold winters*	• *cold winters in east, warmer in west*	• *cold winters in east, warmer in south and west*
• *amber, oil shale*	• *bauxite, lignite, salt*	• *bauxite, oil, lignite*

EYE ON EARTH

Refer to the section titled Climate and Resources on the opposite page. Notice that salt mines have operated in Poland since the 1200s.

Poland's Wieliczka salt mine contains more than 124 miles (200 km) of passages that connect more than 2,000 rooms. The lowest room is 1,073 feet (327 m) below the surface.

There is a tradition among the Wieliczka miners to carve the salt into churches, altars, and large statues. In recent years, increased humidity started to dissolve the carvings. An international team of scientists has made great progress in stopping the deterioration.

Critical Thinking: Once students have read the section on Poland, remind them of the salt carvings. Ask them how these carvings are an expression of Polish culture. *(Possible answer: shows importance of religion)*

internet connect

GO TO: go.hrw.com
KEYWORD: SJ3 CH21
FOR: Web sites about Eastern Europe

Our Amazing Planet

Amber is golden, fossilized tree sap. The beaches along the eastern coast of the Baltic Sea are the world's largest and most famous source of amber. Baltic amber is approximately 40 million years old.

This aerial view of the Danube Delta shows Romania's rich farmland.

internet connect

GO TO: go.hrw.com
KEYWORD: SJ3 CH21
FOR: Web sites about Poland

Physical Features

Eastern Europe stretches southward from the often cold, stormy shores of the Baltic Sea. In the south are the warmer and sunnier beaches along the Adriatic and Black Seas. We can divide the countries of this region into three groups. Poland, the Czech Republic, Slovakia, and Hungary are in the geographical heart of Europe. The Baltic countries are Estonia, Latvia, and Lithuania. Yugoslavia, Bosnia and Herzegovina, Croatia, Slovenia, Macedonia, Romania, Moldova, Bulgaria, and Albania are the Balkan countries.

Landforms Eastern Europe is a region of mountains and plains. The plains of Poland and the Baltic countries are part of the huge Northern European Plain. The Danube River flows through the Great Hungarian Plain, also called the Great Alföld.

The Alps extend from central Europe southeastward into the Balkan Peninsula. Where they run parallel to the Adriatic coast, the mountains are called the Dinaric (duh-NAR-ik) Alps. As the range continues eastward across the peninsula its name changes to the Balkan Mountains. The Carpathian (kahr-PAY-thee-uhn) Mountains stretch from the Czech Republic across southern Poland and Slovakia and into Ukraine. There they curve south and west into Romania. In Romania they are known as the Transylvanian Alps.

Rivers Eastern Europe's most important river for trade and transportation is the Danube. The Danube stretches for 1,771 miles (2,850 km) across nine countries. It begins in Germany's Alps and flows eastward to the Black Sea. Some 300 tributaries flow into the Danube. The river carries and then drops so much silt that its Black Sea delta grows by 80 to 100 feet (24 to 30 m) every year. The river also carries a heavy load of industrial pollution.

✓ **READING CHECK:** *Places and Regions* What are the main physical features in Eastern Europe? areas—Baltics, heartland, Balkans; landforms—mountains, plains; rivers—Danube, tributaries

CLOSE

Ask a student to describe one of the region's countries just by naming the bodies of water or countries that border it, plus one fact about its physical geography. Then have that student call on another student to name the country. Continue until all of the countries have been covered.

REVIEW AND ASSESS

Have students complete the Section Review. Then pair students. Have one student name a country in Eastern Europe and the other name a physical feature or resource found in or bordering that country. Then have students complete Daily Quiz 21.1. **COOPERATIVE LEARNING**

RETEACH

Have students complete Main Idea Activity 21.1. Then organize the class into triads and give each student an outline map of the region. Assign each group member the task of labeling either the plains, mountains, or rivers. Instruct members to exchange maps until all maps are complete.
ENGLISH LANGUAGE LEARNERS, COOPERATIVE LEARNING

EXTEND

Invite interested students to conduct research on ways that the physical geography of the Northern European Plain has affected history. They may want to create an illustrated map that shows the invasions and influences that have swept across this broad, lowland area. **BLOCK SCHEDULING**

Climate and Resources

The eastern half of the region has long, snowy winters and short, rainy summers. Farther south and west, winters are milder and summers become drier. A warm, sunny climate has drawn visitors to the Adriatic coast for centuries.

Eastern Europe's mineral and energy resources include coal, natural gas, oil, iron, lead, silver, sulfur, and zinc. The region's varied resources support many industries. Some areas of the Balkan region and Hungary are major producers of bauxite. Romania has oil. Estonia has deposits of **oil shale**, or layered rock that yields oil when heated. Estonia uses this oil to generate electricity, which is exported to other Baltic countries and Russia. Slovakia and Slovenia mine a soft form of coal called **lignite**. Nevertheless, many countries must import their energy because demand is greater than supply.

For thousands of years, people have traded **amber**, or fossilized tree sap. Amber is found along the Baltic seacoast. Salt mining, which began in Poland in the 1200s, continues in central Poland today.

During the years of Communist rule industrial production was considered more important than the environment. The region suffered serious environmental damage. Air, soil, and water pollution, deforestation, and the destruction of natural resources were widespread. Many Eastern European countries have begun the long and expensive task of cleaning up their environment.

✓ **READING CHECK:** *Environment and Society* What factors affect the location of economic activities in the region? *climate, location of resources*

Do you remember what you learned about climate types? See Chapter 3 to review.

Homework Practice Online
Keyword: SJ3 HP21

Section Review 1

Define and explain: oil shale, lignite, amber

Working with Sketch Maps On a map of Eastern Europe that you draw or that your teacher provides, label the following: Baltic Sea, Adriatic Sea, Black Sea, Danube River, Dinaric Alps, Balkan Mountains, and Carpathian Mountains.

Reading for the Main Idea

1. *Places and Regions* On which three major seas do the countries of Eastern Europe have coasts?

2. *Environment and Society* What types of mineral and energy resources are available in this region? How does this influence individual economies?

Critical Thinking

3. **Making Generalizations and Predictions** Would this region be suitable for agriculture? Why?

4. **Identifying Cause and Effect** How did Communist rule contribute to the pollution problems of this region?

Organizing What You Know

5. **Summarizing** Copy the following graphic organizer. Use it to summarize the physical features, climate, and resources of the heartland, the Baltics, and the Balkans. Then write and answer one question about the region's geography based on the chart.

Region	Physical features	Climate	Resources

Section Review 1

Answers

Define For definitions, see: oil shale, p. 453; lignite, p. 453; amber, p. 453

Working with Sketch Maps Maps will vary, but listed places should be labeled in their approximate locations.

Reading for the Main Idea
1. Adriatic, Baltic, and Black Seas (NGS 4)
2. coal, iron, lead, natural gas, silver, sulfur, zinc, bauxite, oil shale, lignite, amber, salt; wide range of industries, energy shortages, need to import (NGS 4)

Critical Thinking
3. yes; wide plains, plenty of rain
4. increased production and ignored pollution

Organizing What You Know
5. heartland—Northern European Plain, Great Hungarian Plain, Carpathian Mountains; cold winters in east, milder in west; bauxite, lignite, salt; Baltics—Northern European Plain; cold winters; amber, oil shale; Balkans—Dinaric Alps, Balkan Mountains, Transylvanian Alps; cold winters in east, warm in south and west; bauxite, oil, lignite; questions will vary but should focus on geographic distributions

Section 2

Objectives

1. Identify what peoples contributed to the early history of northeastern Europe.

2. Analyze how northeastern Europe's culture was influenced by other cultures.

3. Describe how the political organization of the region has changed since World War II.

LET'S GET STARTED

Copy the following question onto the chalkboard: *What does a country need to do or have to attract tourists? (Possible answers: good hotels and restaurants, historical sites, beautiful scenery, entertainment, reliable transportation)* Tell students that many of the countries of northeastern Europe, particularly the Czech Republic, became popular tourist destinations during the 1990s. Before that time, few people traveled to the region because the governments were communist and tourist facilities were undeveloped. Tell students that in Section 2 they will learn more about the region's ancient and modern history.

Building Vocabulary

Write **Indo-European** on the chalkboard and call on a volunteer to read the text's definition aloud. Point out that the definition only tells about the languages the Indo-European peoples spoke—not what they looked like or what their customs were. Ask what two large regions are represented in the term. *(India and Europe)* Many languages of India are related to European languages.

Section 2 RESOURCES

Reproducible

◆ Guided Reading Strategy 21.2

◆ Graphic Organizer 21

◆ Geography for Life Activities 14 and 16

◆ Map Activity 21

◆ Cultures of the World Activity 3

◆ Creative Strategies for Teaching World Geography, Lesson 11

Technology

◆ One-Stop Planner CD–ROM, Lesson 21.2

◆ Homework Practice Online

◆ HRW Go site

Reinforcement, Review, and Assessment

◆ Section 2 Review, p. 459

◆ Daily Quiz 21.2

◆ Main Idea Activity 21.2

◆ English Audio Summary 21.2

◆ Spanish Audio Summary 21.2

Section 2 The Countries of Northeastern Europe

Read to Discover

1. What peoples contributed to the early history of northeastern Europe?

2. How was northeastern Europe's culture influenced by other cultures?

3. How has the political organization of this region changed since World War II?

Define

Indo-European

Locate

Estonia
Poland
Czech Republic
Slovakia
Hungary

Lithuania
Latvia
Prague
Tallinn
Riga

Warsaw
Vistula River
Bratislava
Budapest

WHY IT MATTERS

Jerzy Giedroye, a Polish editor living in France, helped keep the free exchange of ideas alive in Eastern Europe during Communist rule. Use **CNN fyi.com** or other **current events** sources to discover the efforts of people like Giedroye. Record your findings in your journal.

Musical score by Hungarian Béla Bartók

The Teutonic knights, a German order of soldier monks, brought Christianity and feudalism to northeastern Europe. They built this castle at Malbork, Poland, in the 1200s.

History

Migrants and warring armies have swept across Eastern Europe over the centuries. Each group of people brought its own language, religion, and customs. Together these groups contributed to the mosaic of cultures we see in Eastern Europe today.

Early History Among the region's early peoples were the Balts. The Balts lived on the eastern coast of the Baltic Sea. They spoke **Indo-European** languages. The Indo-European language family includes many languages spoken in Europe. These include Germanic, Baltic, and Slavic languages. More than 3,500 years ago, hunters from the Ural Mountains moved into what is now Estonia. They spoke a very different, non-Indo-European language. The language they spoke provided the early roots of today's Estonian and Finnish languages. Beginning around A.D. 400, a warrior people called the Huns invaded the region from Asia. Later, the Slavs came to the region from the plains north of the Black Sea.

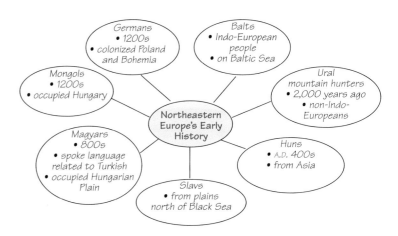

Teaching Objective 1

ALL LEVELS: (Suggested time: 10 min.) Copy the following graphic organizer onto the chalkboard, omitting the italicized answers. Call on students to fill in the outer circles with the names of peoples who contributed to northeastern Europe's early history. Then call on others to supply details about each group to fill in the circles.

Graphic organizer — Northeastern Europe's Early History:
- Germans • *1200s* • *colonized Poland and Bohemia*
- Balts • *Indo-European people* • *on Baltic Sea*
- Mongols • *1200s* • *occupied Hungary*
- Ural mountain hunters • *2,000 years ago* • *non-Indo-Europeans*
- Magyars • *800s* • *spoke language related to Turkish* • *occupied Hungarian Plain*
- Huns • *A.D. 400s* • *from Asia*
- Slavs • *from plains north of Black Sea*

In the 800s the Magyars moved into the Great Hungarian Plain. They spoke a language related to Turkish. In the 1200s the Mongols rode out of Central Asia into Hungary. At the same time German settlers pushed eastward, colonizing Poland and Bohemia—the western region of the present-day Czech Republic.

Emerging Nations Since the Middle Ages, Austria, Russia, Sweden, and the German state of Prussia have all ruled parts of Eastern Europe. After World War I ended in 1918, a new map of Eastern Europe was drawn. The peace treaty created two new countries: Yugoslavia and Czechoslovakia. Czechoslovakia included the old regions of Bohemia, Moravia, and Slovakia. At about the same time, Poland, Lithuania, Latvia, and Estonia also became independent countries.

✓ **READING CHECK:** *Human Systems* What peoples contributed to the region's early history? Balts, hunters from Ural Mountains, Huns, Slavs, Magyars, Mongols, Germans

Culture

The culture and festivals of this region show the influence of the many peoples who contributed to its history. As in Scandinavia, Latvians celebrate a midsummer festival. The festival marks the summer solstice, the year's longest day. Poles celebrate major Roman Catholic festivals. Many of these have become symbols of the Polish nation. The annual pilgrimage, or journey, to the shrine of the Black Madonna of Częstochowa (chen-stuh-KOH-vuh) is an example.

Traditional Foods The food of the region reflects German, Russian, and Scandinavian influences. As in northern Europe, potatoes and sausages are important in the diets of Poland and the Baltic countries. Although the region has only limited access to the sea, the fish of lakes and rivers are often the center of a meal. These fish often include trout and carp. Many foods are preserved to last through the long winter. These include pickles, fruits in syrup, dried or smoked hams and sausages, and cured fish.

The Arts, Literature, and Science
Northeastern Europe has made major contributions to the arts, literature, and sciences. For example, Frédéric Chopin (1810–1849) was a famous Polish pianist and composer. Marie Curie (1867–1934), one of the first female physicists, was also born in Poland. The writer Franz Kafka (1883–1924) was born to Jewish parents in Prague (PRAHG), the

Hungarian dancers perform in traditional dress.

Interpreting the Visual Record
How does this Hungarian costume compare to those you have seen from other countries?

Cultural Kaleidoscope
Czech Texans

During the 1850s affordable land attracted Czech settlers to the fertile low hills of southeastern Texas. The newcomers missed the great cathedrals of their homeland. As soon as they could afford to do so, the immigrants built churches with high vaulted ceilings, stained glass windows, and religious statuary. Interior walls were elaborately painted with angels, clouds, trailing ivy, flowers, and symbols. One surviving ceiling features 66 types of flowers, vines, and shrubs.

Today, services and festivals held at the painted churches of Fayette County draw thousands of visitors, Czech and non-Czech alike. At its Veterans Day ceremony, airplanes drop flower petals on one of the churchyards to honor church members who died in World War II.

Critical Thinking: How did Czech architecture in Texas reflect architectural traditions from the homeland?

Answer: The Czechs built churches in Texas that reflected the great cathedrals of their homeland.

◀ **Visual Record Answer**

Answers will vary according to students' experiences.

455

Teaching Objective 2

ALL LEVELS: (Suggested time: 30 min.) Organize the class into small groups. Provide each group with colored markers and a sheet of butcher paper. Instruct the students to create a mural showing the cultural traits of the region. Ask them also to label their pictures, explaining the origins of the various contributions. Display the murals around the classroom.
ENGLISH LANGUAGE LEARNERS, COOPERATIVE LEARNING

Jean Eldredge of Altamonte Springs, Florida, suggests the following activity to teach students about northeastern Europe: Have students choose a country in the region and write five complete sentences about it, using the material in Section 2, the unit's atlas, and any other resources available in your classroom. Collect the papers. Call on a volunteer to read one student's statements. Have the other students guess which country is being described. Limit the number of guesses. You may want to offer extra-credit points for correct identifications. Repeat the process until all the countries have been covered. This activity can be used as an initial teaching activity, to check on student progress, or to review for a test.

DAILY LIFE

Folk music is an important part of Baltic cultures. Many folk artists from the region now record for the world market.

The *kokle* is a native Latvian instrument. It is played flat on the musician's lap. The strings are plucked. Because the *kokle* can be adapted to a wide range of musical styles, musicians often include non-Latvian songs in their performances.

Modern *kokles* have 30 strings and can be several feet long. Earlier versions had between 5 and 13 strings and were hollowed from a single block of wood. A soundboard was then attached. No two instruments were exactly alike.

Critical Thinking: Why might there be an increase in folk music's popularity in Latvia and the other Baltic countries?

Answer: Students might suggest that since the countries became independent from the Soviet Union they have had more freedom to celebrate their unique cultures.

internet connect

GO TO: go.hrw.com
KEYWORD: SJ3 CH21
FOR: Web sites about Baltic culture

Chart Answer ▶

Latvia, Lithuania

Northeastern Europe

COUNTRY	POPULATION/ GROWTH RATE	LIFE EXPECTANCY	LITERACY RATE	PER CAPITA GDP
Czech Republic	10,264,212 −0.1%	71, male 78, female	100%	$12,900
Estonia	1,423,316 −0.6%	64, male 76, female	100%	$10,000
Hungary	10,106,017 −0.3%	67, male 76, female	99%	$11,200
Latvia	2,385,231 −0.8%	63, male 75, female	100%	$7,200
Lithuania	3,610,535 −0.3%	63, male 76, female	98%	$7,300
Poland	38,633,912 −0.03%	69, male 78, female	99%	$8,500
Slovakia	5,414,937 0.1%	70, male 78, female	not available	$10,200
United States	281,421,906 0.9%	74, male 80, female	97%	$36,200

Sources: Central Intelligence Agency, *The World Factbook 2001*; U.S. Census Bureau

Interpreting the Chart Based on the data in the table, which two countries have the lowest levels of economic development?

This suspension bridge spans the Western Dvina River in Riga, the capital of Latvia.

present-day capital of the Czech Republic. Astronomer Nicolaus Copernicus (1473–1543) was born in Toruń (TAWR-oon), a city in north-central Poland. He set forth the theory that the Sun—not Earth—is the center of the universe.

✓ **READING CHECK:** *Human Systems* How is the region's culture a reflection of its past and location? includes festivals and foods introduced by invaders, migrants, and neighbors

Northeastern Europe Today

Estonia, Latvia, and Lithuania lie on the flat plain by the eastern Baltic Sea. Once part of the Russian Empire, the Baltic countries gained their independence after World War I ended in 1918. However, they were taken over by the Soviet Union in 1940 and placed under Communist rule. The Soviet Union collapsed in 1991. Since then, the countries of northeastern Europe have been moving from communism to capitalism and democracy.

Estonia A long history of Russian control is reflected in Estonia today. Nearly 30 percent of Estonia's population is ethnic Russian. Russia remains one of Estonia's most important trading partners. However, Estonia is also building economic ties to other countries, particularly Finland. Ethnic Estonians have close cultural ties to Finland. In fact, the Estonian language is related to Finnish. Also, most people in both countries are Lutherans. Ferries link the Estonian capital of Tallinn (TA-luhn) with Helsinki, Finland's capital.

LEVEL 1: (Suggested time: 30 min.) Have students create covers for a special edition of a magazine titled News from the Northeast. In advance, duplicate a sheet of paper with the magazine title and date printed on it. (You may want to use the magazine cover template available on the go.hrw.com Web site at Keyword: SK3 Teacher.) Tell students that the special edition highlights the many changes that have occurred in northeastern Europe in the last 100 years or so. Have students think about what images and selected article titles would best represent the concept and re-create them for their own magazine covers.

LEVELS 2 AND 3: (Suggested time: 45 min.) Have students write an article for the News from the Northeast magazine.

▶**ASSIGNMENT:** Collect the articles from the Levels 2 and 3 lesson. Duplicate several, removing students' names, and distribute them to the class. Then instruct students to write a letter to the magazine's editor commenting on or disputing a point or fact in one of the articles.

CONNECTING TO *Literature*

Toy robot

While he wrote many books, Czech writer Karel Capek is probably best known for his play R.U.R. This play added the word robot *to the English language. The Czech word* robota *means "drudgery" or forced labor. The term is given to the artificial workers that Rossum's Universal Robots factory make to free humans from drudgery. Eventually, the Robots develop feelings and revolt. The play is science fiction. Here, Harry Domin, the factory's manager, explains the origin of the Robots to visitor Helena Glory.*

ROBOT ROBOTA

Domin: "Well, any one who has looked into human anatomy will have seen at once that man is too complicated, and that a good engineer could make him more simply. So young Rossum began to overhaul anatomy and tried to see what could be left out or simplified. . . . [He] said to himself: 'A man is something that feels happy, plays the piano, likes going for a walk, and in fact, wants to do a whole lot of things that are really unnecessary. . . .

But a working machine must not play the piano, must not feel happy, must not do a whole lot of other things. A gasoline motor must not have tassels or ornaments, Miss Glory. And to manufacture artificial workers is the same thing as to manufacture gasoline motors. The process must be of the simplest, and the product of the best from a practical point of view. . . .

Young Rossum . . . rejected everything that did not contribute directly to the progress of work—everything that makes man more expensive. In fact, he rejected man and made the Robot. My dear Miss Glory, the Robots are not people. Mechanically they are more perfect than we are, they have an enormously developed intelligence, but they have no soul."

Analyzing Primary Sources
1. Why does Rossum design the Robots without human qualities?
2. How has Karel Capel's play influenced other cultures?

As countries in northeastern Europe have gained freedom from communism, television and radio have been released from government control.

There are few comedies or dramas on television stations in the Baltic countries. Private groups have little experience creating television programs. Broadcasting is still dominated by political discussions and news programs.

Activity: Organize students into small groups and have them create a premise for a television comedy or drama set in any of the Baltic countries. The groups should share their ideas with the class.

Connecting to Literature
Answers
1. because he wished to create a machine that was simple, inexpensive, and that only contributed to the progress of work
2. by contributing the word *robot* to the English language

Latvia Latvia is the second largest of the Baltic countries. Its population has the highest percentage of ethnic minorities. Some 57 percent of the population is Latvian. About 30 percent of the people are Russian. The capital, Riga (REE-guh), has more than 1 million people. It is the largest urban area in the three Baltic countries. Like Estonia, Latvia also has experienced strong Scandinavian and Russian influences. As well as having been part of the Russian Empire, part of the country once was ruled by Sweden. Another tie between Latvia, Estonia, and the Scandinavian countries is religion. Traditionally most people in these countries are Lutheran. In addition, Sweden and Finland are important trading partners of Latvia today.

Teaching Objectives 1-3

LEVEL 1: (Suggested time: 45 min.) Pair students and assign each pair one of the region's countries. Using the text and other sources, have each pair summarize the major ethnic groups, cultural traits, and important facts about its country. Then have students organize their information into a brief oral report. You may want to require a visual aid as part of the presentation. **COOPERATIVE LEARNING**

LEVELS 2 AND 3: (Suggested time: 45 min.) Have students use the information they compiled for the Level 1 lesson to prepare a lesson about the selected country for an elementary classroom. Encourage them to use comparisons with familiar concepts so the younger children could understand the material. You may want to have students who complete their work early

design a Web page to be used in an elementary school curriculum on Eastern Europe. The Web page should highlight one aspect of the country students reported on for the Level 1 lesson.

CLOSE

Ask: What historical events do all the countries in this region share? *(influenced by communism and the Soviet Union)* Which countries seem to have developed the most since the fall of the Soviet Union? *(Poland, Hungary, the Czech Republic)* the least? *(Slovakia)*

National Geography Standard 11

Human Systems The fall of communism in Eastern Europe has led to social improvements as well as changes in government. However, not everyone has experienced the benefits of these improvements equally. In Hungary, Slovakia, the Czech Republic, and Poland, those who have reaped the greatest benefits are generally former Communist Party officials, government workers, and private business owners. State company employees, women, retirees, and the working class have gained less.

Activity: Have students conduct research on the growth of new political parties in Eastern Europe since the fall of communism. Ask students to identify which countries are still dominated by former communists and which have developed new political parties.

Visual Record Answer ▶

possible answers: lined with statues, has pedestrians instead of vehicular traffic

458

This Polish teenager wears traditional dress during a local festival.

Prague's Charles Bridge is lined with historical statues.

Interpreting the Visual Record How does this bridge compare to other bridges you have seen?

Lithuania Lithuania is the largest and southernmost Baltic country. Its capital is Vilnius (VIL-nee-uhs). Lithuania's population has the smallest percentage of ethnic minorities. More than 80 percent of the population is Lithuanian. Nearly 9 percent is Russian, while 7 percent is Polish. Lithuania has ancient ties to Poland. For more than 200 years, until 1795, they were one country. Roman Catholicism is the main religion in both Lithuania and Poland today. As in the other Baltic countries, agriculture and production of basic consumer goods are important parts of Lithuania's economy.

Poland Poland is northeastern Europe's largest and most populous country. The total population of Poland is about the same as that of Spain. The country was divided among its neighbors in the 1700s. Poland regained its independence shortly after World War I. After World War II the Soviet Union established a Communist government to rule the country.

In 1989 the Communists finally allowed free elections. Many businesses now are owned by people in the private sector rather than by the government. The country has also strengthened its ties with Western countries. In 1999 Poland, the Czech Republic, and Hungary joined the North Atlantic Treaty Organization (NATO).

Warsaw, the capital, has long been the cultural, political, and historical center of Polish life. More than 2 million people live in the urban area. The city lies on the Vistula River in central Poland. This location has made Warsaw the center of the national transportation and communications networks as well.

The Former Czechoslovakia Czechoslovakia became an independent country after World War I. Until that time, its lands had been part of the Austro-Hungarian Empire. Then shortly before World War II, it fell under German rule. After the war the Communists, with the support of the Soviet Union, gained control of the government. As in Poland, the Communists lost power in 1989. In 1993 Czechoslovakia peacefully split into two countries. The western part became the Czech Republic. The eastern part became Slovakia. This peaceful split helped the Czechs and Slovaks avoid the ethnic problems that have troubled other countries in the region.

The Czech Republic The Czech Republic experienced economic growth in the early 1990s. Most of the country's businesses are completely or in part privately owned. However, some Czechs worry that the government remains too involved in the economy. As in Poland, a variety of political parties compete in free elections. Czech lands have coal and other important

Have students complete the Section Review. Then organize the class into small groups and have students write one quiz question each on the history, culture, and characteristics of the region. Each question should be written on a slip of paper, folded, and placed in a pile. Students should take turns reading questions to the group while the other members answer. Then have students complete Daily Quiz 21.2. **COOPERATIVE LEARNING**

RETEACH

Have students complete Main Idea Activity 21.2. Then pair students and have pairs create a page for a travel magazine featuring a country in northeastern Europe. Students may divide up tasks according to their abilities.

The page should include a descriptive title, a map or flag, and a short paragraph summarizing the history and culture of the country. Display and discuss the magazine pages. **ENGLISH LANGUAGE LEARNERS, COOPERATIVE LEARNING**

EXTEND

The czardas (CHAHR-dash) is the national dance of Hungary. This couples dance begins with a slow section but becomes fast and lively. Many versions of the czardas have been developed. Have interested students conduct research on how to perform a basic version of the dance. You may wish to have volunteers perform the dance in class. **BLOCK SCHEDULING**

---

mineral resources that are used in industry. Much of the country's industry is located in and around Prague, the capital. The city is located on the Vltava River. More than 1.2 million people live there. Prague has beautiful medieval buildings. It also has one of Europe's oldest universities.

Slovakia Slovakia is more rugged and rural, with incomes lower than in the Czech Republic. The move toward a freer political system has been slow. However, progress has been made. Bratislava (BRAH-tyee-slah-vah), the capital, is located on the Danube River. The city is the country's most important industrial area and cultural center. Many rural Slovaks move to Bratislava looking for better-paying jobs. Most of the country's population is Slovak. However, ethnic Hungarians account for more than 10 percent of Slovakia's population.

Hungary Hungary separated from the Austro-Hungarian Empire at the end of World War I. Following World War II, a Communist government came to power. A revolt against the government was put down by the Soviet Union in 1956. The Communists ruled until 1989.

Today the country has close ties with the rest of Europe. In fact, most of Hungary's trade is with members of the European Union. During the Communist era, the government experimented with giving some businesses the freedom to act on their own. For example, it allowed local farm managers to make key business decisions. These managers kept farming methods modern, chose their crops, and marketed their products. Today, farm products from Hungary's fertile plains are important exports. Much of the country's manufacturing is located in and around the capital, Budapest (BOO-duh-pest). Budapest is Hungary's largest city. Nearly 20 percent of the population lives there.

✓ **READING CHECK:** *Human Systems* How have the governments and economies of the region been affected by recent history? *moved from communism to democracy and private ownership following collapse of Soviet Union*

▲ The Danube River flows through Budapest, Hungary.

Interpreting the Visual Record Why might Hungary's capital have grown up along a river?

go.hrw.com **Homework Practice Online**
Keyword: SJ3 HP21

Section Review 2

Define and explain: Indo-European

Working with Sketch Maps On the map you drew in Section 1, label the countries of the region, Prague, Tallinn, Riga, Warsaw, Vistula River, Bratislava, and Budapest.

Reading for the Main Idea

1. *Human Systems* How did invasions and migrations help shape the region?

2. *Places and Regions* What has the region contributed to the arts?

Critical Thinking

3. **Drawing Conclusions** What influence did the Soviet Union have on the region?

4. **Summarizing** What social changes have taken place here since the early 1990s?

Organizing What You Know

5. **Sequencing** Copy the following graphic organizer. Use it to show the history of the Baltics since 1900.

|————————|————————|————————|
1900   1945   1999

Section Review 2

Answers

Define For definition, see: Indo-European, p. 454

Working with Sketch Maps Maps will vary, but listed places should be labeled in their approximate locations. The Czech Republic, Hungary, and Slovakia do not have seacoasts.

Reading for the Main Idea

1. Balts, hunting people from Ural Mountains, Huns, Slavs, Magyars, Mongols, and Germans all moved into the area, bringing their own languages and traditions. (NGS 9)

2. music—Frédéric Chopin, Béla Bartók; literature—Franz Kafka, Karel Capek (NGS 10)

Critical Thinking

3. imposed communism

4. Change from communism to capitalism and democracy has allowed free elections, greater political and economic freedom.

Organizing What You Know

5. after World War I (1918)—became independent; during World War II (1940)—taken over by Soviet Union; after 1991—changing from communism to democracy

◄ **Visual Record Answer**

Rivers provide reliable sources of water and serve as trade routes.

Section 3

Objectives

1. Describe the early history of southeastern Europe.
2. Explain how other cultures have influenced southeastern Europe.
3. Summarize the changes in the region's governments during the 1980s and 1990s.

LET'S GET STARTED

Copy the following question onto the chalkboard: *How do you identify yourself in terms of your country, state, region, or ethnic group? Write down more than one term. (Examples: American, Texan, Southerner, Puerto Rican, Hispanic, African American)* Discuss responses. Then ask how students would feel if suddenly they had to define themselves as Canadians or as a member of a different ethnic group. Explain that some people in southeastern Europe have had to make similar changes. Tell students that in Section 3 they will learn more about the countries of southeastern Europe.

Building Vocabulary

Write the key terms on the chalkboard. Point out that the people who call themselves **Roma** were once known as Gypsies, when their origins were thought to be in Egypt. If your class has studied Canada, ask students if they recall another group of people who are referred to now by the name they call themselves. *(Inuit)* If they have not covered Canada, lead a discussion about the terms *Hispanic, Chicano, African American*, or others.

Section 3 RESOURCES

Reproducible

- Guided Reading Strategy 21.3
- Geography for Life Activity 21
- Cultures of the World Activity 3
- Creative Strategies for Teaching World Geography, Lesson 11

Technology

- One-Stop Planner CD–ROM, Lesson 21.3
- Homework Practice Online
- HRW Go site

Reinforcement, Review, and Assessment

- Section 3 Review, p. 465
- Daily Quiz 21.3
- Main Idea Activity 21.3
- English Audio Summary 21.3
- Spanish Audio Summary 21.3

Visual Record Answer ▶

Greek

Section 3

The Countries of Southeastern Europe

Read to Discover

1. How did Southeastern Europe's early history help shape its modern societies?
2. How does culture both link and divide the region?
3. How has the region's past contributed to current conflicts?

Define

Roma

Locate

Bulgaria	Albania	Zagreb
Romania	Yugoslavia	Ljubljana
Croatia	Kosovo	Skopje
Slovenia	Montenegro	Bucharest
Serbia	Macedonia	Moldova
Bosnia and	Belgrade	Chişinău
Herzegovina	Podgorica	Sofia
	Sarajevo	Tiranë

WHY IT MATTERS

Bosnia and the other republics of the former Yugoslovia have experienced ethnic violence since independence. Use CNNfyi.com or other current events sources to check on current conditions in the region. Record your findings in your journal.

Dolls in traditional Croatian dress

These ancient ruins in southern Albania date to the 500s B.C.

Interpreting the Visual Record What cultural influence does this building show?

History

Along with neighboring Greece, this was the first region of Europe to adopt agriculture. From here farming moved up the Danube River valley into central and western Europe. Early farmers and metalworkers in the south may have spoken languages related to Albanian. Albanian is an Indo-European language.

Early History Around 750–600 B.C. the ancient Greeks founded colonies on the Black Sea coast. The area they settled is now Bulgaria and Romania. Later, the Romans conquered most of the area from the Adriatic Sea to the Danube River and across into Romania. When the Roman Empire divided into west and east, much of the Balkans and Greece became part of the Eastern Roman Empire. This eastern region eventually became known as the Byzantine Empire. Under Byzantine rule, many people of the Balkans became Orthodox Christians.

🌐 **Teaching Objective 1**

LEVEL 1: (Suggested time: 20 min.) Pair students and have each pair create a flowchart to show how the ancient Greeks, the Roman Empire, the Ottoman Empire, the Austro-Hungarian Empire, and World War I affected the boundaries and religions of the countries of southeastern Europe. **ENGLISH LANGUAGE LEARNERS, COOPERATIVE LEARNING**

LEVELS 2 AND 3: (Suggested time: 30 min.) Focus students' attention on the photographs of buildings and other structures in Section 3. Ask if they have heard the phrase "if these walls could talk." Discuss the meaning of the phrase. Then have students use the information from the Level 1 lesson to write what the walls of one of those structures would say about southeastern Europe's history if they could indeed talk. You may want to invite some students to dramatize their "wall soliloquies."

Kingdoms and Empires Many of today's southeastern European countries first appear as kingdoms between A.D. 800 and 1400. The Ottoman Turks conquered the region and ruled until the 1800s. The Ottomans, who were Muslims, tolerated other religious faiths. However, many peoples, such as the Bosnians and Albanians, converted to Islam. As the Ottoman Empire began to weaken in the late 1800s, the Austro-Hungarians took control of Croatia and Slovenia. They imposed Roman Catholicism.

Slav Nationalism The Russians, meanwhile, were fighting the Turks for control of the Black Sea. The Russians encouraged Slavs in the Balkans to revolt against the Turks. The Russians appealed to Slavic nationalism— to the Slav's sense of loyalty to their country. The Serbs did revolt in 1815 and became self-governing in 1817. By 1878 Bulgaria and Romania were also self-governing.

The Austro-Hungarians responded to Slavic nationalism by occupying additional territories. Those territories included the regions of Bosnia and Herzegovina. To stop the Serbs from expanding to the Adriatic coast, European powers made Albania an independent kingdom.

In August 1914 a Serb nationalist shot and killed the heir to the Austro-Hungarian throne. Austria declared war on Serbia. Russia came to Serbia's defense. These actions sparked World War I. All of Europe's great powers became involved. The United States entered the war in 1917.

Creation of Yugoslavia At the end of World War I Austria-Hungary was broken apart. Austria was reduced to a small territory. Hungary became a separate country but lost its eastern province to Romania. Romania also gained additional lands from Russia. Albania remained independent. The peace settlement created Yugoslavia. *Yugoslavia* means "land of the southern Slavs." Yugoslavia brought the region's Serbs, Bosnians, Croatians, Macedonians, Montenegrins, and Slovenes together into one country. Each ethnic group had its own republic within Yugoslavia. Some Bosnians and other people in Serbia were Muslims. Most Serbs were Orthodox Christians, and the Slovenes and Croats were Roman Catholics. These ethnic and religious differences created problems that eventually led to civil war in the 1990s.

✔️ **READING CHECK:** ⟨ *Human Systems* ⟩ How is southeastern Europe's religious and ethnic makeup a reflection of its past? Colonizers and conquerors introduced different religions.

This bridge at Mostar, Bosnia, was built during the 1600s. This photograph was taken in 1982.

▼

▲

This photograph shows Mostar after civil war in the 1990s.

Interpreting the Visual Record **What differences can you find in the two photos?**

461

Teaching Objective 2

ALL LEVELS: (Suggested time: 45 min.) Organize the class into small groups. Ask each group to plan a feast for a major Roman Catholic, Orthodox Christian, or Muslim holiday. They should include the history of the religion in southeastern Europe as background. Have students use the information in the text as a starting point for creating their menu and conduct additional research as necessary. Then have the groups present their menus and historical summaries to the class. **COOPERATIVE LEARNING**

Teaching Objective 3

ALL LEVELS: (Suggested time: 25 min.) Copy the following graphic organizer onto the chalkboard, omitting the italicized answers. Call on volunteers to fill in the charts. Then lead a discussion on how the various groups have affected changes in the region's governments during the 1980s and 1990s. **ENGLISH LANGUAGE LEARNERS**

Ethnic and Religious Groups of Southeastern Europe

	Major ethnic or religious group	Important minority group
Yugoslavia	*Orthodox Christians*	*Albanian Muslims*
Bosnia and Herzegovina	*Muslims*	*Catholic Croats, Orthodox Serbs*
Croatia	*Roman Catholics*	*Orthodox Christian Serbs*
Slovenia	*Roman Catholics*	
Macedonia	*Orthodox Christians*	*Albanian Muslims*
Romania	*Romanian*	*Roma, Hungarians*
Moldova	*(diverse)*	
Bulgaria	*Bulgarians*	*Turks, Macedonians*
Albania	*Muslims*	

HUMAN SYSTEMS

During the conflict in Kosovo in the late 1990s, American teenagers used e-mail to keep in touch with teens in the war-torn region. The Kosovar youths shared their concerns and fears. The American teens distributed those messages to the news media and to charity groups in order to spread the word about the war's effects.

Critical Thinking: How has access to e-mail affected the world?

Answer: It has improved communication around the world and has allowed people to communicate almost instantaneously.

▲
Ethnic Albanians worship at a mosque in Pristina, Serbia.

The Danube Delta, on the Romanian coast of the Black Sea, is part of a unique ecosystem. Most of the Romanian caviar-producing sturgeon are caught in these waters. Caviar is made from the salted eggs from three types of sturgeon fish. Caviar is considered a delicacy and can cost as much as $50 per ounce.

Culture

The Balkans are the most diverse region of Europe in terms of language, ethnicity, and religion. It is the largest European region to have once been ruled by a Muslim power. It has also been a zone of conflict between eastern and western Christianity. The three main Indo-European language branches—Romance (from Latin), Germanic, and Slavic—are all found here, as well as other branches like Albanian. Non-Indo-European languages like Hungarian and Turkish are also spoken here.

Balkan diets combine the foods of the Hungarians and the Slavs with those of the Mediterranean Greeks, Turks, and Italians. In Greek and Turkish cuisines, yogurt and soft cheeses are an important part of most meals, as are fresh fruits, nuts, and vegetables. Roast goat or lamb are the favorite meats for a celebration.

In the Balkans Bosnian and ethnic Albanian Muslims celebrate the feasts of Islam. Christian holidays—Christmas and Easter—are celebrated on one day by Catholics and on another by Orthodox Christians. Holidays in memory of ancient battles and modern liberation days are sources of conflict between ethnic groups.

✔ **READING CHECK:** *Places and Regions* Why is religion an important issue in southeastern Europe? *source of conflict within and between countries*

Southeastern Europe Today

Like other southeastern European countries, Yugoslavia was occupied by Germany in World War II. A Communist government under Josip Broz Tito took over after the war. Tito's strong central government prevented ethnic conflict. After Tito died in 1980, Yugoslavia's Communist government held the republics together. Then in 1991 the republics of Slovenia, Croatia, Bosnia and Herzegovina, and Macedonia began to break away. Years of bloody civil war followed. Today the region struggles with the violence and with rebuilding economies left weak by years of Communist-government control.

Yugoslavia The republics of Serbia and Montenegro remain united and have kept the name of Yugoslavia. Belgrade is the capital of both Serbia and Yugoslavia. It is located on the Danube River. The capital of Montenegro is Podgorica (PAWD-gawr-eet-sah). The Serbian government supported ethnic Serbs fighting in civil wars in Croatia and in Bosnia and Herzegovina in the early 1990s. Tensions between ethnic groups also have been a problem within Serbia. About 65 percent of the people in Serbia and Montenegro are Orthodox Christians. In the

►**ASSIGNMENT:** Copy the following sentence onto the chalkboard: *Times are always changing in southeastern Europe.* Have students copy the sentence into their notebooks. Ask students to write a paragraph supporting or disputing this statement. *(To agree with the statement, students might cite the many changes in governmental control over the countries in the region. Those who dispute the statement might state that constant change and ethnic or religious conflict are so common in southeastern Europe that times do not really change.)*

Teaching Objectives 1–3

LEVEL 1: (Suggested time: 40 min.) Give each student two or three index cards. Assign or have each student select a country of southeastern Europe. Be sure that each country is selected by at least one student. Instruct students to provide basic information about the country (relative location, history, culture, economy, and government) on one side of the card and to draw a simple map of the country on the other side. Have students exchange cards to evaluate them.

---

southern Serbian province of Kosovo, the majority of people are ethnic Albanian and Muslim. Many of the Albanians want independence. Conflict between Serbs and Albanians led to civil war in the late 1990s. In 1999 the United States, other Western countries, and Russia sent troops to keep the peace.

Bosnia and Herzegovina Bosnia and Herzegovina generally are referred to as Bosnia. Some 40 percent of Bosnians are Muslims, but large numbers of Roman Catholic Croats and Orthodox Christian Serbs also live there. Following independence, a bloody civil war broke out between these groups as they struggled for control of territory. During the fighting the once beautiful capital of Sarajevo (sar-uh-YAY-voh) was heavily damaged.

Croatia Croatia's capital is Zagreb (ZAH-greb). Most of the people of Croatia are Roman Catholic. In the early 1990s, Serbs made up about 12 percent of the population. In 1991 the ethnic Serbs living in Croatia claimed part of the country for Serbia. This resulted in heavy fighting. By the end of 1995 an agreement was reached and a sense of stability returned to the country. Many Serbs left the country.

Slovenia Slovenia is a former Austrian territory. It looks to Western European countries for much of its trade. Most people in Slovenia are Roman Catholic, and few ethnic minorities live there. Partly because of the small number of ethnic minorities, little fighting occurred after Slovenia declared independence from Yugoslavia. The major center of industry is Ljubljana (lee-oo-blee-AH-nuh), the country's capital.

These Muslim refugees are walking to Travnik, Bosnia, with the assistance of UN troops from Britain in the 1990s.

Interpreting the Visual Record **What effect might the movement of refugees have on a region?**

▼

Slovenia's capital, Ljubljana, lies on the Sava River.

▼

National Geography Standard 10

Human Systems In the 1380s, Muslim Turkish armies began raiding the region that is now Bosnia and Herzegovina. The Turks had gained complete control of the area within about 100 years.

Unlike in most other parts of Europe conquered by the Ottoman Empire, a large percentage of Bosnians converted to Islam. There were several reasons for this. First, Muslims had a higher legal status than Christians. Second, Sarajevo and Mostar were mainly Muslim, and those who wanted to participate fully in city life had to convert. Also, the Bosnian Catholic Church was relatively weak.

Discussion: Lead a discussion about the goals, needs, and ideas that people of different religions might have in common.

▲ **Visual Record Answer**

possible answer: ethnic conflict, overcrowding, political ramifications, new cultural traditions introduced in new countries

LEVELS 2 AND 3: (Suggested time: 45 min.) Have students use the cards they created for the Level 1 lesson to design a card trading game. The game should have a goal, rules, and a point system. For example, the goal might be to acquire wealth by collecting the countries with the strongest economies. As an example of the point system, countries with coastlines along the Mediterranean might be worth more points than landlocked countries. Have students play the game and summarize what they learned about the countries for which they did not create a card.

Read the Links to Our Lives feature at the beginning of the chapter to the class. Call on volunteers to add a detail or to explain each point further. Then ask students to propose a single sentence to answer the question "Why should we study Eastern Europe?" *(Possible answer: Events in Eastern Europe can affect people everywhere.)*

Section Review 3

Answers

Define For definitions, see: Roma, p. 464

Working with Sketch Maps Maps will vary, but listed places should be labeled in their approximate locations. Slovenia, Croatia, Bosnia and Herzegovina, Macedonia, Serbia, and Montenegro

Reading for the Main Idea

1. Invasions and foreign control have led to multiple religious and ethnic groups living in the region. (NGS 9)

2. conflict among ethnic, religious groups (NGS 13)

Critical Thinking

3. created after World War I in response to Slav nationalism

4. independent countries without Communist governments; political and economic troubles; developing market economies

Organizing What You Know

5. languages—Romance, Germanic, Slavic, and Non-Indo-European languages; foods—yogurt, soft cheeses, fresh fruits, nuts, vegetables, goat, lamb; celebrations—Catholic and Orthodox Christian holidays, Islamic feasts, national holidays

Chart Answer

23 percent

464

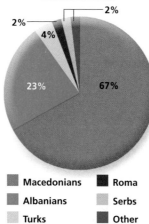

Ethnic Groups in Macedonia

2%
2%
4%
23%
67%

◼ Macedonians ◼ Roma
◼ Albanians ◼ Serbs
◼ Turks ◼ Other

Source: Central Intelligence Agency, *The World Factbook 2001*

In 2001 ethnic Albanian rebels launched months of fighting in hopes of gaining more rights for Macedonia's Albanian minority.

Interpreting the Chart What percentage of Macedonia's people are Albanian?

Macedonia When Macedonia declared its independence from Yugoslavia, Greece immediately objected to the country's new name. Macedonia is also the name of a province in northern Greece that has historical ties to the republic. Greece feared that Macedonia might try to take over the province.

Greece responded by refusing to trade with Macedonia until the mid-1990s. This slowed Macedonia's movement from the command—or government-controlled—economy it had under Communist rule to a market economy in which consumers help to determine what is to be produced by buying or not buying certain goods and services. Despite its rocky start, in recent years Macedonia has made progress in establishing free markets.

Romania A Communist government took power in Romania at the end of World War II. Then in 1989 the Communist government was overthrown during bloody fighting. Change, however, has been slow. Bucharest, the capital, is the biggest industrial center. Today more people work in agriculture than in any other part of the economy. Nearly 90 percent of the country's population is ethnic Romanian. **Roma**, or Gypsies as they were once known, make up almost 2 percent of the population. They are descended from people who may have lived in northern India and began migrating centuries ago. Most of the rest of Romania's population are ethnic Hungarian.

Moldova Throughout history control of Moldova has shifted many times. It has been dominated by Turks, Polish princes, Austria, Hungary, Russia, and Romania. Not surprisingly, the country's population reflects this diverse past. Moldova declared its independence in

Turkish Roma girls

Have students complete the Section Review. Then put the names of all the countries on slips of paper in a hat and have each student draw a name. (You will need to repeat country names.) Ask students to write a sentence that begins "If I lived in this country, I would . . . " and complete it with a detail from Section 3 about the country they have drawn. Have other students guess which country is being discussed. Then have students complete Daily Quiz 21.3.

RETEACH

Have students complete Main Idea Activity 21.3. Then, pair students and provide each pair with an outline map of the region. Have students take turns finding two facts about each country on the map.
ENGLISH LANGUAGE LEARNERS, COOPERATIVE LEARNING

EXTEND

Have interested students create maps of southeastern Europe to show how boundaries have changed, areas where different ethnic and religious groups have settled, and areas where conflicts have been most severe. Let the students who created the maps lead a discussion on conclusions that may be drawn from the maps. **BLOCK SCHEDULING**

1991 from the Soviet Union. However, the country suffers from difficult economic and political problems. About 40 percent of the country's labor force works in agriculture. Chişinău (kee-shee-NOW), the major industrial center of the country, is also Moldova's capital.

Bulgaria Mountainous Bulgaria has progressed slowly since the fall of communism. However, a market economy is growing gradually, and the people have more freedoms. Most industries are located near Sofia (SOH-fee-uh), the capital and largest city. About 9 percent of Bulgaria's people are ethnic Turks.

Albania Albania is one of Europe's poorest countries. The capital, Tiranë (ti-RAH-nuh), has a population of about 270,000. About 70 percent of Albanians are Muslim. Albania's Communist government feuded with the Communist governments in the Soviet Union and, later, in China. As a result, Albania became isolated. Since the fall of its harsh Communist government in the 1990s, the country has tried to move toward both democracy and a free market system.

✓ READING CHECK: **Human Systems** What problems does the region face, and how are they reflections of its Communist past? ethnic violence— Communist government prevented conflict; economic conditions—years of government control left economies weak; democratic reforms—Communist government limited freedoms

Southeastern Europe

COUNTRY	POPULATION/ GROWTH RATE	LIFE EXPECTANCY	LITERACY RATE	PER CAPITA GDP
Albania	3,510,484 0.9%	69, male 75, female	93%	$3,000
Bosnia and Herzegovina	3,922,205 1.4%	69, male 74, female	not available	$1,700
Bulgaria	7,707,495 −1.1%	68, male 75, female	98%	$6,200
Croatia	4,334,142 1.5%	70, male 78, female	97%	$5,800
Macedonia	2,046,209 0.4%	72, male 76, female	not available	$4,400
Moldova	4,431,570 0.05%	60, male 69, female	96%	$2,500
Romania	22,364,022 −0.2%	66, male 74, female	97%	$5,900
Yugoslavia (Serbia and Montenegro)	10,677,290 −0.3%	70, male 76, female	93%	$2,300
United States	281,421,906 0.9%	74, male 80, female	97%	$36,200

Sources: Central Intelligence Agency, *The World Factbook 2001;* U.S. Census Bureau

Interpreting the Chart Based on the table, which southeastern European country has the highest level of economic development?

go.hrw.com **Homework Practice Online**
Keyword: SJ3 HP21

Section Review 3

Define and explain: Roma

Working with Sketch Maps On the map you drew in Section 2, label the region's countries and their capitals. They are listed at the beginning of the section. In a box in the margin, identify the countries that once made up Yugoslavia.

Reading for the Main Idea

1. *Human Systems* How has the region's history influenced its religious and ethnic makeup?
2. *Human Systems* What events and factors have contributed to problems in Bosnia and other countries in the region since independence?

Critical Thinking

3. **Summarizing** How was Yugoslavia created?
4. **Analyzing Information** How are the region's governments and economies changing?

Organizing What You Know

5. **Categorizing** Copy the following graphic organizer. Use it to identify languages, foods, and celebrations in the region.

```
         Foods
Languages
              Celebrations
     Southeastern
     European culture
```

CHAPTER 21

Review
ANSWERS

Building Vocabulary
For definitions, see: oil shale, p. 453; lignite, p. 453; amber, p. 453; Indo-European, p. 454; Roma, p. 464

Reviewing the Main Ideas

1. Northern European Plain, Great Hungarian Plain, Dinaric Alps, Balkan Mountains, Carpathian Mountains, Transylvanian Alps (NGS 4)

2. Balts, hunting people from the Ural Mountains, Huns, Slavs, Magyars, Mongols, Germans; language, customs (NGS 9)

3. air, soil, and water pollution; deforestation, destruction of natural resources (NGS 14)

4. communism; moving to democracy and a market economy; collapse of the Soviet Union in 1991 (NGS 9)

5. Slovenia, Croatia, Bosnia and Herzegovina, Macedonia; ethnic identity and religion (NGS 13)

465

RETEACH

Supply pairs of students with poster board or butcher paper. Help students create time lines showing major events in Eastern Europe. Have students annotate and illustrate the time lines with details about how the events affected culture. Display the time lines around the classroom.
ENGLISH LANGUAGE LEARNERS, COOPERATIVE LEARNING

PORTFOLIO EXTENSIONS

1. Although cultural conflict is part of Eastern Europe's history, so is cooperation. Have students work in groups to write constitutions for "Cromavania"—an imaginary Eastern European country where people live together peacefully. When the groups have completed their constitutions, lead a discussion on the strong points of each. Have students place their constitutions and a summary of what they learned in their portfolios.

2. Have students work in groups to write a series of short skits with the republics that have been part of Yugoslavia as characters. Each skit should depict a different period in the region's history. The countries should express how they feel at each stage in Yugoslavia's history. Have each group perform its skit. Place scripts in student portfolios.

Review
ANSWERS

Understanding Environment and Society

Answers may vary according to the resources available to the student. Students may find that COMECON was founded in 1949. The purpose of the organization was to aid its members in economic integration. The Soviet Union provided fuel and raw materials. In turn, Eastern Europe supplied the Soviet Union with finished machinery and goods. Economies are moving from command to market economies. Use Rubric 29, Presentations, to evaluate student work.

Thinking Critically

1. Possible answer: Balkan diets combine the foods of the Hungarians and the Slavs with those of the Mediterranean Greeks, Turks, and Italians.

2. Danube River; there are several cities located along its route.

3. its location on the Vistula River in central Poland; perhaps less connected to the rest of Poland, more open to attack or Western influence

4. Yugoslavia—violent; Czechoslovakia—peaceful

Reviewing What You Know

Building Vocabulary

On a separate sheet of paper, write sentences to define each of the following words.

1. oil shale
2. lignite
3. amber
4. Indo-European
5. Roma

Reviewing the Main Ideas

1. **Places and Regions** What are the major landforms of Eastern Europe?

2. **Human Systems** What groups influenced the culture of Eastern Europe? How can these influences be seen in modern society?

3. **Environment and Society** What were some environmental effects of Communist economic policies in Eastern Europe?

4. **Human Systems** What political and economic systems were most common in Eastern Europe before the 1990s? How and why have these systems changed?

5. **Human Systems** List the countries that broke away from Yugoslavia in the early 1990s. What are the greatest sources of tension in these countries?

Understanding Environment and Society

Economic Geography

During the Communist era, the Council for Economic Assistance, or COMECON, played an important role in planning the economies of countries in the region. Create a presentation comparing the practices of this organization to the practices now in place in the region. Consider the following:

- How COMECON organized the distribution of goods and services.
- How countries in the region now organize distribution.

Thinking Critically

1. **Drawing Inferences and Conclusions** How has Eastern Europe's location influenced the diets of the region's people?

2. **Analyzing Information** What is Eastern Europe's most important river for transportation and trade? How can you tell that the river is important to economic development?

3. **Identifying Cause and Effect** What geographic factors help make Warsaw the transportation and communication center of Poland? Imagine that Warsaw is located along the Baltic coast of Poland or near the German border. How might Warsaw have developed differently?

4. **Comparing/Contrasting** Compare and contrast the breakups of Yugoslavia and Czechoslovakia.

5. **Summarizing** How has political change affected the economies of Eastern European countries?

FOOD FESTIVAL

Kolaches are popular Czech or Polish pastries. For a shortcut version, use frozen bread dough. Or, make a sweetened yeast dough from scratch. After the dough has risen, roll it out, cut it into circles 2–3 inches across, and indent the center of each. Add a filling of fruit preserves, cottage cheese with egg and sugar, or a sweetened poppyseed paste. Let rise again. Sprinkle with a streusel topping of sugar, cinnamon, butter, and a little flour. Bake at 375° for 15–20 minutes. There are many *kolache* recipes on the Internet. *Kolacky* and *kolachke* are alternate spellings.

CHAPTER 21 REVIEW AND ASSESSMENT RESOURCES

Reproducible
- Readings in World Geography, History, and Culture 37, 38, and 39
- Critical Thinking Activity 21
- Vocabulary Activity 21

Technology
- Chapter 21 Test Generator (on the One-Stop Planner)
- HRW Go site

- Audio CD Program, Chapter 21

Reinforcement, Review, and Assessment
- Chapter 21 Review, pp. 466–67
- Chapter 21 Tutorial for Students, Parents, Mentors, and Peers

- Chapter 21 Test
- Chapter 21 Test for English Language Learners and Special-Needs Students
- Unit 5 Test
- Unit 5 Test for English Language Learners and Special-Needs Students

Building Social Studies Skills

Map ACTIVITY

On a separate sheet of paper, match the letters on the map with their correct labels.

Baltic Sea
Adriatic Sea
Black Sea
Danube River
Dinaric Alps

Balkan Mountains
Carpathian Mountains

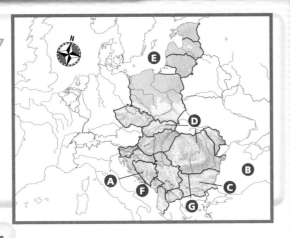

Mental Mapping Skills ACTIVITY

Using the chapter map or a globe as a guide, draw a freehand map of Eastern Europe and label the following:

Bosnia and Herzegovina
Croatia
Czech Republic
Estonia

Hungary
Macedonia
Poland
Yugoslavia

WRITING ACTIVITY

Imagine that you are a teenager living in Romania and want to write a family memoir of life in Romania. Include accounts of life for your grandparents under strict Soviet rule and life for your parents during the Soviet Union's breakup. Also describe your life in free Romania. Be sure to use standard grammar, spelling, sentence structure, and punctuation.

Alternative Assessment

Portfolio ACTIVITY

Learning About Your Local Geography Cooperative Project Ask international agencies or search the Internet for help in contacting a boy or girl in Eastern Europe. As a group, write a letter to the teen, telling about your daily lives.

internet connect

Internet Activity: go.hrw.com
KEYWORD: SJ3 GT21

Choose a topic to explore Eastern Europe:
- Investigate the conflicts in the Balkans.
- Take a virtual tour of Eastern Europe.
- Learn about Baltic amber.

5. Change to capitalism and market economies is happening at the same time as the change from communism to democracy.

Map Activity
A. Dinaric Alps
B. Black Sea
C. Balkan Mountains
D. Carpathian Mountains
E. Baltic Sea
F. Adriatic Sea
G. Danube River

Mental Mapping Skills Activity
Maps will vary, but listed places should be labeled in their approximate locations.

Writing Activity
Answers will vary, but students should discuss all three generations. Details should be consistent with text information. Use Rubric 40, Writing to Describe, to evaluate student work.

Portfolio Activity
Students might tell about their community, families, school, sports, or other topics. Use Rubric 25, Personal Letters, to evaluate student work.

internet connect

GO TO: go.hrw.com
KEYWORD: SJ3 Teacher
FOR: a guide to using the Internet in your classroom

Analyzing an Economic System

Ask students to list the 15 current member countries of the European Union *(Austria, Belgium, Denmark, Finland, France, Germany, Greece, Ireland, Italy, Luxembourg, the Netherlands, Portugal, Spain, Sweden, and the United Kingdom).* Point out that each country had to meet specific requirements before it was admitted into the union. Organize the class into pairs. Assign one EU country to each pair or call on some pairs to volunteer for more than one country. Have students conduct research on the requirements their assigned country had to meet to become a member of the EU and what changes the country had to make. Ask the pairs to compile lists of the requirements and changes.

GEOGRAPHY SIDELIGHT

As it is working to unite the economic and political aspects of the member countries, the EU is also supporting the preservation of particular ethnic cultures. For example, almost 50 million Europeans speak minority languages such as Wendish, Frisian, or Basque. The European Union supports several organizations that try to ensure the survival of these languages.

The EU has an ambitious range of goals. If successful, it will have eased intercountry commerce among its member countries and strengthened the region's political and social systems while maintaining the unique culture of each member country.

Critical Thinking: How is the EU working to link its member countries?

Answer: by easing intercountry commerce, strengthening the region's political and social systems

► This Focus On Government feature addresses National Geography Standards 6, 11, and 13.

FOCUS ON GOVERNMENT

The European Union

What if . . . ? Imagine you are traveling from Texas to Minnesota. You have to go through a border checkpoint in Oklahoma to prove your Texas identity. The guard charges a tax on the cookies you are bringing to a friend in Minnesota. Buying gas presents more problems. You try to pay with Texas dollars, but the attendant just looks at you. You discover that they speak "Kansonian" in Kansas and use Kansas coins. All this would make traveling from one place to another much more difficult.

The European Union Fortunately, that was just an imaginary situation. However, it is similar to what might happen while traveling across Europe. European countries have different languages, currencies, laws, and cultures. For example, someone from France has different customs than someone from Ireland.

However, many Europeans also share common interests. For example, they are interested in peace in the region. They also have a common interest in Europe's economic success.

A shared belief in economic and political cooperation has resulted in the creation of the European Union (EU). The EU has 15 countries that are members. They are: Austria, Belgium, Denmark, Finland, France, Germany, Great Britain, Greece, Ireland, Italy, Luxembourg, the Netherlands, Portugal, Spain, and Sweden.

The Beginnings of a Unified Europe
Proposals for an economically integrated Europe first came about in the 1950s. After World War II, the countries of Europe had many economic problems. A plan was made to unify the coal and steel production of some countries. In 1957 France, Germany, Italy, Belgium, the Netherlands, and Luxembourg formed the European Economic Community (EEC). The name was later shortened to simply the European Community (EC). The goal of the EC was to combine each country's economy into a single market. Having one market would make trading among them easier. Eventually, more countries became interested in joining the EC. In 1973 Britain, Denmark, and Ireland joined. In the 1980s Greece, Portugal, and Spain joined.

The Eurostar train carries passengers from London to Paris. These two cities are only about 200 miles (322 km) apart. However, they have different cultures and ways of life.

Going Further: Thinking Critically

Have students identify the issues and benefits awaiting the member countries of the European Union during the early phase of unification. Then have students conduct research on the economic issues that faced the United States as the country was evolving from a group of colonies into a unified country. As a class, compare and contrast the U.S. experience with those arising now for the European countries. Have students prepare a report that warns European member countries of potential pitfalls, recommends how to overcome them, and lists likely benefits.

The European Union (EU)

The 15 EU countries produce a wide range of exports and are one of the world's richest markets.

The flag of the EU features 12 gold stars on a blue background. The EU's currency, the euro, replaced the currencies of most EU countries.

In the early 1990s, a meeting was held in Maastricht, the Netherlands, to discuss the future of the EC. This resulted in the Maastricht Treaty. The treaty officially changed the EC's name from the European Community to the European Union (EU).

The Future Some people believe the EU is laying the foundation for a greater sense of European identity. A European Court of Justice has been set up to enforce EU rules. According to some experts, this is helping to build common European beliefs, responsibilities, and rights.

The introduction of a common currency, the euro, will continue in the future. Currently, each country has its own form of money. For example, France uses the Franc. Italy uses the lira. On January 1, 2002, the euro became a common form of money for most EU countries. With the exception of Denmark, Sweden, and the United Kingdom, the euro replaced the currencies of member countries.

Another goal of the EU is to extend membership to other countries.

The EU has resulted in many important changes in Europe. Cooperation between member countries has increased. Trade has also increased. EU members have adopted a common currency and common economic laws. The EU is creating a more unified Europe. Some people even believe that the EU might someday lead to a "United States of Europe."

Understanding What You Read

1. What was the first step toward European economic unity?
2. What is the euro? How will it affect the other currencies of the EU countries?

Understanding What You Read

Answers

1. The first step toward European economic unity was a proposal that came about in 1950, following World War II.
2. The euro is the common currency of most of the European Union. It replaced the currencies of most of the EU countries.

Going Further: Thinking Critically

Edge cities are defined as "nodal concentrations of retail and office space that are situated on the outer fringes of metropolitan areas, typically near major highway intersections." Essentially, edge cities are major commercial centers that grow up near major cities. They differ from residential suburbs in that they contain much more office, retail, and hotel space than the communities made up mainly of houses and apartments. Tysons Corner, outside Washington, D.C., in Fairfax County, Virginia, is a typical example. Edge cities are particularly common near the rapidly growing cities of the Sunbelt.

Have students research the development of an edge city in your state or region. Organize the class into groups and have them pursue specific aspects of the topic. Here are some suggestions for the different groups' investigations:

- Contact the administration of the nearby large city for information on how the edge city has affected it.
- Acquire old and recent maps of the area.
- Find old and new photos of the development.
- Find information on how the edge city has affected utility services.
- Ask highway department officials how traffic patterns have changed.
- Investigate changes in property values and taxes in the tax assessor's office.
- Search newspaper files for news stories about the development.
- Research the effect on employment and unemployment patterns.

When the research is complete, have the groups compile their information in a large flowchart.

PRACTICING THE SKILL

1. Students should compare their definitions with dictionary definitions. They should note that the differences among the three types of settlements is primarily size; a city is larger than a town, which is larger than a village. Students should note how their definitions are different from the dictionary's.

2. Students should note the apparent age of their settlement and about how many people live there. Students should describe some jobs in their settlement and how their settlement is connected to others, such as by highway or rail. Students might note unique features, such as buildings, parks, and cultural attractions.

3. Students might suggest capital cities, mountain cities, river cities, port cities, colonial cities, modern cities, tourist cities, or other types.

➤ This GeoSkills feature addresses National Geography Standards 4, 6, 12, and 17.

Building Skills for Life: Analyzing Settlement Patterns

There are many different kinds of human settlements. Some people live in villages where they farm and raise animals. Others live in small towns or cities and work in factories or offices. Geographers analyze these settlement patterns. They are interested in how settlements affect people's lives.

All settlements are unique. Even neighboring villages are different. One village might have better soil than its neighbors. Another village might be closer to a main road or highway. Geographers are interested in the unique qualities of human settlements.

Geographers also study different types of settlements. For example, many European settlements could be considered medieval cities. Medieval cities are about 500–1,500 years old. They usually have walls around them and buildings made of stone and wood. Medieval cities also have tall churches and narrow, winding streets.

▲ This illustration shows a German medieval city in the 1400s.

Analyzing settlement patterns is important. It helps us learn about people and environments. For example, the architecture of a city might give us clues about the culture, history, and technology of the people who live there.

You can ask questions about individual villages, towns, and cities to learn about settlement patterns. What kinds of activities are going on? How are the streets arranged? What kinds of transportation do people use? You can also ask questions about groups of settlements. How are they connected? Do they trade with each other? Are some settlements bigger or older than others? Why is this so?

THE SKILL

1. How do you think a city, a town, and a village are different from each other? Write down your own definition of each word on a piece of paper. Then look them up in a dictionary and write down the dictionary's definition. Were your definitions different?

2. Analyze the settlement where you live. How old is it? How many people live there? What kinds of jobs do people have? How is it connected to other settlements? How is it unique?

3. Besides medieval cities, what other types of cities can you think of? Make a list of three other possible types.

HANDS on GEOGRAPHY

One type of settlement is called a planned city. A planned city is carefully designed before it is built. Each part fits into an overall plan. For example, the size and arrangement of streets and buildings might be planned.

There are many planned cities in the world. Some examples are Brasília, Brazil; Chandigarh, India; and Washington, D.C. Many other cities have certain parts that are planned, such as individual neighborhoods. These neighborhoods are sometimes called planned communities.

Suppose you were asked to plan a city. How would you do it? On a separate sheet of paper, create your own planned city. These guidelines will help you get started.

1. First, decide what the physical environment will be like. Is the city on the coast, on a river, or somewhere else? Are there hills, lakes, or other physical features in the area?

2. Decide what to include in your city. Most cities have a downtown, different neighborhoods, and roads or highways that connect areas together. Many cities also have parks, museums, and an airport.

3. Plan the arrangement of your city. Where will the roads and highways go? Will the airport be close to downtown? Try to arrange the different parts of your city so that they fit together logically.

4. Draw a map of your planned city. Be sure to include a title, scale, and orientation.

Some people think the city plan for Brasília looks like a bird, a bow and arrow, or an airplane. **Interpreting the Visual Record What do you notice about the arrangement of Brasília's streets?**

Lab Report

1. How was your plan influenced by the physical environment you chose?

2. How do you think planned cities are different from cities that are not planned?

3. What problems might people have when they try to plan an entire city?

Lab Report

Answers

1. Students might note that streets would have to fit around physical features such as rivers or mountains. They may also note that the environment affected the businesses in their cities. For example, a port city might have docks and businesses related to fishing, travel, or recreation located near the waterfront.

2. Students might suggest that planned cities are more efficient, have stronger economies, or have a higher standard of living. They might also argue that planned cities have less character or "atmosphere."

3. Students might suggest the difficulty of accommodating different populations, such as locating the airport near downtown for easy business travel or on the outskirts so that noise pollution bothers fewer people. Students may mention other problems, such as the difficulty of predicting growth.

◄ **Visual Record Answer**

possible answers: wide, symmetrical arrangement; attractive pattern

471

RUSSIA AND NORTHERN EURASIA

Direct students' attention to the photographs on these pages. Point out that the bear is a symbol of Russia. Ask what other animal symbols students have seen and for which countries they stand. *(Possible answers: bald eagle—United States; kangaroo, koala—Australia; panda, dragon, tiger—China)* Ask why animal symbols seem to be popular. *(Possible answers: portray strength and power, not associated with any one ethnic group or political party, refer to the country's natural heritage)*

Refer to the photo of the Buryat people. Ask how they compare to the Russian citizens and politicians that are typically seen on news broadcasts. *(Possible answers: more colorful clothing, facial features similar to those of Chinese people)* Point out that the Buryat and some other ethnic groups in Russia have more in common with the peoples of Mongolia and China than they do with Russians who live in cities thousands of miles to the west.

Point out the domes on St. Basil's Cathedral in Moscow. Ask students to suggest why they are shaped like onions. *(so snow does not accumulate)* The cathedral was built in the 1550s to commemorate a major military victory.

UNIT OBJECTIVES

1. Describe the physical features of Russia and northern Eurasia and link the region's natural environment with its economic development.
2. Examine the progression of the region's history, from early periods through communism to the breakup of the Soviet Union.
3. Analyze the relationships among the region's ethnic and culture groups.
4. Use special-purpose maps to analyze relationships among climate, population patterns, and economic activities in Russia and northern Eurasia.
5. Identify the challenges that lie ahead for the governments and peoples of Russia and northern Eurasia.

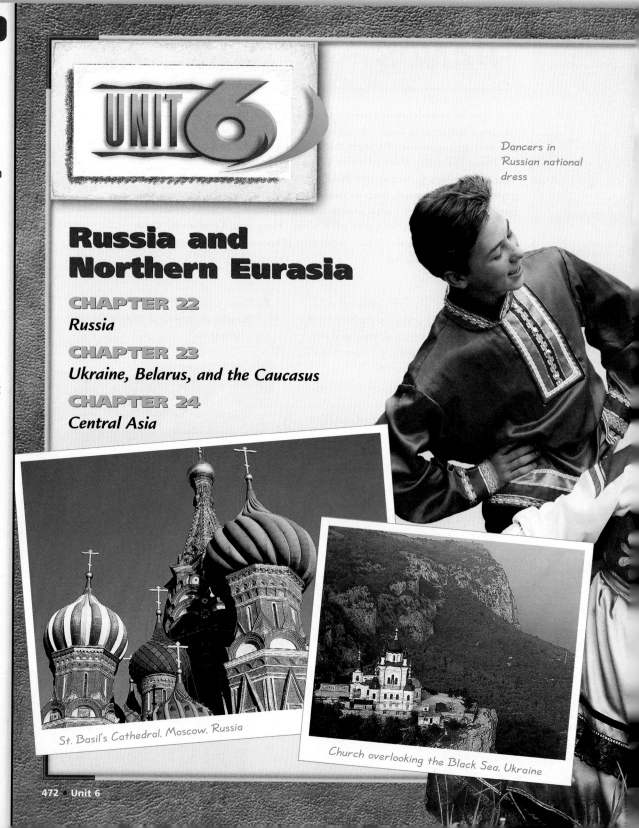

Russia and Northern Eurasia

CHAPTER 22
Russia

CHAPTER 23
Ukraine, Belarus, and the Caucasus

CHAPTER 24
Central Asia

Dancers in Russian national dress

St. Basil's Cathedral, Moscow, Russia

Church overlooking the Black Sea, Ukraine

472

Notes FROM THE field

Journalists in Russia

Journalists Gary Matoso and Lisa Dickey traveled more than 5,000 miles across Russia. They wrote this account of their visit with Buyanto Tsydypov. He is a Buryat farmer who lives in the Lake Baikal area. The Buryats are one of Russia's many minority ethnic groups. **WHAT DO YOU THINK?** *If you visited a Buryat family, what would you like to see or ask?*

"**Y**ou came to us like thunder out of the clear blue sky," said our host. The surprise of our visit did not, however, keep him from greeting us warmly.

Buyanto brought us to a special place of prayer. High on a hillside, a yellow wooden frame holds a row of tall, narrow sticks. On the end of each stick, Buddhist prayer cloths flutter in the biting autumn wind.

In times of trouble and thanks, Buryats come to tie their prayer cloths—called *khimorin*—to the sticks and make their offerings to the gods. Buyanto builds a small fire. He unfolds an aqua-blue *khimorin* to show the drawings.

"All around are the Buddhist gods," he says, "and at the bottom we have written our names and the names of others we are praying for."

He fans the flames slowly with the cloth, purifying it with sacred smoke. After a time he moves to the top of the hill where he ties the *khimorin* to one of the sticks.

Buryat people. Lake Baikal area. Russia

MORE FROM THE FIELD

In the early 1930s the Soviets began enforcing a policy of state ownership of farmland. Harsh measures were used— land confiscations, arrests, and deportations to prison camps. Soviet farm policies resulted in a famine during 1932–33 in which millions of people died.

Since the fall of the Soviet Union, the Russian government has given land grants and tax breaks to private farmers. They decide for themselves what crops to plant and are not required to sell to state agencies. Buyanto Tsydypov, the Buryat farmer described on this page, is one such private farmer. Many people in his village admire his independence and determination.

Critical Thinking: What type of economic system did the Soviet Union have?

Answer: command

Understanding Primary Sources

1. How do you know that Buyanto Tsydypov was surprised to meet the two American journalists?

2. What is a *khimorin*?

Understanding Primary Sources
Answers

1. He compares their arrival to thunder out of a clear blue sky.

2. prayer cloth

Brown bear

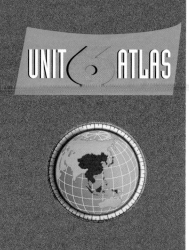

UNIT 6 ATLAS

In this unit, students will learn about the land and people of Russia and northern Eurasia—a vast region that is undergoing many political and economic changes.

Huge open plains and deep evergreen forests dominate Russia's landscape. The population is concentrated in the western plains, where the best farmland and the largest cities are located. Harsh climates keep Siberia and Russia's Far East thinly populated. Development of the region's rich resources may increase settlement there.

For most of the country's history, Russia's leaders have limited citizens' personal freedoms. During the more-than-70 years of communist rule, the Soviet Union became a superpower. Since the collapse of the Soviet Union, Russia has struggled with rising crime, corruption, and unemployment.

The other countries of northern Eurasia were also part of the Soviet Union. They are now independent republics. Their economies are emerging, with varying degrees of success. Distinctive religious and ethnic groups live in the area.

PEOPLE IN THE PROFILE

Note that the elevation profile crosses the West Siberian Plain. The Khanty are one of the many indigenous groups that live in this region of subarctic forest and bogs. They have adapted to the harsh environment. Winters are very cold, and the summer skies are filled with mosquitoes.

The Khanty, who settle in extended family groups, live mainly by fishing, hunting, and reindeer herding. They make almost everything they use, including their shelters, fishing boats, winter clothing, and containers. Arts and crafts include colorful birchbark baskets.

Russia's oil wealth poses problems for the Khanty and other Siberian peoples. Oil spills and pollution threaten the wetlands. Raised roads trap water, causing floods and ruining the land's ability to function as pasture for reindeer. Fires caused by worker carelessness and oil-soaked debris are common. Acid rain has damaged huge areas of land. The Khanty have responded to these threats by organizing to protect their land and way of life from further decline.

Critical Thinking: To what harsh conditions have the Khanty adapted?

Answer: very cold winters, summer skies filled with mosquitoes

UNIT 6 ATLAS The World in Spatial Terms

Russia and Northern Eurasia

Elevation Profile

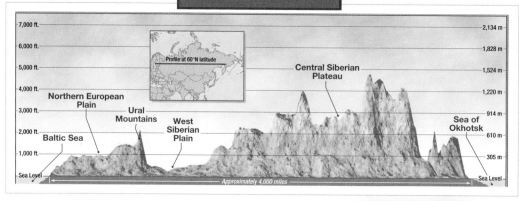

The United States and Russia and Northern Eurasia
Comparing Sizes

GEOSTATS:

Russia

- World's largest country in area: 6,659,328 sq. mi. (17,075,200 sq km)

- World's sixth-largest population: 145,470,197 (July 2001 estimate)

- World's largest lake: Caspian Sea—143,244 sq. mi. (371,002 sq km)

- World's deepest lake: Lake Baikal—5,715 ft. (1,742 m)

- Largest number of time zones: 11

- Highest mountain in Europe: Mount Elbrus—18,510 ft. (5,642 m)

USING THE PHYSICAL MAP

Focus students' attention on the **physical map** of Russia and northern Eurasia on this page. What is the first thing that students notice about Russia? *(Many students will note its large size.)* Challenge students to estimate the distance between Russia's eastern and western borders. *(about 6,000 miles or 9,650 km)*

Note that the Northern European Plain is a physical feature of western Russia and that the eastern portion of the country is called Siberia. Ask students what physical feature appears to separate European Russia from Siberia. *(Ural Mountains)*

Russia and Northern Eurasia: Physical

UNIT 6 ATLAS

internet connect

ONLINE ATLAS
GO TO: go.hrw.com
KEYWORD: SJ3 MAPSU6
FOR: Web links to online maps of the region

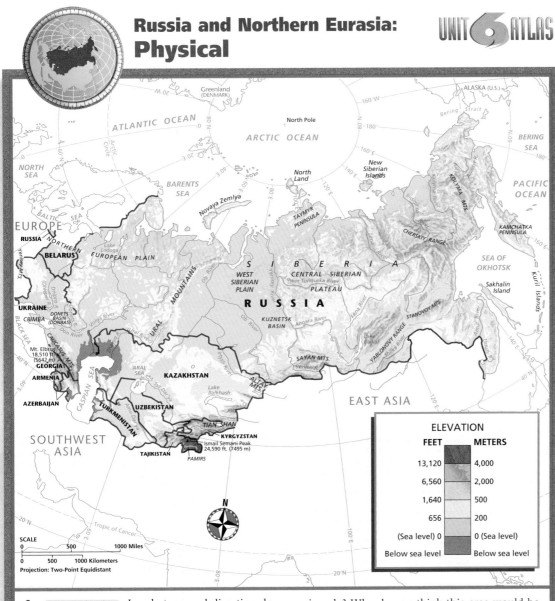

ELEVATION

FEET	METERS
13,120	4,000
6,560	2,000
1,640	500
656	200
(Sea level) 0	0 (Sea level)
Below sea level	Below sea level

SCALE
0 — 500 — 1000 Miles
0 — 500 — 1000 Kilometers
Projection: Two-Point Equidistant

1. *Places and Regions* In what general direction do the great rivers of Siberia flow?

2. *Places and Regions* Which countries have areas that are below sea level?

Critical Thinking

3. Drawing Inferences and Conclusions Russia has often been invaded by other countries. Which part of Russia might be easy to invade? Why do you think this area would be a good invasion route?

4. Comparing Northern Russia appears to have many good harbors. Compare this map to the **climate map** of the region. Why have few harbors been developed on Russia's north coast?

Physical Map

Answers

1. north
2. Azerbaijan, Russia, Kazakhstan, and Turkmenistan

Critical Thinking

3. western Russia; because it is an open plain
4. because the water there is frozen much of the year

475

USING THE POLITICAL MAP

Refer students to the **political map** of Russia and northern Eurasia on this page. Tell the class that all the countries represented in color on the map used to be part of the Soviet Union, but that the Soviet Union broke up in 1991. New independent countries emerged.

Ask students to use the **physical** and **land use and resources maps** to predict how the breakup of the Soviet Union may have affected Russia's economy and transportation links. *(lost access to valuable mineral and energy resources, reduced access to the Caspian and Black Seas)*

internet connect

ONLINE ATLAS
GO TO: go.hrw.com
KEYWORD: SJ3 MAPSU6

Your Classroom Time Line

These are the major dates and time periods for this unit. Have students enter them on the time line you created earlier. You may want to watch for these dates as students progress through the unit.

600s B.C. Greeks establish trading colonies along the Black Sea.

500s B.C. The Persian Empire controls the Caucasus.

100s B.C. Chinese trade and military expeditions begin moving into Central Asia.

A.D. 400 The Georgian language has its own alphabet by this date.

Political Map

Answers

1. no

Critical Thinking

2. less than 100 mi. (161 km); about 780 mi. (1,255 km)

3. All of the irregular curved borders appear to follow natural features.

4. The boundary crosses a wide, flat desert. No physical features, such as rivers, that might define a natural boundary appear on the maps.

Russia and Northern Eurasia: Political

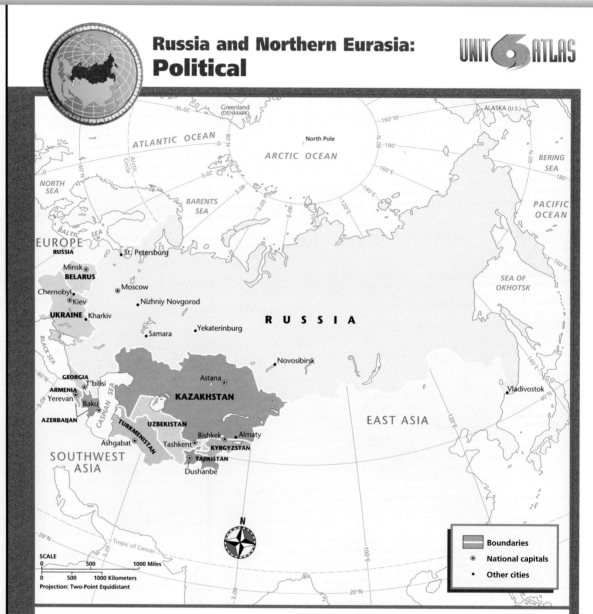

1. (*Places and Regions*) Compare this map to the **physical map** of the region. Do any physical features define a border between Russia and the countries of Europe?

Critical Thinking

2. Analyzing Information About how far apart are Russia and the Alaskan mainland? the Russian mainland and the North Pole?

3. Analyzing Information Which borders separating the countries south and southeast of Russia appear to follow natural features?

4. Drawing Inferences and Conclusions Compare this map to the **physical** and **climate maps** of the region. Why do you think the boundary between Kazakhstan and northwestern Uzbekistan is two straight lines?

Have students examine the **climate map** of Russia and northern Eurasia on this page. Ask them to name the climates of the largest areas of Russia. *(subarctic, tundra)* Have them compare the climate map to the **population map** on the next page. Ask students to answer the following questions: In which climate region is the population density highest? *(humid continental)* In what three types of climate is the population density lowest? *(tundra, subarctic, desert)*

Russia and Northern Eurasia:
Climate

UNIT 6 ATLAS

CLIMATE
- Desert
- Steppe
- Mediterranean
- Humid subtropical
- Humid continental
- Subarctic
- Tundra
- Highland

SCALE
0   500   1000 Miles
0   500   1000 Kilometers
Projection: Two-Point Equidistant

1. *(Places and Regions)* Compare this map to the **political map**. Which is the only country that has a humid subtropical climate?

2. *(Physical Systems)* Which climate types stretch across Russia from Europe to the Pacific?

3. *(Places and Regions)* Compare this map to the **political map**. Which country has only a humid continental climate?

Critical Thinking

4. Drawing Inferences and Conclusions Compare this map to the **land use and resources map**. Why do you think nomadic herding is common east of the Caspian Sea?

5. Comparing Compare this to the **political map**. In which countries would you expect to find the highest mountains? Why?

Your Classroom Time Line, (continued)

500s Turkic-speaking nomads from northern Asia spread through Central Asia.

700s Arab armies take over much of Central Asia.

800s The Rus establish Kiev as their capital.

1200s Tatars sweep across the steppes.

late 1400s Muscovy wins control over parts of Russia from the Mongols.

late 1400s The Kremlin's red brick walls and towers are built.

1547 Ivan IV crowns himself czar of all Russia.

early 1700s Peter the Great expands the Russian Empire.

Climate Map
Answers
1. Georgia
2. tundra and subarctic
3. Belarus

Critical Thinking
4. The land is flat, and the climate is dry; herd animals can live on land too dry for farming.
5. Kyrgyzstan and Tajikistan; because they have large areas of highland climates

477

USING THE POPULATION MAP

Direct your students' attention to the **population map** on this page. Ask them to write a question, with answer, about the population patterns they see on the map. *(Possible answer: Most people live in the western part of the region; the eastern part is sparsely populated.)* Have students exchange their questions with another student and then answer the exchanged questions. Then ask students to suggest reasons for the patterns of population density. They might consult the **climate map** for assistance.

Your Classroom Time Line (continued)

1700s Russian fur traders establish settlements along North America's Pacific coast.

late 1800s The Russian Empire begins to decline.

1867 Russia sells Alaska to the United States.

1891 Construction of the Trans-Siberian Railroad begins.

1893 Composer Peter Tchaikovsky dies.

1917 Czar Nicholas II abdicates his throne.

1917 The Russian Revolution begins.

1918 Writer Aleksandr Solzhenitsyn is born.

1922 The Soviet Union is established.

1922 Georgia, Armenia, and Azerbaijan unite to oppose Soviet rule.

Population Map
Answers
1. Russia, Ukraine
2. Russia

Critical Thinking
3. that the land is fairly flat and that farming is common in western Russia
4. Charts, graphs, databases, and models should accurately reflect population figures for cities on the map.

Russia and Northern Eurasia: Population

UNIT 6 ATLAS

POPULATION DENSITY

Persons per sq. mile	Persons per sq km
520	200
260	100
130	50
25	10
3	1
0	0

● Metropolitan areas with more than 2 million inhabitants

○ Metropolitan areas with 1 million to 2 million inhabitants

SCALE
0   500   1000 Miles
0   500   1000 Kilometers
Projection: Two-Point Equidistant

1. **Places and Regions** Which countries have a large area with more than 260 people per square mile and cities of more than 2 million people?

2. **Places and Regions** Which country has areas in the north where no one lives?

Critical Thinking

3. **Making Generalizations and Predictions** What can you assume about landforms and farming in western Russia just by looking at the **population map**? Check the **physical** and **land use and resources maps** to be sure.

4. **Analyzing Information** Use the map on this page to create a chart, graph, database, or model of population centers in Russia and northern Eurasia.

Direct students' attention to the **land use and resources map** on this page. Then have them compare it to the **population map**. Ask: What area is rich in natural resources but sparsely populated? *(Siberia)* Challenge students to use information from this and the other maps to suggest why Russia may have trouble making use of Siberia's resources. *(Possible answers: vast distance from population centers, difficulties in transporting goods across the region)*

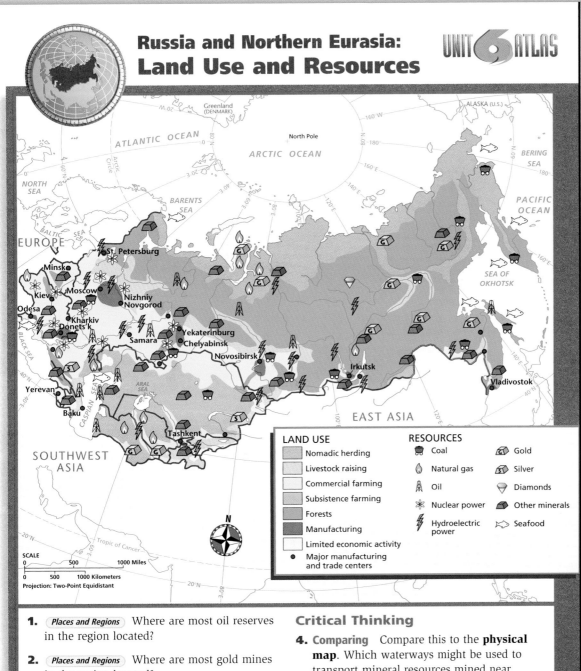

Russia and Northern Eurasia: Land Use and Resources

UNIT 6 ATLAS

LAND USE
- Nomadic herding
- Livestock raising
- Commercial farming
- Subsistence farming
- Forests
- Manufacturing
- Limited economic activity
- Major manufacturing and trade centers

RESOURCES
- Coal
- Natural gas
- Oil
- Nuclear power
- Hydroelectric power
- Gold
- Silver
- Diamonds
- Other minerals
- Seafood

SCALE
0 — 500 — 1000 Miles
0 — 500 — 1000 Kilometers
Projection: Two-Point Equidistant

1. *Places and Regions* Where are most oil reserves in the region located?

2. *Places and Regions* Where are most gold mines in the region located?

3. *Places and Regions* Which country has diamonds? Which countries in the region have silver deposits?

Critical Thinking

4. Comparing Compare this to the **physical map**. Which waterways might be used to transport mineral resources mined near Irkutsk to manufacturing and trade centers?

5. Analyzing Information Use the map on this page to create a chart, graph, database, or model of economic activities in the region.

Your Classroom Time Line (continued)

1939–45 World War II is fought.

late 1940s The Cold War begins.

1957 The Soviets launch Sputnik.

1989 The Baikal–Amur Mainline (BAM) railway is completed.

1986 The Chernobyl nuclear reactor disaster occurs.

1991 The Soviet Union collapses.

1990s Many civil wars erupt in the former Soviet republics.

1991 Russia and the other former Soviet republics form the Commonwealth of Independent States (CIS).

1995 In the Russian Far East, an earthquake causes severe damage on Sakhalin Island.

Land Use and Resources Map

Answers

1. in the Caspian Sea area and in Siberia

2. eastern Siberia

3. Russia; Russia and Kazakhstan

Critical Thinking

4. Angara River and Lake Baikal

5. Charts, graphs, databases, and models should accurately reflect information from the map.

RUSSIA AND NORTHERN EURASIA

LEVEL 1: (Suggested time: 25 min.) Explain that the unemployment rate shows the percentage of people who could work but cannot find a job. The U.S. unemployment rate of 4 percent meant that 4 percent of the people who were of age to work—that is, older than 16 and younger than 64—and could work did not have jobs when the data was collected.

Have students create a two-column chart with these headings: Less than 10 percent unemployment and More than 10 percent unemployment. Ask students to place each of the unit's countries except Turkmenistan in the correct column.

LEVEL 2: (Suggested time: 30 min.) After the collapse of the Soviet Union, many northern Eurasian countries were left with unstable economies and high unemployment and underemployment rates. To demonstrate a 20 percent unemployment rate, have students compute 20 percent of the class. *(Multiply the total number of students by 0.2.)* Have 20 percent of the class stand on one side of the room while the remaining students stand on the other.

Define the term "underemployment." *(Underemployed workers can only find jobs that require less skill than they have, or pay less money than their*

UNITED STATES OF AMERICA

CAPITAL:
Washington, D.C.
AREA:
3,717,792 sq. mi.
(9,629,091 sq km)
POPULATION:
281,421,906
MONEY:
U.S. dollar (US$)
LANGUAGES:
English, Spanish (spoken by a large minority)
UNEMPLOYMENT:
4 percent

Russia and Northern Eurasia

ARMENIA

CAPITAL:
Yerevan
AREA:
11,506 sq. mi. (29,800 sq km)
POPULATION:
3,336,100
MONEY:
dram (AMD)
LANGUAGES:
Armenian, Russian
UNEMPLOYMENT:
20 percent

BELARUS

CAPITAL:
Minsk
AREA:
80,154 sq. mi.
(207,600 sq km)
POPULATION:
10,350,194
MONEY:
Belarusian rubel (BYB/BYR)
LANGUAGES:
Byelorussian, Russian
UNEMPLOYMENT:
2.1 percent (and many underemployed workers)

AZERBAIJAN

CAPITAL:
Baku
AREA:
33,436 sq. mi.
(86,600 sq km)
POPULATION:
7,771,092
MONEY:
manat (AZM)
LANGUAGES:
Azeri, Russian, Armenian
UNEMPLOYMENT:
20 percent

GEORGIA

CAPITAL:
T'bilisi
AREA:
26,911 sq. mi.
(69,700 sq km)
POPULATION:
4,989,285
MONEY:
lari (GEL)
LANGUAGES:
Georgian (official), Russian, Armenian, Azeri
UNEMPLOYMENT:
14.9 percent

KAZAKHSTAN

CAPITAL:
Astana
AREA:
1,049,150 sq. mi.
(2,717,300 sq km)
POPULATION:
16,731,303
MONEY:
tenge (KZT)
LANGUAGES:
Kazakh, Russian
UNEMPLOYMENT:
13.7 percent

Geese flock to this pasture in Ukraine

KYRGYZSTAN

CAPITAL:
Bishkek
AREA:
76,641 sq. mi.
(198,500 sq km)
POPULATION:
4,753,003
MONEY:
Kyrgyzstani som (KGS)
LANGUAGES:
Kirghiz, Russian
UNEMPLOYMENT:
6 percent

Countries not drawn to scale.

skills are generally worth, or both.) Separate out another 20 percent from the employed group of students to represent the underemployed. Ask: Why might a country have a high rate of underemployment? *(Workers are likely to accept any job available, whether or not it meets their skill level or salary needs.)* When students have sat down, have them summarize their observations.

LEVEL 3: (Suggested time: 45 min.) Explain to students that during the Soviet era, the economies of the areas under Soviet control were closely interconnected. When the Soviet Union broke up, many countries did not have enough resources to sustain their populations independently. Have students use the **land use and resources map** to make connections between resources and unemployment rates.

Organize the class into small groups. Have each group brainstorm and create a list of possible solutions to the problems of unemployment and underemployment in the independent republics. Then have each group present its list of solutions to the class.

UNIT 6 ATLAS

Mosque in Uzbekistan

UKRAINE

CAPITAL: Kiev

AREA: 233,089 sq. mi. (603,700 sq km)

POPULATION: 48,760,474

MONEY: hryvna

LANGUAGES: Ukranian, Russian, Romanian, Polish, Hungarian

UNEMPLOYMENT: 4.3 percent officially registered (and many unregistered or underemployed)

RUSSIA

CAPITAL: Moscow

AREA: 6,592,735 sq. mi. (17,075,200 sq km)

POPULATION: 145,470,197

MONEY: Russian ruble (RUR)

LANGUAGES: Russian

UNEMPLOYMENT: 10.5 percent (and many underemployed workers)

UZBEKISTAN

CAPITAL: Tashkent

AREA: 172,741 sq. mi. (447,400 sq km)

POPULATION: 25,155,064

MONEY: Uzbekistani sum (UZS)

LANGUAGES: Uzbek, Russian, Tajik

UNEMPLOYMENT: 10 percent (and many underemployed)

TAJIKISTAN

CAPITAL: Dushanbe

AREA: 55,251 sq. mi. (143,100 sq km)

POPULATION: 6,578,681

MONEY: somoni (SM)

LANGUAGES: Tajik (official), Russian

UNEMPLOYMENT: 5.7 percent (and many underemployed workers)

internet connect

COUNTRY STATISTICS
GO TO: go.hrw.com
KEYWORD: SJ3 FACTSU6
FOR: more facts about Russia and northern Eurasia

TURKMENISTAN

CAPITAL: Ashgabat

AREA: 188,455 sq. mi. (488,100 sq km)

POPULATION: 4,603,244

MONEY: Turkmen manat (TMM)

LANGUAGES: Turkmen, Uzbek, Russian

UNEMPLOYMENT: data not available

Sources: Central Intelligence Agency, *The World Factbook 2001*; *The World Almanac and Book of Facts 2001*; pop. figures are 2001 estimates.

Objectives	Pacing Guide	Reproducible Resources
SECTION 1		
Physical Geography (pp. 483–86) 1. Identify the physical features of Russia. 2. Identify the climates and vegetation found in Russia. 3. Describe the natural resources of Russia.	Regular .5 day **Block Scheduling** .5 day *Block Scheduling Handbook, Chapter 22*	**RS** Guided Reading Strategy 22.1 **E** Creative Strategies for Teaching World Geography, Lessons 12 and 13
SECTION 2		
History and Culture (pp. 487–92) 1. Describe Russia's early history. 2. Describe how the Russian empire grew and then fell. 3. Describe the former Soviet Union. 4. Describe what Russia is like today.	Regular 2 days **Block Scheduling** .5 day *Block Scheduling Handbook, Chapter 22*	**RS** Guided Reading Strategy 22.2 **RS** Graphic Organizer 22 **E** Cultures of the World Activity 4 **SM** Geography for Life Activity 22 **IC** Interdisciplinary Activities for Middle Grades 17, 18, 20 **SM** Map Activity 22
SECTION 3		
The Russian Heartland (pp. 493–95) 1. Describe why European Russia is considered the country's heartland. 2. Identify the characteristics of the four major regions of European Russia.	Regular 1 day **Block Scheduling** .5 day *Block Scheduling Handbook, Chapter 22*	**RS** Guided Reading Strategy 22.3
SECTION 4		
Siberia (pp. 496–98) 1. Describe the human geography of Siberia. 2. Identify the economic features of the region. 3. Describe how Lake Baikal has been threatened by pollution.	Regular 1 day **Block Scheduling** .5 day *Block Scheduling Handbook, Chapter 22*	**RS** Guided Reading Strategy 22.4
SECTION 5		
The Russian Far East (pp. 499–501) 1. Describe how the Russian Far East's climate affects agriculture in the region. 2. Identify the major resources and cities of the region. 3. Identify the island regions that are part of the Russian Far East.	Regular 1 day **Block Scheduling** .5 day *Block Scheduling Handbook, Chapter 22*	**RS** Guided Reading Strategy 22.5

Chapter Resource Key

RS Reading Support
IC Interdisciplinary Connections
E Enrichment
SM Skills Mastery
A Assessment
REV Review

ELL Reinforcement and English Language Learners
 Transparencies
 CD–ROM
 Music

 Video
 Internet
 Holt Presentation Maker Using Microsoft® Powerpoint®

 One-Stop Planner CD–ROM

See the *One-Stop Planner* for a complete list of additional resources for students and teachers.

One-Stop Planner CD–ROM

It's easy to plan lessons, select resources, and print out materials for your students when you use the *One-Stop Planner CD–ROM with Test Generator.*

<table>
<tr><td colspan="2">Technology Resources</td><td colspan="2">Review, Reinforcement, and Assessment Resources</td></tr>
<tr>
<td></td>
<td>One-Stop Planner CD–ROM, Lesson 22.1

Geography and Cultures Visual Resources with Teaching Activities 31–36

ARGWorld CD–ROM: Planning in Russia and Its Neighbors

Homework Practice Online

HRW Go site</td>
<td>ELL
REV
A
ELL
ELL</td>
<td>Main Idea Activity 22.1
Section 1 Review, p. 354
Daily Quiz 22.1
English Audio Summary 22.1
Spanish Audio Summary 22.1</td>
</tr>
<tr>
<td></td>
<td>One-Stop Planner CD–ROM, Lesson 22.2

Homework Practice Online

HRW Go site</td>
<td>ELL
REV
A
ELL
ELL</td>
<td>Main Idea Activity 22.2
Section 2 Review, p. 360
Daily Quiz 22.2
English Audio Summary 22.2
Spanish Audio Summary 22.2</td>
</tr>
<tr>
<td></td>
<td>One-Stop Planner CD–ROM, Lesson 22.3

ARGWorld CD–ROM: Land Values in Post-Soviet Moscow

Homework Practice Online

HRW Go site</td>
<td>ELL
REV
A
ELL
ELL</td>
<td>Main Idea Activity 22.3
Section 3 Review, p. 363
Daily Quiz 22.3
English Audio Summary 22.3
Spanish Audio Summary 22.3</td>
</tr>
<tr>
<td></td>
<td>One-Stop Planner CD–ROM, Lesson 22.4

Homework Practice Online

HRW Go site</td>
<td>ELL
REV
A
ELL
ELL</td>
<td>Main Idea Activity 22.4
Section 4 Review, p. 366
Daily Quiz 22.4
English Audio Summary 22.4
Spanish Audio Summary 22.4</td>
</tr>
<tr>
<td></td>
<td>One-Stop Planner CD–ROM, Lesson 22.5

Geography and Cultures Visual Resources with Teaching Activities 24–29

Homework Practice Online

HRW Go site</td>
<td>ELL
REV
A
ELL
ELL</td>
<td>Main Idea Activity 22.5
Section 5 Review, p. 369
Daily Quiz 22.5
English Audio Summary 22.5
Spanish Audio Summary 22.5</td>
</tr>
</table>

internet connect

HRW ONLINE RESOURCES

<u>GO TO: go.hrw.com</u>
Then type in a keyword.

TEACHER HOME PAGE
KEYWORD: SJ3 TEACHER

CHAPTER INTERNET ACTIVITIES
KEYWORD: SJ3 GT22

Choose an activity to:
• take a trip on the Trans-Siberian Railroad.
• examine the breakup of the Soviet Union.
• view the cultural treasures of Russia.

CHAPTER ENRICHMENT LINKS
KEYWORD: SJ3 CH22

CHAPTER MAPS
KEYWORD: SJ3 MAPS22

ONLINE ASSESSMENT
Homework Practice
KEYWORD: SJ3 HP22
Standardized Test Prep Online
KEYWORD: SJ3 STP22
Rubrics
KEYWORD: SS Rubrics

COUNTRY INFORMATION
KEYWORD: SJ3 Almanac

CONTENT UPDATES
KEYWORD: SS Content Updates

HOLT PRESENTATION MAKER
KEYWORD: SJ3 PPT22

ONLINE READING SUPPORT
KEYWORD: SS Strategies

CURRENT EVENTS
KEYWORD: S3 Current Events

Meeting Individual Needs

Ability Levels

Level 1 Basic-level activities designed for all students encountering new material

Level 2 Intermediate-level activities designed for average students

Level 3 Challenging activities designed for honors and gifted-and-talented students

English Language Learners Activities that address the needs of students with Limited English Proficiency

Chapter Review and Assessment

E Readings in World Geography, History, and Culture 40, 41, 42

SM Critical Thinking Activity 22

REV Chapter 22 Review, pp. 370–71

REV Chapter 22 Tutorial for Students, Parents, Mentors, and Peers

ELL Vocabulary Activity 22

A Chapter 22 Test

Chapter 22 Test Generator (on the One-Stop Planner)

Audio CD Program, Chapter 22

A Chapter 22 Test for English Language Learners and Special-Needs Students

LAUNCH INTO LEARNING

Tell students that Russia occupies more land area than any other country in the world. Then ask them to consider the advantages and disadvantages of such enormous size. *(Students may mention as advantages the possibility of many natural resources and access to trade with many other countries. Disadvantages might include difficulties in communication, defense, distribution of goods and services, and maintaining a sense of unity.)*

Section 1

Objectives

1. Identify the physical features of Russia.
2. Name the climates and vegetation that are found in Russia.
3. List Russia's natural resources.

LINKS TO OUR LIVES

You may wish to highlight to students these reasons why we should know more about Russia:

▶ Russia occupies more land area than any other country on Earth. It also has one of the largest populations.

▶ For many years, the Soviet Union and the United States were enemies. Now economic and cultural connections between the countries are increasing.

▶ As one of the few countries with nuclear weapons, Russia could pose a threat to U.S. national security.

▶ Russia continues to struggle with economic, environmental, and political problems. Because of Russia's size and importance, these problems can affect people around the world.

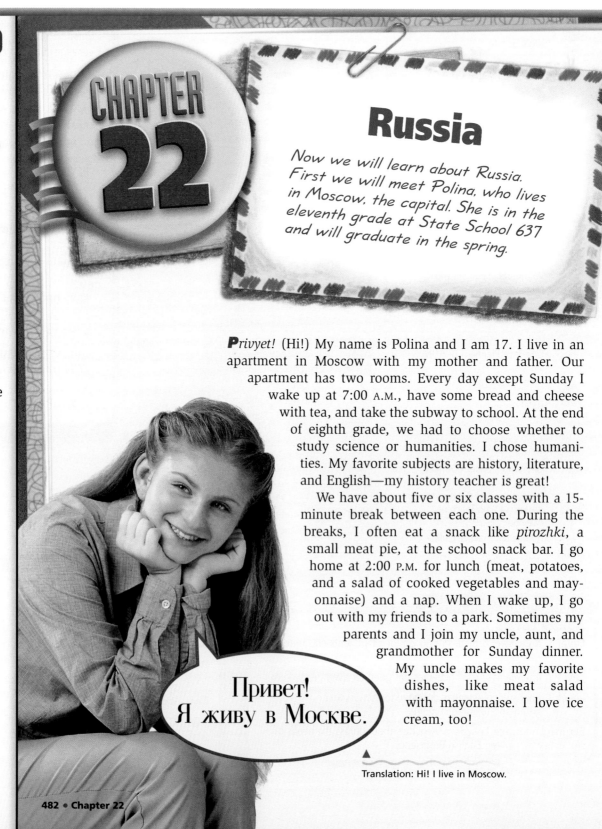

CHAPTER 22

Russia

Now we will learn about Russia. First we will meet Polina, who lives in Moscow, the capital. She is in the eleventh grade at State School 637 and will graduate in the spring.

*P*rivyet! (Hi!) My name is Polina and I am 17. I live in an apartment in Moscow with my mother and father. Our apartment has two rooms. Every day except Sunday I wake up at 7:00 A.M., have some bread and cheese with tea, and take the subway to school. At the end of eighth grade, we had to choose whether to study science or humanities. I chose humanities. My favorite subjects are history, literature, and English—my history teacher is great!

We have about five or six classes with a 15-minute break between each one. During the breaks, I often eat a snack like *pirozhki*, a small meat pie, at the school snack bar. I go home at 2:00 P.M. for lunch (meat, potatoes, and a salad of cooked vegetables and mayonnaise) and a nap. When I wake up, I go out with my friends to a park. Sometimes my parents and I join my uncle, aunt, and grandmother for Sunday dinner. My uncle makes my favorite dishes, like meat salad with mayonnaise. I love ice cream, too!

Привет!
Я живу в Москве.

Translation: Hi! I live in Moscow.

LET'S GET STARTED

Copy the following instructions onto the chalkboard: *Look at the physical-political map of Russia in your textbook. Choose a place on the map and write a sentence describing what you think that place might look like.* Discuss student responses in terms of the landscapes found across Russia. Tell students that in Section 1 they will learn more about the physical geography of Russia.

Using the Physical-Political Map

Have students examine the map on this page. Call on students to describe the general physical characteristics of Russia. *(more mountains in eastern half, flat lowlands, long rivers, one large lake)* Ask if they are already familiar with any of the features on the map. *(Possible answers: Moscow, Siberia, Arctic Circle, Arctic Ocean)* Ask students to predict the general climate of Russia. *(cold)*

Building Vocabulary

Write the key terms on the chalkboard. Tell students that **taiga** and **steppe** come from Russian words. You may want to remind students that "steppe" is also a climate type. Call on volunteers to find the definitions of the Define terms in the section text and to write sentences using them.

Section 1 — Physical Geography

Read to Discover

1. What are the physical features of Russia?
2. What climates and vegetation are found in Russia?
3. What natural resources does Russia have?

Define

taiga
steppe

Locate

Arctic Ocean
Caucasus Mountains
Caspian Sea
Ural Mountains
West Siberian Plain
Central Siberian Plateau
Kamchatka Peninsula
Kuril Islands
Volga River
Baltic Sea

WHY IT MATTERS

Like many nations, Russia is concerned about environmental issues. Use CNNfyi.com and other current events sources to investigate environmental concerns there. Record your findings in your journal.

A Siberian sable

Russia: Physical-Political

ELEVATION

	FEET	METERS
⊛ National capital	13,120	4,000
• Other cities	6,560	2,000
	1,640	500
	656	200
	(Sea level) 0	0 (Sea level)
	Below sea level	Below sea level

SCALE
0   500   1000 Miles
0   500   1000 Kilometers
Projection: Two-Point Equidistant

Size comparison of Russia to the contiguous United States

Section 1 RESOURCES

Reproducible
◆ Block Scheduling Handbook, Chapter 22
◆ Guided Reading Strategy 22.1
◆ Creative Strategies for Teaching World Geography, Lessons 12 and 13

Technology
◆ One-Stop Planner CD–ROM, Lesson 22.1
◆ Homework Practice Online
◆ Geography and Cultures Visual Resources with Teaching Activities 31–36
◆ *ARGWorld* CD–ROM: Planning in Russia and Its Neighbors
◆ HRW Go site

Reinforcement, Review, and Assessment
◆ Section 1 Review, p. 486
◆ Daily Quiz 22.1
◆ Main Idea Activity 22.1
◆ English Audio Summary 22.1
◆ Spanish Audio Summary 22.1

Teaching Objectives 1–3

LEVEL 1: (Suggested time: 15 min.) Copy the following graphic organizer onto the chalkboard, omitting the italicized answers. Call on volunteers to fill in the circle's quarters with Russia's major landforms, rivers, climate and vegetation types, and natural resources.
ENGLISH LANGUAGE LEARNERS

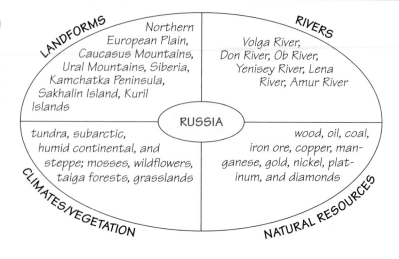

LANDFORMS
Northern European Plain, Caucasus Mountains, Ural Mountains, Siberia, Kamchatka Peninsula, Sakhalin Island, Kuril Islands

RIVERS
Volga River, Don River, Ob River, Yenisey River, Lena River, Amur River

RUSSIA

CLIMATES/VEGETATION
tundra, subarctic, humid continental, and steppe; mosses, wildflowers, taiga forests, grasslands

NATURAL RESOURCES
wood, oil, coal, iron ore, copper, manganese, gold, nickel, platinum, and diamonds

---

EYE ON EARTH

In addition to being the world's largest lake, the Caspian Sea is also one of the greatest salt lakes. Stretching 746 miles (1,200 km) long and 270 miles (434 km) wide, the Caspian Sea covers an area of about 143,550 square miles (371,795 sq km)—an area almost larger than Japan. Scientific studies have shown that until recent times, geologically speaking, the Caspian Sea was actually linked to the Atlantic Ocean through the Sea of Azov, the Black Sea, and the Mediterranean Sea.

Activity: Have students use the physical-political map to find which bodies of water once connected the Caspian Sea to the Atlantic Ocean.

internet connect
GO TO: go.hrw.com
KEYWORD: SJ3 CH22
FOR: Web sites about the Caspian Sea

Visual Record Answer ▶

that Siberia has a cold climate and barren, rugged mountains

internet connect
GO TO: go.hrw.com
KEYWORD: SJ3 CH22
FOR: Web sites about Russia

▲
A train chugs through the cold Siberian countryside.
Interpreting the Visual Record What does this photograph tell you about the physical features and climate of Siberia?

Our Amazing Planet

The coldest temperature ever recorded outside of Antarctica in the last 100 years was noted on February 6, 1933, in eastern Siberia: −90° F (−68°C).

Physical Features

Russia was by far the largest republic of what was called the Union of Soviet Socialist Republics, or the Soviet Union. Russia is the largest country in the world. It stretches 6,000 miles (9,654 km), from Eastern Europe to the Bering Sea and Pacific Ocean.

The Land Much of western, or European, Russia is part of the Northern European Plain. This is the country's heartland, where most Russians live. To the north are the Barents Sea and the Arctic Ocean. Far to the south are the Caucasus (KAW-kuh-suhs) Mountains. There Europe's highest peak, Mount Elbrus, rises to 18,510 feet (5,642 m). The Caucasus Mountains stretch from the Black Sea to the Caspian (KAS-pee-uhn) Sea. The Caspian is the largest inland body of water in the world.

East of the Northern European Plain is a long range of eroded low mountains and hills. These are called the Ural (YOOHR-uhl) Mountains. The Urals divide Europe from Asia. They stretch from the Arctic coast in the north to Kazakhstan in the south. The highest peak in the Urals rises to just 6,214 feet (1,894 m).

East of the Urals lies a vast region known as Siberia. Much of Siberia is divided between the West Siberian Plain and the Central Siberian Plateau. The West Siberian Plain is a large, flat area with many marshes. The Central Siberian Plateau lies to the east. It is a land of elevated plains and valleys.

A series of high mountain ranges runs through southern and eastern Siberia. The Kamchatka (kuhm-CHAHT-kuh) Peninsula, Sakhalin (sah-kah-LEEN) Island, and the Kuril (KYOOHR-eel) Islands surround the Sea of Okhotsk (uh-KAWTSK). These are in the Russian Far East. The rugged Kamchatka Peninsula and the Kurils have active volcanoes. Earthquakes and volcanic eruptions are common. The Kurils separate the Sea of Okhotsk from the Pacific Ocean.

Rivers Some of the world's longest rivers flow through Russia. These include the Volga (VAHL-guh) and Don Rivers in European Russia. The Ob (AWB), Yenisey (yi-ni-SAY), Lena (LEE-nuh), and Amur (ah-MOOHR) Rivers are located in Siberia and the Russian Far East. The Amur forms part of Russia's border with China.

The Volga is Europe's longest river. Its course and length make it an important transportation route. It flows southward for 2,293 miles (3,689 km) across the Northern European Plain to the Caspian Sea. Barges can travel by canal from the Volga to the Don River. The Don empties into the Black Sea. Canals also connect the Volga to rivers that drain into the Baltic Sea far to the northwest.

In Siberia, the Ob, Yenisey, and Lena Rivers all flow thousands of miles northward. Eventually, they reach Russia's Arctic coast. These and other Siberian rivers that drain into the Arctic Ocean freeze in winter. In spring, these rivers thaw first in the south. Downstream in

LEVEL 2: (Suggested time: 20 min.) Ask students to imagine that they are taking part in a summer camp exchange program in Russia. First, students should choose a rural area where their camp is located. Then tell students to write a letter home to their families. They should describe the land, weather conditions, vegetation, and the outdoor recreational activities they are enjoying. Students should also mention some of the natural resources that are located near their camp.

LEVEL 3: (Suggested time: 20 min.) Have students write an outline for a documentary film script about the physical geography of Russia. Outlines should indicate the various physical features, climates, vegetation types, and resources found in Russia. Ask students to write descriptions of the scenes that the film would include.

the north, however, the rivers remain frozen much longer. As a result, ice jams there block water from the melting ice and snow. This causes annual floods in areas along the rivers.

✓ **READING CHECK:** *Places and Regions* What are the major physical features of Russia? Northern European Plain, Caucasus Mountains, Caspian Sea, Ural Mountains, Siberia, Kamchatka Peninsula, Sakhalin Island, Kuril Islands, several long rivers, including the Volga

Climate and Vegetation

Nearly all of Russia is located at high northern latitudes. The country has tundra, subarctic, humid continental, and steppe climates. Because there are no high mountain barriers, cold Arctic winds sweep across much of the country in winter. Winters are long and cold. Ice blocks most seaports until spring. However, the winters are surprisingly dry in much of Russia. This is because the interior is far from ocean moisture.

Winters are particularly severe throughout Siberia. Temperatures often drop below –40°F (–40°C). Although they are short, Siberian summers can be hot. Temperatures can rise to 100°F (38°C).

Vegetation varies with climates from north to south. Very cold temperatures and permafrost in the far north keep trees from taking root. Mosses, wildflowers, and other tundra vegetation grow there.

The vast **taiga** (TY-guh), a forest of mostly evergreen trees, grows south of the tundra. The trees there include spruce, fir, and pine. In European Russia and in the Far East are deciduous forests. Many temperate forests in European Russia have been cleared for farms and cities.

Wide grasslands known as the **steppe** (STEP) stretch from Ukraine across southern Russia to Kazakhstan. Much of the steppe is used for growing crops and grazing livestock.

✓ **READING CHECK:** *Physical Systems* How does Russia's location affect its climate? Arctic winds blow, creating winters that are very cold.

BUILD on WHAT You Know

Do you remember what you learned about tundra climates? See Chapter 3 to review.

Camels graze on open land in southern Siberia near Mongolia.

Russia

National Geography Standard 8

Physical Systems Tundra plants are well adapted to Russia's harsh climate. Vegetation in Russia's tundra climate is limited to lichens and mosses, small shrubs, and grasses. Most plants grow low to the ground to reduce the effects of cold and wind, but may have massive root systems to store nutrients. Other plants have waxy or hairy surfaces that protect them from freezing winds.

Critical Thinking: What adaptations do humans make to survive cold environments like the tundra? How do the adaptations made by humans compare to those made by plants?

Answer: Students might mention that like plants, humans cover themselves with insulating materials, try to keep abundant supplies of food, and try to minimize exposure to wind and cold.

Refer students to the chapter map and the time zone map in the textbook's Skills Handbook. Call out physical features and ask students what time it is where that feature is located when it is 3 P.M. in Moscow. *(Examples: western Siberian lowlands: 5 P.M.; White Sea: 3 P.M.; mouth of the Lena River: 9 P.M.)*

RETEACH

Have students complete Main Idea Activity 22.1. Then have students work in groups to create large sketch maps of Russia. Students should take turns labeling physical features, climates, or locations of natural resources on their maps. **ENGLISH LANGUAGE LEARNERS, COOPERATIVE LEARNING**

REVIEW AND ASSESS

Have students complete the Section Review. Then pair students. Tell each student to create five fill-in-the-blank questions based on the material in Section 1. Then have students take turns asking their questions. Students should copy the questions and answers into their notes. Then have students complete Daily Quiz 22.1. **COOPERATIVE LEARNING**

EXTEND

Have interested students conduct research on the Tunguska Event of 1908, when a huge explosion of mysterious origin leveled about 1,200 square miles of the Siberian wilderness. Ask students to identify when plant and animal life returned. **BLOCK SCHEDULING**

Section Review 1

Answers

Define For definitions, see: taiga, p. 485; steppe, p. 485

Working with Sketch Maps Maps will vary, but listed places should be labeled in their approximate locations. The Volga River is Europe's longest river. It flows across the Northern European Plain to the Caspian.

Reading for the Main Idea

1. the Ural Mountains (NGS 4)
2. by canals to the Don and other rivers that empty into Baltic and Black Seas (NGS 4)
3. severe, with temperatures as low as −40° F (−40° C), and surprisingly dry; answers will vary, but students might mention various human adaptations to the cold, such as more indoor activities. (NGS 15)

Critical Thinking

4. by providing Russia with materials to export as well as limiting its need for imported goods

Organizing What You Know

5. climates—tundra, subarctic, humid continental, and steppe; vegetation—mosses, wildflowers, tundra vegetation, evergreen trees, deciduous forests, and grasslands; resources—oil, coal, iron ore, copper, manganese, gold, platinum, nickel, diamonds, forests

486

▲ A blast furnace is used to process nickel in Siberia. Nickel is just one of Russia's many natural resources.

Resources

Russia has enormous energy, mineral, and forest resources. However, those resources have been poorly managed. For example, much of the forest west of the Urals has been cut down. Now wood products must be brought long distances from Siberia. Still, the taiga provides a vast supply of trees for wood and paper pulp.

Russia has long been a major oil producer. However, many of its oil deposits are far from cities, markets, and ports. Coal is also plentiful. More than a dozen metals are available in large quantities. Russia also is a major diamond producer. Many valuable mineral deposits in remote Siberia have not yet been mined.

✓ **READING CHECK:** *Places and Regions* How might the location of its oil deposits prevent Russia from taking full advantage of this resource? It makes transporting to other markets difficult.

go.hrw.com **Homework Practice Online** Keyword: SJ3 HP22

Section Review 1

Define and explain: taiga, steppe

Working with Sketch Maps On a map of Russia that you sketch or that your teacher provides, label the following: Arctic Ocean, Caucasus Mountains, Caspian Sea, Ural Mountains, West Siberian Plain, Central Siberian Plateau, Kamchatka Peninsula, Kuril Islands, Volga River, and Baltic Sea.

Reading for the Main Idea

1. *Places and Regions* What low mountain range in central Russia divides Europe from Asia?
2. *Places and Regions* How is the Volga River linked to the Baltic and Black Seas?
3. *Environment and Society* What are winters like in much of Russia? How might they affect people?

Critical Thinking

4. **Making Generalizations and Predictions** How might Russia's natural resources make the country more prosperous?

Organizing What You Know

5. **Categorizing** Copy the following graphic organizer. Use it to list the climates, vegetation, and resources of Russia.

Climates	Vegetation	Resources

Section 2

Objectives

1. Outline the major events in Russia's early history.
2. Describe the growth and decline of the Russian Empire.
3. Report on the Soviet Union.
4. Identify characteristics of present-day Russia.

FOCUS

LET'S GET STARTED

Arrange the classroom so that the desks are in two large groups with one aisle down the middle. Ask a volunteer to walk from one corner of the room to the opposite corner by the easiest route. *(Most students will use the aisle.)* Then, explain to the class that the great Russian steppe, or plain, was like a flat aisle or hallway that created easy access to the Russian interior for both invaders and immigrants. Tell students that in Section 2 they will learn more about those groups as they study the history and culture of Russia.

Building Vocabulary

Write the Define terms on the board. Tell students that **czar** comes from a Latin word, *caesar*, which means "emperor." Point out that **abdicate** comes from two Latin word parts: *ab-* ("away") and *dicare* ("to proclaim"). Have students suggest other words related to **allies**. Call on volunteers to define **superpowers** and **consumer goods** based on the terms' component words. Ask students to contrast the **Cold War** with other wars they have studied. Check all suggestions against definitions.

Section 2 — History and Culture

Read to Discover

1. What was Russia's early history like?
2. How did the Russian Empire grow and then fall?
3. What was the Soviet Union?
4. What is Russia like today?

Define

czar
abdicated
allies
superpowers
Cold War
consumer goods

Locate

Moscow

WHY IT MATTERS

Russia is well known for its ballet companies. Use CNNfyi.com or other current events sources to discover more about Russia's culture and its international recognition in the field of dance and other arts. Record your findings in your journal.

Russian caviar, blini (pancakes), and smoked salmon

Section 2 RESOURCES

Reproducible
- Guided Reading Strategy 22.2
- Graphic Organizer 22
- Geography for Life Activity 22
- Interdisciplinary Activity for the Middle Grades 17, 18, 20
- Map Activity 22
- Cultures of the World Activity 4

Technology
- One-Stop Planner CD–ROM, Lesson 22.2
- Homework Practice Online
- HRW Go site

Reinforcement, Review, and Assessment
- Section 2 Review, p. 492
- Daily Quiz 22.2
- Main Idea Activity 22.2
- English Audio Summary 22.2
- Spanish Audio Summary 22.2

Early Russia

The roots of the Russian nation lie deep in the grassy plains of the steppe. For thousands of years, people moved across the steppe bringing new languages, religions, and ways of life.

Early Migrations Slavic peoples have lived in Russia for thousands of years. In the A.D. 800s, Viking traders from Scandinavia helped shape the first Russian state among the Slavs. These Vikings called themselves Rus (ROOS). The word *Russia* comes from their name. The state they created was centered on Kiev. Today Kiev is the capital of Ukraine.

In the following centuries, missionaries from southeastern Europe brought Orthodox Christianity and a form of the Greek alphabet to Russia. Today the Russian language is written in this Cyrillic alphabet.

Mongols After about 200 years, Kiev's power began to decline. In the 1200s, Mongol invaders called Tatars swept out of Central Asia across the steppe. The Mongols conquered Kiev and added much of the region to their vast empire.

The Mongols demanded taxes but ruled the region through local leaders. Over time, these local leaders established various states. The strongest of these was Muscovy, north of Kiev. Its chief city was Moscow.

▲
This painting from the mid-1400s shows a battle between soldiers of two early Russian states.

✓ **READING CHECK:** *Human Systems* What was the effect of Viking traders on Russia? They helped shape the first Russian state, centered on Kiev, now the capital of Ukraine.

Teaching Objective 1

ALL LEVELS: (Suggested time: 20 min.) Have students draw a time line of the major events in Russian history up to the Russian Revolution in 1917. You may want to have students color code the entries based on the ethnic or religious group involved. Display the time lines around the classroom. **ENGLISH LANGUAGE LEARNERS, COOPERATIVE LEARNING**

Teaching Objective 2

ALL LEVELS: (Suggested time: 20 min.) Copy the following graphic organizer onto the chalkboard, omitting the italicized answers. Use it to help students understand the rise and fall of the Russian Empire. Call on volunteers to fill in details to connect the events. **ENGLISH LANGUAGE LEARNERS**

RISE AND FALL OF THE RUSSIAN EMPIRE

RISE

expansion and growth

BEGINNING — *Ivan IV crowned czar*

FALL — *food shortages, economic woes, defeat in war*

Linking Past to Present

Invasions of

Russia Russia has been invaded several times during its long history. Each invasion eventually failed. In (1707–09) Sweden's king Charles XII invaded Russia, only to be defeated by brutal winter weather, burned land and crops, and the czar's army. In 1812, Napoléon I of France invaded Russia with about 400,000 men. Only 30,000 or so survived the battles, cold, disease, hunger, and attacks by Russian soldiers and citizens. In 1941 Nazi Germany invaded the Soviet Union. Again, cold was a major factor in the Nazis' defeat.

Critical Thinking: What roles might Russia's size and landforms have played in these defeats?

Answer: Students might mention that the size made supply lines extremely long and thus difficult to maintain. Armies had to cover such great distances that severe weather could trap unprepared troops before they could return home. The featureless steppe offered little shelter.

Map Answer ▶

between 1801 and 1945

History of Russian Expansion

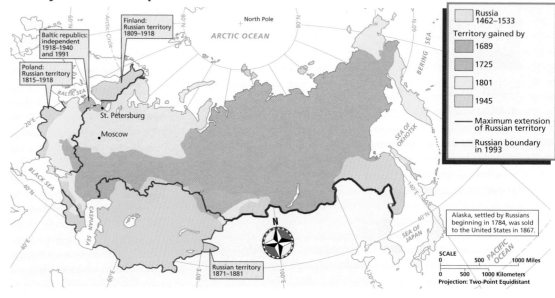

The colors in this map show land taken by the Russian Empire and the Soviet Union over time.

Interpreting the Map When was the period of Russia's greatest expansion?

▲ Ivan the Terrible became grand prince of Moscow in 1533. He was just three years old. He ruled Russia from 1547 to his death in 1584.

The Russian Empire

In the 1400s Muscovy won control over parts of Russia from the Mongols. In 1547 Muscovy's ruler, Ivan IV—known as Ivan the Terrible—crowned himself **czar** (ZAHR) of all Russia. The word *czar* comes from the Latin word *Caesar* and means "emperor."

Expansion Over more than 300 years, czars like Peter the Great (1672–1725) expanded the Russian empire. By the early 1700s the empire stretched from the Baltic to the Pacific.

Russian fur traders crossed the Bering Strait in the 1700s and 1800s. They established colonies along the North American west coast. Those colonies stretched from coastal Alaska to California. Russia sold Alaska to the United States in 1867. Around the same time, Russia expanded into Central Asia.

Decline The Russian Empire's power began to decline in the late 1800s. Industry grew slowly, so Russia remained largely agricultural. Most people were poor farmers. Far fewer were the rich, factory workers, or craftspeople. Food shortages, economic problems, and defeat in war further weakened the empire in the early 1900s.

In 1917, during World War I, the czar **abdicated**, or gave up his throne. Later in 1917 the Bolshevik Party, led by Vladimir Lenin, overthrew the government. This event is known as the Russian Revolution.

✔ **READING CHECK:** *Human Systems* What conflict brought a change of government to Russia? the Russian Revolution in which the Bolshevik Party overthrew the government

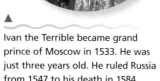

Teaching Objective 3

LEVEL 1: (Suggested time: 20 min.) Pair students and have them design a political poster for the Soviet Union. Posters should include references to the Russian Revolution, past leaders, and Cold War elements concerning competition between the Soviet Union and the United States. **ENGLISH LANGUAGE LEARNERS, COOPERATIVE LEARNING**

LEVEL 2: (Suggested time: 30 min.) Have students create editorial cartoons referring to the Soviet Union during the Cold War from an American viewpoint. Ask them to explain how their cartoons show Americans' frame of reference regarding the Cold War. **ENGLISH LANGUAGE LEARNERS**

LEVEL 3: (Suggested time: 30 min.) Have students play the role of an American journalist living in the Soviet Union just after its collapse in 1991. Tell them to write a newspaper article describing the conditions and factors leading up to the collapse of the Soviet Union, doing additional research as needed. Have each student write an appropriate headline for his or her article and share it with the class.

The Soviet Union

The Bolsheviks, or Communists, established the Soviet Union in 1922. Most of the various territories of the Russian Empire became republics within the Soviet Union.

Under Lenin and his successor, Joseph Stalin, the Communists took over all industries and farms. Religious practices were discouraged. The Communists outlawed all other political parties. Many opponents were imprisoned, forced to leave the country, or even killed.

The Soviet leaders established a command economy, in which industries were controlled by the government. At first these industries grew dramatically. However, over time the lack of competition made them inefficient and wasteful. The quality of many products was poor. Government-run farms failed to produce enough food to feed the population. By the late 1950s the Soviet Union had to import large amounts of grain.

Cold War The Soviet Union in the 1950s was still recovering from World War II. The country had been a major battleground in the war. The United States and the Soviet Union had been **allies**, or friends, in the fight against Germany. After the war the two **superpowers**, or powerful countries, became rivals. This bitter rivalry became known as the **Cold War**. The Cold War lasted from the 1940s to the early 1990s. The Soviet Union and the United States built huge military forces, including nuclear weapons. The two countries never formally went to war with each other. However, they supported allies in small wars around the world.

Collapse of the Soviet Union The costs of the Cold War eventually became too much for the Soviet Union. The Soviet government spent more and more money on military goods. **Consumer goods** became expensive and in short supply. Consumer goods are products used at home and in everyday life. The last Soviet leader, Mikhail Gorbachev, tried to bring about changes to help the economy. He also promoted a policy allowing more open discussion of the country's problems. However, the various Soviet republics pushed for independence. Finally, in 1991 the Soviet Union collapsed. The huge country split into 15 republics.

In late 1991 Russia and most of the other former Soviet republics formed the Commonwealth of Independent States, or CIS. The CIS does not have a strong central government. Instead, it provides a way for the former Soviet republics to address shared problems.

✓ **READING CHECK:**
Human Systems What was the Cold War, and how did it eventually cause the Soviet Union's collapse? rivalry between Soviet Union, U.S.; military costs eventually ruined Soviet economy

Tourists can visit the czar's Summer Palace in St. Petersburg.
Interpreting the Visual Record How do you think the rich lifestyle of the czars helped the Bolsheviks gain support?

A man lights candles in front of portraits of the last czar, Nicholas II, and his wife. Russians are divided over what kind of government their country should have today.
Interpreting the Visual Record Why do you think some Russians might wish to have a czar again?

ENVIRONMENT AND SOCIETY

Russia is faced with many environmental problems that it has inherited from the era of Soviet rule. During the 1900s, the Soviet government expanded settlement into Siberia with little concern for environmental impacts. The size and richness of the land made it seem that there was no limit to the resources or to the land on which to dump wastes.

Now Russia is left with large areas that have polluted air, soil, or water. The Russian economy is troubled, and Russia has limited resources to devote to leftover environmental damage.

Discussion: Where should environmental issues, such as cleaning up pollution and toxic waste, rank in national priorities? Have students discuss the issue from different points of view. *(Students may suggest that such issues are important but that they decline in priority in the face of an economic crisis. Others may feel environmental issues should take first priority because life itself depends on a healthy environment.)*

▲ **Visual Record Answers**

by alienating the common people from the czar

◄

in hopes of regaining their cultural heritage

➤**ASSIGNMENT:** Have students design a tombstone or write an obituary for the Soviet Union. You may want to provide students with examples of obituaries from the daily newspaper to use as models for their work. Display tombstones or obituaries around the classroom.

Teaching Objective 4

LEVELS 1 AND 2: (Suggested time: 20 min.) Have students use their textbook to create a chart on modern-day Russia. The chart should include information on the country's ethnic make up, religions, foods and festivals, arts and sciences, and government. When all students have completed their charts, ask volunteers to share their charts with the class.
ENGLISH LANGUAGE LEARNERS

LEVEL 3: (Suggested time: 30 min.) Pair students and instruct them to create a chart with two columns, one labeled "Russia Today" and the other "United States." Using their textbooks, students should select a fact about life in Russia and write it in the first column. Then they should compare or contrast that fact with life in the United States in the second column. Tell them to be sure to include government, economy, and culture in their charts. **COOPERATIVE LEARNING**

HUMAN SYSTEMS

The Tatars, also known as the Tartars, are an ethnic group living in various parts of Russia. One branch of this group is the Crimean Tatars. They have a unique history. The Crimean Tatars formed their own Soviet republic in 1921, but this republic was dissolved in 1945 when they were accused of helping the Germans in World War II. Most of the Crimean Tatars were deported to other parts of the Soviet Union and forbidden to speak their native language. In 1956 they regained their civil rights but were still refused the right to move back to the Crimea region. It was not until the breakup of the Soviet Union in 1991 that Crimean Tatars were allowed back into that region. Today there are about 270,000 Tatars in the Crimea.

Critical Thinking: Why might a government outlaw a specific language?

Answer: to try to break up a certain group of people

Connecting to Literature Answers
1. She is living in a Soviet republic where there is little information about the United States.
2. a wonderful place with many consumer goods

CONNECTING TO *Literature*

AUNT RIMMA'S TREAT

The former Soviet Union was composed of many republics, which are now independent countries. Nina Gabrielyan's The Lilac Dressing Gown *is told from the point of view of an Armenian girl living in Moscow before the Soviet breakup.*

Aunt Rimma. . . came to visit and gave me a pink caramel which I, naturally, popped straight in my mouth. "Don't swallow it," Aunt Rimma says in an odd sort of voice. "You're not supposed to swallow it, only chew it." "Why," I ask, puzzled by her solemn tone. "It's chewing gum," she says with pride in her eyes. "Chewing gum?" I don't know what she means. "American chewing gum," Aunt Rimma explains. "Mentor's sister sent it to us from America." "Oh, from America? Is that where the capitalists are? What is she doing there?" "She's living there," says Aunt Rimma, condescending to my foolishness.

But I am not as foolish as I used to be. I know that Armenians live in Armenia. Our country is very big and includes many republics: Armenia, Georgia, Azerbaijan, Tajikistan, Uzbekistan, Ukraine, Belorussia [Belarus], the Caucasus and Transcaucasia. All this together is the Soviet Union. Americans . . . live in America. Clearly, Mentor's sister cannot possibly be American. . . . Rimma goes on boasting: "Oh! the underwear they have there! . . . And the children's clothes!" I begin to feel a bit envious. . . . Nobody in our house has anyone living in America, but Aunt Rimma does. My envy becomes unbearable. So I decide to slay our boastful neighbor on the spot: "Well we have cockroaches! This big! Lots and lots of them!'"

Analyzing Primary Sources
1. How does the Armenian girl's frame of reference affect her view of the United States?
2. What does Aunt Rimma seem to think the United States is like?

Russia Today

Russia has been making a transition from communism to democracy and a free market economy since 1991. Change has been slow, and the country faces difficult challenges.

People and Religion More than 146 million people live in Russia today. More than 80 percent are ethnic Russians. The largest of Russia's many minority groups are Ukrainians and Tatars. These Tatars are the descendants of the early Mongol invaders of Russia.

Teaching Objectives 1–4

ALL LEVELS: (Suggested time: 30 min.) Organize the class into groups. Have each group design a poster for a film about one of the major eras of Russian history. Each poster should include the film title, a phrase to describe the action, and an illustration. Ask students to present and explain their posters to the class. **ENGLISH LANGUAGE LEARNERS, COOPERATIVE LEARNING**

Marcia Clevenger of Charleston, West Virginia, suggests the following activity to help students learn about Russia: Have students create a pop-up book or picture book about Russia. Students should cut up old magazines, use Internet pictures, or draw illustrations to tell a story about life in Russia. Ask volunteers to present their books to the class.

In the past, the government encouraged ethnic Russians to settle in areas of Russia far from Moscow. They were encouraged to move to places where other ethnic groups were in the majority. Today, many non-Russian peoples in those areas resent the domination of ethnic Russians. Some non-Russians want independence from Moscow. At times this has led to violence and even war, as in Chechnya in southern Russia.

Since 1991 a greater degree of religious expression has been allowed in Russia. Russian Orthodox Christianity is becoming popular again. Cathedrals have been repaired, and their onion-shaped domes have been covered in gold leaf and brilliant colors. Muslims around the Caspian Sea and the southern Urals are reviving Islamic practices.

Food and Festivals Bread is an important part of the Russian diet. It is eaten with every meal. It may be a rich, dark bread made from rye and wheat flour or a firm white bread. As in other northern countries, the growing season is short and winter is long. Therefore, the diet includes many canned and preserved foods, such as sausages, smoked fish, cheese, and vegetable and fruit preserves.

Black caviar, one of the world's most expensive delicacies, comes from Russia. The fish eggs that make up black caviar come from sturgeon. Sturgeon are fish found in the Caspian Sea.

The anniversary of the 1917 Russian Revolution was an important holiday during the Soviet era. Today the Orthodox Christian holidays of Christmas and Easter are again becoming popular in Russia. Special holiday foods include milk puddings and cheesecakes.

The Arts and Sciences Russia has given the world great works of art, literature, and music. For example, you might know *The Nutcracker*, a ballet danced to music composed by Peter Tchaikovsky (1840–93). It is a popular production in many countries.

Russia

COUNTRY	POPULATION/ GROWTH RATE	LIFE EXPECTANCY	LITERACY RATE	PER CAPITA GDP
Russia	145,470,197 −0.4%	62, male 73, female	98%	$7,700
United States	281,421,906 0.9%	74, male 80, female	97%	$36,200

Sources: Central Intelligence Agency, *The World Factbook 2001;* U.S. Census Bureau

Interpreting the Chart How many times greater is the U.S. population than the Russian population?

Across the Curriculum
ART

Fabergé Eggs For centuries Russians have exchanged eggs at Easter. In 1884 the czar gave his wife a special jeweled egg made by master craftsman Peter Carl Fabergé. It opened to reveal a smaller gold egg, which in turn contained a tiny golden chicken and glittering crown. The czarina was so delighted that the Fabergé egg became an annual tradition. Each was a small triumph of the jeweler's art.

Critical Thinking: How do Fabergé eggs reflect Russian traditions?

Answer: Their creation began with the czar's gift to his wife, reflecting a tradition of exchanging eggs at Easter.

▲ **Chart Answer**

1.9 times greater

Ballet dancers perform Peter Tchaikovsky's *Swan Lake* at the Mariinsky Theater in St. Petersburg.

491

CLOSE

Remind students that the United States and the Soviet Union were bitter enemies for many years. Ask students to suggest reasons why the Cold War never exploded into World War III. *(Students might mention the threat of mutual destruction by nuclear weapons.)*

REVIEW AND ASSESS

Have students complete the Section Review. Then pair students and have each student write five major events in Russian history, in random order. Students should exchange lists with their partners and place the events in chronological order. Then have students complete Daily Quiz 22.2.
COOPERATIVE LEARNING

RETEACH

Have students complete Main Idea Activity 22.2. Then have students work in groups to create time lines of major events in Russian history.
ENGLISH LANGUAGE LEARNERS, COOPERATIVE LEARNING

EXTEND

Have interested students conduct research on the current Russian government and another government of their choice. Ask them to create a chart comparing the two governments. **BLOCK SCHEDULING**

Section Review 2

Answers

Define For definitions, see: czar, p. 488; abdicated, p. 488; allies, p. 357; superpowers, p. 489; Cold War, p. 489; consumer goods, p. 489

Working with Sketch Maps Moscow should be labeled in its approximate location. Kiev was the first Russian state among the Slavs.

Reading for the Main Idea

1. traders, who called themselves Rus and organized the first Russian state (NGS 4)

2. Its leaders overthrew the Russian government in 1917. (NGS 13)

3. struggling economy, corruption, competing political groups (NGS 4)

Critical Thinking

4. Russian Empire—food shortages, economic problems, and defeat in war; Soviet Union—lack of consumer goods, too much money spent on military goods

Organizing What You Know

5. Answers will vary but should include information from the section.

Many Russian writers are known for how they capture the emotions of characters in their works. Some writers, such as Aleksandr Solzhenitsyn (1918–), have written about Russia under communism.

Russian scientists also have made important contributions to their professions. For example, in 1957 the Soviet Union launched *Sputnik*. It was the first artificial satellite in space. Today U.S. and Russian engineers are working together on space projects. These include building a large space station and planning for a mission to Mars.

Government Like the U.S. government, the Russian Federation is governed by an elected president and a legislature called the Federal Assembly. The Federal Assembly includes representatives of regions and republics within the Federation. Non-Russians are numerous or in the majority in many of those regions and republics.

The government faces tough challenges. One is improving the country's struggling economy. Many government-owned companies have been sold to the private sector. However, financial problems and corruption have made people cautious about investing in those companies.

Corruption is a serious problem. A few people have used their connections with powerful government officials to become rich. In addition, many Russians avoid paying taxes. This means the government has less money for salaries and services. Agreement on solutions to these problems has been hard.

Republics of the Russian Federation

Adygea	Karachay-Cherkessia
Alania	Karelia
Bashkortostan	Khakassia
Buryatia	Komi
Chechnya	Mari El
Chuvashia	Mordvinia
Dagestan	Sakha
Gorno-Altay	Tatarstan
Ingushetia	Tuva
Kabardino-Balkaria	Udmurtia
Kalmykia	

✓ **READING CHECK:** *Human Systems*
What are the people and culture of Russia like today? 80 percent ethnic Russian; free practice of religion; bread and preserved foods important; many great works of art, literature, and ballet, and scientific contributions

Section Review 2

Define and explain: czar, abdicated, allies, superpowers, Cold War, consumer goods

Working with Sketch Maps On the map you created in Section 1, label Moscow. In the margin, explain the role Kiev played in Russia's early history.

Reading for the Main Idea

1. *Places and Regions* How did Russia get its name?

2. *Human Systems* What was the Bolshevik Party?

3. *Places and Regions* What are some of the challenges that Russia faces today?

Critical Thinking

4. **Comparing** Compare the factors that led to the decline of the Russian Empire and the Soviet Union. List the factors for each.

Organizing What You Know

5. **Summarizing** Copy the following graphic organizer. Use it to identify important features of Russia's ethnic population, religion, food, and arts and sciences.

Russian people and culture

Section 3

Objectives

1. Explain why European Russia is considered the heartland of the country.
2. Describe the four major regions of European Russia.

FOCUS

🎧 LET'S GET STARTED

Copy the following "equation" and question onto the chalkboard: *St. Petersburg → Petrograd → Leningrad → St. Petersburg. What do you think this means?* Tell students that "-burg" and "-grad" mean "city" and that the city now called St. Petersburg has changed names several times. Ask students to speculate what might have prompted the name change. *(changes in government or rulers)* Tell students that in Section 3 they will learn more about St. Petersburg and other cities of Russia's heartland.

Building Vocabulary

Write the terms **light industry** and **heavy industry** on the chalkboard. Based on what students know about industry, ask them to speculate what the adjective before each of these terms does to the meaning of the term. *(Light and heavy generally reflect the weight of the product created by the industry.)* Have a student locate and read aloud a definition of **smelters**. Call on a volunteer to describe the relationship between smelters and heavy industry. *(Smelters process metal ores that are primarily used in heavy industry.)*

Section 3 — The Russian Heartland

Read to Discover

1. Why is European Russia considered the country's heartland?
2. What are the characteristics of the four regions of European Russia?

Define

light industry
heavy industry
smelters

Locate

St. Petersburg
Nizhniy Novgorod
Astrakhan
Yekaterinburg
Chelyabinsk
Magnitogorsk

WHY IT MATTERS

Following the fall of Communism, some Russian cities' landscapes began to change. In the larger cities like Moscow there are newer buildings and more restaurants. Use **CNNfyi.com** or other **current events** sources to find information about Moscow and other large Russian cities. Record your findings in your journal.

Jeweled box made by Peter Carl Fabergé

Section 3 — RESOURCES

Reproducible
◆ Guided Reading Strategy 22.3

Technology
◆ One-Stop Planner CD–ROM, Lesson 22.3
◆ Homework Practice Online
◆ *ARGWorld* CD–ROM: Land Values in Post-Soviet Moscow
◆ HRW Go site

Reinforcement, Review, and Assessment
◆ Section 3 Review, p. 495
◆ Daily Quiz 22.3
◆ Main Idea Activity 22.3
◆ English Audio Summary 22.3
◆ Spanish Audio Summary 22.3

The Heartland

The European section of Russia is the country's heartland. The Russian nation expanded outward from there. It is home to the bulk of the Russian population. The national capital and large industrial cities are also located there.

The plains of European Russia make up the country's most productive farming region. Farmers focus mainly on growing grains and raising livestock. Small gardens near cities provide fresh fruits and vegetables for summer markets.

The Russian heartland can be divided into four major regions. These four are the Moscow region, the St. Petersburg region, the Volga region, and the Urals region.

✓ **READING CHECK:** *Places and Regions* Why is European Russia the country's heartland? The bulk of the population, the national capital, and large industrial cities are there.

The Moscow Region

Moscow is Russia's capital and largest city. More than 9 million people live there. In addition to being Russia's political center, Moscow is the country's center for transportation and communication. Roads, railroads, and air routes link the capital to all points in Russia.

At Moscow's heart is the Kremlin. The Kremlin's red brick walls and towers were built in the late 1400s. The government offices, beautiful palaces, and gold-domed churches within its walls are popular tourist attractions.

Twenty towers, like the one in the lower left, are spaced along the Kremlin's walls.

Interpreting the Visual Record What was the advantage of locating government buildings and palaces within the walls of one central location?

▼

◄ Visual Record Answer

for protection from attack

Teaching Objectives 1–2

ALL LEVELS: (Suggested time: 15 min.) Copy the following graphic organizer onto the chalkboard, omitting the italicized answers. Call on students to fill in the chart with descriptive characteristics that make this region Russia's heartland. **ENGLISH LANGUAGE LEARNERS**

REGIONS OF THE RUSSIAN HEARTLAND			
Moscow	St. Petersburg	Volga	Ural
• *capital and largest city* • *huge industrial area*	• *second-largest city* • *major seaport* • *important centers of learning*	• *major shipping route* • *many factories and industries*	• *many mineral resources*

Teaching Objective 2

ALL LEVELS: (Suggested time: 30 min.) Organize the class into four groups, assigning each group a region of the Russian heartland. Have students create a "Welcome to ___" billboard to be placed at the region's border. Students should emphasize positive aspects of their region. Display the billboards around the classroom. **ENGLISH LANGUAGE LEARNERS, COOPERATIVE LEARNING**

Across the Curriculum

SCIENCE

The Pulkovo Observatory Among the many learning institutions of the St. Petersburg region is the Pulkovo Observatory. The observatory's 15-inch (38 cm) refracting telescope was the world's largest when it was built in 1839. Known for its quality of observations, the observatory doubled the size of its refracting telescope to 30 inches (76 cm) in 1878. Although it was destroyed during World War II, the Pulkovo Observatory was rebuilt in 1951.

Critical Thinking: Why might an observatory be destroyed during a war?

Answer: The enemy might have targeted all scientific sites in an effort to destroy technological advancements and to reduce potential surveillance.

Vendors sell religious art and other crafts at a sidewalk market in Moscow.
Interpreting the Visual Record What do the items in this market suggest about the status of religion in Russia since the communist era?

Moscow is part of a huge industrial area. This area also includes the city of Nizhniy Novgorod, called Gorky during the communist era. About one third of Russia's population lives in this region.

The Soviet government encouraged the development of **light industry**, rather than **heavy industry**, around Moscow. Light industry focuses on the production of lightweight goods, such as clothing. Heavy industry usually involves manufacturing based on metals. It causes more pollution than light industry. The region also has advanced-technology and electronics industries.

The St. Petersburg Region

Northwest of Moscow is St. Petersburg, Russia's second-largest city and a major Baltic seaport. More than 5 million people live there. St. Petersburg was Russia's capital and home to the czars for more than 200 years. This changed in 1918. Palaces and other grand buildings constructed under the czars are tourist attractions today. St. Petersburg was known as Leningrad during the communist era. Much of the city was heavily damaged during World War II.

The surrounding area has few natural resources. Still, St. Petersburg's harbor, canals, and rail connections make the city a major center for trade. Important universities and research institutions are located there. The region also has important industries.

✓ **READING CHECK:** *Human Systems*
Why are Moscow and St. Petersburg such large cities? centers of politics, transportation, communication, industry, trade, education

The Mariinsky Theater of Opera and Ballet is one of St. Petersburg's most beautiful buildings. It was called the Kirov during the communist era.

Visual Record Answer ▶

It is practiced freely.

CLOSE

Invite students to write their names in Russian style. To create a middle name, boys add -*ovich*, meaning "son of," and girls add -*ovna*, meaning "daughter of," to the father's first name. Girls also add -*a* to the last name. If the last name ends in -*sky* or -*ski*, girls use -*skaya* instead of the -*a*. So Yuri and Helena, the son and daughter of Viktor Maksimov, would be named Yuri Viktorovich Maksimov and Helena Viktorovna Maksimova.

REVIEW AND ASSESS

Have students complete the Section Review. Then pair students. Ask each student to write five fill-in-the-blank questions on the section material and to exchange papers with his or her partner. Then have students complete Daily Quiz 22.3. **COOPERATIVE LEARNING**

RETEACH

Have students complete Main Idea Activity 22.3. Then organize the class into four groups, one for each of the regions in the Russian heartland. Have students draw pictures that show characteristics of their region and place them on a wall map or sketch map of the region. Discuss the picture maps. **ENGLISH LANGUAGE LEARNERS, COOPERATIVE LEARNING**

EXTEND

Have interested students conduct research and write a report on the founding of St. Petersburg by Peter the Great. Ask them to include the difficulties created by the region's physical geography. **BLOCK SCHEDULING**

The Volga Region

The Volga region stretches along the middle part of the Volga River. The Volga is often more like a chain of lakes. It is a major shipping route for goods produced in the region. Hydroelectric power plants and nearby deposits of coal and oil are important sources of energy.

During World War II, many factories were moved to the Volga region. This was done to keep them safe from German invaders. Today the region is famous for its factories that produce goods such as motor vehicles, chemicals, and food products. Russian caviar comes from a fishery based at the old city of Astrakhan on the Caspian Sea.

The Urals Region

Mining has long been important in the Ural Mountains region. Nearly every important mineral except oil has been discovered there. Copper and iron **smelters** are still important. Smelters are factories that process copper, iron, and other metal ores.

Many large cities in the Urals started as commercial centers for mining districts. The Soviet government also moved factories to the region during World War II. Important cities include Yekaterinburg (yi-kah-ti-reem-BOOHRK) (formerly Sverdlovsk), Chelyabinsk (chel-YAH-buhnsk), and Magnitogorsk (muhg-nee-tuh-GAWRSK). Now these cities manufacture machinery and metal goods.

✓ **READING CHECK:** *Places and Regions* What industries are important in the Volga and Urals regions? automobile, chemical, food, mining, machinery, metal goods

A fisher gathers sturgeon in a small shipboard pool in the Volga region. The eggs for making caviar are taken from the female sturgeon. Then the fish is released back into the water.

Russia

Homework Practice Online
go.hrw.com
Keyword: SJ3 HP22

Section Review 3

Define and explain: light industry, heavy industry, smelters

Working with Sketch Maps On the map you created in Section 2, label St. Petersburg, Nizhniy Novgorod, Astrakhan, Yekaterinburg, Chelyabinsk, and Magnitogorsk. In the margin of your map, write a short caption explaining the significance of Moscow and St. Petersburg.

Reading for the Main Idea

1. (*Places and Regions*) Why might so many people settle in Russia's heartland?

2. (*Places and Regions*) Where did the Soviet government move factories during World War II?

Critical Thinking

3. **Drawing Inferences and Conclusions** Why do you think the Soviet government encouraged the development of light industry around Moscow?

4. **Finding the Main Idea** What role has the region's physical geography played in the development of European Russia's economy?

Organizing What You Know

5. **Contrasting** Use this graphic organizer to identify European Russia's four regions. Write one feature that makes each region different from the other three.

European Russia

Section Review 3

Answers

Define For definitions, see: light industry, p. 494; heavy industry, p. 494; smelters, p. 495

Working with Sketch Maps Maps will vary, but listed places should be labeled in their approximate locations. Moscow is Russia's capital and largest city. St. Petersburg is Russia's second-largest city and a major Baltic seaport.

Reading for the Main Idea
1. because it is home to the national capital and large industrial cities; contains the country's most productive farmland (NGS 4)
2. Volga and Urals (NGS 4)

Critical Thinking
3. because it causes less pollution
4. The region's natural resources allowed it to develop an industrial economy.

Organizing What You Know
5. Answers will vary but could include the following: Moscow—center of transportation and communication; St. Petersburg—major seaport; Volga—known for heavy-machine industries and giant factories; Urals—important mining area.

Objectives

1. Describe the human geography of Siberia.
2. Identify the economic features of the region.
3. Analyze how Lake Baikal has been threatened by pollution.

LET'S GET STARTED

Copy the following question onto the chalkboard: *Where are the coldest places in the world?* If students do not name Siberia, refer them to the "Our Amazing Planet" feature in Section 1. You may want to point out Verkhoyansk, in Siberia, where the temperature was recorded, on a more detailed wall map of Russia. Tell students that in Section 4 they will learn more about Siberia.

Building Vocabulary

Write **habitation fog** on the chalkboard. Ask students to infer the meaning of the term based on what they know about the words "habitat" or "inhabit" and "fog." *(Habitation fogs occur over cities—places people inhabit.)* To confirm the meaning, call on a volunteer to find and read the term's definition aloud.

Section 4 RESOURCES

Reproducible
◆ Guided Reading Strategy 22.4

Technology
◆ One-Stop Planner CD–ROM, Lesson 22.4
◆ Homework Practice Online
◆ HRW Go site

Reinforcement, Review, and Assessment
◆ Section 4 Review, p. 498
◆ Daily Quiz 22.4
◆ Main Idea Activity 22.4
◆ English Audio Summary 22.4
◆ Spanish Audio Summary 22.4

Visual Record Answer ▶

subarctic

Section 4

Siberia

Read to Discover

1. What is the human geography of Siberia like?
2. What are the economic features of the region?
3. How has Lake Baikal been threatened by pollution?

WHY IT MATTERS

It takes more than a week by train to cross Russia. The Trans-Siberian Railroad and the Baikal-Amur Mainline let travelers see more of the country. Use CNNfyi.com or other **current events** sources to learn more about these rail systems. Record your findings in your journal.

Define

habitation fog

Locate

Siberia
Trans-Siberian Railroad
Baikal-Amur Mainline
Kuznetsk Basin
Ob River
Yenisey River
Novosibirsk
Lake Baikal

Russian matryoshka *nesting doll*

A Sleeping Land

East of European Russia, across the Ural Mountains, is Siberia. Siberia is enormous. It covers more than 5 million square miles (12.95 million sq. km) of northern Asia. It extends all the way to the Pacific Ocean. That is nearly 1.5 times the area of the United States! To the north of Siberia is the Arctic Ocean. To the south are the Central Asian countries, Mongolia, and China.

Many people think of Siberia as simply a vast, frozen wasteland. In fact, in the Tatar language, *Siberia* means "Sleeping Land." In many ways, this image is accurate. Siberian winters are long, dark, and severe. Often there is little snow, but the land is frozen for months. During winter, **habitation fog** hangs over cities. A habitation fog is a fog caused by fumes and smoke from cities. During the cold Siberian winter, this fog is trapped over cities.

Siberia has lured Russian adventurers for more than 400 years. It continues to do so today. This vast region has a great wealth of natural resources. Developing those resources may be a key to transforming Russia into an economic success.

Reindeer graze around a winter camp in northern Siberia.

Interpreting the Visual Record **What climate does this area appear to have, tundra or subarctic?**

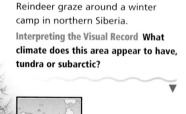

Teaching Objectives 1–2

ALL LEVELS: (Suggested time: 20 min.) Copy the following graphic organizer onto the chalkboard, omitting the italicized answers. Have students list human geography and economic reasons for or against living in Siberia. **ENGLISH LANGAGE LEARNERS**

Living in Siberia	
Reasons for:	Reasons against:
great wealth and natural resources, Trans-Siberian Railroad and the BAM, higher wages	*severe climate, sparsely populated, little industry*

Teaching Objective 3

ALL LEVELS: (Suggested time: 30 min.) Pair students and have them write contracts for new commercial development at or near Lake Baikal. Contracts should start with a review of Lake Baikal's special qualities, include a review of past problems, and conclude with requirements for operating a business that will not harm the lake or its plants and animals. **ENGLISH LANGUAGE LEARNERS, COOPERATIVE LEARNING**

People Siberia is sparsely populated. In fact, large areas have no human population at all. Most of the people live in cities in western and southern parts of the region.

Ethnic Russians make up most of the population. However, minority groups have lived there since long before Russians began to expand into Siberia.

Settlements Russian settlement in Siberia generally follows the route of the Trans-Siberian Railroad. Construction of this railway started in 1891. When it was completed, it linked Moscow and Vladivostok, a port on the Sea of Japan.

Russia's Trans-Siberian Railroad is the longest single rail line in the world. It is more than 5,700 miles (9,171 km) long. For many Siberian towns, the railroad provides the only transportation link to the outside world. Another important railway is the Baikal-Amur Mainline (BAM), which crosses many mountain ranges and rivers in eastern Siberia.

✓ **READING CHECK:** *Places and Regions* Where is Russian settlement located in Siberia, and why do you think this is the case? *along the Trans-Siberian Railroad; because it provides a means of transportation and communication in this vast land*

Siberia's Economy

The Soviet government built the Baikal-Amur Mainline so that raw materials from Siberia could be easily transported to other places. Abundant natural resources form the foundation of Siberia's economy. They are also important to the development of Russia's struggling economy. Siberia's natural resources include timber, mineral ores, diamonds, and coal, oil, and natural gas deposits.

Although Siberia has rich natural resources, it contains a small percentage of Russia's industry. The harsh climate and difficult terrain have discouraged settlement. Many people would rather live in European Russia, even though wages may be higher in Siberia.

Lumbering and mining are the most important Siberian industries. Large coal deposits are mined in the Kuznetsk Basin, or the Kuzbas. The Kuzbas is located in southwestern Siberia between the Ob and Yenisey Rivers. It is one of Siberia's most important industrial regions.

Siberia's largest city, Novosibirsk, is located near the Kuznetsk Basin. The city's name means "New Siberia." About 1.5 million people live there. It is located about halfway between Moscow and Vladivostok on the Trans-Siberian Railroad. Novosibirsk is Siberia's manufacturing and transportation center.

✓ **READING CHECK:** *Environment and Society* How do Siberia's natural resources influence the economies of Siberia and Russia? *They are the foundation of Siberia's economy and could help develop Russia's economy.*

▲ The Omsk (AWMSK) Cathedral in Omsk, Siberia, provides an example of Russian architecture. Omsk was founded in the early 1700s.

▲ A worker repairs an oil rig in Siberia.

Interpreting the Visual Record How is this worker protected from the cold Siberian climate?

The *Rossiya* train takes 153 hours and 49 minutes to travel from Moscow to Vladivostok on the Trans-Siberian Railroad. The famous railroad required 12 years and more than 70,000 workers to complete. It opened Siberia to settlement and provided access to the region's vast natural resources for development.

The number of Russians using the Trans-Siberian Railroad declined about 50 percent between 1991 and 1997. Even so, trains carry half of all passenger traffic in Russia, compared to less than 1 percent in the United States.

Critical Thinking: How did the Trans-Siberian Railroad make economic development of the region possible?

Answer: It opened the region to settlement and gave access to the natural resources there.

internet connect

GO TO: go.hrw.com
KEYWORD: SJ3 CH22
FOR: Web sites about the Trans-Siberian Railroad

◄ **Visual Record Answer**

with heavy clothing

CLOSE

Remind students that Siberia alone is 1.5 times the size of the United States. Lead a discussion on the various problems Russia faces governing such a large area.

REVIEW AND ASSESS

Have students complete the Section Review. Then have each student create a web diagram with "Siberia" in the center. When they have finished, call on volunteers to reproduce their webs on the chalkboard. Then have students complete Daily Quiz 22.4.

RETEACH

Have students complete Main Idea Activity 22.4. Then organize the class into small groups and have groups create a script for an educational television special on Siberia. The script should cover the physical, human, economic, and environmental geography of the region. Have students add pictures and other visual aids if possible. **ENGLISH LANGUAGE LEARNERS, COOPERATIVE LEARNING**

EXTEND

Have interested students conduct research on the rivers of Siberia, the problems caused by their freezing and thawing, and what life is like on board the rivers' freight barges. They may want to compile their findings into a job description and résumé for a barge worker. **BLOCK SCHEDULING**

Section Review 4

Answers

Define For definition, see: habitation fog, p. 496

Working with Sketch Maps Maps will vary, but listed places should be labeled in their approximate locations.

Reading for the Main Idea

1. west—Ural Mountains; east—Pacific Ocean; south—Central Asian republics, Mongolia, and China; north—Arctic Ocean (NGS 4)

2. in cities in the western and southern parts along the Trans-Siberian Railroad; because it is a means of transportation and communication (NGS 9)

3. The harsh climate and difficult terrain have discouraged settlement there. (NGS 9)

Critical Thinking

4. Answers will vary, but students should include reasons to justify their responses.

Organizing What You Know

5. natural resources—timber, mineral ores, diamonds, coal, oil, and natural gas; major industries—lumbering and mining

Visual Record Answer ▶

cause health problems for the plants and animals

498

The scenery around Lake Baikal is breathtaking. The lake is seven times as deep as the Grand Canyon.

Interpreting the Visual Record
How would pollution affect this lake and the plants and animals that live there?

Our Amazing Planet

Lake Baikal covers less area than do three of the Great Lakes: Superior, Huron, and Michigan. Still, Baikal is so deep that it contains about one fifth of all the world's freshwater!

Lake Baikal

Some people have worried that economic development in Siberia threatens the region's natural environment. One focus of concern has been Lake Baikal (by-KAHL), the "Jewel of Siberia."

Baikal is located north of Mongolia. It is the world's deepest lake. In fact, it holds as much water as all of North America's Great Lakes. The scenic lake and its surrounding area are home to many kinds of plants and animals. Some, such as the world's only freshwater seal, are endangered.

For decades people have worried about pollution from a nearby paper factory and other development. They feared that pollution threatened the species that live in and around the lake. In recent years scientists and others have proposed plans that allow some economic development while protecting the environment.

✓ **READING CHECK:** *Environment and Society* How has human activity affected Lake Baikal? The paper factory and other developments have posed an environmental threat.

Homework Practice Online
Keyword: SJ3 HP22

Section Review 4

Define and explain: habitation fog

Working with Sketch Maps On the map you created in Section 3, label Siberia, Trans-Siberian Railroad, Baikal-Amur Mainline, Kuznetsk Basin, Ob River, Yenisey River, Novosibirsk, and Lake Baikal.

Reading for the Main Idea

1. *Places and Regions* What are the boundaries of Siberia?

2. *Human Systems* Where do most people in Siberia live? Why?

3. *Places and Regions* Why does this huge region with many natural resources have little industry?

Critical Thinking

4. **Making Generalizations and Predictions** Do you think Russians should be more concerned about rapid economic development or protecting the environment? Why?

Organizing What You Know

5. **Categorizing** Use this organizer to list the region's resources and industries that use them.

Natural Resources	⇨	Major Industries

Objectives

1. Explain the relationship between climate and agriculture in the Russian Far East.
2. List the major resources and cities of the Russian Far East.
3. Identify the islands of the Russian Far East.

LET'S GET STARTED

Ask students to predict what the economy of the Russian Far East might be like based solely on its location. *(Students might suggest that shipping, fishing, and lumbering are important there.)* Tell students that in Section 5 they will learn more about the economy of the Russian Far East.

Building Vocabulary

Write **icebreakers** on the chalkboard. Tell students that when a group of people gets together for the first time, such as in a classroom or at a club meeting, the leader of the group often starts with an activity called an icebreaker. Ask students if they have ever been part of such an activity and to describe its purpose. *(Possible answer: to break up the cold social atmosphere so that people can better communicate and work together)* Ask students to relate that description to the ships called icebreakers.

Section 5 — The Russian Far East

Read to Discover

1. How does the Russian Far East's climate affect agriculture in the region?
2. What are the major resources and cities of the region?
3. What island regions are part of the Russian Far East?

Define

icebreakers

Locate

Sea of Okhotsk
Sea of Japan
Amur River
Khabarovsk
Vladivostok
Sakhalin Island

A Russian figurine

WHY IT MATTERS

Because of conflict over the Kuril Islands, Russia and Japan did not sign a peace agreement to end World War II. Use **CNNfyi.com** or other **current events** sources to find information on this controversy and other political concerns. Record your findings in your journal.

Reproducible
◆ Guided Reading Strategy 22.5

Technology
◆ One-Stop Planner CD–ROM, Lesson 22.5
◆ Homework Practice Online
◆ HRW Go site

Reinforcement, Review, and Assessment
◆ Section 5 Review, p. 501
◆ Daily Quiz 22.5
◆ Main Idea Activity 22.5
◆ English Audio Summary 22.5
◆ Spanish Audio Summary 22.5

Agriculture

Off the eastern coast of Siberia are the Sea of Okhotsk and the Sea of Japan. Their coastal areas and islands make up a region known as the Russian Far East.

The Russian Far East has a less severe climate than the rest of Siberia. Summer weather is mild enough for some successful farming. Farms produce many goods, including wheat, sugar beets, sunflowers, meat, and dairy products. However, the region cannot produce enough food for itself. As a result, food must also be imported.

Fishing and hunting are important in the region. There are many kinds of animals, including deer, seals, rare Siberian tigers, and sables. Sable fur is used to make expensive clothing.

✔ **READING CHECK:** *Environment and Society* How does scarcity of food affect the Russian Far East? It forces people in the region to import food.

◀

The Siberian tiger is endangered. The few remaining of these large cats roam parts of the Russian Far East. They are also found in northern China and on the Korean Peninsula.

Teaching Objective 1

ALL LEVELS: (Suggested time: 30 min.) Have students draw picture postcards detailing the relationship between agriculture and climate in the Russian Far East. **ENGLISH LANGUAGE LEARNERS**

Teaching Objectives 2–3

ALL LEVELS: (Suggested time: 15 min.) Copy the following graphic organizer onto the chalkboard, omitting the italicized answers. Call on volunteers to come fill in the diagram. **ENGLISH LANGUAGE LEARNERS**

Resources — *forests, coal, oil, geothermal energy* — **RUSSIAN FAR EAST** — Cities — *Khabarovsk and Vladivostok*

Islands — *Sakhalin Island and Kuril Islands*

Section Review 5

Answers

Define For definition, see: icebreakers, p. 500

Working with Sketch Maps Maps will vary, but listed places should be labeled in their approximate locations. Russia and Japan dispute possession of Sakhalin Island and the Kuril Islands.

Reading for the Main Idea

1. far less severe (NGS 4)

2. wheat, sugar beets, sunflowers, meat, dairy products; coal, oil, and geothermal energy (NGS 4)

Critical Thinking

3. It remains a major naval base and home port for a large fishing fleet.

4. The natural resources around these islands make them attractive to both countries.

Organizing What You Know

5. Answers will vary but could include the following: Khabarovsk—its location where the Trans-Siberian Railroad crosses the Amur River made it ideal for processing forest and mineral resources; Vladivostok—its location on the Sea of Japan made it ideal as a naval base and fishing port.

BUILD on WHAT You Know

Do you remember what you learned about plate tectonics? See Chapter 2 to review.

Historical monuments and old architecture compete for attention in Vladivostok.

Economy

Like the rest of Siberia, the Russian Far East has a wealth of natural resources. These resources have supported the growth of industrial cities and ports in the region.

Resources Much of the Russian Far East remains forested. The region's minerals are only beginning to be developed. Lumbering, machine manufacturing, woodworking, and metalworking are the major industries there.

The region also has important energy resources, including coal and oil. Another resource is geothermal energy. This resource is available because of the region's tectonic activity. Two active volcanic mountain ranges run the length of the Kamchatka Peninsula. Russia's first geothermal electric-power station was built on this peninsula.

Cities Industry and the Trans-Siberian Railroad aided the growth of cities in the Russian Far East. Two of those cities are Khabarovsk (kuh-BAHR-uhfsk) and Vladivostok (vla-duh-vuh-STAHK).

More than 600,000 people live in Khabarovsk, which was founded in 1858. It is located where the Trans-Siberian Railroad crosses the Amur River. This location makes Khabarovsk ideal for processing forest and mineral resources from the region.

Vladivostok is slightly larger than Khabarovsk. *Vladivostok* means "Lord of the East" in Russian. The city was established in 1860 on the coast of the Sea of Japan. Today it lies at the eastern end of the Trans-Siberian Railroad.

Vladivostok is a major naval base and the home port for a large fishing fleet. **Icebreakers** must keep the city's harbor open in winter. An icebreaker is a ship that can break up the ice of frozen waterways. This allows other ships to pass through them.

Ask students why residents of the Russian Far East might feel closer to Japan and other Pacific nations than to European Russia. *(Possible answers: physically closer to Japan, trade relationships with Pacific countries, isolation created by Siberia)* Ask students to predict if this isolation will increase or decrease. *(Students may predict that new communication technology will help decrease isolation.)*

Have students complete Main Idea Activity 22.5. Then, organize the class into small groups, assigning resources, cities, agriculture, and islands to different groups. Ask each group to create a visual aid to illustrate the important facts about its topic. **ENGLISH LANGUAGE LEARNERS, COOPERATIVE LEARNING**

Have students complete the Section Review. Then have each student write two true-false questions about agriculture, resources, cities, and islands in the Russian Far East. Pair students and have them quiz each other. Then have students complete Daily Quiz 22.5. **COOPERATIVE LEARNING**

Have interested students conduct research on one of the non-Russian ethnic groups of the Russian Far East. Have them include information on the group's common occupations, history, language, and customs. **BLOCK SCHEDULING**

The Soviet Union considered Vladivostok very important for defense. The city was therefore closed to foreign contacts until the early 1990s. Today it is an important link with China, Japan, the United States, and the rest of the Pacific region.

✓ **READING CHECK:** *Environment and Society* How do the natural resources of the Russian Far East affect its economy? They have supported mining and timber industries, as well as the growth of industrial cities and ports.

Islands

The Russian Far East includes two island areas. Sakhalin is a large island that lies off the eastern coast of Siberia. To the south is the Japanese island of Hokkaido. The Kuril Islands are much smaller. They stretch in an arc from Hokkaido to the Kamchatka Peninsula.

Sakhalin has oil and mineral resources. The waters around the Kurils are important for commercial fishing.

Russia and Japan have argued over who owns these islands since the 1850s. At times they have been divided between Japan and Russia or the Soviet Union. The Soviet Union took control of the islands after World War II. Japan still claims rights to the southernmost islands.

Like other Pacific regions, Sakhalin and the Kurils sometimes experience earthquakes and volcanic eruptions. An earthquake in 1995 caused severe damage on Sakhalin Island, killing nearly 2,000 people.

✓ **READING CHECK:** *Environment and Society* How does the environment of the Kuril Islands and Sakhalin affect people? Earthquakes, such as the one in 1995, and volcanoes pose a constant threat to lives and property.

An old volcano created Crater Bay in the Kuril Islands. The great beauty of the islands is matched by the terrible power of earthquakes and volcanic eruptions in the area.

Interpreting the Visual Record What do you think happened to the volcano that formed Crater Bay?

Section Review 5

Define and explain: icebreakers

Working with Sketch Maps On the map you created in Section 4, label the Sea of Okhotsk, the Sea of Japan, the Amur River, Khabarovsk, Vladivostok, and Sakhalin Island. In the margin, explain which countries dispute possession of Sakhalin Island and the Kuril Islands.

Reading for the Main Idea

1. *Places and Regions* How does the climate of the Russian Far East compare to the climate throughout the rest of Siberia?
2. *Places and Regions* What are the region's major crops and energy resources?

go.hrw.com **Homework Practice Online** Keyword: SJ3 HP22

Critical Thinking

3. **Drawing Inferences and Conclusions** In what ways do you think Vladivostok is "Lord of the East" in Russia today?
4. **Drawing Inferences and Conclusions** Why do you think Sakhalin and the Kuril Islands have been the subject of dispute between Russia and Japan?

Organizing What You Know

5. **Finding the Main Idea** Copy the following graphic organizer. Use it to explain how the location of each city has played a role in its development.

Khabarovsk	Vladivostok

ANSWERS

Building Vocabulary

For definitions, see: taiga, p. 485; steppe, p. 485; czar, p. 488, abdicated, p. 488; allies, p. 489, superpowers, p. 489; Cold War, p. 489; consumer goods, p. 489; light industry, p. 494; heavy industry, p. 494; smelters, p. 495; habitation fog, p. 496; icebreakers, p. 500

Reviewing the Main Ideas

1. Northern European Plain, Caucasus Mountains, Ural Mountains, Siberia, the Kamchatka Peninsula, Sakhalin Island, and the Kuril Islands; in the Kuznetsk Basin (NGS 4)
2. the Ural Mountains (NGS 4)
3. It has an elected executive and a legislative branch, like the United States. (NGS 12)
4. Moscow, St. Petersburg, Volga, and the Urals (NGS 4)
5. to the Volga and Urals regions; because of the threat of German invasion (NGS 4)

◄ **Visual Record Answer**

It became extinct and was filled with water.

501

Have students complete the Chapter 22 Test.

RETEACH

Organize the class into five groups and assign each group a different section from the chapter. Tell the students that each group will create one act for a play titled *Russia—Then, Now, and Forever.* Students should create one or more scenes for their act. The scene can relate to any topic, subject, or plot; however, students should weave facts from their section into the script. Have students perform their plays for the class.
ENGLISH LANGUAGE LEARNERS, COOPERATIVE LEARNING

PORTFOLIO EXTENSIONS

1. Have students create a "family tree" of important rulers and leaders of Russia and the Soviet Union. Some of the most significant are Rurik, Ivan I, Peter the Great, Catherine the Great, Nicholas II, Vladimir Lenin, Joseph Stalin, Nikita Khrushchev, Mikhail Gorbachev, and Boris Yeltsin. Students should use biographies to research each person then summarize how he or she affected Russian or Soviet history and culture.

2. Have students locate a copy of the Cyrillic alphabet. They should create a chart that compares the Cyrillic alphabet to the Latin alphabet. Ask students to report on how some Russian citizens of other ethnic groups are now using the Latin alphabet instead of the Cyrillic.

Review
ANSWERS

Understanding Environment and Society

Students' presentations should indicate an understanding of the Russian steppe as well as the agricultural resources of that region. Use Rubric 29, Presentations, to evaluate student work.

Thinking Critically

1. because it has natural resources for factories and industries

2. command; government-, not individual-, owned industries

3. because the size of the country and movement of goods and resources; built two transcontinental railroads

4. provides a trading port for Siberia and the Russian Far East

5. food shortages, economic problems, defeat in war, and lack of consumer goods; collapse of the Russian Empire and the Soviet Union (NGS 17)

Map Activity

A. Vladivostok

B. Kamchatka Peninsula

C. Moscow

D. Arctic Ocean

Reviewing What You Know

Building Vocabulary

On a separate sheet of paper, write sentences to define each of the following words.

1. taiga
2. steppe
3. czar
4. abdicated
5. allies
6. superpower
7. Cold War
8. consumer goods
9. light industry
10. heavy industry
11. smelters
12. habitation fog
13. icebreakers

Reviewing the Main Ideas

1. (Places and Regions) What are the major physical features of Russia? Where in Siberia are large coal deposits?

2. (Places and Regions) What landform separates Europe from Asia?

3. (Human Systems) How is Russia's government organized, and how does it compare with that of the United States?

4. (Places and Regions) What four major regions make up European Russia?

5. (Places and Regions) Where were Russian factories relocated during World War II and why?

Understanding Environment and Society

Resource Use

The grasslands of the steppe are one of Russia's most valuable agricultural resources. Create a presentation on farming in the steppe. You may want to consider the following:

• The crops that are grown in the Russian steppe.

• The kinds of livestock raised in the region.

• How the climate limits agriculture in the steppe.

Thinking Critically

1. **Finding the Main Idea** In what ways might Siberia be important to making Russia an economic success?

2. **Contrasting** What kind of economic system did the Soviet Union have, and how did it differ from that of the United States?

3. **Drawing Inferences and Conclusions** Why is transportation an issue for Russia? What have

Russians done to ease transportation between European Russia and the Russian Far East?

4. **Analyzing Information** How does Vladivostok's location make it an important link between Russia and the Pacific world?

5. **Identifying Cause and Effect** What problems existed in the Russian Empire and the Soviet Union in the 1900s, and what was their effect?

FOOD FESTIVAL

Here is a basic recipe for borscht, a traditional Russian soup. In a large pot, sauté a chopped onion in 2 tbsp. butter. Stir in 1½ lb. sliced raw red beets, ¼ c. red wine vinegar, 1 tsp. sugar, 2 chopped fresh tomatoes, 1 tsp. salt, and some black pepper. Pour in ½ c. beef stock, cover, and simmer one hour. Pour in 5 c. more of beef stock and ½ lb. shredded cabbage. Bring to a boil. Add ¼ lb. cubed ham, 1 lb. cooked sliced beef, ½ c. chopped parsley, and a bay leaf. Simmer for 30 minutes. Garnish with sour cream.

CHAPTER 22 REVIEW AND ASSESSMENT RESOURCES

Reproducible
- Readings in World Geography, History, and Culture 41, 42, and 43
- Critical Thinking Activity 22
- Vocabulary Activity 22

Technology
- Chapter 22 Test Generator (on the One-Stop Planner)

- Audio CD Program, Chapter 22
- HRW Go site

Reinforcement, Review, and Assessment
- Chapter 22 Review pp. 502–03

- Chapter 22 Tutorial for Students, Parents, Mentors, and Peers
- Chapter 22 Test
- Chapter 22 Test for English Language Learners and Special-Needs Students

Building Social Studies Skills

Map ACTIVITY

On a separate sheet of paper, match the letters on the map with their correct labels.

Arctic Ocean
Caucasus Mountains
Caspian Sea
West Siberian Plain
Central Siberian Plateau

Kamchatka Peninsula
Volga River
Moscow
St. Petersburg
Vladivostok

E. West Siberian Plain
F. Volga River
G. Caucasus Mountains
H. Central Siberian Plateau
I. Caspian Sea
J. St. Petersburg

Mental Mapping Skills Activity
Maps will vary, but listed places should be labeled in their approximate locations.

Writing Activity
Students' descriptions should include the various physical features along the route and an accurate estimation of the distance of the trip. Use Rubric 40, Writing to Describe, to evaluate student work.

Portfolio Activity
Students should develop an accurate list of organizations and projects for their community. Ask students who volunteered to share their experiences with the class. Use Rubric 30, Research, to evaluate student work.

Mental Mapping Skills ACTIVITY

On a separate sheet of paper, draw a freehand map of Russia and label the following:

Baltic Sea
Kuril Islands
Lake Baikal

Sakhalin Island
Siberia
Ural Mountains

WRITING ACTIVITY

Imagine that you are a tour guide on a trip by train from St. Petersburg to Vladivostok. Use the chapter map or a classroom globe to write a one-page description of some of the places people would see along the train's route. How far would you travel? Be sure to use standard grammar, spelling, sentence structure, and punctuation.

Alternative Assessment

Portfolio ACTIVITY

Learning About Your Local Geography Youth Organizations The Baikal-Amur Mainline (BAM) was built partly by youth organizations. Make a list of some projects of youth organizations in your community.

internet connect

Internet Activity: go.hrw.com
KEYWORD: SJ3 GT22

Choose a topic to explore about Russia:
- Take a trip on the Trans-Siberian Railroad.
- Examine the breakup of the Soviet Union.
- View the cultural treasures of Russia.

internet connect

GO TO: go.hrw.com
KEYWORD: SJ3 Teacher
FOR: a guide to using the Internet in your classroom

Russia • 503

503

Ukraine, Belarus, and the Caucasus

Chapter Resource Manager

Objectives	Pacing Guide	Reproducible Resources
SECTION 1		
Physical Geography (pp. 505–07)	**Regular** .5 day	**RS** Guided Reading Strategy 23.1
1. Identify the region's major physical features.	**Block Scheduling** .5 day	**E** Creative Strategies for Teaching World Geography, Lesson 13
2. Identify the climate types and natural resources found in the region.	*Block Scheduling Handbook, Chapter 23*	
SECTION 2		
Ukraine and Belarus (pp. 508–12)	**Regular** 1 day	**RS** Guided Reading Strategy 23.2
1. Identify the groups that have influenced the history of the Ukraine and Belarus.	**Block Scheduling** .5 day	**RS** Graphic Organizer 23
2. Identify some important economic features and environmental concerns of Ukraine.	*Block Scheduling Handbook, Chapter 23*	**E** Cultures of the World Activity 4
3. Describe how the economy of Belarus has developed.		
SECTION 3		
The Caucasus (pp. 513–15)	**Regular** 1 day	**RS** Guided Reading Strategy 23.3
1. Identify the groups that have influenced the early history and culture of the Caucasus.	**Block Scheduling** .5 day	**E** Cultures of the World Activity 4
2. Describe the economy of Georgia.	*Block Scheduling Handbook, Chapter 23*	**SM** Geography for Life Activity 23
3. Describe what Armenia is like today.		**SM** Map Activity 23
4. Describe what Azerbaijan is like today.		

Chapter Resource Key

RS Reading Support

IC Interdisciplinary Connections

E Enrichment

SM Skills Mastery

A Assessment

REV Review

ELL Reinforcement and English Language Learners

 Transparencies

 CD–ROM

 Music

 Video

 Internet

 Holt Presentation Maker Using Microsoft® Powerpoint®

One-Stop Planner CD–ROM

See the *One-Stop Planner* for a complete list of additional resources for students and teachers.

One-Stop Planner CD–ROM

It's easy to plan lessons, select resources, and print out materials for your students when you use the *One-Stop Planner CD–ROM with Test Generator.*

internet connect

HRW ONLINE RESOURCES

<u>GO TO: go.hrw.com</u>
Then type in a keyword.

TEACHER HOME PAGE
KEYWORD: SJ3 TEACHER

CHAPTER INTERNET ACTIVITIES
KEYWORD: SJ3 GT23

Choose an activity to:
• trek through the Caucasus Mountains.
• design Ukrainian Easter eggs.
• investigate the Chernobyl disaster.

CHAPTER ENRICHMENT LINKS
KEYWORD: SJ3 CH23

CHAPTER MAPS
KEYWORD: SJ3 MAPS23

ONLINE ASSESSMENT
Homework Practice
KEYWORD: SJ3 HP23
Standardized Test Prep Online
KEYWORD: SJ3 STP23
Rubrics
KEYWORD: SS Rubrics

COUNTRY INFORMATION
KEYWORD: SJ3 Almanac

CONTENT UPDATES
KEYWORD: SS Content Updates

HOLT PRESENTATION MAKER
KEYWORD: SJ3 PPT23

ONLINE READING SUPPORT
KEYWORD: SS Strategies

CURRENT EVENTS
KEYWORD: S3 Current Events

Technology Resources

 One-Stop Planner CD–ROM, Lesson 23.1

 Geography and Cultures Visual Resources with Teaching Activities 31–35

 Homework Practice Online

HRW Go site

Review, Reinforcement, and Assessment Resources

ELL Main Idea Activity 23.1
REV Section 1 Review, p. 507
A Daily Quiz 23.1
ELL English Audio Summary 23.1
ELL Spanish Audio Summary 23.1

 One-Stop Planner CD–ROM, Lesson 23.2

Homework Practice Online

HRW Go site

ELL Main Idea Activity 23.2
REV Section 2 Review, p. 512
A Daily Quiz 23.2
ELL English Audio Summary 23.2
ELL Spanish Audio Summary 23.2

 One-Stop Planner CD–ROM, Lesson 23.3

Homework Practice Online

HRW Go site

ELL Main Idea Activity 23.3
REV Section 3 Review, p. 515
A Daily Quiz 23.3
ELL English Audio Summary 23.3
ELL Spanish Audio Summary 23.3

Meeting Individual Needs

Ability Levels

Level 1 Basic-level activities designed for all students encountering new material

Level 2 Intermediate-level activities designed for average students

Level 3 Challenging activities designed for honors and gifted-and-talented students

English Language Learners Activities that address the needs of students with Limited English Proficiency

Chapter Review and Assessment

E Readings in World Geography, History, and Culture 43 and 44
SM Critical Thinking Activity 23
REV Chapter 23 Review, pp. 384–85
REV Chapter 23 Tutorial for Students, Parents, Mentors, and Peers
ELL Vocabulary Activity 23
A Chapter 23 Test
 Chapter 23 Test Generator (on the One-Stop Planner)
 Audio CD Program, Chapter 23
A Chapter 23 Test for English Language Learners and Special-Needs Students

CHAPTER 23

Read the following quotation from Russian writer Nikolay Gogol's *Taras Bulba*: "*The farther they penetrated the steppe, the more beautiful it became. . . . Nothing in nature could be finer. The whole surface resembled a golden-green ocean, upon which were sprinkled millions of different flowers. . . . Oh, steppes, how beautiful you are!*" Ask what resources the region described might have. *(arable land)* Tell students that in this chapter they will learn more about the fertile area described, which extends into the countries of Belarus and Ukraine. The countries of the Caucasus Mountains, which have a very different landscape, are also covered in this chapter.

Section 1

Objectives

1. Identify the region's major physical features.
2. Describe the climate types and natural resources found in the region.

LINKS TO OUR LIVES

You might point out to students that there are many reasons why we should know more about Ukraine, Belarus, and the three nations of the Caucasus, these among them:

▶ The nations in this region were part of the Soviet Union. Now they are struggling to adopt freer economic and political practices. Americans are now free to travel to these countries and to invest in them.

▶ These countries possess many natural resources, including deposits of oil, natural gas, metals, and minerals. The United States may wish to establish stronger trade relationships with these countries.

▶ Ukraine's government has the task of ensuring that the damaged Chernobyl nuclear reactor is adequately contained. Inadequate safeguards that result from the country's economic hardships could have far-reaching effects.

CHAPTER 23

Ukraine, Belarus, and the Caucasus

This region consists of plains in the north and mountains in the south. Both of these physical features made this area important to ancient invaders. Before you learn the history of this region, you should meet Ana.

Hi! I am a senior in high school in the city of Tbilisi, Georgia. I live with my parents and my younger sister. I go to school from 9:00 A.M. to 2:00 P.M. and study foreign languages—English and Spanish. I hope to be a journalist. In school the teachers decide which classes everyone must take.

After school I do my homework as fast as possible and then get together with my friends. I come home in the early evening and listen to music, read, or watch television.

We also have great food. My favorite dish is baked chicken with nuts. If you came to Georgia, I would take you to the mountains, to the seaside, and to some hot springs.

We might also go to a festival where you could see Georgians in the country's national dress. Women wear a long red or purple robe with a white head scarf. Men wear a black suit or robe with gold embroidery.

Привіт! Я Анна.

▲
Translation: Hi! I am Ana.

LET'S GET STARTED

Copy the following passage and instructions onto the chalkboard: *Ukraine was considered the breadbasket of the Soviet Union, while Georgia was a popular vacation spot. Name a pair of states in the United States that could fit the same description.* Discuss student responses. *(Possible answers: Nebraska/Florida; Iowa/California; Kansas/Hawaii)* Tell students that in Section 1 they will learn more about the physical geography of Ukraine, Belarus, and the Caucasus nations.

Using the Physical-Political Map

Have students examine the map on this page. Call on individual students to locate the Dnieper River, the Donets River, and the Black Sea. Point out why navigable waterways are important in agricultural and industrial regions such as Ukraine *(for shipping agricultural products, manufactured goods, and mineral resources to markets or processing plants).*

Building Vocabulary

Write the Define term on the chalkboard. Have a student read aloud a dictionary definition of a **reserve**. *(The definition should pertain to the idea of preserving something.)* Ask volunteers to speculate how **nature reserves** might serve the public interest. You may want to ask students to describe nature reserves they have visited or have seen on television programs.

Section 1 — Physical Geography

Read to Discover

1. What are the region's major physical features?
2. What climate types and natural resources are found in the region?

Define

nature reserves

Locate

Black Sea
Caucasus Mountains

Caspian Sea
Pripyat Marshes
Carpathian Mountains
Crimean Peninsula
Sea of Azov

Mount Elbrus
Dnieper River
Donets Basin

WHY IT MATTERS

Ukraine is trying to create a nature reserve to protect its natural environment. Use **CNNfyi.com** or other **current events** sources to find information about how other countries are trying to protect their environments. Record your findings in your journal.

A gold pig from Kiev

Section 1 — RESOURCES

Reproducible
- Block Scheduling Handbook, Chapter 23
- Guided Reading Strategy 23.1
- Creative Strategies for Teaching World Geography, Lesson 13

Technology
- One-Stop Planner CD–ROM, Lesson 23.1
- Homework Practice Online
- Geography and Cultures Visual Resources with Teaching Activities 31–35
- HRW Go site

Reinforcement, Review, and Assessment
- Section 1 Review, p. 507
- Daily Quiz 23.1
- Main Idea Activity 23.1
- English Audio Summary 23.1
- Spanish Audio Summary 23.1

Ukraine, Belarus, and the Caucasus: Physical-Political

SCALE

0 — 200 — 400 Miles

0 — 200 — 400 Kilometers

Projection: Two-Point Equidistant

NORTHERN EUROPEAN PLAIN

EUROPE

CARPATHIAN MOUNTAINS

Vitsyebsk

BELARUS

Minsk

Pripyat (Pinsk) Marshes

Homyel'

Chernobyl

Kiev

UKRAINE

Kharkiv

Dnipropetrovs'k

Kryvyy Rih

Donets'k

DONETS BASIN

Odessa

CRIMEAN PENINSULA

Sevastopol'

Yalta

SEA OF AZOV

RUSSIA

BLACK SEA

Mount Elbrus 18,510 ft. (5642 m)

CAUCASUS MOUNTAINS

GEORGIA

T'bilisi

ARMENIA

Yerevan

AZERBAIJAN

Baku

AZERBAIJAN

Naxçivan

CASPIAN SEA

AEGEAN SEA

Size comparison of Ukraine, Belarus, and the Caucasus to the contiguous United States

ELEVATION

FEET		METERS
13,120		4,000
6,560		2,000
1,640		500
656		200
(Sea level) 0		0 (Sea level)
Below sea level		Below sea level

⊛ National capitals
• Other cities

Teaching Objective 1

ALL LEVELS: (Suggested time: 30 min.) Tell students to imagine that they are tourists visiting Ukraine, Belarus, and the Caucasus countries. Then have students write a postcard or letter that describes the region's physical features to a friend back home. Ask volunteers to read their postcards or letters to the class. **ENGLISH LANGUAGE LEARNERS**

Teaching Objective 2

ALL LEVELS: (Suggested time: 20 min.) Copy the following graphic organizer onto the chalkboard, omitting the italicized answers. Have students complete the organizer, placing the resource that all three have in common in the center. Then pair students and have each pair draw a map that depicts the region's various climates. Have students create keys for their maps. **ENGLISH LANGUAGE LEARNERS, COOPERATIVE LEARNING**

NATURAL RESOURCES

coal, iron — Ukraine

(no specific resources listed) — Belarus

farmland

oil, gas, copper, manganese, iron, other metals — Caucasus Countries

---

Linking Past to Present

Coal Mining in the Donets Basin Note in the section on resources that the Donets Basin is an important coal-mining area. Coal was first discovered there in 1721. However, coal mining became a significant industry only after 1869, when the first railway reached the region. Donets Basin coal mining reached its peak importance by 1913, when the region produced 87 percent of Russian coal. The main part of the coal field covers nearly 9,000 square miles (23,300 sq km) in Ukraine and southwestern Russia, an area slightly smaller than the state of Vermont.

Activity: Have students conduct further research on how the coal mines of the Donets Basin have affected the political and economic history of the region. Ask them to create a time line showing major developments.

Visual Record Answer ▶

volcanic action

506

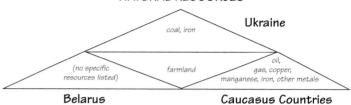

internet connect

GO TO: go.hrw.com
KEYWORD: SJ3 CH23
FOR: Web sites about Ukraine, Belarus, and the Caucasus

Snow-capped Mount Elbrus is located along the border between Georgia and Russia. The surrounding Caucasus Mountains lie along the dividing line between Europe and Asia.

Interpreting the Visual Record What physical processes do you think may have formed the mountains in this region of earthquakes?

Russia
● Mt. Elbrus
Georgia
Turkey

Physical Features

The countries of Ukraine (yoo-KRAYN) and Belarus (byay-luh-ROOS) border western Russia. Belarus is landlocked. Ukraine lies on the Black Sea. Georgia, Armenia (ahr-MEE-nee-uh), and Azerbaijan (a-zuhr-by-JAHN) lie in a rugged region called the Caucasus (KAW-kuh-suhs). It is named for the area's Caucasus Mountains. The Caucasus region is located between the Black Sea and the Caspian Sea.

Landforms Most of Ukraine and Belarus lie in a region of plains. The Northern European Plain sweeps across northern Belarus. The Pripyat (PRI-pyuht) Marshes, also called the Pinsk Marshes, are found in the south. The Carpathian Mountains run through part of western Ukraine. The Crimean (kry-MEE-uhn) Peninsula lies in southern Ukraine. The southern Crimean is very rugged and has high mountains. It separates the Black Sea from the Sea of Azov (uh-ZAWF).

In the north along the Caucasus's border with Russia is a wide mountain range. The region's and Europe's highest peak, Mount Elbrus (el-BROOS), is located here. As you can see on the chapter map, the land drops below sea level along the shore of the Caspian Sea. South of the Caucasus is a rugged, mountainous plateau. Earthquakes often occur in this region.

Rivers One of Europe's major rivers, the Dnieper (NEE-puhr), flows south through Belarus and Ukraine. Ships can travel much of its length. Dams and reservoirs on the Dnieper River provide hydroelectric power and water for irrigation.

Vegetation Mixed forests were once widespread in the central part of the region. Farther south, the forests opened onto the grasslands of the steppe. Today, farmland has replaced much of the original vegetation.

Ukraine is trying to preserve its natural environments and has created several **nature reserves**. These are areas the government has set aside to protect animals, plants, soil, and water.

✔ **READING CHECK:** *Places and Regions* What are the region's major physical features? Caucasus Mountains, Northern European Plain, Pripyat Marshes, Carpathian Mountains, Crimean Peninsula, Black Sea, Caspian Sea, Sea of Azov, Dnieper River

CLOSE

Ask students to describe some things they might do for fun if they visited this region. *(Possible answers include skiing Mount Elbrus, taking a boat ride on the Dnieper River, visiting a nature reserve in Ukraine, going to the beach in Georgia, and so on.)*

REVIEW AND ASSESS

Have students complete the Section Review. Then organize the class into triads and have each triad write eight quiz questions that match a country to its landform, climate, or natural features. Have groups exchange their quizzes and then solve them. Then have students complete Daily Quiz 23.1.
COOPERATIVE LEARNING

RETEACH

Have students complete Main Idea Activity 23.1. Then have students illustrate the postcards or letters they created earlier with appropriate physical features, climate type(s), and natural resources.
ENGLISH LANGUAGE LEARNERS

EXTEND

Have interested students conduct further research on the plants and animals of the Pripyat Marshes of Belarus. Then have students create a publicity campaign to raise awareness of the region's unique characteristics and the forces that endanger it. **BLOCK SCHEDULING**

Climate

Like much of western Russia, the northern two thirds of Ukraine and Belarus have a humid continental climate. Winters are cold. Summers are warm but short. Southern Ukraine has a steppe climate. Unlike the rest of the country, the Crimean Peninsula has a Mediterranean climate. There are several different climates in the Caucasus. Georgia's coast has a mild climate similar to the Carolinas in the United States. Azerbaijan contains mainly a steppe climate. Because it is so mountainous, Armenia's climate changes with elevation.

✓ **READING CHECK:** *Places and Regions* What climate types are found in this area? humid continental, steppe, Mediterranean

Resources

Rich farmlands are Ukraine's greatest natural resource. Farming is also important in Belarus. Lowland areas of the Caucasus have rich soil and good conditions for farming.

The Donets (duh-NYETS) Basin in southeastern Ukraine is a rich coal-mining area. Kryvyy Rih (kri-VI RIK) is the site of a huge open-pit iron-ore mine. The region's most important mineral resources are Azerbaijan's large and valuable oil and gas deposits. These are found under the shallow Caspian Sea. Copper, manganese, iron, and other metals are also present in the Caucasus.

✓ **READING CHECK:** *Environment and Society* How have this region's natural resources affected economic development? Rich farmlands, minerals, metals, and oil and gas support the economy.

BUILD on WHAT You Know

Do you remember what you learned about steppe climates? See Chapter 3 to review.

go.hrw.com **Homework Practice Online**
Keyword: SJ3 HP23

Section Review 1

Define and explain: nature reserves

Working with Sketch Maps On a map of Europe that you draw or that your teacher provides, label the following: Black Sea, Caucasus Mountains, Caspian Sea, Pripyat Marshes, Carpathian Mountains, Crimean Peninsula, Sea of Azov, Mount Elbrus, Dnieper River, and Donets Basin. Where in the region is a major coal-mining area?

Reading for the Main Idea

1. *Places and Regions* What three seas are found in this region?

2. *Places and Regions* What creates variation in Armenia's climate?

Critical Thinking

3. **Drawing Inferences and Conclusions** Why has so much farming developed in Ukraine, Belarus, and the Caucasus?

4. **Drawing Inferences and Conclusions** How do you think heavy mining in this region could create pollution?

Organizing What You Know

5. **Categorizing** Copy the following graphic organizer. Use it to describe the region's physical features, climates, and resources.

	Physical features	Climate	Resources
Belarus			
Caucasus			
Ukraine			

Section Review 1

Answers

Define For definition, see: nature reserves, p. 506

Working with Sketch Maps Maps will vary, but listed places should be labeled in their approximate locations. The Donets Basin is a major coal-mining area.

Reading for the Main Idea
1. Black Sea, Sea of Azov, Caspian Sea (NGS 4)
2. elevation (NGS 4)

Critical Thinking
3. because of the areas' rich soil and good weather conditions
4. metals being washed into water supply

Organizing What You Know
5. Belarus—Northern European Plain, Pinsk Marshes, Dnieper River; humid continental; farmland; Caucasus region—Black Sea, Caspian Sea, Caucasus Mountains, Mount Elbrus; steppe climate, mild coastal climate; rich soil, copper, manganese, iron, oil, and gas; Ukraine—plains, steppe, Carpathian Mountains, Crimea, Black Sea, Sea of Azov, Dnieper River; humid continental, steppe, Mediterranean; farmland, coal, iron

Section 2

Objectives

1. Identify the groups that influenced the history of Ukraine and Belarus.
2. Discuss some important economic features and environmental concerns of Ukraine.
3. Trace the development of Belarus's economy.

Section 2 — Ukraine and Belarus

Read to Discover

1. Which groups have influenced the history of Ukraine and Belarus?
2. What are some important economic features and environmental concerns of Ukraine?
3. How has the economy of Belarus developed?

Define
serfs
Cossacks
soviet

Locate
Ukraine
Belarus
Kiev
Chernobyl
Minsk

WHY IT MATTERS

Energy created by nuclear power plants is important to the United States, Ukraine, and Belarus. Go to **CNNfyi.com** or other **current events** sources to find information about nuclear energy. Record your findings in your journal.

A hand-painted Ukrainian egg

гео·гра·фия

▲
These are the syllables for the Russian word for geography, written in the Cyrillic alphabet.

History and Government

About 600 B.C. the Greeks established trading colonies along the coast of the Black Sea. Much later—during the A.D. 400s—the Slavs began to move into what is now Ukraine and Belarus. Today, most people in this region speak closely related Slavic languages.

Vikings and Christians In the 800s Vikings took the city of Kiev. Located on the Dnieper River, it became the capital of the Vikings' trading empire. Today, this old city is Ukraine's capital. In the 900s the Byzantine, or Greek Orthodox, Church sent missionaries to teach the Ukrainians and Belorussians about Christianity. These missionaries introduced the Cyrillic alphabet.

St. Sophia Cathedral in Kiev was built in the 1000s. It was one of the earliest Orthodox cathedrals in this area. Religious images decorate the dome's interior.

◄

508

🌍 Teaching Objective 1

LEVEL 1: (Suggested time: 20 min.) Pair students and have each pair create a collage that represents the various groups that have influenced the history of Ukraine and Belarus. Display collages around the classroom. **ENGLISH LANGUAGE LEARNERS, COOPERATIVE LEARNING**

LEVELS 2 AND 3: (Suggested time: 20 min.) Tell students to imagine that they are historians who specialize in the history of Ukraine and Belarus. Pair students and have each pair write an outline for a book on the various groups that have influenced the history of Ukraine and Belarus. **COOPERATIVE LEARNING**

Mongols and Cossacks A grandson of Genghis Khan led the Mongol horsemen that conquered Ukraine in the 1200s. They destroyed most of the towns and cities there, including Kiev.

Later, northern Ukraine and Belarus came under the control of Lithuanians and Poles. Under foreign rule, Ukrainian and Belorussian **serfs** suffered. Serfs were people who were bound to the land and worked for a lord. In return, the lords provided the serfs with military protection and other services. Some Russian and Ukrainian serfs left the farms and formed bands of nomadic horsemen. Known as **Cossacks**, they lived on the Ukrainian frontier.

The Russian Empire North and east of Belarus, a new state arose around Moscow. This Russian kingdom of Muscovy won independence from the Mongols in the late 1400s. The new state set out to expand its borders. By the 1800s all of modern Belarus and Ukraine were under Moscow's rule. Now the Cossacks served the armies of the Russian czar. However, conditions did not improve for the Ukrainian and Belorussian serfs and peasants.

Soviet Republics The Russian Revolution ended the rule of the czars in 1917. Ukraine and Belarus became republics of the Soviet Union in 1922. Although each had its own governing **soviet**, or council, Communist leaders in Moscow made all major decisions.

Ukraine was especially important as the Soviet Union's richest farming region. On the other hand, Belarus became a major industrial center. It produced heavy machinery for the Soviet Union. While Ukraine and Belarus were part of the Soviet Union, the Ukrainian and Belorussian languages were discouraged. Practicing a religion was also discouraged.

After World War II economic development continued in Ukraine and Belarus. Factories and power plants were built with little concern for the safety of nearby residents.

This watercolor on rice paper depicts Kublai Khan. He was the founder of the vast Mongol empire in the 1200s. The Mongols conquered large areas of Asia and Europe, including Ukraine.

Kiev remained an important cultural and industrial center during the Soviet era. Parts of the city were destroyed during World War II and had to be rebuilt. Today tree-lined streets greet shoppers in the central city.

ENVIRONMENT AND SOCIETY

Although the Soviet era left parts of Ukraine badly polluted, there are also large natural areas that are protected from industry.

Ukrainians have long supported conservation. Ukraine's first nature reserve began as a private wildlife refuge in 1875. This reserve covers about 27,400 acres (11,000 hectares) and protects a portion of virgin steppe. A successful breeding program for endangered species has been established there. Two of the species are onagers, or wild donkeys, and ostriches.

The Black Sea Nature Reserve, established in 1927, includes protected areas of the sea. The Danube Water Meadows was established for the scientific study and protection of the Danube River's tidewater plant life. The Ukrainian Steppe Reserve, consisting of three separate sections, preserves three special kinds of steppe: meadow steppe, black-earth steppe, and stony steppe. Other reserves protect forest-steppe woodlands, marshes, forests, and mountains.

Activity: Have students conduct additional research and create a collage of pictures of animals found in Ukraine's nature reserves.

Teaching Objective 2

LEVEL 1: (Suggested time: 30 min.) Tell students to imagine that they are compiling an almanac about contemporary Ukraine. Pair students and have each pair compile a list of important economic features and environmental concerns of Ukraine. *(Lists should mention Ukraine's rich soil and the importance of agriculture; it is the world's largest producer of sugar beets; it is a top steel producer; Ukraine experienced rapid industrial growth under the Soviets; the world's worst nuclear disaster occurred at Chernobyl in 1986.)* Ask volunteers to present their lists to the class.
ENGLISH LANGUAGE LEARNERS, COOPERATIVE LEARNING

LEVELS 2 AND 3: (Suggested time: 30 min.) Have students use the lists they created for the *Level 1* activity as the basis for recommendations to Ukraine's government for developing the country's economy and protecting its environment.

TEACHER TO TEACHER

Joanne Sadler of Buffalo, New York, suggests this activity to teach about the geography and culture of Ukraine and Belarus. Organize students into groups, and have groups study either Ukraine or Belarus. Have students write and illustrate another section for this textbook, to be titled Ukraine (or Belarus), Close-Up and in Detail.

DAILY LIFE

The Ukrainian custom of the first haircutting, *postryzhyny*, marks a child's first birthday. The child's whole family gathers to participate in the event and to share a feast.

Guests contribute coins that are collected and placed in a soup bowl. Each of the child's godparents then takes a turn cutting the child's hair as the guests observe. First a lock is cut from the front of the child's head, then from the back, and then from each side. These cuttings are taken from four areas on the head, which represent the four directions of the compass.

After the haircutting, liquor is poured into the bowl to cover the coins. The baby's feet are then dipped into the bowl. This ritual symbolizes that the child will never be controlled by alcohol or money in years to come. The coins are then dried and saved for the child.

Activity: Have students recall rites of passage in their lives such as religious confirmations or graduations and explain the relationship between these rituals and their culture.

Chart Answer ▶

the United States

Near the end of World War II, Soviet, American, and British leaders met at Livadia Palace in Yalta, Ukraine. There they planned the defeat and occupation of Germany.

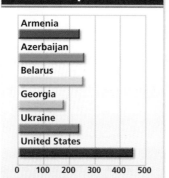

Patients per Doctor

	0	100	200	300	400	500
Armenia						
Azerbaijan						
Belarus						
Georgia						
Ukraine						
United States						

Source: *The World Book Encyclopedia of People and Places*

Interpreting the Graph In what country is the number of patients per doctor greatest?

End of Soviet Rule When the Soviet Union collapsed in 1991, Belarus and Ukraine declared independence. Each now has a president and a prime minister. Both countries still have economic problems. Ukraine has also had disagreements with Russia over control of the Crimean Peninsula and the Black Sea naval fleet.

✓ **READING CHECK:** *Human Systems* Which groups have influenced the history of Ukraine and Belarus? Greeks, Slavs, Vikings, Christians, Mongols, Lithuanians, Poles

Ukraine

Ethnic Ukrainians make up about 75 percent of Ukraine's population. The largest minority group in the country is Russian. There are other ties between Ukraine and Russia. For example, the Ukrainian and Russian languages are closely related. In addition, both countries use the Cyrillic alphabet.

Economy Ukraine has a good climate for growing crops and some of the world's richest soil. As a result, agriculture is important to its economy. Ukraine is the world's largest producer of sugar beets. Ukraine's food-processing industry makes sugar from the sugar beets. Farmers also grow fruits, potatoes, vegetables, and wheat. Grain is made into flour for baked goods and pasta. Livestock is also raised. Ukraine is one of the world's top steel producers. Ukrainian factories make automobiles, railroad cars, ships, and trucks.

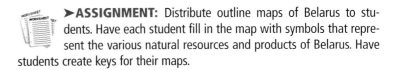

Teaching Objective 3

ALL LEVELS: (Suggested time: 15 min.) Copy the graphic organizer onto the chalkboard, omitting the italicized answers. Use it to help students understand the development of Belarus's economy. Have each student complete the organizer. Ask volunteers to share their answers with the class. Conclude by leading a discussion about Belarus's culture, natural resources, and products. **ENGLISH LANGUAGE LEARNERS**

►**ASSIGNMENT:** Distribute outline maps of Belarus to students. Have each student fill in the map with symbols that represent the various natural resources and products of Belarus. Have students create keys for their maps.

Economic Development in Belarus

- *WWII destroyed most of Belarus's agriculture and industry.*
- *Belarus received the worst of the radiation fallout from Chernobyl.*
- *The country has resisted economic reforms.*
- *Belarus has limited mineral resources.*

⇨ ⇨ ⇨ *slow economic growth*

CONNECTING TO Science

A combine used during July harvest

Wheat: From Field to Consumer

Wheat is one of Ukraine's most important farm products. The illustration below shows how wheat is processed for use by consumers.

- The head of the wheat plant contains the wheat kernels, wrapped in husks. The kernel includes the bran or seed coat, the endosperm, and the germ from which new wheat plants grow.

- Whole wheat flour contains all the parts of the kernel. White flour is produced by grinding only the endosperm. Vitamins are added to some white flour to replace vitamins found in the bran and germ.

- People use wheat to make breads, pastas, and breakfast foods. Wheat by-products are used in many other foods.

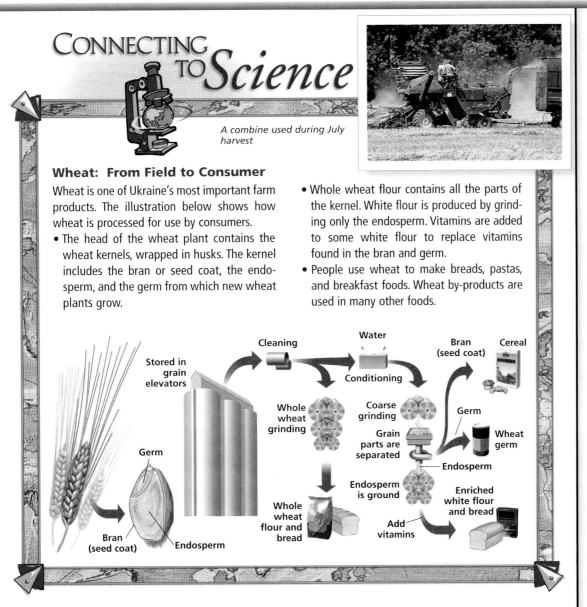

Environment During the Soviet period, Ukraine experienced rapid industrial growth. There were few pollution controls, however. In 1986 at the town of Chernobyl, the world's worst nuclear-reactor disaster occurred. Radiation spread across Ukraine and parts of northern Europe. People near the accident died. Others are still suffering from cancer. Many Ukrainians now want to reduce their country's dependence on nuclear power. This has been hard because the country has not developed enough alternative sources of power.

✓ **READING CHECK:** *Environment and Society* How has the scarcity of alternative sources of power affected Ukraine? It has forced a reliance upon nuclear power.

GLOBAL PERSPECTIVES

Some effects of the Chernobyl nuclear explosion only became apparent years after the accident. The resulting radioactive contamination of the region's food supply and environment has hurt human and animal health and the region's economies.

Some 10 years after the accident, children's cancer rates in parts of Belarus, Russia, and Ukraine were as much as 30 times higher than before the accident. Also, scientists have noted that genetic damage in humans and animals has been passed on to the next generation.

The contaminated part of Ukraine and Belarus was estimated to cover an area the size of England, Wales, and Northern Ireland. Moreover, Belarus has spent up to 20 percent of its annual budget to cope with Chernobyl's aftermath.

internet connect

GO TO: go.hrw.com
KEYWORD: SJ3 CH23
FOR: Web sites about Chernobyl

Chart Answer (p. 512)

lower; answers will vary but might mention nuclear fallout, poverty, diet, or other causes.

Ask students to name some cultural and historical similarities between Ukraine and Belarus. *(Students should mention similarities in languages, use of a Cyrillic alphabet, and Soviet rule.)* Then ask students how the accident at Chernobyl has affected the region. *(The accident has affected people's health and has slowed down economic development.)*

Have students complete Main Idea Activity 23.2. Then organize the class into small groups; have each group use the Define and Locate terms and a few other terms from this section to create a crossword puzzle with clues. Have groups exchange their puzzles and then solve them.
ENGLISH LANGUAGE LEARNERS, COOPERATIVE LEARNING

After students complete the Section Review, have them write four quiz questions related to the history and culture of Ukraine and Belarus. Collect the questions and use them to quiz the class before students complete Daily Quiz 23.2.

Have students research the Mongol conquest of the Ukraine or the role of the Cossacks in Russian and Ukrainian history. Ask students to prepare a five-minute oral report based on their research. **BLOCK SCHEDULING**

Section Review 2

Answers

Define For definitions, see: serfs, p. 509; Cossacks, p. 509; soviet, p. 509

Working with Sketch Maps Maps will vary, but listed places should be labeled in their approximate locations.

Reading for the Main Idea
1. trading colonies, Slavic languages, city of Kiev, Christianity, Cyrillic alphabet (NGS 9)
2. Ukrainian, Belorussian, Russian (NGS 9)

Critical Thinking
3. Ukraine and Belarus declared independence but have economic problems. Ukraine and Russia have had disagreements about political and military issues.
4. contaminated region, killed people, led to cancer

Organizing What You Know
5. 600 B.C.—Greeks establish colonies; A.D. 400s—Slavs move into region; 800s—Kiev established; 900s—Christianity, Cyrillic alphabet brought in; 1200s—Mongols invade; late 1400s—Muscovy wins independence; 1800s—region ruled by Moscow; 1917—Russian Revolution, beginning of Soviet rule; 1991—Soviet Union collapses

Belarus and Ukraine

COUNTRY	POPULATION/ GROWTH RATE	LIFE EXPECTANCY	LITERACY RATE	PER CAPITA GDP
Belarus	10,350,194 −0.2%	62, male 75, female	98%	$7,500
Ukraine	48,760,474 −0.9%	61, male 72, female	98%	$3,850
United States	281,421,906 0.9%	73, male 80, female	97%	$36,200

Sources: Central Intelligence Agency, *The World Factbook 2001*; U.S. Census Bureau

Interpreting the Chart How does life expectancy in the region compare to that of the United States? Why do you think this is the case?

Belarus

The people of Belarus are known as Belorussians, which means "white Russians." Ethnically they are closely related to Russians. Their language is also very similar to Russian.

Culture Ethnic Belorussians make up about 75 percent of the country's population. Russians are the second-largest ethnic group. Both Belorussian and Russian are official languages. Belorussian also uses the Cyrillic alphabet. Minsk, the capital of Belarus, is the administrative center of the Commonwealth of Independent States.

Economy Belarus has faced many difficulties. Fighting in World War II destroyed most of the agriculture and industry in the country. Belarus also received the worst of the radiation fallout from the Chernobyl nuclear disaster, which contaminated the country's farm products and water. Many people developed health problems as a result. Another problem has been slow economic progress since the collapse of the Soviet Union. Belarus has resisted economic changes made by other former Soviet republics.

There are various resources in Belarus, however. The country has a large reserve of potash, which is used for fertilizer. Belarus leads the world in the production of peat, a source of fuel found in the damp marshes. Mining and manufacturing are important to the economy. Flax, one of the country's main crops, is grown for fiber and seed. Cattle and pigs are also raised. Nearly one third of Belarus is covered by forests that produce wood and paper products.

✔ **READING CHECK:** *Human Systems* How has the economy of Belarus developed? slowly, because of the collapse of the Soviet Union and resistance to economic change

go.hrw.com
Homework Practice Online
Keyword: SJ3 HP23

Section Review 2

Define and explain: serfs, Cossacks, soviet

Working with Sketch Maps On your map from Section 1, label Ukraine, Belarus, Kiev, Chernobyl, and Minsk.

Reading for the Main Idea
1. *Human Systems* What contributions were made by early groups that settled in this region?
2. *Human Systems* What ethnic groups and languages are found in this region today?

Critical Thinking
3. **Finding the Main Idea** How did the end of Soviet rule affect Ukraine and Belarus?
4. **Summarizing** How has the nuclear disaster at Chernobyl affected the region?

Organizing What You Know
5. **Sequencing** Copy the time line below. Use it to trace the region's history from the A.D. 900s to today.

A.D. 900 ———————————— Today

Section 3

Objectives

1. List the groups that influenced the early history and culture of the Caucasus.

2. Discuss Georgia's economy.

3. Describe today's Armenia.

4. Discuss what Azerbaijan is like today.

FOCUS

LET'S GET STARTED

Copy the following scenario onto the chalkboard: *Imagine that you live in a small country that has recently gained independence from a large country. Your country is struggling to change in ways that bring more freedom to your citizens. Independence has been difficult, however. Many people have concluded life was better under the large country's rule. Do you agree? Write down your response.* Discuss responses. Tell students that in Section 3 they will learn about three countries—Georgia, Armenia, and Azerbaijan—where this scenario is a reality.

Building Vocabulary

Have volunteers read aloud the definitions of **homogeneous** and **agrarian** from the glossary. Then ask students to apply these terms to describe various societies or economies they have already studied.

Section 3

The Caucasus

Read to Discover

1. What groups influenced the early history and culture of the Caucasus?

2. What is the economy of Georgia like?

3. What is Armenia like today?

4. What is Azerbaijan like today?

Define

homogeneous
agrarian

Locate

Georgia
Armenia
Azerbaijan

Cover of The Knight in Panther's Skin

WHY IT MATTERS

Each of the countries in this section has been involved in a war since the collapse of the Soviet Union in 1991. Use **CNNfyi.com** or other **current events** sources to find information about the reasons for this unrest. Record your findings in your journal.

Section 3 RESOURCES

Reproducible
◆ Guided Reading Strategy 23.3
◆ Cultures of the World Activity 4
◆ Geography for Life Activity 23
◆ Map Activity 23

Technology
◆ One-Stop Planner CD–ROM, Lesson 23.3
◆ Homework Practice Online
◆ HRW Go site

Reinforcement, Review, and Assessment
◆ Section 3 Review, p. 515
◆ Daily Quiz 23.3
◆ Main Idea Activity 23.3
◆ English Audio Summary 23.3
◆ Spanish Audio Summary 23.3

History

In the 500s B.C. the Caucasus region was controlled by the Persian Empire. Later it was brought under the influence of the Byzantine Empire and was introduced to Christianity. About A.D. 650, Muslim invaders cut the region off from Christian Europe. By the late 1400s other Muslims, the Ottoman Turks, ruled a vast empire to the south and west. Much of Armenia eventually came under the rule of that empire.

Modern Era During the 1800s Russia took over eastern Armenia, much of Azerbaijan, and Georgia. The Ottoman Turks continued to rule western Armenia. Many Armenians spread throughout the Ottoman Empire. However, they were not treated well. Their desire for more independence led to the massacre of thousands of Armenians. Hundreds of thousands died while being forced to leave Turkey during World War I. Some fled to Russian Armenia.

After the war Armenia, Azerbaijan, and Georgia were briefly independent. By 1922 they had become part of the Soviet Union. They again became independent when the Soviet Union collapsed in 1991.

This wall painting is one of many at the ancient Erebuni Citadel in Yerevan, Armenia's capital. The fortress was probably built in the 800s B.C. by one of Armenia's earliest peoples, the Urartians.

▼

513

Teaching Objectives 1–4

ALL LEVELS: (Suggested time: 30 min.) Copy the following graphic organizer onto the chalkboard, omitting the italicized answers. Use it to help students understand the shared history and culture of the Caucasus countries and the economic and cultural differences between these countries. Have students work in pairs to complete the organizer.
ENGLISH LANGUAGE LEARNERS, COOPERATIVE LEARNING

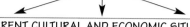

SHARED HISTORY AND CULTURE

- *ruled by ancient Persian Empire*
- *under Byzantine influence*
- *Caucasus isolated from*
- *Christian groups in Europe*
- *united in 1922 to oppose Soviet rule; ruled by Soviet Union*

CURRENT CULTURAL AND ECONOMIC SITUATION

GEORGIA	ARMENIA	AZERBAIJAN
• *Georgian language* • *shortage of good farmland* • *tea, citrus fruits, vineyards, and tourism important to economy* • *imports most of its energy*	• *Armenian language predominant* • *varied industry includes mining, carpets, clothes, and footwear* • *agriculture important*	• *Turkish language predominant* • *agrarian society* • *oil, natural gas, cotton, and fishing important to economy*

Section Review 3

Answers

Define For definitions, see: homogeneous, p. 515; agrarian, p. 515

Working with Sketch Maps Maps will vary, but listed places should be labeled in their approximate locations. Proximity to Asia has led to Turkish, Persian, and Muslim influences. Geographic and cultural isolation from the rest of Europe is a result of the Caucasus Mountains.

Reading for the Main Idea

1. Persians, Byzantine Empire, Muslims, Ottoman Turks (NGS 9)

2. the Soviet Union (NGS 13)

Critical Thinking

3. The massive 1988 earthquake destroyed much of the country's industry.

4. around farming, oil

Organizing What You Know

5. Students should discuss physical, historical, and cultural differences. They might mention that the countries share a history of Soviet domination and reliance on agriculture.

Visual Record Answer ▲

cucumbers and other vegetables

►

possible answer: because the building is sacred to the people

▲

This Georgian family's breakfast includes local specialties such as *khachapuri*—bread made with goat cheese.
Interpreting the Visual Record
What other agricultural products do you see on the table?

The Orthodox Christian Agartsya Monastery was built in Armenia in the 1200s.
Interpreting the Visual Record
Why do you think this building's exterior is so well preserved?

►

Government Each country has an elected parliament, president, and prime minister. In the early 1990s there was civil war in Georgia. Armenia and Azerbaijan were also involved in a war during this time. Ethnic minorities in each country want independence. Disagreements about oil and gas rights may cause more regional conflicts in the future.

✓ **READING CHECK:** *Human Systems* How has conflict among cultures been a problem in this region? Ethnic minorities want independence; disagreements about oil and gas rights may cause problems.

Georgia

Georgia is a small country located between the high Caucasus Mountains and the Black Sea. It has a population of about 5 million. About 70 percent of the people are ethnic Georgians. The official language, Georgian, has its own alphabet. This alphabet was used as early as A.D. 400.

As in all the former Soviet republics, independence and economic reforms have been difficult. Georgia has also suffered from civil war. By the late 1990s the conflicts were fewer but not resolved.

Georgia has little good farmland. Tea and citrus fruits are the major crops. Vineyards are an important part of Georgian agriculture. Fish, livestock, and poultry contribute to the economy. Tourism on the Black Sea has also helped the economy. Because its only energy resource is hydropower, Georgia imports most of its energy supplies.

✓ **READING CHECK:** *Human Systems* In what way has scarcity of energy resources affected Georgia's economy? It has made Georgia dependent on other nations for energy imports.

Armenia

Armenia is a little smaller than Maryland. It lies just east of Turkey. It has fewer than 4 million people and is not as diverse as other countries

CLOSE

Have students use the information in the graphic organizer to compose a few lines of rhyming verse about one of the countries.

REVIEW AND ASSESS

Have students complete the Section Review. Then organize students into groups. Have each group create eight flash cards using information in the section and exchange these flash cards with another group. Have students within each group use the flash cards to quiz each other. Then have students complete Daily Quiz 23.3. **COOPERATIVE LEARNING**

RETEACH

Have students complete Main Idea Activity 23.3. Then ask students to outline the section. **ENGLISH LANGUAGE LEARNERS**

EXTEND

Have interested students conduct research on the cultural history of a resource or product of one of the Caucasus countries (for example, carpets in Armenia, vineyards in Georgia, or caviar in Azerbaijan). Then have each student deliver a short oral report to the class about his or her chosen subject. **BLOCK SCHEDULING**

in the Caucasus. Almost all the people are Armenian, belong to the Armenian Orthodox Church, and speak Armenian.

Armenia's progress toward economic reform has not been easy. In 1988 a massive earthquake destroyed nearly one third of its industry. Armenia's industry today is varied. It includes mining and the production of carpets, clothing, and footwear.

Agriculture accounts for about 40 percent of Armenia's gross domestic product. High-quality grapes and fruits are important. Beef and dairy cattle and sheep are raised on mountain pastures.

✓ **READING CHECK:** *Environment and Society*
How did the 1998 earthquake affect the people of Armenia?
destroyed nearly one third of Armenian industry

The Caucasus

COUNTRY	POPULATION/ GROWTH RATE	LIFE EXPECTANCY	LITERACY RATE	PER CAPITA GDP
Armenia	3,336,100 −0.2%	62, male 71, female	99%	$3,000
Azerbaijan	7,771,092 0.3%	59, male 68, female	97%	$3,000
Georgia	4,989,285 −0.6%	61, male 68, female	99%	$4,600
United States	281,421,906 0.9%	74, male 80, female	97%	$36,200

Sources: Central Intelligence Agency, *The World Factbook 2001*; U.S. Census Bureau

Interpreting the Chart Which country has the lowest per capita GDP in the region? Why might this be the case?

Azerbaijan

Azerbaijan has nearly 8 million people. Its population is becoming ethnically more **homogeneous**, or the same. The Azeri, who speak a Turkic language, make up about 90 percent of the population.

Azerbaijan has few industries except for oil production. It is mostly an **agrarian** society. An agrarian society is organized around farming. The country's main resources are cotton, natural gas, and oil. Baku, the national capital, is the center of a large oil-refining industry. Oil is the most important part of Azerbaijan's economy. Fishing is also important because of the sturgeon of the Caspian Sea.

✓ **READING CHECK:** *Human Systems* What are some cultural traits of the people of Azerbaijan? They are mostly ethnically homogenous and speak a Turkic language.

Section Review 3

Define and explain: homogeneous, agrarian

Working with Sketch Maps On the map you created for Section 2, label Georgia, Armenia, and Azerbaijan. How has the location of this region helped and hindered its growth?

Reading for the Main Idea

1. *Human Systems* Which groups influenced the early history of the Caucasus?

2. *Human Systems* Which country controlled the Caucasus during most of the 1900s?

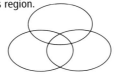
go.
hrw
.com
Homework Practice Online
Keyword: SJ3 HP23

Critical Thinking

3. **Analyzing Information** Why has economic reform been difficult in Armenia?

4. **Finding the Main Idea** How is Azerbaijan's economy organized?

Organizing What You Know

5. **Comparing/Contrasting** Copy the following graphic organizer. Use it to show the similarities and differences among the countries of the Caucasus region.

Review ANSWERS

Building Vocabulary
For definitions, see nature reserves, p. 506; serfs, p. 509; Cossacks, p. 509; soviet, p. 509; homogeneous, p. 515; agrarian, p. 515

Reviewing the Main Ideas

1. Ukraine—steel, coal, agriculture; Belarus—mining, manufacturing, wood and paper products; Georgia—fish, livestock, poultry, tourism; Armenia—agriculture, mining, carpets, clothes, footwear; Azerbaijan—cotton, oil, natural gas, agriculture, fishing (NGS 12)

2. on the Dnieper River; Viking traders (NGS 12)

3. the Greeks and Slavs (NGS 12)

4. Ukraine became rich farming region; Belarus became major industrial center; industrialization polluted some areas. (NGS 14)

5. destroyed much of Armenia's industry and economy (NGS 15)

▲ **Chart Answer**

Azerbaijan; answers will vary, but might mention because there is little industry.

ASSESS

Have students complete the Chapter 23 Test.

RETEACH

Organize students into groups and have each group choose one of the five countries discussed in the chapter. Then have the group prepare a monologue that might be spoken by a senior citizen of their chosen country. The monologue should tell the story of the person's life, and selected facts about the person's country. Have one student from each group read or recite the group's monologue to the class. **COOPERATIVE LEARNING**

PORTFOLIO EXTENSIONS

1. Organize students into small groups. Have each group write a new national anthem for Ukraine or Belarus in which the lyrics pertain to their chosen country's geographical features and natural resources. Have volunteers perform their anthems for the class. Have students place their national anthems in their portfolios.

2. Tell students to imagine that they work for the tourist bureau of one of the Caucasus countries. Organize students into small groups and have each group choose a country. Then have each group prepare a three-minute radio commercial that advertises its chosen country's cultural and geographical features to tourists. Have groups perform their radio commercials for the class. Have students place their commercials in their portfolios.

Review ANSWERS

Understanding Environment and Society

Answers will vary, but the information should address oil and natural gas as it relates to the Caucasus region. Use Rubric 29, Presentations, to evaluate student work.

Thinking Critically

1. Students might suggest that ethnic rivalries can intensify conflicts.
2. Their status as Soviet Republics and proximity to Russia made them enemy targets.
3. Answers will vary. Students should use information from the chapter to justify their opinions.
4. Possible answer: They are physically isolated, have different physical features, and different ethnic groups.
5. Possible answer: Their climates and resources differ, therefore their economies differ.

 **Reviewing What You Know**

Building Vocabulary

On a separate sheet of paper, write sentences to define each of the following words.

1. nature reserves
2. serfs
3. Cossacks
4. soviet
5. homogeneous
6. agrarian

Reviewing the Main Ideas

1. (Human Systems) Which industries have traditionally been very important to the economies of the countries covered in this chapter?
2. (Human Systems) What is the relative location of Kiev? Which group of people founded Kiev?
3. (Human Systems) What peoples were apparently the first to settle in Belarus and Ukraine?
4. (Environment and Society) How did industrialization under Soviet rule affect these regions?
5. (Environment and Society) What effect did the earthquake in 1988 have on Armenia's economy?

Understanding Environment and Society

Resource Use

Prepare a chart and presentation on oil and natural gas production in the Caucasus. In preparing your presentation, consider the following:

• When and where oil and natural gas were discovered.
• The importance of oil and natural gas production to the former Soviet Union.
• How oil is transported from there to international markets today.

When you have finished your model and presentation, write a five-question quiz, with answers, about your chart to challenge fellow students.

Thinking Critically

1. **Drawing Inferences and Conclusions** How might ethnic diversity affect relations among countries?
2. **Analyzing Information** How did the location of Ukraine and Belarus contribute to their devastation during World War II?
3. **Analyzing Information** Of the countries covered in this chapter, which do you think was the most important to the former Soviet Union? Why do you think this was so?
4. **Summarizing** Why did the countries of the Caucasus develop so differently from Russia, Ukraine, and Belarus?
5. **Finding the Main Idea** Why are the economies of each of the Caucasus countries so different from one another?

FOOD FESTIVAL

Compote is a popular dessert in Ukraine. It is sweetened stewed fruit, cooked carefully to keep the fruit as whole as possible. This compote is a more modern version of compote as it does not use lemons, which are hard to get in many areas of Ukraine. To make the compote, wash and drain 1 pound of fresh strawberries or raspberries. Combine ¾ cup of sugar and 1 cup of water and bring it to a boil. Pour the boiling syrup over the fruit and let it stand for several hours before eating.

CHAPTER 23 REVIEW AND ASSESSMENT RESOURCES

Reproducible
◆ Readings in World Geography, History, and Culture 43 and 44
◆ Vocabulary Activity 23

Technology
◆ Chapter 23 Test Generator (on the One-Stop Planner)

◆ HRW Go site
◆ Audio CD Program, Chapter 23

Reinforcement, Review, and Assessment
◆ Chapter 23 Review, pp. 516–17

◆ Chapter 23 Tutorial for Students, Parents, Mentors, and Peers
◆ Chapter 23 Test
◆ Chapter 23 Test for English Language Learners and Special-Needs Students

Building Social Studies Skills

Map ACTIVITY

On a separate sheet of paper, match the letters on the map with their correct labels.

Caucasus Mountains
Pripyat Marshes
Carpathian Mountains
Crimean Peninsula
Mount Elbrus
Donets Basin
Chernobyl

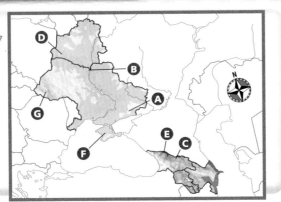

Mental Mapping Skills ACTIVITY

On a separate sheet of paper, draw a freehand map of Ukraine, Belarus, and the Caucasus. Include Russia and Turkey for location reference. Make a map key and label the following:

Armenia
Azerbaijan
Belarus
Black Sea
Caspian Sea
Dnieper River
Georgia
Russia
Turkey
Ukraine

WRITING ACTIVITY

Choose one of the countries covered in this chapter to research. Write a report about your chosen country's struggle to establish stability since 1991. Include information about the country's government and economic reforms. Describe the social, political, and economic problems the country has faced. Be sure to use standard grammar, sentence structure, spelling, and punctuation.

Map Activity
A. Donets Basin
B. Chernobyl
C. Caucasus Mountains
D. Pripyat Marshes
E. Mount Elbrus
F. Crimean Peninsula
G. Carpathian Mountains

Mental Mapping Skills Activity
Maps will vary, but listed places should be labeled in their approximate locations.

Writing Activity
Answers will vary but should consider the social, political, and economic problems faced by the country. Use Rubric 30, Research, to evaluate student work.

Portfolio Activity
Answers will vary based on the region. Students should include all facets of production of the crop, and display the information in a flowchart. Use Rubric 7, Charts, to evaluate student work.

Alternative Assessment

Portfolio ACTIVITY

Learning About Your Local Geography
Cooperative Project What is an important crop in your region of the United States? Work with a partner to create an illustrated flowchart that shows how this crop is made ready for consumers.

internet connect

Internet Activity: go.hrw.com
KEYWORD: SJ3 GT23

Choose a topic to explore about Ukraine, Belarus, and the Caucasus.
• Trek through the Caucasus Mountains.
• Design Ukrainian Easter eggs.
• Investigate the Chernobyl disaster.

internet connect

GO TO: go.hrw.com
KEYWORD: SJ3 Teacher
FOR: a guide to using the Internet in your classroom

Central Asia
Chapter Resource Manager

CHAPTER 24

Objectives	Pacing Guide	Reproducible Resources
SECTION 1		
Physical Geography (pp. 519–21)	**Regular** .5 day	**RS** Guided Reading Strategy 24.1
1. Identify the main landforms and climates of Central Asia.	**Block Scheduling** .5 day	**SM** Geography for Life Activity 24
2. Identify the resources that are important to Central Asia.	*Block Scheduling Handbook, Chapter 24*	**SM** Map Activity 24
SECTION 2		
History and Culture (pp. 522–24)	**Regular** 1 day	**RS** Guided Reading Strategy 24.2
1. Describe how trade and invasions affected the history of Central Asia.	**Block Scheduling** .5 day	**RS** Graphic Organizer 24
2. Describe political and economic conditions in Central Asia today.	*Block Scheduling Handbook, Chapter 24*	**E** Cultures of the World Activity 4
		IC Interdisciplinary Activity for the Middle Grades 24
SECTION 3		
The Countries of Central Asia (pp. 525–27)	**Regular** 1 day	**RS** Guided Reading Strategy 24.3
1. Identify aspects of culture in Kazakhstan.	**Block Scheduling** .5 day	
2. Describe Kyrgyz culture.	*Block Scheduling Handbook, Chapter 24*	
3. Explain politics in Tajikistan.		
4. Identify art forms in Turkmenistan.		
5. Descrube Uzbekistan's population.		

Chapter Resource Key

RS Reading Support

IC Interdisciplinary Connections

E Enrichment

SM Skills Mastery

A Assessment

REV Review

ELL Reinforcement and English Language Learners

 Transparencies

 CD-ROM

 Music

 Video

 Internet

 Holt Presentation Maker Using Microsoft® Powerpoint®

 One-Stop Planner CD-ROM

See the *One-Stop Planner* for a complete list of additional resources for students and teachers.

One-Stop Planner CD–ROM

It's easy to plan lessons, select resources, and print out materials for your students when you use the *One-Stop Planner CD–ROM with Test Generator.*

internet connect

HRW ONLINE RESOURCES

GO TO: go.hrw.com
Then type in a keyword.

TEACHER HOME PAGE
KEYWORD: SJ3 TEACHER

CHAPTER INTERNET ACTIVITIES
KEYWORD: SJ3 GT24

Choose an activity to:
• study the climate of Central Asia.
• travel along the historic Silk Road.
• learn about nomads and caravans.

CHAPTER ENRICHMENT LINKS
KEYWORD: SJ3 CH24

CHAPTER MAPS
KEYWORD: SJ3 MAPS24

ONLINE ASSESSMENT
Homework Practice
KEYWORD: SJ3 HP24
Standardized Test Prep Online
KEYWORD: SJ3 STP24
Rubrics
KEYWORD: SS Rubrics

COUNTRY INFORMATION
KEYWORD: SJ3 Almanac

CONTENT UPDATES
KEYWORD: SS Content Updates

HOLT PRESENTATION MAKER
KEYWORD: SJ3 PPT24

ONLINE READING SUPPORT
KEYWORD: SS Strategies

CURRENT EVENTS
KEYWORD: S3 Current Events

Technology Resources

- One-Stop Planner CD–ROM, Lesson 24.1
- *ARGWorld* CD–ROM: An Inland Sea in Central Asia
- Geography and Cultures Visual Resources with Teaching Activities 31–35
- Homework Practice Online
- HRW Go site

Review, Reinforcement, and Assessment Resources

ELL	Main Idea Activity 24.1
REV	Section 1 Review, p. 521
A	Daily Quiz 24.1
ELL	English Audio Summary 24.1
ELL	Spanish Audio Summary 24.1

- One-Stop Planner CD–ROM, Lesson 24.2
- Homework Practice Online
- HRW Go site

ELL	Main Idea Activity 24.2
REV	Section 2 Review, p. 524
A	Daily Quiz 24.2
ELL	English Audio Summary 24.2
ELL	Spanish Audio Summary 24.2

- One-Stop Planner CD–ROM, Lesson 24.3
- Homework Practice Online
- HRW Go site

ELL	Main Idea Activity 24.3
REV	Section 3 Review, p. 527
A	Daily Quiz 24.3
ELL	English Audio Summary 24.3
ELL	Spanish Audio Summary 24.3

Meeting Individual Needs

Ability Levels

Level 1 Basic-level activities designed for all students encountering new material

Level 2 Intermediate-level activities designed for average students

Level 3 Challenging activities designed for honors and gifted-and-talented students

English Language Learners Activities that address the needs of students with Limited English Proficiency

Chapter Review and Assessment

IC	Interdisciplinary Activities for the Middle Grades 13, 14, 15
E	Readings in World Geography, History, and Culture 34, 35, 36
SM	Critical Thinking Activity 24
REV	Chapter 24 Review, pp. 530–31
REV	Chapter 24 Tutorial for Students, Parents, Mentors, and Peers
ELL	Vocabulary Activity 24
A	Chapter 24 Test
	Chapter 24 Test Generator (on the One-Stop Planner)
	Audio CD Program, Chapter 24
A	Chapter 24 Test for English Language Learners and Special-Needs Students

CHAPTER 24

Read to the class the following description by traveler Marco Polo: "For twelve days the course is along this elevated plain, which is named Pamer [Pamir Plateau]. . . . So great is the height of the mountains, that no birds are to be seen near their summits; and however extraordinary it may be thought, it was affirmed, that from the keenness of the air, fires when lighted do not give the same heat as in lower situations, nor produce the same effect in cooking food." Explain to students that Polo is describing the thinness of air at higher elevations. Ask students to name other areas where this effect might be observed. *(Possible answers: high in the Rockies, the Sierra Nevada, the Andes, or the Alps)* Tell students they will learn more about the physical and cultural geography of Central Asia in this chapter.

Section 1

Objectives

1. Identify the main landforms and climates of Central Asia.
2. Describe the resources that are important to Central Asia.

LINKS TO OUR LIVES

You may want to remind students that even a region as remote as Central Asia connects with us in several ways:

▶ The region has large oil and gas deposits, which the United States may want to use.

▶ People in the region are struggling with the shift from communism to more open governments.

▶ Damage from Soviet nuclear tests in the area may last far into the future.

▶ The region is opening to tourism. It offers some of the most spectacular scenery in the world.

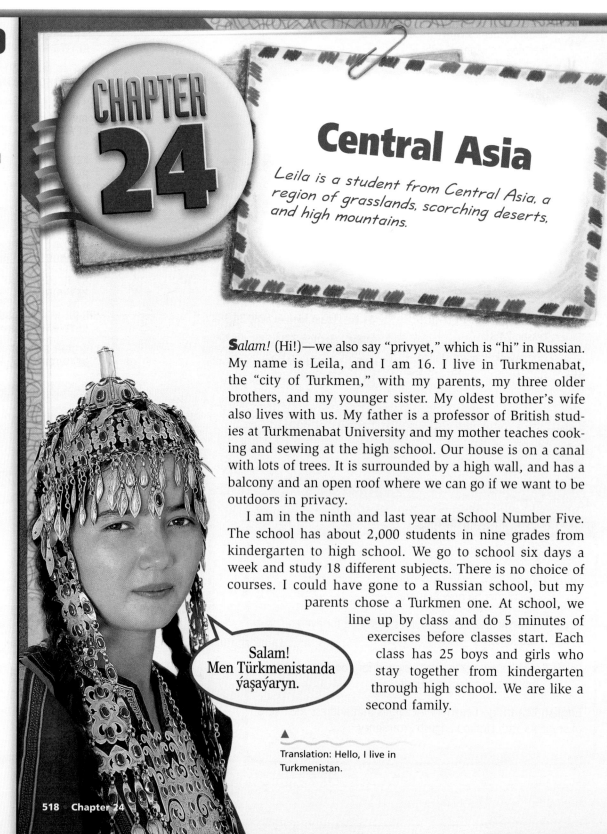

CHAPTER 24

Central Asia

Leila is a student from Central Asia, a region of grasslands, scorching deserts, and high mountains.

Salam! (Hi!)—we also say "privyet," which is "hi" in Russian. My name is Leila, and I am 16. I live in Turkmenabat, the "city of Turkmen," with my parents, my three older brothers, and my younger sister. My oldest brother's wife also lives with us. My father is a professor of British studies at Turkmenabat University and my mother teaches cooking and sewing at the high school. Our house is on a canal with lots of trees. It is surrounded by a high wall, and has a balcony and an open roof where we can go if we want to be outdoors in privacy.

I am in the ninth and last year at School Number Five. The school has about 2,000 students in nine grades from kindergarten to high school. We go to school six days a week and study 18 different subjects. There is no choice of courses. I could have gone to a Russian school, but my parents chose a Turkmen one. At school, we line up by class and do 5 minutes of exercises before classes start. Each class has 25 boys and girls who stay together from kindergarten through high school. We are like a second family.

Salam!
Men Türkmenistanda ýaşaýaryn.

Translation: Hello, I live in Turkmenistan.

LET'S GET STARTED

Copy the following instructions onto the chalkboard: *Look at the map on this page. Estimate the distance between Semey and Tashkent. How many major cities are there between Semey and Mary?* Have students share their responses. *(The distance is about 750 miles. There is one city—Bukhara.)* Then ask what this information tells us about the urban density of Kazakhstan. *(Students might suggest that Kazakhstan has a low urban density.)* Tell students that in Section 1 they will learn more about the physical geography of Central Asia.

Using the Physical-Political Map

Have students examine the map on this page. Ask volunteers to describe the boundaries of the Central Asian countries and to name the nations that border the region. *(Afghanistan, China, Iran, and Russia)* Ask students why Russia might want to stay on good terms with the countries of Central Asia. *(Possible answer: to maintain a buffer zone between itself and other countries)*

Building Vocabulary

Copy the following terms onto the chalkboard: **landlocked** and **oasis**. Have volunteers read aloud the definitions from the text's glossary. Then ask students to explain what these terms might tell us about Central Asia's need for water.

Section 1 Physical Geography

Read to Discover

1. What are the main landforms and climates of Central Asia?
2. What resources are important to Central Asia?

Define

landlocked
oasis

Locate

Pamirs
Tian Shan
Aral Sea
Kara-Kum
Kyzyl Kum
Syr Dar'ya
Amu Dar'ya
Fergana Valley

WHY IT MATTERS

The countries of Uzbekistan, Kazakhstan, and Turkmenistan have large oil and natural gas reserves. Use **CNNfyi.com** or other **current events** sources to find examples of efforts to develop these resources. Record your findings in your journal.

A Bactrian camel and rider

Section 1 RESOURCES

Reproducible
- Block Scheduling Handbook, Chapter 24
- Guided Reading Strategy 24.1
- Geography for Life Activity 24
- Map Activity 24

Technology
- One-Stop Planner CD–ROM, Lesson 24.1
- Homework Practice Online
- Geography and Cultures Visual Resources with Teaching Activities 31–35
- *ARGWorld* CD–ROM: An Inland Sea in Central Asia
- HRW Go site

Reinforcement, Review, and Assessment
- Section 1 Review, p. 521
- Daily Quiz 24.1
- Main Idea Activity 24.1
- English Audio Summary 24.1
- Spanish Audio Summary 24.1

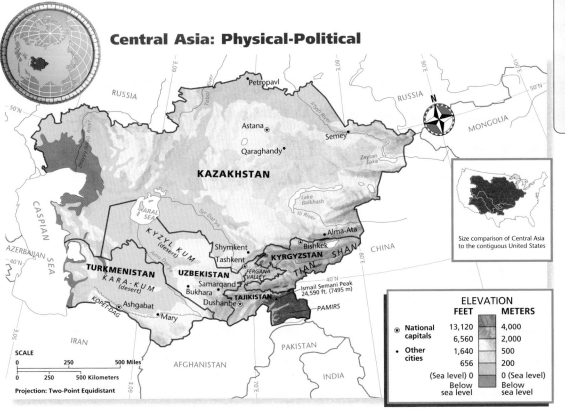

Central Asia: Physical-Political

RUSSIA

Petropavl

Astana

Semey

Qaraghandy

Zaysan Lake

KAZAKHSTAN

Lake Balkhash

Ili River

RUSSIA

MONGOLIA

CASPIAN SEA

ARAL SEA

Syr Dar'ya

KYZYL KUM (desert)

Alma-Ata

Shymkent

Bishkek

KYRGYZSTAN

TIAN SHAN

CHINA

AZERBAIJAN

Amu Dar'ya

Tashkent

FERGANA VALLEY

TURKMENISTAN

KARA-KUM (desert)

UZBEKISTAN

Samarqand

Bukhara

TAJIKISTAN

Dushanbe

Ismail Semani Peak 24,590 ft. (7495 m)

PAMIRS

KOPET-DAG

Ashgabat

Mary

IRAN

AFGHANISTAN

PAKISTAN

INDIA

Size comparison of Central Asia to the contiguous United States

SCALE
0   250   500 Miles
0   250   500 Kilometers
Projection: Two-Point Equidistant

ELEVATION

	FEET	METERS
National capitals	13,120	4,000
	6,560	2,000
Other cities	1,640	500
	656	200
	(Sea level) 0	0 (Sea level)
	Below sea level	Below sea level

Teaching Objectives 1–2

ALL LEVELS: (Suggested time: 20 min.) Copy the following graphic organizer onto the chalkboard, omitting the italicized answers. Use it to help students identify the main landforms, climates, and resources of Central Asia. Pair students and have each pair complete the organizer.
ENGLISH LANGUAGE LEARNERS, COOPERATIVE LEARNING

THE PHYSICAL GEOGRAPHY OF CENTRAL ASIA

Landforms	*mountains—the Pamirs, and Tian Shan; plains; low plateaus; deserts—Kara-Kum and Kyzyl Kum*
Climates	*steppe, desert, and highland*
Resources	*water—Syr Dar'ya and Amu Dar'ya; the Fergana Valley; oil, gas, gold, copper, uranium, zinc, lead, and coal*

➤**ASSIGNMENT:** Have students write a short story about an adventure in the mountains or deserts of Central Asia. Ask them to use elements in the Section 1 illustrations and Our Amazing Planet as ideas for the plot.

EYE ON EARTH

Lake Sarez in eastern Tajikistan was created in 1911 when a massive earthquake caused part of a mountain to collapse, blocking the Murgab River. Today the lake stretches about 37 miles (60 km) behind the dam created by the landslide. Downstream, waters of the Murgab flow into the Pyandzh River which becomes the Amu Dar'ya.

If another earthquake were to cause the dam to collapse, a huge wall of water would destroy entire villages and threaten the lives of thousands of people. The dam's isolation in the Pamirs and the poverty of Central Asian countries make it difficult to find a solution to this potentially disastrous problem.

Discussion: Ask students to consider how wealthy nations could help Central Asian and other poor countries meet the challenges posed by problems such as Lake Sarez.

☑ internet connect

GO TO: go.hrw.com
KEYWORD: SJ3 CH24
FOR: Web sites about the Pamirs

Visual Record Answer ▶

Their heavy fur coats protect them from the cold.

520

Mountain climbers make camp before attempting to scale a peak in the Pamirs.

Our Amazing Planet

With temperatures above 122°F (50°C), it is not surprising that the creatures that live in the Kara-Kum are a tough group. Over a thousand species live there, including cobras, scorpions, tarantulas, and monitor lizards. These lizards can grow to more than 5 feet (1.5m) long.

Lynxes can still be found in the mountains of Central Asia.
Interpreting the Visual Record How are these lynxes well suited to the environment in which they live?

Landforms and Climate

This huge, **landlocked** region is to the east of the Caspian Sea. Landlocked means the region does not border an ocean. The region lies north of the Pamirs (puh-MIRZ) and Tian Shan (TYEN SHAHN) mountain ranges.

Diverse Landforms As the name suggests, Central Asia lies in the middle of the largest continent. Plains and low plateaus cover much of this area. Around the Caspian Sea the land is as low as 95 feet (29 m) below sea level. However, the region includes high mountain ranges along the borders with China and Afghanistan.

Arid Lands Central Asia is a region of mainly steppe, desert, and highland climates. Summers are hot, with a short growing season. Winters are cold. Rainfall is sparse. However, north of the Aral (AR-uhl) Sea rainfall is heavy enough for steppe vegetation. Here farmers can grow crops using rain, rather than irrigation, as their water source. South and east of the Aral Sea lie two deserts. One is the Kara-Kum (kahr-uh-KOOM) in Turkmenistan. The other is the Kyzyl Kum (ki-ZIL KOOM) in Uzbekistan and Kazakhstan. Both deserts contain several **oasis** settlements where a spring or well provides water.

✔ **READING CHECK:** *Places and Regions* What are the landforms and climates of Central Asia? plains, low plateaus, high mountain ranges; steppe, desert, highland

Resources

The main water sources in southern Central Asia are the Syr Dar'ya (sir duhr-YAH) and Amu Dar'ya (uh-MOO duhr-YAH) rivers. These rivers flow down from the Pamirs and then across dry plains. Farmers have used them for irrigation for thousands of years. When it first flows down from the mountains, the Syr Dar'ya passes through the

CLOSE

Tell students that the mountains in Kyrgyzstan, particularly the Tian Shan—also known as the Celestial Mountains—are considered by some to be as beautiful as the mountains in Nepal and Switzerland. Have students share memories and impressions of mountains they have visited.

REVIEW AND ASSESS

Have students complete the Section Review. Then organize students into triads. Have each triad write eight matching questions that link landforms, climate, or natural resources to a particular country in Central Asia. Have groups exchange their questions and answer them. Then have students complete Daily Quiz 24.1. **COOPERATIVE LEARNING**

RETEACH

Have students complete Main Idea Activity 24.1. Then have students draw storyboards for a television commercial attempting to attract tourists to Central Asia. Commercials should highlight the region's landscape, climate, and resources. **ENGLISH LANGUAGE LEARNERS**

EXTEND

Tell students about the Torugart Pass, an overland route from Bishkek, Kyrgyzstan, through the Tian Shan to China. The pass was until recently closed to international travelers. Have interested students conduct research on and map the overland trade routes to China.
BLOCK SCHEDULING

Fergana Valley. This large valley is divided among Uzbekistan, Kyrgyzstan, and Tajikistan. As the river flows toward the Aral Sea, irrigated fields line its banks.

During the Soviet period, the region's population grew rapidly. Also, the Soviets encouraged farmers to grow cotton. This crop grows well in Central Asia's sunny climate. However, growing cotton uses a lot of water. Increased use of water has caused the Aral Sea to shrink.

A Dying Sea Today, almost no water from the Syr Dar'ya or Amu Dar'ya reaches the Aral Sea. The rivers' waters are used up by human activity. The effect on the Aral Sea has been devastating. It has lost more than 60 percent of its water since 1960. Its level has dropped 50 feet (15 m) and is still dropping. Towns that were once fishing ports are now dozens of miles from the shore. Winds sweep the dry seafloor, blowing dust, salt, and pesticides hundreds of miles.

Mineral Resources The Central Asian countries' best economic opportunity is in their fossil fuels. Uzbekistan, Kazakhstan, and Turkmenistan all have huge oil and natural gas reserves. However, transporting the oil and gas to other countries is a problem. Economic and political turmoil in some surrounding countries has made it difficult to build pipelines.

Several Central Asian countries are also rich in other minerals. They have deposits of gold, copper, uranium, zinc, and lead. Kazakhstan has vast amounts of coal. Rivers in Kyrgyzstan and Tajikistan could be used to create hydroelectric power.

✓ **READING CHECK:** *Environment and Society* How has human activity affected the Aral Sea? It has used up the water from the rivers that used to flow into the sea and has depleted the sea's water.

GO TO: go.hrw.com
KEYWORD: SJ3 CH24
FOR: Web sites about Central Asia

This boat sits rusting on what was once part of the Aral Sea. The sea's once thriving fishing industry has been destroyed.

Homework Practice Online
Keyword: SJ3 HP24

Section Review 1

Define and explain: landlocked, oasis

Working with Sketch Maps On a map of Central Asia that you draw or that your teacher provides, label the following: Pamirs, Tian Shan, Aral Sea, Kara-Kum, Kyzyl Kum, Syr Dar'ya, Amu Dar'ya, and Fergana Valley.

Reading for the Main Idea

1. *Environment and Society* What has caused the drying up of the Aral Sea?

2. *Places and Regions* What mineral resources does Central Asia have?

Critical Thinking

3. **Analyzing Information** Why did the Soviets encourage Central Asian farmers to grow cotton?

4. **Finding the Main Idea** What factors make it hard for the Central Asian countries to export oil and gas?

Organizing What You Know

5. **Sequencing** Copy the following graphic organizer. Use it to describe the courses of the Syr Dar'ya and Amu Dar'ya, including human activities that use water.

Melting snows in the Pamirs ⇨ [ ] ⇨ Aral Sea

Section Review 1

Answers

Define For definitions, see: landlocked, p. 520; oasis, p. 520

Working with Sketch Maps Maps will vary, but listed places should be labeled in their approximate location.

Reading for the Main Idea

1. The rivers' waters are used up by human activity upstream. (NGS 14)

2. oil, natural gas, gold, copper, uranium, zinc, lead, and coal (NGS 4)

Critical Thinking

3. probably because cotton grows well in Central Asia

4. Economic and political turmoil in some surrounding countries has made it difficult to build pipelines in the region.

Organizing What You Know

5. irrigation of crops

Section 2

Objectives

1. Explain how trade and invasions affected Central Asia's history.
2. Describe the political and economic conditions in Central Asia today.

LET'S GET STARTED

Copy the following instructions onto the chalkboard: *Write down three things about life as a nomad. Use knowledge you may have of the Plains Indians, touring performers, migrant workers, and so on.* Discuss responses. Tell students that in Section 2 they will learn more about the nomadic peoples of Central Asia.

Building Vocabulary

Write the Define terms **nomads** and **caravans** on the chalkboard. Have volunteers read aloud the definitions from the text's glossary. Tell students that the word *nomad* comes from the Greek language and that the word *caravan* is Persian in origin. Point out that Greece and Persia are two of the major cultural influences on Central Asia.

Section 2 RESOURCES

Reproducible
- Guided Reading Strategy 24.2
- Graphic Organizer 24
- Cultures of the World Activity 4

Technology
- One-Stop Planner CD–ROM, Lesson 24.2
- Homework Practice Online
- HRW Go site

Reinforcement, Review, and Assessment
- Section 2 Review, p. 524
- Daily Quiz 24.2
- Main Idea Activity 24.2
- English Audio Summary 24.2
- Spanish Audio Summary 24.2

Visual Record Answer ▶

Students might mention the mosques and minarets.

Section 2 — History and Culture

Read to Discover

1. How did trade and invasions affect the history of Central Asia?
2. What are political and economic conditions like in Central Asia today?

WHY IT MATTERS

Even though the Soviet Union has collapsed, traces of its influence remain in Central Asia, particularly in government. Use **CNNfyi.com** or other **current events** sources to find examples of political events in these countries. Record your findings in your journal.

Define
nomads
caravans

An ancient Kyrgyz stone figure

Bukhara, in Uzbekistan, was once a powerful and wealthy trading center of Central Asia.

Interpreting the Visual Record
What architectural features can you see that distinguish Bukhara as an Islamic city?

▼

History

For centuries, Central Asians have made a living by raising horses, cattle, sheep, and goats. Many of these herders lived as **nomads**, people who often move from place to place. Other people became farmers around rivers and oases.

Trade At one time, the best land route between China and the eastern Mediterranean ran through Central Asia. Merchants traveled in large groups, called **caravans**, for protection. The goods they carried included silk and spices. As a result, this route came to be called the Silk Road. Cities along the road became centers of wealth and culture.

Central Asia's situation changed after Europeans discovered they could sail to East Asia through the Indian Ocean. As a result, trade through Central Asia declined. The region became isolated and poor.

TEACH

Teaching Objectives 1–2

ALL LEVELS: (Suggested time: 20 min.) Copy the following graphic organizer onto the chalkboard, omitting the italicized answers. Use it to help students understand how trade and invasions affected Central Asia's history and its current political and economic conditions. Have volunteers fill in the organizer on the chalkboard. Tell students to copy the organizer into their notebooks. **ENGLISH LANGUAGE LEARNERS**

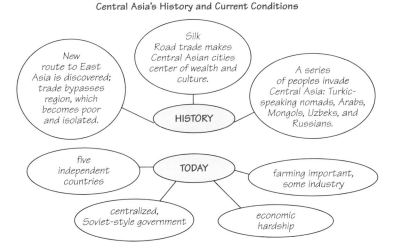

New route to East Asia is discovered; trade bypasses region, which becomes poor and isolated.

Silk Road trade makes Central Asian cities center of wealth and culture.

A series of peoples invade Central Asia: Turkic-speaking nomads, Arabs, Mongols, Uzbeks, and Russians.

HISTORY

five independent countries

TODAY

farming important, some industry

centralized, Soviet-style government

economic hardship

CONNECTING TO History

Silk processing in modern Uzbekistan

The Silk Road

The Silk Road stretched 5,000 miles (8,000 km) across Central Asia from China to the Mediterranean Sea. Along this route passed merchants, armies, and diplomats. These people forged links between East and West.

The facts of the Silk Road are still wrapped in mystery. Chinese trade and military expeditions probably began moving into Central Asia in the 100s B.C. Chinese trade goods soon were making their way to eastern Mediterranean ports.

Over the next several centuries, trade in silk, spices, jewels, and other luxury goods increased. Great caravans of camels and oxcarts traveled the Silk Road in both directions. They crossed the harsh deserts and mountains of Central Asia. Cities like Samarqand and Bukhara grew rich from the trade. In the process, ideas and technology also moved between Europe and Asia.

Travel along the Silk Road was hazardous. Bandits often robbed the caravans. Some travelers lost their way in the desert and died. In addition, religious and political turmoil occasionally disrupted travel.

Understanding What You Read

1. What was the Silk Road?
2. Why was the Silk Road important?

Trade center

Silk Road

Invasions and the Soviet Era About A.D. 500, Turkic-speaking nomads from northern Asia spread through Central Asia. In the 700s Arab armies took over much of the region, bringing Islam. In the 1200s the armies of Mongol leaders conquered Central Asia. Later, another Turkic people, the Uzbeks, took over parts of the region. In the 1800s the Russian Empire conquered Central Asia.

After the Russian Revolution, the Soviet government set up five republics in Central Asia. The Soviets encouraged ethnic Russians to move to this area and made the nomads settle on collective ranches or farms. Religion was discouraged. Russian became the language of government and business. The government set up schools and hospitals. Women were allowed to work outside the home.

✓ READING CHECK: *Human Systems* What type of government system did the five republics set up by the Soviet Union have? one in which the government controlled many aspects of life

GLOBAL PERSPECTIVES

Between 1949 and 1989 the Soviet government conducted more than 400 nuclear tests above, on, or below the surface of an area in Kazakhstan called the Polygon. Nearby residents—some of whom could see mushroom clouds from their homes—were told that the tests were important scientific research. They were not informed of the dangers.

Radiation has since devastated area residents. Thousands have died from cancer, including rare forms of the disease. As many as 100,000 people who received the heaviest doses of radiation have passed genetic defects on to their children and grandchildren. Many babies in the area are born with health problems.

Critical Thinking: What responsibility does Kazakhstan's new government have to its citizens who were hurt by Soviet nuclear testing?

Answer: Some students might mention that the country has limited responsibility but should nonetheless help its citizens.

Connecting to History
Answers

1. a trade route linking China to the Mediterranean Sea

2. linked East and West; helped ideas move between Europe and Asia

523

Section Review 2

Answers

Define For definitions, see: nomads, p. 522; caravans, p. 522

Working with Sketch Maps Places should be labeled in their approximate locations.

Reading for the Main Idea

1. farming, herding, trading (NGS 14)

2. Possible answers: Turkic-speaking nomads, Arab armies, Mongols, Uzbeks, the Russian Empire (NGS 13)

3. Soviets set up five republics, made nomads settle on collective ranches or farms, discouraged religion, made Russian official language, established schools. (NGS 12)

Critical Thinking

4. reaction against long Soviet rule; desire to be more Western

Organizing What You Know

5. primary—growing crops, mining metals, raising livestock, drilling for oil; secondary—making cloth, food products, chemicals from oil; manufacturing tractors

Visual Record Answer ▶

Students might mention the desks or the students' uniforms.

524

BUILD on WHAT You Know

Do you remember what you learned about acculturation? See Chapter 5 to review.

A Kyrgyz teacher conducts class.
Interpreting the Visual Record
How is this class similar to yours?

▼

Section Review 2

Define and explain: nomads, caravans

Working with Sketch Maps On the map you created in Section 1, draw and label the five Central Asian countries.

Reading for the Main Idea

1. *Environment and Society* How have the people of Central Asia made a living over the centuries?

2. *Human Systems* What are four groups that invaded Central Asia?

3. *Human Systems* How did Soviet rule change Central Asia?

Central Asia Today

The five republics became independent countries when the Soviet Union broke up in 1991. All have strong economic ties to Russia. Ethnic Russians still live in every country in the region. However, all five countries are switching from the Cyrillic alphabet to the Latin alphabet. The Cyrillic alphabet had been imposed on them by the Soviet Union. The Latin alphabet is used in most Western European languages, including English, and in Turkey.

Government All of these new countries have declared themselves to be democracies. However, they are not very free or democratic. Each is ruled by a strong central government that limits opposition and criticism.

Economy Some of the Central Asian countries have oil and gas reserves that may someday make them rich. For now, though, all are suffering economic hardship. Causes of the hardships include outdated equipment, lack of funds, and poor transportation links.

Farming is important in the Central Asian economies. Crops include cotton, wheat, barley, fruits, vegetables, almonds, tobacco, and rice. Central Asians raise cattle, sheep, horses, goats, and camels. They also raise silkworms to make silk thread.

Industry in Central Asia includes food processing, wool textiles, mining, and oil drilling. Oil-rich Turkmenistan and Kazakhstan also process oil into other products. Kazakhstan and Uzbekistan make heavy equipment such as tractors.

✓ **READING CHECK:** *Human Systems* How do political freedoms in the region compare to those of the United States? *Citizens' criticism of or opposition to the government is limited, whereas in the United States people can speak freely.*

go.hrw.com **Homework Practice Online**
Keyword: SJ3 HP24

Critical Thinking

4. **Drawing Inferences and Conclusions** What does the switch to the Latin alphabet suggest about the Central Asian countries?

Organizing What You Know

5. **Categorizing** Copy the following graphic organizer. Use it to categorize economic activities in Central Asia. Place the following items in the chart: making cloth, growing crops, mining metals, making food products, raising livestock, making chemicals from oil, drilling for oil, and manufacturing tractors.

Primary industries	Secondary industries

Section 3

Objectives

1. Identify some important aspects of culture in Kazakhstan.
2. Explain how Kyrgyz culture reflects nomadic traditions.
3. Discuss the political violence of recent years in Tajikistan.
4. Describe two important art forms in Turkmenistan.
5. Analyze how Uzbekistan's population is significant.

FOCUS

LET'S GET STARTED

Copy the following instructions onto the chalkboard: *Imagine that a student from Central Asia has joined our class for the next year. Write down four customs or other facets of our lives you believe that he or she should see or experience.* Ask volunteers to share their lists with the class. Tell students that in Section 3 they will learn more about the individual countries of Central Asia.

Building Vocabulary

Write the terms **yurt** and **mosques** on the chalkboard. Have volunteers locate and read aloud the definitions from the text's glossary. Ask students to suggest structures they have seen or heard of with similar shapes or functions. *(Possible answers: church, temple, tepee, or trailer)*

Section 3: The Countries of Central Asia

Read to Discover

1. What are some important aspects of culture in Kazakhstan?
2. How does Kyrgyz culture reflect nomadic traditions?
3. Why have politics in Tajikistan in recent years been marked by violence?
4. What are two important art forms in Turkmenistan?
5. How is Uzbekistan's population significant?

Define
yurt
mosques

Locate
Tashkent
Samarqand

WHY IT MATTERS

Since the collapse of the Soviet Union, religious freedom is more common in Central Asia. Use **CNN fyi.com** or other **current events** sources to find examples of religious and ethnic differences in the countries of Central Asia. Record your findings in your journal.

A warrior's armor from Kazakhstan

internet connect

GO TO: go.hrw.com
KEYWORD: SJ3 CH19
FOR: Web sites about carpets

Kazakhstan

Of the Central Asian nations, Kazakhstan was the first to be conquered by Russia. Russian influence remains strong there. About one third of Kazakhstan's people are ethnic Russians. Kazakh and Russian are both official languages. Many ethnic Kazakhs grow up speaking Russian at home and have to learn Kazakh in school.

Kazakhstanis celebrate the New Year twice—on January 1 and again on Nauruz, the start of the Persian calendar's year. Nauruz falls on the spring equinox.

Food in Central Asia combines influences from Southwest Asia and China. Rice, yogurt, and grilled meat are common ingredients. One Kazakh specialty is smoked horsemeat sausage with cold noodles.

✓ **READING CHECK:** *Human Systems* How has Kazakhstan been influenced by Russia? was first to be conquered by Russia; about one third of the people are ethnic Russians; Russian is commonly spoken

Kyrgyzstan

Kyrgyzstan has many mountains, and the people live mostly in valleys. People in the southern part of the country generally share cultural ties with Uzbekistan. People in northern areas are more linked to nomadic cultures and to Kazakhstan.

A woman in Uzbekistan grills meat on skewers.

Teaching Objectives 1–5

ALL LEVELS: (Suggested time: 30 min.) Copy the following graphic organizer onto the chalkboard, omitting the italicized answers. Point out to students that each circle represents one of the Central Asian countries. The lines indicate that the countries have much in common. Refer students to the Read to Discover questions. Call on students to suggest phrases that answer the questions and that highlight the countries' distinct traits. Write the phrases in the circles. **ENGLISH LANGUAGE LEARNERS**

Distinctive Traits of Central Asian Countries

Kazakhstan
- *Russian influence strong*
- *Kazakh and Russian official languages*
- *combined influences in food*

Turkmenistan
- *English—second official language*
- *Islamic principles taught in schools*
- *important art forms—carpets, poetry*

Kyrgyzstan
- *in north—linked to nomadic cultures*
- *clan membership still important*
- *black and white felt hats to show class status*
- *yurts—movable round house*

Tajikistan
- *civil war in mid-1990s: Soviet-style government against reformers*
- *language related to Persian*
- *Persian literature part of heritage*

Uzbekistan
- *largest population in Central Asia*
- *Uzbek—official language*
- *Tashkent and Samarkand— Silk Road cities*
- *traditional art— embroidering with gold*

Section Review 3

Answers

Define For definitions, see: yurt, p. 526; mosques, p. 526

Working with Sketch Maps Maps should correctly depict the locations of the two cities.

Reading for the Main Idea

1. Kazakhstan; about one third of people there are ethnic Russians (NGS 4)

2. type of government— Soviet-style or reform (NGS 13)

3. It has experienced a revival. (NGS 10)

Critical Thinking

4. Answers might include yurts, carpets.

Organizing What You Know

5. Soviet—communist; Russian; Cyrillic; Islam discouraged; today—so-called democracies but actually centralized, Soviet-style; in some places, Russian and local language, like Kazakh; Latin; government supports revival

Chart Answer ▶

It is much greater.

Visual Record Answer ▶

Students might say that carpets can be transported to any location.

Ethnic Makeup of Kazakhstan and Uzbekistan

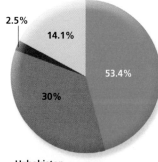

Kazakhstan

2.5%
14.1%
53.4%
30%

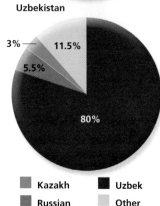

Uzbekistan

3%
11.5%
5.5%
80%

■ Kazakh ■ Uzbek
■ Russian ■ Other

Source: Central Intelligence Agency, *The World Factbook 2001*

Interpreting the Chart How does the number of Russians in Kazakhstan compare to that in Uzbekistan?

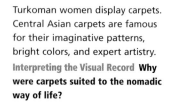

Turkoman women display carpets. Central Asian carpets are famous for their imaginative patterns, bright colors, and expert artistry.

Interpreting the Visual Record Why were carpets suited to the nomadic way of life?

▶

The word *kyrgyz* means "forty clans." Clan membership is still important in Krygyz social, political, and economic life. Many Kyrgyz men wear black and white felt hats that show their clan status.

Nomadic traditions are still important to many Kyrgyz. The **yurt** is a movable round house of wool felt mats over a wood frame. Today the yurt is a symbol of the nomadic heritage. Even people who live in cities may put up yurts for weddings and funerals.

✓ **READING CHECK:** *Human Systems* In what ways do the Kyrgyz continue traditions of their past? by emphasizing clan membership and constructing yurts

Tajikistan

In the mid-1990s Tajikistan experienced a civil war. The Soviet-style government fought against a mixed group of reformers, some of whom demanded democracy. Others called for government by Islamic law. A peace agreement was signed in 1996, but tensions remain high.

The other major Central Asian languages are related to Turkish. However, the Tajik language is related to Persian. Tajiks consider the great literature written in Persian to be part of their cultural heritage.

✓ **READING CHECK:** *Human Systems* What has happened in politics in recent years in Tajikistan? In the mid-1990s there was a civil war between reformers and Soviet-style government; tensions remain high.

Turkmenistan

The major first language of Turkmenistan is Turkmen. In 1993 Turkmenistan adopted English, rather than Russian, as its second official language. However, some schools teach in Russian, Uzbek, or Kazakh.

Islam has experienced a revival in Central Asia since the breakup of the Soviet Union. Many new **mosques**, or Islamic houses of worship, are being built and old ones are being restored. Donations from other Islamic countries, such as Saudi Arabia and Iran, have helped

CLOSE

Ask students whether they can note any similarities between cultural practices in the United States and those in Central Asia.

REVIEW AND ASSESS

Have students complete the Section Review. Then organize students into groups of four. Have each group write five multiple-choice questions about the section. Have groups trade questions and then answer them. Then have students complete Daily Quiz 24.3. **COOPERATIVE LEARNING**

RETEACH

Have students complete Main Idea Activity 24.3. Then have students design flags for each Central Asian nation, including symbols for peoples, customs, and religion. **ENGLISH LANGUAGE LEARNERS**

EXTEND

Have interested students conduct research on the resurgence of barter as a means of doing business in the Central Asian republics. Have students develop their own barter system for basic goods. **BLOCK SCHEDULING**

---

these efforts. The government of Turkmenistan supports this revival and has ordered schools to teach Islamic principles. However, like the other states in the region, Turkmenistan's government views Islam with some caution. It does not want Islam to become a political movement.

Historically, the nomadic life required that all possessions be portable. Decorative carpets were the essential furniture of a nomad's home. They are still perhaps the most famous artistic craft of Turkmenistan. Like others in Central Asia, the people of Turkmenistan also have an ancient tradition of poetry.

✓ **READING CHECK:** *Human Systems* What are two forms of art in Turkmenistan, and how do they reflect its cultural traditions? decorative carpets, poetry; ancient forms of artistic expression

Central Asia

COUNTRY	POPULATION/ GROWTH RATE	LIFE EXPECTANCY	LITERACY RATE	PER CAPITA GDP
Kazakhstan	16,731,303 .03%	58, male 69, female	98%	$5,000
Kyrgyzstan	4,753,003 1.4%	59, male 68, female	97%	$2,700
Tajikistan	6,578,681 2.1%	61, male 67, female	98%	$1,140
Turkmenistan	4,603,244 1.9%	57, male 65, female	98%	$4,300
Uzbekistan	25,155,064 1.6%	60, male 68, female	99%	$2,400
United States	281,421,906 0.9%	74, male 80, female	97%	$36,200

Sources: Central Intelligence Agency, *The World Factbook 2001;* U.S. Census Bureau

Interpreting the Chart **Which country has the lowest per capita GDP in the region?**

Uzbekistan

Uzbekistan has the largest population of the Central Asian countries—about 24 million people. Uzbek is the official language. People are required to study Uzbek to be eligible for citizenship.

Tashkent and Samarqand are ancient Silk Road cities in Uzbekistan. They are famous for their mosques and Islamic monuments. Uzbeks are also known for their art of embroidering fabric with gold.

✓ **READING CHECK:** *Human Systems* What is one of an Uzbekistan citizen's responsibilities? to study the Uzbek language

go.hrw.com **Homework Practice Online**
Keyword: SJ3 HP24

Section Review 3

Define and explain: yurt, mosques

Working with Sketch Maps On the map you created in Section 2, label Tashkent and Samarqand.

Reading for the Main Idea

1. *Places and Regions* In which Central Asian nation is the influence of Russia strongest? Why is this true?

2. *Human Systems* What were the two sides in Tajikistan's civil war fighting for?

3. *Human Systems* What is the role of Islam in the region today?

Critical Thinking

4. **Finding the Main Idea** What are two customs or artistic crafts of modern Central Asia that are connected to the nomadic lifestyle?

Organizing What You Know

5. **Contrasting** Copy the following graphic organizer. Use it to describe the conditions in Central Asia during the Soviet era and today.

	Soviet era	Today
Type of government		
Official language		
Alphabet		
Government attitude toward Islam		

Review ANSWERS

Building Vocabulary

For definitions, see: landlocked, p. 520; oasis, p. 520; nomads, p. 522; caravans, p. 522; yurt, p. 526; mosques, p. 526

Reviewing the Main Ideas

1. steppe, desert, and highland

2. Much of the fishing industry has been destroyed, and dust from the sea floor endangers people's health. (NGS 4)

3. Soviets required that nomads settle on collective farms and ranches, discouraged the practice of religion, and made Russian the official language. (NGS 15)

4. They have strong economic ties, and many ethnic Russians still live throughout the region. (NGS 13)

5. Russian, Kazakh, Tajik, English, Uzbek, and other local languages (NGS 4)

▲ **Chart Answer**

Tajikistan

527

Setting the Scene

In areas like Central Asia where climate and soil are not favorable for agricultural development, pastoral nomadism is a lifestyle that allows people to feed themselves. The Kazakhs have traditionally supported themselves by raising camels, cattle, horses, and sheep. In order to feed their herds and to ensure that fresh pastures are always available, the Kazakhs have developed a complex seasonal migration system. This migratory lifestyle is reflected in the kinds of homes they build—yurts. During the Soviet rule of Kazakhstan, new systems of agriculture requiring fixed settlements were introduced into the area. These farms became obstacles to the seasonal movement of animals.

Building a Case

After students have read the case study, ask them the following questions: Who are pastoral nomads? *(people who move around during the year to maintain herds of livestock)* What are some of the animals that pastoral nomads herd? *(cattle, horses, sheep, goats, yaks, reindeer, and camels)* Why must animals in desert regions be kept moving throughout the year? *(to prevent overgrazing and provide fresh grass shoots)* How far do some Kazakh families move in a year? *(500 miles or 805 km)* How do the kinds of houses built by Kazahks reflect their nomadic lifestyle? *(Yurts are impermanent structures that are designed to be easily moved.)*

HISTORICAL GEOGRAPHY

An example of a current conflict between farmers and nomads is the fighting between the Hutu and Tutsi in eastern Africa. The Hutu first settled the area that is now Rwanda and Burundi from about 500 B.C. to the year 1000. Hutu life centered on small-scale agriculture. During the 1300s and 1400s the Tutsi entered the region. They were pastoral nomads who depended on large herds of big-horned cattle. The Tutsi soon gained control over the area, gave up their nomadic lifestyle and became "lords" over Hutu farmers. Despite their small numbers, the Tutsi dominated the Hutu economically and politically for almost 600 years. In the late 1990s violence erupted between the Hutu and the Tutsi. In 1994 the United Nations sent peace-keepers to stabilize the region.

Activity: Have students conduct research on the current relationship between the Hutu and the Tutsi. Ask students to find out how successful UN forces were at bringing peace to the region. Have students report their findings in the form of a newsmagazine article.

➤ This Case Study feature addresses National Geography Standards 3, 4, 8, 9, and 15.

CASE STUDY

KAZAKHS: PASTORAL NOMADS OF CENTRAL ASIA

Nomads are people who move around from place to place during the year. Nomads usually move when the seasons change so that they will have enough food to eat. Herding, hunting, gathering, and fishing are all ways that different nomadic groups get their food.

Nomads that herd animals are called pastoral nomads. Their way of life depends on the seasonal movement of their herds. Pastoral nomads may herd cattle, horses, sheep, goats, yaks, reindeer, camels, or other animals. Instead of keeping their animals inside fenced pastures, pastoral nomads let them graze on open fields. However, they must make sure the animals do not overgraze and damage the pastureland. To do this, they keep their animals moving throughout the year. Some pastoral nomads live in steppe or desert environments. These nomads often have to move their animals very long distances between winter and summer pastures.

The Kazakhs of Central Asia are an example of a pastoral nomadic group. They have herded horses, sheep, goats, and cattle for hundreds of years. Because they move so much, the Kazakhs do not have permanent homes. They bring their homes with them when they travel to new places.

A Kazakh nomad keeps a watchful eye over a herd of horses.
▼

Drawing Conclusions

Focus discussion on the challenges confronted by the pastoral nomads in Kazakhstan. Ask students when Russians and eastern Europeans began settling the area. *(late 1800s)* How was their food production system different? *(They were farmers, not herders.)* How did this system conflict with the nomadic way of life? *(It blocked access to pasture lands.)* Here is some additional information to share with students: Between 1906 and 1912, Russian settlers established more than 500,000 farms. Many Kazakhs were displaced. In the 1950s and 1960s Soviet officials tried to settle the remaining nomads and expand cultivation. Some 60 percent of Kazakhstan's pasture lands were planted with crops. Today, 84 percent of pastures are devoted to cattle and sheep. Point out that this suggests the area may be more suited to herding than farming.

Going Further: Thinking Critically

The Kazakhs are only one example of nomads. Other examples include the Inuit of the Arctic, Plains Indians of North America, !Kung of the Kalahari Desert, Ona of Tierra del Fuego, Fulani of West Africa, Bedouins of the Arabian Desert, Masai of East Africa, Chukchi of Siberia, Lapps of Finland, and Tuareg of the Sahara. Have students work in small groups to conduct research on a nomadic group. Ask them to answer these questions: What is the people's main food source? What is the climate and soil like where they live? What environmental pressures do they experience? What kind of houses do they build? Have they had conflicts with settled farming peoples? If so, what were the issues? What was the outcome? You may want to have each group create a puppet show to present its findings.

◄ Yurts are carefully stitched together by hand. When it is time to move to new pastures, they are carried from place to place on horseback or on small wagons.

The Kazakhs live in tent-like structures called yurts. Yurts are circular structures made of bent poles covered with thick felt. Yurts can be easily taken apart and moved. They are perfect homes for the Kazakhs' nomadic lifestyle.

During the year, a Kazakh family may move its herds of sheep, horses, and cattle as far as 500 miles (805 km). For one Kazakh family, each year is divided into four different parts. The family spends the first part of the year in winter grazing areas. Then, in early spring, they move to areas with fresh grass shoots. When these spring grasses are gone, the family moves their animals to summer pastures. In the fall, the animals are kept for six weeks in autumn pastures. Finally, the herds are taken back to their winter pastures. Each year, the cycle is repeated.

The nomadic lifestyle of the Kazakhs has changed, however. In the early 1800s people from Russia and eastern Europe began to move into the region. These people were farmers. They started planting crops in areas that the Kazakhs used for pasture. This made it more difficult for the Kazakhs to move their animals during the year. Later, when Kazakhstan was part of the Soviet Union, government officials encouraged the Kazakhs to settle in villages and cities. Many Kazakhs still move their animals during the year. However, tending crops has also become an important way to get food.

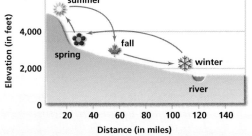

Seasonal Movement of a Kazakh Family

▲ During the year, a Kazakh family may move its herds to several different pasture areas as the seasons change.

Interpreting the Graph Why do you think animals are moved to higher elevations during the summer and to lower elevations during the winter?

Understanding What You Read

1. Why do some pastoral nomads have to travel such great distances?
2. How has the nomadic lifestyle of the Kazakhs changed during the last 100 years?

ASSESS

Have students complete the Chapter 24 Test.

RETEACH

Point out that making carpets and writing poetry are two forms of art highly valued in Central Asia, particularly in Turkmenistan. Organize the class into two groups, with one group creating a carpet design and the other writing a poem. Have students meet within their groups to decide on a overall concept and to volunteer for certain topics. Topics will include landforms, mineral resources, invasions, and other subjects listed in the chapter. Then let individual students work on their topics, either as designs or verses. Finally,

have students combine their contributions into one carpet design and one poem. **ENGLISH LANGUAGE LEARNERS, COOPERATIVE LEARNING**

PORTFOLIO EXTENSIONS

1. Have students build model yurts with sticks, felt, and cloth. Point out that to be useful as movable structures, yurts must be constructed so they can be folded and loaded onto a wagon or horse. Have students photograph their yurts for their portfolios.

2. Have students conduct research on calligraphy, which is important to Uzbekistan, and use it to render the Latin and Cyrillic alphabets in pen and ink. Ask students to write their names in both alphabets. You might also challenge students to find a sentence or phrase in the Uzbek language and write it in calligraphic script.

Review
ANSWERS

Understanding Environment and Society

Students may find that beginning in the 1960s the water level was reduced because water was being diverted for irrigation. Governmental agencies have suggested options for saving the sea, including reducing the amount of cotton being grown, more efficient irrigation methods, or shifting from an agricultural economy. Students may find that consequences such as habitat destruction, species loss, and negative health effects may worsen if the sea's water level continues to drop. Use Rubric 29, Presentations, to evaluate student work.

Thinking Critically

1. Soviets imposed Russian, Cyrillic alphabet on Central Asia; after Soviet collapse, some nations returned to local languages, adopted Latin alphabet.

2. Answers may vary, but carpets and yurts—two art forms—can be moved around.

3. because a growing population needed more cotton; caused the Aral Sea to shrink

Reviewing What You Know

Building Vocabulary

On a separate sheet of paper, write sentences to define each of the following words.

1. landlocked
2. oasis
3. nomads
4. caravans
5. yurt
6. mosques

Reviewing the Main Ideas

1. (*Places and Regions*) What types of climates are most common in Central Asia?

2. (*Environment and Society*) What problems have resulted from the shrinking of the Aral Sea?

3. (*Human Systems*) How did Soviet rule change Central Asians' way of life?

4. (*Human Systems*) What kinds of ties do the Central Asian countries have to Russia today?

5. (*Places and Regions*) What are the various languages spoken in Central Asia?

Understanding Environment and Society

Aral Sea in Danger

The rapid disappearance of the Aral Sea is a serious concern in Central Asia. Research and create a presentation on the Aral Sea. You may want to think about the following:

- Actions that could be taken to preserve the Aral Sea.
- What could be done to help slow the dropping of the sea's water level.
- Possible consequences if the level of the Aral Sea continues to drop.

Include a bibliography of the sources you used.

Thinking Critically

1. **Summarizing** How have politics influenced language and the alphabet used in Central Asia?

2. **Finding the Main Idea** How do the artistic crafts of Central Asia reflect the nomadic lifestyle?

3. **Analyzing Information** Why did the Soviets encourage cotton farming in the region? What were the environmental consequences?

4. **Finding the Main Idea** What obstacles are making it hard for the Central Asian countries to export their oil?

5. **Summarizing** What are some reasons the Central Asian countries have experienced slow economic growth since independence? How are they trying to improve the situation?

FOOD FESTIVAL

Here is an easy recipe for Kazakh Rice Salad. You may want to ask students to contribute the ingredients. Then mix up the salad in class. This recipe serves six. Double or triple it to serve everyone. Combine these ingredients in a large bowl: 1 c. cooked brown rice, cold; 1 c. cooked buckwheat, cold; ½ c. dates, chopped; 3 cloves garlic, minced; 2 tbs. fresh ginger, minced; ¾ c. cashews or almonds, crushed; ½ c. rice wine vinegar; ½ c. extra virgin olive oil; ¾ c. fresh cilantro, chopped.

CHAPTER 24 REVIEW AND ASSESSMENT RESOURCES

Reproducible
- ◆ Readings in World Geography, History, and Culture 45 and 46
- ◆ Critical Thinking Activity 24
- ◆ Vocabulary Activity 24

Technology
- ◆ Chapter 24 Test Generator (on the One-Stop Planner)

- ◆ HRW Go site
- ◆ Audio CD Program, Chapter 24

Reinforcement, Review, and Assessment
- ◆ Chapter 24 Review, pp. 530–31
- ◆ Chapter 24 Tutorial for Students, Parents,

Mentors, and Peers
- ◆ Chapter 24 Test
- ◆ Chapter 24 Test for English Language Learners and Special-Needs Students
- ◆ Unit 6 Test
- ◆ Unit 6 Test for English Language Learners and Special-Needs Students

---

Building Social Studies Skills

Map ACTIVITY

On a separate sheet of paper, match the letters on the map with their correct labels.

Caspian Sea	Kara-Kum
Pamirs	Kyzyl Kum
Tian Shan	Tashkent
Aral Sea	Samarqand

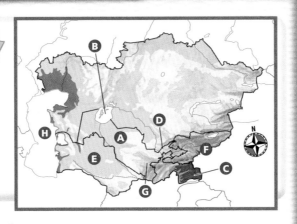

Mental Mapping Skills ACTIVITY

On a separate sheet of paper, draw a freehand map of Central Asia. Make a key for your map and label the following:

Amu Dar'ya	Syr Dar'ya
Aral Sea	Tajikistan
Kazakhstan	Turkmenistan
Kyrgyzstan	Uzbekistan

WRITING ACTIVITY

Imagine that you are a caravan trader traveling along the Silk Road during the 1200s. Write a journal entry describing your journey from the Mediterranean Sea through Central Asia. Be sure to use standard grammar, spelling, sentence structure, and punctuation.

---

4. economic, political turmoil has made it difficult to build pipelines
5. outdated equipment, lack of funds, poor transportation links; some nations are trying to diversify their economic products.

Map Activity
A. Kyzyl Kum E. Kara-Kum
B. Aral Sea F. Tian Shan
C. Pamirs G. Samarqand
D. Tashkent H. Caspian Sea

Mental Mapping Skills Activity
Maps will vary, but listed places should be labeled in their approximate locations.

Writing Activity
Information could include difficult travel due to unfamiliar terrain, varied climate, warring factions. Use Rubric 15, Journals, to evaluate work.

Portfolio Activity
Students' findings will vary, but they should include information about a local hero. Use Rubric 4, Biographies, to evaluate student work.

Alternative Assessment

Portfolio ACTIVITY

Learning About Your Local Geography

History The Mongol conqueror Genghis Khan is a hero to many Central Asians. Use biographies or interviews with residents to find out about a person who is special to your area. Report your findings.

internet connect
Internet Activity: go.hrw.com
KEYWORD: SJ3 GT24

Choose a topic to explore about Central Asia:
- Study the climate of Central Asia.
- Travel along the historic Silk Road.
- Learn about nomads and caravans.

internet connect
GO TO: go.hrw.com
KEYWORD: SJ3 Teacher
FOR: a guide to using the Internet in your classroom

Recalling Concepts

Review with students what they learned in the preceding unit about the political and economic relationships between the countries of Central Asia and the former Soviet Union. *(Students should mention that these countries were Soviet republics, that the economies of these countries were part of the Soviet economy, and that the Soviet Union set up schools and hospitals and made Russian the language of government and business.)* Point out that these countries were part of the Soviet Union for approximately 70 years.

Remind students that the cultural and ethnic differences between the people of these countries and of Russia contributed to the drive by Central Asians to gain independence from the Soviet Union. On a wall map show students the relative location of Kazakhstan, Uzbekistan, Turkmenistan, Azerbaijan, Kyrgyzstan, and Tajikistan between Russia and the countries of Southwest Asia.

FOCUS ON CULTURE

Facing the Past and Present

Patterns of trade and culture can change quickly in our modern world. For example, the United States used to trade primarily with Europe. Most immigrants to the United States also came from Europe. Today, the American connection to Europe has faded. The United States now trades more with Japan and other Pacific Rim countries than with Europe. New ideas, new technology, and immigrants to the United States come from all around the world.

Central Asia Since the breakup of the Soviet Union, similar changes have taken place in Central Asia. In the past, Central Asia had many ties to the Soviet Union. For example, the economies of the two regions were linked. Central Asia exported cotton and oil to Russia and to countries in Eastern Europe. In exchange, Central Asia received a variety of manufactured goods. The Soviet Union also heavily influenced the culture of Central Asia. Many Central Asians learned to speak Russian.

Looking South Today, Central Asia's links to the former Soviet Union have weakened. At the same time, its ancient ties to Southwest Asia have grown stronger. The Silk Road once linked Central Asian cities to Southwest Asian ports on the Mediterranean. Now the peoples of Central Asia are looking southward once again. New links are forming between Central Asia and Turkey. Many people in Central Asia are traditionally Turkic in culture and language. Turkey's business leaders are working to expand their industries in Central Asia. Also, regular air travel from Turkey to cities in Central Asia is now possible as well.

Religion also links both Central Asia and Southwest Asia. Islam was first introduced into Central Asia in the A.D. 700s. It eventually became the region's dominant religion. However, Islam declined during the Soviet era. Missionaries from Arab countries and Iran are now working to strengthen this connection. Iran is also spending millions of dollars to build roads and rail lines to Central Asia.

◄

These children are learning about Islam in Dushanbe, Tajikistan. Although the former Communist government discouraged the practice of religion, today Islam flourishes in the independent Central Asian republics.

Going Further: Thinking Critically

Direct students' attention to the map on this page. Have them name the countries that share each language group. Then ask the following questions: What language is common in a very small area of Southwest Asia and Central Asia? *(Greek)* What is the main language group of Central Asia? *(Turkic)* What language groups are most common in Southwest Asia?

(Iranic and Semitic) Then organize the class into three groups and assign each group one of the three dominant language groups. Have each group conduct research on the history of the language group in Central Asia or Southwest Asia.

Language Groups of Southwest and Central Asia

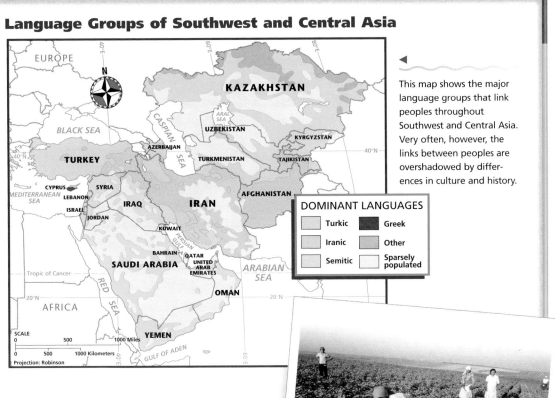

This map shows the major language groups that link peoples throughout Southwest and Central Asia. Very often, however, the links between peoples are overshadowed by differences in culture and history.

DOMINANT LANGUAGES

- Turkic
- Iranic
- Semitic
- Greek
- Other
- Sparsely populated

Central Asia and Southwest Asia share a similar climate, environment, and way of life. Both regions are dry, and water conservation and irrigation are important. Many people in both regions grow cotton and herd animals. In addition, both Central Asia and Southwest Asia are dealing with changes caused by the growing influence of Western culture. Some people are worried that compact discs, videotapes, and satellite television from the West threaten traditional beliefs and ways of life. Shared fears of cultural loss may bring Central Asia and Southwest Asia closer together.

Defining the Region As the world changes, geographers must reexamine this and other regions of the world. Will geographers decide to include the countries of Central Asia in the region of Southwest Asia? Will Russia regain control of Central Asia? The geographers are watching and waiting.

Many people in Central and Southwest Asia grow cotton, such as here in Uzbekistan.

Understanding What You Read

1. What ties did Central Asia have to the Soviet Union in the past?
2. Why are ties between Central Asia and Southwest Asia growing today?

Going Further: Thinking Critically

Have students investigate ways in which their peers are involved in environmental movements. Organize students into groups. Have each group find an example of young people's activism in environmental issues. Students may select an individual or an organization to profile. If a group has selected an individual, ask its members to prepare a résumé for that person. If a group selected an organization, ask its members to prepare a brochure about the organization. Have students include answers to as many of the following questions as possible:

• What is the individual's or organization's main goal?

• Did a certain specific incident or issue inspire the activist(s) to get involved?

• How might other students become involved in this effort?

• Are there aspects of the work done by the organization or individual that might appeal especially to young people?

• How is work done by the organization or individual publicized?

• What is the most important accomplishment achieved so far by the individual or organization?

Have each group present its work to the class.

PRACTICING THE SKILL

1. Students should describe a local environmental problem fully. They should relate the environmental problem to their community.

2. Students should present options for solving the problem.

3. Students should list advantages and disadvantages of options.

4. Students should choose an option, create a plan with the option, explain their plan, and present it.

Visual Record Answer ▶

Possible answer: by containing the oil within a flexible barrier and removing it manually from the surface

➤ This GeoSkills feature addresses National Geography Standards 4 and 14.

GeoSKILLS

Building Skills for Life: Addressing Environmental Problems

The natural environment is the world around us. It includes the air, animals, land, plants, and water. Many people today are concerned about the environment. They are called environmentalists. Environmentalists are worried that human activities are damaging the environment. Environmental problems include air, land, and water pollution, global warming, deforestation, plant and animal extinction, and soil erosion.

People all over the world are working to solve these environmental problems. The governments of many countries are trying to work together to protect the environment. International organizations like the United Nations are also addressing environmental issues.

▲

An oil spill in northwestern Russia caused serious environmental damage in 1995.

Interpreting the Visual Record
Can you see how these people are cleaning up the oil spill?

THE SKILL

1. Gather Information. Create a plan to present to the city council for solving a local environmental problem. Select a problem and research it using databases or other reference materials. How does it affect people's lives and your community's culture or economy?

2. List and Consider Options. After reviewing the information, list and consider options for solving this environmental problem.

3. Consider Advantages and Disadvantages. Now consider the advantages and disadvantages of taking each option. Ask yourself questions like, "How will solving this environmental problem affect business in the area?" Record your answers.

4. Choose, Implement, and Evaluate a Solution. After considering the advantages and disadvantages, you should create your plan. Be sure to make your proposal clear. You will need to explain the reasoning behind the choices you made in your plan.

HANDS on GEOGRAPHY

The countries of the former Soviet Union face some of the worst environmental problems in the world. For more than 50 years, the region's environment was polluted with nuclear waste and toxic chemicals. Today, environmental problems in this region include air, land, and water pollution.

One place that was seriously polluted was the Russian city of Chelyabinsk. Some people have called Chelyabinsk the most polluted place on Earth. The passage below describes some of the environmental problems in Chelyabinsk. Read the passage and then answer the Lab Report questions.

Chelyabinsk was one of the former Soviet Union's main military production centers. A factory near Chelyabinsk produced nuclear weapons. Over the years, nuclear waste from this factory polluted a very large area. A huge amount of nuclear waste was dumped into the Techa River. Many people in the region used this river as their main source of water. They also ate fish from the river.

In the 1950s many deaths and health problems resulted from pollution in the Techa River. Because it was so polluted, the Soviet government evacuated 22 villages along the river. In 1957 a nuclear accident in the region released twice as much radiation as the Chernobyl accident in 1986. However, the accident near Chelyabinsk was kept secret. About 10,000 people were evacuated. The severe environmental problems in the Chelyabinsk region led to dramatic increases in birth defects and cancer rates.

The village of Mitlino was evacuated after a nuclear accident in 1957.

Lab Report

1. How did environmental problems near Chelyabinsk affect people who lived in the region?

2. What might be done to address environmental problems in the Chelyabinsk region?

3. How can a geographical perspective help to solve these problems?

Lab Report

Answers

1. The environmental problems in Chelyabinsk caused many deaths and health problems, including birth defects and cancer.

2. Students might suggest cleaning up the nuclear waste in Chelyabinsk and the Techa River. Students may also suggest further evacuating the area until it is safe for people to live there.

3. Physical geographic studies can help locate sources of pollution and demonstrate the effects of pollution on air, land, soil, and water. Cultural geographic studies can show the effects of pollution on people. These studies can propose ways to prevent similar problems in the future.

☑ internet connect

GO TO: **go.hrw.com**
KEYWORD: **SJ3 CH24**
FOR: **Web sites about children and pollution**

GAZETTEER

Phonetic Respelling and Pronunciation Guide

Many of the key terms in this textbook have been respelled to help you pronounce them. The letter combinations used in the respelling throughout the narrative are explained in this phonetic respelling and pronunciation guide. The guide is adapted from Webster's Tenth New College Dictionary, Merriam-Webster's New Geographical Dictionary, and Merriam-Webster's New Biographical Dictionary.

MARK	AS IN	RESPELLING	EXAMPLE
a	alphabet	a	*AL-fuh-bet
ā	Asia	ay	AY-zhuh
ä	cart, top	ah	KAHRT, TAHP
e	let, ten	e	LET, TEN
ē	even, leaf	ee	EE-vuhn, LEEF
i	it, tip, British	i	IT, TIP, BRIT-ish
ī	site, buy, Ohio	y	SYT, BY, oh-HY-oh
	iris	eye	EYE-ris
k	card	k	KAHRD
ō	over, rainbow	oh	OH-vuhr, RAYN-boh
ù	book, wood	ooh	BOOHK, WOOHD
ò	all, orchid	aw	AWL, AWR-kid
òi	foil, coin	oy	FOYL, KOYN
àu	out	ow	OWT
ə	cup, butter	uh	KUHP, BUHT-uhr
ü	rule, food	oo	ROOL, FOOD
yü	few	yoo	FYOO
zh	vision	zh	VIZH-uhn

A syllable printed in small capital letters receives heavier emphasis than the other syllable(s) in a word.

Acapulco (17°N 100°W) city on the southwestern coast of Mexico, 293

Adriatic Sea sea between Italy and the Balkan Peninsula, 117, 451

Aegean (ee-JEE-uhn) **Sea** sea between Greece and Turkey, 114, 117, 393

Africa second-largest continent; surrounded by the Atlantic Ocean, Indian Ocean, and Mediterranean Sea, A2–A3

Albania country in Eastern Europe on the Adriatic Sea, 451

Alberta province in Canada, 261

Aleutian Islands volcanic islands extending from Alaska into the Pacific Ocean, 217

Alps major mountain system in south-central Europe, 117

Altiplano broad, high plateau in Peru and Bolivia, 361

Amazon River major river in South America, 343

Amsterdam (52°N 5°E) capital of the Netherlands, 411

Amu Dar'ya (uh-MOO duhr-YAH) river in Central Asia that drains into the Aral Sea, 519

Amur (ah-MOOHR) **River** river in northeast Asia forming part of the border between Russia and China, 483

Andes (AN-deez) great mountain range in South America, 111, 343

Andorra European microstate in the Pyrenees mountains, A15

Andorra la Vella (43°N 2°E) capital of Andorra, A15

Antarctica continent around the South Pole, A22

Antarctic Circle line of latitude located at 66.5° south of the equator; parallel beyond which no sunlight shines on the June solstice (first day of winter in the Southern Hemisphere), A4–A5

Antigua and Barbuda island country in the Caribbean, 311

Antwerp (51°N 4°E) major port city in Belgium, 411

Apennines (A-puh-nynz) mountain range in Italy, 393

Appalachian Mountains mountain system in eastern North America, 217, 235

Aral (AR-uhl) **Sea** inland sea between Kazakhstan and Uzbekistan, 519

Arctic Circle line of latitude located at 66.5° north of the equator; the parallel beyond which no sunlight shines on the December solstice (first day of winter in the Northern Hemisphere), A4–A5

Arctic Ocean ocean north of the Arctic Circle; world's fourth-largest ocean, A2–A3

Argentina second-largest country in South America, 343

Armenia country in the Caucasus region of Asia; former Soviet republic, 505

Ashgabat (formerly Ashkhabad) (40°N 58°E) capital of Turkmenistan, 519

Asia world's largest continent; located between Europe and the Pacific Ocean, A3

Astana (51°N 71°E) capital of Kazakhstan, 519

Astrakhan (46°N 48°E) old port city on the Volga River in Russia, 483

Asunción (25°S 58°W) capital of Paraguay, 343

Atacama Desert desert in northern Chile, 361

Athens (38°N 24°E) capital and largest city in Greece, 117

Atlanta (34°N 84°W) capital and largest city in the U.S. state of Georgia, 217, 240

Atlantic Ocean ocean between the continents of North and South America and the continents of Europe and Africa; world's second-largest ocean, A2

Australia only country occupying an entire continent, located between the Indian Ocean and the Pacific Ocean, A3

Austria country in west-central Europe south of Germany, 411

Azerbaijan country in the Caucasus region of Asia; former Soviet republic, 505

Bahamas island country in the Atlantic Ocean southwest of Florida, 311

Baja California peninsula in northwestern Mexico, 293

Baku (40°N 50°E) capital of Azerbaijan, 505

Balkan Mountains mountain range that rises in Bulgaria, 451

Baltic Sea body of water east of the North Sea and Scandinavia, 431

Baltimore (39°N 77°W) city in Maryland on the western shore of Chesapeake Bay, 217, 235

Barbados island country in the Caribbean, 311

Barcelona (41°N 2°E) Mediterranean port city and Spain's second-largest city, 393

Basel (48°N 8°E) city in northern Switzerland on the Rhine River, 411

Basseterre (17°N 63°W) capital of St. Kitts and Nevis, 311

Bay of Biscay body of water off the western coast of France and the northern coast of Spain, 411

Belarus country located north of Ukraine; former Soviet republic, 505

Belém (1°S 48°W) port city in northern Brazil, 343

Belfast (55°N 6°W) capital and largest city of Northern Ireland, 431

Belgium country between France and Germany in west-central Europe, 411

Belgrade (45°N 21°E) capital of Serbia and Yugoslavia on the Danube River, 451

Belize country in Central America bordering Mexico and Guatemala, 311

Belmopan (17°N 89°W) capital of Belize, 311

Bergen (60°N 5°E) seaport city in southwestern Norway, 431

Berkshire Hills hilly region of western Massachusetts, 235

Berlin (53°N 13°E) capital of Germany, 411

Bern (47°N 7°E) capital of Switzerland, 411

Birmingham (52°N 2°W) major manufacturing center of south-central Great Britain, 431

Bishkek (43°N 75°E) capital of Kyrgyzstan, 519

Black Sea sea between Europe and Asia, A14, 107, 114

Blue Ridge Mountains southern region of the Appalachians, 240

Bogotá (5°N 74°W) capital and largest city of Colombia, 327

Bolivia landlocked South American country, 361

Bombay see Mumbai.

Bonn (51°N 7°E) city in western Germany; replaced by Berlin as the capital of reunified Germany, 411

Bosnia and Herzegovina country in Eastern Europe between Serbia and Croatia, 451

Boston (42°N 71°W) capital and largest city of Massachusetts, 217

Brasília (16°S 48°W) capital of Brazil, 343

Bratislava (48°N 17°E) capital of Slovakia, 451

Brazil largest country in South America, 343

Brazilian Highlands regions of old, eroded mountains in southeastern Brazil, 343

Brazilian Plateau area of upland plains in southern Brazil, 343

Bridgetown (13°N 60°W) capital of Barbados, 311

British Columbia province on the Pacific coast of Canada, 261

British Isles island group consisting of Great Britain and Ireland, A15

Brittany region in northwestern France, 411

Brussels (51°N 4°E) capital of Belgium, 411

Bucharest (44°N 26°E) capital of Romania, 451

Budapest (48°N 19°E) capital of Hungary, 451

Buenos Aires (34°S 59°W) capital of Argentina, 343

Bulgaria country on the Balkan Peninsula in Eastern Europe, 451

Calgary (51°N 114°W) city in the western Canadian province of Alberta, 261

Callao (kah-YAH-oh) (12°S 77°W) port city in Peru, 361

Campeche (20°N 91°W) city in Mexico on the west coast of the Yucatán Peninsula, 293

Canada country occupying most of northern North America, 261

Canadian Shield major landform region in central Canada along Hudson Bay, 261

Cancún (21°N 87°W) resort city in Mexico on the Yucatán Peninsula, 293

Cantabrian (kan-TAY-bree-uhn) **Mountains** mountains in northwestern Spain, 393

Cape Cod peninsula offthe coast of southern New England, 235

Cape Horn (56°S 67°W) cape in southern Chile; southernmost point of South America, 361

Caracas (kuh-RAHK-uhs) (11°N 67°W) capital of Venezuela, 327

Cardiff (52°N 3°W) capital and largest city of Wales, 431

Caribbean Sea arm of the Atlantic Ocean between North and South America, A10, 311

Carpathian Mountains mountain system in Eastern Europe, 451

Cascade Range mountain range in the Northwestern United States, 217, 253

Caspian Sea large inland salt lake between Europe and Asia, A16, 351

Castries (14°N 61°W) capital of St. Lucia, 311

Cauca River river in western Colombia, 327

Caucasus Mountains mountain range between the Black Sea and the Caspian Sea, 483

Cayenne (5°N 52°W) capital of French Guiana, 327

Central America narrow southern portion of the North American continent, 311

Central Siberian Plateau upland plains and valleys between the Yenisey and Lena Rivers in Russia, 483

Central Valley narrow plain between the Sierra Nevada and Coast Ranges, 253

Chelyabinsk (chel-YAH-buhnsk) (55°N 61°E) manufacturing city in the Urals region of Russia, 483

Chernobyl (51°N 30°E) city in north-central Ukraine; site of a major nuclear accident in 1986, 505

Chesapeake Bay largest estuary on the Atlantic Coast, 235

Chicago (42°N 88°W) major city on Lake Michigan in northern Illinois, 217, 244

Chile country in South America, 361

China country in East Asia; most populous country in the world, 139

Chișinău (formerly Kishinev) (47°N 29°E) capital of Moldova, 451

Ciudad Juárez (syoo-thahth HWAHR-es) (32°N 106°W) city in northern Mexico near El Paso, 293

Coastal Plain North American landform region stretching along the Atlantic Ocean and Gulf of Mexico, 217

Coast Ranges rugged coastline along the Pacific, 253

Cologne (51°N 7°E) manufacturing and commercial city along the Rhine River in Germany, 411

Colombia country in northern South America, 327

Colorado Plateau uplifted area of horizontal rock layers in the western United States, 217

Columbia Basin region of dry basins and mountains east of the Cascades, 253

Columbia River river that drains the Columbia Basin in the northwestern United States, 217

Copenhagen (56°N 12°E) seaport and capital of Denmark, 431

Córdoba (30°S 64°W) large city in Argentina northwest of Buenos Aires, 343

Cork (52°N 8°W) seaport city in southern Ireland, 431

Costa Rica country in Central America, 311

Crater Lake lake in Oregon; deepest in the United States, 253

Crete largest of the islands of Greece, 114

Crimean Peninsula peninsula in Ukraine that juts southward into the Black Sea, 505

Croatia Eastern European country and former Yugoslav republic, 451

Cuba country and largest island in Caribbean, 311

Cumberland Plateau landform in the Coastal Plain area of the South, 240

Cuzco (14°S 72°W) city southwest of Lima, Peru; former capital of the Inca Empire, 361

Czech Republic Eastern European country and the western part of the former country in Czechoslovakia, 451

Dallas (33°N 97°W) city in Northern Texas, 240

Danube River major river in Europe that flows into the Black Sea in Romania, 411

Death Valley east of the Sierra Nevada; the lowest point in all of North America, 248

Denmark country in northern Europe, 431

Detroit (42°N 83°W) major industrial city in Michigan, 217, 247

Devil's Island (5°N 53°W) French island off the coast of French Guiana in South America, 327

Dinaric Alps mountains extending inland from the Adriatic coast to the Balkan Peninsula, 451

Dnieper River major river in Ukraine, 505

Dominica Caribbean island country, 311

Dominican Republic country occupying the eastern part of Hispaniola in the Caribbean, 311

Donets Basin industrial region in eastern Ukraine, 505

Douro River river on the Iberian Peninsula that flows into the Atlantic Ocean in Portugal, 393

Dublin (53°N 6°W) capital of the republic of Ireland, 431

Dushanbe (39°N 69°E) capital of Tajikistan, 519

E

Ebro River river in Spain that flows into the Mediterranean Sea, 393

Ecuador country in western South America, 361

Edmonton (54°N 113°W) provincial capital of Alberta, Canada, 261

Egypt country in North Africa located east of Libya, 107

El Salvador country on the Pacific side of Central America, 311

England southern part of Great Britain and part of the United Kingdom in northern Europe, 431

English Channel channel separating Great Britain from the European continent, 411

equator the imaginary line of latitude that lies halfway between the North and South Poles and circles the globe, A4–A5

Essen (51°N 7°E) industrial city in western Germany, 411

Estonia country located on the Baltic Sea; former Soviet republic, 411, 451

Euphrates River major river in southwestern Asia, 107

Europe continent between the Ural Mountains and the Atlantic Ocean, A3

Everglades large wetland area in Florida, 240

Fergana Valley fertile valley in Uzbekistan, Kyrgyzstan, and Tajikistan, 519

Finland country in northern Europe located between Sweden, Norway, and Russia, 431

Flanders northern coastal part of Belgium where Flemish is the dominant language, 411

Florence (44°N 11°E) city on the Arno River in central Italy, 393

France country in west-central Europe, 411

Frankfurt (50°N 9°E) main city of Germany's Rhineland region, 411

French Guiana French territory in northern South America, 327

G

Galway (53°N 9°W) city in western Ireland, 431

Geneva (46°N 6°E) city in southwestern Switzerland, 411

Genoa (44°N 10°E) seaport city in northwestern Italy, 393

Georgetown (8°N 58°W) capital of Guyana, 327

Georgia (Eurasia) country in the Caucasus region; former Soviet republic, 505

Germany country in west-central Europe located between Poland and the Benelux countries, 411

Glasgow (56°N 4°W) city in Scotland, United Kingdom, 431

Göteberg (58°N 12°E) seaport city in southwestern Sweden, 431

Gran Chaco (grahn CHAH-koh) dry plains region in Paraguay, Bolivia, and northern Argentina, 343

Great Basin dry region in the western United States, 217

Great Bear Lake lake in the Northwest Territories of Canada, 261

Great Britain major island of the United Kingdom, 431

Greater Antilles larger islands of the West Indies in the Caribbean Sea, 311

Great Lakes largest freshwater lake system in the world; located in North America, 217

Great Plains plains region in the central United States, 217, 248

Great Slave Lake lake in the Northwest Territories of Canada, 261

Great Smoky Mountains southern mountain range in the Appalachians, 240

Greece country in southern Europe located at the southern end of the Balkan Peninsula, 114, 393

Greenland self-governing province of Denmark between the North Atlantic and Arctic Oceans, 431

Green Mountains major range of the Appalachian Mountains in Vermont, 235

Grenada Caribbean island country, 311

Guadalajara (21°N 103°W) industrial city in west-central Mexico, 219

Guadalquivir (gwah-thahl-kee-VEER) **River** important river in southern Spain, 393

Guatemala most populous country in Central America, 311

Guatemala City (15°N 91°W) capital of Guatemala, 311

Guayaquil (gwy-ah-KEEL) (2°S 80°W) port city in Ecuador, 361

Guiana Highlands elevated region in northeastern South America, 327

Gulf of Bothnia part of the Baltic Sea west of Finland, 431

Gulf of California part of the Pacific Ocean east of Baja California, Mexico, 293

Gulf of Mexico gulf of the Atlantic Ocean between Florida, Texas, and Mexico, 293

Gulf of St. Lawrence gulf between New Brunswick and Newfoundland Island in North America, 261

Guyana (gy-AH-nuh) country in South America, 327

Haiti country occupying the western third of the Caribbean island of Hispaniola, 311

Halifax (45°N 64°W) provincial capital of Nova Scotia, Canada, 261

Hamburg (54°N 10°E) seaport on the Elbe River in northwestern Germany, 411

Havana (23°N 82°W) capital of Cuba, 311

Hawaii U.S. Pacific state consisting of a chain of eight large islands and more than 100 smaller islands, 253

Helsinki (60°N 25°E) capital of Finland, 431

Hispaniola large Caribbean island divided into the countries of Haiti and the Dominican Republic, 311

Honduras country in Central America, 311

Houston (30°N 95°W) major port and largest city in Texas, 217, 240

Hudson Bay large bay in Canada, 261

Hungary country in Eastern Europe between Romania and Austria, 451

Iberian Peninsula peninsula in southwestern Europe occupied by Spain and Portugal, 393

Iceland island country between the North Atlantic and Arctic Oceans, 431

India country in South Asia, 139

Indian Ocean world's third-largest ocean; located east of Africa, south of Asia, west of Australia, and north of Antarctica, A3

Interior Plains vast area between the Appalachians and the Rocky Mountains in North America, 217, 219, 244

Iquitos (4°S 73°W) city in northeastern Peru on the Amazon River, 361

Ireland country west of Great Britain in the British Isles, 431

Irish Sea sea between Great Britain and Ireland, 431

Israel country in southwestern Asia, A5

Italy country in southern Europe, 393

Jamaica island country in the Caribbean Sea, 311

Japan country in East Asia consisting of four major islands and more than 3,000 smaller islands, A17

Jeruselem (32°N 35°E) capital of Israel, A5

Jutland Peninsula peninsula in northern Europe made up of Denmark and part of northern Germany, 431

Kamchatka Peninsula peninsula along Russia's northeastern coast, 483

Kara-kum (kahr-uh-KOOM) desert region in Turkmenistan, 519

Kazakhstan country in Central Asia; former Soviet republic, 519

Khabarovsk (kuh-BAHR-uhfsk) (49°N 135°E) city in southeastern Russia on the Amur River, 483

Kiev (50°N 31°E) capital of Ukraine, 505

Kingston (18°N 77°W) capital of Jamaica, 311

Kingstown (13°N 61°W) capital of St. Vincent and the Grenadines, 311

Kjølen (CHUHL-uhn) **Mountains** mountain range in the Scandinavian Peninsula, 431

Korea peninsula on the east coast of Asia, A16

Kosovo province in southern Serbia, 451

Kourou (5°N 53°W) city in French Guiana, 327

Kuril (KYOOHR-eel) **Islands** Russian islands northeast of the island of Hokkaido, Japan, 483

Kuznetsk Basin (Kuzbas) industrial region in central Russia, 483

Kyrgyzstan (kir-gi-STAN) country in Central Asia; former Soviet republic, 519

Kyzyl Kum (ki-zil KOOM) desert region in Uzbekistan and Kazakhstan, 519

Labrador region in the territory of Newfoundland, Canada, 261

Lake Baikal (by-KAHL) world's deepest freshwater lake; located north of the Gobi in Russia, 483

Lake Maracaibo (mah-rah-KY-buh) extension of the Gulf of Venezuela in South America, 327

Lake Nicaragua lake in southwestern Nicaragua, 311

Lake Poopó (poh-oh-POH) lake in western Bolivia, 361

Lake Titicaca lake between Bolivia and Peru at an elevation of 12,500 feet (3,810 m), 361

La Paz (17°S 68°W) administrative capital and principal industrial city of Bolivia with an elevation of 12,001 feet (3,658 m); highest capital in the world, 361

Lapland region extending across northern Finland, Sweden, and Norway, 431

Las Vegas (36°N 115°W) city in southern Nevada, 217, 248

Latvia country on the Baltic Sea; former Soviet republic, 451

Lesser Antilles chain of volcanic islands in the eastern Caribbean Sea, 311

Liechtenstein microstate in west-central Europe located between Switzerland and Austria, 411

Lima (12°S 77°W) capital of Peru, 361

Lisbon (39°N 9°W) capital and largest city of Portugal, 393

Lithuania European country on the Baltic Sea; former Soviet republic, 451

Ljubljana (lee-oo-blee-AH-nuh) (46°N 14°E) capital of Slovenia, 451
London (52°N 0°) capital of the United Kingdom, 431
Longfellow Mountains major range of the Appalachian Mountains in Maine, 235
Los Angeles, California (34°N 118°W) major city in California, 253
Luxembourg small European country bordered by France, Germany, and Belgium, 411
Luxembourg (50°N 7°E) capital of Luxembourg, 411

Macedonia Balkan country; former Yugoslav republic, 451
Machu Picchu (13°S 73°W) ancient Inca city in the Andes of Peru, 361
Madrid (40°N 4°W) capital of Spain, 393
Magdalena River river in Colombia that flows into the Caribbean Sea, 327
Magnitogorsk (53°N 59°E) manufacturing city of the Urals region of Russia, 483
Malta island country in southern Europe located in the Mediterranean Sea between Sicily and North Africa, 384
Managua (12°N 86°W) capital of Nicaragua, 311
Manaus (3°S 60°W) city in Brazil on the Amazon River, 343
Manchester (53°N 2°W) major commercial city in west-central Great Britain, 431
Manitoba prairie province in central Canada, 261
Marseille (43°N 5°E) seaport in France on the Mediterranean Sea, 411
Martha's Vineyard island off the coast of southern New England, 235
Mato Grosso Plateau highland region in southwestern Brazil, 343
Mazatlán (23°N 106°W) seaport city in western Mexico, 293
Mediterranean Sea sea surrounded by Europe, Asia, and Africa, 114, 117
Mexican Plateau large, high plateau in central Mexico, 293
Mexico country in North America, 293
Mexico City (19°N 99°W) capital of Mexico, 293
Miami (26°N 80°W) city in southern Florida, 217, 240
Milan (45°N 9°E) city in northern Italy, 393
Minsk (54°N 28°E) capital of Belarus, 505
Mississippi Delta low swampy area of the Mississippi formed by a buildup of sediment, 240
Mississippi River major river in the central United States, 217
Moldova Eastern European country located between Romania and Ukraine; former Soviet republic, 451
Monaco (44°N 8°E) European microstate bordered by France, 411
Montenegro See Yugoslavia.
Monterrey (26°N 100°W) major industrial center in northeastern Mexico, 293
Montevideo (mawn-tay-bee-THAY-oh) (35°S 56°W) capital of Uruguay, 343
Montreal (46°N 74°W) financial and industrial city in Quebec, Canada, 261
Moscow (56°N 38°E) capital of Russia, 483
Mount Elbrus (43°N 42°E) highest European peak (18,510 ft.; 5,642 m); located in the Caucasus Mountains, 505
Mount Orizaba (19°N 97°W) volcanic mountain (18,700 ft.; 5,700m) southeast of Mexico City; highest point in Mexico, 293

Mount Whitney highest peak in the 48 contiguous United States, 253
Munich (MYOO-nik) (48°N 12°E) major city and manufacturing center in southern Germany, 411

Nantucket island off the coast of southern New England, 235
Naples (41°N 14°E) major seaport in southern Italy, 393
Nassau (25°N 77°W) capital of the Bahamas, 311
Netherlands country in west-central Europe, 411
New Brunswick province in eastern Canada, 261
Newfoundland eastern province in Canada including Labrador and the island of Newfoundland, 261
New Orleans (30°N 90°W) major Gulf port city in Louisiana located on the Mississippi River, 217, 240
New York Middle Atlantic state in the northeastern United States, 217, 235
Nicaragua country in Central America, 311
Nice (44°N 7°E) city in the southeastern coast in France, 411
Nile River world's longest river (4,187 miles; 6,737 km); flows into the Mediterranean Sea in Egypt, 107
Nizhniy Novgorod (Gorky), Russia (56°N 44°E) city on the Volga River east of Moscow, 483
North America continent including Canada, the United States, Mexico, Central America, and the Caribbean Islands, A2
Northern European Plain broad coastal plain from the Atlantic coast of France into Russia, 411
Northern Ireland the six northern counties of Ireland that remain part of the United Kingdom; also called Ulster, 431
North Pole the northern point of Earth's axis, A22
North Sea major sea between Great Britain, Denmark, and the Scandinavian Peninsula, 411
Northwest Highlands region of rugged hills and low mountains in Europe, including parts of the British Isles, northwestern France, the Iberian Peninsula, and the Scandinavian Peninsula, 383
Northwest Territories division of a northern region of Canada, 261
Norway European country located on the Scandinavian Peninsula, 431
Nova Scotia province in eastern Canada, 261
Novosibirsk (55°N 83°E) industrial center in Siberia, Russia, 483
Nunavut Native American territory of northern Canada, 261
Nuuk (Godthab) (64°N 52°W) capital of Greenland, 431

Ob River large river system that drains Russia and Siberia, 483
Okefenokee Swamp large wetland area in Florida, 240
Ontario province in central Canada, 261
Orinoco River major river system in South America, 327
Oslo (60°N 11°E) capital of Norway, 431
Ottawa (45°N 76°W) capital of Canada; located in Ontario, 261
Ozark Plateau rugged, hilly region located mainly in Arkansas, 240

Pacific Ocean Earth's largest ocean; located between North and South America and Asia and Australia, A2–A3

Pamirs mountain area mainly in Tajikistan in Central Asia, 519

Panama country in Central America, 311

Panama Canal canal allowing shipping between the Pacific Ocean and the Caribbean Sea; located in central Panama, 311

Panama City (9°N 80°W) capital of Panama, 311

Paraguay country in South America, 343

Paraguay River river that divides Paraguay into two separate regions, 343

Paramaribo (6°N 55°W) capital of Suriname in South America, 327

Paraná River major river system in southeastern South America, 343

Paris (49°N 2°E) capital of France, 411

Patagonia arid region of dry plains and windswept plateaus in southern Argentina, 343

Pearl Harbor U.S. naval base in Hawaii; attacked by Japan in 1941, 253

Peloponnesus (pe-luh-puh-NEE-suhs) peninsula forming the southern part of the mainland of Greece, 393

Peru country in South America, 361

Philadelphia (40°N 75°W) important port and industrial center in Pennsylvania in the northeastern United States, 217, 235

Phoenix (34°N 112°W) capital of Arizona, 217, 248

Podgorica (PAWD-gawr-ett-sah) capital of Montenegro, 462

Poland country in Eastern Europe located east of Germany, 451

Po River river in northern Italy, 393

Port-au-Prince (pohr-toh-PRINS) (19°N 72°W) capital of Haiti, 311

Portland (46°N 123°W) seaport and largest city in Oregon, 217, 253

Port-of-Spain (11°N 61°W) capital of Trinidad and Tobago, 311

Portugal country in southern Europe located on the Iberian Peninsula, 393

Prague (50°N 14°E) capital of the Czech Republic, 451

Prince Edward Island province in eastern Canada, 261

Pripyat Marshes (PRI-pyuht) marshlands in southern Belarus and northwest Ukraine, 505

Puerto Rico U.S. commonwealth in the Greater Antilles in the Caribbean Sea, 311

Puget Sound Lowland lowland area of Washington state, 253

Pyrenees (PIR-uh-neez) mountain range along the border of France and Spain, 411

Quebec province in eastern Canada, 261

Quebec (47°N 71°W) provincial capital of Quebec, Canada, 261

Quito (0° 79°W) capital of Ecuador, 361

Red Sea sea between the Arabian Peninsula and northeastern Africa, 107

Reykjavik (RAYK-yuh-veek) (64°N 22°W) capital of Iceland, 431

Rhine River major river in Western Europe, 411

Riga (57°N 24°E) capital of Latvia, 451

Río Bravo Mexican name for the river between Texas and Mexico, 293

Rio de Janeiro (23°S 43°W) major port in southeastern Brazil, 343

Río de la Plata estuary between Argentina and Uruguay in South America, 343

Rocky Mountains major mountain range in western North America, 217

Romania country in Eastern Europe, 451

Rome (42°N 13°E) capital of Italy, 393

Rosario (roh-SAHR-ee-oh) (33°S 61°W) city in eastern Argentina, 343

Roseau (15°N 61°W) capital of Dominica in the Caribbean, 311

Russia world's largest country, stretching from Europe and the Baltic Sea to eastern Asia and the coast of the Bering Sea, 483

St. George's (12°N 62°W) capital of Grenada in the Caribbean Sea, 311

St. John's (17°N 62°W) capital of Antigua and Barbuda in the Caribbean Sea, 311

St. Kitts and Nevis Caribbean country in the Lesser Antilles, 311

St. Lawrence River major river linking the Great Lakes with the Gulf of St. Lawrence and the Atlantic Ocean in southeastern Canada, 261

St. Lucia Caribbean country in the Lesser Antilles, 311

St. Petersburg (formerly Leningrad; called Petrograd 1914 to 1924) (60°N 30°E) Russia's second largest city and former capital, 483

St. Vincent and the Grenadines Caribbean country in the Lesser Antilles, 311

Sakhalin Island Russian island north of Japan, 483

Salvador (13°S 38°W) seaport city of eastern Brazil, 343

Salzburg state of central Austria, 411

Samarqand (40°N 67°E) city in southeastern Uzbekistan, 519

San Andreas Fault point in California where the Pacific and North American Plates meet, 253

San Diego (33°N 117°W) California's third largest urban area, 217

San Francisco (38°N 122°W) California's second largest urban area, 217

San José (10°N 84°W) capital of Costa Rica, 311

San Juan (19°N 66°W) capital of Puerto Rico, 311

San Marino microstate in southern Europe surrounded by Italy, 393

San Salvador (14°N 89°W) capital of El Salvador, 311

Santa Cruz (18°S 63°W) city in south central Bolivia, 361

Santiago (33°S 71°W) capital of Chile, 361

Santo Domingo (19°N 70°W) capital of the Dominican Republic, 311

São Francisco River river in eastern Brazil, 343

São Paulo (24°S 47°W) Brazil's largest city, 343

Sarajevo (sar-uh-YAY-voh) (44°N 18°E) capital of Bosnia and Herzegovina, 451

Saskatchewan province in central Canada, 261

Scandinavian Peninsula peninsula of northern Europe occupied by Norway and Sweden, 431

Scotland northern part of the island of Great Britain, 431

Sea of Azov sea in Ukraine connected to and north of the Black Sea, 505

Sea of Japan body of water separating Japan from mainland Asia, 483

Sea of Okhotsk inlet of the Pacific Ocean on the eastern coast of Russia, 483

Seattle (48°N 122°W) largest city in the U.S. Pacific Northwest located in Washington, 217, 252

Seine River river that flows through Paris in northern France, 411

Serbia See Yugoslavia.

Shannon River river in Ireland; longest river in the British Isles, 431

Siberia vast region of Russia extending from the Ural Mountains to the Pacific Ocean, 483

Sicily island region of Italy, 393

Sierra Madre Occidental mountain range in western Mexico, 293

Sierra Madre Oriental mountain range in eastern Mexico, 293

Sierra Nevada located in eastern California; one of the longest and highest mountain ranges in the United States, 217, 253

Skopje (SKAW-pye) (42°N 21°E) capital of Macedonia, 451

Slovakia country in Eastern Europe; formerly the eastern part of Czechoslovakia, 451

Slovenia country in Eastern Europe; former Yugoslav republic, 451

Sofia (43°N 23°E) capital of Bulgaria, 451

South Pole the southern point of Earth's axis, A22

Spain country in southern Europe occupying most of the Iberian Peninsula, 393

Stockholm (59°N 18°E) capital of Sweden, 431

Strait of Gibraltar (juh-BRAWL-tuhr) strait between the Iberian Peninsula and North Africa that links the Mediterranean Sea to the Atlantic Ocean, 393

Strait of Magellan strait in South America connecting the South Atlantic with the South Pacific, 361

Sucre (19°S 65°W) constitutional capital of Bolivia, 361

Suriname (soohr-uh-NAH-muh) country in northern South America, 327

Susquehanna River one of several major rivers that cut through the Appalachians, 235

Sweden country in northern Europe, 431

Switzerland country in west-central Europe located between Germany, France, Austria, and Italy, 411

Syr Dar'ya (sir duhr-YAH) river draining the Pamirs in Central Asia, 519

Tagus River longest river on the Iberian Peninsula in southern Europe, 393

Taiwan (TY-WAHN) island country located off the southeastern coast of China, A17

Tajikistan (tah-ji-ki-STAN) country in Central Asia; former Soviet republic, 419

Tampico (22°N 98°W) Gulf of Mexico seaport in central-eastern Mexico, 293

Tashkent (41°N 69°E) capital of Uzbekistan, 519

T'bilisi (42°N 45°E) capital of Georgia in the Caucasus region, 505

Tegucigalpa (14°N 87°W) capital of Honduras, 311

Thessaloníki (41°N 23°E) city in Greece, 393

Tian Shan (TIEN SHAHN) high mountain range separating northwestern China from Russia and some Central Asian republics, 519

Tiber River river that flows through Rome in central Italy, 393

Tierra del Fuego group of islands at the southern tip of South America, 343

Tigris River major river in southwestern Asia, 107

Tijuana (33°N 117°W) city in northwestern Mexico, 293

Tiranë (ti-RAH-nuh) (42°N 20°E) capital of Albania, 451

Toronto (44°N 79°W) capital of the province of Ontario, Canada, 261

Trinidad and Tobago Caribbean country in the Lesser Antilles, 311

Tropic of Cancer parallel 23.5° north of the equator; parallel on the globe at which the Sun's most direct rays strike the earth during the June solstice (first day of summer in the Northern Hemisphere), A4–A5

Tropic of Capricorn parallel 23.5° south of the equator; parallel on the globe at which the Sun's most direct rays strike the earth during the December solstice (first day of summer in the Southern Hemisphere), A4–A5

Turin (45°N 8°E) city in northern Italy, 393

Turkey country of the eastern Mediterranean occupying Anotolia and a corner of southeastern Europe, A5

Turkmenistan country in Central Asia; former Soviet republic, 519

Ukraine country located between Russia and Eastern Europe; former Soviet republic, 505

United Kingdom country in northern Europe occupying most of the British Isles; Great Britain and Northern Ireland, 431

United States North American country located between Canada and Mexico, 217

Ural Mountains mountain range in west central Russia that divides Asia from Europe, 483

Uruguay country on the northern side of the Río de la Plata between Brazil and Argentina in South America, 343

Uzbekistan country in Central Asia; former Soviet republic, 519

Vaduz (47°N 10°E) capital of Liechtenstein, A15

Valletta (36°N 14°E) capital of Malta, 384

Valparaíso (33°S 72°W) Pacific port for the national capital of Santiago, Chile, 361

Vancouver (49°N 123°W) Pacific port in Canada, 261

Vatican City (42°N 12°E) European microstate surrounded by Rome, Italy, 393

Venezuela country in northern South America, 327

Vienna (48°N 16°E) capital of Austria, 411

Vietnam country in Southeast Asia, A17

Vilnius (55°N 25°E) capital of Lithuania, 451

Virgin Islands island chain lying just east of Puerto Rico in the Caribbean Sea, 311

Vistula River river flowing through Warsaw, Poland, to the Baltic Sea, 451

Vladivostok (43°N 132°E) chief seaport of the Russian Far East, 483

Volga River Europe's longest river; located in west central Russia, 483

Wales part of the United Kingdom occupying a western portion of Great Britain, 431

Wallonia region in southern Belgium, 411

Warsaw (52°N 21°E) capital of Poland, 451

Washington, D.C. (39°N 77°W) U.S. capital; located between Virginia and Maryland on the Potomac River, 217, 235

West Siberian Plain region with many marshes east of the Urals in Russia, 483

White Mountains major range of the Appalachian Mountains in New Hampshire, 235

Willamette Valley lowland area of Oregon, 253

Windsor (42°N 83°W) industrial city across from Detroit, Michigan, in the Canadian province of Ontario, 261

Winnipeg (50°N 97°W) provincial capital of Manitoba in central Canada, 261

Yekaterinburg (formerly Sverdlovsk) (57°N 61°E) city in the Urals region in Russia, 483

Yenisey (yi-ni-SAY) major river in central Russia, 483

Yerevan (40°N 45°E) capital of Armenia, 505

Yucatán Peninsula peninsula in southeastern Mexico, 293

Yugoslavia former Eastern European country of six republics; now including only the republics of Serbia and Montenegro, 451

Yukon Territory Canadian territory bordering Alaska, 261

Zagreb (46°N 16°E) capital of Croatia, 451

Zurich (47°N 9°E) Switzerland's largest city, 411

GLOSSARY

Phonetic Respelling and Pronunciation Guide

Many of the key terms in this textbook have been respelled to help you pronounce them. The letter combinations used in the respelling throughout the narrative are explained in this phonetic respelling and pronunciation guide. The guide is adapted from *Webster's Tenth New College Dictionary, Merriam-Webster's New Geographical Dictionary,* and *Merriam-Webster's New Biographical Dictionary.*

MARK	AS IN	RESPELLING	EXAMPLE
a	alphabet	a	*AL-fuh-bet
ā	Asia	ay	AY-zhuh
ä	cart, top	ah	KAHRT, TAHP
e	let, ten	e	LET, TEN
ē	even, leaf	ee	EE-vuhn, LEEF
i	it, tip, British	i	IT, TIP, BRIT-ish
ī	site, buy, Ohio	y	SYT, BY, oh-HY-oh
	iris	eye	EYE-ris
k	card	k	KAHRD
ō	over, rainbow	oh	OH-vuhr, RAYN-boh
ù	book, wood	ooh	BOOHK, WOOHD
ò	all, orchid	aw	AWL, AWR-kid
òi	foil, coin	oy	FOYL, KOYN
àu	out	ow	OWT
ə	cup, butter	uh	KUHP, BUHT-uhr
ü	rule, food	oo	ROOL, FOOD
yü	few	yoo	FYOO
zh	vision	zh	VIZH-uhn

*A syllable printed in small capital letters receives heavier emphasis than the other syllable(s) in a word.

A

abdicated Gave up the throne, **488**

abolitionist Someone who wants to end slavery, **227**

absolute authority Monarch's power to make all the governing decisions, **141**

absolute location The exact spot on Earth where something is found, often stated in latitude and longitude, **7**

acculturation The process of cultural changes that result from long-term contact with another society, **77**

acid rain A type of polluted rain, produced when pollution from smokestacks combines with water vapor, **64**

Age of Exploration A period when Europeans were eager to find new and shorter sea routes so that they could trade with India and China, **138**

aggression Warlike action, such as an invasion or an attack, **187**

agrarian A society organized around farming, **515**

air pressure The weight of the air, measured by a barometer, **39**

alliance A formal agreement or treaty among nations formed to advance common interests or causes, **152**

allies Friendly countries that support one another against enemies, **489**

alluvial fan A fan-shaped landform created by deposits of sediment at the base of a mountain, **32**

amber Fossilized tree sap, **453**

Antarctic Circle The line of latitude located at 66.5° south of the equator, **20**

anti-Semitism Hatred of Jews, **190**

aqueducts Artificial channels for carrying water, **62, 120**

aquifers Underground, water-bearing layers of rock, sand, or gravel, **62**

arable Suitable for growing crops, **576**

archipelago (ahr-kuh-PE-luh-goh) A large group of islands, **312**

Arctic Circle The line of latitude located at 66.5° north of the equator, **20**

arid Dry with little rainfall, **46**

armistice An agreement to stop fighting, **176**

arms race Competition to create and have more advanced weapons, **194**

atmosphere The layer of gases that surrounds Earth, **22**

axis An imaginary line that runs from the North Pole through Earth's center to the South Pole, **19**

B

badlands Rugged areas of soft rock that have been eroded by wind and water into small gullies and have little vegetation or soil, **249**

balance of power A way of keeping peace when no one nation or group of nations is more powerful than the others, **157**

bankrupt Having no money, **179**

barrier islands Long, narrow, sandy islands separated from the mainland, **241**

basins Regions surrounded by mountains or other higher land, **219**

bauxite The most important aluminum ore, **313**

bloc A group of nations united under a common idea or for a common purpose, **193**

bog Soft ground that is soaked with water, **442**

cacao (kuh-KOW) A small tree on which cocoa beans grow, **316**

caldera A large depression formed after a major eruption and collapse of a volcanic mountain, **255**

calypso A type of music with origins in Trinidad and Tobago, **320**

cantons Political and administrative districts in Switzerland, **425**

capital The money and tools needed to make a product, **160**

capitalism Economic system in which private individuals control the factors of production, **163**

caravans Groups of people who travel together for protection, **522**

cardamom A spice used in Asian foods, **317**

carrying capacity The maximum number of a species that can be supported by an area, **89**

cartography The art and science of mapmaking, **13**

cash crops Crops produced primarily to sell rather than for the farmer to eat, **305**

cassava (kuh-SAH-vuh) A tropical plant with starchy roots, **332**

cathedrals Huge churches sometimes decorated with elaborate stained-glass windows, **128**

caudillos (kow-THEE-yohs) Military leaders who ruled Venezuela in the 1800s and 1900s, **335**

center-pivot irrigation A method of irrigation which uses long sprinkler systems mounted on huge wheels which rotate slowly, irrigating the area within a circle, **251**

chancellor Germany's head of government, or prime minister, **419**

chinampas (chuh-NAM-puhs) The name the Aztecs gave to raised fields on which they grew crops, **298**

chinooks Strong, dry winds that blow from the Rocky Mountains onto the Great Plains in the United States during drought periods, **250**

chivalry A code of behavior including bravery, fairness, loyalty, and integrity, **127**

city-states Self-governing cities, such as those of ancient Greece, **116, 396**

civil war A conflict between two or more groups within a country, **317**

civilization A highly complex culture with growing cities and economic activity, **80**

clergy Officials of the church, including the priests, bishops, and pope, **128**

climate The weather conditions in an area over a long period of time, **37**

climatology The field of tracking Earth's larger atmospheric systems, **13**

cloud forest A high-elevation, very wet tropical forest where low clouds are common, **313**

coalition governments Governments in which several political parties join together to run a country, **401**

Cold War The rivalry between the United States and the Soviet Union that lasted from the 1940s to the early 1990s, **489**

collective farms Large farms owned and controlled by the central government, **181**

colony Territory controlled by people from a foreign land, **139**

command economy An economy in which the government owns most of the industries and makes most of the economic decisions, **85**

commercial agriculture A type of farming in which farmers produce food for sale, **80**

commonwealth A self-governing territory associated with another country, **323**

communism An economic and political system in which the government owns or controls almost all of the means of production, industries, wages, and prices, **85, 192**

condensation The process by which water changes from a gas into tiny liquid droplets, **24**

conquistadores (kahn-kees-tuh-DAWR-ez) Spanish conquerors during the era of colonization in the Americas, **299**

constitution A document that outlines basic laws that govern a nation, **142**

constitutional monarchy A government with a monarch as head of state and a parliament or other legislature that makes the laws, **435**

consumer goods Products used at home and in everyday life, **489**

contiguous Units, such as states, that connect to or border each other, **217**

Continental Divide The crest of the Rocky Mountains that divides North America's rivers into those that flow eastward and those that flow westward, **219**

continental shelf The gently sloping underwater land surrounding each continent, **25**

continents Earth's large landmasses, **29**

cooperatives Organizations owned by their members and operated for their mutual benefit, **321**

cordillera (kawr-duhl-YER-uh) A mountain system made up of parallel ranges, **328**

core The inner, solid part of Earth, **28**

cork The bark stripped from a certain type of oak tree and often used as stoppers and insulation, **407**

Corn Belt The corn-growing region in the Midwest from central Ohio to central Nebraska, **245**

cosmopolitan Having many foreign influences, **424**

Cossacks Nomadic horsemen who once lived on the Ukrainian frontier, **509**

Counter-Reformation Attempt by the Catholic Church, following the Reformation, to return the Church to an emphasis on spiritual matters, **136**

coup (KOO) A sudden overthrow of a government by a small group of people, **369**

creoles American-born descendents of Europeans in Spanish South America, **368**

crop rotation A system of growing different crops on the same land over a period of years, **59**

Crusades A long series of battles starting in 1096 between the Christians of Europe and the Muslims to gain control of Palestine, **128**

crust The outer, solid layer of Earth, **28**

culture A learned system of shared beliefs and ways of doing things that guide a person's daily behavior, **75**

culture region Area of the world in which people share certain culture traits, **75**

culture traits Elements of culture, **75**

currents Giant streams of ocean water that move from warm to cold or from cold to warm areas, **41**

czar (ZAHR) Emperor of the Russian Empire, **488**

Dairy Belt Area including Wisconsin and most of Minnesota and Michigan which produces milk, cheese, and dairy products, **245**

deforestation The destruction or loss of forest area, **61**

deltas Landforms created by the deposits of sediment at the mouths of rivers, **32**

democracy A political system in which a country's people elect their leaders and rule by majority, **86**

desalinization The process in which the salt is taken out of seawater, **63**

desertification The long-term process of losing soil fertility and plant life, **60**

developed countries Industrialized countries that have strong secondary, tertiary, and quaternary industries, **84**

developing countries Countries in different stages of moving toward development, **84**

dialect A variation of a language, **406**

dictator One who rules a country with complete authority, **180, 316**

diffusion The movement of ideas or behaviors from one cultural region to another, **9**

direct democracy Government in which citizens take part in making all decisions, **116**

diversify To produce a variety of things, **242**

division of labor Organization of society in which each person performs a specific job, **106**

domestication The growing of a plant or taming of an animal by a people for their own use, **79**

dominion A territory or area of influence, **266**

droughts Periods when little rain falls and crops are damaged, **245**

earthquakes Sudden, violent movement along a fracture in the Earth's crust, **29**

ecology The study of connections among different forms of life, **51**

economic nationalism Putting the economic interests of one's own country above the interests of other countries, **185**

ecosystem All of the plants and animals in an area together with the nonliving parts of their environment, **54**

ecotourism The process of using an area's natural environment to attract tourists, **318**

ejidos (e-HEE-thohs) Lands owned and worked by groups of Mexican Indians, **300**

El Dorado (el duh-RAH-doh) "The Golden One," a legend of the early Chibcha people of Colombia, **330**

El Niño An ocean and weather pattern in the Pacific Ocean in which ocean waters become warmer, **364**

Emancipation Proclamation Lincoln's document that freed the slaves, **227**

emigrate Leave one's country to move to another, **166**

empire A system in which a central power controls a number of territories, **299**

encomienda A system in which Spanish monarchs gave land to Spanish colonists in the Americas; landowners could force Indians living there to work the land, **351**

Enlightenment An era of new ideas from the mid-1600s through the 1700s, **147**

entrepreneurs People who use their money and talents to start a business, **85**

epidemic Widespread outbreak, often referring to a disease, **299**

equinoxes The two days of the year when the Sun's rays strike the equator directly, **21**

erosion The movement by water, ice, or wind of rocky materials to another location, **32**

estuary A partially enclosed body of water where salty seawater and freshwater mix, **238, 345**

ethnic groups Cultural groups of people who share learned beliefs and practices, **75**

evaporation The process by which heated water becomes water vapor and rises into the air, **24**

extinct Something that dies out completely; is no longer present, **51**

F

factors of production The natural resources, money, labor, and capital needed for business operation, **85, 160**

factory A large building to house workers and their equipment, **160**

famine A great shortage of food, **440**

fascism A political movement that puts the needs of the nation above the needs of the individual, **180**

fault A fractured surface in Earth's crust where a mass of rock is in motion, **31**

favelas (fah-VE-lahs) Huge slums that surround some Brazilian cities, **349**

feudalism A system after the 900s under which most of Europe was organized and governed by local leaders based on land and service, **127**

fief A grant of land, **127**

fjords (fee-AWRDS) Narrow, deep inlets of the sea set between high, rocky cliffs, **432**

floodplain A landform of level ground built by sediment deposited by a river or stream, **32**

food chain A series of organisms in which energy is passed along, **51**

fossil fuels Nonrenewable resources formed from the remains of ancient plants and animals, **68**

free enterprise An economic system in which people, not government, decide what to make, sell, or buy, **85**

front The unstable weather where a large amount of warm air meets a large amount of cold air, **40**

frontier Unsettled land, **224**

G

gauchos (GOW-chohz) Argentine cowboys, **351**

genocide The planned killing of a race of people, **190**

geography The study of Earth's physical and cultural features, **3**

geothermal energy A renewable energy resource produced from the heat of Earth's interior, **70**

geysers Hot springs that shoot hot water and steam into the air, **446**

glaciers Large, slow-moving sheets or rivers of ice, **33**

glen A Scottish term for a valley, **437**

global warming A slow increase in Earth's average temperature, **64**

globalization Process in which connections around the world increase and cultures around the world share similar practices, **199**

Good Neighbor Policy Franklin D. Roosevelt's plan for the United States to cooperate with Latin American countries without interfering with their governments, **185**

Great Depression Period after the Stock Market crashed in 1929 when worldwide business slowed down, banks closed, prices and wages dropped, and many people were out of work, **179**

greenhouse effect The process by which Earth's atmosphere traps heat, **38**

gross domestic product The value of all goods and services produced within a country, **83**

gross national product The value of all goods and services that a country produces in one year within or outside the country, **83**

groundwater The water from rainfall, rivers, lakes, and melting snow that seeps into the ground, **25**

guerrilla An armed person who takes part in irregular warfare, such as raids, **321**

H

habitation fog A fog caused by fumes and smoke trapped over Siberian cities by very cold weather, **496**

haciendas (hah-see-EN-duhs) Huge farmlands granted by the Spanish monarch to favored people in Spain's colonies, **300**

heavy industry Industry that usually involves manufacturing based on metals, **494**

hieroglyphics A form of ancient writing that uses pictures and symbols, **112**

history A written record of human civilization, **107**

Holocaust The mass murder of millions of Jews and other people by the Nazis in World War II, **190, 419**

hominid An early human like creature, **103**

homogeneous Sharing the same characteristics, such as ethnicity, **515**

human geography The study of people, past or present, **11**

humanists Scholars of the Renaissance who studied history, poetry, grammar, and other subjects taught in ancient Greece and Rome, **132**

humus Decayed plant and animal matter, **55**

hurricanes Tropical storms that bring violent winds, heavy rain, and high seas, **48**

hydroelectric power A renewable energy resource produced from dams that harness the energy of falling water to power generators, **70**

I

icebreakers Ships that can break up the ice of frozen waterways, allowing other ships to pass through them, **500**

impressionism A form of art that developed in France in the late 1800s and early 1900s, **417**

indentured servants People who agree to work for a certain period of time, often in exchange for travel expenses, **337**

indigo (IN-di-goh) A plant used to make a deep blue dye, **334**

individualism A belief in the political and economic independence of individuals, **148**

Indo-European A language family that includes many languages of Europe, such as Germanic, Baltic, and Slavic languages, **454**

Industrial Revolution Period that lasted through the 1700s and 1800s when advances in industry, business, transportation, and communications changed people's lives in almost every way, **159**

inflation The rise in prices that occurs when currency loses its buying power, **304**

Inuit North American Eskimos, **273**

irrigation A system of bringing water from rivers to fields through ditches and canals to water crops, **105**

isthmus A neck of land connecting two larger land areas, **28**

junta (HOOHN-tuh) A small group of military officers who rule a country after seizing power, **373**

knight A nobleman who serves as a professional warrior, **127**

land bridges Strips of dry land between continents caused by sea levels dropping, **104**

landforms The shapes of land on Earth's surface, **28**

landlocked Completely surrounded by land, with no direct access to the ocean, **356, 520**

lava Magma that has broken through the crust to Earth's surface, **29**

levees Large walls, usually made of dirt, built along rivers to prevent flooding, **10**

light industry Industry that focuses on the production of lightweight goods, such as clothing, **494**

lignite A soft form of coal, **453**

limited government Government in which government leaders are held accountable by citizens through their constitutions and democratic processes, **86**

limited monarchy Monarchy in which the powers of the king were limited by law, **141**

literacy The ability to read and write, **166**

Llaneros (lah-NE-rohs) Cowboys of the Venezuelan Llanos, **336**

Llanos (LAH-nohs) A plains region in eastern Columbia and western Venezuela, **328**

lochs Scottish lakes located in valleys carved by glaciers, **432**

loess (LES) Fine, windblown soil that is good for farming, **413**

Loyalists Colonists loyal to Great Britain, **152**

magma Melted rock in the upper mantle of Earth, **28**

mainland A region or country's main landmass, **394**

mantle The liquid layer that surrounds Earth's core, **28**

maquiladoras (mah-kee-lah-DORH-ahs) Foreign-owned factories located along Mexico's northern border with the United States, **307**

maritime On or near the sea, such as Canada's Maritime Provinces, **270**

market economy An economy in which consumers help determine what is to be produced by buying or not buying certain goods and services, **85**

mass production System of producing large numbers of identical items, **163**

medieval Refers to the period from the collapse of the Roman Empire to about 1500, **415**

megalopolis A string of cities that have grown together, **236**

mercantilism Economic theory using colonies to increase a nation's wealth by gaining access to labor and natural resources, **139**

Mercosur A trade organization that includes Argentina, Brazil, Paraguay, Uruguay, and two associate members (Bolivia and Chile), **354**

merengue The national music and dance of the Dominican Republic, **320**

mestizos (me-STEE-zohs) People of mixed European and American Indian ancestry, **300**

metallic minerals Shiny minerals, like gold and iron, that can conduct heat and electricity, **65**

meteorology The field of forecasting and reporting rainfall, temperatures, and other atmospheric conditions, **13**

Métis (may-TEES) People of mixed European and Canadian Indian ancestry in Canada, **267**

middle class Class of skilled workers between the upper class and poor and unskilled workers, **129, 164**

migrated To move from one region or climate to another, **110**

militarism Use of strong armies and the threat of force to gain power, **173**

minerals Inorganic substances that make up Earth's crust, occur naturally, are solids in crystalline form, and have a definite chemical composition, **65**

missions Spanish church outposts established during the colonial era, particularly in the Americas, **300**

monarchy A territory ruled by a king who has total power to govern, **141**

monsoon The seasonal shift of air flow and rainfall, which brings alternating wet and dry seasons, **45**

Moors Muslim North Africans, **405**

moraine A ridge of rocks, gravel, and sand piled up by a glacier, **237**

mosaics (moh-ZAY-iks) Pictures created from tiny pieces of colored stone, **398**

mosques Islamic houses of worship, **526**

mulattoes (muh-LA-tohs) People of mixed European and African ancestry, **300**

multicultural A mixture of different cultures within the same country or community, **75**

national parks Large scenic areas of natural beauty preserved by the United States government for public use, **252**

nationalism The demand for self-rule and a strong feeling of loyalty to one's nation, **167, 426**

NATO North Atlantic Treaty Organization, a military alliance of various European countries, the United States, and Canada, **194, 415**

nature reserves Areas a government has set aside to protect animals, plants, soil, and water, **506**

navigable Water routes that are deep enough and wide enough to be used by ships, **412**

neutral Not taking a side in a dispute or conflict, **444**

New Deal Franklin D. Roosevelt's program to help end the Great Depression, **179**

newsprint Cheap paper used mainly for newspapers, **263**

nobles People who were born into wealthy, powerful families, **127**

nomads People who often move from place to place, **522**

nonmetallic minerals Minerals that lack the characteristics of metal, **66**

nonrenewable resources Resources, such as coal and oil, that cannot be replaced by Earth's natural processes or that are replaced slowly, **65**

North Atlantic Drift A warm ocean current that brings mild temperatures and rain to parts of northern Europe, **433**

nutrients Substances promoting growth, **52**

oasis A place in the desert where a spring or well provides water, **520**

oil shale Layered rock that yields oil when heated, **453**

oppression Cruel and unjust use of power against others, **154**

orbit The path an object makes around a central object, such as a planet around the Sun, **17**

ozone A form of oxygen in the atmosphere that helps protect Earth from harmful solar radiation, **22**

Pampas A wide, grassy plains region in central Argentina, **344**

Pangaea (pan-GEE-uh) Earth's single, original super-continent from which today's continents were separated, **31**

pardos Venezuelans of mixed African, European, and South American Indian ancestry, **336**

Parliament An English assembly made up of nobles, clergy, and common people which had the power to pass and enforce laws, **141**

partition To divide a land area into smaller parts, **195**

Patriots Colonists who wanted independence from British rule, **152**

peat Matter made from dead plants, usually mosses, **442**

peninsula Land bordered by water on three sides, **28**

permafrost The layer of soil that stays frozen all year in tundra climate regions, **49**

perspective Point of view based on a person's experience and personal understanding, **3**

Peru Current A cold ocean current off the coast of western South America, **363**

petroleum An oily liquid that can be refined into gasoline and other fuels and oils, **68**

photosynthesis The process by which plants convert sunlight into chemical energy, **51**

physical geography The study of Earth's natural landscape and physical systems, including the atmosphere, **11**

pidgin languages Simple languages that help people who speak different languages understand each other, **337**

pioneers Settlers who lead the way into new areas, **225**

plain A nearly flat area on Earth's surface, **28**

plantains A type of banana used in cooking, **322**

plantations Large farms that grow mainly one crop to sell, **223**

plant communities Groups of plants that live in the same area, **53**

plant succession The gradual process by which one group of plants replaces another, **54**

plateau An elevated flatland on Earth's surface, **28**

plate tectonics The theory that Earth's surface is divided into several major, slowly moving plates or pieces, **28**

police state A country in which the government has total control over the people using the police, **180**

pope The bishop of Rome and the head of the Roman Catholic Church, **401**

popular sovereignty Governmental principle based on just laws and on a government created by and subject to the will of the people, **149**

population density The average number of people living within a square mile or square kilometer, **81**

potash A mineral used to make fertilizer, **263**

precipitation The process by which water falls back to Earth, **24**

prehistory A time before written records, **103**

primary industries Economic activities that directly involve natural resources or raw materials, such as farming and mining, **83**

Protestants Christians who broke away from the Catholic Church during the Reformation, **135**

provinces Administrative divisions of a country, **266**

Pueblo A group of North American Indians who built houses with adobe, **111**

pulp Softened wood fibers used to make paper, **263**

Puritan Member of a group of Protestants who rebelled against the Church of England, **141**

quaternary industries Economic activities that include specialized skills or knowledge and work mostly with information, **83**

Quechua (KE-chuh-wuh) The language of South America's Inca; still spoken in the region, **370**

quinoa (KEEN-wah) A native plant of South America's Andean region that yields nutritious seeds, **365**

quipu (KEE-poo) Complicated system of knots tied on strings of various colors, used by the Inca of South America to record information, **113, 366**

race A group of people who share inherited physical or biological traits, **76**

rain shadow The dry area on the leeward side of a mountain or mountain range, **44**

ratified An approval (of the Constitution by the states), **225**

reactionaries People who want to return to an earlier political system, **158**

reason Logical thinking, **148**

refineries Factories where crude oil is processed, **68**

reforestation The planting of trees in places where forests have been cut down, **61**

reform Making something better by removing its faults, **168**

Reformation A movement in Europe to reform Christianity in the 1500s, **135, 418**

refugees People who flee to another country, usually for economic or political reasons, **321**

reggae A type of music with origins in Jamaica, **320**

regionalism The stronger connection to one's region than to one's country, **269**

Reign of Terror Period from 1789 to 1794 when thousands of people died at the guillotine in France, **156**

relative location The position of a place in relation to another place, **7**

Renaissance (re-nuh-SAHNS) French word meaning "rebirth" and referring to a new era of learning that began in Europe in the 1300s, **131, 401**

renewable resources Resources, such as soils and forests, that can be replaced by Earth's natural processes, **59**

republic Government in which voters elect leaders to run the state, **118**

Restoration The reign of Charles II in English history in which the monarchy was restored to power, **142**

revolution One complete orbit around the Sun, **19**

revolutionaries Rebels demanding self-rule, **184**

Roma An ethnic group also known as Gypsies who are descended from people who may have migrated from India to Europe long ago, **464**

rotation One complete spin of Earth on its axis, **19**

rural An area of open land that is often used for farming, **4**

Santeria A religion, with origins in Cuba, that mixes West African religions and traditions with those of Roman Catholicism, **320**

satellite A body that orbits a larger body, **18**

scarcity Situation that occurs when demand is greater than supply, **87**

Scientific Revolution The period during the 1500s and 1600s when mathematics and scientific instruments were used to learn more about the natural world, **137**

secede To separate from, **227**

secondary industries Economic activities that change raw materials created by primary industries into finished products, **83**

second-growth forests The trees that cover an area after the original forest has been cut, **238**

secularism Playing down the importance of religion, **148**

sediment Small bits of mud, sand, or gravel which collect at the river's mouth, **241**

selvas The thick tropical rain forests of eastern Ecuador, eastern Peru, and northern Bolivia, **363**

semiarid Relatively dry with small amounts of rain, **62**

serfs People who were bound to the land and worked for a lord, **128, 509**

sinkholes A steep-sided depression formed when the roof of a cave collapses, **294**

sirocco (suh-RAH-koh) A hot, dry wind from North Africa that blows across the Mediterranean to Europe, **395**

slash-and-burn agriculture A type of agriculture in which forests are cut and burned to clear land for planting, **307**

smelters Factories that process metal ores, **495**

smog A mixture of smoke, chemicals, and fog, **306**

social contract Idea that government should be based on an agreement made by the people, **149**

soil exhaustion The loss of soil nutrients needed by plants, **346**

solar energy A renewable energy resource produced from the Sun's heat and light, **71**

solar system The Sun and the objects that move about it, including planets, moons, and asteroids, **17**

solstice The days when the Sun's vertical rays are farthest from the equator, **21**

soviet A council of Communists who governed republics and other places in the Soviet Union, **509**

spatial perspective Point of view based on looking at where something is and why it is there, **3**

steppe (STEP) A wide, flat grasslands region that stretches from Ukraine across southern Russia to Kazakhstan, **485**

steppe climate A dry climate type generally found between desert and wet climate regions, **46**

stock market An organization through which shares of stock in companies are bought and sold, **178**

strait A narrow passageway that connects two large bodies of water, **362**

strip mining A kind of coal mining in parts of the Great Plains which strips away soil and rock, **252**

subduction The movement of one of Earth's heavier tectonic plates underneath a lighter tectonic plate, **29**

subregions Small areas of a region, **8**

subsistence agriculture A type of farming in which farmers grow just enough food to provide for themselves and their own families, **80**

suburbs Areas just outside or near a city, **167**

suffragettes Women who fought for all women's right to vote, **168**

superpowers Powerful countries, **489**

symbol A word, shape, color, flag, or other sign that stands for something else, **77**

taiga (TY-guh) A forest of evergreen trees growing south of the tundra of Russia, **485**

tepees Tents that are usually made of animal skins, **110**

tepuís (tay-PWEEZ) Layers of sandstone that have resisted erosion atop plateaus in the Guiana Highlands, **328**

terraces Horizontal ridges built into the slopes of steep hillsides to prevent soil loss and aid farming, **60**

tertiary industries Economic activities that handle goods that are ready to be sold to consumers, **83**

textiles Cloth products, **435**

traditional economy Economy based on custom and tradition, **85**

tributary Any smaller stream or river that flows into a larger stream or river, **25**

Tropic of Cancer The line of latitude that is 23.5° north of the equator, **21**

Tropic of Capricorn The line of latitude that is 23.5° south of the equator, **21**

tundra climate A cold region with low rainfall, lying generally between subarctic and polar climate regions, **49**

typhoons Tropical storms that bring violent winds, heavy rain, and high seas, **48**

U-boats German submarines from World War I, **174**

uninhabitable Not capable of supporting human settlement, **446**

unlimited governments Governments that have total control over their citizens, **86**

urban An area that contains a city, **4**

vassals People who held land from a feudal lord and received protection in return for service to the lord, especially in battle, **127**

vernacular Everyday speech which varies from place to place, **130**

viceroy The governor of a colony, **368**

water cycle The circulation of water from Earth's surface to the atmosphere and back, **23**

water vapor The gaseous form of water, **23**

weather The condition of the atmosphere at a given place and time, **37**

weathering The process of breaking rocks into smaller pieces through heat, water, or other means, **31**

wetlands Land areas that are flooded for at least part of the year, **241**

Wheat Belt The wheat-growing area in the United States which stretches across the Dakotas, Montana, Nebraska, Kansas, Oklahoma, Colorado, and Texas, **251**

working class Unskilled and semi-skilled workers with low-paying jobs, **165**

yurt A movable round house of wool felt mats over a wood frame, **526**

SPANISH GLOSSARY

Phonetic Respelling and Pronunciation Guide

Many of the key terms in this textbook have been respelled to help you pronounce them. The letter combinations used in the respelling throughout the narrative are explained in this phonetic respelling and pronunciation guide. The guide is adapted from *Webster's Tenth New College Dictionary, Merriam-Webster's New Geographical Dictionary,* and *Merriam-Webster's New Biographical Dictionary.*

MARK	AS IN	RESPELLING	EXAMPLE
a	alphabet	a	*AL-fuh-bet
ā	Asia	ay	AY-zhuh
ä	cart, top	ah	KAHRT, TAHP
e	let, ten	e	LET, TEN
ē	even, leaf	ee	EE-vuhn, LEEF
i	it, tip, British	i	IT, TIP, BRIT-ish
ī	site, buy, Ohio	y	SYT, BY, oh-HY-oh
	iris	eye	EYE-ris
k	card	k	KAHRD
ō	over, rainbow	oh	OH-vuhr, RAYN-boh
u̇	book, wood	ooh	BOOHK, WOOHD
ȯ	all, orchid	aw	AWL, AWR-kid
ȯi	foil, coin	oy	FOYL, KOYN
au̇	out	ow	OWT
ə	cup, butter	uh	KUHP, BUHT-uhr
ü	rule, food	oo	ROOL, FOOD
yü	few	yoo	FYOO
zh	vision	zh	VIZH-uhn

*A syllable printed in small capital letters receives heavier emphasis than the other syllable(s) in a word.

abdicated/abdicar Renunciar al trono, **488**

abolitionist/abolicionista Persona que desea terminar con la esclavitud, **227**

absolute authority/autoridad absoluta Poder de un rey o reina para tomar todas las decisiones de gobierno, **141**

absolute location/posición exacta Lugar exacto de la tierra donde se localiza un punto, por lo general definido en términos de latitud y longitud, **7**

acculturation/aculturación Proceso de asimilación de una cultura a largo plazo por el contacto con otra sociedad, **77**

acid rain/lluvia ácida Tipo de lluvia contaminada que se produce cuando partículas de contaminación del aire se combinan con el vapor de agua de la atmósfera, **64**

Age of Exploration/Edad de la exploración Periodo en que los europeos estaban ansiosos por hallar rutas nuevas y más cortas para comerciar con la India y China, **138**

aggression/agresión Acción militar, **187**

agrarian/agrario Sociedad basada en la agricultura, **515**

air pressure/presión atmosférica Peso del aire, se mide con un barómetro, **39**

alliance/alianza Acuerdo entre diferentes países para respaldarse los temas de interès y las causas, **152**

allies/aliados Países que se apoyan entre sí para defenderse de sus enemigos, **489**

alluvial fan/abanico aluvial Accidente geográfico en forma de abanico que se origina por la acumulación de sedimentos en la base de una montaña, **32**

amber/ámbar Savia de árbol fosilizada, **453**

Antarctic Circle/círculo antártico Meridiano localizado a 66.5° al sur del ecuador, **20**

anti-Semitism/antisemitismo Sentimiento de rechazo hacia los judíos, **190**

aqueducts/acueductos Canales artificiales usados para transportar agua, **62, 120**

aquifers/acuíferos Capas subterráneas de roca, arena y grava en las que se almacena el agua, **62**

arable/cultivable Tierra con características que favorecen el cultivo, **576**

archipelago/archpiélago Grupo grande de islas, **312**

Arctic Circle/círculo ártico Meridiano localizado a 66.5° al norte del ecuador, **20**

arid/árido Territorio done la lluvia es muy escasa, **46**

armistice/armisticio Tregua, **176**

arms race/carrera armamentista Competencia entre países para producir y tener armas más avanzadas, **194**

atmosphere/atmósfera Capa de gases que rodea a la tierra, **22**

axis/eje Línea imaginaria que corre del polo norte al polo sur, pasando por el centro de la Tierra, **19**

badlands/tierras de baldío Terrenos irregulares de roca blanda y escasa vegetación o poco suelo, **249**

balance of power/equilibrio de poder Condición que surge cuando varios países o alianzas mantienen niveles tan similares de poder evitar guerras, **157**

bankrupt/bancarrota Sin dinero, **179**

barrier islands/islas de barrera Islas costeras formadas por depósitos de arema arrastrada por las mareas y las corrientes de aguas poco profundas, **241**

basins/cuencas Regiones rodeadas por montañas u otras tierras altas, **219**

bauxite/bauxita El mineral con contenido de aluminio más importante, **313**

bloc/bloque Grupo de naciones unidas por una idea o un propósito común, **193**

bog/ciénaga Tierra suave, humedecida por el agua, **442**

cacao/cacas Árbol pequeño que produce los granos de cacao, **316**

caldera/caldera Depresión grande formada por la erupción y explosión de un volcán, **255**

calypso/calipso Tipo de música originado en Trinidad y Tobago, **320**

cantons/cantones Distritos políticos y administrativos de Suiza, **425**

capital/capital Dinero ganado, ahorrado e invertido para conseguir ganancias, **160**

capitalism/capitalismo Sistema económico en el que los negocios, las industrias y los recursos son de propiedad privada, **163**

caravans/caravanas Grupos de personas que viajan juntas por razones de seguridad, **522**

cardamom/cardamomo Especia que se usa en Asia para condimentar alimentos, **317**

carrying capacity/capacidad de carga Número máximo de especies que puede haber en una zona determinada, **89**

cartography/cartografía Arte y ciencia de la elaboración de mapas, **13**

cash crops/cultivos para la venta Cultivos producidos para su venta y no para consumo del agricultor, **305**

cassava/mandioca Planta tropical de raíces almidonadas, **332**

cathedrals/catedrales Iglesias grandes, **128**

caudillos/caudillos Líderes militares que gobernaron Venezuela en los siglos XIX y XX, **335**

center-pivot irrigation/irrigación de pivote central Tipo de riego que usa aspersores montadios en grandes ruedas giratorias, **251**

chancellor/canciller jefe de gobierno o primer ministro alemán, **419**

chinampas/**chinampas** Nombre dado por los aztecas a los campos elevados que usaban como tierras de cultivo, **298**

chinooks/chinooks Vientos cálidos y fuertes que soplan de las montañas Rocosas hacia las Grandes Planicies, **250**

chivalry/caballerosidad Código o sistema medieval de cavellería, **127**

city-states/ciudades estado Ciudades con un sistema de autogobierno, como en la antigua Grecia, **116, 396**

civil war/guerra civil conflicto entre dos o más grupos dentro de un país, **317**

civilization/civilización Cultura altamente compleja con grandes ciudades y abundante actividad económica, **80**

clergy/clérigo Oficiante de la Iglesia, **128**

climate/clima condiciones meteorológicas registradas en un periodo largo, **37**

climatology/climatología Registro de los sistemas atmosféricos de la Tierra, **13**

cloud forest/bosque nuboso Bosque tropical de gran elevación y humedad donde los bancos de nubes son muy comunes, **313**

coalition governments/gobiernos de coalición Gobiernos en los que la administración del país es regida por varios partidos políticos a la vez, **401**

Cold War/guerra fría Rivalidad entre Estados Unidos y la Unión Soviética que se extendió de la década de 1940 a la década de 1990, **489**

collective farms/fincas colectivas Tierras mancomunadas en grandes fincas, donde las personas trabajan juntas como un grupo, **181**

colony/colonia Territorio controlado por personas de otro país, **139**

command economy/economía autoritaria Economía en la que el gobierno es propietario de la mayor parte de las industrias y toma la mayoría de las decisiones en materia de economía, **85**

commercial agriculture/agricultura comercial Tipo de agricultura cuya producción es exclusiva para la venta, **80**

commonwealth/mancomunidad Territorio autogobernado que mantiene una sociedad con otro país, **323**

communism/comunismo Sistema politica y económico en el cual los gobiernos poseen los medios de producción y controlan el planeamiento de la economia, **85, 192**

condensation/condensación Proceso mediante el cual el agua cambia de estado gaseoso y forma pequeñas gotas, **24**

conquistadores/conquistadores Españoles que participaron en la colonización de América, **299**

constitution/constitución Documento que contiene las leyes y principios básicos que gobiernan una nación, **142**

constitutional monarchy/monarquía constitucional Gobierno que cuenta con un monarca como jefe de estado y un parlamento o grupo legislador similar para la aprobación de leyes, **435**

consumer goods/blenes de consumo Productos usados en la vida cotidiana, **489**

contiguous/contiguo Unidades de territorio (como los estados) que colindan entre sí, **217**

Continental Divide/divisoria continental Cordillera que divide los ríos de Estados Unidos en dos partes: los que fluyen al este y los que fluyen al oeste, **219**

continental shelf/plataforma continental zona costera de pendiente suave que bordea a todos los continentes, **25**

continents/continentes Grandes masas de territorio sobre la Tierra, **29**

cooperatives/cooperativas Organizaciones creadas por los propietarios de una empresa y operados para beneficio propio, **321**

cordillera/cordillera Sistema montañoso de cordilleras paralelas, **328**

core/núcleo Parte sólida del interior de la Tierra, **28**

cork/corcho Corteza extraída de cierto tipo de roble, usada principalmente como material de bloqueo y aislante, **407**

Corn Belt/región maicera Región del medio oeste de Estados Unidos, de Ohio a Iowa, cuya actividad agrícola se basa en el cultivo del maíz, **245**

cosmopolitan/cosmopolita Que tiene influencia de muchas culturas, **424**

Cossacks/cosacos Arrieros nómadas que habitaban en la región fronteriza de Ucrania, **509**

Counter-Reformation/Contrareforma Intento de la Iglesia Católica, luego de la Reforma, por devolver a la Iglesia a un énfasis en asuntos espirituales, **136**

coup/golpe de estado Ataque repentino de un grupo reducido de personas para derrocar a un gobierno, **000**

creoles/criollos Personas de descendencia europea nacidas en la América colonial, **368**

crop rotation/rotación de cultivos Sistema agrícola en el que se siembran productos differentes en periodos específicos, **59**

Crusades/Cruzadas Expediciones hecas por los cristianos para recuperar la Tierra Santa de los musulmanes, **128**

crust/coteza Capa sólida de la superficie de la Tierra, **28**

culture/cultura Sistema de creencias y costumbres comunes que guía la conducta cotidiana de las personas, **75**

culture region/región cultural Región del mundo en la que se comparten ciertos rasgos culturales, **75**

culture traits/rasgos culturales Características de una cultura, **75**

currents/corrientes Enormes corrientes del océano que transportan agua tibia a las regiones frías y viceversa, **41**

czar/zar Emperador ruso, **488**

Dairy Belt/región lechera Región del medio oeste de Estados Unidos, al norte de la franja del maíz, donde la elaboración de productos lácteos es una importante actividad económica, **245**

deforestation/deforestación Destrucción o pérdida de un área boscosa, **61**

deltas/deltas Formaciones creadas por la acumulación de sedimentos en las desembocadura de los ríos, **32**

democracy/democracia Sistema político en el que l a población elige a sus líderes mediante el voto de mayoría, **86**

desalinization/desalinización Proceso mediante el cual se extrae la sal del agua de mar, **63**

desertification/desertificación Proceso a largo plazo en el que el suelo pierde su fertilidad y vegetación, **60**

developed countries/países desarrollados Países industrializados que cuentan con industrias primarias, secundarias, terciarias y cuaternarias, **84**

developing countries/países en vaís de desarrollo países que se encuentran en alguna etapa de su proceso de desarrollo, **84**

dialect/dialecto Variación de un idioma, **406**

dictator/dictadore Persona que ejercen total autoridad sobre un gobierno, **180, 316**

diffusion/difusión Extensión de ideas o conducta de una región cultura a otra, **9**

direct democracy/democracia directa Forma de democracia en la que todos los ciudadanos participan directamente en la toma de decisiones, **116**

diversify/diversificar En la agirgultura, se refiere a la siembra de varios productos y uno solo, **242**

division of labor/división de labores Característa de las civilizaciones en la cual diferentes personas realizan diferentes trabajos, **106**

domestication/domesticación Cuidado de una planta o animal para uso personal, **79**

dominion/dominio Territorio en el que se ejerce una influencia, **266**

droughts/sequías Periodos en los que los cultivos sufren daños debido a la escasez de lluvia, **245**

earthquakes/terremotos Movimientos repentinos y fuertes que se producen en las fisuras de la superficie de la tierra, **29**

ecology/ecología Estudio de las relacions entre las diversas formas de vida del planeta, **51**

economic nationalism/nacionalismo económico Política económica que usa una nación para tratar de mejorar su economía, por medio de limites en el comercio, **185**

ecosystem/ecosistema Conjunto de plantas y animales que habitan una región, junto con los elementos no vivos de ese entorno, **54**

ecotourism/ecoturismo Uso de regiones naturales para atraer visitantes, **318**

ejidos/ejidos Territorios de cultivo propiedad de los indígenas de México, **300**

El Dorado/El Dorado leyenda que habla de los chibchas, antiguos habitantes de Colombia, **330**

El Niño/El Niño Patrón oceánico y climatológico del océano Pacífico que elevó la temperatura del agua en dicho océano, **364**

Emancipation Proclamation/Proclama de Emancipación Decreto emitido por el presidente Abraham Lincoln en 1863 que acabó efectivamente con la esclavitud, a final de la Guerra Civil, **227**

emigrate/emigrar Dejar un país para ir a otro, **166**

empire/imperio Sistema cuyo gobierno central controla diversos territorios, **299**

encomienda/encomienda Sistema mediante el cual los monarcas españoles cedían territorios de América a los colonizadores de su país, quienes obligaban a los indígenas de esas dichas tierras a trabajar para ganarse el susteno, **351**

Enlightenment/Ilustración Período en los años 1700, cuando los filósofos creían que podían aplicar el método cientifico y el uso de la razón para explicar de manera lógica la naturaleza humana, **147**

entrepeneurs/empresarios Personas que usan su dinero y su talento para iniciar un negocio, **85**

epidemic/epidemia Expansión vasta, por general de una enfermedad, **299**

equinoxes/equinoccios Los dos días del año en que los rayos de sol caen directamente sobre el ecuador, **21**

erosion/erosión Desplazamiento de agua, hielo, viento o minerales a otro lugar, **32**

estuary/estuario Cuerpo de agua parcialmente cerrado en el que el agua de mar se combina con ague dulce, **238, 345**

ethnic groups/grupos étnicos Grupos culturales que comparten creencias y prácticas comunes, **75**

evaporation/evaporación Proceso mediante el cual el agua se convierte en vapor y se eleva en el aire, **24**

extinct/extinto Que deja de existir por completo, **51**

factors of production/factores de producción Los recursos naturales, el dinero, el trabajo y los empresarios que se necesitan para las operaciones de negocios, **85, 160**

factory/factoría Fábrica o conjunto de fábricas donde se elabora o produce algún producto, **160**

famine/hambruna Gran escasez de alimeno, **440**

fascism/fascismo Teoría politica que demanda la creación de un gobierno fuerte encabezado por un solo individuo donde el estado sea más importante que el individuo, **180**

fault/falla Fractura de la superficie de la tierra que causa el movimiento de grandes masas de rocas, **31**

favelas/favelas Grandes poblaciones localizadas en los alrededores de algunas ciudades brasileñas, **349**

feudalism/feudalismo Sistema de gobierno local basado en la concesión de tierras como pago por lealtad, ayuda militar y otros servicios, **127**

fief/feudo Concesión de tierras de un amo a su vasayo, **127**

fjords/fiordos Grietas estrechas y profundas localizadas entre altos acantilados donde se acumula el agua de mar, **432**

floodplain/llanura aluvial Especie de plataforma a nivel de la tierra, formada por la acumulación de los sedimentos de una corriente de agua, **32**

food chain/cadena alimenticia Serie de organismos que proveen alimento y energía unos o otros, **51**

fossil fuels/combustibles fósiles Recursos no renovables formados por restos muy antiguos de plantas y animales, **68**

free enterprise/libre empresa Sistema económico en el que las personas, y no el gobierno, deciden qué productos fabrican, venden y compran, **85**

front/frente Inestabilidad climatológica en la que una gran masa de aire tibio choca con una gran masa de aire frío, **40**

frontier/región fronteriza Terreno despoblado, **224**

gauchos/gauchos Arrieros argentinos, **351**

genocide/genocidio Aniquilamiento intencional de un pueblo, **190**

geography/geografía Estudio de las características físicas y culturales de la Tierra, **3**

geothermal energy/energía geotérmica Fuente energética no removable producida por el calor del interior de la tierra, **70**

geysers/géiseres Manantiales que lanzan chorros de agua caliente y vapor a gran altura, **446**

glaciers/glaciares Grandes bloques de hielo que se desplazan con lenitud sobre el agua, **33**

glen/glen Término de origen escocés que se sinónimo de valle, **437**

global warming/calentamiento global Aumento lento y constante de la temperatura de la Tierra, **64**

globalization/globalización Proceso mediante el que as comunicaciones alrededor del mundo se han incrementado haciendo a las culturas más parecidas, **199**

Good Neighbor Policy/Política de Buen Vecino Política esta dounidense durante los año 30, que promulga la cooperación entre las naciones de América Latina, **185**

Great Depression/Gran Depresión Depresión mundial a principios de los años 1930, cuando los salarios cayeron, la actividad comercial bajó y hubo mucho desempleo, **179**

greenhouse effect/efecto invernadero Proceso mediante el cual la atmósfera terrestre atrapa el calor de su superficie, **38**

gross domestic product/producto interno bruto Valor de todos los bienes y servicios producidos en un país, **83**

gross national product/producto nacional bruto Valor de todos los bienes y servicios producidos en un año por un país, dentro o fuera de sus límites, **83**

groundwater/agua subterránea Agua de lluva, ríos, lagos y nieve derretida que se filtra al subsuelo, **25**

guerrilla/guerrillero Persona armada que participa en una lucha armada irregular (los ataques sorpresa, por ejemplo), **321**

habitation fog/humo residente Especie de niebla producida por el humo atrapado en la atmósfera de las cuidades siberianas debido al intenso frío, **496**

haciendas/haciendas Granjas de gran tamaño cedidas por los monarcas españoles a los colonizadores de América, **300**

heavy industry/industra pesada Industria basada en la manufactura de metales, **494**

hieroglyphics Forma antigua de escritura con imágenes y símbolos usados para registrar información, **112**

history/historia Registro escrito de la civilización humana, **107**

Holocaust/haulocausto Asesinato masivo de millones de judíos y personas de otros grupos a manos. de los nazis durante la Segunda Guerra Mundial, **190, 419**

hominid/homínido Primera criatura similares al hombre, **103**

homogeneous/homogéneo Agrupamiento que comparte ciertas características, como el origin étnico, **515**

human geography/geografía humana estudio del pasado y presente de la humanidad, **11**

humanists/humanistas Filósofos del Renacimiento que hacían énfasis en la individualidad, los logros personales y la razón, **132**

humus/humus Materia vegetal o animal en descomposición, **55**

hurricanes/huracanes Tormentas tropicales con intensos vientos, fuertes lluvias y altas mareas, **48**

hydroelectric power/energía hidroeléctrica Fuente energética renovable producida en generadores impulsados por caídas de agua, **70**

icebreakers/rompehielos Barcos que rompen la capa de hielo que se forma en la superficie de algunos cuerpos de agua para permitir el paso de otras embarcaciones, **500**

impressionism/impresionismo Forma de arte desarrollada en Francia a finales del siglo XIX y principios del siglo XX, **417**

indentured servants/trabajadores por contrato Personas que trabajan por un tiempo determinado, en la mayoría de los casos a cambio de gastos de viaje, **337**

indigo/índigo Planta que se usa para fabricar un tinte de color azul oscuro, **334**

individualism/individualismo Creencia en la independencia económica y política de los individuos, **148**

Indo-European/Indoeuropeo Familia que incluye muchos idiomas europeos como el germánico, el báltico y los dialectos eslavos, **454**

Industrial Revolution/Revolución Industrial Cambios producidos a principios de los años 1700, cuando las maquinarias empezaban a hacer mucho del trabajo que las personas tenían que hacer antes, **159**

inflation/inflación Aumento de los precios que ocurre cuando la moneda de un país pierde poder adquisitivo, **304**

Inuit/inuit Tribu esquimal de América del norte, **273**

irrigation/riego Proceso mediante el cual el agua se hace llegar a los cultivos de manera artificial, **105**

isthmus/istmo Franja de tierra que conecta dos áreas de mayor tamaño, **28**

junta/junta Grupo de oficiales militares que asumen el control de un país al derrocar al poder anterior, **373**

knight/caballero Guerrero profesional noble, **127**

land bridges/puentes de terreno Franjas de terreno seco que conecta grandes masas de tierra, **104**

landforms/accidentes geográficos Forma de la tierra en differentes partes de la superficie, **28**

landlocked/sin salida al mar Zona rodeada de agua por completo y sin acceso directo al océano, **356, 520**

lava/lava Magma que emerge del interior de la tierra por un orificio de la corteza, **29**

levees/diques Paredes altas, por lo general de tierra, que se construyen a la orilla de un río para prevenir inundaciones, **10**

light industry/indutria ligera Industria que se enfoca en la manufactura de objetos ligeros como la ropa, **494**

lignite/lignita Tipo de carbón suave, **453**

limited government/gobierno limitado Gobierno en el que los ciudadanos hacen responsables a los dirigentes por medio de la constitución y los procesos democráticos, **86**

limited monarchy/monarquía limitada Sistema de gobierno dirigido por una reina o un rey que no tiene el control absoluto de un país, **141**

literacy/alfabetismo Capacidad de leer y escribir, **166**

Llaneros/**llaneros** Vaqueros de los llanos de Venezuela, **336**

Llanos/llanos Planicies localizadas al este de Colombia y al oeste de Venezuela, **328**

lochs/lagos Lagos escoceses enclavados en valles labrados por los glaciales, **432**

loess/limo Suelo fino de arenisca, excelente para la agricultura, **413**

Loyalists/Loyalistas, Leales Colonistas americanos que se oponían a independización de Intalterra, **152**

magma/magma Roca fundida que se localiza en el manto superior de la tierra, **28**

mainland/región continental Región donde se localiza la mayor porción de terreno de un país, **394**

mantle/manto Capa líquida que rodea al centro de la Tierra, **28**

maquiladoras/maquiladoras Fábricas extranjeras establecidas en la frontera de México con Estados Unidos, **307**

maritime/marítimo En o cerca del mar, como las provincias marítimas de Canada, **270**

market economy/economía de mercado Tipo de economía en la qué los consumidores ayudan a determinar qué productos se fabrican al comprar o rechazar ciertos bienes y servicios, **85**

mass production/producción en serie Sistema de producción de grandes cantidades de productos idénticos, **163**

medieval/medieval Periodo de colapso del imperio romano, aproximadamente en el año 1,500 de nuestra era, **415**

megalopolis/megalópolis Enorme zona urbana que abarca una serie de ciudades que se han desarrollado juntas, **236**

mercantilism/mercantilismo Creación y conservación de riquezas mediante un control minucioso de intercambios comerciales, **139**

Mercosur/Mercosur Organización comercial en la que participan Argentina, Brasil, Paraguay, Uruguay y dos países asociados (Bolivia y Chile), **354**

merengue/merengue Tipo de música y baile nacional en la República Dominicana, **320**

mestizos/mestizos Personas cuyo origen combina las razas europeas y las razas indígenas de América, **300**

metallic minerals/minerales metálicos Minerales brillantes, como el oro y el hierro, que conducen el calor y la electricidad, **65**

meteorology/meteorología Predicción y registro de lluvias, temperaturas y otras condiciones atmosféricas, **13**

Métis/Métis Personas cuyo origin combina las razas europeas y las razas indígenas de Canadá, **267**

middle class/clase media Clase formada por comerciantes patrones de pequeña y mediana industria y profesiones liberales. Está entre la clase noble y la clase campesina en la Edad Media, **129, 164**

migrated/migrar Trasladarse de un lugar a otro, **110**

militarism/militarismo Uso de armamento pesado y amenazas para obtener poder, **173**

minerals/minerales Sustancias inorgánicas que conforman la corteza de la tierra; en su medio natural, aparecen en forma cristalina y tienen una composición química definida, **65**

missions/misiones Puestos españoles de evangelización establecidos en la época colonial, especialmente en América, **300**

monarchy/monarquía Sistema de gobierno dirigido por un rey o una reina, **141**

monsoon/monzón Cambio de corrientes aire y lluvias que produce temporadas alternadas de humedad y sequía, **45**

Moors/moros Musulmanes del norte de África, **405**

moraine/morena Cresta de rocas, grava y arena levantada por un glaciar, **237**

mosaics/mosaicos Imágenes creadas con pequeños fragmentos de piedras coloreadas, **398**

mosques/mezquitas Casas de adoración islámica, **526**

mulattoes/mulatos Personas cuyo origen combina las razas europeas y las razas indígenas de África, **300**

multicultural/multicultural Mezcla de culturas en un mismo país o comunidad, **75**

nationalism/nacionalismo Demanda de autogo-bierno y fuerte sentimiento de lealtad hacia una nación, **167, 426**

national parks/parques nacionales Terreno escénico grande preservado por un gobierno para uso público, **252**

NATO/OTAN (Organización del Tratado del Atlántico Norte); alianza militar formada por varios países europeos, Estados Unidos y Canadá, **194, 415**

nature reserves/reservas naturales Zonas asignadas por el gobierno para la protección de animales, plantas, suelo y agua, **506**

navigable/navegable Rutas acuáticas de profunidad suficiente para la navegación de barcos, **412**

neutral/neutral Que no toma ningún partido en una disputa o conflicto, **444**

New Deal/New Deal Programa del presidente Franklin D. Roosevelt en el que el gobierno federal estableció un amplio programa de obras públicas para crear empleo y conceder dinero a cada estado para sus necesidades, **179**

newsprint/papel periódico Papel económico usado para imprimir publicaciones periódicas, **263**

nobles/nobles Personas que nacen entre familias ricas y poderosas, **127**

nomads/nómadas Personas que se mudan frecuentemente de un lugar a otro, **522**

nonmetallic minerals/minerales no metálicos Minerales que no tienen las características de los metales, **66**

nonrenewable resources/recursos no renova-bles Recursos, como el carbón mineral y petróleo, que no pueden reemplazarse a corto plazo por medios naturales, **65**

North Atlantic Drift/Corriente del Atlántico Norte corriente de aguas tibias que aumenta la temperatura y genera lluvias en el norte de Europa, **433**

nutrients/nutrientes sustancias que favorecen el crecimiento, **52**

oasis/oasis Lugar del desierto donde un manantial proporciona una fuente natural de agua, **520**

oil shale/pizarra petrolífera Capa de roca que al calentarse produce petróleo, **453**

oppression/opresión Uso de poder cruel e injusto contra otros, **154**

orbit/órbita Trayectoria que sigue un objeto que gira alrededor de otro objeto (como la Tierra alrededor del Sol), **17**

ozone/ozono Forma del oxígeno en la atmósfera que ayuda a proteger a la Tierra de los daños que produce la radiación solar, **22**

Pampas/pampas Región extensa cubierta de hierba en la zona central de Argentina, **344**

Pangaea/Pangaea Supercontinente original y único del que se separaron los continentes actuales, **31**

pardos/**pardos** Venezolanos descendientes de la unión entre africanos, europeos e indígenas sudamericanos, **336**

Parliament/Parlamento Órgano británico de legislación, **141**

partition/repartir Separar un terreno formando grupos más pequeños, **195**

Patriots/Patriotas Colonistas americanos que favorecieron con la independencia de Inglaterra, **152**

peat/turba Sustancia formada por plantas muertas, por lo general musgos, **442**

peninsula/península Tierra rodeada de agua en tres lados, **28**

permafrost/permafrost Capa de suelo que permanece congelada todo el año en las regiones con clima de la tundra, **49**

perspective/perspectiva Punto de vista basado en la experiencia y la comprensión de una persona, **3**

Peru Current/corriente de Perú Corriente oceánica fría del litoral oeste de América del Sur, **363**

petroleum/petróleo Líquido graso que al refinarse produce gasolina y otros combustibles y aceites, **68**

photosynthesis/fotosíntesis Proceso por el que las plantas convierten la luz solar en energía quimica, **51**

physical geography/geografía física Estudio del paisaje natural y los sistemas físicos de la Tierra, entre ellos la atmósfera, **11**

pidgin languages/lenguas francas Lenguajes sencillos que ayuden a entenderse a personas que hablan idiomas diferentes, **337**

pioneers/pioneros Primeras personas que llegan a poblar una región, **225**

plain/planicie Área casi plana de la superficie terrestre, **28**

plantains/banano Tipo de plátano que se usa para cocinar, **322**

plantations/plantaciones Granjas muy grandes en las que se produce un solo tipo de cultivo para vender, **223**

plant communities/comunidades de plantas Grupos de plantas que viven en la misma zona, **53**

plant succession/sucesión de plantas Proceso gradual por el que un grupo de plantas reemplaza a otro, **54**

plateau/meseta Terreno plano y elevado sobre la superficie terrestre, **28**

plate tectonics/tectónica de placas Teoría de que la superficie terrestre está dividida en varias placas enormes que se mueven lentamente, **28**

police state/estado totalitario País en que el gobierno tiene un control total sobre la vida de las personas, **180**

pope/papa Obisco de Roma y líder de la Iglesia católica romana, **401**

popular sovereignty/soberanía popular Principio gubernamental basado en leyes justas, y en un gobierno creado y sujeto a la voluntad del pueblo, **149**

population density/densidad de población Número promedio de personas que viven en una milla cuadrada o un kilómetro cuadrado, **81**

potash/potasa Mineral que se usa para hacer fertilizantes, **263**

precipitation/precipitación Proceso por el que el agua vuelve de regreso a la Tierra, **24**

prehistory/prehistoria Tiempo antiguo del que no se conservan registros escritos, **103**

primary industries/industrias primarias Actividades económicas que involucran directamente recursos naturales o materia prima, tales como la agricultura y la minería, **83**

Protestants/protestantes Reformistas que protestaban por la realización de ciertas prácticas de la Iglesia Católica, **135**

provinces/provincias Divisiones administrativas de un país, **266**

pueblo/pueblo grupo de personas que vivian en asientamentos permanentes en el suroests de Estados Unidos, **111**

pulp/pulpa Fibras reblandecidas de madera para hacer papel, **263**

Puritan/Puritano Persona que aspiraban a una doctrina más pura que la propuesta por la Iglesia Católica Inglesa, **141**

stock market/mercado de valores Organización mediante la cual se venden y se compran partes de compañías, **178**

strait/estrecho Paso angosto que une dos grandes cuerpos de agua, **362**

strip mining/minería a cielo abierto Tipo de minería en que se retiran la tierra y las rocas para extraer carbón y otros recursos que están bajo la superficie terrestre, **252**

subduction/subducción Movimiento en el que una placa tectónica terrestre más gruesa se sumerge debajo de una más delgada, **29**

subregions/subregiones Áreas pequeñas de una región, **8**

subsistence agriculture/agricultura de subsistencia Tipo de agricultura en que los campesinos siembran sólo lo necesario para mantenerse a ellos mismos y a sus familias, **80**

suburbs/suburbios Áreas residenciales en las afueras de la ciudad, **167**

suffragettes/sufragistas Mujeres que lucharon por el derecho de las mujeres de votar, **168**

superpowers/superpotencias Países poderosos, **489**

symbol/símbolo Palabra, forma, color, estadarte o cualquier otra cosa que se use en representación de algo, **77**

taiga/taiga Bosque de árboles siempre verdes que existen en el sur de la tundra en Rusia, **485**

tepees/tepees Tiendas de forma cónica, hechas de piel de búfalo, **110**

tepuís/tepuís Capas de roca arenisca resistentes a la erosión en las mesetas de los altiplanos de las Guyanas, **328**

terraces/terrazas Crestas horizontales que se construyen sobre las laderas de las colinas para prevenir la pérdida de suelo y favorecer la agricultura, **60**

tertiary industries/industrias terciarias Actividades económicas que trabajan con productos listos para vender a los consumidores, **83**

textiles/textiles Productos para fabricar ropa, **435**

traditional economy/economía tradicional Economía basada en las costumbres y las tradiciones, **85**

tributary/tributario Cualquier corriente pequeña o río que fluye hacia un río o una corriente más grande, **25**

Tropic of Cancer/Trópico de Cáncer Línea de latitud que está a 23.5 grados al norte del ecuador, **21**

Tropic of Capricorn/Trópico de Capricornio Línea de latitud que está a 23.5 grados al sur del ecuador, **21**

tundra climate/clima de la tundra Región fría de lluvias escasas, que por lo genera se encuentra entre los climas de las regiones subártica y polar, **49**

typhoons/tifones Tormentas tropicales de vientos violentos, fuertes lluvias y altas marejadas, **48**

U-boats/U-boats Submarinos alemanes usados en la Primer Guerra Mundial, **174**

uninhabitable/inhabitable Que no es propicio para el establecimiento de seres humanos, **446**

unlimited governments/gobiernos ilimitados Gobiernos que tienen un control total sobre sus ciudadanos, **86**

urban/urbano Área en que se encuentra una ciudad, **4**

vassals/vassalos Personas a la que un amo le concedía tierras, como pago por sus servicios, **127**

vernacular/vernácular Lenguaje doméstico, natino, propio de un país, **130**

viceroy/virrey Gobernador de una colonia, **368**

water cycle/ciclo del agua Circulación del agua del la superficie de la Tierra a la atmósfera y su regreso, **23**

water vapor/vapor de agua Estado gaseoso del agua, **23**

weather/tiempo Condiciones de la atmósfera en un tiempo y un lugar determinados, **37**

weathering/desgaste Proceso de desintegración de las rocas en pedazos pequeños por la acción del calor, el agua y otros medios, **31**

wetlands/terreno pantanoso Paisaje cubierto de agua al menos una parte del año, **241**

Wheat Belt/región triguera Zona de la región de las Grandes Planicies en Estados Unidos en la que la actividad principal es el cultivo del trigo, **251**

working class/clase trabajadora Personas capacitadas y poco capacitadas con empleos de salarios bajos, **165**

yurt/yurta Tienda redonda y portátil de lana tejida que se coloca sobre una armazón de madera, **526**

INDEX

abdicated, 488
abolitionist, 222, 227
absolute authority, 137, 141
absolute location, 7, *g7*
Acapulco, Mexico, *m293*, 305
acculturation, 77
acid rain, 64
Acropolis, *p397*
Adriatic Sea, 117, *m451*, 452
Aegean Sea, 114, *m114*, *m117*, 393
Age of Exploration, 138–40, *p138*, *m138–39*, *p139*, *g140*
age structure diagrams, S11, *gS11*
aggression, 187
agrarian, 515
agriculture: Aztec method, 298; cash crops, 305; civilization and, 80, 106; commercial, 80; corn, *g513*, 245, 332; crop rotation, 59; development of, 79–80; domestication and, 79; environment and, 80; Stone age, 105–06; Industrial Revolution and, 159, *p159*; plantations, 222–23; seawater agriculture, 376–77; slash-and-burn, 307; subsistence, 80; types of, 80
air: as natural resource, 63–64
air pressure, 39–40, *g40*
al Qaeda, 202–03, 231
Alabama, 214, *m217*; climate, 220–21, 242; economy, 243; natural resources, 221; physical features, 218–20, *m219*, 241–32; statistics, 104
Alaska, *p49*, *m253*, *p256*; climate, 256; economy, 257; natural resources, 221; physical features, 255; statistics, 104
Alaska pipeline, *p69*
Albania, *m451*, 460–62, 465, *g465*, statistics, 388
Alberta, Canada, 215, *m261*, 267, 273
Aleutian Islands, *m217*, 220
Alexander the Great, 117, *p117*, 396
alliance, 151, 152, 173
allies, 188, 489
alluvial fan, 32
Alpine countries, *m411*, 412, 427, *g427*. *See also* Austria; Switzerland
Alps, 117, *m117*, 394, 412, *p412*
Altiplano, *m343*, 362
Amazon River, 344, 362
amber, 452, 453
American Indians, 222, *p222*, 224,

226. *See also* specific peoples
American Revolution, 151–52, *p152*, 225; effects of American Independence, 153; impact on North American migration, 266; Stamp Act, *p151*
Amsterdam, the Netherlands, *m411*, 424
Amu Dar'ya, *m519*, 520
Amundsen-Scott South Pole Station, Antarctica, *p50*
Amur River, *m483*, 499
Andes, *p31*, *m111*, 328, *g328*, *m343*, 344, 362
Andorra, 388
Angkor *See* Cambodia
Angles, 434
Anne of Green Gables (Montgomery), 271
annexed, 187
Antarctica, *p50*
Antarctic Circle, 20, *g20*
Antigua and Barbuda, 288, *m311*, *g320*, 322
anti-Semitism, 187, 189, 190
Antwerp, Belgium, *m411*, 424
Apache, 222
Apennines, *m393*, 394
Appalachian Mountains, *m217*, 218, *p218*, *m219*, 262
Appian Way, Italy, *p400*
aqueducts, 62, 114, 119, 120, *p120*, 400
aquifers, 62
arable, 576
Aral Sea, *m519*, 520, 521
archeology, 97
archipelago, 312
Arctic Circle, 20, *g20*
Arctic Ocean, 484
Argentina, *m343*; climate, 345; history, 351–52, *p351*, *p352*; in modern times, 353–54; people and culture, 352, *p352*; physical features, 344–45; resources, 346; statistics, 288, *g352*; vegetation and wildlife, 345
arid, 46
Aristotle, 398
Arizona, *m248*; climate, 250; economy, 251–52; landforms, 249; natural resources, 221; physical features, 218–20, *m219*, 249–50; statistics, 214
Arkansas, *m240*; climate, 220–21, 242; economy, 243; natural resources, 221; physical features, 218–20, 241; statistics, 214

Armenia, 480, *m505*, 513–15
armistice, 173, 176
arms race, 192, 194
art: French impressionists, 417; Norwegian stave churches, 445; recycled art, 66; Renaissance, 132
Astrakhan, Russia, *m483*, 495
Asunción, Paraguay, *m343*, 357
Atacama Desert, *m361*, 363
Athens Greece, *m114*, 116, *m117*, 397, *p397*, 399
Atlanta Georgia, *m217*, *m240*, 243
Atlantic South America, *m343*; Argentina, 351–54; Brazil, 347–50; climate, 345; culture, 342; Paraguay, 356–57; physical features 344–45; resources, 346; Uruguay, 355–56; vegetation, 345; wildlife, 345; *See also* specific countries
atmosphere, 22, *g22*
Austria, *m411*; culture, 426–27; economy, 427; history, 425–26; statistics, 388, *g427*
Austrian Empire, 426. *See also* Austria
Austro-Hungarian Empire, 426, 459. *See also* Austria
avalanche, *p13*
axis, 19, 188
Azerbaijan, 480, *m505*, 513–14, 515
Aztec, 112, *p112*, 113, *p182*, 298, *p298*

badlands, 248, 249
Bahamas, 288, *m311*, 312, *g321*, 322
Baikal-Amur Mainline (BAM), 497
Baja California, *m293*, 294
balance of power, 151, 157
Balkan countries, *m451*, 452, 461, 462
Balkan Mountains, 452
Balkan Peninsula, *m451*, 452
Baltic countries, 452
Baltic Sea, *m431*, 432, 452, 484
Baltimore, Maryland, *m217*, *m235*, 236
bananas, *g317*, 329, 332
Banff National Park, Alberta, *p262*, 267
Bangladesh, *p11*
bankrupt, 178, 179
Barbados, 288, *m311*, *g320*, 322
Barcelona, Spain, *m393*, 407
bar graphs, S10, *gS10*
barrier islands, 240, 241
Basel, Switzerland, *m411*, 427
basins, 219

ACKNOWLEDGMENTS

For permission to reprint copyrighted material, grateful acknowledgment is made to the following sources:

Casa Juan Diego: Adapted from *Immigrants Risk All: Cry for Argentina* by Ana Maria from *Houston Catholic Worker*, July/August 1997, accessed February 2, 2000, at http://www.cjd.org/stories/risk.html.

Doubleday, a division of Random House, Inc.: From "Marriage Is a Private Affair" from *Girls at War and Other Stories* by Chinua Achebe. Copyright © 1972, 1973 by Chinua Achebe.

FocalPoint f/8: From "October 5—Galtai" from "Daily Chronicles" and from "Buddhist Prayer Ceremony" from "Road Stories" by Gary Matoso and Lisa Dickey from *The Russian Chronicles* from *FocalPoint f/8*, accessed October 14, 1999, at http://www.f8.com/FP/Russia.

Glas Publishers (Russia): From "The Lilac Dressing Gown" by Nina Gabrielyan, translated by Joanne Turnbull from *A Will and a Way: New Russian Writing*, edited by Natasha Perova and Arch Tait. Copyright © 1996 by Glas: New Russian Writing.

HarperCollins Publishers, Inc.: From *My Days* by R. K. Narayan. Copyright © 1973, 1974 by R. K. Narayan.

James Li, M.D.: From "Africa" by James Li, M.D from *eMedicine*, accessed October 11, 1999, at http://www.emedicine.com/emerg/topic726.htm. Copyright © 1999 by James Li.

Ms. Magazine: From "Foresters Without Diplomas" by Wangari Maathai from *Ms.*, vol. 1, no. 5, March/April 1991. Copyright © 1991 by *Ms.* Magazine.

Penguin Books Ltd.: From *The Epic of Gilgamesh*, translated by N. K. Sandars (Penguin Classics 1960, Third Edition, 1972). Copyright © 1960, 1964, 1972 by N. K. Sandars.

Puffin Books, a division of Penguin Putnam Inc.: From *Sadako and the Thousand Paper Cranes* by Eleanor Coerr. Copyright © 1977 by Elizabeth Coerr.

Sandy Wiseman: From "Photo-Journey Through a Costa Rican Rainforest" by Sandy Wiseman from *EcoFuture™ PlanetKeepers*, accessed October 14, 1999, at http://www.ecofuture.org/ecofuture/pk/
pkar9512.html. Copyright © 1999 by Sandy Wiseman.

Writer's House, Inc. c/o The Permissions Company: From *For Love Alone* by Christina Stead. Copyright © 1944 by Harcourt, Inc.; copyright renewed © 1972 by Christina Stead.

SOURCES CITED:

From "The Aztecs in 1519" from *The Discovery and Conquest of Mexico* by Bernal Díaz. Published by Routledge and Kegan Paul, London, 1938.

ART CREDITS

Abbreviated as follows: (t) top, (b) bottom, (l) left, (r) right, (c) center.

Unit flags of United States created by One Mile Up, Inc. Unit flags of Canadian provinces created by EyeWire, Inc. Other flags, country silhouettes, feature maps and atlas maps created by MapQuest.com., Inc. All other illustrations, unless otherwise noted, contributed by Holt, Rinehart and Winston.

Study Skills: Page S2, MapQuest.com, Inc.; S3, MapQuest.com, Inc.; S4 (cl, bl), MapQuest.com, Inc.; S5, MapQuest.com, Inc.; S6 (tr), MapQuest.com, Inc.; S7, MapQuest.com, Inc.; S8 (tr), MapQuest.com, Inc.; S9, MapQuest.com, Inc.; S10 (tc, br), Leslie Kell; S11 (cr), Leslie Kell; S11 (b), Ortelius Design; S12 (b), Leslie Kell; S12 (tr), Rosa + Wesley; S13, Uhl Studios, Inc.; S14 (br), MapQuest.com, Inc.; S15 (bl), MapQuest.com, Inc.

Chapter 1: Page 5 (cl), MapQuest.com, Inc.; 6 (bl), MapQuest.com, Inc.; 7 (tl, tr), MapQuest.com, Inc.; 8 (l, bc), MapQuest.com, Inc.; 9 (tr), MapQuest.com, Inc.; 10 (tc), MapQuest.com, Inc.; 11 (bc), MapQuest.com, Inc.; 15 (tr), MapQuest.com, Inc.

Chapter 2: Page 17 (b), Uhl Studios, Inc.; 18 (b), Uhl Studios, Inc.; 19 (bl), Uhl Studios, Inc.; 20 (b), Uhl Studios, Inc.; 22 (tl), Robert Hynes; 23 (cr), MapQuest.com, Inc.; 24 (tr), Uhl Studios, Inc.; 25 (tr, br), Uhl Studios, Inc.; 26 (b), MapQuest.com, Inc.; 28 (b), Uhl Studios, Inc.; 29 (tl), MapQuest.com, Inc.; 29 (c, br), Uhl Studios, Inc.; 31 (tr), MapQuest.com, Inc.; 32 (tr), Ortelius Design; 35 (tr), MapQuest.com, Inc.

Chapter 3: Page 38 (tl), Uhl Studios, Inc.; 38 (cl), MapQuest.com, Inc.; 39 (bl), MapQuest.com, Inc.; 40 (tr), Uhl Studios, Inc.; 42 (cr), Rosa + Wesley; 42 (b), MapQuest.com, Inc.; 43 (t), Rosa + Wesley; 45 (tl), Uhl Studios, Inc.; 46 (br), MapQuest.com, Inc.; 47 (all), MapQuest.com, Inc.; 48 (tl, bc), MapQuest.com, Inc.; 49 (tr), MapQuest.com, Inc.; 50 (tl), MapQuest.com, Inc.; 52 (tr), Robert Hynes; 55 (tr), Uhl Studios, Inc.; 57 (tr), MapQuest.com, Inc.

Chapter 4: Page 60 (tc), MapQuest.com, Inc.; 62 (bl), MapQuest.com, Inc.; 63 (tr, br), MapQuest.com, Inc.; 65 (br), Leslie Kell; 67 (t), MapQuest.com, Inc.; 69 (tl, br), MapQuest.com, Inc.; 70 (tl), MapQuest.com, Inc.; 73 (tr), MapQuest.com, Inc.

Chapter 5: Page 76 (t, bl), MapQuest.com, Inc.; 77 (tr), MapQuest.com, Inc.; 78 (bl), MapQuest.com, Inc.; 81 (cr), MapQuest.com, Inc.; 83 (br), MapQuest.com, Inc.; 86 (c), Rosa + Wesley; 88 (cl), MapQuest.com, Inc.; 89 (tc), MapQuest.com, Inc.; 91 (tr), MapQuest.com, Inc.; 93 (tl), MapQuest.com, Inc.; 98 (t, bl), Ortelius Design.

Chapter 6: Page 106 (t) MapQuest.com, Inc.; 107 (b) MapQuest.com, Inc.; 109 (b) MapQuest.com, Inc.; 111 (b) MapQuest.com, Inc.; 112 (t) Nenad Jakesevic; 114 (b) MapQuest.com, Inc.; 117 (br) MapQuest.com, Inc.; 119 (b) MapQuest.com, Inc.; 123 (tr) MapQuest.com, Inc.

Chapter 7: Page 126 (br) MapQuest.com, Inc.; 128 (cl) Nenad Jakesevic; 138–139 (tr) MapQuest.com, Inc.; 145 (tr) MapQuest.com, Inc.

Chapter 8: Page 153 MapQuest.com, Inc.; 171 MapQuest.com, Inc.; 171 MapQuest.com, Inc.

Chapter 9: Page 174 MapQuest.com, Inc.; 175 (tr) Craig Attebery; 193 (bl) MapQuest.com, Inc.; 197 MapQuest.com, Inc.; 201 (tr) MapQuest.com, Inc.; 203 (tr) MapQuest.com, Inc.; 208 (bl), Ortelius Design; 208 (t), Ortelius Design; 209 (all), MapQuest.com, Inc.; 210 (all), MapQuest.com, Inc.; 211 (all), MapQuest.com, Inc.; 212 (all), MapQuest.com, Inc.; 213 (all), MapQuest.com, Inc.

Chapter 10: Page 217 (b), MapQuest.com, Inc.; 219 (t), MapQuest.com, Inc.; 219 (br), MapQuest.com, Inc.; 220 (bl), MapQuest.com, Inc.; 224 (bl) MapQuest.com, Inc.; 226 (tr) MapQuest.com, Inc.; 233 (tr) MapQuest.com, Inc.; 233 MapQuest.com, Inc.

Chapter 11: Page 237 MapQuest.com, Inc.; 238 Phototake; 239 MapQuest.com, Inc.; 240 (b) MapQuest.com, Inc.; 243 MapQuest.com, Inc.; 244 MapQuest.com, Inc.; 246 (tr) The John Edwards Group; 246 (b) MapQuest.com, Inc.; 247 MapQuest.com, Inc.; 248 (b) MapQuest.com, Inc.; 249 MapQuest.com, Inc.; 249 MapQuest.com, Inc.; 250 MapQuest.com, Inc.; 253 (b) MapQuest.com, Inc.; 255 MapQuest.com, Inc.; 255 MapQuest.com, Inc.; 257 MapQuest.com, Inc.; 259 (tr) MapQuest.com, Inc.; 259 MapQuest.com, Inc.

Chapter 12: Page 261 (b), MapQuest.com, Inc.; 262 (cl), MapQuest.com, Inc.; 263 (tr), MapQuest.com, Inc.; 266 (tr), MapQuest.com, Inc.; 266 (tl), Leslie Kell; 266 (bl), MapQuest.com, Inc.; 267 (cr), MapQuest.com, Inc.; 267 (br), Rosa + Wesley; 268 (cl), MapQuest.com, Inc.; 270 (tl), MapQuest.com, Inc.; 272 (tl), MapQuest.com, Inc.; 275 (tr), MapQuest.com, Inc.; 277 (tr), MapQuest.com, Inc.; 182 (t, bl), Ortelius Design.

Chapter 13: Page 293 (b), MapQuest.com, Inc.; 295 (tc), MapQuest.com, Inc.; 295 (t,br), MapQuest.com, Inc.; 298 (br), MapQuest.com, Inc.; 300 (bl), MapQuest.com; 302 (tr), Leslie Kell; 303 (t), MapQuest.com, Inc.; 304 (cl), Rosa + Wesley; 305 (bl); MapQuest.com; 306 (bl), MapQuest.com, Inc.; 309 (tr), MapQuest.com, Inc.

Chapter 14: Page 311 (b), MapQuest.com, Inc.; 312 (bl), MapQuest.com, Inc.; 314 (cl), MapQuest.com, Inc.; 315 (cr), MapQuest.com, Inc.; 317 (br), Leslie Kell; 318 (tl), Rosa + Wesley; 319 (br), MapQuest.com, Inc.; 320 (tl), Rosa + Wesley; 321 (tr, bl), MapQuest.com, Inc.; 323 (tr), MapQuest.com, Inc.; 325 (tr), MapQuest.com, Inc.

Chapter 15: Page 327 (b), MapQuest.com, Inc.; 328 (b), Uhl Studios, Inc.; 330 (bc), MapQuest.com, Inc.; 331 (tr), Rosa + Wesley; 331 (cr), MapQuest.com, Inc.; 333 (tc), MapQuest.com, Inc.; 333 (cr), Leslie Kell; 335 (bl), MapQuest.com, Inc.; 335 (br), MapQuest.com, Inc.; 336 (cl), Rosa + Wesley; 337 (cr), MapQuest.com, Inc.; 338 (tl), Rosa + Wesley; 338 (bc), MapQuest.com, Inc.; 339 (tl), MapQuest.com, Inc.; 341 (tr), MapQuest.com, Inc.

Chapter 16: Page 343 (b), MapQuest.com, Inc.; 344 (bl), MapQuest.com, Inc.; 349 (tr), MapQuest.com, Inc.; 349 (br), Rosa + Wesley; 350 (tl), Leslie Kell; 352 (tl,br), MapQuest.com, Inc.; 354 (tc), MapQuest.com, Inc.; 354 (cl), Rosa + Wesley; 355 (c), MapQuest.com, Inc.; 356 (tc), MapQuest.com, Inc.; 356 (bl), Rosa + Wesley; 359 (tr), MapQuest.com, Inc.

Chapter 17: Page 361 (b), MapQuest.com, Inc.; 363 (cr), MapQuest.com, Inc.; 366 (tl), MapQuest.com, Inc.; 368 (tl), MapQuest.com, Inc.; 370 (bl), MapQuest.com, Inc.; 371 (tr), MapQuest.com, Inc.; 372 (tl), Rosa + Wesley; 373 (tr), Leslie Kell; 375 (tr), MapQuest.com, Inc.; 380 (t, bl), Ortelius Design.

Chapter 18: Page 393 (b), MapQuest.com, Inc.; 394 (bl), MapQuest.com, Inc.; 395 (tr), MapQuest.com, Inc.; 395 (cr), Leslie Kell; 396 (br), MapQuest.com, Inc.; 397 (tc), MapQuest.com, Inc.; 399 (tr), Rosa + Wesley; 400 (cl), MapQuest.com, Inc.; 401 (tc), MapQuest.com, Inc.; 402 (bl), MapQuest.com, Inc.; 403 (tr), Rosa + Wesley; 405 (br), MapQuest.com, Inc.; 405 (c), MapQuest.com, Inc.; 406 (tl), Rosa + Wesley; 407 (tl), MapQuest.com, Inc.; 409 (tr), MapQuest.com, Inc.

Chapter 19: Page 411 (b), MapQuest.com, Inc.; 412 (tl, bc), MapQuest.com, Inc.; 413 (tr), MapQuest.com, Inc.; 416 (cr), MapQuest.com, Inc.; 417 (cl), Rosa + Wesley; 418 (bl), MapQuest.com, Inc.; 419 (cr), MapQuest.com, Inc.; 420 (tl), Leslie Kell; 420 (c), MapQuest.com, Inc.; 421 (cr), Rosa + Wesley; 422 (cl), MapQuest.com, Inc.; 424 (cl), Rosa + Wesley; 426 (tl), MapQuest.com, Inc.; 427 (cr), Rosa + Wesley; 429 (tr), MapQuest.com, Inc.

Chapter 20: Page 431 (b), MapQuest.com, Inc.; 432 (bl), MapQuest.com, Inc.; 434 (cl), MapQuest.com, Inc.; 435 (c), MapQuest.com, Inc.; 436 (bl), Rosa + Wesley; 439, Nenad Jakesevic; 437 (cr), Leslie Kell; 440 (bl), MapQuest.com, Inc.; 442 (tl), Rosa + Wesley; 442 (cl), MapQuest.com, Inc.; 443 (br), Rosa + Wesley; 445 (tl, bc), MapQuest.com, Inc.; 436 (tl), MapQuest.com, Inc.; 437 (cr), MapQuest.com, Inc.; 439 (tr), MapQuest.com, Inc.

Chapter 21: Page 441 (b), MapQuest.com, Inc.; 442 (bl), MapQuest.com, Inc.; 443 (tr), MapQuest.com, Inc.; 446 (tl), Rosa + Wesley; 446 (cl), MapQuest.com, Inc.; 448 (bc), MapQuest.com, Inc.; 449 (tr), MapQuest.com, Inc.; 450 (bl), MapQuest.com, Inc.; 451 (cr), MapQuest.com, Inc.; 452 (tc), MapQuest.com, Inc.; 453 (tr, cr), MapQuest.com, Inc.; 456 (tl), Leslie Kell; 456 (bl), MapQuest.com, Inc.; 455 (tr), Rosa + Wesley; 457 (tr), MapQuest.com, Inc.; 459 (tl), MapQuest.com, Inc.; 474 (t, bl), Ortelius Design.

Chapter 22: Page 483 (b), MapQuest.com, Inc.; 485 (br), MapQuest.com, Inc.; 488 (t), MapQuest.com, Inc.; 489 (tr, bc), MapQuest.com, Inc.; 491 (tr, cr), Rosa + Wesley; 491 (br), MapQuest.com, Inc.; 494 (br), MapQuest.com, Inc.; 494 (tl, bl), MapQuest.com, Inc.; 495 (tr), MapQuest.com, Inc.; 496 (bl), MapQuest.com, Inc.; 497 (tl), MapQuest.com, Inc.; 498 (tl), MapQuest.com, Inc.; 500 (bl), MapQuest.com, Inc.; 501 (tr), MapQuest.com, Inc.; 503 (tr), MapQuest.com, Inc.

Chapter 23: Page 505 (b), MapQuest.com, Inc.; 506 (bl), MapQuest.com, Inc.; 509 (bl), MapQuest.com, Inc.; 510 (tl), MapQuest.com, Inc.; 511 (c), Uhl Studios, Inc.; 512 (tl), Leslie Kell; 512 (cl), Rosa + Wesley; 514 (bc), MapQuest.com, Inc.; 515 (tr), Rosa + Wesley; 517 (tr), MapQuest.com, Inc.

Chapter 24: Page 519 (b), MapQuest.com, Inc.; 521 (cr), MapQuest.com, Inc.; 522 (cl), MapQuest.com, Inc.; 523 (cr), MapQuest.com, Inc.; 524 (l), Leslie Kell; 526 (bc), MapQuest.com, Inc.; 527 (tr), Rosa + Wesley; 529 (cr), Leslie Kell; 531 (tr), MapQuest.com, Inc.; 533 (t), MapQuest.com, Inc.

PHOTO CREDITS

Cover and Title Page: (child image) AlaskaStock Images; (bkgd) Image Copyright ©2003 PhotoDisc, Inc./HRW

Table of Contents: iv (bl), © SuperStock; ix (bl) "Werner Forman Archive Art Resource, NY"; ix (tr), © Stone/Chad Ehlers; v (tl) "S. Sherbell/SABA Press Photos, Inc."; vi (bl) "Walters Art Gallery, Baltimore, Courtesy of Robert Forbes Collection"; vi © From the Collections of Henry Ford Museum and Greenfield Village; vii (tl) "Bob Daemmrich/The Image Works, Woodstock, NY"; vii (br), © Marc Muench/CORBIS; viii (tl), © Nazima Kowall/CORBIS; viii (br) Stone/Richard A. Cooke III; x (tl), © The Stock Market/Don Mason; x (br), © Leo de Wys/Siegfried Tauquer; xi (b), © Index Stock Photography; xi (tl), © The Stock Market/H. P. Merten; xi (tr), © Claus Meyer/Black Star/PNI; xii (tl), © Travelpix/FPG International LLC; xii (b) Steve Raymer/NGS Image Collection; xiii (bl), © The State Russian Museum/CORBIS; xiv (tl) Alain Le Garsml/Panos Pictures; xv (br) "Image Copyright © 2002 PhotoDisc, Inc./HRW"; xviii (b), © Matt Glass/CORBIS-Sygma; xxiii, CORBIS Images/HRW; xxiv (tl), David Young-Wolf/PNI; xxiv (b), Klaus Lahnstein/Getty Images; S20 (b), © Nik Wheeler/CORBIS; 21 (b), Global Hydrology and Climate Center, NASA Marshall Space Flight Center.; 21 (t), Global Hydrology and Climate Center, NASA Marshall Space Flight Center.; S21 (cl), Jeffrey Aaronson/Network Aspen; S21 (c), Index Stock Imagery, Inc.; S21 (tr), Amit Bhargava/ Newsmakers/Getty Images; S22 (l), Norman Owen Tomalin/Bruce Coleman, Inc.; S22 (cl), © Fritz Polking/Peter Arnold, Inc.; S22 (tl), Larry Kolvoord Photography; S22 (tr), Runk/Schoenberger/Grant Heilman Photography; S22 (tr), Wolfgang Kaehler Photography; S27 (b), Rosenback/ZEFA/Index Stock Imagery, Inc.; **Unit 1:** 1 (b), Photo © Transdia/Panoramic Images, Chicago 1998; 1 (cl), © Joe Viesti/The Viesti Collection; 1 (tr), Francois Gohier/Photo Researchers, Inc.; 1 (t), © Norbert Wu/www.norbertwu.com; 1 (br), Steven David Miller/Animals Animals/Earth Scenes; **Chapter 1**: 2 (cl), © Stone/Philip & Karen Smith; 2 (bl), Sam Dudgeon/HRW Photo; 2 (tr), CORBIS; 2 (c), © Joseph Sohm; ChromoSohm Inc./CORBIS; 3 (bl), © Stone/Ken McVey; 3 (tr), British Library, London, Great Britain/ Art Resource, NY; 4, Luca Turi/AP/Wide World Photos; 5 (cl), The Stock Market/Jose Fuste Raga; 5 (cr), NASA; 6 (cl), Robert Caputo/Aurora; 6 (tr), © Stone/Robert Frerck; 7-8 © Bob Daemmrich; 8 (b), K.D. Frankel/Bilderberg/Aurora; 9 (t,br), © Wolfgang Kaehler; 10, © Nik Wheeler/CORBIS; 11 (cr), © Ilene Perlman/Stock, Boston; 11 (tr), © Archivo Iconografico S.A./CORBIS; 13 (cr), © Chris Rainier/CORBIS; **Chapter 2**: 16 (c), © Roger Ressmeyer/CORBIS; 16 (tl), Digital Stock Corp./HRW; 16 (t), Steve Winter/ Black Star; 16 (cl), T.A. Wiewandt/DRK Photo; 16 (bl), Scala/Art Resource, NY; 17, Monticello. Photo: Edward Owen; 19, © Stone/Earth Imaging; 20 (tl), courtesy of NASA, Marshall Space Flight Center; 21 (t,b), Global Hydrology and Climate Center, NASA Marshall Space Flight Center.; 23 (t), Image copyright ©2002 PhotoDisc, Inc.; 23 (b), © Gerald French/FPG International LLC; 24, © Tom Bean/CORBIS; 26 (t), © Norbert Wu/www.norbertwu.com; 26 (b), Boll/ Liaison Agency; 27, © Ed Kashi; 28, © Gary Braasch/ CORBIS; 30, © Stone/Andrew Rafkind; 31, © Galen Rowell/CORBIS; 32, © Yann Arhtus-Bertrand/CORBIS; **Chapter 3**: 36 (c), © Stone/Glenn Christianson; 36 (tr), © Stone/A. Witte/C. Mahaney; 36 (br), © Stone/ Brian Stablyk; 36 (b), © 1999 Michael DeYoung/ AlaskaStock.com; 36 (tl), Corbis Images; 37, Thomas A. Wiewandt, Ph.D., Wild Horizons; 38, © Layne Kennedy/CORBIS; 39, © Chinch Gryniewicz; Ecoscene/CORBIS; 40, © SuperStock; 41, © Flip Nicklin/Minden Pictures; 43 (cr), NASA; 44 (bl), Image copyright ©1998 PhotoDisc, Inc./HRW; 44 (t), Paul Seheult; Eye Ubiquitous/CORBIS; 45 (br), © Stone/Martin Puddy; 46 © Laurence Parent; 48 (tl), © Gail Mooney/CORBIS; 48 (b), © DiMaggio/Kalish/ Peter Arnold, Inc.; 49, © Darrell Gulin/CORBIS; 50, © Galen Rowell/CORBIS; 51 (c), © Darrell Gulin/CORBIS; 51 (t), Image Copyright ©2002 PhotoDisc, Inc.; 53 (t), Robert Maier; 54